Compound Semiconductors 1994

Other titles in the Series

The Institute of Physics Conference Series regularly features papers presented at important conferences and symposia highlighting new developments in physics and related fields. Recent titles include:

96: **Gallium Arsenide and Related Compounds 1988, papers presented at the 15th International Symposium, Atlanta, GA, USA**
J Harris

106: **Gallium Arsenide and Related Compounds 1989, papers presented at the 16th International Symposium, Karuizawa, Japan**
T Ikoma and H Watanabe

112: **Gallium Arsenide and Related Compounds 1990, papers presented at the 17th International Symposium, Jersey, Channel Islands**
K E Singer

120: **Gallium Arsenide and Related Compounds 1991, papers presented at the 18th International Symposium, Seattle, WA, USA**
G B Stringfellow

129: **Gallium Arsenide and Related Compounds 1992, papers presented at the 19th International Symposium, Karuizawa, Japan**
T Ikegami, F Hasegawa and Y Takeda

136: **Gallium Arsenide and Related Compounds 1993, papers presented at the 20th International Symposium, Freiburg im Braunschweig, Germany**
H S Rupprecht and G Weimann

Compound Semiconductors 1994

Proceedings of the Twenty-First International Symposium on Compound Semiconductors held in San Diego, California, 18–22 September 1994

Edited by Herb Goronkin and Umesh Mishra

Previous symposia have been published as Gallium Arsenide and Related Compounds in the Institute of Physics Conference Series

Institute of Physics Conference Series Number 141
Institute of Physics Publishing, Bristol and Philadelphia

Copyright ©1995 by IOP Publishing Ltd and individual contributors. All rights reserved. No part of this publication may be reproduced, stored in a retrieval system or transmitted in any form or by any means, electronic, mechanical, photocopying, recording or otherwise, without the written permission of the publisher, except as stated below. Single photocopies of single articles may be made for private study or research. Illustrations and short extracts from the text of individual contributions may be copied provided that the source is acknowledged, the permission of the authors is obtained and IOP Publishing Ltd is notified. Multiple copying is permitted in accordance with the terms of licences issued by the Copyright Licensing Agency under the terms of its agreement with the Committee of Vice-Chancellors and Principals. Authorization to photocopy items for internal or personal use, or the internal or personal use of specific clients in the USA, is granted by IOP Publishing Ltd to libraries and other users registered with the Copyright Clearance Center (CCC) Transactional Reporting Service, provided that the base fee of $19.50 per copy is paid directly to CCC, 27 Congress Street, Salem, MA 01970, USA.
0305-2346/95 $19.50+.00

CODEN IPHSAC 141 932 (1995)

British Library Cataloguing in Publication Data

A catalogue record for this book is available from the British Library.

ISBN 0-7503-0226-7

Library of Congress Cataloging-in-Publication Data are available

This work relates to Department of Navy Grant N00014-94-1-1070 issued by the Office of Naval Research. The United States Government has a royalty-free license throughout the world in all copyrightable material contained herein.

Published by Institute of Physics Publishing, wholly owned by The Institute of Physics, London
Institute of Physics Publishing, Techno House, Redcliffe Way, Bristol BS1 6NX, UK
US Editorial Office: Institute of Physics Publishing, The Public Ledger Building, Suite 1035, Independence Square, Philadelphia, PA 19106, USA

Printed in the UK by Galliard (Printers) Ltd, Great Yarmouth, Norfolk

International Symposium on Compound Semiconductors Award and Heinrich Welker Gold Medal

The International Symposium on Compound Semiconductors Award was initiated in 1976; the recipients are selected by the International Symposium on Compound Semiconductors Award Committee for outstanding research in the area of III–V compound semiconductors. The Award consists of $2500 and a plaque citing the recipient's contribution to the field. The Award is accompanied by the Heinrich Welker Gold Medal, established by Siemens AG, Munich, in honour of the foremost pioneer in III–V semiconductor development.

The winners of the GaAs Symposium Award and the Heinrich Welker Medal are:

Nick Holonyak	1976, for developing the first practical light-emitting diodes
Cyril Hilsum	1978, for contributions in the fields of transferred electron logic devices (TELDs) and GaAs MESFETs
Hisayoshi Yanai	1980, for his work on TELDs, GaAs MESFETs and ICs, and laser diode modulation with TELDs
Gerald L Pearson	1981, for research and teaching in compound semiconductors physics and device technology
Herbert Kroemer	1982, for his work on hot-electron effects, Gunn oscillators and III–V heterostructure devices
Izuo Hayashi	1984, for development and understanding of room temperature operation of DH lasers
Heinz Beneking	1985, for his contributions to III–V semiconductor technology and novel devices
Alfred Y Cho	1986, for pioneering work on molecular beam epitaxy and his contribution to III–V semiconductor research
Zhores I Alferov	1987, for outstanding contributions in theory, technology and devices, especially epitaxy and laser diodes

Jerry Woodall	1988, for introducing the III–V alloy AlGaAs and fundamental contributions to III–V physics
Don Shaw	1989, for pioneering work on epitaxial crystal growth by chemical vapour deposition
George S Stillmann	1990, for the characterization of high-purity GaAs and developing avalanche photodetectors
Lester F Eastman	1991, in recognition of his dedicated work in the field, especially on ballistic electron transport, δ-doping, buffer layer technique, and AlInAs/GaInAs heterostructures
Harry C Gatos	1992, for his contribution to science and technology of GaAs and related compounds, particularly in relating growth parameters, composition and structure to electronic properties
James A Turner	1993, for pioneering the development of GaAs MESFETs, MMICs, circuit fabrication and analytical techniques

The recipient of the 1994 International Symposium on Compound Semiconductors Welker Medal was Federico Capasso. He received his Doctor of Physics degree, *summa cum laude*, from the University of Rome, Italy, in 1973. He joined AT&T Bell Laboratories first as a visiting scientist in 1976 and then as member of the technical staff in 1977. Since 1987 he has been head of the Quantum Phenomena and Device Research department.

Federico Capasso's work on bandgap engineering of semiconductor devices and heterostructures has opened up new areas of research in electronics and optoelectronics and led him to the discovery of many new phenomena in artificially structured semiconductors. In a series of pioneering contributions he and his collaborators showed how, with appropriate combinations of bandgap and doping profiles, one can engineer the band diagram of semiconductor heterostructures in a nearly arbitrary and continuous way and thus tailor their electronic and optical properties for specific applications. This led him to the invention and demonstration of avalanche photodiodes (APDs) with enhanced ionization rates ratio, including multiquantum well APDs and the conception of a solid-state photomultiplier. Federico Capasso and his collaborators also made important contributions to heterojunction bipolar transistors including the first demonstration of a graded gap base bipolar and the measurement of the drift velocity of minority electrons in p-type graded gap semiconductors. His pioneering research on quantum effect devices includes the invention and realization of resonant tunnelling bipolar transistors and of multistate transistors. He and his collaborators showed experimentally that the complexity of many analog and digital circuits can be greatly reduced using these functional devices. His research on superlattice devices led him to the invention of new photoconductors based on the effective mass filtering effect, the first observation of negative differential resistance by Bragg reflections, predicted by Esaki and Tsu in 1970, and to the observation of sequential resonant tunnelling through multiquantum well superlattices. Building on the latter phenomenon he started a research programme on unipolar semiconductor lasers; recently he and his collaborators demonstrated the first unipolar intersubband laser (quantum cascade laser), following a 25-year worldwide quest. In their work on optical properties of quantum structures Capasso and his collaborators designed and demonstrated new coupled-quantum well semiconductors with giant intersubband nonlinear susceptibilities in the infrared and large linear Stark effects for modulator applications, Fabry–Pérot electron filters and structures with Bragg-confined continuum states. He also pioneered the tailoring of heterojunction band discontinuities using doping interface dipoles with acceptor and donor planes.

He has co-authored about 200 papers, given over 100 invited talks at conferences and holds 21 US patents and 45 foreign patents. He is co-editor of four volumes, including the book *Physics of Quantum Electron Devices* (Springer). He is a member of the editorial board of *Semiconductor Science and Technology* and *Il Nuovo Cimento*, he co-chaired two conferences and served on the programme committees of over 25 conferences and on several advisory committees and panels.

He is a recipient of the 1993 New York Academy of Sciences Award, the 1991 IEEE David Sarnoff Award, the 1984 AT&T Bell Laboratories Distinguished Member of Technical Staff Award and the 1984 Award of Excellence of the Society for Technical Communication. He is a Fellow of the IEEE, the Optical Society of America, the American Physical Society, the American Association for the Advancement of Science and SPIE. He is listed in *Who's Who in America*, *Who's Who in Engineering*, *Who's Who in the East*, *Who's Who in New Jersey* and *American Men and Women of Science*.

Young Scientist Award

The International Advisory Committee of the International Symposium on Compound Semiconductors has established a Young Scientist Award to recognize technical achievements in the field of compound semiconductors by a scientist under the age of forty. The Award consists of a financial reward and a plaque citing the recipient's contributions.

The Young Scientist Award recipients are:

1986	Russel D Dupuis	for work in the development of organometallic vapour phase epitaxy of compound semiconductors.
1987	Naoki Yokoyama	for contributions to self-aligned gate technology for GaAs MESFETs and ICs and the resonant tunnelling hot-electron transistor.
1989	Russel Fischer	for demonstration of state of the art performance, at DC and microwave frequencies, of MESFETs, HEMTs and HBTs using (AlGa)As on Si.
1990	Yasuhiko Arakawa	for pioneering work on low-dimensional semiconductor lasers, showing the superior performance of quantum wire and quantum box devices.
1992	Umesh K Mishra	for pioneering and outstanding work on AlInAs–GaInAs HEMTs and HBTs.
1993	Young-Kai Chen	for significant advancements in the fields of high-speed III–V electronic and optoelectronic devices.

The recipient of the 1994 Young Scientist Award was Michael A Haase. He was born in Illinois in 1959. He received his BSc and MSc degrees in Electrical Engineering from the University of Illinois at Urbana-Champaign, where he also earned his PhD in 1988 after developing an internal photoemission technique for measuring band offsets in semiconductor heterostructures. While at Illinois he also studied high-field electron transport in III–Vs, and did MOCVD crystal growth which led to the first use of CC_l4 for C-doping of GaAs.

After graduate school he joined 3M where, in addition to research on novel guided-wave optical modulators in the III–Vs, he was responsible for device development in the wide bandgap II–VI group. That work resulted in blue LEDs and ZnSe-based electro-optic modulators. In 1991, Michael and his co-workers developed the world's first blue-green laser diode, for which they were awarded the 1993 Rank Prize for Optoelectronics. Michael is now the leader of the II–VI laser diode effort at 3M. He is a member of the IEEE and APS, and has co-authored over 40 technical publications and 8 US patents.

Preface

The *21st International Symposium on Compound Semiconductors* was held in San Diego, California from 18 to 22 September 1994. This was the first of the Symposia to carry the new name which was adopted by the International Advisory Committee to signify both the wider scope and underlying common baseline of technology in today's field of compound semiconductors. A total of 184 papers including five plenary session papers were selected from over 300 submissions from 16 countries for presentation at the conference.

The field of compound semiconductors has changed significantly since the first *International Symposium on GaAs* was held in 1966. By the time the third symposium was held in 1970, the scope was expanded to include new work on GaP, InP and mixed crystals of III–V compounds. Correspondingly, the symposium was renamed *Gallium Arsenide and Related Compounds*. As the need to improve the performance and expand the capability of electronic and photonic devices continually drives our research into new materials and fabrication/characterization technologies, the symposium has continued to modify its scope to remain current. However, the charter remains unchanged. The *21st International Symposium on Compound Semiconductors* continues to provide a forum for the myriad activities involved in the conception, fabrication, application and analysis of compound semiconductor devices. Here, researchers have the unique opportunity to discuss subjects with participants from many disciplines and come away from the symposium with perspectives on their own work that could not be so readily obtained in other ways.

The chapters are organized by subject. You will find that there are substantial cross-links between them, so we hope you will roam through this digest and utilize the many fine papers in your research.

We would like to thank the Technical Programme Committee for reviewing the abstracts, constructing a high-quality symposium and chairing the sessions. We would like to also thank Janie Lee (University of California at Santa Barbara) for her valuable help.

Herb Goronkin
Motorola
Technical Programme Chair

Umesh Mishra
University of California at Santa Barbara
Associate Programme Chair

The organizers of the Symposium gratefully acknowledge the support provided by our sponsors:

Motorola
Office of Naval Research
ARPA
Siemens
EPI
EMCORE
Bellcore
Sumitomo Electric
Aixtron Semiconductor Technologies, Inc.

Organizing Committee
Vassilis Keramidas (*Conference Chair*), Herb Goronkin (*Technical Programme Chair*), Umesh Mishra (*Associate Technical Programme Chair*), Gerald B Stringfellow (*Treasurer*), Mark Reed (*Secretary*), James Harbison (*Publicity*), Greg Stillman (*Awards Committee Chair*), Alfred Y Cho (*member*), James S Harris (*member*), Don Wolford (*member*)

Programme Committee
Herb Goronkin (*Technical Programme Chair*), Umesh Mishra (*Associate Technical Programme Chair*), Gunter Weimann (*Europe Programme Coordinator*), Naoki Yokoyama (*Asia Programme Coordinator*), Khalid Ismail (*Middle East Programme Coordinator*)

Subcommittee on Nanoelectronics and Nanophotonics
Pierre Petroff (*Chair*), Evelyn Hu, Toshiaki Ikoma, El-Hang Lee, Mark Reed, Hiroyuki Sakaki, Saied Tehrani

Subcommittee on Heterostructure Transistors
Peter Asbeck (*Chair*), Michael Adlerstein, Jon Abrokwah, Craig Farley, Tony Ma, Ali Khatibzadeh, Bumman Kim, David Rensch, Dwight Streit

Subcommittee on High-Power, High-Temperature Devices
Michael Spencer (*Chair*), Kenneth Davis, Mike Melloch, Calvin Carter, Yoon Soo Park, Jin-Koo Rhee, Chris Clark

Subcommittee on Simulation and Modelling
Mark Lundstrom (*Chair*), Bill Frensley, Karl Hess, Steve Laux, Soong Hak Lee, Michael Littlejohn, Serge Luryi, James Luscombe, Christine Maziar, Mark Pinto, Wolfgang Porod, Jasprit Singh

Subcommittee on Epitaxy and In-situ Processing
George Maracas (*Chair*), Al Cho, Ralph Dawson, Jim Harris, Vassilis Keramidas, Bob Kolbas, Suk-Ki Min, Christ Palmstrom, Gerry Stringfellow, Charles Tu, Ray Tsui, Jerry Woodall

Subcommittee on OEICs
Henryk Temkin (*Chair*), S E Swirhun, N K Dutta, Young Se Kwon, Jing-Ming Xu, S Chandrasekhar

Subcommittee on Characterization
Greg Stillman (*Chair*), Dave Aspnes, Randy Feenstra, Guy D Gilliland, Eugene Haller, Jim Harbison, Tung Shung Kuan, Craig Taylor, Wen Wang, Don Wolford, Jong-Chun Woo

Subcommittee on Visible Emitters
George Craford (*Chair*), Doyeol Ahn, Robert Gunshor, Dave Welsch, Asif Khan, Mike Haas, John A Edmond

International Advisory Committee
A Christou, L R Dawson, L F Eastman, T Ikoma, J Magarshack, T Nagahara, H S Rupprecht, D Shaw, G E Stillman, T Sugano, M Uenohara, B L H Wilson

Contents

International Symposium on Compound Semiconductors Award and Heinrich Welker Gold Medal   v

Young Scientist Award   ix

Preface   xii

Chapter 1: Plenary Papers

Progress towards high temperature, high power SiC devices
P G Neudeck   1

Ballistic electron transport and superconductivity in mesoscopic Nb-(InAs/AlSb) quantum well heterostructures
H Kroemer, C Nguyen and E L Hu   7

Chapter 2: Epitaxy

Optimization of InAsSb/InGaAs strained-layer superlattice growth by metal-organic chemical vapor deposition for use in infrared emitters
R M Biefeld, K C Baucom, D M Follstaedt and S R Kurtz   13

Growth of aluminum antimonide and gallium antimonide by MOMBE
K Shiralagi, R Tsui, D Cronk, G Kramer and N D Theodore   17

Low-pressure-MOVPE growth and characterization of InAs/AlSb/GaSb-heterostructures
M Behet, P Schneider, D Moulin, H Hamadeh, J Woitok and K Heime   23

Real time monitoring of III-V alloy composition and real time control of quantum well thickness in MBE by multi-wavelength ellipsometry
C-H Kuo, S Anand, R Droopad, D L Mathine, G N Maracas, B Johs, P He, J A Woollam and T Levola   29

In-situ determination of optical constants for growth control of AlAs/InGaAs RTDs using spectroscopic ellipsometry
F G Celii, Y-C Kao, W M Duncan and B Johs   35

In-situ monitoring and control of MOCVD growth using multiwavelength ellipsometry
S Pittal, B Johs, P He, J A Woollam, S D Murthy and I B Bhat   41

Strained InAs/(AlGaIn)As/InP wells for 1.5–2.5 μm laser applications grown by virtual surfactant MBE
K H Ploog and E Tournié   51

MBE growth of AlGaAs using Sb as a surfactant
R Kaspi, D C Reynolds, K R Evans and E N Taylor   57

Growth of InP, GaP, and GaInP by chemical beam epitaxy using alternative sources
H H Ryu, C W Kim, L P Sadwick, G B Stringfellow, R W Gedridge Jr and A C Jones   63

Chemical beam epitaxy of InGaAs/GaAs multiple quantum wells using cracked or uncracked tris-dimethylaminoarsenic
H K Dong, N Y Li and C W Tu   69

Strain characterization of AsH_3 induced exchange reactions in InP grown by OMVPE
A R Clawson and C M Hanson   75

Comparative investigation of electrical and optical characteristics of $Al_xGa_{1-x}As$/GaAs structures deposited by LP-MOVPE and MBE
H Hardtdegen, M Hollfelder, Chr Ungermanns, T Raafat, R Carius, A Förster, J Lange and H Lüth   81

Graded $In_{x_0 \geq x \geq 0.5}Ga_{1-x}As$/InP buffer layers on GaAs prepared by molecular beam epitaxy
K Häusler and K Eberl   87

Heteroepitaxial growth of InSb on Si by plasma-assisted Epitaxy
H Ohba and T Hariu   93

Growth parameters for metastable $GaP_{1-x}N_x$ alloys in MOVPE
S Miyoshi, H Yaguchi, K Onabe, Y Shiraki and R Ito   97

Properties of GaN films grown on a-Plane (11$\bar{2}$0) and c-Plane (0001) sapphire
K Doverspike, L B Rowland, D K Gaskill, S C Binari, J A Freitas Jr, W Qian and M Skowronski   101

Epitaxial growth and structural, optical properties of cubic GaN on (100) and (111) GaAs grown by metalorganic chemical vapor deposition
C-H Hong, K Wang and D Pavlidis   107

Growth and characterization of p/n type GaN grown at reduced substrate temperatures by plasma enhanced (PE-) MOCVD
L D Zhu, P E Norris, J Zhao, R Singh, R Molnar, O Razumovsky and T Moustakas   113

Large area growth of GaN thin films in a multi-wafer rotating disk reactor
H Liu, S Liang, T Salagaj, J Damian, C S Chern, R A Stall, C-Y Hwang and W E Mayo   119

Heteroepitaxial growth of GaN on GaAs by ECR plasma assisted MBE
H Lee, D B Oberman, W Götz and J S Harris Jr   125

MBE growth of GaN with ECR plasma and hydrogen azide
D B Oberman, H Lee, W K Götz and J S Harris Jr   131

Growth mechanics of GaN by Monte Carlo modeling
K Wang, J Singh and D Pavlidis   137

InAs epitaxial nanocrystal growth on Se-terminated GaAs(001)
Y Watanabe, F Maeda and M Oshima   143

Artificial control of heterojunction band discontinuities by two delta dopings
Y Hashimoto, N Sakamoto, K Agawa, T Saito and T Ikoma   149

Flow modulation epitaxy of short period GaInAs/InP superlattices
D T Emerson, L Whittingham and J R Shealy   155

Strain compensation in InGaP/InGaAs quantum wells with improved interfaces grown by gas-source molecular beam epitaxy
C H Yan and C W Tu   161

$(InAs)_1(GaAs)_4$ short period strained layer superlattices grown by molecular beam epitaxy
J H Lewis, M G Spencer, J A Griffin, D P Zhang, F Grunthaner and T George   167

Growth and characteristics of high optical quality lattice matched GaInAsP layers and GaInAsP/GaInAs quantum wells on InP by MBE using solid phosphorus and arsenic valved cracking cells
J N Baillargeon, A Y Cho, F A Thiel, P J Pearah and K Y Cheng   173

A theoretical and experimental analysis of non-planar epitaxy using shadow masked MOCVD growth
E A Armour, S Z Sun and S D Hersee   177

Selectivly grown vertical sub-100 nm dual-gate GaAs FETs
W Langen, H Hardtdegen, H Lüth and P Kordoš   183

Selective regrowth of highly resistive InP current blocking layers by a low pressure metalorganic vapor phase epitaxy
D K Oh, H M Kim, C Park, J S Kim, S W Lee, H R Choo, H M Kim, H M Park and S-C Park   189

Double heterostructure produced by ordering in GaInP
L C Su, I H Ho and G B Stringfellow   195

Reflectance anisotropy and the ordering mechanism in GaInP at high growth temperatures
J S Luo and J M Olson   201

Advances in correlating the unusual optical properties of $Ga_{0.52}In_{0.48}P$ to the microstructure
M C DeLong, C E Ingefield, P C Taylor, L C Su, I H Ho, T C Hsu, G B Stringfellow, K A Bertness and J M Olson   207

Composition and structure analyses on spontaneously formed AlGaAs superlattice-like structures on (100) GaAs
S S Cha, Y K Shin, H I Jeon, Z S Piao, K Y Lim, E-K Suh, Y H Lee, D K Kim, B T Lee and H J Lee — 213

Hyperabrupt Si-doping profile in MOVPE growth of varactor diodes
F Bugge, P Heymann, E Richter and M Weyers — 217

Chapter 3: Characterization

Emission spectra and lifetimes of near surface GaInAs/GaAs quantum wells
J Dreybrodt, F Daiminger, J P Reithmaier and A Forchel — 221

Photoinduced near-surface electric field screening in GaAs/AlGaAs MQW structures
V N Astratov, O Z Karimov and Yu A Vlasov — 227

Photocurrent spectroscopy of 2-dimensional excitons in 5-nm wide InGaAs/InAlAs multi quantum wells
N Kotera and K Tanaka — 233

Theoretical and experimental study of polarization dependent photoluminescence in $In_{0.2}Ga_{0.8}As$/GaAs quantum wires
J Shuttlewood, K Ko, P Bhattacharya, S W Pang, I Vurgaftman, J Singh and T Brock — 237

Photoluminescence determination of valence-band symmetry and Auger-1 threshold energy in compressed InAsSb layers
S R Kurtz, R M Biefeld and L R Dawson — 243

Electron transport, negative differential resistance and domain formation in weakly coupled quantum wells
A Shakouri, Y Xu, I Gravé and A Yariv — 247

Minority carrier mobility and density of states in SiGe HBTs determined by a new method
A Gruhle, A Schüppen, H Kibbel and U König — 253

Activation and diffusion of shallow impurities in compound semiconductors
W Walukiewicz — 259

Characterization of carbon doped GaAs layers grown by chemical beam epitaxy
R Driad, F Alexandre, J L Benchimol, R Rahbi, B Pajot, B Jusserand, B Sermage, G LeRoux, M Juhel, J Wagner and P Launay — 265

Characterization of inactive carbon in GaAs
A J Moll, J W Ager III, K M Yu, W Walukiewicz and E E Haller — 269

Annealing of nitrogen-doped ZnSe at high pressures: toward suppression of native defect formation
A L Chen, W Walukiewicz, E E Haller, H Luo, G Karczewski and J Furdyna — 275

Electron emission at Si_3N_4-GaAs interfaces prepared with H_2, Ar and H_2+Ar plasma pretreatments
Q H Wang, M I Bowser and J G Swanson — 281

Local vibrational mode spectroscopy of beryllium- and zinc-hydrogen complexes in GaP
M D McCluskey, E E Haller, J Walker and N M Johnson — 287

High quality etched/regrown GaAs/GaAs interfaces formed by an all in-situ Cl_2 etching process
D S L Mui, T A Strand, B J Thibeault, L A Coldren, P M Petroff and E L Hu — 291

Electrical and structural characterization of MBE GaAs layers grown at temperatures between 200 and 600 °C
P Kordoš, J Betko, A Förster, S Kuklovský, Ch Dieker and F Rüders — 295

Electrical characterisation of GaAs layers grown by molecular beam epitaxy at low temperature
J K Luo, H Thomas, D V Morgan, D Westwood, R H Williams and D Theron — 301

Influence of CH_4/H_2 reactive ion etching on the electrical and optical properties of AlGaAs/GaAs and pseudomorphic AlGaAs/InGaAs/GaAs heterostructures
C M van Es, T J Eijkemans, J H Wolter, R Pereira, M Van Hove and M Van Rossum — 307

Structural and magnetotransport properties of InGaAs/InAlAs heterostructures grown on linearly-graded Al(InGa)As buffers on GaAs
R S Goldman, J Chen, K L Kavanagh, H H Wieder, V M Robbins and J N Miller — 313

Si/SiGe modulation doped structures grown by gas source molecular beam epitaxy
A Matsumura, R S Prasad, T J Thornton, J M Fernández, M H Xie, X Zhang, J Zhang and B A Joyce — 319

Effects of annealing using plasma-CVD SiN film as a cap on 2DEG mobility
S Nakata, M Yamamoto and T Mizutani — 323

Low ion energy ECR etching of InP using Cl_2/N_2 mixture
S Miyakuni, R Hattori, K Sato and O Ishihara — 329

Surface structure of GaSb and AlSb grown by molecular beam epitaxy
B Brar, D Leonard and J H English — 335

Thermal stability of acceptor Si in δ-doped GaAs grown on GaAs(111)A by molecular beam epitaxy
M Hirai, H Ohnishi, K Fujita and T Watanabe — 339

Al/Ga atom-exchange during AlAs/GaAs heterointerface formation in alternating source supply
T Saitoh, M Tamura and J E Palmer — 345

Excitons as a probe of interfacial defects in GaAs/AlAs short-period structures
F T Bacalzo, L P Fu, G D Gilliland, A Antonelli, R Chen, K K Bajaj, D J Wolford and J Klem — 351

Heterointerface microroughness analyzed by PL and PLE of quasi-2D GaAs-AlGaAs single quantum well
S J Rhee, H S Ko, Y M Kim, W S Kim, D W Kim and J C Woo — 355

Characterization of in-situ variable-energy focused ion beam/MBE MQW structures
D J Bone, H Lee, K Williams, J S Harris Jr and R F W Pease — 359

Thickness and strain effects on optical constants of thin epitaxial layers
C M Herzinger, P G Snyder, F G Celii and Y-C Kao — 363

Characterization of layered InGaAsP/InP structures by means of time resolved transient grating technique
J Vaitkus, S Juodkazis, M Petrauskas and M Willander — 369

Chapter 4: High-Power, High-Temperature Semiconductor Devices

Advances in silicon carbide device processing and substrate fabrication for high power microwave and high temperature electronics
C D Brandt, A K Agarwal, G Augustine, A A Burk, R C Clarke, R C Glass, H M Hobgood, J P McHugh, P G McMullin, R R Siergiej, T J Smith, S Sriram, M C Driver and R H Hopkins — 373

Silicon carbide substrates and power devices
J W Palmour, V F Tsvetkov, L A Lipkin and C H Carter Jr — 377

Low leakage, high performance GaAs-based high temperature electronics
L P Sadwick, R J Crofts, Y H Feng, M Sokolich and R J Hwu — 383

SiC microwave power MESFET's and JFET's
C E Weitzel, J W Palmour, C H Carter Jr, K J Nordquist, K Moore and S Allen — 389

Cell design and peripheral logic for nonvolatile random access memories in 6H-SiC
W Xie, G M Johnson, Y Wang, J A Cooper Jr, J W Palmour, L A Lipkin, M R Melloch and C H Carter Jr — 395

In situ processing of silicon carbide
S Ahmed, C J Barbero and T W Sigmon — 399

The lifetime limiting defect in SiC
N T Son, E Sörman, W M Chen, O Kordina, B Monemar and E Janzén — 405

Deep levels in undoped, bulk 6H-SiC and associated impurities: application of optical admittance spectroscopy to SiC
W C Mitchel, M Roth, S R Smith, A O Evwaraye and J Solomon — 411

Optically detected cyclotron resonance study of high-purity 6H-SiC CVD-layers
N T Son, O Kordina, A O Konstantinov, W M Chen, E Sörman, B Monemar and E Janzén — 415

GaN/AlGaN field effect transistors for high temperature applications
M S Shur, A Khan, B Gelmont, R J Trew and M W Shin   419

Some aspects of GaN electron transport properties
D K Gaskill, K Doverspike, L B Rowland and D L Rode   425

X-ray differential diffractometry applied to GaN grown on SiC
I P Nikitina and V A Dmitriev   431

Step-controlled epitaxial growth of α-SiC and application to high-voltage Schottky rectifiers
T Kimoto, A Itoh, H Akita, T Urushidani, S Jang and H Matsunami   437

TEM study of low temperature CVD silicon carbide films grown on on-axis 6H-SiC substrates
K Fekade, M G Spencer, K Irvine, A K Ballal, D P Besabathina and L G Salamanca-Riba   443

Electrical characterization of the thermally oxidized SiO_2/SiC interface
J N Shenoy, L A Lipkin, G L Chindalore, J Pan, J A Cooper Jr, J W Palmour and M R Melloch   449

Monolithic digital integrated circuits in 6H-SiC
W Xie, J Pan, J A Cooper Jr and M R Melloch   455

DC, microwave and high temperature characteristics of GaN FET structures
S C Binari, L B Rowland, G Kelner, W Krupper, H B Dietrich, K Doverspike and D K Gaskill   459

Chapter 5: Visible Emitters and OEICs

Physics and simulation of InGaAsP/InP lasers
R F Kazarinov   463

Frequency limitation of quantum wire lasers
S Tiwari   471

Fundamental limits for linearity of cable TV lasers
V B Gorfinkel and S Luryi   477

Mechanical stress in AlGaAs ridge lasers: its measurement and effect on the optical near field
P W Epperlein, G Hunziker, K Dätwyler, U Deutsch, H P Dietrich and D J Webb   483

Highly efficient light-emitting diodes with microcavities
E F Schubert and N E J Hunt   489

SiC-A^3N alloys and wide band gap nitrides grown on SiC substrates
V A Dmitriev, K G Irvine, J A Edmond, G E Bulman, C H Carter Jr, A S Zubrilov, I P Nikitina, V Nikolaev, A I Babanin, Yu V Melnik, E V Kalinina and V E Sizov   497

Effect of atomic hydrogen on 670nm AlGaInP visible laser
W-J Choi, J-H Chang, W-T Choi, S-H Kim, J-S Kim, S-J Leem and T-K Yoo   503

Threshold current density reduction by annealing in 630–650nm GaInP strained single quantum well lasers
I Nomura and K Kishino   507

Compositional control of (Zn,Mg)(S,Se) epilayers grown by MBE for II/VI blue green laser diodes
M D Ringle, D C Grillo, J Han, R L Gunshor, G C Hua and A V Nurmikko   513

Dark defects in II-VI blue-green laser diodes
G C Hua, D C Grillo, J Han, M D Ringle, Y Fan, R L Gunshor, M Hovinen, A V Nurmikko and N Otsuka   519

Stimulated emission from GaN grown on SiC
A S Zubrilov, V I Nikolaev, V A Dmitriev, K G Irvine, J A Edmond and C H Carter Jr   525

High speed optoelectronics
R Nagarajan, K Giboney, R Yu, D Tauber, J Bowers and M Rodwell   531

Silicon recrystallization effects in mirror coatings of high-power 980-nm InGaAs/AlGaAs lasers
P W Epperlein and M Gasser   537

Record low resistance vertical cavity surface emitting lasers grown by molecular beam epitaxy using hybrid mirror approach
M Hong, D Vakhshoori, J P Mannaerts and F A Thiel   543

Effects of impurity-free and impurity-induced disordering (IID) on the optical properties of GaAs/(Al,Ga)As distributed Bragg reflectors
P D Floyd and J L Merz   547

Period deviations in distributed Bragg reflectors: x-ray diffraction and optical reflectivity measurements
K Matney, M S Goorsky, J Tartaglia, R Rai and C Parsons   553

0.85-μm 8×8 bottom-surface-emitting laser diode arrays grown on AlGaAs substrates by MOCVD
Y Kohama, Y Ohiso, C Amano and T Kurokawa   559

Vertical-cavity surface-emitting lasers with buried lateral current confinement
S Rochus, T Röhr, H Kratzer, G Böhm, W Klein, G Tränkle and G Weimann   563

MBE growth of low threshold laser structures employing short-period superlattices
H Riechert, D Bernklau, A Milde, M Schuster, H Cerva, C Hoyler and H-D Wolf   567

Monolithically integrated optical receivers and transmitters
D T Nichols, W S Hobson, P R Berger, N K Dutta, P R Smith, J Lopata, D L Sivco and A Y Cho   573

Monolithically integrated microwave OEICs by selective OMVPE
H Kanber, W W Lam, S X Bar, C Beckham and P E Norris   579

Optical refractometry with GaAs/AlGaAs interferometers
H P Zappe   585

Lateral npn-phototransistors with high gain and high spatial resolution fabricated by focused laser beam induced Zn doping of GaAs/AlGaAs quantum wells
P Baumgartner, C Engel, G Abstreiter, G Böhm, G Tränkle and G Weimann   591

Ga_2O_3 films for insulator/III-V semiconductor interfaces
M Passlack, M Hong, E F Schubert, J P Mannaerts, W S Hobson, N Moriya, J Lopata and G J Zydzik   597

MOCVD GaInAsSb heterostructure materials for 2-4 μm photodetectors
G Wei, R Peng and W Wu   603

InGaAs/AlGaAs quantum well infrared photodetectors with 3-5μm response
L C Lenchyshyn, H C Liu, M Buchanan and Z R Wasilewski   607

Planar, low loss InGaAsP/InP photoelastic waveguide devices
X S Jiang, Q Z Liu, L S Yu, B Zhu, Z F Guan, S A Pappert, P K L Yu and S S Lau   613

Chapter 6: Heterojunction Transistors

A planar self-aligned GaAlAs-GaAs HBT technology achieved by CBE selective base and collector contacts regrowth
P Launay, R Driad, F Alexandre, Ph Legay and A M Duchenois   619

Experimental investigation of low frequency noise properties of AlGaAs/GaAs and GaInP/GaAs heterojunction bipolar transistors
J-H Shin, J-W Lee, Y-S Seo, Y-S Kim and B Kim   625

Effect of carrier recombination at the emitter-base heterojunction on the performance of GaInP/GaAs and AlGaAs heterojunction bipolar transistors
Z H Lu, A Majerfeld, L W Yang and P D Wright   629

Energy transport modeling of HBTs considering composition-, doping- and energy-dependence of transport parameters
A Nakatani and K Horio   633

Measurement of the electron ionization coefficient at low electric fields in heterojunction bipolar transistors
C Canali, S Chandrasekhar, F Capasso, R A Hamm, R Malik, C Forzan, A Neviani, L Vendrame and E Zanoni   639

Electron transport mechanisms in abrupt- and graded-base/collector AlInAs/GaInAs/InP double heterostructure bipolar transistors
H J De Los Santos, M Hafizi, T Liu and D B Rensch   645

Contactless HBT equivalent circuit analysis using the modulation frequency
dependence of photoreflectance
W Krystek, H Qiang, F H Pollak, D C Streit and M Wojtowicz — 651

Threshold tunable MESFET and ultra-fast static RAM
*P J Zampardi, S M Beccue, R L Pierson, W J Ho, J Hu, M F Chang, A Sailor and
K C Wang* — 657

An n-HJFET-pnp HBT process for complementary circuit applications
*P Parikh, K Kızıloğlu, M Mondry, P Chavarkar, B Keller, S Denbaars and
U Mishra* — 663

Heterostructure insulated-gate-FET with improved gate barrier characteristics
R Westphalen, K-M Lipka, J Schneider and E Kohn — 667

Strain-symmetrized $In_xGa_{1-x}As/In_yAl_{1-y}As$ HEMTs with extremely high 2DEG
densities and mobilities
W Klein, G Böhm, H Heiß, S Kraus, D Xu, R Semerad, G Tränkle and G Weimann — 673

Power combining of double barrier resonant tunnelling diodes at W-band
R E Miles, D P Steenson, R D Pollard, J M Chamberlain and M Henini — 679

Selective molecular beam epitaxy for multifunction microwave integrated circuits
*D C Streit, D K Umemoto, T R Block, A-C Han, M Wojtowicz, K Kobayashi and
A K Oki* — 685

High performance AlGaAs/InGaAs pseudomorphic HEMTs after epitaxial lift-off
*Y Baeyens, C Brys, J De Boeck, W De Raedt, B Nauwelaers, G Borghs,
P Demeester and M Van Rossum* — 689

Electron transport in doped InAs/InGaAs/AlInAs quantum wells and the effect on
FET performance
J K Zahurak, A A Iliadis, S A Rishton and W T Masselink — 693

Very low contact resistance to n^+-InP grown using $SiBr_4$
M T Fresina, S L Jackson and G E Stillman — 699

Ultra-high-speed HEMTs: progress towards the next-generation
L D Nguyen — 703

External gate fringing capacitance model for optimizing Y-shaped, recess-etched
gate FET structure
M Fukaishi, M Tokushima, S Wada, N Matsuno and H Hida — 707

Electron cyclotron resonance microwave plasma enhanced SiGe oxidation and
MOS transistors
P-W Li, E S Yang, Y F Yang and X Li — 711

Non-alloyed contacts to GaAs using non-stoichiometric GaAs
M P Patkar, J M Woodall, T P Chin, M S Lundstrom and M R Melloch — 717

Development of refractory NiGe-based ohmic contacts to n-type GaAs
T Oku, H Wakimoto, M Furumai, H Ishikawa, H R Kawata, A Otsuki and M Murakami — 721

Rapid thermal annealing effects on the structural integrity of InAlAs/GaAsSb HIGFET epilayers on InP
K G Merkel, C L A Cerny, R T Lareau, V M Bright, P W Yu and F L Schuermeyer — 727

A novel area-variable varactor diode
D-W Kim, J-H Son, S-C Hong and Y-S Kwon — 733

Fabrication and performance of $In_{0.53}Ga_{0.47}As/AlAs$ resonant tunneling diodes overgrown on GaAs/AlGaAs heterojunction bipolar transistors
K B Nichols, E R Brown, M J Manfra, O B McMahon, P J Zampardi, R L Pierson and K C Wang — 737

Chapter 7: Simulation and Modelling

Two-dimensional self-consistent modeling of InP/GaInAsP lateral current injection lasers
D A Suda, T Makino and J M Xu — 743

SimWindows: a new simulator for studying quantum well optoelectronic devices
D W Winston and R E Hayes — 747

Development of an intelligent heterostructure material parameter database system
W R Frensley, R C Bowen, C L Fernando and M E Mason — 751

Steady-state dynamic and spectral performance of microcavity surface-emitting strained quantum-well lasers
I Vurgaftman and J Singh — 759

Table-based Monto Carlo simulation of electron, phonon, and photon dynamics in quantum well lasers
Md A Alam and M S Lundstrom — 765

Tight-binding calculation of linear optical properties of $In_{0.53}Ga_{0.47}As$ alloys and heterostructures
V Sankaran, K W Kim and G J Iafrate — 769

Resonant tunneling devices: effect of scattering
S Datta, G Klimeck, R K Lake and M P Anantram — 775

Quantum cellular automata: computing with quantum dot molecules
P D Tougaw and C S Lent — 781

Quantum scattering states in open two-dimensional electronic systems and the local density of states
H K Harbury and W Porod — 787

Monte Carlo investigation of three-dimensional effects in sub-micron GaAs MESFETs
S S Pennathur and S M Goodnick — 793

Evaluation of electron ionization threshold energies for $Al_xGa_{1-x}As$ on a hypercube
V Chandramouli and C M Maziar — 797

Ensemble Monte Carlo calculation of the hole initiated impact ionization rate in bulk GaAs and silicon using a k-dependent, numerical transition rate formulation
I H Oguzman, Y Wang, J Kolnik and K F Brennan — 803

Analysis of heterojunction MOSFET structure for deep-submicron scaling
S A Hareland, A F Tasch and C M Maziar — 807

Full scale simulation of quantum devices with interface disorder and randomness
J P Leburton, D Jovanovic and I Adesida — 813

Chapter 8: Nanoelectronics and Nanophotonics

Progress in self-assembled quantum dots of $In_xGa_{1-x}As$ on GaAs
D Leonard, G Meideros-Ribeiro, H Drexler, K Pond, W Hansen, J P Kotthaus and P M Petroff — 819

Controlled nucleation of InAs clusters on (100) GaAs substrates by electron beam lithography
J W Sleight, R E Welser, L J Guido, M Amman and M A Reed — 825

Quantum interference effects in finite antidot lattices
R Schuster, K Ensslin, D Wharam, V Dolgopolov, J P Kotthaus, G Böhm, W Klein, G Tränkle and G Weimann — 831

Interference of ballistic quantum electron waves in electronic stub tuners
P Debray, J Blanchet, R Akis, P Vasilopoulos and J Nagle — 835

Efficient modulation of carrier density in novel double quantum well structure by low optical pump power
M Rüfenacht, H Akiyama, S Tsujino, Y Kadoya and S H Sakaki — 841

Resonance tunneling in semi-metal/semiconductor, ErAs/(Al,Ga)As, heterostructures
K Zhang, D Brehmer, S J Allen Jr and C J Palmstrom — 845

Hole charging effects in Sb-based resonant tunneling structures
J N Schulman and D H Chow — 851

Resonant interband tunneling quantum functional device
S Tehrani, J Shen, H Goronkin, G Kramer, R Tsui and T X Zhu — 855

Investigation of magneto-tunneling processes in InAs/AlSb/GaSb based resonant interband tunneling structures
H Obloh, J Schmitz and J D Ralson — 861

DC and transient simulation of resonant tunneling devices in NDR-SPICE
P Mazumder, J P Sun, S Mohan and G I Haddad 867

Resonant tunneling through zero-dimensional Si-impurity states in GaAs/AlAs single-barrier structures
H Fukuyama and T Waho 873

Quantum dot single-photon and electron-hole turnstile devices
A Imamoḡlu and Y Yamamoto 879

Chapter 9: An International Perspective on Nanoelectronics and Nanophotonics

Overview and status of the quantum functional devices project
S Okayama, I Ishida, N Aoi and S Maeda 885

Overview and status of the European Framework ESPRIT programmes
S P Beaumont 891

Keyword index 897

Author index 905

Progress towards high temperature, high power SiC devices

Philip G. Neudeck

NASA Lewis Research Center, M.S. 77-1, 21000 Brookpark Road, Cleveland, OH 44135 USA

Abstract. Silicon carbide's demonstrated ability to function under extreme high-temperature, high-power, and/or high-radiation conditions is expected to enable significant enhancements to a far-ranging variety of applications and systems. However, improvements in crystal growth and device fabrication processes are needed before SiC-based devices and circuits can be scaled-up and reliably incorporated into electronic systems. This paper surveys the present status of SiC-based semiconductor electronics within the context of identifying areas where technological maturation is most needed, and speculating on the prospects for resolution of crucial technological obstacles. Recent achievements include the monolithic realization of SiC integrated circuit operational amplifiers and digital logic circuits, as well as significant improvements to epitaxial and bulk crystal growth processes that will impact the overall viability of this rapidly emerging technology.

1. Introduction

Silicon carbide (SiC) based semiconductor electronic devices and circuits are presently being developed for advantageous use in high-temperature, high-power, and/or high-radiation conditions under which conventional semiconductors cannot adequately perform. Silicon carbide's demonstrated ability to function under extreme high-temperature, high-power, and/or high-radiation conditions is expected to enable significant improvements to a far-ranging variety of applications and systems. These range from improved high-voltage switching [1,2] for energy savings in public electric power distribution and electric vehicles to more powerful microwave electronics for radar and communications [3] to sensors and controls for cleaner-burning more fuel-efficient jet aircraft and automobile engines [4,5]. However, there are many crucial crystal growth and device fabrication issues that must be addressed before SiC-based devices and circuits are ready for scale-up and reliable incorporation into electronic systems. This paper surveys recent progress towards the realization of high temperature and/or high power SiC devices and circuits within the context of identifying specific performance-limiting areas where technological maturation is most needed. The prospects for resolution of crucial technological obstacles are also discussed.

Silicon carbide occurs in many different crystal structures (called polytypes) with each crystal structure having its own unique electrical and optical properties. The electrical properties of the more common SiC polytypes are compared to the properties of silicon and GaAs in Table 1. In many device applications, SiC's exceptionally high breakdown field (> 5 times that of Si), wide bandgap energy (> 2 times that of Si), high carrier saturation velocity (> 2 times that of Si), and high thermal conductivity (> 3 times that of Si) could enable substantial performance gains, greatly overcoming non-trivial low-field carrier mobility disadvantages. In the particularly attractive area of power devices, Bhatnagar and Baliga [2] indicate that SiC power MOSFET's and Schottky diode rectifiers would operate over higher voltage and temperature ranges, have superior switching characteristics, and yet have die sizes nearly 20 times smaller than correspondingly rated silicon-based devices.

2. Crystal Growth

Although some of silicon carbide's advantageous electrical properties have been known for decades, until recently there was a lack of wafers with reproducible SiC of sufficient electrical quality to realize advantageous devices and circuits. Efforts to solve

Table 1. Comparison of selected semiconductor room temperature physical properties.

	Si	GaAs	6H-SiC	4H-SiC	3C-SiC
Bandgap (eV)	1.1	1.42	3.0	3.2	2.3
Breakdown Field @ 10^{17} cm^{-3} (MV/cm)	0.6	0.6	3.2	3	> 1.5
Electron Mobility @ 10^{16} cm^{-3} (cm^2/ V-s)	1100	6000	370	800	750
Saturated Electron Drift Velocity (cm/s)	10^7	8×10^6	2×10^7	2×10^7	2.5×10^7
Thermal Conductivity (W/cm-K)	1.5	0.5	4.9	4.9	5.0
Hole Mobility @ 10^{16} cm^{-3} (cm^2/ V-s)	420	320	90	115	40
Commercial Wafers	12"	6"	1.375"	1.375"	None

the material shortage problem through the heteroepitaxial growth of 3C-SiC on large-area substrate materials (primarily silicon wafers) have not proven successful to date, as the resulting SiC material still contains too many defects to be useful. Only with the development of the modified Lely seeded sublimation growth technique have acceptably large and reproducible single-crystal SiC wafers of usable electrical quality become available [6,7]. 1-inch 6H-SiC wafers first became commercially available in 1989 [8], and the vast majority of silicon carbide semiconductor device technology development has taken place since that time.

Of the numerous polytypic forms of silicon carbide, 4H- and 6H-SiC electronic devices presently exhibit the most promise due to the availability and quality of reproducible single-crystal wafers in these polytypes. The size of commercially available 4H- and 6H-SiC wafers has recently been increased from 1 inch to 1.375 inches in diameter, and further up-scaling to 2-inch and 3-inch wafer sizes is eventually expected. Although only one U.S. company is presently selling greater than 1-inch SiC wafers on the open market [8], at least three other companies are producing similar SiC wafers on a regular basis for internal purposes. Having been introduced commercially as recently as 1993, the 4H-SiC wafers are presently more expensive and slightly less developed than 6H-SiC wafers. However, 4H-SiC's substantially higher carrier mobility [9] should make it the polytype of choice for most SiC electronic devices, provided that all other device processing, performance, and cost-related issues play out as being roughly equal between the two polytypes. Furthermore, the inherent mobility anisotropy that degrades conduction parallel to the crystallographic c-axis in 6H-SiC [9,10] will particularly favor 4H-SiC for vertical power devices. The emergence of higher mobility 4H-SiC has largely overshadowed significant progress made in obtaining greatly improved 3C-SiC through heteroepitaxy on low-tilt-angle 6H-SiC substrates [11]. If on-going work ever solves the crystallographic defect problems associated with the heteroepitaxial growth of 3C-SiC on large-area silicon substrates, fabrication line compatibility and economic advantages would probably push 3C-SiC to the forefront.

The controlled growth of high-quality epilayers is a key issue in the realization of SiC electronics, especially given the fact that present-day commercial SiC wafers exhibit bulk resistivities no higher than 10 Ω-cm. Homoepitaxial growth of n-type (nitrogen-doped) or p-type (aluminum-doped) epilayers is primarily accomplished using chemical vapor deposition (CVD) [12]. Recently, a major advancement which greatly enhances the range and control of in-situ doping of SiC during CVD growth was reported by Larkin et. al. [13,14]. This technique, called site-competition epitaxy, has enabled reproducible doping concentrations low enough to enable the fabrication of the first 2 kV SiC rectifiers ever reported [15], as well as doping concentrations high enough that a wide variety of contact metals form ohmic contacts to SiC in their as-deposited state [16]. Despite this recent accomplishment, further reductions in background epilayer doping concentrations will be needed before experimental SiC devices will be capable of 5 to 10 kV standoff voltages. Improvements in epilayer uniformity and surface morphology will also be needed as SiC upscales from prototype devices towards production integrated circuits, but it is anticipated that technological maturation via refined CVD reactor designs and growth conditions will address these problems.

3. Discrete Devices

A variety of small-area prototype SiC devices have been reported in the literature in recent years [17-19], and some have already made their way into the marketplace. Blue light emitting diodes were the first silicon carbide based devices to reach high volume sales, while small signal diodes and JFET's rated to 350 °C and ultraviolet-sensitive photodiodes are more recently introduced commercial products [8]. For the most part, early SiC devices have been produced using nonoptimized device designs and fabrication procedures. Only limited investigations into fundamental SiC device processing techniques, such as contact metallization, ion implantation, surface passivation, oxidation, and etching, have been carried out to date [17-19].

In spite of the lack of optimized fabrication processes, some highly encouraging results, only some of which are specifically mentioned below, have been obtained from a variety of prototype SiC devices. The experimentally realized power performance of prototype X-band SiC MESFET's on highly parasitic low-resistivity substrates nevertheless exceeds the theoretical maximum power output density attainable in GaAs MESFET's at 1 GHz [20,21]. The first microwave MESFET's fabricated on high resistivity 6H-SiC substrates attained a measured RF gain of 8.5 dB at 10 GHz and an f_{max} of 25 GHz [22]. Despite the low mobility of 6H-SiC in the vertical (c-axis) direction, Kimoto et. al. [23] demonstrated high voltage (500 - 1100 V) 6H-SiC Schottky rectifiers whose specific on-resistances were more than 10 times smaller than the theoretical minimum on-resistances attainable in silicon Schottky diodes. Operation of SiC p-n junction diodes, MOSFET's, MESFET's, JFET's, BJT's, and thyristors at temperatures above 300 °C (and in some cases as high as 650 °C) has been well-established (Fig. 1) [17-19]. When these unpackaged devices are operated in atmospheric environments at temperatures near 600 °C, chemical degradation of the contact metallizations

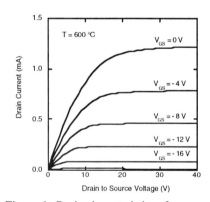

Figure 1. Drain characteristics of a 10 μm x 90 μm 6H-SiC buried-gate JFET at 600 °C.

restricts the functional lifetime to less than a few hours [24], but much longer 600 °C contact lifetimes have been demonstrated in inert non-oxidizing environments [25]. Clearly, reliable interconnection, passivation, and packaging technologies remain to be developed and proven before SiC devices can become truly useful in extreme high-temperature environments.

Given the extreme usefulness and success of MOSFET-based electronics in silicon, it is naturally desirable to implement high-performance inversion channel MOSFET's in SiC. Research results to date indicate that the quality of the SiO_2 formed by thermal oxidation of n-type 6H-SiC is comparable to oxides used for silicon MOSFET's [26]. However, thermal oxides grown on p-type 6H-SiC are generally poorer, exhibiting higher fixed charge and interface state densities. This leads to low inversion channel electron mobilities that seriously degrade the performance of n-channel MOSFET's [27]. However, on-going work towards improving the electrical quality of oxides on p-type SiC [28,29] offers some encouragement that the predicted advantages of inversion channel SiC N-MOSFET's (especially vertical power SiC MOSFET's [2]) might be realized. Since SiC devices will be operating at higher electric fields and temperatures than their silicon-based counterparts, challenging oxide and other surface passivation reliability issues will undoubtedly be faced as the technology progresses forward.

Though many of the SiC devices described above exhibit very promising area-normalized performances, micropipe defects present in the SiC wafers (and propagate into subsequently grown homoepilayers) prevent small-area prototype power device results from being scaled-up into useful large-area (> 1 mm^2), multi-amp power devices [30]. The micropipe defects generally lead to junction breakdown at electric fields well below the known critical-field. Figures 2 (top view) and 3 (cross-sectional view) show optical micrographs of localized microplasmas associated with premature reverse-bias failure at micropipes in 6H-SiC devices. The origin of these defects is still very much a topic of current debate and research [31-34], but their density has been steadily decreasing at a roughly twofold rate every year to a present-day minimum density of 55 per square centimeter [35]. The density of dislocation defects has been measured on the order of 10,000 cm^{-2}, but these defects are apparently not nearly as detrimental (as evidenced by SiC prototype device performances) as micropipes. The fact that areas larger than 1 mm^2 within SiC Lely-platelet crystals (which are not considered suitable for mass production) have been observed to be totally free of dislocations and micropipes [31] suggests that these defects are perhaps preventable.

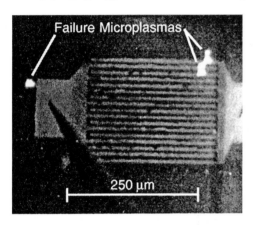
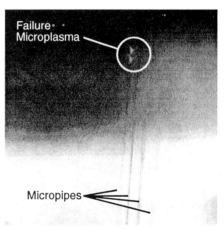

Figure 2. Top view of highly localized microplasmas observed in the near-dark on a probing station as a 6H-SiC transistor fails due to the presence of micropipes.

Figure 3. Cross-sectional view (~ 400X) of localized failure microplasma in micropipe running through a 6H-SiC p-n junction. After Ref. [30].

4. Integrated Circuits

Although they have a severe impact on high-field power devices, the micropipes appear to be less of a problem for signal-level electronics where devices are operated at much lower electric fields. This is evidenced by the recent achievement of Brown and co-workers at General Electric, who successfully fabricated the first complete monolithic integrated SiC operational amplifier chips [36]. The 1 mm x 2 mm op-amp chip shown in Figure 4 exhibited yields far higher than could be expected if micropipes were fatal to the active devices in this low-voltage circuit. Based on highly conservative 7 μm design rules, the circuit demonstrated 49 to 54 dB gains and bandwidths of 724 kHz to 269 kHz as temperature increased from 25 °C to 300 °C. The chip was based on depletion mode n-channel MOSFET technology, which alleviated the present difficulties associated with p-type SiC oxides needed to fabricate inversion channel N-MOSFET's [37].

An indication that the p-type oxide problems can be overcome somewhat can be found in the work of Xie and co-workers at Purdue University [38], who recently demonstrated SiC digital integrated circuits based on inversion channel N-MOSFET's. Basic digital logic gates, latches, flip-flops, binary counters, and half adder circuits with up to a dozen transistors were fabricated and successfully operated over the temperature range from 25 °C to 300 °C. These circuits are envisioned as first-generation prototypes for on-chip peripheral logic to drive 1-transistor non-volatile random access memory (NVRAM) arrays [39], but clearly demonstrate the present feasibility of small SiC digital integrated circuits. It is anticipated that increasingly larger digital circuits will be demonstrated in the near future, but the functional sizes and yields of such circuits will probably be influenced by the aforementioned crystal growth uniformity and defect issues.

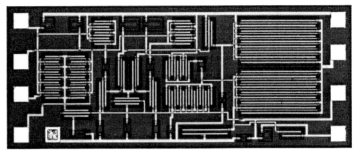

Figure 4. Photo micrograph of SiC MOSFET operational amplifier chip. The chip size is 1 mm x 2 mm. After Ref. [36]. (Courtesy of D. M. Brown, General Electric Company)

5. Conclusion

Although the advantageous properties of SiC have been known for decades, it was largely an enabling technical breakthrough in crystal growth that made mass production of useful SiC semiconductor devices and circuits seem possible. This, coupled with an acknowledged growing need for high temperature electronics, has led to SiC's accelerated development over the last half-decade. Although progress to date has yielded a few products and highly encouraging prototype results, some crucial technical obstacles remain to be solved before SiC can achieve its true potential. It is of paramount importance that crystal growth continue to improve, as larger wafers with far lower defect densities and improved epilayer doping and thickness control will be required for the majority of envisioned SiC electronic products. When combined with the continued maturation of device processing and high temperature packaging technologies, an increasingly capable variety of SiC devices and circuits will evolve to meet the system demands for hostile-environment electronics.

References

[1] Hingorani N G and Stahlkopf K E 1993 *Scientific American* **269** (5) 78-85
[2] Bhatnagar M and Baliga B J 1993 *IEEE Trans. Electron Devices* **40** 645-655
[3] Trew R J, Yan J-B and Mock P M 1991 *Proc. IEEE* **79** 598-620
[4] Nieberding W C and Powell J A 1982 *IEEE Trans. on Industrial Elect.* **29** 103-106
[5] Przybylko S J 1993 *American Institute of Aeronautics and Astronautics Report* AIAA 93-2581
[6] Tairov Y M and Tsvetkov V F 1978 *J. Cryst. Growth* **43** 209-212
[7] Barrett D L, Seidensticker R G, Gaida W, Hopkins R H and Choyke W J 1991 *J. Cryst. Growth* **109** 17-23
[8] Cree Research, Inc. (Durham, NC)
[9] Schaffer W J, Negley G H, Irvine K G and Palmour J W 1994 to appear in Diamond, SiC, and Nitride Wide-Bandgap Semiconductors *Materials Research Society Symposia Proceedings* (Pittsburgh, PA: Materials Research Society)

[10] Lomakina G A 1973 Silicon Carbide-1973 (Columbia, SC: University of South Carolina Press) 520-526
[11] Neudeck P G, Larkin D J, Starr J E, Powell J A, Salupo C S and Matus L G 1994 *IEEE Trans. Electron Devices* **41** 826-835
[12] Powell J A, Petit J B and Matus L G 1991 *Trans. 1st Int. High Temperature Electronics Conf.* (Sandia National Laboratories: Albuquerque, NM) 192-197
[13] Larkin D J, Neudeck P G, Powell J A and Matus L G 1994 to appear in *Appl. Phys. Lett.*
[14] Larkin D J, Neudeck P G, Powell J A and Matus L G 1994 Silicon Carbide and Related Materials *Institute of Physics Conference Series* **137** (Bristol, United Kingdom: IOP Publishing) 51-54
[15] Neudeck P G, Larkin D J, Powell J A, Matus L G and Salupo C S 1994 *Appl. Phys. Lett.* **64** 1386-1388
[16] Petit J B, Neudeck P G, Salupo C S, Larkin D J and Powell J A 1994 Silicon Carbide and Related Materials *Institute of Physics Conference Series* **137** (Bristol, United Kingdom: IOP Publishing) 679-682
[17] Davis R F, Kelner G, Shur M, Palmour J W and Edmond J A 1991 *Proc. IEEE* **79** 677-701
[18] Spencer M G, Devaty R P, Edmond J A, Khan M A, Kaplan R and Rahman M 1994 Silicon Carbide and Related Materials *Institute of Physics Conference Series* **137** (Bristol, United Kingdom: IOP Publishing) 465-690
[19] King D B and Thome F V 1994 Sessions X & XI*Trans. 2nd. Int. High Temperature Electronics Conf.* (Charlotte, NC) X-1 - XI-28
[20] Sriram S et al 1994 Silicon Carbide and Related Materials *Institute of Physics Conference Series* **137** (Bristol, United Kingdom: IOP Publishing) 491-494
[21] Weitzel C E, Palmour J W, Carter C H Jr. and Nordquist K J 1994 *this volume*
[22] Sriram S et al 1994 *52nd Annual IEEE Device Research Conference* (Boulder, CO)
[23] Kimoto T, Itoh A, Urushidani T, Jang S and Matsunami H 1994 *this volume*
[24] Neudeck P G, Petit J B and Salupo C S 1994 *Trans. 2nd. Int. High Temperature Electronic Conf.* (Charlotte, NC) X-23-28
[25] Crofton J, Williams J R, Bozack M J and Barnes P A 1994 Silicon Carbide and Related Materials *Institure of Physics Conference Series* **137** (Bristol, United Kingdom: IOP Publishing) 719-722
[26] Brown D M, Ghezzo M, Kretchmer J, Downey E, Pimbley J and Palmour J 1994 *IEEE Trans. Electron Devices* **41** 618-620
[27] Sheppard S T, Melloch M R and Cooper J A Jr. 1994 *IEEE Trans. Electron Devices* **41** 1257-1264
[28] Ouisse T, Becourt N, Templier F and Jaussaud C 1993 *J. Appl. Phys.* **75** 604-607
[29] Shenoy J, Lipkin L A, Chindalore G, Pan J, Cooper J A Jr., Palmour J W and Melloch M R 1994 *this volume*
[30] Neudeck P G and Powell J A 1994 *IEEE Electron Device Lett.* **15** 63-65
[31] Fazi C, Dudley M, Wang S and Ghezzo M 1994 Silicon Carbide and Related Materials *Institute of Physics Conference Series* **137** (Bristol, United Kingdom: IOP Publishing) 487-490
[32] Wang S, Dudley M, Carter C Jr., Asbury D and Fazi C 1993 Applications of Synchrotron Radiation Techniques to Materials Science *Materials Research Society Symposia Proceedings* **307** (Pittsburgh: Materials Research Society) 249-254
[33] Stein R A 1993 *Physica B* **185** 211-216
[34] Yang J-W 1993 Case Western Reserve University Ph. D. thesis (Cleveland, OH)
[35] Palmour J W 1994 personal communication
[36] Brown D M, Ghezzo M, Kretchmer J, Krishnamurthy V, Michon G and Gate G 1994 *Trans. 2nd. Int. High Temperature Electronics Conf.* (Charlotte, NC) XI-17-22
[37] Krishnamurthy V, Brown D M, Ghezzo M, Kretchmer J, Hennessy W, Downey E and Michon G 1994 Silicon Carbide and Related Materials *Institute of Physics Conference Series* **137** (Bristol, United Kingdom: IOP Publishing) 483-486
[38] Xie W, Pan J, Cooper J A Jr. and Melloch M R 1994 *this volume*
[39] Xie W, Johnson G M, Yang Y, Cooper J A Jr., Palmour J W, Lipkin L A, Melloch M R and Carter C H Jr. 1994 *this volume*

Ballistic Electron Transport and Superconductivity in Mesoscopic Nb-(InAs/AlSb) Quantum Well Heterostructures.

Herbert Kroemer, Chanh Nguyen*, and Evelyn L. Hu

ECE Department, University of California
Santa Barbara, CA 93106

Abstract. When superconductors and semiconductors are brought into barrier-free contact, the electrons can interact across the interface and the transport properties on the semiconductor side of the interface can get drastically modified. Under favorable conditions, the entire structure may become superconducting. The modification of the semiconductor properties is due to so-called Andreev Reflections. The preferred semiconductor for such experiments is InAs, especially in the form of modulation-doped quantum wells with AlSb barriers.

1. Introduction

Throughout their history, the fields of semiconductors and of superconductors have developed in parallel, with little overlap or interaction between those two solid-state disciplines. During the last few years, a small but increasing group of researchers have crossed over the boundary between the two fields, which has led to great mutual stimulation, and the area of overlap between the two fields is rapidly becoming an area of intense research activity. The purpose of this paper, written by a participant coming from the semiconductor side of the boundary, is to introduce his fellow semiconductor researchers to these developments.

When a superconductor and a semiconductor are brought together into atomically intimate contact, with an interface that is free from intervening oxides and/or contaminants, and which does not form an electron-blocking Schottky barrier, the electrons in the two materials can interact with each other in ways that can drastically alter the current flow through what may be called "super-semi-super double heterostructures." Under favorable conditions, superconductivity may be induced in the semiconductor itself. The key point is that the interface must be of such a quality that information about the *phase* of the Cooper pair wave functions is preserved in the electron transfer across the semiconductor.

Although transport across SpSm interfaces has been studied using other semiconductor systems [1-3], our own work since 1989 indicates that the effects of interest are most pronounced in structures in which the semiconductor is InAs, specifically, InAs in the form of quantum wells with AlSb barriers [4-8], in planar structures with a basic cross-sectional geometry as shown in Fig. 1. The present paper therefore restricts itself to structures of this kind, in all cases with Nb electrodes.

---

* Now at Hughes Research Laboratories, Malibu, CA 90265

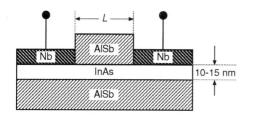

Fig. 1. Cross-sectional geometry of the basic InAs-AlSb quantum well structures employed in this work.

The principal reason why InAs *should* be the most suitable semiconductor is that the Fermi level at metal-to-InAs interfaces is pinned inside the InAs conduction band, leading to contacts without significant Schottky barriers impeding the free coupling of electrons between the two materials. This advantage was already pointed out by Clark et al. in their proposal for a hybrid Josephson FET's (JOFET's) [9]. In fact, much of the present work on SpSm structures — certainly our own — was probably stimulated by that work.

A second advantage of InAs is that it has higher electron mobilities than any other III-V compound except InSb. However, in bulk InAs, or homo-epitaxial InAs layers, this advantage can be utilized only to a limited extent, because the high electron concentrations required for induced superconductivity require doping levels that drastically reduced the low-temperature mobilities through impurity scattering. It is for this reason that in all of our own work we have used the InAs in the form of modulation-doped narrow (10-15nm) InAs quantum wells with AlSb barriers. Because of the very high barriers (1.35eV [10]) at the InAs-AlSb interface these quantum wells are very deep, which makes it possible to achieve very high electron concentrations inside the wells while maintaining high electron mobilities [11]. The work reported here employed MBE-grown wells 15nm wide, with electron sheet concentrations of about $8\times10^{12}\,\mathrm{cm}^{-2}$, and low-temperature (< 10K) electron mobilities of about $8\times10^4\,\mathrm{cm}^2\mathrm{V}^{-1}\mathrm{s}^{-1}$.

Fig. 2 shows the *I-V* characteristic of a device with the basic geometry of Fig. 1, with gap of 0.6µm. at two temperatures, 3.9K and 2.9K. The device width was 50µm. Readers familiar with Josephson junctions will immediately recognize the characteristics as "classical" Josephson characteristics (especially at the lower temperature), albeit with an unusually large critical current (20µA/µm), and unusually large $I_c R_n$ product ($\approx$ 1mV). Furthermore, such structures always retain strong non-linearities with a greatly enhanced zero-bias conductance even at temperatures above the onset of superconductivity ($\approx$3.9K in the present case), decreasing with increasing temperature, but remaining visible all the way up to until the Nb electrodes themselves go normal.

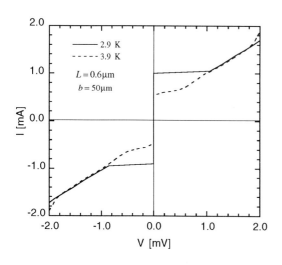

Fig. 2. Josephson-type *I-V* characteristics of a device with 0.6µm electrode separation, at two temperatures

2. Andreev Reflections

2.1. Basic Concept

The work of Clark et al. cited above assumed that the mechanism of supercurrent transport through the semiconductor is what we would like to call the *conventional* proximity effect, basically a form of Cooper pair tunneling through the normal part of the structure, as discussed in many textbooks. However, in the last few years it has become clear that in the kind of SpSm structures of interest here the proximity effect is dramatically enhanced by a phenomenon called *Andreev reflections* (AR's), a concept originated thirty years ago by Andreev [12], but receiving major attention only during the last few years, especially amongst those studying super-semi interactions.

Consider a semi-super interface between a degenerately doped semiconductor and a superconductor. As shown in Fig. 3, a superconducting energy gap has opened up on the superconductor side. If now an electron with an energy ε *above* the Fermi level (but still inside this gap) is incident on the interface from the semiconductor side, the absence of single-particle states within the gap prevents that electron from entering the superconductor as a *single* electron, and one might expect this electron to be reflected, and the electrical resistance to current flow across the interface actually to increase at the onset of superconductivity in the metal.

However, a single electron may pair up with a second electron at the same energy ε *below* the Fermi level, forming a Cooper pair, which *can* enter the superconductor, causing a doubling of the current compared to that in the absence of superconductivity, rather than a reduction. As the electron below the Fermi level is removed from the semiconductor, it leaves behind a hole below the surface of the Fermi sea. The generally accepted jargon associated with this phenomenon is to say that the incident electron is *reflected as a hole*, and the process itself is called an *Andreev reflection* (AR).

In a semiconductor with a large mean free path for the electrons ($\approx 3\mu m$ in our structures), the Andreev hole left behind at the interface has a large mean free path itself, roughly equal to that of the original electron, and theory shows that the hole travels back into the semiconductor along a trajectory that essentially re-traces the trajectory of the original incident electron. If its mean free path is sufficiently large, the hole will eventually reach the opposite superconducting electrode. In the absence of any bias across the structure, the energy of the hole is still within the superconducting gap on that side. Such a

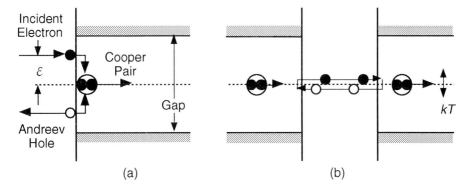

Fig. 3. Andreev reflections. (a): Basic concept. (b): Quasi-persistent current flow by multiple Andreev reflections.

hole cannot enter the superconductor, but it can be annihilated by breaking up a Cooper pair inside the other superconductor: One of the electrons of the pair annihilates the hole, the other electron takes up the annihilation energy, and is injected into the semiconductor as a ballistic electron *above* the Fermi level, at an equal to that of the initial electron. This process, illustrated in Fig. 3b, can evidently be repeated: The Andreev reflections act as what we might call a "Cooper pair pump" annihilating Cooper pairs on one side, and re-constituting them on the other. If *all* reflections of electrons and holes were Andreev reflections rather than "ordinary" reflections, and if there were no other kinds of scattering processes, the result would be a persistent current. However, perturbations are always present, and we would expect simply an enhancement of the conductivity by a factor equal to the number of ballistic traverses before an unfavorable reflection or collision event randomize either the electron or the hole flow in this chain reaction.

2.2. *Phase-Coherent Andreev Reflections and Persistent Currents*

In order to understand how multiple AR's can lead to a Josephson junction behavior with true supercurrents, it is necessary to consider the role of the *phase* of the electron and hole wave functions, neglected so far. In order for the state represented in Fig. 3b to be a stationary state, the round trip phase change must be a multiple of 2π, just as the ordinary bound states in a conventional GaAs-(Al,Ga)As are states for which the round trip phase change is such a multiple of 2π. The big difference for an AR stationary state is that the phase on one of the two traverses is carried by an electron, on the other traverse by a hole, but this does not change the requirement that the *total* round-trip phase change be a multiple of 2π. In fact, with regard to the electrons and holes in the semiconductor portion of the structure, the stationary states may indeed be viewed as a new kind of bound state, albeit one that actually carries current, in contrast to the current-less conventional bound states in a GaAs-(Al,Ga)As quantum well.

That in itself does not yet lead to a supercurrent through the structure: for every bound state, such as the state shown in Fig. 3b carrying Cooper pairs from the left to the right, there will also be a time-reversed loop with the same round trip phase shift, but with the opposite direction of the current flow. In thermodynamic equilibrium, both of these states have the same occupation probability, hence their current contributions cancel, and no *net* current will flow.

However, in order for two superconductors to be in true thermal equilibrium, it is necessary that no current would flow between them even if they were connected directly or through a conventional Josephson junction. Now it is a fundamental fact of superconductivity that in the presence of a non-zero phase difference between the two superconductors (non-zero here and in the following means modulo 2π), a current *will* flow between them when they are connected; true thermal equilibrium requires that the phases of their Cooper pair wave function be the same, or differ by a multiple of 2π.

It now turns out that any phase difference between the two superconductors also removes the degeneracy of the two stationary states of each pair of Andreev bound states. Just as the round-trip phase change in a GaAs-(Al,Ga)As quantum well contains a contribution from the penetration into the quantum well barriers (this is what causes the bound state energies to depend on the height of the barriers), the round trip phase change for an Andreev loop depends on the phase of the Cooper pairs in the two barriers, more specifically, on the phase *difference* between the two superconductors—the absolute phases do not matter. If there is zero phase difference, the two states of each pair of states have the same energy, and hence the same occupation probability. But when a non-zero phase

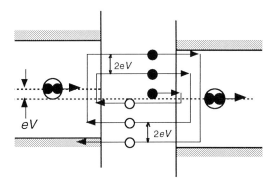

Fig. 4. Andreev "spiral" in the presence of a non-zero bias voltage.

difference is somehow introduced, the two states will move slightly in energy to re-establish their round-trip phase shifts of multiples of 2π. These energy shifts will be in opposite direction for the two states, that is, their degeneracy gets removed. This in turn leads to unequal occupation probabilities, and hence to non-canceling currents flowing between the two superconductors.

2.3 Andreev Reflection Effects at Non-Zero Bias: Sub-Gap Structure in dI/dV

Throughout the above discussion we had assumed that there is no bias voltage applied between the two superconducting electrodes. In the presence of a non-zero bias, the Andreev loop of Fig. 4b ceases to be a closed loop, and develops into a spiral, as shown in Fig. 4. The number of electron/hole traverses then becomes necessarily finite, decreasing with increasing bias voltage.

Whenever, with increasing bias, the number of traverses decreases by one, the differential conductance also decreases in quasi-steplike fashion, with the steps occurring at voltages roughly equal to the integer fractions of the superconducting gap voltage. They are readily visible in Fig. 5, which shows the differential conductance at 4.2K as a function of voltage for one of our samples with a relatively large 1µm electrode spacing, and hence no superconductivity in the temperature range investigated (>1.4K). (The sample *does* show a pronounced enhancement of the zero-bias conductivity).

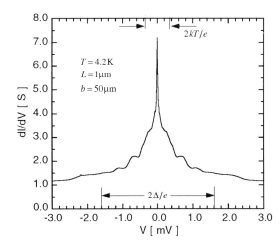

Fig. 5. Zero-bias spike, and sub-harmonic gap structure in the differential conductance dI/dV vs. voltage at 4.2K for a device with 1µm electrode spacing.

The existence of such a sub-harmonic-gap structure in the dI/dV characteristic is generally considered the key "fingerprint" evidence for the presence of pronounced multiple AR's [13], and it is in fact the most direct evidence that the electron transport in the kind of Sp-Sm-Sp structures discussed here is governed by multiple AR's.

2.4. Recent Advances

Our presentation has only scratched the surface of the field of Adreev-reflection-mediated current transport through Sp-Sm-Sp double heterostructures, concentrating on the discussion of the general principles of AR's rather than on presenting the most recent device results.

Most of our most recent work involved not single-gap devices as in Fig. 1, but series-connected arrays, made by laser holography, of large numbers (≥ 300) of nominally identical devices. This design has a number of advantages, for example, working with large number of series-connected devices improves the voltage sensitivity of the measurements in the range of near-vanishing resistance, at temperatures just above the onset of superconductivity, and under the influence of small magnetic fields — see [8].

3. Acknowledgements

This work was supported in part by the Office of Naval Research and in part by the National Science Foundation, the latter through the NSF Science and Technology Center for Quantized Electronic Structures, Grant DMR 91-20007, and through the NSF Materials Research Laboratory Program, Award DMR 912-3048. One of us (C.N.) acknowledges the financial support from the UCSB Vice Chancellor's Fellowship for Advanced Research on Quantized Structures.

References

[1] Takayanagi H and Kawakami T 1985 *Phys. Rev. Lett.* **54** 2449-2452.
[2] Ivanov Z, Claeson T, and Andersson T 1987 *Jpn. J. Appl. Phys.* **26 Supplement 3** DP31-32. (Proc. 18th Internat. Conf. on Low Temperature Physics, Kyoto, 1987)
[3] Nishino T, Hatano M, Hasegawa H, Kure T, and Murai F 1990 *Phys. Rev. B* **41** 7274-7276.
[4] Nguyen C, Werking J, Kroemer H, and Hu E L 1990 *Appl. Phys. Lett.* **57** 87-89.
[5] Nguyen C, Kroemer H, and Hu E L 1992 *Phys. Rev. Lett.* **69** 2847-2850.
[6] Kroemer H, Nguyen C, and Hu E L 1994 *Solid-State Electron.* **37** 1021-1025.
[7] Nguyen C, Kroemer H, and Hu E L 1994 *Appl. Phys. Lett.* **65** 103-105.
[8] Kroemer H, Nguyen C, Hu E L, Yuh E L, Thomas M, and Wong K C 1994 *Physica B* to be published. (Proc. NATO Advanced Research Workshop on Mesoscopic Superconductivity, Karlsruhe, 1994)
[9] Clark T D, Prance R J, and Grassie A D C 1980 *J. Appl. Phys.* **51** 2736-2743.
[10] Nakagawa A, Kroemer H, and English J H 1989 *Appl. Phys. Lett.* **54** 1893-1895.
[11] Nguyen C, Brar B, Bolognesi C B, Pekarik J J, Kroemer H, and English J H 1993 *J. Electron. Mat.* **22** 255-258.
[12] Andreev A F 1964 *Sov. Phys. JETP* **19** 1228-1231.
[13] Klapwijk T M, Blonder G E, and Tinkham M 1982 *Physica B+C* **109&110** 1657-1664.

Optimization of InAsSb/InGaAs Strained-Layer Superlattice Growth by Metal-Organic Chemical Vapor Deposition For Use In Infrared Emitters

R. M. Biefeld, K. C. Baucom, D. M. Follstaedt, and S. R. Kurtz
Sandia National Laboratory, Albuquerque, NM 87185-0601

Abstract. We have prepared InAsSb/InGaAs strained-layer superlattices (SLSs) by metal-organic chemical vapor deposition using a variety of growth conditions. The presence of an InGaAsSb interface layer was indicated by x-ray diffraction. This interface effect was minimized by optimizing the purge times, reactant flows, and growth conditions. The optimized growth conditions involved the use of low pressure, short purge times between the growth of the layers, and no reactant flow during the purges. Electron diffraction indicates that CuPt-type compositional ordering occurs in $InAs_{1-x}Sb_x$ alloys and SLSs which explains an observed bandgap reduction from previously accepted alloy values.

1. Introduction

Recent work by Menna et al. on a metal-organic chemical vapor deposition (MOCVD)-grown 3.06 μm diode laser yielded a maximum operating temperature of 35 K and threshold current densities of 200 - 330 A/cm^2 [1]. These results indicate the potential of this system for making infrared devices and the need for improved devices able to operate in this wavelength range at higher temperature[1]. We are continuing to explore the growth of $InAs_{1-x}Sb_x/In_{1-y}Ga_yAs$ SLS's for their possible use in mid-wave, 2-6 μm infrared optoelectronic and heterojunction devices. This system was chosen because the compositions span the 2-6 μm wavelength range and they can be grown lattice matched to InAs with type I band offsets. Our previous studies of the InAsSb ternary system have demonstrated accurate composition control through the use of a thermodynamic model [2,3]. We have made high quality infrared detectors in our laboratory from doped strained-layer superlattices (SLS's) grown by MOCVD in the Sb rich end of this ternary[2,3]. We have focused our recent efforts on the growth of $InAs_{1-x}Sb_x/In_{1-y}Ga_yAs$ heterostructures on InAs with emphasis on the growth and characterization of the InAsSb layers due to the importance of this material in the active devices [4,5]. We reported detailed growth conditions for $InAs_{1-x}Sb_x/In_{1-y}Ga_yAs$ SLS's and $InAs_{1-x}Sb_x$ alloys [4]. We have also reported on an optically pumped laser stripe constructed from these strained heterostructures with emission at 4 μm and the growth and characterization of a 4 μm infrared photodiode, both of which operated up to ≈100 K [5]. The efficiency of emitters in the 2-6 μm range is usually limited by Auger processes [4,5]. The increased electron-hole effective mass ratio of biaxially compressed InAsSb reduces the Auger transition rates and increases the infrared emission. In the present work, we determined the effects of growth conditions, purge times, and reactant flow sequences on the x-ray diffraction patterns of these SLS's from a series of experiments. Herein we discuss the details of the growth conditions and characterization results and the applicability to infrared devices of As rich $InAs_{1-x}Sb_x$ materials.

2. Experimental

This investigation was carried out in a previously described horizontal MOCVD system [6]. The sources of In, Sb and As were trimethylindium (TMIn), trimethylantimony (TMSb), trimethylgallium (TMG), and 100 percent arsine (AsH$_3$), respectively. Hydrogen was used as the carrier gas at a total flow of 8 SLM. The V/III ratio was varied from 2.7 to 10.3 over a temperature range of 475-525 °C, at total growth pressures of 200 to 660 Torr and growth rates of 0.75 to 3.0 µm/h. The group V molar fraction of TMSb in the vapor phase [nTMSb/(nTMSb + nAsH$_3$)] was varied from 0.06 to 0.09.

The growth was performed on (001) InAs substrates. InAs was cleaned by degreasing in solvents and deionized water. It was then etched for 20 to 30 seconds in a 50:1 mixture of sulfuric acid and hydrogen peroxide, rinsed with deionized water and blown dry with nitrogen.

Sb compositions, x, reported for the InAs$_{1-x}$Sb$_x$ layers were determined by x-ray diffraction using a Cu x-ray source and a four crystal Si monochromator. The (004) reflection was used to measure the lattice constant normal to the growth plane and the (115) or (335) reflections were used to determine the in-plane lattice constant [6]. In this way the composition determination is corrected for partial strain relaxation by misfit dislocations. The simulations were done using the Bede RADS program.

The optical properties of these SLS's were determined by infrared photoluminescence and absorption measurements. Infrared photoluminescence was measured at 14 K using a double-modulation technique which provides high sensitivity, reduces sample heating, and eliminates the blackbody background from infrared emission spectra [7].

Cross-sectional specimens of some samples were prepared for transmission electron microscopy (TEM) by cleaving along (110) and epoxying two alloy surfaces together. TEM examination was done using a Philips CM20T (200 keV) microscope. The specimens were examined in a <110> direction perpendicular to the [001] growth direction. Both transmission electron diffraction (TED) patterns and TEM images were obtained.

3. Results And Discussion

The growth results for the As rich end of the InAs$_{1-x}$Sb$_x$ ternary are similar to those described previously for the Sb rich end of the ternary [2]. The growth rate was found to be proportional to the TMIn and the TMGa flow into the reaction chamber and independent of the TMSb and AsH$_3$ flow. The observed trends for the effects of input vapor concentrations on the resulting solid composition can be described by a thermodynamic model [2]. The model predicts that the more stable III/V compound will control the composition. For the InAsSb system when the vapor phase III/V ratio < 1, As is preferentially incorporated into the solid and the solid-vapor distribution coefficient of Sb (k_{Sb}) is < 1. This is because InAs has a lower free energy of formation than InSb at 475-525 °C. For III/V ratios close to one, k_{Sb} approaches one and for III/V $\geq$ 1, k_{Sb} = 1. When III/V $\geq$ 1, all of the As and Sb will be incorporated into the solid. Some slight deviations from the predicted behavior of the thermodynamic model are observed for the present results. For some samples grown at 475 °C, k_{Sb} appears to be $\geq$ 1. This can be explained by the incomplete decomposition of AsH$_3$ at temperatures below 500 °C [2,8]

An x-ray diffraction pattern of the (004) reflection of a selected InAs$_{1-x}$Sb$_x$/In$_{1-y}$Ga$_y$As SLS (CVD1263) grown on InAs is shown in Figure 1. The pattern shows the intense InAs peak and as well as the satellites due to the SLS. This sample was grown at atmospheric pressure with 30 second purges between layers with uninterrupted AsH$_3$ flow. The calculated pattern shown in curve C is for an InAs$_{0.9}$Sb$_{0.1}$/ In$_{0.91}$Ga$_{0.09}$As SLS with 100 Å/ 130 Å layer thicknesses. The slight asymmetry and the width of the peaks in the observed pattern is most likely due to the existence of an interface layer between the layers in the SLS. High resolution TEM confirms the existence of an interface layer in similar samples. Several simulations were tried in an effort to match the observed spectrum. If the thickness and composition of both the InGaAs and InAsSb layers were varied then the pattern became uniformly broad. If only the Sb composition and layer thickness was varied in the InAsSb layer but the sum of the thickness of both layers remained constant then the pattern remained symmetric and sharp. If the compositions of the layers including two In$_{1-x}$Ga$_x$As$_{1-y}$Sb$_y$ interface layers were varied but the total period remained constant the

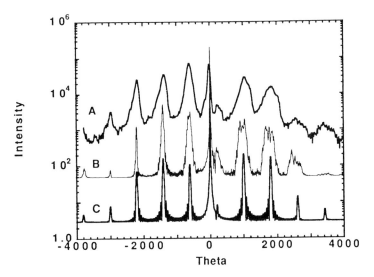

Figure 1. Curve A is the observed x-ray diffraction pattern of the (004) reflection of CVD1263. It is compared to calculated patterns in the lower curves to illustrate the improved match when interface layers are included in the x-ray simulation B when compared to a simulation with no interface layers shown in C.

simulation produced uniformly broadened peaks. The best match was obtained by varying the composition and thickness of the InAsSb layer and the interface layers through the structure with no variations for the InGaAs layer. This simulation is also illustrated in Figure 1, curve B as explained below.

In Figure 1, curve B, the calculated pattern is for a structure that consists of 100 layers grouped in five segments with different interface layer compositions and thicknesses. The structure contains the basic period of

$$In_{1-x}Ga_xAs_{1-y}Sb_y/InAs_{1-z}Sb_z/In_{1-x}Ga_xAs_{1-y}Sb_y/In_{1-w}Ga_wAs,$$

where the two interface layers are identical. The $In_{1-w}Ga_wAs$ layers are invariant throughout the structure with compositions of $w = 0.09$, and thicknesses of 110 Å each. The thickness of the $InAs_{1-z}Sb_z$ layers are all equal but the composition varies from $z = 0.06$ to 0.08 in 4 steps of 0.005. Both the composition and the thickness of the interface layers change with the layer thickness varying from 2 to 10 Å in steps of 2 Å and the Sb composition changing from $z = 0.05$ to 0.09 while the Ga composition remains unchanged at $x = 0.1$. The order of the layers makes no difference to the calculated pattern. This simulated pattern, curve B, is a good match to the observed pattern, curve A, and is support for the existence of an interface layer as the explanation for the width and asymmetry observed in the x-ray diffraction pattern.

A variety of growth conditions were explored in an effort to improve the sharpness of the x-ray diffraction pattern. The optimum x-ray diffraction pattern was obtained at 200 Torr with a 10 second growth interrupt at 475 °C, a V/III ratio of 7, a growth rate of 1.8 μm/hour, and no reactants flowing during the interrupt. An x-ray diffraction pattern of an SLS grown using the optimum growth technique was simulated without using interface layers and a very good match between the observed and calculated pattern was obtained. This improvement is due to the minimization of exposure of the InAsSb surface to AsH3 and/or As. Extended AsH3/As exposures are known to cause interface roughening in the

InGaAsP/InP and InAs/GaSb systems [9,10]. The presence of Ga in these layers can not be explained by the alteration in the purging times and gas flows since no TMGa was used during the purges. The presence of Ga in the interfaces must be due to either the roughness of the surfaces which enables mixing of Ga into the InAsSb layers at the interface or the diffusion of Ga into the InAsSb during growth.

The bandgap of the unstrained, $InAs_{0.9}Sb_{0.1}$ alloy was determined to be approximately 270 meV from optical studies of bulk ternary alloy and subsequent studies of SLSs and quantum wells. The accepted value for $InAs_{0.9}Sb_{0.1}$ is $\approx$ 330 meV [8]. Throughout our studies of As-rich, InAsSb (5-50% Sb), the optically determined bandgap of InAsSb alloys was smaller than accepted values [5,11]. This InAsSb bandgap anomaly was observed in both MOCVD and MBE grown samples.

The TED patterns for $InAs_{1-x}Sb_x$ (x=0.07-0.14) alloys and an InGaAs/InAsSb SLS show bright reflections of the [110] zone pattern of the zinc-blende lattice as well as non-zinc-blende reflections at half the distance between (000) and {111} spots. These weak reflections indicate that compositional ordering of the Cu-Pt type is occurring on the {111}B planes of the group-V sublattice which is the same type as that which was previously observed for $InAs_{0.6}Sb_{0.4}$ alloys and SLSs [11]. Compositional ordering of InAsSb should result in bandgap reduction as seen above even for low Sb concentrations [11,12].

In conclusion, we have established the growth conditions for $InAs_{1-x}Sb_x/In_{1-y}Ga_yAs$ SLS's. The vapor-solid distribution coefficient for Sb can be described by a thermodynamic model. The PL peak energies of the SLS's and alloys grown under the conditions of this study are lower than those previously reported. The lower energy anomaly in these SLSs is probably due to ordering which is observed in $InAs_{1-x}Sb_x$ alloys and SLSs. We anticipate that with these improvements in material quality, we should soon realize a reduction of Auger rates and higher temperature operation of midwave emitters with biaxially compressed InAsSb.

Acknowledgments

We wish to thank J. A. Bur and M. W. Pelczynski for providing technical assistance. This work was supported by the US DOE under Contract No. DE-AC04-94AL85000.

References

[1] R. J. Menna, D. R. Capewell, R. U. Martinelli, P. K. York, and R. E. Enstrom, Appl. Phys. Lett. 59, (1991) 2127.
[2] R. M. Biefeld, J. Crystal Growth 75 (1986) 255.
[3] R. M. Biefeld, S. R. Kurtz, and S. A. Casalnuovo, J. Crystal Growth 124 (1992) 401.
[4] R. M. Biefeld, K. C. Baucom, S. R. Kurtz, and D. M. Follstaedt, Mat. Res. Soc. Symp. Proc., Physics and Applications of Defects in Advanced Semiconductors, 325, 493 (1994).
[5] S. R. Kurtz and R. M. Biefeld, L. R. Dawson, K. C. Baucom, and A. J. Howard, Appl. Phys. Lett. 64, 812 (1994).
[6] R. M. Biefeld, C. R. Hills and S. R. Lee, J. Crystal Growth 91 (1988) 515.
[7] S. R. Kurtz and R. M. Biefeld, Phys. Rev. B 44, 1143 (1991).
[8] Z. M. Fang, K. Y. Ma, D. H. Jaw, R. M. Cohen, and G. B. Stringfellow, J. Appl. Phys. 67, 7034 (1990).
[9] G. Landgren, J. Wallin, and S. Pellegrino, J. Electronic Mater., 21 (1992) 105.
[10] D. H. Chow, R. H. Miles, J. R. Soderstrom, and T. C. McGill, J. Vac. Sci. Technol. B8 (1990) 710.
[11] S. R. Kurtz, L. R. Dawson, R. M. Biefeld, D. M. Follstaedt, and B. L. Doyle, Phys. Rev. B 46, 1909 (1992).
[12] D. M. Follstaedt, R. M. Biefeld, S. R. Kurtz, and K. C. Baucom submitted to J. Electronic Materials

Growth of Aluminum Antimonide and Gallium Antimonide by MOMBE

K. SHIRALAGI[†], R. Tsui[†], D. Cronk[†], G. Kramer[†], and N. D. Theodore*

[†]Motorola Inc., Phoenix Corporate Research Laboratories, 2100 E. Elliot Road, Tempe, AZ 85284
*Motorola Inc., Materials Research & Strategic Technologies, 2200 W. Broadway, Mesa, AZ 85201

Abstract: The growth of AlSb and GaSb by MOMBE (utilizing trimethylaminealane and triethylgallium) has been studied. We have explored the effects of temperature and V/III flux ratio on the growth rates using RHEED and cross-sectional TEM studies. Both compounds exhibit a very strong decrease in growth rate with decreasing temperature and/or increasing antimony flux. Conditions under which device quality material can be obtained in the 500 °C temperature range will be reported. Results are also compared with those for antimonide growth in MBE and arsenide growth in CBE to develop a better understanding of the reaction kinetics.

1. Introduction

A number of novel devices such as heterojunction-FETs (Werking *et al.* 1992) and resonant tunneling diodes (Söderström *et al.* 1989, Tehrani *et al.* 1994) have been demonstrated using the InAs/(Al,Ga)Sb material system. A wide range of band gap energies (0.3 to 1.7 eV), large conduction band discontinuities, and high electron mobility in InAs are some of the attractive features of this material system. These quantum structures, consisting of very thin, nearly lattice matched layers, are typically grown by conventional MBE using elemental Al and Ga as the group III sources for the antimonides. MOMBE combines many of the advantages of MBE and MOVPE, but relatively little has been reported in the literature on the growth of these compounds by this technique, which allows quick and accurate flux changes and provides easier composition control for the growth of ternary alloys.

In this study, the layers were grown on GaAs substrates. A superlattice buffer layer is used to reduce the density of dislocations caused by the large lattice mismatch ($\approx 8\%$) and to obtain strong intensity oscillations in reflection high energy electron diffraction (RHEED) measurements. The growth of antimonides using solid Sb, triethylgallium (TEG) and trimethylaminealane (TMAA) was investigated. In the past, triisobutylaluminum (TIBA) was chosen over trimethylaluminum because the strong Al-C bonds in the latter severely degraded layer quality in MOVPE (Tromson-Carli *et al.* 1981). Later, TIBA was found to be unsuitable for use in MOMBE due to its low vapor pressure and thermal instability. More recently, TMAA has been gaining popularity as a cleaner alternate source, especially for the growth of AlGaAs (Abernathy and Bohling 1992). TMAA exhibits a vapor pressure of ≈ 1 torr at room temperature and does not contain C-Al bonds that enhance carbon incorporation in the layers.

TMAA was also used for the growth of AlGaSb and was shown to be promising for growth at high temperatures, with the growth behavior reported to be similar to that of TEG as used in the growth of GaSb (Okuno et al. 1993). The presence of InAs in many of the novel heterostructures results in an upper limit for the growth temperature of ≈ 500 °C for the active region of the device. However, previously published results on the MOMBE of AlSb (Okuno et al. 1993) reported a very large decrease in growth rate for a small temperature change in the 500 °C range, thus making MOMBE less attractive as a growth technique for these applications. In this study, we have optimized the growth of AlSb at more reasonable growth rates even at lower substrate temperatures to obtain good quality material. Results of growth studies on AlSb and GaSb in the 500°C-600°C temperature range at varying V/III ratios are reported. These results are also compared with GaAs/AlAs growth rates and the growth of antimonides using elemental Al and Ga. The InAs for device structures was grown using trimethylindium (TMI) and arsine, the details of which will be reported elsewhere.

2. Experimental procedure

The growth was performed in a Fisons V90H MOMBE evacuated by a diffusion pump with a liquid nitrogen cold trap. TEG and TMI were introduced into the growth chamber through one of the injectors kept at 70°C and TMAA through another injector. A set of control valves were used to maintain constant pressure of the alkyls before injection. TMI and TEG have been used extensively in MOMBE and MOVPE in the past with good success (Jones 1993). In this study TMAA was kept at a bath temperature of about 25 °C, with a slightly higher temperature for the line through which the vapors are passed before being injected into the chamber. Elemental Al and Ga were also used to grow AlSb and GaSb for comparison purposes. Growth was performed on (100) oriented semiinsulating GaAs substrates. After oxide blow-off and the growth of a thin GaAs smoothing layer, the substrate was cooled to the required antimonide growth temperature and a 20 period AlSb/GaSb superlattice was grown, followed by 0.5 µm of AlSb. Temperature measurements were done with an infrared pyrometer which was calibrated in the 500 to 600 °C range by using the InSb melting point as well as the temperatures for oxide blow-off and the transition of the RHEED pattern from an As-stable (2x4) to a C(4x4) reconstruction for GaAs. Growth rates were obtained by RHEED intensity oscillations and by cross-sectional transmission electron microscopy on a structure consisting of alternating layers of GaSb and AlSb grown under different conditions.

3. Results and discussion

Alkyl Al and Ga arrival rates were chosen to result in a growth rate equal to 0.5 monolayer (ML)/s of AlAs and GaAs, respectively, in the 500 °C temperature range. Similar growth rates were used while growing with elemental Al and Ga sources. An As-stable GaAs surface, subsequent to the oxide blow-off and the growth of a thin GaAs buffer layer, shows a (2x4) RHEED reconstruction. The same surface under an Sb flux shows a diffused pattern before the commencement of AlSb growth. Upon the initiation of AlSb growth, the pattern turns spotty in a couple of seconds and within 30 seconds, the Sb-stable (1x3) pattern begins to appear and progressively becomes clearer. The pattern becomes well defined in about a minute, quite similar to the observations for conventional MBE. With alkly Ga, however, the reconstruction shows a diffused (1x3) pattern even after the growth of GaSb for several minutes, and the layer starts to become hazy even when grown under a wide range of V/III flux ratios and temperatures. GaSb with excellent surface morphology could only be obtained when grown on top of a starting AlSb layer a few hundred angstroms thick. The presence of the AlSb initiates migration of alkyl Ga molecules to step edges, resulting in a layer-by-layer growth. The alkyl Ga has a short residence time in the presence of excess Sb to begin with, and it starts to form islands on top of GaAs, which do not coalesce due to short migration lengths as in the case of InAs grown on GaAs under high As flux (Kubiak *et al.* 1984).

All layers in this study were grown under group V-stable conditions with (1x3) reconstruction patterns since attempts at growing under group III-rich conditions yield droplets, similar to observations in the growth of other III-V compounds. Under group III-stable conditions (V/III ratios < 1.0), AlSb changes to a (4x2) reconstruction. The minimum Sb flux needed to maintain group V stable conditions is determined by measuring the time it takes for an Al-rich surface in the absence of any Sb flux to recover back to a Sb-stable condition when Sb is supplied once again to the surface. Unfortunately, this approach does not work well for GaSb. Here, the Sb-stable (1x3) pattern becomes more diffused as the Sb flux is reduced and no clear (4x2) pattern is observed when growth becomes Ga-rich. Surface haze due to the formation of droplets confirms this Sb-deficient state.

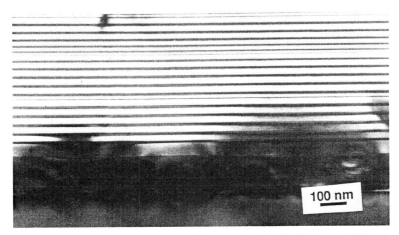

Fig. 1. Cross sectional TEM picture of three stacks of GaSb/AlSb layers; V/III ratio of 1.0, 1.5 and 3.0 was used for each stack from bottom up. The middle stack consists of 5 pairs with temperature decreasing from 580°C in steps of 20°C for each pair in the growth direction, and the other two with temperature increasing in steps of 20°C.

RHEED intensity oscillations could be obtained for AlSb and GaSb grown by MBE and AlSb by CBE, for layers grown on the buffer. Oscillations were examined in a temperature range of 480 to 600 °C and over a wide range of Sb over-pressure. Below 480 °C, very low growth rates are seen that cannot be determined accurately with our RHEED measuring setup, so in the case of MOMBE they were confirmed by growing alternate layers of GaSb and AlSb, 400 Å and 200 Å thick, respectively, as shown in Fig. 1. The cross-sectional TEM picture shows at the bottom a superlattice and buffer structure grown on top of a GaAs substrate, followed by three stacks of alternating AlSb/GaSb layers, with each stack grown with a different Sb over-pressure. A thin marker layer helps locate each stack easily. Within each stack, each pair of GaSb (dark layer) and AlSb (light layer) was grown at one substrate temperature. The middle stack was grown with the temperature decreasing from nominally 580 °C to 500 °C in 20 °C steps for each pair, while the top and bottom stacks were grown with the substrate temperature increasing from the bottom up. Some qualitative information can be easily seen form the picture. The high density of dislocations is quickly reduced after the growth of the buffer. The bottom stack, grown at the lowest Sb over pressure (V/III ≈ 1.0), is the thickest while the top one, grown at the highest Sb flux (V/III ≈ 3.0), is the thinnest. Even within each stack, layers grown at higher temperatures are thinner and this effect is stronger with higher Sb flux (top stack). AlSb layer thicknesses are close to the target thickness of 200 Å (except in the top stack where they are significantly thinner), but GaSb layers are much thinner even though the target thickness was 400 Å.

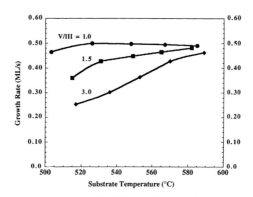

Fig. 2. Growth rate of MOMBE AlSb as a function of substrate temperature for increasing V/III ratios as marked. Arrival rate of Al corresponds to 0.5 ML/s.

Growth rates based on RHEED and TEM measurements are plotted in Fig. 2 for the growth of AlSb by MOMBE. The rate approaches that of AlAs (0.5 ML/s) for temperatures greater than 580 °C, independent of the Sb flux. When the V/III ratio is close to unity, the AlSb growth rate remains nearly constant over a wide temperature range. Layers with mirror like surface as seen by the unaided eye can be obtained under all these growth conditions. The surface morphology is similar to that of AlSb grown using elemental Al. Only when grown at a high V/III ratio does the rate decrease rapidly as shown and this results in an activation energy of about 14 Kcal/ mol in agreement with previously published results (Okuno *et al.* 1993). Under these conditions, the excess Sb on the surface was believed to suppress the decomposition of alkyl Al, resulting in enhanced desorption from the surface and a reduced growth rate. However, with low Sb over-pressure (V/III ≈ 1.0), AlSb growth proceeds at the arrival rate of Al molecules. Under limited Sb supply, the growth rate agrees with that of AlAs, so the case is similar to where excess As readily desorbs from the surface and site blocking does not occur. When grown with elemental Al, no significant variation in AlSb growth rate is observed for a wide range of temperatures and Sb over-pressure. Here the presence of excess Sb does not instigate any Al desorption (due to the high sticking coefficient of elemental Al), resulting in a constant growth rate (0.5 ML/s in this case) over the entire range.

Fig. 3 is a plot of the AlSb growth rate as a function of V/III ratio at a TMAA arrival rate corresponding to about 0.5 ML/s. The V/III ratio has a more pronounced effect at lower temperature than at temperatures close to 600 °C. This may be attributed to the more efficient desorption at higher temperatures of excess Sb, resulting in lesser surface coverage and a growth rate approaching that defined by the Al arrival rate. Thus with low V/III ratios, AlSb can be grown at higher rates even at temperatures close to 500 °C as needed for structures with In containing compounds. Owing to the instability of TMAA, anomalous growth rates may be seen when the vapors are left unused at high gas line temperatures for long periods of time before being fed into the injector.

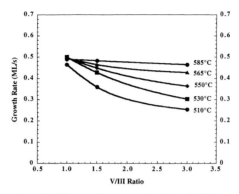

Fig. 3. MOMBE AlSb growth rate as a function of V/III ratio for various substrate temperatures as marked and for a fixed group III arrival rate.

GaSb growth rate variations plotted in Fig. 4 show trends similar to that of AlSb. As a comparison, the growth rate of GaAs in CBE remains fairly constant in the 500-550 °C temperature range and shows a gradual fall-off beyond that. However, the GaSb growth rate is less than that of GaAs (0.5 ML/s) even at high temperatures. At low temperatures, ≈ 500 °C, growth rates are a very strong function of Sb flux and higher rates can be achieved at lower V/III ratios. Growth rate reduction due to the presence of excess Sb prohibiting Ga decomposition has been explained in the past by means of a theoretical model and compared with that of GaAs (Kaneko *et al.* 1992). Fig. 4 shows that by reducing the effective excess Sb coverage, higher growth rates are possible. Of course, higher growth rates can also be achieved by increasing the TEG arrival rate and that effectively reduces the V/III ratio and also prevents the sharp growth rate fall off at lower temperatures. No such variation is seen when GaSb is grown with elemental Ga; the growth rate remains constant at 0.5 ML/s in the entire temperature range of interest.

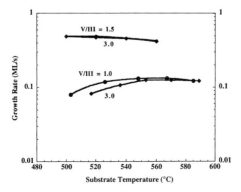

Fig. 4. MOMBE GaSb growth rate as a function of substrate temperature for two V/III ratios (bottom curves). Corresponding GaAs growth rate with arsine are shown for two V/III ratios (top curves). All data correspond to a group III arrival rate of 0.5 ML/s.

AlSb bulk layers were grown directly on GaAs substrates whereas GaSb layers were grown on a thin AlSb nucleating layer for the measurement of electrical properties. 1 μm thick AlSb layers exhibit a room temperature hole concentration of 3-5 x 10^{16} /cm^3 and a mobility of 150 cm^2/Vs, comparable to a solid source Al grown layer. GaSb yielded similar carrier concentrations with a mobility of 400 cm^2/Vs. InAs/(Al,Ga)Sb resonant interband tunneling diodes which consisted of 25 Å thick barriers of AlSb (grown by MOMBE) on either side of a 65 Å thick GaSb quantum well (grown with elemental Ga) yielded respectable peak-to-valley current ratios of 5-8 at room temperature. These layers were grown at 500 °C with growth rates obtained from results of the present study.

4. Conclusions

The growth of AlSb and GaSb by MOMBE on GaAs substrates has been investigated. Growth rates were determined by RHEED intensity oscillation and cross-sectional TEM measurements. Both compounds show very highly temperature-dependent growth rates under high V/III ratios, in strong contrast to the growth of GaAs/AlAs using alkyl sources or GaSb/AlSb using elemental sources. However, when grown with low V/III ratios, higher growth rates comparable to those obtained at high temperatures and layers with good surface morphology can both be obtained even at 500 °C. In the case of GaSb, a thin AlSb starting layer is necessary for nucleation on GaAs substrates. The data provide growth conditions for quantum structures consisting of InAs layers and antimonides by MOMBE.

5. Acknowledgment

This work was partially performed under the management of FED (the R&D Association for Future Electron Devices) as a part of the R&D of Basic Technology for Future Industries supported by NEDO (New Energy and Industrial Technology Development Organization, Japan).

6. References

Abernathy C R and Bohling D A 1992 *J. Crystal Growth* **120** 195
Asahi H, Kaneko T, Okuno Y and Gondo S 1991*J. Crystal Growth* **107** 1009
Jones 1993 *J. Crystal Growth* **129** 728
Kaneko T, Asahi H and Gondo S 1992 *J. Crystal Growth* **120** 39
Kubiak R A A, Parker E H C, Newstead S, and Harris J 1984, *Appl. Phys. A* **35** 61
Okuno Y, Asahi H, Liu X.F, Inoue K., Itani Y, Asami K, and Gondo S 1993 *J. Crystal Growth* **127** 143
Söderström J R, Chow D H, and McGill T C 1989 *Appl. Phys. Lett.* **55** 109
Tehrani S, Shen J, Goronkin H, Kramer G, Hoogstra M, and Zhu T 1994 in *Proc. 20th International Symp. on GaAs and Related Compounds, Freiburg, 1993* (IOP, Bristol), p. 209
Tromson - Carli A, Gibart P, Bernard C 1981 *J. Crystal Growth* **55** 125
Werking J D, Bolognesi C R, Chang L D, Nguyen C, Hu E L, and Kroemer H 1992 *IEEE Electron Dev. Lett.* **EDL13** 164

Low-pressure-MOVPE growth and characterization of InAs/AlSb/GaSb-heterostructures

M. Behet, P. Schneider, D. Moulin, H. Hamadeh, J. Woitok*, K. Heime

Institut für Halbleitertechnik, RWTH Aachen, Templergraben 55, 52056 Aachen, Germany
*I. Physikalisches Institut, RWTH Aachen, Templergraben 55, 52056 Aachen, Germany

Abstract. This contribution presents for the first time results on the deposition and characterization of InAs/AlSb/GaSb-heterostructures that were grown by MOVPE at low total pressure (p_{tot} = 20 hPa). A comparative investigation on the growth kinetics of the basic binary compounds revealed that only the combination of the metalorganics TEGa, TEAl, TMIn and TESb exhibited a well matched reactivity for the deposition of these heterostructures. X-ray diffraction and photoluminescence measurements on GaSb/AlSb- and InAs/AlSb-multi quantum well (MQW) structures proved the good structural and optical quality of the layer stacks. Finally, a symmetrical nin-InAs/AlSb DBRT-structure (double barrier resonant tunneling) was fabricated exhibiting regions of negative differential resistance in the I/V-characteristic.

1. Introduction

Among III-V compound semiconductors, InAs/AlSb/GaSb constitutes a closely lattice matched material system, yet significantly differing in band parameters and transport properties [1]. For example, combining the high intrinsic electron mobility in InAs with the largest conduction band offset (1.35 eV) of all III-V-heterostructures in InAs/AlSb-layer structures opens new perspectives for high speed electronic applications. The unique features of InAs/AlSb/GaSb-heterostructures have already resulted in a number of elegant device concepts and applications, such as double barrier resonant (interband) tunneling structures [2,3], n-, p-channel HFET [4,5] and strained layer superlattices for infrared applications [6].

In spite of these possible applications Sb-based semiconductors have received far less attention than their As- and P-analogues. This is most probably due to the fact that the device processing technology for Sb-containing semiconductors has not been developed to the sophisticated level which has become a standard for As- and P-based alloys. Furthermore, the absence of semiinsulating GaSb and InAs substrates complicates the on-wafer device isolation.

A possible solution could be the use of semiinsulating GaAs or InP substrates. This of course necessitates the deposition of thick buffer layers to accomodate the large lattice mismatch.

Up to now, all studies on InAs/AlSb/GaSb-based structures and devices have been performed on MBE-samples. Due to the MOVPE-specific problems like suitable Al-precursors and the necessity of higher deposition temperatures compared with MBE InAs/AlSb/GaSb-structures have not been realized by MOVPE so far. Our paper presents for the first time results on the deposition and characterization of InAs/AlSb/GaSb-heterostructures which were grown by MOVPE at low total pressure (p_{tot} = 20 hPa). A prerequisite for the growth of high quality heterostructures is the ability to deposit the binary bulk materials reproducibly at a set of compatible growth parameters. Therefore, we first briefly summarize experimental results on the growth kinetics and characterization of the binary semiconductors GaSb, AlSb and InAs. Afterwards, X-ray diffraction and photoluminescence measurements on GaSb/AlSb- and InAs/AlSb-MQW structures will be presented. Basic problems like the gas switching sequence in InAs/AlSb-MQW structures will also be treated. Finally the feasibility for the device quality of the MOVPE-grown structures is shown by a nin-InAs/AlSb-DBRT structure.

2. Experimental

A standard LP-MOVPE system with a horizontal quartz reactor was used for the epitaxial growth. Starting materials for the group III components were TEGa, TMIn, TEAl and for group V components TESb, AsH_3 and PH_3. Palladium-diffused hydrogen was used as carrier gas. The growth experiments were carried out at a total pressure of 20 hPa and a carrier gas velocity of 1.2 m/s. GaSb and InAs substrates were used for deposition. Before growth an annealing treatment of the substrate at 870 K for 5 minutes under TESb (GaSb substrate) or AsH_3 (InAs substrate) ambient was carried out in order to remove any remaining oxide layers.

Layer thicknesses of the bulk materials were determined by measuring the weight of the samples before and after growth (accuracy : ± 0.1 μm). The crystallinity and period thickness of the MQW structures was verified by double crystal X-ray diffraction, the optical quality of the samples by low temperature photoluminescence measurements.

3. Results and discussion

First, we performed a comparative study on the growth kinetics of InAs, GaSb and AlSb with methyl- and ethyl-group III precursors to determine the preferential combination for the LP-MOVPE deposition of these binary semiconductors. For the growth of InAs/AlSb/GaSb-heterostructures it is essential to utilize precursors with similar reactivity. This is a prerequisite to evaluate a set of compatible deposition parameters.

Fig. 1 shows the growth rate of InAs, GaSb and AlSb as a function of the deposition temperature for the use of TEGa, TEAl, TMIn, TESb and AsH_3. We found that the ethyl-precursors exhibit a well matched reactivity for GaSb and AlSb. A kinetically controlled growth regime was determined for temperatures lower than 870 K. At higher temperatures the deposition rate saturates indicative of diffusion limited growth. Diffusion limitation is also present for InAs over the whole displayed temperature range. Kinetically controlled growth for InAs occurs for deposition temperatures lower than 700 K [7].

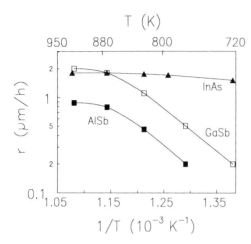

Fig. 1: *Temperature dependence of InAs, GaSb and AlSb growth rate (parameters: p(TMIn) = 0.29 Pa, p(AsH$_3$) = 57.5 Pa, p(TEGa) = 0.19 Pa, p(TEAl) = 0.08 Pa, p(TESb) = 1.3 Pa)*

For Sb-based semiconductors the V/III ratio is another key parameter because both a deficiency and an excess of antimony on the surface leads to a strong degradation of the material quality. With respect to good surface morphology we determined optimum V/III ratios of 7 for GaSb and 16 for AlSb at a growth temperature of 870 K. For lower growth temperatures the V/III ratios for GaSb and AlSb have to be increased to maintain the good material quality. After determination of the growth parameter windows for compatible deposition of all three semiconductors, finetuning of the deposition parameters lead to an optimization of the electrical and optical quality of the material.

A GaSb/AlSb-single quantum well structure (type I band alignment) consisting of 18, 12 and 8 nm GaSb wells separated by 5 nm AlSb barriers was investigated by photoluminescence measurements (fig. 2). On top of the quantum well structure a 5 nm thick GaSb cap-layer was deposited to prevent the Al-containing layers from oxidation.

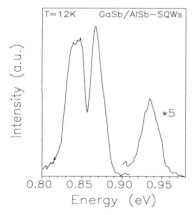

Fig. 2: *Photoluminescence spectra of GaSb/AlSb-SQWs (d(AlSb) = 5 nm)*

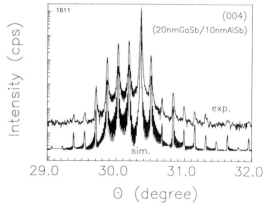

Fig. 3: *Cu-Kα1 diffraction pattern of GaSb/AlSb-MQWs (20/10)nm, 10 periodes*

The observed transition energies of the recombination processes in the QW-structure shift with decreasing GaSb-well width to higher energies as expected from theory. Additionally, for GaSb-well thicknesses smaller than approximately 10 nm a strong decrease of the photoluminescence intensity was found. This intensity drop can be explained by the fact that the first Γ state in the GaSb-well can be easily pushed above the first L state by quantum-confinement (L-valley minimum of GaSb is only 63 meV located above the Γ-minimum). As a result the GaSb quantum well ground state becomes the L state for sufficiently small well widths.

GaSb/AlSb-MQW structures (10 periodes) with different GaSb-well and AlSb-barrier thicknesses were grown at a temperature of 870 K. The compositional and structural quality of the MQW-stacks was characterized by double crystal X-ray diffraction measurements. Fig. 3 displays a typical Cu-Kα1 diffraction pattern of a GaSb/AlSb-MQW structure with a nominal period thickness of 30 nm (20 nm GaSb/10 nm AlSb). The presence of a large number of high-order satellite peaks and their narrow peak width demonstrates the high structural quality of the sample. Additionally, the asymmetry of the MQW-diffraction pattern with respect to the GaSb substrate peak reflects the relevance of strain-induced effects in this material system. A simulation of the experimental data lead to a periode thickness of 31.9 nm.

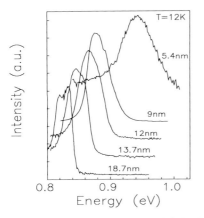

Fig. 4: Photoluminescence spectra of GaSb/AlSb-MQWs with different GaSb-well widths (d(AlSb) = 15 nm)

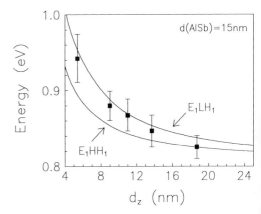

Fig. 5: Dependence of photoluminescence peak energy on GaSb-well width (calculation without strain-effects)

Fig. 4 shows photoluminescence spectra (T=12K) of GaSb/AlSb-MQWs (10 periodes) with a constant AlSb-barrier thickness of 15 nm and the denoted GaSb-well thicknesses (determined by X-ray diffraction). The recombination energy shifts with decreasing well width to higher energy. In fig. 5 the dependence of the MQW peak energy on the GaSb-well width (squares) and a quantum mechanical calculation (lines) based on the transfer matrix method of the expected transitions is plotted. On the first sight our experimental data does not correspond to the low energetic E_1HH_1-transition but to the higher energetic E_1LH_1-transition. Further calculations including strain effects (e.g. resulting from partially relaxed AlSb-barriers) did not support the experimentally suggested light and heavy valence subband reversal. However, from Raman and X-ray diffraction measurements there are hints that the well and barrier material are not the pure binary GaSb and AlSb but ternary AlGaSb with low Al- and Ga-content in the well and barrier, respectively. An explanation for this unexpected result could be a memory effect from evaporated material that was priorly deposited on the susceptor or the liner-tube. A low Al-content in the well material could indeed explain the higher E_1HH_1-transition energies because of the higher band gap of the ternary AlGaSb compared with the pure binary GaSb. Further investigations are being performed to clarify these results.

The next step towards the deposition of InAs/AlSb/GaSb-heterostructures was the investigation of InAs/AlSb layer sequences (band II alignment). In this material system both the group III and the group V constituents are exchanged at the interface. Hence, two types of heterointerfaces ("InSb-like" or "AlAs-like") which significantly affect transport properties can be generated by the appropiate switching sequence of the reactive gases. Generally, the "InSb-like" interface is preferable because of the much better transport properties in two dimensional structures compared to "AlAs-like" interfaces [8].

Our MOVPE-InAs/AlSb-heterostructures were deposited at a temperature of 790 K which is very similar to MBE growth conditions. We found that the switching sequence from InAs to AlSb and vice versa is a critical point for the deposition of InAs/AlSb layer structures with good crystalline quality. If an overlap for the group V gases AsH_3 and TESb was applied for switching from InAs to AlSb and vice versa, the crystalline and morphological quality was insufficient. MQW structures of high structural quality could simply be obtained by introducing a hydrogen purging time of several seconds between InAs and AlSb growth. Fig. 6 displays a X-ray diffraction measurement of an InAs/AlSb-MQW structure (10 periods) consisting of nominally 12 nm InAs and 8 nm AlSb.

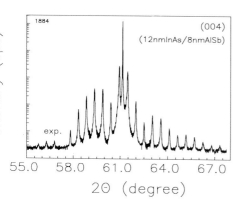

Fig. 6: $Cu\text{-}K\alpha 1$ diffraction pattern of InAs/AlSb-MQWs (12nm/8nm), 10 periodes

In this case hydrogen purging times of 8 sec. for the group V and 3 sec. for the group III gases were used. Many narrow high order satellite peaks with fringe pattern in between can be observed. This data set proves the high quality of the InAs/AlSb-MQW stack with respect to layer composition and thickness. The asymmetry in the MQW-diffraction pattern reflects the relevance of strain in this structure which is introduced by the 1.35 % lattice mismatch between InAs and AlSb. Raman-, TEM- and electrical measurements are in progress to gain further insight in the nature of the heterointerface.

Nevertheless, we have fabricated the first MOVPE-grown symmetrical nin InAs/AlSb-DBRT structure (see fig. 7) to demonstrate the high quality of the material. The device structures were grown on S-doped InAs substrates. For n-doping of the InAs buffer and InAs top contact layer H_2S was used as dopand source. We defined circular devices with 100 μm diameter by using standard photolithography. Ohmic top and bottom contacts were made using Pd/Au. A $HCl/CH_3COOH/H_2O_2$ (6:4:1) solution was used for wet chemical mesa etching.

Fig. 8 shows the DC I/V-characteristics at 300 K for the layer sequence described in fig. 7. Two distinct regions of NDR (negative differential resistance) can be observed for positive and negative polarity. Figures of merit for a resonant tunneling device are the PVC ratio (peak to valley current) and the current density ΔI. In our case the PVC ratio is approximately 1.5 and the current density $\Delta I \approx 2$ kA/cm^2. These values are promising if one considers that this was not an optimized InAs/AlSb-DBRT structure.

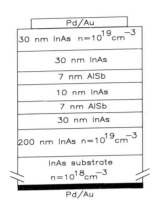 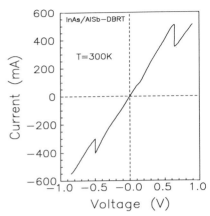

Fig. 7: Layer sequence of a nin-InAs/AlSb DBRT structure

Fig. 8: I/V-characteristics of a nin-InAs/AlSb DBRT structure (layer sequence see fig. 7)

The fact that the I/V-curve is not strongly symmetric is probably due to some asymmetry in the structure, e.g. smearing out of one of the AlSb barriers. Further optimization of the switching sequences would probably enhance the device performance a lot because the quality of resonant tunneling devices is naturally very sensitve to the abruptness of the heterointerfaces.

4. Summary and conclusion

We demonstrated for the first time the possibility of depositing InAs/AlSb/GaSb-heterostructures by LP-MOVPE. High-quality GaSb/AlSb- and InAs/AlSb-MQW structures were grown and characterized by photoluminescence and X-ray diffraction measurements. In addition to that, promising preliminary data on resonant tunneling in InAs/AlSb-DBRT structures was presented. Further systematic investigations on the nature of the InAs/AlSb-heterointerface are being performed to improve e.g. DBRT and n-HFET device performance.

5. References

[1] Esaki L, Chang L L and Mendez E E 1981 *J. Appl. Phys.* **20** L529
[2] Luo L F, Beresford R and Wang W I 1988 *Appl. Phys. Lett.* **53** 2320
[3] Luo L F, Beresford R and Wang W I 1989 *Appl. Phys. Lett.* **55** 2023
[4] Luo L F, Longenbach K F and Wang W I 1990 *IEEE Electron Device Lett.* **11** 567
[5] Bolognesi C R, Caine E J and Kroemer H 1994 *IEEE Electron Device Lett.* **15** 16
[6] Zhang Y, Baruch N and Wang W I 1993 *Appl. Phys. Lett.* **63** 1068
[7] Behet M, Stoll B and Heime K 1992 *J. Crystal Growth* **124** 389
[8] Tuttle G, Kroemer H and English J H 1989 *J. Appl. Phys.* **65** 5239

# Real Time Monitoring of III-V Alloy Composition and Real Time Control of Quantum Well Thickness in MBE by Multi-Wavelength Ellipsometry

Chau-Hong Kuo, Satish Anand, Ravi Droopad, David L. Mathine and George N. Maracas; Arizona State University EE Department, Tempe, AZ 85287-5706, B. Johs, P. He, and J. A. Woollam, J. A. Woollam Co., Inc., 650 J St.. #39, Lincoln, NE 69508, Tapani Levola, DCA Instruments, Telekatu 14, 20360 Turku, FINLAND

Abstract. Real time measurement of AlGaAs alloy composition and epitaxial layer thickness control in molecular beam epitaxy (MBE) is demonstrated by multiwavelength ellipsometry (MWE). MWE uses fewer (44) discrete wavelengths than a fully spectroscopic system but is considerably faster (spectra can be obtained 5 times per second) making it compatible with real-time growth process monitoring and control. In order to obtain epitaxial layer property measurements under substrate rotation, a substrate manipulator having angular stability of 0.02 degrees was developed. Growth tracking and extrapolation of layer thickness were used to control the thickness of quantum wells and distributed Bragg reflectors in a Fabry Perot cavity. A comparison of 75Å and 100Å quantum well structures grown with and without MWE control is made. Growth control of a Fabry Perot cavity consisting of distributed Bragg reflector stacks is also demonstrated.

1. Introduction

One approach to real-time control of epitaxial crystal growth is to monitor the properties of the growing film and generate a signal that can be used to adjust the appropriate control variable. Spectroscopic ellipsometry (SE) is used here as a sensor that can measure the real and imaginary parts of the index of refraction of an epitaxial layer in an MBE growth reactor. These indices can be used to extract layer temperature, alloy composition and thickness. If measured in real time (i.e. layer properties are tracked), growth rate and variations in the aforementioned material parameters can also be measured.

2. MBE system considerations

Two major requirements for implementing SE control on a molecular beam epitaxy (MBE) system include a) the ability to record and analyze spectroscopic data on time scales faster than the growth is occurring (~1 monolayers (ML)/sec) and b) having a substrate manipulator that maintains the angle of incidence (typically 75°) under rotation to less than 0.1 degrees [1,2]. The latter is required because it obviates the need to use angle of incidence as a data fitting parameter, thus reducing analysis time and fitting errors.

2.1 Stable substrate manipulator

In an MBE system, substrate rotation (~1 revolution per second) is essential for obtaining films with high thickness and alloy composition uniformity across the substrate. In polarization sensitive measurements such as SE, small deviations in the sample position and tilt affect the confidence of the extracted layer thickness and composition. To obtain accurate information from the rotating epitaxial layer, the plane of incidence (which is normal to the substrate surface) and angle of incidence must be kept constant within 0.1 degree [1]. To achieve this, the stability of the substrate position and of the substrate normal with respect to the incoming light beam must be maintained. This is shown schematically in figure 1. If the surface normal is not collinear with the manipulator axis, then the reflected light spot will precess under rotation and produce a time-varying angle of incidence. The position of the substrate along the manipulator axis can also change because of thermal expansion and contraction. This effect moves the reflected spot vertically a distance d in the plane of incidence and displaces it on the output polarization optics and optical detector. Because the thermal load on the manipulator varies with epitaxial structure (e.g., different for a bipolar transistor and a laser) and between successive runs on similar structures, the thermal history is different from sample to sample and thus produces fairly unpredictable variations in the thermal expansion distance d versus time. Hence, the optical data can become unreproducible.

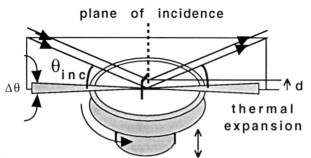

Figure 1. Schematic of a light beam incident onto an MBE substrate manipulator. The vertical position of the substrate is affected by the thermal expansion of the manipulator shaft. Angular deviations are caused by wobble in the manipulator bearings and tilted mounting of the substrate in the holder.

Commercially available MBE manipulators have not been designed to meet the angular stability specification for in-situ SE. An ultra-stable manipulator has been designed in conjunction with DCA Instruments that minimizes the substrate wobble. A wobble of less than 0.02 degree was achieved by designing the manipulator with stable bearings and so that an external adjustment of substrate tilt is possible. This adjustment allows the substrate axis to be tilted to coincide with the manipulator axis. As a result, imperfections in the bearings and sample mount (indium-free mounting rings) can be compensated. The thermal stability of the substrate was also improved by choosing titanium for the manipulator shaft material. No measurable variation due to thermal expansion was observed using a titanium shaft as compared to a stainless steel shaft that exhibited a substrate position dependent on its thermal history.

2.2 SE description

MBE process monitoring and control by SE require ellipsometric spectra to be recorded and analyzed approximately two times per second (growth rates approximately 1ML/sec). This has been achieved using an ellipsometer designed in conjunction with the J.Woollam Co. The rotating analyzer instrument is capable of recording 44 wavelengths in the spectral range of $415 < \lambda < 750$ nm at a maximum rate of 25 spectra per second. SE data fitting is performed in real-time and parameters transferred to the MBE control computer via serial link for process feedback. For the following thickness control experiments, the SE computer was used to directly control the MBE shutters via an RS232 line.

3. Thickness control by spectroscopic ellipsometry

3.1 Quantum wells

Tracking of thickness and composition (and other parameters) by SE is performed by monitoring spectra and fitting the data to a structure model consisting of target layer thicknesses and compositions approximately two times per second. After each fit, the structure model is updated with the underlying layer parameters held constant. Thus the only variables are the thickness and composition of the surface layer that is growing which is called the virtual substrate [3] method. A measure of the growth rate is directly obtained in this way. The shutter closing time is calculated based upon the target thickness, present thickness and present growth rate. This extrapolation technique reduces the frequency at which the structure model needs to be updated.

Control of quantum well thickness was attempted on a multiple quantum well structure consisting of five 100Å GaAs wells and five 75Å wells separated by 500Å of AlGaAs. The structure was grown on semi-insulating GaAs, all wells had 200Å $Al_{0.3}Ga_{0.7}As$ barriers and the entire structure was capped with 50Å of GaAs. One sample was grown with closed-loop SE feedback control of the shutters while a second sample was grown using growth rates measured by SE and timed shutter sequences (no feedback control). Figure 2 shows the 77K photoluminescence (PL) of the two test structures. The peaks correspond to the n=1 electron to heavy hole transitions in the GaAs wells. Thicknesses shown above the peaks were calculated using a self-consistent Schrödinger-Poisson solver using a barrier composition of x=0.3 that was independently measured by PL. The "timed" quantum well thicknesses were calculated to be 72Å and 91Å while the "SE control" thicknesses were 75Å and 94Å for the nominally 75Å and 100Å wells respectively. This demonstrates that SE closed-loop feedback can control multiple quantum well thickness to an accuracy of a few percent.

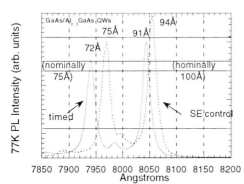

Figure 2. 77K photoluminescence of two MQW samples. One sample was grown using nominal growth rates (timed) and the other was grown under SE shutter control.

Accurate control of epitaxial layer thickness requires a-priori knowledge of the complex index of refraction versus wavelength at the growth temperature. There are two ways to obtain this data: standard optical constant reference tables or values measured just before growth. A database of optical constants of GaAs [4] and AlGaAs [5] have been measured up to 700°C in the wavelength range 250nm < λ < 1000nm. This data has a temperature uncertainty of approximately 2°C and a composition uncertainty of x=±0.01. Temperature and composition can be fitted in real time using this data. The second method involves measuring optical constants just before the growth in the growth reactor at the growth temperature. While this method has the advantage that a large database of materials constants is not necessary, it has an inherent calibration problem

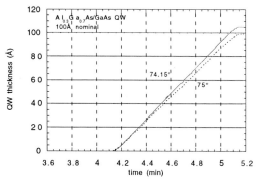

Figure 3. Thickness control of a GaAs/Al$_{0.3}$Ga$_{0.7}$As quantum well by spectroscopic ellipsometry.

in that the substrate temperature uncertainty is large and an independent measure of the alloy composition is lacking. An additional problem is that the optical constants are extracted using an angle of incidence that may not be known to less than 0.1 degree. The angular uncertainty introduces errors in thickness and composition extraction as illustrated in figure 3.

Figure 3 illustrates the growth of a single, nominally 100Å GaAs/Al$_{0.3}$Ga$_{0.7}$As quantum well using SE thickness control. The sample was grown at by solid source MBE at 600°C with a growth rate of nominally 0.6μm/hr. At 4.15 minutes the Ga shutter was opened under SE control and growth of the GaAs well begun. At approximately 5.1 minutes (100Å) the Ga shutter was closed under SE control. Two curves are shown on the graph corresponding to incident angles θ=75° and θ=74.15°. Optical constants for the GaAs and AlGaAs at the growth temperature were obtained just before the growth run assuming an angle of incidence of 75°. 75Å quantum wells were grown in a similar manner. It is seen that the QW thickness calculated using θ=74.15° is 6Å larger than that for 75°. 74.15° was the actual angle of incidence obtained by fitting the 600°C (growth temperature) optical constants from [4] to the spectrum measured just before the growth run in the MBE chamber. The GaAs growth rate was measured to be 1.766 and 1.844 ML/sec for θ=75° and 74.15° respectively. Table 1 lists the angle of incidence and the corresponding growth rate and quantum well thicknesses for the 100Å and 75Å wells. The effect of using an approximate angle of incidence to measure optical constants and subsequently control thickness is apparent.

Angle of incidence (°)	GaAs growth rate (ML/sec)	thickness 100(Å) well	thickness 75(Å) well
75.0	1.766	98.90	75.57
74.80	1.779	100.57	76.72
74.60	1.805	101.43	77.82
74.40	1.831	103.54	78.93
74.15	1.844	105.49	79.87

Table 1. GaAs quantum well growth rate and thickness versus angle of incidence. 74.15° was the actual angle of incidence while 75° was assumed for the measurements.

3.2 Fabry-Perot Cavity

SE control of MBE growth was extended to a Fabry-Perot (FP) cavity. Such structures consist of a cavity having an optical thickness corresponding to a particular optical mode placed in between two dielectric mirrors. This particular structure was used to test SE control of thick epitaxial layers because the position of the FP mode is a very sensitive measure of the cavity thickness. Distributed Bragg reflectors (DBRs) consist of alternating λ/4 high and low index of refraction layers forming mirrors that have a high reflectance wavelength band. High reflectance indicates high thickness and composition uniformity among the DBR layers. Such structures are used for fabricating vertical cavity surface emitting lasers (VCSELs) and resonant cavity FP electro-optic modulators for example.

The FP cavity design for this experiment consisted of ten period bottom and top DBRs having AlAs/GaAs thicknesses of 829.9Å and 695.8Å respectively. The GaAs 1λ FP cavity was designed to have a mode at 975nm so the thickness of the GaAs was nominally 2783.3Å. The entire structure was grown on an n^+ GaAs substrate. Figure 4 shows the result of growing the FP cavity under SE control.

In this experiment all the optical constants used were from the optical constant database measured at ASU. The angle of incidence was measured before the run by fitting to the GaAs optical constants. AlAs and GaAs optical constants at 600°C (from the database) were used during the growth and the substrate was rotated for uniformity. In figure 4, the calculated normal incidence reflectance for the 975 nm design is shown along with that calculated using the layer thicknesses and compositions measured by SE during the growth run. The two agree. The FP mode measured by normal incidence reflectance was observed at 974.7nm which is 0.3 nm from the design value. This corresponds to a thickness error of 0.3% or 8.6Å in the 2783Å cavity from the design value.

One reason for this discrepancy is that during the growth run, it was assumed that the temperature of the sample did not change and thus the 650°C optical constants were valid. A change in temperature produces a change in optical constants which produce an error in measured optical thickness. Because the SE assumes the optical constants to be correct, then the calculated normal incidence reflectance agrees with the design spectrum. This case is analogous to the quantum well thickness error arising from an error in angle. Fitting for temperature would thus increase the accuracy in placing the FP mode at the design wavelength. Such control would be possible if data around the semiconductor critical point energies could be measured in real time and included in the data fitting. This would require a fast ellipsometer which can measure data between 250nm and 1000nm.

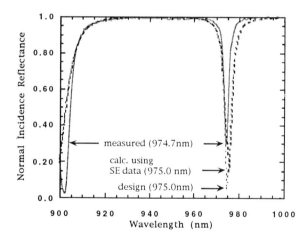

Figure 4. Normal incidence reflectance of a 975.0nm (design) Fabry Perot cavity consisting of 10 period AlAs/GaAs DBRs. The wavelength calculated from SE measured thicknesses and compositions was 975.0nm. Superimposed on the calculation is the measured curve showing a FP mode at 974.7nm.

4. Conclusions

Spectroscopic ellipsometry has been used to control the thickness of multiple quantum wells and Fabry-Perot optical cavities. In both cases, thickness control to within a few monolayers was achieved. This was possible by using a fast 44 wavelength ellipsometer and an ultra-stable substrate manipulator capable of maintaining the angle of incidence under rotation to 0.02°. Variations between design and actual layer properties introduced by angle of incidence and temperature errors were also discussed.

Acknowledgments

This work was supported by ARPA/ULTRA under contract No. N00014-92-J-1931.

References

[1] G. N. Maracas, J. L. Edwards, K. Shiralagi, K. Y. Choi, R. Droopad, B. Johs and J. A. Woollam, J. Vac. Sci. Tech. A 10(4), 1832-1839 (1992).
[2] G.N. Maracas, J.L. Edwards, K. Shiralagi, K.Y. Choi, R. Droopad, B. Johs and J.A. Woollam, J. Vac. Sci. Tech. A 10(4) (1992)
[3] D. E. Aspnes, W. E. Quinn, S. Gregory, Appl. Phys. Lett. 56 (1990) 2569
[4] C. Kuo, S. Anand, R. Droopad, K.Y. Choi and G.N. Maracas, NA MBE Conference, Stanford, CA (1993) JVST B, 12(2), 1214-1216 (1994)
[5] C. Kuo, R. Droopad and G.N. Maracas, in press

In Situ Determination of Optical Constants for Growth Control of AlAs/InGaAs RTDs using Spectroscopic Ellipsometry[*]

Francis G. Celii, Yung-Chung Kao and Walter M. Duncan

Corporate Research & Development/Technology
M/S 147, Texas Instruments, P. O. Box 655936, Dallas, TX 75265

Blaine Johs

J. A. Woollam Co., 650 J St., Suite 39, Lincoln, NE 68508

Abstract. We have determined the dielectric functions at 450 °C for InGaAs, InP, and strained InAs and AlAs, using *in situ* spectroscopic ellipsometry (SE). The SE data were synchronously acquired from rotating wafers during MBE growth. Multiple spectra representing different thicknesses of the growing layer were fit simultaneously to obtain self-consistent optical constants and layer thicknesses. The optical constants for InGaAs and InP show the expected decrease of critical point energies when compared with room temperature values. Optical functions for both psuedomorphic and relaxed InAs and AlAs layers were obtained in conjunction with previous results of *in situ* laser light scattering measurements of layer relaxation. Initial results on *in situ* monitoring of AlAs/InGaAs resonant-tunneling diode (RTD) growth using the optical constants are also reported.

1. Introduction

The use of *in situ* sensors is a promising approach to improving the precision and reproducibility of epitaxial growth.[1,2] A variety of sensor-based techniques, including reflection mass spectrometry (REMS),[3,4,5,6] optical flux monitoring,[7] and spectroscopic ellipsometry (SE)[8,9] have been applied for both open and closed-loop control of molecular beam epitaxial (MBE) growth. Quantum devices, such as resonant-tunneling diodes (RTDs), make challenging vehicles for demonstrating sensor-based growth improvements because they are particularly sensitive to variations in layer thickness.[10,11] Closed-loop control of RTD growth based on mass spectrometry has been demonstrated.[12]

The AlAs/InGaAs RTD structure requires precise growth control because of the presence of thin (20 Å), highly-strained AlAs and InAs layers (-3.5% and +3.2% mismatch with InP, respectively). SE has advantages for real-time, on-wafer growth control, capable of simultaneous acquisition of layer composition and thickness with monolayer precision;[8,13] however, the optical constants at growth temperature needed for analysis of *in situ* SE data are currently available only for AlGaAs alloys.[14,15,16] We report here the dielectric functions at 450 °C for InP, lattice-matched $In_{0.53}Ga_{0.47}As$, and strained InAs and AlAs, needed for SE-based growth control of AlAs/InGaAs RTDs.

---

[*] Work partially supported by ARPA under contract MDA972-93-H-0005.

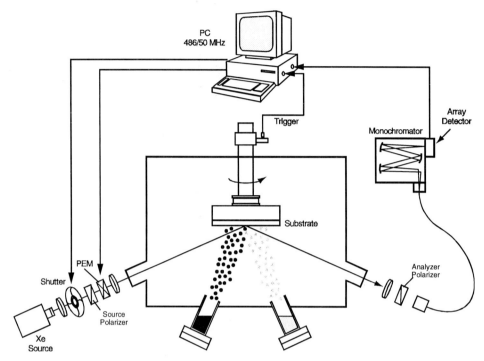

Figure 1. Schematic of experimental apparatus, showing MBE system with rotating wafer which triggers acquisition of the phase-modulated SE system.

2. Experimental

Experiments were performed on a Vacuum Generators 4" V90 solid-source MBE system, equipped with ports to allow SE optical access at 75° from the surface normal (Fig. 1). The 2" InP substrates (Nippon Mining, (100)) were used as-received and mounted using an indium-free holder. Surface oxide was desorbed by heating to 560 °C for two minutes under As_4 flux. Buffer layers consisted of at least 1000 Å of lattice-matched $In_{0.53}Ga_{0.47}As$ (hereafter referred to as InGaAs), verified using x-ray diffraction The InGaAs, InAs and AlAs layers were deposited at growth rates of 2.2, 1.2, and 1.0 Å/s, respectively, under a V/III flux ratio of ~3, and at a nominal 450 °C substrate temperature (obtained using an uncorrected one-color pyrometer with bandpass at 940 nm).

SE measurements were made using a rapid-scan, phase-modulated spectroscopic ellipsometer, developed at Texas Instruments and described elsewhere,[17,18] during MBE growth on rotating InP wafers. Briefly, the collimated output from a 150 W Xe source was linearly polarized before passing through a photoelastic modulator (PEM), lens and strain-free quartz window into the MBE chamber (Fig. 1). The light reflected from the growth surface was collected, passed through a second fixed polarizer and 0.32 m monochromator, and detected using a 46-element Si diode array. The Ψ and Δ values were extracted from the DC, first and second harmonic components of the 50 kHz modulation imposed by the PEM, and determined using fast waveform digitization and Fourier transform analysis. Waveform acquisition was triggered synchronously with wafer rotation, which alleviated the effects of substrate wobble. For each spectrum, we typically averaged twenty waveforms at each wavelength in a total time of under 2 sec. The *in situ* SE data was analyzed post-growth, with techniques described below, using the WVASE software package.

3. Results and Discussion

3.1. In situ SE data acquisition

Sample data for growth of InGaAs on InP are shown in Fig. 2. Part of the raw SE data (the Ψ component), obtained at 46 discrete wavelengths and sampled once each rotation period (10 rpm), are shown for the oxide-desorbed InP substrate and at three InGaAs layer thicknesses during the growth (Fig. 2a). The spectral representation is appropriate for real-time monitoring applications, where thickness or composition values are obtained from point-by-point fits. For post-growth data analysis, a powerful technique is multi-sample analysis (MSA),[19] in which the spectral data from numerous thicknesses are fit simultaneously. With MSA, kinetic plots are a more natural representation of the data (Fig. 2b). Similar measurements were made for InAs and AlAs layers on InGaAs/InP. The critical layer thicknesses of these strained layers (25-35 Å[20]) was exceeded such that optical functions could be obtained for both psuedomorphic and relaxed films through judicious choice of data regions for analysis.

3.2. SE data analysis: InP and InGaAs

The optical functions of both InP and InGaAs can be obtained from growth of a single layer of InGaAs on InP. Results of such analysis are shown in Fig. 2b, which shows a kinetic plot of a portion (7 out of 46 wavelengths) of the observed SE data along with the multi-sample analysis fit to that data. (The same fit is shown in the spectral representation of Fig. 2a.) Analysis of the InGaAs data at various growth times shows a constant growth rate, indicating shutter transient effects[6] to be negligible. Dielectric functions extracted from this fit are shown in Fig. 3 and compared with room temperature values. The InGaAs optical functions (Fig. 3b) are distinct from those of InGaAs deposited on relaxed AlAs and InAs layers (not shown) due to (at least) the significant surface morphology caused by misfit dislocations and/or islanding in the latter case.

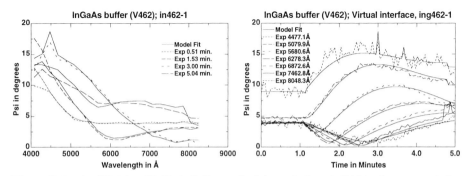

Figure 2. *In situ* SE data of InGaAs/InP growth: (a) spectral, and; (b) kinetic representation.

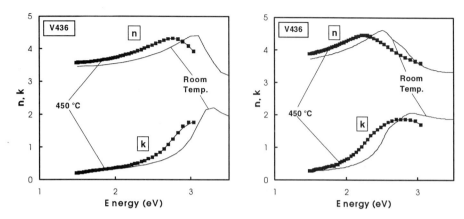

Figure 3. Derived n and k values for (a) InP; (b) InGaAs at 450 °C and room temperature.

3.3. SE data analysis: Strained InAs and AlAs on InGaAs/InP

Optical functions were obtained for thin (pseudomorphic) and thick (relaxed) strained layers of AlAs and InAs on InGaAs/InP, summarized in Figs. 4 and 5, respectively. SE data of thick layers (100 Å) were selectively analyzed in various growth regions to determine the effect of strain relaxation on the optical functions. The optical functions for strained AlAs and InAs were able to fit the thin-layer SE spectra, but failed to fit the latter-time growth data, corresponding to thicker, relaxed films (Figs. 4a and 5a). Optical functions for strained and relaxed AlAs and InAs are collected in Figs. 4b and 5b.

3.4. RTD Growth Monitoring

Growth-temperature optical functions were employed for post-growth analysis of *in situ* SE data from RTD growths. Preliminary analysis shows the expected correlation between RTD

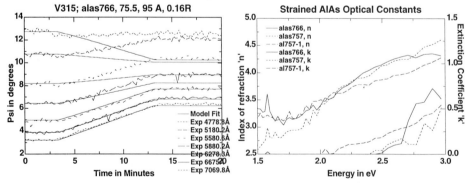

Figure 4. (a) Observed and calculated SE data for AlAs growth (3-13 min) on InGaAs/InP. The fit data uses strained optical constants (alas766, from early growth times); (b) Optical functions of strained (solid and dotted lines) and relaxed (dashed) AlAs. Relaxed films show higher absorption (k value) between 2.0 and 2.7 eV.

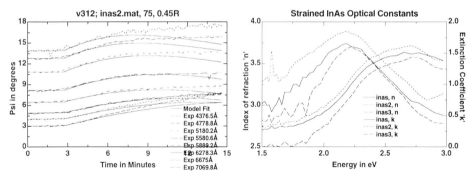

Figure 5. (a) Observed and calculated SE data for InAs growth (3-15 min) on InGaAs/InP. Strained optical constants (inas2) were used in the fit; (b) Optical functions of strained (dotted line) and relaxed (solid and dashed lines) InAs.

current density and AlAs barrier thickness, determined using either SE or REMS. The SE data also indicate the presence of thickness asymmetry between the first and second barriers in each RTD, which is consistent with the observed asymmetry in RTD peak current.

4. Conclusions

Optical constants for the AlAs/InGaAs RTD material system have been determined using *in situ* SE monitoring and post-growth, multi-sample analysis. Knowledge of these constants will enable SE-based growth control to improve reproducibility of quantum devices such as RTDs.

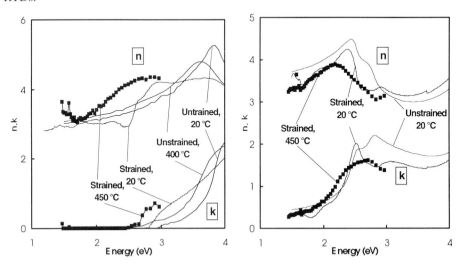

Figure 6. Growth-temperature dielectric functions for pseudomorphic (a) AlAs and (b) InAs on InGaAs/InP (squares), compared with unstrained and strained films at 450 °C and room temperature.

5. Acknowledgments

We thank Chyi Sheng and Matt Woolsey for software support, O. Faye Phillips for data management, Randy Thomason and Frank Stovall for the MBE operation, and Andy Purdes for insightful discussions. We also acknowledge the contributions of Craig Herzinger, who developed some of the SE analysis techniques implemented in the WVASE software.

6. References

[1] Barna G G, Loewenstein L M, Henck S A, Chapados P, Brankner K J, Gale R J, Mozumder P K, Butler S W and Stefani J A 1994 *Sol. State Technol.* **37** 47-53
[2] Celii F G, Kao Y-C and Purdes A J 1993 *Proc. of 25th Government Microcircuit Applications Conference* 511-5
[3] SpringThorpe A J and Mandeville P 1988 *J. Vac. Sci. Technol.* **B6** 754-7
[4] Tsao J Y, Brennan T M, Klem J F and Hammons B E 1989 *Appl. Phys. Lett.* **55** 777-9
[5] Evans K R, Kaspi R, Cooley W T, Jones C R and Solomon J S 1993 *Mat. Res. Soc. Symp. Proc.* **300** 495-500
[6] Celii F G, Kao Y-C, Beam E A III, Duncan W M and Moise T S 1993 *J. Vac. Sci. Technol. B* **11** 1018-22
[7] Chalmers S A and Killeen K P 1993 *Appl. Phys. Lett.* **63** 3131-3
[8] Aspnes D E, Quinn W E, Tamargo M C, Pudensi M A A, Schwarz S A, Brasil M J S P, Nahory R E and Gregory S 1992 *Appl. Phys. Lett.* **60** 1244-6
[9] Maracas G N, Edwards J L, Shiralagi K, Choi K Y, Droopad R, Johs B and Woollam J A 1992 *J. Vac. Sci. Technol. A* **10** 1832-9
[10] Seabaugh A C, Luscombe J H, and Randall J N 1993 *Future Electron. Dev. J.* **3** Suppl. 1 9-20
[11] Seabaugh A C, Taddiken A H, Beam E A III, Randall J N, Kao Y-C and Newell B 1993 *Electon. Lett.* **29** 1802-4
[12] Celii F G, Harton T B, Kao Y-C and Moise T S 1994 *Appl. Phys. Lett.* submitted for publication
[13] Celii F G, Duncan W M and Kao Y-C 1994 *J. Electron. Mater.* submitted for publication.
[14] Kuo C H, Anand S, Droopad R, Choi K Y and Maracas G 1994 *J. Vac. Sci. Technol. B* **12** 1214-6
[15] Yao H, Snyder P G, Stair K and Bird T 1992 *Mat. Res. Soc. Symp. Proc.* **242** 481-6
[16] Kuo C H, Anand S, Droopad R, Choi K Y, Maracas G N, Johs B, He P, and Woollam, J A 1994 unpublished
[17] Duncan W M and Henck S A 1993 *Appl. Surf. Sci.* **63** 9-16
[18] Henck S A, Duncan W M, Lowenstein L M and Butler S W 1993 *J. Vac. Sci. Technol. A* **11** 1179-85
[19] Herzinger C M, Yao H, Snyder P G, Celii F G, Kao Y-C, Johs B.and Woollam J A *J. Appl Phys.* submitted for publication
[20] Celii F G, Kao Y-C, Beam E A III and Files-Sesler L A 1993 unpublished

In-situ monitoring and control of MOCVD growth using multiwavelength ellipsometry

Shakil Pittal, Blaine Johs, Ping He, and John A. Woollam
J. A. Woollam Co., Inc. 650 J street, suite 39, Lincoln NE 68508 USA
S. Dakshina Murthy and Ishwara B. Bhat
Electrical, Computer and Systems Engineering,
Rensselaer Polytechnic Institute, Troy, New York 12180 USA

Abstract
In this paper we describe real-time growth control of MOCVD grown HgCdTe using a novel and low cost in-situ multiwavelength ellipsometer. This ellipsometer acquired in-situ data in real-time at 44 wavelengths in the spectral range of 415-750 nm. The in-situ ellipsometer was mounted on an MOCVD HgCdTe reactor, and a library of optical constants as a function of nominal composition of HgCdTe was extracted from the ellipsometric data acquired during the growth of HgCdTe multilayer growth with varying alloy composition. This library of optical constants was later used by the ellipsometer in the composition control of HgCdTe. Physical model parameters such as composition were determined in real-time from the analysis of the acquired ellipsometric data, and were passed to a feedback control program (external to the ellipsometer program). Alloy compositions of 0.25 and 0.3 were stabilized by the feedback control program by regulating the H_2 flow through a dimethyl cadmium (DMCd) source. When an intentional temperature disturbance was introduced during the control experiment, the program re-adjusted the H_2 flow to maintain the as-grown composition.

1. Introduction

II-VI semiconductors, especially HgCdTe are extremely important for military optoelectronic applications such as imaging and guidance systems. There are efforts worldwide to grow high quality HgCdTe by MBE[1] and MOCVD[2]. In spite of the significant efforts invested in the growth of HgCdTe, the current device yields are extremely low and costs unacceptably high, due mainly to lack of control of material quality and reproducibility during crystal growth. This increasing demand to bring about improved material quality, device yields and costs requires an in-situ, non destructive technique for monitoring and process control.

Ellipsometry is an optical technique used to measure the change in polarization state of light, which is very sensitive to physical properties, such as film thickness, composition, and optical constants. Ex-situ spectroscopic ellipsometry has long been an established technique for analyzing bulk materials and thin films[3]. Combined with sophisticated mathematical modeling capabilities, in-situ ellipsometry is emerging as an excellent technique to monitor and control film deposition processes[4-7].

In this paper we describe the application of multiwavelength ellipsometry in monitoring and controlling composition of HgCdTe growth by MOCVD.

2. Instrumentation and data analysis

A novel and compact multiwavelength ellipsometer was retro-fitted to an MOCVD reactor. The details of the ellipsometer and the MOCVD reactor are described elsewhere[7]. This ellipsometer acquires in-situ ellipsometric data in real-time at 44 wavelengths in the spectral range of 415-750 nm. The maximum data acquisition rate is 25 measurements per second, with typical acquisition rate of 1 measurement (44-wavelength data) in 2-3 seconds. The ellipsometer data acquisition and analysis are integrated into a single computer program. In addition, physical model parameters such as thickness and composition are determined in real-time from the analysis of the acquired ellipsometric data, and are passed to a feedback control program (external to the ellipsometer program). The external program tunes the process controller accordingly to achieve a desired growth. The ellipsometer calibration[8] and various in-situ ellipsometric data analysis techniques[7] were described earlier.

The growth of HgCdTe was carried out in a standard atmospheric pressure horizontal flow reactor. The reactor is made of quartz tube (3 inch diameter) with two optical ports for the ellipsometer. These optical ports were continuously purged by flowing H_2 to keep windows free of any deposition. HgCdTe was grown by simultaneous pyrolysis of precursors DMCd, DIPTe and elemental mercury on a CdTe buffer layer grown on a GaAs substrate at 350°C. These precursors were carried into the reactor by H_2 gas.

3. Library of optical constants of HgCdTe at growth temperature

The goal of the experiment was to perform composition control of the HgCdTe grown at 350° C. This required a library of optical constants of HgCdTe of various compositions at the growth temperature of 350°C. To obtain this library, a multilayer structure containing a series of different compositions, x in $Hg_{1-x}Cd_xTe$, was grown on a CdTe buffer on a GaAs substrate. In addition, real-time ellipsometric psi and delta data were simultaneously obtained during the multilayer growth. The composition of HgCdTe in the multilayer structure was varied (from Cd rich composition to Hg rich composition) by changing the hydrogen flow rate through the cadmium organometallic source in steps of 5 from 30 to 5 sccm. In order to obtain optical contrast between the layers, the H_2 flow rates were not changed in a descending order, but were alternated between a high and low flow rate (i.e. 15, 10, 25, 5 and 30 sccm). We intended to obtain x values evenly spread out between 1 and 0. However, we could not determine the actual compositions after growth either from FTIR or SIMS measurements as the layers were not thick enough. But, from the comparison of the solved optical constants of these layers with the known optical constants of HgCdTe at room temperature[9], it appeared that these compositions were Hg rich. Therefore, we arbitrarily assigned nominal values to the compositions grown at 30, 25, 20, 15, 10, and 5 sccm H_2 flow rates as 0.3, 0.25, 0.2, 0.15, 0.1, 0.08, and 0.0 respectively. The real-time, in-situ ellipsometric data are as shown in figure 1.

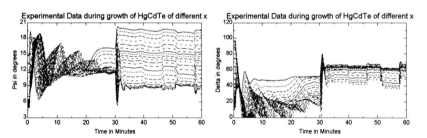

Figure 1. Real-time ellipsometric psi and delta data during the growth of a multilayer stack containing layers of HgCdTe of various compositions. Every other wavelength of the 44 wavelengths in 415-750 nm is shown. t=0 to t=11.7 -CdTe growth; t=11.7 to t=30-heat reactor walls and mercury; t=30 to t=40-HgCdTe growth at 15 sccm H_2 flow rate; t=40 to t=46-10 sccm; t=46 to t=51- 25 sccm; t=51 to t=57- 5 sccm; t=57 to t=60- 30 sccm.

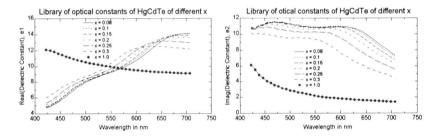

Figure 2. Library of optical constants (represented as dielectric function) of HgCdTe at 350°C of nominal compositions 0.08, 0.1, 0.15, 0.2, 0.25, 0.3 and 1.0.

The optical constants of the CdTe (x =1) were obtained by selecting ellipsometry data at multiple instants during the CdTe growth and treating the data as if they were obtained from different models. This approach, called multi-model analysis, assumed that the optical constants of CdTe at each instant (model) were coupled together. The ellipsometer program then solved for the CdTe optical constants and layer thicknesses. The advantage of this multi-model approach is that optical constants can be obtained in the presence of surface roughness on the film.

Obtaining optical constants of HgCdTe by the multi-model analysis was complicated due to the presence of interdiffusion. We simplified the analysis by assuming that each layer in the multilayer structure became optically thick after a certain time during the growth, and the pseudo-optical constants at that time represented the optical constants at that composition. Interdiffusion and surface roughness were ignored in this simplified analysis. Figure 2 shows the extracted library of optical constants at various compositions of HgCdTe.

4. HgCdTe composition control

Having obtained a library of optical constants of HgCdTe at the growth temperature of 350°C, we implemented a feedback control experiment to monitor and control the composition of the growing HgCdTe. For this an analog output line was run from the computer to the 0-5 volts input of a mass flow controller (MFC). The MFC was used to regulate the flow of hydrogen gas carried through the DMCd source. A qualitative relation between the MFC voltage and the composition was obtained. The ellipsometer program acquired and analyzed the real-time data, and passed the model fit parameter (composition in this case) to a control program external to the ellipsometer program. The control program adjusted the MFC voltage to achieve and maintain a target composition using a simple proportional algorithm. The time constant representing the system response was set to a reasonable value. The ellipsometer data analysis used the "virtual interface" approach[7]. A constant growth rate of 3 Å/second was assumed.

Figure 3 shows the instantaneous composition of HgCdTe and feedback voltage at the MFC. Initially, the MFC was set to grow a composition of 0.35 . The control was then given to the computer to grow a composition of 0.25. The control program accordingly adjusted the mass flow controller to grow the composition of 0.25, and in less than two minutes it achieved the target composition of 0.25 and maintained it to within ±0.01. At time t=10 minutes, the target composition was changed to 0.3. The feedback response changed again, this time, to grow the composition of 0.3, and obtained a composition of 0.29±0.02 in about

7 minutes. From t =0 to t=27, the substrate temperature was held constant at 350°C. At t=27 minutes, a disturbance in the temperature of 5°C was given, which caused the composition of the growing material to change from 0.29 to 0.34 [10]. The control program accordingly adjusted the MFC to compensate for the temperature disturbance to maintain the composition at 0.29. The compositions were later confirmed from the SIMS analysis of the post-deposited film, and the actual compositions were found to be 0.25 and 0.34 for the nominal values of 0.25 and 0.30 respectively.

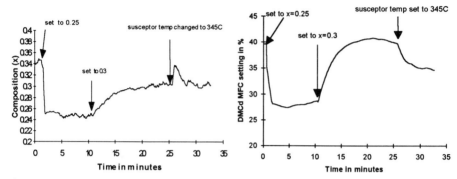

Figure 3. Instantaneous composition of HgCdTe and feedback voltage (in percentage of 5 volts) during the composition control experiment.

5. Conclusions

We have demonstrated the use of multiwavelength ellipsometry in monitoring and control of the composition of HgCdTe. Although the experiment was performed using the library of optical constants of nominal HgCdTe compositions, the library can be easily related to the actual compositions once they are determined by using some independent technique such as FTIR or SIMS. The feedback control algorithm can further be optimized by implementing a more advanced control algorithm and determining an accurate time constant of the system.

Acknowledgments

This work was sponsored by DARPA Contract DAAH01-92-C-R191 and U.S. Army Contract DAAB07-92-C-K755.

References

[1] R. Hartley, M. Folkard, D. Carr, P. Orders, D. Rees, V. Kumar, G. Shen, T. Steele, H. Buskes and J. Lee 1992 *J. Vac. Sci. Technol.* B10 (4) 1410-1414
[2] I. Bhat, N. Taskar and S. Ghandhi 1986 *J. Vac. Sci. Technol.* A4 p 2230
[3] D. E. Aspnes 1993 *Thin Solid Films* 233 1-8
[4] R. W. Collins 1990 *Rev. Sci. Instrum.* 61 2029-2062
[5] D. Aspnes, W. Quinn and S. Gregory 1990 *Appl. Phys. Lett.* 57 p. 2707
[6] B. Johs, J. Edwards, K. Shiralagi, R. Droopad, K. Choi, G. Maracas, D. Meyer, G. Cooney and J. Woollam 1992 *Materials Research Society Symp.* Proc. 222 p.75
[7] B. Johs, D. Doerr, S. Pittal, I. Bhat and S. Dakshinamurthy 1993 *Thin Solid Films* 233 293-296
[8] B. Johs 1993 *Thin Solid Films* 234 395-398
[9] Edward Palik 1991 *Handbook of Optical Constants of Solid II* (Academic Press, INC.)
[10}J. B. Mullin and S. J. C. Irvine 1982 *J. Vac. Sci. Technol.* 21 p 178

Strain Effect On AlAs/InGaAs/InAs RTD Growth and Electrical Characteristics

Y.C. Kao, A.J. Katz, F.G. Celii, T.S. Moise, E.A. Beam III, and H.Y. Liu

Corporate Research & Development,
Texas Instruments, P.O. Box 655936, MS 147, Dallas, TX 75265

Abstract. We have directly linked the interface roughness and electrical characteristics for pseudomorphic AlAs/InGaAs/InAs resonant tunneling diodes (RTDs). For strained InAs growth at low growth rate (~ 0.01 nm/s), enhanced surface roughening has been observed by *in situ* laser light scattering and *ex situ* transmission electron microscopy (TEM); these data are consistent with Monte Carlo simulations, which indicate increased surface islanding at lower growth rates and higher temperatures in strained systems. The RTD current-voltage curves are also sensitive to the interface roughness with peak-to-valley ratio sharply decreasing at lower growth rates.

1. Introduction

Pseudomorphic AlAs/In$_{.53}$Ga$_{.47}$As/InAs resonant tunneling diode (RTD) heterostructures have been used as the basic building blocks in advanced circuits such as multi-state memory [1] and compressed functionality integrated circuits.[2-4] Through the use of pseudomorphic InAs notch quantum wells (+3.2% lattice mismatch to InP) and AlAs barriers (-3.5% lattice mismatch to InP), peak-to-valley current ratios exceeding 50 have been achieved [5]. However, the RTD electrical characteristics, in particular the valley current, are sensitive to both the structural characteristics [6] of the device (i.e. thickness, composition, etc.) and the epitaxial growth parameters (i.e. growth rate, temperature), as we show in this paper. Specifically, we attribute the change induced by growth rate and temperature to a 3-dimensional (3D) islanding, or roughening, of the InAs compressive strained layer.

Surface roughening in strained epitaxial system has received considerable attention recently.[7] Certain types of strained epilayers (eg. InAs) tend to form 3D islands during the initial stages of epitaxy. The surface morphology depends strongly on the layer strain: as the strain increases, the system grows in the form of islands, as thees reduce the elastic energy in the film.[8] In this case, surface diffusion is critical for island formation. Surface roughening, even prior to dislocation formation, has been shown theoretically to affect RTD electrical characteristics.[9] Elastic scattering of electrons at rough interfaces provides a mechanism to alter the forward electron momentum and enhance off resonance electron transmission. In this paper, we provide a direct experimental link between interfacial roughness and the electrical properties, specifically the valley current, of a resonant tunneling diode.

In the RTD growth, by changing the growth conditions of the InAs-notched quantum wells, we systematically vary the roughness of the InAs/InGaAs and InGaAs/AlAs interfaces. Monte-Carlo simulations suggest that slower growth rates and higher temperatures promote island formation in strained epitaxial systems. In fact, we find experimentally that the growth rate is the effective parameter for control of interfacial roughness. Variations in interfacial roughness are directly observed through application of LLS (laser light scattering), TEM (transmission electron microscopy), and AFM (atomic force microscopy) techniques. We demonstrate the effect of interfacial roughness on RTD electrical characteristics by measuring I-V curves. Severe degradation in the peak-to-valley ratio is found with increasing roughness. Our results establish a direct link between morphology in strained heterostructures and electrical characteristics of RTDs.

2. Experimental

Samples were prepared in a VG V90-MBE system on 2-inch, epi-ready InP(100) wafers. Following oxide desorption at 550C for 2 min., an n+ buffer layer of 500 nm $In_{.53}Ga_{.47}As$, (hereafter referred to as InGaAs), was first grown on all samples. Double barrier AlAs/In(Ga)As/AlAs RTD were then deposited. The double barrier structure consisted of two AlAs barriers of 2 nm separated by a 4.5 nm thick InGaAs quantum well (QW) with an InAs notch (varied from 2 to 3.5 nm). Our standard RTD structure contains 2 nm InAs.[1] All growth was conducted at a nominal temperature of 450C with the substrate rotating at 20 rpm. Growth interruptions were inserted before and after each AlAs barier growth.

To quantify the effects of lattice relaxation on the current-voltage characteristics of the RTD, we have developed a vertically-integrated resonant-tunneling diode (VIRTD) heterostructure in which 4 RTDs are grown consecutively and are separated by 200 nm of n^+ InGaAs.[6] The entire stack is capped by 50 nm of n^+ (1×10^{18} cm^{-3}) InGaAs and 100 nm of n^{++} (5×10^{18} cm^{-3}) InGaAs to minimize the contact resistance. A schematic cross-section of the fabricated VIRTD is shown in Figure 1.

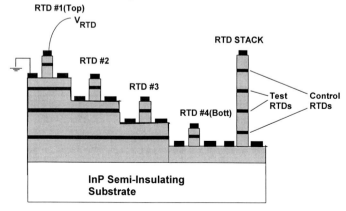

Fig. 1 The cross-sectional schematic of a fabricated VIRTD

Using this stacked RTD approach, we perform a series of experiments within a single epitaxial growth which enables us to explore perturbations in the current-voltage characteristics caused by the change of a single growth variable. For example, to investigate the dependence of lattice relaxation on growth rate, we altered the growth rate of the strained InAs layer within the two middle QWs by 50% and 90%, respectively. The growth rate for the top and bottom QWs were held constant as experimental controls.

The LLS experimental apparatus has been described in detail previously.[10] Briefly, a 633 nm 5 mW HeNe laser reflected off the wafer at normal incidence. Diffuse reflectance caused by non-specular surface morphology was collected at 65° from the surface normal and detected using a photomultiplier tube (Burle 4832). With this geometry and laser wavelength, the detection configuration is particularly sensitive to surface features with correlation length of 0.3 μm (K=20 μm^{-1}). Use of a narrow (3nm) bandpass filter and optical chopping with lock-in detection eliminated various noise sources and background drift. *In situ* reflection mass spectrometry (REMS) was used as a process monitor of effusion cell flux.[11] Structural defects were verified with cross-sectional TEM and surface morphology was investigated with AFM.

3. Interfacial Roughness: Modeling

We conducted Monte Carlo simulations to investigate how variations in growth conditions affect surface islanding using a model developed by Orr *et al.*[7] The Orr model incorporates the lattice elastic energy of both substrate and overlayer film, and the surface energy of the film. Particles on the surface diffuse according to a solid-on-solid model, where the probability of hopping to a neighboring site is controlled by the difference in strain and surface energies. Formation of islands is determined by a characteristic diffusion length, which is a function of such factors as growth rate, temperature, and the relative magnitudes of elastic and surface energies. Island formation is enhanced for higher temperatures and lower growth rates (conditions that increase surface diffusion) and is suppressed for lower temperatures and higher growth rates.

Our simulations indicate that an order of magnitude change in growth rate or surface diffusion rate will strongly influence island formation. Figure 2 shows surface roughness, which we define in terms of the rms height fluctuations on the surface, as a function of growth rate in arbitrary units for a lattice misfit of 11% between the substrate and the overlayer. We observe that roughness decreases as growth rate increases because of the decrease in adatom diffusion length at higher growth rates. Figure 3 displays surface roughness as a function of surface diffusion rate, which depends, in turn, on substrate temperature. As expected, surface roughness increases with higher diffusion rates.

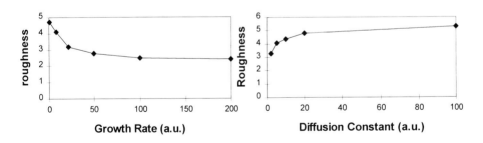

Fig.2. Roughness dependence on growth rate. Roughness is defined in terms of height fluctuations: Roughness = $<(h-<h>)^2>$

Fig.3. Roughness dependence on diffusion constant. Roughness is defined as height fluctuation in Figure 2.

4. Interfacial Roughness: Experiment

We have systematically varied the growth conditions for the InGaAs/InAs-notched QW within a VIRTD sample. In these experiments, we varied the QW growth rate by as much as a factor of 20 while holding the substrate temperature unchanged. In separate experiment, we found that only small changes in islanding were observed over the acceptable range of temperatures (350C to 480C) for epitaxial growth.

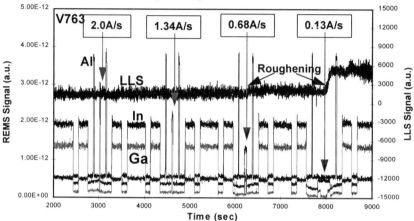

Fig. 4 The REMs and LLS spectra of VIRTD growth prepared under various growth rates. Notice the different growth rate can be seen from the In and Ga REMS signals. Roughening causes an increase in LLS intensity.

Figure 4 shows a composite plot of REMS and LLS data for a VIRTD sample grown at 460 °C with the following InAs QW growth rates: RTD#4(first grown): 0.2 nm/s; RTD#3 0.134 nm/s; RTD#2: 0.068A/s; RTD#1: 0.013 A/s. The barrier and quantum well growth regimes for the four RTDs are indicated by the REMS data clearly. The LLS scattering data indicate no roughening during growth of the first two RTDs. During growth of the third and fourth RTDs, the magnitude of the LLS signal increases as the result of enhanced interfacial roughening (correlation between increases in LLS scattering and surface roughening are described in detail in [11]). Because the amplitude of the LLS signal does not continue to increase beyond the growth of the quantum well, we conclude that islanding has taken place without formation of dislocations, which usually result in much larger and continuing increases of the LLS signal.

Figure 5 shows the TEM micrographs of two RTD QWs grown with InAs growth rates of 0.013 nm/s and 0.134 nm/s. Island formation or roughening is clearly indicated at the Fig. 5(a) top Alas/InGaAs/InAs interface. Because the top RTD barrier layer is too thin to allow for flattening of the interface, the top barrier layer follows the roughened contour of the InGaAs/InAs interface. Model simulations [9] indicate that roughness at the AlAs/InGaAs (barrier/quantum well) interface significantly enhances the RTD valley current. AFM scans confirm the TEM results. The scans were taken on two terminated-RTDs surface prepared without the top AlAs barrier. They show the average peak to valley depth is 5 nm for the one prepared with lower InAs growth rate; while the depth is around 1nm for the one grown at high InAs growth rate.

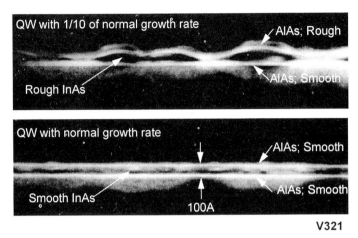

Fig. 5. TEM micrographs of two RTD layers grown with different InAs growth rates: (a) 0.013 nm/s, (b) 0.13 nm/s. The bottom RTD shows smooth interfaces, while the top AlAs/InGaAs/InAs heterostructure is islanded, although the first AlAs under the QW remains flat.

Using straightforward arguments [12], we show that at the lower growth rates we expect growth to occur under equilibrium conditions, where islanding is most favored. We first define the roughness scale in terms of the wavelength, λ, of the maximally unstable perturbative surface mode. The equilibrium growth regime occur when the diffusion length, L (defined by $L=\sqrt{(D\tau)}$, where D is the diffusion constant and τ is the time to deposit a monolayer), is larger than the characteristic island size set by λ. For parameter values appropriate for our material system, a λ of 9 nm is calculated.[12] We next calculate the diffusion length L for our experimental growth conditions to determine whether we are in the equilibrium or nonequilibrium growth regime. For Ga adatoms, the diffusion constant $D = D_o(\exp{-(E_a/T)})$, where the activation energy E_a is 1.36 eV and the constant D_o is 10^{-3} cm^2/sec. [6] Using a monolayer thickness of 0.3 nm, growth temperature of 450 °C, and a growth rate of 0.2 nm/sec, we calculate $D = 3.3 \times 10^{-13}$ cm^2/sec and $L = 8.1$ nm, which is smaller than the characteristic wavelength calculated above. For T = 450 C and a growth rate of 0.013 nm/sec, we find $L = 27.5$ nm, which should be well into the equilibrium regime where islanding is optimized.

5. RTD Electrical Characteristics

Figure 6 shows a set of four I-V curves for a 4-RTD stacked set grown at 460 °C with the following InAs growth rates: RTD#4 (first grown): 0.13 nm/s; RTD#3: 0.013 nm/s; RTD#2: 0.065 nm/s; RTD#1(Top): 0.13 nm/s. The first and fourth RTDs serve as controls. A systematic increase in the valley current is observed as the growth rate is decreased from 0.13 nm/s to 0.013 nm/s. At the lowest growth rate, negative differential resistance, the characteristic RTD signature, has disappeared. These data indicate the large impact of growth rate (and, therefore, surface roughening) on I-V characteristics. Similar trends have been observed in other repeat experiments.

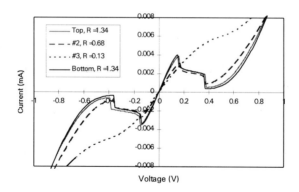

Fig. 6. I-V curves for a 4-peak VIRTD as a function of QW growth rate. The peak to valley ratio drops to < 1.0 for the RTD grown with the lowest QW growth rate.

6. Conclusions

We have varied the quantum well growth rates during growth of pseudomorphic RTDs to control the degree of surface roughening. Data from LLS, TEM, and AFM show enhanced surface roughening at lower growth rates; these data are consistent with Monte Carlo simulations, which indicate increased surface islanding at lower growth rates and higher temperatures in strained systems. We have measured the I-V curves of RTDs as a function of growth rate and have found that the peak-to-valley ratio sharply decreases at the lower growth rates. We believe our data provide the most direct link to date between interfacial roughness in heterostructures and electrical characteristics of physical devices.

References

[1] Seabaugh A C, Kao Y C, and Yuan H T 1992 *IEEE Electron. Lett.* **13**, 479
[2] Seabaugh A C, Taddiken A H, Beam, III E A, Randall J N, and Kao Y C 1993 *Electron. Lett.* **29**, 1802
[3] Imamura K, Takatsu M, Mori T, Adachihara T, Ohnishi H, Muto S, and Yokoyana N 1992 *IEEE Tran. Electron. Dev.* **39**, 2707
[4] Seabaugh A C, Taddiken A H, Beam, III E A, Randall J N, Kao Y C, and Newell B 1993 *IEDM*, **93**, 419
[5] Smet J, Broekaert T P E, Fonstat C G 1991 *J. Appl.Phys.* **71**, 2475
[6] Moise T S, Kao Y C, Broekaert T P E, Luscombe J H, and Celii F G, "Sensitivity Analysis of Pseudomorphic InGaAs/AlAs Resonant Tunneling Diodes," in preparation.
[7] Orr B G, Kessler D, Snyder C W, and Sander L 1992 *Europhys. Lett.* **19**, 33
[8] Berger P, Chang K, Bhattacharya P K, Singh J, and Bajaj K K 1988 *Appl. Phys. Lett.* **53**, 684
[9] Roblin P and Liou W 1993 *Phys. Rev. B* **47**, 2146
[10] Celii F G, Kao Y C, Liu H Y, Files-Sesler L A, and Beam, III E A 1993 *J. Vac. Sci. Technol. B*, **11(3)**, 1014
[11] Celii F G, Kao Y C, Beam, III E A, Duncan W M, and Moise T S 1993 *J. Vac. Sci. Technol. B*, **11(3)**, 1018
[12] Srolovitz D J 1989 *Acta Metall.* **37**, 621

Strained InAs/(AlGaIn)As/InP quantum wells for 1.5 - 2.5 μm laser applications grown by virtual surfactant MBE

Klaus H. Ploog and Eric Tournié*

Paul-Drude-Institut für Festkörperelektronik, D-10117 Berlin, Germany

Abstract. Strained InAs quantum wells confined by $Al_xGa_{0.48-x}In_{52}As$ barriers on InP substrate (3.2 % mismatch) were fabricated by a new growth technique which prevents islanding of the highly strained InAs layers. Under In-rich conditions (virtual surfactant MBE) strain relaxation is suppressed for kinetic reasons, and $InAs/Al_xGa_{0.48-x}In_{52}As$ interfaces of high structural perfection were obtained even far beyond the critical InAs layer thickness. Spontaneous emission from InAs single quantum wells up to 2.4 μm at 300 K and stimulated emission from electrically pumped laser structures has been achieved.

1. Introduction

The understanding of compressive or tensile strain to substantially improve the performance of optoelectronic InP-based devices operating in the 1.3 - 1.6 μm wavelength range [1 - 3] has led to a renewed interest in strained $Ga_xIn_{1-x}As$ quantum well (QW) structures on InP substrate which emit in the spectral range of 1.6 - 3.0 μm [4 - 9]. Single-mode semiconductor lasers emitting in this mid-IR region of the spectrum are needed for applications of LIDAR (light detection and ranging) using atmospheric transmission windows and remote sensing of atmospheric gases and for optical communication systems based on low-loss fluoride fibers. The alloy system $Ga_xIn_{1-x}As_ySb_{1-y}/Al_xGa_{1-x}As_ySb_{1-y}$ lattice-matched to GaSb substrate has been used successfully for lasers operating in this wavelength range [10, 11]. However, the GaSb-based alloy system has the disadvantages of GaSb compounds being difficult to etch, having low thermal conductivity and dissociating at low temperature [12], which is detrimental to the development of advanced laser structures. The superior quality of InP substrates as compared to GaSb substrates and the elaborate growth and processing technology of $Ga_xIn_{1-x}P_yAs_{1-y}$ alloys makes this materials system attractive for optoelectronic devices which require lateral confinement of the active layers. In addition, the high compressive strain in lattice-mismatched $Ga_xIn_{1-x}As$ QW structures on InP substrate results in a sharp peak of the highest in-plane valence-band dispersion at the center of the zone. This is favorable for reducing undesired non-radiative processes, such as Auger recombination or intervalence band absorption, which represent the major part of the threshold current [1 - 3].

Strained InAs QWs confined by $Al_xGa_{0.48-x}In_{0.52}As$ barriers lattice-matched to InP substrates have theoretically been credited a high potential for laser application [13]. Due to the 3.2 % difference in lattice constants, the InAs binary compound experiences a large amount of strain when grown in registry with the underlying InP substrate. This strain results in a room-temperature (RT) cut-off wavelength of λ_c = 3.0 μm for InAs, whereas the corresponding value for the unstrained $Ga_{0.47}In_{0.53}As$ ternary alloy (lattice-matched to InP) is λ_c = 1.7 μm. It follows that the RT cut-off wavelength for strained InAs QWs confined by $Ga_{0.47}In_{0.53}As$ barriers can theoretically be tuned between 1.7 < λ_c < 2.5 μm by adjusting the

QW width between 0 and 25 monolayers (ML) [14].

In this paper we demonstrate that the strained-InAs/Al$_x$Ga$_{0.48-x}$In$_{0.52}$As materials system is indeed qualified for emission above 1.7 μm at room temperature and we report the operation of separate-confinement-heterostructure (SCH) lasers made from this system.

2. Growth procedure

The samples of the present study are strained single InAs QWs embedded in a Ga$_{0.47}$In$_{0.53}$As matrix which are grown on n$^+$-InP(100) substrates by solid-source MBE in a Riber 32 P system. Previous thermodynamic considerations [15] have indicated that above a certain thickness a strained layer forms islands to relax the strain, which leads to samples of inferior structural quality. To avoid islanding during MBE growth of InAs on Ga$_{0.47}$In$_{0.53}$As (3.2 % lattice mismatch) we had to develop a modified growth procedure, called virtual-surfactant mediated (VSM) MBE [16]. In VSM-MBE the InAs QW is grown under In-stable conditions instead of the conventional As-stable ones, which can be monitored in-situ by reflection high-energy electron diffraction (RHEED). Unter these conditions kinetic restrictions are imposed to the sticking species which can then not migrate on the surface to find proper sites for island formation. As a result, strained-InAs/Ga$_{0.47}$In$_{0.53}$As interfaces of superior structural perfection are obtained and the critical thickness of the strained InAs layer before defects are generated is considerably increased [16].

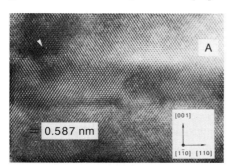

Fig. 1   HRTEM images of strained InAs embedded in Ga$_{0.47}$In$_{0.53}$As; (A) 20 ML InAs grown under As-stable conditions, (B) 23 ML InAs grown under In-stable conditions. The white arrow indicates a dislocation.

3. Structural and optical properties

In Fig. 1 we display the HRTEM (high resolution transmission electron microscopy) lattice images of two Ga$_{0.47}$In$_{0.53}$As/InAs/Ga$_{0.47}$In$_{0.53}$As heterostructures on (100)InP substrate grown under As-stable (A) and In-stable conditions (B). The sample A with 20 ML InAs grown under As-stable conditions contains dislocations, in particular at the upper interface, whereas no dislocations are detectable in the image taken from sample B (23 ML InAs grown under In-stable conditions) as well as in other images taken from this sample covering an area as large as 600 lattice planes. Although 23 ML thick, any strain relaxation of the InAs layer is negligible and the defect density is thus extremely low. The striking result is that the growth mode of the InAs film has a drastic effect on the structural integrity of the upper InAs/Ga$_{0.47}$In$_{0.53}$As interface and hence on the optical and electrical properties of the InAs QW heterostructure.

In Fig. 2 we depict the low-temperature (6 K) photoluminescence (PL) spectra obtained from several strained InAs single QWs embedded in $Ga_{0.47}In_{0.53}As$ of different well widths ranging from 2 to 23 ML. All samples were grown under the favorable In-stable conditions, and their emission spans the 1.6 - 2.2 µm wavelength range. The vertical arrows indicate the positions of the transitions between the confined electron state e_1 and the confined heavy-hole state hh_1 calculated with the envelope function model assuming fully strained InAs QWs and using the parameters given in Ref. [17]. Excellent agreement between calculated and experimental values is obtained except for the thin 2 ML and 5 ML wells which are more sensitive to interface fluctuations. The PL linewidth first decreases when increasing the well width, and then increases again above 13 ML. This behavior is due to the fact that for wider wells (i) the QW width fluctuations have less influence and (ii) the wavefunction is more localized in the QW, i. e. the ternary-alloy broadening from the barrier material is reduced (the broader line of the 8 ML QW reveals a lower quality sample). The linewidth increase for the wider wells, together with the slight decay of the PL efficiency, is indicative of a minor structural deterioration, as confirmed by X-ray diffraction. Nevertheless, the excellent agreement between experimental and calculated peak positions indicates that even the widest InAs QWs are elastically strained, as also observed by HRTEM [16]. Note that we have measured the narrowest PL lines ever reported for InAs QWs grown on InP substrates [18], which reveals the superior quality of our samples.

To further evaluate the potential of single InAs QWs embedded in $Ga_{0.47}In_{0.53}As$ for light-emitting devices, we have studied the temperature dependence of the luminescence. We have obtained PL up to 300 K from all samples fabricated under In-stable conditions. As an example we show in Fig. 3 the 300 K spectrum taken from the 23 ML InAs QW sample. Although the QW line is rather broad, its peak position (2.38 µm) fits well with the calculated value indicated by the vertical arrow. Any structural relaxation of this QW would give a red-shift by as much as 0.28 µm. This wavelength of 2.38 µm is, to our knowledge, the longest wavelength achieved up to now from quantum wells grown on InP substrates.

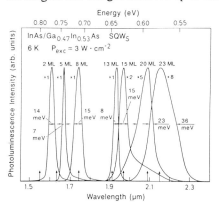
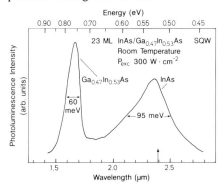

Fig. 2 Low-temperature PL spectra taken from strained InAs single QWs embedded in $Ga_{0.47}In_{0.53}As$ with well widths as indicated. Excitation was by the 647.1 nm line of a Kr⁺ laser.

Fig. 3 Room-temperature PL spectrum taken from the 23 ML wide strained InAs single QW embedded in $Ga_{0.47}In_{0.53}As$ using the 647.1 nm line of a Kr⁺ laser for excitation.

4. Laser characteristics

Electrically pumped laser diodes were fabricated from a 10 ML InAs single QW clad by stepped $Al_xGa_{0.48-x}In_{0.52}As$ barriers, as schematically indicated in Fig. 4. In Fig. 5 we show the radiative recombination channels I, II and III which are possible in the studied laser structure under different excitation conditions and which will be discussed later.

The output power versus current injection for a typical broad-area (200 x 500 μm^2) laser diode under pulsed current injection (800 ns pulse length, 1 kHz repition rate) at 80 K is shown in Fig. 6. The inset in the figure displays a spectrum taken from the same laser diode at $J = 1.1\ J_{th}$. The spectrum is centered at 1.86 μm with about ten clearly-resolved longitudinal modes. This is the first observation of laser emission from InAs QWs grown on InP substrates. The peak position well fits the value of the e_1-hh_1 transition calculated for a 10 ML InAs/Ga$_{0.47}$In$_{0.53}$As QW (1.845 μm). Planar narrow-stripe devices (7μm x 480 μm)

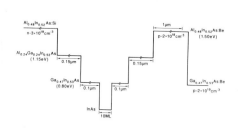

Fig. 4   Layer sequence of strained InAs single QW laser structure indicated by the variation of the conduction band edge (energy gaps are 80 K values).

Fig. 5   Radiative recombination channels in the 10 ML InAs QW laser structure (e_1 = 174 meV, hh_1 = 38 meV at 80 K).

were also fabricated from the same epitaxial wafer. In this case, the threshold current intensity at 80 K is 40 mA (J_{th} = 1.2 kA/cm^2), which allow continuous wave (cw) operation to be achieved. Quantitative measurements of the light emitted in cw operation under 100

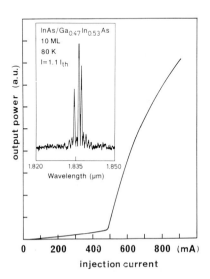

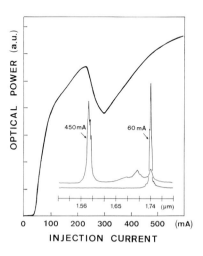

Fig. 6   Output power versus drive current measured at 80 K from a broad-area (200 x 500 μm^2) laser diode. The inset shows a spectrum taken at $I = 1.1\ I_{th}$.

Fig. 7   Output power vs drive current measured at 80 K from a narrow-stripe diode (7 x 480 μm^2). The inset shows two spectra taken at 60 and 450 mA current injection, respectively.

mA current-injection yield an output power of 5 mW and a differential quantum efficiency of ≈ 15 %, both per uncoated facet.

The lasing characteristics of the present diodes strongly depend on the stripe width, the injection current and the heat-sink temperature. Figure 7 shows the output-power vs drive-current measured at 80 K from a narrow-stripe device. Two spectra taken under pulsed current injection of 60 mA and 450 mA are displayed in the inset of the figure. Several features have to be noted. First, under low injection (60 mA), the spectrum is centered at 1.74 μm instead of the 1.84 μm obtained with a broad-area diode (Fig. 6), although both devices were made from the same epitaxial wafer. Second, under high injection the main laser emission occurs at 1.57 μm, i. e. at the $Ga_{0.47}In_{0.53}As$ barrier wavelength. Finally, we have also noted that when increasing the temperature above 80 K the threshold current density increases as $\exp(T/T_o)$ with a low characteristic temperature of $T_o \approx 30$ K for the broad-area diodes and $T_o \approx 35 - 40$ K for the narro-stripe devices. Above 110 K, the main laser mode in both cases corresponds to the $Ga_{0.47}In_{0.53}As$ barrier emission.

To understand these a priori puzzling findings one has to consider the energy band diagram of the 10 ML $InAs/Ga_{0.47}In_{0.53}As$ QW displayed in Fig. 5. The strained InAs well is rather shallow and the heavy-hole and electron levels lie only 42 meV and 86 meV below the barrier band-edges, respectively. In addition, InAs has a low *in-plane* heavy-hole effective mass which leads to rapid filling of the hole subband as previously demonstrated by PL spectroscopy [17]. As for the laser structure, reducing the stripe width leads to large lateral losses [19] and the high gain required to overcome them leads then to band filling. In our case, the broad-area diode lases on the e_1-hh_1 QW transition (Fig. 5 arrow I). For a narrow stripe under low injection (60 mA) we observe a laser emission at 1.74 μm which coincides with the transition between the QW electron level e_1 and the $Ga_{0.47}In_{0.53}As$ valence band-edge (Fig. 5: arrow II). Finally, when increasing the pump current the emission from the well saturates (Fig. 6), before gain is achieved on the $Ga_{0.47}In_{0.53}As$ barrier levels (Fig. 5: arrow III).

5. Conclusion

We have fabricated strained InAs quantum wells confined by $Al_xGa_{0.48-x}In_{0.52}As$ barriers on InP substrate (3.2 % mismatch) by a new growth technique, called virtual-surfactant MBE, which prevents islanding of the highly strained InAs layers. Under such In-rich MBE growth conditions, strain relaxation is retarded for kinetic reasons, and $InAs/Al_xGa_{0.48-x}In_{0.52}As$ interfaces of high structural perfection were obtained even far beyond the critical InAs layer thickness. We have achieved spontaneous emission from strained InAs single quantum wells up to 2.4 μm at room temperature, and we have observed stimulated emission at 80 K from a 10 ML InAs quantum well SCH laser diode using electrical pumping. However, the poor carrier confinement in the InAs quantum well of the present structure results in a rapid filling of the quantum-well levels which is detrimental to the device performance. Work is underway in our laboratory to improve the carrier confinement.

Acknowledgements. We gratefully acknowledge the very stimulating cooperation with the many of our collegues throughout this project who are listed in Refs. [8, 16, 17, 18].
Part of this work was sponsored by the Bundesministerium für Forschung und Technologie of the Federal Republic of Germany and by the GSI Darmstadt.

References

* present address: C.N.R.S., Centre de Recherche sur l'Hétéroépitaxie et ses Applications, F-06560 Valbonne-Sophia-Antipolis, France

[1] Adams A R 1986 *Electron. Lett.* **22** 249 - 250
[2] Yablonovitch E and Kane E O 1986 *J. Lightwave Technol.* **LT-4** 504 - 506 and 961
[3] Thijs P J A, Tiemeijer L F, Kuindersma P I, Binsma J J M and Van Dongen T 1991 *IEEE J. Quantum Electron.* **QE-27** 1426 - 1438
[4] Bour D P, Mortinelli R U, Enstrom R E, Stewart T R, DiGuiseppe N G, Hawrylo F Z and Cooper D B 1992 *Electron. Lett.* **28** 37 - 39
[5] Forouhar S, Larsson A, Ksendzov A, Lang R J, Tothill N and Scott M D 1992 *Electron. Lett.* **28** 945 - 947
[6] Forouhar S, Ksendzov A, Larsson A and Temkin H 1992 *Electron. Lett.* **28** 1431 - 1432; 1993 *Electron. Lett.* **29** 574 - 576
[7] Major J S, Nam D W, Osinski J S and Welch D F 1993 *IEEE Photonics Technol. Lett.* **5** 594 - 596
[8] Tournié E, Grunberg P, Fouillant C, Kadret S, Boissier G, Baranov A, Joullié A, Gaumont-Goarin E and Ploog K H 1993 *Electron. Lett.* **29** 1255 - 1257
[9] Martinelli R U, Menna R J, Triano A, Harvey MG and Olsen G H 1994 *Electron. Lett.* **30** 324 - 326
[10] Bochkarev A E, Dolginov L M, Darkin A E, Eliseev P G and Sverdlov B N 1988 *Sov. J. Quantum Electron.* **18** 1362 - 1363
[11] Eglash S J and Choi H K 1990 *Appl. Phys. Lett.* **57** 1292 - 1294; 1991 *IEEE J. Quantum Electron.* **QE-27** 1555 - 1559
[12] Joullié A, Jia Hua F, Karouta F, Mani H and Alibert C 1985 *Proc. SPIE-Int. Soc. Opt. Eng.* **587** 46 - 57
[13] Yablonovitch E and Kane E O 1988 *J. Lightwave Technol.* **LT-6** 1292 - 1299
[14] Tournié E, Ploog K H and Alibert C 1992 *Appl. Phys. Lett.* **61** 2808 - 2810
[15] Bauer E 1958 *Z. Kristallogr.* **110** 372 - 380
[16] Tournié E and Ploog K H 1993 *Appl. Phys. Lett.* **62** 858 - 860
     Tournié E, Brand O, Ploog K H and Hohenstein M 1993 *Appl. Phys.* A **56** 91 - 94
     Tournié E and Ploog K H 1994 *J. Cryst. Growth* **135** 97 - 112
[17] Tournié E, Brandt O and Ploog K H 1993 *Appl. Phys.* A **56** 109 - 111
[18] Tournié E, Schönherr H P, Ploog K H, Giannini C and Tapfer L 1992 *Appl. Phys. Lett.* **61** 846 - 848
[19] Shieh C, Mantz J, Lee H, Ackley D and Engelmann R 1989 *Appl. Phys. Lett.* **54** 2521 - 2523

MBE Growth of AlGaAs Using Sb as a Surfactant

R. Kaspi and D. C. Reynolds

University Research Center, Wright State University, Dayton, OH 45435, USA

K. R. Evans and E. N. Taylor

Solid State Electronics Directorate (WL/ELR), Wright Laboratory, Wright-Patterson AFB, OH 45433-7323, USA

Abstract. We have employed a surface segregating population of Sb as an isoelectronic surfactant during solid-source molecular beam epitaxy (MBE) of AlGaAs layers. A steady-state population of Sb was maintained at the AlGaAs growth surface by providing a continuous Sb_2 flux to compensate for loss due to thermal desorption. A significant improvement in the optical quality of AlGaAs layers was observed by photoluminescence. Sharper GaAs QW PL lines also indicate smoother inverted (GaAs on AlGaAs) interfaces when the Sb surfactant is employed. These improvements may be attributed to a reduced incorporation of impurities and point defects, and/or improved surface diffusion kinetics during AlGaAs MBE.

1. Introduction

In recent years, there has been an emerging interest in the use of surfactants to improve film growth during molecular beam epitaxy (MBE). Most notably, a single surface segregating monolayer of Sb [1,2], As [3], or Te [4] has been shown to alter the growth mode of Ge on Si (100) by suppressing island growth, promoting two-dimensional growth, and thereby improving the crystalline quality of the Ge overlayer. Other recent examples include layer-by-layer growth of Cu on Ru using O [5], delayed onset of relaxation of InGaAs on GaAs using Te [6], and improved (111)B-GaAs homoepitaxy using In [7] as surfactants. The mechanism behind surfactant mediated growth is not well understood, however, passivation of dangling bonds at the surface, the modification of the surface energy, as well as a change in the surface diffusion kinetics of the parent species are all considered to be important factors. The primary criteria for a possible surfactant are that it will segregate to the surface of the growing film without being diluted into the layers, and that its surface population can be maintained during the course of the growth.

While high quality $Al_xGa_{1-x}As$ layers are of extreme technological importance, the growth of alloys with $x > 0.2$ by MBE is often characterized by rougher surface morphology [8] and weaker photoluminescence signal relative to GaAs [9]. This is generally attributed to the lower surface mobility of Al compared to Ga adatoms at the usual AlGaAs growth temperatures [10], as well as elevated background impurity levels due to the gettering abilty of Al [11]. In addition, a "forbidden range" of intermediate growth temperatures is commonly identified in which faceted island growth occurs [12,13]. While the mechanisms leading to the observations stated above are still in debate, we consider here the utility of using a surfactant to improve the bulk optical properties and surface morphology of AlGaAs layers grown by solid-source MBE. Co-evaporation of Sb_2 under conditions where the Sb sticking coefficient (σ^{Sb}) is negligible is shown to be a convenient way to satisfy the criteria for surfactant growth. Preliminary results described below indicate a significant improvement in the optical quality of $Al_{0.22}Ga_{0.76}As$ layers as well as a smoother inverted (GaAs on AlGaAs) interface as observed by photoluminescence analysis.

2. Sb as surfactant on AlGaAs

The mechanisms determining the incorporation rates of As_2 and Sb_2 during MBE growth of (001) oriented GaAsSb and AlGaAsSb layers are complex due to the competition among these species to chemisorb onto the same surface sites. However, As_2 is known to incorporate preferentially over Sb_2 in these alloys as the substrate temperature T_s is increased [14,15]. The temperature at which σ^{Sb} becomes negligible will depend on the arsenic/antimony flux ratio, the overall V/III ratio as well as the film growth rate. Based on the reported values of the variation of σ^{Sb} with T [14,15], and the onset of Ga re-evaporation [16], there will be a range of T_s in which σ^{Sb} will be essentially zero while no loss of Ga to thermal desorption will occur (i.e.; roughly between 650-700 °C) . In this parameter range, therefore, it is possible to co-evaporate Sb, yet effectively deposit an AlGaAs alloy film. Moreover, the incident Sb-flux will maintain a steady-state surface Sb population by continuously replenishing Sb lost from the surface by thermal desorption. This surface Sb layer can behave as a surfactant during AlGaAs MBE.

3. Experimental

Samples containing AlGaAs layers were deposited with and without Sb_2 co-evaporation in Varian GEN-II MBE system on 2-inch diameter semi-insulating radiatively heated (001) oriented GaAs substrates. Both group-V tetrameric sources were cracked to provide an incident As_2 flux $J(As_2)$, and an incident Sb_2 flux $J(Sb_2)$. The thermocouple measurement of T_s was calibrated against the substrate oxide desorption temperature (~580 °C) and the temperature at which Ga re-evaporation during AlGaAs growth becomes significant (700 °C). The run-to-run variation in T_s is estimated to be < 5 °C whereas the absolute value of T_s is estimated to be +/- 15 °C.

Nominally undoped 1 μm thick $Al_{0.22}Ga_{0.76}As$ layers were deposited using $J(As_2)$ ~ 7 x 10 Torr beam equivalent pressure (BEP) at T_s ~ 690 °C in which $J(Sb_2)$ was varied. The growth rate in each case was 0.58 μm/hour. The change in the incident Sb_2 flux for each Sb-oven temperature setpoint was separately measured by comparing the amu/q = 121 peak intensity from a mass spectrometer positioned in line-of sight of the substrate while the entire incident Sb_2 flux was reflected from a static GaAs surface at T_s ~ 690 °C. Absolute flux values were obtained by comparing these peak intensities to a drop in the intensity measured when a known incident flux of Ga is used to deposit GaSb at T_s ~ 500 °C, and an equivalent amount of Sb is consumed.

In addition, AlGaAs/GaAs quantum well structures were grown in order to probe the inverted interface smoothness by photoluminescence (PL). Each of these nominally undoped structures consisted of a 5000 Å $Al_{0.24}Ga_{0.76}As$ layer grown under conditions described above, 250 Å GaAs quantum well, a 250 Å $Al_{0.24}Ga_{0.76}As$ barrier, and a 250 Å GaAs cap layer. A 30 growth interruption at the inverted interface was used to deplete all of the remaining Sb at the surface and to reduce T_s to 600 °C, which was maintained for the remainder of the growth.

For PL analysis, all samples were excited with an Ar^+-ion laser using an exciting power density of 400 mW/cm^2, and were analyzed at 2 K using a modified Bausch & Lomb 4-m grating spectrograph. In-situ surface reconstruction transition time measurements were made using reflection high energy electron diffraction (RHEED) along the [-110] azimuth.

4. Results

Two peaks were mainly observed in the near bandgap photoluminescence spectra from unintentionally doped p-type $Al_{0.22}Ga_{0.78}As$ layers. The higher energy peak is attributed to bound exciton (BE) recombinations, while the lower energy peak includes contributions from donor-acceptor pair recombination and conduction band to acceptor recombination, collectively referred to as (D,A). The films were p-type and, based on the energy separation, the acceptor was identified as carbon. A factor of merit of the optical quality of AlGaAs is the (BE)/(D,A) intensity ratio at low temperatures. Typically, the control samples grown using conventional MBE at T_s= 690 °C exhibited spectra in which the (D,A) intensity was dominant over (BE). In contrast, a dramatic increase in the (BE) intensity relative to that of the (D,A) was observed in samples grown using Sb as a surfactant. Figure 1 compares PL spectra from a 1 μm thick layer grown using $J(Sb_2) \sim 4.9 \times 10^{13}$ mol cm^{-2} s^{-1}, to that grown without Sb. The (BE)/(D,A) intensity ratio increased from about 0.3 to over 12, along with a five-fold increase in the integrated intensity.

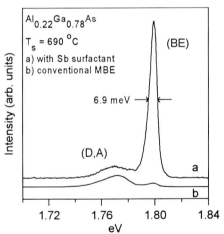

Figure 1. Near bandgap PL spectra from 1 μm thick $Al_{0.22}Ga_{0.78}As$ samples grown with (a), and without (b) Sb surfactant. The spectra are vertically shifted for clarity.

The $J(Sb_2)$ employed in this study ranged from $\sim 8 \times 10^{12}$ mol cm^{-2} s^{-1} to $\sim 1.05 \times 10^{14}$ mol cm^{-2} s^{-1}. The highest (BE)/(D,A) ratio and integrated intensity, and the narrowest (BE) peak (2.6 meV FWHM) was observed in the sample grown using an intermediate $J(Sb_2)$ of 1.9×10^{13} mol cm^{-2} s^{-1}. The peak energies remained essentially unchanged for the entire set of samples, thus incorporation of Sb into the layers was well below 1 at. % under the growth conditions employed.

In-situ observations of the film surface under the same growth conditions as described above were made using RHEED. The AlGaAs surface exhibited a streaky (2x1) surface reconstruction pattern with a faint half-order streak when no Sb$_2$ was employed. Intensity features indicative of three-dimensional growth were not observed even after 1000 Å of deposition. This would indicate that the growth conditions were not within the "forbidden range" despite the fact that a (2x4) reconstruction was not observed. Within a few seconds after the Sb shutter was opened during growth, a surface phase transition from a (2x1) to a (2x4) reconstruction symmetry was observed with the four-fold symmetry along the [-110] azimuth. When the Sb shutter was closed during deposition, a clear transition back to the (2x1) reconstruction shortly followed. Figure 2 shows the "Sb-on (2x4)" to "Sb-off (2x1)" transition time as a function of the J(Sb$_2$) employed. An increase in the transition time from less than 1s to nearly 4 s was observed when J(Sb$_2$) was varied from 8 x 10^{12} mol cm^{-2}s^{-1} to 1.05 x 10^{14} mol cm^{-2} s^{-1} at T$_s$ = 690 °C. This trend can be explained by an increasing steady-state surface riding Sb population as J(Sb$_2$) is increased. At higher initial surface coverages, more time is needed for the Sb to be depleted by thermal desorption so that the (2x4) to (2x1) transition can occur. Shown in the same plot is the PL (BE(/D,A) ratio as a function of J(Sb$_2$).

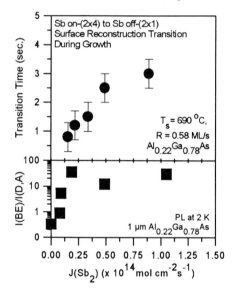

Figure 2. (2x1) to (2x4) RHEED transition time and PL (BE)/(D,A) intensity versus incident Sb$_2$ flux.

Specular RHEED beam intensity measurements at T$_s$ = 690 °C indicate an enhancement in both the amplitude and the persistence of growth oscillations in the presence of Sb at the surface. Structures incorporating an inverted interface were examined by PL to verify surface smoothening as a result of the Sb surfactant. Figure 3 compares 2 K PL spectra from a 250 Å GaAs quantum well layer deposited on a conventionally grown 0.5 µm thick $Al_{0.24}Ga_{0.76}As$ layer with that grown using $J(Sb_2) = 1.9 \times 10^{13}$ mol cm^{-2} s^{-1}. Both layers were grown at T_s = 690 °C. The latter case results in a reduction of the FWHM from 5.1 meV to 2.9 meV, accompanied by an eight-fold increase in intensity.

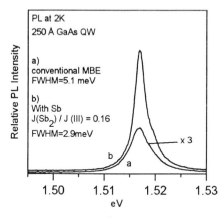

Figure 3. PL spectra from 250 Å GaAs QW grown on 0.5 µm thick AlGaAs deposited with (a), and without (b) Sb surfactant

4. Discussion

The mechanisms resulting in the improvement of AlGaAs layers when the Sb surfactant is present are not clear at this time. The simultaneous increase in the integrated PL intensity and the (BE)/(D,A) ratio suggests that carbon incorporation is reduced by the presence of Sb at the surface. Also, the narrowing of the (BE) peak as $J(Sb_2)$ is increased suggests that the free exciton recombination is enhanced in these layers. A reduction of nonradiative recombination centers in the presence of an Sb_2 flux may be due to the passivation of electron traps such as As-vacancies or As-vacancy containing complexes by a small amount of incorporated antimony. Trap gettering by 0.2-1 atomic % Sb in GaAs has been reported to reduce electron traps related to As vacancies [17]. Alternatively, the effect of the surfactant on the growth processes may lead to a lower intrinsic trap density in the films without necessitating passivation.

Isoelectronic doping of GaAs films grown by metalorganic vapor phase epitaxy has recently been shown to result in improved surface morphology and reduced dislocation density, attributed to the elastic interaction of the Sb with Ga and As vacancies [18]. A similar mechanism may be present in the AlGaAs layers studied here. However, incorporation of Sb into the layers may not be a prerequisite to yield a smooth surface. A smoother growth front may also be due to increased surface diffusion kinetics in the presence of Sb , especially for Al adatoms that are far less mobile than Ga at T_s= 690 °C. The Al-Sb bond, being relatively weaker than the Al-As bond, may help decrease the overall diffusion activation barrier for Al and result in the enhancement of layer-by-layer growth as evidenced by RHEED oscillations. One suggested mechanism by which roughness is introduced onto the AlGaAs surface is the difficulty the adatoms may encounter in moving down the edges of islands at intermediate growth temperatures [19]. One could speculate that the Sb at the surface preferentially occupies these more stable sites at the step edges, and by changing the local energy configuration, facilitates the descent of Ga or Al down the step edge. This would lead to a smoother inverted interface as observed. Another possible mechanism for surface roughening has been attributed to the excessive gettering of ambient oxygen by AlGaAs which then segregates and accumulates at the film surface [11]. One possible role of the Sb surfactant is that it may shield the underlying AlGaAs surface from impurities. During thermal desorption, Sb may help pull oxygen, carbon or other impurities, that would otherwise incorporate, away from the surface, thus leading to a smoother surface. A common feature of all of these scenarios is that any increase in the surface Sb population in the low-coverage regime is likely to enhance the beneficial effect. This consistent with the data shown presented in Figure 2.

In conclusion, we have shown that it is possible to maintain surfactant mediated MBE growth of AlGaAs by co-evaporation of Sb_2 in the appropriate temperature range. We have also demonstrated that growth with the surfactant results in an improvement of the optical quality as well as a smoother inverted interface.

Acknowledgments

We thankfully acknowledge D.C. Look (WSU) for many stimulating discussions. Authors R.K and D.C.R. were supported by USAF Contract No. F33615-91-C-1765, and all of their work was performed at the Solid State Electronics Directorate, Wright Laboratory. This work was also partially funded by AFOSR.

References

[1] Osten H J, Klatt J, Lippert G and Bugiel E 1993 *J. Cryst. Growth* **127** 396-400
[2] Copel M and Tromp R M 1991 *Appl. Phys. Lett.* **58** 2648-2651
[3] Copel M, Reuter M C and Tromp R M 1989 *Phys. Rev. Lett.* **62** 632-635
[4] Higuchi S and Nakanishi Y 1991 *Surface Sci.* **254** L465-467
[5] Wolter H, Schmidt M and Wandelt K 1993 *Surface Sci.* **298** 173-186
[6] Grandjean N, Massies J, Delamarre C, Wang L P, Dubon A and Laval J Y 1993 *Appl Phys. Lett.* **63** 66-68
[7] Ilg M, Eisler D, Lange C and Ploog K 1993 *Appl. Phys.* **A56** 397-399
[8] Weisbuch C, Dingle R, Gossard A C and Wiegmann W 1980 *J. Vac. Sci. Technol.* **17** 1128-1132
[9] Pavesi L and Guzzi M 1994 *J. Appl Phys.* **75** 4779-4842
[10] Alexandre F, Goldstein L, Leroux G, Joncour M C, Thibierge H and Rao E V K 1985 *J. Vac. Sci. Technol.* **B3** 950-955
[11] Chand N, Chu S N G and Geva M 1991 *Appl. Phys. Lett.* **59** 2874-2876
[12] Morkoç H, Drummond T J, Kopp W and Fisher R 1982 *J. Electrochem Soc.* **129** 824-828
[13] Lee H, Nouri N, Colvard C and Ackley D 1989 *J. Cryst. Growth* **95** 292-295
[14] Klem J, Fisher R, Drummaond T J Morkoç H and Cho A Y 1983 *Electron. Lett.* **19** 453-455
[15] Evans K R, Stutz C E, Yu P W, Wie C R 1990 *J. Vac. Sci. Technol.* **B8** 271-275
[16] Van Hove J M and Cohen P I 1985 *Appl. Phys. Lett.* **47** 726-728
[17] Li A, Kim HK, Jeong JC, Wong, D, Zhao JH, Fang Z-Q, Schlesinger TE and Milnes AG 1989 *J. Crystal Growth* **95** 296-300
[18] Yakimova R, Paskova T and Ivanov I 1993 *J. Crystal Growth* **129** 143-148
[19] Dabiran A M, Nair S K, He H D, Chen K M and Cohen P I 1993 *Surface Sci.* **298** 384-391

Growth of InP, GaP, and GaInP by Chemical Beam Epitaxy using Alternative Sources

H.H. Ryu[1], C.W. Kim[1], L.P. Sadwick[2], G.B. Stringfellow[1,2], R. W. Gedridge, Jr.[3], and A.C. Jones[4]

[1]Department of Material Science and Engineering, The University of Utah, Salt Lake City, Utah 84112; [2]Department of Electrical Engineering, The University of Utah, Salt Lake City, Utah 84112; [3]Naval Air Warfare Center Weapons Division, China Lake, CA 93555; [4]EpiChem, Ltd., Power Road, Bromborough, Wirral, Merseyside L623QF, United Kingdom

Abstract. This paper reports the growth of indium phosphide (InP), gallium phosphide (GaP), and gallium indium phosphide (GaInP) by chemical beam epitaxy (CBE) using alternative group III and V precursors. For the work reported here, ethyldimethylindium (EDMIn) and triisopropylgallium (TIPGa) were used as the In and Ga precursors, respectively, in conjunction with three alternative phosphorous (P)-sources: tertiarybutylphosphine (TBP), bisphosphinoethane (BPE), and trisdimethylaminophosphine (TDMAP). We also report for the first time the growth of GaInP by CBE without precracking the group V source using TIPGa, EDMIn, and TDMAP.

1. Introduction

In the chemical beam epitaxy (CBE) technique, the choice of the source precursors for growth is extremely important because of their influence on background impurity concentration and contamination, growth temperature, etc. Given this fact, a sizable effort has been devoted to developing new sources. Traditionally, phosphine (PH_3), triethylgallium (TEGa), and trimethylindium (TMIn) or triethylindium (TEIn) were the most commonly used sources in CBE.

PH_3 is highly toxic and safer alternatives need to be found. Tertiarybutylphosphine (TBP), bisphosphinoethane (BPE), and trisdimethylaminophosphine (TDMAP) are all liquids with toxicity levels substantially lower than for PH_3. However, to date, relatively little has been published on the growth of III-V compounds by CBE using these alternative sources [1-9]. TBP, BPE, and TDMAP are 50% pyrolyzed at 540, 500, and 480 °C, respectively [9], in an ersatz reactor with a residence time of approximately 1.5 seconds.

The goal of the present work was to grow high quality InP, GaP, and GaInP without the need for a group V cracker cell. As discussed below this was not possible for InP using TBP, BPE, or TDMAP; however, growth of GaP and GaInP without thermally precracking was realized using TDMAP as the phosphorous source.

The growth of InP using the alternative precursors will be discussed first followed by a discussion of the growth of GaP with the main focus of this paper being the growth of GaInP. Due to a limited quantity of BPE, results pertaining to the growth of GaP and GaInP using BPE will not be discussed in detail in this paper.

Triisopropylgallium (TIPGa) is known to have a lower pyrolysis temperature than TEGa because of the weaker Ga-alkyl bonds in TIPGa [6]. This lower pyrolysis temperature can lead to lower growth temperatures which is highly desirable. For these reasons, TIPGa was chosen as the Ga precursor. It has been reported that the effective vapor pressure of TMIn varies with time due to the tendency to crystallize in the bubbler [10] and that TEIn tends to decompose in the cylinder during storage [11]. Due to these shortcomings of conventional In sources, ethyldimethylindium (EDMIn) [10,12] was used as the indium precursor in the present work.

2. Experimental Considerations

Epitaxial growth of InP, GaP and GaInP was performed in a custom-designed CBE chamber equipped with diffusion pumps and gas handling systems for the group III and V sources. Thermally precracked TBP was injected into the growth chamber and controlled by a conventional mass flow controller (MFC). The flow rates of BPE and TDMAP were controlled with a closed-loop pressure measurement system. The TIPGa and EDMIn vapors were introduced through pressure-controlled servo-valves and pyrolyzed on the substrate. No carrier gas was used for any of the group III or group V precursors. Group V flow rates were in the range of 1 to 6 sccm; group III flow rates were in the range of 0.03 to 0.09 sccm. Typical group V cracker cell temperatures were in the range of 700 to 850°C with the exception of TDMAP for which the cracker cell was held at room temperature during all growth runs. Typical growth rates for GaP, InP, and GaInP using the above-mentioned precursors were in the range of 0.5 to 1.5 µm/hr.

3. Growth of InP using TBP, BPE, and TDMAP

All InP epilayers were grown on device grade semi-insulating iron (Fe)-doped (100) InP substrates. Without thermal precracking of TBP, no InP epitaxy was observed for V/III ratios up to 100 and growth temperatures up to 525°C. Therefore, it was concluded that it was not possible to grow InP using TBP as the group V source without first thermally precracking TBP. Cracker cell temperatures below 750°C produced epitaxial InP that had inferior materials, electrical and photoluminescence (PL) properties. Details of the dependence of the InP epilayer quality on the cracker cell temperature using TBP will be presented elsewhere.

Although excellent morphology was obtained at growth temperatures as low as 440°C, both the electrical and PL measurements indicated that the InP epilayers had large carrier concentrations ($\approx 10^{17}$ cm^{-3}), low mobilities, µ, (≈ 1800 cm^2/Vs), and PL spectra dominated by donor to acceptor (D-A) transitions. For growth temperatures above 460°C, the D-A peak was extremely small and strong bound exciton emission was observed in the PL spectra with a corresponding decrease in the carrier concentration. To date, we have grown InP using TBP with room temperature carrier concentrations of 10^{15} cm^{-3} and values of 17 K PL full width at half maximum (FWHM) as low as 3 meV at an optimum growth temperature of 500°C and cracker cell temperature of 900°C with a V/III ratio of 80.

Table 1 lists typical Hall and PL properties of InP grown using TBP as a function of substrate growth temperature, Tg, at a fixed V/III ratio of 40. Table 2 lists the Hall and PL properties for InP grown with TBP as a function of V/III ratio with Tg fixed at 500°C. At growth temperatures above 460°C PL FWHM values were typically in the range of 7 to 9 meV. The growth rate of InP using TBP is independent of growth temperature in the range of 400 to 500°C. From the results presented in Tables 1 and 2, it can be seen that higher CBE growth temperatures and higher TBP/EDMIn ratios produce InP with better quality

Table 1. Hall and PL properties of InP grown by CBE using TBP as a function of Tg

Tg (°C)	µ (cm^2/Vs)	n (cm^{-3})	17K PL peak energy (eV)	PL FWHM (meV)
440	1808	1.3 x10^{17}	1.38	16.5
460	1868	8.8 x10^{16}	1.41	9.2
475	2085	2.1 x10^{16}	1.41	6.7
490	1921	3.0 x10^{16}	1.41	8.2
500	2614	2.0 x10^{16}	1.41	8.7

Table 2. Hall and PL properties of InP grown by CBE using TBP as a function of V/III ratio

V/III	µ (cm^2/Vs)	n (cm^{-3})	17K PL peak energy (eV)	PL FWHM (meV)
10	2277	1.6 x10^{17}	1.41	9.3
15	2447	7.7 x10^{16}	1.41	9.2
20	2300	9.1 x10^{16}	1.41	9.0
30	2403	4.1 x10^{16}	1.41	7.9
40	2740	2.8 x10^{16}	1.41	7.9

electrical and optical properties. With our present cracker cell design, we have performed some preliminary growths at higher V/III ratios (up to 80) and found that the carrier concentration decreases to 10^{15} cm^{-3} with a 17K PL FWHM of 3 eV. At substrate temperatures of 520°C, surface morphology worsens and the electrical properties degrade. Figure 1 shows a plot of growth rate as a function of cracker cell temperature for a fixed EDMIn flow of 0.088 sccm and fixed V/III ratio of 40 at a growth temperature of 480°C for InP grown using TBP. The decrease in growth rate with increasing cracker cell temperature could be due to either site blocking by radicals generated from the t-butyl radical breakup or desorption of In due to reactions with the t-butyl radicals at the growth surface.

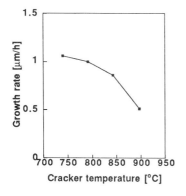
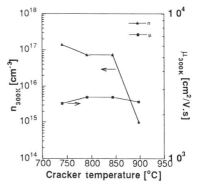

Figure 1. Growth rate as a function of cracker cell temperature for InP grown using TBP.

Figure 2. Carrier Concentration and mobility as a function of cracker cell temperature for InP grown using TBP.

Table 3. Hall properties of InP grown by CBE using BPE as a function of approximate cracker cell temperature, Tc.

Tc (°C)	μ (cm^2/Vs)	n (cm^{-3})
724	3262	1.5 x10^{16}
760	3900	6.3 x10^{15}
800	3235	2.3 x10^{16}

Figure 2 shows the carrier concentration and mobility for the same set of samples used to construct the plot shown in Figure 1. As can be seen from Figure 2, the carrier concentration decreases with increasing cracker cell temperature while the mobility remains relatively flat and even undergoes a slight decrease with increasing cracker cell temperature. The mobility behavior with cracker cell temperature may be due to increased carbon incorporation in the epilayer at higher cracker cell temperatures.

Based on the same methodology used above for TBP, it was not possible for us to grow InP by CBE without thermally precracking BPE. Depending on the growth parameters, typical BPE-grown InP epilayer mobilities were in the range of 3000 to 4000 cm^2/Vs with corresponding background doping levels in the mid 10^{15} to low 10^{16} cm^{-3}. Table 3 lists the electrical properties of BPE-grown InP samples as a function of approximate cracker cell temperature at a growth temperature of 485°C, a EDMIn flow rate of 0.053 sccm and a V/III ratio of 12. The cracker cell temperature is approximate as a detailed calibration was not performed on the particular cracker cell (which was a different one than the cracker cell used to obtain the TBP data above) used to obtain the data in Table 3. The surface morphologies of these samples were excellent with values of 4°K PL FWHM typically less than 5 meV. Further details on the growth of InP using BPE can be found elsewhere [4,5,9].

After repeated unsuccessful attempts, it was concluded that it was not possible to grow InP by CBE without first precracking TDMAP. This inability to grow InP without first precracking TDMAP is consistent with the results of Abernathy et. al. [8]. As discussed below, growth of GaP and GaInP is observed without precracking the TDMAP. We postulate that the surface residence time is longer on a GaP surface than an InP surface presumably due to the larger bond strength of Ga-P compared to that of In-P.

4. Growth of GaP using TBP, BPE, and TDMAP

Although there is a relatively large lattice mismatch, semi-insulating (100) GaAs was used as the substrate for the growth of all GaP samples discussed in this work. Due to the indirect bandgap of GaP, no attempt was made to perform PL measurements.

GaP could be grown with TBP at a cracker cell temperature of approximately 800°C and TIPGa at substrate temperatures as low as 325°C. The best surface morphology was obtained at a V/III ratio of 13 at T_g = 350°C and a V/III ratio of 21 at T_g = 425°C [9]. As the growth temperature increased, the growth rate increased. This behavior is similar to the result obtained by Garcia et al. [7] who used TEGa and PH$_3$. Typical growth rates were in the range of 1 to 2 µm/hr.

Using TDMAP, we were able to grow GaP without thermally precracking at substrate temperatures as low as 480°C. Sharp and strong x-ray peaks were observed for GaP samples grown at 480 and 500°C [9]. At 480°C, the growth rate of GaP was 0.48 µm/hr while at 500°C, the growth rate was 0.66 µm/hr. At both temperatures, the flow rates of TIPGa and TDMAP were 0.07 and 0.52 sccm, respectively.

As mentioned in Section 3, we postulate that GaP can be grown without precracking while InP cannot because of the longer residence time on the GaP surface due to the larger Ga-P bond strength. TDMAP is thus able to crack before desorbing on a GaP surface while it desorbs before cracking on an InP surface. Similar GaP results using TDMAP were obtained by Abernathy et al. [8]

5. Growth of GaInP using TBP, BPE, and TDMAP

GaInP was grown on semi-insulating (100) GaAs substrates using EDMIn and TIPGa with thermally precracked TBP at substrate temperatures ranging from 440 to 500°C and V/III ratios from 25 to 60 with TBP flow rates ranging from 3.5 to 8.5 sccm. Intense PL emission was observed for GaInP grown using TBP. It was further observed that, for a given V/III ratio, as the growth temperature increased, the PL FWHM monotonically decreased. The PL FWHM was less sensitive to the V/III ratio although, for a given growth temperature, a trend toward smaller values of FWHM was seen as the V/III increased. The lowest PL FWHM for our CBE grown GaInP using TBP was approximately 40 meV. Recently Garcia et al. [7] have reported GaInP material growth using TBP with a value of PL FWHM of 40 meV.

GaInP samples were grown using BPE with intense 17K PL emission at 1.982 eV with a FWHM of 36 meV. As with TBP, it appears that BPE is too stable to crack on the heated substrate surface alone and that a thermal cracker cell is required for growth of GaInP by CBE to occur.

For the first time, the growth of GaInP by CBE on GaAs (100) substrates without precracking the group V source was accomplished using TDMAP. All published papers to date for the growth of InP or GaInP by CBE have used cracker cells to precrack the group V source [1-7,13-16]. This is due to the large V-radical bond strengths for the group V sources, such as PH_3 and TBP, so that decomposition of the source does not occur in a "single bounce" on the heated substrate at temperatures convenient for growth. Before epitaxial growth, the substrate was heated to the desired growth temperature under TDMAP to thermally desorb the oxide. The group III flow was initiated to start growth. Typical chamber pressures during growth were in the range of 10^{-5} to low 10^{-4} torr. GaInP was grown on GaAs substrates at 500°C without precracking the TDMAP. The flow rates of TIPGa, EDMIn and TDMAP were 0.05, 0.04 and 1.1 sccm, respectively. This gave an input V/III ratio of 12. Poor morphology was observed with lower V/III ratios. The growth rate of GaInP is proportional to the total group III flow rate. The GaInP composition was determined from x-ray diffraction measurements. Several Intentionally mismatched $Ga_{0.43}In_{0.57}P$ epilayers to the GaAs substrate were grown to demonstrate that In-rich GaInP can be grown by CBE without the need for precracking TDMAP. The 17K PL peak energy for a $Ga_{0.43}In_{0.57}P$ layer was 1.8 eV which was about 100 meV lower than expected [17]. It was also observed from electron diffraction [18] that there was some degree of Cu-Pt ordering, which might account for the low PL peak energy [17]. In this work, FWHM values as low as 34 meV were measured. Table 5 lists typical growth efficiencies and growth rates for GaInP grown using uncracked TDMAP. As can be seen from Table 5, The growth rate is proportional to the total group III flow amount and the growth efficiency is extremely high for GaInP epilayers grown using uncracked TDMAP.

Table 5. Relation between the growth rate of GaInP, growth efficiency and the total group III flow amount. The TDMAP flow rate was 1.1 sccm.

TIPGa+EDMIn Flow (sccm)	Growth Rate (μm/hr)	Growth efficiency (μm/mol)
0.06	0.38	7.54×10^5
0.09	0.53	7.13×10^5
0.11	0.70	8.09×10^5

7. Conclusions

In conclusion we report on the growth of InP, GaP, and GaInP by CBE using three alternative phosphorus sources: TBP, BPE, and TDMAP. All three sources were capable of growing high quality InP, GaP, and GaInP epilayers. For example, PL FWHM as low as 3 meV and carrier concentrations of 10^{15} cm^{-3} were obtained from TBP-grown InP. It was necessary to thermally precrack the phosphorus source to grow InP using TBP, BPE, and TDMAP and to grow GaP and GaInP using TBP and BPE.

We report several "firsts" for the growth of GaInP: 1) no precracking of the P source (TDMAP); 2) use of novel P-precursor TDMAP; and 3) use of Ga and In precursors EDMIn plus TIPGa. Good morphology and strong PL were observed. For approximately lattice-matched GaInP, the FWHM of the PL peak was typically less than 40 meV. Growth of GaP without precracking was also achieved. Further work is under way to optimize growth conditions and reduce the unintentional doping levels in the InP and GaInP epilayers.

8. Acknowledgments

This research was supported by the U.S. Army Office of Research under grant number DAAL 03-91-G-0153 and by the National Science Foundation under grant number ECS-9113992.

References

[1] Ritter D, Panish M B, Hamm R A, Gershoni D and Brener I 1990 Appl. Phys. Lett. **56** 1448-1450
[2] Beam III E A, Henderson T S, Seabaugh A C and Yang J Y 1992 J. Crystal Growth **116** 436-446
[3] Hincelin G, Zahzouh M, Mellet R and Pougnet A M 1992 J. Crystal Growth **120** 119-123
[4] Chin A, Martin P, Das U, Mazurowski J and Ballingall J 1992 Appl. Phys. Lett. **61** 2099-2101
[5] Chin A, Martin P, Das U, Mazurowski J and Ballingall J 1992 J Vac. Sci. Technol. B **11** 847-85
[6] Jones A C 1993 J. Cryst. Growth **129** 728-773
[7] Garcia J Ch, Regreny Ph, Delage S L, Blanck H and Hirtz J P 1993 J. Crystal Growth **127** 255-257
[8] Abernathy C A, Bohling D A, Muhr G T and Wisk P W 1994 MRS Spring Meeting Symposium E in press
[9] Sadwick L P, Kim C W, Ryu H H, Stringfellow G B, Gedridge, Jr. R W and Jones A C 1994 MRS Spring Meeting Symposium E in press
[10] Knauf J, Schmitz D, Strauch G, Jurgensen H and Heyen M 1988 J. Crystal Growth **93**, 34-40
[11] Stringfellow G B Organometallic Vapor Phase Epitaxy:Theory and Practice (Academic Press, Boston, 1989).
[12] Fry K L, Kuo C P, Larsen C A, Cohen R M and Stringfellow G B 1986 J. Electron Mat. **15**, 91-96
[13] Garcia J Ch, Maurel Ph, Bove Ph and Hirtz JP 1991 J. Appl. Phys. **69** 3297-3302
[14] Tsang W T Appl. Phys. Lett. **45** 1234-1236
[15] Ozasa K, Yuri M, Tanaka M and Matsunami H 1989 J. Appl. Phys. **65** 2711-2716
[16] Abernathy C R, Wisk P W, Ren F, Pearton S J, Jones A C and Rushworth S A 1993 J. Appl. Phys. **73** 2283-2287
[17] Su L C, Ho I H and Stringfellow G B 1994 J. Crystal Growth to be published
[18] Soh S H, Ryu H H, Sadwick L P and Stringfellow G B 1994 (unpublished results)

Chemical beam epitaxy of InGaAs/GaAs multiple quantum wells using cracked or uncracked tris-dimethylaminoarsenic

H K Dong, N Y Li, and C W Tu

Department of Electrical and Computer Engineering, University of California at San Diego, La Jolla, California 92093-0407, USA

Abstract: InGaAs/GaAs multiple quantum wells have been grown by chemical beam epitaxy (CBE) using triethylgallium, trimethylindium, and cracked or uncracked tris-dimethylaminoarsenic (TDMAAs). Good InGaAs/GaAs interfaces can be achieved for samples grown with cracked TDMAAs in the substrate temperature range studied, 440-550 °C, as evidenced by satellite peaks in x-ray rocking curves. For samples grown with uncracked TDMAAs at high temperatures, however, no satellite peaks were observed. The indium composition in the quantum wells differs for these two cases even with the same substrate temperature. The discrepancies may come from the different arsenic species on the surface.

1. Introduction

In the past few years, there have been many attempts to replace the highly toxic and high-pressured AsH_3 with novel organometallic arsenic reagents. There are mainly three requirements for new precursors to be used in metalorganic molecular beam epitaxy (MOMBE) or chemical beam epitaxy (CBE), i.e., low but reasonable vapor pressure, low cracking temperature, and replacement of the As-H bond (which is believed to introduce the extremely toxic function) by other ligands. Tris-dimethylaminoarsenic (TDMAAs), $As[N(CH_3)_2]_3$, with As directly bonded to N, has been proposed as an alternative source to arsine[1-8]. Since there are no As-H and As-C direct bonds, one can expect TDMAAs to have lower toxicity and lower carbon incorporation. TDMAAs has been successfully used in MOMBE or CBE [1,2] and metalorganic chemical vapor deposition (MOCVD) [3,4] growth of GaAs. However, no results have been published so far on the growth of InGaAs on a GaAs substrate using TDMAAs. Pseudomorphic, strained InGaAs/GaAs quantum well structures are important for their applications in high-speed transistors and optoelectronic devices.

In this paper, we report the CBE growth and characterization of InGaAs/GaAs multiple quantum wells (MQWs) using trimethylindium (TMIn), triethylgallium (TEGa), and TDMAAs. In order to obtain more understanding on the CBE growth mechanism, comparisons are made for MQW structures grown by cracked ($T_{crack}=300°C$) and uncracked TDMAAs.

2. Experimental Procedures

Experiments were performed in a modified Perkin Elmer 425B MBE system, equipped with gas lines for TEGa, TMIn, and TDMAAs. TEGa and TMIn were introduced into the growth chamber without any carrier gas through vapor source mass flow controllers. TDMAAs, with a 20°C vapor pressure of ~1.35 Torr, was carried by hydrogen and injected into the chamber through a leak valve. The TDMAAs flow was determined by the hydrogen flow rate, which was controlled by a mass flow controller. Semi-insulating (100) GaAs substrates were chemically etched and thermally cleaned at about 600°C. The reflection high-energy electron diffraction (RHEED) image was monitored by a video-camera system. The intensity oscillation data were extracted and analyzed from a defined region of a digitized image that covered the specular beam.

$In_xGa_{1-x}As$/GaAs MQW structures (10 periods) were grown using cracked (T_{crack}=300°C) or uncracked TDMAAs at different substrate temperatures (440-550°C). During growth, the flow rates of TEGa and TMIn were kept at 0.4 and 0.06 sccm, respectively, and the hydrogen flow rate for TDMAAs was 2 sccm. The growth was interrupted for 10 seconds between wells and barriers with TDMAAs supplied constantly. The well composition and thickness were determined from x-ray rocking curve (XRC) measurements and simulations using the dynamical theory. The results were confirmed by low-temperature photoluminescence (PL) measurements and simulations based on the envelope-function model.

3. Results and Discussion

Good InGaAs/GaAs interfaces can be achieved for cracked TDMAAs samples in the whole substrate temperature range studied, as indicated by the sharp satellite peaks in the XRC graph of an InGaAs/GaAs MQW, as shown in Fig. 1(a). For uncracked TDMAAs samples, however, when the substrate temperature is high (525 or 550°C), no satellite peaks were observed in the XRC graph, as shown in Fig. 1(b). The dissimilar behavior could be due to the different surface chemical reactions during InGaAs growth with or without pre-cracking TDMAAs. In order to obtain good interfaces for InGaAs/GaAs MQWs using uncracked TDMAAs, the substrate temperature must be in the range of 440-500°C.

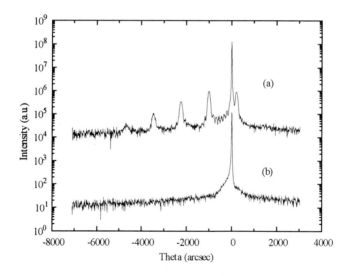

Fig. 1 X-ray rocking curves for InGaAs/GaAs MQWs grown at 550°C with (a) cracked TDMAAs and (b) uncracked TDMAAs

Fig. 2 shows the full width at half maximum (FWHM) of PL exciton peaks as a function of substrate temperature. Solid squares in the figure indicate the results using cracked TDMAAs, and the FWHM decreases as the substrate temperature increases. Therefore, higher substrate temperature gives better optical property. The best growth temperature in our experiment is determined to be 550°C, where we can obtain a FWHM of about 5.7 meV while maintaining a good InGaAs/GaAs interface. Open circles in the plot represent the results using uncracked TDMAAs. The FWHM in this case has a minimum value at 525°C, where good InGaAs/GaAs interfaces can not be obtained. Therefore, we choose the preferred growth temperature for uncracked TDMAAs samples to be between 480 and 500°C, where the FWHM is about 6.0-7.0 meV and the interfaces are good.

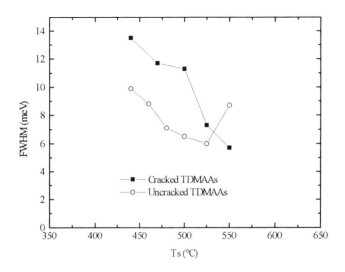

Fig. 2 FWHM of PL exciton peaks as a function of substrate temperature

From the XRC data, computer simulations using the dynamical theory were conducted and the InGaAs well composition, well thickness, and the GaAs barrier thickness were deduced. These values are used in the envelope-function model to calculate the exciton energy of the MQW structure, which agrees with the PL peak very well.

Fig. 3(a) shows the indium composition in the quantum wells as a function of substrate temperature with cracked and uncracked TDMAAs. For uncracked TDMAAs samples grown at 525 and 550°C, composition and thickness can not be determined since the XRCs are too broad to be compared with simulations. The indium composition decreases monotonically as the substrate temperature increases from 440°C to 500°C for uncracked TDMAAs samples, whereas for cracked TDMAAs samples, the indium composition has a minimum at 500°C. Fig. 3(b) shows the InGaAs well thickness as a function of substrate temperature. The InGaAs growth rate has a peak value at 480°C for uncracked TDMAAs and 510°C for cracked TDMAAs. The dotted line in Fig. 3(a) represents the corresponding GaAs growth rate (with the same TEGa flow rate) as a function of substrate temperature. The GaAs growth rate increases as the substrate temperature increases, and becomes saturated above 500°C, indicating a complete decomposition of TEGa. Moreover, the complete decomposition of TMIn happens below 440°C. Therefore, the incomplete decomposition of TEGa in the low substrate temperature range results in a high indium composition and low InGaAs growth rate. In the high substrate temperature range, surface segregation of indium during growth and lower sticking coefficient of TEGa molecules on an indium-covered surface are believed to be the reason that the InGaAs growth rate becomes lower than that of GaAs with the same TEGa flow rate [9]. As the substrate temperature increases, more TEGa will desorb from the surface, and hence result in a lower InGaAs growth rate. The higher indium composition at high substrate temperatures comes from the more severe indium segregation. This is verified in the cracked TDMAAs case.

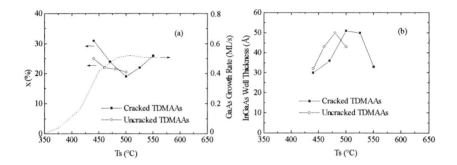

Fig. 3 (a) Indium composition, and (b) InGaAs well thickness as a function of substrate temperature with cracked or uncracked TDMAAs.

By cracking TDMAAs, the arsenic species involved in the surface chemical reactions could be different from those of the uncracked TDMAAs. The arsenic surface adsorption behavior can greatly influence the InGaAs growth. In order to better understand the adsorption behavior, a similar study to Liang and Tu[10] was performed. It is known that the RHEED specular-beam intensity corresponds to the smoothness of the surface. During growth interruption, at high substrate temperatures, the RHEED intensity decreased after the TDMAAs shutter was closed because the desorption of the adsorbed arsenic species roughened the surface, as shown in Fig. 4(a). However, at lower substrate temperatures, the RHEED intensity increased after closing the TDMAAs shutter since the desorption of the excess accumulated arsenic species from the surface created a smoother surface, as shown in Fig. 4(b).

Since the intensity increases or decreases exponentially, indicating a first-order process, the time constant, t_0, can be determined. The desorption rate constant, k, is then $1/t_0$. Fig. 5 shows the desorption rate constant of cracked and uncracked TDMAAs from GaAs as a function of the reciprocal substrate temperature. A discontinuity of the desorption rate constant is observed for both cracked and uncracked TDMAAs at about 520°C. Above this temperature, the adsorbed arsenic species start to desorb from GaAs surface. In the low temperature range (T_s<520°C), the activation energy is 46 kcal/mol for both cracked and uncracked TDMAAs. In the high temperature range (T_s>540°C), the activation energy is 51 kcal/mol for cracked TDMAAs and 64 kcal/mol for uncracked TDMAAs. This tells us that different arsenic species are generated from cracked and uncracked TDMAAs at high substrate temperatures. The different arsenic species will then affect the InGaAs growth and result in different InGaAs/GaAs interfaces as have been seen in Fig. 1. At low substrate temperatures, uncracked TDMAAs has higher desorption rate constants than cracked TDMAAs, and accordingly, less incorporation sites are blocked. Therefore, uncracked TDMAAs samples have a higher InGaAs growth rate than cracked TDMAAs samples.

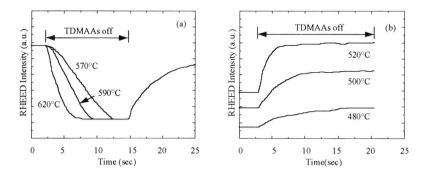

Fig. 4 The RHEED specular-beam intensity of GaAs as a function time when the TDMAAs (uncracked) shutter closed at (a) high temperature range and (b) low temperature range, respectively.

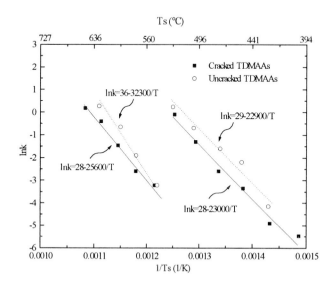

Fig. 5 The natural logarithm of the desorption rate constants of cracked and uncracked TDMAAs as a function of the reciprocal substrate temperature.

4. Conclusions

InGaAs/GaAs MQW structures have been successfully grown with both cracked and uncracked TDMAAs. Good InGaAs/GaAs interfaces and optical property can be achieved. The indium composition in the quantum wells and the InGaAs growth rate are different for these two cases even with the same substrate temperature. The discrepancies may come from the different arsenic species on the surface from cracked and uncracked TDMAAs, as have been shown by the RHEED study of TDMAAs desorption from a GaAs surface.

Acknowledgment

This work is supported by the Air Force Wright Laboratory. The authors wish to thank W. S. Wong and C. H. Yan for their assistance. They are also grateful to the generous donation of the MBE system by the Rockwell International Science Center, Thousand Oaks, California.

References

[1] Abernathy C R, Wisk P W, Bohling D A and Muhr G T 1992 Appl. Phys. Lett. **60** 2421.
[2] Dong H K, Li N Y and Tu C W 1994 Mater. Res. Soc. Symp. Proc. **340** 173.
[3] Zimmer M H, Hövel R, Brysch W and Brauers A 1991 J. Cryst. Growth **107** 348.
[4] Zimmermann G, Protzmann H, Marschner T, Zsebök O, Stolz W, Göbel E O, Gimmnich P, Lorberth J, Filz T, Kurpas P and Richter W 1993 J. Cryst. Growth **129** 37.
[5] Fujii K, Suemune I, Koui T and Yamanishi M 1992 Appl. Phys. Lett. **60** 1498.
[6] Fujii K, Suemune I and Yamanishi M 1992 Appl. Phys. Lett. **61** 2577.
[7] Koui T, Suemune I, Miyakoshi K, Fujii K and Yamanishi M 1992 Jpn. J. Appl. Phys. **31** L1272.
[8] Salim S, Lu J P, Jensen K F and Bohling D A 1992 J. Cryst. Growth **124** 16.
[9] Iimura Y, Nagata K, Aoyagi Y and Namba S 1990 J. Cryst. Growth **105** 230.
[10] Liang B W and Tu C W 1992 J. Appl. Phys. **72** 2806.

Strain Characterization of AsH3 Induced Exchange Reactions in InP Grown by OMVPE

A. R. Clawson [†] and C. M. Hanson [§],

[†] Univ. California San Diego, ECE Dept-0407, 9500 Gilman Dr., La Jolla, CA 92093-0407
[§] NCCOSC RDTE Div 555, 49285 Bennett St., Rm 111, San Diego, CA 92152-5790

Abstract. Behavior of As-P intermixing in heterojunctions of As-compounds grown on InP (001) has been inferred from strain of very thin mismatched layers inserted periodically in InP to form multilayer superlattice structures. Strain components are observed for both As-P exchange in the InP and from inserted layer growth. The data suggest that for As-compounds grown on an InP surface the availability of As for both layer growth and As-P exchange is rate-limited, most likely by surface kinetics, and an optimum growth rate occurs for minimizing the As-P exchange. There is also a fixed InAs strain component inherent to interfaces of As-compounds on InP. Monolayer thickness layers of AlP or GaP grown on the InP to change the surface chemical bonds are shown to reduce the As-P exchange somewhat but they do not stabilize the InP surface against exposure to AsH3. H2 and PH3-exposures of the As-terminated surface show that P-As exchange to desorb the As is slow compared to As adsorption on InP by AsH3 exposure.

1. Introduction

Compositional intermixing at InGaAs/InP heterojunctions has long been recognized as a limitation to achieving perfectly abrupt interfaces desired for device structures. It is generally concluded that intermixing on the group III sublattice is either small or negligible, but intermixing on the group V sublattice is significant[1,2]. Furthermore, the As-P exchange is different for the two different InGaAs/InP interfaces with resulting differences in heterojunction characteristics.

In this investigation we provide insight into the As-P intermixing behavior of As-compounds grown on InP substrates by analyzing the strain in periodic multilayer structures specially configured to reveal As-P exchange. The structures consist primarily of InP with very thin mismatched binary compounds inserted at ~170 Å spacing. A nominal "superlattice" of 30 periods is grown and analyzed by x-ray diffraction (XRD) to obtain a characteristic superlattice pattern. The separation of the superlattice zero order peak from the substrate (004) peak shows the change of average lattice dimensions due to elastic accommodation of different size atoms within the thickness of a 170 Å thick single period.

2. Experimental

The OMVPE growth was performed in a small scale horizontal reactor with a radiantly heated graphite susceptor at temperature of 650°C and pressure of 20 Torr on InP {001} substrates. Transient-free gas switching was achieved with a Thomas Swann Epifold low dead-space manifold using vent/run switching of established gas flows, keeping the gas flow to the chamber constant, and balancing the pressure between vent and run manifolds. Growth conditions to achieve reproducible growth rates were determined from previous studies[3,4].

Table 1 III-V OMVPE Growth Rates for This Study

III-V	⊥ Growth Rate (Å/second)	Measurement Technique
InP	2.24	from InP superlattice satellite peak spacing.
InAs	2.03	from InAs monolayer strain in GaAs superlattice.
GaAs	2.18	from GaAs superlattice satellite peak spacing.
GaP	2.52	from GaP monolayer strain in InP superlattice.
AlAs	2.34	from AlP = 2.19 Å/S assuming same Al atom per second rate.
AlP	2.19	from AlP monolayer strain in InP superlattice.

To analyze the strain it is important to accurately know growth rates to determine layer thicknesses. We have previously shown that fractional monolayer growth rates are the same as rates for thick layers, and that the strains of superlattices generally are predictable from the inserted layer thickness and mismatch[5]. Correction for the strain expansion or contraction of the layer's perpendicular lattice dimension is based on the Poisson ratio as outlined by Hornstra and Bartels[6]. We use the growth rates of these earlier studies as summarized in Table 1 to grow layers of predetermined thickness and strain. The measured superlattice strain is related to the individual strain components over the thickness of one period as:

$$(\Delta a_\perp/a)_{SL} \times d_{SL} = (\Delta a_\perp/a)_{ML} \times d_{ML} + (\ unknown\ ) \quad (1)$$
$$i.\ e., [\text{superlattice strain}] = [\text{inserted layer strain}] + [\text{As-P intermixing strain}]$$

where $(\Delta a_\perp/a)_{SL}$ is the XRD measured superlattice strain, d_{SL} = 170 Å is the measured period thickness, $(\Delta a_\perp/a)_{ML}$ is the perpendicular strain of the elastically deformed thin "monolayer" calculated from its lattice parameter and elastic coefficients, and d_{ML} is the inserted monolayer thickness. The unknowns are the strain and thickness of the As replacement of P, which is easiest to describe as an equivalent thickness of InAs. Other possible strain components are considered insignificant. In particular, strains from group III sublattice intermixing contribute nothing to the superlattice strain[5], and data are presented below that show the rate of P replacement of As is comparatively slow thus we ignore its contribution. The superlattice strains resulting from thin As-compound layers inserted in InP are shown in Fig. 1. Three sets of data are shown: InAs layer growth, GaAs layer growth and InAs(P) formed by

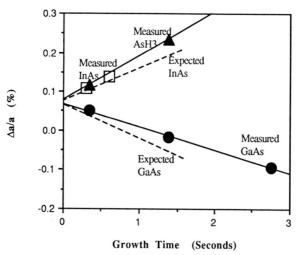

Figure 1. Strain in 30-period InP superlattices formed by periodic insertion of monolayer equivalent thicknesses of InAs (open squares) or GaAs (closed circles), or from AsH3 exposure of InP (closed triangles) at 170 Å intervals.

periodic exposure of InP to AsH3. The strains for each of the three kinds of layers increased with growth time, however, in each case the extrapolation to zero time shows an offset compressive strain component, $\sigma_0 \equiv \frac{\Delta a_\perp}{a}(t=0)$, equivalent to ~0.6 to 0.7 monolayers of InAs. σ_0 is a very rapidly formed strain attributed to a combination of In-As bonds formed when As preferentially replaces P on surface sites when InP is exposed to AsH3 and As surface atoms that are relatively unaffected by subsequent exposure to PH3.

An important feature of the data is evident by comparing the slopes of the measured strains of InAs and GaAs layer growth (solid lines) with their expected slopes (dashed lines). The difference between measured and expected slopes shows an additional time dependent compressive strain component, $\sigma(t) \equiv \frac{\Delta a_\perp}{a}(\text{meas}) - \frac{\Delta a_\perp}{a}(\text{expected})$, which can only be attributed to formation of an As rich compound, InAs in our simplified view. It is significant that the rate of excess As-compound formation and therefore the quantity of As-compound causing the total strains for InAs and GaAs growths is the same and is also identical to the amount of InAs formed by exposure to only AsH3. The implication is that the amount of available As to supply these growths is rate-limited to the same fixed amount, and that there are two mechanisms consuming the As: the layer growth preferentially uses whatever As necessary to achieve compound stoichiometry and the balance of As contributes to As-P exchange. In the absence of growth all the available As contributes to As-P exchange.

The As rate-limiting reaction step is attributed to the surface chemistry kinetics rather than to gas phase reactions. In Fig. 2 the strain is compared for AsH3-exposed InP superlattices that are identical except for a 1.9 times increase in the concentration of AsH3 used. The resultant strain dependencies on time are identical, thus the amount of As available to incorporation on the surface is unchanged; excess AsH3 vapor has no effect. Only the σ_0 component has increased slightly.

There are several predictions for As-P exchange which can be made from this As rate-limited interpretation of the data in Fig. 1. One is that As-P exchange would be minimized if all the As is consumed in the layer growth. Another is that if a growth rate is established which exceeds the As available to maintain stoichiometry, it would result in a Group III rich surface which might result in liquid formation with consequent roughening and possible

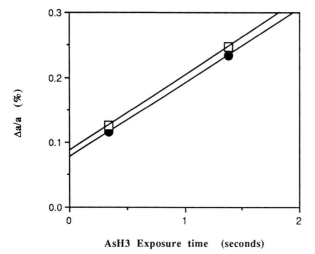

Figure 2. Strain of InAsP formed from AsH3 exposure of InP at 170 Å intervals for 30 periods comparing effects of AsH3 vapor concentration. Closed circles are for 0.008 mole fraction AsH3 and open squares are for 0.0152 mole fraction AsH3.

compositional intermixing when InP growth is resumed. There looks to be an optimum growth rate for As-compounds grown on InP.

To further explore the initial exposure of InP to AsH3 the InP surface was capped with monolayers of AlP or GaP to change the surface atom bond strength as a means of stabilizing the InP surface against As-P exchange, as has been reported for OMMBE growth of InAlAs on InP[7]. Results shown in Fig. 3 compare strains of identical AsH3 exposed superlattices with and without a cap layer of AlP or GaP. The strains of AlP-only and GaP-only in InP superlattices were determined for the zero AsH3 exposure time, then structures with identical AlP or GaP monolayers were grown with additional AsH3 exposure. The tensile strains from the AlP and GaP of -0.240% and -0.271% respectively, are fixed contributions to the total strain. The data of the strains with AsH3 exposure extrapolate to a σ_0 that is identical to the offset without the cap layer. The absence of any change in σ_0 in the presence of AlP or GaP suggests that the interface displacement of surface P by As is the same. AlP or GaP monolayers were also grown following the AsH3 exposure and again no change is seen in σ_0. On the other hand, $\sigma(t)$ is sensitive to AlP and GaP cap layers on the InP surface, and for both shows a decrease in the compressive strain when the cap layers are grown prior to AsH3 exposure, while there is little or no effect when the cap layer are grown after the AsH3 exposure.

To further assess the behavior of the As-terminated surface the desorption of As during exposure to H2-only or H2+PH3 atmospheres was studied. Fig. 4 shows the results of a 1.38 second AsH3 exposure of InP followed by an exposure of the As-terminated surface to varying periods of H2 or PH3 prior to resuming InP growth of the superlattice. The only strain which could be introduced in these structures would be compressive, from InAs. Exposure to H2 shows negligible desorption of As. This relates to the apparent InGaAs surface stability that allows use of H2 interrupts at this interface to improve interface abruptness[8,9]. The interrupt allows time to fully deplete the very reactive arsine vapor with no degradation of the surface prior to starting InP growth. The decrease of strain when the As-terminated surface is exposed to PH3 indicates removal of As and its replacement by P. The rate of P-As exchange of the As-terminated surface in the presence of phosphine is at a much slower rate than the rate of As-P exchange in the presence of arsine. There is a distinct change in the rate of As desorption when the strain is reduced to the equivalence of ~2/3

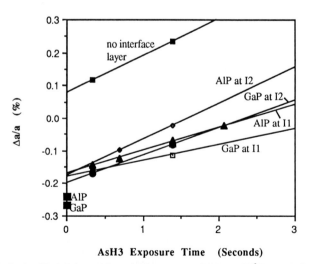

Figure 3 Strain of InAsP formed from AsH3 exposure of InP at 170 Å intervals for 30 periods showing effects of adding an interface layer on the InP of ~1 monolayer AlP or ~1 monolayer GaP prior to AsH3 exposure (interface I1) or following the AsH3 exposure (interface I2).

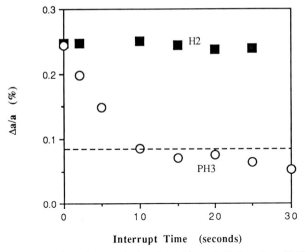

Figure 4. Strain in InP 30 period superlattices formed by periodic insertion of 1.38 second AsH3 exposure followed by an As-desorption growth interrupt in H2 or H2+PH3 at 170 Å intervals. The dotted line is the expected strain for inclusion of the ~2/3 monolayer InAs in the superlattice equivalent to σ_0 from figure 1.

monolayer InAs suggesting that the As is more easily removed when in excess of 2/3 monolayer of InAs.

Using the data of Fig. 4 along with Fig. 1, both σ_0 and $\sigma(t)$ might be interpreted in terms of surface reconstruction. For As-compounds formed on InP from AsH3 the occurrence of σ_0 equivalent to ~2/3 monolayer InAs seems to be a fundamental feature. This suggests that a stable surface on InP exposed to AsH3 has a single layer surface reconstruction with only partial occupancy of the surface sites. The offset strain, σ_0, is envisioned as forming partly when As substitutes for P on the surface at the start of As-compound growth and partly when the InP growth resumes on the As-terminated surface. The time-dependent strain, $\sigma(t)$, is envisioned as resulting from a multilayer accumulation of adsorbed As in excess of the semiconductor surface layer analogous to the multiple dimer layers seen for c(4x4)/d(4x4) GaAs surface reconstruction. This excess could then release into the next InP layer as the surface adjusts to the observed equilibrium InP single layer (2x4) reconstruction[10].

3. Conclusions

Superlattices formed by periodic insertion of strained layers into an unstrained binary compound can be used to elucidate characteristics of group V exchange across heterojunction interfaces. Excess compressive strain when As-compounds are grown on InP is attributed to As replacement of P, and the excess InAs strain component can be separated from the expected strain of the inserted thin layer. Strains of thin As-compound layers grown in InP show that during growth the amount of As available for reaction either in layer growth or in As-P exchange is rate-limited by surface kinetics, and an optimum growth rate should provide stoichiometric layer growth with minimum As-P exchange. Modification of the InP surface chemical bond strength by growing monolayers of AlP or GaP has some effect on reducing As-P exchange but does not stabilize the InP surface against AsH3 exposure for the OMVPE growth conditions. For As-terminated surfaces exposed to PH3 the P replacement of As is a slow reaction compared to As replacement of P under AsH3 exposure. Indication of a stable ~2/3 coverage As surface layer on InP suggests that a part monolayer of InAs at each interface is inherent to growth of As-compounds in InP, and additional excess As from accumulation on the As-compound surface can carryover into subsequent InP growth.

Acknowledgements

This work was supported by the Office of Naval Research, Program Element 62234N, Microelectronics Project RS34M40.

References

[1]. J. P. Wittgreffe, M. J. Yates, S. D. Perrin and P. C. Spurdens, *J. Crystal Growth* 130, 51-58 (1993)
[2]. R. Meyer, M. Hollfelder, H Hardtdegen, B. Lengeler and H. Lüth, *J. Crystal Growth* 124, 583-588 (1992)
[3]. A. R. Clawson, T. T. Vu, S. A. Pappert and C. M. Hanson, *J. Crystal Growth* 124, 536-540 (1992)
[4]. A. R. Clawson, X. Jiang, P. K. L. Yu, C. M. Hanson and T. T. Vu, *J. Electron. Mater.* 22, 423 (1993)
[5]. A. R. Clawson and C. M. Hanson, presented at the Electronic Materials Conference, June 22-24, 1994, Boulder, CO; submitted to *J. Electron. Mater.*
[6]. J. Hornstra and W. J. Bartels, *J. Crystal Growth* 44, 513-519 (1978)
[7]. W. E. Quinn, M. C. Tamargo, M. J. S. P. Brazil, R. E. Nahory and H. H. Farrell, *J. Vac. Sci. Technol.* B 10, 978-981 (1992)
[8]. N. J. Long, A. G. Norman, A. K. Petford-Long, B. R. Butler, C. G. Cureton, G. R. Booker and E. J. Thrush, *Microsc. Semicond. Mater. Conf.*, Oxford, 25-28 March 1991 (Inst. Phys. Conf. Ser. No. 117; section 2) p. 69-74
[9]. A. Kohl, A. Mesquida Küsters, S. Brittner, K. Heime, J. Finders., D. Gnoth, J. Geurts and J. Woitok, *Proc. 6th Int'l. Conf. on Indium Phosphide and Related Materials*, March 27-31, 1994, Santa Barbara, CA (IEEE Cat.No. 94CH3369-6) p. 151-154
[10]. J. Jönsson, F. Reinhardt, K. Ploska, M. Zorn, W. Richter and J.-Th. Zettler, *Proc. 6th Int'l. Conf. on Indium Phosphide and Related Materials*, March 27-31, 1994, Santa Barbara, CA (IEEE Cat.No. 94CH3369-6)p.53-56

Comparative investigation of electrical and optical characteristics of $Al_xGa_{1-x}As/GaAs$ structures deposited by LP-MOVPE and MBE

Hilde Hardtdegen, M. Hollfelder, Chr. Ungermanns, T. Raafat, R. Carius, A. Förster, J. Lange and H. Lüth

Institut für Schicht- und Ionentechnik, Forschungszentrum Jülich, D-52425 Jülich, Germany

Abstract: A comparison of electrical and optical characteristics of (AlGa)As and GaAs bulk material as well as (AlGa)As/GaAs heterostructures deposited by LP-MOVPE and MBE at growth conditions employed for high frequency device structural growth is presented. Characterization of bulk GaAs layers shows that electrical and optical quality is comparable. Bulk (AlGa)As layers, however, differ in their shallow and deep level impurity content - their concentration is lower for MOVPE samples. Characterization of heterostructures documents that the interfaces are much rougher for MOVPE samples than for MBE samples.

1. Introduction

In the past many studies have dealt with the optimization of the MBE (molecular beam epitaxy) and MOVPE (low pressure metalorganic vapor phase epitaxy) growth processes for bulk $Al_xGa_{1-x}As$ layers. The conditions under which optimum results are obtained - a high deposition temperature of over 700°C - is, however, detrimental to high frequency device structures since diffusion takes place and interfaces are not perfectly abrupt Therefore the usual deposition temperature for device structures ($T_D \leq 650°C$) is a compromise between the growth temperature for high quality bulk material ($T_D \geq 700°C$) and that for sharp interfaces ($T_D \leq 600°C$). The aim of this contribution is *not* to compare the optical and electrical characteristics of the best bulk layers that can be obtained, but *rather* to understand the differences in optical and electrical properties of MBE and LP-MOVPE AlGaAs/GaAs structures deposited under the conditions used for high frequency device structural growth.

From the manufacturing point of view, LP-MOVPE belongs to one of the most attractive processes for mass production of device structures. Nevertheless (AlGa)As/GaAs device structures such as HBTs (hetero bipolar transistors) and HEMTs (high electron mobilty transistors) are industrially deposited by MBE, since the purity and crystal perfection of (AlGa)As deposited by MOVPE was insufficient so far. The metalorganic Al-source TMAl employed is responsible for the poor purity of MOVPE (AlGa)As samples. The source is contaminated with silicon and oxygen and is so stable that it decomposes incompletely: Therefore besides Si and O also carbon is incorporated into the layers. The MBE process is, in contrast, well known for its high purity AlGaAs layers, since only elemental sources are employed. Production of layers with high crystal perfection, homogeneity and sharp interfaces together with the possibilty of in situ-growth control are additional advantages of the MBE method.

Recently (AlGa)As layers were deposited by LP-MOVPE using a new approach (Hardtdegen 1994): nitrogen instead of hydrogen was employed as the carrier gas together with dimethylethylaminealane (DMEAAl, $(CH_3)_2(C_2H_5)N-AlH_3$) and triethylgallium (TEGa)

as metalorganic sources and AsH_3. This approach led to greater crystal perfection and homogeneity due to the carrier gas selected and less impurity incorporation due to the high purity of the Al source and the absence of Al-C bonds in the Al source compound. It is now possible for the first time to compare the electrical and optical characteristics of (AlGa)As/GaAs deposited by both MBE and LP-MOVPE.

2. Experimental

The LP-MOVPE experiments were carried out in a horizontal reactor (Aixtron) at 20 mbar and a gas velocity of 0.9 m/s using the source materials TEGa, DMEAAl and AsH_3. The ambient, N_2, was purified using a getter column. SiH_4 was used as the source for n-type doping. The MBE samples were deposited in a standard solid source Varian Mod-Gen II system using silicon as the dopant

The characterization of bulk layers was carried out by van der Pauw-Hall measurements at 77 and 300 K to determine the conductivity and free carrier concentration, by photoluminescence spectroscopy (PL) to evaluate the crystal quality and by secondary ion mass spectrometry (SIMS) to investigate impurities.

Heterostructures were characterized by van der Pauw-Hall studies on modulation doped two dimensional electron gas (2 DEG) structures in the temperature range from 4 to 300 K to investigate differences in dominant scattering mechanisms. Interface roughness was studied optically by photoluminescence spectroscopy on quantum well structures and electrically by the evaluation of AlAs/GaAs resonant tunneling diodes (RTD), whose peak to valley ratio responds sensitively to the interface quality.

3. Results and discussion

3.1. 2 DEG structures

The starting point of our investigations was a 2 DEG structure designed for obtaining high mobilities. The growth parameter deposition temperature and the structural parameters such as sheet carrier concentration and spacer thickness were optimized with respect to obtaining high electron mobility for each growth technique separately. Only a slightly different growth temperature was found - 650 for MOVPE and 630°C for MBE. All other optimum structural parameters happened to be the same. The Si-doped $Al_{0.3}Ga_{0.7}As$ layer was separated from the GaAs channel by a 40 nm thick $Al_{0.3}Ga_{0.7}As$ spacer. The optimum sheet carrier concentration for this structure was determined to be $n_s \leq 3\times10^{11}$ cm^{-2}. Temperature dependent Hall measurements were done on these samples (Fig. 1). From room temperature to about 70 K the mobility of the samples is similar. Below this temperature the curves have a different slope. The increase in mobilty is higher for the MBE samples than for the MOVPE sample. At 4 K a value of 1.9×10^6 cm^2/Vs was found for the MBE sample - a value which is state of the art when no special optimization procedures are carried out concerning vacuum etc. For the MOVPE sample, however, a mobilty of 0.65×10^6 cm^2/Vs was obtained, which is a factor of 3 lower than for the MBE sample, but which is one of the highest values ever obtained in MOVPE. Since three different scattering mechanisms are dominant in this lower temperature range (Walukiewicz 1984) - residual impurity scattering, remote impurity scattering and scattering due to interface roughness - either more impurities are incorporated into the layers or rougher interfaces are obtained during MOVPE growth compared to MBE growth. The purity of bulk AlGaAs and GaAs material and/or the quality of the (AlGa)As/GaAs heterointerface is not only important for obtaining high electron mobility at low temperature in 2 DEG structures, it also is of great importance for high frequency devices such as HEMT`s, HBT`s, RTD`s and MESFET`s. At the growth conditions used for 2 DEG structural growth first the electrical and optical quality of bulk GaAs was under investigation. Under the same growth conditions (i.e. conditions used for 2 DEG structural growth), then bulk AlGaAs was characterized and at last heterostructures were deposited and studied to evaluate the interface roughness.

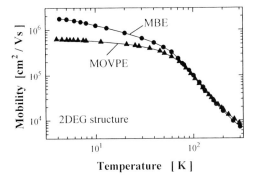

Fig. 1: Dependence of 2 DEG mobility on temperature for MOVPE and MBE deposited modulation doped $Al_{0.3}Ga_{0.7}As$/GaAs heterostructures with 40 nm spacer thickness and a sheet carrier concentration of $3.0 \times 10^{11} cm^{-2}$

3.2. Comparison of bulk material

Pure GaAs is not only very important for high mobility 2 DEG structures and GaAs/(AlGa)As HEMTs (due to residual impurity scattering, see above), it also has a big influence on MESFET characteristics since this device uses GaAs as the active layer. To investigate the impurities incorporated in GaAs, 6 µm thick undoped GaAs layers were deposited on (100) semiinsulating GaAs substrates oriented 2° off towards the nearest ⟨110⟩ plane. Photoluminescence experiments on these samples document, that there is a great difference of the acceptor to donor ratio between GaAs grown by MOVPE and material deposited by MBE. The intensity ratio of the acceptor bound exciton transition to the donor bound exciton transition is a factor 3.5 higher for the MBE sample. Electrical characterization by means of Hall effect and C-V measurements shows that the major difference between both samples is the type of the majority carriers: the MBE layer was p-type with a carrier concentration of $2 \times 10^{14} cm^{-3}$, the MOVPE sample was n-type with the same carrier concentration. The influence of the residual majority carrier type on the 4 K mobility of 2 DEG structures with the same sheet carrier concentration, however, is not fully understood. The FWHM of the donor bound exciton transition of 0.13 meV for MOVPE GaAs is comparable to that of the MBE sample. All in all the difference in crystal purity and quality found was not that big that it should have an influence on device characteristics.

To study the (remote) impurity incorporation in $Al_{0.3}Ga_{0.7}As$, 1 µm thick undoped layers were deposited on semiinsulating GaAs substrates, oriented 2° off (100) towards the nearest ⟨110⟩ plane. Electrical characterization of the layers using the Hall effect was not successful since the layers were highly resistive, i.e. the free carrier concentration in both cases was so low that the layer thickness deposited did not exceed the depletion zone thickness. Photoluminescence studies at 2K were done on these samples. Fig.2 shows the spectra obtained for the MOVPE and the MBE sample. The MOVPE sample exhibits a very strong peak at 1.95 eV with a FWHM of 7.4 meV, which is assigned to excitonic transitions. Temperature dependent as well as excitation and energy dependent PL-studies reveal, that 2 excitonic transitions are observed: the free exciton transition with a FWHM of 4.3 meV and the bound exciton transition (which shows up as a shoulder in Fig. 2) with a FWHM of 3.5 meV. Only the superposition of both signals is seen in Fig. 2, the dominant signal being the free exciton transition for the MOVPE sample. This feature is not usually observed in (AlGa)As with 30% Al content and can only be detected in (AlGa)As of high purity. At a 30 meV lower energy an extremely weak transition is observed, that can be identified as the donor acceptor transition, the acceptor being carbon. The MBE sample shows a factor of 250 weaker luminescence. Its dominant transition is the donor acceptor transition, which in part can also be attributed to carbon. The bound exciton transition has a FWHM of 8.0 meV, no free exciton transition was observed. The larger FWHM of the bound exciton transition for MBE (AlGa)As together with the absence of the free exciton transition, the lower luminescence intensity and a factor of 185 lower ratio of the bound exciton to the donor acceptor transition document the higher impurity content and inferiority of crystal quality in comparison to MOVPE material. For heterostructural devices with vertical current transport, highly pure (AlGa)As is important. Here our new MOVPE growth process has an advantage over the standard MBE process. Room temperature photoluminescence measurements were also done on these samples. The luminescence intensity for the MBE sample was 40 times weaker than for the MOVPE sample. Again, the lower luminescence intensity in the case of MBE AlGaAs is an indication for a higher concentration of non-radiative recombination centers and therefore for the higher quality of the MOVPE material. All in all it has been shown that at 650°C MOVPE (AlGa)As is of excellent quality. Remote impurity scattering in $Al_xGa_{1-x}As$ is not the limiting factor for the 2 DEG electron mobility in MOVPE structures at low temperatures.

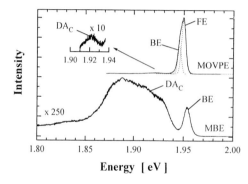

Fig. 2: 2 K photoluminescence spectra for 1 μm thick $Al_{0.3}Ga_{0.7}As$ layers deposited by MOVPE and MBE at 650 and 630°C, respectively. A lower PL-intensity (factor of 250), higher FWHM of the bound exciton (BE) transition (a factor of 1.6) and a lower ratio of the bound exciton to donor acceptor (DA) transition (a factor of 185) are measured for the MBE sample; the free exciton transition (FE) is additionally observed for the MOVPE sample.

Comparative SIMS studies on the (AlGa)As samples were performed. The oxygen and carbon concentration in the layers could not be determined for both layers mentioned above since the levels were below the detection limit. There is, however, a difference in metallic impurities: even though both samples were deposited on undoped GaAs wafers, chrome was detected in the MBE deposited sample in a quantatity of $3 \times 10^{14} cm^{-3}$, which is a factor of 10 higher than for the MOVPE sample. Chrome is a deep level impurity in GaAs and (AlGa)As. The incorporation of chrome could be one reason for the lower PL intensity observed for MBE samples. However, this does not seem to be detrimental to obtaining high electron mobilities in 2 DEG structures at low temperatures. Deep level transient spectroscopy studies on (AlGa)As layers deposited by both growth techniques still need to be carried out to be able to quantify the deep level impurity content and their nature.

3.3. Investigation of interface roughness

Using photoluminescence spectroscopy at 2 K a single quantum well structure was examined. The 7 nm thick GaAs well was embedded in 70 nm thick $Al_{0.3}Ga_{0.7}As$ layers. The energetical position of the PL-signal originating from the well is correlated to the quantum well thickness. The FWHM is dependent on the abruptness of the GaAs/(AlGa)As and (AlGa)As/GaAs interfaces. A FWHM of about 1 meV (as obtained by MBE samples) indicates that the interfaces are "optically smooth". This does not mean that they realy are microscopically smooth since first of all the signal represents an average over a lateral range of the laser spot (100μm). Second, samples with smaller step widths than the exciton radius (the optical probe) of about 100 Å tend to suggest smooth interfaces (Reynolds 1985). If two signals are found and the energetical separation is appropriate, this feature can be explained by monolayer splitting (Christen 1990). For samples deposited by both deposition techniques only one signal from the well is observed. No monolayer splitting was detected. The FWHM of the MOVPE sample was 3.1 meV under the conditions used for 2 DEG structures. The broadening in comparison to MBE samples points to problems at the interface. This could be the case since the growth process in terms of growth interruptions was not yet optimized with respect to interface smoothness. Although the analysis of FWHM is not unambiguous, the evidence is strong that smoother interfaces are obtained by MBE growth. The ability to obtain smoother interfaces will be one of the most important tasks for future investigation and optimization for the MOVPE growth process. Smoother interfaces should then lead to higher mobilty in 2 DEG structures and higher conductivity in HEMT's.

Interface roughness on a smaller microscopic scale, where PL studies fail (step widths < 100 Å, Chevoir 1993) can be detected electrically using resonant tunneling diodes. Their I-V characteristics have one peculiarity: a negative differential resistance region both in the negative and positive voltage range. The valley current of these RTD's depends sensitively on scattering processes at the interfaces, which give rise to a non-resonant tunneling current contribution. This effect reduces the PVR (peak to valley ratio). Our RTD's were fabricated on highly Si-doped GaAs (100) wafers with the following layer sequence (Förster 1994): 1 μm GaAs buffer layer, 10 nm GaAs (n = 1 x 10^{17}cm^{-3}), 10 nm GaAs (n = 1 x 10^{16}cm^{-3}), 5 nm GaAs spacer, 6 monolayers (ML) AlAs barrier, 5 nm GaAs spacer, 10 nm GaAs (n = 1 x 10^{16}cm^{-3}), 10 nm GaAs (n = 1 x 10^{17}cm^{-3}) and finally a 0.5 μm thick highly-doped n+ (n = 5 x 10^{18}cm^{-3}) GaAs top layer. The diodes were prepared in the way described by Förster (1994). The I-V characteristics for both the MBE and the MOVPE deposited diodes are presented in Fig. 3. Using diodes of the same size deposited at the growth conditions used for 2 DEG structures a PVR of 2.4 was obtained for the MOVPE diode whereas more than double this value (PVR = 5.0) was measured for MBE diodes. The higher PVR of the MBE diode is another indication that the interface roughness is higher for MOVPE diodes not only on a larger but also on a microscopically smaller scale and documents the superiority of the MBE growth technique at present for devices that are extremely sensitive to interface roughness.

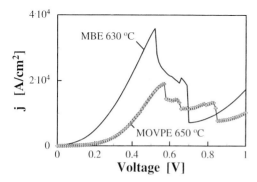

Fig. 3: I-V characteristics of AlAs/GaAs resonant tunneling diode structures deposited by MOVPE at 650°C and MBE at 630°C containing 5 nm of GaAs well surrounded by 6 monolayers of AlAs barriers. The PVR is 2.4 and 5.0 for the MOVPE and MBE samples, respectively.

3.4. Comparison of homogeneity

One precondition for industrial mass production of electronic devices is the uniformity of the layers deposited. The homogeneity in thickness, composition and conductivity should be within the range of ± 1%. This precondition is easily fulfilled by the MBE deposition technique since the deposition of highly homogeneous layers has always been a special strongpoint of MBE. More effort has to be made with MOVPE. Earlier investigations have shown (Hardtdegen 1994), that the use of the carrier gas nitrogen greatly improves the homogeneity of MOVPE layers. This new approach together with substrate rotation should lead to the best results. Macroscopic thickness homogeneity of GaAs and compositional homogeneity of $Al_{0.3}Ga_{0.7}As$ was studied by PL spectroscopy using the quantum well structure mentioned above and a bulk layer, respectively. The uniformity was within a standard deviation of ± 1 % on a full two inch wafer. The thickness homogeneity of (AlGa)As, which was detected by surface profiling on selectively etched samples, proved to be also within a standard deviation of ± 1 %. The biggest challenge is to meet the conductive homogeneity value. Here the homogeneity for a Si-doped GaAs layer even was well within the 1 % limit i.e. ± 0.6 % on a full two inch wafer. In conclusion all requirements for homogeneous growth have now also been fulfilled for the MOVPE growth technique.

4. Conclusion

In this study we have compared the optical and electrical characteristics of LP-MOVPE and MBE (AlGa)As/GaAs structures deposited at growth conditions used for device structures. The 2 DEG electron mobility at low temperature was a factor of 3 lower for MOVPE structures than for MBE structures. The residual impurity concentration in GaAs is the same for both growth techniques, the type of majority carriers is, however different (p-type for MBE and n-type for MOVPE samples). The remote impurity concentration in $Al_{0.3}Ga_{0.7}As$ is lower for the MOVPE samples. At the growth conditions employed the optical quality of MOVPE (AlGa)As is superior to that of the MBE. Studies on interface roughness indicate, that the factor of 3 lower 2 DEG mobilities at low temperature for MOVPE samples is due to scattering at the interface. The control of layer homogeneity in terms of thickness, composition and conductivity is comparable for both deposition techniques and meets the requirements for device fabrication.

Acknowledgements

The authors would like to thank W. Michels (AIXTRON) for conductive homogeneity studies, H. Holzbrecher (Zentralabteilung für chemische Analysen, Forschungszentrum Jülich) for SIMS studies, P. Kordoš for fruitful discussions and K. Wirtz for technical assistance.

References

Chevoir F and Vinter B 1993 Phys. Rev. B 47 7260
Christen J 1990 "Characterization of Semiconductor Interfaces with Atomic Scale Resolution by Luminescence" in Festkörperprobleme/Advances in Solid State Pysics 20 251, Queisser H J, ed
Förster A, Lange J, Gerthsen D, Dieker Ch and Lüth H 1994 J. Phys. D: Appl. Phys. 27, 175
Hardtdegen H, Ungermanns Chr, Hollfelder M, Raafat T, Carius R, Hasenöhrl St and Lüth H 1994, Proceedings of the Seventh International Conference on Metalorganic Vapor Phase Epitaxy, Yokohama, J. Cryst. Growth, in press
Reynolds D C, Bajaj K K and Litton C W, Yu P W, Singh J, Maselink W T, Fischer R and Moroc H 1985 Appl. Phys Lett. 46 51
Walukiewicz W, Ruda H E, Lagowski J and Gatos H C 1984 Phys. Rev. B 30; 4571

Graded $In_{x_0 \leq x \leq 0.5}Ga_{1-x}As/InP$ buffer layers on GaAs prepared by molecular beam epitaxy

K. Häusler and K. Eberl

Max-Planck-Institut für Festkörperforschung
D-70569 Stuttgart, Germany

Abstract. Compositionally graded $In_{x_0 \leq x \leq 0.5}Ga_{1-x}As$ buffer layers are prepared to relieve the misfit strain between GaAs substrate and $In_{0.5}Ga_{0.5}As/In_{0.5}Al_{0.5}As$ modulation doped heterostructure. Different linearly graded buffer layers, with the initial In mole fraction x_0 varying between 0 and 0.5 are investigated by X-ray diffraction, atomic force microscopy and Hall measurements. The best compromise in terms of surface smoothness and electron mobility is achieved for $x_0 = 0.18$. An additional InP buffer layer between graded $In_{0.18 \leq x \leq 0.5}Ga_{1-x}As$ layer and modulation doped heterostructure provides a significant improvement of the electron mobility and increases the photoluminescence intensity. The highest electron mobilities, which are achieved for $In_{0.5}Ga_{0.5}As$ quantum wells, are 11,050 cm^2/Vs at room temperature and 74,830 cm^2/Vs at 77K.

1. Introduction

The heteroepitaxy of lattice mismatched semiconductor crystals, such as for example SiGe/Si, ZnSe/GaAs or InGaAs/GaAs, has attracted considerably interest in recent years. Many investgations of these materials have been made concerning device applications and strain relaxation processes [1-4]. Strain relaxation of lattice mismatched epilayers occurs by the formation of misfit dislocations when the layer thickness exceeds the critical thickness for plastic relaxation. Strain relaxed buffer layers with a thickness well above the critical value are applied for semiconductor devices which have a high lattice mismatch of several per cent.

Significant work has been focused on the $In_{0.5}Ga_{0.5}As$ buffer layers with the lattice constant close to InP. Uniform $In_{0.5}Ga_{0.5}As$ layers on GaAs with several μm thickness, have a high density of misfit defects on the surface of typically 10^9 cm^{-2}, which results in poor crystal quality [5]. The application of compositionally graded buffer layers provides a much lower defect density of less than 10^6 cm^{-2} [6]. In addition, these strain relaxed graded layers show a characteristic surface roughness ('cross hatching').

In this contribution we report about graded $In_{x_0 \leq x \leq 0.5}Ga_{1-x}As$ buffer layers which are prepared by molecular beam epitaxy on (001) GaAs. The graded buffer layers have a thickness of less than 1 μm. Different buffer layers, with the initial In mole fraction x_0 varying between 0 and 0.5 are investigated by X-ray diffraction, Hall measurements, photoluminescence (PL), atomic force microscopy (AFM) and transmission electron microscopy (TEM). The introduction of a relaxed InP buffer layer on top of the graded $In_{0.18\leq x\leq 0.5}Ga_{1-x}As$ layer is investigated with regard to the optical and electronic properties of modulation doped heterostructures.

2. Crystal growth

The samples are prepared using solid source molecular beam exitaxy (MBE). The GaAs substrates are (001) oriented with a miscut of about 0.1°. The graded InGaAs layers are grown at a low substrate temperature ($T_{sub}=320°C$), which is measured with a thermocouple refering to the oxide desorbtion temperature of the GaAs substrate (580°). The low growth temperature results in two-dimensional growth as observed with reflecting high energy electron diffraction (RHEED), because the island formation (Stransky-Krastanov mode) is suppressed. However, samples grown at low temperature indicate a microscopic surface roughness. In order to diminish this surface roughness the deposition was interrupted after every 200 monolayers, and subsequently the samples were annealed at 420°C. This annealing steps lead to a smoothening of the surface as observed by RHEED. The graded InGaAs buffer layers are grown by ramping the In mole fraction linearly from x_o up to 0.5 with a grading of 0.5/µm. This is performed by increasing the In-cell temperature and at the same time decreasing the Ga-cell temperature to achieve a constant growth rate of about one monolayer per second.

Table 1 shows a list of samples with different uniform $In_{0.5}Ga_{0.5}As$, $In_{0.5}Al_{0.5}As$, or InP buffer layers on top of the graded layer. The substrate temperature for the growth of uniform InGaAs and InAlAs layers is 380°C. The InP layers are deposited at $T_{sub} = 420°C$ using a special P_2 source which is based on GaP decomposition. Finally, a n-type modulation doped $In_{0.5}Al_{0.5}As$ layer ($N_{Si}= 3..5 \times 10^{18} cm^{-3}$) followed by an $In_{0.5}Ga_{0.5}As$ cap layer is grown on top of the buffer layer for applying Hall measurements. In this way different samples with strain relaxed buffers are prepared which have either a single modulation doped heterostructure or a quantum well. The samples I to V listed in Table 1 have the same layer structure, but different values of the initial In mole fraction x_o in the graded buffer layer. These samples have an uniform 1µm thick $In_{0.5}Ga_{0.5}As$ layer and a modulation doped heterostructure. In addition, samples (A), (B), and (C) are prepared with an InP layer on top of the graded $In_{0.18\leq x\leq 0.5}Ga_{1-x}As$ layer followed by a modulation doped heterostructure or a quantum well structure, respectively.

Table 1: Nominal layer composition of the samples

sample I, II, III, IV, V	sample (A)	sample (B)	sample (C)
$In_{xo\leq x\leq 0.5}Ga_{1-x}As$ $x_o=0/0.12/0.18/0.24/0.5$	$In_{0.18\leq x\leq 0.5}Ga_{1-x}As$	$In_{0.18\leq x\leq 0.5}Ga_{1-x}As$	$In_{0.18\leq x\leq 0.5}Ga_{1-x}As$
	1µm InP	200nm InP	1µm InP
1µm $In_{0.5}Ga_{0.47}As$	1µm $In_{0.5}Ga_{0.5}As$	200nm $In_{0.5}Al_{0.5}As$	1µm $In_{0.5}Al_{0.5}As$
		10nm $In_{0.5}Ga_{0.5}As$	10nm $In_{0.5}Ga_{0.5}As$
5nm $In_{0.5}Al_{0.5}As$	5nm $In_{0.5}Al_{0.5}As$	5nm $In_{0.5}Al_{0.5}As$	5nm $In_{0.5}Al_{0.5}As$
30nm $In_{0.5}Al_{0.5}As$:Si	30nm $In_{0.5}Al_{0.5}As$:Si	30nm $In_{0.5}Al_{0.5}As$:Si	30nm $In_{0.5}Al_{0.5}As$:Si
5nm $In_{0.5}Ga_{0.5}As$	5nm $In_{0.5}Ga_{0.5}As$	5nm $In_{0.5}Ga_{0.5}As$	5nm $In_{0.5}Ga_{0.5}As$

3. Results and discussion

Investigations of the electronic properties of samples I to V, which have different initial In mole fraction x_0 of the graded buffer, indicate the highest electron mobility for sample III with $x_0 = 0.18$. In addition, the surface roughness is reduced as x_0 increases from 0 to 0.5 as observed by AFM. For example, the averaged roughness amplitude is about 12 nm for sample I with $x_0 = 0$, but it is 3 nm for sample III ($x_0 = 0.18$). Furthermore, TEM micrographs of sample I and sample III show that the arrangement of misfit dislocations is similar for both samples and it is typical for linearly graded buffer layers. In contrast, the TEM micrographs obtained from sample V whithout graded buffer, indicate a high density of threading dislocations which diminishes the electronic and crystal quality (see also Refs. [11] and [12]).

Table 2 summarizes the X-ray data for the samples I to V. Listed are the full width at half maximum (fwhm) of the main InGaAs peak, the tilt angle of the homogenous $In_{0.5}Ga_{0.5}As$ cap layer against the substrate, the tilt direction, the substrate miscut, and the direction of the substrate miscut. For the samples I to IV, which have a graded layer, the minimum fwhm is measured for $x_0 = 0.18$. The sample without grading (sample V) can not be directly compared to the samples with graded buffer, since there is an asymmetrical broadening due to the compositional grading itself.

Strain relaxation in lattice mismatched heterostructures may cause a tilting between epilayer and substrate. Such tilting can be measured by X-ray diffraction [8]. Fig. 1 shows the 004-reflection from sample I, which is measured along the four in plane <110> directions. The angle between the [110] direction within the (001) surface plane and the beam direction is indicated by $\omega = 0$, 90°, 180°, 270°. The small peak close to the dotted line at zero originates from the substrate. The little shift of about ±300 arcsec around this dotted line originates from the substrate miscut. The surface of the substrate was precisly alligned (±50 arcsec) using a laser beam reflection on the substrate. The broader peak on the left side of the substrate reflection originates mainly from the $In_{0.5}Ga_{0.5}As$ layer. This peak shifts by about 1900 arcsec as a function of the rotation angle. This is due to the tilt of the $In_{0.5}Ga_{0.5}As$ layer against the substrate.

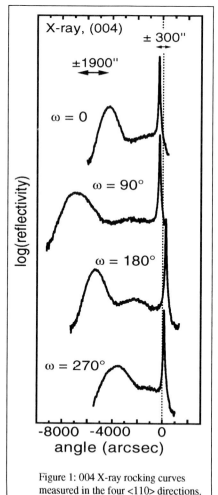

Figure 1: 004 X-ray rocking curves measured in the four <110> directions.

The measurement allows to determine the lattice tilt angle θ_t using the following equation: $\theta = \theta_B + \theta_t \cos(\omega - \omega_0)$, where θ is the angle of the 004-reflection from the uniform InGaAs layer, which is due to the broad shifted peaks in Fig. 1. Furthermore, θ_B is the Bragg angle and ω_0 the angle between tilt direction and the [110] direction.

Table 2: X-ray data of samples I to V.

sample	I	II	III	IV	V
Initial In-content x_0	0	0.12	0.18	0.24	0.50
X-ray fwhm (arcsec)	1700	3100	1200	1500	1000
Tilt angle (arcsec)	1900	560	290	290	≈ 0
tilt direction (ω_0)	≈[$\bar{1}$10] (292°)	≈[$\bar{1}\bar{1}$0] (162°)	≈[$\bar{1}$10] (261°)	≈[$\bar{1}\bar{1}$0] (190°)	---
substrate miscut (")	250	390	250	170	---
miscut direction (ω_0)	≈[1$\bar{1}$0] (41°)	≈[110] (345°)	≈[1$\bar{1}$0] (107°)	≈[1$\bar{1}$0] (77°)	---

The tilt direction is defined as the projection of the [001] epilayer vector into the (001) substrate plane. The tilt angle decreases from 1900 arcsec to about zero as x_0 increases from 0 to 0.5. The substrates, which have been used, have a miscut of about 0.1° (360 arcsec). The data listed in Table 2 are always taken from one specific position of the sample close to the center of the wafer. The magnitude and the direction of miscut are listed in Table 2. For the samples I and II with $x_0 = 0$ and 0.12 the tilt angle is significantly larger than the substrate miscut. In addition, the direction of the epilayer tilt is not clearly correlated to the miscut direction of the substrate.

The lattice tilt is caused by preferential glide or nucleation of dislocations which have the same burgers vector. Investigations on strain relaxation in graded and uniform SiGe buffer layers on miscut Si substrates have been reported, recently [8]. In this case the glide limited relaxation in uniform layers causes a small lattice tilt compared to the large tilt in graded layers where the relaxation is nucleation limited. For the relaxed SiGe layers the direction of tilt is the same as the miscut direction and the tilt magnitude is smaller than the miscut. However, these results are not directly comparable to the data for lattice tilting of our samples I to IV, because we find no obvious correlation between substrate miscut and epilayer tilt and also the magnitude of tilt is pronounced larger than the miscut.

Table 3 shows the electron mobility obtained from Hall measurements which were performed at 300 K and at 77 K. The electron mobility and the carrier density data for the samples I to V are discussed in Ref. [11]. The highest electron mobility among these samples,

Table 3: Electronic properties of samples III, (A), (B), and (C)

sample	III	(A)	(B)	(C)
μ_{300K} (cm^2/Vs)	9300	9849	11050	10690
n_{300K} (10^{12}cm^{-2})	2.847	4.756	1.872	2.512
μ_{77K} (cm^2/Vs)	32430	51490	56530	74830
n_{77K} (10^{12}cm^{-2})	2.041	2.878	1.692	1.814

which is 9300 cm^2/Vs at 300 K and 33000 cm^2/Vs at 77 K, was measured for sample 3 with $x_0=0.18$. Starting out with the 0.7µm thick graded In$_{0.18 \leq x \leq 0.5}$Ga$_{1-x}$As buffer layer we deposited a relaxed InP overlayer for further improvement of the crystal quality. The InP layer can be grown at higher substrate temperatures in a two dimensional growth mode. Generally relaxed InP buffer layers on GaAs have lower defect densities than homogenious In$_{0.5}$Ga$_{0.5}$As layers, which have the same lattice constant as InP [10,12]. This is probably due to faster dislocation movement and lower barrier for dislocation nucleation in InP. The introduction of the InP layer in the samples A, B, and C, which are listed in Table III, reduces the micrscopic surface roughness as observed with AFM. The averaged vertical amplitude is 3 nm for sample III, but it is 1nm for sample (C) which was grown with an additional InP layer.

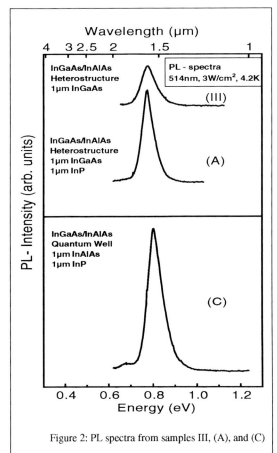

Figure 2: PL spectra from samples III, (A), and (C)

The composition of the samples (A), (B), and (C) is described in Table 1. The electron mobility and the carrier density data for these sampes are listed in Table 3. The introduction of a 1µm thick InP layer between the graded layer and the uniform In$_{0.5}$Ga$_{0.5}$As layer (sample A) increases the mobility to 9900 cm^2/Vs at 300 K and 51,000 cm^2/Vs at 77 K.

Even better results are obtained for the samples with a modulation doped 10 nm thick In$_{0.5}$Ga$_{0.5}$As quantum well (QW). The mobility is about 11,000 cm^2/Vs at room temperature for sample (B) and (C). At 77 K a mobility of 74,830 cm^2/Vs is achieved for sample (C) which has a thicker InP/In$_{0.5}$Al$_{0.5}$As buffer layer on top of the graded layer. To our knowledge this is the highest mobility reported for strain relaxed In$_{0.5}$Ga$_{0.5}$As layers on GaAs substrates. The electronic data for the strain relaxed layers of sample (A), (B) and (C) are comparable to the results obtained from lattice matched In$_{0.53}$Ga$_{0.47}$As epilayers on InP substrates [9].

The nomial Si doping level within the 30 nm thick In$_{0.5}$Al$_{0.5}$As layer is about 3x10^{18} cm^{-3} except for sample (A) where the Si doping was 5x10^{18} cm^{-3}. The carrier density for sample (A) is 4.756x10^{12} cm^{-2} at room temperature and 2.878x10^{12} cm^{-2} at 77 K. Illumination of the samples at 77 K increases both electron density and electron mobility by less than 10% indicating a low density of DX centers.

Figure 2 shows the PL spectra of samples III, (A), and (C). The peaks at about 0.8 eV originate from the In$_{0.5}$Ga$_{0.5}$As layers. The PL measurements were performed at 4.2 K using an Ar-laser with a 514 nm wavelength and an excitation power of 3 W/cm^2. The PL spectra were measured applying a Ge-detector which provides a spectral range between 1µm and

2μm. The PL signal from sample (C) is shifted towards higher energy by about 40meV relative to the signal from sample (A). This energy shift is attributed to the subband separation in the 100 Å thick QW and to slightly variations in the InGaAs composition.

The peak intensity from sample III is more than a factor 2 smaller than that of sample (A) which has an additional 1μm thick InP layer between graded buffer and $In_{0.5}Ga_{0.5}As$ layer. Also, the fwhm of sample (A) is smaller (62meV) than that of sample III, which has a fwhm of 80 meV. The higher intensity and the smaller fwhm of the PL signal from sample (A) indicate an improved crystal quality of the relaxed InGaAs layer. The PL signal measured from the 100Å thick quantum well of sample (C) has a fwhm of 71meV. The PL intensity of this sample is about a factor 1.5 higher than that of sample (B) (not shown in Fig 2), which has a fwhm of 92meV. These data demonstrate the low density of dislocations in the surface region and the excellent crystal quality of sample (C), which is consistent with the results obtained from Hall measurements.

4. Conclusions

In summary we have shown that the introduction of a misfit step with $x_0 = 0.18$ in the graded $In_{x_0 \leq x \leq 0.53}Ga_{1-x}As$ buffer layer provides a smoother surface and a higher electron mobility than a linearly graded buffer without such step. In addition, X-ray measurements show that the epilayer in linearly graded buffer layers exhibit a pronounced tilting which magnitude is reduced as x_0 increases. The fwhm of the X-ray reflections obtained from the $In_{0.5}Ga_{0.5}As$ layer on top of the graded layer is the smallest for $x_0 = 0.18$. A further enhancement of the electron mobility, the luminescence intensity, and the surface smoothness is achieved by the introduction of an additional InP layer between graded buffer and uniform layer.

We would like to thank C. Lange for AFM investigations and A. Trampert, and T. Shitara for expert help and stimulating discussion.

References

[1] K. Inoue, J. C. Hamand, and T. Matsuno, J. Cryst. Growth **111,** 313 (1991)
[2] J. Chen, J. M. Fernandez and H. H. Wieder, Semicond. Sci. Technol. 8, 315 (1993)
[3. M. J. Eckenstedt, T. G. Andersson, and S. M. Wang, Phys. Rev. B **48,** 5289 (1993)
[4] . Krishnamoorthy, P. Ribas, and R. M. Park, Appl. Phys. Lett. 58, 2000 (1990)
[5] D. I. Westwood, D. A. Woolf, A. Vilà, A. Cornet, and J. R. Morante, J. Appl. Phys. **74,** 1731 (1993)
[6] F. K. LeGoues, B. S. Meyerson, J. F. Morar, and P. D. Kirchner, J. Appl. Phys. **71,** 4230 (1992).
[7] D. E. Jesson and S. J. Pennycook, Phys. Rev. Lett. **71,** 1744 (1993)
[8] F. K. LeGoues, P. M. Mooney, and J. O. Chu, Appl. Phys. Lett. **62,** 140 (1993)
[9] E. Tournié, L. Tapfer, T. Bever and K. Ploog, J. Appl Phys **71,** 1790 (1992)
[10] S. Z. Chang, S. C. Lee, H. P. Shiao, W. Lin, and Y. K. Tu, Appl. Phys. Lett. **63,** 2417 (1993)
[11] K. Häusler, K. Eberl, and W. Sigle, submitted to Semicond. Sci. Technol.
[12] K. Eberl, K. Häusler, T. Shitara, Y. Kershaw, and W. Sigle, Proceed. on 'Interface formation and dynamics in layered structures', Toronto 1994, to be publ. in Scanning Microscopy.

Heteroepitaxial Growth of InSb on Si by Plasma-Assisted Epitaxy

H. Ohba and T. Hariu
Department of Electronic Engineering, Tohoku University, Sendai 980-77, Japan

ABSTRACT: Improved direct epitaxial growth of InSb on Si by plasma-assisted epitaxy has been achieved by employing As-termination of Si surface after hydrogen-plasma treatment and low-temperature growth of initial layer of a few monolayers. The As-terminated Si surface is much more stable than hydrogen terminated surface. The initial layer grown at 240 °C, strained with longitudinal lattice mismatch around 12-13 % compared with 19 % between bulk InSb and Si, should be useful to relax large mismatch.

1. INTRODUCTION

Heteroepitaxial growth of compound semiconductors on Si has been intensively investigated with partial success in view of possible realization of OEIC and galvanomagnetic IC combined with Si IC. InSb has the highest electron mobility and the narrowest bandgap among III-V compound semiconductors, and is attractive for these applications. Epitaxial layers of InSb have been typically grown on GaAs (Ohshima et al 1989 and the references therein), but active layers should be directly grown on Si for monolithic integration of these devices. It has been found, however, that InSb on Si with large lattice mismatch 19 % is a much more difficult system than GaAs on Si because the epitaxial growth is much more sensitive to surface cleanliness and structures of Si substrate and InSb tends easily to poly-crystallize. In this paper we describe improved direct epitaxial growth of InSb on Si by plasma-assisted epitaxy (PAE) (Takenaka et al 1979), and emphasize the importance of As-termination of Si surface after hydrogen-plasma treatment and also of an initial strained layer of a few monolayers grown at a lower temperature in order to suppress three dimensional islands growth.

2. EXPERIMENTAL

PAE, in which enhanced chemical reactivity and surface migration of atoms are supplied through discharging plasma, has been studied for low temperature epitaxial growth. It has been also confirmed that hydrogen plasma is effective for surface cleaning at a low temperature with removal of native oxide layers on the Si substrate (Gao et al 1987).

A similar PAE apparatus as described elsewhere (Matsushita et al 1984) was used in the present growth. Hydrogen gas was used as a discharging gas. 6-nine purity elemental In and Sb were evaporated by resistive heating and supplied through hydrogen plasma toward the substrate. Discharging plasma was excited by rf power at 13.56 MHz through inductive coupling.

(111)Si was employed for substrate, which were chemically etched by the conventional RCA method prior to loading into the PAE chamber. The pressure of hydrogen gas was kept at 0.02 Torr. Si surface was exposed to hydrogen plasma at 600 °C prior to epitaxial growth for the removal of native oxide. Two-step growth was used, in which a thin InSb layer is grown at a low temperature (240-300°C) with the low growth rate (0.1-0.3μm/h). Subsequently, the substrate is heated to 380-450 °C and InSb thicker layer was grown with relatively high growth rate (0.7 μm/h).

3. RESULTS and DISCUSSIONS

The surface cleaning is prerequisite for successive epitaxial growth. Hydrogen termination of Si has not given robust surface sufficient for the present purpose because it is sensitive to such surface structure as steps and is not stable against residual oxygen at relatively low temperature. Instead we employed hydrogen plasma treatment at 600 °C to establish the removal of native oxide (Gao et al 1987) and then exposed to arsenic vapor in plasma during cooling, in view of the fact that As-Si bond is much stronger than H-Si bond. Fig. 1 shows the RHEED pattern of As-terminated Si surface which was exposed to air for several minutes. It was confirmed that the 1x1 structure of As-terminated surface is stable even in air. After the above surface cleaning with and without arsenic vapor exposure, InSb films were grown. The RHEED patterns of grown films are shown in Fig.2. On the Si surface terminated with arsenic, epitaxial films grew [Fig.2 (a)], in contrast to the polycrystalline InSb [Fig.2 (b)] grown on the surface without exposure to arsenic, probably due to partial oxidation of the surface by residual oxygen. These observations indicate that As-termination is effective for surface passivation against native oxide growth prior to epitaxial growth.

Then the importance of control of the initial stage in the two-step growth of InSb on this As-stabilized Si is discussed. Fig.3 shows the RHEED patterns and SEM micrographs of initial layers grown at 270 °C with a growth rate of 0.3 μm/h. Ring-like patterns of polycrystalline InSb were observed at an early stage of initial growth of several monolayers and InSb grains grew larger with thickness. It was found that this tendency is more remarkable at higher temperature. Then a very thin initial layer with a few monolayers

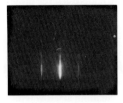

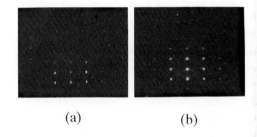

(a) (b)

FIG.1 RHEED pattern of As-terminated Si surface after exposure to air. The direction of electron beam is [$\bar{2}$11].

FIG.2 RHEED patterns of InSb films grown on Si cleaned by hydrogen plasma (a) with and (b) without arsenic vapor exposure.

which is not poly-crystallized should be used for the initial layer in the two-step growth.

Then, we observed the initial stage within 4 monolayers by lowering the growth rate down to 0.1 μm/h. Fig.4 shows the RHEED patterns and SEM micrographs of the InSb initial layers grown at temperature (a) 240 °C and (b) 270 °C up to the thickness around 4 monolayers before they are poly-crystallized. Although the three dimensional island growth is suppressed at lower temperature, these layers are still not continuous (Volmer-Weber growth mode) and are strained with longitudinal lattice mismatch around 12-13 %, compared with 19 % between bulk InSb and Si. The RHEED patterns of the films grown successively at 450 °C on these initial layers are shown in Fig.5. The film deposited on the initial layer grown at 240 °C has better crystal quality with less microtwins compared with that grown at 270 °C. We believe this type of strained initial layer of a few monolayers grown at lower temperature with almost flat surface is useful generally to relax the large lattice mismatch. The electrical property of the former film with thickness 1.5 μm ($n=1.8 \times 10^{17}$ cm^{-3}, $\mu_n=12,000$ cm^2/Vs) is also better than that of the latter film with the same thickness ($n=4.2 \times 10^{17}$ cm^{-3}, $\mu_n=6,700$ cm^2/Vs). It was confirmed that the epitaxial layers were successfully and reproducibly grown at 380 °C with much improved crystallographic and electrical properties on the above type of initial layer grown at lower temperature.

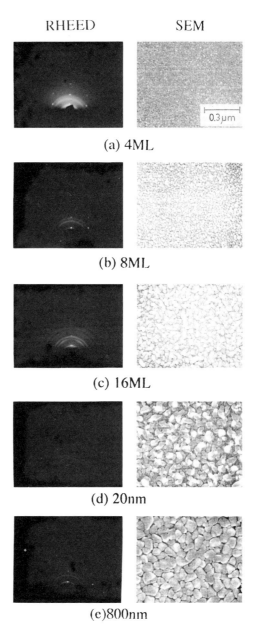

FIG.3 Observation of initial layer grown at 270 °C with a growth rate of 0.3 μm/h.

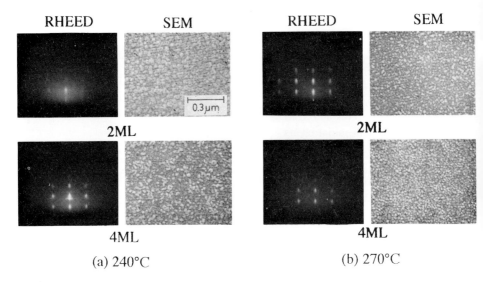

FIG.4 RHEED patterns and SEM photographs of the InSb initial layers grown at temperature (a) 240 °C and (b) 270 °C up to the thickness around 2-4 monolayers.

4. CONCLUSION

The hydrogen plasma treatment at 600 °C and exposure to arsenic vapor during cooling were employed for the cleaning of Si surface and we confirmed As-terminated surface is stable even in air. Detailed observations of the initial stage of the InSb growth revealed that the initial layer tends to poly-crystallize easily within several monolayers. The thinner initial layer grown at 240 °C, strained with longitudinal lattice mismatch around 12-13 % compared with 19 % between bulk InSb and Si, should be useful to relax large mismatch.

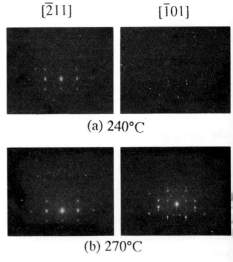

FIG.5 RHEED patterns of InSb epitaxial layers successively grown on initial layers which were grown at (a) 240 °C and (b) 270 °C.

REFERENCES

Gao Q Z, Hariu T and Ono S, 1987 Jpn. J. Appl. Phys. **26** L1576
Matsushita K, Sato Y, Hariu T and Ono S, 1984 J. Vac. Sci. Jpn. **27** 569
Ohshima T, Yamauchi S and Hariu T, 1989 Jpn. J. Appl. Phys. **28** L13
Takenaka K, Hariu T and Shibata Y, 1979 Jpn. J. Appl. Phys. **19** 765

Growth parameters for metastable GaP$_{1-x}$N$_x$ alloys in MOVPE

S. Miyoshi, H. Yaguchi, K. Onabe, Y. Shiraki[*], and R. Ito

Department of Applied Physics, The University of Tokyo,
7-3-1 Hongo, Bunkyo-ku, Tokyo 113, Japan

[*]Research Center for Advanced Science and Technology (RCAST),
The University of Tokyo, 4-6-1 Komaba, Meguro-ku, Tokyo 153, Japan

Abstract: In order to clarify what determines the solid composition (x) in the metalorganic vapor phase epitaxy (MOVPE) of metastable GaP$_{1-x}$N$_x$ alloys, we applied a growth interruption technique in this alloy growth. In our growth procedure, different from conventional MOVPE, the cycle of ~1 ML growth and growth interruption (only Ga precursor supply is turned off) is repeated. As increasing the growth interruption time, x decreases. This indicates that nitrogen atoms desorb from the film surface and that phosphorus atoms occupy the vacant sites during the growth interruption. The time constant of this substitution process is obtained to be 0.71, 0.47 and 0.39 sec at 655, 670 and 685°C, respectively. This result can explain the growth temperature dependence and the growth rate dependence of x in the conventional MOVPE of this alloy system. It also gives a guide to growing the alloys with large x.

1. Introduction

The GaP$_{1-x}$N$_x$ alloy is potentially a candidate for blue or ultraviolet optical devices, because its bandgap should change from 2.3 eV (GaP) up to 3.4 eV (GaN). Due to the large miscibility gap caused by the difference in the lattice structure and in the lattice constant between GaP and GaN, the growth of this alloy system had been considered to be very difficult. The recent progress of some crystal growth procedures using supersaturation circumstances such as molecular beam epitaxy (MBE) and metalorganic vapor phase epitaxy (MOVPE) has enabled the growth of this alloy. There have been two demonstrations of GaP$_{1-x}$N$_x$ alloy growth, with x < 0.08 by MBE [1] and x > 0.91 by chemical vapor deposition [2]. We have also succeeded in the alloy growth by MOVPE [3]. However, the maximum x obtained in the MOVPE is limited to a very low value (x < 0.04). In order to grow alloys with larger x, it is necessary to clarify the key factor which determines the solid composition x. It is expected that growth interruption will change the solid composition, because it has a large effect on the growth process.

In the present study, we applied a growth interruption technique to the MOVPE growth of GaP$_{1-x}$N$_x$ alloys. The cycle of ~1 ML growth and the growth interruption were repeated 1500 times to grow the film. The growth interruption time (t_i) dependence of x was studied. It was shown that the nitrogen desorption was considered to be a key factor which determined x.

2. Experimental procedure

Samples were grown on nominally on-axis (100) GaP substrates with a conventional low-pressure (60 Torr) MOVPE system with H_2 carrier [3]. The Ga, P, and N precursors were trimethylgallium (TMG), PH_3, and dimethylhydrazine (DMHy), respectively. After the growth of a 0.3 μm-thick GaP buffer layer at 750°C, the substrate temperature was lowered to the growth temperature (T_g), 655 ~ 685°C, and then DMHy and PH_3 were supplied. In order to grow the alloys, the ~1ML growth and the growth interruption were repeated 1500 times by turning the TMG flow on and off (which is shown in the inset of Fig. 1). The flow of DMHy and that of PH_3 during the growth interruption were kept at the same as those during the growth. The TMG flow rate was 2.0 - 4.0 sccm (= 4.5 - 9.0 μmol / min). The growth rate was 1.2 ML / sec with f_{TMG} = 2.0 sccm. The relative molar flow of DMHy in the total group V source gas was 0.625. The V/III ratio was 80 with f_{TMG} = 2.0 sccm. All the samples had a mirrorlike surface. We assumed Vegard's law between the solid composition x and the lattice constant, and determined x from the lattice constant of the alloy measured by the double crystal x-ray rocking curve of (511) reflection. The lattice constant of GaP and that of cubic GaN were taken from refs. [5] and [6], respectively. All the samples had a zincblende structure. The solid composition (x) of the alloy was 0.0022 - 0.021.

3. Results and discussion

Figure 1 shows the growth interruption time (t_i) dependence of the solid composition (x) of the $GaP_{1-x}N_x$ alloys grown at 655°C. As t_i increases, x decreases. Although not shown in the figure, the full width at half maximum (FWHM) of the x-ray diffraction rocking curve of the samples grown with a growth interruption is as narrow as that grown without growth interruption (about 80 arcsec). This shows that the growth interruption affects the solid composition of only the surface layer. (If inner regions are also affected, there should be a composition gradient along the growth direction, thus the FWHM of the x-ray diffraction should be broader at large t_i.) Therefore, it is clear that some surface reaction which decreases x is occurring at the film surface. Considering the fact that the equilibrium pressure of nitrogen is very high [4], we suspect that nitrogen atoms desorb from the film surface during the growth interruption. This process decreases x because phosphorus atoms will occupy the vacant sites

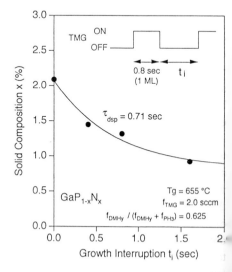

Fig. 1. Growth interruption time (t_i) dependence of solid composition x for $GaP_{1-x}N_x$ alloys grown at 655°C.

where nitrogen atoms are desorbed. Then the solid composition at $t_i = \infty$ will be determined by a thermal equilibrium between nitrogen supply and desorption. Figure 1 shows that the solid composition x at $t_i = 0$ is larger than that at $t_i > 0$. There are two possible explanations for this. One is that the partial pressure of the P and N reactants changes as changing the TMG partial pressure (p_{TMG}) through the law of mass action. Jou et al. have reported that in the MOVPE of GaP$_{1-x}$Sb$_x$ system, x increases as p_{TMG} increases [7]. They explained it by the change of the equilibrium constant of PH$_3$ = (1/2)P$_2$ + (3/2)H$_2$ reaction with changing the TMG flow rate. The same effect might occur in our growth, though it is impossible at present to investigate the effect quantitatively because we have no data on the thermochemical properties of DMHy. The other explanation is that TMG and DMHy form an adduct. Then the Ga and N atoms will arrive at the film surface and react simultaneously, thus the N atoms are taken into the alloy preferentially.

By considering the nitrogen desorption and phosphorus substitution process, the t_i dependence of x shown in Fig. 1 is explained quantitatively. As x is very small, we assume that if a nitrogen atom desorbs from the film surface, the vacant site is immediately occupied by a phosphorus atom. Then the following rate equation holds:

$$dn_s/dt = G - n_s/\tau_{dsp}, \qquad (1)$$

where n_s is the nitrogen content at the growing surface ($\propto x$), G is nitrogen supply from the vapor phase, and τ_{dsp} is the time constant of this nitrogen desorption. Applying eq.(1) to the experimental data, the solid line of Fig. 1 is obtained. The time constant τ_{dsp} is determined to be 0.71 sec. We have done the same growth study at 670°C. The t_i dependence of x is shown in Fig. 2. At 670°C, τ_{dsp} is obtained to be 0.47 sec. As increasing T_g, the nitrogen desorption rate increases, thus τ_{dsp} gets shorter. Though not shown in the figure, τ_{dsp} is obtained to be 0.39 sec at 685°C.

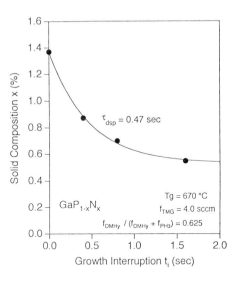

Fig. 2. Growth interruption time (t_i) dependence of solid composition x for GaP$_{1-x}$N$_x$ alloys grown at 670°C.

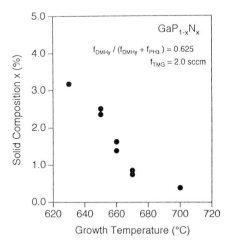

Fig. 3. Growth temperature dependence of solid composition x for GaP$_{1-x}$N$_x$ alloys grown by conventional MOVPE, i.e., without growth interruption.

This increasing nitrogen desorption as increasing T_g also explains the T_g dependence of x in the conventional (without growth interruption) MOVPE of this alloy. The experimental results are shown in Fig. 3. The increasing nitrogen desorption (which may occur not only during growth interruption but also during growth) decreases x as increasing T_g. The nitrogen desorption process can also be applied to explain the growth rate dependence of x [3]. As the growth rate increases, x increases. It is considered that at a high growth rate, the film surface is covered with the next growing surface before the nitrogen desorption from the film surface occurs sufficiently, or, that as increasing TMG flow rate ($\propto$ growth rate), the number of TMG-DMHy adducts increases and thus x increases.

These results indicate that, though we have no direct evidence, nitrogen desorption from the film surface is a key factor which determines the solid composition in the MOVPE of $GaP_{1-x}N_x$ alloys. It is clear that in order to grow the alloys with larger x, low growth temperature and high growth rate is effective. However, these growth conditions usually deteriorate the crystal quality. It is thus important to balance the need for large x and the crystal quality.

4. Conclusion

We applied a growth interruption technique in the MOVPE of metastable $GaP_{1-x}N_x$ alloys. The cycle of ~1ML growth (TMG supply on) and growth interruption (TMG off) is repeated. As increasing the growth interruption time, x decreases. This indicates that nitrogen atoms desorb from the film surface and that phosphorus atoms occupy the vacant sites during the growth interruption. The time constant of this substitution process gets shorter as increasing growth temperature. This process can explain the growth temperature dependence and the growth rate dependence of the solid composition in the conventional MOVPE of this alloy system. It also gives a guide to growing the alloys with large x.

Acknowledgements

The authors would like to express their sincere gratitude to W. Pan, K. Ota, A. Shima, K. Takemasa and X. Zhang for assistance in MOVPE growth and to T. Osada and S. Fukatsu for useful discussion, and acknowledge S. Ohtake for his technical support. We greatly appreciate Sumitomo Chemical Co., Ltd. for supplying the TMG source.

References

[1] Baillargeon J N, Cheng K Y, Hofler G E, Pearah P J and Hsieh K C 1992 *Appl. Phys. Lett.* **60** 2540-2
[2] Igarashi O 1992 *Jpn. J. Appl. Phys.* **31** 3791-3
[3] Miyoshi S, Yaguchi H, Onabe K, Ito R and Shiraki Y 1993 *Appl. Phys. Lett.* **63** 3506-8
[4] Thurmond C D and Logan R A 1972 *J. Electrochem. Soc.* **119** 622-6
[5] Straumanis M E, Krumme J -P and Rubenstein M 1967 *J. Electrochem. Soc.* **114** 640-1
[6] Miyoshi S, Onabe K, Ohkouchi N, Yaguchi H, Ito R, Fukatsu S and Shiraki Y 1992 *J. Cryst. Growth* **124** 439-42
[7] Jou M J and Stringfellow G B 1989 *J. Cryst. Growth* **98** 679-89

Properties of GaN Films Grown on A-plane (11$\bar{2}$0) and C-Plane (0001) Sapphire

K.Doverspike, L.B.Rowland, D.K.Gaskill, S.C.Binari, J.A.Freitas, Jr.*, W.Qian**, and M.Skowronski**

Laboratory for Advanced Material Synthesis, Naval Research Laboratory, Washington, DC 20375-5347;
* Sachs-Freeman Assoc., Landover, MD 20785-5396
** Carnegie Mellon University, Pittsburgh, PA 15213

Abstract. This paper presents a comparative study of the properties of GaN, using an AlN buffer layer, as a function of sapphire orientation (a-plane *vs* c-plane). Results are presented for varying the AlN buffer layer thickness. The electron Hall mobilities of the GaN films grown on a-plane were, in general, higher than that of the corresponding growth on c-plane sapphire. Because electron mobilities obtained on a-plane sapphire are large (> 300 cm^{2}V^{-1}s^{-1}) over a wide range of buffer layer thicknesses (115 Å to 375 Å), growth on a-plane may be more desirable for mass production of GaN-based devices.

1. Introduction

In the last several years, considerable progress has been made in growing high quality GaN by using either a GaN or AlN buffer layer[1,2]. Previous results in our group and others indicate that properties of the GaN films grown by organometallic vapor phase epitaxy (OMVPE) are sensitive to the conditions under which the buffer layer is grown, e.g., buffer layer thickness[3,4] and buffer layer growth temperature[5]. Current understanding is limited as to whether this sensitivity to the buffer layer growth conditions is the result of various impurities or defects being incorporated in the buffer layer under certain conditions, or whether the morphology of the buffer surface, which will subsequently affect the nucleation of the GaN film, is the dominant factor. While the basal plane (c-plane) of sapphire (0001) is the most commonly used substrate for the growth of wurtzite GaN, there have been a few reports of GaN films grown on a-plane (11$\bar{2}$0) sapphire[6-9]. These have focused mainly on the epitaxial relationship to the substrate, the crystalline quality, and the surface morphology of the resulting GaN layers. GaN films grown on either c-plane or a-plane sapphire are oriented with the GaN c-plane (0001) parallel to the substrate. Growing on a-plane, therefore, offers an excellent opportunity to investigate the role sapphire orientation may have on the subsequent nucleation of the GaN film and its properties. The role of the sapphire substrate surface is also unclear, since the quality of the surface polish could affect not only the nucleation of the buffer layer, but also the GaN film nucleation. In an effort to investigate the role of the sapphire substrate surface, a-plane sapphire was chosen, because it is believed that the a-plane sapphire is easier to polish than the c-plane [10], which may result in a higher quality polish on the sapphire. In this study, using AlN as the buffer layer, GaN layers were grown on both c-plane and a-plane sapphire by reduced-pressure OMVPE. For each of the experiments described, simultaneous growth on a-plane and c-plane was performed. The thickness of the AlN buffer layer was varied while the conditions for the GaN film grown on the buffer remained unchanged.

2. Experimental

An inductively-heated, water cooled, vertical OMVPE reactor operated at 57 torr was used for the growth[11]. The AlN buffer layer was deposited at 450°C using 1.5 μmol/min triethylaluminum (TEAl), 2.5 SLM NH_3, and 3.5 SLM H_2. The growth rate of the buffer layer was determined independently by growing a thick film using the buffer layer growth conditions and viewing the cross-section using scanning electron microscopy. After the buffer layers were deposited, they were annealed in 2.5 SLM NH_3 and 3.5 SLM H_2 for 10 min. at 1025°C. The GaN films grown on the buffer layers were approximately 3μm thick with a growth rate of 1.6 μm/hr. They were formed using 25.7 μmol/min trimethylgallium (TMGa), 2.5 SLM NH_3, and 3.5 SLM H_2. Sample thickness was measured using interference fringes obtained by infrared reflectance spectroscopy. For the buffer layer thickness study, a growth temperature of 1040°C was used for the GaN film. Surface morphology was examined using Nomarski optical microscopy. Double crystal x-ray diffractometry using Cu Kα radiation was used to assess film crystallinity. Van der Pauw Hall measurements (2240 G) were performed on clover leaf geometries using In contacts. Capacitance-voltage (C-V) measurements were also performed using Pt/Au Schottky barriers. Plan-view transmission electron microscopy (TEM) specimens were prepared by mechanical thinning from the substrate side to about 30 μm followed by 5 kV Ar^+ ion milling at 77K. The TEM observation was carried out in a Philips EM420, operating at 120 keV. The 325nm line of a He-Cd laser was used as the excitation source for the low temperature (6K) photoluminescence (PL) experiments.

3. Results and Discussion

The lowest full width half maximum (FWHM), when using c-plane or a-plane sapphire, occurs with an AlN buffer thickness of about 160 to 175 Å as seen in Figure 1. Plan-view bright field TEM images of samples grown on a-plane and c-plane are shown in figures 2a and 2b respectively. The FWHM of the samples shown in Figure 2 were approximately 300 arc sec, while the 300 K room temperature electron Hall mobilities were 367 and 325 $cm^2v^{-1}s^{-1}$ for growth on a-plane and c-plane sapphire, respectively. The TEM micrographs are virtually identical, and the main defects are dislocations with Burgers vectors of <1120> type propagating along the c-axis. The dislocation density in both c-plane and a-plane samples is about $1.5 \times 10^{10} cm^{-2}$. Therefore, for the optimized buffer thickness, these a-plane and c-plane samples do not show a large difference in either the crystallinity, the number, or types of defects as seen by TEM.

Photoluminescence was also used to characterize the material. Figure 3 shows a PL spectrum of a typical GaN film grown on an AlN buffer. PL spectra of films grown on c-plane or a-plane sapphire were similar with the the band edge emission being the dominant feature. Using higher spectral resolution, the band edge emission peak width at half maximum has been as narrow as 2.7 meV, which is indicative of high quality material. As can be seen in Figure 3, there is only a small indication of the deep level 2.2 eV band which is commonly observed in GaN films and may be associated with compensation in the GaN films involving deep donor states and shallow acceptor states[12].

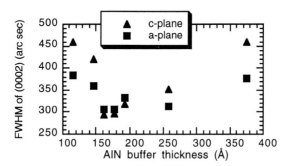

Figure 1. Crystallinity of GaN film as measured by the FWHM of the (0002) peak as a function of the AlN buffer layer thickness for GaN growth on both c-plane and a-plane sapphire.

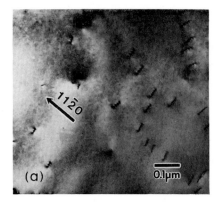 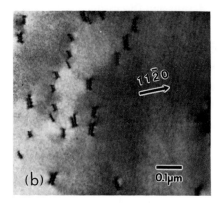

Figure 2 (a) Plan-view bright field image (g=1120), taken from GaN film grown on a-plane (b) GaN grown on c-plane.

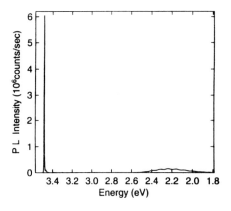

Figure 3. Photoluminescence spectrum (6K) of GaN grown on an AlN buffer on a-plane sapphire.

Hall measurements performed on the films where the AlN buffer layer thickness was varied, yielded electron concentrations between 1.5×10^{17} and 4.5×10^{17} cm^{-3}. Capacitance-voltage measurements were also done in order to obtain electron concentrations as a function of depth. The near surface carrier concentrations obtained by C-V measurements were in agreement with that obtained from Hall measurements, and also verified that the net carrier concentration was uniform, in the region probed by C-V measurements (within 0.30 μm of the top surface). No trend was observed in the electron concentrations as a function of buffer layer thickness for growth on either c-plane or a-plane. There was, however, some interesting trends found in the mobilities as a function of buffer layer thickness (Figure 4). For all the films measured in this thickness study, the mobility of the GaN deposited on a-plane sapphire was higher than that deposited on c-plane, while the buffer layer thickness yielding the highest mobility was similar for the two orientations; 162 Å for c-plane and 175 Å for a-plane. Thus, growth on a-plane sapphire accommodated a wide range of buffer thicknesses (115 Å to 375 Å) while still yielding GaN films with high mobilities (> 300 cm^{2}v^{-1}s^{-1}).

It is very intriguing that the mobility of GaN grown on a-plane is fairly insensitive to buffer layer thickness, since growth on c-plane sapphire shows a very strong dependence between the mobility of the GaN and the buffer layer thickness[3,4]. Assuming the a-plane sapphire does indeed have a higher quality surface polish, this may lead to a buffer layer whose properties or characteristics are fairly uniform and are independent of the buffer layer thickness. Therefore, a wide range of buffer layer thicknesses could be used and still lead to GaN films of sufficiently high quality (> 300 cm^{2}v^{-1}s^{-1}). For growth on c-plane sapphire, however, the buffer layer properties vary significantly with thickness. One possible explanation may be the decreased quality of the surface polish which would effect the manner in which the buffer nucleates on the sapphire and the subsequent GaN nucleation. Only when the buffer layer thickness is optimized on c-plane does the GaN nucleation on the buffer layer, and the subsequent film properties, compare with the GaN deposited on a-plane sapphire.

Aside from the surface polish, growing on a-plane sapphire will also have an effect on the lattice mismatch between the sapphire and GaN. If the rectangular unit cell (12.97 Å x 8.23 Å) of a-plane sapphire is doubled, then the sapphire can now easily accommodate several unit cells of GaN basal planes[9]. This would then lower the lattice mismatch to 1.6 % along the (0001) of sapphire and 0.6 % along the ($1\bar{1}00$) axis of the sapphire. It is unclear at this time whether the differences in the GaN films grown on c-plane versus a-plane sapphire are due to the surface polish or the orientation.

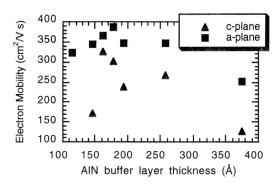

Figure 4. Electron mobility (300K) of GaN film as a function of the AlN buffer thickness.

4. Summary

In this study, GaN was grown on an AlN buffer layer on both c-plane (0001) and a-plane (11$\bar{2}$0) sapphire by reduced-pressure OMVPE. The thickness of the buffer layer was varied and optimized to approximately 160 to 175Å, where the films showed the highest crystalline quality (measured by the FWHM of the (0002) peak) and also the highest electron Hall mobilities. Not only were the electron Hall mobilities of the GaN films grown on a-plane higher than the corresponding growth on c-plane sapphire, but were obtainable over a wide range of buffer thicknesses (115 Å to 375 Å).

5. Acknowledgments

One of us (LBR) acknowledges support from a NRC-NRL Cooperative Research Associateship.

6. References

[1] Amano H, Sawaki N, Akasaki I and Toyoda Y 1986 *Appl. Phys. Lett.* **48** 353

[2] Nakamura S 1991 *Jpn. J. Appl. Phys.* **30** L1705

[3] Kuznia J N, Khan M A, Olson D T, Kaplan R and Freitas, J.A.,Jr. 1993 *J. Appl. Phys.* **73** 4700-2

[4] Doverspike K, Rowland L B, Gaskill D K and Freitas J A, Jr. To be published in *J. Electronic Mater.*

[5] Rowland L B, Doverspike K, Gaskill D K and Freitas Jr. J A 1994 *Mater. Res. Soc. Symp. Proc.* 339

[6] Wickenden D K, Faulkner K R, Brander R W and Isherwood B J 1971 *J. Cryst. Growth* **9** 158-64

[7] Amano H, Hiramatsu K, Kito M, Sawaki N and Akasaki I 1988 *J. Cryst. Growth* **93** 79-82

[8] Amano H, Hiramatsu K and Akasaki I 1988 *Jpn. J. Appl. Phys.* **27** L1384-6

[9] Moustakas T D, Lei T and Molnar R J 1993 *Physica B* **185** 36-49

[10] Private commumication with K.Heikkinen, Union Carbide.

[11] Rowland L B, Doverspike K, Giordana A, Fatemi M, Gaskill D K, Skowronski M and Freitas Jr J A 1993 *Inst. Phys. Conf. Ser.* **137** 429-32

[12] Glaser E R, Kennedy T A, Crookham H C, Freitas J A, Jr., Khan M A, Olson D T and Kuznia J N 1993 *Appl. Phys. Lett.* **63** 2673-5

Epitaxial growth and structural, optical properties of cubic GaN on (100) and (111) GaAs grown by metalorganic chemical vapor deposition

Chang-Hee Hong, Kun Wang, and Dimitris Pavlidis

Department of Electrical Engineering & Computer Science, The University of Michigan, Ann Arbor, MI48109, U.S.A

Abstract: GaN films were grown on (100) and (111)A or (111)B GaAs substrates by low pressure metalorganic chemical vapor deposition (MOCVD) and their structural and optical properties were investigated by X-ray diffraction, PL and Raman spectroscopy. The growth of cubic GaN on (100) GaAs was strongly influenced by the growth temperature. However cubic GaN on (111)GaAs showed no dependence on growth temperature in terms of phase formation. The PL and Raman characteristics evidence the quality of the grown films and are very similar for both (100) and (111) GaAs substrates.

1 Introduction

GaN is a wide bandgap material with great potential for optoelectronic device applications from the blue to the ultraviolet wavelengths, high-power and high-temperature devices[1,2]. GaN can be crystallized in either hexagonal(wurtzite) or cubic (zincblende) structure depending on the substrate symmetry and growth conditions. In certain cases both structures may co-exist because of the small difference in energy of formation between them[3,4]. High-quality wurtzitic GaN has been grown successfully on a variety of substrates, in particular on the basal plane of sapphires. However, it is believed that cubic structures possess in principle superior electronic properties for device applications[5,6]. Furthermore, it is desirable to use cubic substrates such as silicon or GaAs for the growth of GaN since this could lead to reduction of interfacial defects and impurities as well as, integration of GaN with Si or GaAs-based devices.

Since the first realization of cubic GaN grown on GaAs substrates[7], only few additional attempts have been reported for growing cubic GaN on GaAs. Various techniques have been used which include (i) MOCVD[8][9], (ii) Gas source MBE using dimetylhydrazine(DMHy) as the source of nitrogen [10] and (iii) plasma-assisted MBE[11]. Recently, Miyoshi et al [8] have reported high quality cubic GaN grown on (100) GaAs at 650°C using DMHy, and showed Raman peaks characteristic of cubic symmetry. However, there is no comprehensive investigation of GaN on GaAs by MOCVD using NH_3 as a nitrogen source. In this paper, we report the growth and the structural, as well as, optical properties of GaN on (100) and (111) GaAs substrates by low pressure MOCVD using NH_3 as the nitrogen source.

Work supported by ONR Contract No: N00014-92-J-1552

2. MOCVD growth of GaN on GaAs substrates

The GaN films were grown in our low pressure (60 Torr) MOCVD system using H_2 carrier gas. Various GaAs substrate types were used including (100), (111)A and (111)B orientations. The substrates were epi-ready or chemically cleaned in a solution of $H_2SO_4:H_2O_2:H_2O = 5:1:1$. They were loaded in a horizontal reactor and annealed in H_2 and AsH_3 at 670°C for 5 minutes to remove the surface oxide. After cooling down to the desired growth temperature, NH_3 alone was supplied for 10 minutes so that surface nitridation could take place. TMG and NH_3 were then supplied simultaneously and GaN films were grown. The samples were grown at a fixed V/III ratio of 3300 and substrate temperatures ranging from 530°C to 700°C. The NH_3 flow was 1000 sccm. The TMG flow rate was fixed at 13.4 µmol/min for all the experiments reported here. The H_2 supply rate was 2 slm. Typical growth rates were of the order of 0.5µm/hr at 600°C and the total layer thickness of the samples was typically 0.5µm. Some thick (1.65µm) films were also grown for comparison. The structural properties of the GaN films were characterized by X-ray diffraction(XRD) and their optical properties were investigated by photoluminescenec and Raman spectroscopy.

3. Structural and optical characterization of GaN films grown on GaAs

XRD characterization took place using a single-axis goniometer of θ-2θ geometry with Cu Kα line(λ=1.537Å) and a goniometer of four-circle geometry with Mo Kα line (0.709Å). For the photoluminescence measurements a frequency-doubled picosecond DCM dye laser was used to provide a quasi-continuous excitation source at 325 nm. Doubling was performed outside the dye laser cavity using an angle-tuned LiIO3 crystal at room temperature.

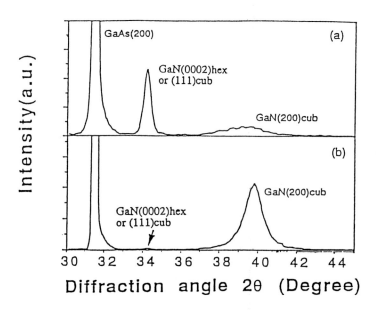

Fig.1a XRD pattern of GaN grown on (100) GaAs: (a) at 530°C; (b)at 600°C.

The luminescence spectrum from the sample was then collected in a back-scattered geometry and focused onto the entrance slits of a 1 meter spectrometer. Photoluminescence(PL) signals were detected with a C31034A-02 GaAs photo multiplier in photon-counting mode, and the spectra were recorded as a function of temperature and excitation intensity. Raman scattering measurements were performed in a backscattering geometry. For excitation, 100 mW of 488 nm radiation from an argon ion laser was focused onto the sample with a 150 mm focal length lens. A small pick-off mirror steered the light onto the sample.

3.1 Results and discussion on strunctural properties of GaN films

The XRD patterns of GaN epilayers grown on (100) GaAs at 530 and 600°C are shown in Figs 1a and 1b respectively. The peak at around $2\theta=39.8°$ is assigned to the (200) diffraction ($(200)_{cub}$) of cubic GaN while the peak at around 34.3° can be related to (0002) hexagonal $((0002)_{hex})$ or (111) cubic $((111)_{cub})$ GaN. One observes that when the growth temperature is below 530°C or above 650°C (not shown in figure) there are two clear XRD peaks, one around $2\theta=39.8°$ and another at 34.3° (see Fig. 1 a). At 600°C (Fig. 1b) one observes a single peak at around 39.8° which is associated again with diffraction from $(200)_{cub}$ GaN. It is obvious that structural changes take place for GaN grown on (100) GaAs when the growth temperature changes. In particular, one observes a change from $(0002)_{hex}$ or $(111)_{cub}$ phase to $(200)_{cub}$ phase GaN as the temperature is raised from 530 to 600°C. There appears consequently to exist a temperature window within which one obtains cubic GaN on (100) GaAs substrates. The lattice constant for cubic GaN estimated from the $(200)_{cub}$ pattern was found to be 4.528Å.

Growth on (111)A and (111)B GaAs at 600°C revealed the same cubic (111) peak of GaN at 34.3°. Experiments with thicker layers showed a reduction of full-width-half-maximum (FWHM) and a FWHM of 12mins was recorded for 1.65µm thick layers grown non (111)B GaAs.

In contrast to the results obtained on (100) GaAs, GaN grown on (111) GaAs shows only a single phase. Figure 2 shows the main peaks of typical θ-2θ diffraction spectra for a GaN film grown on (111) GaAs substrate.

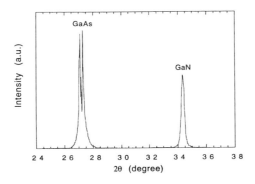

Fig.2 XRD pattern of GaN on (111) GaAs substrate

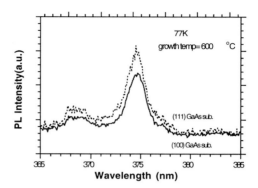

Fig.3 Photoluminescence (PL) spectra of GaN on (100) and (111) GaAs

The diffraction peak at about $2\theta=27.1°$ is related to the (111) GaAs substrate and the diffraction peak at $2\theta=34.4°$ can be associated with cubic (111) planes from the GaN films. In fact, it is very difficult to distinguish between (111) cubic plane and (0002) hexagonal plane structures only from conventional XRD characterization because the d-spacing of XRD in (0002) hexagonal and (111) cubic lattices is the same. A four-circle x-ray diffractometer, which is capable of distinguishing the stacking sequence, has therefore been used to verify the phase information [12]. The results of the stacking sequence of GaN on (111) GaAs suggest that the material grown is indeed of (111) cubic phase. The lattice constant calculated from XRD measurements of the (111) GaN films was 4.52Å, which is in good agreement with previously reported values.

3.2 Results and discussion on Optical properties of GaN films grown on GaAs

Fig.3 shows the photoluminescence (PL) spectra of cubic GaN grown on (100) and (111) GaAs substrates. The spectra were measured at a temperature of 77K and the excitation power was $38 W/cm^2$. Both spectra show similar features. The peak at 3.31 eV is assigned to donor-acceptor (DA) pair recombination and the peak at 3.366 eV is assigned to band edge emission. The satellites around the peak of 3.31 eV are believed to be phonon replicas. The best FWHM of the 3,366eV peak was 5.3meV for spectra obtained at 6.5K. A detailed analysis of the PL characteristics will be published elsewhere[13].

Fig.4 shows typical z(xx)z and z(xy)z Raman spectra from c-GaN films grown on (100) and (111) GaAs. As one can see, both show similar characteristics. The spectra indicate a strong peak at 736 cm^{-1}, a small shoulder at 768 cm^{-1}, and a weak continuum extending from 700 cm^{-1} to 500 cm^{-1} with a second peak at 560 cm^{-1}. The results shown in this figure agree with previously published data on cubic GaN [14] but contrast Raman characteristics reported for wurtzitic GaN films [15]. The peaks at 736 cm^{-1} and 560 cm^{-1} are consistent with cubic longitudinal (LO) and transverse (TO) phonon energies. The small shoulder at 768 cm^{-1} is assumed to be defect-related, although it is close in energy to quoted wurtzitic LO phonon energies [16].

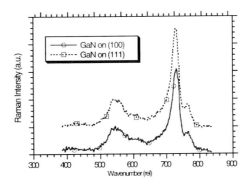

Fig. 4 Raman spectra of GaN on (100) and (111) GaAs

More details on the analysis of the GaN Raman characteristics on (100) GaAs have been reported in [17] and a more complete study of their features on (111) GaAs substrates will be reported elsewhere.

4. Conclusion

GaN films were grown by low pressure MOVPE using TMGa and NH3 and showed cubic characteristics on both (100) and (111) GaAs substrates. The cubic versus hexagonal phase formation of GaN on (100) substrates was found to depend on the growth temperature. The growth on (111) substrates revealed a significant improvement of controllability of the GaN cubic phase when compared with growth on (100) GaAs. Minimum FWHM of 12mins was obtained for thick (1.65µm) GaN films grown on (111)B GaAs. PL and Raman characteristization of the GaN films revealed similar characteristics for growth on both (100) and (111) GaAs. Band edge emission in the PL spectra occured at 3.366eV and DA pair recombination was revealed by a peak at 3.31eV. A minimum FWHM of 5.3meV was recorded at 6.5K. The Raman spectra peaks at 736 cm^{-1} and 560 cm^{-1} were consistent with cubic LO and TO phonon energies

Acknowledgments

The authors would like to thank Max Yoder of ONR for his support and encouragmernt, Chuck Taylor and Prof. Roy Clarke for help in four circle XRD characterization and Steve Brown and Prof. Steve Rand for help in PL and Raman characterization.

References

1. R. F. Davis, Proc. IEEE **79**, 702 (1991).
2. S. Strite and H. Morkoç, J. Vac. Sci. Technol. B **10**, 1237 (1992).
3. J. I. Pankove, S. Bloom, and G. Harbeke, RCA Rev., 36 (1975) 163.
4. B. J. Min, C. T. Chan, and K. M. Ho, Phys. Rev, B 45 (1992) 1159.
5. J. I. Pankove, MRS Symp. Proc., 162 (1990) 515.
6. K. Das and D. K. Ferry, Solid-State Electron.,19 (1976) 851.
7. M. Mizuta, S. Fujieda, Y. Matsumoto, and T. Kawamura, Jpn. J. Appl.Phys., 25 (1986) L945.
8. S. Miyoshi,K. Onabe, N. Ohkouchi, H. Yaguchi, R. Ito, S. Fukatsu, and Y. Shiraki, J.Crystal Growth, 124 (1992) 439.
9. D. Pavlidis, C. H. Hong and K. Wang, 18th WOCDICE, Cork, Ireland , May 1994
10. H. Okumura, S. Misawa , S. Yoshida, and E. Sakuma, J. Crystal Growth, 120 (1992) 114.
11. S. Strite, J. Ruan, Z, Li, A. Salvador, H. Chen, D.J. Smith,
12. C.H. Hong, D. Pavlidis, C. Taylor and R. Clarke to be published
13. C.H. Hong, D. Pavlidis, S. Brown and S. Rand to be published
14. S. Murugkar, R. Merlin, T. Lei, and T. D. Moustakas, Abstract from the American Physical Society Meeting, March 1992.
15. C.-J. Sun and M. Razeghi, Appl. Phys. Lett. **63**, 973 (1993).
16. D. D. Manchon, Jr., A. S. Barker, Jr., P. J. Dean, and R. B. Zetterstrom, Solid State Comm. **8**, 1227 (1970).
17. S. Brown, S. Rand, C.H. Hong and D. Pavlidis, Proc. of MRS, Symp.D, San Fransisco (1994)

Growth and Characterization of P/N Type GaN Grown at Reduced Substrate Temperatures by Plasma Enhanced (PE-) MOCVD

L.D. Zhu, P.E. Norris and J. Zhao

NZ Applied Technologies, 12 Whitney Avenue, Cambridge, MA 02139 USA

R. Singh, R. Molnar, O.Razumovsky and T.Moustakas

Boston University Dept. of Electrical Engineering, 44 Cummington St., Boston, MA 02215 USA

Abstract. GaN epitaxial layers have been grown on c- and r-plane sapphire substrates using Plasma Enhanced Metalorganic Chemical Vapor Deposition (PE-MOCVD). The growth temperature for single crystalline GaN films was reduced to 620-850°C in comparison to the conventional MOCVD growth temperature of 1050°C. X-ray diffraction studies showed the films to be highly oriented single crystalline phase. For the first time, p-type GaN layers with strong blue-green luminescence have been realized using PE-MOCVD and Mg-doping. This demonstrates the utility of the PE-MOCVD approach for GaN materials and device technology.

1. Introduction

The need to extend the operating wavelength of optoelectronic devices to the shorter blue and ultra violet spectral region, as well as the interest in developing high temperature electronics, has motivated the present intensive research activity on the wide bandgap semiconductors GaN (and related III-nitride compounds), SiC and Diamond[1][2]. Of these wide bandgap materials, the group III-Nitride compounds GaN, AlN, InN and their alloys are direct bandgap materials promising efficient optoelectronic devices with operating wavelengths covering the range from red to the UV. Recent progress in p-type doping and improvement of crystal quality in these materials by so-called 'Two-Flow' MOCVD method, has led to the fabrication of bright blue LEDs[3][4].

Naturally, wide band gap semiconductors are also candidates for high temperature, high power and high frequency FETs. In an assessment of materials for power FET applications, Baliga's figure of merit (BFOM)[5] can be used:

$$BFOM = \varepsilon\mu(E_g)^3 \qquad (1)$$

Values of BFOM, which gauges power handling capability of a FET (conduction/unit area), have been calculated for α-SiC and α-GaN and compared to Si, as shown in Table 1.

Table 1. Baliga's figure of merit for Si, SiC and GaN [normalized to Si]

	μ	ε	$\varepsilon\mu/[\varepsilon\mu(Si)]$	E_g (eV)	$E_g/[E_g(Si)]^3$	$BFOM/(BFOM)_{Si}$
Si	1300	11.4	1.00	1.1	1.00	1.0
α-SiC	260	9.7	0.17	2.9	18.32	3.1
α-GaN	1300	9.5	0.83	3.4	29.53	24.6

It is clear from the table that GaN, due to its wide bandgap and expected high electron mobility, is more than one order of magnitude superior to silicon in FET power handling capability.

We have grown GaN on sapphire substrates by remote microwave plasma-enhanced (PE-)MOCVD. It is believed that activation/dissociation of N_2 or NH_3 will significantly reduce the growth temperature[6],[7]. This is because nitrogen is activated and/or ionized by microwave rather than thermal energy and the kinetic energy of the nitrogen species helps to reduce the thermal energy necessary for the atoms to migrate on the growing crystal surface. Besides, low temperature growth should improve the stoichiometry of these refractory compound semiconductors with high nitrogen dissociation pressures by easing the problems associated with nitrogen vacancies. Low temperature growth is also believed to reduce impurity contamination.

2. Low Temperature Growth of GaN by Plasma Enhanced MOCVD

GaN growth was carried out in a vertical chamber, low pressure MOCVD system. A 2.45GHz, remote microwave power source was used to activate a 3%, ultra pure hydrogen in nitrogen mixture. The remote microwave source was used to avoid damage to the layer from the energetic plasma species. The triethylgallium (TEG) source was transported separately from the H_2/N_2 plasma line. C-plane(0001) and r-plane(1$\bar{1}$02) sapphire were used as substrates. Prior to growth, the wafers were degreased, etched, rinsed and then blown dry.

GaN growth was initiated by thermal treatment of the substrate, followed by a plasma treatment. In most of the runs, a 400-600Å thick GaN buffer layer was grown at about 600°C. GaN epitaxial layers were grown over a temperature range of 620-850°C, which is still much lower than the conventional MOCVD growth temperature of 1050°C. For nominally undoped GaN growth, the TEG transport rate was chosen to be 5.7 μmole/m, while for p-doped GaN growth, the TEG flow rate was decreased to 0.5 μmole/m. Bis(cyclopentadienyl)magnesium (CP_2Mg) was used for p-type doping. The Mg-doped GaN was thermally annealed in-situ or ex-situ at 720°C.

It is meaningful to compare the PE-MOCVD growth conditions with other growth methods in terms of GaN phase diagram. Newman et al [8] discussed the GaN phase stability for the case of NH_3 and N_2 reactants, as a function of group V reactant partial pressure and substrate temperature. Our use of an activated N_2/H_2 mixture could apply to both cases. Due to the large kinetic barrier to decomposition, GaN actually undergoes congruent sublimation rather than decomposition.

We found that PE-MOCVD is performed closer to the region of the phase diagam where GaN is stable, as compared with other activated processes [e.g. reactive sputtering, plasma-assisted MBE or ECR-assisted CVD]. This is due to the higher group V partial pressure possible in the PE-MOCVD process. In fact, if we restrict ourselves to epitaxial growth processes at 850°C and below, it is the only one even near phase stability.

3. Experimental results

3.1 X-ray diffraction studies

The crystalline quality of the GaN films were characterized by X-ray diffraction measurements (XRD). Fig.1 shows an X-ray diffraction θ-2θ plot of a GaN layer grown on (0001) Al_2O_3. The only XRD peaks observed are (002)GaN and (004)GaN reflection peaks indicating that the film is single crystalline phase. Furthermore, the peak positions at 34.6° and 73.0° indicate that the GaN has a wurtzitic structure, as expected on c-plane Al_2O_3.

An X-ray φ-scan, shown in Fig. 2, was performed to evaluate the degree of in-plane ordering on a macroscopic scale, that is, grains which may have their c-axis aligned with that of the substrate, but may possibly be misoriented in the plane of the substrate surface. The (10$\bar{1}$4) plane reflections in the φ-scan show a 60° periodicity, characteristic of the wurtzitic symmetry around c-axis. The absence of any intermediate or randomly occurring peaks indicates that the GaN is well ordered in-plane and no misoriented domains exist. The φ-scan of the c-plane Al_2O_3 substrate shows its 3-fold symmetry around the c-axis, and the peaks are

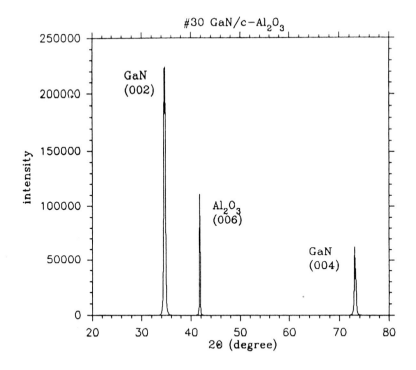

Fig. 1. XRD Θ-2Θ Plot of Wurtzitic GaN/(0001)Al$_2$O$_3$ Grown by PE-MOCVD

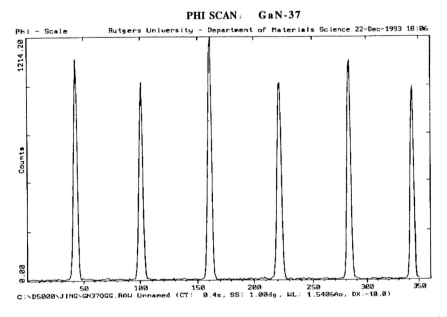

Fig. 2 X-ray Φ-scan of GaN/(0001) Al$_2$O$_3$ with 60° periodicity, indicating a high degree of in-plane ordering

displaced 30° relative to the GaN (10$\bar{1}$4) peaks as expected, indicative of an atomic epitaxial relationship between GaN layer and substrate. The X-ray φ-scan measurement on a series of GaN samples show that single crystalline GaN films can be grown with the growth temperatures from 620°C to 850°C by the PE-MOCVD.

Rocking curve measurements were performed on a series of GaN samples with both nitriding at 755°C and GaN buffers grown at 630-640°C. For growth between 800-850°C rocking curve FWHM values were between 31-32 arc min. It should be noted that these thin films are all sub-micron in thickness and therefore have broadened rocking curves. The dependence of the FWHM on layer thickness has been studied and found to decrease strongly in the thickness range from 2-5 μm[9]. It is expected that thicker PE-MOCVD GaN films will have greatly reduced FWHMs.

The surface morphology of the GaN layers were examined by scanning electron microscopy. The GaN layers on c-plane Al_2O_3 substrates showed an essentially featureless surface at 1000x magnification. However, at 30,000x magnification a fine, submicron surface texture is apparent. These results are similar to those of MBE grown GaN layers[10].

3.2 77K PL measurements

Photoluminescence (PL) measurements were carried out at 77K on undoped and Mg-doped GaN films using a pulsed nitrogen laser (337.1nm). Fig. 3 shows the results for an undoped GaN layer grown at 850°C on a (0001)Al_2O_3 substrate. The 77K PL is dominated by a strong, single peak at 3.45eV which is attributed to exciton emission of the wurtzitic GaN[11]. The only other peaks are small and probably the phonon replicas of the unresolved donor-acceptor recombination[11]. The most important features of this spectrum are the unresolvable donor-acceptor pair emission and the absence of the defects and impurity related deep level luminescence, implying good crystal quality and reasonable purity.

One further point should be made with respect to N_2 vs. NH_3 sources. The contamination levels for various impurities in ultra pure N_2 and NH_3 have recently been compared[12]. The overall purity level of ultra high purity N_2 is nearly 50 times higher than

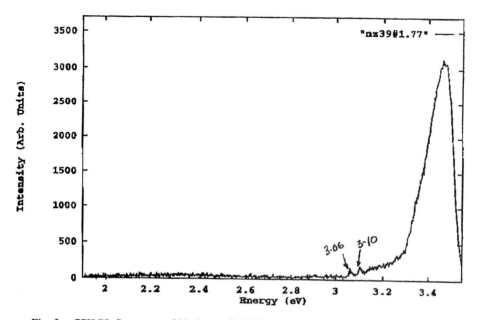

Fig. 3. 77K PL Spectrum of Undoped GaN/(0001) Al_2O_3 Grown by PE-MOCVD

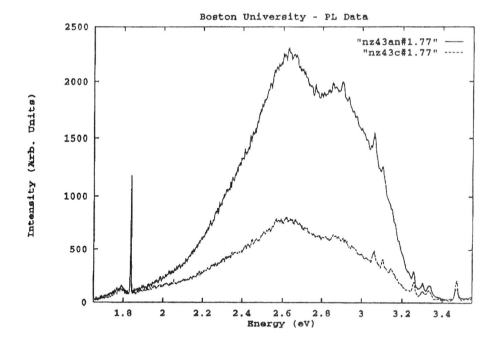

Fig. 4. 77K PL Spectrum of Mg-doped GaN Grown by PE-MOCVD at 645°C, the dotted curve is GN-43 [as-grown] ; solid curve is GN-43an [annealed]

the best available NH$_3$. Therefore, the N$_2$ plasma process used in this work offers the potential for higher purity GaN than those using NH$_3$.

Two 77K PL spectra from a Mg-doped GaN layer grown by PE-MOCVD are shown in Fig. 4. The wafer was grown at 645°C with a nitridation pretreatment but no low temperature GaN buffer layer because the growth temperature is already low. The bottom PL spectrum [dotted curve] is for an as-grown Mg-doped sample, GN-43. The upper PL spectrum [solid curve] is for an annealed sample cut from the same wafer. The annealing was done at 740° C in a nitrogen environment. It is readily apparent that annealing increases PL intensity by about three times without changing the spectral shape. This implies that annealing reduces non-radiative recombination without altering the radiative recombination mechanism.

Magnesium doping at 650°C produced a strong blue-green light emission which appeared very bright even at room temperature. This strong luminescence signal peaked at 2.60eV with a prominent shoulder at around 2.85eV. Several other samples also showed a second, lower-energy shoulder at 2.3eV. The latter two peaks are consistent with the observation by Dingle on Mg-doped VPE GaN[11], but to our knowledge the 2.60eV peak was not reported. It is speculated that Mg-doping in GaN grown at low temperature by PE-MOCVD is responsible for this radiative Mg-complex center.

3.3 Hall Effect Measurements

Hall effect measurements, using the van der Pauw method, were performed on GaN grown on c-plane Al$_2$O$_3$. A magnetic field of 0.7T was used and both current and field were reversed in the course of the measurement sequence. The measurements were performed at 300K. Undoped GaN was n-type with electron concentration in the mid-10^{17} to mid-10^{19} cm^{-3} range. This has been previously observed and has been associated with N$_2$ vacancies or oxygen incorporation. This includes samples with a wide variety of PL characteristics, including GN-39, which showed only excitonic PL and no deep levels. No apparent correlation was observed between optical and electrical properties for undoped samples.

The mobilities measured in this study were relatively low, with the highest values being from 18-80 cm^2/v-s at 300K. We believe that this is due to excess scattering from compensated residual impurities such as carbon due to inadequate N$^+$ or H$^+$ concentration. All Mg-doped samples were measured to determine carrier type and concentration. The initial run with Mg-doping, GN-42 [on r-plane Al$_2$O$_3$], resulted in resistivity of 6.5×10^5 ohm-cm, and a much reduced n-type carrier concentration of $\approx 10^{13}$ cm^{-3} after growth. After N$_2$ plasma annealing at 720°C for 30 min, resistivity decreased substantially, as indicated in Table 2.

Table 2.   Electrical Properties of Mg-doped GaN on Al$_2$O$_3$

Sample	Al$_2$O$_3$ Orient.	Resistivity [ohm-cm.]	Carrier Conc. [cm^{-3}]	Cond. Type	Mg-carr. flow [sccm]
GN-42	r-plane	6.5E+05	≈1E+13	n	5
GN-42an	"	1.1E+01	1.0E+17	n	5
GN-43	c-plane	high	-	-	7
GN-43an	"	1.0E+03	1.0E+15	n	7
GN-46	"	5.9E+01	1.3E+18	p	30

The Mg-doping level was increased, by increasing the Mg-carrier gas flow to 30 sccm, in run GN-46, p-type GaN with a low mobility of .08 cm^2/v-sec. was obtained. The low mobility is believed to be associated with compensated hole conduction. To the best of our knowledge, this is the first observation of p-type conductivity in GaN using a PE-MOCVD process and demonstrates the feasibility of using PE-MOCVD for the growth of GaN-based heterostructures.

Acknowledgments

The authors wish to thank Dr. Ping Lu of Rutgers University Dept. of Mechanics and Materials Science for the X-ray measurement. The work was supported by the Air Force Office of Scientific Research, Dr. Gerald F. Witt contract monitor.

References

[1] R.F. Davis, Z. Sitar, B.E. Williams, H.S. Kong, H.J. Kim, J.W. Palmour, J.A. Edmond, J. Ryu, J.T. Glass, and C.H. Carter Jr., Matr. Sci. Eng. B 1, 77(1988)
[2] S. Strite and H. Morkoc, J. Vac. Sci. Technol. B 10(4), 1237(1992)
[3] S. Nakamura, Y. Harada, and M. Senoh, Appl. Phys. Lett. 58, 2021(1991)
[4] Shuji Nakamura, Takashi Mukai, and Masayuki Senoh, Appl. Phys. Lett. 64, 1687(1994)
[5] B.J. Baliga, IEEE Electron. Dev. Lett. 10, 455(1989)
[6] H. Nomura, S. Meikle, Y. Nakanishi, and Y. Hatanaka,, J. Appl. Phys. 69, 990(1991)
[7] S.W. Choi, K.J.Bachmann, and G. Lukovsky, J. Matr. Res. 8, 847(1993)
[8] N. Newman, J. Ross, and M. Rubin, Appl. Phys. Lett. 62, 1243(1993)
[9] C.J. Sun and M. Razeghi, Appl. Phys. Lett. 63, 973(1993)
[10] T.D. Moustakas, T. Lei, and R. Molnar, Physica B, 185, 36(1993)
[11] M. Illegems and R. Dingle, J. Appl. Phys. 44, 4234(1973)
[12] W. Kern, Handbook of Semiconductor Wafer Cleaning Technology, Noyes Publications, Park Ridge, N.J., (1993), pp88-89

Large Area Growth of GaN Thin Films in a Multi-Wafer Rotating Disk Reactor

H. Liu*, S. Liang, T. Salagaj, J. Damian, C.S. Chern, and R. A. Stall
EMCORE Corporation, 35 Elizabeth Avenue, Somerset, NJ 08873

C-Y Hwang and W. E. Mayo
Department of Material Science and Engineering, Rutgers University, Piscataway, NJ 08855

Abstract. The rapid advancement in the area of III-V Nitride based materials and devices has stimulated great interest in the search of a high volume production technology. We have utilized multi-wafer MOCVD method utilizing high speed rotation to grow GaN thin films on (0001) sapphire substrates. GaN films with good thickness uniformity (±2%) and surface morphology were obtained. Typical room temperature background electron concentration is in the mid-10^{17} cm^{-3} with mobility of 200 cm^2/V-s. Room temperature photoluminescence showed a strong 356nm bandedge emission and a much weaker deep level band peaking at 575 nm. At 77K, two weak peaks at 364 nm and 375 nm were observed, which may be related to the compensation centers in the films.

1. Introduction

III-V nitrides are wide bandgap semiconductor materials that posses many unique properties. their high thermal and physical strength along with low electrical leakage and radiation hardness make these materials suitable for fabricating high temperature and high power electronic devices and circuits. Also, their direct bandgap ranging from 2.2 eV to 6.0 eV are promising for optical device applications in the full visible and ultraviolet (UV) spectra. Recently, high brightness light emitting diode (LED) emitting at 350 nm with 1.5 mW optical power has been made available commercially [1]. This opens up new opportunities in the applications of full color displays, road signs, etc., as well as, future potential for high density optical storage system when short wavelength (<50 nm) laser diode can be made from these materials.

Compared with the much more matured GaAs- or InP-based technology, III-V nitrides are still in their preliminary development stage and many basic material issues require engineering solutions. Namely, these include the lack of lattice matched substrates, nitrogen vacancies, p-type doing, etc. Research into these problematic areas has been increasingly active, and much progress has been resulted in the last few years [2,3,4,5]. Very high quality GaN films with room temperature electron mobility up to 200 cm^2/V-s have been reported by Nakamura et. al. [2]. They also reported p-type GaN with hole concentration as high as 3 x 10^{18}cm^{-3} using their Two Flow-MOCVD reactor [6]. These results have stimulated great interest in the search for a high throughput growth process and equipment for future production of III-V nitride based materials and devices.

In this paper we will describe a multi-wafer MOCVD growth process utilizing a high speed rotating disk to achieve high quality GaN films grown on (0001) Al$_2$O$_3$ substrates.

* On leave from EMCORE

2. High speed rotating disk MOCVD reactor

The gas dynamics of the high speed, low pressure, rotating disk reactor allows for the growth of uniform epitaxial layers over a large area [7,8]. Theoretical studies of this type of CVD reactor along with the flow visualization experiments have been reported [9,10]. A simplified schematic of a typical rotating disk reactor is shown in Fig.1. During the deposition process, the disk (wafer carrier) rotates at a high speed (500 - 1000 rpm) and a 'top flow' which is a mixture of reactant gases with a much higher main hydrogen flow is introduced from the top flange. The group III and nitrogen precursors mixed with hydrogen are transported into their own divided zones with flow distributions regulated by individual needle valves before each zone. The group III and nitrogen precursors do not mix before entering the chamber so the chemical reaction inside the top flange can be avoided. The distribution of 'top flow' in each zone is adjustable. This along with the wafer rotation results in highly uniform films. The high speed rotation also provides additional advantages over conventional MOCVD processes. Due to the relatively high speed between the gas flow and the wafer surface, the thermodynamic boundary layer thickness is much reduced, which results in higher growth rates while maintaining uniformities (thickness, doping, composition). These attributes are especially useful for the growth of GaN since the generation of gas flow recirculation due to the much higher gas viscosity (high ammonia flow rate) and the likely premature reactions (high growth temperature) between metalorganic precursors and ammonia can be effectively reduced, while maintaining high growth rate by using low growth pressure in conjunction with high speed rotation.

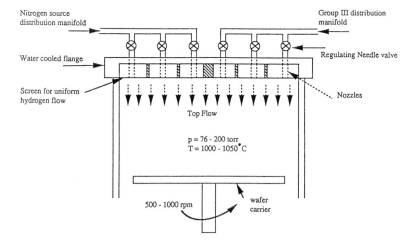

Fig. 1. A simplified schematic of high speed rotating disk MOCVD reactor for III-V Nitrides.

3. Experimental procedure

An EMCORE GS3300 MOCVD reactor with a process throughput of six 2" wafers per growth run was used. However, during this experimental period we have only used 1"-dia. size substrates. Sapphire substrates with (0001) orientation were first cleaned with solvents and followed by etching in 1:1 HNO_3: H_2O_2 solution, DI water rinse and blow dry with UHP nitrogen. The substrate was then loaded in the load-lock which was pumped down to 10^{-6} torr before transfered to the growth chamber. The sources used are trimethylgallium (TMG) and ammonia (NH_3, 100%). Growth experiments were performed under a reactor pressure 200 torr and a wafer carrier rotation speed at 1000 rpm. The growth process started with 10-minute substrate annealing at 1050 °C under H_2 ambient. The substrate temperature was then reduced to 540 °C and a 250 Å thick GaN buffer layer was grown. Immediately after the buffer layer growth, the substrate temperature was ramped up to 1030 °C for the GaN epitaxial growth. During the temperature ramping, NH_3 continued flowing to the growth chamber to avoid decomposition of the GaN buffer layer. The growth rate of the buffer layer and the epitaxial GaN layer were 1 μm/hr and 2.5 μm/hr, respectively. The resulted GaN film thickness was 2.5 μm. All samples were evaluated by Surface profilometry, Nomarski optical microscopy, Scanning Electron Microscopy (SEM), double crystal x-ray rocking curve (DCXRC), Hall measurement, and photoluminescence (PL).

4. Results

Highly uniform GaN films were obtained with thickness uniformity ±2% across 1" wafers. The surface morphology of the films were evaluated by both Nomarski optical microscopy and SEM, as shown in Fig. 2a and 2b. Some faint texture can be observed under low magnification (50X). The roughness is less than 100 Å measured by DEKTAK surface profilometer. At high magnification, as shown in Fig. 2b, the surface morphology is very smooth.

50 μm
(a)

2 μm
(b)

Fig. 2 (a) Surface morphology of a GaN film observed by Normarski optical microscopy, and (b) Same surface observed by SEM.

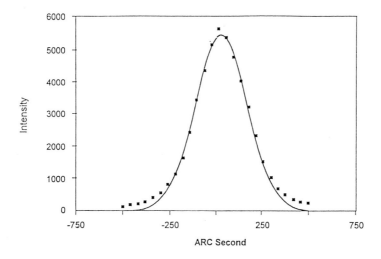

Fig. 3. Typical DCXRC of GaN films grown by high speed rotating disk MOCVD method.

DCXRC is a useful method to evaluate the crystallinity of thin films. A typical DCXRC result of the GaN films we grew is shown in Fig. 3. The FWHM is about 330 arc second. The electrical transport properties was measured by Hall effect using Van der Paul method. All the as-grown GaN samples exhibit n-type conduction with room temperature electron concentration in the mid-10^{17} cm^{-3} range and mobility of about 200 cm^2/V-s. However, the electron mobility decreased slightly when the samples were cooled down to 77K, indicating compensation. The origin of the compensation centers is not clear yet. Research into possible causes such as source purity, growth condition, etc., are currently under investigation.

The optical properties of the GaN films were evaluated by PL spectroscopy. At room temperature, as shown in Fig. 4a, band edge emission at 356 nm dominates the spectrum and a much weaker and broader peak at 575 nm is observed. A shoulder on the longer wavelength side of the band edge emission is also observed, indicating the presence of shallow donors or acceptors. The deep level emission at 575 nm has been commonly observed by other research groups [11,12] and its origin is still an open area to be further studied. At 77K, as shown in Fig. 4b, the ratio of the intensity between the band edge and the deep level emission increases significantly such that the deep level emission is not observable under the scale of Fig. 4b. The shallow energy states at 77K can be distinquished into two major components, peaking at 364 nm and 375 nm respectively. These impurities or defect levels may be related to the aforementioned compensating centers since no such shallow states were observed in some of the reported PL results obtained from GaN films with no compensation. More efforts are needed to further clarify the origin of these energy states.

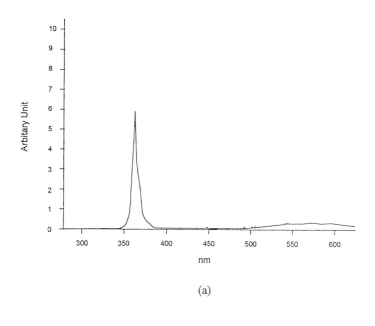

(a)

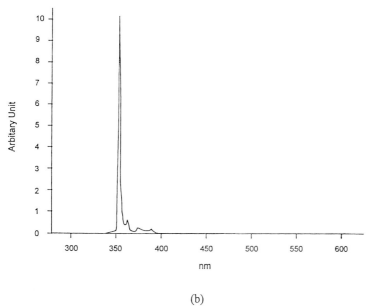

(b)

Fig. 4 Photoluminescence of GaN films measured at (a) room temperature, and (b) 77K

5. Summary

Highly uniform GaN thin films grown on (0001) sapphire substrates were obtained using a high speed rotating disk MOCVD reactor with high wafer throughput. Surface morphology is good with roughness less than 100 Å measured by surface profilometry. Material properties measured by DCXRC, Hall effect and PL indicate the GaN films grown by this method are of high quality. More studies are needed to explore the origins of the deep level emission, electrical compensation and the shallow energy states observed by PL measurements.

6. Acknowledgment

This work is supported by BMDO/IST under the contract N00014-93-C-0269, and is managed by Mr. Max Yoder. We also appreciate the support of surface profilometry measurement from Dr. Robert Anderson and Dr. Brian McDerott at Rockwell International Science Center, and photoluminescence measurements from Professor Dennis Wickenden at Johns Hopkins University.

References:

[1] Nakamura S., Mukai T., and Senoh M., 1994 Appl Phys. Lett. **64** (13), 1687-1689.
[2] Nakumura S., Mukai T., and Senoh M., 1992 J. Appl. Phys. **71** (11), 5543-5549.
[3] Amano H., Kito M., Hiramatsu K., and Akasaki I., 1989 Jpn. J. Appl. Phys. **28**, **L2112-L2114**.
[4] Detchprohm T., Hiramatsu K., Sawaki N., andAkasai I., 1994 J. Crystal Growth 137, 170-174.
[5] Rubin M., Newman N., Chan J.S., Fu T.C., and Ross J.T., 1994 Appl. Phys. Lett. **64** (1), 64-66.
[7] Evans G.H. and Grief R., 1987 J. Heat Transfer **109**, 928.
[8] Evans G.H. and Grief R., 1987 Numerical Heat Transfer **12**, 243
[9] Coltrin M.E., Kee R.J., and Evans G.H., 1989 J. Electrochem. Soc. **136**, 819 - 829.
[10] Breiland W., and Evans, G.H., 1991 J. Electrochem. Soc. **138**, 1806 - 1816.
[11] Akasaki I., andAmano H., 1990 SPIE **1361**, 138-149.
[12] Khan M.A., Olson D.T., and Kuzinia J. N., Carlo W.E., Freitas J.A., Jr., 1993 J. Appl. Phys. **74**, 5901-5903.

Heteroepitaxial Growth of GaN on GaAs by ECR plasma assisted MBE

Heon Lee, David B. Oberman, Werner Götz and James S. Harris Jr.
Solid State Electronics Laboratory, Stanford University, Stanford, CA 94305

Abstract. Heteroepitaxial GaN was grown on (100) GaAs by plasma assisted molecular beam epitaxy. The effect of GaAs surface treatment and low temperature grown buffer layers are investigated. Optimized growth conditions are obtained using X-ray rocking curve and Raman Spectroscopy. High resolution cross-sectional TEM photographs show the GaN films are defective near the interface and have columnar domains. Only the (0002) peaks of wurtzite GaN are observed in X-ray Diffraction patterns and the FWHM of (0002) GaN peak is about 720 seconds. We investigate the effect of heat treatment of GaN films at trmperatures between 600°C to 1000°C for 30 minutes. It is observed that the crystalline quality of the GaN films is relatively little changed by post-growth heat treatment.

1. Introduction

GaN and other III-V nitride semiconductors (AlN, InN and their alloys) have direct wide bandgaps and reasonable electron mobility. They thus have great potential for the fabrication of optoelectronic devices operating in the blue to UV region and electronic devices operating under harsh environments and high temperatures. Some of their potential applications have already been realized. Two new commercial products using these materials are Nichia's blue light emitting diode[1] and APA Optic's UV sensors.

Despite the attractive features of these materials, they have been almost unexplored until recently and there are a number of materials problems which have not been solved and understood yet. These problems are a) the absence of a substrate that has a good lattice and thermal expansion match to any of these nitrides, b) the nature of structural and intrinsic electronic defects, c) the nature of both intentional and unintentional impurity species and d) the very high equilibrium pressure of nitrogen over GaN[2], so that GaN must be grown under non-equilibrium conditions. The lattice mismatch between the nitride film and the substrate causes misfit dislocations, stacking faults and other crystalline defects. Complete incorporation of nitrogen atoms into the growing film is also very difficult, especially at the typical growth temperature.

One of the key issues for MBE is the nitrogen source. We have chosen an electron cyclotron resonant (ECR) plasma source to provide chemically reactive nitrogen ions and radicals. In this ECR-MBE technique, the nitride film is grown by the reaction of elemental Ga with an activated atomic nitrogen species produced by dissociation of nitrogen gas in which is a trade-off between the ECR plasma discharge and the plasma energy. One of the major issues for this technique is the ion damage in the film. Compared to the MOCVD technique, ECR assisted MBE technique has advantages including lower growth temperature, availability of in-situ monitoring tools and absence of hydrogen, which is believed to passivate p-type dopants[3, 4], and enhanced surface diffusion of adatoms on the growing film due to the lower growth pressure.

2. Experimental

The growth system used in this study is a modified Varian GEN II MBE system with a Wavemat^{TM} electron cyclotron resonance (ECR) plasma source. Because the V/III ratio may be very important, this source was positioned in the center of the source flange (normal "pyrometer port") in order to obtain a uniform flux of activated nitrogen over the entire substrate surface. Group III elements are evaporated from standard effusion cells. During the growth, the nitrogen pressure is typically a few times of 10^{-4} torr, which is still in the molecular flow regime.

Semi-insulating (100) GaAs wafers were chemically cleaned, prior to loading. Wafers were baked at 500°C for an hour in the loading chamber to drive off water and then transferred to the growth chamber. The substrate was exposed to the nitrogen plasma at 740°C for 30 minutes for native oxide removal and surface nitridation. A 400 Å thick GaN or AlN buffer layer was grown at 400°C and the substrate temperature was then raised to the normal growth temperature (650 - 740°C). After the growth, some films were heat-treated in a quartz tube furnace under N_2 ambient between 600°C and 1000°C for 30 minutes to investigate the stability and potential improvement in crystalline quality of the GaN films. The typical growth conditions for the GaN films grown with ECR plasma are described in Table 1.

substrate temperature	650 - 740°C
nitrogen flow rate	10 - 20 scam
Ga beam flux	$2 - 3 \times 10^{-7}$ torr
Plasma power	150 - 200 W
chamber pressure	1.9×10^{-4} torr

Table 1. Typical growth conditions for GaN films

The growth rate was about 0.15 - 0.2 μm/hr, and was not appreciably changed by the range of growth parameters in Table 1. Some of the films showed ring-type patterns from ion damage due to the toroidal plasma.

3. Results and Discussion

Figure 1-(a), a high resolution cross-sectional TEM photograph of the GaN film, shows the GaN grown on (100) GaAs is wurtzitic. Near the interface, the film contained a very high density of dislocations and stacking faults. A small cluster of zincblende GaN was observed, but the GaN film showed mainly wurtzite stacking sequence. Figure 1-(b), cross-sectional TEM photograph at lower magnification, shows that the GaN film consists of columnar domains and that each domain is relatively free of defects.

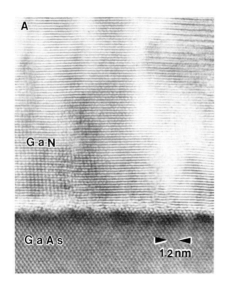

Fig. 1. Cross-sectional TEM photograph of GaN film on GaAs
(a) HRTEM (b) at lower magnification

A typical X-ray diffraction pattern is shown in Figure 2. The X-ray diffraction pattern shows only (0002) peaks of wurtzitic GaN with a FWHM of 720 seconds. An example of Raman spectrum obtained in the back scattering configuration along the z-axis is shown in Figure 3. The longitudinal A_1 mode at 730 cm^{-1} and the E_2 mode at 565 cm^{-1} are both observed, in correspondence with selection rules for wurtzitic GaN[5]. These results shows the GaN film is oriented along the z-axis.

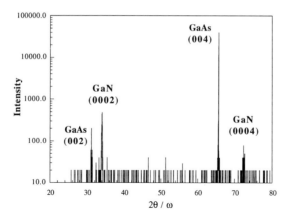

Fig. 2. X-ray Diffraction Pattern of GaN film on GaAs

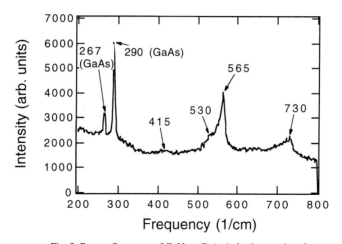

Fig. 3. Raman Spectrum of GaN on GaAs in backscattering along z-axis

To verify the composition of the film, we analyzed the film by X-ray Photoelectron Spectroscopy (XPS). The top 100Å of the surface was sputter etched in vacuum prior to the analysis to remove surface contaminants. The XPS spectra revealed that the GaN film contained about 7.4% oxygen and the N to Ga ratio was 0.933, indicating the presence of nitrogen vacancies. No obvious Si peak was observed, which means Si content should be less than 0.1 atomic %.

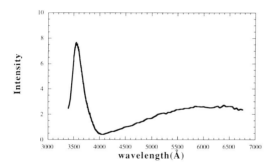

Fig. 4. Photoluminescence Spectrum of GaN on GaAs at 77°K

The optical properties of GaN were investigated using low temperature photoluminescence. A 325nm HeCd laser (10mW power) was used to excite electrons-hole pairs. As shown in Figure 4, there is also very broad band luminescence around 550nm which is likely due to several deep level intrinsic defects, impurities or defect-impurity complexes. The sharp bandedge luminescence is somewhat surprising in light of the high oxygen concentration shown by XPS.

Post-growth heat treatment was carried out under N_2 ambient at various temperature in order to investigate the crystalline quality of GaN films. These samples were measured by X-ray diffraction after annealing and the FWHM of the GaN (0002) rocking curves are shown in Figure 5. Unfortunately, there is only slight improvement of the film crystal quality.

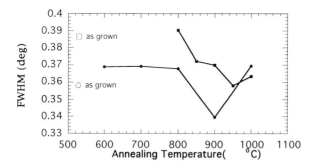

Fig. 5. Effect of annealing temperature on FWHM of GaN (0002) peak

4. Conclusions

GaN films grown on (100) GaAs by ECR assisted MBE showed the wurtzite structure and consisted of columnar domains. The activated atomic nitrogen flux limits the growth rate to a very low level (0.15 - 0.2 µm/hr). The plasma also gives rise to a fairly high concentration of silicon and oxygen contamination to the film due to sputtering of the SiO_2 plasma cup. Only (0002) wurtzite peaks were observed by X-ray diffraction. XPS studies revealed a considerable amount of oxygen impurities in the film and the absence of a 1 to 1 stoichiometry in the film composition.

Acknowledgment

This Study was supported by ONR/ARPA under the contract No. N00014-93-1-1375 and by CNOM under the contract number N00014-92-J-1903.

Reference

1. S. Nakamura et. al., Applied Physics Letters, 64, 1687(1994)
2. J. Karpinski, J. Crystal Growth, 66, 1(1984)
3. Pearton, S.J., et al. Electronics Letters, vol.30, no.6, 527(1994)
4. J. van Vechten, Jap. J. Appl. Phys., Part 1, 31, 3662(1992)
5. T. Kozawa et. al., J. Appl. Phys., 75, 1098(1994)

MBE Growth of GaN with ECR Plasma and Hydrogen Azide

D. B. Oberman, H. Lee, W. K. Götz, J.S. Harris, Jr.

Solid State Electronics Laboratory, Stanford University, Stanford, CA 94305

Abstract. GaN films were grown on (0001) sapphire by molecular beam epitaxy (MBE). Two sources of activated nitrogen were investigated: an electron cyclotron resonance (ECR) plasma, and hydrogen azide (HN_3). With the ECR plasma source, typical growth rates were ~0.1 µm/hr. Films grown in this manner showed significant surface damage from ions, and very little if any photoluminescence. With HN_3, growth rates were ~0.25 µm/hr. Azide-grown films showed smooth surfaces, and sharp band-to-band photoluminescence. This is the first reported use of HN3 to grow III-V nitrides by MBE, and it shows great promise as a nitrogen source.

1. Introduction

The growth of III-V nitrides poses particular difficulties for molecular beam epitaxy (MBE). Due to the high equilibrium partial pressure of nitrogen over these materials at typical growth temperatures, a very active source of nitrogen is needed in order to achieve film growth while simultaneously maintaining the ultra-high vacuum conditions necessary for MBE [1]. In this study we compare two different sources of active nitrogen- a plasma, in which nitrogen gas is physically exicted, and a highly reactive nitrogen-containing compound, hydrogen azide.

2. Growth with ECR Plasma

2.1. Experimental Procedure

All samples in this study were grown in a modified Varian GEN II MBE machine. Standard MBE effusion cells were used for the Group III elements. Both a nitrogen plasma and hydrogen azide gas were used as nitrogen sources. The plasma was generated by a Wavemat™ electron cyclotron resonance (ECR) plasma source which sits in the center of the source flange, normal to the substrate (Figure 1).

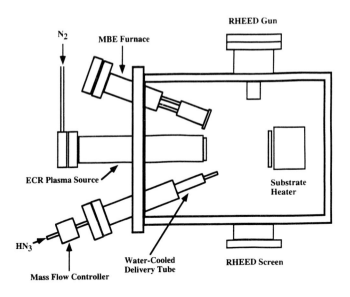

Figure 1. MBE chamber with ECR plasma source.

Sapphire wafers with the (0001) orientation were degreased in solvents, placed into a 160 °C mixture of H_2SO_4: H_3PO_4 (3:1) for 15 minutes, and rinsed in deionized water for 10 minutes. After being blown dry with nitrogen gas, they were indium-bonded to molybdenum block substrate holders. The wafers were then loaded into a load chamber attached to the MBE system. After the load chamber was pumped down to ~10^{-6} torr, the wafers were baked at 500 °C for one hour in order to drive off water.

Wafers were then loaded into the MBE chamber. Prior to growh, they were heated to 900 °C to remove impurities from the surface. They were then cooled to 400 °C. The plasma was started, and wafers were exposed to the plasma for 30 minutes in order to nitridize the surface. A low-temperature (400 °C) buffer layer of ~200 Å was then grown. Conditions for the remainder of the growth were varied according to Table 1 below.

Substrate temperature	600 °C - 860 °C
Nitrogen flow rate	10 - 20 sccm
Plasma power	150 - 250 W
Ga beam equivalent pressure	1 - 3 x 10^{-7} torr
Plasma pressure	> 10^{-4} torr
Growth time	6 - 12 hrs

Table 1. Growth conditions with ECR Plasma

2.2. Experimental Results

Films grown with the ECR plasma were found to be single crystal GaN, with a high concentration of defects. A typical X-ray diffraction spectrum, with an FWHM of ~20 arcminutes, is shown below in Figure 2. Growth rates were determined by profilometry to be approximately 0.1 μm/hr. Many of the films grown with the ECR plasma displayed significant ion damage. In particular, several showed ring-shaped damage patterns corresponding to the toroidal plasma.

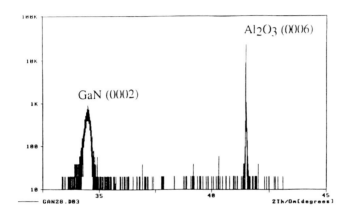

Figure 2. X-ray diffraction spectrum of GaN on Al_2O_3

3. Growth with Hydrogen Azide

3.1. Hydrogen Azide (HN₃)

Hydrogen azide (HN_3), the most reactive of the hydronitrogens, is particularly well-suited to growing nitrides, as it decomposes at relatively low temperatures (~300 °C) to form an N_2 molecule and a metastable HN (nitrene) radical with two dangling bonds. This radical provides a ready source of active nitrogen. Hydrogen azide also has the lowest hydrogen-to-nitrogen ratio of any hydronitrogen; this could have important consequences, as hydrogen is believed to passivate p-type dopants in GaN [2].

Hydrogen azide must be handled with great care, as it is both highly toxic and explosive. It is most prone to explode when condensed. For that reason, it cannot be pumped with cryopumps; turbopumps must be used instead. When stored at low pressure (~100 torr), hydrogen azide has been shown to be stable for months [3]. Due to its corrosive nature, HN3 can only be used with compatible materials, such as stainless steel and Viton™.

As hydrogen azide is not commercially available, it must be synthesized by the user in the laboratory. An apparatus was constructed for generating, storing, and dispensing HN3 (Figure 3). For safety reasons, the entire apparatus sits inside a Semi-Gas™ gas cabinet. To form HN3, 20 grams of sodium azide (NaN3) are mixed with 1 kg of stearic acid (CH3(CH2)16COOH) in a stainless steel reaction chamber. The chamber is then evacuated and heated to 100 °C. After pumping on the chamber for 30 minutes to remove impurities, the hydrogen azide gas evolved by the reactin is collected in an evacuated stainless steel cylinder until it reaches a pressure of 100 torr. The reaction is allowed to go to completion, and the remainder of the HN3 gas that is generated is pumped away into the exhaust.

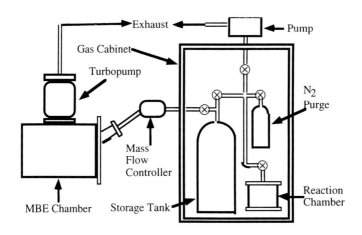

Figure 3. Hydrogen azide generation, storage, and dispensing system.

3.2. Experimental Procedure

Sapphire substrates were prepared and loaded as described in the previous section. They were heated to 900 °C for 30 minutes and then cooled to 400 °C. HN3 was then introduced into the MBE chamber through a water-cooled delivery tube in one of the furnace ports. Each substrate sat in the HN3 flux for 30 minutes in order to nitridize the surface of the sapphire and form a thin layer of AlN. The gallium shutter was then opened. After a low-temperature (400 °C) GaN buffer layer of ~200 Å was grown, the substrate was heated to the final growth temperature. After 8 hours of growth, the substrate was cooled and the Ga and HN_3 fluxes were shut off. Growth parameters are listed in Table 2.

Substrate temperature	600 °C - 860 °C
Ga beam equivalent pressure	$1 - 3 \times 10^{-7}$ torr
HN_3 flow rate	1 - 3 sccm
HN_3 beam equivalent pressure	$1 - 5 \times 10^{-5}$ torr
Growth time	8 hours

Table 2. Growth conditions with Hydrogen Azide

3.3. Experimental Results

Gallium nitride films grown with hydrogen azide were much smoother and more uniform than those grown with an ECR plasma. They displayed none of the surface damage that the plasma-grown films showed. Growth rates were measured to be approximately 0.25 µm/hr, or almost three times those attained with the plasma source.

Photoluminescence (PL) measurements were peformred at 77 K using a 10 mW He-Cd laser operating at 325 nm. Films grown with hydrogen azide showed a strong PL peak at 360 nm, corresonding to a band-to-band transition, and very little deep level luminescence. A typical spectrum is shown in Figure 4.

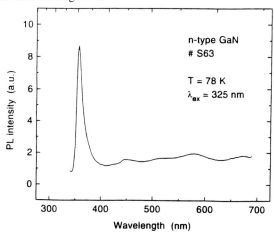

Figure 4. Photoluminescence of GaN on Al_2O_3 Grown with HN_3

4. Conclusions

A new nitrogen source, hydrogen azide, has been successfully used to grow gallium nitride films by MBE. To our knowledge, this is the first reported use of HN3 to grow III-V nitrides by MBE. Films grown with hydrogen azide show better surface morphology and much stronger photoluminescence than films grown with a plasma source, and growth rates are almost three times higher. Hydrogen azide is thus an extremely promising nitrogen source for use in the MBE growth of III-V nitrides.

5. Acknowledgments

This work was supported by ONR/ARPA under contract no. N00014-93-1-1375 and by CNOM under contract no. N00014-92-J-1903.

6. References

[1] Newman N et al 1993 *Applied Physics Letters* **62** 1242-4
[2] van Vechten J 1992 *Jap. J. Appl. Phys., Part 1* **31** 3662-3
[3] Dows D et al 1954 *J. Chem. Phys.* **23** 1258-60

Growth Mechanics of GaN by Monte Carlo Modeling

Kun Wang, Jasprit Singh and Dimitris Pavlidis

Department of Electrical Engineering and Computer Science, The University of Michigan, Ann Arbor, MI 48109-2122, USA

Abstract The growth mechanics of GaN are studied by Monte Carlo modeling. An atomistic model has been developed for this purpose and MBE-like growth conditions were assumed. The growth kinetics suggest that GaN growth surfaces are likely to be Ga stabilized. The impact of substrate temperature, Ga flux and N/Ga ratio on growth front quality is analyzed. The results indicate that there is a temperature window and a N/Ga ratio window in which a good growth front could be obtained while preserving a reasonable growth rate. 2D and 3D growth front contours evaluated under various growth conditions are used to analyze the growth of GaN.

1 Introduction

Recent experimental studies of GaN resulted in very encouraging results in terms of crystallographic and optical characteristics, as well as, application of this material to the realization of devices such as Light-Emitting-Diodes[1]. In spite of this progress there is a lack of detailed understanding of the mechanics involved in GaN growth and the exact role of growth parameters in determining the quality of the resulting material. This paper addresses these issues by Monte Carlo modeling.

The thermodynamics, kinetics, and the microscopic details of incorporation of atoms play very important roles in controlling the quality of growth[2]. To fully understand these important issues, an atomistic understanding of GaN growth must be developed. In this paper, a model for nitrogen incorporation in molecular-beam-epitaxy-like (MBE-like) growth of GaN is presented. A theoretical study based on Monte Carlo techniques is carried out. The importance of group V element to group III element (V/III) flux ratios, growth temperatures, and Ga flux on the growth quality of GaN are examined.

2 Theoretical Model

A first attempt in simulating the growth of GaN can be made by considering molecular beam epitaxy (MBE)-like conditions. Electron-cyclotron-resonance MBE (ECR-MBE) like conditions were chosen for our simulations because of the relatively simple processes involved in this case. In ECR MBE, the ECR source is used to produce the atomic nitrogen and Gallium comes from the conventional Knudsen-effusion-cell[3]. (001) GaAs has been selected as the substrate for the simulation of the growth reported here due to the importance of cubic GaN in device applications. The calculated trends are, however, of more general nature and may be employed as general guidelines for understanding GaN growth.

Work supported by ONR contract No: N00014-92-J1552

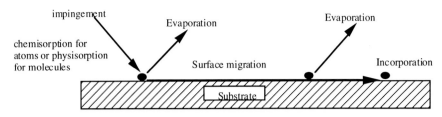

FIG 1 Schematic representation of "MBE-like" GaN growth processes

The lattice gas model using Monte Carlo techniques has been quite successful in understanding the atomistic nature of the growth of III-V semiconductors[2], such as GaAs, AlAs, and InAs, and II-VI systems as in HgTe-CdTe[4]. With some important modifications discussed below, we follow this formalism for the GaN system.

The dynamics of crystal growth from the vapor could be simulated by four basic events: impingement, surface reaction, surface migration, and evaporation. In Figure 1, we show a schematic representation of the MBE growth process for GaN, in which we assure that atomic species are chemisorbed while molecular species are physisorbed on the surface[5]. The kinetic processes are described by the evaporation and diffusion rates. The evaporation rate R_e^i is taken to be of an Arrhenius form:

$$R_e^i = R_{oe}^i \exp(-E_{evap}^i / k_B T) \qquad (1)$$

where i is the site for the cation or anion under question; R_{oe} is a prefactor for evaporation and E_{evap}^i is the activation energy of the atom in the site i.

Migration rates are taken to be in an Arrhenius form as well:

$$R_d^i = R_{od}^i \exp[-(E_{tot}^i - \Delta) / k_B T] \qquad (2)$$

where R_{od} is a prefactor; E_{tot}^i is the total energy at site i; Δ is a parameter which is related to the energy adjustment for a particular migration process and ($E_{tot}^i - \Delta$) is the activation barrier for migrations. Since nitrogen atoms have very high probability of migration, the limiting kinetics controlling the growth are the cation(gallium) surface kinetics; a more detailed discussion about this will follow later on in this section. Migration rates were therefore considered only for Ga. Four migration processes were considered: (a) hopping on the same surface layer in a direction which is defined by the intercept of the surface and orbital planes ; (b) hopping on the same surface layer in a direction which is perpendicular to the intercept line of the surface and orbital planes; (c) hopping to the lower layer; (d) hopping to the upper layer. The energy barrier for hopping to occur by mechanism (a) is equal to breaking half of its bonds with the atoms in the layers below it. The energy barrier for process (b) is equal to breaking 6 out of the 8 existing bonds in case of a kink site. In general, the energy barriers depend on the number of neighbors that each atom has. Figure 2 shows intralayer hopping, where $E_\parallel$ is the energy adjustment for intralayer hopping.

Since the nitrogen evaporation rate is extremely high, particularly at the high temperature usually required for growth, the residence time of a nitrogen atom is much smaller than that of Ga[6]. Furthermore, experimental evidence from growth of GaN indicates that the sticking coefficient of Ga is small (0.25) but not zero, and any attempt to increase the Ga flux leads to a Ga rich film[7]. An important point that emerges from these results is that the growing surface of GaN may very likely be cation-stabilized. This would then result in non-stoichiometric GaN film which could account for the often

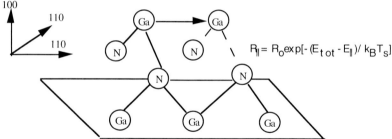

FIG 2 Hopping in the same layer ($E_{\parallel}$ is the activation barrier)

observed high unintended carrier concentrations. Furthermore, the incorporation of Ga appears to be limited by the supply of nitrogen radicals at the growth surface. It could consequently be expected that the incorporation of Ga and N are related. Accordingly, a tentative model for N incorporation could be drawn and is discussed below.

When the nitrogen atom impinges, it could reach thermodynamic equilibrium with the surface in a short time. During the residence time of the absorbed nitrogen atom, only if at the same time a Ga atom comes to its top, and forms a bond which "traps" the nitrogen atom, the nitrogen atom would be incorporated; otherwise the nitrogen atom would re-evaporate very rapidly. As a result, a tetrahedral binding structure will be formed in the [100] direction, where each Ga traps two nitrogen atoms. Another possibility would of course be N incorporation by the formation of surface N-N bonds. These are, however, much weaker than the Ga-N bonds and it is therefore the latter that are favored. Overall, we can consequently visualize this as a process where Ga keeps the N down on the surface.

Let us next examine the incorporation probability of nitrogen since it is crucial for increasing the efficiency of GaN growth simulation. The residence time of impinging nitrogen atoms is inversely proportional to the evaporation rate. The higher the evaporation rate, the shorter the residence time, and the smaller the probability that a nitrogen atom would be trapped. In addition, the more nitrogen atoms impinge, the higher are the chances for the nitrogen atoms to meet gallium atoms, and consequently the probability that they could be trapped by gallium atoms increases. The incorporation of nitrogen can consequently be described by a surface site occupation probability function $P_N(E_{tot})$. It will depend on both the site residence time and the available flux. For each of the growing sites i, this is defined through thermodynamics by:

$$P_N(E_{tot}) = J_N / R_e(E_{tot}) \qquad (3)$$

where J_N is the nitrogen flux, and $R_e(E_{tot})$ is the evaporation rate for the site with energy E_{tot}. For the site under question for which the probability of N incorporation is evaluated, a random number is generated. This is then used to describe whether or not the N atom is incorporated during the chemisorption of a Ga atom; N was considered to be incorporated only if P_N was larger than the randomly generated number.

Ga incorporation depends on the availability of two nitrogen atoms on the surface. The formation of two bonds between the Ga and the two N atoms ensures Ga incorporation. Thus incorporation is limited by the supply of nitrogen radicals at the grown surface. The nitrogen incorporation probability function was thus used to evaluate the Ga incorporation.

The above probability function was also used to evaluate Ga migration. Since Ga forms two bonds with the N atoms of the layer below it, migration of Ga requires that two such bonds are available at the site which it will move to. To estimate this availability we employed the probability function of Eq.3.

Based on the above considerations, a special Monte Carlo simulation was carried out. A 30x30 site area with periodic boundary condition is chosen for the simulation and the results obtained are discussed in the following section. The computing time depends on the substrate temperature chosen for the simulation.

3 Results of growth simulation and Discussion

3.1 The impact of N/Ga flux ratio on the growth rate

First, the relation between the growth rate and V/III ratio, as well as, the substrate temperature was evaluated. Fig.3 shows that the growth rate increases with the V/III ratio, reaching a saturation value which is determined by the Ga flux (not shown in Fig.) and is independent of temperature.

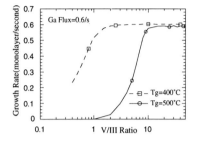

FIG 3 Growth rate vs. V/III ratio (Ga flux=0.6/s)

The V/III ratio at which this saturation value is reached is determined by the substrate temperature. Furthermore, since the growth rate is determined by the Ga flux, an increase of V/III ratio is accompanied by an enhancement in the Ga incorporation. It also shows front that a critical V/III ratio is required at each growth temperature for the growth rate to become significant.

3.2 The impact of the substrate temperature on the growth front

Figs.4 a and b, show the plots of [nth layer coverage(dq_n)]/[total coverage(dq_{tot})] as a function of growth time for constant Ga flux of 0.61/s. Here, q_n is the number of atoms contained in the n-th layer. Similarly, q_{tot} is the number of atoms contained in all the grown layers; this refers to all layers completed and uncompleted from the point of view of growth. Thus, dq_n/dq_{tot} represents the ratio of the additional incorporated atoms in the n-th layer with respect to the total incorporated atoms within a time interval dt. The significance of the dq_n/dq_{tot} ratio is in fact very similar to that obtained from RHEED oscillations. "Smooth" growth takes place when layer-by-layer growth occurs. In case of true layer-by-layer growth, the growth front depicted in Fig.4 should show no overlap between successive layers n. It is obvious that the larger the overlap, the rougher the growth front is and that eventually at extreme cases a three-dimensional front will be observed. As one can see, at a given V/III flux ratio, and gallium flux, the growth

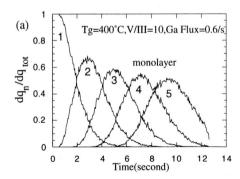
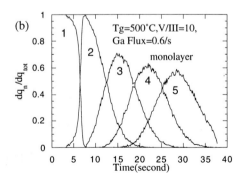

FIG 4 Ratio of nth layer coverage over total coverage vs growth temperature a(T=400°C, V/III=10. Ga flux=0.61/s); b(T=500°C, V/III=10. Ga flux=0.61/s)

is better at higher temperatures where surface migration is high. Surface atoms move in this case rapidly to kink sites and edge stepsand growth takes place by a layer-by-layer mechanism. The accompanying increased evaporation rate of Ga at high temperatures does not in this case play an important role since migration is the dominant mechanism. In other words, the high surface migration occurring at elevated temperatures improves the quality of the growth front. However, as the temperature increases, the growth rate decreases rapidly. Furthermore, as the temperature goes beyond certain point, entropy controlled effects will cause a poor quality of film due to high defect densities. From this point of view, one can say that there is a temperature "window" in which a good growth front can be obtained with reasonably high growth rates.

3.3 The impact of V/III flux ratio on the growth front:

For the study of the impact of V/III flux ratio on the growth front, the growth temperature was kept constant, and the Ga flux was kept at 0.6/sec. The contours are shown in Fig.5 ,a and b, respectively for V/III=1,000 and V/III=10,000. As one sees, the growth front becomes rougher at high V/III ratios. The growth front of the V/III=1000 case corresponds to 3 MLs, a situation which is close to a layer-by-layer growth mechanism. When the V/III ratio increases to say 10,000, the growth front corresponds to 6 MLs. The growth rate is about 2 seconds per monolayer for the V/III=10000 case, and becomes 5 seconds per monolayer for the V/III=1000 case. A low V/III ratio corresponds to a good growth front. However, if the V/III ratio is too low, the growth rate becomes too low for practical applications. The larger growth rate at high V/III ratios can be understood by the model described in Sec. II. One could see in that section that an increase of group V atoms increases the probability of nitrogen to meet gallium atoms and be trapped by them. The degradation of growth front at high V/III ratios can be explained as follows. As we have seen, the increasing of V/III ratio results in increased incorporation of gallium and nitrogen.

(a) (b)

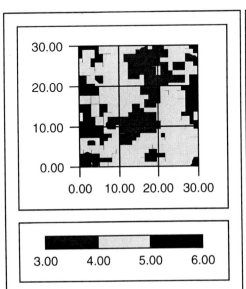
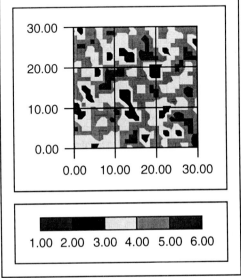

FIG 5 Growth front contours[T=600°C]: (a) V/III=1,000; (b) V/III=10,000; The legend on the botttom gives the number of surface monolayers.

In this condition, the growth may happen at free sites, which leads to forming island structures. As a result, the growth becomes 3-D and the growth front becomes rougher.

3.4 The impact of Ga and N flux on the growth rate and the roughness of the growth front

Simulations of the dependence of growth rate on Ga flux were made at a constant temperature of 600°C. The results show that the growth rate is dependent on the Ga flux. As the Ga flux increases the growth rate increases. However, as the Ga flux increases, the growth front becomes rougher. This indicates that as the gallium incorporation increases, 3-D growth starts taking place.

The impact of N flux was finally discussed in section 3.3 where the role of V/III ratio was examined. The results indicate that for given Ga flux, the higher the N flux is, the rougher the growth front.

4 Conclusions

The growth processes of GaN were simulated by Monte Carlo approaches. The growth rate and the quality of the growth front are determined by the substrate temperature, Ga flux, and V/III ratio. At a given temperature and V/III ratio, the lower the Ga flux, the better the growth front. At a given temperature, the lower the V/III ratio, the better the growth front, but the lower the growth rate. There is a temperature "window" in which a good growth front can be obtained with reasonably high growth rates. As-grown GaN may be associated with a cation stabilized surface. This could account for the high intrinsic carrier concentration often observed in this material. Experimental verification of the suggested Ga-stabilized surface conditions would shed the light on the growth mechanisms of GaN.

Acknowledgments

The authors would like to thank Max Yoder of ONR for his support and encouragmernt and Chang-Hee Hong, Kyushik Hong and Marcel Tutt for helpful discussions.

References

1 S. Strite and H. Morkoc, J. Vac. Sci. Technol. B 10(4),1237(1992)

2 J. Singh, Rev. Solid State Sci., 4, 785(1990)

3 C.R. Eddy Jr,and T.D.Moustakas, J. Appl. Phys. 73(1),448(1993)

4 J. Singh and J. Arias, J. Vac. Sci. Technol.,A7,2562(1989)

5 J. R. Arthur, Surf. Sci, 43, 1999 (1974)

6 C.D. Thurmond, J. Electrochem. Soc. 119,622(1972)

7 S. Strite, et al, J. Vac. Sci. Technol. B9 1924(1991)

InAs epitaxial nanocrystal growth on Se-terminated GaAs(001)

Yoshio Watanabe, Fumihiko Maeda, and Masaharu Oshima

NTT Interdisciplinary Research Laboratories
3-9-11 Midori-cho, Musashino-shi, Tokyo 180, Japan

ABSTRACT: We studied InAs epitaxial growth by molecular beam epitaxy on a Se-terminated GaAs surface as well as on an As-stabilized GaAs surface by *in situ* synchrotron radiation photoelectron spectroscopy (SRPES), *in situ* reflection high-energy electron diffraction, high-resolution scanning electron microscopy, and cross-sectional transmission electron microscopy to evaluate the electronic and structural properties of the InAs epitaxial nanocrystals. While an InAs layer grew on the As-stabilized GaAs substrate in the Stranski-Krastanov growth mode, InAs nanocrystal islands formed on the Se-terminated GaAs surfaces at the very early growth stages. InAs nanocrystals with a lateral width of about 70-80 nm, a density of about 2×10^9 cm^{-2}, and a height of about 20 nm were reproducibly grown on Se-terminated GaAs. The SRPES deconvoluted results indicate that although band bending is reduced by the Se treatment, it reappears at the heterointerface between InAs nanocrystals and the Se-treated GaAs surface during InAs deposition due to the lattice mismatch.

1. INTRODUCTION

Semiconductor nanocrystals grown on semiconducting substrates are expected to exhibit quasi-zero-dimensional quantum effects, and are attractive materials for advanced optoelectronic devices such as extremely efficient semiconductor lasers. There have been several reports on fabricating this type of nanocrystal without using photolithography, dry etching and regrowth. These methods involve growing fractional monolayers on tilted substrates (Brandt *et al* 1991), growing a lattice-mismatched epilayer by hydride vapor phase epitaxy (Ahopelto *et al* 1992), by molecular beam epitaxy (MBE) (Leonard *et al* 1993), or by metalorganic chemical vapor deposition (Notzel *et al* 1994), and growing lattice-matched GaAs microcrystals on sulfur-terminated AlGaAs by utilizing droplet formation with sequentially supplied Ga and As molecular beams, that is, a migration-enhanced epitaxy (MEE) mode (Koguchi *et al* 1991). We have recently demonstrated the inertness of Se-terminated GaAs surfaces against impinging Sb atoms at more than 200°C by using *in situ* synchrotron radiation photoelectron spectroscopy (SRPES) and reported evidence for a two-dimensional arrangement of InSb nanocrystals on a Se-terminated, terraced GaAs substrate by using a conventional MBE growth mode (Watanabe *et al* 1994a). Furthermore, the results of *in situ* SRPES showed that the MBE-grown InSb nanocrystals are stoichiometric, whereas MEE-grown ones are not (Watanabe *et al* 1994b).

In this work, we studied InAs epitaxial growth by MBE on a Se-terminated GaAs surface as well as on an As-stabilized GaAs surface by *in situ* SRPES, *in situ* reflection high-energy electron diffraction (RHEED), high-resolution scanning electron microscopy (HRSEM), and cross-sectional transmission electron microscopy (XTEM) to evaluate the electronic and structural properties of the InAs epitaxial nanocrystals. Furthermore, InSb growth was examined for reference by *in situ* SRPES because, in this heterostructure system, where the group-III and -V atoms are different from the atoms of the substrate materials, the chemical states of the InSb epitaxial crystals can be selectively evaluated from the SRPES spectra.

2. EXPERIMENTAL

The MBE growth and Se-treatment were performed in a growth chamber, and the *in situ* SRPES was performed in an analysis chamber located at beamline BL-1A in the *Photon Factory* at the National Laboratory for High Energy Physics. The two chambers were connected to each other through an ultra-high vacuum. The samples used here were Si-doped *n*-type GaAs(001) wafers with a carrier density of 1×10^{18} cm^{-3}. The sample chemical treatment, surface cleaning, GaAs buffer-layer growth, and Se treatment were previously described in detail (Watanabe *et al* 1993). Two kinds of samples were prepared by (i) depositing nominally 3-ML-thick InAs on As-stabilized GaAs at 200°C by MBE, and (ii) depositing nominally 3-ML-thick InAs on Se-terminated GaAs at 200°C by MBE. A fine streaky, As-stabilized 2x4 RHEED pattern was observed after GaAs epitaxial buffer layer growth. After the Se treatment, the surface structure changed to show a 2x1 RHEED pattern, which implies a Se-terminated GaAs surface (Scimeca *et al* 1992).

The synchrotron radiation photon energy was set at 90.0 eV to obtain surface-sensitive information about the Se 3d, As 3d, Sb 4d, Ga 3d, and In 4d core levels. The advantages of using synchrotron radiation over conventional x-ray photoelectron spectroscopy to analyze In chemical states are that the In 4d photoionization cross section increases by over a factor of 100 as the incident photon energy is changed from 1486.6 eV (Al Kα) to 90.0 eV and that the electron mean free path decreases from about 20 to 5 Å. In addition to providing greater overall sensitivity as well as greater surface sensitivity, the synchrotron radiation is a bright source providing a photon flux of about 10^{10} to 10^{11} photons/s. Total energy resolution was determined to be about 0.3 eV from the observed broadening of the Au Fermi edge.

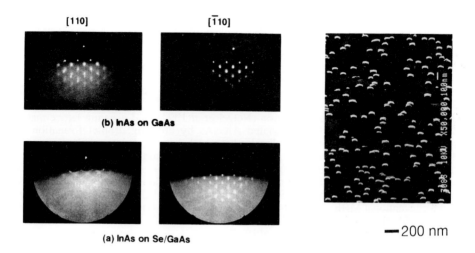

Fig. 1 RHEED patterns for the InAs-grown samples on GaAs with and without Se treatment.

Fig. 2 HRSEM photograph for the InAs-grown sample on Se-terminated GaAs.

3. RESULTS AND DISCUSSION

In order to clarify the difference in the InAs epitaxial growth modes between the Se-terminated GaAs surface and As-stabilized one, their RHEED patterns were compared. Figure 1 shows the RHEED patterns for both samples. Even after InAs growth on the Se-terminated GaAs surface at 200°C with a nominal thickness of 3 ML, the half-order streak pattern along [110] can still be observed, accompanied by new strong streak patterns along [$\bar{1}$11] and [1$\bar{1}$1], whereas the [$\bar{1}$10] RHEED pattern shows only a spotty feature, indicating that the surface morphology has an asymmetric feature with ($\bar{1}$11) and (1$\bar{1}$1) facets, and that crystal islanding takes place. These results provide evidence of the formation of InAs nanocrystals on Se-terminated GaAs at the very early growth stages. In contrast, when the As-stabilized GaAs substrate was used, spotty patterns along [110] and [$\bar{1}$10] were observed, suggesting that an InAs layer grows on the As-stabilized GaAs surface in the Stranski-Krastanov growth mode (Munekata *et al* 1987). InAs nanocrystal formation on Se-terminated GaAs is confirmed by HRSEM micrographs and XTEM images. Figure 2 shows the HRSEM micrograph for the InAs-grown sample on the Se-terminated GaAs substrate. Many rectangular-shaped InAs islands with a lateral width of 70-80 nm can be seen and the density is estimated to be about 2×10^9 cm^{-2}. XTEM images (figure 3) indicate that these InAs nanocrystals have a smooth terrace with a height estimated to be about 20 nm.

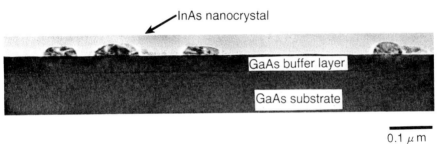

Fig. 3 XTEM image for the InAs-grown sample on Se-terminated GaAs.

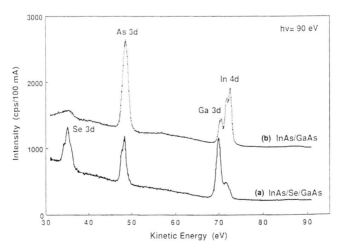

Fig. 4 Core-level SRPES spectra for the InAs-grown samples on GaAs with and without Se treatment.

In order to examine the electronic properties of the InAs nanocrystals on the Se-terminated GaAs surface as well as those of the InAs layer on the As-stabilized one, the core-level SRPES spectra were measured. Figure 4 shows the core-level SRPES spectra for both samples, where the intensities are normalized by the synchrotron radiation ring current. For the sample without Se treatment, the peak intensity of In 4d is stronger than that of Ga 3d, whereas for the Se-treated sample the peak intensity of In 4d is very weak. This difference in the ratios of Ga 3d and In 4d is also observed with InSb growth, as shown in figure 5. These results are consistent with InAs or InSb island formation on the Se-terminated GaAs surface, by considering that the height of InAs or InSb islands is larger than the photoelectron mean free path of In 4d SRPES photoelectrons, which is estimated to be about 5 Å. Thus, these comparative results support the interpretation that, while an InAs layer grows on the As-stabilized GaAs substrate in the Stranski-Krastanov growth mode, InAs islands form on the Se-terminated GaAs surfaces at the very early growth stages.

From the results in figure 4, we obtained deconvoluted results for the Ga 3d and In 4d core-level SRPES spectra. These are shown in figure 6. There are two peaks in the In 4d spectrum for the Se-treated sample, which are assigned as the In-Se bonding states and the In-As bonding states. The In-Se bonding states also exist for the sample without Se treatment. This is because Se treatment and MBE growth were performed in the same chamber, so Se-contamination could not be avoided. Indeed, a very small, broad peak in the Se 3d spectrum at a kinetic energy of around 35 eV can be seen in figure 4 for the sample without Se treatment. To evaluate the validity of this assignment, we simulated these peak intensities. Figure 7 shows a model for this calculation.

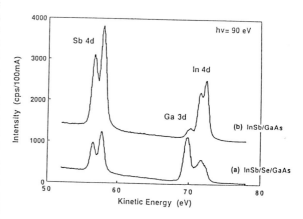

Fig. 5 Core-level SRPES spectra for the InSb-grown samples on GaAs with and without Se treatment.

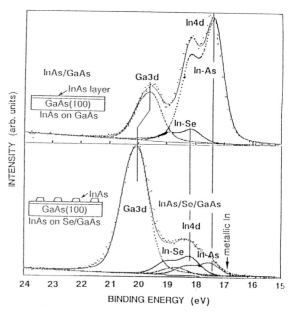

Fig. 6 Ga 3d and In 4d spectra with the deconvoluted results for the InAs-grown samples on GaAs with and without Se treatment.

Since the height of InAs islands is larger than the photoelectron mean free path, we assumed that the photoelectrons from the In $4d$ core-level assigned as the In-Se bonding states come from the InAs-uncovered region. The peak intensities in In $4d$ caused by the In-As bonding states and In-Se ones, I_{InAs} and I_{InSe}, are expressed by the following equations,

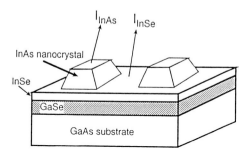

Fig. 7 Model for the In $4d$ peak intensity calculation.

$$I_{InAs} \propto \sigma \times \int_0^{d_{InAs}} \exp(-z/\lambda) dz, \quad (1)$$

$$I_{InSe} \propto (1-\sigma) \times \int_0^{d_{InSe}} \exp(-z/\lambda) dz. \quad (2)$$

Here σ, d_{InAs}, d_{InSe}, and λ are the lateral area percentage of InAs islands, the height of the InAs islands, the effective thickness of the InSe bonding states, and the photoelectron mean free path, respectively, where the values of σ and d_{InAs} are estimated from the results of HRSEM and XTEM observations. The experimental value of the ratio of I_{InAs} to I_{InSe} is 0.69. On the other hand, the calculated value of this ratio is 0.67 if we assume that d_{InSe} is 0.9 Å, which means that the reconstructed 2×1 Se-terminated surface partially changes to a 1×1 surface due to the In-Se bonding. Indeed, a slight 2×1 RHEED pattern can be seen even after InAs deposition as shown in figure 1. This calculation result supports the validity of the assignment. It is noted that the Ga $3d$ peak position shifts to the higher binding energy side, which is induced by Se treatment (Simeca et al 1992). On the other hand, no shift occurs in the In $4d$ peak position. This behavior can be explained by using the energy band diagrams as shown in figure 8. For the Se-treated sample, the Ga $3d$ spectrum comprises photoelectrons coming from the non-InAs-covered substrate surface area. Thus, the Ga $3d$ peak shift can be observed accompanied by a reduction of band bending due to Se treatment. However, no change in the In $4d$ peak occurred. Therefore, band bending takes place again during InAs deposition, as a result of the residual strain and/or generation of dislocation at the heterointerface due to the lattice mismatch between InAs and GaAs.

4. CONCLUSION

We have shown InAs nanocrystal formation on a Se-terminated GaAs surface compared with using an As-stabilized GaAs substrate and with InSb nanocrystal formation by using SRPES, RHEED, HRSEM, and XTEM. While an InAs layer grows on the As-stabilized GaAs substrate in the Stranski-Krastanov growth mode, InAs islands form on the Se-terminated GaAs surfaces at the very early growth stages. InAs nanocrystals with a lateral width of about 70-80 nm, a density of about 2×10^9 cm^{-2}, and a height of about 20 nm are reproducibly grown on Se-terminated GaAs. The SRPES deconvoluted results indicate that, although band bending is reduced by Se treatment, it reoccurs at the heterointerface between InAs nanocrystals and the Se-treated GaAs surface during InAs deposition, as a result of the residual strain and/or generation of dislocations at the heterointerface due to the lattice mismatch.

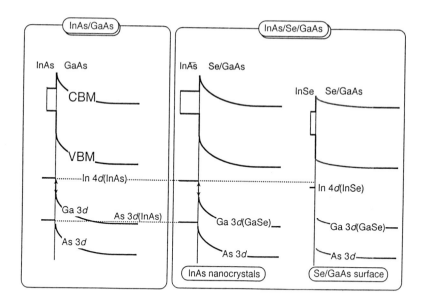

Fig. 8 Schematic energy band diagrams at the heterointerfaces for the InAs-grown samples on GaAs with and without Se treatment.

References

Ahopelto J, Yamaguchi A A, Nishi K, Usui A, Sakaki H, Mochizuki Y 1992 *Extended Abstracts the 1992 International Conference on Solid State Devices & Materials*, Tsukuba, p. 281 (Business Center for Academic Societies Japan, Tokyo, 1992).
Brandt O, Tapfer L, Ploog K, Bierwolf R, Hohenstein M, Phillipp F, Lage H, Heberle A 1991 Phys. Rev. B **44**, 8043.
Koguchi N, Takahashi S, Chikyow T 1991 J. Crystal Growth **111**, 688.
Leonard D, Krishnamurthy M, Reaves C M, Denbaars S P, Petroff P M 1993 Appl. Phys. Lett. **63**, 3203.
Munekata H, Chang L L, Woronick S C, Kao Y H 1987 J. Crystal Growth **81**, 237.
Notzel R, Temmyo J, Tamamura T 1994 Nature **369**, 131.
Scimeca T, Watanabe Y, Berrigan R, Oshima M 1992 Phys. Rev. B **46**, 10201.
Watanabe Y, Scimeca T, Maeda F, Oshima M 1993 *Extended Abstracts of the 1993 International Conference on Solid State Devices & Materials*, Makuhari, p. 116 (Business Center for Academic Societies Japan, Tokyo, 1993).
Watanabe Y, Maeda F, Oshima M 1994 *Springer Proceedings of Physics Series, "Nanostructure and Quantum Effects"* (in press).
Watanabe Y, Maeda F, Oshima M 1994 *3rd International Symposium on Atomic Layer Epitaxy and Related Surface Processes*, Sendai (1994) p. 236. Watanabe Y, Maeda F, Oshima M 1994 Appl. Surf. Sci. (in press)

Artificial control of heterojunction band discontinuities by two delta dopings

Y Hashimoto*, N Sakamoto, K Agawa**, T Saito and T Ikoma***

Institute of Industrial Science, University of Tokyo, 7-22-1 Roppongi, Minato-ku, Tokyo 106, Japan

Abstract. The energy band diagram at a heterojunction is controlled by p- and n-type δ-dopings inserted very near the junction. GaAs/AlAs heterojunctions with two Si δ-dopings grown by molecular beam epitaxy were studied by *in situ* x-ray photoemission spectroscopy. When the Si δ-doping layers are inserted, the energy difference between Ga $3d$ and Al $2p$ levels increases by 0.5 eV indicating that the valence band discontinuity is enlarged by 0.7 eV.

1. Introduction

Artificial control of heterojunction band discontinuity is an innovative technique which gives more freedoms in designing heterostructure devices such as high electron mobility transistors, semiconductor lasers, and resonant tunneling devices [1-5]. Roles of a Si layer inserted at a GaAs/AlAs heterointerface has been studied as a candidate to realize the control of band discontinuities [1-4]. However, we have found that one has to precisely control the atomic sites of the inserted Si atoms to achieve a control of heterojunction band discontinuities by a Si insertion layer at a GaAs/AlAs heterointerface [1-2]. This is very difficult under the present experimental conditions [6].

When one inserts a p-type layer and an n-type layer at both sides of the interface, a dipole is formed between the doped layers to vary the band diagram [5]. A combination of p- and n-type doping sheets inserted at a heterojunction sets an artificial interface dipole between the two doping sheets. When the spacing between the two layers is very thin, the effects of the doping sheets on the electron and/or hole conduction are modified as if the band discontinuities are controlled.

Figure 1 shows an energy band diagram at a GaAs/AlAs interface with p- and n-type Si δ-dopings. The spacing between the p- and n-type layers should be as thin as a few nanometers, which electrons transmit through the tunneling. To effectively control the band discontinuities, a high doping density ($\sim 10^{12}$ cm^{-2}) and a good confinement of dopants (~a few nm) are needed. Moreover one has to balance the p- and n-type doping density so as to confine the band bending in the thin layer between the doping sheets. Then, the band diagram is modified as shown in Fig. 1. $\Delta E_v(o)$ is the intrinsic valence band discontinuity. ΔE_v(effective) shows the effective valence band discontinuity, which governs the hole transmission.

Table I summarizes the above two methods to control band discontinuities and their crucial points. To control the energy band diagram with use of a Si insertion layer at a GaAs/AlAs heterointerface, one must place Si atoms at a Ga site and the next As site as shown by (a). However, the Si atoms tend to spread over 3 atomic layers or more as shown in (b) [6]. Then, the band discontinuities are not effectively controlled [1]. It is rather easier to use the two δ-dopings to control the band discontinuities.

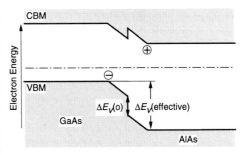

FIG. 1. An energy band diagram at a GaAs/AlAs interface with p- and n-type Si δ-dopings. A ⊕ (⊖) symbol indicates Si donors (acceptors).

In this work, we report the variation in the energy band diagram at a GaAs/AlAs heterointerface with p- and n- type Si δ-dopings grown by molecular beam epitaxy (MBE) and characterized by *in situ* x-ray photoemission spectroscopy (XPS). It will be shown that we can successfully control the band diagram and increase the valence band discontinuity by 0.7 eV.

TABLE I. A comparison of methods proposed to control heterojunction band discontinuities.

Idea	Single layer insertion	Two δ-doping sheets
Crucial points	(a) Ga site Si, As site Si; ΔE_v change = 1.59 eV (theory). Si occupied sites must be precisely controlled. (b) Ga site Si, As site Si; ΔE_v change = −0.22 eV (theory). When the Si atoms distribute over three layers, the interface dipole is not effectively formed.	Spacing between δ-dopings. n-type δ-doping, p-type δ-doping. (1) Effective δ-dopings have been realized. (2) The Si occupied sites is well controlled by the growth parameters. (3) A spacing between δ-dopings can be lowered to be a few nanometers.
legends: ● Si, ○ Ga, ◍ Al, ○ As		
Applicability	Poor	Good

Further, we confine the interface dipole due to the Si δ-dopings within 50 Å, which is much thinner than in the samples studied by Capasso et al. [5] and enables us to set an abrupt change in the band diagram.

2. Application to GaAs/AlAs (311)A system

GaAs (311)A is an attractive surface on which we can grow a p-type GaAs layer doped with Si by MBE. Recently, we have found that the conduction type of a Si-doped MBE-grown GaAs (311)A sample can be controlled by the growth temperature, T_g; i.e. the Si δ-doping sheet is n-type when $T_g < 480$ °C and it is p-type when $T_g > 480$ °C [7]. This fact indicates that both p- and n-type doping sheets can be formed using only Si as a dopant. We use a GaAs/AlAs (311)A structure to realize a control of the energy band discontinuities [8]. Further, this structure has the following two advantages. (1) Si is a stable and reliable dopant. (2) When one inserts only Si, the Si segregation effect is mostly canceled, because we can control the doping type by T_g including segregation and/or diffusion species.

3. Samples and XPS measurements

XPS measurements were performed on the following GaAs/AlAs structures including the Si δ-dopings. Each of the samples consists of a GaAs buffer layer, a p-type Si δ-doping layer (0.05 ML-thick), a GaAs layer (25 Å), an AlAs layer (25 Å), an n-type Si δ-doping layer (0.1 ML-thick), and an AlAs layer ($t = 30\sim60$ Å) successively grown on an n-type GaAs (311)A substrate. The growth temperature was 600 °C for the GaAs, bottom Si layer and AlAs intermediate layer and 400 °C for the top Si layer and the top AlAs layer.

A sample was studied by in situ XPS immediately after the growth. Energy regions which cover the Ga 3d and Al 2p levels were scanned for 1~3 days. From each of the Ga 3d spectra a background signal due to the AlAs layer was subtracted by using a spectrum from bulk AlAs [9].

Table II summarizes the measured energy difference, ΔE_{CL}, between the Ga 3d and Al 2p levels. ΔE_{CL} increases by 0.5 eV when the Si δ-dopings are inserted. This change shows that the energy band diagram is controlled.

TABLE II. Measured energy differences (in eV).

	ΔE_{CL}	Al 2p FWHM	Ga 3d FWHM	ΔE_v(effective)	E_{BB}
Sample I ($t = 30$ Å)	54.86	1.03	1.22	1.18	0.23
Sample II ($t = 60$ Å)	54.69	1.02	1.07[a]	1.18	0.46
AlAs/GaAs[b]	54.38	0.97	1.03	0.44	0.00
AlAs/Si/GaAs[c]	54.0	1.2	1.03	0.44	0.7

[a] An ambiguity of the line width is due to the large background signal from the 85 Å-thick AlAs overgrown layer.
[b] Reference [10]. ΔE_{CL} values are updated. Since the electron spectrometer has changed its gain function, 54.41 eV in Ref [10] is converted to be 54.38 eV in the present experimental setups.
[c] Reference [1]. Data obtained on the AlAs (30 Å)/Si (0.5 Å)/GaAs (100) structures.

4. Discussions

In order to deduce the band discontinuities from the values determined by XPS, we have to consider the effects of the XPS probing depth ($\lambda_e \sim 25$ Å) [1]. Figure 2 shows the energy band diagrams of GaAs/AlAs junctions with and without Si δ-doping. As shown in Fig. 2 (a), at a GaAs/AlAs junction the valence band discontinuity is given by [10],

$$\Delta E_v = \Delta E_{CL} + E_{v\text{-Ga }3d} - E_{v\text{-Al }2p}$$
$$= \Delta E_{CL} - 53.94 \text{ eV}. \qquad (1)$$

Here, $E_{v\text{-Ga }3d}$ and $E_{v\text{-Al }2p}$ denote the binding energies of Ga 3d and Al 2p core levels measured from the valence band maxima (VBM) in GaAs and AlAs, respectively. It should be noted that the band should be flat for Eq. (1) to be valid. When the δ-doping layers are inserted the band diagram changes as shown in Figs. 2 (b) and (c). ΔE_{CL} is, then, shown by,

$$\Delta E_{CL} = E_{\text{Ga }3d}(\text{average}) - E_{\text{Al }2p}(\text{average})$$
$$= E_v^{\text{GaAs}}(\text{average}) - E_v^{\text{AlAs}}(\text{average}) + 53.94 \text{ eV}. \qquad (2)$$

Here, $E_{\text{Ga }3d}(\text{average})$ and $E_{\text{Al }2p}(\text{average})$ denote the energies of the Ga 3d and Al 2p levels, respectively, which are averaged over the XPS probing depth. $E_v^{\text{GaAs}}(\text{average})$ and $E_v^{\text{AlAs}}(\text{average})$ are the averaged VBM of GaAs and AlAs, respectively. ΔE_{CL} is thus affected by the dipole potential between the Si layers as well as a band bending in the AlAs layer over the top Si layer.

When the interface dipole is formed between the doped layers and no band bending is left above the top Si layer as shown in Fig. 1, ΔE_{CL} is almost independent of the top AlAs layer thickness t. However, this value changes when t increases from 30 Å to 60 Å. The band diagram including the interface dipole potential between the δ-doping layer and the band bending above the top δ-doping layer is estimated under the following assumptions.

(1) The dipole potential, Δ, between the δ-doping layers and hence the effective valence band discontinuity is the same for $t = 30$ Å and $t = 60$ Å.
(2) The band bending in the above AlAs layer is proportional to t.

Then, the Ga 3d and Al 2p binding energies, $E_{\text{Ga }3d}$ and $E_{\text{Al }2p}$, are shown as a function of the depth d from the sample surface as,

$$E_{\text{Ga }3d} = E_{\text{Ga }3d}^{d=50 \text{ Å}+t} \qquad (d > 50 \text{ Å}+t)$$

$$= E_{\text{Ga }3d}^{d=50 \text{ Å}+t} - \frac{\Delta}{2} \times \frac{(50 \text{ Å}+t)-d}{25 \text{ Å}} \qquad (50 \text{ Å}+t \geq d \geq t+25 \text{ Å}), \qquad (3)$$

$$E_{\text{Al }2p} = E_{\text{Ga }3d}^{d=25 \text{ Å}+t} - 54.38 \text{ eV} - \frac{\Delta}{2} \times \frac{(25 \text{ Å}+t)-d}{25 \text{ Å}} \qquad (t+25 \text{ Å} > d \geq t)$$

$$= E_{\text{Al }2p}^{d=t} + E_{BB} \times \frac{t-d}{t} \qquad (0 \geq d \geq t), \qquad (4)$$

$$E_{BB}(t = 60 \text{ Å}) = 2 \times E_{BB}(t = 30 \text{ Å}). \qquad (5)$$

By substituting the values given by Eqs. (3-5) into the exponential weighting function [1], $E_v^{\text{GaAs}}(\text{average})$ and $E_v^{\text{AlAs}}(\text{average})$ are obtained. The ΔE_{CL}'s for

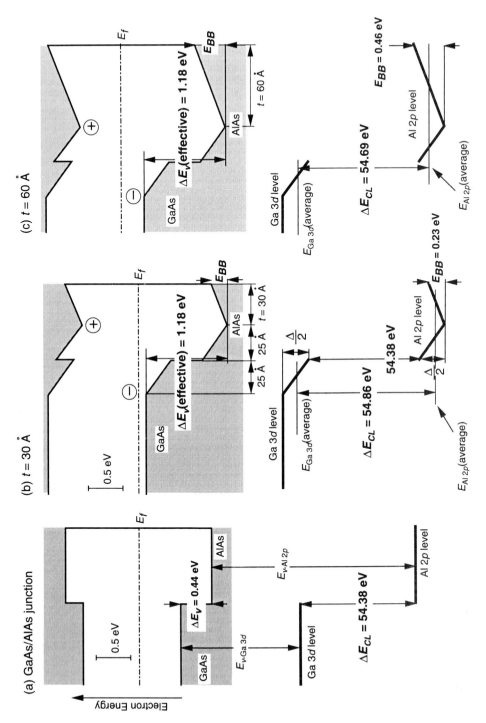

FIG. 2. Energy band diagrams of the valence band maximum, the conduction band minimum, and core levels, which are studied by the XPS measurements.

$t = 30$ Å and $t = 60$ Å are shown as functions of Δ and the band bending above the top Si doped layer, E_{BB}. By solving the equations we have $\Delta = 0.74$ eV and $E_{BB} = 0.23$ eV ($t = 30$ Å). Then, the energy band diagram shown in Fig. 2 are drawn. It indicates that the effective valence band discontinuity is enlarged by 0.7 eV due to the Si layers and that the n-type doping density is larger than the p-type doping density. This might be due to the difference in the activation efficiency of Si donor and acceptor as well as Si diffusion/segregation effects. The observed peak line width of the Ga $3d$ and Al $2p$ levels confirms the validity of the above determination. The Al $2p$ line width is relatively narrow as compared with the AlAs/Si/GaAs samples [1], in which the band bending in the AlAs layer is as large as 0.7 eV. The Ga $3d$ width is broadened because XPS probes mainly the top few nanometer of the GaAs layer where the interface dipole is formed.

These experimental observations demonstrate that the insertion of two δ-dopings very near the junction can effectively control the heterojunction band discontinuities, although it requires us to balance the p- and n-type doping density.

5. Conclusions

GaAs/AlAs (311)A heterojunctions with Si p- and n-type δ-dopings were grown by molecular beam epitaxy and studied by *in situ* x-ray photoemission spectroscopy. The energy difference between the Ga $3d$ and Al $2p$ levels increases by 0.5 eV when the Si layers are inserted. This change indicates that the effective valence band discontinuity is enlarged by 0.7 eV when the doping densities are 0.05 ML (p-type) and 0.1 ML (n-type) and their separation is 50 Å. The energy band diagram is thus controlled by using p- and n-type δ-dopings inserted very near the heterointerface.

Acknowledgment. This work was supported by the Grant-in-Aid for Scientific Research on Priority Areas Supported by Ministry of Education, Science and Culture, "Quantum Coherent Electronics".

References

* Present address: Department of Electrical and Electronic Engineering, Faculty of Engineering, Shinshu University, 500 Wakasato, Nagano 380, Japan.
** Present address: Toshiba Corporation, 1-1-1 Shibaura, Minato-ku, Tokyo 105-01, Japan.
*** Present address: Texas Instruments Tsukuba Research and Development Center Ltd., 17 Miyukigaoka, Tsukuba, Ibaraki 305, Japan.

[1] Hashimoto Y, Tanaka G and Ikoma T, J. Vac. Sci. Technol. B **12**, 125 (1994); Hashimoto Y, Hirakawa K, Tanaka G and Ikoma T, Proc. Int. Conf. Phys. Semicond. (1992); Hashimoto Y, Saito T, Hirakawa K and Ikoma T, Inst. Phys. Conf. Ser. **129**, 259 (1993).
[2] Akazawa M, Hasegawa H, Tomozawa H and Fujikura H, Jpn. J. Appl. Phys. **31**, L1012 (1992).
[3] Muños A, Chetty M and Martin R M, Phys. Rev. B **41**, 2976 (1990).
[4] Sorba L, Bratina G, Ceccone G, Antonini A, Walker J F, Micovic M and Franciosi A, Phys. Rev. B **43**, 2450 (1991).
[5] Capasso F, Mohammed M and Cho A Y, J. Vac. Sci. Technol. B **3**, 1245 (1985).
[6] Chambers S A and Trans T T, Phys. Rev. B **47**, 13023 (1993).
[7] Agawa K, Hirakawa K, Sakamoto N, Hashimoto Y and Ikoma T, Appl. Phys. Lett. **65** (1994) to be published.
[8] Saito T, Hashimoto Y and Ikoma T, Solid-state Electronics **37**, 743 (1994).
[9] Hirakawa K, Hashimoto Y and Ikoma T, Surf. Sci. **267**, 166 (1992).
[10] Hirakawa K, Hashimoto Y and Ikoma T, Appl. Phys. Lett, **57**, 2555 (1990).

Flow Modulation Epitaxy of Short Period GaInAs/InP Superlattices

D T Emerson, K L Whittingham, and J R Shealy

OMVPE Facility, School of Electrical Engineering
Cornell University, Ithaca, N.Y. 14850 USA

Abstract. Flow Modulation Epitaxial growth of short period GaInAs/InP superlattices was investigated. Superlattice quality was assessed by Double Crystal X-Ray Diffraction, room and low temperature Photoluminescence, Raman Scattering, and Atomic Force Microscopy. Growth of bulk GaInAs, single GaInAs quantum wells, and GaInAs/InP superlattices (including results on the shortest period, high quality GaInAs/InP superlattices prepared to date by OMVPE) are presented. Finally, a qualitative model relating accumulated interface roughness and preferential Group III incorporation during GaInAs nucleation is presented.

I. Introduction

GaInAsP lattice matched InP is an important materials system for long wavelength optoelectronic devices. There has been a great deal of research into the mixed crystal, but few reports on quaternary replacement by GaInAs/InP short period superlattices [1]. Because heterostructures in this system require the exchange of both Group III and Group V constituents across one or more interfaces while maintaining lattice matched growth, they are among the most difficult interfaces to form by Organometallic Vapor Phase Epitaxy (OMVPE). The role of the Group V sources is especially significant because of the nonlinear nature of the As/P incorporation ratio [2] and the necessity of having at least one of the Group V sources in the reaction cell at all times to preserve the growth surface.

Most studies on the interfaces in this system have involved single quantum wells [3][4] or large period superlattices [5]. These reports have often concentrated on optimizing the growth of thin wells and not thin barriers and have generally relied on one primary characterization technique (e.g. Photoluminescence or X-Ray Diffraction). Studies of this kind do not address the effect of the thick InP layers on smoothing the growth surface [6] or the question of interface repeatability, nor do they seek to reconcile the sometimes disagreeing results of different characterization techniques. In contrast, this study includes the growth of structures with both thick and thin wells (as thin as approximately 10 Å) and barriers (as thin as approximately 20 Å) and employs a variety of characterization techniques including Double Crystal X-Ray Diffraction (DCXRD), Raman Scattering, room and low temperature Photoluminescence (PL), and Atomic Force Microscopy (AFM).

II. Experimental

GaInAs/InP superlattices were grown at 600°C using Flow Modulation Epitaxy (FME) at low pressure (76 torr) in a vertical barrel OMVPE system [7]. Before each run the reaction cell was cleaned via a hot wall HCl vapor etch to remove deposits from the previous run. The sources were triethylgallium (TEG), trimethylindium (TMIn), arsine (AsH_3), and phosphine (PH_3). The TMIn flow was regulated with an ultrasonic analyzer. The V/III ratios and growth rates were approximately 30 and 0.5 µm/hr for the GaInAs and 275 and 0.25 µm/hr for the InP. Two substrates, (100) ±0.5° S-doped InP and (100) semi-insulating Fe-doped misoriented 2±0.5° toward the nearest (110), were loaded side by side on the susceptor during each experiment.

Symetric (004) and asymmetric {115} DCXRD was used to determine superlattice period and lattice parameters in directions both parallel and perpendicular to the growth surface.

Most of the samples described here were coherent, so *superlattice lattice parameter* will be used to describe the perpendicular lattice constant. Room (300K) and low temperature (1K) PL were used to assess optical quality and determine the superlattice bandgap. Raman scattering results supplemented the structural information provided by the DCXRD, especially the composition of the wells and the extent of layer intermixing. Finally, surface morphology of selected samples was assessed by AFM.

Samples consisted of single layers, quantum wells, and superlattices. Superlattice periods ranged from 30 Å to 260 Å, GaInAs thickness from 10 Å to 130 Å, and InP thickness from 20 Å to 130 Å. Several interface formation techniques, depicted in Figure 1, were used in order to simulate switching procedures characteristic of a large number of OMVPE reactors. Referring to the figure, all growth stops were fixed at 10 sec. For Interface I, the H_2 interrupt time was varied from 0.5-2 sec; for Interface II, the GaInAsP overlap time was varied from 0.1-5 sec; and for Interface III, the hydride overlap was fixed at 0.1 sec while the surface conversion time was varied from 0.375-0.75 sec.

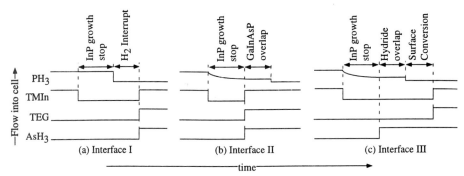

Figure 1 Interface formation techniques used in this study. The InP to GaInAs switching and flow sequences are depicted in the figure. Interfaces in the same sample were formed symmetrically, so the GaInAs to InP sequence is described by interchanging the Group III and Group V sources accordingly.

III. Results & Discussion

Superlattices with nominal 250 Å periods were produced using bulk lattice matching flows. DCXRD spectra of samples grown on on-axis substrates utilizing Interface I (H_2 purge time of 2 sec) and Interface II (GaInAsP overlap times of 0.1 and 5 sec) are shown in Figure 2 and are denoted a,b, and c, respectively. The lattice parameters of the samples are consistently smaller than that of the substrate (0th order diffraction feature on high angle side of substrate). Since these samples were produced with reactant flows that resulted in thick, lattice matched GaInAs, this indicates a change in lattice matching conditions for thick and thin layers of GaInAs. Contrasting the structural quality of the samples, Interface I resulted in broad diffraction features and only two orders of

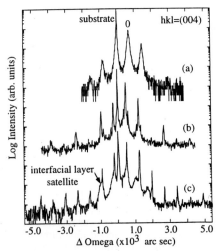

Figure 2 DCXRD for three different interface formation techniques (nominal period - 250 Å)

superlattice satellites, demonstrative of poor crystal quality and/or rough interfaces. On the other hand, Interface II resulted in more intense, narrower, and more numerous satellites. Furthermore, even though the 5 sec GaInAsP overlap time resulted in an additional set of satellites related to the intentionally introduced GaInAsP transition layers, the interlayers were not detrimental to structural quality as is evidenced by the comparable diffraction results. Optical quality as determined by intensity and spectral width of 1K and 300K PL of the samples prepared with Interface II and 0.1 sec and 5 sec GaInAsP overlap times also compared favorably, confirming the difficulty in evaluating the effect of interface formation technique on sample quality in structures with thick layers.

As the period was decreased, the samples produced with Interface I were consistently of poorer quality than those produced with Interface II. Deterioration of the growth surface, even for H_2 purge times as short as 0.5 seconds, indicates rapid transport of residual group V species from the reaction cell. Combined with the 250 Å period results, this demonstrates that one of the group V sources is needed in the reaction cell at all times to preserve the growth surface. Following the lattice parameter trend begun with the 250 Å period samples, the superlattice lattice parameter decreased with GaInAs layer thickness as the period was reduced. Furthermore, as the period was decreased below approximately 60 Å, samples with a large number of periods (>50) produced with bulk calibration reactant flows were characterized by poor optical and structural quality and inferior surface morphology relative to the larger period samples.

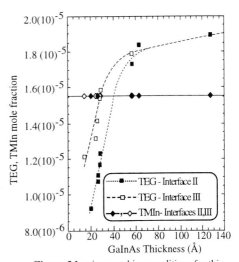

Figure 3 Lattice matching conditions for thin GaInAs layers.

To preserve lattice matched growth and restore sample quality, the group III mole fraction ratio was adjusted as a function of the GaInAs layer thickness for both Interfaces II and III as shown in Figure 3. This effect was repeatable over the duration of the study, with the possibility of systematic errors eliminated by repeating several experiments for the shorter period (30-120 Å) structures and by periodically verifying the bulk lattice matching conditions. Notably, even after the lattice matching conditions were established, samples grown on the off-axis substrates were of poorer quality than those grown on the on-axis substrates, indicating a fundamental difference between growth on the on- and off-axis substrates. The change in lattice matching conditions with GaInAs thickness and interface formation technique indicates that growth of thin GaInAs with bulk calibration data does not necessarily result in lattice matched structures and provides a possible explanation for the large dispersion in PL data for thin GaInAs single quantum wells (e.g. 1.0 to 1.2 eV at 1K for 10 Å GaInAs/InP wells) [4].

Identification of the lattice matching trend enabled the synthesis of a quaternary replacement with a wide range of bandgaps as shown in Figure 4. Kronig-Penney calculations for expected ground state electron to heavy hole transition energies are included for comparison. Abrupt interfaces, parabolic bands, and equal well and barrier thicknesses are assumed, with paramaters as used in reference [4]. Bandgaps and band offsets are adjusted for the strain [4]. The shaded areas reflect the width of the minibands. This study also resulted in the shortest period high quality GaInAs/InP superlattices with a large number of periods produced by OMVPE to date. Room and low temperature PL and a DCXRD spectrum of a representative sample with a 54 Å period repeated 100 times are

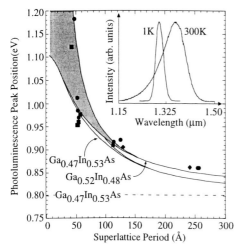
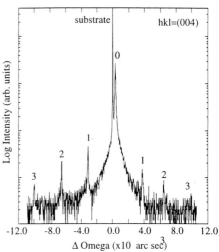

Figure 4: GaInAs/InP superlattice 1K PL peak position. Symbols correspond to Interface I (♦), Interface II (●), and Interface III (■). Inset and Kronig-Penney Calculations are described in the text. GaInAs well compositions are as indicated.

Figure 5: DCXRD spectrum centered around InP (004) reflection. Superlattice satellite orders are enumerated.

shown in the Figure 4 inset and in Figure 5, respectively. This result is in conflict with an earlier report claiming that such a structure could not be grown by OMVPE [6]. Single quantum wells were also prepared with Interfaces II and III; both bulk lattice matching flows and thin layer lattice matching flows determined from the superlattice study were used. All samples were characterized by strong 1K PL; however, the dispersion of the thin well emission energy emphasized the importance of determining lattice matching conditions for thin layers.

The surface morphology of the large period structures was smooth with the dominant features being monolayer terraces separated by approximately 0.5 μm as shown in the AFM image of the surface of a 250 Å period superlattice in Figure 6a. This is consistent with the morphology of thick InP and GaInAs layers grown under the same conditions. As the period was decreased, the dominant features were depressions as shown in the image of a 120 Å period superlattice in Figure 6b. When the period was further reduced to the shortest period structure investigated, 30 Å, the depressions became larger in both width

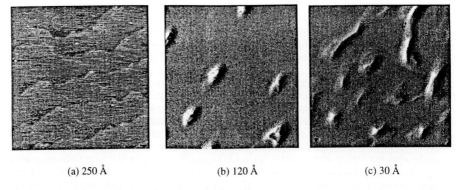

(a) 250 Å   (b) 120 Å   (c) 30 Å

Figure 6 Superlattice surface morphology. Field size is 2 μm by 2 μm. Superlattices are synthesized with Interface II. Superlattice period is as indicated.

and depth. In the image shown in Figure 6c, the depressions were approximately 60 Å deep, twice the superlattice period. The surface morphology trend seen here can be used as an indication of the individual interface roughness, with the interface roughness increasing with decreasing period. Furthermore, growth of similar samples which differed only in the number of periods revealed that the interface roughness is cumulative.

Raman spectra for structures with GaInAs layer thicknesses ranging from 1 μm to 10 Å are shown in Figure 7; Interface II (0.1 sec overlap) was used for all samples shown. The dominant spectral features are identified in the figure caption. Modes (i-iii) are associated with the GaInAs layers while modes (iv-v) are related to the InP layers. Mode (vi) is discussed below. The TO phonons were identified by varying the polarization of the incident and scattered light. The increase in TO intensity in the shorter period samples indicates deteriorating sample quality and increasing interface roughness. Samples produced with Interface III showed similar features. Bulk GaInAs and the two largest period samples have virtually identical spectra; the similarity between these superlattice and the bulk spectra, the absence of a GaP-like mode, and the small shift of the InP-like LO phonon from its bulk value indicate layer intermixing on the order of monolayers [8]. However, as the period is reduced, the Raman spectra diverge considerably from the spectrum of the bulk reference sample. The increase in

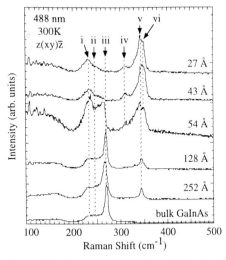

Figure 7 Raman scattering from GaInAs/InP superlattices. Superlattice period is indicated on the right side of the figure. Mode identification is as follows: (i) InAs-like LO, (ii) & (iii) GaAs-like TO & LO, (iv) & (v) InP-like TO & LO, and (vi) GaP-like LO. Vertical, dotted lines are intended to indicate approximate mode position.

the strength of the InP-like modes relative to the GaInAs-like modes could be casued by a resonance apparent only in the shorter period structures or by an additional mode superimposed on the InP-like region of the spectra (location of mode (vi) consistent with vibrational frequency of GaP-like LO phonon [9]). The intensity changes within the GaInAs-like region of the spectra may result from a resonance condition, the presence of Ga-deficient wells (not likely based on DCXRD and PL data and the position of the proposed GaP-like phonon), or increased interaction of In and As atoms at the interfaces as the interfaces roughen and become less abrupt.

Growth by FME of GaInAs/InP superlattices and single quantum wells has been found to be sensitive to the interface formation technique, substrate orientation, superlattice period, and number of periods. DCXRD results demonstrated the decrease in superlattice lattice parameter with period. AFM measurements and surface morphology trends showed increasing interface roughness with decreasing GaInAs thickness and accumulating with the number of periods. The Raman scattering indicated a possible incorporation of P in the GaInAs in short period structures, increasing interface roughness with decreasing period, and relative changes in the reactant incorporation ratio with period. The lattice parameter trend could be caused by incorporation of residual P in the GaInAs; however, the amount of residual PH_3 in the reaction cell is known to be small (see discussion of Interface I) and P does not incorporate in GaInAs efficiently at 600°C [2]. Furthermore, if this was the only cause of the lattice matching trend, the superlattice lattice parameter should decrease by exposing the GaInAs surface to PH_3 for increasing amounts of time; the opposite effect was seen in Interface III experiements in which all parameters except the surface conversion time were held constant. Instead, we believe that the observed variation in the lattice matching conditions is caused by preferential Ga incorporation in the GaInAs near the interfaces which in turn produces a rough growth surface. This effect is not significant in structures with thick InP layers because InP smooths the growth surface and prevents the accumulation of interface roughness [6]. The change in the Ga/In incorporation ratio near the interfaces is likely to occur due to the stronger Ga-group V bond compared to the In-group V bond [10]. In the early stages of GaInAs nucleation on InP, preferential Ga deposition occurs because the sticking coefficient of Ga on P is larger than that of In on P. Similarly, during the growth interruption following GaInAs deposition, preferential In desorption is likely because the sticking coefficient of Ga on As is larger than that of In on As. The strain associated with the Ga-rich layer at either interface would explain the surface roughening and the resulting preferential Ga incorporation would explain the necessity of reducing the Ga flux for short periods where thin, Ga-rich layers would decrease the lattice constant by the largest amount.

IV. Summary

In summary, bulk GaInAs growth calibration data may be used to grow high quality thick period (> 200 Å) superlattices, short period superlattices with a relatively small number of interfaces, and single quantum wells, but not good quality short period superlattices with a large number of interfaces (> 50). As the GaInAs layer thickness is decreased from 1 μm to 10 Å , the Ga mole fraction must be likewise decreased relative to the In to preserve lattice matched growth. We have discussed the plausibility of P incorporation in the GaInAs as well as enhanced Ga incorporation and surface roughening during the GaInAs growth to explain this trend. Regardless of which effect is dominant in causing the change in lattice matching conditions as the superlattice period is decreased, we have provided an explanation for the large dispersion in PL data for thin wells and defined a repeatible process for successfully synthesizing superlattice quaternary replacements. Finally, we have demonstrated the necessity of using several characterization techniques to investigate complicated epitaxial structures such as those prepared here.

V. Acknowledgements

The authors wish to thank B.P. Butterfield, M.J. Cook, and M.J. Matragrano for technical assistance. This work was supported by JSEP (grant No. F49620-90-C-0039) and ARPA (contract No. MDA97290C0058) Optoelectronics Technology Center.

VI. References

[1] Dotor M L, Huertas P, Golmayo D, and Briones F 1993 *Appl. Phys. Lett.* **62** 891-3
[2] Samuelson L, Omling P, and Grimmeiss H G 1983 *J. Crys. Growth* **61** 425-6
[3] Wang T Y, Reihlen E H, Jen H R, and Stringfellow G B 1989 *J. Appl. Phys.* **66** 5376-5383
[4] Wang T Y and Stringfellow G B 1990 *J. Appl. Phys.* **67** 344-352
[5] Landgren G, Wallin J, and Pellegrino S 1992 *J. Elec. Mat.* **21** 105-108
 Clawson A R, Vu T T, Pappert S A, and Hanson C M 1992 *J. Crys. Growth* **124** 536-540
[6] Bhat R, Koza M A, Hwang D M, Kash K, Caneau C, and Nahory R E 1991 *J. Crys. Growth* **110** 353-362
[7] Pitts B L, Matragrano M J, Emerson D T, Sun B, Ast D T, and Shealy J R 1993 *Proc. of 20th Int. Sym. on GaAs and Related Compounds*
[8] Asahi H, Kohara T, Soni R K, Asami K, Emura S, and Gonda S 1993 *J Crys. Growth* **127** 194-198.
[9] Mozume T 1994 *Mat. Res. Soc. Symp. Proc.* **324** 285-290
[10] Sato M and Horikoshi Y 1991 *J. Appl. Phys.* **69** 7697-7701

Inst. Phys. Conf. Ser. No 141: Chapter 2
Paper presented at Int. Symp. Compound Semicond., San Diego, 18–22 September 1994
© 1995 IOP Publishing Ltd

Strain compensation in InGaP/InGaAs quantum wells with improved interfaces grown by gas-source molecular beam epitaxy

C H Yan* and C W Tu

Department of Electrical and Computer Engineering
University of California at San Diego La Jolla, California 92093-0407 U.S.A.

Abstract: We discuss in detail how the driving force for dislocation generation changes in quantum-well structures with and without strain compensation. The criteria for designing a structurally stable, strain-compensated stack during and after growth are to make the average strain close to zero in the whole structure and to keep the thickness of each layer below its critical layer thickness. With improved interfaces, the photoluminescence of an InGaP/InGaAs single quantum well with strain compensation shows a much stronger and narrower peak than the one without strain compensation. For a strain compensated InGaP/InGaAs multiple quantum well structure, the double crystal x-ray rocking curve exhibits much narrower satellite peaks.

1. Introduction

The strain in strained semiconductor quantum wells offers an additional degree of freedom in tailoring electronic properties, which presents a considerable potential for developing novel and improved electronic and optoelectronic devices[1,2]. The device performance may be maximized by using a larger strain. which, however, requires a thinner critical thickness (h_c). For certain devices, such as field-effect transistors, the channel layer must be kept relatively thick to minimize the effect of size quantization and reduced carrier confinement. According to the Matthews-Blakeslee's model[3] for the $In_xGa_{1-x}As/GaAs$ system, when x equals 0.2, h_c is about 65Å; when x is 0.3, h_c is about 38Å. Therefore, it is difficult to realize an $In_{0.3}Ga_{0.7}As/GaAs$ strained quantum well with a well thickness between 100Å and 150Å. Even though h_c of the second and the upper wells can be increased by using wider barriers to dilute the effect of strain, it can not be varied too much and h_c of the first well remains the same. Lowering the growth temperature can also increase h_c, because dislocation multiplication will decrease at low temperatures[4], but the material quality may be inferior.

Another method for increasing h_c is strain compensation. Previously, several authors have discussed the possibilities of strain compensation[5] based on the Matthews-Blakeslee's model[3]. Basically, tensile and compressive strain in alternating layers allow synthesis of a strained quantum well stack with a considerably reduced net strain, which results in a low residual effective stress for misfit dislocation injection during and after growth. Strain compensation holds the promise of reduced defect densities with an enhanced structural stability, and it should permit the growth of highly strained quantum wells for electronic and optoelectronic devices.

---

* Permanent address: Institute of Semiconductors, Chinese Academy of Sciences, Beijing, China.

In this paper, we compare strain compensated and uncompensated InGaP/InGaAs single quantum well (SQW) and multiple quantum well (MQW) structures.

2. Theory

A partially compensated stack can be grown easily by using wells and barriers with opposite strain. The net strain (ε_n) is then reduced. For an N-period MQW structure it can be expressed as follows:

$$\varepsilon_n = \frac{Nh_b\varepsilon_b + (N-1)h_w\varepsilon_w}{Nh_b + (N-1)h_w}$$

where h'_s and ε'_s are layer thickness and strain, respectively; b means barrier layer and w means well layer. We can choose the strain and the thickness of the wells and the barriers by making $\varepsilon_n = 0$ to obtain a strain-compensated structure. However, $\varepsilon_n = 0$ means only that the whole stack is stable after growth; it can not tell if the stack is stable during the growth. If the strain in the first several layers is already relaxed, we can not obtain a pseudomorphic structure even though ε_n is equal to zero. Therefore, another important condition in the design of a strain-compensated structure is that the thickness of each layer must be kept within its h_c.

According to the model of Houghton et al., which is based on the Matthews and Blakeslee's energy balance model, the driving force (τ_{DP}) for the propagation of a dislocation pair in a multilayer structure is[5]:

$$\tau_{DP} = A\left\{\mu_{xy}\left|\frac{Nhx+(N-1)Hy}{L}\right| - \frac{B}{L}\left[\ln\frac{\beta}{b}(L+Z) + \ln\left(\frac{\beta}{b}Z\right) + 2\ln\left(\frac{L}{L+Z}\right)\right]\right\}$$

$$A = \frac{2(1+v)}{(1-v)}\cos\psi\cos\lambda$$

$$B = \frac{\mu_x\mu_y}{\mu_x+\mu_y}\frac{b(1-v\cos^2\theta)}{4\pi(1+v)\cos\lambda}$$

where N is the period of alternating layers with opposite strain x and y; h and H are well and barrier thicknesses, respectively; ψ is the angle between the strained interface normal and the slip plane; θ is the angle between the dislocation line and its Burgers vector; λ is the angle between the Burgers vector and the direction in the interface normal to the dislocation line; μ_x is the shear modulus of the layer x; μ_{xy} is the average shear modulus for the multilayer; v is Poison's ratio ~ 0.3; β is the core parameter; $b = 4$Å is the Burgers vector of the misfit dislocation; $L=Nh+(N-1)H$ is the height of the multilayer cut by the threading arm of the misfit dislocation pair; and Z is the thickness of the top layer.

If τ_{DP} is negative, then there is no tendency for strain relaxation, and the whole structure is thermodynamically stable. If τ_{DP} is positive, then the strained structure is unstable. When $\tau_{DP} = 0$, it is the critical condition at which h_c can be determined. Using the driving force equation, we can design a stable structure during and after growth by choosing appropriate barrier-layer parameters for a given well strain and thickness. Therefore, with strain compensation an infinite period of QWs can be grown.

The driving force in InGaP/InGaAs SQW structure with and without strain compensation is shown in Fig. 1. Curve (a) is the driving force for dislocation generation in a strained $In_{0.38}Ga_{0.62}P$ single layer on a GaAs substrate. Its h_c is about 145Å determined at $\tau_{DP}=0$. For an $In_{0.48}Ga_{0.52}P(50Å)/In_{0.2}Ga_{0.8}As(50Å)/In_{0.48}Ga_{0.52}P(50Å)$ SQW structure without strain compensation, shown in curve (b), the $In_{0.48}Ga_{0.52}P$ layer has a lattice-mismatch of about 1×10^{-3}, based on the actual structure grown, as described later, and the strain in the $In_{0.2}Ga_{0.8}As$ layer is about 1.4×10^{-2}. This results in an average strain of about 5.4×10^{-3}. For a strain-compensated $In_{0.38}Ga_{0.62}P(50Å)/In_{0.2}Ga_{0.8}As(50Å)/In_{0.38}Ga_{0.62}P(50Å)$ SQW, curve (c), the barrier strain is about -7.8×10^{-3}, and the average strain in the whole stack is about 5×10^{-4}. Since the thickness of the InGaAs and InGaP layers are all smaller than their h_c's, the driving force is less than zero during and after growth for both SQWs. After the growth of InGaAs and InGaP, the driving force in the strain-compensated structure is less than that of the uncompensated structure, i.e., the strain compensated structure should be more stable.

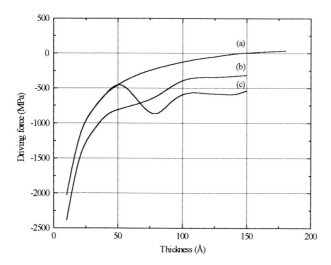

Fig. 1 Driving force for dislocation generation as a function of layer thickness in (a) a single strained layer $In_{0.38}Ga_{0.62}P$ on GaAs substrate, (b) an $In_{0.48}Ga_{0.52}P(50Å)/In_{0.2}Ga_{0.8}As(50Å)/In_{0.48}Ga_{0.52}P(50Å)$ SQW without strain compensation, and (c) an $In_{0.38}Ga_{0.62}P(50Å)/In_{0.2}Ga_{0.8}As(50Å)/In_{0.38}Ga_{0.62}P(50Å)$ SQW with strain compensation

3. Experiment

Materials were grown in a gas-source molecular beam epitaxy (GSMBE) system modified from a Varian GEN II MBE system, equipped with two 2200 l/s cryopumps. Elemental indium and gallium were used, and arsenic and phosphorus dimers (As_2 and P_2) were obtained by cracking pure arsine and phosphine, respectively, at 1000°C.

Strained QW structures, with and without strain compensation, were grown consecutively on semi-insulating (100) GaAs substrates at 500°C. The typical SQW structure used was $In_xGa_{1-x}P(50Å)/In_{0.2}Ga_{0.8}As(50Å)/GaAs(20Å)/In_xGa_{1-x}P(50Å)$, and the MQW structure was 10 periods of $In_yGa_{1-y}P/GaAs(9Å)/In_{0.1}Ga_{0.9}As/GaAs(9Å)$. The reason for using the GaAs interfacial layer is described in the next section. The chamber pressure during growth was $3\times10^{-6} \sim 6\times10^{-6}$ Torr. The total hydride flow rate was 2~3SCCM. The lattice-mismatch of InGaP on GaAs (less than $\pm$ 10^{-3}) for strain-uncompensated structures was determined by double-crystal x-ray rocking curves (DCXR) from bulk epitaxial layers. The quality of SQW and MQW was evaluated by photoluminescence (PL) and DCXR, respectively.

4. Results and discussion

During growth, we noticed that the interface of InGaP on InGaAs was always worse than that of InGaAs on InGaP and both interfaces of GaAs-InGaP, because the reflection high energy electron diffraction (RHEED) pattern changed from a streaky to a spotty pattern. Since the interface between GaAs and InGaP is good, a thin GaAs layer was inserted between InGaAs and InGaP to prevent P-As substitution around In atoms, resulting in an improved InGaAs-GaAs-InGaP interface dramatically.

With the improved interface, the SQW structure with strain compensation exhibited a better PL quality as shown in Fig. 2. Fig. 2(a) shows the PL of an $In_{0.48}Ga_{0.52}As/In_{0.2}Ga_{0.8}As/GaAs/In_{0.48}Ga_{0.52}P$ SQW structure, without strain compensation, but with a 20Å GaAs interfacial layer. The peak near 8300Å is due to GaAs; that at 9127Å is from the SQW; and that at 9381Å may be a defect- or impurity-related transition. Fig. 2(b) shows the PL spectrum of a strain-compensated $In_{0.38}Ga_{0.62}P/In_{0.2}Ga_{0.8}As/GaAs/In_{0.38}Ga_{0.62}P$ SQW. No peak near 9380Å is observed in this case. The beneficial effect of strain compensation is evident. The PL intensity of the strain-compensated SQW is more than twice that of the SQW without strain compensation, and the line width is slightly narrower.

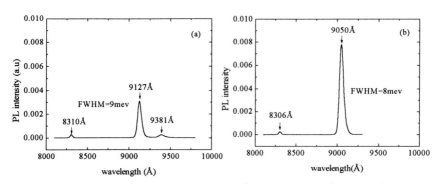

Fig. 2 Strain compensation effect in $In_xGa_{1-x}P(50Å)/In_{0.2}Ga_{0.8}As(50Å)/GaAs(20Å)In_xGa_{1-x}P(50Å)$ SQWs from 10K photoluminescence spectra. (a) x=0.48 and (b) x=0.38

The DCXR results of InGaP/InGaAs MQWs are shown in Figs. 3 and 4. Curve (a) in Fig. 3 is from a 10-period $In_{0.54}Ga_{0.46}P(100Å)/GaAs(9Å)/In_{0.10}Ga_{0.90}As(90Å)/GaAs(9Å)$

MQW structure without strain compensation, whereas curve (a) in Fig. 4 is from a 10-period $In_{0.42}Ga_{0.58}P(105Å)/GaAs(9Å)/In_{0.11}Ga_{0.89}As(89Å)/GaAs(9Å)$ MQW structure with strain compensation. Both curve (b) in Figs. 3 and 4 are dynamical-theory simulations. Peak A is from the 3000Å $GaAs_{0.96}P_{0.04}$ buffer layer The 4% P in the buffer layer is due to the fact we used only one cracker for both arsine and phosphine. The zeroth order satellite peak of Fig. 3(a) is at -403 arcsec, i.e., the net strain in the MQW stack is about 5.1×10^{-3}, whereas the zeroth order peak of Fig. 4(a) is at -158 arcsec, i.e., the net strain is about 6.5×10^{-4}. The full width at half maximum (FWHM) of the first-order peak in Fig. 3(a) is about 109 arcsec, whereas that in Fig. 4(a) is 88 arcsec. The strain-compensated structure with a reduced net strain shows a better crystalline quality than the uncompensated MQW structure.

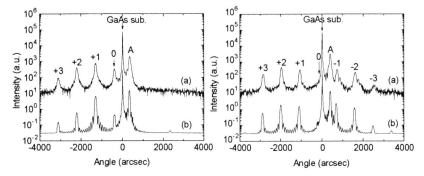

Fig. 3 (a) An x-ray rocking curve from a 10-period InGaP/InGaAs MQW structure without strain compensation. (b) A dynamical-theory simulation; peak A is from the 3000Å $GaAs_{0.96}P_{0.04}$ buffer layer

Fig. 4 (a) An x-ray rocking curve from a 10-period InGaP/InGaAs MQW structure with strain compensation. (b) A dynamical-theory simulation; peak A is from the 3000Å $GaAs_{0.96}P_{0.04}$ buffer layer

Summary

The driving force for misfit dislocation generation in strained InGaP/InGaAs/InGaP SQW structure with and without strain compensation was compared. To design a strain-compensated structure that is stable during and after growth, first, the average strain in the whole stack must be kept close to zero; and second, the thickness of each layer must be less than h_c. The strain-compensated SQW shows a better PL quality than one without strain compensation, and the strain compensated MQW structure exhibits much narrower satellite peaks than one without strain compensation.

Acknowledgment

The authors wish to acknowledge the Air Force Wright Laboratory (F. Schuermeyer and C. Cerny) and Hughes Aircraft Co. (C. S. Wu) for their support of this work.

References

[1] Thijs P J and Dongen T V 1989 Electron. Lett. **25** 1735
[2] Koren U, Oron M, Yong M G, Miller B I, Demiguel J L, Raybon G and Chien M 1990 Electron. Lett. **26** 465
[3] Matthews J W and Blakeslee A E 1974 J. Crystal Growth **27** 118
[4] Teng D, Mandeville P and Eastman L F 1994 J. Crystal Growth **135** 36
[5] Houghton D C, Davies M and Dion M 1994 Appl. Phys. Lett. **64** 505
[6] Anan Takayoshi, Sugou Shigeo, Nishi Kenichi and Ichihashi Toshinari 1993 Appl. Phys. Lett. **63** 1047

$(InAs)_1/(GaAs)_4$ Short Period Strained Layer Superlattices Grown by Molecular Beam Epitaxy

J.H. Lewis, M.G. Spencer, J.A. Griffin, and D.P. Zhang
Materials Science Research Center of Excellence, Howard University, School of Engineering
Washington, DC 20059

F. Grunthaner, and T. George, Jet Propulsion Laboratory, Pasadena, CA 91109

Abstract. We present a study of the effects of growth conditions on the transport properties of $(InAs)_1/(GaAs)_4$ short period strained layer superlattices (SPSLSL's) grown on (100) GaAs substrates. The superlattices were grown at substrate temperatures of 420, 480, and 520°C, and under various arsenic beam equivalent pressures (BEP). Reflection high energy electron diffraction (RHEED) observations along the [110] azimuth were made to determine the growth mode [i.e. two-dimensional (2D) or three-dimensional growth (3D)], and set the growth conditions. Under optimum growth conditions, RHEED oscillations of an 80-period $(InAs)_1/(GaAs)_4$ superlattice were obtained. Two-dimensional electron gas (2DEG) mobilities of 7,600 $cm^2/V \cdot s$ at 300°K and 43,000 $cm^2/V \cdot s$ at 77°K have been obtained for high electron mobility transistor (HEMT) structures fabricated on 8-period $(InAs)_1/(GaAs)_4$ superlattices. To our knowledge, this 77°K value represents the highest mobility obtained for $(InAs)_1/(GaAs)_4$ SPSLSL superlattices.

I. Introduction

InAs/GaAs strain layer superlattices are of interest for use in high speed devices[1,2] and quantum well lasers[3]. Toyoshima[2] et. al reported an $(InAs)_1/(GaAs)_4$ SPSLSL HEMT channel grown at 540°C with a 77°K electron mobility of 25,000 $cm^2/V \cdot s$. In this work, we investigate the effects of growth conditions on the electron transport properties of $(InAs)_1/(GaAs)_4 x N$ superlattices, where N is the number of periods.

II. Experimental

SPSLSL were grown on nominally (100) GaAs substrates. In and Ga deposition rates were set by RHEED oscillation techniques. The arsenic pressure was provided by a cracker cell which was set to 635°C. At this setting the primary growth species produced is As_4. The InAs and GaAs growth rates for this study were 0.18 ML/s and 1.0 ML/s respectively. The In and Ga fluxes were adjusted to provide complete monolayers. The superlattice was grown under constant As pressure and growth interrupts were performed after the deposition of each InAs monolayer and at the end of each period. During the growth of the SPSLSL RHEED oscillations were recorded using a computer controlled image capture system. To study the 2DEG mobility of the $(InAs)_1/(GaAs)_4xN$ superlattices, HEMT structures were grown. A typical structure was as follows: a 5000Å undoped GaAs buffer, undoped InAs/GaAs superlattice channel, a 60Å undoped $Al_{0.3}Ga_{0.7}As$ spacer layer, a 300Å n-type $Al_{0.3}Ga_{0.7}As$ supply layer ($n=2x10^{18}$ cm^{-3}), and a 200Å n$^+$ GaAs cap. This structure produced a 2DEG carrier concentration of approximately $1.5x10^{12}$ cm^{-2}. A HEMT consisting of an $In_{0.2}Ga_{0.8}As$ alloy channel was grown for comparison. The SPSLSL growths were performed as a function of the arsenic pressure as well as the substrate temperature.

III. Result and Discussion

Superlattices grown at 420°C exhibited the strongest RHEED oscillations provided that the As pressure was sufficient to maintain an As-stable surface. The As$_4$ pressure was determined by finding the minimum pressure at which As-stable growth of GaAs could be maintained. Figure 1 shows the RHEED oscillations for a $(InAs)_1/(GaAs)_4x80$ superlattice grown at 420°C and with an As beam equivalent pressure of $5x10^{-6}$ Torr.

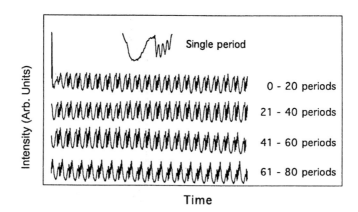

Figure 1. RHEED oscillations for an $(InAs)_1/(GaAs)_4$x80 superlattice.

A streaked RHEED pattern was observed during the growth of all 80 periods. TEM micrographs confirm the existence of an 80-period superlattice. Superlattices grown at reduced As pressures resulted in a metal-stable surface after the deposition of only a few monolayers. The As requirements at higher substrate temperatures differ from those at 420°C. The minimum As pressure required to maintain As-stable growth of GaAs at 420°C was sufficient to maintain an As-stable surface during growth of the superlattice. However, the RHEED pattern for superlattices grown under these conditions would transition from a streaked to a spotty pattern after growth of only a few monolayers, indicating 3D growth, and thereby preventing further intensity oscillations. The degree of spotting was inversely related to the As pressure. To maintain a 2D growth mode it was necessary to increase the As pressure. Figure 2 shows the RHEED oscillations for an $(InAs)_1/(GaAs)_4 \times 8$ superlattice grown at a substrate temperature of 520°C and an As BEP = 2.0×10^{-5} Torr.

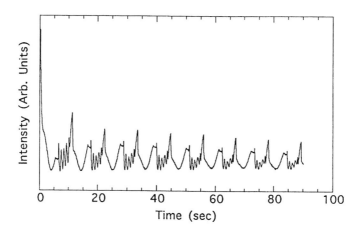

Figure 2. RHEED oscillations for an $(InAs)_1/(GaAs)_4 \times 8$ superlattice.

Shown in Figure 3 is the dependence of the 2DEG Hall mobility versus As_4 pressure for SPSLSL's grown at 420, 480, and 520°C. Figure 3a shows a constant mobility for superlattices grown at As pressures greater than 5×10^{-6} Torr. RHEED observations under these conditions indicate As-stable 2D growth of the superlattice. The reduction in 2DEG mobility seen at lower As pressures corresponded with the observation of metal-stable growth. Figure 3b shows a similar dependence of 2DEG mobility on As_4 pressure at 480°C. The mobility is essentially constant under high As pressures, and as the As_4 pressure is reduced the mobility decreases. However, at 480°C the As pressure at which the mobility begins to decrease does not correspond with the observation of metal-stable growth. Streaked RHEED patterns were observed during growth of the superlattice for samples that produced a high 2DEG mobility. Reducing the As pressure resulted in the observation of spotty RHEED patterns during the InAs cycle. The reduction in mobility corresponds with the observation of 3D growth of InAs. Figure 3c shows a constant 2DEG mobility for superlattices grown at 520°C with As pressures of 1.3 to 6.8×10^{-5} Torr.

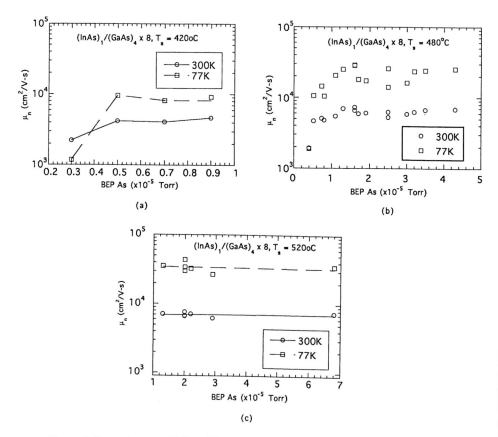

Figure 3. Dependence of Hall mobility on As beam equivalent pressure and growth temperature. (a) 420°C, (b) 480°C, (c) 520°C.

The 2DEG mobilities obtained for $(InAs)_1/(GaAs)_4 \times 8$ superlattice channel HEMT's grown at different temperatures is shown in Figure 4. The 2DEG mobility was found to be directly proportional to the substrate temperature. The highest 2DEG mobility for the superlattice grown at 520°C was 7600 and 43,000 cm²/V·s at 300°K and 77°K respectively. The 2DEG mobility for a corresponding alloy in our system was 6900 and 24,000 cm²/V·s at 300°K and 77°K respectively. This represents an 80% increase in the 2DEG mobility of the superlattice over the alloy at 77°K.

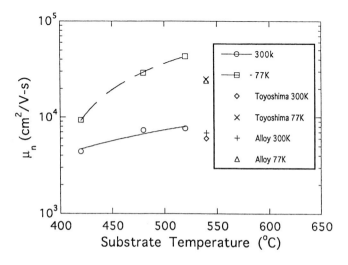

Figure 4. 2DEG mobilities vs growth temperature for $(InAs)_1/(GaAs)_4 \times 8$ HEMTs.

Comparison of the variation of RHEED oscillation intensity and 2DEG mobility with temperature shows that there is a clear trade off between maximum superlattice thickness and 2DEG mobility. To grow superlattices that exceed the critical thickness, it is necessary to go to lower substrate temperatures. At these lower temperatures the mobility is degraded. To study the effects of the number of periods on 2DEG mobility, a substrate temperature of 480°C was chosen. Figure 5 shows the 2DEG mobility versus N for an $(InAs)_1/(GaAs)_4 \times N$ channel HEMT. The 2DEG mobilities at 300°K and 77°K decreased by only 15% and 23% respectively for a superlattice with a thickness nearly double that of the critical thickness of a corresponding alloy.

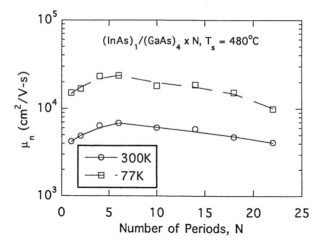

Figure 5. 2DEG mobilities vs N (periods) for $(InAs)_1/(GaAs)_4 \times 8$ HEMTs

IV. Summary

High mobility SPSLSL channels have been demonstrated. The mobility of the material is directly related to substrate temperature. Provided arsenic pressure exceeds a critical value it is possible to grow material whose mobilities exceeds that of the corresponding alloy. Further, it was demonstrated that it is possible to exceed the Matthews Blakeslee limit and still maintain good mobilities, particularly at lower growth temperatures.

V. References

[1] T. Yao, Jpn. J. Appl. Phys. 22 L680 (1983).
[2] H. Toyoshima, K. Onda, E. Mizuki, N. Samopo, M. Kuzuhara, T. Itoh, A. Okamoto, T. Anan, T. Ichihashi, J. Appl. Phys. 69 3941 (1991).
[3] N.K. Dutta, N. Chand, J. Lopata, Appl. Phys. Lett. 61 7 (1992).

Growth and Characteristics of High Optical Quality Lattice Matched GaInAsP Layers and GaInAsP/GaInAs Quantum Wells on InP by MBE using Solid Phosphorus and Arsenic Valved Cracking Cells

J. N. Baillargeon, A. Y. Cho, and F. A. Thiel
AT&T Bell Laboratories, 600 Mountain Avenue, Murray Hill, NJ 07974

P. J. Pearah, and K. Y. Cheng
University of Illinois, ECE Department, Urbana, IL 61801

Abstract. High optical quality lattice matched GaInAsP layers and GaInAsP/GaInAs quantum wells on InP were grown by solid source MBE. Separate valved cracking cells, one utilizing *in-situ* generated white phosphorus and the other arsenic, were used to achieve a column V stoichiometric reproducibility of <1% for quaternary layers. Photoluminescence emission at 300K from the quaternary layers displayed spectral widths as narrow as 40.6 meV at $\lambda \sim 1.43$ μm. GaInAs QW emission ($\lambda \sim 1.55$ μm) at 77K displayed spectra as narrow as 4.3 meV for a 75Å width. The data show that the quality of GaInAsP/GaInAs grown on InP with all solid sources is as good as that produced by any other method.

1. Introduction

Growth of phosphorus-based III-V semiconductor compounds by MBE is generally performed with highly toxic hydride and metalorganic sources.[1] A less hazardous growth process, such as that utilizing solid phosphorus and arsenic sources, is more desirable because of reduced environmental and safety pressures. Solid arsenic has been used successfully for the growth of III-V compounds because only one primary form of arsenic exists within MBE operating pressure and temperature ranges. Solid phosphorus though has several primary allotropes that have different vapor pressures for the same vapor species. As an effusion cell source, red phosphorus has not resulted in a reproducible stoichiometric growth. This is due to the inherent difficulty in conducting growth with P_4, as opposed to P_2, and the uncontrolled generation of the white allotrope from the red source inside the cell. The inability to control the transformation rate of the red phosphorus source to white phosphorus within the cell results in an uncontrollable beam flux. Growth of lattice matched compounds such as GaInAsP on InP with solid sources is an impractical goal unless a constant phosphorus flux can be maintained. Separate valved cracking cells designed specifically for phosphorus and arsenic allows for such precise control of the beam fluxes. The phosphorus cell used in these growth experiments is fundamentally different from that used previously to grow InAsP/InP structures.[2]

2. Experimental

The growth system was a Riber 2300. The valved cracking cells are a Riber KPC40 for phosphorus and a EPI RB500P for arsenic. The arsenic flux is generated from As_4 vapor supplied by the arsenic cell oven which is held constant during growth at 380 °C. The phosphorus flux is generated from P_4 vapor supplied by an *in-situ* generated white phosphorus source. The temperature of the white phosphorus during growth was 25 °C. White phosphorus was derived *in-situ* by sublimation of the red phosphorus at 360 °C and the condensation of the P_4 vapor at ~70 °C. Generation of white phosphorus, using the above parameters, for a period of 5 hours enables the Riber KPC40 cell to operate 33 hours while supplying a beam equivalent pressure (BEP) of 1×10^{-6} Torr. The cracking zone temperature during growth was 920 °C for both cells. The performance characteristics of the phosphorus cell has been detailed elsewhere.[3]

Growth of the lattice matched GaInAsP layers and GaInAsP/GaInAs quantum wells (QWs) were performed on on-axis (100) Fe-doped InP substrates at temperatures ranging from 480 and 500 °C. The GaInAsP alloy used in this study was grown using a P_2 BEP of 3×10^{-6} Torr and As_2 BEP of 2.3×10^{-6} Torr. The growth rate of this alloy was 1.5 μm/h. QWs were produced by closing the In and Ga shutters, switching off the phosphorus flux, and pausing for 10s prior to opening the In and Ga shutters. At the completion of the QW the In and Ga shutters were again closed and the phosphorus beam was switched back into the chamber for 10s prior to initiating barrier layer growth. The pause at each interface was inserted to allow for equilibration of the P_2 beam flux. The arsenic flux was not modulated or switched during growth to permit assessment of the P_2 beam stability, derived from the *in-situ* white phosphorus source, on the optical properties of the layers.

3. Results

Data depicting three representative 300K photoluminescence spectra of eight 1 μm thick $Ga_{0.30}In_{0.70}As_{0.68}P_{0.32}$ samples is shown in Fig. 1. All samples, having been produced with the same growth parameters, show a maximum (sample-to-sample) deviation in the peak emission wavelength of 130Å ($\Delta\lambda/\lambda \sim 0.9\%$). In terms of peak emission energy, this equates to a maximum spread which is < 10 meV (at room temperature). The measured sample-to-sample deviation in lattice constant for the samples was $|\Delta a/a| < 1 \times 10^{-3}$. This corresponds to a maximum variation of <1% in the arsenic/phosphorus stoichiometry. The 300K spectral full-width-at-half-maximum (FWHM) for seven of the samples were between ~41 - 44 meV; the eighth sample measured 63 meV owing to a change in substrate temperature during growth. The 300K peak emission intensities were also very similar for all samples.

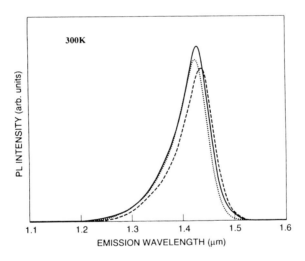

Fig. 1 The 300K photoluminescence intensity spectra for three of the eight $Ga_{0.30}In_{0.70}As_{0.68}P_{0.32}$ samples. The samples represent the maximum, minimum and mid-range emission energy spectra recorded for the eight epitaxial layers.

The 77K photoluminescence intensity spectrum, measured with a LN_2 cooled Ge detector, of a sample with four QWs is displayed in Fig. 2. The structure consists of a 1 μm thick $Ga_{0.30}In_{0.70}As_{0.68}P_{0.32}$ buffer layer followed by four $Ga_{0.30}In_{0.70}As$ wells with thickness of 100Å, 75Å, 50Å and 25Å, each separated by a 200Å $Ga_{0.30}In_{0.70}As_{0.68}P_{0.32}$ barrier layer.

The FWHMs measured at 77K are 10.3 meV, 4.3 meV, 5.7 meV and 10.8 meV, corresponding to 100Å, 75Å, 50Å and 25Å well widths, respectively.

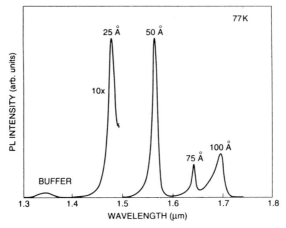

Fig. 2 The 77K photoluminescence intensity emission spectrum of a sample with four $Ga_{0.30}In_{0.70}As$ quantum wells with $Ga_{0.30}In_{0.70}As_{0.68}P_{0.32}$ barriers.

The spectral width of all the QWs are quite narrow indicating little phosphorus has incorporated into either of the QWs during growth. This is directly attributable to the rapid on/off switching of the P_2 beam from the cell.[4] The narrow QW widths also suggest that the As/P interfaces undergo no significant degradation during the 10s pause, independent of a P_2 flux. Somewhat longer pause times of 15s did not result in significant broadening of the QW emission spectra.

The results presented demonstrate that precise control of the As_2 and P_2 beam fluxes are possible using valved cracking cells that are of the proper design. Its technological importance as an eventual replacement for the more toxic hydrides and metalorganic sources in MBE appears promising.

References

[1] M. B. Panish 1993 *Gas Source Molecular Beam Epitaxy* (Berlin, Springer-Verlag).
[2] J. P. R. David et al 1993, *Fourth Intern. InP and Related Materials Conference* (Paris).
[3] J. N. Baillargeon, A. Y. Cho, and R. J. Fischer, to be published in 1994 *J. Vac. Sci. Technology B*, Nov-Dec.

Inst. Phys. Conf. Ser. No 141: Chapter 2
Paper presented at Int. Symp. Compound Semicond., San Diego, 18–22 September 1994
© *1995 IOP Publishing Ltd*

A Theoretical and Experimental Analysis of Non-Planar Epitaxy Using Shadow Masked MOCVD Growth

E.A. Armour, S.Z. Sun, and S.D. Hersee

University of New Mexico, Center for High Technology Materials, Albuquerque, NM 87131 USA

Abstract. Shadow masked growth is unique in its ability to deposit smoothly-varying non-planar epitaxial layers that are free from macroscopic facets. It is proposed that increased lateral gas phase diffusion and the absence of facets, are due to the recombination of surface species with organic radicals that are temporarily 'trapped' in the shadow mask channel. This kinetic mechanism is supported by a finite element analysis model that accurately predicts the shape of the non-planar growth profiles as a function of the shadow mask geometry. This work also shows that in conventional non-planar growth, it is vapor phase diffusion rather than surface diffusion, that is primarily responsible for facet formation.

1. Introduction

Non-planar epitaxial growth allows the creation of optoelectronic device structures with novel optical waveguiding and electronic confinement properties [1-5]. Non-planar epitaxy generally exhibits facetting and the anisotropy of growth rate, composition and doping on different facets can be used to advantage. However, there are applications where a high quality, smoothly varying, non-facetted epilayer is more appropriate. Such applications include multiwavelength laser arrays [6], wide spectrum LED's [7], integral lenses on vertical cavity lasers and a continuous antiguide laser [8].

Vapor phase crystal growth through a shadow mask window allows the growth of non-planar epitaxial layers, that vary smoothly in thickness and are free of macroscopic facets [9-10]. Furthermore, this technique appears to be controllable and reproducible. Originally proposed many years ago [11-13], the progress of shadow masked growth has been limited by the absence of a complete understanding of its physical basis and by the lack of a model to accurately predict its behavior. In this paper, we show that if we take into account methyl radicals that are temporarily trapped under the shadow mask, we can accurately describe the non-planar growth behavior using a simple 2-D finite element analysis.

2. Shadow Mask Growth Process

Shadow masked growth [3] is conducted through a window in an overhanging mask layer, that is suspended above the growth surface by a thick spacer layer as shown in Fig. 1. In the shadow masking process the first epitaxial growth is terminated by an $Al_{0.6}Ga_{0.4}As$ layer (thickness h=3-10 μm) and an undoped GaAs mask layer (thickness 1-2 μm). Selective wet-chemical etching is then used to open a stripe window (width W=15-150 μm) in the GaAs mask layer and to undercut the GaAs mask layer, forming the shadow mask channel.

Our MOCVD shadow mask growths were conducted at 725 ° C, 100 Torr, with a V/III ratio of 50 and using AsH_3, trimethylgallium, and trimethylaluminum. The growth rate was nominally 500 Å/min, except for the thick AlGaAs spacer layers used in the mask, where 1000 Å/min was used. In general multi-layer stacks, consisting of alternating periods of AlGaAs

(50nm) and AlAs (59nm), were grown as these allowed the evolution of the non-planar features to be observed.

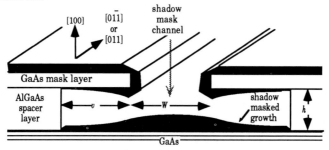

Figure 1. Perspective view of shadow mask geometry. The shadow mask geometry contains a window of width W etched in an overhanging GaAs mask layer (overhang distance v) which is suspended above a thick $Al_{0.6}Ga_{0.4}As$ spacer layer of height h.

3. Characteristics of Shadow Masked Growth

Shadow mask growths were made for stripes oriented along the [011] and [0$\bar{1}$1] directions. Both stripe orientations showed multiple facets on the overhanging GaAs mask, however the epilayers within the channel exhibit smoothly varying non-planar profiles absent of macroscopic facetting (Figure 2). The shape of this non-planar growth was independent of the stripe orientation and the gas flow direction. For the [011] direction stripes, the mask edge prior to shadow mask growth terminated in (111)B planes, which are non-growth planes under standard growth conditions. During the shadow mask growth, a variety of different facets were formed on the mask, as shown in Fig. 2. In addition, lateral growth under the mask terminated in a (111)B plane. For [0$\bar{1}$1]-oriented stripes, the mask edges were (111)A facets prior to growth and these extended during the shadow mask growth. In this case lateral growth under the mask terminated in a (111)A facet.

3.1 Kinetic Processes of Shadow Mask Growth

The standard vapor phase epitaxial model [15] consists of several kinetic events which may occur both in series and in parallel. A schematic of the general kinetic processes is shown in Fig. 3. In general, vapor phase epitaxy consists of the forward processes (a) diffusion from the vapor phase down to the adatom (surface) phase, (b) surface migration within the adatom phase, and (c) eventual incorporation into the crystal phase. These forward processes are accompanied by their respective reverse processes (d) desorption from the surface phase back into the vapor phase and (e) dissociation from the crystal phase back into the surface adatom phase.

MOCVD growth over non-planar features traditionally exhibits facet formation that extends laterally over distances of tens of microns. Large growth rate differences on these facets indicate that there is lateral diffusion of the reactant species. These gradients may exist either in the gas phase (a) or within the adatom (surface) phase (b). Most previous studies have concluded that these gradients are formed in the gas phase [2,16,17], and that the surface diffusion length is quite small. The existence of neighboring facetted and non-facetted regions within the shadow mask geometry, provides a unique opportunity to address this issue.

We have developed a finite element analysis model that calculates the growth profiles based upon simplified cases of the kinetic model. In all cases, dissociation (e) is ignored. For the computations, the shadow mask geometry is divided into a square mesh and a finite difference approximation is used to solve the individual steady-state diffusion equations (LaPlace's equation). The growth rate is assumed to be directly proportional to the flux of source species incident upon the growth surface.

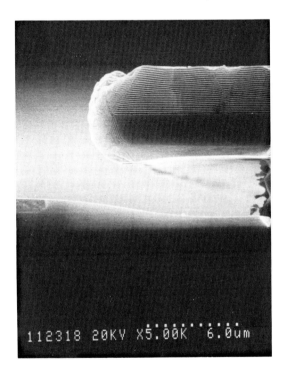

Figure 2. SEM micrograph of an [011]-oriented channel. The shadow mask growth consisted of alternating periods of $Al_{0.15}Ga_{0.85}As$ and AlAs. The growth exhibits a smoothly varying thickness profile within the shadow mask channel that is terminated in a (111)B non-growth plane.

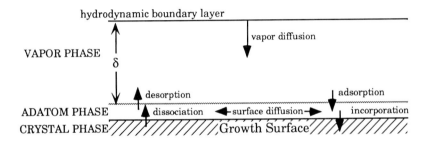

Figure 3. Kinetic processes in vapor phase epitaxy.

4. Finite Element Analysis of Shadow Masked Growth

Three different sets of boundary conditions were used in the finite element model. In all cases the growth profiles (Figs. 4 and 5) are normalized to the center thickness to remove the influence of the boundary height. This is a reasonable assumption under normal growth conditions where the boundary layer is significantly larger than the spacer height h, and changes in the boundary layer height have neglible effects on the normalized growth profile. The experimental data shown in Figs. 4 and 5 was obtained by high resolution SEM measurements.

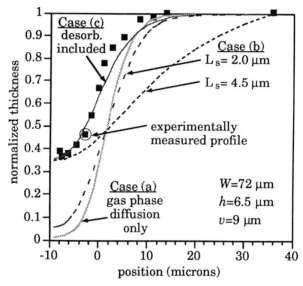

Figure 4. Normalized Finite element analysis profiles, comparing case (a) the model containing only vapor phase diffusion (thick black line), (b) vapor phase diffusion coupled to surface migration with surface diffusion lengths $L_s = 2.0$ µm (thin dotted) and $L_s = 4.5$ µm (thick dotted), (c) vapor phase diffusion with desorption (thin black) and experimentally measured profile (■). For both the simulations and the experimental measurements, the stripe width was 72 µm, and the spacer height 6.5 µm. Negative distance values indicate positions underneath the shadow mask overhang.

For the first case (a), only vapor phase diffusion has been considered. Here a constant concentration of reactants is assumed at the top of the boundary layer and vapor phase diffusion occurs to the growth surface. Once on the surface as adatoms, the reactants are promptly incorporated without diffusing along the surface. This case is similar to that examined by DeVlamynck et al. [14] and is represented as curve (a) in Fig. 4. This case predicts an abrupt decrease in growth thickness underneath the shadow mask, that is not observed experimentally.

In the second case (b), vapor phase diffusion and surface diffusion are allowed to occur and it is assumed that once reactants reach the surface they do not desorb. Again there is poor agreement between the predicted and measured profile. It is necessary to assume a surface diffusion length of 4.5 µm to achieve the experimentally observed scale of lateral diffusion but this gives a gradual growth profile that does not correlate with the measured profile.

For the third case (c), desorption from the surface (back into the vapor phase) was allowed, with a constant desorption rate which is directly proportional to the concentration gradient at the surface. Surface migration was not allowed. In this case the model accurately predicts the growth profile shapes, giving enhanced lateral diffusion while maintaining abrupt thickness reductions for thin spacer layers. Case (c) accurately predicts the non-planar profile shape for a wide range of shadow mask geometries, overhang distances v, spacer thicknesses h, and window widths W for a given growth condition and a fixed desorption factor.

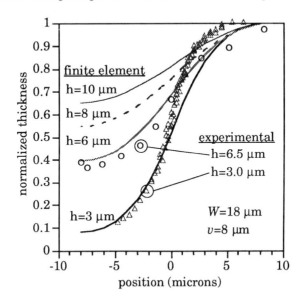

Figure 5. Comparison of different spacer heights using the finite element analysis model of case (c). The finite element models (solid lines) accurately fit the experimentally measured shadow mask growth profiles (width $W=18$ μm, spacer heights $h=3.0$ μm (Δ) and $h=6.5$ μm (O)) for a wide variety of geometries.

5. A Kinetic Model of Shadow Masked Growth

In shadow masked growth, the simultaneous presence of facetted growth on the upper mask layer and non-facetted growth in the channel, indicates that surface diffusion (which will be the same for both regions) is not dominating the lateral thickness variation. We conclude that vapor phase diffusion is therefore dominant. However, vapor phase diffusion alone (cases (a) and (b)) does not predict the extensive lateral diffusion that is observed experimentally. The most accurate prediction occurs when 98 % of reactants that originally enter the adatom phase, are allowed to desorbed back into the vapor phase. We propose that this high desorption rate is reasonable if we consider the unique geometry of shadow masked growth. When the metalorganic precursors decompose they will liberate methyl radicals. This is consistent with the work of Creighton et al. [18,19], who showed that the primary hydrocarbon products from the decomposition of TMGa on (100) GaAs are methyl radicals. In the shadow mask channel (and especially beneath the mask) these radicals will be temporarily "trapped" as their only escape route is through the mask opening. We propose that these radicals recombine with the surface group III species, returning them to the vapor phase where further lateral diffusion can occur. The existence of high concentrations of methyl radicals also explains the absence of facetting. If a facet begins to form, for example in a local region where the growth rate is higher, this will create a local high concentration of radicals that will enhance desorption and locally reduce the growth rate.

6. Conclusions

MOCVD growth through a shadow mask is unique in that it yields non-planar epitaxial layers which are free from macroscopic facets. We propose a kinetic growth mechanism to explain this observation, that is based upon the recombination of group III surface species with gas phase methyl radicals. This model explains the absence of facetting and when incorporated into a finite element analysis model, allows accurate prediction of the non-planar profile shape for a wide range of shadow mask geometries. The ability to accurately predict the profile shape will allow shadow masking to be used in wider array of applications and we anticipate many more novel device structures based on this improved understanding of the shadow mask growth technique.

7. Acknowledgements

The authors would like to acknowledge Dr. Piet Demeester, who provided us with details of his shadow mask process. The authors also wish to thank Kang Zheng, Jeff Ramer, and Dr. Saket Chadda for their helpful discussions. This work was supported by the Air Force Office of Scientific Research.

8. References

[1] Hersee S D, Barbier E, and Blondeau R 1986 *J. Crystal Growth* **77** 310
[2] Dzurko K M, Hummel S G, Menu E P, and Dapkus P D 1990 *J. Electron Mat.* **19** 1367
[3] Demeester P, Van Daele P, and Baets R 1988 *J. Appl. Phys.* **63** 2284
[4] Frateschi N C, Osinski J S, Beyler C A, and Dapkus P D 1992 *IEEE Photon. Tech. Lett.* **4** 209
[5] Bhat R, Kapon E, Hwang D M, Koza M A, and Yun C P 1988 *J. Crystal Growth* **93** 850
[6] Coudenys G, Moerman I, Zhu Y, Van Daele P, and Demeester P 1992 *IEEE Photon. Tech. Lett.* **4** 524
[7] Vermier G, Buydens L, Van Daele P, and Demeester P 1992 *Electron. Lett.* **28** 903
[8] Armour E A and Hersee S D 1994 *U.S. patent application* submitted May 31
[9] Demeester P, Buydens L, Moerman I, Lootens D, and Van Daele P 1991 *J. Crystal Growth* **107** 161
[10] Armour E A, Sun S Z, Zheng K, and Hersee S D, 1994 submitted for publication *J Appl. Phys.*
[11] Burnham R D 1984 *U.S. patent* #4,447,904
[12] Burnham R D 1984 *U.S. patent* #4,448,797
[13] Fekete D, Burnham R D, Scifres D R, Streifer W, and Yingling R D 1981 *Appl. Phys. Lett.* **38** 607
[14] De Vlamynck K, Coudenys G, and Demeester P 1991 *Appl. Phys. Lett.* **59** 3145
[15] Ohtsuka M and Suzuki A 1993 *J. Appl. Phys.* **73** 7358
[16] Gibbon M, Stagg J P, Cureton C G, Thrush E J, Jones C J, Mallard R E, Pritchar R E, Collis N, and Chew A 1993 *Semicond. Sci. Tech.* **8** 998
[17] Kayser O 1991 *J Crystal Growth* **107** 989
[18] Creighton J R 1991 *J. Vac. Sci. Tech. A* **9** 2895
[19] Zhu X Y, White J M, and Creighton J R 1992 *J. Vac. Sci. Tech. A* **10** 316

Selectively Grown Vertical Sub-100 nm Dual-Gate GaAs FETs

W. Langen, H. Hardtdegen, H. Lüth and P. Kordoš

Institut für Schicht- und Ionentechnik (ISI),
Forschungszentrum Jülich GmbH,
P.O. Box 1913, D-52425 Jülich, Germany

Abstract: A selectively grown vertical dual-gate GaAs FET was fabricated. In contrast to lateral devices the gate length is given by the thickness of an evaporated metal layer. It can be easily controlled in the sub-100 nm region. The two gates are separated by a SiO_2 layer and likewise embedded in SiO_2. A source-drain distance smaller than 1 μm is possible. The DC - IV characteristic shows a clear pentode-like behaviour with an output conductance of 0.3 mS for a total channel length of 0.25 mm. To study the optimization of the device performance, the channel width was systematically varied from 0.8 μm to 0.3 μm.

1. Introduction

Dual-gate GaAs FETs have been widely investigated and have been applied in mixers and gain-controlled amplifiers (Darling 1989). In comparison to single gate FETs dual-gate FETs show a smaller output conductance, a higher power gain and a reduced feedback. Some theoretical and experimental investigations have shown that a reduction of device geometry, in particular the gate length and the source-drain distance, would further improve these characteristics (Allamando et al. 1982). Using the concept of permeable base transistors, Vojak et al. (1984) constructed a device with two metallic gates to control the source-drain current. However, the output conductance did not show the desired characteristics of a dual-gate FET. This might be due to its geometric configuration, which is not like that of a dual-gate FET. It might also be due to the overgrowing of the gate structure with GaAs. Adachi et al. (1985) suggested that voids and crystal defects near the sensitive gate region may decrease the current density.

In contrast to overgrown or lateral devices, we constructed a *selectively* grown *vertical* dual-gate transistor where the gate length is given by the thickness of an evaporated metal layer, which can be easily controlled in the sub-100 nm region. The two gates are separated by a SiO_2 layer and likewise embedded in SiO_2 (see Fig. 1). Since the thickness of the SiO_2 can be very small, the total thickness of this layer structure and thus the source-drain distance can be smaller than 1 μm. This creates the possibility of investigating short channel effects. If the ratio of channel width to channel height is too small, selective epitaxy has to be used to fill up the channels. Selective epitaxy, instead of overgrowing the structure, results in a good GaAs morphology (Langen et al. 1994).

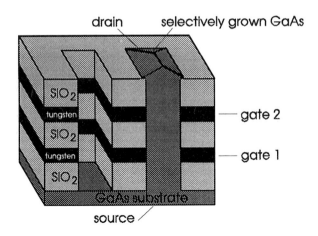

Fig. 1   Device cross-section of the selectively grown dual gate FET.

2. Fabrication

Dual-gate structure preparation starts by successively depositing SiO_2, tungsten, SiO_2, tungsten and SiO_2 on a GaAs (100) wafer. The tungsten with a thickness between 30 nm and 120 nm is evaporated and the SiO_2 with a thickness between 200 nm and 300 nm is deposited on the wafer by plasma-enhanced chemical vapour deposition. The channel width is defined by electron beam writing. We succeeded in constructing devices with a channel width of as little as 300 nm. A layer of 20 nm titanium and 30 nm chromium is evaporated on the wafer and partially removed using a Lift Off process. This metallic layer serves as a mask for the following reactive ion etching (RIE), where vertical trenches are defined. The SiO_2 is etched with CHF_3 and the tungsten with SF_6. The etch process is controlled by a laser interferometer which indicates when a given layer has been removed. Then the etch gas has to be switched from CHF_3 to SF_6 and back again. When the substrate has been reached the metallic mask can be totally removed in dilute HF acid. Vertical trenches of the layer structure can be achieved by means of RIE. The etched trenches are now filled up with n-doped GaAs using selective LP-MOVPE. An ohmic contact to the source and drain is realized by the deposition and annealing of a standard Ni/AuGe/Ni system. The tungsten gates, which are still buried under SiO_2, are partially opened by RIE, using the drain metallization as a self-aligned mask for the upper gate.

3. Selective epitaxy

Selective growth is carried out in an air-cooled horizontal low pressure reactor. N_2 is used as a carrier gas. The samples are grown at a reactor pressure of 20 hPa (Hardtdegen et al., 1992). The growth temperature is 650 ° C, the gas velocity is 0.5 m/s and the V/III - ratio is 50. An $n^+/n/n^+$ doping profile is used in the channels. The highly doped regions $N_D = 3 * 10^{18}$ cm^{-3} reduce the contact resistivity to source and drain and the lower channel doping $N_D = 3 - 8 * 10^{16}$ cm^{-3} gives a good Schottky contact to the gate. Because of the technological steps prior to epitaxy, the contaminations have to be removed without attacking the tungsten or the SiO_2. We use dilute HF to remove metallic contaminations such as titanium. We use sulphuric acid to remove the oxide from the GaAs. The trenches, which were etched by RIE, have a thickness of about

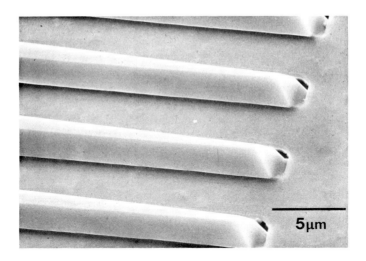

Fig. 2   SEM picture of the RIE-etched trenches filled with GaAs. The trenches are orientated in the [001] direction.

0.5 µm to 1 µm. They can be filled with GaAs by using selective epitaxy without any deposition of polycrystalline GaAs on the SiO_2 surface. A SEM picture of a transistor structure filled with selectively grown GaAs is shown in Fig. 2. The channels could not be filled by a non-selective process. The lateral growth on the SiO_2 surface would then close the channels before they were completely filled with GaAs grown in the vertical direction.

The GaAs growth rate depends on the orientation of the trenches relative to the substrate. The growth of GaAs produces facets. This is due to the different growth rate of GaAs along the crystallographic planes. The emerging facets are characterized by a relatively low growth rate. Because of the facetting, the GaAs has to be grown slightly above the SiO_2 surface. This has to be done to achieve contact between the GaAs and the side walls of the trenches. The electrical measurements presented in this paper were performed on devices where the trenches are orientated 15 ° off from the [0$\bar{1}$1] towards the [011] direction. In this direction, a close contact between the GaAs and the channel side wall can be achieved. However, at the end of the trenches a gap may remain between the grown GaAs and the $SiO_2/W/SiO_2/W/SiO_2$ gate structure. This is due to the different crystallographic direction of the grown GaAs at the end of the trenches. In order to avoid short circuits, these gaps can either be filled with a dielectric material or the metallic area can be aligned for the ohmic contacts so that they do not cover the gaps.

The growth rate of the GaAs in the trenches is further influenced by the ratio of the SiO_2 to the GaAs surface on the wafer. The growth rate on a patterned substrate is higher than the growth rate on a pure GaAs wafer. GaAs growth species migrate on the SiO_2 surface (Hiruma et al., 1990), which can lead to rough facets. To prevent these migrating species from reaching the channels, we surrounded the device trenches with GaAs areas. These GaAs areas are of no significance for the electrical properties of the device, but lead to good growth control and morphology.

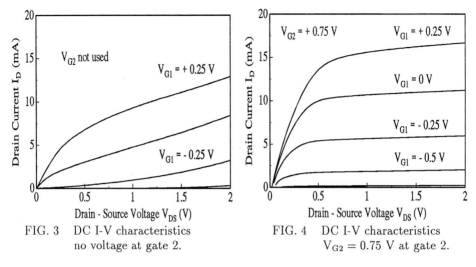

FIG. 3　DC I-V characteristics no voltage at gate 2.

FIG. 4　DC I-V characteristics $V_{G2} = 0.75$ V at gate 2.

4. Electrical Characterization

We obtained a high current flow from source to drain ($j_{DS} = 0.5$ mA/μm^2). We attribute this to the high crystal quality achieved by using selective epitaxy instead of overgrowing the gate structure. The effect of the second gate, which can be supplied with an independent voltage, on the DC-current voltage characteristic can be seen by comparing Figures 3 and 4. The gate nearer to the drain is called the screen gate (G_2) and the gate nearer to the source is called the control gate (G_1) following the designation used with lateral devices. In Figure 3 the screen gate is not supplied with a voltage and the typical characteristic of a vertical device is obtained, e.g. a high output conductance. Figure 4 gives an output characteristic of a constant voltage of $V_{G2} = + 0.75$ V applied to the screen gate. A clear pentode-like behaviour with an output conductance of 0.3 mS for a total channel length of 0.25 mm can be observed. To our knowledge, this value is one of the smallest values for vertical FETs ever attained. This demonstrates the feasibility of a vertical dual-gate FET with very short dimensions. The low output conductance is due to the constant potential at the screen gate.

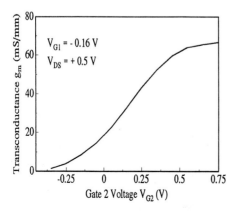

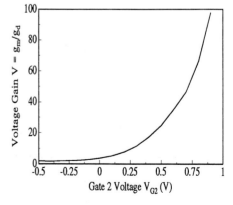

FIG. 5　Transconductance as a function of V_{G2}.

FIG. 6　Voltage gain as a function of V_{G2}.

Table 1. Electrical parameters as a function of the channel width.
$U_{DS} = 2V$, $U_{G2} = 0.75V$, $U_{G1} = 0V$

channel width µm	transconductance mS/mm	output conductance mS/mm	g_m/g_d ratio
0.3	69	0.5	138
0.4	63	8	8
0.5	54	25	2
0.6	32	33	1
0.7	13	142	0.1
0.8	8	152	0.006

This gate shields the drain potential from the gate next to the source. The shielding mechanism can be seen in the formation of a charge domain between the control gate and the screen gate (Dollfus P. and Hesto P., 1993).

The transconductance of the vertical dual gate FET is shown in Fig. 5. The values are normalized to the sum of the total channel length. This is the length of a single channel multiplied by the number of channels. The transconductance can be controlled by a voltage V_{G2} applied to the second gate. This can be used for an amplifier with a variable gain. If V_{G2} is negative, the channel can be pinched off, so that the transconductance tends to zero. But if a positive voltage V_{G2} is applied, the channel opens more and more and the transconductance rises until saturation. The output conductance is also a function of V_{G2}, and it decreases if the voltage becomes positive. This behaviour of the transconductance and the output conductance is advantageous for obtaining a high voltage gain V. It is given by the ratio of g_m and g_d (see Fig. 6).

To study the influence of the channel width on the electrical parameters, we varied the channel width from 0.8 µm to 0.3 µm. The results are summarized in Table 1. We fabricated these transistors in the same epitaxy. It follows that the channel height (1040 nm), the cleaning process and the doping profile (channel doping $N_D = 3 * 10^{16} cm^{-3}$)

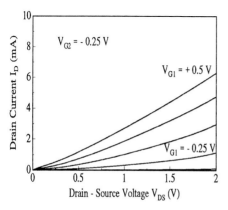

FIG. 7   DC I-V characteristics
$V_{G2} = -0.25$ V at gate 2.

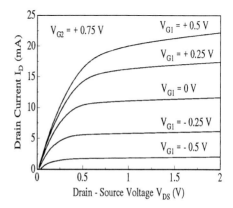

FIG. 8   DC I-V characteristics
$V_{G2} = 0.75$ V at gate 2.

are the same for all devices. We found that the transconductance increases with decreasing channel width. This behaviour is well known for lateral transistors with very short gates and has proved true in this case as well. The output conductance decreases with decreasing channel width and in the case of smaller channel width, the screening effect of the second gate becomes more effective. Through the combination of these effects, the voltage gain V can be raised from V=0.1 for a channel width of 0.7 μm to V=138 for a channel width of 0.3 μm. By properly adjusting the constant potential applied to the second gate the DC - current voltage characteristic can be switched from a triode-like to a pentode-like behaviour. Simultaneously, the source drain current increases with increasing gate 2 voltage, as can be seen by comparing Fig. 7 and Fig. 8. This is due to barrier lowering, which is caused by the change from a negative to a positive potential in the screen gate region. The triode-like behaviour, which also occurs in conventional vertical transistors with a single gate, is associated with a negative voltage V_{G2}. When the voltage V_{G2} is changed to a positive value, the shape of the IV - curve changes increasingly to a pentode-like behaviour. As a result, only a positive second gate voltage is able to effectively reduce the output conductance. The smallest values for the output conductance can be reached when the difference $V_{G2} - V_{G1}$ is greatest (see Fig. 8). A limitation on the growing difference between V_{G2} and V_{G1} is, of course, the leakage current in the metal/semiconductor/metal diode.

5. Conclusion

We have demonstrated that a second gate used in vertical devices with a $SiO_2/W/SiO_2/W/SiO_2$ gate structure can be very useful to avoid some undesirable short channel effects. In particular, the output conductance can be decreased and thus the voltage gain increased. Selective epitaxy instead of overgrowing leads to a high current density. For narrow and deep channels, selective epitaxy is necessary to fill up the channels. The output characteristic can be switched from a triode-like to a pentode-like behaviour by a potential applied to the screen gate.

Acknowledgments

The authors would like to thank A. v.d. Hardt and M. Nonn for Electron Beam Lithography, and T. Arrafat and C. Wirtz for performing the selective epitaxy.

References

Adachi S., Ando S., Asai H. and Susa N., IEEE Electron Device Letters, 1985, Vol. 6, pp. 264-266

Allamando E., Salmer G., Bouhess M. and Constant E., Electronics Letters, 1982, Vol. 18, pp. 791-793

Darling R. B., IEEE Trans. on Microwave Theory a. Techn., 1989, Vol. 37, pp. 1351-1360,

Dollfus P. and Hesto P., Solid-State Electronics, 1993, Vol. 36, pp. 711 - 715

Hardtdegen H., Hollfelder M., Meyer R., Carius R., Münder H., Frohnhoff S., Szynka D. and Lüth H., J. Crystal Growth 1992, Vol. 124, pp. 420-426

Hiruma K., Haga T., and Miyazaki M., J. Crystal Growth 1990, Vol. 102, pp. 717-724

Langen W., Raafat T.M., Hardtdegen H., v.d. Hart A., Meyer F., Lüth H. and Kordoš P., Proc. 24th ESSDERC, 1994, pp. 635-638

Vojak B.A., McClelland R.W., Lincoln G.A., Calawa A.R., Flanders D.C., and Geis M.W., IEEE Electron Device Letters, 1984, Vol. 5, pp. 270-272

> # Selective Regrowth of Highly Resistive InP Current Blocking Layers by a Low Pressure Metalorganic Vapor Phase Epitaxy

Dae Kon Oh, Hyung Mun Kim, Chongdae Park, Jeong Soo Kim, Seung Won Lee, HeungRo Choo, Hong Man Kim, Hyung Moo Park, and Sin-Chong Park

Semiconductor Technology Division, Electronics and Telecommunications Research Institute, Yusong P.O. Box 106, Taejon 305-600, Republic of Korea

Abstract. For the fabrication of planar buried hetero structure (BH) lasers, low pressure metal organic vapor phase epitaxy (MOVPE) has been used to regrow semi-insulating (SI) InP on various etched mesa structures. Excellent mesa side wall coverage without void and mask over-growth was obtained using a optimized non-reentrant etched mesa shape and a mask overhang length less than 1.2 µm. The lasers with highly resistive ($> 10^9$ Ωcm) SI InP current blocking layer exhibit a low (~ 10 mA) threshold current at room temperature. Comparison of laser leakage performance between different current blocking layers, i.e., SI only, SI/n, SI/n/SI/n, and p/n/SI/n, has been carried out. As a result, SI/n structure shows the lowest leakage performance in the forward lasing voltage region.

1. Introduction

The formation of current blocking layer is an important step in the fabrication of high bit rate InGaAsP/InP BH lasers. Even though a few investigations have been reported, a current blocking layer is still in interest for the best laser performance. Most BH lasers use p-n junction for the current blocking layer [1,2]. However, the reverse-biased current blocking junction results in a slow modulation speed because of large leakage current and parasitic capacitance [3]. More recently, lasers with SI current blocking layers [4,5] have provided advantages of high performance lasing characteristics, i.e., their low parasitic capacitance. These SI layers themselves have been found to be highly resistive because of their ability to compensate for shallow donor background levels by deep acceptor electron-trap levels. Since the negative charges of electrons trapped by deep level centers prevent further injection of electrons from n-layer into the SI layer, the layer becomes highly resistive. However, when a SI layer contacts a p-layer, "double injection" [6], which both electrons and holes are injected into SI layers, can occur. Therefore, we must focus on how to block double injection in the SI layers for the fabrication of high speed BH laser with minimal leakage current. In addition, when MOVPE is used, planarization of the final surface in the current blocking layer regrowth is also one of the important issues [7].

In this paper, we will describe an InGaAsP/InP planar BH lasers with SI-InP current blocking layers by vertical geometry low pressure MOVPE, and report on the comparison of leakage performance between various SI current blocking structures.

2. Experimental

The two MOVPE reactor systems were used in this work. One is the high speed susceptor rotational vertical type for the growth of 1.55 μm MQW double hetero structure (DH) layer, and the other is the vertical quartz reactor with 2 rpm susceptor rotation for SI re growth and 3rd p-layer growth. The 1.55 μm MQW DH layer was grown on a (100) oriented S-doped InP substrate. This hetero structure consists of a 0.1 μm n-InGaAsP (1.24 μm band-gap) wave guide layer (Si doped, n=1×10^{18} cm^{-3}), a strained MQW (1.55 μm wavelength, 8 periods) active layer, a 0.1 μm p-InGaAsP (1.24 μm band-gap) wave guide layer, and a 0.2 μm p-InP clad layer (Zn doped, p=5×10^{17} cm^{-3}).

A mesa stripe was etched in the <110> direction using SiNx mask. The reactive ion etching (RIE) is used to form a mesa to 2.5 μm high and the chemical etching (HBr:H_2O_2:H_2O) adjusted it to 3.0 μm high and the active region to 2.0 μm wide for the optimal overhang length. The selective non-planar growth of current blocking layers was performed with the MOVPE growth parameters as shown in Table 1. After the removal of SiNx mask and InGaAs cap layer, 2 μm thick p-InP clad layer and 0.3 μm thick p-InGaAs ohmic layer were grown. SiNx film was deposited by plasma enhanced chemical vapor deposition, and then 5 μm stripes were opened. A Ti/Pt/Au p-contact was deposited and sintered at 425 °C for 30 sec using RTA. The back of the n-InP substrate was thinned to about 100 μm and a Cr/Au n-contact was evaporated. Finally, the Au layer was plated on p-contact metal to aid the die and th wire bonding.

3. Results and discussions

3.1. MOVPE grown Fe-doped SI InP

The SI-InP with the Fe concentrations between 7×10^{16} cm^{-3} and 6×10^{17} cm^{-3} was obtained by varying the input ferrocene mole fraction between 6×10^{-6} and 1.2×10^{-4}. The resistivity of SI-InP was ranged between 9×10^8 and 2.2×10^9 Ωcm at room temperature. The highest resistivity value was obtained at 6×10^{-5} Fe/In gas mole fraction with no evidence of crystal defects including Fe-P precipitates by transmission electron microscope analysis. The current-voltage characteristics of Fe-doped InP are shown in Fig. 1. The sample with n-SI-n type consists of Ti/Pt/Au contact(80×80 μm^2) and 2.6 μm thick SI layer. Resistivity of the sample was calculated to be 2.2×10^9 Ωcm. A breakdown field was approximately 9×10^4 Vcm^{-1}, which was consistent with the previous work [14]. Figure 1 shows that Fe-doped InP is ohmic for a bias voltage less than 3V. The increasing steepness of slope above 3V suggests the onset of space-charge current mechanism, as discussed by Macrander [15].

Table 1. Growth Parameters for Highly Resistive Fe-Doped SI-InP

Growth Temperature = 650 °C
Growth Pressure = 76 Torr
Ferrocene, [Fe(C_5H_5)$_2$] / Tri Methyl Indium, [(CH_3)$_3$In] = 5.9 x 10^{-5}
Total Gas Flow = 5 slpm
Growth Rate = 5.25 μm/hr
V/III = [PH_3] / [(CH_3)$_3$In] = 130

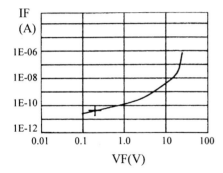

Fig.1. I-V Characteristics of SI-InP

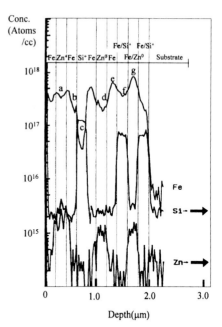

Fig. 2. SIMS Profile of Fe Atom in Si/Fe/Zn Doped Layer.

3.2 Diffusion Behavior of Zn/Fe/Si

The atomic inter-diffusion between Si and Fe, as well as, between Zn and Fe was observed by SIMS measurement. Figure 2 represents the SIMS doping profile of a sample comprising two different Zn doping regions (Zn^+; 2×10^{18} cm^{-3}, Zn^0; 5×10^{17} cm^{-3}), the high Si doping region (Si^+; 2×10^{18} cm^{-3}), and Fe/Zn^0 and Fe/Si^+ codoped regions. The Fe density in this particular structure was kept above the solubility limit which is assumed to be 7×10^{16} cm^{-3} at the growth temperature of 650 °C [16]. The Fe doping profile shown in Fig. 2 suggests the following: (i) the Fe impurities strongly diffuse to the Zn^+ doped region(mark "a") by zinc stimulated out-diffusion of iron as reported by E.W.A. Young group [14]; (ii) the Si doped layer does not allow the Fe inter-diffusion, because of its apparent difference with Zn doped layer, as marked by "c" and "a" in the figure; (iii) the low zinc doped InP (Zn^0) shows reduced out-diffusion behavior of iron compared to the Zn^+ region as marked by "d" and "a", with the similar trend of Zn out-diffusion to SI layer; (iv) the Fe dopants in Fe/Zn^0 codoped region have been gathered from the neighbored Fe/Si codoped InP layer (marks "g" and "f"), where as the Zn did not move to Fe/Si codoped layer due to Si barriers.

3.3 Selectively regrown profiles around etched mesas

For the fabrication of BH lasers, the non-planar selective regrowth of current blocking layer was performed around an etched mesa structure by low pressure MOVPE. The regrowth profile is sensitive to the parameters such as the mesa side wall shape and the extent of the etch mask overhang. To understand the growth mechanism on mesa, InP/InGaAs multi-pairs were grown and etched by stain etchant to see the growth profile contour.

Table 2. Etch Procedures

Sample Label	Step	Constituents	Substrate	Side Wall Type (height of neck from top)
a	1	HBr:H_2O_2:H_2O	MQW HD w/o InGaAs cap	reentrant (0.2 μm)
b	1	HBr:H_2O_2:H_2O	InP[S] bulk sub.	reentrant (1.1 μm)
c	1	RIE(CH_4:H_2)	MQW DH with InGaAs cap	reentrant
	2	HCl:HNO_3:H_2O		(0.4 and 0.8 μm)
d	1	RIE(CH_4:H_2)	InP[S] bulk sub.	vertical
	2	BOE		
e-h	1	HBr:H_2O_2:H_2O	MQW DH with InGaAs cap	non-reentrant

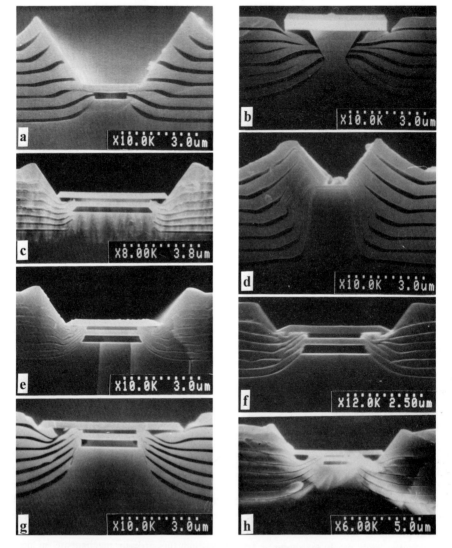

Fig. 3. Crossectional SEM Pictures of Regrowth Profile on Different Mesa Shapes, The Direction of All Mesa Stripe is <011>.

The etched mesas capped with SiNx were produced using various common wet chemical and/or dry etching technique. Figure 3 shows SEM cross sections of various sample regrown under low pressure (0.1 atm.) in vertical geometry reactor. The etch procedures labeled (a) - (e), shown in Table 2, resulted in general mesa shapes. For these samples, all the stripe orientations were in the <110> direction on the wafer surface and most of the etched mesas have mask undercut to prevent over growth. Some regrowth experiments were performed directly on an etched InP substrate patterned with SiNx stripe, and others were performed on the etched MQW DH layers grown by MOVPE. In the case of mask with overhang, when the growth reaches the height of the mesa, almost flat surface is obtained. As the growth extends beyond the height of the mesa, a shoulder appears at the edges of the growing layers due to lateral migration of reactants from the top of the mask. Depending on the height of the neck from the surface in the reentrant side wall shape (shown in label "a" and "b"), the growth profile shows a little difference due to no growth on the long side wall of (111)A reverse mesa plane. The vertical side wall allows the lateral and the facet oriented growths besides the planar growth, so that the over-growth may be more probable as the growth exceeds the height of the mask. The non-reentrant mesa (labeled "e" - "h" in Fig. 3) gave the planarized surface without lateral growth and mask over-growth as the most promising candidate for BH lasers.

The regrowth of wet etched mesas in the <110> directions with non-reentrant side walls was made with the variation of overhang length. Each of them consisted of etched mesas of overhang lengths 0.8, 1.2, 1.8, and 2.1 μm as labeled (e)-(h) in Fig. 3. The planar surfaces were obtained in all these samples, differing from the results of atmospheric pressure growth [17], which shows the anomalous facet growth over the mask in the short overhang mask. A remarkable difference of growth profile on overhang length was only the formation of void under the mask shown in overhang above 1.8 μm. This may be interpreted as reactant deficiency under the mask with long overhang length. Therefore, the optimized mask overhang length should be considered in the cases of various reactor pressures and etched mesa shapes.

3.4. A comparison of leakage performance of lasers with the different current blocking structures

The MQW planar BH lasers of 3 μm active width were fabricated with SI current blocking layers on the non-reentrant mesa side wall. The different current blocking structures, i.e., SI only, SI/n, SI/n/SI/n, and p/n/SI/n, were tested for the best lasing characteristics, especially the lowest leakage performance. In Fig. 4, the current-voltage characteristics of these lasers show the differences in relative leakage performances at several forward voltage regions. The leakage current increases at the lasing region in the order of SI/n, SI/n/SI/n, p/n/SI/n, and SI-only structure above 0.9 V, where as at the voltage region of below threshold point, it increases in the order of p/n/SI/n, SI/n/SI/n, SI/n, and SI-only between 0.9 and 0.6 V. The remarkable current increasing rate in the laser, with SI-only current blocking layer, may be mainly due to the "double injection" in the p-i-n structure. From the comparison of leakage performance, the best choice was SI/n structure, even though it has a room for improvements in SI/n/SI/n and p/n/SI/n structures. The lasing performance of typical SI/n structure was shown in Fig. 5, representing the 12 mA threshold current and 22% slope efficiency at 2 mW in the 1.55 μm peak wavelength.

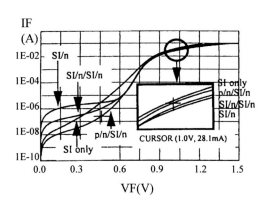
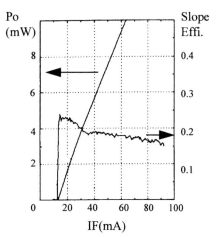

Fig. 4. I-V Characteristics of BH Lasers with Different Current Blocking Structures

Fig. 5. Lasing Characteristics of Planar BH Laser with a SI/n Current Blocking Structure.

4. Summary

We have grown current blocking layers, including Fe-doped InP, selectively on the mesa etched MQW DH epi layer by low pressure MOVPE. The resistivity of Fe-doped SI-InP was 2.2×10^9 Ωcm at the 6×10^{-5} Fe/In gas mole fraction. From the understanding of growth profile dependence on the shape of side wall and the overhang length, the planarised regrown current blocking layers were obtained at the overhang below 1.2 μm in the non-reentrant mesa shape. Using above results, we fabricated the 1.55 μm planar BH laser with SI/n current blocking layer, which shows the lowest leakage performance among SI-only, SI/n, SI/n/SI/n, and p/n/SI/n current blocking structures.

References

[1] A. W. Nelson, W. J. Devlin, R. E. Hobbs, C. G. D. Lenton, and S. Wong 1985 Electron. Lett. **21** 888
[2] H. Ishikawa, H. Imai, T. Tanahashi, and K. Takahei 1982 IEEE J. Quantum Electron. **QE-18** 1704
[3] Richard A. Linke 1984 J. Light. Tech. **LT-2** 40-43
[4] P. J. Williams, A. P. Webb, I. H. Goodridge, and A. C. Carter 1986 Electron. Lett. **22** 472
[5] B. I. Miller, U. Koren, and R. J. Capik 1986 Electron. Lett. **22** 947
[6] S. Asada, S. Sugou, K. Kasahara, and S. Kumashiro 1989 IEEE J. Quantum Electron. **25** 1362-8
[7] J. L. Zilko, B. P. Segner, U. K. Chakrabarti, R. A. Logan, J. Lopata, D. L. Van Haren, J. A. Long, and V. R. McCray 1991 J. Cryst. Growth **109** 264-271
[8] E. W. A. Young and G. M. Fontijn 1990 Appl. Phys. Lett **56** 146-7
[9] A. T. Macrander, J. A. Long, V. G. Riggs, A. F. Bloemeke, and W. D. Jonston, Jr. 1984 Appl. Phys. Lett. **45** 1297-8
[10] D. Franke, P. Harde, P. Wolfram, and N. Grote 1990 J. Cryst. Growth **100** 309-312
[11] T. Sanada, K. Nakai, K. Wakao, M. Kuno, and S. Yamakoshi 1987 Appl. Phys. Lett. **51** 1054-6

Double heterostructure produced by ordering in GaInP

L.C. Su, I.H. Ho, and G.B. Stringfellow

Departments of Materials Science and Engineering and Electrical Engineering, University of Utah, Salt Lake City, UT 84112

Abstract. Ordering has been studied in $Ga_xIn_{1-x}P$ with $x \approx 0.52$ grown by organometallic vapor phase epitaxy at a rate of 0.5 µm/hr. The GaAs substrates were misoriented from the (001) plane toward the $[1\bar{1}0]$ direction by angles of 0, 3, 6, and 9°. Growth temperature is a key factor in determining both the rate of the ordering process, occurring at the surface during growth, and the annealing process, occurring in the layer after growth. Results are reported for growth at temperatures of 520°C. These layers are only weakly ordered. The lowest 10-K PL peak energy of 1.936 eV, about 60 meV below that of highly disordered $Ga_{0.52}In_{0.48}P$, occurs for a misorientation angle of 9°. The degree of order, judged from the PL peak energy, generally increases with increasing misorientation angle. The results indicate that the annealing process is slow at 520°C, but so is the ordering process. A double heterostructure was produced by growing sequential layers at temperatures of 740°C (disordered), 620 °C (highly ordered), 520°C (slightly ordered).

1. Introduction

Atomic-scale ordering is a phenomenon observed in essentially all III/V alloys[1]. It is an important phenomenon for materials used in photonic devices since it is known to produce a reduction in the band gap energy[2]. In fact, the use of order/disorder structures in GaInP has contributed to the production of the highest efficiency solar cells reported to date[3].

Ordering is believed to be driven by the thermodynamic stability of the Cu-Pt ordered structure, with ordering on {111} planes, at the (001) surface reconstructed to produce [110] group V dimer rows[1]. The Cu-Pt structure is not stable in the bulk. Thus, high temperature annealing results in a decrease in the degree of order[4]. Kinetic factors also appear to be important since, for example, high growth rates of 12 µm/hr produce material with very little ordering[5]. The use of low growth temperatures also appears to decrease the rate of the ordering process occurring at the surface during growth[6-7].

Surface steps produced by intentional misorientation of the nominal (001) substrate are also important. For low growth temperatures (e.g. 570°C) [110] steps on the surface apparently assist the formation of the Cu-Pt structure at the surface during growth[7]. However, for larger misorientation angles and/or higher growth temperatures the increase in the annealing process occurring below the surface during growth caused by the [110] steps is more significant.

This paper consists of two parts. The first involves the extension of our earlier systematic studies[7-9] of the effects of growth parameters on the extent and type of ordering observed during organometallic vapor phase (OMVPE) epitaxial growth of GaInP lattice matched to the GaAs substrates. This paper reports results of ordering at a growth temperature of 520°C, the lowest reported for OMVPE growth. The second part of the paper involves the use of temperature to produce a disorder/order/disorder double heterostructure by using a different growth temperature for each of the three $Ga_{0.52}In_{0.48}P$ layers.

2. Experimental

The GaInP epitaxial layers were grown by OMVPE on semi-insulating GaAs substrates misoriented by angles of 0, 3, 6, and 9° (± 0.5°) in the [$\bar{1}$10] direction, to produce various densities of [110]-oriented steps on the surface. Substrate preparation consisted of degreasing followed by a 5 min. etch in a 1% Bromine in methanol solution. A horizontal, atmospheric pressure OMVPE reactor was used. The source materials were trimethylgallium (TMGa at -9°C), trimethylindium (TMIn at 25°C), and phosphine. The carrier gas was Pd-diffused hydrogen with a flow rate of 4 slm. The growth temperature was constant at 520°C for all layers except the double heterostructure, where the layers were grown at 740, 620, and 520°C, with a 5 min. interruption between layers. The input phosphine partial pressure was 2.3 torr. The growth rate was constant at 0.5 μm/hr. The GaInP layers were typically approximately 0.7 μm thick. Before beginning the GaInP growth, a 0.15 μm GaAs buffer layer was deposited using TMGa and arsine to improve the quality of the GaInP layers. Good morphology GaInP layers could be obtained only when the GaAs layer was grown at a temperature higher than 520°C. For the layers described here a GaAs growth temperature of 670°C was used.

The solid composition of the GaInP layers was measured by x-ray diffraction using Cu K_α radiation. The 10K-PL was excited with the 488 nm line of an Ar^+ laser. The emission was dispersed using a Spex Model 1870 monochromator and detected using a Hamamatsu R1104 head-on photomultiplier tube. [110] cross-sectional transmission electron microscope (TEM) samples were prepared by cleaving two facets, glued face to face and mechanically polished, followed by Ar-ion milling to electron transparency at 77 K. The transmission electron diffraction (TED) patterns and TEM images were obtained using a JEOL 200 CX scanning transmission electron microscope operated at 200 KV.

3. Results and Discussion

For the optimized growth conditions, the surface morphologies of the epitaxial layers for all values of the misorientation angle, ϑ_m, were mirror-like to the naked eye and virtually featureless when viewed by interference contrast optical microscopy. A very slight cross-hatch pattern was observed for some samples, even though the results of the x-ray diffraction measurements indicates that they are lattice-matched, i.e., the values of x for these alloys are 0.52±0.005.

The 10-K PL spectra for samples grown on substrates with misorientation angles of 0-9° to produce [110] steps on the surface are shown in Fig. 1. For all four samples, the PL intensity was comparable to measurements on samples grown at higher temperatures. However, a major difference is that the PL consists of two peaks. The lower-energy peak dominates for the excitation intensity of 10 mW (30 W/cm^2) used to obtain these spectra. Increasing the excitation intensity resulted in a blue shift of approximately 6 meV/decade for the lower energy peak and no discernible shift in the high energy peak. The higher energy peak becomes more dominant at higher excitation intensities, indicating that it is due to band-edge recombination while the lower energy peak is due to a

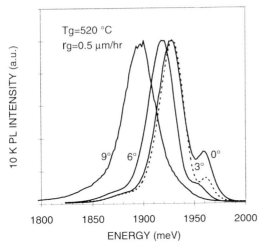

Fig. 1: 10-K PL spectra obtained for layers grown at 520°C. Substrate misorientation angles are indicated. Power is 10 mW.

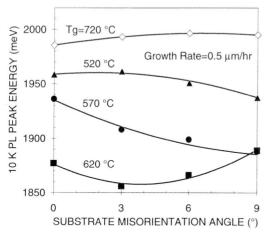

Fig. 2: 10-K PL peak energy versus substrate misorientation for various temperatures. Only the high energy peak is included for the 520°C samples.

recombination process involving an impurity. The peak separation of approximately 40 meV, suggests involvement of a residual acceptor impurity.

The PL peak energy is plotted versus misorientation angle in Fig. 2, with the substrate temperature during growth as the second parameter. Data are taken from the present study as well as from references 7-10. For the samples grown at 520°C, the peaks were deconvoluted, by assuming Gaussian peak shapes, to obtain the peak energies. The data plotted here show that the samples grown at 720 and 620°C, with a misorientation angle of 3°, represent the materials with the highest and lowest bandgap energies, respectively. For high growth temperatures the low degree of order is probably due to the rapidity of the annealing process that disorders the material below the surface during growth. In this case, the misorientation angle has little effect.

For growth at 520°C, the degree of order is low, although the samples are not as disordered as those grown at 720°C. The clearly observed decrease in the degree of order with decreasing growth temperature from 570 to 520°C suggests that annealing is not responsible for the low degree of order. More likely, the decrease in order is due to a decrease in the rate of the ordering process occurring at the surface during growth. As the rate of the surface ordering process decreases, the time for the surface atoms to rearrange becomes insufficient to segregate the In and Ga atoms into alternating rows before the arrangement is frozen by being covered by the next layer. This explains the decrease in the degree of order as the temperature is decreased from 570 to 520°C. It also explains the decrease in order with increasing growth rate[5]. It is seen that for growth temperatures of both 520 and 570°C the PL peak energy

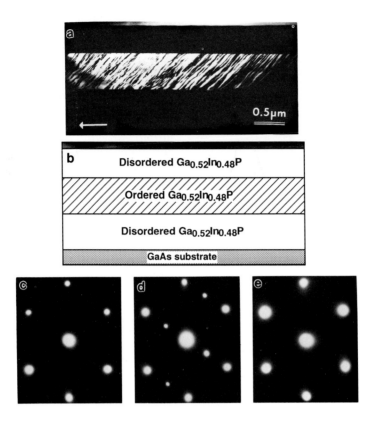

Fig. 3: Schematic diagram of disorder/order/disorder double heterostructure. Also shown are the TEM image and the TED patterns for the three layers individually.

shrinks as the substrate misorientation angle increases. This suggests that in the absence of annealing, the presence of [110] steps assists the surface ordering process.

Using the knowledge of the effect of substrate temperature on the degree of order, a disorder/order/disorder double heterostructure was produced by growing on a substrate misoriented by 3° to produce [110] steps. The first layer was grown at 740°C (disordered), the second layer at 620°C (highly ordered), and the third layer at 520°C (slightly ordered). This is expected to result in a small bandgap layer sandwiched between higher bandgap layers. A schematic illustration of the structure is shown for comparison with the dark-field TEM image in Fig. 3. The dark field image shows that only the second layer is ordered. The TED patterns shown at the bottom confirm that only the 620°C layer is ordered. The PL spectra at several excitation intensities from the double heterostructure are shown in Fig. 4. By comparison with our previous results, the low energy peak is determined to originate from the ordered layer grown at 620°C. Comparison with the spectra seen in Fig. 1 indicates that the two higher energy peaks originate in the top layer grown at 520°C. The peak from the layer grown at 740°C is expected to appear at

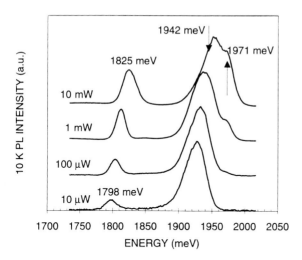

Fig. 4: PL spectra for the double heterostructure shown in Fig. 4 for various excitation powers.

an energy of 1.995 eV[10], as indicated in Fig. 2. It is apparently missing. This is probably because this layer is deeply buried in the structure. This is supported by the PL spectra from a similar single heterostructure consisting of layers grown at 740 and 620°C: The peak from the 740°C layer is much weaker than the peak from the top, 620°C layer[10]. The dependence of the peak positions on excitation intensity confirms these assignments.

4. Conclusions

Natural ordering of GaInP to form the Cu-Pt structure has been studied for OMVPE growth at a very low growth temperature of 520°C. Epitaxial layers with excellent morphology and strong 10-K PL were produced at 520°C with a growth rate of 0.5 µm/hr. Nominally (001) substrates, misoriented by angles of 0, 3, 6, and 9° toward the [1̄10] direction in the lattice were used. The 10K PL spectra consist of two peaks separated by approximately 40 meV. With increasing excitation energy the higher energy peak becomes relatively stronger. The lower energy peak shifts to higher energy with increased excitation intensity while the higher energy peak is stationary. Together, this behavior suggests that the low energy peak is due to recombination involving unintentional impurities; thus, the term "peak energy" refers here to the high energy peak. The PL peak energies of 1.94-1.96 eV suggest that the samples are more ordered than for growth at 720°C, where the peak energy is 1.995 eV. However, the order parameter is much smaller than for samples grown at 620°C, with $\vartheta_m = 3°$, where the peak occurs at approximately 1.855 eV. The effect of growth temperature on degree of order was exploited to produce a disorder/order/disorder double heterostructure. The layers were grown by sequentially varying the growth temperature from 740°C for the first layer to 620°C for the middle, highly ordered layer, to 520°C for the top layer.

ACKNOWLEDGMENTS

This research was supported by the Department of Energy, Basic Energy Sciences Division.

REFERENCES

1) Stringfellow G B 1993 in Common Themes and Mechanisms of Epitaxial Growth, (Materials Research Society, Pittsburgh) 35-46.
2) Wei S H and Zunger A 1990 Appl. Phys. Lett. **56** 662; Su L C Pu S T Stringfellow G B Christen J Selber H and Bimberg D 1993 Appl. Phys. Lett. **62** 3496.
3) Bertness K A Kurtz S R Friedman D J Kibbler A E Kramer C and Olson J M 1994 Appl. Phys. Lett. **65** 989..
4) Gavrilovic P Dabkowski F P Meehan K Williams J E Stutius W Hsieh K C Holonyak N Shahid M A and Mahajan S 1988 J. Crystal Growth **93** 426.
5) Cao D S Kimbal A W Chen G S Fry K L and Stringfellow G B 1988 J. Appl. Phys. **66** 5384.
6) Kurtz S R Olson J M Arent DJ Kibbler A E and Bertness K A 1993 in Common Themes and Mechanisms of Epitaxial Growth (Materials Research Society, Pittsburgh) 83-88.
7) Su L C Ho I H and Stringfellow G B 1994 J. Appl. Phys. (to be published).
8) Su L C Ho I H and Stringfellow G B 1994 J. Appl. Phys. **75** 5135.
9) Su L C Ho I H Kobayashi N and Stringfellow G B 1994 J. Crystal Growth, (to be published).
10) Su L C Ho I H and Stringfellow G B 1994 Appl. Phys. Lett. (to be published).

Reflectance Anisotropy and the Ordering Mechanism in GaInP at High Growth Temperatures

J.S. Luo and J.M. Olson

National Renewable Energy Laboratory
Golden, CO 80401

Abstract. We use reflectance difference spectroscopy (RDS) to investigate the mechanism of ordering in GaInP. The RD spectrum of GaInP includes a peak at E_0 that scales with the order parameter and decreases significantly at high and low growth temperatures. However, we also observe *in all GaInP grown by metal organic chemical vapor deposition (MOCVD)* a prominent peak at 3.2 eV (absorption depth of 20 nm) that is bulk-induced. This peak does not scale with E_0 but decreases only slightly with growth temperature from 650° to 770 °C. These observations support a model that requires rapid subsurface Ga/In interdiffusion to explain the diminished ordering observed at high growth temperatures.

1. Introduction

The CuPt-type ordered structure is consistently observed in MOCVD-grown GaInP [1-3]. It has been generally accepted that the ordering occurs on or near the surface and, as growth proceeds, is locked into a metastable state by slow, bulk Ga/In interdiffusion. The ordering causes the band gap of the GaInP at constant composition to shift to lower energy. Therefore, using band gap as a measure of order, experimentally one finds that the degree of order, under certain conditions, goes through a maximum for growth temperatures around 670 °C [4]. There is general agreement that, for growth temperatures below 670 °C, the surface ordering is limited by low surface mobility of the Ga and In adatoms. Therefore, at low growth temperatures ordering is usually enhanced by a lower growth rate or a surface with a higher B-type step density. However, at least two models have been proposed to explain the pronounced increase (decrease) of the band gap (bulk ordering) for growth temperatures above 670 °C. In the first model, the degree of surface ordering decreases abruptly as the growth temperature is increased. This could be an intrinsic property similar to a first-order phase transition, or it could be extrinsic caused by, for example, a loss of phosphorus from the surface [5]. This loss of P destroys the surface P-P dimerization at a B-type surface step that is believed to be important for an ordered surface arrangement of Ga and In [5]. In an alternate model, the degree of surface ordering does not decrease abruptly with temperature, but at higher growth temperatures the kinetic barrier that locks the ordered structure into a metastable state begins to weaken and the order in the bulk is reduced by rapid interdiffusion of Ga and In [6]. Again, it is known experimentally that real bulk interdiffusion even at growth temperatures as high as 750 °C is too slow to cause this effect [7], so this model suggests the existence of a near-surface region, where, due to injection of vacancies from the surface, a higher rate of interdiffusion is expected. There is no direct evidence for the existence of a near-surface transition layer in GaInP, but enhanced interdiffusion of GaAs/AlGaAs superlattices near a free surface has been detected by Kim et al. [8] using chemical lattice imaging. This model is consistent with the experimental observation that, at high growth temperatures, this subsurface disordering effect is pronounced for layers grown at slow rates.

In this study, we use RDS to investigate the parameters that affect the ordering mechanism in GaInP. We will show that RDS observations support the model which requires rapid subsurface Ga/In interdiffusion to explain the diminished ordering observed at high growth temperatures. The data are also consistent with the model proposed for the low-temperature kinetics-limited regime.

2. Experimental

All of the samples were grown by MOCVD at atmospheric pressure using trimethylindium, trimethylgallium, and phosphine in a Pd-purified hydrogen carrier gas. The substrates were n-GaAs misoriented 2° to 4° in the (111)B direction from (100). The RDS apparatus is mounted on the MOCVD reactor chamber with near normal optical access to the substrate through a stain-free quartz window. The RDS intensity is measured as $2 \times (R_{1\bar{1}0} - R_{110})/(R_{1\bar{1}0} + R_{110})$ where $R_{1\bar{1}0}$ and R_{110} are reflectances in the [1$\bar{1}$0] and [110] directions, respectively. Further details of the RDS system are described elsewhere [9,10].

3. Results and discussions

RDS is sensitive to both surface and bulk optical anisotropy. As such, a typical RD spectrum for MOCVD-grown GaInP (Fig. 1) is rich with prominent bulk-related peaks at E_0 and E_1, and surface-related features between 2.0 and 2.6 eV (inset of Fig. 1) that can be reversibly altered by annealing the surface in H_2 or a PH_3 and H_2 mixture, or by exposing the sample to air. Details can be found in elsewhere [9,11]. From theoretical modeling studies[12], we propose that the feature at E_1 in Fig. 1 is composed of two peaks: a strong positive peak near 3.2 eV and a weaker negative peak near 3.0 eV. The intensity of the E_1 peak is then taken as the peak-to-peak sum of these spectral features. This has the added benefit of reducing baseline errors in the vicinity of E_1.

The penetration depth for photons with energies near E_0 is of the order of several μm, and that at E_1 is estimated from spectral ellipsometry measurements by Kato et al. [13] to be about 15 nm. Hence, RDS features at E_0 and E_1 provide complementary information about the degree of long range bulk ordering far below and near the growing surface, respectively. This is important for our analysis of the ordering mechanism.

In Fig. 2, we show a series of *in situ* RD spectra for a sequence of layers grown at temperatures from 730 to 770 °C on a single GaAs substrate. It clearly shows that, as the growth temperature increases, E_0 blue shifts and decreases in intensity. These are both indications that the degree of ordering in the GaInP *bulk* is decreasing with increasing growth temperature. On the other hand, except for an apparent baseline shift, the intensity of the E_1 peak decreases only slightly with increasing growth temperature. These intensity data for E_0 and E_1 (combined with similar data from other spectra not shown) are presented graphically in Fig. 3 as a function of growth temperature for a fixed growth rate and phosphine input partial pressure. For temperatures below 730 °C, E_1 tracks E_0 fairly well. Below 650 °C (which we call region I), where the ordering process is limited by surface diffusion, the intensity of both peaks decreases with decreasing growth temperature. For growth temperatures between 650° and 730 °C (region II), the E_1 and E_0 features decrease slightly with increasing growth temperature and track with the surface order parameter (indicated with a dash-dot line in Fig. 3) calculated by Osorio et al. [14]. Above 730 °C (region III), however, the E_0 peak intensity falls off rapidly with growth temperature and no longer follows E_1 or the calculated surface order parameter.

Kinetic versus thermodynamic effects are explored by studying the time dependent behavior of the system. In Fig. 4, we compare the response at E_1 $\Delta R = \left(R_{1\bar{1}0} - R_{110} \right)_{E_1}$ as a

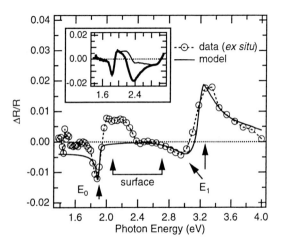

Fig. 1 Typical RD spectrum of ordered GaInP measured at 22 °C. The circles are experimental data and the dashed line is a guide to the eye. The solid line is a model fit obtained from our recent work [12]. The RD spectra in the inset are measured at 85°C after two different surface treatments: annealed in H_2(thin line), and 1% PH_3 in H_2 (thick line) at 400 °C for several minutes.

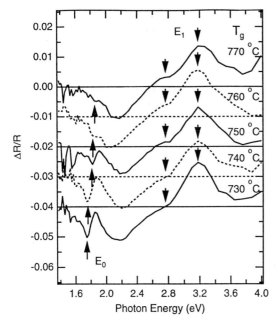

Fig. 2 RD spectra of GaInP grown at T_g between 730° and 770 °C. The spectra are measured at 300 °C. All samples have a layer thickness of about 500 Å. The arrows are indicators of RDS features at E_0 or E_1. The inflection of the feature at E_1 appears to be offset at higher growth temperature. The offset is associated with surface roughening. Surface roughening is usually observed for GaInP grown at high temperature [5].

function of the relative growth rate (0, 0.25, 0.5, and 1, where 1 corresponds to a growth rate of 0.1 μm/min) at growth temperatures in the three ordering regions of Fig. 3. In Figs. 4(a), 4(c), and 4(d), the initial signal with $\Delta R \approx 0$ is from the uncoated GaAs substrate and gives a good measure of the baseline. In Fig. 4(a), the growth temperature is well within region I of parameter space where slow adatom surface mobility supposedly limits the degree of near surface and bulk ordering and slower grow rates promote more ordering. Here, we see that the signal at E_1 is strongest for no growth and decreases monotonically with growth rate. The growth-to-no-growth transient *for low growth rates* is relatively fast; for the highest growth rates studied it is on the order of 10 seconds. If the next layer in a no-growth-to-growth transient is grown at a rate different than that of the preceding layer, then the transient is modulated by interference effects. This is a direct result of the fact that the growth rate affects the degree of order and therefore the optical properties of the GaInP[6].

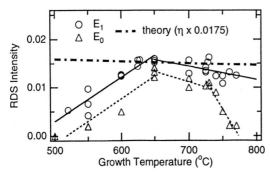

Fig. 3 RDS intensity at E_0 and E_1 vs. growth temperature. The thin dashed and solid lines are guides to the eye. The dash-dotted line is from ref. [14] scaled to fit the plot. A detailed description of the method for determining the RDS intensities at E_0 and E_1 is given elsewhere [12].

For intermediate growth temperatures (region II), the degree of ordering and therefore the optical properties of GaInP are relatively insensitive to growth temperature. This is a region where the surface is near thermodynamic equilibrium. This is confirmed by the data in Fig. 4(b). The E_1 signal is virtually independent of growth rate, and only a small interference effect is detected in the no-growth-to-growth transient.

At higher temperatures (region III), two new effects begin to evolve. Under steady-state growth conditions, E_1 now decreases slightly with decreasing growth rate. This is expected if 1) the probe depth is larger than the proposed subsurface region where enhanced interdiffusion of Ga and In occurs, and 2) a slower growth rate yields a residence time within this subsurface region, which is comparable to the Ga/In interdiffusion time. This is plausible and supports the subsurface disordering model. However, the other new effect observed at these higher growth temperatures is that, under non steady-state no-growth conditions, E_1 exhibits a nonsteady behavior with a time constant of several minutes or longer (depending on growth temperature and PH_3 partial pressure [12]). During this same time, the surface becomes morphologically unstable and a nonzero, sloping RDS baseline

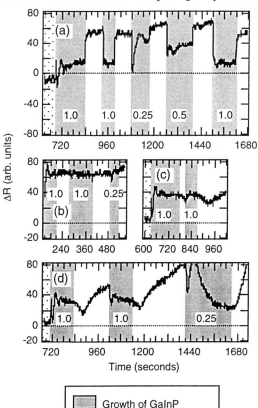

Fig. 4 Time-scan RDS of GaInP at various growth temperatures and growth rates. Growth temperatures T_g are (a) 550 °C, (b) 650 °C, (c) 730 °C, and (d) 750 °C. The values in the shaded areas are associated with growth rates and are described in the text. The monitoring light wavelength is 410 nm. Essentially, the RDS feature around E_1 is measured. Due to a redshift with increasing temperature, monitoring at different positions of the E_1 feature is expected.

develops due to anisotropic scattering from the resulting rough surface. Undoubtedly, surface roughness will ultimately have an effect on the surface ordering process, but it is not clear at this time why the surface roughens or whether it is the sole cause of the loss bulk ordering. However, the small interference modulation between the first two layers grown at 0.1 µm/min in Fig. 4(d), a feature which one can not explain by surface roughening, appears to be associated with a change of the *near-surface* optical properties (caused by near-surface disordering) of the first layer.

4. Conclusion

We have observed two ordering-induced bulk features in the RD spectrum of ordered GaInP: one is associated with bulk ordering and the other with near-surface or surface ordering. The variation of the two RDS features with growth temperature supports the model that disordering at high growth temperatures is caused mostly by rapid near-surface Ga/In interffusion. RDS observations are also consistent with the accepted model for ordering at low growth temperatures.

References

[1] Gomyo A., Suzuki T., Kobayashi K., Kawata S., Hino I. and Yuasa T., Appl Phys Lett **50**, 673 (1987).

[2] Kondow M., Kakibayashi H. and Minagawa S., J of Cryst Growth **99**, 291 (1988).

[3] Friedman D. J., Zhu J. G., Kibbler A. E., Olson J. M. and Moreland J Appl Phys Lett **63**, 1774 (1993).

[4] Kurtz S. R., Olson J. M., Arent D. J., Kibbler A. E. and Bertness K. A., (Mater. Res. Soc. Proc., 1993), vol. **312**, pp 83.

[5] Kurtz S. R., Arent D. J., Bertness K. A. and Olson J. M., (Mater. Res. Soc. Proc., 1994), vol. **340**.

[6] Kurtz S. R., Olson J. M. and Kibbler A., Appl Phys Lett **57**, 1922 (1990).

[7] Gavrilovic P., Dabkowski F. P., Meehan K., Williams J. E., Stutius W., Hsieh K. C., Holonyak N., Shahid M. A. and Mahajan S., J Cryst Growth **93**, 426 (1988).

[8] Kim Y., Ourmazd A., Malik R. J. and Rentschler J. A., (Mater Res Soc Proc, 1989), vol. 159, pp 351.

[9] Luo J. S., Olson J. M., Bertness K. A., Raikh M. E. and Tsiper E. V., Accepted by J Vac Sci Tech.

[10] Aspnes D. E., Harbison J. P., Studna A. A. and Florez L. T., J Vac Sci Technol **A6**, 1327 (1988).

[11] Luo J. S., Olson J. M., Kurtz S. R., Arent D. J., Bertness K. A., Raikh M. E. and Tsiper E. V., Submitted to Phys. Rev. B.

[12] Luo J. S. and Olson J. M., to be published.

[13] Kato H., Adachi S., Nakanishi H. and Ohtsuka K., Jpn J Appl Phys **33**, 186 (1994).

[14] Osorio R., Bernard J. E., Froyen S. and Zunger A., Phys Rev **B45**, 11173 (1992).

Advances in correlating the unusual optical properties of $Ga_{0.52}In_{0.48}P$ to the microstructure

M. C. DeLong, C. E. Ingefield and P. C. Taylor

Department of Physics, University of Utah, Salt Lake City, UT 84112

L. C. Su, I. H. Ho, T. C. Hsu and G. B. Stringfellow

Department of Materials Science and Engineering, University of Utah, Salt Lake City, UT 84112

K. A. Bertness and J. M. Olson

National Renewable Energy Laboratory, Golden, CO 80401

Abstract. The microstructure in partially ordered samples of $Ga_{0.52}In_{0.48}P$ depends critically on the growth parameters and the orientation of the substrate. Ordering in $Ga_{0.52}In_{0.48}P$ produces several unusual optical properties, including a photoluminescence whose energy and lifetime depend strongly on the intensity of the exciting light. This PL process is correlated with the appearance of two variants (regions partially ordered along different {111} B directions) of ordering. These correlations are established using samples produced in two different laboratories.

1. Introduction

It is well established that the partially ordered samples of $Ga_{0.52}In_{0.48}P$, in which alternate (111) planes are preferentially occupied by Ga and In, have several unusual optical properties [1,2,3,4]. Perhaps the most prominent example of these unusual optical properties is the reduction of the optical band gap in the partially ordered samples. There also exists a splitting of the valence band, the appearance of an absorption that "tails" into the gap, and a photoluminescence (PL) emission whose peak energy and radiative lifetime both depend strongly on the intensity of the exciting light. In what follows we shall call this unusual PL property the moving PL emission process. Although unusual features occur in the PL spectra in many partially ordered samples of $Ga_{0.52}In_{0.48}P$, we shall document that rapidly moving emission occurs only in materials that contain two variants of the ordering, i. e., regions partially ordered along different {111} B directions. In some two-variant samples, rates for the shift of the PL peak energy with excitation intensity can be as large as 18 meV per decade of change in excitation intensity, and PL lifetimes can be as long as 10 ms at low excitation intensities [3,4]. These long lifetimes have been used as evidence that the PL process occurs between carriers that are spatially separated [2,4].

2. Experimental details

The four samples investigated in this study were grown by atmospheric pressure organometallic vapor phase epitaxy (OMVPE) using standard techniques as described elsewhere [5]. Pairs of samples were grown under essentially identical conditions in reactors at NREL and Utah. The first pair of samples was grown on substrates misoriented $6°$ toward {111} B under conditions where only one ordered variant is observed. We shall call these samples single-variant samples. The Utah sample was grown at 670 C with a growth rate of about 1 µm/h and a V/III ratio in the source gases of 160. The NREL sample was grown at 700 C with a growth rate of 5.5 µm/h and a V/III ratio of 80. Transmission electron diffraction (TED) measurements indicate that both samples are strongly ordered with only one variant. Transmission electron microscopy (TEM) photographs show that these samples are relatively homogeneous within the domains.

The second pair of samples was grown on substrates oriented exactly on {100} under conditions where two variants of ordered domains are known to occur. We shall call these samples two-variant samples. Both the Utah and NREL samples were grown at 670 C at a growth rate of 0.5 µm/h and a V/III ratio of 510. TED measurements show the presence of two variants, and a streaking of the superspots along the growth direction indicates a distribution of small domain dimensions in this direction in both of these samples. TEM photographs show small, platelet-like domains with one dimension as small as 20 to 30 Å. These domains are organized into *macro-domains* whose dimensions are much larger (on the order of 1000 Å). Within these macro-domains the boundaries between the individual domains are probably *anti-phase boundaries* where the ordered cation planes change from indium-rich to gallium-rich, and vice versa, over a few lattice spacings.

The continuous wave (cw) PL and PL excitation spectra were obtained at 5 K using experimental techniques also described elsewhere [4,5]. Time resolved PL spectra were taken using an acousto-optic modulator to produce a pulsed source and a digital, signal-averaging oscilloscope to collect the PL data.

3. Results

A typical PL spectrum for the Utah single-variant sample is shown in Fig. 1. [The spectrum for the NREL single-variant sample is effectively identical in all important aspects.] This spectrum contains two PL peaks. The lower energy peak changes in shape slightly with increasing excitation intensity. The higher- and lower-energy PL peaks are consistent with excitonic recombination and an unresolved combination of band-acceptor and donor-acceptor-pair recombination, respectively. These recombination processes, including the subtle changes that occur with excitation intensity (see Fig. 1) are very similar to those that often occur in homogeneous, binary III-V compounds. The position of the excitonic PL peak in this sample at about 1.92 eV clearly indicates that the material is ordered. In $Ga_{0.52}In_{0.48}P$ where the cation sublattice is a random distribution of Ga and In, the excitonic PL at 5 K occurs at 2.0 eV [6].

The PL behavior in the two-variant samples is markedly different. Typical PL spectra are shown as a function of the power density of the exciting light in Fig. 2 for the NREL two-variant sample. The behavior of the PL shown in Fig. 2 is very unusual. The major PL peak shifts continuously through a manifold of states at a rate of more than 10 meV per decade of excitation power density for a total shift of more than 70 meV for power densities varying between approximately 3 $\mu W/cm^2$ and 300 W/cm^2. The situation is even more striking in Fig. 3 where similar data are shown for the Utah two-variant sample. In this case the major PL peak shifts by about 18 meV per decade of incident excitation power density, and the total shift

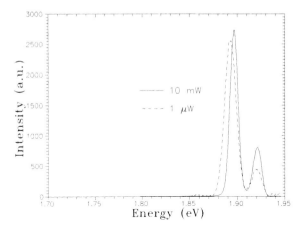

Figure 1. Dependence of the PL on exciting light intensity in the Utah sample of single-variant, partially-ordered $Ga_{0.52}In_{0.48}P$. The laser powers listed as 10 mW and 1 μW correspond to power densities at the sample of 300 mW/cm^2 and 30 μW/cm^2, respectively. See text for details.

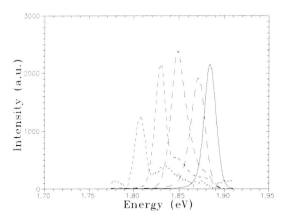

Figure 2. Dependence of the PL on exciting light intensity in the NREL sample of two-variant, partially-ordered $Ga_{0.52}In_{0.48}P$. For the PL peaks from low energy to high energy, the laser power densities at the sample are 3 μW/cm^2, 300 μW/cm^2, 30 mW/cm^2, 3 W/cm^2, and 300 W/cm^2, respectively. See text for details

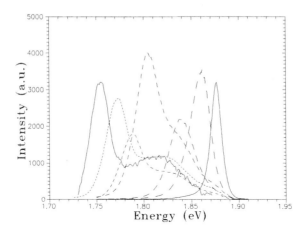

Figure 3. Dependence of the PL on exciting light intensity in the Utah sample of two-variant, partially-ordered $Ga_{0.52}In_{0.48}P$. For the PL peaks from low energy to high energy, the laser power densities at the sample are 3 $\mu W/cm^2$, 30 $\mu W/cm^2$, 300 $\mu W/cm^2$, 3 mW/cm^2 300 mW/cm^2, 3 W/cm^2, and 300 W/cm^2, respectively. See text for details

from a power density of approximately 3 $\mu W/cm^2$ to 300 W/cm^2 is over 120 meV. The same data are replotted on a semi-logarithmic scale in Fig. 4. This figure emphasizes the fact that there appears to be a high energy cut-off to the PL emission that is fairly independent of the excitation intensity. This feature is probably related to the "band-tails" that are observed for these samples in PL excitation experiments [5,7].

There are at least two PL peaks in the spectra of the two-variant samples [3,4]. In this paper we concentrate only on the dominant PL peak that moves dramatically with the intensity of the exciting light. A careful examination of the data in Figs. 2 and 3 shows that there are clearly quantitative differences between the two samples that were made under nominally identical growth conditions. As mentioned above, in the case of the one-variant samples, only one of which is shown in Fig. 1, the optical properties are both qualitatively and quantitatively similar despite the fact that the growth conditions were not identical. It is clear that the two-variant samples that are highly ordered are much more sensitive to precise conditions of growth. We speculate that this situation results because the optical properties, such as the rapidly moving PL emission, are dominated by interfaces separating domains ordered in different directions. From the TEM results it is clear that the microstructures are much more complicated in these two-variant samples than they appear to be in the single-variant samples.

The moving PL emission in the two-variant samples is also a very slow process [4]. When excited by pulsed laser excitation, this component of the PL exhibits a slow rise and a slow decay, both of which that may be on the order of ms at low power densities. The PL rise and decay rates both increase roughly in proportion to the intensity of the exciting light, and the decay is highly non-exponential as is evident from the data shown in Fig. 5 for the Utah two-variant sample. The data in Fig. 5 were taken using an exciting light intensity of 300 $\mu W/cm^2$ and the PL was detected at 1.77 eV, which is near the position of the peak of the PL at this

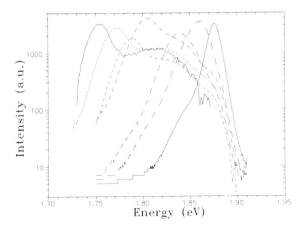

Figure 4. Semi-logarithmic plot of the data of Fig. 3. Note the high energy "cut-off" of the PL intensities. See text for details.

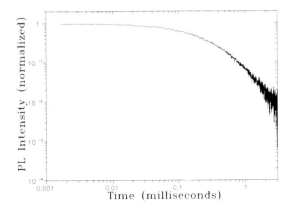

Figure 5. Time decay of PL in Utah sample of two-variant, partially-ordered $Ga_{0.52}In_{0.48}P$. The excitation power density is 300 $\mu W/cm^2$ and the PL is detected at 1.77 eV. See text for details.

excitation power as shown in Fig. 3. Note that the PL decay is well approximated by a power law at times greater than about 0.2 ms. If the excitation intensity is decreased by an order of magnitude and the PL detected at the same energy, then the decay curve is shifted one decade to longer times but is otherwise unchanged in shape on a log-log plot. If the PL is detected at a different energy, then the decay curve exhibits a slightly different shape, which also can be scaled in time for different excitation power densities. These time resolved data are consistent with earlier cw data [4] that indicated a decay of the moving PL emission that was essentially independent of PL energy at a given excitation power density, but that decreased by approximately an order of magnitude for each decade increase in excitation power density. Although a detailed microscopic interpretation of these results is not currently available, it is encouraging to note that there is general reproducibility between samples made in different laboratories, even for the two-variant materials where the microstructure depends critically on the growth conditions.

4. Summary

Both the microstructure and the optical properties of single-variant, partially-ordered films of $Ga_{0.52}In_{0.48}P$ are highly reproducible between laboratories. At the growth temperatures employed these samples exhibit relatively large ordered domains and the optical properties are similar to those observed in disordered (random placement of Ga and In on the cation sublattice) $Ga_{0.52}In_{0.48}P$ except for the reduction in the optical band gap. In the case of two-variant samples, the reproducibility is qualitatively good, but there are quantitative differences in the optical properties that probably result from detailed differences in the microstructure. These samples exhibit a PL whose peak energies and lifetimes vary rapidly with the power density of the exciting light. This unusual behavior may be related to the interfaces between domains that are partially ordered along different {111} B directions.

Acknowledgements

This research was supported by the Office of Naval Research, the National Renewable Energy Laboratory, and the Department of Energy.

References

[1] Gomyo A, Kobayashi K, Kawata S, Hino I, Suzuki T and Yuasa T 1986 *J. Cryst. Growth* **77** 367-73.
[2] Fouquet J E, Robbins V M, Rosner S J and Blum O 1990 *Appl. Phys. Lett.* **57** 1566-8.
[3] DeLong M C, Taylor P C and Olson J M 1990 *Appl. Phys. Lett.* **57** 620-2.
[4] DeLong M C, Ohlsen W D, Viohl I, Taylor P C and Olson J 1991 *J. Appl. Phys.* **70** 2780-7.
[5] DeLong M C, Mowbray D J, Hogg R A, Skolnick M S, Hopkinson M, David J P R, Taylor P C, Kurtz S R and Olson J M 1993 *J. Appl. Phys.* **73** 5163-72.
[6] DeLong M C, Mowbray D J, Hogg R A, Skolnick M S, Williams J E, Meehan K, Kurtz S R,Olson J M, Wu M C and Hopkinson M 1994 *AIP Conf. Proc.* **306** 529-32.
[7] DeLong M C, Hogg R A, Mowbray D J, Skolnick M S, Taylor P C, Olson J M, Kurtz S R and Kibbler A E 1992 *AIP Conf. Proc.* **268** 332-7.

Composition and structure analyses on spontaneously formed AlGaAs superlattice-like structures on (100) GaAs

S. S. Cha, Y. K. Shin, H. I. Jeon, Z. S. Piao, K. Y. Lim,
E.-K. Suh, Y. H. Lee, D. K. Kim*, B. T. Lee*, and H. J. Lee

Semiconductor Physics Research Center and
Department of Physics
Jeonbuk National University, Jeonju 560-756, KOREA

*Department of Metallurgical Engineering
Chonnam National University, Kwangju 500-757, KOREA

Abstract. $Al_xGa_{1-x}As$ superlattice-like structures are spontaneously formed during the nominally homogeneous epitaxial layer growth on 2° off (100) GaAs substrate by the metalorganic chemical vapor deposition. The structures are observed in the range of nominal Al gas composition range of x_g = 0.1~0.3. The solid composition x values constituting the well are shown to be constant ~0.035 regardless of the x_g value. Meanwhile the barrier composition increases with x_g and exhibits the maximum value at x_g = 0.25. The well width decreases with increasing growth temperature Tg and increasing x_g, and the barrier width increases with Tg. The period is almost independent of Tg and increases with decreasing x_g.

1. Introduction

The natural superlattices (NS), which can spontaneously formed during the growth of nominally homogeneous epitaxial layers of some III-V ternary alloys, are well evidenced in the materials with miscibility gap [1]. Even though AlGaAs alloys do not have miscibility gap, in bulk materials Pefroff et al.[2] have observed the spontaneously formed multilayer structure in MBE-grown AlGaAs films on (110) oriented GaAs substrates. Kuan et al.[3] also have reported monolayer superlattices of AlGaAs grown by both methods of MBE and MOCVD.

In this report we present NS structures in MOCVD-grown AlGaAs alloys, which are quite different form from those reported so far.

2. Experiment

AlGaAs epilayer were grown by low-pressure MOCVD system using TMAl, TMGa, and AsH₃ (20% in H₂). The layers were grown on (100) GaAs 2° off toward (110) using a constant V/III ratio, 100. The reactor pressure was kept atmospheric, and bubbler pressures of TMGa and TMAl were 1,100 and 800 Torr, respectively. Typical growth rate is 10.5Å/sec and the epilayer thicknesses are $2 \sim 4 \mu$m.

The layer characteristics were examined as functions of Al gas composition x_g and growth temperature T_g. Various methods such as double crystal X-ray diffraction (DCXD), Raman scattering (RS), transmission electron microscopy (TEM) and photoluminescence (PL) were used in analyzing the layers.

3. Results

The present analyses exhibit that superlattice-like structures are spontaneously formed with clearly defined interfaces modulated along the growth direction. These structures are quite different from those reported so far, no plate-like structures being observed in the present layers. The [110] cross-section dark field micrographs were obtained by TEM(JEOL 2000EX) and one of the typical results is shown in Fig. 1.

Figure 1. TEM [110] cross-section micrograph of AlGaAs natural superlattice-like structures with $x_g = 0.20$. Bright region indicates barrier.

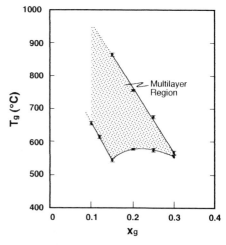

Figure 2. Multilayer formation as a function of x_g and T_g.

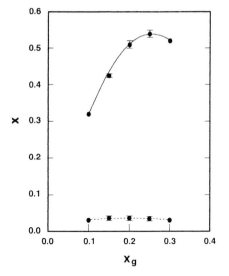
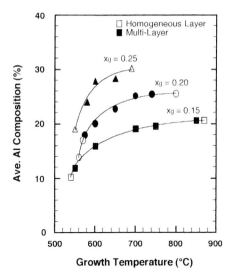

Figure 3. Composition variation with nominal gas composition.

Figure 4. The average Al composition variation with growth temperature.

Two phases of AlGaAs layers are clearly observed in a series of alternating layers parallel to the substrate surface. These structures are observed in a very wide ranges of x_g and T_g as shown in Fig. 2.

The structures and compositions were determined by comprehensive analyses using DCXD, RS and TEM all together. The epilayers are composed of alternate layers with different Al concentration when x_g is 0.1 to 0.3. The lower x values constituting the well of the superlattice are in good agreement with the PL data and show almost constant, ~0.035, regardless of the x_g value. Meanwhile the barrier composition increases with x_g up to $x_g = 0.25$: the Al concentration varies from 0.32 to 0.54 and the variation is shown in Fig. 3. The weighting averaged Al compositions show growth-temperature-dependency and increase with T_g as shown in Fig. 4.

The interfaces between consecutive layers are planar and compositionally abrupt. Fairly good periodic structures were obtained for the x_g values 0.15~0.25. One period thickness increases with decreasing x_g and is almost independent of growth temperature. But the well and barrier widths vary with x_g and T_g as shown in Figs. 5 and 6. The well width decrease with increasing T_g and vice versa in the barrier width.

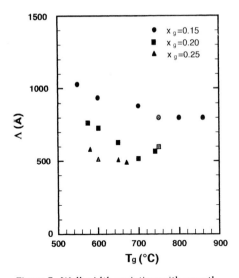

Figure 5. Well width variation with growth temperature.

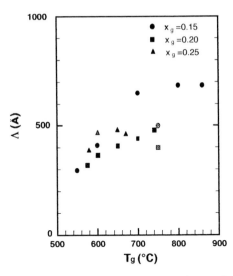

Figure 6. Barrier width variation with growth temperature.

4. Summary

We observed for the first time the superlattice structures of AlGaAs spontaneously grown on (100) GaAs substrate during the homogeneous epitaxial growth. The structure and composition of the superlattice were studied as functions of the Al gas composition and the growth temperature.

This work wark was supported by the Korea Science and Engineering Foundation through Semiconductor Physics Research Center, Jeonbuk National University.

References

[1] Norman, AG et al. 1993 *Semiconductor Science and Technology* **8** S9-S15
[2] Petroff, PM et al. 1982 *Phys. Rev. Lett.* **48** 170-3
[3] Kuan, TS et al. 1985 *Phys. Rev. Lett.* **54** 201-4

Hyperabrupt Si-doping profile in MOVPE growth of varactor diodes

F. Bugge, P. Heymann, E. Richter, and M. Weyers

Ferdinand-Braun-Institut für Höchstfrequenztechnik, Rudower Chaussee 5,
D-12489 Berlin, Fed. Rep. Germany

To obtain very short and steep electrical pulses a transmission line containing nonlinear varactor diodes can be used. Such diodes require a precisely controlled doping profile with a relatively lowly doped graded layer on top of a highly doped buried contact layer.
  We have grown GaAs:Si varactor diodes by MOVPE using disilane as the dopant source. A wide range of growth conditions was used to investigate the influence of these parameters on the doping mechanism and the obtained doping profiles. To avoid a high background doping level after the growth of an N^+ buried layer an annealing step at 825°C was necessary. This annealing step allows for the subsequent growth of layers with background doping levels below 2 E15 cm^{-3} and a very sharp doping front as required for the varactor diode.

1. Introduction

A MMIC nonlinear transmission line (NLTL) periodically loaded by varactor diodes can be used as a source of picosecond electrical transients. The most promising applications are frequency multipliers with several hundred gigahertz output frequency and pulse generators that can drive very fast sampling diodes enabling 300 GHz instrument bandwidths.

  The characteristics of the NLTL depend mainly on the geometry and the doping profile of the diodes. They must have a strongly varying depletion capacitance, small parasitic elements and a breakdown voltage sufficient to support the wave amplitudes of several volts. This requires a highly doped buried N^+-layer on a semiinsulating GaAs substrate which serves as the diode cathode connection with a resistance as low as possible. The cathode is covered by a hyperabrupt n-doped active layer of about 450 nm thickness (fig. 1). A steep doping profile results in a strong voltage dependence of the capacitance and a reduced number of diodes necessary for optimum performance of the NLTL.

  The growth of a n^--layer on top of a highly doped N^+-layer by MOVPE has been studied and optimized with respect to the desired well controlled hyperabrupt doping profile.

2. Experimental procedure

MOVPE growth of the GaAs layer structure was performed in a horizontal reactor at a total pressure of 70 hPa and at growth temperatures between 600°C and 825°C. Starting materials were pure arsine and trimethyl gallium (TMGa). Disilane diluted in hydrogen was used for n - type doping. The structures were grown on undoped, semi-insulating (100) GaAs

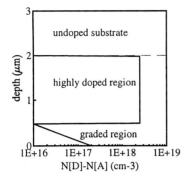
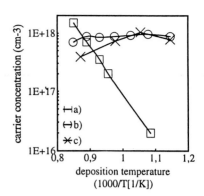

Fig.1. Schematic of desired doping profile for varactor diode

Fig. 2. Arrhenius plot of carrier concentration (at 300 K) in dependence on growth temperature
a) silane [2]   b) disilane [2]
c) disilane (this study) p_0 = 6.9 E-5 Pa

"epiready" substrates (LEC-grown) without additional substrate preparation and on Si-doped (100) GaAs (Bridgman-grown) after standard chemical cleaning.

Deposition temperature, doping flow and Si_2H_6/TMGa mole fraction ratio were varied in the fabrication of both doped and undoped layers to study the influence of these parameters on the doping mechanism and the obtained doping profiles. Carrier concentration profiles have been measured by electrochemical capacitance-voltage profiling (ECV) in a Polaron profiler using tiron as the electrolyte. The spatial resolution in this method is given by the triple Debye length due to free-electron leakage and is about 130 nm for the undoped ($[N_D-N_A] \approx 1$ E16 cm^{-3}) and about 10 nm for the highly doped ($[N_D-N_A] \approx 2$-3 E18 cm^{-3}) region. To allow for an evaluation of the lowly doped part of the active region an additional undoped layer of 400 nm has been inserted on top of the N$^+$-layer in some cases.

For the fabrication of varactor diodes Schottky contacts are made with a Ti/Pt/Au lift-off process and ohmic contacts to the N$^+$-layer are formed by a recess etch through the n-layer. The diode area is defined by an H$^+$-implantation converting the surrounding areas into semiinsulating material. By this process varactor dimensions of down to 10 μm^2 are obtained.

3. Results and discussions

For the growth of a varactor diode a group VI donor offers a higher doping level and thus a smaller resistance of the buried N$^+$ cathode layer. However, selenium as the group VI donor employed in MOVPE of GaAs suffers from memory effects due to the strong adsorption of H$_2$Se on the steel tubing [1]. This makes the desired steep doping profile difficult to achieve. Thus we used Si as the dopant in our experiments which offers the additional advantage of a lower diffusion coefficient.

Disilane was chosen as the dopant source, not only due to its higher doping efficiency in comparison to silane. Disilane has also been reported to show a Si incorporation behaviour which is nearly independent of temperature in the usual range of growth temperatures [2]. In our own experiments this could not be completely verified. Fig. 2 shows the Arrhenius plot

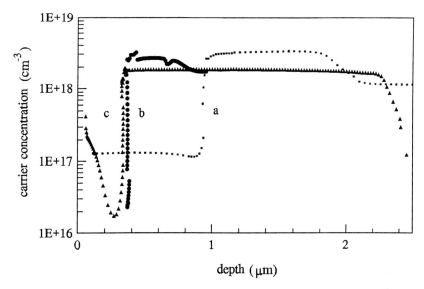

Fig.3. Doping profiles of different structures
a) continous growth; b) annealed after growth of N$^+$-layer; c) varactor diode

of the obtained carrier concentration. At our growth conditions the incorporation efficiency increases with increasing the temperature from 600°C to 675 °C and then again decreases from 675°C to 825°C. SIMS measurements show that it is the Si concentration that changes with temperature. It is approximately 10 % higher than the measured carrier concentrations in fig. 2. The observed temperature dependence of Si incorporation from disilane shows that also growth parameters like linear flow velocity (3 m/s) or V/III ratio (210) affect the doping process.

First attempts of growing the desired structure showed that on top of the buried N$^+$-layer a background doping of 1-2 E17 cm^{-3} was obtained even up to thicknesses of 1 µm of the intentionally undoped layer (fig. 3a). Comparison of ECV- and SIMS profiles showed that at a net carrier concentration of 3.5 E18 cm^{-3} the Si concentration was 1.2 E19 cm^{-3} and outdiffusion of the not electrically active Si on interstitial sites was suspected as the reason for the high doping level in the top layer. However, reduction of the Si concentration in the N$^+$-layer to ≈ 1 E18 cm^{-3} (i.e. below the level of saturation of Si incorporation as a substititional donor) resulted only in a slight reduction of the background doping level.

The desired steep gradient can be obtained if after the growth of the highly doped layer an annealing step is carried out. After annealing at 825°C for 20 min under 210 Pa of arsine the following layer shows a background doping level of below 2 E15 cm^{-3} (fig. 3b). The doping level could not be precisely evaluated since the depletion zone extends down to the front of the N$^+$-layer. This annealing step also is effective if the Si concentration is already above the saturation level. During the annealing step different processes can take place. Si incorporated in GaAs deposits on the quartz liner tube and the graphite susceptor can be desorbed or converted into a stable configuration thus eliminating memory effects. Such deposits can act as an only slowly depleting Si source giving rise to a constant background doping level over a considerable layer thickness as seen in fig. 3a. Also Si segregating on the growing surface can be desorbed at high temperatures in MOVPE. However, a constant background doping level of above 1 E17 cm^{-3} due to Si segregation appears to be rather unlikely since δ-doping has successfully been demonstrated [3]. Finally

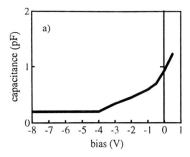 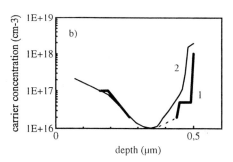

Fig. 4. a) Capacitance versus bias voltage extracted from microwave measurements in the frequency range of 1 to 50 GHz
b) 1 - carrier concentration profile calculated from capacitance values in a)
2 - ECV profile (for comparison)

the annealing step can affect the vacancy concentration in the N^+-layer. The enhanced carrier concentration on top of the contact layer seen in fig. 3b points to an enhancement of the Si activation by switching from interstitial sites to Ga lattice sites. This should reduce the tendency for Si outdiffusion into the undoped layer. The importance of each of these processes still has to be clarified.

Using this growth interruption and annealing scheme layer structures with a hyperabrupt doping profile have successfully been grown. Fig. 3c shows the ECV profile of such a structure. To obtain the desired steep increase in the active layer the disilane partial pressure has been ramped from 1.75 E-5 Pa to 1.4 E-4 Pa while the TMGa partial pressure was reduced from 4 Pa to 1 Pa. Fig. 4a shows the capacitance of the diode in dependence of the bias voltage. The doping profile calculated from these data is in excellent agreement with the desired profile (fig. 4b). For the low concentration region between 0.25 µm and 0.45 µm a precise quantitative evaluation of the doping level from the microwave measurements is not possible. It can only be stated that the doping level is below 2 E16 cm^{-3}. For comparison also the profile obtained by ECV is shown.

4. Conclusions

The results of MOVPE growth for varactor diodes with a hyperabrupt doping profile are presented. The growth of 1-2 µm thick N^+-doped layers results in a high background doping level, which can be suppressed by an annealing step at 825°C. Annealing reduces the background doping to values lower than 2 E15 cm^{-3} and yields a steep transition between the N^+-layer and the undoped region. This way the doping profile necessary for varactor diodes has been realized and first NLTL are under fabrication.

References

[1] Lewis C R, Dietze W T and Ludowise M J 1983 *J. Electronic Materials* 12 507-524
[2] Kuech T F, Veuhoff E and Meyerson B S 1984 *J. Crystal Growth* 68 48-53
[3] Li G, Jagadish C, Clark A, Larsen C A and Hauser N 1993 *J. Applied Physics* 74 2131-2133

Emission spectra and lifetimes of near surface GaInAs/GaAs quantum wells

J. Dreybrodt, F. Daiminger, J.P. Reithmaier and A. Forchel

Technische Physik, University of Würzburg, Am Hubland, 97074 Würzburg, Germany

Abstract. Quantum wells with very thin top barrier layers in the range of several nanometers are strongly influenced by the surface. We have studied the time resolved photoluminescence of GaInAs quantum wells as a function of the distance to the surface and of the quantum well width to probe the strength of the coupling. For top barrier thicknesses below 5 nm a strong emission line shift to higher energies is observed which is due to the influence of the high vacuum potential at the surface. The excitonic lifetime decreases strongly for top barrier thicknesses less than 10 nm. Calculations for the barrier layer dependent probability to find electrons within a 5 nm thick trapping layer at the surface reproduce the measured data very well.

1. Introduction

Influences of the surface on the properties of quantum well (QW) structures have become of considerable interest during the last years [1]. QWs allow to study the influence of the surface on a several nanometer thick volume within the semiconductor [2]. By changing the distance of the QW to the surface the strength of the interaction is variable. This allows a systematical investigation of the underlying physical processes.

We report on the optical properties of GaInAs/GaAs QWs as a function of the GaAs top barrier thickness and the well width. We observe with decreasing top barrier thickness a shift of the emission line to higher energies and a strong broadening. Simultaneously the emission intensity and lifetime are strongly reduced. A model taking into account the properties of the surface (vacuum potential and surface states) and a finite top barrier thickness is in good agreement with the observed behaviour.

2. Experiment

Sets of QWs with various top barrier thicknesses were prepared by a wet chemical etch process. We used 5 and 15 nm thick QWs grown by molecular beam epitaxy

(MBE) with Indium contents of 20% and 13%, respectively. The growth temperature was 580 °C for GaAs and 520 °C for GaInAs. For the etch process we used a highly diluted H_2SO_4:H_2O_2:H_2O etchant (ratio 1:10:6000) with an etch rate of 0.25 nm/s. This allows us to control the top barrier thickness within an accuracy of ± 1 nm (further details in Ref. [3]).

We performed continous wave (cw) and time resolved photoluminescence (PL) spectroscopy at liquid helium temperature. The samples were excited for the cw experiments by an Ar-ion laser (λ=514 nm) with a density of about 10 W/cm^2. A Ti:Sapphire laser (λ=827 nm) with a pulse length of 2 ps and a power peak density of 70 kW/cm was used for the time resolved measurements. The emission signal was detected after dispersion by a 32 cm-monochromator by a CCD- or a streak camera.

3. Photoluminescence Results

Figure 1 shows the PL spectra of the etched 5 nm $Ga_{0.8}In_{0.2}As$ QW for top barrier thicknesses of 16, 4, 1 and 0 nm. We observe a distinct PL line shift to higher energies with reduced top barrier thickness. Simultaneously the line broadens and the intensity decreases about three orders of magnitude. The PL spectra at the bottom shows within the etch accuracy the emission of a surface QW, where the GaAs top barrier layer is completely removed. Here the low energy discontinuity of the semiconductor is replaced by the high potential of the vacuum. We determine a blue shift of 33 meV in comparison to the unetched reference sample with a 16 nm thick top barrier layer.

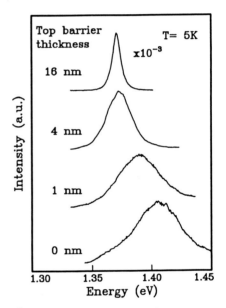

Figure 1. 5K PL spectra from a 5 nm $Ga_{0.8}In_{0.2}As$ QW for top barrier thicknesses of 16, 4, 1 and 0 nm.

4. Model and Discussion

In figure 2 the energy shift of the PL lines of the 5 and 15 nm wide QWs is plotted versus the top barrier thickness. The blue shift for the 5 nm QW starts already at about 5 nm top barrier thickness. The large energy shift up to the GaAs band edge is due to the reduction of the GaInAs layer indicated as negative values of the top barrier thickness in the figure. The 15 nm QW shows even for the surface QW a weak energy shift. Only when the thickness of the GaInAs layer is reduced we observe a QW width related shift.

The shift of the emission lines can be modeled when the high vacuum potential is taken into account. The solid lines in figure 2 display calculations assuming a finite top barrier thickness and a 5 eV potential [4] at the surface. In the vacuum we take the free electron mass as effective mass. The calculations are in good agreement with the experimental data. They explain the energy shift for non zero top barrier thicknesses for the 5 nm QW as well as the significantly smaller shift for the 15 nm QW. The reason for the difference is the low quantization energy of the 15 nm QW. This causes a lower sensitivity of the energy eigenstate on the replacement of the GaAs top barrier layer by the vacuum potential.

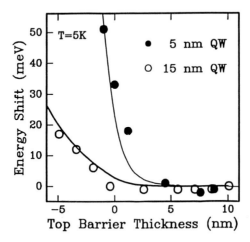

Figure 2. PL line energy shift as function of the top barrier thickness for etch series of a 5 nm $Ga_{0.8}In_{0.2}As$ QW and a 15 nm $Ga_{0.87}In_{0.13}As$ QW. The solid lines represent calculations assuming a finite GaAs top barrier layer with a 5 eV vacuum potential at the surface. A negative top barrier thickness corresponds to a reduction of the GaInAs layer.

Figure 3 shows the electron wave functions corresponding to the energy eigenstates of the 5 nm QW with 0, 1, 5 and 10 nm top barrier thickness. The maximum of the wave function shifts into the semiconductor, when the top barrier thickness is reduced from 5 to 0 nm. This is due to the repelling of the wave function into the semiconductor by the high vacuum potential. At the surface the wave function decreases very quickly, because of the large discontinuity and the free electron mass in the vacuum. These effects cause an enhancement of the confinement resulting in an enlarged quantization energy, which is observed in the PL spectrum as a blue shift. Above 5 nm thick barrier layers the wave function does not change significantly. This leads to constant transition energies.

Due to the penetration of the wave function into the GaAs top barrier layer there is a nonzero probability to find carriers at the surface. The surface is characterized besides the vacuum potential by a high density of midgap surface states [5]. Therefore carriers in near surface QWs have two main recombination channels: Radiative recombination in the QW and nonradiative recombination at the surface. The measured lifetime τ is the sum of both recombination processes [6].

The capture process of carriers for example by impurities, defects and acceptors can be characterized by a capture cross section. The theory by Lax et. al. [7] describes the recombination process by a trap as a capture of carriers into a potential followed by a relaxation through bound states via phonon emission. Therefore we express the lifetime for the nonradiative recombination at the surface by the probability P to find

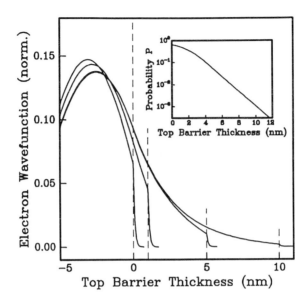

Figure 3. Electron wave functions for a 5 nm $Ga_{0.8}In_{0.2}As$ QW with a 0, 1, 5 and 10 nm thick top barrier layer. The dashed lines mark the surface position for each wave function. Note the strong decrease of the wave function in the vacuum and the shift of the maximum due to the reduction of the top barrier thickness. The inset shows calculations for the probability to find an electron within a 4.5 nm thick trapping layer at the surface. The probability saturates when the trapping layer is shifted into the GaInAs layer.

an electron within a trapping layer of the thickness L_t multiplied by the capture time τ_C for surface states. We obtain for the lifetime τ

$$\frac{1}{\tau} = \frac{1}{\tau_R} + P\frac{1}{\tau_C} \qquad (1)$$

where τ_R is the life time for radiative recombination in the QW.

The inset in figure 3 shows the calculated probability to find an electron within an 4.5 nm thick trapping layer at the surface in dependence of the top barrier thickness. The probability P increases strongly with reduced top barrier thickness. The value is 10^{-3} for a 10 nm top barrier layer and 0.6 for a surface QW. For top barrier layer thicknesses lower than 4.5 nm the trapping layer is shifted into the optically active well region. This causes a saturation of the probability P as shown in the inset. We observe this saturation effect in the top barrier thickness dependence of the PL intensities. The 5 and 15 nm thick QWs show after a strong decrease of about three orders of magnitude a low constant intensity below 5 nm top barrier thickness. This is consistent with the value choosen for the trapping layer thickness in the calculations.

Figure 4 displays the measured lifetimes for the 5 and 15 nm QW versus the top barrier thickness. We observe for both QWs a strong decrease of the lifetime beginning at a certain value. The onset of the decrease scales with the QW width and starts for the 15 nm QW at lower top barrier layers. The slope of the decrease is for both samples similar.

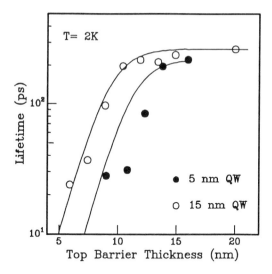

Figure 4. Lifetime as a function of the top barrier thickness for 5 nm and 15 nm wide QWs. The solid lines represent calculations assuming nonradiative recombination within a 4.5 nm trapping layer at the surface determined by the probability of the electrons within this layer.

The solid lines represents calculations according to Eq. (1). We used for the radiative lifetimes τ_R the measured values for the as grown samples with thick top barrier layers. A capture time τ_C of 165 fs was used, which is on the time scale for electron-phonon scattering processes [8] [9]. The trapping layer thickness was 4.5 nm.

The model fits the measured data with one set of parameters very well. The difference between both curves is explained by the different probabilities to find the electrons within the trapping layer. The slope of the decrease is determined by the exponential decrease of the wave function in the top barrier layer. It is similar for both samples, because the important parameter is the energetic distance between the electron energy level and the GaAs barrier.

5. Summary

We have investigated the influence of the surface on the optical properties of GaInAs QWs located at different distances to the surface. The emission line shifts to higher energies with reduced top barrier thickness. The blue shift scales with the QW width and reaches 33 meV for a 5 nm $Ga_{0.8}In_{0.2}As$ surface QW without GaAs top barrier layer. The PL line of a 15 nm thick QW shows only a blue shift when the optically active layer itself is reduced. This behaviour is in good agreement with calculations assuming a finite top barrier layer with a 5 eV vacuum potential at the surface.

Time resolved measurements have shown that the excitonic lifetimes are already influenced for top barrier thicknesses below around 10 nm. The life time decreases strongly below a certain value of the top barrier thickness, which is for the 15 nm QW lower than for the 5 nm QW. A model taking into account nonradiative recombination of electrons within a 4.5 nm thick trapping layer at the surface describes the observed behaviour very good. The dominating parameter is the probability to find electrons within this trapping layer.

Acknowledgements

We would like to thank F. Faller for advice in questions of the sample growth. The financial support by the Deutsche Forschungsgemeinschaft is gratefully acknowleged.

References

[1] J.M. Moison, K. Elcess, F. Houzay, J.Y. Marzin, J.M. Gerard, F. Barthe, M. Bensoussan, Phys. Rev. B **41**, 12945 (1990).

[2] S. Fafard, E. Fortin and A.P. Roth, Phys. Rev. B **45**, 13769 (1992).

[3] J. Dreybrodt, A. Forchel, and J.P. Reithmaier, Phys. Rev. B **41**, 14741 (1993).

[4] J.M. Moison, C. Guille, M. Van Rompay, F. Barthe, F. Houzay, and M. Bensoussan, Phys. Rev. B **39**, 1772 (1989).

[5] T. Sawada, K. Numata, S. Tohdoh, T. Saitoh, and H. Hasegawa, Jpn. J. Appl. Phys. **32**, 511 (1993).

[6] W. Shockley, *Electrons and Holes in Semiconductors* (van Nostrand, New York, 1950), p. 318.

[7] M. Lax, Phys. Rev. **119**, 1502 (1960).

[8] J.A. Kash, J.C. Tsang, and J.M. Hvam, Phys. Rev. Lett. **54**, 2151 (1985).

[9] M.C. Tatham, J.F. Ryan, C.T. Foxon, Phys. Rev. Lett. **63**, 1637 (1989).

Photoinduced near-surface electric field screening in GaAs/AlGaAs MQW structures.

V.N.Astratov, O.Z.Karimov, Yu.A.Vlasov.

A.F. Ioffe Physical-Technical Institute, 194021 St.-Petersburg, Russia.

Abstract. The paper is devoted to the study of interaction mechanisms between quantized states and neighbouring surfaces. The surface/well interaction was detected by photoluminescence (PL) of quantum wells (QW) located at nanometric distances from the structure surface. The redshift and the strong quenching of QW PL line were observed with thinning of the top barrier by wet etching. The mechanism accounting for the band bending and quantum confined Stark effect is maintained. In order to study field screening processes the QW PL was investigated in a wide temperature and excitation power ranges. Under strong pumping the indication on field-induced charge build-up in near-surface QW was found due to observation of PL line broadening effect. It is shown that the latter process can lead to the transition of near-surface field screening pattern from linear to pronounced step-like shape.

1. Introduction.

The interaction of states confined in quantum wells (QW) with nearby surfaces and interfaces are of major significance to the optical and electrical properties of III-V based nanostructures. In particular it defines the radiative properties of ultrasmall objects such as etched quantum wires and dots [1]. A relative new and promising possibility of studying the surface/well interaction is offered by the optical spectroscopy of near-surface QWs [2-7]. This method involves nanoscale variation of the surface/well separation by epitaxial growth or etching of the top barrier layer. The interaction with the surface is detected by the photoluminescence (PL) from a near-surface QW considered as a kind of local optical probe of the surface influence. The latest spectroscopic study of the near-surface QWs has revealed a complex pattern of interactions with various kinds of surfaces, the major models proposed for their description being: (i) The influence of the potential discontinuity at the surface given by the electron affinity of the semiconductor [3]; (ii) Tunnelling coupling with surface states [2]; (iii) The ambipolar tunnelling model [4] accounting for the formation of photoinduced dipole across the top barrier due to asymmetric tunnelling conditions for electrons and holes. These models are characterized by relatively small distances' scale (<150A) of the surface/well interaction determined by the tunnelling length.

The paper presented concerns the interaction with wet-etched and sulphur-passivated surface in GaAs/AlGaAs system. Due to observation of long-range interaction (>300A) we maintained the mechanism caused by near-surface band bending and quantum-confined Stark effect (QCSE) [5-7]. We observed that the redshift under pumping is accompanied by a strong broadening of PL line. These results indicate the possibility of field-induced charge build-up in QWs and formation of step-like character of nonequilibrium field distribution.

2. Experimental details.

The experimental study of the PL of near-surface QW as a function of the distance from the surface is usually provided by the set of the samples with different top barrier thickness d. The variation of the distance d can be performed by MBE growth or by etching on the different depth of the set of samples cuted from the same wafer. However this techniques gives rize to the noncontrollable divergence as of the surface properties defined in different experiments, so the properties of the samples in a given set. This fact leads to significant scattering of the experimental data making the interpretation difficult. In the paper presented we propose the new technique that allows to overcome these difficulties. It is based on using the gradient 'wedge-like' MQW sample.

The MQW gradient structures were grown by MBE without substrate rotation. The average width of QWs numbered opposite to the growth order were consequently: 1-13A, 2-20A, 3-30A, 4-38A, 5-53A, 6-77A. The QWs were separated by AlGaAs (x=0.3) barriers with 200A average width. The MQW region was sandwiched between two AlGaAs (x=0.4) layers 2200A thick. Due to variation of the growth rate along the nonrotating substrate the widths of all the QWs and the barriers smoothly changed. Thus the structure possesses a pronounced 'wedge-like' geometry shown on the inset of Fig. 1. which is characterized by the increasing angle formed by the plane of the substrate and planes of different interfaces sequentially grown far and far away from the substrate. The parameters of the structure were controlled by local optical measurements. The uniformity of the barriers Al percentage was checked by well resolved barrier exciton peak in reflectance spectra. The thickness gradients of the barriers were controlled by laser reflectometry, and that of the GaAs layers by PL of corresponding QWs. Fig. 1 presents the set of low-temperature (T=2K) PL spectra measured with excitation by a focused laser spot (~0.1mm diameter) at different points (A-C) along the gradient direction. It can be seen that the PL lines corresponding to QWs of different width display the blueshift from spectra A to C that is naturally explained by the thinning of the structure. The changing of the QW width reached 3-4%/mm along the gradient direction. Note that a given energetic position of PL peaks is strictly corresponds to a given point on the structure that permits hte identification of the excitation spot position directly from the PL data. The results from both methods (PL and reflectometry) were found to be in a good agreement and showed equal relative changing of widths of the barriers and wells in accordance with described above 'wedge-like' geometry.

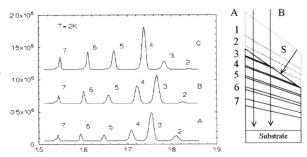

Fig. 1 *PL spectra measured at different positions of laser spot on the surface of as-grown 'wedge-like' structure.* **Inset.** *Schematic view of 'wedge-like' sample (not to scale). The etched part shown by dotted lines. The arrows marked A, B represent different positions of excitation spot corresponding to different spectra.*

In order to decrease the distance d from QW to the surface the wet etching in highly diluted sulphuric-peroxide etchant was used. During etching the surface was transferred to the depth S deep into the structure being parallel to its initial plane and crossing different QWs (see inset Fig. 1). The surface was passivated after etching by Na_2S deposition to enhance the PL efficiency according to the method [14]. The continuous variation in nanoscale of the distance from QWs to the surface can be easily realised in PL experiments by the shift of the position of excitation spot along the sample. The surface properties in thus obtained sample are fixed along it, because they were defined during one cycle of etching-passivation treatment and only one parameter to be changed - the distance from QW to the surface. Thus the proposed technique permits to avoid the noncontrollable diversity of the samples and their surface properties simplifying the interpretation of PL

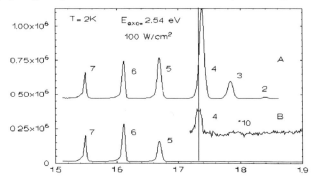

Fig. 2 *PL spectra of the 'wedge-like' sample measured before (a) and after etching (b) from the same point. T=2K, excitation intensity $100W/cm^2$.*

data.

3. Results and discussion.

The PL was excited by a focused beam of 488 nm line of Ar-laser. The PL spectra were measured with a double monochromator and a standard photon counting. The excitation power was 100W/cm^2 (except the experiments on PL pumping dependences discussed in section 3.3).

Figure 2,a shows the typical PL spectrum from nonetched sample. The spectrum shown on Fig. 2,b was measured from the same point on the sample after wet-etching. Note, that the spectral position of PL peaks from deeply embedded QWs (marked 6-7) can be used for control the coincidence of the position of excitation spot before and after etching because these wells are insensitive to the surface influence. It can be seen from Fig. 2,b that the etching of the structure results in disappearance of PL lines 1 - 3 that can be naturally attributed to the etching away of corresponding QWs. The comparison of the spectra of Fig. 2 shows that the approaching of the surface to the QW (here the QW 4) is accompanied by the two main spectroscopic effects: the redshift of corresponding PL line maximum and drastic quenching of its intensity. For the detailed study of these effects we perform experiments with variable distance from QW to the surface and variable intensity of exciting light.

3.1. PL dependence on the distance between QW and the surface.

In order to vary the distance d from near-surface QW to the surface the set of PL spectra was measured as a function of sequential shift of excitation spot position along the gradient direction. Taking into account described above 'wedge-like' sample geometry it can be shown that the changing of distance d corresponding to the different positions of excitation spot can be estimated as $\Delta d = \Delta L_z / L_z \times S$. Here $\Delta L_z / L_z$ -relative change of the sample thickness wich can be found from the PL lines of deeply embedded QWs. As a reference point with zero distance the position of laser spot was chosen corresponding to the spectrum with PL line of near-surface QW quenched down to 10^{-4} relative to that in spectrum before etching.

The observed effects of the surface influence on PL line of near-surface QWs with different widths as a function of the distance from the surface d are presented in Fig. 3. The following peculiarities should be noted: the large distances (>300A) of the interaction with the surface and well pronounced increase of the redshift for the thicker QWs. They can not be explained by the models accounting for the tunnelling to surface states [2-4], which are characterized by smaller interaction distances determined by the tunnelling length (~100A) and by opposite dependence of the redshift on QW width. Note also, that observed effects are disagreed with the model [3] of quasi-infinite barrier at the surface of the structure because it leads to the blueshift of near-surface QW PL line in opposition to the observed behaviour.

We interpret the experimental data in the framework of

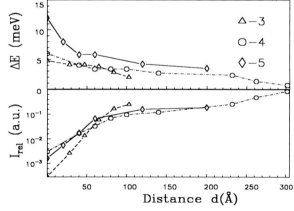

Fig. 3 *The redshift ΔE of PL line of near-surface QW's (upper plot) and its relative intensity I_{rel} normalized on intensity of PL line of deeply buried QW 7 (down plot) as a function of the distance from the surface d. The average widths of QW's are: 3 - 30A, 4 - 40A, 5 - 50A.*

the mechanism, based on band bending and quantum confined Stark effect (QCSE). Near semiconductor surface the built-in electric field exists due to pinning of the Fermi level at the energetic position of charged surface states. The electric field in depleted layer distorts the QW potential that leads to the redshift of PL line of near-surface QW due to QCSE [8]. Therefore the surface/well interaction in this mechanism expands on distances determined by the Schottky length reaching under photoexcitation several hundreds of angstroms that is significantly larger than characteristic tunnelling length and is in accordance with observed interaction distances. This mechanism can also describe the behaviour of the redshift on QW width due to QCSE dependence $E \sim L_z^4$. Moreover the influence of built-in field on near-surface QW can also explain the effect of PL quenching by the depletion of near-surface region as well as by the field induced tunnelling of carriers out of the QW through the tilted barrier. Note, that the latter process is sufficient for the explanation of the observed magnitude of quenching (similar values of quenching due to field induced tunnelling were observed in experiments [9] with external electric fields $\sim 10^5$ V/cm).

3.2. The field profile in MQW structure.

Despite the fact that the surface field induced nature of the observed effects seems to be clear the real spatial distribution of the electric field in MQW structure needs further investigation. For the bulk semiconductor under the above band gap photoexcitation the theoretical description of the screening processes is provided in the framework of the drift-diffusion model (see for instance [11]). The electron-hole pairs generation in the depleted region should be accompanied by the separation of the charges. The minority photocarriers drift to the surface under the influence of the electric field created by the charged surface states partially compensating them. This leads to the flattening of the bands that reduces the potential barrier for the diffusion and thermionic emission flows of the majority carriers to the surface. In the steady state the surface directed currents of the majority and manotity carriers are balanced and recombine via surface states. This model well describes photoinduced band flattening effect and decreasing of the band bending at low temperatures [13] observed in GaAs.

The specific properties of MQW structures are connected with the possibility of photoinduced built-in charge accumulation within QWs. In the undoped MQW structures under consideration these processes can be accompanied by drastic changes of the field distribution. The charging of the QWs can be caused by ifferent mechanisms. So in small distances scale (d<150 A) the QW charging can be explained on the basis of recently proposed ambipolar tunnelling model [4] taking into account the nominally different tunnelling probabilities for two types of carriers to reach the surface. In the steady state the ambipolar tunnelling regime is established due to formation of a photoinduced dipole electric field across the top barrier. It should be noted however that in larger distances scale (d>150 A) corresponding to our experimental situation the tunnelling interaction with the surface states exponentially falls with d and other factors of QW charging are more probable.

Under the influence of electric field the probabilities for electrons and holes to be captured in the well (or tunnel from it) become different thus providing the filling of QW by the one sign carriers. These charging processes are stronger under the above barrier than under the below barrier excitation. Such mechanisms were observed in experiments with the application of external electric field to DBRT structure [12]. In the structures with near-surface QWs these mechanisms also can be expected under condition $L_{tunn} < d < L_{sh}$, where L_{tunn} - tunnelling length, $L_{SH} = \sqrt{2 \cdot \dfrac{\varepsilon \varepsilon_0 \varphi_S}{eN}}$ - Schottky length, ε- dielectric permittivity, ε_0- static dielectric constant, N - concentration of ionized centers in the barrier, φ_S- surface band bending. The theory of field screening in this situation is not developed at present thus we shall only briefly mention two main ultimate cases.

a. Weak QW charging. At low photoexcitation the concentration n_S of the built-in charge in QWs is expected to be less than space charge N in the barriers ($n_S \ll N \cdot b$, where b - the barrier width) and the screening is provided mainly by the formation of a depleted layer in the barrier material in accordance with drift-diffusion model. The QWs itselves can be considered as a kind of local optical probes of the field strength due to QCSE nondisturbing the field distribution.

b. Strong QW charging. At high photoexcitation the built-in charge in the near surface QW can exceed the space charge in the barriers ($n_S > N \cdot b$). The charge built-up in near-surface QW should lead to significant screening of electric field within the QW. The field spatial distribution should be nonlinear with pronounced steps at QWs due to partial screening by 2D carrier gas - see Fig. 4, corresponding to p-type material. The electrons populate states from the bottom of conduction band up to Fermi energy $E_F = \hbar^2 \cdot n_s / 4\pi m^*$, where m^* -effective electron mass. The band filling should be accompanied by the asymmetric PL line broadening [12]. In this case the redshift reflects the magnitude of electric field averaged within the QW.

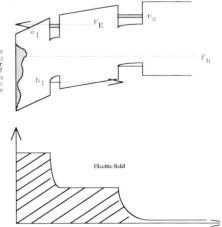

Fig. 4 *QWs of different width in near-surface space charge region under strong above band gap photo-excitation. The arrows shows the field induced tunneling of the carriers out of the QW.*

Our experimental data may be analysed basing on the assumption that field screening is not strongly affected by the presence of QWs (case *a*). The field spatial distribution was calculated (see also [5-7]) from the redshift data $E(d)$ of Fig. 3 by modelling the QCSE within the variational approach [8]. The calculated data fits well to the linear dependence, however the estimation of the space charge leads to too high concentration $\sim 2 \cdot 10^{17} \text{cm}^{-3}$ that is larger than the total concentration of defects in our samples. This fact points out the possibility of QW charging under pumping.

3.3. PL dependence on the intensity of exciting light.

For detailed study of QW charging we investigated the PL pumping dependences on the planar MQW structure with relatively narrow PL lines instead of 'wedge-like' nonuniform sample. The GaAs/AlGaAs (x=0.3) MQW structure was grown by MBE on semi-insulating substrate and contained the set of QWs (40 nm, 30 nm, 20 nm, 10 nm, 5 nm, 2.5 nm - growth order) separated by the 20 nm barriers' layers. The top barrier layer (20 nm) was capped by the 5 nm GaAs. The distance from the surface was decreased by the etching in sulphuric-peroxide mixture. The etching depth was controlled by laser reflectometry with the accuracy 2 nm. The temperature of T=77K was chosen for PL experiments in order to enhance the magnitude of near-surface electric field [13] and stimulating QW charging processes. In the PL spectra the redshift and quenching of PL lines of near-surface QWs were revealed. In accordance with discussed above experiments on 'wedge-like' structures these effects occur at large distances from the surface (up to 50 nm). Figure 5 presents the PL pumping dependences of PL line from QW with L_z=50A in as-grown

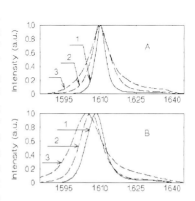

Fig. 5 *Normalised PL spectra (QW width 50A) of the planar sample measured at T=77K before (a) and after (b) etching at different excitation intensities: 1- 0.1W/cm2, 2-2W/cm2, 3-100W/cm2.*

(A) and etched (B) samples. Note that the distance d from the QW to the surface in as-grown sample was about 480A and that in etched sample 200A. As it can be seen from the Fig. 5,a the QW PL line of as-grown sample undergoes weak broadening and its postition remains practically unchanged with pumping. Fig. 5,b reveals strong broadening of PL line with increasing of the excitation level in the spectra of etched sample accompanying by the increase of the redshift. Figure 6 illustrates the dependencies of the full width at half maximum (FWHM) (a) and the PL line positions (b) as a function of pumping power in both the as-grown and etched samples.

The observed broadening of PL line allows to suppose the QW band filling by the free carriers. The magnitude of broadening more probably corresponds to the accumulation of the electron gas in the near-surface QW. It should be noted that the shape of PL line with increasing pumping became asymmetric in the high energy range that is typical for the PL spectra of DBRT structures [12] with charge build-up in QW. The FWHM - more than 14 meV, see Fig. 6 - can be used for the estimation of charge concentration $n_s \sim 10^{11} cm^{-2}$ according to method [12] taking into account FWHM in as-grown sample 5 meV. Therefore the value n_s/b at high photoexcitation 100 W/cm^{-2} is comparable with the upper limit of the concentration $N \sim 10^{16} cm^{-3}$ determined by unwanted backgrownd doping of the barriers. In other words the field screening across the barrier layer is comparable with the screening within the QW - the situation intermediate between the ultimate cases a) and b) discussed in Sect. 3.2. At lower excitation levels the screening pattern approaches the linear Schottky law typical for the bulk semiconductors.

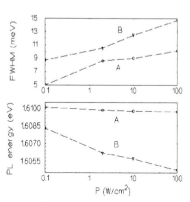

Fig. 6 *The FWHM (upper plot) and the redshift (down plot) 0f QW PL line of as-grown (a) ana etched (b) sample as a function of pumping power P.*

Acknowledgments. The authors are grateful to Prof.A.A.Kaplyanskii for the encouragement and to Dr.I.E.Itskevich, Profs. V.D.Kulakovskii, I.V.Kukushkin and V.B.Timofeev for critical comments and useful discussions.

References

1. P.D.Wang, C.M.Sotomayor Torres, H.Benisty, C.Weisbuch and S.P.Beaumont, Appl.Phys.Lett. **61**, 946 (1992).
2. J.-M.Moison, K.Elcess, F.Houzay, J.-Y.Marzin, J.-M.Gerard, F.Barthe and M.Bensoussan, Phys.Rev. **B 41** , 12945 (1990)
3. J.Dreybrodt, A.Forchel and J.P.Reithmaier to be published in Phys.Rev.B (1994).
4. B.Bonanni, M.Capizzi, V.Emiliani, A.Frova, C.Presilla, M.Colocci, M.Gurioli, Y.Chang,I.Tan and J.L.Merz to be published in J.Appl.Phys.(1994).
5. V.N.Astratov and Yu.A.Vlasov, Semiconductors **27(7)**, 606 (1993).
6. V.N.Astratov and Yu.A.Vlasov, Journal de Physique **2**, 3, 277 (1993).
7. V.N.Astratov and Yu.A.Vlasov, Material Science Forum,**143-147**,599-603 (1993)
8. G.Bastard, E.E.Mendez and L.Esaki, Phys.Rev.**B 28**, 3241 (1983).
9. Y.Horicoshi, A.Fisher and K.Ploog, Phys.Rev.**B 31**, 7859 (1985).
10. H.H.Weider, J.Vac.Sci.Technol. **17**, 1009 (1980).
11. M.H.Hecht, J.Vac.Sci.Technol.**B 8**, 1018 (1990).
12. M.S.Skolnick, D.G.Hayes, P.E.Simmonds, A.W.Higgs, G.W.Smith, H.J.Hutchinson, C.R.Whitehouse, L.Eaves, M.Henini, O.H.Hughes, M.L.Leadbeater and D.P.Halliday, Phys.Rev. **B41**, 1074 (1990).
13. C.R.Lu, J.R.Anderson, W.T.Beard and R.A.Wilson, Superlatt. and Microstr., **8**, 155 (1990).

Photocurrent spectroscopy of 2-dimensional excitons in 5-nm wide InGaAs/InAlAs multi quantum wells

N. Kotera and K. Tanaka

Kyushu Institute of Technology, Iizuka, Fukuoka 820, Japan

Abstract. The excitons confined in a narrow quantum well of InGaAs/InAlAs MQW structures lattice-matched to InP have been observed at 77 K by photocurrent spectroscopy experiments. Distinctive 6 peaks were successfully assigned to be two series of consecutive 3 heavy-hole subbands. The assignment was based on the square-law dependence of the peak energy on the heavy-hole quantum number. Theoretical calculation using InGaAs heavy-hole effective mass of $0.46m_0$ and bandgap difference, ΔE_v, of 0.28 eV fits well with experimental results.

1. Introduction

A number of observations of 2-dimensional excitons confined in InGaAs/InAlAs MQW's lattice-matched to InP substrates were reported so far. Most observations were made using a specimen of wider well widths than 10 nm. Quantitative analysis of the band parameters has never been tried because of the small number of peaks.

In this paper, specimens of MBE-grown p-i-n photodiodes including 33 undoped MQW's were measured. The well (InGaAs) width, L_z, was 5 nm and the barrier (InAlAs) width was 10 nm. Many wells were uncoupled. Thus, a simple particle-in-a-box model with a finite barrier height was used for energy level calculation. The effective mass of the barrier material was also taken into account. By comparing experiments and theory, heavy-hole effective mass in the well normal to the well-barrier interface can be analyzed quantitatively.

2. Experimental

A specimen was irradiated by light from a halogen lamp behind a monochromator through a glass window of a cryostat. A specimen was placed on a cooled base metal in vacuum and a probing needle was pressed onto the contact layer of a p-i-n photodiode. Dc reverse bias of -1 V was applied to the junction. Photocurrents perpendicular to the well interface were measured at 77 K. The electric field in the quantum well ranged from 30 to 100 kV/cm approximately and pure allowed- and forbidden- optical transitions cannot be observed experimentally.

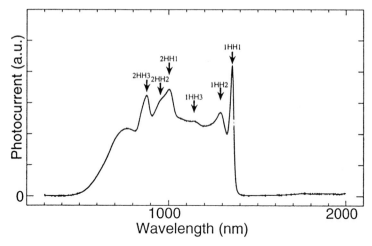

Fig.1 A photocurrent spectrum of a specimen (V-749) at 77 K. A notation, nHHℓ, stands for exciton formation by an electron of n-th conduction subband and a heavy hole of ℓ-th valence subband.

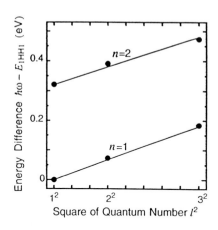

Fig.2 The dependence of peak energies on the square of quantum number in the heavy hole subbands (the square law).

Fig.3 Calculated subband energy vs. heavy-hole effective mass of InGaAs (well). Solid lines indicate $E_{HH}(\ell)$. A dotted line indicates the energy difference, $E_{HH}(3)$-$E_{HH}(1)$. The horizontal bar indicates the experimental value.

3. Results and discussions

Photocurrent spectra as a function of wavelength were plotted as shown in Fig. 1. The peak heights reflect the exciton absorption intensity. If excitons are formed by an electron of n-th conduction subband and a heavy hole of ℓ-th valence subband, the resonant absorption occurs if the irradiated photon energy, $\hbar\omega$, is equal to the sum of the InGaAs bandgap energy, E_G, the conduction subband energy at the bottom, $E_e(n)$, and the heavy-hole subband energy at the top, $E_{HH}(\ell)$. The notation, HH, stands for the heavy hole. Then, the resonant photon energy, $E_{nHH\ell}$, is equal to $E_G+E_e(n)+E_{HH}(\ell)$. The exciton formation energy was assumed negligible.

In Fig. 1, sharp and clear 6 peaks were observed. The strongest peak in Fig. 1 was assumed to be the fundamental absorption corresponding to the energy, E_{1HH1}. Based on the theoretical calculation using a finite well model, it was found that the square law, $E_{HH}(\ell) \approx \ell^2 E_{HH}(1)$, was applicable to our systems. The energy difference, $\hbar\omega - E_{1HH1}$, was plotted as a function of the prospective quantum numbers, $\ell=1, 2, 3$, in Fig. 2 and the square-law dependence was verified experimentally. Double series of consecutive 3 peaks in Fig. 1 might be caused by the ground($n = 1$) and the higher($n = 2$) conduction subbands. The gradient of this curve will give the effective mass of related holes. If the mass value is close to that of the bulk heavy hole, it will be concluded that the contribution of light hole is scarce in this experiment.

Theoretical calculation was made by varying the effective mass of InGaAs heavy-hole. As seen in Fig. 3, three to four $E_{HH}(\ell)$ levels were found when the bandgap

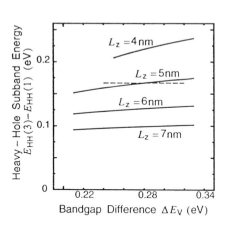
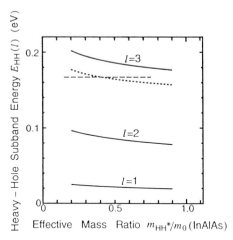

Fig.4 Calculated the 3rd heavy-hole subband energy as a function of the bandgap difference at the valence band. Parameters were the well width.

Fig.5 Calculated heavy-hole subband energy vs. heavy-hole effective mass of InAlAs (barrier). Solid lines indicate $E_{HH}(\ell)$. A dotted line indicates the energy difference, $E_{HH}(3)$-$E_{HH}(1)$. The horizontal bar indicates the experimental value.

difference of the valence bands, ΔE_v, was 0.28 eV and well width, L_z, was 5 nm. The observed $E_{HH}(3)$-$E_{HH}(1)$ of 169 meV inserted in the figure gives the m^*_{HH} of $0.46m_0$ which agrees with the bulk cyclotron effective mass of InGaAs [1]. This validates the peak assignment shown in Fig. 1. The parameters used in this calculation were as follows: $m^*_c(\text{InGaAs})=0.041m_0$, $m^*_c(\text{InAlAs})=0.075m_0$, $\Delta E_c=0.47$ eV, $m^*_{HH}(\text{InAlAs}) = 0.41m_0$ and $\Delta E_v = 0.28$ eV. The m_0 is the free electron mass in vacuum. Using these parameters, a theoretical square-law curve comparable to Fig. 2 was calculated. Agreement between theory and experiment was excellent.

Similar calculation varying the bandgap difference, ΔE_v, was made between 0.21 eV and 0.33 eV as shown in Fig. 4. If 169 ± 10 meV was the experimental value, ΔE_v of 0.28 ± 0.04 eV could be obtained. If the well width, L_z, was varied in the calculation, the experiments showed the L_z must be 5 ± 0.5 nm when the heavy hole mass was assumed.

Finally, calculation varying the heavy hole mass of the barrier(InAlAs) was made as shown in Fig. 5. Comparison with experiments gives the approximate effective mass of $0.41m_0$. The energy depends slightly on this effective mass.

4. Conclusion

Photocurrent spectroscopy of the 2-dimensional electron and hole system was utilized to characterize the subband structures. The square-law dependence of the peak energy on the heavy-hole quantum number was clearly observed, which validates the peak assignment. A calculation based on a simple finite-square-well model fits well with experiments. The analysis offers a probable parameter set including effective masses as follows; L_z=5 nm, $m^*_{HH}(\text{InGaAs})=0.46m_0$, $\Delta E_v=0.28$ eV, and $m^*_{HH}(\text{InAlAs})=0.41m_0$.

The authors appreciate Mr. Hitoshi Nakamura of Central Research Laboratory, Hitachi Ltd. for the fabrication of p-i-n photodiodes.

References

[1] Alavi K, Aggarwal R L, and Groves S H 1980 Phys. Rev. B21 (3) 1311-1315

Theoretical and experimental study of polarization-dependent photoluminescence in $In_{0.2}Ga_{0.8}As/GaAs$ quantum wires

J. Shuttlewood, K. Ko, P. Bhattacharya, S. W. Pang, I. Vurgaftman, J. Singh and T. Brock

Solid State Electronics Laboratory, Department of Electrical Engineering and Computer Science, University of Michigan, Ann Arbor, MI 48109-2122

Abstract. We report a theoretical and experimental study of polarization-dependent photoluminescence in quantum wires. The experimental samples were made by electron beam lithography, dry etching with an electron cyclotron resonance source, wet chemical etching and molecular beam epitaxial regrowth. The luminescence from the wires at 9K is blue-shifted by 6 meV from that of the quantum wells and both are of comparable intensity, taking the fill factor into account. The ratio of the peak intensities for polarizations parallel and perpendicular to the wire axis is $\sim$2.5.

1. Introduction

The altered density of states due to reduced dimensionality in semiconductor heterostructures can be exploited to enhance the performance of electronic and optoelectronic devices. This is one of the motivations behind the increased interest in structures such as quantum dots and wires [1,2,3]. Interface disorder effects in quantum wire structures may cause localization of electronic states along the length of the wire with serious consequences on the optical and electronic properties [4]. The effect of increasing interfacial disorder is thus to broaden the bandedge singularities and weaken the polarization dependence of optical transitions. The study of the polarization dependence of the luminescence from a quantum wire can therefore be a good probe of the quality of quantum wires. More specifically, the polarization dependence of the luminescence can provide direct feedback to the various processing steps involved in fabricating the quantum wires.

2. Theoretical Study of Polarization-Dependent Luminescence

The theoretical formulation for the study of polarization-dependent optical properties of disordered quantum wires has been described by Tanaka et al [5]. The essential features are reiterated here. The calculations have been done for the GaAs/AlGaAs system. The conduction band states in this system are well described by an isotropic effective mass and can thus be described by a single band Schrodinger equation. To determine the valence bandstructure the Kohn-Luttinger formalism is used to include the heavy hole (HH)-light hole (LH) band mixing. In the perfect wire, the confinement potential V(x, y, z) changes abruptly from that of the well material (say zero) to that of the barrier material (the band discontinuity) across the interfacial plans. In the disordered wire, we represent the wire cross-section by three regions: i) the perfect well regions; ii) the perfect barrier region; and iii) the disordered region which has a certain width. The disordered region, we assume is made up of GaAs islands and $Al_{0.3}Ga_{0.7}As$ islands randomly intermixed together. The islands are described by their height Δx which is expected to be a few monolayers, and their lateral extends ΔL_z and ΔL_x. Once the dimensions of the disordered islands are decided upon, we use the Monte Carlo technique to randomly create the appropriate potential fluctuations in the disordered region [6]. For the studies presented here we assume that the islands are only of GaAs and $Al_{0.3}Ga_{0.7}As$ kind and are distributed with equal probabilities in the interfacial region.

For band-to-band transitions, we note that in the 3-dimensionally confined structure, k is not a good quantum number. The absorption coefficient for polarization along ε is given by,

$$\alpha(\hbar\omega) = \frac{4\eta^2 e^2 \hbar}{\eta cm^2 \hbar\omega V} \sum_{i,j} |\varepsilon \cdot P_{ij}|^2 x\delta\left(E_i^e - E_j^h - \hbar\omega\right) \qquad (1)$$

where the indices i and j represent the electron and hole states, V is the volume of the structure over which the electronic state is confined, η is the refractive index. The momentum matrix element is given by,

$$P_{ij} = \sum_{v=1}^{4} \int \phi_i^c(x,y,z)\phi_j^v(x,y,z)\,dxdydz < u_c |p| u_v > \qquad (2)$$

where ϕ_i^c (x, y, z) and ϕ_j^v (x, y, z) are the electron and hole envelope functions and $< u_c |p| u_v >$ is the momentum matrix element between the conduction and valence states.

The valence band state is the 4-band state evaluated by our solution in the confined structure. The polarization dependent effects are obtained from the relative mixtures of the p_x, p_y, p_z states in the valence band solutions [7]. Since the conduction band states are s-type, the only non-zero momentum matrix elements are,

$$< s |p_x| p_x > = < s |p_y| p_y > = < s |p_z| p_z > = P_{cv} \qquad (3)$$

The values of P_{cv} are listed for several semiconductors by Lawaetz [8]. We have calculated the density of states in perfect and disordered wires and it is found that the effect of disorder is to broaden the bandedge singularities in the density of states due to localization of electron and hole states. As a result, the relative intensities of x, y, z polarization transitions change. Here the z-direction represents the wire axis and x and y

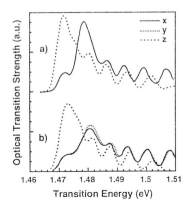

Figure 1: Polarization dependence of optical transitions in a 150Å x 150Å x 1000Å wire; a) perfect wire; b) wire with 40Å x 20Å x 120Å islands placed randomly on two interfaces.

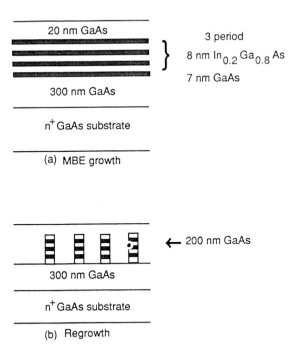

Figure 2: Schematic of the quantum well and wire structures (a) grown and (b) regrown by MBE.

represent the cross-section. The net results of increasing interfacial disorder is to weaken the polarization dependence. This is shown in Figure 1.

3. Experimental

The samples were grown by MBE on (100) Si-doped GaAs substrates. They consisted (Fig. 2) of a 300 nm GaAs layer (undoped), 3 undoped quantum wells made up of 7 nm of GaAs and 8 nm of $In_{0.2}Ga_{0.8}As$, and 20 nm undoped GaAs cap layer. Electron beam lithography was used for direct writing of the quantum wires on a bilayer consisting of 2%, 496K polymethylmethacrylate (PMMA) in anisol spun at 5 krpm for 60 sec and prebaked for 30 min at 200°C; followed by 2%, 950K PMMA in anisol spun on 5 krpm for 60 sec and prebaked for 30 min at 200°C. These result in a total resist thickness of ~100 nm. A subsequent liftoff process using 25 nm Ni formed the mask for dry etching. The quantum wire array for photoluminescence (PL) measurements consisted of wires which were 40–70 nm in width with a 200 nm pitch. The fill factor for this array is 25%. Free-standing quantum wires were then defined by dry etching using an electron cyclotron resonance (ECR) source and a rf-powered electrode [9,10]. The GaAs capping layer and the 3 quantum wells were etched using a 30% Cl_2 in Ar mixture with an etch rate of 130 nm/min. The total etch depth was 100 nm. A short wet etch (~5 nm) in $NH_4OH:H_2O_2:H_2O=4:1:2000$ was used to remove the process-induced damage. Buried quantum wires were realized by MBE regrowth. The regrown structure consisted of a 200 nm undoped GaAs layer. A 10 sec rapid thermal anneal (RTA) at 800°C was performed to improve the luminescence of the quantum wires.

Photoluminescence (PL) measurements at low (9K) and higher temperatures were made using the 488 nm output of an Ar^+ laser as the excitation source and a 1.0 m scanning spectrometer. Only low-temperature PL is discussed in this letter, since the polarization dependence is observed most clearly at low sample temperatures. Several areas were left unetched to serve as control (quantum well) regions. Figure 3 shows the PL spectra of the as grown $In_{0.2}Ga_{0.8}As$ quantum wells. A strong bound exciton peak is observed at 1.350 eV. The PL spectrum of a 50 nm quantum wire array after etching and regrowth is also shown in Figure 3. The intensity of the luminescence is lowered by a factor of 5 compared to the quantum well luminescence, but when the fill factor is taken into account the intensities are nearly identical. The peak in the quantum wire is observed at 1.356 eV, and it is blue-shifted from the quantum well peak by 6 meV. These results indicate that we have achieved 1D quantum confinement in the wires without excessive interfacial disorder.

The polarization dependence of the PL at 9K was measured by placing a Glan-Thomson prism analyzer between the sample and the spectrometer. Figure 4 shows the PL spectra of the quantum wires for luminescence parallel and perpendicular to the wires. It is evident that the luminescence intensity parallel to the wires is about 2.5 times greater than the intensity perpendicular to the wires, as predicted from theoretical calculations. These polarization-dependent characteristics, which were not observed in the quantum wells, clearly demonstrate the 1D quantum confinement of carriers in the wires, and the fact that in this particular sample interfacial disorder is quite small. However, it may be noted that the peak intensities for the two polarizations differ by

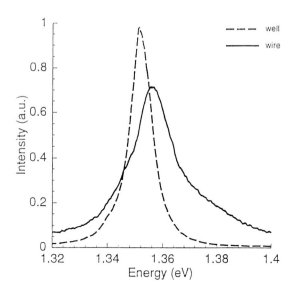

Figure 3: Low-temperature (9 K) photoluminescence from the quantum wells and 50 nm buried (by regrowth) quantum wires. The luminescence from the wires takes into account the fill factor of 4.

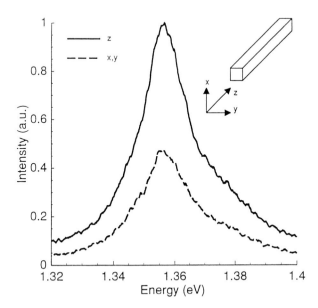

Figure 4: 9K polarization-dependent spectra measured for 50 nm buried wires.

a factor of 2.5, whereas theoretical calculations predict a difference of a factor of 3–4. Hence, further improvements are possible, and we find that wet etching after dry etching and before regrowth is a critical step.

4. Conclusion

In conclusion, we have shown the fabrication of 20–70 nm quantum wires using electron beam lithography, dry etching with an ECR source, and MBE regrowth with strongly polarization-dependent luminescence and a blue shift of 6 meV. We believe that this fabrication method shows great potential for monolithic optoelectronic structures. The observed polarization dependence is in agreement with theoretical calculations.

The work was support by the Army Research Office (URI program) under Grant DAAL03-92-G-0109.

References

[1] Arakawa Y and Sakaki H 1982 *Appl. Phys. Lett.* **40**, 939

[2] Arakawa Y, Vahala K and Yariv A 1984 *Appl. Phys. Lett.* **45** 950

[3] Asada M, Miyamoto Y and Suematsu Y 1986 *J. Quantum Electron.* **QE-22** 1915

[4] Singh J, Arakawa Y and Bhattacharya P 1992 *IEEE Photonics Tech. Lett.* **4** 835

[5] Tanaka T, Singh J, Arakawa Y and Bhattacharya P 1993 *Appl. Phys. Lett.* **62** 756

[6] Singh J 1991 *Appl. Phys. Lett.* **59** 3142

[7] Hong S and Singh J 1987 *Superlatt. and Microstructures* **3** 645

[8] Lawaetz R 1971 *Phys. Rev.* **B4** 3400

[9] Ko K and Pang S 1993 *J. Vac. Sci. Technol.* **B11** 2275

[10] Ko K and Pang S 1994 *J. Electrochem. Soc.* **141** 255

Photoluminescence Determination of Valence-Band Symmetry and Auger-1 Threshold Energy in Compressed InAsSb Layers

S. R. Kurtz, R. M. Biefeld, and L. R. Dawson
Sandia National Laboratories, Albuquerque, NM 87185

Abstract: InAsSb/InGaAs strained-layer superlattices (SLSs) and InAsSb quantum wells, both with biaxially compressed InAsSb layers, were characterized using magneto-photoluminescence and compared with unstrained InAsSb and InAs alloys. In heterostructures with biaxially compressed InAsSb, the holes exhibited a decrease in effective mass, approaching that of the electrons. Correcting the data for the magneto-exciton binding energy, we obtain electron-hole reduced mass values in the range, μ=0.010-0.015, for the InAsSb heterostructures, whereas μ=0.026 and μ=0.023 for unstrained InAsSb and InAs alloys respectively. In the 2-dimensional limit, a large increase in the Auger-1 threshold energy accompanies this strain-induced change in valence-band symmetry. Correspondingly, the activation energy for nonradiative recombination in the SLSs displayed a marked increase compared with that of the unstrained alloys

1. Introduction

Frequently, the Auger-1 process (i.e. An electron and hole recombine by scattering a second electron up into the conduction band.) dominates radiative recombination in narrow bandgap III-V semiconductors, and as a result, the wavelength of diode lasers operating at room temperature has been limited to $\leq$ 2.1-2.3 µm.[1] In a biaxially compressed III-V layer, the |3/2,±3/2> hole ground state can increase the in-plane, electron-hole effective mass ratio (m^*_e/m^*_h) over that found in bulk material. In the 2-dimensional limit, this effective mass ratio will result in an increased threshold energy for Auger-1.[2-4] Therefore, midwave infrared (2-6 µm) emitters with biaxially compressed, InAsSb active regions may exhibit improved performance and higher temperature operation.[5] In this paper, we will describe the optical characterization of InAsSb/InGaAs strained-layer superlattices (SLSs) and InAsSb quantum wells with InAs barriers. We present evidence indicating that a large increase in Auger-1 threshold energy results from the strain-induced valence-band symmetry of the InAsSb layer.

2. Experimental Details

InAsSb and InGaAs alloys and heterostructures were grown by metal-organic chemical vapor deposition (MOCVD) on InAs substrates. SLS and ternary compositions, layer thicknesses, and lattice constants were determined from both (004) and (115) or (335) x-ray rocking curves. In this study, photoluminescence spectra for an InAsSb/InGaAs SLS and InAsSb quantum wells with InAs barriers are compared with those for unstrained InAs$_{0.93}$Sb$_{0.07}$ and InAs alloys. The SLS characterized in this study was an InAs$_{0.91}$Sb$_{0.09}$ / In$_{0.87}$Ga$_{0.13}$As SLS (90 Å / 130 Å layer thicknesses), nominally lattice matched to the InAs substrate. The multiple quantum well sample consisted of 318, 159, 106, and 53 Å thick quantum wells of InAs$_{0.91}$Sb$_{0.09}$, separated by 500 Å thick, InAs barriers. The sum of the four quantum well thicknesses is less than the critical layer thickness ($\approx$ 1000 Å), and the quantum wells are psuedomorphic.

Throughout our studies of As-rich, InAsSb (5-50% Sb), the bandgaps of our InAsSb alloys were smaller than accepted values.[6] Electron diffraction results indicate compositional ordering and phase separation may occur in the As-rich, InAsSb grown at low temperatures by vapor phase epitaxy.[7] As demonstrated in the following discussion, the InAsSb material displays "single phase, random-alloy-like" optical properties except for the bandgap anomaly, and we assume that the strain imposed by the InAs heterostructure dominates any internal strain occurring within domains of the InAsSb.

Midwave infrared photoluminescence was measured by operating an FTIR spectrometer in a double-modulation mode. In the magneto-photoluminescence experiments, a fluoride optical fiber was used to transmit infrared light in the magnet cryostat. The photoluminescent light was collected by the fiber and analyzed with the FTIR equipped with an InSb photodiode. All measurements were made in the Faraday configuration with the magnetic field parallel to the growth, (001), direction of the sample.

3. Magneto-Photoluminescence

Low temperature, zero-field photoluminescence spectra for the multiple quantum well, SLS, unstrained InAsSb, and InAs samples are shown in Figure 1. For each sample, the photoluminescence spectra consists of a single peak with a linewidth of approximately 10 meV. Emission from single quantum wells is clearly resolved in Fig 1(c), with the quantum well thickness indicated in the figure. The photoluminescence energies of the thick (318 Å) quantum well and the InAsSb alloy provide estimates of the bandgap energies of the MOCVD alloys, and both samples exhibit the bandgap anomaly. A large quantum size shift is observed for the quantum wells. Analysis of quantum size effects for InAsSb/InGaAs and InAsSb/InAs heterostructures indicates that both have type I band offsets. The bandgap and conduction band offset are sensitive to ordering or phase separation occurring in the InAsSb; the valence band offset is insensitive to these effects.

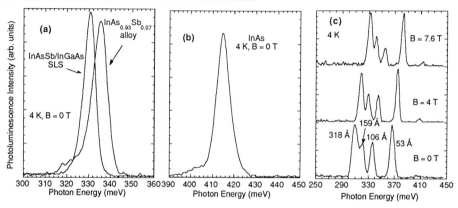

Figure 1 - Photoluminescence spectra of (a) $InAs_{0.91}Sb_{0.09}$ / $In_{0.87}Ga_{0.13}As$ SLS and $InAs_{0.93}Sb_{0.07}$ alloy, (b) InAs, and (c) $InAs_{0.91}Sb_{0.09}$/InAs quantum wells.

The magnetic field dependence of the photoluminescence energies for each of these samples is shown in Figure 2. For each photoluminescence line, the reduced mass, μ, corresponding to the slope obtained from the free electron-hole approximation

($\frac{dE}{dB} = \mu_B \left(\frac{1}{m_e^*} + \frac{1}{m_h^*} \right) = \frac{\mu_B}{\mu}$; μ_B is the Bohr magneton), is indicated in the figure. The reduced masses for the unstrained alloys (μ_{InAsSb}= 0.037 and μ_{InAs}=0.031, in units of free electron mass) are significantly larger than those observed in the strained heterostructures (μ=0.015-0.024) due to the low in-plane hole mass associated with the heavy-hole ($|3/2,\pm 3/2>$) ground state of the biaxially compressed, InAsSb layers.

Excitonic behavior is revealed in the magneto-photoluminescence results. For all samples, the photoluminescence peak energy is insensitive to magnetic field for B< 2T, characteristic of a diamagnetic exciton, and in the linear region observed at higher fields, the reduced mass values obtained from the free carrier approximation are consistently too large, due in part to the binding energy of the exciton. Also, note the increase in reduced mass (or decrease in slope) as the quantum wells become thinner (Fig. 2(b)).

Using measured parameters for InAs and semi-empirical expressions for nonparabolicity and magneto-exciton energies, we can estimate exciton binding energies and correct the reduced mass values.[8] Assuming a 3-dimensional exciton, we obtain corrected reduced mass values, µ=0.010, 0.026, and 0.023, for the 318 Å thick quantum well, unstrained InAsSb alloy, and InAs, respectively. The exciton binding energies for the 318 Å well and the alloys were 1.0 meV and 1.8 meV, respectively. Examining the thinnest (53 Å) quantum well, we estimate an exciton binding energy of 3-4 meV for a band minimum reduced mass, µ=0.010, and assuming a 2-dimensional exciton. The change in slope in Fig 2(b), associated with quantum confinement, indicates an increase in exciton binding energy of the correct magnitude.[8]

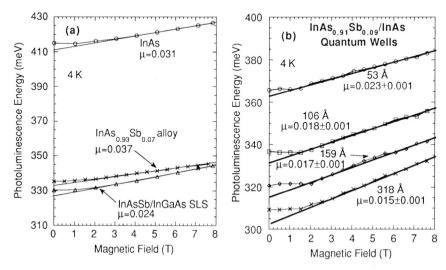

Figure 2 - Magnetic field-induced shift of the photoluminescence energy for (a) the InAsSb/InGaAs SLS and the InAsSb and InAs alloys, and (b) the InAsSb/InAs quantum wells.

4. Modification of Auger-1 Threshold Energies

In the 2-dimensional limit, a large increase in the Auger-1 threshold energy accompanies this strain-induced change in valence-band symmetry. We examined the temperature dependence of the photoluminescence intensity for SLS and bulk, $InAs_{0.93}Sb_{0.07}$ and InAs samples. (see Figure 3) At > 100 K, the radiative efficiency decreases exponentially for all samples, and the activation energy for nonradiative recombination (ΔE) in the SLS displayed a marked increase compared with that of the unstrained alloys ($\Delta E=0.26\, E_{gap}$ vs $\Delta E=0.06\, E_{gap}$). Other InAsSb/InGaAs SLSs that we examined also displayed large activation energies. Due to variations in optical alignment and sample doping and defects, the differences in the relative radiative efficiency between samples may not be accurate.

Approaching room temperature, Auger-1 will be the dominant nonradiative process. The radiative efficiency for extrinsic, n-type material is

$$\eta = \frac{\tau_A}{\tau_R} \propto (1/n_0) \cdot \exp\left(\frac{\Delta E}{kT}\right) \quad (1) \quad \text{and} \quad \Delta E = \left(\frac{m_e^*/m_h^*}{1+(m_e^*/m_h^*)}\right) \cdot E_{gap} \quad (2),$$

where n_0 is the electron density, $\tau_R \propto (T/n_0)$ is the radiative lifetime, and $\tau_A \propto (T/n_0^2)\exp(\Delta E/kT)$ is the Auger-1 lifetime (in 2 dimensions).[9] Eq. 1 is based on simple energy-momentum conservation for isotropic, parabolic bands, and the expression for activation energy, ΔE (ΔE = Auger threshold - Bandgap), is valid in 2 or 3-dimensions.[2] In the 2-dimensional limit, the SLS in-plane effective masses determine ΔE,[3,4] and therefore, the SLS activation energy is larger than that of the unstrained alloys due to the decreased hole mass in biaxially compressed layers of the SLS.

Electron-hole effective mass ratios for each sample are compared with theoretical values obtained from k·p calculations and with experimental values obtained from magneto-photoluminescence and temperature dependence studies. In each case there is good agreement between theoretical and experimentally determined mass ratios. The effective mass ratios (m^*_e / m^*_h) determined by the 3 methods for the SLS were in the range, 0.3-0.5, and the effective mass ratios for the unstrained alloys were 0.06±0.02.[9] Comparing the SLS with the alloys, the SLS consistently displayed the predicted effects of the valence band under biaxial compression.

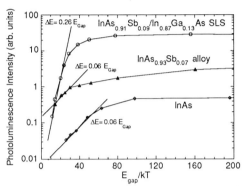

Figure 3 - Temperature dependence of the photoluminescence intensity (radiative efficiency) for the InAsSb/InGaAs SLS and InAsSb and InAs alloys. Activation energies (ΔE) are indicated in the figure.

5. Conclusions

Using infrared photoluminescence and magneto-photoluminescence, we have examined the electronic properties of a series of heterostructures with biaxially compressed InAsSb. Holes are confined to the heavy hole (small in-plane effective mass) ground state of the biaxially compressed InAsSb layers, and the band offsets for these heterostructure are type I. Analysis of the magneto-excitonic behavior indicates that the hole mass may be quite small ($\mu=0.010$), and exciton binding energy increases with quantum confinement. The photoluminescence efficiency versus temperature revealed that an increased activation energy for nonradiative recombination accompanies the decreased hole mass in the SLS. Photoluminescence efficiency activation energies for the SLS and unstrained alloys agree with estimated Auger-1 threshold energies, with the activation energy for the SLS approaching the Auger-1 value in the 2-dimensional limit. Although the material properties of the MOCVD-grown InAsSb are non-ideal, these heterostructures display Auger-1 threshold energies, effective masses, band offsets, and "random alloy-like" properties that are desirable for active regions in midwave infrared diode lasers.

J. A. Bur, K. C. Baucom, and M. W. Pelczynski provided technical assistance. This work was performed at Sandia National Laboratories, supported by the U. S. Department of Energy under contract No. DE-AC04-94AL85000.

References
[1] Eglash S J and Choi H K 1990 Appl. Phys. Lett. **57** 1292; 1992 Appl. Phys. Lett. **61** 1154.
[2] Dutta N K 1983 J. Appl. Phys. **54** 1236
[3] Adams A R 1986 Elect. Lett. **22** 249
[4] Yablonovich E and Kane E O 1988 IEEE J. Lightwave Tech. **6** 1292
[5] Kurtz S R, Biefeld R M, Dawson L R, Baucom K C, and Howard A J 1994 Appl. Phys. Lett. **64** 812
[6] Fang Z M, Ma K Y, Jaw D H, Cohen R M, and Stringfellow G B 1990 J. Appl. Phys. **67** 7034, and references therein.
[7] Follstaedt D M, Biefeld R M, Baucom K C, and Kurtz S R (submitted to J. Elect. Mat.)
[8] Kurtz S R and Biefeld R M (submitted to Appl. Phys. Lett.)
[9] Kurtz S R, Biefeld R M, and Dawson L R (submitted for publication)

Electron transport, negative differential resistance and domain formation in weakly coupled quantum wells

Ali SHAKOURI, Yuanjian XU, Ilan GRAVÉ, Amnon YARIV

Department of Applied Physics, 128-95
California Institute of Technology
Pasadena, California 91125

Abstract. Using photocurrent spectroscopy and dark current measurements, different electron transport mechanisms in multistack multiquantum well detectors are investigated. In particular we will report on the first observation of sequential resonant tunneling induced negative differential resistance in very weakly coupled multiquantum wells (separated by 44nm barriers).

1. Introduction

Quantum well infrared photodetectors (QWIP) have been the subject of a considerable research effort [1,2]. Recently we reported on a new infrared intersubband detector, capable of working in number of modes; its features included tunability and multi-color operation [3,4]. Analysis of the photocurrent spectrum and the dark current enables one to study different transport mechanisms of optically or thermally excited electrons in these structures. In the following sections, we will present design and characterization of the device and will discuss the evidence for sequential resonant tunneling induced electric field domain formation in the multiquantum well region.

2. Design and characterization of the multistack infrared detector

The structure was grown by molecular beam epitaxy on a semi-insulating GaAs substrate. The multiquantum well region, clad by two n-doped contact layers, consisted of three stacks of 25 quantum wells each; the first 25 wells (called stack (a)) were 3.9 nm wide and were separated by $Al_{0.38}Ga_{0.62}As$ barriers; the second stack (b) consisted of 4.4 nm wide wells with $Al_{0.30}Ga_{0.70}As$ barriers; the last stack (c) had 5.0 nm wide wells and $Al_{0.24}Ga_{0.76}As$ barriers. All the barriers were 44 nm wide; the wells and the contacts

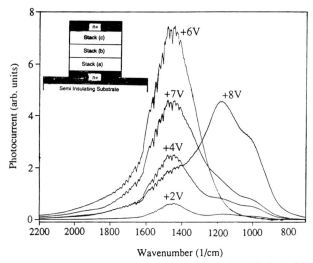

Figure 1. Spectral photoresponse for few values of applied positive voltage. Note the switching in peaks at an applied voltage around 8.0V. The responsivity, at the peak of $1140 cm^{-1}$ and the applied voltage of 8.0V, is 0.75 A/W. The units are the same for both Figure 2 and Figure 1. The inset shows the device structure.

were uniformly doped with Si to $n = 4 \times 10^{18} cm^{-3}$ (a schematic of the device is drawn in the inset of figure 1). The absorption spectrum at zero field and room temperature, obeying the intersubband selection rule, showed a peak at $1364\ cm^{-1}$ due to the 3.9nm wells and a stronger peak at $964\ cm^{-1}$ which is the composite contribution of the two other stacks of quantum wells, which, individually, have absorption peaking at 1080 and $920\ cm^{-1}$ [3]. These results agree with our design values; our calculations anticipated absorpion peaks at room temperature at 1335,1052, and $880\ cm^{-1}$, respectively. The blue shift in the experimental values versus the calculated ones can be explained by the omission of the exchange interaction from our calculations; in these heavily doped samples, the correction supplied by many-body effects is noticeable. the existence of the two absorption peaks that merge into a wide and strong peak was also experimentally verified by analyzing the absorption of a few additional MBE grown control wafers, which were designed to include, each time, only two of the stacks described above.

Devices were processed out of the grown wafer and prepared as etched mesas, $200\mu m$ in diameter. Figure 1 displays the photocurrent spectroscopy of a device at a temperature of 10K, for positive values of applied voltage. It is seen that, for low applied fields ($\leq 5V$), the stack (a) of 3.9nm wells, closer to the substrate, provides most of the photocurrent at the appropriate excitation energies around the peak of 1450 cm^{-1}. When the bias is increased to 6V, the contribution of stack (a) increases while the contribution of the other two stacks *disappears*. Increasing the bias even further (to 7V) reduces the photocurrent peak at 1450 cm^{-1}, but the other two stacks start to contribute again. When the bias reaches a threshold of 8V, a sharp transition takes place and the responsivity peak *switches* to 1190 cm^{-1}. It is apparent that stack (b) and (c) are now responsible for most of the photocurrent, while the contribution from the stack (a) has

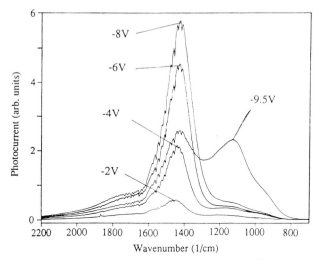

Figure 2. Spectral photoresponse for few values of applied negative voltage for the three-stack quantum well infrared photodetector. Note the broadening in the spectral response below $-9.0V$.

sensibly decreased.

If we apply a negative bias to the detector [see figure 2], again, at low voltages, the photocurrent is due mostly to electrons excited in the stack (a). The responsivity increases with the applied voltage, but its magnitude is always less than that corresponding to the same forward bias; in addition, one observes that the photocurrent peak around 1450 cm^{-1} is $\approx$ 50% broader in the forward bias mode. When the bias is increased to more negative values (-9V), the responsivity extends to lower energies, showing increasing contributions from the stack (b): stack (a) still contributes equally, in contrast to the sharp reduction in response experienced in the opposite polarity of the applied electric field.

These features in the photocurrent, which are observed at a temperature of 10K, persist at higher temperatures. In the reverse bias direction, one observes the same general behavior also at 77K. In the forward polarity, the switching of the photoresponse from the higher energy peak to the lower one is observed up to a temperature of 60K; the critical voltage at which the switching occurs increases slightly with the temperature.

3. Current-voltage characteristics

The I-V curves of the device at different temperatures are shown in Figure 3. One can note a strong asymmetry between the two polarities. A fine structure in the plateau of the I-V curves, corresponding to regions of negative differential resistance, was observed. For example the T=10K curve has oscillations in the voltage range of -10.0 to -7.0 volts and +4.0 to +7.0 volts. In each of these intervals 24±2 oscillations with a period of 0.128±0.008 volts were measured. Figure 4 shows an expansion of the forward bias plateau region. One can see in the inset, the oscillations of the differential resistance of

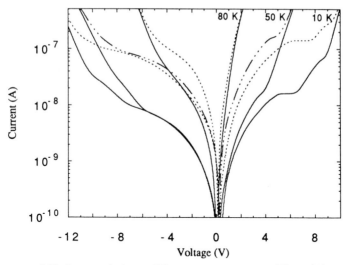

Figure 3. I-V characteristic at different temperatures. The solid curves are without illumination and the dashed ones are with illumination of a black body source. Note the important contribution of photo-assisted transport to the total current, specially at low biases.

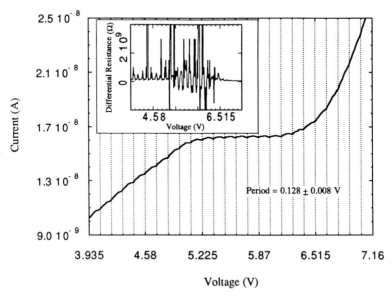

Figure 4. An expansion of the [+4,+7] range in the 10K I-V curve (without illumination). One can see 24 oscillations with a period of 0.128±0.008 volts. The inset shows oscillations in the differential resistance of the device.

the device. This measurement gives a most important clue to the possible origin of photocurrent peaks disappearing and switching, since these negative differential resistance oscillations are the signature of the formation and expansion of a high field domain along the sample multiquantum well region [5,6,7,8,9,10].

Basically, when a series of quantum wells are under applied bias, a uniform distribution of electric field is not *stable* because all of the quantum wells will be out of resonance i.e. none of the energy levels of pairs of adjacent wells will be aligned. Instead, the system will settle into a configuration in which the electric field profile includes high and low field regions. In the high field region we have ground level to excited level sequential resonant tunneling, and in the low electric field region ground level to ground level tunneling. Transport within each domain is thus resonant, while at the boundary between the two regions it is generally non-resonant. This boundary then acts as a bottleneck that limits the current. Charge accumulation or depletion at this boundary takes place because of the change in the slope of electric fields, as required by Poisson's equation. An increase in the bias will cause more quantum wells to enter the high field domain region, and this is reflected by the oscillatory behavior in the I-V curve.

4. Discussion and conclusion

Under illumination, the light is absorbed in all the quantum wells but only photoexcited carriers which are in a region with a high electric field can be swept out of the quantum well and contribute to the photocurrent. Those in low field region have a high probability of being recaptured by their own well, contributing negligibly to the photocurrent. The observed oscillations in the I-V curve are in the same voltage range where photocurrent peak disappearance and switching occurs. This is an indication that electric field domain formation can be responsible for the observed behavior in the photocurrent spectrum. On the other hand, the low bias behavior of this device did *not* show any negative differential resistance, and different contribution of the three stacks to the photocurrent can be explained using Liu's equivalent circuit analysis [11].

The presence of plateaux in the I-V characteristic is consistent with the current being limited by the boundary between high and low field domains (see e.g. refs.[6, 9]). The period of oscillation (0.128±0.008 volts) is close to the separation between ground state and the excited state in stack (c) (123 meV). However, for these bound-to-continuum detectors where the excited state is near the quantum well edge (typically 10 meV above the barrier), it is possible that sequential resonant tunneling does not occur to the states which are maximally localized in the well region (corresponding to the absorption peak).

In conclusion, We presented evidence of sequential resonant tunneling induced negative differential resistance in very weakly coupled quantum wells. In the design of QWIPs, which have long barriers to reduce the dark current, inter-well coupling is usually neglected. Taking into account transport through sequential resonant tunneling, can be used to design detectors with improved performances.

This work was supported by DARPA and the Office of Naval Research.

References

[1] J.S. Smith, L.C. Chiu, S. Margalit, A. Yariv, and, A.Y. Cho, 'A new infrared detector using electron emission from multiple quantum wells', *J. Vac. Sci. Technol. B*, **1**,376 (1983).

[2] B.F. Levine, 'Quantum well infrared photodetectors', *J. Appl. Phys.*, **74**(8),R1 (1993).

[3] I. Gravé, A. Shakouri, N. Kuze, and A. Yariv, 'Voltage-controlled tunable GaAs/AlGaAs multistack quantum well infrared detector', *Appl. Phys. Lett.*, **60**,2362 (1992).

[4] A. Shakouri, I. Gravé, Y. Xu, A. Ghaffari, and A. Yariv, 'Control of electric field domain formation in multiquantum well structures', *Appl. Phys. Lett.*, **63**,1101 (1993). Note that the results presented in the paper are a little different from the ones reported in this reference. These two devices were processed from different places of the same grown wafer. The device performances are very sensitive to quantum well parameters and specially to the defect concentration.

[5] L. Esaki, and L.L. Chang, 'New transport phenomenon in a semiconductor superlattice', *Phys. Rev. Lett.*, **33**,495 (1974)

[6] K.K. Choi, B.F. Levine, R.J. Malik, J. Walker and, C.G. Bethea, 'Periodic negative conductance by sequential resonant tunneling through and expanding high-field superlattice domain', *Phys. Rev. B* , **35**,4172 (1987)

[7] R.F. Kazarinov and, R.A. Suris, 'Electric and electromagnetic properties of semiconductors with a superlattice', *Sov. Phys. Semicond.*, **6**, 120 (1972); *ibid*, 'Possibility of the amplification of electromagnetic waves in a semiconductor with superlattice', *Sov. Phys. Semicond.*, **5**,707 (1971)

[8] F. Capasso, K. Mohammad, and A.Y. Cho, 'Sequential resonant tunneling through a multiquantum well superlattice', *Appl. Phys. Lett.*, **48**, 478 (1986)

[9] H.T. Grahn, H. Schneider and K. von Klitzing, 'Optical studies of electric field domains in GaAs/AlGaAs superlattices', *Phys. Rev. B*, **41**, 2890 (1990); H.T. Grahn, R.J. Haug, W. Muller, and K. Ploog, 'Electric field domains in semiconductor superlattices; a novel system for tunneling between 2D systems', *Phys. Rev. Lett.*, **67**,1618 (1991)

[10] B. Laikhtman and, D. Miller, 'Theory of current-voltage instabilities in superlattices', *Phys. Rev. B*, **48**, 5395 (1993); B. Laikhtman 'Current-voltage instabilities in superlattices', *Phys. Rev. B*, **44**,11260 (1991)

[11] H.C. Liu, J. Li, Z.R. Wasilewski, M. Buchanan, P.H. Wilson, M. Lamm and J.G. Simmons, 'A three-color voltage tunable quantum well intersubband photodetector for long wavelength infrared', in *Quantum well intersubband transition physics and devices*, eds. H.C. Liu et al., 123 (1994).

Minority Carrier Mobility and Density of States in SiGe HBTs Determined by a New Method

A. Gruhle, A. Schüppen, H. Kibbel and U. König

Daimler-Benz Research, W. Runge Str. 11, D-89081 Ulm, Germany

Optimization of SiGe HBTs calls for an exact knowledge of both mobility and density of the minority carriers in the SiGe base. The anisotropy of the strained SiGe layers excludes usual extraction methods. Only the collector saturation current J_s can provide the unknown parameters minority carrier mobility μ_n, the density of states $N_c N_v$, and the bandgap narrowing BGN. However, it will only give the product of all three. We show how $\mu_n N_c N_v$ can be determined independently of ΔE_g and BGN by plotting $d/dT \log(J_s)$ against $\log(J_s)$, therefore being insensitive to errors in the Ge content. Highly precise (0.1°C) measurements revealed that with increasing Ge from 0% to 26% the $\mu_n N_c N_v$ product decreases by a factor of 3 for moderate (10^{19}cm^{-3}) base dopings but remains almost constant in the case of 10^{20}cm^{-3} bases. One possible explanation is an increasing minority mobility with germanium content.

1. Introduction

The key advantage of SiGe heterojunction bipolar transistors (HBTs) over conventional Si bipolar transistors is a thin and heavily doped base. The low base resistance and the short transit time leads to high cutoff frequencies above 100GHz /1/ and low noise performance /2/. For further device improvement an exact knowledge of both mobility and density of the minority carriers in the base is necessary. Usual extraction methods /3/ that determine these two parameters simultaneously in silicon fail in the case of strained SiGe layers due to the anisotropy, i.e. the in-plane carrier transport differs considerably from perpendicular currents. The only way to obtain data on minority carrier transport through strained SiGe layers are therefore measurements on HBTs itself, i.e. the variation of the collector saturation current density J_s with Ge-content, base doping level and temperature. The expression for J_s

$$J_s = q\, \alpha\, \mu_n\, V_T\, G_B^{-1}\, n_i^2(Si)\, \exp(\Delta E_g/kT)\, \exp(BGN) \qquad (1)$$

contains the product of minority carrier mobility μ_n, the ratio of the density of states $\alpha = N_c N_v(SiGe) / N_c N_v(Si)$ between Si and SiGe, the base Gummel number $G_B = \int N_A$, the intrinsic carrier concentration n_i^2 of silicon, the bandgap reduction ΔE_g and the apparent bandgap narrowing BGN /5/. Several attempts have been made to determine one or more of these parameters at different Ge contents. However, in practice measurement errors introduce large uncertainties: (i) Determination of the Gummel number from base sheet resistance measurements gives a wrong value because Hall factor and (in-plane) mobility are unknown in SiGe. (ii) The precision of the Ge percentage from RBS or X-ray rocking is limited to +/- 0.5%. (iii) The measurement of J_s itself is very sensitive to leakage currents or non-ideality (e.g. from out-diffusion) and requires an extremely accurate temperature measurement ,e.g. 0.1°C. A plot of $\log(J_s)$ against $1/T$ therefore usually shows a large spread of the data points leading to uncertainties of a factor of two or more of the above parameters.

This paper describes how J_s can be measured with high precision and it introduces a new analysis method that allows to separate some of the above parameters. Before analyzing the measured J_s values it is convenient to use a modified collector current J_c by eliminating the known parameters of equ.(1). Dividing equ.(1) through n_i^2 has the advantage that the log(J_s(1/T))-curves become less steep. The remaining slope only reflects E_g, BGN and a possible temperature variation of the other parameters. In silicon the relationship for n_i is /4,5/

$$n_i^2 = 9.61 \times 10^{23} \, T^3 \, \exp(-E_g(T)/kT) \tag{2}$$

where $E_g(T)$ is the temperature-dependent bandgap /4/. There are different models for n_i in the literature, e.g. /6/ , but they differ by a few meV only. In addition, it is the same formula which is used for all HBTs , so it cannot introduce an error when looking at the relative differences between the devices.

Finally we eliminate G_B^{-1} by using a high precision SIMS value (+/-5%) assuming complete dopant activation (MBE with elementary boron source). The modified current J_c is then

$$J_c = \alpha \, \mu_n \, \exp(\Delta E_g/kT) \, \exp(BGN) \;=\; \alpha \, \mu_n \, \exp(\Delta E_{tot}/kT) \tag{3}$$

2. Sample Preparation

Device fabrication starts with 4 inch n+ substrates . After an RCA clean the wafers are loaded into the MBE system where the complete HBT layer structure is grown without interruption. Growth starts after a 900°C flash-off with the 100-200nm thick collector layer doped $1 \times 10^{17} cm^{-3}$ by secondary ion implantation of Sb. The 40nm thick SiGe base is grown by coevaporation and doped from an elementary boron effusion cell. The Ge content varies between 0% and 26% at four different doping levels of $3 \times 10^{18} cm^{-3}$, $1 \times 10^{19} cm^{-3}$, $3 \times 10^{19} cm^{-3}$ and $1 \times 10^{20} cm^{-3}$. Undoped spacer layers on both the collector (10nm) and emitter side (5nm) avoid outdiffusion of the boron out of the base. This is extremely important to be sure that no parasitic conduction band barriers are formed which strongly degrade device performance /7/. The emitter is 70nm thick doped 1×10^{18} with Sb followed by a 230nm thick n+ emitter contact layer.

After unloading the wafers from the MBE system transistors are fabricated using a double-mesa technology. Access to the base layer is done with a selective KOH etch that stops on SiGe. Typical device sizes were $5\text{-}20 \times 10^{-4} cm^2$. Samples were prepared by bonding the transistors into hermetically sealed ceramic chip carriers. The packages were immersed in a temperature bath which was controlled to an accuracy of 0.1°C. To prevent leakage currents through the liquid in the bath the packages and wires were embedded in silicone plastic.

3. Measurements

First measurements revealed that the Gummel plots obtained from an HP4145B parameter analyzer were insufficient for the required precision in our cases: The voltage resolution limit of 1mV of the HP4145B leads to an uncertainty of V_{be} of the same magnitude. (In addition, from time to time the self calibration feature causes shifts of almost 1mV). At 95°C this produces a current error of 3% which corresponds to an unacceptable temperature drift of more than 1°C.

The second point is the fine tuning of V_{bc} which is necessary when measuring I_c in the pA range. This will compensate unavoidable leakage currents, e.g. due to thermoelectric voltages. Depending on the parasitic resistance between base and collector the fine tuning voltage is sometimes only fractions of a mV which cannot be resolved by the HP4145B. For

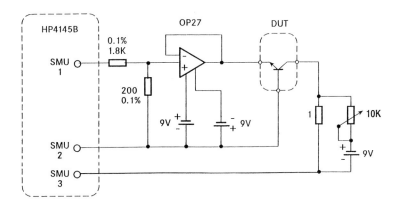

Fig.1: Schematic diagram for high precision Gummel plot measurements

this reason the measurement setup was modified according to Fig.1. The base-emitter voltage is devided by 10 and buffered through the high precision operational amplifier (<<1mV offset). This increases the voltage resolution by a factor of 10. (The base current certainly increases, however it is unimportant for our measurements). The fine tuning of V_{bc} is accomplished by the variable 10KΩ resistor.

Fig.2 shows the Gummel plot of the collector current of an HBT at 4 different temperatures and the corresponding ideality factors scaled in °C. This assumes an ideality of 1.0 which in fact is almost true for HBTs with high base dopings /7/. Usually several different transistors were measured at the same time and their ideality factors indicated the same temperature within an error of 0.1°C. Actually, temperature measurements using the ideality

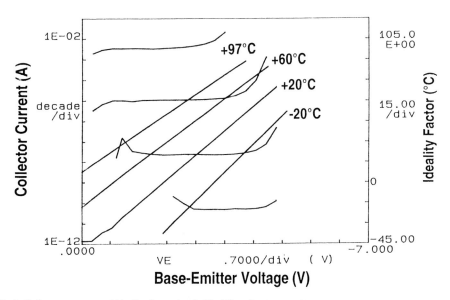

Fig.2: Collector current and ideality factor (scaled in °C) at four temperatures

factor were so precise that they revealed the nonlinearities of a digital thermocouple thermometer and up to 0.5°C deviation of mercury thermometers. Care was taken that only HBT samples were evaluated that showed a plateau in the ideality curve within a current range of at least 2 decades. In some cases the transistors were measured in inverse operation because of better characteristics.

Unfortunately a complete series of HBT samples with low ($3 \times 10^{18} cm^{-3}$) base doping had to be excluded from the evaluation because of the following reasons: The base thickness modulation increases the ideality factor /7/ (typically 1°-2°C above the actual temperature). Therefore the correct temperature has to be measured independently. This is no problem, however the extrapolation of the J_c-curve will give erroneous results, particularly as the optimum V_{be} (in the middle of the plateau) changes with temperature. Much more serious is the fact that the BE and BC space charge layers vary the base Gummel number considerably (up to 50% reduction). The exact value is difficult to estimate.

About 20 samples were measured between -20°C and 95°C at 4 or 5 different temperatures. Plotting $\log(J_c)$ versus 1/T gives a curve which was approximated by a straight line. The slope can be used to calculate ΔE_{tot}. The distance between the measured points and the curve (on the y-axis, because curves are relatively flat) corresponded to less than 0.2°C error. Samples with larger deviations were not considered.

4. Results and Discussion

Fig. 3 is a plot of $\log (J_{c300K})$ versus the Ge content, Fig.4 plots the value of ΔE_{tot} in the same way. Despite the precise J_s-measurements the spread of the data is considerable and no tendency concerning the three different base dopings can be found beside the fact that there seems to be no BGN effect for high Ge concentrations (in pure Si the expected difference is 30meV /5/). Due to the uncertainty in the germanium content each point actually has a horizontal error bar of +/- 0.5%, equal to the resolution of the x-ray rocking (the SIMS value is even less accurate). Vertical error bars are about as large as the symbols reflecting +/- 5% spread of the G_B-value. (The dashed curve in Fig.4 is the theoretical People-Bean-fit which assumes no BGN.)

A way of eliminating the Ge-percentage is to plot ΔE_{tot} against $\log J_c$ as in Fig. 5. The slope of the curves is now independent of errors in ΔE_g, i.e. uncertainties in the germanium content will shift the measurement points up and down the curve. A clear difference between the different base dopings can now be observed. The dashed lines indicate the ideal slope, i.e. if $\alpha \mu_n$ in equ.(3) remained constant. This is obviously not the case, as can be seen even more clearly in Fig.6, where $\alpha \mu_n$ has been plotted vs. the germanium content.

The value for μ_n in the case of the BJTs (0% Ge) where $\alpha=1$ seems to be rather low, references /3/ and /5/ suggest 200-250 cm^2/Vs. This is because μ_n was assumed to be independent of temperature in equ.(1). If a 1/T-relationship /5/ is used it has the effect of an additional ΔE_{tot}-slope of about -25meV which in turn will multiply the $\alpha \mu_n$ -values of Fig.6 by 2.7 leading in fact to the correct BJT-values (we did not use the 1/T-law in equ.(1) because of the discrepancy between references /3/ and /5/ concerning the temperature dependence of μ_n). Note that the bandgap narrowing is not visible in Fig.6 because BGN is part of ΔE_{tot}. (Unfortunately there was no BJT-sample with $1 \times 10^{19} cm^{-3}$ available).

The curves in Fig.6 may be interpreted in several ways: If the decrease of the 1×10^{19}-curve comes from the change in α, it is in fact the predicted ratio due to band splitting of about 0.3 /8/. However, what happens at higher dopings? Is α then constant or does an increase of the mobility μ_n compensate the loss? Another influence could be the apparent

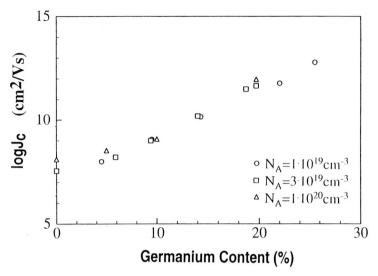

Fig.3: Plot of $logJ_c$ vs. germanium content

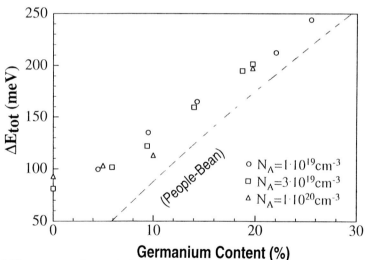

Fig.4: Plot of ΔE_{tot} vs. germanium content

bandgap narrowing. The Fermi-Dirac correction makes the BGN temperature-dependent, especially at high dopings and small N_v-values.

As a conclusion we have shown that BGN and ΔE_g may be separated from $\alpha\mu_n$. However, more independent information about α or μ_n is necessary to explain the observed behaviour.

5. Acknowledgement

The authors thank W. Schäfer and A. Windmüller for their help with the temperature measurements.

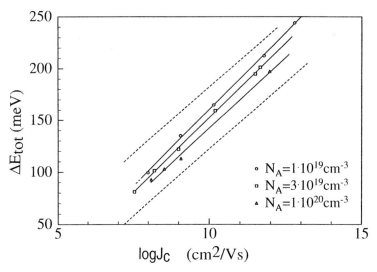

Fig.5: Plot of ΔE_{tot} vs. $\log J_c$

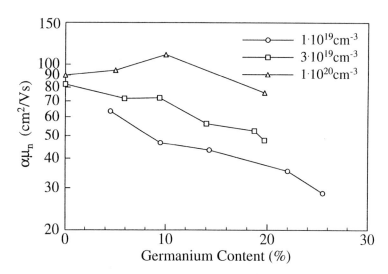

Fig.6: Plot of $\alpha\mu_n$ vs. germanium content

6. References

/1/ A. Schüppen et al., Electron. Lett. 30(14), 1187-1188, (1994)
/2/ H. Schumacher et al., Proc. 1994 MTT-Symp., 1167-1170, (1994)
/3/ I.-Y. Leu, A. Neugroschel, IEEE ED-40(10), 1872-1874, (1993)
/4/ S.M. Sze, Physics of Semiconductor Devices, Wiley (1981)
/5/ D.B.M. Klaassen et al., Sol. St. Electr. 35(2), 125-129, (1992)
/6/ M.A. Green, J. Appl. Phys. 67, 2944 (1990)
/7/ A. Gruhle, IEEE ED-41(2), 198-203, (1994)
/8/ E. Prinz, IEDM 89, 639-642, (1989)

Activation and diffusion of shallow impurities in compound semiconductors

W. Walukiewicz

Center for Advanced Materials, Materials Sciences Division, Lawrence Berkeley Laboratory, University of California, Berkeley CA 94720 USA

Abstract. The formation of native defects is discussed within the framework of the recently proposed amphoteric native defect model. The basis of the model is the existence of an energy reference which is common for compound semiconductors and can be used to determine the differences in the Fermi level dependent part of the defect formation energy in various semiconductors and at semiconductor heterointerfaces. The model is used to explain several closely related phenomena including reduced electrical activity of shallow donors and acceptors, doping induced diffusion of the impurities and doping enhanced superlattice intermixing.

1. Introduction

Efficient doping is one of the principal requirements for most of the practical applications of semiconductor materials. It was recognized very early that many semiconductors are very difficult to dope and that in some instances either n- or p-type of doping cannot be achieved at all. This problem seems to be especially critical for wide gap compound semiconductors which exhibit a very clear propensity to one type of doping. Group III-V phosphides and nitrates are in general good n-type materials, whereas arsenides and antimonides can be easily doped with acceptors. Among the group II-VI semiconductors all selenides and sulfides show an inclination to n-type doping while p-type doping is very difficult to achieve in these materials.

The reduced electrical activity of intentionally introduced dopants has been widely studied and is a hotly debated issue. Several different models were put forward to explain this phenomenon [1]-[6]. It was proposed that in some instances the maximum doping levels resulted from a limited solubility of a given dopant [6] or from doping induced structural transformation of the dopant site [5]. Other models tend to emphasize the role of native defects which compensate intentionally introduced dopants[1]-[4]. The latter models were especially successful in explaining the difficulties encountered in the doping of wide gap semiconductors. It has been argued that a change of the doping from p- to n-type in a wide gap semiconductor is accompanied by a large shift of the Fermi energy position resulting in a change of the formation energy and concentration of compensating, charged defects [2], [3].

In this paper we will show that the reduced electrical activity as well as enhanced dopant and self-diffusion can be understood from a single, unifying point of view. The key point of our approach is the introduction of the Fermi level stabilization energy as a common energy reference that can be used to calculate the differences in the formation energy of compensating native defects in various semiconductors. It will be shown how, within this concept, one can explain experimentally observed trends in dopant activation, dopant diffusion and dopant induced superlattice intermixing in compound semiconductors.

2. Formation Energy of Defects

Under equilibrium conditions the concentration of a defect depends on its Gibbs free energy

$$G = E + pV^* - TS \qquad (1)$$

where E is the internal energy, p is the pressure, V^* is the defect activation volume and S is the entropy. The pressure dependent term pV^* in Eq. 1 plays a important role only in materials strained due to either lattice mismatch [7] or application of external hydrostatic pressure [8]. In homogenous semiconductors under ambient pressure this term can be safely neglected. In semiconductors, typical entropies range from 5 k_B to 10 k_B and should be very similar for the same type of defects in different semiconductor materials. Since here we are mostly interested in comparing behavior of the same type of defect in various systems we can assume that the entropy term is constant.

The formation energy of a charged defect depends on the location of the Fermi energy with respect to the defect charge transition states. In general the location of the charge transition states are not well known, therefore one has to resort to another energy reference to measure the changes in the Fermi energy position. In standard, phenomenological models of defect formation the Fermi energy is usually measured with respect to the Fermi energy in intrinsic material. It has been shown, however, that since the location of this energy reference is determined by the low density of states conduction and valence band edges it has no relationship to the highly localized states of native defects.

A more convenient and much more appropriate energy reference to calculate the formation energy of native defects has been proposed recently [9]. The Fermi level stabilization energy E_{FS} is defined by the location of the Fermi energy in a heavily damaged semiconductor crystal. It has been found that this energy reference is almost constant for all compound semiconductors when measured with respect to the vacuum level. For the Fermi energy $E_F > E_{FS}$ ($E_F < E_{FS}$) acceptor (donor) -like defects are predominantly formed [9]. The formation energy of a highly localized defect can then be written as,

$$E_f = E_{f0} \pm s(E_F - E_{FS}) \qquad (2)$$

where E_{f0} is the formation energy of the defect for $E_F = E_{FS}$ and s the charge state of the defect. It can be easily seen from Eq. 2 that the formation energy is greatly reduced for the defects which can support multiple charges. The extent of the possible reduction of the defect formation energy, given by the second term in Eq. 2, depends on the semiconductor band gap. Also it is quite evident that doping of a semiconductor affects the formation energy and also abundances of native defects. The concept of the Fermi level stabilization energy has been applied to understand a variety of defect related phenomena in semiconductors [3], [9], [10]. The following will give a few examples of the application of this concept to the understanding of the dopant activation and doping enhanced diffusion.

3. Dopant activation and diffusion in n-type semiconductors

Heavily doped n-type GaAs is an example of a most extensively studied system where both the dopant activation as well as dopant and self diffusion are directly affected by the location of the Fermi energy [3], [4], [11]. In this case the dominant defects are triply negatively charged gallium vacancies (V_{Ga}^{3-}). In GaAs $E_{FS} = E_c - 0.8$ eV, therefore in material with a free electron concentration in mid 10^{18} cm^{-3} $E_F = E_c$ and according to Eq. 2 the formation energy of V_{Ga} is reduced by about 2.4 eV. This large reduction in the formation energy leads to a dramatic increase of the concentration of V_{Ga} acceptors which compensate intentionally introduced donors and lead to increased diffusion.

Results of calculations of the electrical activity of donors in GaAs are shown in Fig. 1 [3]. It is seen that for the donor concentration up to the threshold of about 3×10^{18} cm^{-3}, practically all the donors are electrically active. At doping levels exceeding the threshold value the free electron concentration tends to saturate. The saturation is a consequence of an increased incorporation of compensating V_{Ga}. It is also seen that the calculations are in good agreement with the experimental data obtained for various donors [3].

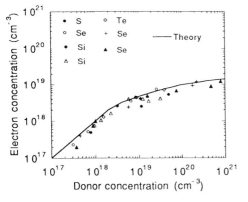

Fig. 1. Free electron vs. donor concentration for a different donor species in GaAs. The solid line represents the theoretical calculations assuming that gallium vacancies are the compensating acceptors.

The increased incorporation of V_{Ga} at high doping levels also has a profound effect on self-diffusion. One of the most spectacular and easy to observe manifestations of self-diffusion is the so called superlattice intermixing. Annealing of a GaAs/AlAs superlattice at high temperatures leads to the diffusion of group III atoms and to a destruction of the superlattice [12]. It has been demonstrated that interdiffusion is strongly enhanced by n-type doping and that very rapid intermixing of GaAs/AlGaAs superlattice is observed for doping levels exceeding the threshold value of 3×10^{18} cm^{-3}. Such behavior can be understood assuming that gallium vacancies are responsible for the interdiffusion.

The calculated diffusion coefficient for group III atoms in the GaAs/AlGaAs superlattice is shown in Fig. 2 [3]. It was assumed that the diffusivity is proportional to the concentration of gallium vacancies $[V_{Ga}]$. At doping levels $N_D < 3 \times 10^{18}$ cm^{-3} $[V_{Ga}] \sim N_D^3 \sim n^3$ for N_D above the threshold value $[V_{Ga}] \sim N_D \sim n^3$. Again the results of the calculations seem to be in reasonably good agreement with available experimental data for Si doped superlattices [12]. The onset of a fast GaAs/AlGaAs superlattice intermixing at threshold carrier concentration of about 3×10^{18} cm^{-3} has also been confirmed in an experiment in which Si ions were implanted into a superlattice [13]. Total intermixing was observed only in the implanted region where the free carrier concentration exceeded the threshold value.

A number of experiments on the diffusion in n-type doped GaAs [14] and on selectively Si doped GaAs/AlGaAs heterostructures [15] have shown that for doping levels higher than the threshold value a rapid redistribution of Si atoms is observed. Thus in MOVPE grown GaAs, a large enhancement of the diffusion rate of Si atoms was observed when the doping level was changed from 1.7×10^{18} to 4×10^{18} cm^{-3} [15]. It has also been found that annealing under Ga deficient conditions leads to further acceleration of the diffusion process.

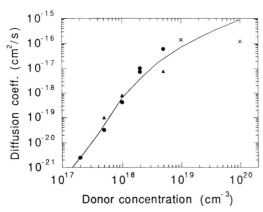

Fig. 2. Interdiffusion coefficient for Si doped superlattice. The theoretically calculated curve was obtained assuming that the interdiffusion is proportional to the concentration of gallium vacancies.

All these experimental and theoretical results provide very convincing support for the early assertions [3] that the Fermi energy induced reduction of the formation energy of V_{Ga} defects is responsible for the low activation efficiency of donor dopants and explains dopant and self-diffusion related phenomena in heavily doped n-GaAs. Although much less work has been done on other semiconductors it has been demonstrated that there is a clear correlation between the location of E_{FS} and activation efficiency of donor and acceptor impurities. InP with $E_{FS} = E_c - 0.4$ eV can be easily doped with donors and concentrations of free electrons approaching 10^{20} cm^{-3} have been reported while p-type doping is limited to the mid 10^{18} cm^{-3} level. An interesting case is represented by $In_{0.53}Ga_{0.47}As$ a ternary compound in which $E_{FS} = E_C - 0.25$ eV. This material can be heavily doped with donors to levels much higher than in GaAs. Also the value of $n = 1.3 \times 10^{19}$ cm^{-3} has been found as the threshold electron concentration for the onset of the intermixing of InGaAs/InAlAs superlattice [16]. This is a much higher concentration than the threshold value of $n = 3 \times 10^{18}$ cm^{-3} found for GaAs. However it should be noted that in both cases that the Fermi energy for the threshold concentrations is located at about 0.8 eV above EFS. This again confirms that the formation of the defects responsible for the intermixing is controlled by the Fermi energy as given by Eq. 2.

4. Dopant activation and diffusion in p-type semiconductors

Numerous studies of group II acceptor doping indicate that the free hole concentration tends to saturate near 10^{20} cm^{-3} in GaAs and at a much lower concentration of about $(3 \text{ to } 5) \times 10^{18}$ cm^{-3} in InP[17]. These concentrations correspond to Fermi energy positions at $E_V - 0.2$ eV in GaAs and $E_V + 0.16$ eV in InP. With the known valence band offset of 0.35 eV we find that in both materials the Fermi level is located at a constant position of about 0.8 eV below E_{FS}. This shows that even without any knowledge of the nature of the compensating defects the location of the valence band relative to E_{FS} can be used to estimate the maximum hole concentrations that can be attained in a given material. Thus from the known valence band offset of 0.5 eV between GaAs and AlAs one could expect much lower activation of group II acceptors in a material with higher Al content. This has been confirmed by studies of Be doped GaAs/AlAs superlattices. It has been found that Be implanted into $GaAs/Al_{0.3}Ga_{0.7}As$ structures segregates into GaAs [18]. The concentration of Be in AlGaAs was 4 times smaller than that in the GaAs layers. It should be noted that the redistribution occurred very rapidly during 3 s rapid thermal annealing at 860 °C. Even more drastic segregation effects were observed during epitaxial growth of Be doped GaAs/AlAs superlattices [19]. It has been found that although uniform doping at the level of 2×10^{19} cm^{-3} was attempted, the maximum Be concentration in AlAs was only 5×10^{17} cm^{-3}. All the excess Be segregated into the GaAs layers.

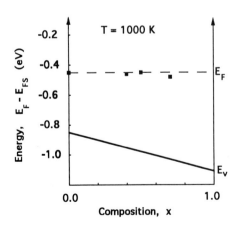

Fig. 3. Location of the Fermi energy for the maximum hole concentration in Zn doped $In_{0.5}(Ga_{1-x}Al_x)_{0.5}P$ measured with respect to Fermi level stabilization energy. The position of the valence band edge E_V is also shown.

Similar behavior of group II acceptors was also observed in $In_{0.5}(Ga_{1-x}Al_x)_{0.5}P$ quaternaries lattice matched to GaAs. The electrical activity of Zn acceptors was found to decrease by almost one order of magnitude with an Al content increasing from 0 to 70 % [20]. Fig. 3 shows the location of the Fermi energy, E_F, calculated for the maximum hole concentration as a function of Al content at the growth temperature of about 1000 K. It is quite evident that E_F remains constant and is located at about $E_{FS} - 0.45$ eV.

We will show here that this behavior of the acceptors can be understood in terms of a balance between substitutional acceptors and interstitial donors. Doping of an $A_{III}B_V$ compound with group II acceptors (F) proceeds via the reaction,

$$F_i^{r+} + A_A \leftrightarrow F_A^- + A_i^{s+} + (r-s+1)h \quad (3)$$

where h is a hole at the Fermi energy and the superscripts represent the charge state.

Experiments on group II acceptor diffusion indicate that $(r+s+1)=2$ which means that 2 holes are formed at the Fermi energy when the reaction 3 proceeds from the left to the right hand side. The energy required for the reaction (3) is,

$$E_f = E_f(F_A) + E_f(A_i) + 2(E_{FS}-E_F) - E_f(F_i) \qquad (4)$$

where $E_f(.)$ represents the formation energy of a given species for $E_F = E_{FS}$.

At elevated temperatures when the defects are in a thermal equilibrium the substitutional to interstitial concentration ratio is given by [21],

$$[F_A]/[F_i] = \exp(-E_f/k_BT) = \exp[2(E_F - E_{SI})/k_BT] \qquad (5)$$

where

$$E_{SI} = [E_f(F_A) + E_f(A_i) + 2E_{FS} - E_f(F_i)]/2 \qquad (6)$$

Here E_{SI} plays the role of an energy reference for the reaction (2). With increasing doping level the Fermi energy shifts down towards E_{SI} leading to a higher concentration of the interstitials donors which compensate substitutional acceptors. This process results in a saturation of the free hole concentration. The experimental results on the GaAlAs and InGaAlAs systems indicate that over the whole Al composition range E_{SI} is located at the same energy relative to E_{FS}.

According to the substitutional-interstitial model [22] the reaction (3) is also responsible for the diffusion of group II acceptors in III-V semiconductors and for diffusion enhanced superlattice intermixing [11]. Again these processes are controlled by the location of the valence band edges with respect to E_{FS} and the onset for the diffusion correlates with the saturation of electrical activity of the acceptors.

The problem of the low electrical activity of dopants is especially critical in II-VI semiconductors. A great deal of effort has been directed towards improving the electrical activity of acceptors in ZnSe and related ZnMgSSe compounds lattice matched to GaAs substrates. It has been shown in a recent study that an increase in the Mg and S contents leads to a larger band gap and also results in a rapid deterioration of the activation efficiency of N acceptors [23]. A maximum free hole concentration of 6×10^{17} cm^{-3} has been achieved in ZnSe. However in $Zn_{0.8}Mg_{0.2}S_{0.26}Se_{0.74}$ with an energy gap of 3.15 eV the free hole concentration is limited to 2.5×10^{15} cm^{-3}. This behavior of N acceptors has been explained within the concept of the amphoteric native defect model [24]. It has been argued that increasing the S content leads to a downward shift of the valence band edge measured with respect to E_{FS}. Consequently, according to reaction (3), this will result in a lower maximum free hole concentration. The Fermi energy calculated for the maximum free hole concentration is shown in Fig. 4. It is seen from the figure that, similarly as in the case of InGaAlP, the Fermi energy remains constant relative to E_{FS}. This indicates that the concept of the common energy reference can also be applied to group II-VI compounds.

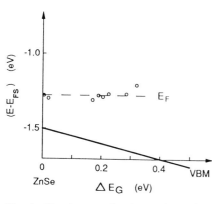

Fig. 4. Fermi energy for the maximum hole concentration as a function of the change in energy gap in N doped $Zn_xMg_{1-x}S_ySe_{1-y}$ lattice matched to GaAs. The energy is measured relative to E_{FS}. Also shown is the location of the valence band maximum, VBM.

Although in discussing the deactivation of acceptors in p-type semiconductors we have considered the interstitials as the compensating donor species it is evident that a similar result will be obtained if the compensation is realized by any highly localized donor for which a charge transition state is pinned to the Fermi level stabilization energy. In particular the same concept should be applicable to donors resulting from structural transformation in which either acceptor-host or host-host bonds are broken [25].

Acknowledgments

The author would like to thank E.E. Haller for many useful discussions. This work was supported by the Director, Office of Energy Research, Office of Basic Energy Sciences, Materials Science Division of the U.S. Department of Energy under Contract No. DE-AC03-76SF00098.

References

[1]   Longini R L and Greene R F 1956 *Phys. Rev.* **102** 992-999
[2]   Mandel G 1964 *Phys. Rev.* **134** A1073-1079
[3]   Walukiewicz W 1989 *Appl. Phys. Lett.* **54** 2094-2096
[4]   Zhang S B and Northrup J E 1991 *Phys. Rev. Lett.* **67** 2339-2342
[5]   Chadi D J and Chang K J 1989 *Appl. Phys. Lett.* **55** 575-577
[6]   Laks D B, Van de Walle C G, Neumark G F, Blochl P E and Pantelides S T 1992 *Phys. Rev. B* **45** 10965-10978
[7]   Baumann F H, Huang J-H, Rentschler J A, Chang T Y and Ourmazd A, *Phys. Rev. Lett.* **73** 448-451
[8]   Chen A L, Walukiewicz W, Haller E E, Luo H, Karczewski G and Furdyna J, this conference proceedings
[9]   Walukiewicz W 1988 *J. Vac. Sci. Technol.* **B6** 1257-1262
[10] Walukiewicz W 1987 *J. Vac. Sci. Technol.* **B5** 1062-1067
[11] Tan T Y, You H-M and Gösele U 1993 *Appl. Phys.* **A56** 249-258
[12] Mei P, Schwarz S A, Venkatesan T, Schwartz C L, Harbison J P, Florez L, Theodore N D and Carter C B 1988 *Appl. Phys. Lett.* **53** 2650-2652
[13] Ishida K J, Matsui K, Fukunaga T, Kobayashi J, Morita T, Miyauchi E and Nakashima H 1987 *Appl. Phys. Lett.* **51** 109-111
[14] Deppe D G, Holonyak N Jr., Plano W E, Robbins V M, Dallesasse J M, Hsieh K C and Baker J E 1988 *J. Appl. Phys.* **64** 1838-1844
[15] Veuhoff E, Baumeister H and Treichler R 1988 *J. Cryst. Growth* **93** 650-655
[16] Miyazawa T, Kawamura T and Mikami O 1988 *J. Appl. Phys.* **27** L1731-L1733
[17] See e.g. Chan L Y, Yu K M, Ben-tzur M, Haller E E, Jaklevic J M and Walukiewicz W 1991 *J. Appl. Phys.* **69** 2998-3006
[18] Humer-Hager T, Treichler R, Wurzinger P, Tews H and Zwicknagl P 1989 *J. Appl. Phys.* **66** 181-186
[19] Kopf R F, Schubert E F, Downey S W and Emerson A B 1992 *Appl. Phys. Lett.* **61** 1820-1822
[20] Nishikawa Y, Tsuburai Y, Nozaki C, Ohba Y, Kokubun Y and Kinoshita H 1988 *Appl. Phys. Lett.* **53** 2182-2184
[21] Walukiewicz W 1994 *Proc. 17th Intern. Conf. on Defects in Semiconductors, Mats. Sci. Forum* Vols. **143-147** 519-530 (Zurich: Trans Tech Publications)
[22] Gösele U and Morehead F 1981 *J. Appl. Phys.* **52** 4617-4619
[23] Okuyama H, Kishita Y, Miyajima T, Ishibashi A and Akimoto K 1994 *Appl. Phys. Lett.* **64** 904-906
[24] Kondo K, Okuyama H and Ishibashi A 1994 *Appl. Phys. Lett.* **64** 3434-3436
[25] Chadi D J, to be published in the proceedings of the 22nd Intern. Conf. on the Physics of Semiconductors, Vancouver, August 14-19, 1994

Characterization Of Carbon Doped GaAs Layers Grown By Chemical Beam Epitaxy

R.Driad, F.Alexandre, J.L.Benchimol, R.Rahbi*, B.Pajot*, B.Jusserand, B.Sermage, G.Le Roux, M.Juhel, J.Wagner⁺ and P.Launay,

FRANCE TELECOM, Centre National d'Etudes des Télécommunications, Paris B
Laboratoire de Bagneux, 196 avenue Henri Ravéra, BP 107, 92225 Bagneux, France.
*Groupe de Physique des Solides, Université de Paris-VII, Tour 23, 75251 Paris Cedex 05, France
⁺Fraunhofer-Institut für Angewandte Festkörperphysik, Tullastrasse 72, D-79108 Freiburg, Germany

Abstract. The thermal stability of carbon doped GaAs layers grown by chemical beam epitaxy (CBE) with different doping levels was investigated. The influence of annealing temperature and doping level on minority carrier lifetime, hole concentration and lattice parameter was observed and discussed. In order to determine the origin of the variations of the electrical, optical and structural properties of the GaAs:C epitaxial layers after thermal annealing, we have performed secondary ion mass spectrometry, Raman spectroscopy and infrared absorption measurements on the as-grown and annealed GaAs:C epilayers.

1. Introduction

Carbon has attracted considerable interest as a p-type dopant for GaAs, mainly because of its extremely high incorporation level ($> 5 \times 10^{20}$ cm^{-3}) [1] and small diffusion coefficient as compared to Be or Zn. From an application view point, such a heavily doped p-type GaAs layer is suitable for the base layer to improve high frequency performances of heterojunction bipolar transistors (HBTs). These applications require the C-dopant to be stable under high current operation or after technological thermal treatments.
However, it has been reported, that the fraction of electrically active carbon in as-grown and annealed GaAs:C epilayers varies depending on the carbon concentration, growth method, carbon sources used and annealing conditions [2,3]. Moreover, a drastic degradation of the current gain, induced by a regrowth process, was observed in selectively regrown HBTs [4].
In this work, we report a systematic study of the thermal stability of carbon doped GaAs layers (1×10^{18} - 3×10^{20} cm^{-3}) grown by chemical beam epitaxy (CBE), by correlating their electrical, optical and structural properties before and after annealing. These results are discussed and compared to carbon doped GaAs layers grown by metalorganic vapor phase epitaxy (MOVPE).

2. Experimental procedure

The C-doped GaAs epilayers were grown by CBE on 2-inch diameter semi-insulating GaAs (001) substrates. The source materials for As, Ga and C dopant were trimethylgallium (TMG) and arsine (AsH$_3$) at growth temperatures in the range 500 to 650°C. Unlike conventional CBE process where hydrogen is used as carrier gas, our experiments were performed

with no carrier gas, because hydrogen reacts with alkyls, and reduces the carbon incorporation. Typical thicknesses of the epilayer varies in the range 0.5-2.5 µm.
The post growth annealing of the samples covered with silicon nitride cap layers was carried out at temperatures ranging from 450 to 800°C for 5 to 60 min under a nitrogen gas flow in an ordinary furnace with an open quartz tube.

The minority carrier lifetime was evaluated using a time-resolved photoluminescence technique using the 730 nm line of a mode-locked Ti:Sapphire laser. The Hall effect measurements of the carrier concentration were conducted using the Van der Pauw method. The absolute concentration of C and H atoms was measured by secondary ion mass spectrometry using a primary beam of Cs^+ ions. The variations of the lattice deformation of the samples was investigated with a double-crystal X-ray diffractometer using the (004) Bragg reflection and $CuK\alpha$ radiation. Infrared absorption measurements were made at liquid He temperature with a Bomem DA8 Fourier-Transform spectrometer. Raman experiments were performed at 77 K using the 457.96 nm line of an Ar^+ ion laser at a power density of 200 $W.cm^{-2}$.

3. Results and discussions

The photoluminescence decay measurements were performed on double heterojunctions (DH), where a 200 nm thick GaAs:C active layer is confined between 250 nm of undoped $Al_{0.3}Ga_{0.7}As$. Figure 1 shows the degradation ratio of the minority carrier lifetime, measured in the GaAs:C layers, before (τ_{eo}) and after (τ_e) thermal treatments for two doping levels as a function of annealing temperature. It can be observed from Fig.1, that the electron lifetime decreases after annealing and this degradation effect is more pronounced at high C concentration and high annealing temperatures.

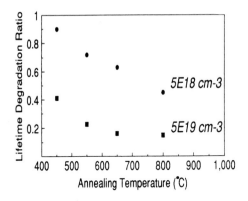
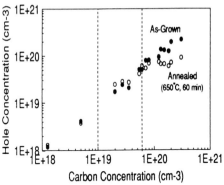

Fig.1 : Minority carrier lifetime degradation ratio (τ_e/τ_{eo}) as a function of the annealing temperature.

Fig.2 : Hole concentration vs total atomic carbon concentration for as-grown (●) and annealed (o) GaAs:C epilayers.

Hall effect measurements were performed in order to determine the active concentration of carbon before and after a 650°C - 60 min annealing treatment (Figure 2). For low C-doping levels (in the $1-10 \times 10^{18}$ cm^{-3} range), the hole concentration remains constant after annealing. Between 1×10^{19} and 6×10^{19} cm^{-3} a slight increase of the hole concentration was observed, corresponding to a reactivation of C atoms, associated to the out-diffusion of H, as will be shown from SIMS measurements. Hydrogen is known to compensate electrically active carbon to form neutral C-H complexes [5]. Above 6×10^{19} cm^{-3} C-doping level, the hole concentration was found to decrease, indicating the possible occurrence of a self compensating phenomena. The same behavior was also observed in the three doping regions for the MOVPE grown epilayers.

From SIMS measurements, the concentration of incorporated H during the growth of C-doped GaAs layers was found to increase with the C-doping level (Figure 3). Moreover, it can be observed that the amount of H incorporated in the MOVPE layers is much higher than that obtained from the CBE grown layers. Fig.3 also shows a decrease of the hydrogen level in the epilayers after a 650°C-60 min anneal. Since incorporation of C from TMG induces incorporation of C-H complexes [5], the decrease of the H concentration corresponds to a breakdown of the C-H bonds. These results are in agreement with the observed increase of the hole concentration after annealing, in the medium doping range, attributed to the reactivation of the carbon atoms.

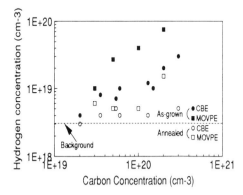 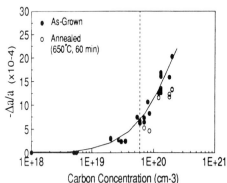

Fig.3 : Hydrogen concentration vs Carbon concentration for as-grown/annealed (650°C, 60 min) GaAs:C layers.

Fig.4 : Relative lattice contraction parameter vs atomic carbon concentration for as-grown and annealed (650°C, 60 min) GaAs:C samples.

Figure 4 shows the effect of annealing on the lattice deformation in the C doped GaAs epilayers. The CBE grown layers are still elastically strained for C concentrations $< 6\times10^{19}$ cm^{-3}. Whereas, a slight increase of the lattice contraction parameter ($-\Delta a/a$) was observed for an MOVPE layer doped to 3×10^{19} cm^{-3}. This increased lattice mismatch was related to the removal of hydrogen [5] after the thermal treatment. A significant decrease was observed in ($-\Delta a/a$) for the more heavily doped samples ($p > 1\times10^{19}$ cm^{-3}) after annealing. The reduction in ($-\Delta a/a$) eliminates the self-compensating hypothesis of C site switching from C_{As} to C_{Ga}, because of the very close values of the atomic radii of Ga and As. Höfler et al [6] have attributed the lattice relaxation observed at high doping levels in MOCVD layers to a displacement of carbon from substitutional to interstitial sites. An alternative explanation for the observed change of $\Delta a/a$ could be attributed to the generation of misfit dislocations after annealing.

Because hydrogen is known to compensate electrically active carbon to form neutral C-H complexes in GaAs:C layers grown from MOVPE and MOMBE [5], we have performed local vibrational mode (LVM) infrared absorption measurements on the as-grown and annealed epitaxial layers.
Infrared absorption spectra of an as-grown MOVPE GaAs:C epitaxial layer doped to 1×10^{20} cm^{-3} is shown in Fig.5. The dominant feature observed at 2637 cm^{-1} is assigned to the H-stretching vibration of the H-passivated C_{As} acceptor (C_{As}-H) [7]. The intensity of this C_{As}-H band was found to disappear after annealing (650°C, 60 min) in all MOVPE layers. These results confirm the reactivation of C, observed from Hall effect measurements.
However, no C-H complexes were detected for both as-grown and annealed CBE layers, in spite of the slight increase of the hole concentration observed for medium doping range. This was confirmed by Raman scattering : only a very weak (C-H) line is observed, which disappears after annealing. This suggests that only weak hydrogen passivation of the C acceptors occurs in the CBE grown layers as compared to MOVPE ones. This was attributed to the growth conditions used in this study, which allow little incorporation of hydrogen as shown from SIMS measurements (Fig.3).

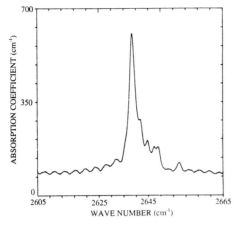
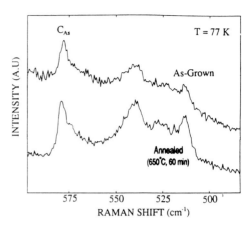

Fig.5 : IR absorption spectra at 6 K of the (C-H) related complexes for a 1×10^{20} cm^{-3} GaAs:C grown by MOVPE.

Fig.6 : Low temperature Raman spectra for as-grown and annealed (650°C, 60 min) GaAs:C sample doped to 1.8×10^{20} cm^{-3} grown by CBE.

Figure 6 displays the as-grown/annealed (650°C, 60 min) Raman spectra of a highly C doped GaAs layer (1.8×10^{20} cm^{-3}) grown by CBE. The LVM of substitutional C on As site (C_{As}) is revealed at 578 cm^{-1}. The broad band centred at 540 cm^{-1} corresponds to the intrinsic second-order phonon scattering [8]. We observed on this sample (see Fig.6) that the intensity of the C_{As} LVM (normalized to the intensity of the intrinsic second-order phonon scattering at 540 cm^{-1}) decreases after annealing. The decrease in C_{As} intensity corresponds to a change location of the C atoms after annealing. This displacement of C should result in the formation of recombination centers (As vacancies) and/or nonradiative recombination centers (precipitates or interstitial carbons).

4. Conclusion

In summary, we have studied the thermal stability of heavily doped GaAs:C films grown by CBE. We have shown that degradation effects are more drastic as the annealing temperature and C-doping level increase. Our data also show that the concentration of substitutional carbon (C_{As}) decreases after annealing. The strong decrease in minority carrier lifetime indicates that annealing results in the formation of recombination centers and/or lattice defects, especially for highly doped samples (> 6×10^{19} cm^{-3}). The observed compensation could be attributed to the presence of intrinsic defects such as As vacancies and/or to C site changes from the substitutional acceptor to the interstitial which forms a non radiative recombination center. Finally, these results show that high C-doping levels and high temperature treatments should be avoided during device processing.

References:

[1]- Abernathy C.R, Pearton S.J, Caruso R, Ren F and Kovalchik J 1989 Appl.Phys.Lett. **55** 1750-1752
[2]- Hobson W.S, Pearton S.J, Kozuch D.M, and Stavola M 1992 Appl.Phys.Lett. **60** 3259-3261
[3]- Hanna M.C, Lu Z.H and Majerfeld A 1991 Appl.Phys.Lett. **58** 164-166
[4]- Zerguine D, Launay P, Alexandre F, Benchimol J.L and Etrillard J 1993 Elect.Lett. **29** 1349-1350
[5]- Kozuch D.M, Stavola M, Pearton S.J, Abernathy C.R and Hobson W.S 1993 J.Appl.Phys. **73** 3716-3724
[6]- Höfler G.E, Höfler H.J, Holonyak N and Hsieh K.C 1992 J.Appl.Phys. **72** 5318-5324
[7]- Clerjaud B, Gendron F, Krause M and Ulrici W 1990 Phys.Rev.Lett. **65** 1800-1803
[8]- Wagner J, Maier M, Lauterbach Th and Bachem K.H 1992 Phys.Rev.B **45** 9120-9125

Characterization of inactive carbon in GaAs

A.J. Moll*, J.W. Ager, III, Kin Man Yu, W. Walukiewicz, and E.E. Haller*

Center for Advanced Materials, Lawrence Berkeley Laboratory, Berkeley, CA 94720
and
*Department of Materials Science and Mineral Engineering, University of California, Berkeley, CA 94720

Abstract. Independent of the doping technique, the electrical efficiency of C doped GaAs is less than 100%. We have shown that Raman spectroscopy provides direct evidence for the formation of C precipitates in GaAs. GaAs samples doped with C and annealed at various temperatures exhibit two broad peaks in the Raman spectra at 1585 cm^{-1} and 1355 cm^{-1} which are characteristic of two sp^2 bonded C atoms. Photoluminescence of annealed C doped GaAs indicates the presence of compensating defects as well. Both precipitation of C and compensation due to native defects play a role in the reduction of the electrical activity of C doped GaAs.

1. Introduction

Carbon is one of the technologically most important acceptor dopants in GaAs. Its diffusion coefficient is at least one order of magnitude lower than those of the other common p-type dopants in GaAs providing better device stability. A number of research groups have grown epitaxial layers with ultra high concentrations (>10^{21} cm^{-3}) of free holes using C doping.[1,2] However, at high concentrations, the doping in these layers is not thermally stable.[3,4] Annealing above 650°C results in a rapid decrease in the free hole concentration. Independent of the original C concentration or the growth technique, the hole concentration tends toward an ultimate limit of approximately 5×10^{19} cm^{-3}. In addition to the reduction in free carrier concentration, the mobility of the epitaxial layers and the strain in the epitaxial layers also decreases.

Doping with C by ion implantation into GaAs is also problematic. Above a bulk concentration of approximately 5×10^{18} cm^{-3}, the concentration of free holes in the valence band is less than 10% of the concentration of C atoms implanted into the substrate.[5] Co-implantation of Ga increases the activation significantly but hole concentrations greater than 5×10^{19} cm^{-3} cannot be attained with implantation.[6]

Various explanations for the inactive C in GaAs have been presented. They include self-compensation, interstitial C, precipitation, and compensation by point defects. No experimental evidence directly supporting any of these mechanisms has been published. The discussion below will concentrate on two specific mechanisms which account for the majority of the inactive C in both heavily C doped epitaxial layers which have been annealed and in implanted layers. These two mechanisms are precipitation of C and compensation of C acceptors by native defects. Other mechanisms which have been suggested include C$_{Ga}$ donors, lone C interstitials, C complexes, and misfit dislocations. No conclusive experimental evidence exists to support any of these mechanisms. Although, they may occur in either epitaxial layers or implanted layers, the data presented here suggests that they do not play a major role in reducing the electrical activity of C.

Table 1. Doping parameters and presence of Raman peaks for samples from this study.

Doping Technique	Doping Parameters	Anneal Conditions	Raman Peaks
None		None	NO
None		950 °C, 10 s	NO
Implant	C: 1×10^{15} cm^{-2}, 80 keV	None	NO
Implant	C: 1×10^{15} cm^{-2}, 80 keV	950 °C, 10 s	YES
Implant	Zn: 2×10^{15} cm^{-2}, 180 keV	950 °C, 10 s	NO
MOMBE	C: 6×10^{20} cm^{-3}	None	NO
MOMBE	C: 6×10^{20} cm^{-3}	850 °C, 3 hrs	YES

2. Carbon precipitation

GaAs substrates were solvent cleaned, etched, and implanted with singly ionized C ions under various conditions as described in Table 1. Following implantation, the samples were rapid thermally annealed or furnace annealed under conditions listed in Table 1. Carbon doped GaAs epitaxial layers were grown by MOMBE at 400°C by procedures reported elsewhere.[7] The hole concentration as measured by Hall effect is approximately 6×10^{20} cm^{-3}. A 1 cm^2 piece was furnace annealed at 850 °C for 3 hours.

In Raman spectroscopy of perfect crystals, the only phonons which are observed are zone center phonons due to the requirement of k conservation. When disorder is introduced the conservation of k is relaxed and other phonons with a large density of states are seen. The Raman spectra of amorphous carbon (a-C) films consists of a "G-band" centered between 1540 cm^{-1} and 1590 cm^{-1}, and a "D-band" centered between 1340 cm^{-1} and 1390 cm^{-1}.[8] The G-band corresponds to the zone center phonon (allowed phonon of graphite) which shifts slightly with the introduction of disorder. The D-band is a disorder allowed phonon which corresponds to a peak in the phonon density of states of sp^2-bonded C. Appearance of the G- and D- bands in a Raman spectrum indicate that "bulk" sp^2-bonded C is present.

Typical Raman spectra for various C-implanted GaAs and C-doped GaAs epitaxial layers are shown Fig. 1. The broad features centered near 1585 cm^{-1} and 1355 cm^{-1} are assigned to sp^2-bonded carbon. All of the samples where these peaks appeared are annealed. These two peaks in the Raman spectra can be unambiguously identified as arising from C precipitates in the doped layer.[9] Only bulk-like sp^2 bonded C can display such a Raman spectrum.

The detection limit for sp^2 bonded C in these experiments is estimated to be 0.25 monolayers of graphitic C which corresponds to an areal C density of 2.5×10^{14} cm^{-2}. This detection limit was determined from the Raman spectrum of single crystal graphite.

Raman spectroscopy is a surface sensitive technique with probe depth in GaAs of 55 nm. C contamination of the surface is the principal difficulty in analyzing the Raman spectra in these experiments. To study the effect of C contamination on the interpretation of the Raman results, several additional samples were characterized as described in Table 1. A piece of the GaAs substrate was processed and rapid thermally annealed in the same manner as the C implanted samples. A second piece of the same substrate was implanted with Zn and annealed using the same annealing schedule. Neither of these samples exhibited any Raman

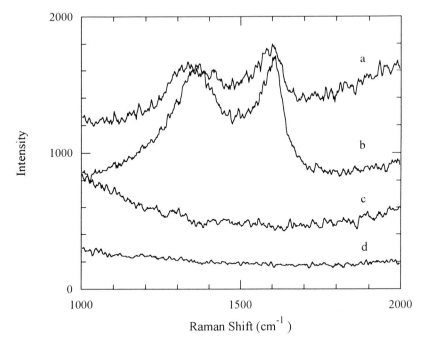

Fig. 1. Raman spectra of carbon precipitates in GaAs for the following samples: a) C implanted and annealed, b) MOMBE-grown and annealed, c) implanted and not annealed, and d) annealed substrate.

peaks in the region from 1300 to 1600 cm^{-1}. Also, a sample which had been implanted with C but not annealed did not exhibit any peaks in this region. However, all the samples which had been doped with carbon and annealed exhibited two Raman peaks centered near 1585 cm^{-1} and 1360 cm^{-1}.

As mentioned above, the most difficult part of this experiment is eliminating C contamination during the annealing. To completely eliminate the possibility that the Raman signal was caused by C contamination, GaAs samples were implanted with ^{13}C using the same implant conditions described above. The Raman spectra for samples implanted with ^{12}C and ^{13}C are shown in Fig. 2. In the samples implanted with ^{13}C the two peaks are shifted to lower frequency by 4% ($\sqrt{12/13}$) as expected. Therefore, the Raman peaks can only be a result of the C clustering of implanted C and not from environmental contamination.

The size of the C precipitates can be estimated from the Raman spectra by comparison to amorphous C films which have been studied extensively.[8] The intensity ratio of the D- and G- bands, the I_d/I_g ratio is a function of sp^2 domain size in disordered graphite. An inverse relationship has been found between domain size, L_a, and I_d/I_g ratio in microcrystalline graphite by Tuinstra and Koenig:[10]

$$L_a(nm) = 4.4 \ (I_d/I_g)^{-1}.$$

(The Raman spectrum of single crystal graphite is a single "G-band" peak at 1580 cm^{-1}, i.e., $I_d/I_g = 0$). The spectra observed in this study are nearly identical to those observed by Dillon[11] from ion-beam films annealed at 900 °C. The I_d/I_g ratios in the C doped samples (computed using the peak heights) are in the range of 0.8 to 1.0. Using the Tuinstra and Koenig relationship, the C precipitates have an average sp^2 domain size of 5 nm.

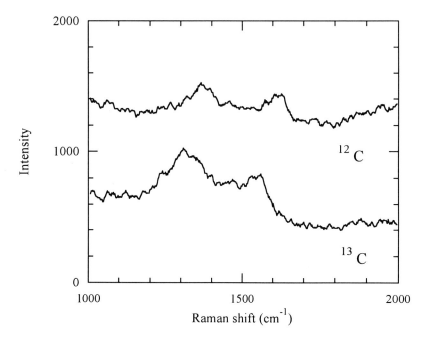

Fig. 2. Raman spectra of C precipitates in GaAs samples implanted with ^{12}C or ^{13}C and annealed.

A simple, rough estimate of the minimum diffusion coefficient required for a C precipitate to form can be calculated. Given the annealing conditions of 950°C for 10 s the diffusion coefficient required for precipitate formation can be estimated by:
$$x = (Dt)^{1/2}.$$
D is then approximately 2×10^{-13} cm^2/s. You et. al.[12] reported a diffusion coefficient of 7.8×10^{-15} cm^2/s for C diffusion at 960 °C. The diffusion coefficient calculated here is more than 10 times higher than that of You. Two possible explanations exist for this effect.

First, diffusion is enhanced either by loss of group V elements at the surface or by radiation damage in the crystal due to the implantation process. To examine the role of the surface in the precipitation of C, two GaAs samples with the same concentration of implanted C were rapid thermally annealed. One sample was capped using the proximity method. The second was not capped. The surface of the second sample was noticeably degraded due to the loss of As. In this sample the C-C peaks were more intense. This result suggests that the loss of the group V element near the surface enhances precipitation.

Another possibility for the larger diffusion coefficient compared to those reported in the literature relates to the method of determining this quantity. In general, diffusion coefficients are determined with the use of SIMS. SIMS experiments do not give any information on the bonding configuration of the C. Therefore, these experiments do not take into account C diffusion to a precipitate within the C doped layer. C in a precipitate has not diffused beyond the original doped layer so this movement is not detected by SIMS.

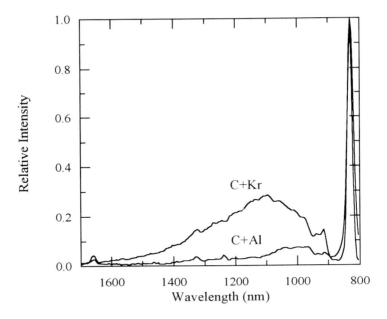

Fig. 3. Photoluminescence spectra for C+Al and C+Kr samples.

3. Compensation by native defects

Precipitation clearly accounts for at least a portion of the inactive C in implanted and epitaxial GaAs:C. The existence of precipitates can account for the reduction in free hole concentration and the decrease in strain in annealed epitaxial layers. However, the decrease in mobility cannot be explained by precipitation. In degenerately doped semiconductors, mobility is inversely proportional to the number of ionized impurities. Precipitation reduces the number of ionized impurities hence the mobility should increase. Evidently, precipitation is not the sole contributor to the decrease in free hole concentration.

In this section the role of native defects in reducing the free hole concentration and mobility in GaAs:C will be examined. In implanted layers, radiation damage provides an abundant source of defects. In annealed epitaxial layers, the precipitation of C can also create defects. C which moves from an As site to a precipitate will create an As vacancy. The accommodation of the precipitate in the lattice can also create defects. A model developed by Walukiewicz[13] indicates that native defects will form complexes and change their electrical character to compensate intentional dopants. In this case, native point defects will have donor character to compensate the C_{As} acceptors.

Experimental evidence for the existence of native defects can be found in the measurement of the mobility and in PL experiments. GaAs epitaxial layers heavily doped with C and annealed exhibit a reduction in hole mobility as well as a reduction in free hole concentration. In C co-implantation experiments described previously[5] the hole mobility was reduced in samples where the stoichiometry of the implanted layer was not maintained during implantation. Samples were co-implanted with Kr or Al and similar hole concentrations were attained in both. However using local vibrational mode spectroscopy the C+Kr sample was found to have nearly twice the concentration of C_{As} as the C+Al sample. The C+Kr sample also had a lower mobility at 77K then the C+Al sample (60 and 106 cm^2/Vs respectively) indicating the presence of compensating defects.

These defects may also be responsible for the increased PL intensity reported for annealed epitaxial layers and also seen in our implantation studies. The PL spectra for C+Kr and C+Al samples are shown in Fig. 3. The C+Kr spectrum is remarkably similar to those shown by Watanabe.[3] Watanabe's PL spectra were recorded with from heavily C-doped MOCVD layers (p=1.3×10^{20} cm^{-3}) annealed at 850°C Both the MOCVD epitaxial layer and the C+Kr implanted layer exhibit a large broad peak centered around 1100 nm in their PL spectra.

4. Conclusions

In conclusion, direct evidence of C precipitates in GaAs has been shown The Raman peaks are clearly not due to C contamination but come from the intentionally doped C as shown by isotope substitution. The precipitates are present in both implanted layers and epitaxially grown layers.

The striking similarity between the results of precipitation experiments and PL experiments on both implanted and epitaxial GaAs:C layers suggest that the limitations in C doping of GaAs are inherent to this system and not particular to the doping technique. At doping levels above 5×10^{19} cm^{-3}, doping of GaAs with C is not thermally stable. The solid solubility of C is exceeded, and the C will precipitate during annealing. The precipitation process decreases the [C$_{As}$] and reduces the strain in the epitaxial layers. Precipitation also occurs in implanted layers. As well as precipitation, the free hole concentration is further reduced in both implanted and epitaxial layers through compensation by native defects.

References

[1]   Hanna, M.C., Lu, Z.H. and Majerfeld, A. 1991 *Appl. Phys. Lett.* **58** 164
[2]   Yamada, T., Tokumitsu, E., Saito, K., Akatsuka, T., Miyauchi, M., Konagai, M. and Takahashi, K. 1989 *J. Cryst. Growth* **95** 145
[3]   Watanabe, K. and Yamazaki, H. 1991 *Appl. Phys. Lett.* **59** 434
[4]   Nozaki, S., Takahashi, K., Shirahama, M., Nagao, K., Shirakashi, J. and Tokumitsu, E. 1993 *Appl. Phys. Lett.* **62** 1913
[5]   Moll, A.J., Ager III, J.W., Yu, K.M., Walukiewicz, W. and Haller, E.E. 1993 *J. Appl. Phys.* **74** 7118
[6]   Moll, A.J., Walukiewicz, W., Yu, K.M., Hansen, W.L. and Haller, E.E. 1992 *Mater. Res. Soc. Proc.* **240** 811
[7]   Konagai, M., Yamada, T., Akatsuka, T., Nozaki, S., Miyake, R., Saito, K., Fukamachi, T., Tokumitsu, E. and Takahashi, K. 1990 *J. Cryst. Growth* **105** 359
[8]   Robertson, J. 1991 *Progress in Solid State Chemistry* **21** 199
[9]   Moll, A.J., Haller, E.E., Ager, I., J.W. and Walukiewicz, W. 1994 *Appl. Phys. Lett.* **65** 1145
[10]   Tuinstra, F. and Koenig, J.L. 1970 *J. of Chem. Phys.* **53** 1126
[11]   Dillon, R.O., Woollam, J.A. and Katkanant, V. 1984 *Phys. Rev. B* **29** 3482
[12]   You, H.M., Tan, T.Y., Gösele, U.M., Lee, S.T., Höfler, G.E., Hsieh, K.C. and Holonyak Jr., N. 1993 *J. Appl. Phys.* **74** 2450
[13]   Walukiewicz, W. 1988 *Mater. Res. Soc. Proc.* **104** 483

Annealing of nitrogen-doped ZnSe at high pressures: Toward suppression of native defect formation

A. L. Chen[1], W. Walukiewicz[1], E. E. Haller[1,2], H. Luo[3], G. Karczewski[3], and J. Furdyna[3]

[1]Center for Advanced Materials, Material Sciences Division, Lawrence Berkeley Laboratory, Berkeley, CA 94720; [2]Department of Material Sciences and Mineral Engineering, University of California, Berkeley, CA 94720; [3]Department of Physics, University of Notre Dame, Notre Dame, IN 46556.

Abstract. Pressure is shown to have a drastic effect on the annealing characteristics of p-type, nitrogen-doped ZnSe. Samples annealed in vacuum show decreased carrier concentrations and simultaneous formation of deep-donor-related luminescence, while samples annealed under pressure show suppression of this compensating donor. The results are interpreted as an increase in the formation energy of the compensating deep donor under pressure. In addition the samples annealed under pressure show emergence of a new, intense, green luminescence band centered at 2.44 eV. The magnitude of the shift of this peak under applied stress suggests that it results from a recombination involving a deep acceptor.

1. Introduction

Processing of semiconductors for device applications often involves annealing at elevated temperatures. Annealing has the effect of removing damage produced by implantation and activating and diffusing dopants. However, in some instances it also has detrimental effects. Undesired impurities such as transition metals often are unintentionally incorporated. Native crystal defects may also be generated. These two types of defects produce deep levels that trap the free carriers.

For the II-VI semiconductors, annealing often decomposes the crystals at temperatures far below those necessary for regrowth of damaged layers. In nitrogen-doped, p-type ZnSe decomposition initially takes the form of native defects that compensate the shallow, acceptor level [1,2]. Thus in order to implant and activate dopants, novel approaches to annealing that suppress the formation of native defects must be investigated.

In this paper we report the first study of using high pressure to suppress the formation of native defects. Compression of the sample is expected to decrease the concentration of defects that have positive activation volumes. In addition pressure shifts the band extrema and the Fermi level so that the electronic component of the formation energy is changed. We investigate by photoluminescence (PL) and by capacitance-voltage

(CV) measurements the difference in samples annealed under high pressure and those annealed in vacuum. Our preliminary findings show that samples annealed in vacuum are compensated by a deep donor that has a characteristic PL peak while samples annealed under high pressure do not show evidence of this peak. The results are discussed along estimates of the pressure dependence of the formation energy of this defect. We also find the appearance of a new, intense, green band in the luminescence spectrum of samples annealed under pressure.

2. Samples and Experimental Procedure

We studied two nitrogen-doped ZnSe samples grown by molecular-beam epitaxy on semi-insulating GaAs substrates. One (sample A) had an active nitrogen concentration (N_a-N_d) of 1.2×10^{17} cm^{-3}; the other (sample B) had N_a-N_d = 5×10^{17} cm^{-3}. These concentrations were determined by CV measurements. The epitaxial layers were about 2 microns thick.

Photoluminescence spectra were taken at 6 K. The emission was spectrally analyzed by a SPEX 1404 double monochromator operated with a spectra resolution of 0.5 meV and detected by a photomultiplier tube. CV measurements were performed on back-to-back Au Schottky barriers evaporated onto freshly etched surfaces. The measurements were done on a Hewlett Packard 4277A LCZ meter.

Samples annealed in vacuum were sealed in clean quartz ampoules and heated in a furnace. Samples annealed under high pressure were cleaved and lapped into small 200x200x80 micron squares. They were then pressurized in an Inconel diamond-anvil cell with liquid nitrogen as the high pressure medium and heated in a standard electrical furnace. The pressure changed irreversibly during the annealing cycle because of thermal expansion of the cell that deformed the gasket. The pressure after the anneal was often less than 50% of its original value. For each anneal performed at high pressure, the initial pressure of the sample is given.

3. Annealing Results: Donor-Acceptor Pair Luminescence and Compensation

Fig. 1 shows representative PL spectra of sample A (a) as-grown, (b) after annealing in vacuum at 400° C, (c) and after annealing at high pressures (38 kbar) at 400° C. The spectrum of the as-grown sample shows the usual acceptor-bound exciton at 2.787 eV and donor-acceptor-pair (DAP) recombination at 2.698 eV. The latter (hereafter referred to as D^sAP) has been attributed to the recombination between a residual shallow donor and the nitrogen acceptor and is followed by a series of LO phonon replicas. When the sample is annealed another DAP series emerges with a zero-phonon-line at 2.683 eV. Pressure [3] and ODMR [1] studies of this peak have shown it to be caused by a deep donor-to-N acceptor recombination (hereafter referred to as D^dAP). This peak also appears in samples heavily doped with nitrogen. Its occurrence has been correlated with compensation of the nitrogen acceptor [2]. The prevailing ODMR evidence suggests that a V_{Se}-related complex is the deep donor [1]. In the sample annealed under pressure, the

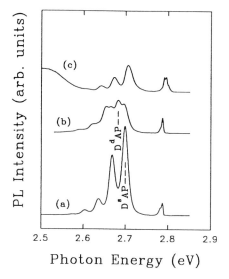

Fig. 1 PL spectra of ZnSe:N (sample A) under different annealing conditions: (a) as-grown, (b) annealed in evacuated ampoule at 400°C for 20 min, (c) annealed at 38 kbar of pressure at 400°C for 20 min.

D^dAP peaks are absent and the D^sAP peaks shift slightly toward higher energy.

To correlate the occurrence of the D^dAP peak with compensation we annealed sample B at various temperatures and analyzed them with both PL and CV measurements (Fig. 2). Although this sample already had evidence of the D^dAP peak before heat treatment, one can clearly see an increase in the D^dAP intensity relative to the D^sAP peak intensity. The measured concentrations of net acceptors are given in the figure. We find a rapid decrease in N_a-N_d when this sample is annealed above 325° C and an increase in D^dAP intensity at these temperatures. Thus the appearance of the D^dAP peak signifies the onset of compensation in annealed samples.

The effect of pressure on annealing can be qualitatively understood by estimating the formation energy of the compensating defect. The concentration of a native defect in a solid is given by the Arrhenius type expression,

$$C \propto \exp\left(\frac{-G}{kT}\right) \quad (1)$$

where

$$G = E + PV^* - TS \quad (2)$$

is the change in the Gibbs free energy that is required to form the defect. Here P is the pressure, V^* is the activation volume, T is the temperature, and S is the change in the entropy. Internal energy, E, has two components

$$E = E_0 - (E_d - E_F) \quad (3)$$

where E_0 is the formation energy of the neutral defect and $-(E_d - E_F)$ is the electronic part [4] of the formation energy that the native donor gains when it gives an electron from

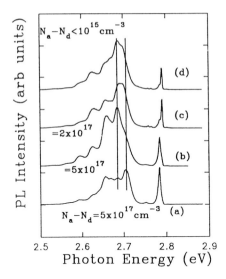

Fig. 2   PL spectra of ZnSe:N (sample B) after various anneals in vacuum: (a) as grown, N_a-N_d= 5×10^{17} cm^{-3}; (b) annealed at 300°C for 20 min, N_a-N_d= 5×10^{17}; (c) annealed at 325°C for 20 min, N_a-N_d= 2×10^{17}; (d) annealed at 335°C for 20 min, N_a-N_d < 5×10^{15}. Net acceptor concentrations were measured by CV technique. Vertical lines indicate positions of deep DAP and shallow DAP peaks.

its donor level (E_d) to the Fermi level (E_F). The activation volume, V^*, for creation of a Se vacancy-interstitial pair is not known. We make a rough estimate of V^* by using the activation volume of V_{Ga} in GaAs, which is reported to be about 20 Å^3 [5]. The PV^* term would then increase the formation energy by about 12 meV/kbar. This increase is partly counteracted by the electronic contribution, which decreases the formation energy. Since E_F can be assumed to remain pinned by the nitrogen acceptors, the pressure dependence of the electronic contribution is then the separation between the deep donor state and the nitrogen acceptor, i.e. the pressure dependence of the D^dAP peak. We previously measured this value to be 4.5 meV/kbar [3]. Thus, overall pressure increases the formation energy of this defect by about 7.5 meV/kbar. At the annealing pressures of about 40 kbar, the increase is 300 meV. At 400°C the concentration of this native defect would be decreased by a factor of about 100.

Our estimate indicates that pressure creates a large increase in the pressure-volume term in the Gibbs free energy and thus suppresses the formation of this compensating native defect. This result is reflected qualitatively in the PL spectra where we witness suppression of the characteristic D^dAP peak when the samples are annealed under pressure.

4. Annealing Results: Broad Green Luminescence

We also find that annealing at high pressures produces an intense broad green luminescence. A representative spectrum is shown in Fig. 3. It is composed of three

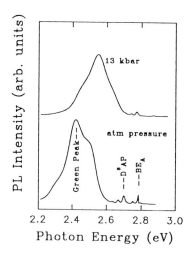

Fig. 3 PL spectra of ZnSe:N after annealing at 20 kbar at 390° C for 40 min. Upper (lower) spectrum was taken at 13 kbar (atm. pressure).

broad peaks at energies of about 2.33, 2.44 and 2.51 eV. We annealed many samples at temperatures between 300 and 400° C and pressures between 10 and 40 kbar and find the following trends in the intensity of this luminescence: (i) the three peaks occur in samples annealed above 300°C, (ii) higher temperatures, higher pressures, and longer anneal times all enhance the peak intensity, and (iii) the peaks are not observed when n-type samples are annealed. It seems that pressure enhances the incorporation of the defect or impurity that is responsible for this luminescence.

To further characterize the peaks we measured their shifts under applied hydrostatic pressure.. A spectrum taken at 13 kbar is also shown in Fig. 3. All three peaks increase at rates that are larger than that of the direct energy gap. The central green peak shifts at 10.2 meV/kbar. Without deconvoluting the peaks, the pressure coefficients of the weaker two peaks cannot be determined accurately. The direct energy gap of ZnSe shifts at 6.6 meV/kbar [3]. It is known that transitions between the conduction band and deep acceptor states in ZnSe have pressure coefficients between 10 and 13 meV/kbar [6]. From this argument we tentatively assign each peak in the series to a deep acceptor incorporated into the crystal. The weaker peak at 2.33 eV may correspond to the well studied "Cu-green" peak which involves a shallow donor and a Cu-related acceptor. We have not been able to find spectra in the literature that even remotely resemble the other two broad features.

5. Conclusions

We have shown that pressure strongly effects the annealing characteristics of p-type, nitrogen-doped ZnSe. It suppresses the formation of one compensating, native defect mainly by increasing the pressure-volume term of the formation energy. However it also introduces additional defects into the crystal that give rise to a characteristic green luminescence band that we tentatively identify as arising from deep acceptor states. It is striking that annealing in vacuum produces deep donors while annealing under pressure creates deep acceptors. Although further measurements of the pressure annealed samples are required in order to determine whether pressure annealing increases the p-type activity, our current results are promising and show that pressure can be used to affect the concentrations of compensating defects.

6. Acknowledgments

This work was supported by the Director, Office of Energy Research, Office of Basic Energy Sciences, Materials Science Division of the U.S. Department of Energy under Contract No. DE-AC03-76SF00098.

References

[1] B.N. Murdin, B.C.Cavenett, C.R. Pidgeon, J. Simpson, I. Hauksson, and K.A. Prior, Appl. Phys. Lett. **63**, 2411 (1993).

[2] K.A. Prior, B. Murdin, C.R. Pidgion, S.Y. Wang, I. Hauksson, J.T. Mullins, G. Horsburgh, B.C. Cavenett, J. of Crystal Growth **128**, 94 (1994).

[3] A.L. Chen, W. Walukiewicz, E.E. Haller, to be published in Appl. Phys. Lett.

[4] W. Walukiewicz, Appl. Phys. Lett. **54**, 2094 (1989).

[5] F.H. Baumann, J-H. Huang, J.A. Rentschler, T.Y. Chang, and A. Ourmazd, Phys. Rev. Lett. **73** 448 (1994).

[6] D.J. Strachan, M. Ming Li, M. Tamargo, and B.A. Weinstein, J. of Crystal Growth **128**, (1994).

Electron emission at Si₃N₄ - GaAs interfaces prepared with H_2, Ar and H_2 +Ar plasma pretreatments

Q.H.Wang, M.I.Bowser and J.G.Swanson

Department of Electronic and Electrical Engineering, King's College London, Strand, London WC2R 2LS

Abstract. A comparison has been made of the effects of predeposition plasma treatments using hydrogen, argon and a mixture of the two using channel current transient spectroscopy. All of the samples exhibited electron emission from an interface state continuum with energies consistent with the interface state band model. When argon and hydrogen were used together two extra processes were observed. One of these was due to an electron trap with an activation energy of 0.05eV, involving states at the remote edge of the depletion region. The second process which had an activation energy of 0.05eV resembled hole emission, but has been attributed to an interfacial polarisation process exhibiting thermally activated relaxation. The necessity for argon and hydrogen suggests that argon had created structural damage permitting the entry of hydrogen atoms to form electrically active complexes in the damaged region.

1. Introduction

GaAs - insulator interfaces have high densities of trap states which cause Fermi level pinning and other anomalous behaviour.[1-7] These states arise from the termination and reconstruction of the GaAs lattice at the surface and from the electronic bonds which are made with the insulating material.[2,7] Plasma pretreatments of this interface might offer a way of improving the interface quality and reduce the trap density thereby unpinning the Fermi level. It has been reported that H_2 treatment might passivate trap states and cause a reduction of the interface state concentration.[8,9] Bowser et al.,[3] had studied GaAs-Si₃N₄ interface which had been subjected to different plasma treatments before the deposition of the insulator. It was found that there was no great reduction in the density of the interface trap states for any of the pretreatments; however, unusual effects were observed for samples with a H_2+Ar pretreatment. A very small depletion capacitance was observed compared with other samples, and the surface potentials calculated using this capacitance and the original doping concentration were unrealistically large, suggesting that the active doping concentration had been reduced near to the surface. In this study, the thermal properties of the trapped charges are investigated using deep level transient spectroscopy (DLTS). The combined effects of H_2 and Ar on the GaAs-Si₃N₄ interface are to be compared with samples that were treated with H_2 and Ar alone.

2. Sample preparation

The MISFET test structures were made from a single <100> sulphur doped n-type VPE epitaxial layer, formed on a semi-insulating substrate with an undoped 1-2μm buffer layer.

The doping concentrations were typically $6 \times 10^{16} cm^{-3}$. Ohmic contacts were formed using AuGeNi annealed at 440°C for 2 minutes in a 5% H_2, 95% N_2 ambient. Prior to insulator deposition, the samples were RF plasma pretreated at 300°C for 3 minutes in an Electrotech PF310 parallel plate PECVD system, pressure 450mtorr, power 40W for Ar and Ar+H_2 treatments and 15W for H_2 treatment. After pretreatment, the deposition reactants were admitted and the plasma re-applied to start deposition of Si_3N_4. The deposition conditions were: SiH_4 2% / N_2 98% (1400sccm) - NH_3 (7sccm), 300°C, 350mtorr, plasma: 190KHz, 40W. The insulator layer thickness was 750Å. Contact windows were formed using CF_4 plasma etching, and gate electrodes by depositing a 500nm Al layer patterned by lift-off.

3. Emission transient observations

The transient observations were made by biasing the Al counter electrode to establish a quiescent surface potential and then applying a positive pulse to fill states within a 50meV range above the quiescent position of the surface Fermi level. Upon the removal of the pulse, excess electrons trapped at the interface or in the bulk caused an expansion of the depletion region proportional to the number of trapped charges. The instantaneous change in width of the depletion region was assessed by measuring the current flowing in the undepleted parallel channel of the FET structure. The current transient was related to the instantaneous excess charge in the interface and analysed using the DLTS method[10].

Ar and H_2 treated samples gave spectra with a single well defined peak. Fig1. The activation energy of the unique mechanism was obtained from an Arrhenius plot. It was measured at different gate biases which positioned the quiescent surface Fermi level within the interface state continuum. Typical activation energy versus bias variations for the Ar and H_2 pretreated samples are shown in Fig.2. The thermal activation energy changed typically from 0.3eV to 0.5eV. This indicates that the range of the interface state continuum which could be probed by the Fermi level was typically 0.20eV. The trend with changing bias was consistent with emission to the conduction band.

Fig.3 shows the DLTS spectrum of a sample subjected to the H_2+Ar treatment. It can be seen that there are two superimposed positive peaks E1 and E2, as well as a negative peak H. The peaks span a temperature range from 100K to 300K. This indicates that there are two majority (electron) trap states and a response which could be assigned to a minority carrier trap. The measured activation energies versus bias for the two majority trap levels are shown in Fig.4, whilst that for the negative peak is shown in the same figure but related to the right

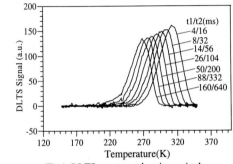

Fig 1. DLTS spectrum showing a single peak for Ar pretreated samples

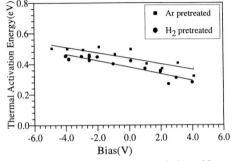

Fig 2. Typical activation energy variation with bias for Ar and H_2 pretreated samples

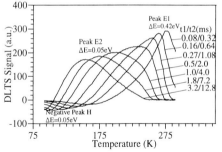
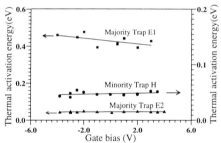

Fig 3. DLTS spectra for H_2 + Ar pretreated sample

Fig 4. Activation energy variation with bias measured from sample with Ar+H_2 pretreatment

hand axis. The trap level E1 in Fig.3 had a zero bias thermal activation energy of 0.42eV which changed with bias indicating that it too was an interface trap state, it had a capture cross-section of the order of $10^{-15} cm^{-2}$ and corresponded with the single response from the H_2 and Ar treated samples. The other trap levels E2 and H demonstrated small bias independent activation energies, indicating that they are related to bulk trap states or other mechanisms which are independent of bias. Even at 100K they still had an effect on the current transients. The capture cross-sections of these two shallow traps were several orders smaller than for E1, typically $10^{-21} cm^{-2}$.

4. Discussions

It has been reported that GaAs-insulator interfaces have a "U"-shaped state continuum with a quiescent surface Fermi level position near to its minimum.[3,7] It is possible to vary the position of the Fermi level over a limited range around the minimum with practical electric fields. The high interface state density limits the Fermi level modulation range to about 0.15eV;[3] this is consistent with the change of activation energy of the interface related response. The quiescent Fermi level position was usually about 0.9eV below the conduction band minimum for H_2 and Ar pretreated samples.[3]

When the surface potential is moved positively states above the quiescent Fermi level tend to fill and then empty when the field is removed or decreased. If the thermal emission were to the conduction band, an activation energy $E_{act} = E_c - E_{fq} - q\Delta\varphi_{sc}$ would be required for thermal emission, where $\Delta\varphi_{sc}$ is the quiescent surface potential change by bias; E_{act} would be not be expected to be less than 0.9eV for the test samples. In this study the emission energy at a particular bias was always much less than the anticipated energy E_{act} whatever the pretreatment. These differences between the anticipated and actual thermal emission energies at zero gate bias were 0.46eV and 0.58eV for the Ar and H_2 treated samples respectively. Similar results have been explained by the interface state band (ISB) model[1,2] which supposed that the interface state density was so large towards the conduction band edge that an interface state band was formed. A trapped charge might then be emitted to the lower edge of the band and then reach the conduction band edge by a series of small energy jumps. The rate limiting step is then the largest step from the trap level to the interface state band lower edge.

The bias dependent response was seen too at about the same energy in the samples that had been treated with Ar and H_2 together. It seems very likely that its origin was also from electron emission from states in an interface continuum. The two bias independent responses could not have been from states at the interface. Their fixed activation energies suggest that very shallow bulk states were involved. An electron trap was observed 0.05eV from the conduction band. The ionisation energy of the sulphur donor in GaAs is only 0.006eV,[11] and it is apparent that the occupation probability of this trap state would have been significantly greater than states at the donor level or in the conduction band. This suggests that mobile electrons from the donor would have been lost into these traps.

Although this might provide an explanation of donor passivation in the bulk of a sample it does not provide a satisfactory explanation for the lower donor concentrations which have been observed using depletion capacitance measurements.[3,8,12] In these measurements charge changes at the edge of a space-charge region are detected as the region expands and contracts. Throughout most of the region in normal samples it is only the ionised donors that are present. In samples containing the 0.05eV traps these would be well above the Fermi level throughout most of the region and would be empty and uncharged. The charge concentration in the region would be therefore unchanged and its width unaffected by the presence of the trap. Only near the edge of the region would the traps be occupied. As the depletion region expands incrementally electrons are expelled at its edge from the conduction band and from the traps too. The latter suffer large changes in population when they move upwards away from the Fermi level which is close to them in the bulk. The overall effect would be to remove the same number of electrons for a given change of surface potential leaving the depletion capacitance unchanged. This suggests that the shallow trap is not responsible for the lower capacitances from which donor passivation had been inferred.

During the filling pulse only shallow states near to the depletion region edge would suffer a significant population change. This shows that the trap levels that were being detected were several thousand angstroms from the surface. Note that this trap was not detected when the same material was treated with either argon or hydrogen alone. The individual roles of argon and hydrogen atoms need to be considered. Argon atoms are 40 times more massive than hydrogen atoms and are capable of delivering much more momentum to a lattice with which it collides. Of the two atomic species it is Ar that is more likely to displace atoms and cause damage to the crystal surface. Rare gas atoms are chemically inert and unlikely to form electronic bonds with other atoms and, as a consequence, are unlikely to create electronic defects, except in as much as the lattice is deformed or damaged. Consistent with this, it was not possible to detect DLTS peaks due to Ar treatment alone.

Hydrogen on the other hand is unlikely to cause displacement damage to the crystal but is known to be electrically active, forming electronic bonds readily. Having a small atomic diameter allows it to penetrate interstitially into a lattice. This would be facilitated by Ar bombardment which could cause damage by creating permanent atomic displacements or by temporary vibrational displacements which would ease the passage of the small hydrogen atoms. That treatment with hydrogen alone was insufficient to produce this trap level is consistent with Ar facilitating its entry to the lattice allowing its electrical activity to be seen relatively deep within the lattice.

The other bulk response was negative and had an activation energy of 0.05eV. It would normally be attributed to hole emission but it is difficult to see how hole injection could have occurred in the MIS samples. It was only seen in the sample that had been treated with argon and hydrogen together. The presence of a negative spectral peak requires that the channel current variation should have contained an exponentially decreasing component with time following the removal of the positive pulse. The depletion region would have had to expand with time. This would be consistent with an increasingly negative electric field at or near to the semiconductor surface.

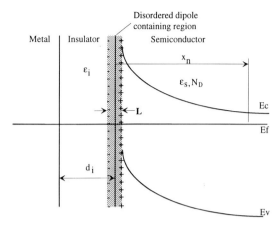

Fig 5. Dipole located at the H$_2$+Ar pretreated interface

An explanation can be formulated in terms of an interfacial polarisation process with a relaxation time activated by temperature.

The Debye model of polarisation[13] assumes that the polarisable species exist in two states of different energy, with a relaxation time controlled by the energy barrier between the two energy levels. It is not possible to say what the polarisable species was in this case. The mechanism cannot be associated with the silicon nitride dielectric because the response was only seen in Ar+H$_2$ treated samples. It could be associated with a damaged interfacial region, though not one created by argon alone, this suggests that the hydrogen atom was again active as a charge species in creating polarisable complexes within the damaged region.

The diagram, Fig. 5 shows an equivalent planar surface dipole of surface moment QL, positioned at the interface. It is polarised during the positive pulse and relaxes to its unpolarised state when the pulse is removed by a thermally activated process. It is unlikely that the polarisation process was caused by a truly planar atomic dipole in the structurally disordered surface region. Nevertheless the representation used here is of a dipole of equivalent moment near to the interface which would yield the observed electrostatic behaviour believed to be caused by charge displacements in a surface disordered region. When a positive pulse was applied, the planar dipole was polarised. The reduction of dipole moment with time following the removal of the positive pulse would then cause the semiconductor electric field to shift negatively causing the depletion region to expand giving rise to a decreasing component of channel current. The time constant of this current component can be expressed as $\tau_p = \tau_0 \exp\left(\frac{\Delta E}{kT}\right)$, where ΔE is the thermal activation energy for the polarisation relaxation. The observation of the negative spectral peak suggests that this activation energy is 0.05eV. The change in channel current associated with this mechanism allowed the change in depletion width to be inferred and related to the magnitude of the interfacial polarisation. The dipole moment per unit area which this yielded was 1.0×10^{-11}C·m^{-1}. Had the dipole had a length of 1nm the equivalent charge density would have been 6×10^{12}cm^{-2}, comparable to the interface state densities observed.

It appears that the joint action of hydrogen and argon was needed to yield these two effects. This is further confirmed by their absence in samples that were not pretreated or were pretreated with nitrogen and ammonia plasmas. This evidence indicates that the deposition process itself had no effect in the production of these phenomena.

5. Conclusions

Samples which were prepared with different plasma pretreatments exhibited electron emission from interfacial states with bias dependent thermal emission energies. The thermal emission energy was significantly less than the separation of the emitting states from the conduction band edge. This behaviour is consistent with the ISB model. For the samples that were pretreated with a combination of Ar and H_2, two other features were seen in the transient spectra. One corresponded to electron emission from a bulk level. It is concluded that these states arose from deep damage caused by argon bombardment combined with the ingress of hydrogen atoms to form an electrically active complex. These states were not observable when either argon or hydrogen were used alone, suggesting that damage facilitated the entry of electrically active hydrogen atoms. This state is not sufficient to explain donor passivation by hydrogen. The other response in the Ar+H_2 treated sample had a negative magnitude and its activation energy was bias independent. It is unlikely that this could have been due to hole emission. The presence of an interfacial electric dipole which polarises under positive gate bias, relaxing by a thermally activated process provides a consistent explanation of this observation. The inferred dipole moment per unit area was $1.0 \times 10^{-11} C \cdot m^{-1}$ corresponding to a state density of $6 \times 10^{12} cm^{-2}$ for a planar dipole length of 1nm. This response also required the presence of both Ar and H_2 suggesting that the polarisation was associated with a damaged region made electrically active by the hydrogen atoms.

References

1. Hasegawa H and Sawada T 1980 IEEE Trans. Electron. Devices.**ED-27**,6,1055
2. Hasegawa H and Sawada T 1983 Thin Solid Films, **103**,119
3. Bowser. M. I and Swanson J G 1988 Int. Symp. on GaAs and Related Compounds, Atlanta, Georgia 405
4. Lee W S and Swanson J G 1983 Thin Solid Films, **103**,177
5. Spicer W E and Eglash S 1982 Thin Solid Films, **89**,447
6. Spicer W E, Weber Z L, Weber E and Newman N 1988 J.Vac. Sci. Tech. B**6**,1245
7. He L, Hasegawa H, Luo J and Ohno H 1988 Applied Surface Sci,**33/44**,1030
8. Pearton S J, Dautremont-Smith W C, Chevallier J and Tu C W and Cummings K D 1986 J. Appl. Phys, **59**,15
9. Pearton S J, Wu C S, Stavola M, Ran F, Lopata L and Dautremont-Smith W C 1987 Appl. Phys. Lett, **51**, 17
10. Lang D V 1974 J. Appl. Phys, **45**, 3014
11. Sze S M 1981 Physics of Semiconductor Devices, 2nd edition, John Wiley & Sons
12. Pan N, Lee B, Bose S S, KimM H, Hughes J S, Stillman J E, Arai K and Nashimoto Y 1987 Appl. Phys. Letts. **50**,1832-34
13. Debye P 1945 Polar Molecules, Dover, New York

Local Vibrational Mode Spectroscopy of Beryllium- and Zinc-Hydrogen Complexes in GaP

M. D. McCluskey and E. E. Haller

Lawrence Berkeley Laboratory and University of California at Berkeley, Berkeley, California 94720

J. Walker and N. M. Johnson

Xerox Palo Alto Research Center, Palo Alto, California 94304

Abstract. Using infrared absorption spectroscopy, we have observed local vibrational modes in GaP:Be and GaP:Zn exposed to a remote hydrogen or deuterium plasma. In GaP:Zn, we attribute the modes at 2379.0 cm^{-1} and 1729.4 cm^{-1} to hydrogen-phosphorus and deuterium-phosphorus bond-stretching modes of complexes adjacent to the zinc acceptors. In GaP:Be, we attribute the modes at 2292.2 cm^{-1} and 1669.8 cm^{-1} to similar complexes adjacent to the beryllium acceptors.

1. Introduction

The subject of hydrogen passivation of defects and impurities in compound semiconductors has attracted a great deal of interest in recent years. Although most studies have focused on GaAs and InP [1,2], significant work has been done on hydrogen passivation in GaP. Singh *et al.* used photoluminescence (PL) spectroscopy to study hydrogen neutralization of donors, acceptors, and the isoelectronic nitrogen trap in GaP [3]. Electrical measurements demonstrated that zinc in GaP is neutralized after exposure to atomic hydrogen [4]. Clerjaud *et al.* observed the C-H and C-D bond-stretching local vibrational modes (LVM's) [5] and the N-H mode [6] in GaP grown by the liquid-encapsulation Czochralski (LEC) technique. LVM's corresponding to hydrogen-defect complexes in LEC-grown GaP have also been observed [7]. Prior to this study, LVM's corresponding to group II acceptor-hydrogen complexes in GaP have not been reported.

2. Experiment

The GaP samples used for this study had a (100) orientation and were approximately 5 mm X 5 mm X 0.3 mm. Prior to zinc diffusion, they were n-type, with a sulfur concentration of around 10^{17} cm^{-3}. To obtain GaP:Zn, GaP samples were placed with 1 g zinc in an evacuated 180 ml quartz ampoule which had been cleaned in HF. The ampoule was placed in a vertical furnace and the GaP samples were diffused for 1 hr at a temperature of 900°C. After completion of the diffusion, the samples were quenched to room temperature by dropping the ampoule into ethylene glycol. Residual zinc on the GaP surfaces was removed by immersion in dilute HCl. A room temperature Hall effect measurement with the Van der

Pauw geometry verified that the samples were p-type after zinc diffusion. Some of the samples were then exposed to monatomic hydrogen or deuterium in a remote plasma system as described in Ref. 8. The hydrogenation temperature was 300°C and the duration of the exposure was 3 hr.

To obtain GaP:Be, GaP samples were implanted with Be ions at energies of 40 keV, 100 keV, and 200 keV, for a total dose of 2.5×10^{15} cm^{-2}, followed by a rapid thermal anneal at 1000°C for 10 s. A room temperature Hall effect measurement indicated a hole concentration of 2.5×10^{15} cm^{-2} prior to hydrogenation.

Infrared absorption spectra were obtained with a Digilab 80-E vacuum Fourier transform spectrometer with a KBr beamsplitter. Spectra were taken at 10 K with an instrumental resolution of 0.25 cm^{-1}. A Ge:Cu photoconductor was used as the detector. GaP:Zn and GaP:Be samples which were not H- or D-plasma exposed were used as reference samples.

3. Results

Spectra recorded with the hydrogenated and deuterated GaP:Zn samples show infrared absorption peaks at 2379.0 cm^{-1} and 1729.4 cm^{-1}, respectively, at a temperature of 10 K (Fig. 1). The isotopic ratio of these frequencies, $r = v_H/v_D$, is 1.3756. Neither peak was seen in GaP:Zn which was not H- or D-plasma exposed. As a further check, we annealed a GaP sample for 1 hr at a temperature of 900°C in an evacuated quartz ampoule but with no zinc present. The sample was then exposed to a hydrogen plasma under the conditions described above. Again, neither peak was seen in the absorption spectrum. GaP:Zn samples which were exposed to a H/D plasma mixture showed both peaks but no new peaks which would have indicated a HD complex. Free carrier absorption prevented measurements below 1300 cm^{-1}.

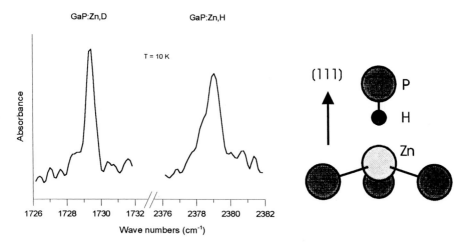

Figure 1. Infrared absorption spectra of deuterated and hydrogenated GaP:Zn and suggested model for H-passivation.

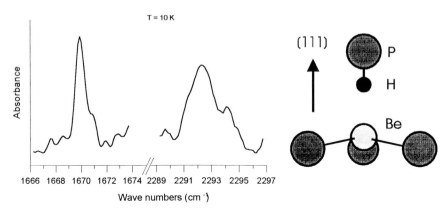

Figure 2. Infrared absorption spectra of deuterated and hydrogenated GaP:Be and suggested model for H-passivation.

By way of comparison, hydrogenated InP:Zn has a bond-stretching mode at 2287.7 cm^{-1} and isotopic ratio r = 1.3744 [9]. The bond-stretching mode has been attributed to a H-P complex oriented along a bond-centered direction, adjacent to the Zn acceptor, with H essentially decoupled from Zn. Since the LVM's and the r-factor for GaP:Zn are similar to the corresponding values for InP:Zn, we assume the structures are the same. The H-P model receives further support from the observation that the LVM frequency is very different from the bond-stretching frequency of Zn-H (1600 cm^{-1}). This is in agreement with pervious studies [10], which determined that in group II acceptor-hydrogen complexes in GaAs, the hydrogen is bound directly to a neighboring arsenic atom.

The hydrogenated and deuterated GaP:Be samples show infrared absorption peaks at 2292.2 cm^{-1} and 1669.8 cm^{-1}, respectively, at a temperature of 10 K (Fig. 2). The isotopic ratio of these frequencies, r = v_H/v_D, is 1.3727. Neither peak was seen in GaP:Be which was not H- or D-plasma exposed. Once again, these values are similar to the corresponding values in hydrogenated InP:Be, which has a H-P bond-stretching mode at 2236.5 cm^{-1} and isotopic ratio r = 1.3714. We therefore assume that the absorption peaks arise from a H-P complex oriented in a bond-centered direction, adjacent to the Be acceptor.

The H-P modes in GaP are higher than the H-P modes in InP; this may be related to the fact that GaP has a smaller lattice constant. In addition, the H-P modes in GaP:Zn are higher than the H-P modes in GaP:Be. The increase in hydrogen bond-stretching frequencies with increasing group II acceptor size has also been observed in InP and GaAs.

The positions and FWHM of the observed peaks are listed in Table I. The FWHM of the D-P peaks at 10 K are approximately twice as small as those of the H-P peaks. This effect has been observed in numerous hydrogen-related complexes in III-V semiconductors and is related to the smaller average vibrational amplitude of the D atom as compared to the H atom. In InP:Zn, for example, the D-P peak at 6 K is narrower than the H-P peak by a factor of 2.9.

Compound	H-P stretch mode		D-P stretch mode		$r=\nu_H/\nu_D$
	Peak (cm^{-1})	FWHM (cm^{-1})	Peak (cm^{-1})	FWHM (cm^{-1})	
GaP:Be	2292.2	1.8	1669.8	0.7	1.3727
InP:Be[a]	2236.5	0.43	1630.9	0.2	1.3714
GaP:Zn	2379.0	1.1	1729.4	0.5	1.3756
InP:Zn[a]	2287.7	0.23	1664.5	0.08	1.3744

[a]Ref 1.

Table 1. Frequencies and FWHM of hydrogen LVM peaks in GaP and InP.

We are indebted to A. G. Elliot for providing the GaP samples. We also wish to acknowledge J. Beeman for constructing our Ge:Cu photoconductor, L. Hsu for his assistance with the Digilab spectrometer, and W. Walukiewicz and J. Wolk for helpful discussions. This work was supported in part by USNSF grant DMR-91 15856 and in part by AFOSR contract F49620-91-C-0082. We acknowledge the use of the facilities at Lawrence Berkeley Laboratory.

References

[1] J. Chevallier, B. Clerjaud, and B. Pajot, in *Semiconductors and Semimetals*, edited by J. I. Pankove and N. M. Johnson (Academic Press, Orlando, FL, 1991), Vol. **34**, p. 447.
[2] E. E. Haller, *Twelfth Record of Alloy and Semiconductor Physics and Electronics Symposium* (1993), to be published.
[3] M. Singh and J. Weber, Appl. Phys. Lett. **54**, 424 (1989).
[4] M. Mizuta, Y. Mochizuki, N. Takadoh, and K. Asakawa, J. Appl. Phys. **66**, 891 (1989).
[5] B. Clerjaud, D. Côte, W.-S. Hahn, and W. Ulrici, Appl. Phys. Lett. **58**, 1860 (1991).
[6] B. Clerjaud, D. Côte, W.-S. Hahn, and D. Wasik, Appl. Phys. Lett. **60**, 2374 (1992).
[7] B. Dischler, F. Fuchs, and H. Seelwind, Physica B **170**, 245 (1991).
[8] N. M. Johnson, in *Semiconductors and Semimetals*, edited by J. I. Pankove and N. M. Johnson (Academic Press, Orlando, FL, 1991), Vol. **34**, Ch. 7.
[9] B. Pajot, J. Chevallier, A. Jalil, and B. Rose, Semicond. Sci. Technol. **4**, 91 (1989).
[10] R. Rahbi, B. Pajot, J. Chevallier, A. Marbeuf, R. C. Logan, and M. Gavand, J. Appl. Phys. **73**, 1723 (1993).

High Quality Etched/Regrown GaAs/GaAs Interfaces Formed by an All In-Situ Cl$_2$ Etching Process

D. S. L. Mui, T. A. Strand, B. J. Thibeault, L. A. Coldren,
P. M. Petroff, and E. L. Hu

Electrical and Computer Department, University of California, Santa Barbara, CA 93106

Abstract. The growth of high quality GaAs/GaAs interfaces on in-situ Cl$_2$ etched (100) GaAs surfaces is demonstrated. Etching is performed on as-grown MBE GaAs surfaces in an integrated ultrahigh vacuum etching chamber by pure chemical gas etching at 140, 180, 210, and 280 °C. Current-voltage characteristics of Schottky diodes formed by this all in-situ process have ideality factor of 1.03 and no measurable current-limiting effect caused by the etched interface (except for the sample etched at 280 °C). Trace amounts (low 10^{10} cm^{-2}) of etch-induced point defects are present at the etched/regrown interface as revealed by CV and DLTS measurements. The defect densities presented in this work are the lowest ever reported on etched GaAs surfaces.

1. Introduction

It is well-established that significant modifications to a GaAs surface, for example, oxidation, causes the surface Fermi level to be pinned around mid-gap. The oxidation process results in the formation of a high density of defects on the surface. To address the problem of oxidation during device processing, we have devised a unique etch and regrowth system for the formation of devices under ultrahigh vacuum (UHV) conditions. We demonstrate in this paper that by the use of the all in-situ processing technique, defect density at etched/regrown interfaces is extremely low and these interfaces do not display any Fermi level pinning effect.

2. Experimental

The experimental setup of the in-situ etching chamber has been described elsewhere [1]. Samples used in this study were all n-type and grown by Molecular Beam Epitaxy (MBE) at 600 °C under As-rich conditions. A heavily doped layer followed by a lightly doped (10^{16} cm^{-3}) buffer was first grown on a heavily doped substrate. The sample was then transferred to an adjacent etching chamber via an UHV transfer tunnel. After the substrate temperature had been stabilized, etching was performed with a Cl$_2$ flow rate of 5 sccm and a chamber pressure of 2×10^{-4} Torr. Chlorine has been extensively used as the etch gas for various dry etching processes of GaAs-based semiconductors [2-4]. After etching, the sample was transferred back to the MBE growth chamber for the subsequent regrowth of a 10^{16} cm^{-3} cap layer. The sample was removed from the growth facility for the thermal evaporation of circular dots of Au, 685 μm in diameter, to form the top Schottky contact. The conductive substrate served as the other electrical contact. In this study, we have compared the electrical properties of etched/regrown interfaces formed with four different etching temperatures - 140, 180, 210, and 280 °C. These samples are labeled A, B, C, and D, respectively. Current-voltage, CV, and DLTS were used for characterizing the electrical properties of these interfaces.

3. Results and Discussion

The presence of a high density of compensating interface defects causes a build up of electrical charges at the interface. This results in the formation of a depletion region and band-bending across the interface. This scenario is depicted in Fig. 1. If the band-bending is larger than a thermal unit, a current blocking barrier will be formed. Not only is this effect easily observable when the diode is forward biased, it is also a good indication on the quality of an interface For a sample doped to 10^{16} cm^{-3}, a band-bending of a thermal unit corresponds to an interface trap density of 1.2×10^{11} cm^{-2}. Only one of the samples used in this study displays a current limiting effect caused by the etched interface. The Schottky plot of the I-V characteristics for samples B and D are shown in Fig. 2. The I-V curves of samples A and C are similar to that of sample B. It is then clear that samples A, B, and C have lower interface trap density then sample D, and Fermi level pinning is absent in all of these samples. Although sample D displays a current blocking effect, the interface trap density (as discussed below) is still low enough that the interface Fermi level is not pinned. In comparison, an ex-situ etched epi/substrate interface displays a larger current blocking effect then that shown in sample D.

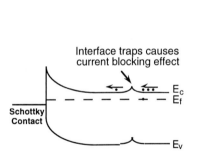

Fig. 1. Energy band diagram depicting the current blocking effect at an interface.

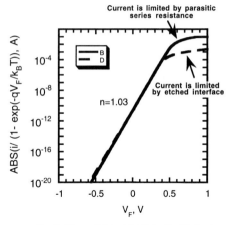

Fig. 2. Comparision of the Schottky-Plots of two etched diodes. Sample B does not show any current limiting effect due to the interface.

To be more quantitative, the CV carrier profile technique is used to determine the total carrier depletion caused by the etched interface. The apparent carrier concentration, N_{app}, as a function of distance can easily be obtained since the total depletion width is a known function of the gate bias. Figure 3 shows the carrier profile for the four samples. It is obvious that carrier depletion across the interface is a strong function of etching-temperatures. It should be pointed out that there are regions in the samples where N_{app} is higher than the background doping concentration. This effect is caused by the response of the interface traps to the applied DC bias (CV stretch out) and does not indicate the presence of carrier accumulation at the interface.

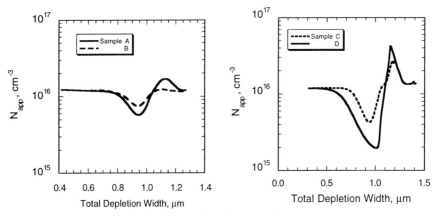

Fig 3. CV carrier profiles of four etched/regrown samples.

The integrated carrier depletion, N_{int}, across the interface for the four samples, A through D, are found to be 1.2, 0.84, 1.6 and 4.0×10^{11} cm^{-2}, respectively. These numbers are the lowest ever reported for any etched GaAs surfaces and interfaces. It can be seen that these results are consistent with our earlier discussion on the I-V characteristics of these samples - the only sample (D) which displays a current limiting effect in the I-V characteristics also shows the largest carrier depletion. In investigating other samples, we have observed a monotonic decrease in N_{int} with decreasing etch depth. For example, we measured an N_{int} of 2.9×10^{10} cm^{-2} in a sample (E) in which 600 Å of GaAs was etched at 180 °C compared to 3000 Å in samples A through D. This etch-depth dependence is not currently understood and is under investigation.

We have performed DLTS measurements on these samples in order to determine what kinds of interface traps are present. The DLTS spectra of sample E, the sample with the lowest N_{int}, is shown in Fig. 4.

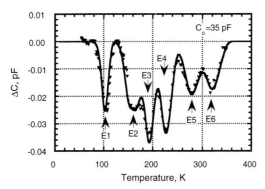

Fig. 4. DLTS spectra of sample E.

The thermal signatures of these discrete traps are very similar to those found in Argon sputtered GaAs bulk samples [5-6], and these traps are known to be non-stoichiometric point defects. The extracted trap activation energies and capture cross-sections from Fig. 4 are listed in Table 1. The facts that these traps are non-stoichiometric defects and N_{int} displays a strong dependence on etch-temperatures suggest that the desorption-rate ratio between GaCl and AsCl plays an important role in the determination of N_{int}.

Table 1. Activation energies and capture cross-sections of etch-induced traps.

Label	E_a, eV	s, cm^2
E1	0.16	0.7×10^{-15}
E2	0.28	7.7×10^{-15}
E3	0.41	1.4×10^{-13}
E4	0.49	1.8×10^{-13}
E5	0.61	1.6×10^{-13}
E6	0.73	4.3×10^{-13}

4. Conclusion

We have demonstrated very high quality GaAs regrowth on etched GaAs surfaces with record low interface trap density. The interface does not exhibit any measurable current blocking effect. The etch induced traps have thermal signatures very similar to those found in Ar sputtered bulk GaAs. Our data suggest that the desorption-rate ratio between GaCl and AsCl plays an important role in the determination of the density of interface traps.

Acknowledgments - The authors would like to thank J. English for his technical assistance. This work is sponsored by the NSF Science and Technology Center for Quantized Electronic Structures (QUEST), #DMR91-20007, and the innovative Science and Technology Division of SDIO via ARO.

References

[1] D. S. L. Mui, T. A. Strand, B. J. Thibeault, L. A. Coldren, P. M. Petroff, and E. L. Hu, *37th International Symposium on Electron, Ion, and Photon Beams*, San Diego, CA, 1993.

[2] J. A. Skidmore, D. L. Green, D. B. Young, J. A. Olsen, E. L. Hu, L. A. Coldren, and P. M. Petroff, *J. Vac. Sci. Technol.*, **B9**, 3516(1991).

[3] E. M. Clausen, Jr., J. P. Harbison, L. T. Florez, and B. Van der Gaag, *J. Vac. Sci. Technol.*, **B8**, 1960(1990).

[4] N. Furuhata, H. Miyamoto, A. Okamoto, and K. Ohata, *J. Appl. Phys.*, **65**, 168(1988).

[5] F. D. Auret, S. A. Goodman, G. Myburg, and W. E. Meyer, *J. Vac. Sci. Technol.*, **b10**, 2366(1992).

[6] G. M. Martin, A. Mitonneau, and A. Mircea, *Elect. Lett.*, **13**, 192(1977).

Electrical and structural characterization of MBE GaAs grown at temperatures between 200 and 600 °C

P. Kordoš[1], J. Betko[2], A. Förster[1], S. Kuklovský[2], Ch. Dieker[1], F. Rüders[1]

[1]Institute of Thin Film and Ion Technology, Research Centre Jülich,
D-52425 Jülich, Germany
[2]Institute of Electrical Engineering, Slovak Academy of Sciences,
SK-84239 Bratislava, Slovakia

Abstract. As-grown and annealed MBE GaAs layers prepared at various temperatures T_g=200-600 °C are analyzed and a strong influence of T_g on the layer properties is demonstrated. In as-grown layers the resistivity extremely increases, by six orders of magnitude, with T_g increased from 200 to 400 °C. An opposite effect, a strong decrease of the resistivity occurs if T_g increases from 450 to 600 °C. Semi-insulating GaAs layers (ρ_{300}~10^7 Ω cm) with regular GaAs lattice constant can be prepared at T_g between 400 and 450 °C without annealing and these layers are thermally stable in their resistivity during annealing (590 °C, 10 min). Photoconductors prepared on 400 °C layers show a higher sensitivity but a longer response time than on typical LTG GaAs layers. Further, the BEP ratio between 7 and 19 has no significant influence on the resistivity of 200 °C layers but for 350 °C grown layers an increase of the resistivity by factor of about 10 with decreasing BEP ratio is observed for both as-grown and annealed layers. Finally, we observed an influence of the annealing time on the resistivity of GaAs layers.

1. Introduction

The growth of GaAs layers by molecular-beam epitaxy (MBE) is normally performed at substrate temperatures of ~600 °C. A material of excellent structural, electrical and optical properties can be obtained. Intentionally undoped layers generally exhibit p-type conductivity with a carrier concentration of 10^{13}-10^{14} cm^{-3} [1]. The deep-trap concentration increases if the growth temperature decreases and below ~450 °C a high resistive GaAs layers will be grown [2]. Recently, it has been shown that even semi-insulating (SI) layers can be prepared at 400 °C [3].

It is known that the growth at extremely low temperatures, T_g ~200 °C, makes it possible to prepare layers attractive for applications at photoconductive switches and field-effect devices [4]. However, low-temperature-grown (LTG) MBE GaAs has a special feature that its properties will change if an annealing is performed. As-grown layers are conductive ($\sim 10^2$ Ω cm) and highly lattice mismatched with ~ 1 % of excess arsenic incorporated mainly as As_{Ga}-related native defects of extremely high concentrations, as high as 10^{20} cm^{-3}. After annealing (usually at 600 °C for 10 min), the native defect concentrations decrease, As-precipitates are formed, the lattice mismatch disappears and the layer becomes high resistive (10^5-10^6 Ω cm). Two models have been proposed to explain the conduction mechanism of LTG GaAs. The first one is based on the action of As-precipitates as buried Schottky barriers which can fully deplete the layer [5]. The second model takes into account the compensation effect of deep As_{Ga}-related donors [6]. In spite of great activity on LTG materials many questions still remain open. Recently, a tendency to study the layer properties grown at higher temperatures than LTG-typical 200 °C exists [3,7]. A systematic study on the GaAs layers as a function of the growth temperature has not been reported.

The aim of the present study is to investigate extensively electrical and structural properties of MBE GaAs layers grown at various temperatures between 200 and 600 °C, i.e. in the whole range between the LTG-typical and the conventional conditions. Precise conductivity measurements in the range of 300-440 K, high resolution x-ray topography (HRXT), transmission electron microscopy (TEM) and electro-optical sampling (EOS) analyses were performed in our study. We investigated properties of as-grown and annealed GaAs layers prepared at the conventional LTG conditions, as well as layer properties as a function of growth and annealing parameters.

2. Experimental procedures

The GaAs layers under investigation in this study were grown in a Varian Mod GEN II MBE system, under As-stabilized conditions, on indium-free-mounted 2-in.-diam (100) SI-GaAs wafers. The growth temperature was adjusted by a calibrated thermocouple in the range between 200 and 600 °C in increments of 50 °C. We will note that all data concerning the growth temperature are only thermocouple readings here. 2 μm thick layers were grown at the growth rate of 1 μm/h and the As_4/Ga beam-equivalent-pressure (BEP) ratio. The layers were predominantly grown at BEP = 19, additionally layers using BEP = 7 and 13 were prepared. No post growth annealing was performed in the MBE chamber. 10x10 mm^2 samples were cleaved from each wafer for extensive electrical and structural characterization. Layers prepared at all growth temperatures were annealed at typical LTG conditions, i.e. T_a = 590 °C, t_a = 10 min, to evaluate their thermal stability. The influence of annealing time, t_a varried between 10 and 60 min, on the layer properties was studied. Rapid thermal processing under local As overpressure was used for annealing.

Precise dc-conductivity measurements (within an error of less than 3 %) [8] were carried out in the temperature range from 300 up to 440 K using a high-impedance system and the van der Pauw method. Nonalloyed contacts were prepared by rubbing in In+Ga at room temperature to avoid the influence of heating before measurements were performed. The I-V characteristics of these contacts showed ohmic behaviour. Before contacting, the samples were treated in HCl for 5 min and consequently rinsed in DI-water to avoid a surface conduction caused by an improper oxide layer. The resistivity of the layer was determined from sheet resistivity measurements before (r_t) and after (r_s) etching away the GaAs layer from the SI GaAs substrate, using the equation

$\rho = d_e(r_t^{-1} - r_s^{-1})^{-1}$, where d_e is the layer thickness [9]. The resistivity of SI GaAs substrates very slightly varried from wafer to wafer, $\rho_{300} \sim (1\text{-}1.5) \times 10^8$ Ω cm. The lattice constant of GaAs layers normal to the layer-substrate interface was determined by measuring the (004) rocking curves using a high-resolution x-ray diffractometer. The lattice mismatch $(a_e - a_s)/a_s$ was evaluated from two separate peaks of the Bragg reflection atributed to the substrate (s) and MBE layer (e). Additionally, TEM micrographs were used to evaluate the structural quality and the formation of As-precipitates. EOS measurements were performed to study the properties of photogenerated carriers.

3. Results and Discussion

3.1. Effect of the growth temperature

The resistivity dependences on the reciprocal temperature for some typical unannealed GaAs layers grown at various temperatures T_g are shown in Fig. 1. The resistivity dependence for SI GaAs substrate is shown for comparison too. The 200 and 250 °C layers represent common LTG material, with low resistivity and dominated hopping conduction above the room temperature. The Hall mobility $\mu_{H(300K)} \leq 0.15$ cm^2/Vs has been obtained in our previous study on LTG GaAs [9]. The lattice mismatch in these layers is high, between 0.11 and 0.13 %, which corresponds to $\simeq$ 0.7-0.9 % of excess As. The lattice mismatch decreases and the resistivity increases with increased T_g from 200 to 350 °C. The 350 °C grown layer is nearly lattice matched to the substrate and the resistivity becomes high, $\sim 10^5$ Ω cm at 300 K. The $\ln\rho = f(T^{-1})$ dependence of 350 °C layer shows a linear region at higher measurement temperatures with the activation energy of 0.68 eV. The value of 0.65 eV, attributed to As$_{Ga}$-related deep donors, has been found before [7]. A further increase of ρ with T_g is observed, as demonstrated on 450 °C layer which shows semi-insulating properties with $\rho_{300} \sim 10^7$ Ω cm. An increase of resistivity of about six orders of magnitude exists between 200 and 450 °C. The activation energy of 0.75 eV is evaluated for the 450 °C sample, which is a typical value for bulk SI GaAs [10]. However, the 550 °C layer is quite conductive, $\rho_{300} \sim 260$ Ω cm. From Hall data p-type conductivity was evaluated, which is expected for conventional growth ($T_g \simeq$ 600 °C) of intentionally undoped MBE GaAs [1].

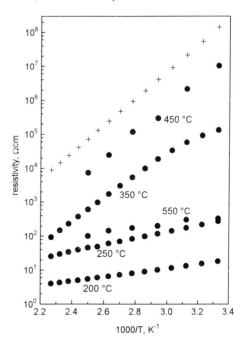

Fig. 1 Resistivity vs inverse measurement temperature for unannealed GaAs layers grown at various temperatures (full dots) and for SI GaAs substrate (crosses).

The influence of the growth temperature on the conduction properties of MBE GaAs is also demonstrated in Fig. 2. The room-temperature resistivity as a function of T_g for layers before and after annealing (590 °C, 10 min) are shown here. Three different regions of ρ_{300} vs T_g dependence can be distinguished. The first, in which a strong increase of as-grown resistivity from about 10 to >10^5 Ω cm with T_g increased from 200 to ~350 °C exists. Layers grown in this temperature range are thermally unstable, the resistivity of all layers increases up to 10^5-10^6 Ωcm after annealing (open dots). This is a typical behaviour of LTG materials. The second region at T_g ~400-450 °C can be characterized by semi-insulating properties of as-grown as well as annealed layers, despite the fact that it is difficult to measure such high resistivities on our samples due to the small difference between sheet resistivities of the layer and substrate. However, from the possible error estimation at performed measurements we have evaluated the lower resistivity limit of 5×10^6 Ω cm for these samples and resitivity of about 2×10^7 Ω cm for

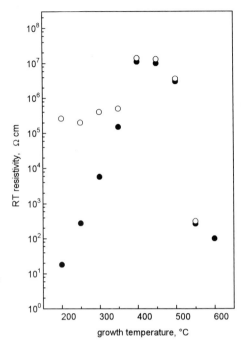

Fig. 2 Room-temperature resistivity vs growth temperature (thermocouple reading only) for as-grown (o) and annealed (o) GaAs layers.

400 °C growth has already been reported [3]. Layers grown at T_g ~450-600 °C show p-type conductivity, are thermally stable and the resistivity decreases extremely, from ~10^7 to ~10^2 Ω cm, to the values expected for ~600 °C growth (assuming p ~10^{14} cm^{-3}, μ_p ~400 cm^2/Vs). It is known that layers grown at 200 °C exhibit very high concentrations of native defects, ~ 10^{19}-10^{20} cm^{-3}, and with increased T_g the defects concentrations decrease to ~10^{14} cm^{-3} at T_g ~600. At the intermediate temperatures, T_g ~400-450 °C, the compensation effect between deep donors and shallow acceptors might be responsible for semi-insulating properties of GaAs. The As-precipitates should not exist in as-grown layers, and we could not find them in as-grown layers either, therefore it is difficult to explain the observed high resistivity of 400-450 °C grown, unannealed GaAs layers by the Schottky barrier model proposed by Warren et al. [5]. According to our results, Look's deep-level model [6] could be used to describe the properties of GaAs layers grown at lower temperatures. On the other hand, some questions are still open and need a more detailed study, for example the different annealing influence on GaAs layers grown below and above 350 °C.

It is known that annealed LTG GaAs, which is highly resistive, can be used as a fast photoconductive switch. We prepared MSM photodetectors on some GaAs layers and as an example two switch responses are shown in Fig. 3. One is for 200 °C grown and 590 °C annealed layer with the resistivity $\rho_{300} = 2.6 \times 10^5$ Ω cm (lower trace, left voltage axis) and the second for 400 °C grown unannealed layer with $\rho_{300} \simeq 10^7$ Ω cm (upper trace, right y-axis). The 400 °C layer exhibit at dc bias of 5 V about 5-times higher sensitivity than the 200 °C layer at 10 V bias. This indicates that in 400 °C layer the carrier mobility should be higher than in 200 °C layer. Recently, Nabet et al [11] have reported on improved dc properties of photodetectors on GaAs layers grown at 350-500 °C and they suppose a higher speed on layers with higher mobility than on typical LTG

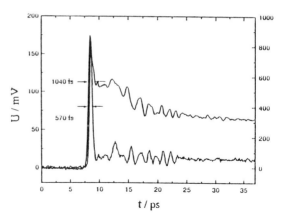

Fig. 3 Photoconductor switch responses of GaAs layers grown at 200 °C (lower trace) and 400 °C (upper trace).

GaAs grown at 200 °C. From our measurements, presented in Fig. 3, it follows that 200 and 400 °C layers show the same rise times but the 400 °C layer exhibit much longer fall time than 200 °C layer. Some slowly recombinating centers must dominate in 400 °C layer. In addition, we have observed different response behaviour on as-grown and annealed layers, both grown at 400 °C, although the layer resistivity of $\sim 10^7$ Ω cm has not changed during annealing.

3.2. Effect of the BEP ratio, annealing time and layer thickness

The BEP of ~ 20 is usually used at the growth of LTG GaAs and its effect on the layer properties is not known. We have grown layers at three different BEP ratios (19, 13 and 7) and measured resistivities on layers grown at 200 and 350 °C are shown in Tab. 1. On the 200 °C layers very slight decrease of the resistivity with decreased BEP ratio is observed for both, as-grown and annealed layers. An opposite effect is observed on 350 °C layers, the resistivity increases with decreased BEP ratio. Practically no influence of annealing on the 350 °C layer resistivity is observed if BEP ratio is 7. However, layers prepared using BEP ratio of 19 are thermally stable only if $T_g \geq 400$ °C (see Fig.2).

Tab. 1 Influence of BEP ratio on the room-temperature resistivity of GaAs layers grown at 200 and 350 °C before and after annealing (590 °C, 10 min).

BEP ratio	$T_g = 200$ °C		$T_g = 350$ °C	
	as-grown	annealed	as-grown	annealed
	resistivity, ρ_{300} [Ω cm]			
19	18	2.6×10^5	1.4×10^5	5.3×10^5
13	17.7	2.4×10^5	1.6×10^5	9×10^5
7	17.5	1.2×10^5	1.8×10^6	2×10^6

Tab. 2 Influence of annealing time on the resistivity of GaAs layers, BEP=19, T_g=200°C, T_a=590°C.

t_a [min]	ρ_{300} [Ω cm]
10	2.6 x 10^5
10 + 20	6.2 x 10^5
30	7.6 x 10^5
30 + 30	1.6 x 10^6

It is common to anneal the LTG layers for the period of only 10 min. An increase of the resistivity with annealing time is demonstrated in Tab. 2. Similar values are obtained if the layer is annealed 30 min in one run or twice (10+20 min). Layers prepared at higher T_g show smaller increase of the resistivity with increased annealing time t_a.

The structural quality of the layers was studied by cross-sectional TEM micrographs. On the 200 °C layers a breakdown of the crystallinity was observed at the thickness of $\sim$1 μm, but even 6 μm thick layers with perfect crystallinity can be grown at 250 °C.

4. Conclusions

The effect of the growth temperature varying between 200 and 600 °C on the properties of MBE GaAs layers has been studied. Thermally stable semi-insulating layers can be grown at $T_g \sim$ 400-450 °C. The influence of the BEP ratio and annealing time on the layer properties is shown. Different photoresponse behaviour of high resistive 200 °C (annealed) and 400 °C (as-grown) layers is demonstrated.

Acknowledgements

The authors wish to thank D. Guggi for x-ray diffraction measurements, D. Gregušová and J. Darmo for assistance with sample annealing and H. Lüth for encouragement. This work was supported, in part, by the German Federal Ministry of Research and Technology (BMFT) and the Slovak Grant Agency for Science.

References

[1] Cho A Y 1983 *Thin Solid Films* **100** 291-317
[2] Murotani T, Shimanoe T and Mitsui S 1978 *J. Cryst. Growth* **45** 302-8
[3] Look D C, Fang Z-Q, Sizelove J R and Stutz C E 1993 *Phys. Rev. Lett.* **70** 465-8
[4] Witt G L 1993 *Mater. Sci. Eng. B* **22** 9-15
[5] Warren A C, Woodall J M, Freeouf J L, Grischkowsky D, McInturff D T, Melloch M R and Otsuka N 1990 *Appl. Phys. Lett.* **57** 1331-3
[6] Look D C, Walters D C, Robinson G D, Sizelove J R, Mier M G and Stutz C E 1993 *J. Appl. Phys.* **74** 306-10
[7] Look D C, Robinson G D, Sizelove J R and Stutz C E 1993 *J. Electron. Mater.* **22** 1425-8
[8] Betko J and Kuklovský S 1994 *Acta Phys. Hung.* **74** 121-7
[9] Betko J, Kordoš P, Kuklovský S, Förster A, Gregušová D and Lüth H 1994 *Mater. Sci. Eng. B,* **23** No. 12 (in press)
[10] Betko J and Merinský K 1986 *Phys. Stat. Solidi (a)* **93** K205-8
[11] Nabet B, Paolella A, Cooke P, Lemuene M L, Moerkirk R P and Liou L-Ch 1994 *Appl. Phys. Lett.* **64** 3151-3

Electrical Characterisation of GaAs Layers Grown by Molecular Beam Epitaxy at Low Temperature

J.K.Luo, H.Thomas, D.V.Morgan, D.Westwood*, R.H.Williams* and D.Theron**

School of Electrical, Electronics and System Engineering, *Department of Physics and Astronomy, Univ. of Wales College of Cardiff, UK.
**IEMN, Univ. of Lille, France

Abstract: The effects of growth and annealing temperatures on the electrical properties of the low temperature (LT-) grown GaAs have been investigated. It was found that the resistivity of the as-grown LT-GaAs layer increased, while the breakdown voltage decreased as the growth temperature was increased, approaching the value of the bulk GaAs. Thermal annealing causes an exponential increases of the resistivity with increasing annealing temperature T_A, giving an activation energy of $E_A \sim 2.1 eV$. The transport in the LT-GaAs layers was found to be dominated by hopping conduction, and additionally by a thermally-activated process at high temperatures. The breakdown voltage V_{BD}, for as-grown LT-GaAs increased with decreasing measurement temperature, but decreased with decreasing temperature following annealing at $T_A > 500°C$. The hopping conduction is held responsible for the observed electrical breakdown properties.

1. Introduction

Gallium Arsenide grown by molecular beam epitaxy (MBE) at low-temperature (LT-) has recently attracted much attention for its unique electrical and optical properties and its great potential in device applications[1-4]. Unlike conventional MBE GaAs, the LT-GaAs grown at 200~300°C is non-stoichiometric owing to the incorporation of excess arsenic, which has a strong dependence on the growth temperature, Tg. The excess arsenic induces a high density of anti-site defects As_{Ga}, of the order $10^{18} \sim 10^{19} cm^{-3}$[5], leading to a defect mini-band and the dominance of hopping conduction[6]. Post-growth annealing at $T_A \sim 600°C$ has been found to cause the precipitation of the excess arsenic[7,8], with the formation of buried Schottky barriers with overlapping depletion regions which results in very high resistivity[5,8]. It has been established that the resistivity of the LT-GaAs is determined by the hopping conduction between As-clusters[9]. In this paper, we report the effect of growth temperature and subsequent annealing temperature on the electrical breakdown properties of the LT-GaAs layers, and the corresponding transport properties.

2. Experimental procedure and sample preparation

The LT-GaAs layers were grown to a thickness of 1μm on SI-GaAs substrates using a VG V80H MBE system. After the removal of the oxide at 630°C, the substrates were cooled to growth temperatures of 200, 250 and 300°C (Samples LT1, LT2 and LT3) respectively, as monitored by a thermocouple. The growth rate was 1μm/hr using As_4 fluxes with a pressure of 10^{-5} mbar.

Thermal annealing was carried out in the temperature range of 350~650°C for intervals of 10minutes in a nitrogen ambient. The LT-GaAs layers were mounted face to face on SI-GaAs substrates to minimise the out-diffusion of arsenic from the surface of the LT-GaAs layer. Au/Ge/Ni was evaporated directly on the LT-GaAs. All samples used here were alloyed at 430°C to form Ohmic contacts. Although the resistivity was found to be changed by more than one order at this alloying process, there was no change of the breakdown properties[9].

The coplanar test structure is shown in Fig.1 inset. Samples were mounted in a Joule-Thomson Refrigerator, and current-voltage characteristics were measured by Keithley source-measure units.

3. Results and discussion

3.1 Electrical properties of the as-grown LT-GaAs

Figure 1 shows log-log I-V plots for Samples LT1, LT2 and LT3. The figure shows essentially Ohmic behaviour for all samples. The resistivity increases with increasing growth temperature up to 300°C, the upper limit of LT-GaAs growth temperature, thereafter the current increased rapidly as the voltage was increased. The plot for a SI-GaAs sample is also shown for comparison, and displays a much higher resistivity. Parallel conduction through SI-GaAs can be ruled out, since the resistivity of the SI-GaAs is more than two orders of the magnitude greater than that of the LT-GaAs. Furthermore, a significant difference in the breakdown characteristics for the SI- and LT-GaAs is seen.

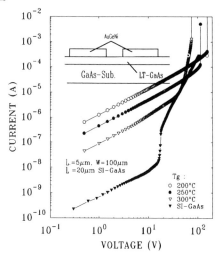

Fig.1 I-V plots for Samples grown at 200, 250, 300°C and SI-Substrate

SI-GaAs I-V properties show an Ohmic behaviour, followed by evidence of trap-filling and space charge limited conduction[10]. No catastrophic breakdown was observed up to 110V for an electrode separation of L=20μm. The electrical field E_T (=V_T/L) corresponding to the sudden increase of current was typically <10kV/cm, in agreement with the results obtained by other authors[11,12]. The low value of E_T for SI-GaAs has previously been explained by surface breakdown[11,12] However, the LT-GaAs layers initially showed Ohmic behaviour followed by a catastrophic breakdown. The electrical breakdown field E_{BD} was found to increase from 100kV/cm to 320kV/cm as the growth temperature was decreased from 300°C to 200°C, approaching the value of the bulk GaAs and more than one order of magnitude higher than that of the SI-GaAs layer. No obvious trap filling process was observed though a high density of defects (>10^{18}cm^{-3}) was known to exist in the LT-GaAs[5]. The high breakdown voltage demonstrates the possibility for novel passivation of the devices using as-grown LT-GaAs.

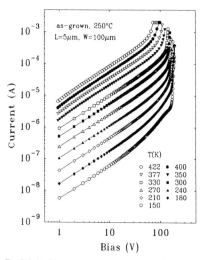

Fig.2 I-V-T characteristics for Sample LT2. The breakdown voltage increases on cooling.

Figure 2 shows the I-V-T characteristics of Sample LT2 grown at 250°C. The response before breakdown is linear at all temperatures and in addition, the breakdown voltage is seen to increase from 55V to 180V as the temperature was decreased from 422K to 150K, contrary to the behaviour of conventional avalanche breakdown in semiconductor junctions which decreased with decreasing measurement temperature[13]. Similar results were obtained from Samples LT1 and LT3. Figure 3 shows the dependence of the normalised breakdown voltage ($V_{BD}(T)/V_{BD}(300K)$) on the measurement temperature for Samples LT2 and LT3, demonstrating that the breakdown voltage increases with decreasing temperature.

3.2 Annealing effect on the electrical breakdown properties

Thermal annealing was found to cause a dramatic change in the electrical properties of the LT-GaAs layers. Figure 4 shows I-V characteristics of the LT-GaAs grown at 250°C with annealing temperature as a variable. Low bias Ohmic behaviour remains unchanged, and the high breakdown electric field is largely preserved after annealing. However, the resistivity increased by more than three orders of magnitude with increasing annealing temperature. After annealing at high temperatures T_A>500°C, samples exhibited non-linear characteristics in the medium bias region as marked by BC and the breakdown characteristics became soft, significantly different from the as-grown materials. The resistivity of the LT-GaAs was found to increase exponentially with increasing annealing temperature, in the form of $\rho \propto \exp(-2.1/kT_A)$ with an activation energy E_A=2.1eV[9].

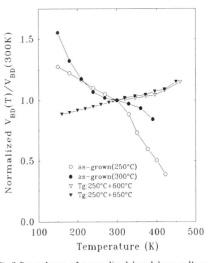

Fig.3 Dependence of normalized breakdown voltage on temperature for as-grown and annealed samples.

It has been shown that the increase of the resistivity of the LT-GaAs is owing to the precipitation of the excess arsenic as As-clusters and the formation of a depletion region surrounding the "metallic" arsenic clusters[7,8]. Therefore, the activation energy obtained is related to that of arsenic precipitation during annealing.

Although annealing does not significantly affect the value of high field breakdown, it changes the breakdown characteristic significantly. Figure 5 shows I-V-T plots for Sample LT2 which has been annealed at 600°C. High temperature annealing causes three changes: firstly, at low temperatures, I-V plots were linear with a gradient of n~1, demonstrating the dominance of Ohmic conduction. The gradient decreases as annealing temperature was increased, implying that the transport is no longer Ohmic at high temperatures. Secondly, the breakdown was abrupt at

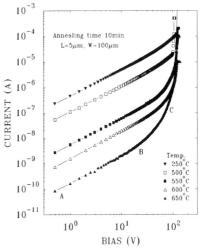

Fig.5 I-V properties of Sample LT2 after annealing at various temperature.

low temperatures, and became soft at higher temperatures. Thirdly, the breakdown voltage was seen to decrease with decreasing measurement temperature, an opposite change to that of as-grown LT-GaAs materials. The dependence of breakdown voltage on measurement temperature has been investigated for all annealed samples, it was found that the dependence of V_{BD} on measurement temperature for samples annealed at T_A<500°C, shows a behaviour similar to that of as-grown samples. After annealing at T_A>500°C, the breakdown voltage V_{BD} decreases with decreasing measurement temperature. The dependence of the normalised breakdown voltage on temperature is also shown in Fig.3 for comparison.

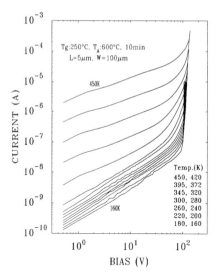

Fig.5 I-V-T characteristics for sample LT2 after annealing at 600°C. V_{BD} decreases on cooling.

3.3. Transport mechanism of the LT-GaAs

The current at a fixed bias has been studied at various measurement temperatures. The Arrhenius plots for Sample LT1 shows a straight line at all temperatures with an activation energy of 0.07eV. For Samples LT2 and LT3, the Arrhenius plots show two regions. In the high temperature region it gives activation energies of 0.17eV to 0.36eV respectively, but shows a similar gradient to that of Sample LT1 at low temperature. The plots of the resistivity against $T^{-1/4}$ revealed straight lines, indicating the dominance of the hopping conduction at low temperatures for sample LT1 at all temperatures and for Samples LT2 and LT3 at low temperatures.

Figure 6 shows the plot of resistivity against $T^{-1/4}$ for Sample LT2 with annealing temperature as a variable. The plot is divided

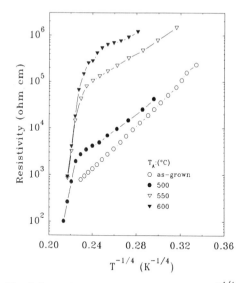

Fig. 6 Dependence of the resistivity on $T^{-1/4}$ for Sample LT2 with T_A as a variable.

into two regions. The straight lines in the low temperature region with low gradients demonstrate the dominance of hopping conduction. In the high temperature region evidence of a thermally-activated process is seen. An Arrhenius plot of the current at a fixed bias at high temperature region reveals a common activation energy in the range of 0.60~0.65eV for samples annealed at T_A>500°C, a value similar to the barrier height obtained by photoemission measurement from GaAs:As photodiodes[14]. From Figs.4 and 5, it is seen that the thermally-activated process is accompanied by the appearance of non-ohmic behaviour and soft breakdown.

4. Arsenic cluster model

The above observed properties of the LT-GaAs can be explained by a unified As-cluster model, provided that the As_{Ga} anti-site defects in as-grown LT-GaAs is an extreme case of the As-clusters. Figure 7 graphically shows the energy band diagrams for as-grown and annealed LT-GaAs. In the case of as-grown LT-GaAs, the high density of As-clusters (anti-site As_{Ga} defects) cause the overlapping of defect wave functions, providing an effective conductive path. Thus hopping conduction dominates the transport. Thermal annealing causes the precipitation of the excess arsenic as As-clusters, forming As/GaAs buried Schottky barriers with overlapping depletion regions. As the precipitation takes place, the cluster size increases with increasing annealing temperature, and it is accompanied by a decrease in density and an increase in the

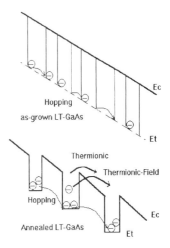

Fig.7 Energy band diagram for as-grown and annealed LT-GaAs materials.

distance between the clusters[7,8]. Subsequently, the hopping conduction decreases and the resistivity of the LT-GaAs increases. The resistivity of the LT-GaAs layer is therefore determined by the hopping conduction between As-clusters. Based on this hopping consideration, we have calculated the As-cluster density from the resistivity obtained[9]. The relationship between cluster density and the annealing temperature was found in the form of $N_T = N_{TO} \exp(-T_A/T_O)$, with $N_{TO} = 3 \times 10^{19} cm^{-3}$ at 200°C, in agreement with that obtained from Infrared absorption[5]. This As-cluster model can also explain the increase of the resistivity of the LT-GaAs with increasing Tg as shown in Fig.1, since the density of the incorporated excess arsenic decreases as the growth temperature was increased.

A high density of defects and hopping conduction is believed to be responsible for the high breakdown voltage of the LT-GaAs layer and the temperature dependence of the breakdown behaviour. Firstly, a higher hopping conductivity minimises the accumulation of injected charge and Ohmic behaviour leads to a linear distribution of the electrical field, preventing the premature breakdown at the contact edge and surface, thus a high breakdown voltage is obtained. Secondly, in the normal case, the scattering probability of free carriers decreases with decreasing temperature, and a lower field is required to cause impact ionisation, so that the breakdown voltage decreases on cooling[13]. In the case of as-grown LT-GaAs, all carriers are effectively captured by defects. No free charge can contribute to the impact ionisation. The hopping conductivity is so high that most injected charge is transported through hopping. Little trapped charge can be released from the defects to the conduction band by thermionic-field emission (or thermionic-emission), especially at high field, and contributes to the impact ionisation. As the temperature decreases, this thermally-activated process is reduced, thus a higher field is necessary and the breakdown voltage is seen to increase on cooling. On the other hand, as the growth temperature is increased, the hopping conduction decreases due to the reduction of excess arsenic. The conduction contribution from thermionic-field emission increases, and the breakdown voltage decreases.

After annealing, the excess of arsenic precipitates as As-clusters and the distance between clusters increases. The hopping probability is reduced, and the conduction attributed to thermionic-field emission increases. More charge is emitted from the defects to the conduction band and contributes to the conduction. Thus the I-V characteristic is no longer Ohmic in behaviour and the contribution of the thermionic-field emission leads to a softer breakdown. In this case, the scattering caused by free charge emitted from the defects is similar to that of conventional impact ionisation, thus the breakdown voltage decreases with decreasing the temperature.

The activation energy corresponding to the high temperature region of Fig.6 was found to be 0.6~0.65eV, a value similar to the barrier height obtained from GaAs:As photodiodes[14]. This provides further evidence for the As-cluster precipitation model.

5. Summary

The effects of thermal annealing on the electrical properties of LT-GaAs grown in the temperature range, Tg of 200~300°C has been investigated. The study has revealed the following results:

1). The resistivity of the LT-GaAs layer decreases as Tg was decreased in the range, Tg:200-300°C, but increases exponentially with annealing temperature, T_A, giving an activation energy of 2.1eV.
2). The corresponding breakdown voltage, V_{BD} decreased with increasing Tg and also decreased as the annealing temperature was increased.
3). The breakdown voltage for as-grown LT-GaAs was found to increase as the temperature was decreased, while that of the annealed samples decreased on cooling.
4). The transport of the LT-GaAs layer is dominated by hopping conduction, and decreased with increasing T_A. Hopping conduction is held responsible for the change of resistivity, the high V_{BD}, and the increase of V_{BD} on cooling for as-grown LT-GaAs.

References:
[1] A.C.Warren, N.Katzenellenbogen, D.Grischkowsky, J.M.Woodall, M.R.Melloch and N. Otsuka; Appl. Phys. Lett.; **58**, 151 (1991).
[2] F.W.Smith, A.R.Calawa, C.L.Chen, M.J.Manfra and L.J.Mahoney; IEEE Electr. Device Lett. **9**, 77, (1988).
[3] C.L.Cheng, F.Smith, B.J.Clifton, J.M.Michael and A.Calawa; IEEE Electr. Device Lett. **12**, 306 (1991).
[4] L.W.Yin, Y.Hwang, J.H.Lee, R.M.Kolbas, R.J.Trew and U.K.Mishra; Electr. Device Lett. **11**, 561 (1990).
[5] M.O.Manasreh, D.C.Look, K.R.Evans and C.E.Stutz; Phys. Rev. **B41**, 10272 (1990).
[6] H.Yamamoto, Z.Q.Fang and D.C.Look; Appl. Phys. lett. **57**, 1537 (1990).
[7] A.C.Warren, J.M.Woodall, J.L.Freeouf, D.Grischkowsky, D.T.Mulnturff, M.R.Melloch and N.Otsuka; Appl. Phys. Lett. 57, 1331 (1990).
[8] M.R.Melloch, D.D.Nolte, N.Otsuka, C.L.Chang and J.M.Woodall; J.Vac. Sci. Technol. **B11**, 795 (1993).
[9] J.K.Luo, H.Thomas, D.V.Morgan and D.Westwood; Appl. Phys. Lett. **64**, 3614 (1994).
[10] P.Mark and A Lampert, 'Current injection in Solids', Academic Press, NY, 1970
[11] T.Kitagawa, H.Hasegawa and H.Ohno; Electr. Lett. **28**, 299 (1985).
[12] S.R.Blight and H.Thomas; J. Appl. Phys. **65**, 215 (1988).
[13] S.M.Sze, 'Physics of Semiconductor Devices' John Wiley & Son, New York.
[14] D.T.McInturff, J.M.Woodall, A.C.Warren, N.Braslau, G.D.Pettit and P.D.Kirchner; Appl. Phys. Lett. **60**, 448 (1992).

Acknowledgement: Investigation supported by the ESPRIT project TAMPETS 6849 of the European Communities, which is in cooporation with the University of Lille, France and the University of Ulm, Germany.

Influence of CH_4/H_2 reactive ion etching on the electrical and optical properties of AlGaAs/GaAs and pseudomorphic AlGaAs/InGaAs/GaAs heterostructures.

C.M. van Es, T.J. Eijkemans, and J.H. Wolter

Cobra, Interuniversity Research Institute
Eindhoven University of Technology, Department of Physics, P.O. Box 513, 5600 MB, Eindhoven, The Netherlands

R. Pereira, M. Van Hove, and M. Van Rossum

Interuniversity Micro-Electronics Center (IMEC), Kapeldreef 75, B-3001, Leuven, Belgium.

Abstract. We investigate the effect of methane/hydrogen (CH_4/H_2) reactive ion etching (RIE) and a subsequent annealing process on AlGaAs/GaAs and pseudomorphic AlGaAs/InGaAs/GaAs heterostructures. We use low temperature Hall, Shubnikov-de Haas, and photoluminescence measurements. We observe that the electron density and mobility of the two-dimensional electron gas in the heterostructure is strongly reduced by the RIE process. After annealing the electron density fully recovers for both types of structures, whereas the electron mobility responds differently. While for the pseudomorphic AlGaAs/InGaAs/GaAs heterostructures thermal annealing restores the electron mobility completely, for the AlGaAs/GaAs heterostructures the electron mobility recovers only to 60% of the original value. This indicates that in the AlGaAs/GaAs heterostructures the structural damage induced by reactive ion etching is not fully removed by thermal annealing. This is confirmed by photoluminescence measurements at low temperatures.

1. Introduction

It is well known that CH_4/H_2 reactive ion etching (RIE) of bulk GaAs and AlGaAs, doped with silicon, leads to a deactivation of the silicon donors. This donor deactivation is associated with the formation of an As-Si-H complex. At present, there is a basic understanding of the processes leading to a deactivation of the shallow and the deep donors. Also the recovery by annealing processes and the relation with the donor density and the Al percentage for a variety of doping materials is basically understood [1-8]. The damage of the surface of bulk material has been studied thoroughly by many investigators. For example, Collot et al. [6,7] investigated the damage by carrying out current-voltage and capacitance-voltage measurements. There are also some papers reporting on the photoluminescence (PL) characteristics after passivation and annealing of the samples [2, 9-12].

RIE is successfully applied in the fabrication of two-dimensional transistor structures. For example, in the fabrication of modulation doped field-effect transistors RIE is applied for gate recessing. Accurate control of the etch depth and its uniformity across the wafer is needed. CH_4/H_2 RIE has proven to be superior to the commonly applied chlorinated gases. Low and controllable etch rates, good uniformity, and smooth surfaces have been obtained by this process [13-15].

For pseudomorphic modulation doped field effect transistors, Pereira et al. [13,16,17] reported that the donors, which had been partly deactivated by the RIE process could be reactivated by an annealing procedure. Also the influence of RIE and subsequent thermal annealing on the PL spectra was studied. These experiments strongly suggest that the material can almost fully recover from the RIE process, if a suitable annealing procedure is applied subsequently.

In this paper we study in detail the influence of RIE and subsequent thermal annealing on the transport and/or optical properties of modulation doped AlGaAs/GaAs and pseudomorphic AlGaAs/InGaAs/GaAs heterostructures. The electrical properties were investigated by Hall effect and Shubnikov-de Haas (SdH) measurements. The PL and SdH measurements were carried out at liquid helium temperature, whereas the Hall measurements were performed in the temperature range of 4.2 K - 300 K. We find that the electron density of the two-dimensional electron gas (2DEG) in the heterostructure is strongly reduced by the RIE process. Also the electron mobility decreases strongly. After annealing the electron density fully recovers for both types of structures. On the other hand with respect to the electron mobility, both types of structures respond differently. While for pseudomorphic AlGaAs/InGaAs/GaAs heterostructures thermal annealing restores the electron mobility completely, for AlGaAs/GaAs heterostructures the electron mobility recovers only to 60% of the original value. This indicates that in the AlGaAs/GaAs heterostructures the structural damage induced by RIE is not fully removed by thermal annealing.

2. Experimental details

Pseudomorphic AlGaAs/InGaAs/GaAs modulation doped heterostructures were grown by molecular beam epitaxy on a semi-insulating GaAs substrate. Starting from the substrate, a GaAs buffer layer, a 13 nm undoped $In_{0.15}Ga_{0.85}As$ channel, a 5 nm undoped $Al_{0.25}Ga_{0.75}As$ spacer layer, a $5*10^{12}$ cm^{-2} Si delta-doped layer, a 30 nm $Al_{0.25}Ga_{0.75}As$:Si ($5*10^{17}$ cm^{-3}) layer, and a 40 nm highly doped GaAs:Si contact layer were grown. Also the AlGaAs/GaAs heterostructures were grown on a semi-insulating GaAs substrate. The layer structure consists of a 4 µm GaAs buffer layer, a 30 nm undoped $Al_{0.33}Ga_{0.67}As$ spacer layer, a 38 nm $Al_{0.33}Ga_{0.67}As$:Si ($1.3*10^{18}$ cm^{-3}) layer, and a 17 nm undoped GaAs cap layer. The same layer structure was grown with a $5*10^{12}$ cm^{-2} Si delta-doped layer incorporated at the interface of the spacer layer and the uniformly doped Si layer.

Hall bars were defined by photolithography. After mesa isolation by wet etching in H_3PO_4:H_2O_2:H_2O (3:1:50), standard AuGeNi ohmic contacts were formed by thermal evaporation and subsequent alloying. PL measurements were performed on unpatterned samples. RIE with CH_4/H_2 was carried out in a commercial parallel plate Plasma Technology PLASMALAB µP system, operating at a radio frequency of 13.56 MHz. The following conditions were used for etching: 15 % CH_4 in H_2, a rf power density of 0.33 W/cm^2, a cathode self-bias of -300 V, a pressure of 35 mTorr, and a temperature of 25 ^{0}C. With these conditions an etch rate of 10 nm/min for GaAs was obtained. The AlGaAs/InGaAs/GaAs structures were exposed to the plasma for 4 min. For the AlGaAs/GaAs structures the exposure time was 1 min. In case of the AlGaAs/InGaAs/GaAs structures the reference sample was wet etched for 1 min in a H_2SO_4:H_2O_2:H_2O (1:8:1000) solution with a calibrated etch rate of 40 nm/min. By removing the highly doped cap layer it was ensured that the etch was deep enough to avoid any second conducting path. From earlier transport measurements on

AlGaAs/GaAs heterostructures we know that wet etching of the undoped cap layer has no effect on the electron density and electron mobility of the 2DEG. For the AlGaAs/GaAs heterostructures, the fact that the thickness of the cap layers of the reference sample and the passivated and annealed samples are different, has no influence on the recovery of the electron density and electron mobility, because RIE only affects the silicon donors in the doped AlGaAs layer. Thermal annealing was carried out for 30 sec between 250 and 550 °C using a halogen rapid thermal annealing system in a forming gas (90 % N_2, 10 % H_2) ambient. This treatment was applied to the samples used for the electrical and the PL measurements.

3. Results and Discussion.

3.1 AlGaAs/InGaAs/GaAs samples

In order to study how the transport properties of the AlGaAs/InGaAs/GaAs heterostructures are affected by plasma etching of the highly doped cap layer, followed by an annealing step, we performed SdH-measurements at liquid helium temperature and Hall-measurements in the temperature range 4.2 K - 300 K. SdH measurements at 4.2 K of the wet etched reference sample (a), as etched sample (b), and the samples annealed at a temperature of 350 °C (c), and 450 °C , with an electron density of 1.70, 0.35, 1.04, and $1.57*10^{16}$ m^{-2} respectively, are shown in Fig. 1. From the SdH data it is clear that for all samples only one subband is occupied, even for the reference sample with the highest sheet carrier density and mobility, because there is no indication of a second oscillation period. This is in agreement with the results of van der Burgt et al [18].

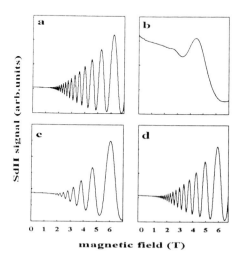 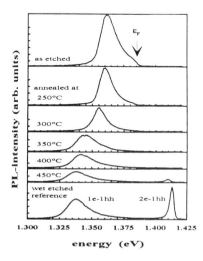

Fig.1 SdH signal of sample H2735 at 4.2 K.

Fig.2 PL spectra of sample H2735. The arrow indicates the energy of the Fermi-level.

In contrast to the PL data (Fig.2), where we clearly observe two subbands in accordance with the self-consistent calculations we have done, the occupation of the second subband obviously is too small to give a second oscillation in the SdH-signal. The measured sheet carrier density and the Hall mobility for the samples as a function of temperature at different annealing temperatures are presented in Fig. 3. From the data it is clear that after the RIE process the sheet carrier density and the mobility of the 2DEG is decreased. With increasing annealing temperature, the electron density and mobility increase again. Within the experimental accuracy at an annealing temperature of 400 °C, there is a total recovery of the photoluminescence energy, the electron density, and the Hall mobility.

The low temperature PL spectra for the reference sample, the passivated sample, and the samples annealed at different temperatures are presented in Fig. 2. For the wet-etched reference sample two peaks are clearly observed at energies 1.338 and 1.413 eV. The lower energy peak is interpreted as emission from the $n=1$ electron subband to the $n=1$ heavy hole subband (1e-1hh), while the higher energy peak is ascribed to the transition from the $n=2$ electron subband to the $n=1$ heavy hole subband (2e-1hh). Note, that in symmetric quantum wells this last transition is forbidden!
After plasma exposure the 2e-1hh peak disappears and only the transition from the $n=1$ electron subband to the $n=1$ heavy hole subband is present. The shift of the luminescence peak to higher energies is due to a shift of the subband structure induced by electric fields. The high energy shoulder is associated with transitions from the Fermi level, in agreement with Colvard et al. [19].

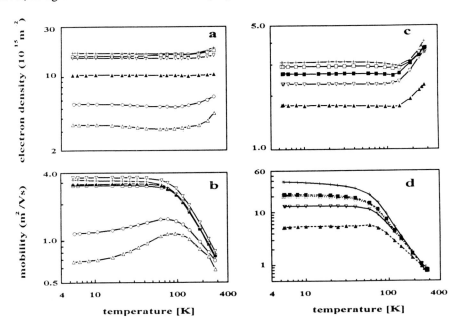

Fig.3 The electron density and the Hall-mobility for the $Al_{0.25}Ga_{0.75}As/In_{0.15}Ga_{0.85}As/GaAs$ sample H2735 (a,b) and the AlGaAs/GaAs sample W24 (c,d) as a function of temperature for the reference sample (+), as etched sample (△), and samples annealed at temperatures of 250 °C (o), 350 °C (▲), 400 °C (▽), 450 °C (□) and 500 °C (■), respectively.

We explain the increase of the PL intensity for as etched samples as follows. Hydrogenation of GaAs and AlGaAs leads to passivation of nonradiative centers as well as shallow and deep levels. According to Mostefaoui et al. [5], the neutralization efficiency of shallow donors and DX centers by hydrogenation in AlGaAs depends on the aluminum fraction. For $x = 0.25$ the deep donors are more effectively passivated than the shallow donors. As a result of the passivation of the nonradiative centers in AlGaAs a strong enhancement of the luminescence occurs, as explained by Pavesi et al. [9]. They concluded that the carrier lifetime and diffusion length is increased. Thus, in case of the multilayer structure, a large number of photoexcited carriers recombine in the two-dimensional channel.

On the other hand the diffusion of hydrogen induced by the dry etching process affects the electrical properties [6] and promotes the generation of new efficient recombination centers [20]. The competition between these two effects is responsible for the observed luminescence intensity of the different samples. The increase of the annealing temperature (Fig. 2) shifts the luminescence peak to its original energy. The spectra in Fig. 2 show that there is a total recovery in the energy of the transitions between 350 and 400 ^{0}C.

3.2 AlGaAs/GaAs heterostructures

For the AlGaAs/GaAs structures it turns out that plasma etching causes a greater damage on these heterostructures than for the AlGaAs/InGaAs/GaAs heterostructures. From Hall measurements in the temperature range 4.2 - 300 K, presented in Fig. 3 for the sample W24, we conclude that the sheet carrier density is totally restored, but that the Hall mobility remains much lower for the annealed samples in comparison with the reference wet-etched sample. For the low temperatures measurements at 4 K there are a lot of problems to make good ohmic contact, because for the passivated and for the samples annealed at low temperatures the electron density in the 2DEG is very small. From the Hall measurements we can conclude that passivation and annealing of the AlGaAs/GaAs samples induces much more destructive damage to the samples than in the case for the pseudomorphic AlGaAs/InGaAs/GaAs heterostructures.

The difference in the recovery of the mobility between the AlGaAs/InGaAs/GaAs and the AlGaAs/GaAs heterostructures can tentatively be explained as follows. The heavily δ-doped layer in the AlGaAs/InGaAs/GaAs structures shields the influence of the plasma treatment. Because of the lower mobility of the AlGaAs/InGaAs/GaAs, due to alloy cluster scattering, the electron scattering induced by structural damage has less influence on the overall mobility for pseudomorphic structures than for the AlGaAs/GaAs heterostructures.

As a first attempt we performed measurements of the sheet carrier density and the Hall mobility of the AlGaAs/GaAs heterostructure with a Si delta-doped layer incorporated in the heterostructure. As preliminary result there is a pronounced enhancement of the recovery of the Hall mobility in the 2DEG from 60 to 75 %. It seems that such a layer prevents further introduction of damage into the semiconductor material. This conclusion is in good agreement with the results of Agarwala et al. [21], using SiCl$_4$/SiF$_4$ plasma on delta-doped GaAs/AlGaAs MODFET structures.

However one should be careful about incorporating a Si delta-doped layer in the heterostructure as one is then possibly introducing a second conducting channel in the heterostructure.

4. Conclusions

RIE affects not only the electron density but also the electron mobility. Both are restored in the case of the AlGaAs/InGaAs/GaAs heterostructures, while in the AlGaAs/GaAs heterostructures the mobility does not fully recover, indicating that some structural damage induced by the RIE process is not removed by thermal annealing.
To explain the different effects in the recovery of the mobility between the AlGaAs/InGaAs/GaAs and the AlGaAs/GaAs heterostructures, we suggest that the heavily doped, δ-doped layer in the AlGaAs/InGaAs/GaAs structures shields the influence of the plasma treatment as follows from the measurements of the AlGaAs/GaAs heterostructure with a Si-delta doped layer incorporated. Moreover the scattering induced by structural damage has more influence on the overall mobility for the AlGaAs/GaAs heterostructures, because of the higher overall mobility of the AlGaAs/GaAs heterostructures.

References

[1] Chevallier J, Dautremont-Smith W C, Tu C W and Pearton S J 1985 *Appl. Phys. Lett.* **47** 108-10
[2] Dautremont-Smith W C, Nabity J C, Swaminathan V, Stavola Michael, Chevallier J, Tu C W and Pearton S J 1986 *Appl. Phys. Lett.* **49** 1098-1100
[3] Pearton S J, Dautremont-Smith W C, Chevallier J, Tu C W and Cummings K D 1986 *J. Appl. Phys.* **59** 2821-7
[4] Nabity J C, Stavola M, Lopata J, Dautremont-Smith W C, Tu C W and Pearton S J 1987 *Appl. Phys. Lett.* **50** 921-3
[5] Mostefaoui R, Chevallier J, Jalil A, Pesant J C, Tu C W and Kopf R 1988 *J.Appl. Phys.* **64** 207-10
[6] Collot P, Gaonach C and Proust N 1989 *Mat. Res. Soc. Symp. Proc.* Vol. **144** 507-12
[7] Collot P and Gaonach C 1990 *Semicond. Sci. Technol.* **5** 237-41
[8] De Wolf I, Van Hove M, Pereira R-G, Van Rossum M, Maes H E and Münder H 1992 *Mat. Res. Soc. Symp. Proc.* Vol. **240** 355-60
[9] Pavesi L, Martelli F, Martin D and Reinhart F K 1989 *Appl. Phys. Lett.* **54** 1522-4
[10] Pavesi L, Martin D and Reinhart F K 1989 *Appl. Phys. Lett.* **55** 475-7
[11] Chen Y-F, Tsai C-S and Chang Y 1990 *Appl. Phys. Lett.* **57** 70-2
[12] Bosacchi A, Franchi S, Vanzetti L, Allegri P, Grilli E, Guzzi M, Zamboni R and Pavesi L 1991 *Physica B* **170** 540-4
[13] Pereira R, Van Hove M, de Potter M, Van Rossum M 1990 *Electron. Lett.* **27** 462-4
[14] Werking J, Schramm J, Nguyen C, Hu E L and Kroemer H 1991 *Appl. Phys. Lett.* **58** 2003-5
[15] Zou G, Pereira R, de Potter M, Van Hove M, De Raedt W and Van Rossum M 1991 *Semicond. Sci. Technol.* **6** 912-5
[16] Pereira R, Van Hove M, De Raedt W, Jansen Ph, Borghs G, Jonckheere R and Van Rossum M 1991 *J. Vac. Sci. Technol. B* **9** 1978-80
[17] Pereira R, Van Hove M, De Raedt W, Van Hoof C, Borghs G, Van Rossum M, Braspenning R H, Eijkemans T J, van Es C M and Wolter J H 1992 *Mat. Res. Soc. Symp. Proc.* Vol. **240** 361-5
[18] Van der Burgt M, Van Esch A, Peeters F M, Van Hove M, Borghs G and Herlach F Yamada Conference XXXIII.
[19] Colvard C, Nouri N, Lee H and Ackley D 1989 *Phys. Rev. B* **39** 8033-8
[20] Capizzi M, Coluzza C, Frankl P, Frova A, Colocci M, Gurioli M, Vinattieri A V and Sacks R N 1991 *Physica B* **170** 561-5
[21] Agarwala S, Tong M, Balleeger D G, Nummila K, Ketterson A A, and Adesida I 1993 *J. Electr. Mat.* **22** 375-81

Structural and Magnetotransport Properties of InGaAs/InAlAs heterostructures grown on linearly-graded Al(InGa)As buffers on GaAs

R.S. Goldman, Jianhui Chen, K.L. Kavanagh, and H.H. Wieder

Department of Electrical and Computer Engineering, University of California at San Diego, La Jolla, CA 92093-0407

V.M. Robbins and J.N. Miller

Hewlett-Packard Laboratories, Palo Alto, CA 94304

Abstract: We have investigated the structural and magnetotransport properties of modulation-doped $In_{0.52}Al_{0.48}As/In_{0.53}Ga_{0.47}As/In_{0.52}Al_{0.48}As$ heterostructures grown by molecular beam epitaxy on GaAs (001) by means of linearly-graded $Al_{0.48}(In_zGa_{1-z})_{0.52}As$ buffer layers. Studies by double-crystal x-ray diffraction and transmission electron microscopy indicate that the linearly-graded buffer is ~ 90% strain-relaxed in both <110> in-plane directions with an epilayer tilt of 0.5° about the [$\bar{1}$10] axis, and threading dislocation or stacking fault defect density approximately $10^7/cm^2$. The peak 1.6K apparent electron mobility, determined from Hall measurements in the [$\bar{1}$10] direction, 1.1 x $10^5 cm^2$/V-s, is comparable to the highest low-temperature mobility reported for a similar heterojunction grown on InP [7], and is consistent with the theoretical alloy-scattering limited mobility for such heterojunctions [8],[9]. Analysis of 1.6K Shubnikov de Haas oscillatory magnetoresistance measurements indicates two-subband occupation of the two-dimensional electron gas.

1. Introduction

$In_xGa_{1-x}As$ alloys grown on GaAs substrates may be employed as quantum wells for the confinement of a two-dimensional electron gas (2DEG) [1]. Modulation-doped InGaAs/InAlAs heterostructures, provide considerable advantages in comparison to GaAs/AlGaAs heterostructures [2], due to their higher electron mobilities, higher conduction band offsets, and reduced deep-level trapping in InAlAs barrier layers [3]. The $In_{0.53}Ga_{0.47}As/In_{0.52}Al_{0.48}As$ heterojunction can be grown lattice matched to InP. In order to overcome the lattice mismatch of this heterojunction relative to GaAs, it is necessary to interpose a compositionally graded buffer layer between the substrate and heterojunction. Ideally, the buffer layer is fully strain-relaxed, prevents dislocations from propagating into the heterojunction, and provides an optimum confinement of charge carriers [4],[5]. These requirements can be met with a buffer which is linearly-graded from $Ga_{0.52}Al_{0.48}As$ to $In_{0.52}Al_{0.48}As$ at 50 atomic % indium per μm. In this paper, we present structural and charge transport properties of modulation doped heterostructures grown on GaAs substrates by means of $Al_{0.48}(In_zGa_{1-z})_{0.52}As$ buffer layers, and compare these to simple modelling schemes.

2. Experimental Procedures

The samples were grown by solid-source molecular beam epitaxy, with As_4 beam equivalent pressure 1×10^{-5} torr and growth rates ~ 1 μm/hr. The substrate temperature was determined in-situ using an optical pyrometer and maintained at 350°C throughout the growth. The targeted sample structure is shown in figure 1. Samples were characterized with transmission electron microscopy (TEM), x-ray rocking curves (XRC), low-field Hall, and high-field Shubnikov de Haas (SdH) oscillatory magnetoresistance measurements. A Philips CM30 operating at 300keV was used for TEM. XRC were measured with a double crystal x-ray diffractometer using $CuK_{\alpha 1}$ radiation monochromated by four Ge (220) crystals. Low-field Hall measurements were performed at 300 and 1.6K with 6-arm Hall bars (200 x 800 μm) aligned along both <110> in-plane directions. 1.6K SdH oscillatory magnetoresistance data was taken as a function of magnetic field from 0 to 8 T. Subband occupation of the 2DEG was determined from a fast Fourier transform (FFT) power spectrum of the SdH oscillations in reciprocal magnetic field. The subband electron concentrations and energy levels were compared with a Schrödinger-Poisson simulation based on the Joyce-Dixon approximation.

50 Å $In_{0.53}Ga_{0.47}As$ (n^+-doped) ($N_d=2 \times 10^{18} cm^{-3}$)
300 Å $In_{0.52}Al_{0.48}As$ (n-doped) ($N_d=1.0 \times 10^{18} cm^{-3}$)
100 Å $In_{0.52}Al_{0.48}As$ (undoped)
250 Å $In_{0.53}Ga_{0.47}As$ (undoped)
2500 Å $In_{0.52}Al_{0.48}As$ (undoped)
1μm linearly-graded $Al_{0.48}(In_zGa_{1-z})_{0.52}As$
SI GaAs substrate

Fig. 1 Schematic of sample structure

3. Results and Discussion

Figure 2 (a) and (b) show cross-sectional and plan-view bright-field TEM images taken with a g=220 two-beam condition. In figure 2(a), the [110] cross-sectional image shows many misfit dislocations in the linearly-graded buffer which have relaxed most of the strain, with few dislocations threading into the active layers. In figure 2(b), the plan-view image shows a relatively low threading dislocation density, $< 10^7$ cm^{-2}. There do not appear to be any <110> oriented misfit dislocations in the vicinity of the 2DEG.

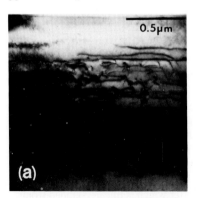
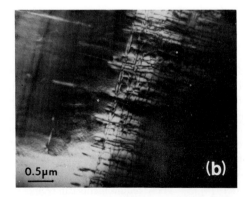

Fig. 2 (a) Cross-sectional transmission electron micrograph, taken at g=220, tilted from [110], showing misfit dislocations in the linearly-graded buffer layer which have relaxed most of the strain and (b) Plan-view transmission electron micrograph, taken at g=220, showing an orthogonal network of <110> misfit dislocations confined in the linearly-graded buffer.

The epilayer tilt, alloy composition, and strain relaxation, were obtained through the measurement and analysis of (004) and (224) x-ray rocking curves [6]. Figure 3 displays the data obtained for the (004) reflections at eight azimuthal angles ranging from ω=0 to 270°. The azimuthal angle, ω, is defined as a clockwise rotation around the [001] growth axis, where ω=0 corresponds to the incident and diffracted x-ray beams parallel to the [110] direction.

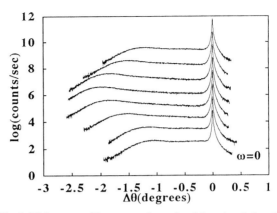

Fig. 3 (004) x-ray rocking curves shown for eight azimuthal angles in 45° increments from ω=0 to 270°.

The variation of the epilayer-substrate peak separation with azimuthal angle ω follows a sinusoidal function, whose amplitude and phase indicate an epilayer tilt of 0.5° about the [$\bar{1}$10] axis. Furthermore, the $In_yAl_{1-y}As$ buffer, nominally y=0.53 ± 0.05, has relaxed 90 ± 10 % in both <110> directions, and the active layers are lattice-matched to the buffer layer, to within the resolution of our present measurements.

The apparent electron concentrations, $[n]_H$, and apparent electron mobilities, μ_H, determined from low-field Hall and resistivity measurements are summarized in Table I. At room temperature, the electron mobility is limited principally by polar optical phonon scattering, and the anisotropy is insignificant in comparison to experimental error. At 1.6K, the primary mobility-limiting scattering mechanisms are

Table I: Apparent electron concentrations, $[n]_H$, and apparent electron mobilities, μ_H, at 300K and 1.6K in both <110> directions, determined from low-field Hall and resistivity measurements taken with a DC current of 10 μA. The 300K and 1.6K data were taken in magnetic fields of 0.1T and 0.15T, repectively. The difference in error bars reflect the differences in experimental uncertainty in the different apparatus.

Transport direction	$[n]_H$ (300K) (10^{12} cm^{-2})	μ_H (300K) (10^4cm^2/V-s)	$[n]_H$ (1.6K) (10^{12}cm^{-2})	μ_H (1.6K) (10^4cm^2/V-s)
[110]	1.2 ± 0.1	1.2 ± 0.1	1.08 ± 0.01	7.47 ± 0.01
[$\bar{1}$10]	1.2 ± 0.1	1.3 ± 0.1	1.05 ± 0.01	11.0 ± 0.01

remote ionized impurity and alloy disorder scattering. The highest electron mobility is equal, to within experimental error, the highest low temperature electron mobility for a similar heterojunction grown on (001) InP [7], and is comparable to the theoretical alloy-scattering limited mobility for such heterojunctions [8],[9]. The mobility anisotropy is 20%. Assuming that the high mobility direction, [$\bar{1}$10], is limited by alloy scattering, then the low mobility direction, [110], is likely to be limited by alloy scattering, and an additional defect-related scattering process. Electron mobility asymmetries in 2DEGs have been attributed to electron scattering from asymmetric distributions of misfit dislocations [10],[11]. However, our TEM images do not indicate the presence of misfit dislocations in the vicinity of the 2DEG. Threading dislocations and stacking faults are observed but their distribution appears to be orientation independent.

Anisotropic properties of GaAs field-effect transistors have been attributed to electron scattering due to the piezoelectric effect [12]. The zincblende lattice has one piezoelectric coupling constant which couples to shear stresses [13]. Since the strain relaxation is nearly symmetric, the in-plane stresses are essentially isotropic, and the in-plane shear stress is negligible. However, the epilayers have tilted by 0.5° with respect to the substrate, about the [$\bar{1}$10] axis. Resolving the epilayer stress tensor into the substrate coordinate system demonstrates that there are shear stresses in the epilayers, which may contribute to the electron mobility asymmetry through the piezoelectric effect.

In addition, a 2DEG is highly degenerate and interfacial contributions to its mobility are expected to be significant. Interface roughness has been suggested as the origin of anisotropic electron mobilities observed in AlGaAs/GaAs 2DEGs [14]. The roughness is postulated to originate from [$\bar{1}$10] elongated islands which evolve during MBE growth. This roughness would only affect electrons travelling perpendicular to the islands. Elongated islands have been observed by scanning tunneling microscopy (STM) under certain growth conditions [15], but disappear at higher substrate temperatures typically used for the growth of AlGaAs/GaAs 2DEGs. Earlier STM studies have indicated a preference for the step-edge alignment along the [$\bar{1}$10] direction during MBE growth of GaAs (001) [16]. This step-edge alignment corresponds with the alignment of As-dimers for the typical 2x4 GaAs (001) surface reconstruction, and may contribute to anisotropic atomic-scale roughness.

The two-dimensional nature of the electron gas was confirmed from 1.6K high-field SdH measurements, which indicate oscillations of the resistivity, ρ_{xx}, and quantum Hall plateaus in the transverse resistivity, ρ_{xy}. Data are shown in figure 4 (a) and (b), for the [110] and [$\bar{1}$10] directions, respectively. The slight beating of the ρ_{xx} oscillations is indicative of the occupancy of a second subband of the 2DEG. The electron concentrations in each subband, $[n]_i$, which were determined from the FFT power spectrum of the ρ_{xx} oscillations in reciprocal magnetic field, are summarized in Table II.

Charge-carrier confinement in the 2DEG was confirmed through a comparison of the total electron concentration, $[n]_t = [n]_0 + [n]_1$, determined from SdH

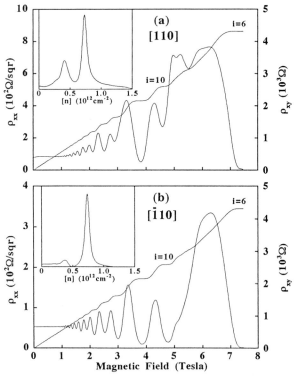

Fig. 4 Magnetic field dependence of the resistivity tensor components, ρ_{xx} and ρ_{xy}, at T=1.6K, in the (a) [110] and (b) [$\bar{1}$10] directions. The insets show the values of [n] determined from the FFT power spectrum of the ρ_{xx} oscillations vs. reciprocal magnetic field, which indicate two-subband occupancy of the 2DEG.

measurements, and the apparent electron concentration, $[n]_H$, determined from Hall measurements. $[n]_t$ is greater than $[n]_H$ by 7 and 2% in the [110] and [$\bar{1}$10] directions, respectively. The slight increase in $[n]_t$ relative to $[n]_H$ may be due to the difference in electron mobilities of the subbands.

Table II: Summary of 1.6K transport parameters in both <110> directions. Electron concentrations, [n], were determined from the FFT power spectrum of SdH oscillations. A classical model for two subband conduction was used to calculate the electron mobility, μ, and corresponding classical relaxation time, $\tau_c = \mu m^*/e$, where m^* is the electron effective mass at the Fermi level, which was corrected for conduction band non-parabolicity [19],[20], and e is the electron charge. The quantum relaxation time, τ_q, was determined following Coleridge [22].

subband	[n] (10^{11} cm^{-2})	μ (10^4cm^2/V-s)	m^*/m_0	τ_c (10^{-12} sec)	τ_q (10^{-13} sec)	τ_c/τ_q
[110]						
E_0	7.3 ± 0.1	8.4 ± 0.1	0.048	2.3 ± 0.1	1.5 ± 0.1	15 ± 0.1
E_1	4.0 ± 0.1	4.4 ± 0.1	0.046	1.1 ± 0.1	1.2 ± 0.1	9 ± 0.1
[$\bar{1}$10]						
E_0	7.3 ± 0.1	11.8 ± 0.1	0.048	3.2 ± 0.1	1.6 ± 0.1	20 ± 0.1
E_1	3.7 ± 0.1	9.8 ± 0.1	0.046	2.3 ± 0.1	1.2 ± 0.1	19 ± 0.1

Subband occupancy and energy level separation were determined theoretically from a Schrödinger-Poisson simulation based on the Joyce-Dixon approximation for the degeneracy of the electron gas. Assuming the targeted structure shown in figure 1, with a supply-layer donor concentration, $N_d = 1.25 \times 10^{18}$ cm^{-3}, we obtained two-subband occupation with a total electron concentration $[n] = 1.11 \times 10^{12}$ cm^{-2}, which is within 2% of the values of $[n]_t$ determined by SdH measurements for both <110> directions. However, the theoretical subband energy level separation is approximately 50% larger than that determined from SdH measurements.

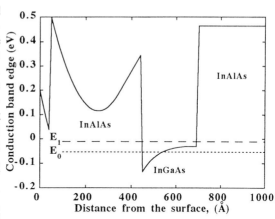

Fig. 5 Schrödinger-Poisson simulation of the heterojunction shown in figure 1, with the Fermi level set at zero energy. E_0 and E_1 correspond to the first and second subbands.

A classical model for two subband conduction [17] was used to calculate the electron mobility, μ_i, in each subband, and the corresponding classical electron relaxation time $\tau_{ci} = \mu_i m^*/e$, where m^* is the electron effective mass at the Fermi level corrected for conduction band non-parabolicity [18],[19], and e is the electron charge. A digital bandpass filter was then used to extract and separate the low-field oscillatory components of the magnetoresistance corresponding to each subband [20],[21],[22]. The quantum scattering time, τ_q, was determined from an analysis of the extrema of the filtered SdH magnetoresistance oscillations [21]. These quantities are also summarized in Table II.

For both subbands, the classical scattering times are larger in the [$\bar{1}$10] than the [110] direction. However, the quantum scattering times are identical to within experimental error for each subband in both <110> directions. The ratio of the scattering times gives us further insight into the electron scattering processes [23]. When $\tau_c/\tau_q =1$, large angle scattering is the most dominant mechanism. As the ratio increases from 1, small-angle scattering processes become more dominant. Since τ_c/τ_q is the smallest in the [110] direction, we expect that large angle scattering has a greater effect on electron transport in the [110] than the [$\bar{1}$10] direction. Furthermore, in the [110] direction, $\tau_{c0}/\tau_{q0} > \tau_{c1}/\tau_{q1}$, indicating that large angle scattering is most significant in the second subband.

4. Conclusions

We have studied the structural and charge transport properties of modulation-doped $In_{0.52}Al_{0.48}As/In_{0.53}Ga_{0.47}As/In_{0.52}Al_{0.48}As$ grown on GaAs substrates by means of linearly graded $Al_{0.48}(In_zGa_{1-z})_{0.52}As$ buffer layers. The buffer layer is essentially strain-relaxed with a low density ($\sim 10^7 cm^{-2}$) of defects propagating into the heterojunction. The 1.6K apparent electron mobility in the [$\bar{1}$10] direction is comparable to the highest low temperature electron mobility reported for a similar heterojunction grown on (001) InP [7] and is consistent with the theoretical alloy-scattering limited mobility for such heterojunctions [8],[9]. We suggest the electron mobility in the [110] direction is limited by an additional defect-related scattering process, which is dominated by large-angle scattering.

References

[1] Kastalsky A, Dingle R, Cheng K Y, and Cho A Y 1982 *Appl. Phys. Lett.* **41** 274.
[2] Mishra U K, Brown A S, and Jensen J F 1989 *Inst. Phys. Conf. Ser.* **106** 605.
[3] Itoh T, Griem T, Wicks G W, and Eastman L F 1985 *Electr. Lett.* **21** 373.
[4] Inoue K, Harmand J C, and Matsuno T 1991 *J. Cryst. Growth* **111** 313.
[5] Mishima T, Tanimoto T, Kudoh M, and Takahama M 1993 *J. Cryst. Growth* **127** 770.
[6] K.M. Matney, G.G. Chu, M.S. Goorsky, R.S. Goldman, and K.L. Kavanagh, unpublished.
[7] Onabe K, Tashiro Y, and Ide Y 1986 *Surface Science* **174** 401.
[8] Walukiewicz W, Ruda H E, Lagowski J, and Gatos H C 1984 *Phys. Rev. B* **30** 4571.
[9] Matsuoka T, Kobayashi E, Taniguchi K, Hamaguchi C, and Sasa S *Jap. J. Appl. Phys.* **29** 2017.
[10] Schweizer T, Kohler K, and Ganser P 1991 *Semicond. Sci. Technol.* **6** 356.
[11] Chen J, Fernandez J M, and Wieder H H 1992, *Mater. Res. Symp. Proc.* **263** 377.
[12] Asbeck P M, Lee C P, and Chang M C F 1984 *IEEE Trans. Electron Devices* **ED-31** 1377.
[13] Nye J F 1979 *Physical Properties of Crystals* (Oxford:Oxford University Press).
[14] Tokura Y, Saku T, Tarucha S, and Horikoshi Y 1992 *Phys. Rev. B* **46** 15558.
[15] Orme C, Johnson M D, Sudijono J L, Leung K T, and Orr B G *Appl. Phys. Lett.* **64** 860.
[16] Pashley M D, Haberern K W, and Gaines J M 1991 *Appl. Phys. Lett* **58** 406.
[17] Ziman J M 1964 *Principles of the Theory of Solids* (Cambridge:Cambridge University Press).
[18] Nicholas R J, Portal J C, Houlber C, Perrier P, and Pearsall T P 1979 *Appl. Phys. Lett.* **34** 492.
[19] Ando T 1982 *J. Phys. Soc. Jpn.* **51** 3893.
[20] Kaiser J F and Reed W A 1978 *Rev. Sci. Instrum.* **49** 1103.
[21] Coleridge P T, Stoner R, and Fletcher R 1989 *Phys. Rev. B* **39** 1120.
[22] Fernandez J M and Wieder H H 1992 *Inst. Phys. Conf. Ser.* **120** 639.
[23] Harrang J P, Higgins R J, Goodall R K, Jay P R, Laviron M, and Delescluse P 1985 *Phys. Rev. B* **32** 8126.

Si/SiGe modulation doped structures grown by gas source molecular beam epitaxy

A. Matsumura [a),b),*], R. S. Prasad [a)], T. J. Thornton [a)],
J. M. Fernández [b)], M. H. Xie [b)], X. Zhang [b)], J. Zhang [b)]
and B. A. Joyce [b)]

a) Dept. of Electrical Engineering and
b) Interdisciplinary Research Centre for Semiconductor Materials
   Imperial College, London SW7, UK
* Permanent address :
   Electronics Research Labs., Nippon Steel Corp.,
   5-10-1 Fuchinobe, Sagamihara-shi, Kanagawa 229, Japan

Abstract. We have grown Si/SiGe modulation doped quantum wells by gas source molecular beam epitaxy using arsenic as an n-type dopant. The layers are characterised by TEM and SIMS analysis, and magnetotransport measurements. The SIMS analysis shows an arsenic concentration in excess of 10^{19} cm^{-3} along with strong surface segregation. Pronounced Shubnikov-de Haas oscillations were observed in the magnetotransport measurements. Mobilities in the two dimensional electron gas were determined to be around 15,000 cm^2/Vs.

1. Introduction

Si/SiGe heterostructures are expected to have widespread commercial applications in future high speed communication systems and have recently been investigated for use in both transport and optoelectronic devices. Modulation doped (MOD) structures are of particular interest for transport device applications due to their unique high speed/low noise properties. At low temperatures these structures have enhanced mobility because of the reduction in ionized impurity scattering and electron mobilities in excess of 100,000 cm^2/Vs have already been obtained[1-3].
   Among growth methods for MOD structures molecular beam epitaxy (MBE) is expected to realize high quality layer structures. Furthermore, by using gas sources, relatively low growth temperatures become available bringing advantages such as a reduction in the level of impurities coming from the system itself. Although MOD Si/SiGe heterostructures have been grown by a variety of techniques such as UHV-CVD[1] and solid source MBE[2,3], the doping source used for n-channel layers has been exclusively phosphorous or antimony. In this paper we shall describe the growth and characterisation of n-channel layers using gas source molecular beam epitaxy (GS-MBE) with arsenic as the source of dopants.

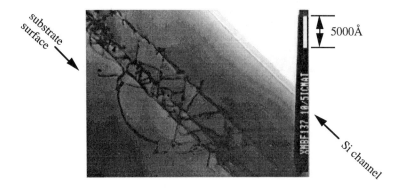

Fig. 1 A cross sectional (TEM) image of the layer structure.

2. Growth of Modulation Doped Structures

In our GS-MBE system the gas sources are Si_2H_6 and GeH_4 for the SiGe layers, and AsH_3 for the arsenic doped n-channel supply layers. The substrate temperature was chosen to allow two dimensional growth and was typically in the range 550°C - 700°C. Typical growth rates were 0.1-1.0 Å/s. To obtain near dislocation free, totally relaxed SiGe layers, either stepwise or linearly graded SiGe buffer layers with Ge concentrations of up to 35% were grown on Si (100) substrates and were terminated by a thick SiGe layer of 30% Ge. The total thickness of the buffer was 1.5 - 2.0 µm. The MOD structures were grown on top of the buffer layers and consisted of a strained Si channel layer, a SiGe spacer layer, an arsenic doped SiGe supply layer and cap layer. Fig. 1 shows a transmission electron microscope (TEM) cross-sectional image of the layer structure. The majority of dislocations are confined below the Si channel indicating the effectiveness of our stepwise buffer layers. The Si channel itself is clearly visible and has sharp interfaces. A similar dislocation behavior was also observed in the case of the linearly graded buffer layers.

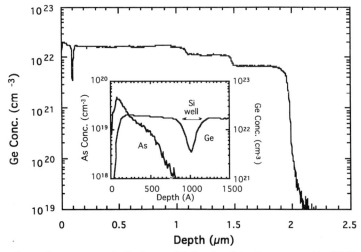

Fig. 2 Germanium and arsenic concentration in a modulation doped structure measured by SIMS. The inset shows an expanded figure around the Si channel.

Secondary ion mass spectroscopy (SIMS) of a typical MOD layer shows the variation of Ge concentration in the stepwise buffer (see Fig. 2). The inset to the Figure shows the very high arsenic concentration in excess of 10^{19} cm^{-3}. The concentration increases towards the surface suggesting strong arsenic segregation. Hall effect measurements of a uniformly doped SiGe layer grown under similar conditions to the MOD heterostructure supply layer indicate an average carrier concentration of 6×10^{18} cm^{-3}. The comparison between the SIMS and Hall measurements indicates a good activation of the arsenic in the SiGe layer.

3. Electrical Properties

To evaluate the electrical properties of the MOD structure, magnetoresistance measurements were made at temperatures down to 0.3 K. A Hall bar pattern was formed by wet-etching and longitudinal (R_{xx}) and Hall resistance (R_{xy}) measurements were made in magnetic fields (B) up to 4 Tesla. Three MOD structures were prepared for these measurements. In Table I, the layer structure and the results of the measurement of each sample are shown. Fig. 3 shows the B-field dependence of R_{xx} and R_{xy} of sample #2 at a temperature of 0.3 K. Pronounced Shubnikov-de Haas (SdH) oscillations can be seen in R_{xx} with corresponding quantum Hall plateaus in R_{xy} both of which are indicative of a high mobility two dimensional electron gas (2DEG) in the Si channel.

The overall sheet carrier density of sample #1 is determined from the gradient of R_{xy} to be 2.4×10^{12} cm^{-2}, while that in the 2DEG is 7×10^{11} cm^{-2} from the period of the SdH oscillations. This discrepancy can be explained by the existence of parallel conduction. A simulation of the conduction band edge profile implies the existence of parallel conduction in the SiGe supply layer close to the surface.

For the case of sample #2 with the thinner supply layer, a Fourier transform analysis of the SdH oscillations shows that two frequency components are present. This suggests the existence of two 2DEGs possibly at both interfaces of the Si channel. The sheet densities determined from the oscillations are approximately 6×10^{11} cm^{-2} and 4×10^{11} cm^{-2}, giving a sum equal to that from the gradient of R_{xy}. This indicates that the parallel conduction in the SiGe supply layer was completely removed in the case of sample #2 due to the thin supply

Table I Structural and electrical properties of MOD samples

(a) Layer structures

Sample No.	Buffer structure	Layer thickness (Å)					arsenic conc. (cm^{-3})
		Si channel	Spacer	Supply layer	SiGe cap	Si cap	
#1	stepwise	200	120	1000	0	100	~ 4×10^{19}
#2	stepwise	200	150	500	100	100	~ 4×10^{19}
#3	linearly graded	100	180	200	200	100	3×10^{18}

(b) Electrical properties

Sample No.	Sheet carrier density (×10^{12} cm^{-2})		2DEG mobility (cm^2/Vs)
	from R_{xy}	from SdH oscillation	
#1	2.4	0.7	~ 16,000
#2	1.0	0.6, 0.4	~ 15,000
#3	0.3	0.3	~ 10,000

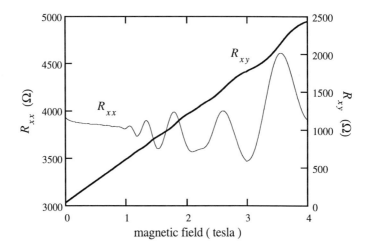

Fig. 3 Magnetoresistance measurements of sample #2.

layer and inserted SiGe cap layer.

A Fourier transform of the SdH oscillations measured in sample #1 shows a single peak which is much broader than those observed for sample #2. This suggests that two 2DEGs may also be present in sample #1 but very similar electron concentrations. However, even if we took this into account the total 2DEG sheet density of 1.4×10^{12} cm^{-2} is well below the overall sheet density, 2.4×10^{12} cm^{-2}. Therefore we can conclude that there definitely exist a parallel conduction path in the case of sample #1.

In the case of sample #3, the sheet densities obtained from the gradient of R_{xy} and the SdH oscillation agree well, indicating the existence of only one 2DEG. This is probably because the 2DEG can exist only at the upper interface of the thinner Si quantum well.

The low temperature electron mobility, μ_e, in the 2DEG of our MOD structure is about 15,000 cm^2/Vs. This value is considered to be limited by interface scattering due to poor morphology as well as background impurity scattering and alloy scattering due to Ge segregation [4] into the Si channel.

4. Conclusions

Modulation doped structures were grown by gas source MBE. Arsenic was used as the n-type dopant and high concentrations in excess of 10^{19} cm^{-3} were obtained with strong segregation. Pronounced Shubnikov-de Haas oscillations were observed in magnetotransport measurements. A sample having a thick supply layer reveals the existence of parallel conduction, while it is completely depleted in the sample of a modified layer structure. The mobilities of the 2DEGs are about 15,000 cm^2/Vs and we believe that the mobilities can be improved greatly by optimising the growth conditions and parameters.

References
[1] Többen D, Schäffler F, Zrenner A and Abstreiter G 1992 *Phys. Rev. B* **46** 4344-4347.
[2] Nelson S F, Ismail K, Jackson T N, Nocera J J and Chu J O 1993 *Appl. Phys. Lett.* **63** 794-796.
[3] Xie Y H, Fitzgerald E A, Monroe D, Silverman P J and Watson G P 1993 *J. Appl. Phys.* **73** 8364-8370.
[4] Ohtani N, Mokler S M, Xie M H, Zhang J and Joyce B A 1993 *Surf. Sci.* **284** 305-314.

Effects of annealing using plasma-CVD SiN film as a cap on 2DEG mobility

S. Nakata, M. Yamamoto and T. Mizutani

NTT LSI Laboratories
3-1 Morinosato Wakamiya, Atsugi-shi, Kanagawa 243-01, Japan

Abstract   We investigated how the mobility, μ, and carrier density, n_s, of AlGaAs/GaAs two-dimensional electron gas (2DEG) are influenced by fabrication processes using plasma-excited chemical vapor deposition (plasma-CVD) SiN film. Annealing with plasma-CVD SiN film as a cap results in a large reduction in μ and n_s. Both, however, are restored by reannealing after removing the SiN film. We further investigated the influence of the same process on a more simplified Si-doped GaAs layer structure through capacitance-voltage (C-V) measurements. Similar degradation and restoration of the carrier density were also observed for Si-doped GaAs. The reannealing temperature dependence shows the degraded carrier density is almost perfectly restored by reannealing at the low temperature of 380 °C for 20 minutes. Reannealing after removing the plasma-CVD SiN film is necessary for fabricating high-quality buried structures.

1. Introduction

Buried quantum wires with lateral heterointerfaces are promising for producing quantum-wire-based quantum-electron devices [1-4]. We fabricated buried wires using a photo-assisted chemical vapor deposition (photo-CVD) SiN film as a mask for etching and as a mask for subsequent MOCVD regrowth [4]. This method enables us to fabricate all kinds of nanometer-sized structures.

Little process-induced damage results from photo-CVD SiN film deposition because of its small photon-assisted energy, as opposed to plasma-excited CVD (plasma-CVD) film deposition. Photo-CVD film, however, does not adhere to the wafer as well as plasma-CVD film. For this reason, it is preferable to use plasma-CVD SiN film as a mask.

In order to obtain high-quality buried wires, we have to clarify how the processes using plasma-CVD SiN film influence the electrical properties of the two-dimensional electron gas (2DEG).

We investigated how the mobility, μ and carrier density, n_s of 2DEG are influenced by fabrication processes using plasma-CVD SiN film. We investigated the effects of annealing using SiN film as a cap. This process simulated MOCVD regrowth, and resulted in a large reduction in μ and n_s. Both were restored, however, by reannealing after removing the SiN film. These experimental results are shown in section 3.

Next, we investigated the influence of the same process on the carrier profile of a more simplified structure of a Si-doped GaAs layer through C-V measurements. We will discuss how the carrier density is reduced by annealing using the plasma-CVD film and how this reduction is restored by reannealing after the film is removed. To determine the lowest temperature required to substantially reduce defects, we investigated the reannealing temperature dependence of carrier density restoration. The results are discussed in section 4.

2. Processes

Schematic representations of the processes used in this study are shown in Fig. 1. We used plasma-CVD and photo-CVD to deposit SiN film and compared the results. For photo-CVD deposition, ultra-violet light from a Hg lamp was used to form vapor phase reactants that enhance the deposition rate. The photo-CVD SiN film was deposited at the substrate temperature, T_s, of 160 °C. Plasma-CVD was done using microwaves with a power of 0.5 kW obtained by applying the radio-frequency voltage and the SiN film was deposited at T_s of 300 °C. Therefore, the large plasma energy added to the thermal energy enhanced the deposition rate.

In process (A) (Fig. 1), SiN was deposited on the wafer to a thickness of 60 nm. After deposition, the film was removed from the wafer with etchant (HF). In process (B), after SiN deposition, the wafer was annealed at 600 °C for 20 minutes. The film was then removed. Process (B) is a simulated process of MOCVD regrowth used for burying wires. In process (C), SiN was deposited, and the wafer was annealed just as in process (B). The SiN film was then removed and the wafer was reannealed at 600 °C for 20 minutes. Ohmic contacts were made by annealing AuGeNi at 380 °C for 2 minutes. μ and n_s were measured using the van der Pauw method at 77 K.

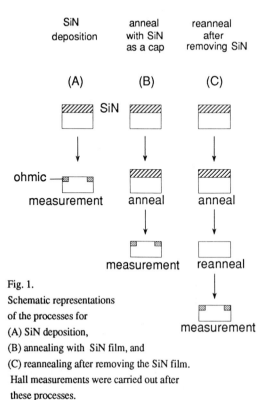

Fig. 1.
Schematic representations of the processes for
(A) SiN deposition,
(B) annealing with SiN film, and
(C) reannealing after removing the SiN film.
Hall measurements were carried out after these processes.

3. Hall measurement

The epitaxial layer used here was a modulation-doped AlGaAs/GaAs heterostructure grown by molecular beam epitaxy (MBE). An undoped GaAs buffer layer (600 nm), an undoped $Al_{0.3}Ga_{0.7}As$ spacer layer (10 nm), a Si-doped $Al_{0.3}Ga_{0.7}As$ layer with a nominal Si concentration of 1.5×10^{18} /cm^3 (50 nm), and a Si-doped GaAs cap layer with a nominal Si concentration of 1.0×10^{18} /cm^3 (10 nm) were succes-

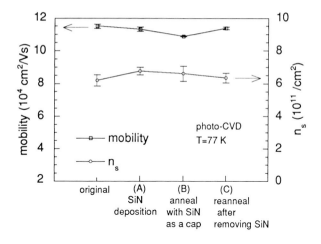

Fig. 2. Mobility and carrier density, n_s, for processes A, B and C shown in Fig. 1 using a photo-CVD SiN film. The original mobility and n_s of 2DEG are also shown.

sively grown on a semi-insulating GaAs substrate. The μ and n_s of the 2DEG in the starting wafer was 1.15×10^5 cm²/Vs and 6.2×10^{11}/cm² at 77 K.

Figure 2 shows the change in μ and n_s after processes (A), (B) and (C) using the photo-CVD SiN film. μ didn't change appreciably after process (A). This indicates a small amount of process-induced damage is caused during the photo-CVD SiN deposition. After process (B), μ showed only a 5 % reduction from its original value, and after process (C), it was almost entirely restored to its original value. n_s also did not change significantly during processes (A), (B) and (C), but rather stayed close to its original value of 6.2×10^{11}/cm².

Figure 3 shows how the μ and n_s are changed by processes (A), (B) and (C) when plasma-CVD SiN film is used. We observed a 8 % reduction in μ after plasma-CVD SiN deposition by process (A). In contrast, we observed a 76% reduction in μ with process (B) from 10.6×10^4 to 2.6×10^4/cm². n_s was also reduced from 6.2×10^{11} to 2.4×10^{11}/cm². The large reduction in μ and n_s after process (B) with the plasma-CVD SiN film stands in contrast to the slight reduction in μ and the almost constant n_s after process (B) using the photo-CVD SiN film.

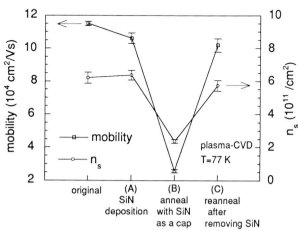

Fig. 3. Mobility and carrier density, n_s, for processes A, B and C shown in Fig. 1 using a plasma-CVD SiN film. The original mobility and n_s of 2DEG are also shown.

The slight reduction in μ after process (A) might be a result of damage during plasma-CVD SiN deposition from the 0.5 kW microwave power source. In order to identify the cause of the significant reduction of μ and n_s after process (B) using plasma-CVD, we investigated the effects of reannealing the wafers after removing the SiN film (process (C)). As shown in Fig. 3, μ and n_s after process (C) are almost

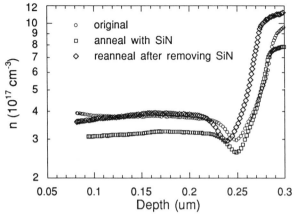

Fig. 4. Carrier profile of the original wafer, annealed wafer with plasma-CVD SiN, and reannealed wafer after removing SiN.

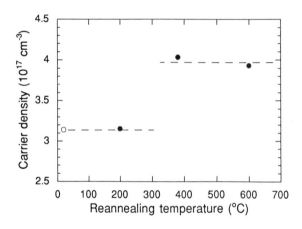

Fig. 5. Reannealing temperature dependence of carrier density. The upper broken line and the lower one represent the original carrier density and the reduced one after annealing with plasma-CVD SiN.

completely restored to their original values. This indicates the increase in defect density caused by process (B) is almost completely eliminated and the reduction of μ and n_s occuring in process (B) is not related to, for example, Si dopant diffusion into the heterointerface channel.

4. C-V measurement

We further investigated the influence of the same process on a more simplified Si-doped GaAs layer structure. The epitaxial layer used here was Si-doped GaAs (300 nm) with a nominal Si concentration of 4×10^{17} cm^{-3} grown by MBE on a Si-doped GaAs substrate with a Si concentration of 1×10^{18} cm^{-3}.

Fig. 4 shows the carrier profile of the Si-doped GaAs wafer obtained from C-V measurements. The circles, squares and diamonds represent respectively the results for the original wafer, the annealed wafer with SiN (process B), and the reannealed wafer after the removal of the SiN (process C). The carrier density of the original wafer at 0.15 μm from the surface (half the epitaxy layer thickness) was 3.8×10^{17} cm^{-3}, which is consistent with the nominal Si concentration. The depletion length was 0.08 μm. The carrier density of the annealed wafer with SiN was reduced to 3.1×10^{17} cm^{-3} at the same depth. The depletion length was increased to 0.09 μm. For the wafer reannealed after removing the SiN, the carrier density was restored to 4.0×10^{17} cm^{-3} and the depletion length was reduced to 0.08 μm. This degradation and restoration of the carrier density is consistent with the Hall measurement results for 2DEG. Thus, the behavior of the depletion length is strongly related to

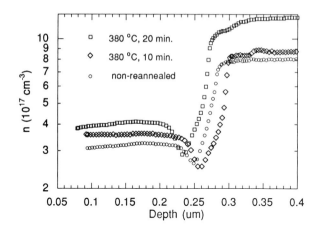

Fig. 6. Carrier profile of wafers reannealed at 380 °C after removing plasma-CVD SiN for 20 minutes and 10 minutes after 600 °C annealing with plasma-CVD SiN. A profile of the non-reannealed wafer is shown for comparison.

that of the carrier density.

Next, we investigated how defects are reduced as a function of the reannealing temperature. After annealing with SiN at 600 °C for 20 minutes, we reannealed the wafer for 20 minutes at 600 °C, 380 °C, and 200 °C after removing the SiN film. We compared the carrier density of these wafers with that of the non-reannealed wafer, as shown in Fig. 5. We plotted the carrier density of these wafers at a 0.15-µm depth, which is a half of the epitaxy layer thickness. The solid and open circles represent the carrier density of the reannealed wafers and that of the non-reannealed wafers. The upper broken line shows the original carrier density and the lower one shows the carrier density of the annealed wafer with the SiN film (Process (B)). Reannealing at 380 °C was found to restore the carrier density to the level of the original wafer, but reannealing at 200 °C didn't restore the carrier density at all. Thus, the lowest temperature at which defects can be reduced substantially is somewhere between these two values.

Next, we investigated how the restoration of the carrier density depends on the reannealing time with the reannealing temperature fixed at 380 °C. We prepared two wafers for reannealing for 10 and 20 minutes after removing the SiN film. The carrier profiles as a function of the reannealing time are shown in Fig. 6 and compared with that of the non-reannealed wafer. The carrier density of the 20 minute reannealed wafer was restored to the original one as described earlier. The carrier density for the 10-minute reannealed wafer, however, was not restored sufficiently. The carrier density of this wafer at 0.15 µm was $3.5 \times 10^{17} cm^{-3}$. This value falls between that of the 20 minute reannealed one and that of the non-reannealed one. For a temperature of 380 °C, 20-minute reannealing is sufficient to restore the value of the original carrier density, but 10-minute reannealing is not.

5. Conclusion

We observed μ and n_s degradation of 2DEG by annealing using plasma-CVD SiN film as a cap. Both of these values, however, could be restored by reannealing after removing the SiN film from the wafer. Similar degradation and restoration of the carrier density were also observed for a Si-doped GaAs layer. Reannealing temperature dependence indicated that the lowest temperature required to reduce defects substantially was somewhere between 200 and 380 ^{0}C. Reannealing after removing the plasma-CVD SiN film is necessary for fabricating high-quality buried structures.

Acknowledgment

We are grateful to T. Enoki and S. Yamahata for helpful advice on SiN film deposition and to K. Wada for valuable discussions. We are also grateful to H. Fushimi for C-V measurement advice and to T. Ishikawa for supplying the high-quality MBE wafers.

References

[1] F. Sols, M. Macucci, U. Ravaioli and K. Hess, 1989 Appl. Phys. Lett. **54** 350
[2] S. Datta, 1989 Superlattices & Microstruct. **6** 83
[3] K. Aihara, M. Yamamoto and T. Mizutani, 1993 Appl. Phys. Lett. **63** 3595
[4] S. Nakata, K. Ikuta, M. Yamamoto and T. Mizutani, 1993 Jpn. J. Appl. Phys. **32** 6258

Low ion energy ECR etching of InP using Cl_2/N_2 mixture

S.Miyakuni, R.Hattori, K.Sato and O.Ishihara

Optoelectronic and Microwave Devices Lab.
Mitsubishi Electric Corporation
4-1 Mizuhara, Itami 664, Japan

Abstract : The low energy (about 30eV) ion etching of InP with excellent etching performances of vertical sidewall profile, smooth surface and high etching rate (more than 1000 Å/min) has been realized by an ECR plasma using a Cl_2/N_2 mixture. Surface analysis by XPS and plasma diagnostics by OES suggest that the mechanism of this etching may be dominated by the two effects; the remarkable reduction of Cl neutral radical density by the reaction of $Cl+N \rightarrow NCl$ and the formation of InN and P_3N_5 due to the reaction between the atomic nitrogen in the Cl_2/N_2 plasma and the InP substrate. This technique appears to be based on a new concept for InP dry etching to control the volatilizing rate of $PCl_x(x=1-5)$ products balancing to the desorption rate of $InCl_x(x=1-3)$ products, in opposition to the conventional method enhancing the desorption of non-volatile $InCl_x$ products comparably to that of volatile PCl_x products by high energy ion bombardment.

1. Introduction

Reactive Ion Etching (RIE) or Reactive Ion Beam Etching (RIBE) technique had been developed for the application of indium-based compounds dry etching for semiconductor processing in this last decade[Franz 1990, Hayes et al. 1989, Contlini 1988], for example to form the facet of laser diode cavity or the emitter mesa of the heterojunction bipolar transistor (HBT). However, InP dry etching technique had been one of the most difficult plasma processing.

With chlorine-based gases, as conventional techniques, for obtaining the smooth surface and anisotropic profile in InP etching it has been necessary to enhance the desorption rate of the non-volatile $InCl_x(x=1-3)$ products in comparable to that of the volatile $PCl_x(x=1-5)$ products by the physical sputtering with high energy ions of 100-1000eV or the temperature rising of the substrate. However, the crystal damage by the high energy ion bombardment causes serious problems for the application to the device processing.

With CH_4- or C_2H_6-based gases, the reaction products of indium and phosphorus can be easily volatilized in contrast with the chlorine-based gases. The InP etching with these gases have been intensively investigated, recently. Pearton et al. pointed out that InP could be etched by the ECR plasma using $CH_4/H_2/Ar$ mixture without rf bias[Pearton et al. 1990]. This may be a result of the ion assisted etching by the higher ion density in the ECR plasma. However, high etching rate and smooth surface could not be yielded.

We previously had studied the low energy ion ECR etching of $In_xGa_{1-x}As(x=0.53)$ using Cl_2/He mixture, and proposed that it was essential for the low energy ion etching of indium-based compounds to reduce the Cl neutral radical density in the processing plasma by diluting Cl_2 with the inert gas such as He[Miyakuni et al.1994]. In this work, we have investigated the dry etching of InP by low energy ion (about 30eV) ECR plasma with Cl_2/N_2 mixture, resulting in vertical profile and smooth etched surface. Furthermore, a qualitative explanation of the smooth etching mechanism has been presented.

2. Experimental

The etching of InP was performed by an electron cyclotron resonance (ECR) plasma etching system, which has both main and sub magnetic coils[Miyakuni et al.1994]. ECR plasma is excited by combination of 2.45GHz microwave and 875G magnetic field. In order to control the energy of incident ions, rf power of 13.56MHz is supplied to the substrate holder.

The etching samples used in this study were 3 inches diameter semi-insulating (100) InP substrates. WSi_x (3000Å in thickness, x=0.2) was used as an etching mask. Gas mixtures of Cl_2/Ar and Cl_2/N_2 were utilized for the etching of InP. The following etching conditions were selected to reduce the Cl neutral radical density. An operating pressure was 0.4mTorr and a total gas flow rate was kept constant to be 10sccm. Microwave and rf power were fixed at 200W and 30W, respectively. Main and sub magnetic current were adjusted to 13A and 25A, respectively. The substrate temperature was raised from 100°C to about 200°C by a sheathed heater installed in the substrate holder.

Optical Emission Spectroscopy (OES) system was utilized to elucidate the relation between the etching property and the Cl radical density in the plasma during etching. Scanning Electron Microscope (SEM) and X-ray Photoelectron Spectroscopy (XPS) analyses were performed to characterize the etched surface morphology or profile and the surface chemical bonding. Furthermore, the Langmuir probe, which is 6mm in diameter, was set above 10mm from the center of the substrate to measure plasma potential and ion flux.

3.Results

A. Characterization of InP etching

Figure 1 shows the InP etching rates as a function of the $Ar/(Cl_2+Ar)$ or $N_2/(Cl_2+N_2)$ mixing ratio. The substrate temperature was kept to about 150°C. The InP etching rate was increased drastically by diluting Cl_2 with a small amount of N_2. With increasing N_2 concentration, the etching rate was decreased rapidly. It should be noted that InP could be etched slightly (about 100 Å/min) even in pure N_2 plasma. On the other hand, in the case of diluting Cl_2 with Ar, the etching rate was decreased gradually with increasing the Ar concentration, and eventually became zero in pure Ar plasma.

Figure 2(a)-2(c) are the SEM micrographs of InP etching profiles etched by pure Cl_2, Cl_2/Ar (=1sccm/9sccm) and Cl_2/N_2 (=3sccm/7sccm),

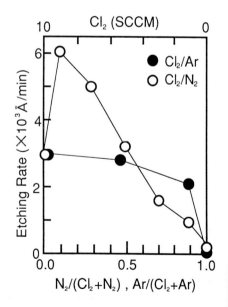

Fig.1. Dependence of the etching rate of InP on $N_2/(Cl_2+N_2)$ or $Ar/(Cl_2+Ar)$ ratio.

respectively.

With pure Cl_2, the etched surface seriously became rough [Fig.2(a)]. The rough etched surface could not be eased by diluting Cl_2 with 90% Ar content [Fig.2(b)]. On the contrary, remarkable smooth surface was obtained by diluting Cl_2 with N_2, and abruptness of the edge profile could be improved with increasing N_2 mixing ratio [Fig.2(c)]. The excellent etching performances of smooth surface and abrupt edge profile could be achieved with $Cl_2(40\%)/N_2(60\%)$ mixture, realizing in high etching rate of more than 1000 Å/min.

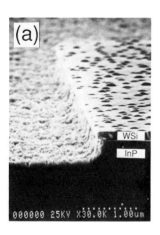

Fig.2. SEM photograph of the InP etched by (a) Cl_2=10sccm, (b) Cl_2/Ar=1sccm/9sccm, (c) Cl_2/N_2=3sccm/7sccm.

B. Plasma diagnostics

Etching plasma was diagnosed by OES system during the etching process. The intensity of Cl atomic line at 725.6nm indicates the Cl radical density in the processing plasma. Figure 3 shows the dependence of the optical emission intensity of Cl on the gas mixing ratio in both cases of Cl_2/Ar and Cl_2/N_2 mixture. Noticable reduction of the emission intensity can be observed in even small amount of N_2 mixing to Cl_2 gas. This may arise from the reaction of Cl+N→NCl, which is caused by the dissociation of Cl_2 and N_2 molecules in the plasma[Colin and Jones 1967]. This result indicates that the Cl neutral radical density is extremely reduced to a large extent by the N_2 addition than by the inert gas such as Ar.

The etching plasma was also inspected through the measurements of plasma potential, dc bias voltage of the substrate induced by rf

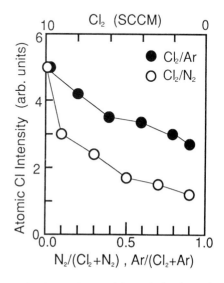

Fig.3. Dependence of the emission intensity from atomic Cl on $N_2/(Cl_2+N_2)$ or $Ar/(Cl_2+Ar)$ ratio.

power and the ion flux, as demonstrated in Figure 4 and 5, respectively. The plasma potential and the dc bias voltage increased with increasing diluting ratio in both cases of Ar and N_2 [Fig.4]. The incident ion energy to the substrate can be estimated to about 20~30eV from dc bias voltage and the plasma potential. The ion flux also increased with further addition of Ar and N_2 [Fig.5]. However, there appeared to be little difference between the ion energy and the ion flux in either type of gas mixture.

C. Chemical analysis of etched surface

The reaction products remaining on the etched surface were analyzed by XPS to examine the chemical reactions in the etching plasma. Figure 6 shows the XPS survey spectra of the etched surfaces demonstrated in Fig.2(a) and Fig.2(c), respectively. (A) is the spectrum from unetched InP control sample. (B) and (C) correspond to the spectra for the surface etched by pure Cl_2, and to that by Cl_2/N_2 (=3sccm/7sccm), respectively. These samples were prepared by the treatment of deionized water rinsing immediately after the etching, in order to remove the free Cl atoms adsorbed on the surfaces, which may conceal some information about chemical bonding states in the near-surface region.

The XPS spectra of In_{3d}, P_{2p}, N_{1s} peaks were analyzed by curve-fitting technique to distinguish the etch product peaks from the InP surface. Few Cl residues were detected on all of samples, which may be washed out by the water rinsing treatment mentioned above. In the case of the etching using Cl_2/N_2 mixture, we found that a large quantity of InN was formed on the InP surface, moreover, a fairly large amount of P_3N_5 and/or $R_2PO(OH)$ products exist, compared with the other samples.

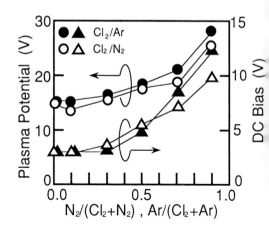

Fig.4. Dependence of the plasma potential and dc bias voltage on $N_2/(Cl_2+N_2)$ or $Ar/(Cl_2+Ar)$ ratio.

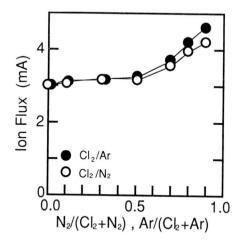

Fig.5. Dependence of the ion flux on $N_2/(Cl_2+N_2)$ or $Ar/(Cl_2+Ar)$ ratio.

It is considered that the possibility for the formation of P_3N_5 is higher than that of $R_2PO(OH)$ [R the organic functional group], which is generated by the water treatment. From these results, wit increasing N_2 mixing ratio in Cl_2/N_2 plasma, the nitridation of InP was deduced to be the dominar surface reaction. Although the vapor pressure of P_3N_5 is not well known, it can be supposed that the volatility of P_3N_5 is much lower than that of PCl_x from its thermal stability point of view.

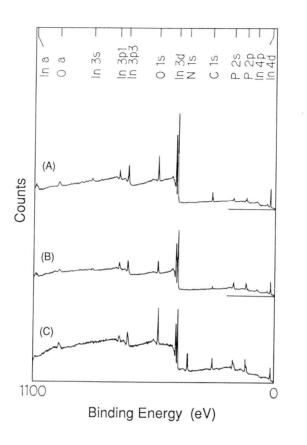

Fig.6. XPS survey spectra from InP control sample (A), etched sample by Cl_2=10sccm (B) and etched sample by Cl_2/N_2=3sccm/7sccm (C).

4. Discussion

The relation between the etching performances and the diagnostics of the etching plasma and moreover the XPS inspections of etch products on the surface have been studied in this work. As a consequence, it was found that the surface roughness and edge profile of InP etching strongly depend on Cl radical density in the etching plasma and the reaction between InP surface and etching gases. The excellent etching performances of smooth surface and vertical sidewall profile can be obtained only under limited condition with low density of Cl radical in the Cl_2/N_2 plasma.

Mechanism of the InP dry etching, described above, with low Cl radical density plasma and low energy (about 30eV) ion bombardment with ECR plasma may be based on a new concept in contrast to the conventional method, enhancing the desorption rate of non-volatile $InCl_x$ products in comparably to that of PCl_x products by high energy ion bombardments. The benefits of utilization of Cl_2/N_2 gas mixture in InP etching with ECR plasma can be attributed to the following two effects, one is the reduction of Cl radical density in the processing plasma, and the other is the nitridation of

In and P. The processing plasma may be in the Cl radical transport limit due to the reduction of Cl radical density, resulting in the effective suppression of PCl_x formation and its desorption. In addition to this, the formation rate of PCl_x can remarkably be reduced by the reaction of non-volatile P_3N_5 formation with increasing N_2 mixing. With these two effects, the selective desorption of PCl_x can be suppressed compared with the other reaction products. Consequently, the smooth etched surface has been realized even in such a low ion energy plasma.

This technique is to control the volatilizing rate of PCl_x products balancing to the desorption rate of $InCl_x$ products with the reduction of Cl radical density and nitridation of In and P. By employing this unique method, InP etching with smooth surface, vertical edge profile and high etching rate can be realized for the first time without high energy ion assists.

5. Conclusion

Low energy (about 30eV) ion etching of InP was performed by an ECR plasma with a Cl_2/N_2 mixture. When the N_2 concentration became higher than 60%, the smooth surface and the vertical sidewall profile were obtained with the high etching rate (more than 1000 Å/min) which has not ever reported in ECR plasma etching. These characteristics are considered to result from the extreme reduction of Cl radical density due to the reaction of $Cl+N \rightarrow NCl$ and the nitridation of In and P in Cl_2/N_2 plasma.

References
Franz G. 1990 J.Electrochem.Soc.137,p.2897
Hayes T R et al. 1989 J.Vac.Sci.Technol.B7,p.1130
Contlini R J. 1988 J.Electrochem.Soc.135,p.929
Pearton S J et al. 1990 Appl.Phys.Lett.56,p.1424
Miyakuni S. 1994 J.Vac.Sci.Technol.B12,p.530
Colin R and Jones W E. 1967 Canad.J.Phys.45,p.301

Surface structure of GaSb and AlSb grown by molecular beam epitaxy

Berinder Brar, Devin Leonard and John H. English
Department of Electrical and Computer Engineering
University of California Santa Barbara, Santa Barbara, CA 93106

> **Abstract.** Atomic force microscopy is employed to obtain images of the surface of GaSb and AlSb epilayers grown on (001) GaAs and GaSb using molecular beam epitaxy. The surface of the lattice-mismatched GaSb epilayers is characterized by mounds that have a spiral step structure. The spirals nucleate at threading screw dislocations present in the lattice-mismatched epilayer. Homoepitaxial layers of GaSb grown on (001) GaSb also exhibit mounds on the surface, however without the accompanying spiral steps observed on the mismatched epilayers. AlSb epilayers are observed to be smooth compared to the GaSb epilayers, and show no evidence of the spiral step structure.

The nearly lattice-matched ($a \approx 6.1$ Å) compound semiconductors GaSb, AlSb and InAs, offer a great deal of flexibility for bandgap engineered devices such as infrared detectors,[1,2] heterostructure field-effect transistors,[3] and resonant tunneling devices.[4,5] The device structures are usually grown on lattice-mismatched (001) GaAs semi-insulating substrates and employ thick buffer layers of GaSb and AlSb to allow the attenuation of threading dislocations. A recent study by Kyutt et. al.[6] shows that the large lattice mismatch between GaSb epilayers and the GaAs substrate ($\Delta a/a \approx 7\%$) is accommodated by the nucleation of a high density of threading screw dislocations 10^{10} cm^{-2} at the hetero-interface, decreasing to approximately 10^8 cm^{-2} after 1 μm of material has been grown. In the present paper, we use an atomic-force microscope to obtain detailed topological information of GaSb and AlSb epilayers grown by molecular beam epitaxy (MBE) in order to investigate their growth in the presence of a high density of threading screw dislocations.

All samples were grown in a Modular Varian GENII MBE machine equipped with elemental group III and V sources. The antimony source was uncracked molecular Sb$_4$. The GaSb and AlSb epilayers were grown at 530 °C and 570 °C respectively, with a nominal growth rate of 1 μm/hr and an Sb beam equivalent pressure of 2×10^6 Torr. A 50 nm AlSb nucleation layer was employed in the GaSb samples to initiate growth on the GaAs substrate. After the growth of the epilayer, the samples were cooled in an Sb beam. The AlSb samples were capped with 6 monolayers of GaSb to protect the surface. After the samples were removed from the MBE machine they were measured in air at room temperature using a Digital Instruments Nanoscope III AFM with a 200 μm cantilever. The sample surfaces remained stable in air, and the images taken immediately after removal from the MBE machine were nominally identical to those taken a few weeks later.

Fig. 1. shows a 5 μm × 5 μm AFM image of the MBE grown heteroepitaxial GaSb surface for the sample grown at 530 °C on a singular (001) GaAs substrate. The bright spots correspond to a vertical dimension of 6 nm compared to the dark areas of the image. The surface consists of a large number of mounds approximately 4 nm high. The largest mounds on the surface have a lateral dimension of $\approx$ 1 μm × 1.5 μm. The mounds on the surface are responsible for the "orange-peel" appearance of the surface when examined under a Nomarski microscope at 100× magnification. The mounds are elongated in the [01$\bar{1}$] direction, and their

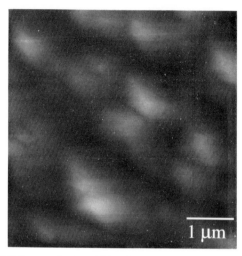

Fig. 1. 5 μm × 5 μm AFM image of a 1 μm thick epilayer of GaSb grown on a (001) GaAs substrate. The bright areas of the image are 6 nm tall compared to the dark areas of the image.

areal density is $\approx 5 \times 10^7$ cm^{-2}. Monolayer-height (0.3 nm) steps are visible on the surface of the mounds in Fig. 1.

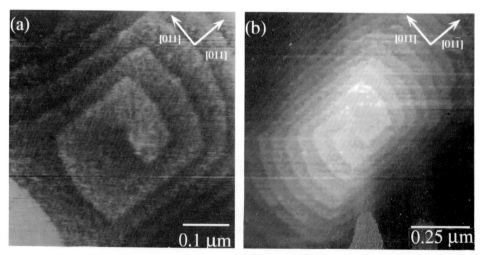

Fig. 2. High-resolution images of the mounds observed in Fig. 1. 2(a) shows the spiral step edge emanating from a screw dislocation at the center of the mound. 2(b) shows a three lobed spiral corresponding to three screw dislocations at the center of the mound.

A higher-resolution image of one of the mounds from Fig. 1 is shown in Fig. 2(a). The detailed structure of the steps can be seen quite clearly. The surface consists of a single step edge that originates from the center of the mound and proceeds outwards in a spiral fashion until it intersects a similar step edge from one of the neighboring mounds. *All* the mounds shown in Fig. 1 display a similar spiral step edge. The spiral growth mode has been studied in GaAs homoepitaxial layers grown by organometallic vapor phase epitaxy,[7] and occurs as a result of screw dislocations on the surface. A screw dislocation has a Burgers vector that points in the direction of the dislocation line and presents a *persistent* step source on the growing surface. In the samples we have studied, we observe growth spirals that have both a clockwise and an anti-clockwise sense. Mounds with one, two, and three spiral lobes originating from the center are also observed. Fig. 2(b) shows a spiral GaSb mound with three anti-clockwise lobes originating from three screw dislocations present near the center of the mound. A more detailed study of spiral growth in mismatched GaSb epilayers will be reported elsewhere.[8]

Fig. 3(a) is an AFM image of a *homo*epitaxial layer of GaSb grown on a singular (001) GaSb substrate. The image also shows the presence of $\approx 2 \times 10^7$ cm^{-2} mounds on the surface. As compared to the mounds on the mismatched GaSb epilayers, the homoepitaxial mounds are not as tall and are larger in lateral size. Fig. 3b is a high-resolution image of a mound on the homoepitaxial layer of GaSb. The step structure associated with the mounds observed on the homoepitaxial lattice-matched GaSb surface is not spiral, but consists of *concentric* rings indicative of a two-dimensional growth mode without the presence of screw dislocations.

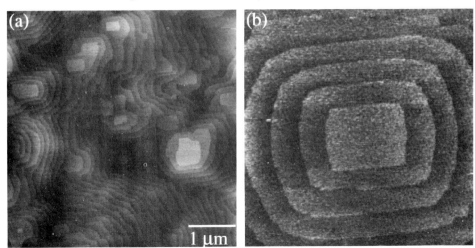

Fig. 3.(a) 5 µm × 5 µm AFM image of a 1 µm thick *homo*epitaxial epilayer of GaSb grown on a (001) GaSb substrate. The bright areas of the image are 6 nm tall compared to the dark areas of the image. 3(b) is a high-resolution image of a mound on the homoepitaxial surface.

In order to study the mismatched AlSb epilayers on (001) GaAs substrates two additional samples were grown . The first AlSb sample consisted of 1 µm thick layer of AlSb with a 6 monolayer thick GaSb cap employed to stabilize the surface in air. Fig. 4(a) is an AFM image of the surface. The surface is smooth compared to the GaSb epilayers, and does not have an "orange-peel" appearance under any magnification of the nomarski optical microscope. The AFM image shows no evidence of the characteristic spiral mounds of the mismatched GaSb epilayers. We believe that the absence of spiral growth in the AlSb epilayers is a direct manifestation of the lower surface-mobility of the Al adatom, which does not allow preferential growth at the persis-

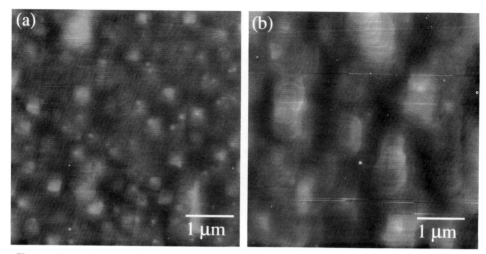

Fig. 4.(a) 5 μm × 5 μm AFM image of a 1 μm thick epilayer of AlSb. 4(b) is an AFM image of a 100 nm thick AlSb layer grown on a 1 μm thick GaSb buffer layer. Both samples were grown on (001) GaAs substrates and employed a a 6 monolayer thick GaSb cap layer.

tent step edge associated with the screw dislocation. The second AlSb sample consisted of a 100 nm thick AlSb layer grown on a 1 μm thick GaSb buffer that provides a template of spiral mounds. After the AlSb layer 6 monolayers of GaSb are deposited to protect the surface. Fig. 4(b) is an AFM image of the second AlSb sample grown on the GaSb buffer. The surface shows mounds with approximately the same density as observed in GaSb epilayers. The few one-lobe spirals that are observed presumably form during the deposition of the thin GaSb caplayer. Most of the mounds do not show any evidence of the spiral growth that was present in the GaSb buffer. We believe that the AlSb layer grows conformally on the GaSb surface thereby reproducing the "orange-peel" structure of the GaSb buffer layer. However, without the high surface mobility of the group III adatom, AlSb is unable to maintain the spiral step structure around the screw dislocations.

We would like to acknowledge fruitful discussions with Dr. Bauser and Dr. Gossard and the support and encouragement of Dr. Kroemer and Dr. Petroff.

References

[1] L. Samoska, B. Brar, and H. Kroemer, Appl. Phys. Lett. **62**, 2539 (1993).
[2] Y. Zhang, N. Baruch, and W. I. Wang, Appl. Phys. Lett. **63**, 1068 (1993).
[3] C. R. Bolognesi, E. J. Caine, and H. Kroemer, IEEE Electron Device Letters **15**, 16 (1994).
[4] D. H. Chow and J. N. Schulman, Appl. Phys. Lett. **64**, 76 (1994).
[5] E. R. Brown, S. J. Eglash, G. W. Turner, C. D. Parker, J. V. Pantano, and D. R. Calawa, IEEE Transactions on Electron Devices **41**, 879 (1994).
[6] R. N. Kyutt, R. Scholz, S. S. Ruvimov, T. S. Argunova, A. A. Budza, S. V. Ivanov, P. S. Kopev, L. M. Sorokin, and M. P. Scheglov, Phys. Solid State **35**, 372 (1993).
[7] C. C. Hsu, Y. C. Lu, J. B. Xu, and I. H. Wilson, Appl. Phys. Lett. **65**, 1959 (1994).
[8] B. Brar and D. Leonard (to be published)

Inst. Phys. Conf. Ser. No 141: Chapter 3
Paper presented at Int. Symp. Compound Semicond., San Diego, 18–22 September 1994
© 1995 IOP Publishing Ltd

Thermal stability of acceptor Si in δ-doped GaAs grown on GaAs(111)A by molecular beam epitaxy

M. Hirai, H. Ohnishi, K. Fujita and T. Watanabe

ATR Optical and Radio Communications Research Laboratories
2-2 Hikaridai, Seika-cho, Soraku-gun, Kyoto 619-02, Japan

Abstract: We report on the thermal stability of Si in δ-doped GaAs layers on GaAs substrates with various orientations. Diffusivity is strongly affected by the off-angle for a misoriented (111)A substrate; the diffusion coefficient for exactly (111)A is the smallest, and becomes larger as the off-angle increases in the [110] or [001] direction. This diffusivity dependence on substrate misorientation can be understood by the step density, depending on the off-angle. Furthermore, measured diffusion parameters in GaAs(111)A are found to differ from those in GaAs(001); the activation energy in GaAs(111)A is larger than that in GaAs(001).

This study shows that acceptor Si is more stable than donor Si. The low diffusivity of Si δ-doped layers on (111)A makes them suitable as p-type heavily doped thin layers in electronic devices.

1. Introduction

Since the characteristics of semiconductor devices depend strongly on the final impurity doping profiles in the electrically active region, a detailed understanding of impurity diffusion is required to control the impurity doping profiles. Impurity diffusion in III–V semiconductors has been studied but many problems remain unresolved. For example, several values have been reported for the diffusion coefficient and activation energy of Si in GaAs [1-4].

Silicon is widely used as the n-type dopant in GaAs growth using molecular beam epitaxy (MBE). It has been reported, however, that in GaAs growth on (n11)A substrates, Si atoms are incorporated as acceptors when n≤3 [5]. We have previously reported that Si acceptors can be heavily doped in GaAs(111)A up to 6×10^{19} cm^{-3} [6] and that Si δ-doped GaAs(111)A is stable compared with δ-doped GaAs(001) [7]. These results are attractive because they show that stable thin p-type layers can be grown by using GaAs(111)A substrates. However, little attention has focused on the effect of substrate misorientation on the diffusivity of Si and the mechanism for diffusion of Si, depending on the occupation site of Si, has not yet been clarified.

In this article we report our investigation on the thermal stability of Si atoms in Si δ-doped GaAs layers on GaAs substrates with various misorientations, and report on the estimation of the diffusion coefficient and activation energy of acceptor Si on (111)A.

2. Experimental procedure

Silicon δ-doped GaAs layers were grown in a Varian Gen II MBE system on semi-insulating GaAs(111)A substrates misoriented 0-55° toward the [110] or [001] direction.

The growth temperature and the growth rate of GaAs were 600 °C and 0.6 μm/h, respectively. The As_4 beam equivalent pressure was 3.0×10^{-5} Torr and the As_4/Ga flux ratio was 7.0. The epitaxial layer consisted of a 200 nm undoped GaAs layer, a Si δ-doped GaAs layer with a doping concentration of $2-4 \times 10^{12}$ cm^{-2} and a 100 nm undoped GaAs cap layer. Post-growth annealing was performed in a rapid thermal annealer (RTA) at 800 °C for 60 sec under N_2 atmosphere. During the anneal, each GaAs sample was placed upside down on a fresh GaAs wafer and another GaAs cover wafer was placed on top in contact with the sample for surface protection. Secondary ion mass spectrometry (SIMS) was used to measure the Si profiles of the Si δ-doped GaAs. Diffusivity was evaluated by taking the full width at half maximum (FWHM) of the Si profile for an as-grown sample and an annealed sample. The SIMS analysis was performed using an O_2^+ primary beam of 3.0 keV impact energy. These conditions were selected to reduce the knock-on effect which is the primary cause limiting the depth resolution of SIMS profiles.

3. Results and discussion

3.1 Substrate misorientation effects

The morphology of each sample was mirrorlike except for (110). The surface features for (110) developed during growth. Figure 1 shows the carrier concentration of as-grown samples. The conductivity type for misoriented (111)A with the off-angle in the [001] direction tends to be n-type except for (111)A misoriented 5° toward the [001] direction and (113)A, while that for misoriented (111)A with the off-angle in the [110] direction tends to be p-type except for (331) and (110). These results roughly agree with the results obtained under the growth temperature of 630°C except for (113)A and (331); the conductivity type for (113)A and (331) are p-type and n-type, respectively. This result indicates that the conductivity type for (113)A and (331) is affected by the growth conditions, as reported previously [8].

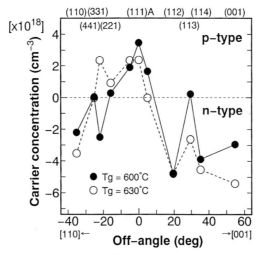

Fig. 1. Carrier concentration versus off-angle of misoriented GaAs(111)A substrate, for a Si δ-doped GaAs layer.

Figure 2 shows the FWHM of Si profiles taken by SIMS for as-grown samples. The FWHM for (110) is not included in this figure because it was very large, due to a poor surface morphology. The FWHM of the Si profile is 7-9 nm for all samples except for (112)A and (001).

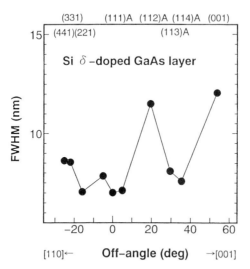

Fig. 2. FWHM of the Si profile in as-grown samples versus off-angle of misoriented GaAs(111)A substrate.

To determine the diffusivity of Si atoms, the diffusion coefficient at each temperature was calculated by means of the following expression[4]:

$$Z_d^2 = 2D_{si}t = (FWHM_A^2 - FWHM_0^2)/4. \qquad (1)$$

Here, Z_d is the diffusion distance, D_{si} is the one-dimensional diffusivity, and t is the anneal time. $FWHM_A$ and $FWHM_0$ denote the FWHM of the Si profile for an annealed sample and an as-grown sample, respectively.

Figure 3 shows the relationship between the diffusion coefficient of Si and off-angle of misoriented (111)A. The diffusion coefficient for (110) is not included in this figure because the FWHM could not be evaluated due to excessive broadening by annealing. The diffusion coefficient for exactly (111)A is the smallest, and becomes larger as the off-angle increases in the [110] or [001] direction expect for (113)A. Furthermore, the diffusion coefficient decreases above (221). This tendency agrees with the result for the growth temperature of 630°C except for (113)A and (331) [8]. These results for (113)A and (331) suggest that the diffusion coefficient is affected by the occupation site of Si atoms because the conductivity type for (113)A and (331) changes to the other type, as shown in Fig. 1.

To understand the reason for the diffusion coefficient dependence on the off-angle of misoriented (111)A, the step density is estimated for each sample. There are several types of steps on misoriented (111)A; e.g. the (001) step induced by misoriented (111)A with the off-angle in the [001] direction, and the (110) step induced by misoriented (111)A with the off-angle in the [110] direction.

Now, we define the terrace length as L. From this, the step density can be calculated as 1/L when using the terrace length. There is a strong correlation between the step density and the diffusion coefficient; the coefficient becomes large as the step density becomes large. This result indicates that the diffusivity of Si is proportional to the step density.

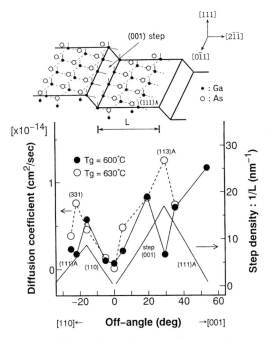

Fig. 3. Diffusion coefficient for Si and step density versus off-angle of misoriented GaAs(111)A substrate.

It is remarkable that the diffusion coefficient is strongly related to the step density of a misoriented substrate. We consider the diffusion coefficient dependence on step density to be related to the occupation site of Si at the step on the misoriented substrate. There are three types of steps, i.e., (001), (110) and (111)A, on misoriented (111)A as shown in Fig. 3. Si atoms are incorporated as an acceptor on the (111)A step while Si atoms are incorporated as a donor on the (001) step or (110) step. We have reported that an acceptor Si is more stable than a donor Si [8]. Therefore, in this case, the diffusivity of Si is enhanced by the presence of the (001) step or (110) step. Initially, the step density of the (001) step or (110) step increases as the off-angle of misoriented (111)A increases in the [110] or [001] direction. As a result, the diffusion coefficient of Si becomes large. But, for larger misorientations, the diffusion coefficient decreases above (221) and (113)A because the (111)A step appears from (113)A and (221) to (001) and (110), respectively, as shown in Fig. 3.

3.2 Diffusion coefficient and activation energy of acceptor Si on (111)A

To determine the diffusion parameters, after being encapsulated with SiN, post-growth annealing was additionally performed (in the RTA at 750–950 °C for 60 sec under N_2 atmosphere). The temperature dependence of the diffusion coefficient of Si δ-doped GaAs layers on (111)A and (001) is shown in Fig. 4. The activation energies of Si diffusion in the (111)A and (001) samples (determined from an Arrhenius plot) are $E_{a(111)A}$ = 2.74 eV and $E_{a(001)}$ = 2.48 eV, respectively. The extrapolated diffusion coefficients at T → ∞ in (111)A and (001) are $D_{0(111)A}$ = 1.14x10^{-2} cm^2/sec and $D_{0(001)}$ = 3.46x10^{-3} cm^2/sec, respectively.

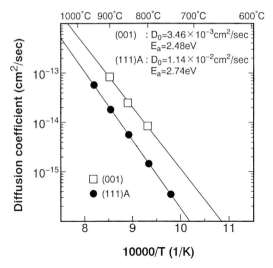

Fig. 4. Arrhenius plot of the diffusion coefficient for Si δ-doped GaAs layers grown on GaAs(111)A and GaAs(001).

As shown in Fig. 4, the diffusion coefficient of the (001) sample is larger than that of the (111)A sample in this temperature range. Furthermore, the activation energy of the (111)A sample is larger than that of the (001) sample. These results indicate that Si δ-doped GaAs layers grown on a (111)A substrate are more stable than the same layers grown on a (001) substrate. Our results for the (001) sample are in agreement with the values of $D_{0(001)} = 4 \times 10^{-4}$ cm^2/sec and $E_{a(001)} = 2.45$ eV, reported by Schubert et al. [3].

We consider that the activation energy of acceptor Si differs from that of donor Si, and that the difference in activation energy between acceptor Si and donor Si is related to the occupation of the neighboring lattice sites, i.e., acceptor Si is surrounded by Ga atoms, while donor Si is surrounded by As atoms. Therefore, acceptor Si has Si-Ga bonds and donor Si has Si-As bonds. Accordingly, for Pauling's model [9], the binding energy of Si-Ga and that of Si-As are 2.13 eV and 1.65eV, respectively. These indicate that acceptor Si is more stable than donor Si because Si-Ga bonds are stronger than Si-As bonds. From the discussion above, the most stable situation for a Si-doped GaAs layer is reached when all of the Si atoms are in As sites.

We think that Si δ-doped GaAs layers grown on GaAs(111)A substrates are suitable for application in electronic devices, such as heterojunction bipolar transistors (HBTs), that require p-type heavily doped thin layers with low diffusivity.

4. Conclusions

In this study, it was found that diffusivity is strongly affected by the off-angle of misoriented (111)A; the diffusion coefficient for exactly (111)A is the smallest, and becomes large as the off-angle increases in the [110] or [001] direction. This diffusivity dependence on the off-angle can be understood by the density of the step, which is caused by substrate misorientation. It was also found that the diffusion parameters in (111)A differ from those in (001); the activation energy in (111)A is larger than that in (001).

Acknowledgments

The authors would like to thank Dr. Hideyuki Inomata for his encouragement throughout this work.

References

[1] Greiner M E and Gibbons J F, *Appl. Phys. Lett.* **44** (1984) 750–752.
[2] Omura E, Wu X S, Vawter G A, Coldren L, Hu E and Merz J L, *Electron. Lett.* **22** (1986) 496–498.
[3] Schubert E F, Stark J B, Chiu T H and Tell B, *Appl. Phys. Lett.* **53** (1988) 293–295.
[4] Cunningham J E, Chiu T H, Jan W and Kuo T Y, *Appl. Phys. Lett.* **59** (1991) 1452–1454.
[5] Wang W I, Mendez E E, Kuan T S and Esaki L, *Appl. Phys. Lett.* **47** (1985) 826–828.
[6] Okano Y, Seto H, Katahama H, Nishine S, Fujimoto I, and Suzuki T, *Japan J. Appl. Phys.* **28** (1989) L151–L154.
[7] Shinoda A, Yamamoto T, Inai M, Takebe T and Watanabe T, *Japan J. Appl. Phys.* **32** (1993) L1374–L1376.
[8] Hirai M, Ohnishi H, Fujita K and Watanabe T, *Appl. Surf. Sci.* (to be published)
[9] Pauling L, *"The Nature of the Chemical Bonds"*, Cornell Univ. Press, New York, (1960) 572.

Al/Ga atom–exchange during AlAs/GaAs heterointerface formation in alternating source supply

T. Saitoh, M. Tamura and J. E. Palmer

Optoelectronics Technology Research Laboratory (OTL)
5-5 Tohkodai, Tsukuba, Ibaraki 300-26, Japan

Abstract. We have investigated the Al/Ga atom–exchange phenomenon at an AlAs/GaAs interface formed by an alternating supply of Al and As_4 in molecular beam epitaxy. The surface composition during interface formation was measured *in situ* using coaxial impact collision ion scattering spectroscopy, and the structure of the formed interface was examined by cross–sectional transmission electron microscope. By varying the amount of Al atoms per cycle in an alternating source supply, we found that an Al/Ga atom–exchange occurs when excess Al atoms which have no bonds with As atoms exist on the surface during interface formation, while an abrupt interface is formed under the As–rich growth condition. This result suggests that the bonding of Al atoms to As atoms on the growing surface is important for obtaining a well–controlled AlAs/GaAs heterointerface without an atom–exchange at the interface.

1. Introduction

Heterointerfaces controlled on the atomic scale are required for achieving higher performance of quantum devices. For heterointerface systems of III–V compound semiconductors, such as AlAs/GaAs and GaAs/InAs, the atom–exchange between the epitaxial layer and the substrate during growth is one of the most essential issues for controlling the abruptness of the interfaces on the atomic scale. We have studied the Ga/In atom–exchange at the GaAs/InAs interface using coaxial impact collision ion scattering spectroscopy (CAICISS), and showed that a Ga/In atom–exchange occurs more frequently under a lower arsenic pressure [1]. This result suggests that the III/V ratio on the surface is an important factor for controlling such an atom–exchange phenomenon at the interface.

In this study, we investigated the Al/Ga atom–exchange at an AlAs/GaAs interface formed by an alternating supply of Al and As_4, as a function of the III/V ratio on the surface during growth. The ratio of Al to As atoms on the GaAs surface was varied by changing the amount of Al atoms deposited per cycle. The processes of Al/Ga atom–exchange are discussed based on the results of real–time observations by CAICISS and structural analyses by cross–sectional transmission electron microscope (XTEM).

2. Experimental

Crystal growth and CAICISS measurements were carried out in a molecular beam epitaxy (MBE) chamber equipped with a CAICISS analyzer [2]. GaAs(001) substrates were first dipped into H_2SO_4, rinsed in deionized water, and introduced to a growth chamber through a loading chamber. The substrates were heated to 620 °C under an As_4 flux to remove native oxides, and subsequently GaAs buffer layers of about 200 nm were grown by conventional MBE at 550 °C. The GaAs surface was then alternately irradiated with Al and As_4 beams. The Al beam was supplied at a rate of 0.1 monolayer (ML)/s. The amount of Al atoms per cycle was varied from 1 to 5 ML.

CAICISS measurements were carried out *in situ* during the growth of AlAs for measuring the surface composition. A pulsed He^+ ion beam of 2 keV was incident on the sample, parallel to the (100) plane with an incident angle of 12° from the surface. Back-scattered He particles (mainly neutrals) were detected by a micro-channel-plate, which was located on the same axis as the incident beam. XTEM observations were performed for examining the structure of the AlAs/GaAs interfaces.

3. Results

First, we compared two different growth processes for 5 ML-AlAs layers on GaAs substrates, in which the AlAs layers were grown by an alternating supply of 5 cycles of 1 ML-Al and As_4 (Sample A) and 1 cycle of 5 ML-Al and As_4 (Sample B). These samples were grown at 550 °C. Figures 1 (a) and (b) show the time-resolved variations in the scattering intensity of a He^+ ion beam from both As and Ga atoms during AlAs/GaAs interface formation in cases of Samples A and B, respectively. In Fig. 1 (a), the scattering intensity decreases during Al supply, and maintains an almost constant value during As_4 supply until the third cycle of the source supply, then the intensity saturates for further growth. The decrease in the scattering intensity observed here is interpreted as the decrease in the scattering from Ga atoms [3]. The value of the saturated intensity for more than 3 ML-AlAs growth is about half of that from the GaAs surface. This result means that Ga atoms are hidden by the AlAs layer and the scattering intensity becomes that due to scattering from only As atoms. The detectable depth in the present CAICISS measurement is about 3 ML [3]. Therefore, the result in Fig.1 (a) indicates that an abrupt AlAs/GaAs interface was formed by layer-by-layer growth of AlAs in Sample A.

In Fig. 1 (b), the scattering intensity decreases while Al atoms are supplied to the surface and rapidly increases immediately after As_4 supply. The value of the scattering intensity after 5 ML-AlAs growth is higher than that of Sample A shown in Fig. 1 (a). This result indicates that Ga atoms still exist at a depth that is detectable by CAICISS. This suggests that Al/Ga atom-exchange occurred between the grown layer and substrate in Sample B. The processes of Al/Ga atom-exchange are discussed in the section 4.

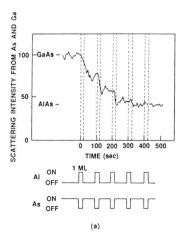
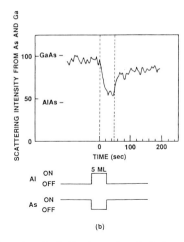

Figure 1 Time-resolved variations in the scattering intensity of a He$^+$ ion beam from both As and Ga atoms during AlAs/GaAs interface formation. Figs. 1 (a) and (b) correspond to the cases where 5 ML-AlAs layers were grown by an alternating supply of 5 cycles of 1 ML-Al and As$_4$ and 1 cycle of 5 ML-Al and As$_4$, respectively.

Figures 2 (a) and (b) show RHEED patterns of the surfaces after the growth of a 5 ML-AlAs layer of Sample A and B, respectively. In the case of Sample A (Fig. 2 (a)), the RHEED pattern does not show a clear surface reconstruction, although a faint twofold pattern is observed for an incidence from the [110] direction. On the other hand, a clear (2x4) pattern can be seen from Sample B (Fig. 2 (b)). This difference in the RHEED patterns also shows difference in the growth processes between Samples A and B. Figures 2 (c) and (d) show RHEED patterns of the surfaces of a 200 nm thick GaAs layer and a 200 nm thick AlAs layer on GaAs substrates grown by conventional MBE at 550 °C, respectively. The RHEED patterns of GaAs and AlAs surfaces show (2x4) and (3x1) reconstructions, respectively. The fact that the (2x4) pattern was observed on the surface of Sample B (Fig. 2 (b)) strongly suggests that an (Al)GaAs layer was formed on the surface, which is in good agreement with the result

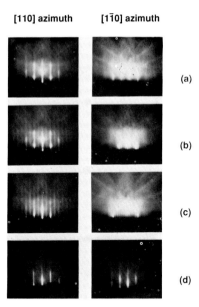

Figure 2 RHEED patterns of the surfaces after 5 ML-AlAs layers were grown on GaAs by the alternating supply of 5 cycles of 1 ML-Al and As$_4$ (a) and 1 cycle of 5 ML-Al and As$_4$ (b). Figs. 2 (c) and (d) show the RHEED patterns of the surfaces of a 200 nm thick GaAs layer and a 200 nm thick AlAs layer on GaAs grown by conventional MBE.

that the contribution from Ga atoms was detected in a CAICISS measurement for Sample B. The reason why the (3x1) pattern was not observed from the surface of Sample A (Fig. 3 (a)) is not clearly understood. We speculate that this is due to the fact that the AlAs film is extremely thin (5 ML).

Next, the structure of AlAs/GaAs interfaces, where AlAs layers were grown by alternating source supply of various amounts of Al atoms, was compared by XTEM. The samples used for XTEM observations were grown at 400 °C. Figures 3 (a), (b), (c) and (d) show the XTEM images of GaAs/AlAs/GaAs structures, where the amount of Al atoms per cycle was 1, 2, 3 and 4 ML, respectively. The number of cycles in the alternating source supply was 12, 6, 4 and 3, respectively, thus the total amount of Al atoms was kept constant at 12 ML for these four samples. The shutter sequences during growth of AlAs layers are also shown in Fig. 3. In Figs. 3 (a) and (b), the AlAs layers are seen as continuous "white" lines with a thickness of about 3.3 nm (12 ML) located between the "black" GaAs regions. Each interface of AlAs/GaAs is very clear, indicating that an abrupt interface was formed under these growth conditions.

On the other hand, in Figs. 3 (c) and (d), black lines can be clearly seen in both AlAs layers. The positions of the most distinct black line in the AlAs layers are at depths of about 1/4 and 1/3 in the AlAs layers from the AlAs/GaAs interface in Figs. 3 (c) and (d), respectively. Considering that the number of cycles in the source supply is 4 and 3, respectively, the black line is located at the positions where

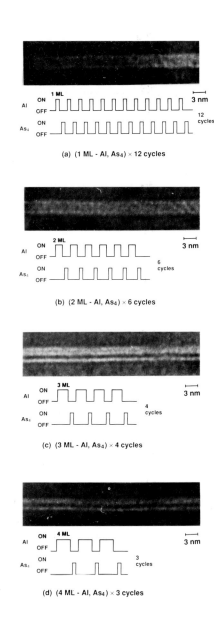

(a) (1 ML - Al, As$_4$) × 12 cycles

(b) (2 ML - Al, As$_4$) × 6 cycles

(c) (3 ML - Al, As$_4$) × 4 cycles

(d) (4 ML - Al, As$_4$) × 3 cycles

Figure 3   Cross-sectional TEM images of GaAs/AlAs/GaAs structures. 12 ML-AlAs layers were grown by the alternating supply of 12 cycles of 1 ML-Al and As$_4$ (a), 6 cycles of 2 ML-Al and As$_4$ (b), 4 cycles of 3 ML-Al and As$_4$ (c) and 3 cycles of 4 ML-Al and As$_4$ (d).

were the sample surfaces just after the first cycle in source supply. Taking into account that the results of the CAICISS and RHEED observations showed the existence of Ga atoms when the AlAs layer was grown by an alternating supply of excess Al atoms and As_4 (Fig. 1 (b) and Fig. 2 (b)), the black lines in the AlAs layers observed in the TEM images are thought to be (Al)GaAs layers. Such (Al)GaAs layers were not observed in cases where the amount of Al atoms per cycle is less than 2 ML (Figs. 3 (a) and (b)). These results indicate that there exists a threshold for the amount of Al that induces an Al/Ga atom−exchange at the interface. The TEM images also show that the Ga atoms did not become uniformly distributed in the grown layer but preferentially located on the surface after growth.

4. Discussion

In the previous section, we showed that Al/Ga atom−exchanges occur under certain growth conditions. Here, we discuss the processes of the Al/Ga atom−exchange during interface formation. In the TEM images shown in Fig. 3, the black lines which are interpreted as being (Al)GaAs layers are formed when the amount of Al atoms per cycle is more than 2 ML. From this result, we found that Al/Ga atom−exchange is promoted when the amount of Al atoms per cycle is more than 2 ML at a growth temperature of 400 °C. Under this growth condition, the RHEED pattern of the GaAs surface before AlAs growth showed the c(4x4) reconstruction. It is known that the c(4x4) reconstructed−GaAs(001) surface is formed with an As−coverage of 1.75 ML [4]. When Al atoms of over 2 ML are supplied on the c(4x4) GaAs surface, there would exist excess Al atoms which would not make bonds with As atoms on the surface. We consider these excess Al atoms are free to exchange with Ga atoms in the substrate, resulting in the promotion of an Al/Ga exchange at the interface.

In Fig. 1 (b), the scattering intensity decreases during the supply of 5 ML−Al atoms, which implies that the AlAs layer grows while supplying Al atoms in spite of no supply of As_4. This suggests that excess Al atoms immediately exchange with Ga atoms in the GaAs substrate and bond with As atoms, resulting in the formation of an AlAs layer on the surface. Morishita et al. observed by micro−probe RHEED that Ga droplets are formed on the surface when excess Al atoms are supplied onto a GaAs(111) surface caused by an Al/Ga atom−exchange [5]. Although there is a difference in the growth conditions between their experiments and ours, the above−mentioned fact concerning the formation of Ga droplets strongly supports our consideration. The rapid increase in the scattering intensity immediately after the supply of As_4 in Fig. 1 (b) indicates that the Ga atoms which already existed on the surface by Al/Ga exchange reacted with As_4 and rapidly formed (Al)GaAs layer on the surface. This can explain the preferential existence of Ga atoms that is seen as black lines in the TEM images shown in Figs. 3 (c) and (d).

5. Conclusion

We investigated the Al/Ga atom-exchange at an AlAs/GaAs interface formed by an alternating source supply. We found that this exchange occurs when excess Al atoms which do not have bonds with As atoms exist on the surface during interface formation. This result suggests that the bonding of Al atoms to As atoms on the growing surface is important for obtaining a well-controlled AlAs/GaAs heterointerface without an atom-exchange at the interface.

Acknowledgement

The authors would like to thank Dr. Y. Katayama and Dr. I. Hayashi for their valuable discussions and continuous encouragement.

References

[1] T. Saitoh, A. Hashimoto and M. Tamura, Appl. Surf. Sci. 60/61, 228 (1992).
[2] A. Hashimoto, N. Sugiyama and M. Tamura, Jpn. J. Appl. Phys. 30, 3755 (1991).
[3] T. Saitoh, J. E. Palmer and M. Tamura, Jpn. J. Appl. Phys. 32, L476 (1993).
[4] D. K. Biegelesen, R. D. Bringans, J. E. Northrup and L. E. Swartz, Phys. Rev. B41, 5701 (1990).
[5] Y. Morishita, S. Goto, Y. Nomura, T. Isu and Y. Katayama, Jpn. J. Appl. Phys. 32, L222 (1993).

Inst. Phys. Conf. Ser. No 141: Chapter 3
Paper presented at Int. Symp. Compound Semicond., San Diego, 18–22 September 1994
© 1995 IOP Publishing Ltd

Excitons as a Probe of Interfacial Defects in GaAs/AlAs Short-Period Structures

F.T. Bacalzo, L.P. Fu, G.D. Gilliland, A. Antonelli, R. Chen, K.K. Bajaj
Emory Univ., Physics Dept., Atlanta, GA 30322
D.J. Wolford
Iowa State Univ., Physics Dept., Ames, IO 50011
J. Klem
Sandia National Laboratories, Albuquerque, NM 87185

We have measured the time- and space-resolved photoluminescence of excitons in type-II GaAs/AlAs structures. We find that these excitons are localized at the heterointerfaces at low temperatures (diffusivity, $D = 10^{-3}$ cm^2/s at 1.8 K) and are highly mobile at higher temperatures ($D = 10$ cm^2/s at 30 K). Thus, photoluminescence lineshapes are indicative of heterointerfacial roughness. We have modeled our transport and kinetics results in terms of a motionally-averaged nonradiative decay at heterointerfacial defects. Comparison of our model with experiment yields nonradiative heterointerfacial defect sheet densities ranging from 10^6 to 10^7 cm^{-2}.

I. Introduction.

Interfacial structure of semiconductor heterostructures is of considerable interest both scientifically and technologically. This microscopic heterointerfacial structure is scientifically interesting from the standpoint of understanding the kinetics and thermodynamics of epitaxial growth and its correlation with the observed optical and electrical properties. Technologically, this interest arises from the incorporation of heterostructures in a plethora of semiconductor devices. Furthermore, the physics of carriers (electrons, holes, or excitons) often involves an interaction between these carriers and a heterointerface, and this interaction may have a profound influence on actual device performance. For these reasons scientists have developed a vast array of experimental techniques for examining heterointerfacial structure. These techniques include: Photoluminescence spectroscopy (PL),[1] Cathodoluminescence,[2] Hall Transport Measurements,[3] Scanning Tunneling Microscopy (STM),[4] and Atomic Force Microscopy (AFM),[5] etc.. However, there is not a one-to-one correspondence between the results of these experimental techniques since they are not all sensitive on the same length scale and do not measure the same quantity. For example, STM and AFM measurements yield microscopic structural information, whereas Hall transport measurements only yield scattering times from which the scale of any heterointerfacial roughness is difficult to determine.[6] PL measurements also allow determination of interfacial roughness, but only on a length scale of the order of the effective size of the recombining species, which for excitons is many lattice sites. In this study we have used the PL emission from excitons in type-II GaAs/AlAs heterostructures to probe the heterointerfacial roughness and defects in these types of structures. Our results provide a correlation between the excitonic transport properties, optical emission properties, and structural roughness and defect properties on a length scale of ~ 100 Å.

In our study[7] we have chosen to examine the heterointerfacial properties of type-II short-period GaAs/AlAs structures in preference to other GaAs/AlAs structures because type-II excitons actually straddle the heterointerface, whereas excitons reside between

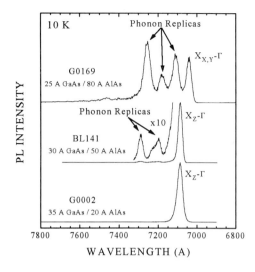

Fig. 1. cw PL spectra taken at 10 K after laser excitation at 4186 A. Spectra were offset vertically for clarity.

heterointerfaces in type-I quantum-wells, and may not even come into contact with a heterointerface in significantly thicker GaAs structures. It is well-known that short-period GaAs/Al$_x$Ga$_{1-x}$As superlattices or quantum wells may be made type-II through appropriate choice of both well and barrier layer thicknesses. In this materials system, a type-II structure manifests itself with the hole residing at the Γ point of the GaAs layers, and the electron residing at the X conduction-band-edges of the Al$_x$Ga$_{1-x}$As layers. Specifically for the (GaAs)$_m$/(AlAs)$_n$ system, this may occur for GaAs-layer thicknesses < 35 Å (m < 13) and AlAs-layer thicknesses > 15 Å (n > 6).[8] Excitons in these type-II systems are thus indirect in real-space and in momentum-space.

II. Experiment.

We have performed PL, PL time-decay, time-resolved PL, and time-resolved PL-imaging experiments in an effort to elucidate the interaction between excitons and heterointerfaces. In order to measure excitonic transport we have resorted to an all-optical analog of the classic Haynes-Shockley experiment, whereby, excitonic PL is spatially and temporally resolved.[9] Our technique relies on confocal laser excitation and imaging of the photoexcited excitons. PL was excited by a cw mode-locked Ti^{3+}:Al$_2$O$_3$ laser which was frequency doubled and pulse-picked to lower the repetition rate. With this experimental arrangement, near diffraction-limited laser spot sizes (~ 3 μm) are achievable, with a temporal and spectral resolution of ~ 50 ps and ~ 0.1 meV, respectively.

The samples used in this study were undoped MBE-prepared GaAs/AlAs superlattices. GaAs layers in all samples are nominally 30 Å thick and the AlAs layers range from 20 to 80 Å. All samples were grown at 600 °C with interrupts of 30 s at each interface, and were grown on top of a 1 μm GaAs buffer layer.

III. Results.

Figure 1 shows the cw PL spectra of 3 samples at 10 K taken under nearly identical conditions. Samples with narrow AlAs layer thicknesses (20 and 50 Å) exhibit strong no-phonon Γ-X$_Z$ transitions and weaker phonon replicas. Samples with thicker AlAs layer thicknesses (> 65 Å) show a weaker no-phonon Γ-X$_{X,Y}$ emission line and more intense phonon replicas relative to the no-phonon line. We find that the no-phonon emission lines do not change appreciably with increasing temperature - peak position is temperature independent, and linewidth increases only moderately. However, PL intensities taken with pulsed excitation (at a repetition rate slower than the longest lifetime) decrease drastically with increasing temperature.

PL time decays of the no-phonon emission are extremely long and nonexponential at low-temperatures and decrease and become rigorously exponential with increasing temperature. We find low-temperature lifetimes (taken from the exponential tail of the PL decays) ranging systematically with AlAs layer thickness from 650 ns to 4 ms. At 30 K all PL decays are similar with lifetimes from 4 to 40 ns.

Time- and spatially-resolved PL imaging measurements versus temperature were performed in order to quantify the transport properties of these type-II excitons. We find diffusivities ranging from 10^{-3} to 100 cm^2/s from 1.8 to 50 K, increasing monotonically with increasing temperature for all samples. These results confirm that these excitons are localized at the heterointerfaces at low temperatures and are thermally activated at higher temperatures.[7] Thus, PL lineshapes are indicative of heterointerfacial roughness induced localization.[10] These kinetic and transport results are summarized in Fig. 2.

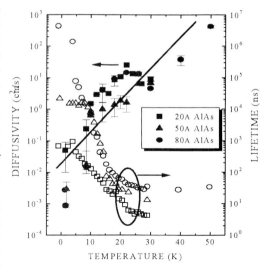

Fig. 2. Diffusivities (Solid Symbols) and Lifetimes (Open Symbols) versus temperature and versus AlAs layer thickness.

IV. Discussion.

These results imply the existence of a strongly temperature-dependent nonradiative decay.[11,12] We have arrived at a microscopic model for motionally-averaged nonradiative decay for these type-II excitons in order to quantify heterointerfacial nonradiative decay.[13] Our model involves the 2-dimensional random walk of excitons along the heterointerfaces. At some point these excitons may encounter a nonradiative defect and be annihilated, emitting phonons. Thus, the transport and kinetic properties become coupled. This model is shown schematically in Fig. 3. Using this model together with the Klein, Cohen, and Sturge theory[14] for radiative decay of localized excitons in a disordered system, we may then compare our theory to our experimental results. Fig. 4 shows our measured and calculated nonradiative decay rates versus diffusivity. From this model we may extract the quantity r_2/r_1. Here $2r_2$ is the average distance between nonradiative defects and $2r_1$ is the capture cross-section. Using excitonic wavefunctions calculated by Cen et al.[15] and assuming point-defects as the nonradiative defects, we may calculate the average distance between defects, and hence their concentration. We find nonradiative defect sheet densities of 4×10^6, 1×10^7, and 1×10^6 cm^{-2} for samples with 20 Å, 50 Å, and 80 Å AlAs layer thicknesses, respectively.

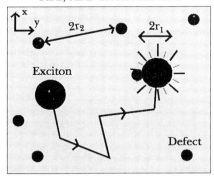

Fig. 3. Microscopic model of motionally-averaged nonradiative decay. Excitons undergo random-walk until they encounter a heterointerfacial defect and subsequently decay nonradiatively.

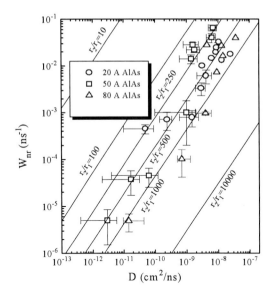

Fig. 4. Comparison of model calculation and experimental results, nonradiative decay rate versus diffusivity.

We have also calculated the diffusion length ($L_d = \sqrt{D\tau}$) from our results. We find diffusion lengths on the order of 3 µm, in excellent agreement with the nonradiative defect densities derived from our model. At this point, we may only speculate as to the nature of these defects. We believe that possible candidates are heterointerfacial vacancies or heterointerfacially incorporated oxygen.[16]

V. Conclusions.

In summary we report here a direct observation of the temperature-dependent, spatial localization of cross-interface excitons in type-II superlattices. Our results suggest the the PL lineshape is indicative of heterointerface roughness induced localization. We have also developed a quantitative model relating the excitonic decay kinetics to their transport. Our model accurately predicts the observed PL decay kinetics using the measured diffusivities. From this model we may characterize the heterointerface quality in terms of the density of nonradiative traps, which for these samples range from 10^6 to 10^7 cm^{-2}. Thus, we have utilized excitons in type-II structures to probe the roughness of heterointerfaces and defects incorporated at heterointerfaces.

Acknowledgements

Supported, in part, by ONR under contract N00014-92-J-1927, AFOSR under grant No. AFOSR-91-0056, and the U.S. DOE under contract DE-AC04-94AL85000.

REFERENCES

1. D. Gammon, B.V. Shanabrook, and D.S. Katzer, Phys. Rev. Lett. **67**, 1547 (1991).
2. D. Bimberg, J. Christen, T. Fukunaga, H. Nakashima, D.E. Mars, and J.N. Miller, J. Vac. Sci. Technol. B **5**, 1191 (1987).
3. J.H. English, A.C. Gossard, H.L. Störmer, and K.W. Baldwin, Appl. Phys. Lett. **50**, 1826 (1987).
4. H.W.M. Salemink, M.B. Johnson, and O. Albrektsen, J. Vac. Sci. Tech. B **12**, 362 (1994).
5. T. Yoshinobu, A. Iwamoto, and H. Iwasaki, Jap. J. Appl. Phys. Part 1 **33**, 383 (1994).
6. H. Sakaki, T. Noda, K. Hirakawa, M. Tanaka, and T. Matsusue, Appl. Phys. Lett. **51**, 1934 (1987).
7. G.D. Gilliland, A. Antonelli, D.J. Wolford, K.K. Bajaj, J. Klem, and J.A. Bradley, Phys. Rev. Lett. **71**, 3717 (1993).
8. E. Finkman, M.D. Sturge, M.H. Meynadier, R.E. Nahory, M.C. Tamargo, D.M. Hwang, and C.C. Chang, J. of Lumin. **39**, 57 (1987).
9. G.D. Gilliland, D.J. Wolford, T.F. Kuech, and J.A. Bradley, Appl. Phys. Lett. **59**, 216 (1991).
10. G.D. Gilliland, D.J. Wolford, J.A. Bradley, and J. Klem, J. Vac. Sci. Tech. B **11**, 1647 (1993).
11. B.A. Wilson, IEEE J. Quantum Elect. **QE-24**, 1763 (1988).
12. J. Ihm, Appl. Phys. Lett. **50**, 1068 (1987).
13. S. Chandrasekhar, Rev. Mod. Physics **15**, 1 (1943).
14. M.V. Klein, M.D. Sturge, and E. Cohen, Phys. Rev. B **25**, 4331 (1982).
15. J. Cen, S.V. Branis, and K.K. Bajaj, Phys. Rev. B **44**, 12848 (1991).
16. M.T. Asom, M. Geva, R.E. Leibenguth, and S.N.G. Chu, Appl. Phys. Lett. **59**, 976 (1991).

Heterointerface microroughness analyzed by PL and PLE of quasi-2D GaAs-AlGaAs single quantum well

S J Rhee, H S Ko, Y M Kim, W S Kim, D W Kim#, and J C Woo

Dept. of Physics, Seoul National University, Seoul 151-742, Korea
Dept. of Physics, Sun Moon University, Asan-Kun, Chung-Nam 337-840, Korea

Abstract. Microroughness on so-called atomically smooth heterointerfaces of narrow quantum wells (QW) has been analyzed by photoluminescence (PL) and PL excitation (PLE) spectroscopies. Different conditions of the normal heterointerfaces were produced by fractional interruption of RHEED oscillation in MBE growth. PL and PLE clearly present the peaks with the monolayer flat interface, while PL of different growth interruption fits well with a theoretical model introducing perturbation due to submonolayer fluctuation. The result shows that flat islands may extend out to a lateral dimension larger than the exciton diameter (Dx). But, the island may be composed of microroughness much smaller than Dx, and the microroughness may be detected as the averaged alloy composition in PL.

1. Introduction

Recently the microscopic structure of the heterointerface has attracted interest in numerous reports related to the so-called monolayer-flat or atomically smooth interface[1-3]. In this work, the results of photoluminescence (PL) and PL excitation (PLE) spectroscopic analysis on quasi-two dimensional single quantum wells (QW) with different normal heterointerface structures is reported. The results show that the microroughness, which extends out much less than an exciton diameter, shifts the ground sublevel spectra by a fraction of the monolayer (ML) difference in QW width (Lz). The submonolayer is detected as if there is an alloy compositional fluctuation at the interface in the excitonic recombination spectroscopy.

2. Experiment

Total of 13 samples with 45° reflection high energy electron diffraction (RHEED) phase separation were prepared with the growth interruption primarily aimed at 13 ML oscillation. In the molecular beam epitaxial growth of GaAs-$Al_{0.25}Ga_{0.75}As$ QW, the microscopic interfacial structure of the normal interface was varied by

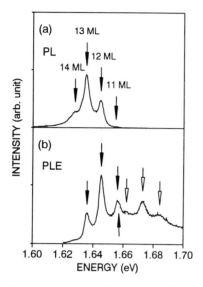

Fig. 1: Typical (a) PL and (b) PLE spectrum of single QW sample

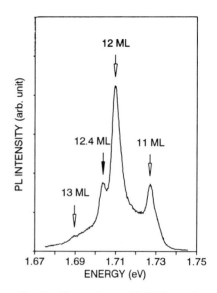

Fig. 2: PL spectrum of MQW sample

interrupting the growth at various phases of RHEED intensity oscillation, while the inverted interface was maintained so as to be atomically smooth. At both interfaces the growth was interrupted for 2 minutes. In addition, a multiquantum well (MQW) sample, which consists of 10 QWs of Lz=12 ML, plus one outermost QW intentionally overgrown by about 0.4 monolayer, i.e. the RHEED phase at the interruption of 135°, was prepared. Each QW was separated by 200Å $Al_{0.5}Ga_{0.5}As$ barriers. The growth temperature was 600°C, and the growth rate was 0.5 μm/hr for GaAs. The PL and PLE were performed at 20 K.

3. Data and Results

In Fig. 1, typical PL and PLE spectra of the sample growth-interrupted at RHEED phase of 45° are shown in (a) and (b), respectively. In PL, four distinct peaks with the most intense 13 ML one and three side peaks from 11, 12 and 14 ML, indicate that the lateral extends are said to be larger than the exciton diameter[1]. In PLE, there are a total of 7 peaks; the solid and open arrows are related to heavy- and light-hole excitons of 11, 12, and 13 ML, respectively, and the rest related to an excited excitonic state. The PL of MQW sample is shown in Fig. 2. The photon energies of PL peaks from 13 samples in steps of 45° RHEED phase are summarized in Fig. 3.

The ground state transition was computed with the sublevel separation using envelope function approximation (EFA) and the estimated low-dimensional confined exciton binding energy[4]. The parameters used in EFA calculation are the same as in Ref. 4. If a rectangular barrier, whose height equals the same

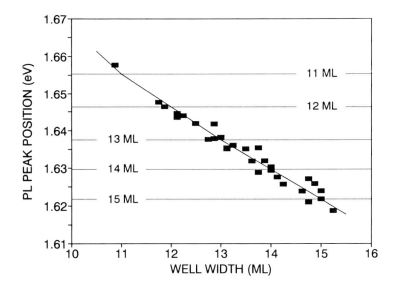

Fig. 3: Summary of PL peak positions of single QW samples. The solid line is theoretical line calculated by EFA and perturbation, and the dotted lines are guides for integer monolayer.

fractions of $Al_{0.25}Ga_{0.75}As$ barrier as that of RHEED phase and whose width of 1 ML, is introduced as perturbation at the normal interface site, the change of ground state transition becomes linear as shown in Fig. 3. Notice that most of the PL peaks lie in the vicinity of the theoretical line. Since the separation of PL peaks of a given sample is almost the same as the energy difference of $\Delta Lz=1$ ML, that is, the spectra are so-called quantized, QWs in this work have large lateral extension of atomically flat interface in the conventional sense. This means that the photon energy should be quantized for Lz of integer ML. However, the observed PL presents a contradicting result with almost linear change for a fraction of ML.

If the perturbation is interpreted as a compositional change of alloy at the interface, the submonolayer shifts of PL peaks which agrees well with the perturbation may be due to the apparent compositional change at the normal interface. When the exciton is used as a probe, as in the case of PL and PLE, the microroughness, roughness smaller than exciton diameter (Dx), will be detected as an average of composition over Dx. Still, a lateral extension larger than Dx, which is often referred to as the island, can be detectable as long as the distribution of microroughness is uniform over Dx. The microroughness should be superimposed on the island, as schematically shown in Fig. 4.

A comprehensive example of the presence of a submonolayer is shown in Fig. 2. There are 4 distinct peaks. The peaks indicated with open arrows are matched with QWs of Lz=11, 12, and 13 ML, while the peak indicated by solid arrow can be identified as Lz=12.4 ML. Here, the exciton binding energy was assumed to be about 10 meV.

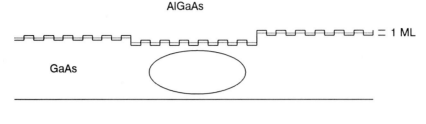

Fig. 4: Schematic diagram of interface configuration of single QW

4. Discussion

In this work, with the excess GaAs of controlled quantities, the microstructure of the heterointerface has been changed, and the limit of spectroscopic analysis was tested. The result shows that microroughness rather than large islands are formed by the excess GaAs, whose presence was observed in chemical lattice imaging[2], and is best explained with "bimodal roughness"[3]. This work also provides supporting evidence for the claim of Warwick that the picture of "atomically smooth island" based on quantized PL and PLE peaks is too naive.
 Conclusively, the PL and PLE analyses of a series of single QWs present the dual features of heterointerface; flat in the scale larger than Dx and rough smaller than Dx. It is indicative that flat islands extend out to the scale larger than the 2-dimensional Dx (around 100Å), but the island surface may consist of the microroughness much smaller than Dx. The microroughness on the lateral island has statistically random distribution at the normal interface and appears as the average of alloy composition over Dx.

References

[1] Fukunaga T, Kobayashi K L I, and Nakashima H 1985 *Surf. Sci.* **174** 71
[2] Ourmazd A, Taylor D W, Cunningham J, and Tu C W 1989 *Phys. Rev. Lett.* **62** 933
[3] Kopf R F, Schubert E F, Harris T D, Becker R S, and Gilmer G H 1993 *J. Appl. Phys.* **74** 6139
[4] Kim D W, Leem Y A, Yoo S D, Woo D H, Lee D H, and Woo J C 1993 *Phys. Rev. B* **47** 2042
[5] Warwick C A, Jan W Y, Ourmazd A, and Harris T D 1990 *Appl. Phys. Lett.* **56** 2666

Characterization of In-Situ Variable-Energy Focused Ion Beam/MBE MQW Structures

D J Bone, H Lee, K Williams, J S Harris, Jr., and R F W Pease

Solid State Laboratory, Stanford University, Stanford, California 94305

Abstract. We characterize surface roughness and quantum well quality in MBE-grown GaAs/AlGaAs MQW structures selectively bombarded *in situ* by a focused beam of gallium ions varying in energy from 25 eV to 30 keV. Photoluminescence intensity from QWs above, below, and at the growth-interrupted interface subjected to Ga+ bombardment decreases as ion energy increases. AFM measurements show that surface roughness also increases with increasing ion energy. Use of 500eV ions can eliminate a QW structure at the interface but leave deeper and proximate regions relatively unscathed.

1. Introduction

Molecular beam epitaxy affords tremendous control on the scale of atomic monolayers vertically, but lateral control has largely been available only by post-growth processing. We use a novel low-energy focused ion beam system[1] to focus low-energy Ga+ ion onto a GaAs wafer in-situ during a suspension of MBE growth, locally impacting the epitaxial structure laterally.

Use of focused Ga+ beams has been reported selectively to damage specific regions, such as those near gates for a HEMT to reduce leakage current and allow independent control of 2DEG regions while allowing high-quality 2DEG formation over the growth-interrupted region[2]. We are using lower-energy ions and considering much shorter distance scales, including affects at the processed interface itself. Low energy ions impart considerably less damage[3]. We use photoluminescence emission from various MQW structures and AFM surface analysis to measure the consequences of ion bombardment at various energies.

2. Growth and Ion Beam Procedures

In-situ processed GaAs wafers are transported to the focused ion beam column during a suspension of MBE growth via a portable vacuum chamber with an on-board battery-powered ion pump maintaining a base pressure below 10^{-9} Torr. Doses $\sim 10^{15}$ cm^{-2} of Ga+ extracted from a liquid-metal-ion source and slowed by a retarding field[1] are applied at normal incidence at room temperature. Upon return to the MBE growth chamber, the samples are thermally cleaned for ten minutes at 700°C under an As flux before additional epitaxial growth. A common structure used to gauge material quality is a MQW structure[4]. A series of GaAs/AlGaAs MQW structures (figure 1) were grown continuously or with growth suspension and processing. For the in-situ sample, growth was interrupted immediately following growth of the 50Å QW, at the bottom of the AlGaAs barrier.

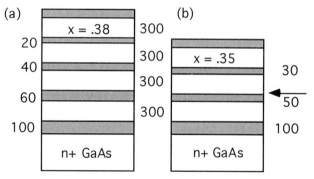

Figure 1: MQW structures used for (a) ex-situ and (b) in-situ 77K photoluminescence measurements. Each structures consists of GaAs QWs of various widths separated by 300Å AlGaAs barriers with x = .38 or x =. 35 respectively. The growth interruption for in-situ processing occurs atop the 50Å GaAs QW.

3. Photoluminescence

Post-growth bombardment of MQW structures has been reported by Green et al[4]. We were interested in ion damage that survives regrowth at temperatures ~620°C, but ex-situ measurements provide an important unannealed baseline. The ex-situ data (Figure 2, left) show extensive diminution of PL intensity, penetrating deeply into the sample. Emission from

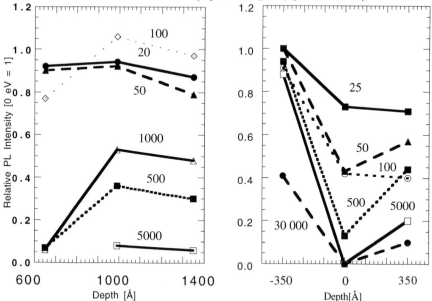

Figure 2: PL emission spectra for a continuously-grown MQW sample bombarded ex-situ (left) and a growth-interrupted and regrown sample (right). Depth indicates position from the ion-bombarded surface, which corresponds to the wafer surface in the ex-situ case and the top of the 50Å QW in the in-situ case The PL signals are normalized to unimplanted control regions. Ion energies in eV.

the 20Å well located close to the surface was not evident over any ion-processed region. None of the wells exhibited emission over regions bombarded with 10,000 eV or 30,000 eV ions. Emission generally decreases as ion energy increases. These results show widespread deleterious effects at a wide variety of depths that scale with increasing ion energy.

All *in-situ* processed samples are returned to the MBE growth chamber for additional thermal cleaning and growth. This step subjects the wafer to substantial "implicit" annealing under an As ambient. Kosugi et al[5] found elevated L-2 trap levels in GaAs at depths greater than 200Å even for low energy ions, but these were greatly ameliorated by annealing at 600°C for ten minutes in a hydrogen ambient.

The top QW, grown following the growth interruption and ion bombardment, emits normally for ion energies of 5000 eV or less (Figure 2, right, depth = -350Å). The 30,000 eV ions reduce emission by about 60%. The peaks also exhibit a shift toward higher wavelength for ion energies of 5000 eV or greater as well as a broadening of 40% in the FWHM, suggesting poor well quality and interface roughening.

Emission from the QW at the growth-interrupted interface was non-existent for ion energies above 500 eV. The top surface (at the time of ion bombardment) was the GaAs well material with the first monolayer after the growth interruption being the bottom of the top AlGaAs barrier. This surface is hence an incredibly sensitive probe of vacuum quality and interface process damage. Interrupting growth on an AlGaAs rather than a GaAs layer or with poor vacuum conditions completely quenches PL emission. Ion bombardment with energies < 500eV simply retards PL emission. No peak shift was indicated as would be the case if a portion of the well were physically etched or if net deposition of Ga took place.

Emission from the QW located 350Å beneath the growth-interrupted layer degraded rapidly for increasing ion energy, though emission was noted even for high-energies. The intensity falloff up to 100 eV is relatively severe, but the additional damage from higher-energy ions is relatively low, in stark contrast to the ex-situ case which exhibited little or now emission at comparable depths. The 300Å deep in-situ results were superior to the 650Å deep ex-situ characteristics at higher energies.

4. Atomic Force Microscopy

Areas of the surface which had been subjected to ion bombardment were examined by atomic force microscopy. We used the standard deviation of surface height as a measure of surface roughness. Lezec et al[6] studied a GaAs surface bombarded with 260 keV Si^{++} ions and annealed for 20 minutes at 655 °C under as As_4 flux and noted hills of 5-10 nm high and 0.1-0.3 μm wide. Our measurements cover the range 50 eV to 30,000 eV and include post-bombardment regrowth of 330Å of epitaxial material, primarily AlGaAs. The control surface, an unimplanted area on the same wafer subjected to the same growth process and growth interruption, exhibited an rms surface roughness of 7Å. The surface roughness remained relatively constant for ion energies of 200eV or lower (Figure 3) but rose substantially for higher energy ions. For example, the surface roughness exceeded that of the control sample by an order of magnitude for 5000 eV ions. Measured values are accurate relative to each other but may be underestimates by as much as 30% judging by monolayer separations visible on height histograms. Kondo et al [7] report a similar roughness for thermal cleaning during a study of surface roughness but smoother surfaces using ECR cleaning.

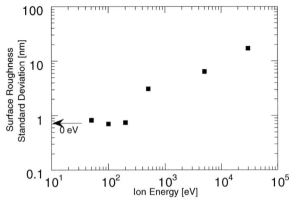

Figure 3: Surface roughness over regions bombarded with Ga ions of various energies and regrowth of 330Å of epitaxial material (primarily AlGaAs).

If selective *in-situ* processing is combined with precise subsequent epitaxy, a smooth post-bombardment surface is desired. The regrown epitaxial surface appears relatively unaffected by low-energy ions ~200 eV but becomes significantly rougher as ion energy increases. The surface roughening observed for ions up to 5000 eV may be tolerable in some structures, as evidenced by PL emission from the overgrown QW.

5. Conclusions

Quality epitaxial material can be regrown by MBE atop regions subjected to bombardment by focused beams of gallium ions. High-energy ions cause extensive damage, even far from the processed interface. Lower ion energies produce less damage. A QW grown at the processed interface continues to emit a PL signal as long as its top surface was bombarded only with ions of energy 500 eV or less. Damage is substantially ameliorated during post-processing regrowth. The roughness of epitaxial material grown atop processed regions is not degraded for ion energies up to ~200s eV. High energy ion bombardment appreciably roughens the surface, creating problems for precise epitaxy over such regions.

6. Acknowledgments

This work was partially supported by ARPA and ONR through contract N000-14-93-1-1375.

References

[1] Narum D and Pease R F W 1988 J. Vac. Sci. Tech. B 6, p. 966; 1988 J. Vac. Sci. Tech B 6, p. 2115
[2] Brown K, Linfield E, Ritchie A, Jones G, Grimshaw M and Pepper M 1994 Apl. Phys. Let. 64 , p. 1827
[3] Gamo K 1993 Beam-Solid Interactions: Fundamentals and Applications, Materials Research Society Symposium Proceedings-279, p. 577
[4] Green D, Hu E, Petroff P, Liberman V, Nooney M and Martin R 1993 J. Vac. Sci. Tech. B 11, p. 2249
[5] Kosugi T, Mimura R, Aihara R, Gamo K and Namba S 1990 Jpn. J. Appl. Phys. 29 Part 1 No. 10, p.2295
[6] Lezec H, Ahopelto J, Usui A and Ochia Y 1993 Jpn. J. Appl. Phys. 32 Part 1 No. 12B p. 6251
[7] Kondo N, Nanishi Y and Fujimoto M 1994 Jpn. J. Appl. Phys. 33 Part 2 No. 1B pp. L 91- L93

Thickness and strain effects on optical constants of thin epitaxial layers

C.M. Herzinger and P.G. Snyder

Center for Microelectronic and Optical Materials Research, and Dept. of Electrical Engineering, University of Nebraska-Lincoln, Lincoln, NE 68588-0511

F.G. Celii and Y.-C. Kao

Texas Instruments, Central Research Lab, P.O. Box 655936, Dallas, TX 75265

D. Chow

Hughes Research Lab, RL-63, 3011 Malibu Cyn Rd, Malibu, CA 90265

B. Johs and J.A. Woollam

J.A. Woollam Co., Inc., 650 J St. Suite 39, Lincoln, NE 68508

Abstract. The dielectric functions of thin (<3 nm) strained well and barrier layers, and of their thin unstrained cap layers, have been investigated by *ex situ* spectroscopic ellipsometry. The E_1, $E_1+\Delta_1$ critical point (cp) structure for the thin layers differs significantly from bulk, in a manner depending on the surrounding material and the layer thickness. Strained InAs wells within $In_{0.53}Ga_{0.47}As$ barriers retain well-resolved cp structure, with a strain-induced increase in Δ_1. Thin strained AlAs and AlSb barriers (between unstrained $In_{0.53}Ga_{0.47}As$ and InAs layers, respectively) lose the E_1, $E_1+\Delta_1$ cp structure entirely. Their unstrained cap layers show a blue-shift of both peaks, consistant with a quantum confinement effect.

1. Introduction

Structures containing very thin well and barrier layers are of interest for applications such as the resonant tunneling diode (RTD). The characteristics of such devices depend strongly on layer thicknesses, which must also be less than the critical thickness (only several nm in highly strained layers). For this work, *ex situ* spectroscopic ellipsometry (SE) has been used to characterize the apparent dielectric functions (ε) of these thin layers. (The reason for qualifying ε with "apparent" is discussed below.) An understanding of the ε of thin layer materials is useful in itself, and is also necessary for quantitative thickness characterization (and control) using real-time SE. One of the most useful spectral regions for characterization is that containing the E_1 and $E_1+\Delta_1$ critical points, because this structure is sensitive to composition, temperature, and strain.

Conventional models for optical analysis consist of a stack of layers with planar interfaces, each layer with a fixed ε independent of thickness or surrounding material type. For a very thin layer, however, the electron wavefunctions involved in an optical transition are not strictly confined to that layer. They extend out of the layer to a distance that depends on the surrounding material type, and which may be comparable to or greater than the layer thickness itself. The conventional multilayer model (which we use here) is an attempt to model wavefunctions which extend relatively far beyond the physical boundaries of the layer, using a quantity (ε) that is defined only within the layer boundaries. For this reason the modeled ε of a layer is a somewhat artificial quantity, that depends on its particular environment: thickness and surrounding material (as well as strain and temperature). Thus we use the term "apparent" dielectric function for these thin layers.

2. Samples and measurements

All samples were grown by molecular beam epitaxy (MBE). The "InAs" group consisted of two samples with structures InGaAs/InAs/InGaAs/InP substrate, with nominal InAs well thicknesses of 2 and 3 nm. The $In_xGa_{1-x}As$ was lattice matched (x=0.53) to InP. Nominal cap and buffer layer thicknesses were 6 and 200 nm, respectively, for both samples. The "AlAs" group contained four samples with structure InGaAs/AlAs/InGaAs/InP substrate. Caps and buffers were nominally 6 and 200 nm in three samples, 2 and 400 nm in the fourth. The AlAs layers ranged from 1.5 to 2.5 nm. Another pair of samples with slightly thicker AlAs layers (5 nm) were studied separately. The "AlSb" group contained four samples with structure InAs/AlSb/InAs/GaAs substrate, where the InAs buffers were 1 micron thick (thus optically opaque over the analyzed spectral range) and relaxed. Caps were all nominally 3 nm, and the AlSb layers ranged from 1.5 to 2.5 nm.

Spectroscopic ellipsometry (SE) measures the change in polarization state of a collimated, totally polarized, quasi-monochromatic light beam after non-normal incidence reflection from the sample surface. The ellipsometric parameters ψ and Δ, measured as functions of photon energy and angle of incidence, are obtained directly from the measured polarization change, and are also related to the sample by [1]

$$\tan(\psi)\, e^{i\Delta} = R_p/R_s$$

where R_p (R_s) is the complex electric field reflection coefficient for light polarized parallel (perpendicular) to the plane of incidence. To fit the measured SE data, R_p and R_s are calculated using a multilayer model and Fresnel reflection and transmission coefficients for each interface, assuming perfectly smooth and abrupt interfaces, and assigning thicknesses and dielectric functions to each layer. The unknown model parameters are then varied to obtain a best-fit to the data. Fitting SE data from several similar samples simultaneously (multiple sample analysis, described elsewhere [2]), improves sensitivity and reduces correlation between model parameters. All samples were measured in an air ambient, so thin native oxide layers (with ε from [3]) were included in the multilayer models.

3. Results

We begin by noting that use of bulk optical constants (which are available for all of the materials contained in these samples) in the multilayer models produces very poor fits to the SE data. If bulk data are used anyway, to try to determine approximate layer thicknesses, the fitted thicknesses are often grossly different from nominal (sometimes even solving to negative values). The ε values presented here were obtained by fitting them to the SE data,

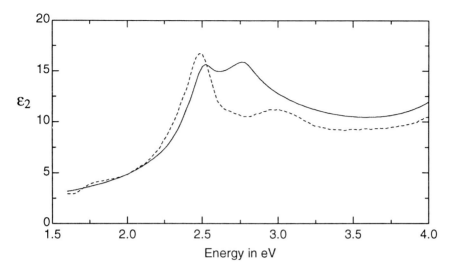

Figure 1. InAs bulk ε_2 (solid) and thin strained layer apparent ε_2 (dashed).

together with the layer thicknesses, using the multiple sample analysis technique. A full description of the analysis and results will be given elsewhere.

Figure 1 shows the apparent ε_2 (imaginary part of ε) solved for thin strained InAs wells, compared with the bulk data [4]. The strained E_1 peak near 2.5 eV is close to the bulk peak, but the $E_1+\Delta_1$ peak is shifted considerably, indicating an increase in Δ_1. This increased splitting is qualitatively consistant with calculations of the effect of strain on bandstructure [5].

Figures 2 and 3 show the apparent ε_2 for thin strained AlAs and AlSb, compared with bulk [6,7]. A surprising effect is observed (which to our knowledge has not been reported by others), namely the E_1, $E_1+\Delta_1$ structure is completely gone. This may be due to strong coupling of the wavefunctions of electrons optically excited into the barrier conduction band, with states in the (lower) conduction bands of the neighboring material. As the barrier thickness increases, its apparent ε should approach that of bulk. Figure 2 includes the ε_2 solved for the thicker (~5 nm) AlAs layers. It contains a peak near the bulk $E_1+\Delta_1$ and a weaker shoulder peak below it at about 3.5 eV, well below the bulk E_1. The peaks are broader than the bulk E_1, $E_1+\Delta_1$ structure, but still distinctly resolved. (These samples may have had imperfect morphology near the surface, since the AlAs thicknesses exceeded the critical layer thickness.) The reappearance of distinctive cp structure in these layers is, we believe, a result of the increased thickness, while the apparent red-shift of E_1 is consistant with calculations of strain effects [5] (indicating that the AlAs layers, though greater than the critical thickness, still contained considerable strain). We have also observed an absence of E_1, $E_1+\Delta_1$ cp structure in a sample containing a thin (~2 nm) *unstrained* AlAs layer, in a GaAs/AlAs/GaAs structure [6]. All of these results lead us to believe that the broadening and loss of E_1, $E_1+\Delta_1$ cp structure in a barrier layer is an effect of the thinness of the layer (and strong coupling to states in neighboring materials with smaller cp energies). Strain is an independent effect which is only observed if the cp structure remains resolvable, as for thin InAs and thicker AlAs layers.

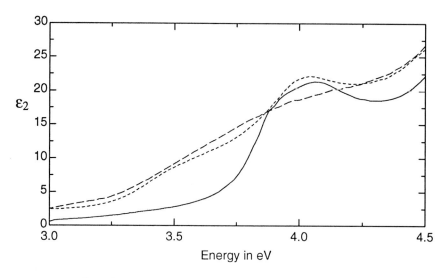

Figure 2. AlAs bulk ε_2 (solid), thin strained layer (1.5-2.5 nm, long dash), and slightly thicker layer (5 nm, short dash) apparent ε_2.

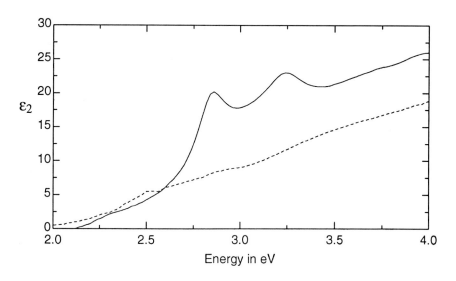

Figure 3. AlSb bulk ε_2 (solid) and thin strained layer apparent ε_2 (dashed).

Just as the results for barriers depend on thickness and the surrounding material, the cap optical constants are likewise affected by the material beneath. In fact, our initial attempts to obtain ε for thin AlAs layers [5] were less successful, because we used bulk data to model the cap layers (resulting in poorer fits, and "bleed through" of cap spectral features into the solved AlAs ε). Good fits to the SE data required fitting the apparent ε of *both* the barrier and cap layers. (This is possible only with multiple sample analysis.)

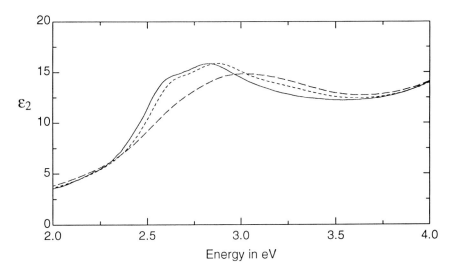

Figure 4. InGaAs (lattice matched to InP) bulk (solid), and apparent ε_2 of a thin cap on AlAs: ~6 nm (short dash) and ~2 nm (long dash).

Figure 4 compares bulk [8] and thin cap apparent ε_2 spectra for the AlAs set of samples. In this set, one of the caps was much thinner than the others, and its ε had to be solved for independently. Both cap spectra are blue-shifted and broadened with respect to bulk, the thinner cap showing the greatest difference. A blue-shift of both peaks is consistant with a quantum confinement effect in the cap. The same effect has been reported for the E_1, $E_1+\Delta_1$ peaks of a GaAs quantum well between $Al_xGa_{1-x}As$ barriers [9,10]. However, to our knowledge no others have reported observing it in a cap layer, which is not a conventional quantum well. The material beneath (AlAs) does form a conventional confinement barrier to electron wavefunctions in the InGaAs, but the material above is native oxide. The same effect was observed for the InAs caps of the AlSb samples. Also, similar blue-shifts were seen for the thin GaAs caps in samples with thick unstrained AlAs layers (these results will be reported elsewhere [6]).

In the case of the AlAs and AlSb samples, the E_1 cp structure of the cap material (InGaAs and InAs, respectively) is lower in energy than that of the barrier. InGaAs and InAs, on the other hand, have nearly the same values of E_1 and $E_1+\Delta_1$, so one would not expect to see quantum confinement effects in the InAs samples. This was indeed the case; the fit did not improve when the cap cp structure was allowed to vary.

The results for single, isolated barrier layers described above suggest that their apparent ε depend strongly on the surrounding material. This further suggests that the apparent ε of a barrier layer within multiple thin layers could be different from that of a single isolated barrier. We have measured one RTD sample, with structure InAs/AlSb/InAs/AlSb/ InAs (GaAs substrate). We attempted to solve for the thicknesses of all the layers, using the thin layer apparent ε obtained for individual InAs wells, AlSb barriers, and InAs caps, but the fit to the SE data was poor. One sample is not enough on which to base a final conclusion, but it suggests that the apparent ε obtained for single thin layers may not be appropriate for layers surrounded by other thin layers. More work needs to be done on characterizing RTD structures.

4. Conclusion

The room temperature dielectric functions of thin (<3 nm) strained InAs, AlAs, and AlSb layers, and their thin unstrained "cap" layers, in the E_1 and $E_1+\Delta_1$ critical point regions, have been investigated using spectroscopic ellipsometric multiple sample analysis. Results confirm that strain has a significant effect on critical point structure in highly strained layers, if the critical point structure is present. They also show that critical point structure can be blue-shifted significantly in thin caps. Even more significantly, in barrier layers the presence of critical point structure itself is apparently a function of layer thickness: it disappears entirely in very thin layers, and reappears as the thickness increases. To the degree (presently unknown) that these effects occur at growth temperatures, they must be understood and accounted for in order to accurately control growth using real-time ellipsometry.

Acknowledgements

Research supported by ARPA consortium agreement number MDA972-93-H-0005.

References

[1] Azzam R M A and Bashara N M 1977 *Ellipsometry and Polarized Light* (New York: North Holland)
[2] McGahan W A, Johs B, and Woollam J A, 1993 *Thin Solid Films* **234** 443-446
[3] Zollner S 1993 *Appl. Phys. Lett.* **63** 2523-2524
[4] InAs bulk data were from our own SE measurements of a thick MBE grown layer, corrected for native oxide. See also Aspnes D E and Studna A A 1983 *Phys. Rev. B* **27** 985-1009
[5] Herzinger C M, Snyder P G, Celii F G, Kao Y-C, Johs B, and Woollam J A 1994 *Proc. Sixth Int. Conf. on Indium Phosphide and Related Materials* (IEEE catalog #94CH3369-6) p 122
[6] AlAs bulk data were from a multiple sample analysis of MBE grown layers, to be published elsewhere. See also Garriga M, Kelly M, and Ploog K 1993 *Thin Solid Films* **233** 123-125
[7] Zollner S, Lin C, Shoenherr E, Boehringer A, and Cardona M 1990 *J. Appl. Phys.* **66** 383-387
[8] $In_{0.53}Ga_{0.47}As$ bulk data were from a multiple sample analysis of MBE grown layers. See also: Aspnes D E and Stocker H J 1982 *J. Vac. Sci. Technol.* **21** 413-416
[9] Erman M, Theeten J B, Frijlink P, Gaillard S, Hia F J, and Alibert C 1984 *J. Appl. Phys.* **56** 3241-3249
[10] Vasquez R P, Kuroda R T, and Madhukar A 1987 *J. Appl. Phys.* **61** 2973-2978

Characterization of layered InGaAsP/InP structures by means of time resolved transient grating technique

J.Vaitkus, S.Juodkazis, M.Petrauskas,* M.Willander

Semiconductor Physics Department, Vilnius University, Sauletekio 9, 2054 Vilnius, LITHUANIA
* Physics Department, Linköping University, S-58183 Linköping, SWEDEN

Abstract. Non-equilibrium charge carriers (NCC) dynamics have been investigated by means of transmission and light induced transient grating techniques in the epitaxial layers of InP, InGaAs, InGaAsP. The values of NCC lifetime and ambipolar diffusion coefficient are determined. These parameters are explained in terms of NCC dynamics under the conditions of epi-layer bleaching, free carrier absorption and NCC separation by internal electric field of the heterojunction.

1. Introduction

Interest in the direct band-gap quaternary semiconductor alloy $In_{1-x}Ga_xAs_yP_{1-y}$ is prompted by its technological use as a light source and detector for 1.3-μm and 1.55-μm fiber optic communication systems. The imperfect behavior of lasers, fabricated from these materials, has been attributed to various loss mechanisms which compete with radiative recombination within the device. Thus, it is important to reveal the mechanism of non-equilibrium charge carriers (NCC) dynamics in highly excited matter. Moreover, there are some different considerations such as NCC cooling and plasma expansion [1,2] or non-linear recombination [3,4] needed to explain picosecond NCC dynamics.

In this paper the transient gratings induced by a short laser pulse were used to investigate the recombination and diffusion of carriers in layered systems composed of InP n and p type, $In_{0.47}Ga_{0.53}As$ and $In_{0.47}Ga_{0.53}As_{0.53}P_{0.47}$ with layers thickness less than NCC diffusion length. In recent paper [5] the parameters of transient grating decay in case of highly excited matter have been measured and in this paper the carrier dynamics has been analyzed. The density of NCC ΔN was up to $2 \cdot 10^{19}$ cm^{-3} in the same layers as in Ref.[5].

2. Experimental

Two structures were grown by Hybrid Vapor Phase Epitaxy at 665°C on (100)-InP substrate. The experimental arrangement of picosecond light induced grating (LIG) measurements is described in [6]. The main point of the method is the temporal probing of light-induced refractive index grating decay. The grating is formed by NCC which are generated in accordance with the interference field of two crossing picosecond pulses. The maximum delay of the probe pulse is 1200 ps with a resolution of 35 ps. Maximum intensity of excitation at λ_{ex}= 1.06 μm is $I_{ex} \approx$ 10 GWcm^{-2}. The comparison of diffraction efficiency experimental kinetics measured at different LIG periods Λ with the results of the numerical simulation of recombination and diffusion [7,8] allows to estimate from $1/\tau = 1/\tau_R + (\Lambda/2\pi)^{-2}D_a$, where τ -LIG erasure time allows to evaluate recombination lifetime τ_R in the range of (0.05-10)ns and ambipolar diffusivity D_a - (1-200)cm^2/s.

Table 1

Samples

Sample layer No	with InGaAsP			with InGaAs		
	material	N, cm^{-3}	d, µm	material	N, cm^{-3}	d, µm
0(substr.)	InP	SI	500	InP	SI	500
1				InP(n)	$5 \cdot 10^{16}$	0.4
2	InP(n)	$5 \cdot 10^{17}$	0.5	InP(n)	$5 \cdot 10^{17}$	0.03
3	InGaAsP	$1 \cdot 10^{15}$	1.0	InGaAs(p)	$5 \cdot 10^{19}$	0.2
4	InP(p)	$5 \cdot 10^{17}$	1.5	InP(n)	$5 \cdot 10^{17}$	0.3

This formula is valid for the case of NCC density independent values of τ_R and D_a, what is fulfilled at our employed NCC density.

The InP wafers homogeneity and behavior of carriers were controlled by mapping of transmission and light self-diffraction efficiency of Nd:YAG laser radiation at 1,06 µm (pulse duration 2 and 12 ns).

3. Results and discussion

Two structure have been investigated. The types of layers and their thickness d are given in Table 1. The most important difference concerns the different environment of mostly narrow band-gap layer where linear absorption of light generates the NCC grating. The excited electrons and holes can be separated by heterojunction barriers that decrease their recombination rate.

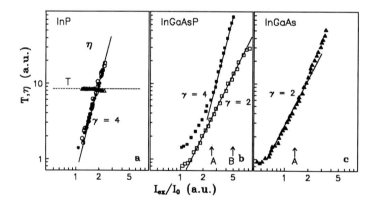

Fig.1. Dependencies of diffraction efficiency (η) and transmission (T) vs excitation intensity for different samples: a) InP, $I_O = 1.25$ MW/cm^2, circle - excitation pulse duration 12 ns, filled circle - 2 ns; b) InGaAsP "covered" layer -filled squares, "uncovered" - squares, $I_O = 0.1$ MW/cm^2, c) InGaAs, $I_O = 1$ MW/cm^2. γ - slope of dependence, A and B arrows sign the employed intensities in experiments.

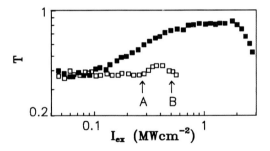

Fig.2.Dependence of transmission vs excitation intensity. Marking as in Fig.1b.

Investigation of light self-diffraction in InP substrate (Fig.1a) showed the lifetime is longer than ten nanosecond and that generation of NCC is caused by the two-photon absorption. The influence of two photon absorption on full transmission of crystal became important at about 2 MW/cm^2. The typical InP substrate was rather homogenous and the variation of light induced diffraction intensity was less than 10% per wafer.

The diffraction efficiency dependence on excitation in InGaAsP layer grown in between InP layers showed the linear generation of carriers. In uncovered layers of InGaAsP (when the upper layer of InP was etched) the two photon absorption was prevailing (Fig.1b). The transmission measurement showed the optical bleaching effect rather strong in the later sample (Fig.2) and this effect was small in other samples. At higher intensities the nonlinear absorption took place. The threshold is near to that in InP. According to it the transient grating in the structure with bleached absorption layer is generated in the layer and substrate of InP. This conclusion is supported by transient grating decay (Fig.3),

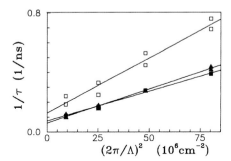

Fig.3. Gratings lifetime vs angle dependence in different samples. Marking the same as in Fig.1.

where the diffusivity and lifetime of carriers are found equal to (4.0± 0.9) cm^2/s and (13.5±1.5) ns, respectively. These data fit well with known data for InP [4,9,10]. It is necessary to stress that the time constant of grating decay in InGaAsP layer covered by InP has to be very similar to it in InP due to the extraction of holes to InP (p) and electrons into InP(n). This double layer diffusivity is near to the same as ambipolar diffusivity in InP. The lifetime of carriers depends on the recombination at and near interface and it is shorter than in the bulk. In the open InGaAsP layer on InP the bleaching is expressed weakly due to redistribution of electrons between InP(n) and InGaAsP. It caused shorter lifetime of electrons (7.9±0.9) ns in comparison with "covered" InGaAsP due to absence of hole location in InP(p) layer what reduces the recombination rate. The increase of electrons density in InP(n) caused the free carriers absorption, that follows the bleaching (if estimate the photon capture by free carrier cross-section to be 10^{-16} cm^2). The diffusivity in this structures is (7.4±0.9) cm^2/s what corresponds to the higher mobility of holes in InGaAsP layer which should be equal to 150 cm^2/Vs (it is close to the literature data [11]).

In the case of InGaAs grown between InP(n) (Fig.2c, Fig.3) the electrons were going to the surrounding InP layers. It has increased the lifetime of non-equilibrium carriers up to (16.7±2.0) ns and set the diffusivity equal to (4.0±0.9) cm^2/s which is controlled by holes and defects in InGaAs layer.

4. Conclusions

The transient light induced grating method was used to investigate the non-equilibrium carrier dynamics in layered InGaAsP structures and to determine recombination lifetime and diffusivity of carriers. The separation of carriers at the interface caused the increase of lifetime, but the electrons and holes interaction caused that the decay of grating was influenced by diffusion of carriers in different layers. The diffusivity was found near to ambipolar value in these materials, etc., (4.0÷7.4) cm^2/s.

5. References

[1] Wiesenfeld M, Taylor A J 1986 Phys.Rev.(b) **34** 8740-8749
[2] Taylor A J, Wiesenfeld J M 1987 Phys.Rev.(b) **35** 2321-2329
[3] Manning R J Fox A M and Marsh J H 1984 Ultrafast Phenomena **5** 210-212
[4] Manning R J, Fox A M and Marsh J H 1984 Electronics Lett. **20** 601
[5] Petrauskas M, Juodkazis S, Netikšis V, Willander M, Quacha A, Hammarlund B 1992 Semicond.Sci.Technol. **7** 1355-1358
[6] Ruckmann I, Kornack J, Kolenda J, Petrauskas M, Ding Y, Smandek B 1992 Phys.Stat.Sol.(b) **70** 353-359
[7] Fox A M, Manning R J, Miller A 1989 J.Appl.Phys. **65** 4287-4298
[8] Eichler H J, Gunter P, Pohl D W 1985 Laser-Induced Dynamics Gratings Chapt.2.3., 3.5; 4 (Springer-Verlag, Berlin)
[9] Hoffman C A, Jarasiunas K, Gerritsen H J and Nurmiko I 1978 Appl. Phys. Lett. **33** 536-538
[10] Bothra S, Tyagi S, Ghandhi S K, Borrego J M 1991 Solid-State Electronics **34** 47-50
[11] Landolt-Borstein Semiconductors 22 (Springer-Verlag)

ially. Growth and fabrication of 2-inch
Advances In Silicon Carbide Device Processing And Substrate Fabrication For High Power Microwave And High Temperature Electronics

C.D. Brandt, A.K. Agarwal, G. Augustine, A.A. Burk, R.C. Clarke,
R.C. Glass, H.M. Hobgood, J.P. McHugh, P.G. McMullin, R.R. Siergiej,
T.J. Smith, S. Sriram, M.C. Driver, and R.H. Hopkins

Westinghouse Science & Technology Center
1310 Beulah Road
Pittsburgh, PA 15235-5098

Abstract. High power density, temperature and radiation tolerant SiC electronics offer an exceptional opportunity to increase the performance and lower the cost of systems ranging from radar transmitters, to aircraft and tank controls, to missile sensors. Growth and fabrication of 2-inch diameter semi-insulating and low resistivity wafers, MESFETs with 25GHz cutoff frequencies and 2-3X the power density of GaAs devices, the world's first SiC static induction transistor, and 300°C analog and digital MOSFET circuits are among recent technological advances.

1.0. Introduction

For more than three decades, scientists have known that electronic devices employing silicon carbide (SiC) semiconductor material had the potential to revolutionize consumer, industrial, and defense electronic products. SiC devices exhibit 10X the voltage capability, 100X the radiation resistance, and 3X the thermal conductivity of devices fabricated in conventional semiconductors such as silicon or gallium arsenide and have the potential to operate at temperatures to 600°C. Westinghouse is investigating the use of SiC electronics in radar, communications, advanced television, electric vehicle controls, avionics and nuclear instrumentation and controls.

Recent technological advances have brought this potential payoff closer to reality. SiC has come a long way since the early 1960s when it was first seriously considered for electronic devices. Since then, methods for growing single crystals have been developed and a variety of devices have been fabricated, including blue LEDs [1], MOSFETs [2,3], and microwave MESFETs [4,5].

In this paper we present an overview of some recent advances in SiC materials and device technology achieved at the Westinghouse Science and Technology Center.

2.0. Substrate Fabrication

2.1. Bulk Crystal Growth

In the early days the only available SiC single crystal material was Lely crystals a few mm in length having uncontrolled thickness and polytype. Now, following the pioneering work of Tairov, Westinghouse has grown boules up to 2 inches in diameter using vapor transport at temperatures of 2200°C. The growth technique [6] utilizes an induction-heated, cold-wall system in which high purity graphite materials constitute the hot-zone of the furnace. Undoped crystals were grown in a 20 Torr high purity Ar ambient and show resistivities of 10^2 to 10^3 ohm-cm over the full crystal length. Low resistivity crystals with resistivities <0.02 ohm-cm have been

produced by controlled nitrogen doping. We have also grown the world's first semi-insulating 6H-SiC boules utilizing vanadium as a deep-level dopant [7]. These 2 inch diameter SI wafers exhibit resistivities in excess of 10^7 ohm-cm. Rapid advances have also been made in the slicing and polishing of wafers using conventional Si wafering technology modified for the increased hardness of SiC and have produced wafers up to 2 inches in diameter [8] as shown in **Figure 1**. These wafers are edge rounded like Si and GaAs in order to reduce breakage during processing. This Si-like wafer fabrication technology now makes SiC device manufacturing more economical.

Figure 1 A Westinghouse 2-inch diameter 6H-SiC boule and polished wafer

SiC suffers from a defect that does not occur in other semiconductors. Small pores or "micropipes" (up to tens of microns in diameter) can form during the growth of SiC and may extend through the entire boule. The origin of these pores is not well understood at this time but their presence in densities of 50 to 1000 cm^{-2} can seriously affect the yield of devices and circuits and is an active area of research.

2.2.   Epitaxial Growth

Significant progress has been made in the epitaxial growth of SiC layers. Epitaxial growths were performed in a water-cooled quartz atmospheric pressure horizontal reactor capable of growth on 3-inch diameter substrates. The growth temperature is approximately 1450°C using silane and propane as the primary reagents. Intentional n-type doping is obtained by admitting controlled amounts of nitrogen, while p-type doping is accomplished using trimethylaluminum.

Depending on the Si/C ratio used during growth [9,10], undoped films generally exhibit background carrier concentrations less than 1×10^{15} cm^{-3}. Intentionally doped films can be grown from 1×10^{15} cm^{-3} to greater than 1×10^{19} cm^{-3}. Film uniformities are improving; 2-inch diameter 6H-SiC epitaxial wafers exhibit thickness and doping uniformities of 7% and <20%, respectively.

3.0. Microwave Devices

3.1. Static Induction Transistors

In the area of power microwave devices, we have fabricated the first static induction transistors (SITs) on 6H-SiC. The structure, similar to a vertical field effect transistor, is composed of vertical pillars formed by reactive ion etching having a source contact on the top of each pillar, gate contacts on the pillar sidewalls, and a common drain contact on the back of the wafer.

Figure 2 shows an SEM of a completed device and the expected triode-like DC characteristics. Gate-drain breakdown voltages in excess of 100V, voltage gains of 6 to 12, and source-drain currents of 30 to 40 mA/mm were measured in these devices. Early RF

 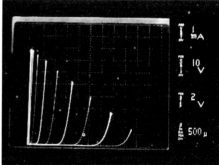

Figure 2   Static Induction Transistor (SIT) fabricated in 6H-SiC. (left) SEM of the 2 micron wide vertical source channels, and (right) the characteristic triode-like DC characteristics.

measurements have shown cutoff frequencies of 800 to 1000 MHz with approximately 6 dB of small signal gain at 300 MHz. Pulsed power testing at 175MHz of a multi-transistor chip yielded more than 30W output power, indicating good combining efficiency.

The SIT device is designed as a discrete power transistor for operation at frequencies up to S-band.

3.2.   MESFETs

MESFETs were fabricated on high resistivity 6H-SiC substrates by electron beam direct write processing [5]. The MESFET epitaxial structure consisted of a 2 micron thick undoped buffer layer ($<1 \times 10^{15}$ cm^{-3}), 0.4 micron thick channel layer (2×10^{17} cm^{-3}), and a 0.1 micron n$^+$ contact layer ($>1 \times 10^{19}$ cm^{-3}). A scanning electron micrograph of a fully air-bridged MESFET with a 0.5 micron long gates and approximately 4 mm of gate periphery is shown in **Figure 3**.

The DC performance of these transistors indicates a 230 mA/mm channel current and breakdown voltages exceeding 70V, making these devices suitable for RF power operation. Small signal RF gain measurements show excellent 8.5 dB gain at 10 GHz with a cutoff frequency of 25 GHz. It is significant to note here that these RF results were obtained with a high drain bias voltage of 40V which clearly demonstrates the simultaneously high voltage and frequency capabilities of SiC devices. Preliminary power measurements have shown 1.3 W/mm at 6 GHz with 8 dB of gain, at least a factor of two higher power output than GaAs MESFETs. This is the highest frequency and microwave power for SiC MESFETs reported to date.

4.0.   High Temperature Electronics

We describe some of our recent efforts to realize a depletion mode 6H-SiC lateral MOSFET for high temperature applications [3]. The device fabrication follows closely to the SiC MESFET fabrication with the exception that before a gate is applied to the device, a thermal oxide is grown on the active channel area. Due to the use of submicron e-beam lithography, channel currents in excess of 200 mA/mm and a record setting transconductance of 12 mS/mm have been obtained from these 6H-SiC MOSFETs. Additionally, simple digital and analog demonstration circuits have been fashioned with these devices.

MOSFET chips were laser scribed and mounted into 40 pin ceramic packages to demonstrate preliminary monolithic SiC circuits. Circuits were fashioned by wire bond connections directly on the die and by off chip connections via the lead package pins. SiC analog

Figure 3 A fully air-bridged 6H-SiC power MESFET having gate lengths of 0.5 micron and a total gate periphery of approximately 4 mm.

circuits such as single-stage transistor amplifiers, depletion-loaded differential pairs, and resistive-loaded differential pairs were tested. Functional SiC digital building blocks such as inverters and NOR gates were also tested.

Proper operation of the digital circuits is achieved from DC to frequencies greater than 1 MHz; however some distortion in the output characteristics is observed at 1 MHz due to parasitic wiring capacitance. These circuits have also been tested from room temperature to 300°C and show little if any degradation at elevated temperatures. More elaborate circuits are currently being fabricated to further test this technology.

5.0. Summary

SiC has made great strides in the last several years. There are still many parameters that need to be optimized in both the material growth and device fabrication but its semiconducting properties offer great promise for power generation and for applications in hostile environments where Si and GaAs cannot operate.

6.0. Acknowledgments

This paper summarizes some of the SiC research at the Westinghouse Science and Technology Center. This work has been funded by the Department of the Air Force under contract F33615-92-C-5912 (Tom Kensky, contract monitor), the Office of Naval Research under contract N00014-92-C-0129 (Y.S. Park, contract monitor), and by internal research and development funds.

References

[1] Edmond JA, Kong HS, and Carter CH 1993 *Physica B* **185** 453
[2] Brown DM, Ghezzo M, Kretchmer J, Krishnamurthy VG, and Michon G 1994 *Transactions of the 2nd International High Temperature Electronics Conference* Charlotte, NC XI-17
[3] Siergiej RR et. al 1994 *Transactions of the 2nd International High Temperature Electronics Conference* Charlotte, NC XI-23
[4] Palmour JW, Edmond JA, Kong HS, and Carter CH 1994 *Silicon Carbide and Related Materials* (London: Institute of Physics) **137** 495-498
[5] Sriram S et. al. 1994 *Electron Device Letters* (submitted)
[6] Barrett DL et. al. 1993 *J. Crystal Growth* **128** 358
[7] Hobgood HM et. al. 1994 *Appl. Phys. Letts.* (submitted)
[8] Hobgood HM, McHugh JP, Greggi J, Hopkins RH, and Skowronski M 1994 *Silicon Carbide and Related Materials* (London: Institute of Physics) **137** 7-12
[9] Burk AA et. al 1994 *Silicon Carbide and Related Materials* (London: Institute of Physics) **137** 29-32
[10] Larkin DJ, Neudeck PG, Powell JA, and Matus LG 1994 *Silicon Carbide and Related Materials* (London: Institute of Physics) **137** 51-54

Silicon Carbide Substrates and Power Devices

John W. Palmour, Valeri F. Tsvetkov, Lori A. Lipkin, and Calvin H. Carter, Jr.

Cree Research, Inc., 2810 Meridian Parkway, Durham, NC 27713

Abstract. Development of high current, high voltage SiC-based power switches relies on the wafer size and defect density of SiC substrates. Thus, the production of 2 inch diameter 6H-SiC, n+ 4H-SiC wafers, and significant advances in the reduction of micropipe defects in bulk crystals will be presented. In addition, two types of power switches, MOSFETs and thyristors, have been demonstrated in SiC. 4H-SiC vertical power MOSFETs capable of blocking 150 V had current densities of 100 A/cm^2 at a drain voltage of 3.3 V, and were successfully operated up to 300°C. 6H-SiC npnp thyristors had blocking voltages of 160 V and current densities of 100 A/cm^2 at 3.0 V, and were operated up to 500°C. The first thyristors ever fabricated in 4H-SiC blocked 210 V and had current densities of 500 A/cm^2 at 3.8 V.

1. Introduction

Silicon carbide has been projected to have tremendous potential for high voltage solid-state power devices with very high voltage and current ratings because of its high electric breakdown field of 4×10^6 V/cm and high thermal conductivity of 4.9 W/cm-K. The high breakdown field allows the use of much higher doping and thinner layers for a given voltage than is required in Si devices, resulting in much lower specific on-resistances for unipolar devices [1]. 4H-SiC shows the best potential for high power operation because the electron mobility in 4H-SiC is almost double that of 6H-SiC. Moreover, 6H-SiC exhibits a high degree of mobility anisotropy, having a mobility in the (0001) direction that is 1/5th that in the basal plane [2], yielding a 10× advantage for 4H-SiC over 6H-SiC for vertical power devices.

High power devices (1-100 A) have large active areas that must withstand high electric fields, so the wafer size and defect density of SiC is a key factor to the commercial success of this technology. This paper presents results on several topics concerning growth of SiC bulk crystals. The development of heavily nitrogen doped 4H crystals up to 10^{20} cm^{-3} for use as low resistivity substrates, the production of 2 inch diameter 6H-SiC wafers and prospects for growing 4H and 6H crystals with diameters > 2 inch will also be discussed. One of the most important defects currently limiting the production of large area power devices in SiC is the "micropipe" defect, which causes device failures at high fields [3]. Advances in the reduction of micropipe defects will be discussed. Some sources of dislocations and micropipes will also be described.

Despite the presence of these micropipe defects, the potential for SiC power devices has now been demonstrated by several types of device structures fabricated in both 6H and 4H-SiC. High voltage rectifiers [4] and power switches such as power MOSFETs and thyristors [5] have shown desirable characteristics. This paper presents results on MOSFETs and thyristors in both 6H-SiC and 4H-SiC which further confirm the potential of this material. Vertical power MOSFETs and thyristors have been demonstrated at both room temperature and high temperatures. All of these devices had relatively small active areas in order to avoid micropipe defects (about 100 cm^{-2}), but show promise for much larger power devices as the material quality is improved.

2. SiC Substrates

The availability of relatively large, high quality wafers of silicon carbide (SiC) for device development has been a major factor in the recent increased interest in this material for electronic and optoelectronic applications. The method used by Cree Research to produce single crystal SiC boules, a modified seeded sublimation process, is described in detail elsewhere [6]. Cree currently produces 35 mm diameter n-type 6H-SiC wafers and 30 mm diameter 4H-SiC wafers. 6H-SiC wafers 50 mm (2 inch) in diameter are in pilot production for blue LEDs and will be in full production by late 1994. An example of one of these 50 mm diameter wafers (lightly nitrogen doped) is shown in Fig. 1(c), along with 30 mm and 35 mm diameter 6H-SiC wafers in Fig.1(a) and (b). We anticipate that 76 mm (3 inch) SiC wafers will be available in 3 years. The increased size to 50 mm and then 76 mm will yield 2× and 4.7× more devices on a wafer, respectively, compared to 35 mm diameter wafers. The crystals can be doped both p (Al-doped) and n-type (N-doped) with resistivities from 0.1 to 40 Ω-cm for p-type and 0.018 to 0.8 Ω-cm for n-type for 6H. Nitrogen doping gives 6H-SiC a green color and aluminum doping gives a blue color. A 30 mm p-type 6H wafer is shown in Fig. 1(d). Although we have produced some p-type 4H in research efforts, production 4H wafers are all n-type (N-doped). A 30 mm 4H wafer is shown in Fig. 1(e). Light nitrogen doping gives 4H-SiC a tan color but heavily doped material is dark brown or opaque. We have recently produced heavily nitrogen doped 4H-SiC wafers as high as $N_d = 1 \times 10^{20}$ cm^{-3} with resistivities as low as 0.0028 Ω-cm. Advances in the production of this polytype are of particular interest for power devices since the electron mobility of 4H-SiC has been measured to be as high as 1000 cm^2/V-s for lightly doped material, parallel to the c-axis [2].

Silicon carbide wafers currently have a total etch pit density of $1-2 \times 10^4$ cm^{-2}. We use a combination of molten salt (KOH) etching and automated digitizing optical microscopy as an in-house quality assurance step in wafer production. Using this method, three primary types of defects are seen. These defects were classified as EP-1, EP-2 and EP-3 by Nakata et al. [7]. The EP-1 defects are the largest and are caused by micropipes, which are often associated with "super" screw dislocations in the crystal [8]. However, there are several mechanisms for micropipe formation including screw dislocations which propagate from the seed, carbon inclusions, and silicon droplets on the growth face. The diameter of the micropipes can vary depending on the growth conditions and formation mechanism but are typically 0.5 to 10 μm in diameter. These defects are fatal for most kinds of devices, especially for high voltage power devices [3]. Silicon carbide wafers have traditionally had rather high densities (400-500 cm^{-2}) of micropipes but recent wafers have micropipe densities as low as 55 cm^{-2} for 4H and 87 cm^{-2} for 6H. A digitized image of an etched 30 mm 6H SiC wafer which has 614 micropipes or 87 cm^{-2} is shown in Fig. 2. It is important to note that the micropipes are not evenly distributed throughout the wafer. Some rather large areas which are free from micropipes can be found, especially in 4H wafers. Figure 3 shows a 0.73 cm^{-2} area of a 4H wafer which only has one micropipe in the lower left corner which corresponds to <1.4 cm^{-2}. We believe that micropipe densities can be rapidly decreased over the next few years and possibly eliminated in 4-5 years.

3. Power Device Structures

The electric breakdown field (E_{max}) of 4H pn junction diodes has been characterized as a function of doping [5] and temperature [9]. A doping of 1.4×10^{16} cm^{-3} yields a breakdown voltage of 1000 V, corresponding to an E_{max} of 2.2×10^6 V/cm. In forward bias, the built-in potential was 2.8 V. In reverse bias, the highest voltage 4H-SiC diode had low leakage rectification to -1130 V. A negative temperature coefficient for the electric breakdown field was observed. The rate of decrease in breakdown field averaged about 500 V/cm-K, which corresponds to a decrease in breakdown voltage of 8% from room temperature to 350°C.

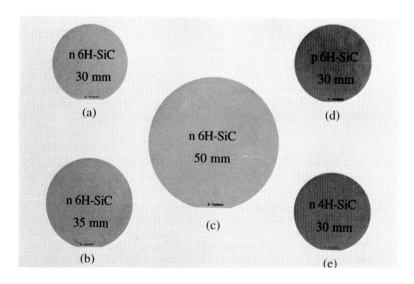

Figure 1: Silicon carbide wafers produced by Cree Research, Inc. (a) 30 mm n-type 6H, (b) 35 mm n-type 6H, (c) 50 mm (2 inch) n-type 6H, (d) 30 mm p-type 6H, (e) 30 mm n-type 4H.

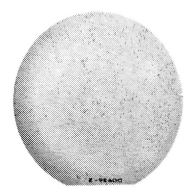

Figure 2: Digitized micrograph of an etched 30 mm 6H-SiC wafer which has 614 micropipes which is 87 cm^{-2}. Note the non-uniform distribution of micropipes.

Figure 3: A micrograph of a 0.73 × 1.0 cm^{-2} area of an etched 4H-SiC wafer. This area has one micropipe in the lower left corner. This corresponds to <1.4 micropipes/cm^2.

3.1 4H-SiC Vertical Power MOSFETs

Silicon carbide vertical power MOSFETs possess a strong advantage over those made in silicon because the high voltage layer (drain-drift region) may use a 10× higher doping level and one-tenth the thickness for a given voltage due to the much higher E_{max} of SiC. Ultimately, this could translate into specific on-resistances that are as low as 1/300th that of equivalent Si devices [1]. The first vertical UMOS power MOSFET structures in both 6H-and 4H-SiC were reported by Palmour et al. [5]. The 4H-SiC MOSFETs have yielded the best results to date. The UMOS design of these devices, shown in Fig. 4, utilized n+ source fingers ion implanted into an epitaxially grown p-type channel layer. The n- drain epilayer was isolated using a reactive ion etched mesa. The active area of this device was 6.7×10^{-4} cm^2 and the gate periphery was 4 mm.

The highest current devices blocked 80 V on the drain and could withstand current densities as high as 368 mA (550 A/cm^2 and 0.92 A/cm of gate periphery) and power densities greater than 10 kW/cm^2. A drain current of 134 mA (200 A/cm^2) was obtained at V_d = 3.5 V, which corresponds to an $R_{ds(on)}$ of 17.5 mΩ-cm^2 at a gate bias of +16 V. The g_{max} for this device was about 40 mS (10 mS/mm). The channel mobility for these lower voltage, higher current devices was about 20 cm^2/V-sec. A higher voltage 4H-SiC UMOS device is shown in Figure 5. This device blocked 150 V and could achieve a drain current of 67 mA (100 A/cm^2) at V_d = 3.3 V and V_g = +18 V. This corresponds to an $R_{ds(on)}$ of 33 mΩ-cm^2. The g_{max} of this device was about 20 mS (5 mS/mm). The channel mobility for this device was about 7-12 cm^2/V-sec. The highest blocking voltage achieved to date with these 4H devices was 180 V, but the devices on that wafer had a higher $R_{ds(on)}$ of 74 mΩ-cm^2.

The 4H-SiC UMOS devices were also characterized as a function of temperature [9]. The drain current and transconductance increased dramatically with a temperature increase from 27°C to 150°C. The drain current increased by a factor of almost two and the transconductance increased by 50%. These increases were primarily due to the decrease in threshold voltage, because the $R_{ds(on)}$ remained very stable. These trends continued up to 300°C. The drain current increased to almost 3× its room temperature value and the transconductance increased by a factor of 2. Unfortunately, these devices had low lifetimes at temperatures above 300°C, exhibiting gate oxide failure during measurement. This problem will require further research.

3.2 4H-SiC Thyristors

The most promising thyristor structure to date has been a npnp device, which allows the use of a low resistivity n-type substrate [5]. These 6H-SiC devices showed forward and reverse voltages of 160 V with no gate current. The forward breakover voltage was reduced to -6 V

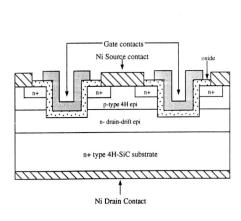

Figure 4: Cross-sectional view of the n-channel 4H-SiC vertical MOSFET device design. The drain-drift layer was mesa isolated.

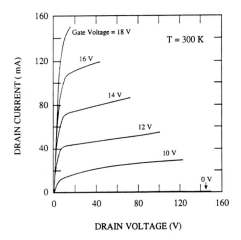

Figure 5: I-V characteristics of a 150 V UMOS 4H-SiC vertical UMOS power MOSFET structure. (A=6.7×10^{-4} cm^2).

with a trigger current of -200 µA. These devices had a relatively low voltage drop of -3.0 V for -105 mA (100 A/cm^2). With the built-in voltage of 2.65 V subtracted, this corresponds to a specific on-resistance of 3.6 mΩ-cm^2.

These 6H-SiC npnp devices were successfully operated at high temperatures [9]. When the device was heated to 350°C, the gate current required for a -7 V breakover dropped to -150 µA. The built-in voltage at 350°C decreased to about -2.0 V, but the bulk resistance increased to 10 mΩ-cm^2 due to decreasing mobilities (for electrons and holes) with increasing temperature. These offsetting effects meant that the voltage drop for 100 A/cm^2 was still about -3.0 V. This thyristor operated well to temperatures as high as 500°C, with a leakage current at -100 V of only -43 µA (4×10^{-2} A/cm^2). The forward breakover voltage with a gate current of -150 µA was reduced to about -3.0 V. The built-in voltage at 500°C was further decreased to -1.8 V and the specific on-resistance was still about 10 mΩ-cm^2, resulting in a voltage drop of -2.75 V for a current density of 100 A/cm^2.

This same structure has now been fabricated for the first time in the 4H-SiC polytype, which has a higher conductivity in the c-direction. The device structure of the 4H-SiC npnp thyristors is shown in Figure 6. They utilized a mesa structure, with all of the doping being done *in situ* during epitaxy. The area of the device was 1.05×10^{-3} cm^2 and the device periphery was terminated using a reactive ion etched mesa. These devices showed a forward blocking voltage of -210 V with no gate current, as shown in Figure 7. The forward breakover voltage was reduced to -3.1 V with a trigger current of -500 µA. The voltage drop for a current of -105 mA (100 A/cm^2) was -3.1 V. This device was subjected to forward currents as high as -525 mA (500 A/cm^2). The voltage drop for this high current density was still less than -3.8 V, as shown in the inset of Fig. 7. With the built-in voltage of 2.85 V subtracted, this thyristor had a specific on-resistance of 1.75 mΩ-cm^2, which compares favorably with the 3.6 mΩ-cm^2 observed for the 6H-SiC 160 V thyristors discussed above. This improved on-resistance results in a lower voltage drop for 500 A/cm^2 for the 4H-SiC devices despite their higher built-in potential of 2.85 V as compared with 2.6 V for 6H-SiC.

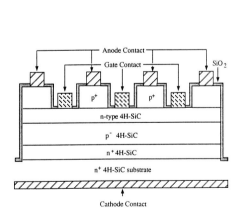

Figure 6: Cross-sectional view of the npnp 4H-SiC thyristor device design. The outer perimeter was mesa isolated.

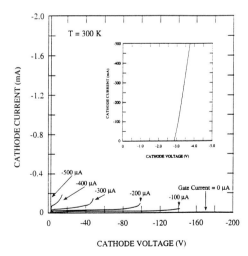

Figure 7: Forward bias I-V characteristics of a 4H-SiC npnp thyristor at 27°C showing 200 V capability. Inset shows the low voltage, high current characteristics. The area is 1.05×10^{-3} cm^{-2}.

4. Conclusions

Significant advances have been made in both the size and quality of both 6H- and 4H-SiC substrates. Fifty mm diameter 6H-SiC substrates are now being transitioned into production quantities. High quality, low resistivity (n+) 4H-SiC wafers can now be produced with micropipe densities as low as 55 cm^{-2} over the entire wafer, with some significantly large areas (0.73 cm^2) that are virtually "micropipe free". Two power switch structures have now been demonstrated in SiC, showing promise for high power, high temperature operation. The level of doping allowed for a given voltage is much higher than in Si and the layer thickness is about 1/10th that required for Si. This is a particularly strong advantage for unipolar power MOSFETs, where it will eventually result in dramatically lower on-resistances than for Si devices of equivalent size, or allow dramatically smaller SiC device sizes for the same on-resistance. Vertical UMOS devices have been fabricated in 4H-SiC with 160 V capability and a specific on-resistance of 33 mΩ-cm^2. These MOSFETs operated well up to 300°C, but are currently limited by oxide breakdown above that temperature. The first reported 4H-SiC npnp thyristors had 210 V capability and had a forward voltage drop of -3.8 V for 500 A/cm^2. 6H-SiC npnp thyristors operated well up to 500°C.

5. Acknowledgments

The SiC substrate development was funded in part by NIST (Award No. 70NAN132H1252), monitored by John Gudas. The 4H-SiC power MOSFET research was funded by BMDO and the Dept. of the Air Force, Wright Laboratories (Contr. F33615-93-C-2340), monitored by C. Severt. The thyristor research was funded by NASA-Lewis Research Center (Contr. NAS3-26927), monitored by G. Schwarze. The authors also wish to thank D. Asbury, S. Macko, and H.S. Kong for their technical assistance and valuable discussions.

References

[1] Bhatnagar M and Baliga B J 1993 *IEEE Trans. on Electron Devices* **40** 645-655
[2] Schaffer W J, Negley G H, Irvine K G, and Palmour J W 1994, to be published in *Diamond, SiC, and Nitride Wide-Bandgap Semiconductors*, C H Carter, Jr, G Gildenblatt, S Nakamura, and R J Nemanich, eds., Mater Res Soc Proc **339**, MRS, Pittsburgh, PA
[3] Neudeck P G and Powell J A 1994 *IEEE Electron Device Lett.* **15** 63-65
[4] Neudeck P G, Larkin D J, Powell J A, Matus L G, and Salupo C S 1994 *Appl Phys Lett* **64** 1386-1388
[5] Palmour J W, Edmond J A, Kong H S, and Carter, Jr, C H 1993 in *Silicon Carbide and Related Materials*, M G Spencer, R P Devaty, J A Edmond, M A Khan, R. Kaplan, and M Rahman, eds, Inst. of Phys. Conf. Series **137**, IOP, Bristol, 499-502
[6] U.S. Patent No. 4,866,005 (September 12, 1989) R.F. Davis, C.H. Carter, Jr. and C.E. Hunter (to North Carolina State University)
[7] Nakata T, Koga K, Matsushita Y, Ueda Y and Niina T 1989, in *Amorphous and Crystalline Silicon Carbide II*, M M Rahman, C Y-W Yang and G L Harris, eds., Springer Proc in Physics **43**, Springer-Verlag, New York, 26
[8] Wang S, Dudley M, Carter, Jr. C, Asbury D, and Fazi C 1993, in *Applications of Synchrotron Radiation Techniques to Materials Science*, D L Perry, N D Shinn, R L Stockbauer, K L D'Amico, and L J Terminello, eds., Mater Res Soc Proc **307**, MRS, Pittsburgh, PA, 249-254
[9] Palmour J W and Lipkin L A 1994 *Trans Second Intern High Temp Electronics Conf - Vol 1*, D B King and F V Thome, eds, Sandia National Labs, Albuquerque, NM, XI-3 -8

Low Leakage, High Performance GaAs-Based High Temperature Electronics

L.P. Sadwick[1], R.J. Crofts[1], Y.H. Feng[1], M. Sokolich[2] and R.J. Hwu[1]

[1]Department of Electrical Engineering, The University of Utah, Salt Lake City, Utah 84112
[2]Hughes Aircraft Company, GaAs Operations; Torrance, California 90509

Abstract. This paper reports on substantial improvements in device performance and reduction of the parasitic and leakage currents in gallium arsenide (GaAs) metal semiconductor field effect transistor (MESFET) devices and circuits at temperatures up to 350°C. These improvements in device performance, which have been verified on over 300 transistors measured on twenty different wafers, are obtained by a judicious choice of substrate material, contact technology, device layout, and device biasing. In addition, these MESFET's also displayed an enhanced resistance to breakdown at elevated temperatures. The methods used to realize these improvements have general applicability and work equally well with both enhancement and depletion n-channel MESFET's.

1. Introduction

High temperature electronics for automobile, aeronautical, geothermal, computer, military, nuclear and space applications [1-9] have received increased attention in the last few years. Wide band-gap semiconductors including silicon carbide (SiC) and gallium nitride (GaN) show great potential for use in high temperature electronic applications. However, for wide band-gap devices, improvements in the process and fabrication technology is required to attain reliable small scale (SSI) and medium scale integration (MSI) circuits. Due to a relatively mature process and fabrication technology, gallium arsenide (GaAs) is an attractive material system for use in high temperature (and high speed) electronics. While GaAs-based devices have been investigated for elevated temperature applications, previous reports on the high temperature characteristics of GaAs metal semiconductor field effect transistors (MESFET's) have indicated problems such as those caused by parasitic currents [10-13].

Recently, there have been several reports on low leakage GaAs MESFET [14] and Heterojunction FET devices [15-18] for high temperature applications. However the techniques used require complex etching and/or multilayer device structures which rule them out for use in low cost, high yield manufacturing and fabrication applications.

2. Fabrication

All of the MESFET's discussed in this paper were fabricated on (100) oriented and semi-insulating (SI) GaAs substrates. To form active regions in the semi-insulating GaAs substrates, single or multiple silicon (^{28}Si) implantation into photoresist coated (selective) and bare (blanket) substrates were performed. Typically, the wafers were degreased and then etched in a solution of NH_4OH-H_2O_2-H_2O. Prior to AuGeNi Ohmic contact deposition, the wafers were cleaned in a NH_4OH-H_2O_2-H_2O etch solution and rinsed in a NH_4OH-H_2O solution. All of the ^{28}Si implants were performed and annealed before the proton bombardment step (for those wafers that received a proton bombardment step). The wafers were proton bombarded at an energy of 100 KeV with a dose of 2 $\times 10^{14}$ cm^2. Annealing of the implants were performed in a furnace under either SiO_2 capping, Si_3N_4 capping or $AsH_3 + H_2$ overpressure at a representative annealing temperature of 850°C for 30 minutes. Typical room temperature leakage currents between Ohmic contacts isolated by proton bombardment or by mesa etching were in the 10^{-9} A range. The MESFET gate metalization was either TiPtAu or refractory TiWN. In the case of TiWN, a self-aligned gate (SAG) process was used in the fabrication of both discrete MESFET's and MESFET-based circuits. It should be noted that the technique used in this work has been applied to devices processed under significantly different technologies and conditions on substrates obtained from a wide variety of vendors. Therefore, the results presented in this paper should be of a very general nature and applicable to all types of GaAs MESFET structures.

3. The thermal leakage reduction technique

The origin of the substrate leakage current at elevated temperatures can be explained as follows: at elevated temperatures, the GaAs SI substrate becomes semiconducting due to thermal generation of carriers and can no longer be viewed or treated as semi-insulating. This elevated temperature semiconducting behavior of the GaAs substrate permits parasitic interactions between the active MESFET and the substrate. Based on comprehensive studies, that have [19] and will be [20] presented in detail elsewhere, the active MESFET/substrate parasitic interactions primarily manifest themselves in source- and drain-to-substrate thermally generated leakage currents. Simply stated, the electrical properties of the source- and drain-to-substrate system is of a rectifying behavior.

The essence of the thermal reduction technique used in this work is the substrate bias effect. Based on the discussion above, a single isolated MESFET with the source at ground potential will be used to illustrate and explain the substrate bias effect. When the substrate is left floating or grounded and a positive bias is applied to the drain, the drain-to-substrate "diode" becomes forward-biased with a resultant thermal "leakage" current that is exponentially dependent on temperature. The key to successful high temperature operation using our technique is to eliminate or substantially reduce the drain-to-substrate leakage current. By treating the substrate as an active element (essentially, the fourth terminal) of the MESFET, it is then possible to turn off the source- and drain-to-substrate "diodes" by applying a voltage that reverse-biases the "diodes". This can be accomplished by applying a voltage to the substrate that is more positive by typically 1 volt than the largest positive voltage applied to the drain. Turning off these "parasitic" diodes thus produces a remarkable reduction in the drain leakage current.

4. Experimental methodology

Figure 1 shows the 25°C drain current (I_d) versus drain voltage (V_{ds}) curves as a function of gate voltage (V_{gs}) of a typical 20x1 µm (i.e., nominal 20 µm width and 1 µm gate length) MESFET used in this work. As can be seen from Figure 1, respectable room temperature MESFET behavior was obtained with an acceptable output resistance (r_o) in saturation of approximately 2 kΩ.

Parasitic leakage current effects were most easily observed by monitoring the changes in the saturation region slope of the I_d-V_{ds} curve (i.e., dI_d/dV_{ds}, or the drain conductance). Ideally, the saturation region dI_d/dV_{ds} should be as small as possible. At some temperature, typically in the range of 175 to 250°C, all standard GaAs MESFET's will experience a sharp rise in leakage currents which eventually dominate MESFET device and circuit behavior. Figure 2 shows the 0 volt substrate, 250°C I_d versus V_{ds} curves for the same MESFET used to obtain the data shown in Figure 1. Based on the data shown in Figures 1 and 2, for Vgs equal to 0.4 volts, the saturation region dI_d/dV_{ds} for the same 20x1 µm depletion

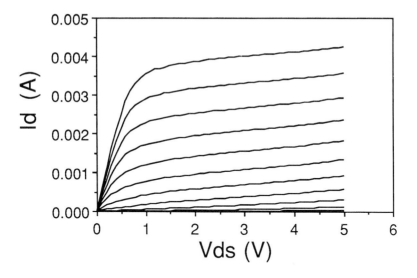

Figure 1. 25°C MESFET Id vs. Vds curves. Vgs ranges from 0.4 to -1.8 volts in 0.2 volt steps. 0 volts applied to substrate.

MESFET without substrate bias increased from 120µS at 25°C to 3.11 mS at 250°C. Increases in the saturation region dI_d/dV_{ds} are undesirable. In contrast, by employing our technique, the saturation region dI_d/dV_{ds} was remarkably stable for temperatures up to

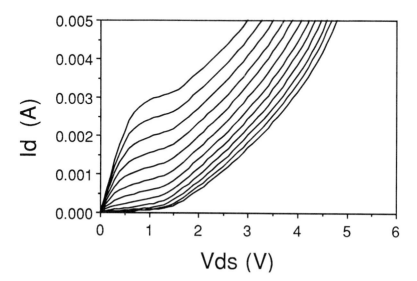

Figure 2. 250°C Id vs. Vds plots of the same MESFET as that of Figure 1. 0 volt substrate bias applied.

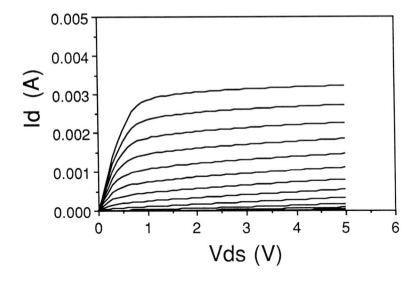

Figure 3. 250°C Id vs. Vds plots of the same MESFET as that of Figure 1. 6 volt substrate bias applied.

300°C. The data shown in Figure 3 was taken under identical conditions to that of Figure 2 except that a bias of 6 volts was applied to the substrate. Under 6 volt substrate bias and for a V_{gs} of 0.4V, the MESFET's saturation region dI_d/dV_{ds} was 116μS at room temperature.

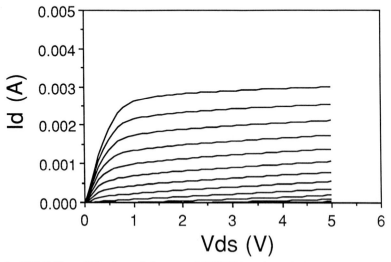

Figure 4. 300°C Id vs. Vds plots of the same MESFET as that of Figure 1. 6 volt substrate bias applied.

This figure of merit reduced to 56 µS at 200°C and remained steady at this value to 300°C. Figure 4 shows the 300°C MESFET's Id versus Vds curves with a 6 volt substrate bias applied. Utilizing the substrate bias technique, the MESFET's were also extremely stable at 350°C. Using this technique, we have also experimentally constructed and demonstrated useful circuits such as differential amplifiers and digital logic with TTL output levels that could operate for prolong periods of time at temperatures over 300°C without any indication of degradation. Without the substrate bias technique, device and circuit operation was not possible above 250°C. Although not discussed in this paper, similar results were also obtained for enhancement MESFET's.

The transconductance characteristics ($g_m = dI_d/dV_{gs}$) were also favorably controlled with a substrate bias of 6 volts at temperatures up to 300°C and beyond. Maximum g_m at 25°C for the 20x1 µm depletion transistor was approximately 179 mS/mm. At 200°C, the g_m value reduced to 141 mS/mm and at 300°C, the g_m value was 119 mS/mm. Figure 5 shows the 300°C transconductance data obtained from two different MESFET's (labelled A and B, respectively) having different nominal turn-on voltages at three different values of V_{ds} (MESFET A: V_{ds} = 1.5, 3, and 4.5 volts; MESFET B: V_{ds} = 1, 2.5, and 4 volts). As can be clearly seen from Figure 5, when using substrate biasing at high temperatures (in this case 6 volts and 300°C, respectively), g_m is well-behaved with no indication of degradation due to parasitic leakage currents.

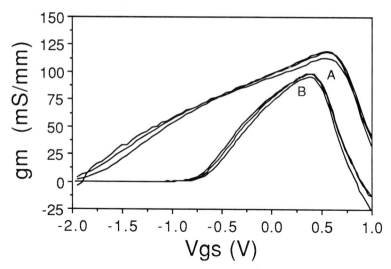

Figure 5. 300°C gm data taken at three values of Vds (MESFET A::V_{ds} = 1.5, 3, and 4.5 volts; MESFET B: V_{ds} = 1, 2.5, and 4 volts) for two different 20x1 µm MESFET's having different nominal turn-on voltages.

5. Conclusions

In conclusion, a novel technique utilizing substrate biasing has been developed which provides reproducible control of the high temperature leakage currents and breakdown of GaAs MESFET devices and circuits. To date digital circuit operation at temperatures up to 350°C with TTL output levels has been realized and functional analog circuits have also been constructed that operate beyond 300°C. We are presently designing small and medium scale high temperature GaAs MESFET digital logic circuits for ambient temperature operation in the temperature range of 200 to 350°C.

References

[1] Evans J L, Romanczuk C S and Bosley L E 1994 Transactions of the Second International High Temperature Conference I3-I8.
[2] Erskine J C, Carter R G, Hearn J A, Fields H L and Himelick J M 1994 Transactions of the Second International High Temperature Conference I9-I18.
[3] Carlin C M and Ray J K 1994 Transactions of the Second International High Temperature Conference I19-I26.
[4] Cloyd S J 1994 Transactions of the Second International High Temperature Conference I27-I28.
[5] Tajima M 1994 Transactions of the Second International High Temperature Conference I29-I34.
[6] Fricke K Hartnagel H L Schutz R Schweeger J and Wurfl J 1989 IEEE Electron Device Lett. **10** 577-579
[7] Canfield P C Lam S C F and Allstot D J 1990 IEEE J. Solid State Circuits **25** 299-306
[8] Bottner T, Fricke K, Goldhorn A, Hartnagel H L, Rappl A, Ritter S and Wurfl J " 1991 Proceedings of the First International High Temperature Conference 77-82
[9] Swonger J W, Gaul S J, and Heedley P L 1991 Proceedings of the First International High Temperature Conference 281-290
[10] Parry T B, Lee D H, Tran T, Sadwick L P and Sokolich M 1991 Proceedings of the First International High Temperature Conference 313-346
[11] Ojala P K, Cooper L S and Shoucair FS 1991 Proceedings of the First International High Temperature Conference 68-73
[12] Wong H, Liang C and Cheung N W 1992 IEEE Trans. Electron Devices **39** 1571-1577
[13] Shoucair F S and Ojala P K 1992 IEEE Trans. Electron Devices **39** 1551-1557
[14] Lee R, Trombley G, Johnson B, Reston R, Havasy C, Mah M and Ito C 1994 Transactions of the Second International High Temperature Conference V3-V8
[15] Papanicolaou N A, Anderson W T, Katzer D S, Jones S H and Jones J R 1994 Transactions of the Second International High Temperature Conference V9-V14
[16] Wilson C D and O'Neill A G 1994 Transactions of the Second International High Temperature Conference V15-V20
[17] Baier S, Nohava J, Jeter R, Carlson R and Hanka S 1994 Transactions of the Second International High Temperature Conference V21-V26
[18] Wurfl J, Janke B, Rooch K H and Thierbach S 1994 Transactions of the Second International High Temperature Conference V33-V38
[19] Sadwick L P 1994 Appl. Phys. Lett. 64 79-81
[20] Sadwick L P, Koniak J, McDonald R M, Sokolich M and Hwu R J 1994 unpublished

SiC Microwave Power MESFET's and JFET's

Charles E. Weitzel *, John W. Palmour †, Calvin H. Carter, Jr. †, Kevin J. Nordquist *, Karen Moore * and Scott Allen †

* Phoenix Corporate Research Laboratories, Motorola, Inc., Tempe, AZ 85284
† Cree Research, Durham, NC 27713

Abstract. In recent years, silicon carbide has received increased attention because of its potential for high power microwave devices. Sub-micron gate length MESFET's fabricated using the 6H and 4H polytypes have achieved f_{max} = 13.9 GHz and 16.3 GHz and f_T = 8.3 GHz and 8.0 GHz, respectively. The 4H-SiC MESFET achieved a maximum power density of 2.8 W/mm @ 1.8 GHz which is the highest ever reported for a SiC MESFET. SiC JFET's are promising for high temperature applications because they should have very low leakage currents. Experimental 6H-SiC JFET's have achieved f_{max} = 4.8 GHz and f_T = 3.0 GHz.

1. Introduction

In recent years, silicon carbide has received increased attention because of its potential for high power microwave devices. The high electric breakdown field of 4 x 10^6 V/cm, high saturated electron drift velocity of 2 x 10^7 cm/sec, and high thermal conductivity of 4.9 W/cm-°K indicate that SiC is a very promising material for high power, high frequency operation. The wide bandgap should also allow SiC FET's to have significant R.F. output power at high temperatures [1]. Until recently the SiC polytype that had received the most attention in the last few years has been 6H-SiC, since it had the best crystal quality of the established polytypes. Submicron MESFET's have been fabricated in 6H-SiC and have demonstrated desirable microwave performance [2-3]. Another polytype that shows even more potential for high power, high frequency operation is 4H-SiC, because the electron mobility in 4H-SiC is about twice as high as that of 6H-SiC [4] while retaining the other desirable features of 6H-SiC.

We report D.C. and S-parameter results obtained with 6H-SiC and 4H-SiC MESFET's. The maximum frequencies of oscillation (f_{max}) of 13.9 GHz and 16.3 GHz calculated from S-parameters are the highest reported to date for 6H-SiC and 4H-SiC MESFET's on conducting substrates, respectively. The measured continuous wave power density of 2.8 W/mm at 1.8 GHz is the highest power density ever reported for a SiC MESFET and two times the highest ever reported for a GaAs MESFET [5]. A 6H-SiC JFET is reported with an f_{max} = 4.8 GHz which is the highest ever reported for a SiC JFET.

2. SiC MESFET's

2.1. Device design and fabrication

The MESFET utilized a two fingered design with each gate finger being 166 µm, for a total gate width of 332 µm. The gate length was 0.5 µm, and the source-gate and gate-drain spacings were 0.2 µm and 0.9 µm, respectively. The MESFET was placed in a coplanar high

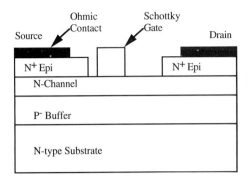
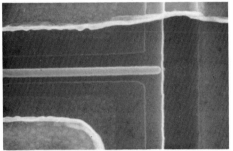

Figure 1. Device cross-section showing material layers and metal contacts.

Figure 2. SEM photomicrograph of MESFET channel region, Lg = 0.5 µm.

frequency probing pattern for on-wafer small signal and power measurements. The cross-sectional device design is shown in Figure 1. It is similar to that reported previously by Palmour, et. al. [3] with the N+ source and drain regions formed by epitaxial growth and the channel recess formed by reactive ion etching. The material structure consisted of an n-type SiC substrate, 7.5 µm p-type isolation epilayer doped about 3×10^{15} cm^{-3}, n-type channel epilayer and 0.15 µm N+ contact layer doped $\geq 2 \times 10^{19}$ cm^{-3}. The MESFET was fabricated using six lithography steps two of which were done on a Cambridge EBML-300 direct write electron beam system. The fabrication flow consisted of mesa isolation etch, N+ epi etch, dielectric deposition, ohmic metal deposition and anneal, Schottky gate metal definition and lift-off, and pad metallization. The source and drain metal contacts were annealed Ni and the pad metallization was Au. An SEM photomicrograph of the channel region of the unpassivated MESFET is shown in Figure 2.

2.2. 6H-SiC D.C. and R.F. performance

The I-V characteristic of a 6H-SiC MESFET fabricated with a channel doping of 2.3×10^{17} cm^{-3} is shown in Figure 3. The drain current density at $V_{ds} = 20$ V, $V_{gs} = 0$ V is about 240 mA/mm with a maximum extrinsic transconductance of 44.7 mS/mm. S-parameters were measured on-wafer from 45 MHz to 26.5 GHz using an HP8510 network analyzer and Cascade Microtech probes. S-parameters from this device were used to calculate current gain (H21), maximum stable gain (MSG), and maximum available gain (MAG) which are plotted in Figure 4. The measurement conditions were Vds = 40 V, I_{ds} = 88.4 mA, V_{gs} = -2.0 V, and I_g = 12.8 µA. The unity current gain cutoff frequency f_T was 8.3 GHz. The f_{max} (13.9 GHz) is the highest reported to date for a 6H-SiC MESFET on a conducting substrate. The small signal gain was 9.4 dB at 5 GHz and 2.7 dB at 10 GHz.

2.3. 4H-SiC D.C. and R.F. performance

The I-V characteristic of a 4H-SiC MESFET fabricated with a channel doping of 3.7×10^{17} cm^{-3} is shown in Figure 5. The drain current density at $V_{ds} = 20$ V, $V_{gs} = 0$ V is about 186 mA/mm with a maximum extrinsic transconductance of 42.7 mS/mm. Measured S-parameters were used to calculate the high frequency gains of the device which are plotted in Figure 6. The measurement conditions were Vds = 40 V, I_{ds} = 92.3 mA, V_{gs} = 0 V, and I_g = 0.1 µA. The unity current gain cutoff frequency f_T was 8.0 GHz. The f_{max} (16.3 GHz) is the highest reported to date for a 4H-SiC MESFET on a conducting substrate. The small signal gain was 11 dB at 5 GHz and 5.1 dB at 10 GHz.

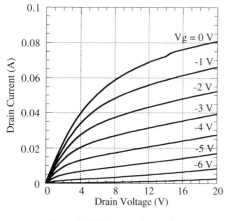
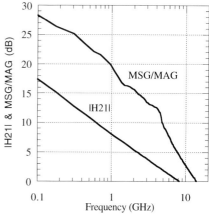

Figure 3. I-V characteristics of 6H-SiC MESFET.

Figure 4. |H21|, MSG, and MAG of 6H-SiC MESFET.

The output power of another 4H-SiC MESFET was measured on-wafer using an ATN LP1 On-Wafer Load Pull System. This device had slightly larger channel dimensions with $L_g = 0.7$ μm, $L_{sg} = 0.3$ μm and $L_{gd} = 0.8$ μm. The channel doping was 1.7×10^{17} cm^{-3}. The unity current gain cutoff frequency and maximum frequency of oscillation of this device were 6.7 GHz and 12.9 GHz, respectively. The power measurement conditions were $V_{ds} = 54$ V, $V_{gs} = -2$ V, and $I_{ds} = 77.4$ mA. The maximum output power 29.72 dBm (0.937 W) was achieved with 6.7 dB gain, 12.7% power added efficiency, and 15.4% drain efficiency using an input power of 23 dBm with the loadpull system tuned for maximum output power. The

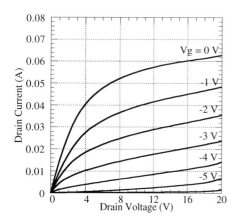
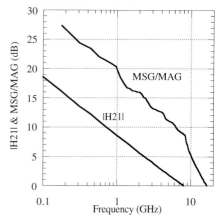

Figure 5. I-V characteristic of 4H-SiC MESFET.

Figure 6. |H21|, MSG, and MAG of 4H-SiC MESFET.

maximum power was limited by the onset of gate breakdown. The maximum power density was 2.8 W/mm which is the highest ever reported for a SiC MESFET. By reducing the drive level to 21 dBm the gain, power added efficiency, and drain efficiency improved to 8.15 dB, 14.8%, and 17% respectively, but the output power decreased to 29.15 dBm (2.5 W/mm). Although the drain efficiency may seem low it should be remembered that for Class A operation the theoretical drain efficiency with the device tuned for maximum output power is only 25%. The device efficiency can be increased by changing the load line in such a way as to reduce the output power level, but increase the power added efficiency. The maximum possible efficiency for Class A operation is 50%.

Although the f_T, f_{max}, and power density are the highest reported for 6H-SiC and 4H-SiC MESFET's on conducting substrates, the present devices are far from optimized. First, the breakdown voltages of these devices are about half of the theoretical values for these doping densities. For the power measurement the maximum operating voltage of the $L_g = 0.7$ μm 4H-SiC MESFET was 54 V, but from a theoretical standpoint, a SiC device with a channel doping of 1.7×10^{17} cm^{-3} should have a breakdown voltage close to 175 V which would allow an operating voltage close to 80 V. The higher operating voltage would significantly increase the power density of the device. Second, the specific contact resistance to the N$^+$ epilayer was 5×10^{-6} Ω-cm^2 and the sum of the metal contact and N$^+$ contact layer resistance to the channel epilayer was 5 Ω-mm. Reduction of these parasitic resistances would significantly increase the extrinsic transconductance of the transistor. Finally, approximately half of the input capacitance of these MESFET's is contributed by the gate pad which presently sits on a dielectric film. The use of a semi-insulating SiC substrate would significantly reduce this parasitic input capacitance. With these improvements a two-fold increase in power density, f_T, and f_{max} is achievable.

3. SiC JFET's

3.1. Device design and fabrication

High frequency SiC JFET's are of interest for high temperature RF applications because much lower gate leakages should be obtained with a pn junction at high temperature than with a Schottky gate. The first 6H-SiC high frequency JFET devices reported had an f_{max} of 1.0 GHz [6]. Subsequent 6H-SiC JFET's have been fabricated using the same processing and cross-sectional design described in ref. [6]. The JFET's utilized the same two fingered device

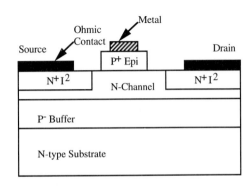

Figure 7. Cross section of 6H-SiC JFET with ion implanted source and drain.

design and the same coplanar probing pattern that was described earlier for the MESFET's. The material structure consisted of an n-type 6H-SiC substrate, 7.5 μm p-type isolation epilayer doped about 4×10^{15} cm^{-3}, 0.23 μm n-type channel epilayer doped 3.1×10^{17} cm^{-3}, and P+ gate epi doped 5×10^{18} cm^{-3}. The epitaxially grown P+ gate layer was subsequently reactive ion etched to form the gate mesa. Source and drain wells were then ion implanted, after which Ni ohmic contacts were deposited, as shown in Fig. 7. The gate contact was then overlayed on top of the gate mesa using the direct write electron beam aligner. The resulting structure was a gate mesa length of 1.3 μm and source-gate and gate-drain spacings of 0.6 μm and 0.7 μm, respectively.

3.2. 6H-SiC JFET D.C. and R.F. performance

The I-V characteristics of a 6H-SiC JFET are shown in Figure 8. The drain current density at Vds = 20 V, Vgs = 0 V is about 220 mA/mm with a maximum extrinsic transconductance of 25.5 mS/mm. S-parameters from this device were used to calculate current gain (H21), maximum stable gain (MSG), and maximum available gain (MAG) which are plotted in Figure 9. The measurement conditions were Vds = 30 V, Ids = 73.3 mA, Vgs = 0 V, and Ig = 1.29 mA. The unity current gain cutoff frequency f_T was 3.0 GHz. The f_{max} was 4.8 GHz, a 4.8X improvement over previous JFET's. This is the highest f_{max} reported to date for a SiC JFET.

Although these devices have shown a marked increase in f_{max}, they are still quite low compared to the MESFET's. However, it should be stressed that the MESFET's have had much more development. Parasitics that are currently limiting the JFET's are gate resistance due to the low hole mobility in the P+ layer, and the relatively long gate length of 1.3 μm compared to the 0.5 μm gate lengths for the MESFET's. The long gate length is currently required because of the need to align the gate metal on top of the gate mesa. Methods of obtaining a self-aligned gate structure are being investigated.

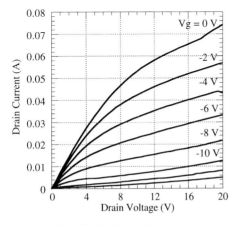

Figure 8. I-V characteristics of 6H-SiC JFET.

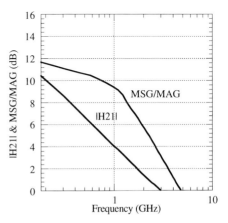

Figure 9. |H21|, MSG, and MAG of 6H-SiC JFET.

4. Acknowledgments

The authors would like to acknowledge the help of S. Macko and H.S. Kong of Cree Research for epitaxial growth, processing engineers and technicians at both Motorola and Cree Research, and the management support at both companies. Thanks also are due to David Wandrei of ATN Microwave, Inc., Billerica, MA 01821 for taking the on-wafer power measurements. This MESFET research was sponsored in part by Dr. Y.S. Park of the Office of Naval Research under Contract # N00014-92-C-0083. The JFET research was sponsored in part the Dept. of the Air Force, Wright Laboratories, Contract No. F33615-93-C-1294, monitored by R. Blumgold.

References

[1] M.W. Shin, G.L. Bilbro, and R.J. Trew, "High Temperature Operation of N-Type 6H-SiC and P-Type Diamond MESFET's," *1993 IEEE/Cornell Conference*, August, 1993, pp. 421-430.

[2] S. Sriram, R.C. Clarke, M.H. Hanes, P.G. McMullin, C.D. Brandt, T.J. Smith, A.A. Burk, Jr., H.M. Hobgood, D.L. Barrett, and R.H. Hopkins, "SiC Microwave Power MESFET's, in *Silicon Carbide and Related Materials*, M.G. Spencer, R.P. Devaty, J.A. Edmond, M. Asif Khan, R. Kaplan, and M. Rahman (Eds.), Inst. Phys. Pub., Bristol, No. 137, 1994, pp. 491-494.

[3] J.W. Palmour, C.H. Carter, Jr., C.E. Weitzel, and K.J. Nordquist, "High Power and High Frequency Silicon Carbide Devices," to be published in *Diamond, SiC and Nitride Wide-Bandgap Semiconductors*, C.H. Carter, Jr., G. Gilden Blatt, S. Nakamura, and R.J. Nemanich (Eds.), Mat. Res. Soc. Proc., Vol. 339, Pittsburgh, PA, 1994.

[4] W.J. Schaffer, H.S. Kong, G.H. Negley, and J.W. Palmour, "Hall Effect and C-V Measurements on Epitaxial 6H- and 4H- SiC," in *Silicon Carbide and Related Materials*, M.G. Spencer, R.P. Devaty, J.A. Edmond, M. Asif Khan, R. Kaplan, and M. Rahman (Eds.), Inst. Phys. Pub., Bristol, No. 137, 1994, pp. 155-159.

[5] H.M. Macksey and F.H. Doerbeck, "GaAs FETs Having High Output Power Per Unit Gate Width," *IEEE Electron Device Letters*, Vol. EDL-2, No. 6, June, 1981, pp. 147-148.

[6] J.W. Palmour, C.E. Weitzel, K. Nordquist, and C.H. Carter, Jr., "Silicon Carbide Microwave FET's," in *Silicon Carbide and Related Materials*, M.G. Spencer, R.P. Devaty, J.A. Edmond, M. Asif Khan, R. Kaplan, and M. Rahman (Eds.), Inst. Phys. Pub., Bristol, No. 137, 1994, pp. 495-498.

… *Inst. Phys. Conf. Ser. No 141: Chapter 4*
Paper presented at Int. Symp. Compound Semicond., San Diego, 18–22 September 1994
© 1995 IOP Publishing Ltd

Cell Design and Peripheral Logic for Nonvolatile Random Access Memories in 6H-SiC

W. Xie*, G. M. Johnson†, Y. Wang*, J. A. Cooper, Jr.*,
J. W. Palmour†, L. A. Lipkin†, M. R. Melloch*, and C. H. Carter, Jr.†

* School of Electrical Engineering, Purdue University, West Lafayette, IN 47907
† Cree Research, Inc., 2810 Meridian Parkway, Suite 176, Durham, NC 27713

Abstract. We recently demonstrated a one-transistor nonvolatile memory cell in 6H-SiC which integrates a bipolar access transistor over a pn junction storage capacitor. Storage times range from about 4 minutes at 350 °C to 120 hours at 210 °C. In this article, we report the first long-term measurements of storage time at room temperature. We also discuss cell design and cell performance, and describe a hybrid test chip in which a SiC cell array is connected to a silicon CMOS peripheral logic chip.

1. Introduction

One-transistor dynamic memory cells implemented in a wide bandgap semiconductor such as 6H-SiC exhibit extremely long storage times at room temperature [1,2]. In silicon, the storage time is limited by thermal generation of minority carriers. Thermal generation is proportional to the intrinsic carrier concentration n_i. At room temperature, the intrinsic carrier concentration of 6H-SiC ($E_G \approx 3.0$ eV) is about 1.6×10^{-6} cm^{-3}, about sixteen orders of magnitude smaller than in silicon. Assuming the density of generation-recombination centers in SiC is similar to the density in silicon, one concludes that the thermal-generation-limited storage times in SiC will be many (sixteen?) orders of magnitude longer than in silicon at room temperature.

These trends have been confirmed by measurements at elevated temperatures [1,2]. The best SiC pn junction capacitors have storage times of approximately 4 minutes at 350 °C, increasing to 120 hours at 210 °C. The storage times are thermally activated with an activation energy of half bandgap (1.5 eV). Extrapolation of these data suggests a room temperature storage time of over one million years! Of course, such numbers are meaningless except to assure us that the mechanisms responsible for the storage time at elevated temperatures will be entirely negligible at room temperature. Other mechanisms having lower activation energies (or non-thermally activated) will dominate as temperature is reduced. To investigate these mechanisms, direct measurements of storage time have been conducted at room temperature.

2. Long-term storage time measurements

Long-term room temperature storage time measurements are conducted as follows: The SiC test chip is mounted in a 24-pin dual in-line package in a shielded dark box. Each storage capacitor is wired to a BNC connector on the box, and is kept shorted by a toggle switch. The recovery transient of a defective storage capacitor is shown in Fig. 1 for illustration purposes. Before t = 0 the capacitor is in equilibrium, and a large capacitance (11.143 pF) is observed. At t = 0, a short bias pulse is applied across the capacitor. This pulse forward biases one of the pn junctions, removing holes from the central p-region.

When the pulse is completed, the bias across the npn capacitor returns to zero, and a smaller capacitance is observed (10.01 pF). The capacitance gradually returns to the equilibrium value as defect-assisted generation supplies holes to the middle p-region. We define the storage time as the time required for the capacitance to return to within 1/e of it's equilibrium value. In the defective capacitor of Fig. 1, the storage time is about 1693 hours (70 days).

Figure 2 shows the recovery transient of a good capacitor on the same test chip. Although the recovery is almost imperceptible on this scale, it is possible to estimate the storage time by expanding the vertical scale. Figure 3 shows that the extrapolated room temperature storage time of this device is approximately 20 years.

3. Cell design considerations

The basic nonvolatile RAM (NVRAM) cell [3] is illustrated in Fig. 4. It can be viewed as an npn bipolar transistor with a floating collector. Charge is stored on the floating collector, which has a large capacitance to the substrate. If the cell is to be used in an array, several new considerations arise. The cell should be as small as possible, but must contain enough charge to provide a measurable voltage on the bit line. It should add as little capacitance as possible to the bit line, while maximizing the storage capacitance. These considerations argue for a small emitter-base junction and a large collector-substrate junction.

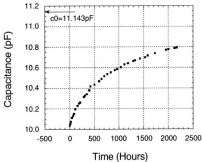

Figure 1. Capacitance transient of a defective storage capacitor at room temperature. The equilibrium capacitance is indicated by the arrow.

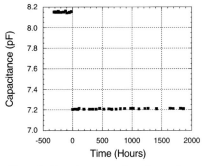

Figure 2. Capacitance transient of a good storage capacitor at room temperature. No detectable recovery is visible at this scale.

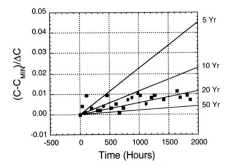

Figure 3. Recovery transient of Fig. 2 on an expanded scale. The initial slope of the transient indicates a storage time of about 20 years.

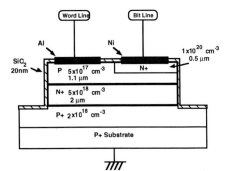

Figure 4. Cross section of a typical SiC bipolar NVRAM cell. The top npn layers form a bipolar transistor with a floating collector.

During readout, the bipolar access transistor must have good efficiency in the inverse-active mode, i.e. when electrons from the floating collector are being injected upward to the reverse-biased emitter. Since electrons must reach the emitter to be detected, it is important to insure that the largest possible fraction of stored electrons actually reaches the emitter. Unfortunately, most of the electrons injected under the base contact will recombine at the base contact and not be collected by the emitter — these electrons are lost. This argues for making the emitter large relative to the base contact.

The most important consideration is the base spreading resistance. Since the hole mobility in p-type SiC is relatively low, considerable voltage drop occurs in the base due to lateral current flow under the emitter. As a result, electron injection from the collector occurs primarily under the emitter periphery. Thus it is desirable to maximize the emitter periphery while minimizing the emitter area (to minimize bit line capacitance). To maximize the stored charge, the floating collector can be extended laterally beyond the boundary of the p-base.

4. Design of a SiC/Si hybrid memory

In order to evaluate various cell designs and test peripheral circuits, the NVRAM is being built first in a hybrid configuration, as shown schematically in Fig. 5. In this implementation, the SiC chip contains the NVRAM cell array, while a silicon CMOS chip contains sense amplifiers, word line drivers, row and column select circuitry, and control logic. The two chips are connected by wire bonding.

The SiC chip consists of a 14x52 NVRAM cell array. Of the 52 rows, 48 contain full-size cells and the remaining 4 contain half-size "dummy" cells. Note that the columns (bit lines) are grouped into 14 pairs, with each pair connecting to one sense amplifier on the CMOS logic chip. A sense amplifier is basically a balanced flip-flop. During the read operation, one side of each sense amplifier is connected to a full-size cell containing stored data, while the opposite side of the sense amplifier is connected to a half-size cell containing a logic "zero". The sense amplifier then falls into the correct logic state depending on the precise balance between the two bit lines. This procedure eliminates common-mode noise and provides an automatic reference level for sensing. The control logic insures that during each read, one row of half-size cells and one row of full-size cells are selected, with the selection rule that if the full-size word line is even numbered, the half-size word line is odd numbered (and vise versa). This places the half-size cell on one bit line in each pair, with the selected full-size cell on the other.

For increased yield, we utilize row and column redundancy. Of the 14 bit line pairs and 48 word lines, we select only the best 8 bit line pairs and best 32 word lines, resulting in a 256-bit NVRAM. The column and row selections are accomplished by fuse programming on the control chip.

5. Conclusions

We have reported new measurements of storage time on one-transistor NVRAM cells in 6H-SiC. Storage times extrapolated from 2000+ hours of direct testing at room temperature suggest a 1/e recovery time of about 20 years. We have also described a hybrid implementation of a 256-bit NVRAM chip, designed to evaluate cell design and sense amplifier performance.

This work is supported by grant No. N00014-93-C-0071 from the Ballistic Missile Defense Organization, Innovative Science and Technology Division, administered by ONR.

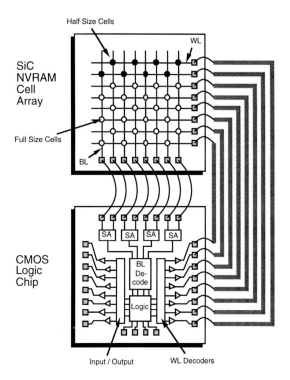

Figure 5. Schematic diagram of the hybrid NVRAM implementation. The SiC chip contains NVRAM cells, while a silicon CMOS chip contains peripheral logic. The bit lines are grouped into pairs, with each pair connected to one sense amplifier on the control chip by direct wire bonding. Word lines are connected via runners on the substrate.

References

[1] J. A. Cooper, Jr., J. W. Palmour, C. T. Gardner, M. R. Melloch, and C. H. Carter, Jr., "Dynamic Charge Storage in 6H-Silicon Carbide: Prospects for High-Speed Nonvolatile RAM's," *IEEE Device Research Conference*, Boston, MA, June 22-24, 1992.

[2] C. T. Gardner, J. A. Cooper, Jr., M. R. Melloch, J. W. Palmour, and C. H. Carter, Jr., "Dynamic Charge Storage in 6H-SiC", *Applied Physics Letters,* Vol. 61, pp.1185-1186, 7 September, 1992.

[3] W. Xie, J. A. Cooper, Jr., M. R. Melloch, J. W. Palmour, and C. H. Carter, Jr., "A Vertically Integrated Bipolar Storage Cell in 6H Silicon Carbide for Nonvolatile Memory Applications," *IEEE Electron Device Letters*, Vol. 15, pp. 212-214, June 1994.

IN SITU PROCESSING OF SILICON CARBIDE

S. Ahmed*, C. J. Barbero* and T. W. Sigmon

Department of Electrical Engineering
Arizona State University
Tempe, AZ 85287-5706

ABSTRACT

Development of doping processes which allow fabrication of highly doped layers *in situ* in SiC is discussed. Comparison of implantation profiles, calculated using an empirical simulator, with experimental results are also reported. Point contact current voltage (PCIV) measurements are presented for dopant profiles obtained using a pulsed *uv*-excimer laser to activate pre-implanted impurities and dope virgin SiC from spin-on-glass dopant sources. Our experimental data shows pulsed excimer laser processing (PELP) has the potential to be an effective, low temperature, selective doping process for the SiC industry and research community.

1. Introduction

Crystalline silicon carbide is emerging as a candidate material for high-power, high-temperature devices. Both figure of merit calculations and theoretical analysis of power devices suggest 6Hα-SiC material properties are superior to either silicon or gallium arsenide for specific applications in power, automotive, aerospace and hostile environment electronics [1,2]. The availability of high-quality, large area crystalline substrates, particularly for 6Hα-SiC, has sparked renewed interest in developing device and process technology for this material system. In particular, doping technology is difficult in SiC, with the material generally being doped during the crystal growth process, or by subsequent growth of doped epitaxial layers [3,4].

Ion implantation and diffusion are two of the most common selective doping techniques used in the semiconductor industry. Efforts at ion implantation has met with limited success due to extremely high temperatures required for annealing the implant damage and activation of the implanted dopant atoms. In an article Vodakov *et al.* reports the diffusion coefficients for common dopants in SiC [5]. Common impurities such as N, Al, Ga have very low diffusivities in SiC. They concluded that very high temperatures (> 2200°C) and long process times are required for the introduction of the above mentioned dopants in SiC using conventional diffusion process.

Nakata *et al.* report the failure of conventional thermal annealing of ion implanted SiC [6]. Although they are able to decrease the sheet resistance of the implanted sample to some extent, the

---

* Department of Electrical Engineering and Applied Physics, Oregon Graduate Institute, Portland, OR 97291.

obtained sheet resistances were not low enough to be useful for device processes. They also mention the inability of a 1500°C thermal anneal to restore the crystallinity of the implanted sample. Recently Brown *et al.* report nitrogen implantation for source-drain formation in a SiC FET [7]. However, the procedure involved hot implantation at 1000°C and subsequent furnace annealing at 1300°C.

Pezoldt *et al.* observe polytype phase transitions in implanted and high-temperature annealed SiC crystals [8]. They find that in implanted and high temperature annealed SiC crystals, partial and complete polytype phase transition may occur. The partial phase transition is attributed to the generation of dislocations and stacking faults along with recrystallization of local defect clusters. From these results it is evident that a low temperature doping technique is essential for commercial viability of SiC devices and monolithic circuits.

With laser processing, however, the substrate is heated to annealing temperatures within a span of a few nanoseconds. The actual annealing is performed in a very short amount of time (50-150 nanoseconds), resulting in a nonequilibrium process. This phenomenon may lead to the formation of nonequilibrium phase and impurity concentrations, preventing the polytype phase transition at high temperatures [9]. Usually for laser annealing of ion implanted dopants, the substrate is melted to a depth where no implantation damage is present, reducing the possibility for local defects and stacking faults to have an adverse effect in the recrystallized surface layer. The following sections discuss the empirical simulator developed for implantation profiles along with our experimental results in the area of pulsed excimer laser processing (PELP) of SiC.

2. Depth Profile Simulator for Ion Implantation in 6Hα-SiC

Knowledge of the depth profile of implanted ions into a material is crucial as this distribution is closely related to the electrical characteristics of the final device. TRIM, a widely used profile simulator, is designed to predict depth profiles in amorphous material. It cannot predict the exponential tails, resulting from channeling of ions, in a crystalline material. In order to minimize the axial and planar channeling occurring during implantation, wafers are usually tilted and rotated with respect to the beam axis during the implantation process. However, our experimental data demonstrates that this step does not completely eliminate ion channeling effects for SiC and an empirical simulator is necessary to accurately predict depth profiles of implanted dopants [10].

In order to develop an empirical simulator for crystalline SiC, commercially available substrates are implanted with nitrogen, arsenic, boron, and aluminum. While N and As are n-type dopants, B and Al contribute to p-type conductivity in silicon carbide. Experimental implantation profiles of 30- to 300-keV N, As, B, and Al in 6Hα-SiC are obtained using secondary ion mass spectrometry (SIMS).

Implanted profiles are then fit with Pearson-IV curves which require knowledge of the first four moments of the distribution. The moments of the impurity distributions are extracted from the experimental data and fit to simple functions of the ion energies. Thus an accurate implantation depth profile simulator, based on experimental data for the common dopants in α-SiC, is developed. This method results in a more accurate implant simulator than is obtained using conventional first principles calculations, primarily due to channeling considerations. Figure 1 shows the depth profiles of 30 and 80 keV nitrogen (N) implantation into SiC. The profiles predicted by the simulator are also shown in figure 1. The equations for the fitted curves for the four dopant atoms are described in reference 11. The four moments, necessary to construct a Pearson IV distribution, for a particular ion and a given energy can be obtained from these equations.

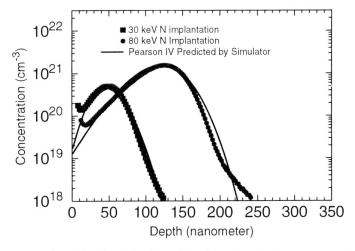

Figure 1: Experimental and simulated depth profiles of implanted nitrogen at two different energies. The doses for the 30 and 80 keV implantations are 2.3×10^{15} cm^{-2} and 1.1×10^{16} cm^{-2} respectively.

3. Activation of Implanted Dopants in 6H-SiC

The objective of a post-implant anneal is the restoration of the crystalline quality of the implanted layers and activation of the implanted dopants. Chou *et al.* previously demonstrated that PELP is an effective method for restoring the crystalline quality of the near surface amorphous layer created by high dose implantation [9]. In our experiments, we use a 308 nm XeCl excimer laser with pulse duration of 27 ns to activate implanted N and Al. The laser beam passes through a spatial homogenizer and is focused down to 3x3 mm^2 spot. Various beam splitters are placed in the path of the beam to obtain the energy required for a particular experiment. The computer controlled laser processing system also has *in situ* characterization capability derived by monitoring the surface reflectivity with a HeNe laser. When a substrate is melted with a high energy laser pulse, the reflectivity changes and is recorded in real time.

To anneal the nitrogen implanted sample, we use an energy fluence of 1.44 J/ cm^2. For the aluminum implanted sample an energy fluence of 1.51 J/ cm^2 is used. In both cases, 5 shots are used and *in situ* reflectivity measurements indicate melting of the SiC surface due to the high energy fluence of the laser. According to Bourdelle *et al.*, the threshold energy fluence for melting of SiC is 0.9 J/ cm^2 [12]. We use higher energies to anneal the implanted dopants. At these energies, the near surface of SiC is melted with a significant redistribution of dopants occurring [9]. This redistribution of the implanted dopants is seen in Figs. 2 and 3. This result confirms our reflectivity measurement observations of the surface of the SiC substrate first being melted with an epitaxial regrowth then taking place. The shape of the resultant dopant profile is controlled by carefully selecting the number and the energy of the laser pulse. More in depth experiments are in progress to investigate the dopant profile and resulting impurity activation in the laser processed SiC substrates.

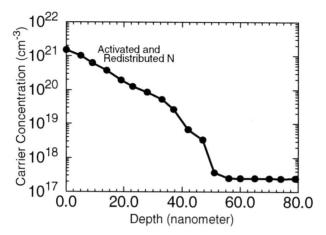

Figure 2: Activated and redistributed N atoms in SiC. An implantation dose of 3×10^{15} cm^{-2} at 30 keV is used.

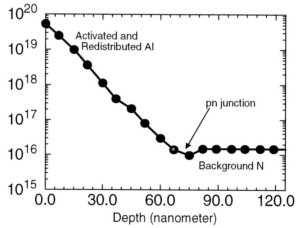

Figure 3: Activated and redistributed Al atoms in SiC. An implantation dose of 3×10^{13} cm^{-2} at 30 keV is used.

4. Spin-on-glass (SOG) Laser Doping of SiC

We have been successful in creating shallow junctions in 6Hα-SiC using a spin-on-glass dopant source. Commercially available n-type substrates are spin coated with spin-on-glass solutions containing the Ga, Al, or B atoms, p-type dopants for SiC. The excimer laser is then used to irradiate the substrates so that the liquid phase diffusion of dopants into SiC occurs. Point Contact Current-Voltage (PCIV) measurements are then performed to extract the carrier concentration of the processed samples [13]. We performed SIMS on selected samples to establish physical evidence for the process [14].

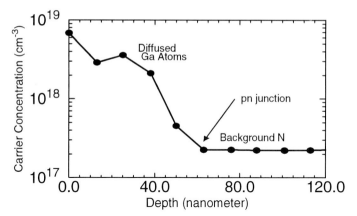

Figure 5: Carrier concentration profile in a SOG laser doped sample. The formation of pn junction is also shown. The sample is irradiated with an energy fluence of 1.45 J/cm^2.

Figure 5 shows the carrier concentration profile in a gallium (Ga) doped sample. A shallow junction is created at a depth of ~60 nm. We have created similar profiles with other p-type dopants (such as Al) in SiC. In Fig. 6, we see the boron concentration profile resulting in a SOG doped sample. Secondary ion mass spectroscopy (SIMS) is performed to obtain the B profile shown in Fig. 6 [14].

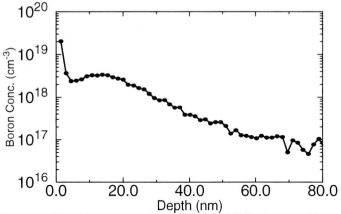

Figure 6: Concentration of boron atoms in a SOG doped SiC substrate. The sample is irradiated with an energy fluence of 1.04 J/cm^2.

5. Conclusion

We demonstrate activation of both n- and p-type dopants implanted into 6H-SiC using a pulsed excimer laser. Initial results suggest full activation of the implanted dopants, making pulsed laser annealing of ion implanted SiC an attractive option as compared to furnace annealing at high temperatures. The SOG laser doping technique shows, we believe for the first time, laser induced diffusion can also be a viable technique to selectively dope SiC. The major advantage of PELP is that it is not only a fast but also a low temperature process. The process can easily be extended to other important poly-types of SiC as well. The process may be performed with *in situ* monitoring of the melt times, resulting in real time knowledge of the resulting profiles.

This effort is supported in part by Wright-Patterson Air Force Materials Laboratory [Contract No. F33615-92-C-5959 (J. King)].

References

[1] R. W. Keyes, "Figure of Merit for semiconductors for high-speed switches," *Proc. IEEE*, vol. 60, p. 225, 1972.

[2] M. Bhatnagar and B. J. Baliga, "Comparison of 6H-SiC, 3C-SiC, and Si for power devices," *IEEE Trans. on Electron Devices*, vol. 40, no. 3, pp. 645-655, March 1993.

[3] Y. C. Wang, R. F. Davis, and J. A. Edmond, "Insitu incorporation of Al and N and p-n junction diode fabrication in alpha(6H)-SiC thin films," *J. of Elec. Materials*, vol. 20, no. 4, pp. 289-294, 1991.

[4] G. Kelner, S. Binari, M. Shur, and J. Palmour, "High temperature operation of α-Silicon Carbide buried-gate junction field-effect transistors," *Electron. Lett.*, vol. 27, no. 12, pp. 1038-1040, June 1990.

[5] Y. A Vodakov, E. N. Mokhov, "Diffusion and Solubility of Impurities in Silicon Carbide", in *The 3rd International Conference on Silicon Carbide*, Edited by R. C. Marshall, J. W. Fault, and C. E. Ryan, pp. 508-519, 1973.

[6] T. Nakata, Y. Mizutani, M. Mikoda, M. Watanabe, T. Takagi, S. Nishino, "Evaluation of Al ion implanted 6H-SiC single crystals," *Nucl. Instr. and Meth.* vol. B74, pp. 131-133, 1993.

[7] D. M. Brown, M. Ghezzo, J. Kretchmer, E. Downey, J. Pimbley, and J. Palmour, "SiC MOS characteristics," *IEEE Trans. on Electron Devices*, vol. 41, no. 4, pp. 618-620, April 1994.

[8] J. Pezoldt, A. A. Kalnin, D. R. Moskwina, W. D. Savelyev, "Polytype transitions in ion implanted silicon carbide," *Nucl. Instr. and Meth.*, vol. B80/81, pp. 943-948, 1993.

[9] S. Y. Chou, Y. Chang, K. H. Weiner, T. W. Sigmon, J. D. Parsons, "Annealing of implantation damage and redistribution of impurities in SiC using a pulsed excimer laser," *Appl. Phys. Lett.*, vol. 56, no. 6, pp. 530-532, February 1990.

[10] S. Ahmed, C. J. Barbero, T. W. Sigmon, J. Erickson, "Boron and aluminum implantation in α-SiC," *Appl. Phys. Lett.*, vol. 65, no. 1, pp. 67-69, July 1994.

[11] S. Ahmed, C. J. Barbero, T. W. Sigmon, J. Erickson, "Empirical simulator for ion implantation in 6Hα-SiC," (to be submitted to *Journal of Applied Physics*)

[12] K. K. Bourdelle, N. G. Chechenin, A. S. Akhmanov, A. Y. Poroikov, and A. V. Suvorov, "Melting and damage production in Silicon Carbide under pulsed laser irradiation," *Phys. Stat. Sol. A*, vol. 121, pp. 399-406, 1990.

[13] Point contact I-V measurements were performed at Solid State Measurements, Pittsburgh, PA.

[14] SIMS measurements were carried out at Charles Evans & Associates, Redwood, CA.

The lifetime limiting defect in SiC

N T Son*, E Sörman, W M Chen, O Kordina, B Monemar and E Janzén

Department of Physics and Measurement Technology, Linköping University, S-581 83 Linköping, Sweden

Abstract. The minority carrier recombination process dominated by a deep center in 4H and 6H SiC has been revealed and studied by optically detected magnetic resonance (ODMR). This center in its paramagnetic charge state has an isotropic g-value close to 2 for both polytypes and is related to the deep photoluminescence band peaking at around 1.7 eV. This center is shown to be a deep level defect with an energy level at about 1.1 eV below the conduction band. This defect provides a very efficient recombination channel, limiting the carrier lifetime. The fact that this defect has been observed in both bulk crystals and epilayers, regardless of their doping type, indicates that it must be a common and basic defect in 4H and 6H SiC.

1. Introduction

Due to its superior electrical, mechanical and chemical properties, SiC has long been of great interest as one of the most promising materials for high temperature, high power and high frequency devices. The progress in applications of SiC-based bipolar high power devices has, however, so far been hindered by the rather short carrier lifetime in the material (typically shorter than 50 ns [1]). This is true even in areas free of crystalline imperfections due to extended defects such as micropipes, and is generally believed to be due to the presence of point defects in the material. Up to now very little is known about deep level defects in SiC, which usually play important roles in carrier recombination. Among them, the dominant recombination center (i.e. the lifetime-limiting defect), despite of its essential importance, is still unknown.

Optical detection of magnetic resonance (ODMR) has in the past decade proven to be a very powerful technique [2], which combines highly sensitive optical spectroscopy with microscopically informative magnetic-resonance techniques. In this work we have carried out a systematic ODMR study of 4H and 6H SiC, in an attempt to reveal and investigate the

---

* Permanent address: Department of Physics, University of Hanoi, 90 Nguyen Trai, Hanoi, Vietnam.

dominant recombination center in the material. In the ODMR experiments, the radiative carrier recombination processes are directly monitored. A magnetic-resonance enhanced carrier recombination via the dominant recombination center will result in a corresponding decrease in carrier recombination via other channels [3]. In other words, the photoluminescence (PL) emission directly related to the dominant recombination channel, if it is partly radiative, will increase upon the magnetic resonance condition. Other PL emissions not related to the dominant recombination center will consequently decrease in intensity.

2. Experimental

The samples studied include bulk crystals (6H SiC) grown by the modified Lely method and epitaxial layers (both 4H and 6H SiC) grown by chemical vapor deposition (CVD). They are either n- or p-type, with a doping level ranging from 10^{17}-10^{18} cm^{-3} to as low as several times 10^{14} cm^{-3} [4]. The radiative carrier recombination processes were monitored via PL emissions from the samples. ODMR measurements were performed with the aid of a modified Bruker ER-200 X-band (9.23 GHz) ESR spectrometer using a TE$_{011}$ cylindrical cavity with optical access in all directions. The sample temperature can be varied from room temperature down to 6 K in an Oxford Instruments continuous flow helium cryostat. The multi-ultraviolet lines (351.1-363.8 nm) of an Argon ion laser were used as excitation source. PL emissions were first filtered by proper optical filters or dispersed by a Jobin-Yvon 0.25-m grating monochromator for spectral studies, and were then collected by a GaAs photomultiplier (for visible PL emissions) or a cooled North-Coast EO-817 Ge detector (for near infrared PL emissions). Signals were synchronously detected with a lock-in amplifier in phase with the amplitude-modulated microwave field. In photo-excitation experiments, a tunable Ti:Sapphire laser was used as a second excitation source.

3. Results and discussion

In all samples studied, an ODMR signal at around B=0.33 T at X-band frequency can be observed via all the PL emissions from the samples. Such an ODMR spectrum from p-type Al-doped 6H SiC detected via the shallow band-edge donor-acceptor PL emission is shown in Fig. 1(a). The negative ODMR signal in this case corresponds to a decrease of the PL intensity. In 6H SiC, the resonance line is rather broad (about 20 Gauss) and probably contains an unresolved hyperfine structure and/or unresolved resonance from different inequivalent lattice sites of the defect. The resonance linewidth in the case of 4H SiC is much narrower (about several Gauss) as illustrated in Fig. 1(b). No definite conclusion on the symmetry and the chemical identity of the defect could be reached from an angular dependence study of the ODMR signal. The averaged g-value was deduced to be 2.01±0.01 from an analysis of the experimental data by a spin Hamiltonian $H=\mu_B BgS$ with the effective electron spin S=1/2 and an assumed isotropic symmetry of the defect.

The spectral dependence study shows that this ODMR signal corresponds to an enhancement in the intensity of the PL band peaking at around 1.7 eV (1.68 eV for 6H-SiC and 1.72 eV for 4H polytype) and a decrease in the intensity of other PL emissions ranging from shallow bound excitons to near infrared deep PL bands. As an example, the spectral

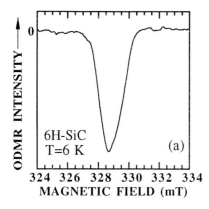 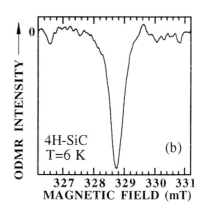

Figure 1: ODMR spectra observed at 6 K (a) from p-type, Al-doped 6H SiC (with the concentration of Al and residual N in the order of 10^{17} cm^{-3}) at ν=9.234 GHz and (b) from n-type, undoped 4H SiC at ν=9.227 GHz.

dependence of the ODMR signal from the same aluminium-doped 6H SiC sample is shown in Fig. 2. It can clearly be seen that the shallow N donor- Al acceptor recombination is decreased at resonance [Fig. 2(a)], accompanied by a corresponding increase in the 1.68 eV PL band [Fig. 2(b)]. Similar spectral dependence of the ODMR signal was observed in 4H SiC (Fig. 3). This experimental observation can be understood as due to competing processes in carrier recombination as schematically shown in Fig. 4: the magnetic resonance at the dominant recombination center (denoted as X in Fig. 4) promotes carrier recombination via this center (monitored by the PL band at 1.7 eV) and consequently reduces carrier recombination via other recombination channels (represented by dashed arrows). We assign the 1.7 eV PL band to the radiative part of a charge transfer process between the X center and acceptors and this is based on the following experimental facts. Firstly, the 1.7 eV PL band may not be related to bound exciton recombination at the defect, since in that case a multi-particle electronic system is involved which would give rise to a more complicated magnetic resonance spectrum. It most likely originates from an inter-center charge transfer process, which can be an efficient process as has been shown in silicon [5]. Secondly, the absence of the magnetic resonance from the predominant shallow N donor in the ODMR spectrum suggests that donors are not involved in the charge transfer process which gives rise to the 1.7 eV band. The observed decrease of the N donor bound exciton PL, in which case it is known that the carrier capture and recombination time is in the order of sub ns for a high N concentration [6], demonstrates that the carrier recombination involving the X center must be extremely efficient.

This dominant recombination center is shown to be a deep level defect, evident from the related deep PL emission and a photo-excitation spectrum of the ODMR signal. The latter reveals the photo-ionization of the center. This photo-ionization corresponds to the removal of the unpaired electron, participating in the microwave-induced magnetic resonance transition, from the defect to the conduction band. This process thus leads to a photo-quenching of the ODMR signal. Preliminary results from such a study in 6H SiC reveals the presence of a

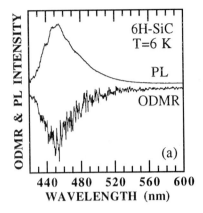
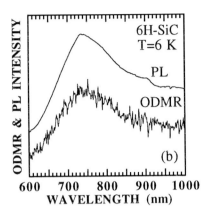

Figure 2: Spectral dependence spectra of the ODMR signal detected at 6 K from the Al-doped 6H SiC sample. PL spectra from the same spectral range are also shown for reference. It is clearly shown that (a) the shallow N donor-Al acceptor pair recombination is decreased at the ODMR condition, accompanied by (b) an corresponding increase in the 1.68 eV PL band.

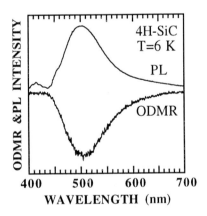

Figure 3: Spectral dependence spectrum of the ODMR signal observed at 6K from a p-type, heavily Sc-doped 4H SiC with residual concentration of N is in the 10^{17}-10^{18} cm^{-3} range. The shallow PL emission was not affected in this case.

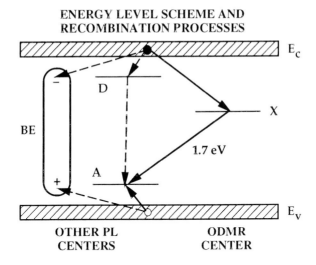

Figure 4: Schematic picture of the energy level and the competing carrier capture and recombination processes. "X" denotes the dominant recombination center, which gives rise to the observed ODMR signal.

threshold for the photo-ionization of the center, which determines the energy level of the defect to be at about 1.1 eV below the bottom of the conduction band. The acceptor level involved in this charge transfer process would then lie at about 0.2-0.3 eV above the top of the valence band, which is close to the energy level positions of the common acceptor impurities in the material. The fact that this ODMR signal has been observed in both 4H and 6H SiC, regardless of their doping type and whether they are bulk or epilayers, indicates that this must be a common and basic defect in SiC, probably involving native defects. This defect provides a very efficient recombination channel, limiting the carrier lifetime.

4. Summary

In summary, we have revealed the dominant carrier recombination center in 4H and 6H SiC. This center relates to the deep PL band peaking at around 1.7 eV which most likely originates from a charge transfer process via acceptors. This defect is shown to be a deep level center with the energy level at about 1.1 eV (for 6H SiC) below the conduction band. The results provide a spectroscopic signature of the dominant carrier recombination center. For the chemical identification of the defect and the establishment of its electronic structure, however, further investigations at a higher microwave frequency are required.

Acknowledgements: Financial support for this work was provided by the Swedish Council for Engineering Sciences (TFR), the Swedish Board for Industrial and Technical Development (NUTEK), the Swedish National

Science Research Council (NFR), the NUTEK/NFR Materials Consortium on Thin Film Growth, the NUTEK/Asea Brown Boveri (ABB) Power Device Program and the Swedish Institute.

References

[1] Naumov A V and Sankin V I 1989 Sov. Phys. Semicond. **23** 630
[2] Cavenett B C 1981 Adv. in Phys. **30** 475
[3] Chen W M and Monemar B 1991 Appl. Phys. **A53** 130
[4] Kordina O, Henry A, Hallin C, Glass R C, Konstantinov A O, Hemmingsson C, Son N T and Janzén E 1994 Proc. MRS Spring Meeting, San Francisco, USA, April 1994, in press
[5] Chen W M, Monemar B, Janzén E and Lindström J L 1991 Phys. Rev. Lett. **67** 1914
[6] Bergman J P, Harris C I, Kordina O, Henry A and Janzén E 1994 accepted for publication in Phys. Rev. B.

Deep Levels in Undoped Bulk 6H-SiC and Associated Impurities: Application of Optical Admittance Spectroscopy to SiC

W.C. Mitchel, Matthew Roth, S. R. Smith*, A.O. Evwaraye**, and J. Solomon*

Wright Laboratory, Materials Directorate, Wright-Patterson Air Force Base, Ohio 45433-7707.

Abstract. Deep levels in bulk 6H-SiC have been studied by Optical Admittance Spectroscopy (OAS), temperature dependent Hall effect, Optical Absorption and Secondary Ion Mass Spectroscopy (SIMS). Temperature dependent Hall effect measurements give activation energies between 0.1 to 0.5 eV. Specific levels are identified by OAS are found in the range .78 to 1.569 eV. Impurities are identified by Optical Absorption and SIMS. Deep levels observed by OAS at 0.78 and 1.59 eV are attributed to vanadium.

1. Introduction

SiC is under development for use in a variety of electronic applications. The physical and electronic properties of silicon carbide make it an ideal choice for high power, high temperature, and radiation resistant devices. In particular, SiC is receiving considerable attention for use in high power microwave field effect transistors, where insulating or semi-insulating substrates are required. However, as-grown, undoped 6H-SiC, the most common polytype, is usually conducting due to residual impurities such as nitrogen, a donor, or boron, an acceptor. Thus the development of semi-insulating SiC requires the introduction of a deep trap near mid gap in sufficient quantities to compensate the dominant residual impurity. At present a suitable dopant has not been identified.

We report here a study of the deep levels in unintentionally doped 6H-SiC primarily by Optical Admittance Spectroscopy (OAS) but also including temperature dependent Hall effect (TDH), optical absorption and SIMS. In order to electrically characterize deep traps in wide band gap semiconductors by thermal techniques such as Deep Level Transient Spectroscopy (DLTS), Admittance Spectroscopy or TDH, temperatures up to several hundred degrees Celsius are required. This requires such experimental modifications as hot stages and refractory metal contacts and leads. A method that has the sensitivity of the capacitance techniques, but that does not require the use of high temperatures would be useful. We have therefore studied the deep levels in 6H-SiC by using Optical Admittance Spectroscopy of Schottky diodes.

Admittance spectroscopy was first introduced by Losee[1] to study deep traps in compound semiconductors. The details of the technique have been worked out by him and others[2,3]. The optical variation of this technique was introduced by Vincent, et al.[2], and further developed by Duenas, et al.[4].

Figure 1 shows a schematic representation of a Schottky diode on n-type material containing one deep level. When the junction is modulated with sinusoidal voltage of frequency ω, charging and discharging processes of the defect level take place around the point where the Fermi level intersects the defect level. Carriers thermally emitted from the defect level lead to additional conductance, G_T and capacitance, C_T. But during optical experiment, the temperature of the junction is at a point where thermal emission of carriers from the defect is negligible. Thus the additional conductance and capacitance are due to optical emission of carriers from the defect level.

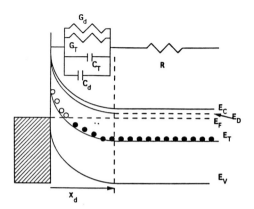

Figure 1. Schematic representation of a metal-semiconductor Schottky diode with one deep level in the semiconductor.

Traps introduce into the measurement a conductance G_T, given by[2]

$$G_T = \frac{e_n \omega^2}{e_n^2 + \omega^2} \frac{N_T}{n} A \left(\frac{\varepsilon q N^+}{2V} \right)^{\frac{1}{2}},$$

and the measured capacitance in the presence of traps is given by

$$C = C_o + C_T = \left(1 + \frac{N_T}{n}\left(\frac{e_n}{e_n^2 + \omega^2}\right)\right) A \left(\frac{\varepsilon q N^+}{2V} \right)^{\frac{1}{2}},$$

where e_n is the emission rate, n is the number of free carriers in the bulk, N_T is the number of deep traps, N^+ is the number of ionized deep centers, w is the measurement frequency, e is the bulk dielectric coefficient, A is the area of the diode, and q is the electronic charge.
Clearly, when N+ is changed by the process of illumination, both C and G_T change, and in the same manner. Hence, the photo-capacitance and the photo-conductance curves have the same shape.
Monochromatic light is focused on the diode and the wavelength of the light is scanned by the use of a monochromater. When light of the proper energy is incident on the diode a carrier is photo excited from the trap level to the conduction band. The presence of these carriers in the conduction band produces a change in the conductance of the diode which can be detected using a capacitance/conductance meter, hence, we see peaks in the conductance corresponding to the energy of the absorbed photons. When the wavelength of the light is not of the proper value to excite carriers, the conductance is lower.

2. Experimental Details

Bulk 6H-SiC samples were obtained from several sources. These included both unintentionally doped and doped material. SIMS measurements were made on selected samples from different sources. Calibration standards were not available so only impurity identification and relative concentrations could be determined. All samples studied were found to contain titanium and vanadium in varying concentrations. Another common but not universal impurity was zirconium. Electrical properties were studied by TDH using van der Pauw squares with either Ni, Ti or Al contacts. Measurements were made up to 400K. Schottky diodes for OAS were fabricated by evaporating Al dots 600mm in diameter onto the other side of the wafer. The capacitance and conductance were measured using an HP4270A Multifrequency LCR Meter, operated in the high resolution mode at a frequency of 100 kHz.

3. Results

Undoped samples were studied by TDH at temperatures up to 400°C to determine the activation energy, E_a, of the material. Three groups of samples were identified by their activation energy, samples with E_a between 70 and 120 meV were dominated by the nitrogen donor (E_a = 90 and 120 eV), another set of samples had E_a close to boron's 300 meV. However, several undoped samples had activation energies in the range 500-650 meV but no specific energy level could be identified. The room temperature resistivities of these samples were found to be up to 10^7 Ω–cm.

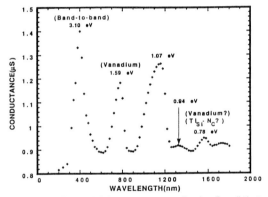

Figure 2. Optical Admittance spectrum showing five distinct peaks. Energy equivalents, and possible sources are noted for each peak.

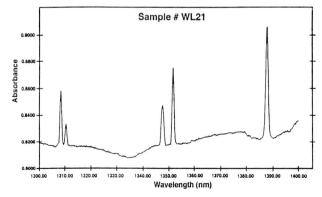

Figure 3. Infrared spectrum for vanadium in 6H-Sic.

Five peaks were identified in optical admittance spectra. A typical spectrum is shown in figure 2. The energies of the transitions are indicated in the figure. The band-to-band transition for 6H-SiC is clearly seen, along with two peaks which are attributable to vanadium, one at 1.59 eV and the other at 0.78 eV. Two other peaks are seen which are most likely attributable to other transition metal impurities. IR absorption measurements, typified by the spectrum in figure 3, confirm the presence of vanadium.

4. Discussion

Of the principal transition metal impurities detected by SIMS, only V is expected to be electrically active on an isolated substitutional site. Ti[6] and Zr are isovalent in SiC. Vanadium is amphoteric in SiC and produces a donor state, $V_{Si}^{4+}(3d^1) \rightarrow V_{Si}^{3+}(3d^0)$ and an acceptor state, $V_{Si}^{4+}(3d^1) \rightarrow V_{Si}^{3+}(3d^2)$. Stiasny et al.[7] have suggested that the donor level of vanadium is at $E_V + 1.45$ eV and the acceptor level is at $E_C - 0.70$ eV. The mid gap location of the vanadium donor in 6H-SiC has been confirmed by photo-ESR experiments[6,8] so we are confident is assigning our 1.59 eV peak to the donor state of V_{Si}. The acceptor state proposed by Stiasny et al. is close to our 0.78 eV level which thus might correspond to that state. Both the 1.59 and 0.78 eV peaks were significantly suppressed in a sample in which the vanadium absorption peaks were not observed. However, another possibility exists. Maier et al. have reported from photo-ESR experiments that the Ti_{Si}-N_C complex has a donor level at 0.6 eV. At present we do not have any suggestions for the identification of the other peaks.

These results suggest that vanadium might be a suitable dopant to compensate residual shallow donors if the Fermi level can be pinned at the deep donor level at 1.59 eV. To date we have not observed samples where this is the case. Further experiments are underway to identify the other peaks and to confirm the identification of the 0.78 eV level.

ACKNOWLEDGMENTS

We would like to acknowledge the work of Mr. Paul Von Richter, Mr. Robert V. Bertke, and Mr. Gerald Landis in preparing the specimens for these experiments. The work of SRS and JS was supported by Air Force contract no. F33615-91-C-5603.
*University of Dayton Research Institute, Dayton, Ohio 45469-0178
**Visiting Scientist, Permanent address, Physics Department, University of Dayton, Dayton, Ohio 45469-2314.

REFERENCES

[1] Losee D L 1975 *J. Appl. Phys.* **46** 2204
[2] Vincent G, Bois D, and Pinard P 1975 *J. Appl. Phys.* **46** 5173
[3] Oldham W D and Naik S S 1972 *Solid State Electron.* **15** 1085
[4] Duenas S, Jaraiz M, Vicente J, Rubio E, Bailon L and Barbolla J 1987 *J. Appl. Phys.* **61** 2541
[5] Evwaraye A O, Smith S R and Mitchel W C 1994 *J. Appl. Phys.* **75** 3472
[6] Maier K, Muller H D and Schneider J 1992 *Mat'ls Sci Forum* **83-87** 1183
[7] Stiasny Th, Helbig R and Stein R A 1992 *Amorphous and Crystalline Silicon Carbide IV, Springer Proceedings in Physics* **71** eds C Y Yang, M M Rahman and G L Harris (Berlin: Springer-Verlag)
[8] Maier K, Schneider J, Wilkening W, Leibenzeder S, Stein R 1992 *Mat'l Sci Eng* **B11** 27

Optically detected cyclotron resonance study of high-purity 6H-SiC CVD-layers

N T Son*, O. Kordina, A.O. Konstantinov, W.M. Chen, E. Sörman, B. Monemar and E. Janzén

Department of Physics and Measurement Technology, Linköping University, S-581 83 Linköping, Sweden

Abstract. We carry out optically detected cyclotron resonance (ODCR) studies at X-band microwave frequency on high-purity 6H-SiC epilayers grown by chemical vapor deposition (CVD). Electron effective mass values $m_\perp^* = (0.42 \pm 0.02)$ m_0 and $m_\parallel^* = (2.0 \pm 0.2)$ m_0 are determined. From the fit of the ODCR line shape, a remarkably high mobility $\mu_\perp \approx 1.1 \times 10^5$ cm^2/Vs is obtained for electrons in the basal plane. The anisotropy of the mobility can be explained by the corresponding anisotropy of the effective masses.

1. Introduction

Among various polytypes, 6H SiC is considered to be very promising for high temperature, high power and high frequency devices. Despite a considerable effort many of the fundamental electronic properties of 6H SiC are not yet well established, mainly due to the difficulty in growing high-quality single crystals. Recently, high purity epitaxial layers with reasonably high carrier mobility became available, providing a possibility to obtain the band structure parameters directly by the cyclotron resonance (CR) technique. Successful CR studies of 3C SiC have been reported recently by Kaplan *et al* [1,2] and Kono *et al* [3]. The effective mass and the band structure of 6H SiC and other polytypes have experimentally been studied by several indirect methods [4-7]. However, no CR study has so far been reported for 6H SiC.

In this work we carry out optically detected cyclotron resonance (ODCR) studies of 6H SiC films grown by the CVD technique. From the observed ODCR spectra, electron effective masses and the scattering time can be directly determined.

---

* Permanent address: Department of Physics, University of Hanoi, 90 Nguyen Trai, Hanoi, Vietnam.

2. Experimental

The 6H SiC films studied in this work were grown by CVD in a hot wall system, on n-type (a few times of 10^{18} cm^{-3}), (0001) Si-face, off axis (~4 degrees off towards the [11$\bar{2}$0] direction), 6H SiC substrates [8]. The thicknesses of the layers are in the range of 15-35 μm and the residual doping concentration, as estimated from PL and CV experiments, is in the mid 10^{14} cm^{-3} range [8]. Samples with higher doping concentration were also studied for comparison. ODCR measurements were performed on a modified Bruker ER-200D X-band (9.23 GHz) ESR spectrometer using a cylindrical TE$_{011}$ cavity fitted with an Oxford-Instruments continuous flow helium cryostat. The multi-ultraviolet lines (351.1-363.8 nm) of an argon ion laser were used as the excitation source and PL emissions were detected with an LN$_2$-cooled Ge detector. During measurements the sample temperature was 6 K.

3. Results and discussion

Figure 1(a) and 1(b) show some typical ODCR spectra observed in uncompensated, n-type 6H SiC layers with residual doping concentrations in the 10^{14}-10^{15} cm^{-3} range. A very sharp peak at B≈0.328 T in these spectra is due to an optically detected magnetic resonance (ODMR) of a defect present in these samples and is beyond the scope of the present paper. An ODCR spectrum with a fairly narrow linewidth peaking at B≈0.14 T with B||c observed in this very low-doped layer (4 × 10^{14} cm^{-3}) is shown in Fig. 1(a).

An excellent fit to the observed ODCR spectrum could be obtained by assuming that the intensity of the ODCR signal is proportional to the absorption of the microwave (MW) power by the crystal, which can be described by

$$P \propto \frac{1+(1+\omega_c^2/\omega^2)\omega^2\tau^2}{[1+(\omega_c^2/\omega^2-1)\omega^2\tau^2]^2+4\omega^2\tau^2} \times \frac{Ne^2\tau}{m^*} \quad (1)$$

for a linearly polarized electric MW field perpendicular to the static magnetic field **B**. Here $\omega_c = eB/m^*$ and ω the MW frequency. N is the free carrier density, m^* the cyclotron effective mass of the carriers and τ is the scattering time. From the fit of the line shape, the parameter $\omega\tau$ was obtained to be $\omega\tau=1.53$ in the case of B||c and $\omega\tau=1.28$ for **B** close to the [1$\bar{1}$00] direction. The corresponding values of the effective scattering time were deduced as $\tau(B||[0001])\approx2.64\times10^{-11}$ s and $\tau(B||[1\bar{1}00])\approx2.20\times10^{-11}$ s. The simulated curve (dashed curve) using the obtained $\omega\tau$ value and equation (1) is plotted in Fig. 1(a).

ODCR measurements with **B** along the [11$\bar{2}$0] direction was also performed to compare with the case of B||[1$\bar{1}$00], to check if the observed ODCR in these two cases are the same, i.e., if the effective mass is isotropic in the basal plane. At the X-band frequency, the two spectra are indeed very similar within the experimental error. It is therefore appropriate to assume that the conduction minima in 6H SiC are ellipsoids with the principal axis along the c-axis. From Raman scattering measurements, Colwell and Klein [6] concluded that the conduction-band minima in 6H SiC lie along the line M-L in the Brillouin zone. Theoretical calculations have also proposed that the conduction-band minima are located on the boundary planes [($1\bar{1}00$) or equivalent planes] of the Brillouin zone, either at a general point along M-L [9] or at the M point [10-12]. Thus, in the case of **B**||c, only a single ODCR peak corresponding to the transverse mass is expected. From the resonance position, this effective mass was determined as $m_\perp^* = (0.42 \pm 0.02)\, m_0$. ODCR measurements with different alignments of **B** with respect to the c-axis were also performed. From the fit to the experimentally observed cyclotron mass: $1/m^* = [(\cos^2\theta/m^{*,\perp 2}) + (\sin^2\theta/m^{*,\perp}m^{*,\|})]^{1/2}$ (where θ denotes the angle between **B** and the c-axis), the longitudinal mass was deduced to be $m_\|^* = (2.0 \pm 0.2)\, m_0$. From these effective masses and τ values, the carrier mobility in the basal plane can be estimated as $\mu_\perp \approx 1.1 \times 10^5$ cm^2/Vs by assuming $\mu \approx e\tau/m^*$. For comparison, samples with a doping level range from 10^{15} to 10^{17} cm^{-3} have also been measured. At X-band frequency, with the doping level of 5×10^{15} cm^{-3}, the resonance peak (in the case of **B**||c) already became very broad with $\omega\tau \approx 1$, corresponding to a mobility of about 60000 cm^2/Vs [Fig. 1(b)]. For the layers with n-type doping level close to 10^{17} cm^{-3} as well as for high-purity compensated layer (10^{14} cm^{-3}, aluminium compensated), no ODCR resonance has been detected due to a lower carrier mobility [Fig. 1(c)].

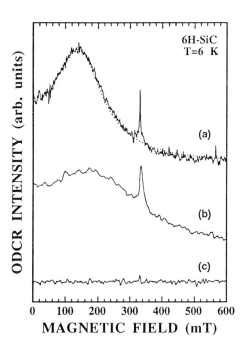

Figure 1: ODCR spectra with **B**||c and MW frequency ν=9.235 GHz observed in 6H SiC CVD-layers with different residual doping concentrations: (a) 4×10^{14} cm^{-3} uncompensated; (b) 5×10^{15} cm^{-3} uncompensated and (c) 3×10^{14} cm^{-3} compensated by Al. The simulated curve (dashed line in (a)) is calculated using equation (1) with parameter $\omega\tau$=1.53.

Recently, Schaffer *et al* [13] measured the carrier mobilities in 6H SiC along the [11$\bar{2}$0] and [0001] directions and found that the mobilities differ by a sizeable factor (about 6 at 80 K) in these two directions. Assuming the scattering time to be more or less isotropic, one would expect to have the ratio $\mu_\perp/\mu_\parallel \approx m_\parallel^*/m_\perp^*$. In our case, this ratio $m_\parallel^*/m_\perp^*$ is about 4.7.

4. Summary

In summary, we have observed optically detected cyclotron resonance at X-band frequency in high-purity 6H SiC epilayers. The electron effective mass values for 6H SiC were obtained as $m_\perp^*=(0.42\pm0.02)$ m_0 and $m_\parallel^*=(2.0\pm0.2)$ m_0. A remarkably high electron mobility at 6 K in the basal plane was determined to be $\mu_\perp \approx 1.1\times10^5$ cm^2/Vs. The anisotropy of the electron mobilities in 6H SiC can be explained by the corresponding anisotropy of the effective masses.

Acknowledgements: Ulf Lindefelt is acknowledged for useful discussions. Financial support for this work was provided by the Swedish Council for Engineering Sciences (TFR), the Swedish Board for Industrial and Technical Development (NUTEK), the Swedish National Science Research Council (NFR), the NUTEK/NFR Material Consortium on Thin Film Growth, the NUTEK/Asea Brown Boveri (ABB) Power Device Program and the Swedish Institute.

References

[1] Kaplan R *et al* 1985 *Solid State Commun.* **55** 67
[2] Kaplan R *et al* 1993 *proceedings of the International Conference on SiC and Related Materials (ICSCRM), Washington, USA, Dec. 1993* in press
[3] Kono J *et al* 1993 *Phys. Rev. B* **48** 10909
[4] Choyke W J and Patrick L 1962 *Phys. Rev.* **127** 1868
[5] Ellis B and Moss T S 1967 *Proc. Soc. A* **299** 383
[6] Colwell P J and Klein M V 1971 *Phys. Rev. B* **6** 498
[7] Suttrop W *et al* 1992 *J. Appl. Phys.* **72** 3708
[8] Kordina O *et al* 1994 *proceedings of the MRS Spring Meeting, San Francisco, USA, April 1994* in press
[9] Herman F *et al* 1969 in *Silicon Carbide*-1968, edited by H.K. Henisch and R. Roy (Pergamon, New York) S167.
[10] Junginger H G and van Hearingen W 1970 *Phys. Status Solidi* **37** 709
[11] Park C H *et al* 1994 *Phys. Rev. B* **49** 4485
[12] Almbladh C O *et al* 1994 unpublished
[13] Schaffer W J *et al* 1994 *proceedings of the MRS Spring Meeting, San Francisco, USA, April 1994* in press

GaN/AlGaN Field Effect Transistors for High Temperature Applications.

M. S. Shur *, A. Khan +, B. Gelmont *, R. J. Trew ++, and M. W. Shin ++

*Department of Electrical Engineering, University of Virginia, Charlottesville, VA 22903

+ APA Optics, APA inc., 2950 N. E. 84th Lane, Blaine, MN 55449 USA

++ Department of Electrical Engineering and Applied Physics, Case Western Reserve University, Cleveland, OH

The results of our analytical and numerical simulations of transport properties of GaN and SiC in a wide range of temperatures and doping levels point out to a relatively modest decrease of the electron mobility with temperature. Based on these results and our experimental data for AlGaN/GaN HFETs, we evaluate and compare the characteristics of AlGaN/GaN heterostructure field effect transistors at elevated temperatures using the Universal Charge Control Model (UCCM) and discuss the temperature dependencies of transconductance, maximum device current, and threshold voltage. Using a large signal computer drift-diffusion model and the harmonic balance method, we then analyze the microwave performance of AlGaN/GaN HFETs. We also present our preliminary experimental data on AlGaN/GaN performance at elevated temperatures We conclude that, in spite of a smaller thermal conductivity than for SiC, the excellent transport properties of GaN at elevated temperatures, a semi-insulating substrate, and a large conduction band discontinuity in the AlN/GaN HFETs make these devices a viable alternative for high temperature microwave and digital applications.

1. Introduction

Recently, we demonstrated a microwave operation of AlN/GaN Heterostructure Field Effect Transistors (HFETs) at room and elevated temperatures (Khan et al (1994)) reaching the cutoff frequency of 17 GHz and the maximum oscillation frequency of 35 GHz. These preliminary results demonstrate an superb potential of these device for microwave and, eventually, for millimeter waver operation. Electronic circuits which operate reliably at elevated temperatures may cut the weight of electronics on board of airplanes and space craft by a factor of two or more (Yoder (1992)).

For a specific application in linear microwave amplifiers, AlN/GaN HFETs will have several advantages including a higher temperature performance, higher power, and a larger dynamic range. In this paper, we consider the potential of the GaN/AlN material system for high temperature electronic applications. Our analysis is based on the theory of the electron transport, measured characteristics of AlGaN/GaN HFETs, and field effect transistor models which allow us to predict the device behavior at elevated temperatures. The predicted performance is much better than that for already fabricated devices where the maximum device transconductance is limited by a very large series resistance. Improved contacts and a more advanced device design should allow us to shrink this gap.

2. Transport properties of GaN and SiC

Monte Carlo simulations predicted a high peak velocity, v_p, high saturation velocity, v_s, and a high electron mobility, μ, in GaN ($v_p = 2.7 \times 10^5$ m/s, $v_s \approx 1.5 \times 10^5$ m/s, and $\mu \approx 1000$ cm^2/Vs at room temperature in GaN doped at 10^{17} cm^{-3} (see Littlejohn et al (1978), Gelmont et al (1993), Shur et al (1994), Joshi (1994)). These values make GaN and GaN based heterostructures quite attractive applications for microwave and millimeter wave devices which may operate in a wide temperature range because of a wide energy gap and fairly good thermal conductivity of this material. (A more accurate mobility calculations are in progress.) Fig. 2 shows the comparison of the calculated dependencies of the mobilities for GaN, 3C-SiC, and 6H-SiC and the measured dependence of the mobility on temperature for 4H-SiC.

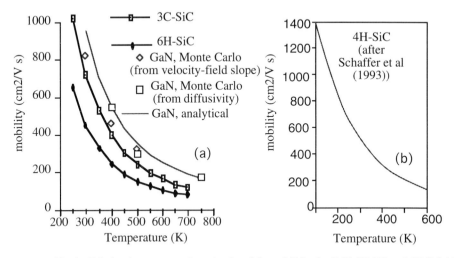

Fig. 1. Calculated temperature dependencies of the mobilities for GaN, 3C-SiC, and 6H-SiC (a) (based on the results presented by Shur et al (1994)) and measured dependence of the mobility on temperature for 4H-SiC (b) (after Schaffer et al (1994)).

6H-SiC is now one of the prime contenders for power, high temperature, and microwave applications. As can be seen from Fig 2, GaN has a big advantage over 6H-SiC and a somewhat smaller advantage compared to 3C-SiC and 4H-SiC at room temperatures and above. This advantage is enhanced by excellent properties of the wide band AlGaN barrier layer which allowed us to implement AlGaN/GaN HFETs.

Fig. 2 shows the velocity-field dependencies in GaN computed at different temperatures by Monte Carlo technique.

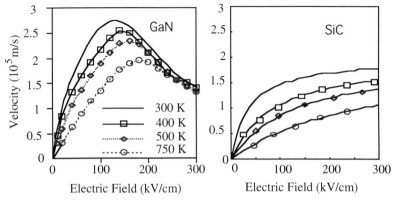

Fig. 2. Electron drift velocity for GaN doped at 10^{17} cm^{-3} and for 6H-SiC. (The GaN curves in Fig. 2 are from Gelmont et al (1993) and Shur et al (1994)).

The electron velocity in very high electric fields (200 to 300 kV/cm) remains practically unchanged with temperature. The peak field in GaN is very high (on the order of 100 kV/cm compared to approximately 3.5 kV/cm) in GaAs. The reason for such a shape of the velocity-field characteristic is a large intervalley separation and a very large energy of polar optical phonons in GaN (nearly three times higher than in GaAs). The peak field is also very large making the peak velocity to be more important in determining the overall device speed compared to GaAs where the peak field is approximately 25 times smaller. All in all, the GaN electron velocity at elevated temperatures is large enough for achieving high transconductance in GaN FETs, especially in short channel devices (since very high electric fields are required for velocity saturation in this material).

Fig. 2 also shows the estimated dependencies of the electron velocity in SiC at the same temperatures. Since no Monte Carlo results are available for SiC at elevated temperatures, we used the usual Trofimenkoff's formula for SiC

$$v(F) = \frac{\mu(T)F}{1+\mu(T)F/v_s} \qquad (1)$$

where $v_s \approx 2 \times 10^5$ m/s is the estimated saturation velocity in SiC and $\mu(T)$ is the low field mobility (see Fig. 1). The comparison of the $v(F)$ dependencies for GaN and 6H-SiC recalls the comparison of the $v(F)$ dependencies for GaAs and Si, respectively. (This approach is similar to that used for SiC by Ruff et al. (1994). It is appropriate for SiC which is an indirect semiconductor (like Si) but not appropriate for GaN which is a direct gap semiconductor (like GaAs).) The relationship between the thermal conductivities for SiC (3.5 W/cm°C) and GaN (1.3 W/cm°C) further supports this analogy.

3. Performance of AlGaN/GaN HFETs at elevated temperatures

Fig. 3 compares the calculated transconductance in the saturation region versus gate bias for three devices: a 0.5 μm gate GaAs MESFET, 0.5 μm gate AlGaAs/GaAs HFET, and 0.25 μm AlGaN/GaN HFET.

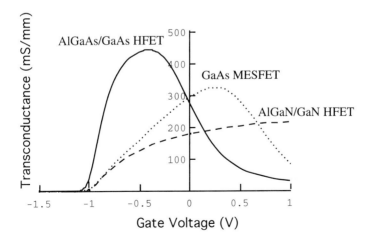

Fig. 3. Computed device transconductance in the saturation region versus gate bias for three devices: 0.5 μm gate GaAs MESFET, 0.5 μm gate AlGaAs/GaAs HFET, and 0.25 μm AlGaN/GaN HFET. The source series resistance per unit width for the MESFET and the AlGaAs/GaAs HFET is 0.3 Ωmm. The source resistance per unit width for the AlGaN/GaN HFET is 2 Ωmm.

The calculations were done using a universal charge control model described by Fjeldly and Shur (1991). This model was incorporated in AIM-Spice (see Lee et al (1993)). (Instructions on downloading a student version of AIM-Spice are available from M. Shur via e-mail (ms8n@virginia.edu)). As can be seen from the figure, the AlGaN/GaN HFET has a big advantage compared to the MESFETs in the device linearity (which is determined by how constant the device transconductance is as a function of the gate bias.) In a MESFET, the gate bias modulates the depletion region width, resulting in an increase of the channel capacitance (and as a consequence, the device transconductance) with a gate bias above threshold. In HFETs, the channel capacitance is determined by the thickness of the wide band gap barrier layer. Ideally, it should be nearly constant. However, in AlGaAs/GaAs HFETs, a relatively small conduction band discontinuity limits the maximum density of the 2d electron gas at approximately 2×10^{12} cm^{-2} and, as a consequence, the transconductance peaks fairly sharply (see Fig. 3).

Fig. 4 shows the computed temperature dependencies of the AlN/GaN HFET transconductance and saturation current with the gate length of 0.25 µm based on the temperature dependence of the electron mobility given in Fig. 1.

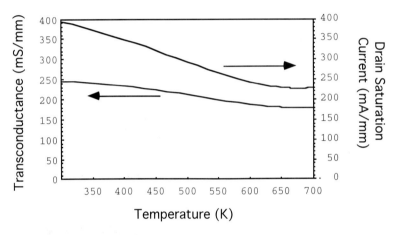

Fig. 4. Calculated transconductance and drain current of AlGaN/GaN HFET in saturation regime versus temperature for gate length 0.25 µm. Gate bias 1 V. Parameters used in this calculation are: threshold voltage, $V_T = -0.5$ V, gate length, $L = 0.25$ µm, AlGaN layer thickness, $d_i = 250$ Å, source and drain series resistances, $R_s = 2$ Ωmm, the maximum sheet electron concentration, $n_s = 2.5 \times 10^{13}$ cm^{-2}, saturation velocity, $v_s = 2 \times 10^5$ m/s.

The reason why there is such a modest predicted decrease in the device transconductance with temperature is that (1) we assumed no significant gate leakage current because of a large conduction band discontinuity and (2) we assumed a large maximum sheet electron concentration for the same reason.

Our experimental data fall much below the expected performance, primarily because of a very large source series resistance and a limited intrinsic gate voltage swing. Fig. 5 a shows the measured and calculated values of the AlGaN transconductance for the devices described by Khan et al (1994). (In this calculation, the value of the electron mobility was scaled down to 500 cm^2/Vs at 300 K.) Whereas the decrease of the device transconductance with temperature was very modest, the AlGaN/GaN HFETs exhibited a substantial threshold voltage shift and a decrease in the output conductance. The variation of the threshold voltage, V_T, was approximately linear with temperature with the temperature coefficient $K_{VT} \approx -0.4$ to -0.5 mV/°C (approximately three times larger than for GaAs based FETs, see Shoucair and Ojala (1992).) This may be related to a large donor activation energy in AlGaN.

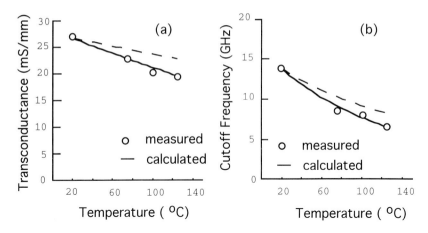

Fig. 5. Measured and calculated dependencies of the maximum transconductance and cutoff frequency on temperature. Dashed curve accounts for the temperature dependence of the low field mobility only. Solid curve also accounts for a small decrease in the intrinsic gate voltage swing (on the order of 0.1% per °C). (From Khan et al (1994a)).

Fig. 5b shows the measured and calculated dependencies of the cutoff frequency on temperature (assuming that the cutoff frequency is determined by the electron transit time). A more accurate description of the device behavior at microwave frequencies, including the calculation of the cutoff frequency requires to consider the device embedded into a microwave circuit. In this work, we used such a device simulator utilizing the harmonic balance method and the analytical FET model described by Khatibzadeh and Trew (1988) (see also Bilbro et al (1994)). The results of this simulation predict an 11 dB linear gain, 25% maximum power added efficiency, and 10 dbm output RF at 10 GHz for an AlGaN/GaN amplifier based on our device. The predicted values of f_{max} and f_t are in excellent agreement with experimental data (see Fig. 6). (The details of this measurement and these calculations will be published elsewhere.)

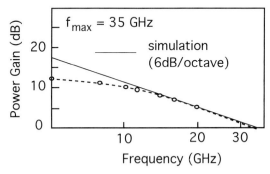

Fig. 6. Measured and calculated small-signal power gain for AlGaN/GaN HFET (f_t = 11 GHz).

4. Conclusions

A high conduction band discontinuity, high saturation velocity, and a relatively high mobility (which modestly decreases with temperature because of a high optical phonon energy) should lead to a superior performance of AlN/GaN HFETs at elevated temperatures at microwave frequencies, comparable to room temperature performance of typical GaAs MESFETs. Our preliminary results on dc and microwave performance of these devices are very encouraging but still fall far below the theoretical expectations because of a high source series resistance and a limited intrinsic gate voltage swing.

5. Acknowledgment

This work has been partially supported by the Office of Naval Research (Project Monitor Max Yoder), and by DCA (Project Monitor Andre Rausch).

References

Bilbro G L Shin M W Trew R J and Clarke R C *Proc. 5th Conf. Silicon Carbide and Related Compounds* (Bristol and Philadelphia: Institute of Physics Publishing) Conference Series Number 137 p 699

Fjeldly T A and Shur M 1991 *Workshop Proc. 21th Euro. Microwave Conf.* (Stuttgart: Institute of Physics Publishing) p. 198

Gelmont B Kim K S and Shur M 1993 *J. Appl. Phys.* 74 1818

Joshi R P and Raha P K 1994 *Proc. 5th Conf. Silicon Carbide and Related Compounds* (Bristol and Philadelphia: Institute of Physics Publishing) Conference Series Number 137 p 687

Khatibzadeh M A and Trew R J 1988 *IEEE Trans Microwave Theory and Tech.* 36 231

Khan M A Kuznia J N Olson D T Schaff W Burm G Shur M S 1994 *Applied Phys. Lett.* 65 1121

Khan M A Kuznia J N Olson D T Schaff W Burm G Shur M S Eppers C 1994a presented at *Device Research Conference* (Boulder: Colorado)

Lee K Shur M Fjeldly T A and Ytterdal T 1993 *Semiconductor Device Modeling for VLSI* (Englewood Cliffs, NJ: Prentice Hall)

Littlejohn M A Hauser J R and Glisson T H 1975 *Appl. Phys. Lett* 26 625

Ruff M Mitlehner H and Helbig R 1994 *IEEE Trans. on Electron Dev.* 41 1040

Schaffer W J Kong H S Negley G H and Palmour J W 1994 *Proc. 5th Conf. Silicon Carbide and Related Compounds* (Bristol and Philadelphia: Institute of Physics Publishing) Conference Series Number 137 p 155

Shoucair F S and Ojala P K 1992 *IEEE Trans. on Electron Dev.* 39 1551

Shur M Gelmont B Saavedra-Munoz C and Kelner G 1994 *Proc. 5th Conf. Silicon Carbide and Related Compounds* (Bristol and Philadelphia: Institute of Physics Publishing) Conference Series Number 137 p 465

Yoder M 1992 *Lecture presented at the Eminent Lecture Series* (Charlottesville: University of Virginia)

Some aspects of GaN electron transport properties

D. K. Gaskill[1], K. Doverspike[1], L.B. Rowland[1], and D. L. Rode[2]

[1]Laboratory for Advanced Material Synthesis, Naval Research Laboratory, Washington, DC 20375
[2]Department of Electrical Engineering, Washington University, St. Louis, MO 63130

Abstract. The present work presents new theoretical calculations, determined from iterative solutions of the Boltzmann equation, on the Hall scattering factor for wurtzite GaN. The effect of compensation on the 300 K mobility is also calculated and compared to the best n-type transport data reported by several groups for both unintentionally doped and intentionally doped samples. These data probably represent transport properties intrinsic to wurtzite GaN on sapphire. In addition, variable-temperature electron transport results from unintentionally doped samples are discussed and compared to previous investigations. In the latter case, the data probably indicate that an important compensation mechanism is due to the parallel transport of charge by impurity conduction.

1. Introduction

Recently, significant advances in the growth of GaN epitaxial layers have resulted in improved material properties which subsequently enabled the successful fabrication of light emitting diodes [1] and RF devices [2,3]. Further, because of the overall improvement in material quality, new light has been shed on our understanding of this substance's electron transport properties. For example, many years ago the reason for the best reported electron transport properties was conjectured to be due to large growth rates [4]. In contrast, the high quality of epilayers currently available, which are grown more slowly than previously, is directly traceable to the improved crystallinity of the samples. Yet, many questions remain to be answered such as: What are the intrinsic transport properties of GaN, and how do these compare with the best reported transport properties? What are the reasons for the high electron concentration frequently reported? And, what is the magnitude of, and reasons for compensation? To help in answering these important questions, this article presents theoretical calculations on wurtzitic GaN transport properties, compares recent experimental data on electron transport with the theoretical results, and shows the effect impurity conduction has on compensation.

The data presented here are for unintentionally doped and Si or Ge doped wurtzite samples grown on (0001) and (11$\bar{2}$0) sapphire substrates. The growth methods used to synthesize the samples were organometallic vapor phase epitaxy (OMVPE)[5] and molecular beam epitaxy (MBE)[6] where details of the sample preparation can be found in the references. Since the Hall effect is the simplest and most widely utilized measurement of transport properties, the transport results quoted here are from Hall measurements of samples with In contacts and using the van der Pauw technique and sometimes the clover-leaf geometry. The magnetic field typically ranged from 0.2 to 0.6 T. The experimental results were derived assuming the Hall scattering factor is unity.

2. Theoretical modeling of electron transport

In the first publication on the theoretical electron transport properties of GaN[7], we showed that the classical Boltzmann equation in the presence of a magnetic field, including Fermi statistics, can be solved in terms of a contraction mapping. A contraction mapping solution ensures existence, uniqueness, and rapid convergence under numerical solution by iteration. Electron scattering mechanisms include polar-optical phonon scattering, deformation potential acoustic phonon scattering, piezoelectric phonon scattering, and ionized impurity scattering.

GaN is allotropic, being capable of synthesis in either the zincblende or the hexagonal wurtzite lattice structure, the latter being the case for the experiments reported here. In previous work[8], we gave the theoretical electron drift mobility of pure hexagonal GaN and showed that it is significantly anisotropic (by a factor of 1.7) when piezoelectric scattering dominates, i.e., for high-purity GaN below about 150 K. At room temperature, where polar-optical phonon scattering is dominant, the mobility is essentially isotropic. For the present state-of-the-art, impurity concentrations are too large for anisotropy to be noticeable. Hence theoretical calculations were carried out only for the c-axis oriented along the crystal growth direction while the electric field is oriented transverse to the c-axis. The GaN band structure is given in terms of the energy band gap (3.39 eV) and the polaron effective mass ($0.218m_O$)[8]. Other material parameters were given previously[8].

We tested the importance of conduction band nonparabolicity, and s and p wave-function admixture, with the following results. At 300 K, the calculated drift and Hall mobilities are decreased by about 1.5 - 3.5% due to these factors for uncompensated electron concentrations of 1×10^{14} - 5×10^{19} cm^{-3}. At 300 K, the calculated electron Hall and drift mobilities of pure intrinsic GaN are 1732 and 1496 cm^{2}V^{-1}s^{-1}, yielding a Hall factor of 1.16. For an electron concentration of about 8×10^{17} cm^{-3} we show in Fig. 1 the calculated dependence of the Hall factor r_B on magnetic field strength B. Clearly, the Hall factor is substantially constant for B less than 5 T for concentrations near where much of the present experimental work was carried out as discussed in the next section. The calculated 300 K Hall mobility for $N_A/N_D = 0$ and 0.4 are shown in Fig. 2. The comparison with the experimental data in Fig. 2 (see next section) indicate that the GaN films grown on sapphire with electron concentration greater than about 5×10^{18} cm^{-3} have a compensation of about 0.4 and those films with lesser electron concentrations are more compensated.

3. Experimental electron transport measurements

Figure 2 shows the best 300 K wurtzite GaN Hall mobility *vs* free electron concentration reported to-date [9-18]. The samples were grown by OMVPE and MBE, with and without buffer layers (consisting of thin layers of AlN or GaN grown under conditions different from the subsequent GaN layer), and were unintentionally (open symbols) and intentionally doped (closed symbols, Si or Ge). The current data span a greater range and have slightly larger mobility for a given electron concentration than data presented previously [4]. Several important observations can be made from this data. (1) The data fall generally in the same pattern with the mobility increasing as the electron concentration decreases showing no signs of saturating. (The spread in values is probably related to insufficient growth optimization and slight differences in measurement techniques.) (2) The dependence of the OMVPE and MBE data on electron concentration are similar. (3) The data for samples with AlN, GaN, or no buffer layers fall into the same general pattern, althought unintentionally doped films grown by OMVPE without a buffer layer generally have higher electron concentrations than those grown with a buffer layer [15]. (4) The mobility for intentionally doped samples is similar to unintentionally doped samples in the region where the results overlap.

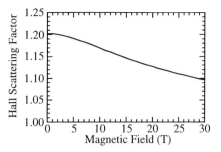
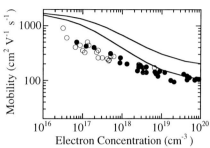

Figure 1 The Hall scattering factor *vs* magnetic field for GaN at 300 K with electron concentration 8.37×10^{17} cm^{-3}.

Figure 2 The best reported 300 K Hall mobility *vs* Hall electron concentration for GaN on sapphire. Open symbols are for unintentionally doped and closed symbols are for Si or Ge doped samples. The top (bottom) theoretical curve is for compensation $N_A/N_D = 0$ (0.4).

Several conclusions can be drawn from the data shown in Fig. 2. The most important is that the data probably represent transport properties intrinsic to wurtzite GaN epilayers on sapphire. This conclusion follows because the data taken were from samples grown under widely varying growth conditions using both atomic and molecular-based growth technologies. Another conclusion is that the compensation of the doped samples is approximately equal to that of the undoped samples in the electron concentration range where the results overlap, that is 7 x 10^{16} to 7 x 10^{18} cm^{-3}. This conclusion derives from the assumption that the 300 K mobility for two samples of similar electron concentration can be used as a gauge of compensation.

The recent improvements in the growth process, i.e., the buffer layer, have enabled the synthesis of GaN with low electron concentration, down to 3 x 10^{16} cm^{-3}. Clearly, there exists an intimate, but poorly understood relationship between growth process improvements and the densities of donors and acceptors in the layer. Some key pieces to this puzzle are the following. Previously, thin epitaxial layers grown without buffer layers had high electron concentrations, whereas those currently grown with buffer layers have lower concentrations. Improvement in crystallinity, as measured by double-crystal x-ray diffraction, has also been reported for samples grown with buffer layers [9]. In addition, this laboratory has shown that lower buffer layer growth temperature, and also (11$\bar{2}$0) sapphire orientation, improves crystallinity and transport properties, respectively [12,19]. Taken together, this evidence suggests that the nucleation of the GaN on the buffer layer has important consequences for the resulting transport properties. Because only the crystallinity or the incorporation of impurities into the GaN film can be affected by the nucleation process, we put forward the following working hypothesis: the presence of defects due to lattice mismatch (and possibly thermal expansion coefficient differences) between the epilayer and the substrate gives rise to donor and acceptor-like states which determine the background electron concentration and the amount of compensation. The use of an optimized buffer layer minimizes these donor and acceptor-like states. The current level of understanding cannot distinguish between defects or impurities (the incorporation of which was enhanced by the defects) as the physical cause for the compensating donor or acceptor-like states.

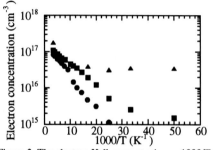
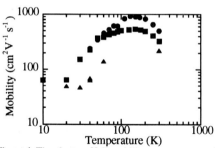

Figure 3 The electron Hall concentration vs 1000/T for samples A (circle), B (square), and C (triangle).

Figure 4 The electron Hall mobility vs temperature for samples A (circle), B (square), and C (triangle).

Figure 3 displays a plot of the electron concentration derived from variable-temperature Hall data vs 1000/T for samples A (circle), B (square), and C (triangle). Details about these and other OMVPE samples grown in this laboratory, can be found in Table I. Consider first the electron concentration of sample A. Below an onset temperature of about 40 K, the concentration data (off the scale of the graph) no longer freezes out but stays approximately constant. The mobility in this region is low and decreases with decreasing temperature. The conductivity in this region is thought to be due to impurity conduction [4]. Two-level Hall analysis has been applied to data of this type by Molnar et al. [20]. The separate contributions to the conductivity showed that the impurity conduction mechanism has a relatively low apparent mobility and high apparent concentration. Between 40 and 300 K, the electron concentration has an activated behavior, with an activation energy of 18 meV. Assuming the film is compensated, twice this activation energy is the donor ionization energy, or 35.9 meV. This is similar to the 34 meV ionization energy reported by Nakamura et al. using a GaN buffer layer [9].

Figure 4 displays a plot of the electron Hall mobility as a function of temperature from the measurements on samples A, B, and C. Sample A has the highest peak mobility, about 1000 $cm^2V^{-1}s^{-1}$, at 125 K. The peak mobility of sample B is also near 125 K. For low temperatures, about 30 to 60 K, the mobility of samples A and B have similar temperature dependencies. Between 40 and 125 K the mobility of samples A can be approximately described by a temperature power law with exponent 1.2. For sample B, the mobility between 20 and 60 K can be approximately described by a temperature power law with exponent 1.7. For higher temperatures, the mobility cannot be described by a simple power law, or activated behavior for either sample. Notice that the mobility of sample C has a similar low-temperature

Table I Sapphire orientation, buffer layer type, x-ray FWHM, electron Hall concentration (N), Hall mobility (μ_H), and resistivity (ρ) at 300 K for samples A-F shown in Figs. 3 - 6. All samples are about 3 µm thick.

Sample	Orientation	Buffer layer	FWHM (arc sec)	N (x 10^{17} cm^{-3})	μ_H ($cm^2V^{-1}s^{-1}$)	ρ (Ω–cm)
A	c-axis	AlN	313	0.837	500	0.149
B	c-axis	AlN	307	1.08	324	0.178
C	c-axis	GaN	378	1.80	301	0.115
D	c-axis	AlN	317	2.32	293	0.0918
E	a-axis	GaN	307	2.60	357	0.0673
F	c-axis	GaN	880	2.28	379	0.0722

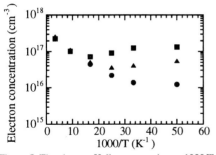

Figure 5 The electron Hall concentration *vs* 1000/T for samples D (circle), E (triangle), and F (square).

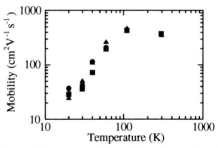

Figure 6 The electron Hall mobility *vs* temperature for samples D (circle), E (triangle), and F (square).

dependence as samples A and B, although the values are displaced relative to samples A and B. This suggests that the same scattering mechanism for samples A and B also play a significant role in sample C.

Returning to Fig. 3, the electron concentration of sample B also displays an activated behavior, but the extracted ionization energy is 22.4 meV. The electron concentration of sample C appears to also display an activated behavior between 60 and 300 K with activation energy, and hence ionization energy, similar to sample B. For the case of sample C, the onset of impurity conduction behavior occurs near 60 K, higher than for samples A and B.

Figures 5 and 6 show the electron concentration and mobility for samples D - F as a function of temperature. The onsets of impurity conduction are about 60, 50, and 30 K for samples F, E, and D, respectively. For these samples a region where the temperature dependence of the electron concentration is activated is observed, although the activated region is narrower than for samples A and B, with activation energies similar to that of sample B. The temperature dependence of the mobility is nearly identical for samples D - F and similar to that of sample C. Thus, samples C - F possess a common low temperature mobility scattering mechanism which is similar to that of samples A and B.

Several observations can be made for the data shown in Figs. 3, 4, 5, and 6. (1) All samples show the impurity conduction behavior, but to varying degrees. (2) Sample resistivity at 300 K is not a reliable indicator of the extent of the impurity conduction. Instead, a sample with 300 K mobility of 500 $cm^2V^{-1}s^{-1}$ had the lowest impurity conduction onset temperature. (3) Because for samples B, C, D, E, and F the mobility peaks (not shown for all samples) at lower values but the samples have nominally the same electron concentration, these samples are probably more compensated than sample A.

Since the data indicate that samples not dominated by impurity conduction, i.e., having lower onset temperatures, are the least compensated, then an important compensation mechanism in the epilayers may be due to the parallel transport of charge by impurity conduction. The data does not rule out that this mechanism may be due to localized transport near the buffer/epilayer interface. Within the context of the working hypothesis, the impurity conduction mechanism exists because the lattice mismatch between the epilayer and sapphire gives rise to the presence of donor and acceptor-like states. Experimentally, the lowest compensated film is one where optimizing the buffer layer results in an epilayer with reduced impurity conduction, that is, the lowest impurity conduction onset temperature.

4. Summary

Theoretical calculations of the Hall scattering factor for wurtzite GaN films as a function of magnetic field strength using numerical contraction mapping solutions to the classical Boltzmann equation has been presented. In addition, the calculated Hall and drift electron mobilities of pure intrinsic GaN were found to be 1732 and 1496 cm^{2}V^{-1}s^{-1}, respectively. The best reported experimental 300 K Hall mobility values as a function of electron concentration for wurtzite GaN on sapphire are shown in Fig. 2 and probably represent transport properties intrinsic to GaN epilayers on sapphire because the results are generally independent of buffer layer or growth technique. The theoretical calculations indicate that the GaN films grown on sapphire with electron concentration greater than about 5 x 10^{18} cm^{-3} have a compensation of about 0.4 and those films with lesser electron concentrations are more compensated. Variable-temperature Hall measurements indicate that an important compensation mechanism is due to the parallel transport of charge by impurity conduction. A working hypothesis is presented whereby the defects generated by the lattice mismatch give rise to donor and acceptor-like states which compensate the film via impurity conduction. For a film dominated the least by impurity conduction, the donor ionization energy was found to be 35.9 meV.

5. Acknowledgments

The authors gratefully acknowledge the data contributions of Dr. Wickenden, Prof. Morkoç, Prof. Amano, Dr. Meyer, and Dr. Hoffman. One of us (LBR) acknowledges a National Research Council/Naval Research Laboratory Cooperative Research Associateship.

6. References

[1] Akasaki I, Amano H, Kito M, and Hiramatsu K 1991 J. Luminescence **48-49**, 666-70
[2] Khan M A, Kuznia J N, Bhattarai A R, and Olson D T 1993 Appl. Phys. Lett. **62**, 1786-7
[3] Binari S C, Rowland L B, Kruppa W, Kelner G, Doverspike K, and Gaskill D K 1994 Electronics Lett., in press
[4] Ilegems M and Montgomery H C 1973 J. Phys. Chem. Solids **34**, 885-95
[5] Rowland L B, Doverspike K, Giordana A, Fatemi M, Gaskill D K, Skowronski M, and Freitas, Jr. J A 1993 Inst. Phys. Conf. Ser. **137**, 429-32
[6] Yoshida S, Misawa S, and Gonda S 1983 Appl. Phys. Lett. **42**, 427-9
[7] Rode D L, 1973 Phys. Stat. Sol. (b) **55**, 687-96
[8] Rode D L, 1973 Semiconductors & Semimetals **10**, 1-89
[9] Nakamura S, 1991 Jpn. J. Appl. Phys. **30**, L1705-7
[10] Wickenden D K and Bryden W A 1993 Inst. Phys. Conf. Ser. **137**, 381
[11] H. Morkoç, private communication.
[12] Rowland L B, Doverspike K, Gaskill D K, and Freitas, Jr. J A 1994 Mater. Res. Soc. Proc., in press
[13] Nakamura S, Mukai T, and Senoh M 1992 J. Appl. Phys. **71**, 5543-9
[14] Nakamura S, Mukai T, and Senoh M 1992 Jpn. J. Appl. Phys. **31**, 2883-8
[15] Khan M A, Skogman R A, Schulze R G, and Gershenzon M 1983 Appl. Phys. Lett. **42**, 430-2
[16] Khan M A, Kuznia J N, Van Hove J M, Olson D T, Krishnankutty S, and Kolbas R M 1991 Appl. Phys. Lett. **58**, 526-7
[17] Nakamura S, Makui T, and Senoh M, 1992 J. Appl. Phys. **71**, 5543-9
[18] Akasaki I and Amano H 1991 Mat. Res. Soc. Symp. Proc. **242**, 383-94
[19] Doverspike K, Rowland L B, Gaskill D K, Binari S C, and Freitas, Jr. J A, 1994 Inst. Phys. Conf. Ser., this volume
[20] Molnar R J, Lei T, and Moustakas T D 1993 Appl. Phys Lett. **62**, 72-4

X-Ray differential diffractometry applied to GaN grown on SiC

Irina P. Nikitina

A F Ioffe Institute and Cree Research EED, 26 Polytechnicheskaya Str., St. Petersburg, 194021 Russia

Vladimir A. Dmitriev

Cree Research, Inc., 2810 Meridian Parkway, Durham, NC, 27713 USA

Abstract. The crystal structure of undoped GaN layers grown on 6H-SiC substrates was investigated by x-ray differential diffractometry. X-ray scattering intensity distributions were obtained from rocking curves employing ω and ω–2Θ scanning geometries. GaN layers tested had a thickness of 1.5 to 6 microns and $N_d - N_a$ concentrations of ~ 10^{16} cm^{-3}. These layers were single crystal and consisted of two sublayers: a thin strained sub-layer near the interface and a relaxed layer above it. The minimum full width at half maximum of ω-scan rocking curves measured for GaN grown on SiC was ~ 1.32 arcmin.

1. Introduction

X-ray diffraction has been used to characterize the crystal quality of GaN by many other researchers [1-7]. Usually the full width at half maximum (W) of x-ray rocking curves indicates the relative crystal quality of the epilayer. Correct comparison of the published data may be made if the x-ray reflection and scanning geometries are known. The best results for GaN layers reported to date are summarized in Table 1. It is important to note that the only accurate indicator of crystal quality is the W value measured in a ω scanning geometry.

2. Samples and measurements

GaN layers were grown on 6H-SiC manufactured by Cree Research, Inc. These layers were grown on the (0001)Si face of the substrate and had a thickness of 1.5 to 6 μm. Concentration of electrically active donor impurities $N_d - N_a$ was in the range of 5×10^{15} - 5×10^{16} cm^{-3}.

Table 1. Full width at half maximum (W) of x-ray rocking curves for GaN layers.

Substrate	Thickness (microns)	Nd-Na (cm^{-3})	Reflection	x-ray scan	W (arcmin)	Reference
Sapphire	4	7x10^{17}	0002	?	1.6	[1]
Sapphire	4	4x10^{16}	0002	?	~5	[1]
Sapphire	1.5	8x10^{18}	0002	?	0.63	[2]
Sapphire	3 - 12	1x10^{17}	0002	ω	1.9	[3]
Silicon	0.7			ω	9.6	[4]
SiC	2.5 - 3.5	2x10^{17}	0002	ω	5.2	[5]
SiC	~0.3	>1x10^{19}	0002	ω	2.4	[6]

The crystal structure of the GaN layers was investigated by x-ray differential diffractometry which allows one to study the distribution of intensity of scattered x-rays near lattice points of the reciprocal lattice. The distribution of intensity of scattered x-rays was measured in two mutually perpendicular directions in reciprocal space (Fig. 1): (1) along the vector H of the reciprocal lattice (radial distribution) and (2) along the direction perpendicular to the vector H (azimuthal distribution).

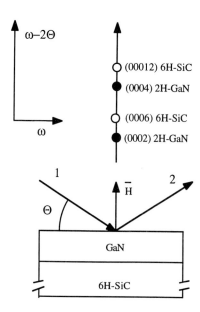

Figure 1. Scheme of reciprocal space and scanning directions: 1 - incident x-ray beam, 2 - scattered x-ray beam.

Information about the radial distribution intensity of scattered x-rays may be obtained from x-ray rocking curves measured using the ω–2Θ scanning geometry (this method is also referred as the Θ–2Θ mode [7]). Information about the azimuthal distribution intensity of scattered x-rays may be obtained from x-ray rocking curves measured using the ω scanning geometry (this method is also referred as the Θ mode [7]). Full width at half maximum for the ω scan and the ω–2Θ scan rocking curves are referred as $W_\perp$ and $W_\parallel$, respectively.

X-ray diffraction measurements were conducted employing double-crystal and triple-crystal spectrometers. High quality 6H-SiC crystals grown by the Lely method served as monochromator and third crystals. Symmetric (0002) and (0004) reflections and asymmetric (11$\bar{2}$4) reflection were studied. X-ray rocking curves in ω geometry were measured using a double-crystal spectrometer with an open slit. X-ray rocking curves in the ω–2Θ geometry were measured using a triple-crystal spectrometer.

3. Results

Results of x-ray measurements showed that all GaN layers grown on SiC were monocrystalline. The W values of ω-scan rocking curves for the (0002) reflection ranged from 79 to 260 arcsec. These results are summarized in Table 2 and in Figures 2-4. Note, Figures 2 and 3 are plotted on a linear vertical scale, while Fig. 4 has a logarithmic vertical scale. The following features of x-ray scattering near (0002) and (0004) points of the reciprocal lattice may be emphasized:

(a) The value of $W_\perp$ was much greater than $W_\parallel$.
(b) The ratio of $W_{\parallel(0004)}$ to $W_{\parallel(0002)}$ for the ω–2Θ rocking curve was close to the ratio of tangents of Bragg angles for corresponding reflections:

$$\frac{W_{\parallel(0004)}}{W_{\parallel(0002)}} \cong \frac{\tan\Theta_{(0004)}}{\tan\Theta_{(0002)}} = 2.37$$

(c) X-ray rocking curves were asymmetric. The asymmetry in the x-ray rocking curve for the (0004) reflection was most pronounced. The asymmetry of the x-ray rocking curve for the (0002) reflection was observed when the data was plotted on a semilogarithmic scale (Fig. 4).

From this x-ray data, the lattice spacing, d, along the c-axis for GaN layers was determined. It was found that all GaN layers had d values smaller than that for a relaxed GaN crystal. For a relaxed GaN crystal Δd/d is 2.89×10^{-2} [8]. The normal deformation ($\varepsilon_\perp$ = Δd/d$_\perp$) of GaN grown on SiC was calculated. To determine the tangent deformation ($\varepsilon_\parallel$), the x-ray rocking curve for the asymmetric reflection (11$\bar{2}$4) was measured. Calculated values of normal and tangent deformation are shown in Table 3.

Table 2. Full width at half maximum of x-ray rocking curves for ω scan ($W_\perp$) and ω–2Θ scan ($W_\parallel$).

Sample	Thickness microns	$W_\perp(0002)$ arcmin	$W_\parallel(0002)$ arcmin	$W_\perp(0004)$ arcmin	$W_\parallel(0004)$ arcmin
1	6	3.68	0.50	2.70	1.08
2	6	2.07	0.58	1.83	0.83
3	6	1.67	0.47	2.80	1.00
4	3.5	2.28	0.58	2.33	0.50
5	3.5	4.05	0.58	3.55	1.17
6	2.2	4.32	0.50	4.20	1.41
7	1.5	1.32	0.30	1.40	0.60
8	1.5	1.35	0.50	1.40	0.78

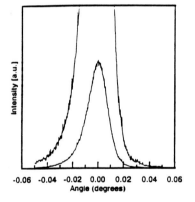

Figure 2. Rocking curve in ω–2Θ scan; (0004) reflection.

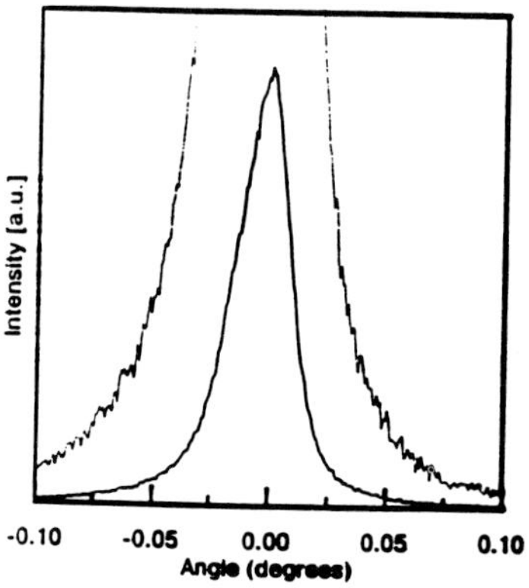

Figure 3. Rocking curve in ω scan; (0004) reflection.

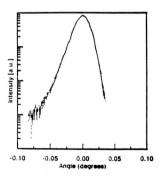

Figure 4. Rocking curve in ω–2Θ scan; (0002) reflection.

Table 3. Calculated values of normal and tangent deformation of GaN layers.

Sample	$\varepsilon_\perp \times 10^2$	$\varepsilon_\parallel \times 10^2$
1	2.728	-
2	2.753	-
3	2.778	2.401
4	2.784	-
5	2.728	-
6	2.816	2.111
7	2.749	-
8	2.733	2.396

4. Discussion

From the above results, an analysis of defect structures in GaN layers was performed. Two possibilities were considered: (1) uniform dislocation distribution in the material and (2) mosaic structures consisting of blocks separated by dislocation accumulations. For case (2), x-ray scattering inside the blocks is coherent. The dislocation density in the blocks is much less than that in the blocks boundaries. The distance between the dislocations in the boundaries is much smaller than the block size.

In the first case, according to [9], a random dislocation distribution leads to x-ray reflection broadening along the vector H, i.e. resulting in broadening of the $\omega-2\Theta$ rocking curve. The absolute value of the broadening is proportional to the tangent of the Bragg angle. In the second case, x-ray scattering is mainly determined by misorientation of the blocks. This results in broadening of the ω rocking curve.

Our results showed that the angular dependence of the radial distribution of scattered x-rays was close to $\tan\Theta$, which indicates that dislocations are randomly distributed. On the other hand, measurements showed that $W_\perp \gg W_\parallel$. This indicates that in the layers there are regions misoriented to each other. Thus, we may conclude that the crystal structure of GaN layers may be described as intermediate between a random dislocation distribution and a mosaic structure. This type of crystal structure was reported for GaSb/GaAs [10], InSb/GaAs [11] and GaAs/Si [12] heteroepitaxial growth.

As was mentioned in the previous section, the resulting x-ray rocking curves have an asymmetric shape. The fact that the rocking curves have an additional broadening at smaller angles indicates that the layers have regions with $d_\perp$ spacing larger than that for the rest of the layer. The relative intensity of this broadening is higher for (0004) reflections indicating that the regions with larger $d_\perp$ spacing are concentrated at the interface. We may conclude that the GaN layers consist of two layers: (1) a thin interface layer with a block structure and high dislocation density and (2) a thick top layer with a random dislocation distribution. Transmission electron microscopy (TEM) proves this model. It was found that the initial GaN layer has a high dislocation density with a thickness of 0.1 - 0.4 micron. This layer is followed by a GaN layer having a dislocation density 2 - 3 orders of magnitude less than the initial layer (details of the TEM results will be published elsewhere).

5. Conclusion

GaN layers were grown on 6H-SiC substrates and were shown to be single crystal with a full width at half maximum of ω-scan rocking curve as low as 1.32 arc min. Based on the results of this x-ray diffraction study, the following model of the defect structure in GaN layers was proposed. GaN grown on SiC substrates consists of two layers, (1) a thin interface layer with a high dislocation density and (2) a top layer with a random dislocation distribution and lower dislocation density.

Acknowledgments. We appreciate the TEM measurements performed by A. Sitnikova and helpful discussion with A. Tsaregorodtsev. This work was supported in part by the Advanced Research Projects Agency and the Office of Naval Research.

References

[1]   Nakamura S 1991 *Jap. J. Appl. Phys.* **30** L1705-7
[2]   Feritta K G, Holmes A L, Neff J G, Ciuba F J, and Dupuis R D 1994 *Technical Program with Abstracts, 1994 Electronic Materials Conference* A36
[3]   Akasaki I, Amano H, Koide Y, Hiramatsu K, Sawaki N 1989 *J. Cryst. Growth* **98** 209-19
[4]   Meng W J, Heremans J, and Cheng Y T 1994 *Inst. Phys. Conf. Ser.* **137** 409-11
[5]   Lin M E, Sverdlov B, Zhou G L, Morkoc H 1993 *Appl. Phys. Lett.* **62** 3479-81
[6]   Weeks T W Jr., Kum D W, Carlson E, Perry W G, Ailey K S, and Davis R F 1994 *Transactions of the Second Int. High Temperature Electronics Conference* (Charlotte, North Carolina) **2** P-209 - 14
[7]   Itoh N and Okamoto K 1988 *J. Appl. Phys.* **63** 1486 - 93
[8]   Porowski S, Jun J, Perlin P, Grzegory I, Teisseyre H, and Suski T 1994 *Inst. Phys. Conf. Ser.* **137** 369-72
[9]   Ryaboshapka K P 1981 *Zavodskaja Laboratoriya* **5** 26-33
[10] Kjutt R N et al. 1993 *Fizika Tverdogo Tela* **35** 3 724-735
[11] Zhang X, Staton-Bevan A E, Pashley D W, Parker S D, Droopad R, Williams R L, Newman R C 1990 *J. Appl. Phys.* **67** 2 800-806
[12] Fang S F, Adomi K, Iyer S, Morkoc H, and Zabel H 1990 *J. Appl. Phys.* **68** 7 R31 - 58

Step-controlled epitaxial growth of α-SiC and application to high-voltage Schottky rectifiers

T.Kimoto, A.Itoh, H.Akita, T.Urushidani[1], S.Jang[2], and H.Matsunami

Department of Electrical Engineering, Kyoto University,
Yoshidahonmachi, Sakyo, Kyoto 606-01, Japan

Abstract: 6H- and 4H-SiC were homoepitaxially grown by step-controlled epitaxy. N-type 6H- and 4H-SiC epilayers with a carrier concentration of $10^{16}cm^{-3}$ showed mobilities of 351 and 724cm^2/Vs at room temperature, 6050 and 11000cm^2/Vs at 77K, respectively. Isothermal capacitance transient spectroscopy (ICTS) measurements revealed that the epilayers have very few electron traps ($<10^{13}cm^{-3}$). Schottky rectifiers with 10μm-thick epilayers demonstrated high breakdown voltages of 600~1100V. Very low specific on-resistances of 1.2~1.7mΩcm^2 were achieved utilizing 4H-SiC which has higher mobilities than 6H-SiC.

1. Introduction

Silicon carbide (SiC) has received increasing attention as a wide bandgap semiconductor material for advanced power devices, owing to its high breakdown field and high thermal conductivity. Theoretical simulation has predicted that high-voltage Schottky rectifiers and metal-oxide-semiconductor field effect transistors (MOSFETs) with extremely low specific on-resistances can be realized using SiC, and SiC devices can replace the present-day Si power devices on account of low power dissipation and the reduced chip sizes [1].

In recent years, the development of 6H-SiC (E_g=2.93eV at 300K) based devices has shown rapid progress, such as high-voltage p-n junction diodes [2], power Schottky rectifiers [3,4], and high-temperature and microwave FETs [5,6]. This progress is brought about by the availability of high-quality 6H-SiC epilayers, which can be obtained by homoepitaxial growth on off-oriented 6H-SiC{0001} substrates (step-controlled epitaxy) [7-10]. However, the systematic characterization of epilayers including deep levels has not been reported. Besides, there have been only a few reports on epitaxial growth of 4H-SiC (E_g=3.26eV at 300K) and its device applications [11,12], in spite of two times higher electron mobilities in 4H-SiC Lely crystals than 6H-SiC [13].

In this paper, we report step-controlled epitaxy of 6H- and 4H-SiC, and their device application. The epilayers are characterized by Hall measurements and isothermal capacitance transient spectroscopy (ICTS). Preliminary results of application to high-voltage Schottky rectifiers are presented.

---

[1]On leave from Fuji Electric R & D., Ltd., Yokosuka, Kanagawa, Japan
[2]On leave from Dongshin University, Naju, Chonnam, Korea

2. Experiments

Substrates used in this study were n-type 6H- or 4H-SiC {0001} grown by the Acheson or a modified Lely methods. Bulk crystal growth of 4H-SiC by a modified Lely method is described elsewhere [11]. The off-orientation of substrates was 4~5° toward <11$\bar{2}$0>. The typical resistivities of substrates were 0.08Ωcm for 6H-SiC and 0.02Ωcm for 4H-SiC. N-type SiC layers were homoepitaxially grown by step-controlled epitaxy [7,9], using a SiH_4-C_3H_8-H_2 system. The growth temperature and growth rate were 1500°C and 2.5μm/h, respectively. The carrier concentration was controlled by in-situ doping with nitrogen using N_2 gas. The typical thickness of grown layers was 10μm.

The electron mobilities of epilayers were measured by the van der Pauw method. The epilayers were electrically isolated from n-type substrates utilizing epitaxially grown p-n junctions. The ICTS measurements were performed using Schottky diode structures in the temperature range of 200~400K.

For the fabrication of power Schottky rectifiers, Ni and Au were used as ohmic and Schottky contacts, respectively. Before Au deposition, the as-grown surfaces were cleaned in organic solvents, heated K_2CO_3, aqua regia, and HF, and then rinsed in deionized water. The size of circular Au Schottky contacts was 120μm in diameter. The characteristics of Schottky rectifiers were measured in air at 293~473K.

3. Results and discussion

3.1 Epitaxial growth and characterization

The use of off-oriented SiC{0001} substrates leads to the replication of the substrate polytypes in epilayers when the growth temperature is high enough to realize step-flow growth [9,10]. By this technique, 6H- and 4H-SiC with featureless surface can be homoepitaxially grown on both off-oriented (0001)Si and (000$\bar{1}$)C faces. No significant differences in surface morphology and growth rate were observed between 6H- and 4H-SiC growth.

Figure 1 shows the temperature dependence of electron mobilities for 6H- and 4H-SiC epilayers with a carrier concentration of 2~3$\times 10^{16}$cm^{-3}. The mobilities at room temperature are 351cm^2/Vs for 6H and 724cm^2/Vs for 4H polytype. The mobilities increase with lowering temperature, and reach up to 6050cm^2/Vs for 6H and 11000cm^2/Vs for 4H at 77K. The mobilities of 6H-SiC is comparable to the previous report [14], and those of 4H-SiC are the highest data ever reported in α-SiC. It has been known that α-SiC has an anisotropy in electron mobility [15,16]. Although we have not measured the anisotropy in detail, the mobilities parallel to the c-axis ($\mu_\parallel$) are reported to be 1/3~1/5 of those perpendicular to the c-axis ($\mu_\perp$) in 6H-SiC, whereas 4H-SiC shows little anisotropy [15,16]. Since two times higher $\mu_\perp$'s were obtained for 4H-SiC epilayers in this study, much larger difference in $\mu_\parallel$ between 6H and 4H can be expected. This difference makes 4H-SiC a more suitable material for power-device applications, because most power devices are fabricated so as to make current flow along the c-axis.

Figure 2 shows typical ICTS spectra obtained from the Schottky diodes of 6H-SiC wafer and epilayer [17]. In the present samples, the donor concentrations of 6H-SiC wafer and epilayer were 9$\times 10^{16}$cm^{-3} and 1$\times 10^{15}$cm^{-3}, respectively. The spectra from the wafer are composed of several overlapping signals which arise from several electron traps with different levels and capture cross sections. As for two large ICTS peaks (peaks 1 and 2 in Fig.2), the trap concentrations were estimated to be 9$\times 10^{15}$cm^{-3} and

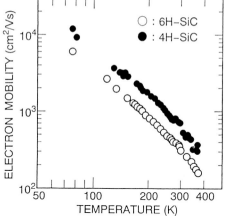
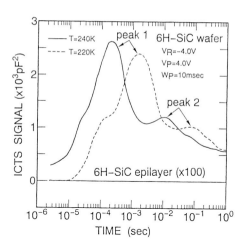

Fig.1 Temperature dependence of electron mobilities for 6H-SiC and 4H-SiC epilayers.

Fig.2 ICTS spectra for 6H-SiC wafer and epilayer. The measurement conditions such as the reverse voltage V_R, the forward pulse voltage V_P, and the pulse width W_P are shown.

$4\times 10^{15} cm^{-3}$ for peaks 1 and 2. The Arrhenius plots of τT^2 (τ : the emission time constant) revealed that the trap levels for peaks 1 and 2 are 0.39eV and 0.43eV from the conduction band, and the capture cross sections are $3\times 10^{-15} cm^2$ and $4\times 10^{-16} cm^2$, respectively. The ICTS spectra of wafers varied from sample to sample, and some wafers showed very small signal of peak 2. However, the peak 1 trap with a concentration of $10^{15} \sim 10^{16} cm^{-3}$ was observed in all the wafers. The origins of these electron traps cannot be identified at present, and further analyses on intentionally doped samples will be required.

On the other hand, epilayers showed very small ICTS signal as shown in Fig.2. Almost no ICTS peaks were observed even if the measurement temperature was varied from 200K to 400K. This result indicates that the epilayers have very few electron traps located above E_c-1.0eV (E_c : the conduction band edge). The trap concentration can be estimated to be less than $10^{13} cm^{-3}$, from the donor concentration of epilayer ($1\times 10^{15} cm^{-3}$) and the detection limit of the measurement system. Thus, the quality of epilayers is adequate for application to majority carrier devices.

3.2 Application to power Schottky rectifiers

At first, the breakdown field was estimated as a function of doping concentration. Estimated breakdown field of 6H-SiC was 1.9×10^6V/cm at $1\times 10^{16} cm^{-3}$, and 4.0×10^6V/cm at $1\times 10^{18} cm^{-3}$. These values are a little lower than those obtained from mesa p-n junction diodes [18]. This is caused by the concentrated electric field at the contact periphery in our Schottky samples. 4H-SiC also showed high breakdown fields, 1.6×10^6V/cm at $1\times 10^{16} cm^{-3}$, and 4.0×10^6V/cm at $1\times 10^{18} cm^{-3}$, which are comparable to those of 6H-SiC.

Figure 3 shows the current density-voltage characteristics of Au/6H- and 4H-SiC Schottky rectifiers at room temperature. Substrates used were off-oriented $(000\bar{1})$C

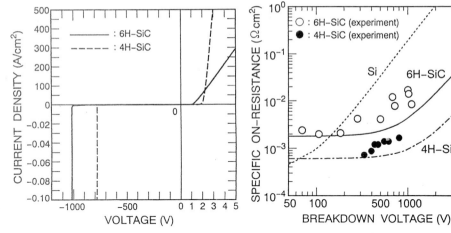

Fig.3 Current density – voltage characteristic of 6H and 4H Schottky rectifiers at room temperature.

Fig.4 Breakdown voltage dependence of specific on-resistance for 6H-, 4H-SiC, and Si Schottky rectifiers. The open and solid circles denote the experimental results obtained for 6H and 4H rectifiers.

wafers. The donor concentration and thickness of epilayers were 3×10^{15}cm^{-3} and 10μm. Very high breakdown voltages of 1020V for 6H and 790V for 4H rectifiers could be achieved. The lower breakdown voltage of the 4H rectifier may be ascribed to crystal imperfection which results from an immature technique in preparing 4H-SiC wafers. The leakage current densities were quite low, 6.8×10^{-5}A/cm^2 for 6H and 6.7×10^{-6}A/cm^2 for 4H rectifiers at -500V, in spite of no surface passivation.

Under the forward bias condition, the current densities follow the thermionic emission equation with ideality factors (n) of 1.08~1.21 at current levels lower than 1A/cm^2. The barrier heights of Au for 6H- and 4H-SiC were estimated to be 1.5eV and 1.8eV, respectively. The 4H rectifier exhibits a little higher turn-on voltage due to its higher barrier height. However, the current density of 4H rectifier increases much more rapidly than that of 6H, indicating the lower on-resistance of 4H rectifier. As a result, the forward voltage drop to deliver 100A/cm^2 becomes lower in 4H (2.31V), and higher in 6H rectifiers (2.54V). A very high current density of 500A/cm^2 can be achieved at a low voltage drop of 3.00V in the 4H rectifier. Although the turn-on voltage of this 4H rectifier is very high, it can be easily reduced by controlling the barrier height.

Figure 4 shows the specific on-resistance (R_{on}) as a function of breakdown voltage (V_B). Open and solid circles denote the R_{on}'s obtained for 6H and 4H rectifiers in the present study. Dashed and solid curves show the theoretical limits for Si and SiC. The theoretical R_{on}'s are given by

$$R_{on} = R_{on}(\text{epi}) + R_{on}(\text{sub}) = \rho_{epi}W_{epi} + \rho_{sub}W_{sub}, \qquad (1)$$

where $R_{on}(\text{epi})$ and $R_{on}(\text{sub})$ are the resistances for unit area of the epilayer and substrate, ρ_{epi} and ρ_{sub} the epilayer and substrate resistivities, W_{epi} and W_{sub} the thicknesses of the epilayer and substrate, respectively. The optimum $R_{on}(\text{epi})$ can be calculated using the following equation [19]

$$R_{on}(\text{epi}) = \frac{4V_B^2}{\epsilon E_B^3 \mu}, \tag{2}$$

where ϵ is the permittivity ($\epsilon=11.9\epsilon_0$ for Si and $9.7\epsilon_0$ for SiC : ϵ_0 is the permittivity in vacuum). $R_{on}(\text{epi})$'s were calculated using breakdown field (E_B) and mobility (μ) data reported for Si [20] and SiC [16,18]. Here, the breakdown field of 4H-SiC was assumed to be the same as that of 6H-SiC. As for SiC, the mobilities parallel to the c-axis [15,16] were employed in the simulation. For calculation of $R_{on}(\text{sub})$, we used a thickness of 300μm and resistivities of 0.01, 0.06, and 0.02Ωcm for Si, 6H- and 4H-SiC, which are commercially available values.

Although theoretical R_{on}'s for low-voltage (<300V) SiC rectifiers are saturated due to the relatively high substrate resistances, SiC devices show much lower R_{on}'s than Si in the high-voltage region. In particular, 4H-SiC is the most promising candidate owing to much higher mobility along the c-axis than 6H-SiC.

The experimentally obtained specific on-resistances for 6H rectifiers with breakdown voltages of 500~1100V are lower than the theoretical limits of Si rectifiers by one order of magnitude. 4H rectifiers realized the further reduction of R_{on} compared with 6H. For example, experimental R_{on}'s of 500V rectifiers are 4.1mΩcm^2 and 1.2mΩcm^2 for 6H- and 4H-SiC. This result is attributed to higher mobility of 4H-SiC. In the high-voltage devices, the experimental R_{on}'s are higher than theoretical values by a factor of 2~3, which may come from the non-optimized device structure. For example, an edge-termination technique is strongly required, because the breakdown voltages obtained in this study were only 40~70% of theoretical values.

The rectifiers showed good rectification at temperatures up to 473K. The R_{on}'s of the SiC rectifiers increased monotonously with temperature, according to T^2 dependence [4]. This increase in R_{on} is caused by the decrease in electron mobility.

The present study demonstrated that SiC (especially 4H polytype) is an excellent candidate for advanced power devices. In order to realize SiC power devices for practical use, however, several problems should be solved, such as improvement of wafer quality and device processing, including structure designing.

4. Conclusions

6H- and 4H-SiC were homoepitaxially grown by step-controlled epitaxy. N-type 6H- and 4H-SiC epilayers with a carrier concentration of 2~3×10^{16}cm^{-3} showed mobilities of 351 and 724cm^2/Vs at room temperature, 6050 and 11000cm^2/Vs at 77K, respectively. The ICTS measurements revealed that the epilayers have very few electron traps (<10^{13}cm^{-3}), although several traps exist in wafers. Schottky rectifiers with 10μm-thick epilayers demonstrated high breakdown voltages of 600~1100V. Very low specific on-resistances of 1.2~1.7mΩcm^2 were achieved utilizing 4H-SiC which has much higher mobilities along the c-axis than 6H-SiC. The on-resistances increased with temperature according to T^2 dependence.

Acknowledgment

This work was partially supported by a Grant-in-Aid for Scientific Research from the Ministry of Education, Science and Culture of Japan. The authors wish to thank Prof. I.Kimura and Dr. Kanno in the Department of Nuclear Engineering, Kyoto University for their facilities in ICTS measurements.

References

[1] Bhatnagar M and Baliga B J, 1993, IEEE Trans. Electron Devices, **40**, 645-655.
[2] Neudeck P G, Larkin D J, Powell J A, Matus L G, and Salupo C S, 1994, Appl. Phys. Lett., **64**, 1386-1388.
[3] Bhatnagar M, McLarty P K, and Baliga B J, 1992, IEEE Electron Device Lett., **13**, 501-503.
[4] Kimoto T, Urushidani T, Kobayashi S, and Matsunami H, 1993, IEEE Electron Device Lett., **14**, 548-550.
[5] Palmour, J W, Edmond J A, Kong H S, and Carter Jr. C H, 1993, Physica B, **185**, 461-465.
[6] Clarke R C, O'Keefe T W, McMullin P G, Smith T J, Sriram S, and Barrett D L, 1992, IEEE Trans. Electron Devices, **39**, 2666-2667.
[7] Kuroda N, Shibahara K, Yoo W S, Nishino S, and Matsunami H, 1987, Ext. Abstr. the 19th Conf. on Solid State Devices and Materials, p.227-230.
[8] Kong H S, Glass J T, and Davis R F, 1988, J. Appl. Phys., **64**, 2672-2679.
[9] Kimoto T, Nishino H, Yoo W S, and Matsunami H, 1993, J. Appl. Phys., **73**, 726-732.
[10] Kimoto T and Matsunami H, 1994, J. Appl. Phys., **75**, 850-859.
[11] Itoh A, Akita H, Kimoto T, and Matsunami H, to be published in Appl. Phys. Lett. **65**.
[12] Palmour J W, Weitzel C E, Nordquist K, and Carter Jr. C H, 1994, *Silicon Carbide and Related Materials*, Spencer M G, Devaty R P, Edmond J A, Khan M A, Kaplan R, and Rahman M, Eds. London:Institute of Physics, p.495-498.
[13] Barrett D L, Campbell R B, 1967, J. Appl. Phys., **38**, 53-55.
[14] Karmann S, Suttrop W, Schöner A, Schadt M, Haberstroh C, Engelbrecht F, Helbig R, Pensl G, Stein R A, and Leibenzeder S, 1992, J. Appl. Phys., **72**, 5437-5442.
[15] Lomakina G A, 1974, *Silicon Carbide 1973*, Marshall R.C, Faust Jr. J W, and Ryan C E, Eds. Columbia:Univ. of South Carolina Press, p.520-526.
[16] Schaffer W J, Negley G H, Irvine K G, and Palmour J W, 1994, to be published in Mat. Res. Soc. Symp. Proc.
[17] Jang S, Kimoto T, and Matsunami H, 1994, Appl. Phys. Lett. **65**, 581-583.
[18] Edmond J A, Waltz D G, Brueckner S, Kong H S, Palmour J W, and Carter Jr. C H, 1991, Trans. 1st Int. High Temp. Electron. Conf., p.207-212.
[19] Baliga B J, 1989, IEEE Electron Device Lett., **10**, 455-457.
[20] Sze S M, *Physics of Semiconductor Devices*, 2nd ed., 1985, New York:Willey-Interscience.

TEM Study of Low Temperature CVD Silicon Carbide Films Grown on on-Axis 6H-SiC Substrates

K. Fekade, M.G. Spencer, K. Irvine

MSRCE, School of Engineering, Howard University, Washington, DC 20059, USA.

A.K. Ballal, D. Prasad Besabathina, and L.G. Salamanca-Riba

Department of Materials & Nuclear Engineering, University of Maryland, College Park, MD 20742, USA.

ABSTRACT

Epitaxial SiC films were grown on 6H-SiC (0001) substrates. The substrates were slightly off axis (<1°). These films were grown on the C- face by the chemical vapor deposition technique. A transmission electron microscopy study was carried out to determine the polytype and crystallinity of the grown films. Both plan-view and cross-sectional samples were observed. In addition, high quality ohmic contacts were produced. The electron microscopy results indicated that films grown both at high (1285°C) and low (1150°C) temperatures resulted in cubic polytype when doped with ammonia (NH_3). However, films doped with Trimethylaluminum (TMA) resulted in 6H poly-type. Ohmic contacts fabricated on the n-type material exhibited ohmic contact resitivities of 2-3×10^{-5} ohm-cm^2.

I. Introduction

It has recently been shown that β-SiC can be grown on α-SiC at temperatures as low as 1150°C.[1-3] Most of the previous investigations consisted of studying the role of surface disturbance, temperature and growth surface in determining the poly type formation. In this paper, we present a transmission electron microscopy (TEM) study of beta silicon carbide grown on 6H substrates using the chemical vapor deposition (CVD) technique, and the role of doping in determining the polytype formation. The ohmic contacts result has been previously published[4], hence will not be discussed here.

II. Experimental

Low pressure chemical vapor deposition (LPCVD) technique was used to grow the samples. The effect of n-type and p-type doping, and low and high growth temperatures, on the polytype formation and crystal structure was investigated. The 6H substrates used in this study were fabricated by the Lely method. Details of the growth of the epitaxial films have been previously published[5]. Growth temperatures as low as 1150°C and as high as 1285°C were used. NH_3 was used as the n-type dopant and trimethylaluminum (TMA) was used as the p-type dopant. The growth time was varied from 5 minutes to 60 minutes, corresponding to film thicknesses in the range 0.25 to 0.66 μm.

The samples for TEM study were prepared by mechanical thinning to about 30-40 μm followed by ion beam thinning. Mechanical thinning was done following the method of Ron Anderson of IBM[6]. Ion milling was done using a Gatan ion mill. Samples were milled at room temperature using argon ions. A JEOL 2000 FX-II TEM operating at 200 KV was used to examine the samples.

III. Results and Discussion

The electron microscopy results indicated that films grown both at high and low temperatures, showed a cubic polytype when doped with NH_3, and Samples doped with TMA grown at both low (1150°C) and high (1285°C) temperatures produced hexagonal polytype. In general, the NH_3 doped samples grown at low temperature were polycrystalline, whereas samples grown at high temperature were single crystal. In the case of the TMA doped samples both temperature ranges resulted in single crystal. The summary of our results is given in Table I.

Table I

Sample #	GTemp	GTime	SiH_4/C_3H_8 Ratio	Doping type	Growth Surface	TEM Result
28	1150	5 min	50/50	TMA	Si-face	Single 6H
28-A	1150	5 min	50/50	TMA	C-face	Poly 6H
63	1153	10 min	50/62	NH_3	C-face	Poly 3C
38	1177	40 min	50/55	none	C-face	Single 6H
38-A	1177	40 min	50/55	TMA	C-face	Single 6H
45	1177	40 min	50/70	NH_3	C-face	Poly 3C
51	1278	10 min	50/48	NH_3	C-face	poly 3C
62	1285	60 min	50/52	NH_3	C-face	single 3C
62-A	1285	10 min	50/52	TMA	C-face	Single 6H
62-B	1285	60 min	50/52	TMA	C-face	Single 6H

Figure 1 is a low magnification TEM micrograph of sample 45. In the case of sample 45 which was grown at 1177°C for 40 minutes, the film was polycrystalline. This sample also showed a columnar structure, as can be seen in Fig. 1, where the columns originated from the interface and extended to the free surface of the film. The larger columns have lengths of ~ 0.35 μm and width of ~ 0.12 μm. The crystallites seem to be randomly oriented with respect to each other. The random orientation was a common observation from the diffraction patterns, bright & dark field images obtained from this sample. We believe that this might be due to a large deviation of the 1:1 silicon to carbon ratio relative to the other samples. The growth mechanism for this sample resembles that of Stranski-Krastanov[7].

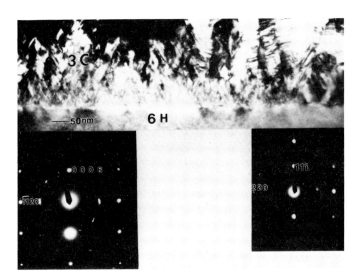

Figure 1: a) Bright field image of 3C/6H SiC interface showing a columnar and polycrystalline film of 3C SiC. b) Diffraction pattern from interface region (right) c) Diffraction pattern from one of the columnar grain (left).

Sample 62 was grown at a higher temperature (1285°C) and doped with NH_3. The film was single crystalline and was a cubic polytype. An epitaxial relationship of the film with the substrate is (111)cubic-SiC//(0001)6H-SiC. and [110]3C//[1120]6H. This result was corroborated by both cross section and plan view images. The interface is fully coherent but the 3C-SiC film is largely faulted and contains twins on (111) planes as can be seen in Fig 2. A cross-sectional high resolution transmission microscope (HRTEM) image of the 3C-SiC film grown on 6H substrate is shown in Figure 3 Some areas of the sample showed twinning on only one of the {111} plane, while other areas twinned in more than one {111} planes. The 3C/6H interface is very smooth as can be seen from Fig. 3.

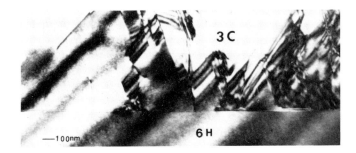

Fig. 2: Bright field image of 3C/6H SiC interface showing the largely faulted area of the 3C grown film.

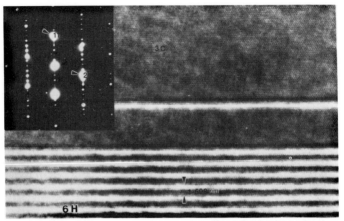

Figure 3: HREM of the 3C/6H interface. Region of microtwin is shown by the arrow head.

The largely twinned area of the 3C-SiC film is shown in Fig. 4. The inset to the micrograph is a diffraction pattern from the area, and shows twinning in two directions.

Samples 28, 38, 38-a, 62-A, and 62-B all exhibited hexagonal polytype films. these films were all single crystalline, some of these films did not show any type of defects. An example is shown in Figure 5, i.e., a bright field image of the interface between the film and the substrate obtained from sample 28. The interface is clean and sharp. The inset in this micrograph shows the diffraction pattern from the film and shows that the film is single crystalline. This sample was grown on Si-face of 6H-SiC for 5 minutes at 1150°C with a 1:1 SiH_4/C_3H_8 ratio, and was doped p-type with TMA.

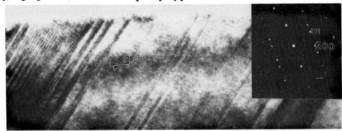

Figure 4: HREM micrograph of 3C-SiC film grown on 6H substrate. The inset shows the diffraction pattern from the area.

Fig 5: Bright field image of the 6H film/6H substrate interface. The inset is the diffraction pattern from the interface.

Even though this film was grown only for 5 minutes it did not become poly-crystalline. This could be because this sample was doped with TMA which makes the film grow as a hexagonal polytype. Also since the substrate is hexagonal the film is grown homoepitaxially and can, therefore, be single crystalline. The inset in this micrograph shows the diffraction pattern from the film, and also shows that the film is single crystalline. Similar observation was reported by S. Nishino et al[8] i.e. when a small amount of TMA was added during the investigation of growth of SiC on AlN/sapphire and on Si(100), the result was single crystalline layer of α-SiC and β-SiC respectively. These results suggests that TMA enhances single crystal formation of SiC, and lowers the epitaxial temperature. However, a sample grown with the exact same condition as sample 28 except on the C-face exhibited hexagonal polytype, but resulted in polycrystalline film as shown in Fig. 6. We believe the polycrystalline film resulted due to nucleation on the terraces of the C-face, coupled with short period and low (1150°C) temperature of growth.

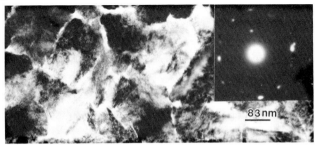

Fig. 6: Bright field image of a poly crystalline film. The inset is the diffraction pattern from this area.

Samples 51 resulted in polycrystalline film with a cubic polytype, even though the dopant in these case was NH_3, and it was grown at high temperature. The diffraction pattern of one of this film is shown in Fig. 7. We believe that this film became polycrystalline due to short period of growth time which did not allow the islands to coalesce to form single crystal.

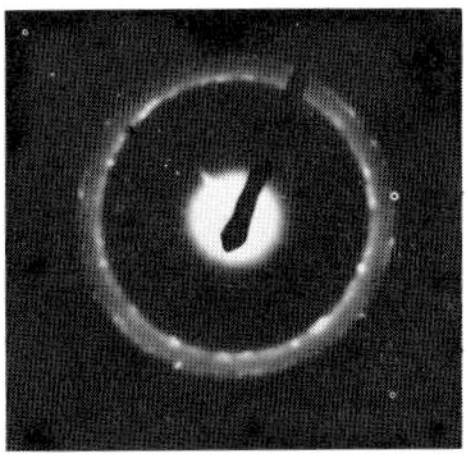

Fig. 7: Diffraction pattern obtained from plan-view sample of 51 showing a polycrystalline film.

The results of this study suggest that the main factor in determining the formation of the cubic polytype is NH_3 doping. It has previously been reported that in bulk growth of SiC, a 6H to 3C transformation occurs at high temperatures and under high pressure in a nitrogen atmosphere[10,11]. It was proposed that the nitrogen diffused into the SiC and gave rise to n-type doping which in turn induced the 6H to 3C transformation at high temperatures. The mechanism for the transformation was not discussed. Pirouz and Yang[12] have recently proposed a model for this transformation based on the mobility of partial dislocations. However, this model was given for polytype transformation due to applied pressure, and did not explicitly state the role of doping.

IV. Conclusion

A comparison study of SiC films grown under different conditions was carried out. The films doped with NH3 resulted in a cubic polytype regardless of the growth temperature and SiH_4/C_3H_8 ratio in the range studied. TMA doping resulted in hexagonal poly-type, and single crystalline films for growth temperatures as low as 1150°C.

V. Acknowledgment

This work was supported by a Grant, #N00014-92-J-1136 from the ofice of Naval Research.

VI. References

1. J. A. Powell, J.B. Petit, J.H. Edgar, I.G. Jenkins and L.G. Matus, J.M. Young, P. Pirouz, W.J. Choyke, L. Clemen and M. Yanagatha, Appl. Phys. Lett 59 (3) p 333 July 15, (1991).

2. H.S. Kong, J. T.. Glass and R.F. Davis, Appl. Phys. Lett. 49, p. 1074, (1986).

3. H. Matsunami, K. Jhibara, N. Kurod, and S. Nishano, In Amorphous and Crystalline Silicon Carbide, edited by G.L.Harris & C.Y.W.Yang, Springer proceeding in physics, Vol. 34 p 34 (Springer , Berlin, Heidelberg, (1989)

4. V.A Dimitriev, K.Irvine, M.Spencer, & G.Kelner App.Phys.lett. Vol 64 (3), p. 318 (1994).

5. K.J. Irvine, M.G. Spencer, and V.A. Dimitriev, Mat. Res. Soc. Symp. Proc. Vol. 281, (1993) Materials Research Society.

6. J.P. Benedict, S.J. Klepris, W.G. Vandygrift and Ron Anderson, Proceedings, EMSA, ed. by Bailey, G.W. p. 720 (1989).

7. M. Ohring, "The Materials Science of Thin Films", Academic Press, Inc., p. 197, (1992).

8. S. Nishino, K. Takahashi, H. Tanaka, and J. Saraie Proc. of SiC and Related, Materials, Vol. 137, p. 63 (1994).

9. K.Takahashi, S. Nishino, and J. Saraie, App. Phys. Lett, Vol.61, No. 17, p. 2081, (1992).

10. N.W. Jepps and T.F. Page, "Polytypic Transformation in SiC in Progress in Crystal Growth and Characterization", edited by P. Krisna, Pergamon Press, Oxford, Vol 9, p. 25 (1983).

11. E.D. Whitney and T.B. Shaffer, "Investigation of The Phase Transformation Between alpha and beta SiC at High Pressure", High Temp. High Press. Vol. 1, p 107, (1969).

12. P. Pirouz and J.W. Yang, Ultramicroscopy, p. 51, 189-214, (1993).

Electrical Characterization of the Thermally Oxidized SiO$_2$/SiC Interface

J. N. Shenoy*, L. A. Lipkin†, G. L. Chindalore*, J. Pan*,
J. A. Cooper, Jr.*, J. W. Palmour†, and M. R. Melloch*

* School of Electrical Engineering, Purdue University, West Lafayette, IN 47907
† Cree Research, Inc., 2810 Meridian Parkway, Suite 176, Durham, NC 27713

Abstract. We use a room temperature photo-CV technique and a high-temperature simultaneous hi-lo CV technique to characterize the thermally oxidized p-type 6H-SiC/SiO$_2$ interface in terms of fixed oxide charge, interface state density, and the density of near-interface oxide traps. Results on both aluminum and boron doped epilayers are reported.

1. Introduction

Silicon carbide (SiC) is a binary compound semiconductor which has remarkable physical and electrical properties. It is one of the hardest materials known. It is extremely resistant to all known chemical etchants at room temperature, and is thermally stable to temperatures above the melting point of silicon. One of the main advantages of SiC over other compound semiconductors is that it can be thermally oxidized to form a SiO$_2$ layer suitable for MOS devices.

In the past several years, many groups have investigated the MOS properties of both n- and p-type 6H-SiC. These interfaces typically exhibit fixed oxide charge Q_f and surface state density D_{it} from 5 to 10 times higher than comparable silicon devices. MOS interfaces on n-type SiC have been observed to have lower Q_f and D_{it} than similar interfaces on p-type SiC [1,2]. This difference has been attributed to the use of aluminum as the p-type dopant [2]. It is known that during oxidation, Al is incorporated into the oxide at about the same density as is present in the SiC [3]. One solution which has been suggested is the use of boron as a p-type dopant [2]. Boron is the standard p-type dopant in silicon, and does not appear to introduce any defects in silicon MOS devices. One purpose of this work is to investigate the MOS properties of boron-doped SiC as compared to aluminum-doped SiC.

The wide bandgap of SiC, about 3 eV for the 6H polytype, has to be taken into account for adopting a characterization methodology for SiC/SiO$_2$ MOS capacitors. The majority of interface states at the SiC/SiO$_2$ interface are energetically far away from the band edges, leading to extremely large emission times at room temperature. In addition, thermal generation of minority carriers is negligible at room temperature [4,5]. Thus conventional CV measurements convey little information about the SiC/SiO$_2$ interface. We use a combination of high-frequency and quasi-static CV measurements at both room temperature and elevated temperature. This combination allows a more complete characterization of the SiC/SiO$_2$ interface in terms of fixed oxide charge, interface state density, and near-interface oxide traps.

2. MOS device fabrication

MOS devices are formed on p-type epilayers grown on the Si-face of p-type 6H-SiC substrates. All wafers and epilayers are supplied by Cree Research, Durham, NC. Epilayers are typically 3 μm thick and are doped in the range of $1-2\times10^{16}$ cm^{-3}.

Two different pre-oxidation cleaning procedures are used. The first cleaning procedure consists of a solvent clean followed by a "piranha" cleaning procedure. The solvent clean consists of successive rinses in acetone / trichloroethane (TCA) / acetone / methanol for 5 minutes, followed by a DI rinse. The wafers are then soaked in 1:1 H_2SO_4:H_2O_2 for 15 minutes, rinsed in DI water, and the oxide etched by a 5 minute soak in buffered HF. The samples are then DI rinsed and blown dried with nitrogen. The RCA clean begins with the above solvent and piranha cleans. The wafers are then boiled in NH_4OH:H_2O_2:DI (1.5:1.5:5) for 15 minutes. This is followed by a DI rinse, oxide strip in buffered HF, and DI rinse. The same procedure is then used with HCl:H_2O_2:DI (1.5:1.5:5).

Oxidation takes place in a conventional resistance-heated quartz-walled furnace tube which is periodically TCA cleaned. All samples reported here were oxidized in wet O_2 at 1150 °C to a thickness of 640 Å (aluminum-doped samples) or 610 Å (boron-doped samples). Following oxidation, all wafers are subjected to a 30 minute in-situ Ar anneal. This Ar anneal is crucial in improving the dielectric integrity of the oxide — samples not receiving this anneal exhibit oxide leakage and do not survive the high temperature characterizations. Two different pull rates are employed. In the "fast pull", wafers are extracted from the 1150 °C Ar anneal in about 100 sec. In the "slow pull", the furnace temperature is gradually lowered to 900 °C (over a period of one hour) before unloading.

After oxidation and annealing, aluminum gates are deposited by thermal evaporation using an alumina-lined tungsten boat, and MOS capacitors are formed by chemically etching the aluminum. All capacitors are circular dots with a diameter of 450 μm. Each dot is surrounded by a guard ring (20 μm wide), and the "open" area between devices is covered by aluminum to form a large area capacitor.

3. Experimental techniques and results

In all capacitance measurements, the metalized area between devices is used as a capacitive connection to the substrate. Measurements are taken using a shielded probe station equipped with a hot chuck. Room temperature, high-frequency CV curves are obtained with an HP-4275 LCR bridge, and illumination in deep depletion is accomplished using the microscope light. We implement the high-temperature hi-lo measurement using a computer controlled simultaneous hi-lo CV measurement system based on the Kiethley 595 quasi-static CV meter, Kiethley 590 CV analyzer, and Kiethley 5951 remote input coupler.

3.1 Room temperature photo-CV measurements

P-type 6H-SiC MOS capacitors typically exhibit a large negative flat band voltage. One component of this is the positive fixed charge Q_f resident at the interface. Added contributions may result from mobile ions in the oxide, charge trapped in the oxide, and charge trapped in interface states at flatband. Without further investigation into the acceptor/donor nature of interface states and into the nature of the charge trapping in the oxides, it is not possible to separate the various contributions. Hence, we treat all contributions as being lumped into an effective fixed charge Q_f that is responsible for the flat band voltage. For these measurements, the flatband voltage is obtained from high frequency CV curves taken at room temperature.

An estimate of the total surface state density N_{it} can be obtained from the interface state ledge in CV curves taken at room temperature. This feature was seen originally in silicon MOS gate-controlled diodes at liquid nitrogen temperature [6], and is also observed in SiC gate-controlled diodes at room temperature [5]. This ledge can also be seen in SiC MOS capacitors at room temperature if photogeneration is used to supply minority carriers in deep depletion. Figure 1 shows an aluminum-doped SiC MOS capacitor oxidized in wet O_2 at 1150 °C using a piranha clean and slow pull. The sample is first swept from accumulation to

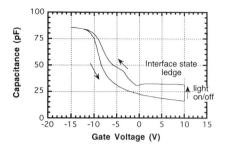
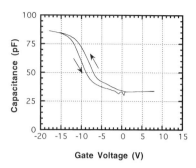

Figure 1. Room temperature CV curves for an Al-doped SiC MOS capacitor oxidized in wet O_2 at 1150 °C using a piranha clean and slow pull.

Figure 2. CV curves for the sample of Fig. 1 obtained at 340 °C. The remaining hysteresis is attributed to traps in the oxide near the interface.

deep depletion in the dark. In deep depletion, the capacitor is illuminated and minority carriers are trapped in interface states. After the illumination is turned off, the bias is swept back from inversion to accumulation. Initially the interface states retain their minority charge, but around midgap, when the majority carrier density becomes appreciable at the interface, the interface states begin to capture majority carriers, resulting in a characteristic "surface state ledge". The hysteresis in the CV curve just below the onset of the ledge corresponds to the total density of surface states that are not able to respond to the CV sweep. At room temperature, this accounts for a majority (perhaps as much as 50–75 %) of the surface states in SiC MOS capacitors, as can be verified by comparison of these numbers with measurements of interface state density D_{it} as a function of band bending (to be discussed later).

As seen in Fig. 1, an additional hysteresis is present in many of our SiC MOS structures in the bias range from accumulation to about midgap. This second hysteresis persists to very high temperatures, as shown in Fig. 2. We attribute this hysteresis to traps in the oxide which communicate with the semiconductor by tunneling, and we designate the density of these "oxide traps" as N_{ot}. Such near-interface charge trapping can also be observed in silicon MOS capacitors, although it is usually significant only after high field stressing or ionizing radiation [7].

Figure 3 shows a boron-doped MOS capacitor oxidized in wet O_2 at 1150 °C using an RCA clean and fast pull. All our boron-doped samples exhibit a small kink in the room temperature CV curves near flat band, as illustrated. This is due to partial freeze-out of acceptors in the semiconductor at room temperature. The ionization energy of the shallow boron level in 6H-SiC is 0.34 eV, compared to 0.21 eV for aluminum [8]. This "freeze-out kink" is also seen in silicon MOS capacitors at low temperatures (40–77 K) [9].

A comparison of Figs. 1 and 3 would suggest that boron-doped MOS capacitors have significantly lower Q_f, N_{it}, and N_{ot} than aluminum-doped capacitors, but this conclusion is incorrect. Note that the boron-doped sample of Fig. 3 was processed with an RCA clean, while the aluminum-doped sample of Fig. 1 was processed with a piranha clean. As we shall see in the next section, the improvement in interface quality is due entirely to the cleaning procedure, not to the use of boron.

3.2 High temperature hi-lo CV measurements

The commonly used techniques for D_{it} measurements on silicon MOS capacitors are the Terman high frequency CV technique, the AC conductance technique, and the hi-lo CV technique. The Terman technique has been applied to SiC MOS capacitors [4,10] but suffers from poor resolution. The AC conductance technique has also been reported on SiC [1], and offers excellent resolution and accuracy. The drawback of this technique is the need to

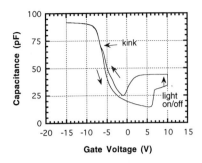
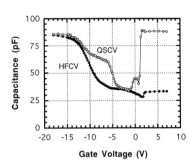

Figure 3. Room temperature CV curves of a boron-doped SiC MOS capacitor oxidized in wet O_2 at 1150 °C with an RCA clean and fast pull.

Figure 4. Simultaneous hi-lo CV curves on the sample of Fig. 1 (Al-doped with a piranha clean and slow pull) measured at 340 °C.

acquire large amounts of data. The frequency dispersion in the conductance is larger for SiC MOS capacitors than for silicon MOS capacitors, so that a larger frequency range is needed for SiC MOS characterization as compared to silicon.

The hi-lo frequency CV method offers a good tradeoff between resolution and measurement complexity. Applying the hi-lo technique to SiC MOS capacitors is complicated by the fact that conventional quasi-static CV sweeps are prone to non-equilibrium effects resulting from injection of excess minority charge into the substrate [11]. We have used an equilibrated quasi-static CV measurement [12], based on the feedback charge technique [13], that allows us to separate the small signal perturbation used to measure the capacitance from the bias sweep itself, unlike in conventional quasi-static CV measurements. We have found 340 °C to be sufficiently high so that most of the interface states respond to the measurement, allowing us to apply the hi-lo analysis and extract D_{it} vs energy over a significant portion of the bandgap.

Figure 4 shows high temperature simultaneous hi-lo CV measurements of the sample of Fig. 1. Using the hi-lo technique we calculate the surface state density D_{it} as a function of energy in the bandgap, as shown in Fig. 5. D_{it} in this sample is relatively constant at about 6×10^{11} eV^{-1} cm^{-2} across the energy range investigated.

Figure 6 compares the surface state density D_{it} of boron- and aluminum-doped samples oxidized in wet O_2 at 1150 °C with an RCA clean and extracted with a fast pull. D_{it} is similar for both samples, ranging from about 2×10^{11} to 3.5×10^{11} eV^{-1} cm^{-2}. We typically find no significant difference between boron- and aluminum-doped samples in terms of either fixed oxide charge density Q_f or surface state density D_{it}.

4. Discussion

Table 1 summarizes our findings on the wet oxidation of both boron- and aluminum-doped 6H-SiC at 1150 °C with a 30 minute in-situ Ar anneal. This temperature and ambient have yielded the best SiC MOS capacitors to date. All samples in this study are doped in the range of $1-2 \times 10^{16}$ cm^{-3}. The results reported in the table are based on only one or two devices of each type, but little variation is seen between different devices across a wafer.

As stated in the last section, results for boron- and aluminum-doped samples are comparable, indicating that the dopant species has little effect on interface quality. Pull rate also appears to have little effect. However, the surface cleaning procedure prior to oxidation has an important effect on the interface quality. For both boron- and aluminum-doped samples, the RCA clean results in lower fixed charge density and surface state density.

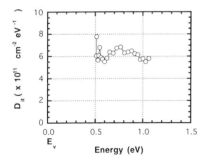
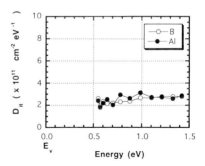

Figure 5. Surface state density versus energy in the bandgap calculated from the curves of Fig. 4 (Al-doped sample with a piranha clean and slow pull)

Figure 6. Comparison of surface state density vs energy for boron- and aluminum-doped SiC MOS capacitors oxidized in wet O_2 at 1150 °C with an RCA clean and fast pull.

The most important result of this work is the observation that the use of boron as a p-type dopant does not produce significant improvement in interface quality compared to aluminum. This is somewhat surprising, since it has been widely assumed that the poor quality of p-type SiC MOS capacitors (as compared to n-type) was due to the incorporation of aluminum at the interface during oxidation. Indeed, aluminum has been shown to be incorporated in the oxide at about the same density as the doping density in the SiC [3]. Since boron produces no degradation in silicon MOS capacitors, it has been suggested that the use of boron would result in better interface quality in p-type SiC.

Although boron does not solve the problems associated with MOS devices on p-type SiC, we note that improvements in material quality and oxidation procedures over the past few years have reduced fixed oxide charge density to the high-10^{11} to low-10^{12} cm^{-2} range and surface state density to the low-10^{11} eV^{-1} cm^{-2} range. These values are acceptable for the fabrication of practical MOS devices in SiC. It is significant that these results have been obtained primarily by optimizing the oxidation procedure itself and employing a high temperature in-situ Ar anneal. No low-temperature post-oxidation anneals have yet been employed. This suggests a promising direction for future research.

5. Conclusions

Thermally oxidized 6H-SiC MOS capacitors are characterized in terms of effective fixed charge density Q_f, surface state density D_{it}, and density of near-interface oxide traps N_{ot}. A combination of room-temperature high-frequency CV measurements and high-temperature hi-lo CV measurements are used to extract interface parameters. We find no significant difference in interface quality between boron- and aluminum-doped samples. We obtain a fixed oxide charge density of 1.4×10^{12} cm^{-2} and a surface state density of 2.5×10^{11} eV^{-1} cm^{-2} in samples which are prepared using the RCA clean, oxidized in wet O_2 at 1150 °C, and annealed in-situ for 30 minute in argon.

This work is supported by the Ballistic Missile Defense Organization, Innovative Science and Technology Division, under grant no. N00014-93-C-0071 (administered by ONR). Additional support is provided by the Semiconductor Research Corporation under grant no. 94-SJ-378.

Oxidation Condition	Dopant	Q_f/q (cm^{-2}) x 10^{12}	N_{it} (cm^{-2}) x 10^{11} (note 1)	D_{it} (cm^{-2} eV^{-1}) x 10^{11} (note 2)	N_{ot} (cm^{-2}) x 10^{11}
piranha clean and fast pull	Aluminum	2.6	12	6.0	10
	Boron	2.6	10	6.0	6.2
piranha clean and slow pull	Aluminum	2.3	14	5.5	5.5
	Boron	2.3	13	5.5	4.6
RCA clean and fast pull	Aluminum	1.4	5.1	2.5	4.1
	Boron	1.4	4.5	2.5	3.4

Table 1. Parameters of boron- and aluminum-doped 6H-SiC MOS capacitors.

Notes: (1) In the calculation of N_{it}, all the hysteresis below the interface state ledge (Fig 1) has been used. Some of this contribution comes from near-interface oxide traps (N_{ot}). N_{ot} measurements are made at 340 °C, and a larger density of these traps may respond at 340 °C than at 23 °C.
(2) The D_{it} value listed is an average over the energy range measured. Variations of ±0.5x10^{11} eV^{-1} cm^{-2} over the energy range are typical (c.f. Figs 5 & 6).

References

[1] T. Ouisse, N. Becourt, C. Jaussad, and F. Templier, *J. Appl. Phys.*, **75**, pp.604-609 (1994).
[2] D. M. Brown, M. Ghezzo, J. Kretchmer, E. Downey, J. Pimbley, and J. Palmour, *IEEE Trans. on Electron Devices*, **41** pp.618-620 (1994).
[3] J. W. Palmour, R. F. Davis, H. S. Kong, S. F. Corcoran, and D. P. Griffis, *J. Electrochem. Soc.*, **136**, pp. 502-507 (1989).
[4] J. W. Sanders, J. Pan, W. Xie, S. T. Sheppard, M. Mathur, J. A. Cooper, Jr., and M. R. Melloch, *IEEE Trans. on Elec. Dev.*, **40**, pp.2130-2131 (1993).
[5] S. T. Sheppard, J. A. Cooper, J r., and M. R. Melloch, *J. Appl. Phys.*, **75**, pp.3205-3207 (1994).
[6] A. Goetzberger and J. C. Irvin,*IEEE Trans. on Elec. Dev*, **15**, pp.1009-1017 (1968).
[7] D. M. Fleetwood, *IEEE Trans. on Nucl. Sci.*, **39**, pp. 269-271 (1992).
[8] W. J. Schaffer, Cree Research, Inc., private communication.
[9] P. V. Gray and D. M. Brown, *Appl. Phys. Lett.* pp. 247-248 (1968).
[10] R. E. Avila, J. J. Kopanski, and C. D. Fung, *Appl. Phys. Lett.*,**49**, pp.334-336 (1986).
[11] P. Neudeck, S. Kang, J. Petit, and M. Tabib-Azar, *J. Appl. Phys.*, **75**, pp.7949-7953 (1994).
[12] J. N. Shenoy, J. A. Cooper, Jr., and M. R. Melloch, submitted to *Elec. Lett.* (1994).
[13] T. J. Mego, *Rev. Sci. Instrum.*, **57**, pp. 2798-2805 (1986).

Monolithic Digital Integrated Circuits in 6H-SiC

W. Xie, J. Pan, J. A. Cooper, Jr., and M. R. Melloch

School of Electrical Engineering, Purdue University, West Lafayette, IN 47907

Abstract. We report the first monolithic digital integrated circuits in 6H-SiC. Non-self-aligned gate MOSFETs are fabricated in p-type epilayers on p+ (Si-face) 6H-SiC substrates. The logic circuits consist of n-channel enhancement mode pull-down networks with n-channel enhancement mode load transistors connected in a non-saturating configuration. We have fabricated logic inverters (beta ratios of 9, 12, and 16), NAND and NOR gates, D-latches, RS flip flops, XNOR gates, binary counters, and half adders. These circuits are operational from room temperature to above 300 °C.

1. Introduction

Silicon carbide (SiC) is a wide bandgap semiconductor (E_G = 3.0 eV for the 6H polytype) having exceptional chemical and thermal stability. Recent improvements in material quality suggest that practical devices and circuits can be developed in this material system. In this paper we report recent progress in the development of monolithic digital integrated circuit technology in 6H-SiC [1,2].

2. Fabrication

Integrated circuits are formed with n-channel MOSFETs having non-self-aligned gates. The MOSFETs are fabricated in 3 μm p-type epilayers (doped with Al at 2×10^{16} cm^{-3}) on p+ (Si-face) 6H-SiC substrates obtained from Cree Research, Durham, NC. Registration marks are produced by RIE in SF_6 using an Al etch mask. Source and drain regions are formed by ion implantation of N through a Ti implant mask. A second implant introduces Al into regions outside the active devices to serve as a channel stop. (This channel stop implant may be unnecessary, but was included here as a precaution.) Both implants are conducted with the SiC sample at 650 °C. The implants are then activated by a 1500 °C 30-min. anneal in Ar. A 50 nm gate oxide is formed by wet oxidation at 1150 °C, followed by a 30 min. in-situ Ar anneal. Source and drain ohmic contacts are E-beam evaporated Ni, which is patterned by liftoff. The p-type ohmic contacts are formed to the channel stop region by evaporated Al, which is also patterned by liftoff. At this point, both ohmic contacts would normally be annealed at high temperature in Ar. However, in the devices reported in this presentation, the contact anneal step was omitted to expedite the processing. Finally, the gate and interconnect metals are formed by evaporated Al, forming non-self-aligned gate MOSFETs.

3. Logic circuit characterization

The basic NMOS logic circuit is illustrated schematically in Fig. 1. In each circuit, the pull-down network is comprised of n-channel enhancement mode MOSFET's. The load, or pull-up, devices are n-channel enhancement mode MOSFET's connected in a non-saturating

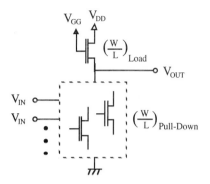

Figure 1. Basic logic circuit consisting of n-channel enhancement-mode MOSFET's. The load transistor is connected in the non-saturating configuration with separate V_{DD} and V_{GG} supplies. The pull-down network (inside dashed box) is configured to implement the desired logic function.

configuration with separate V_{DD} and V_{GG} bias supplies. Threshold voltages are 2.5 V at room temperature, decreasing to about 0.5 V at 304 °C.

Logic inverters with beta ratios of 9, 12, and 16 have been fabricated and characterized. At room temperature, with V_{DD} = 10 V and V_{GG} = 15 V, inverters with a beta ratio of 16 exhibit logic levels of 0.8 V and 10 V, as shown in Fig. 2. Noise margins are approximately 1.5 V for logic zero and 3.5 V for logic one. At 304 °C, with V_{DD} = 5 V and V_{GG} = 6 V, the inverters exhibit logic levels of 0.2 V and 4.9 V (Fig. 2).

We have also implemented more complicated logic circuits, including NAND and NOR gates, D-latches, RS flip-flops, binary counters, XNOR gates, and half adders. All circuits are functional from room temperature to 304 °C. For illustration purposes, we will discuss the binary counter, XNOR gate, and half adder here.

The binary counter, shown in Fig. 3, consists of two pairs of inverters which are connected by transmission gates to form a master-slave data latch. To form a counter, the inverted output (Q-bar) is connected to the data input, so that each positive-going clock edge causes the master latch to go to the opposite state from the current output. This new state is transferred into the slave latch on the negative-going edge of the clock. The result is a divide-by-two counter, as verified by the waveforms in Fig. 3.

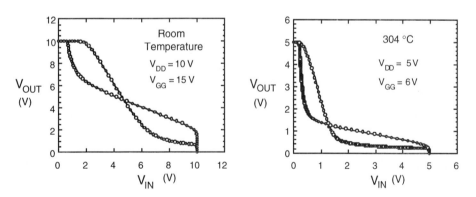

Figure 2. Input-Output characteristics of the logic inverter at room temperature and 304 °C.

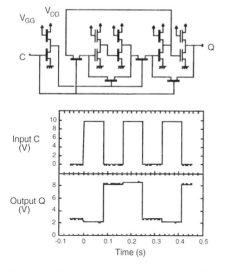
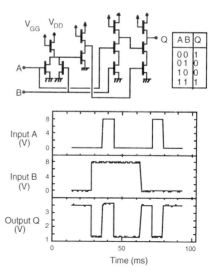

Figure 3. Circuit diagram and waveforms of a SiC binary counter at room temperature.

Figure 4. Circuit diagram and waveforms of a SiC XNOR gate at room temperature.

The XNOR gate, shown in Fig. 4, consists of an inverter, a NOR gate, and two NAND gates. The output is a logic one whenever the two inputs A and B are the same. The truth table is given in the figure, and can be verified by an examination of the waveforms shown. In these circuits, the output transistors are not designed to drive off-chip loads, and the output voltage swing is limited by loading due to the 1 MΩ oscilloscope probe.

The half adder is shown in Fig. 5. It consists of two inverters, a NOR gate, and an AOI (and-or-invert) gate. Outputs are sum "S" and carry "C". The truth table and waveforms are also shown in the figure. The S output does not go all the way to ground when the C output is high because during this state, S is being pulled low by the transistor which is gated by the C output. Since the C output does not go all the way to 10 V due to loading by the scope probe, the transistor gated by the C output is not adequately turned on, and the S output does not pull all the way to ground.

The half adders (and all other circuits) operate properly at temperatures from room temperature to 304°C, the highest temperature tested. Figure 6 shows waveforms of the half adder at 304 °C.

4. Discussion

Since the transistors are not optimized, we have not attempted to determine the maximum switching speed of these circuits. In this test chip, the minimum channel length of the transistors is 5 µm, and all transistors have metal gates which overlap the source and drain. These overlap capacitances severely degrade switching performance. In addition, the source and drain ohmic contacts are not annealed, and source and drain resistances are high. These deficiencies can easily be corrected in future designs.

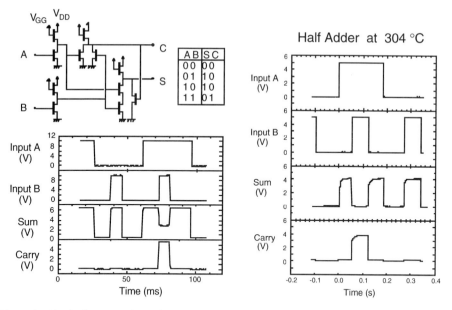

Figure 5. Circuit diagram and waveforms of a SiC half adder at room temperature.

Figure 6. Waveforms of the half adder at 304 °C.

Although no failures were observed at 304 °C, this does not prove that these circuits have high reliability at these temperatures. More extensive testing is required before long-term high-temperature reliability can be claimed.

5. Conclusions

We have reported the first monolithic digital integrated circuits in 6H-SiC. The logic circuits consist of n-channel enhancement mode pull-down networks with n-channel enhancement mode load transistors connected in a non-saturating configuration. We have operated logic inverters, NAND and NOR gates, D-latches, RS flip flops, XNOR gates, binary counters, and half adders from room temperature to above 300 °C.

This work is supported by grant No. N00014-93-C-0071 from the Ballistic Missile Defense Organization, Innovative Science and Technology Division, administered by ONR. The authors acknowledge valuable discussions with J. W. Palmour and C. H. Carter, Jr., of Cree Research, Inc.

References

[1] J. A. Cooper, Jr. and M. R. Melloch, "NMOS Digital Integrated Circuits in 6H Silicon Carbide," *WOCSEMMAD Conf.*, San Francisco, CA, February 21 - 23, 1994.

[2] W. Xie, J. A. Cooper, Jr., and M. R. Melloch, "NMOS Digital Integrated Circuits in 6H Silicon Carbide," *IEEE Device Research Conference*, Boulder, CO, June 20-22, 1994.

DC, Microwave, and High-Temperature Characteristics of GaN FET Structures

S. C. Binari, L. B. Rowland, G. Kelner, W. Kruppa[†], H. B. Dietrich, K. Doverspike, D. K. Gaskill

Naval Research Laboratory, Washington, D. C. 20375
[†]SFA, Inc., Landover, MD 20785

Abstract. We report on the dc, microwave, and high-temperature characteristics of GaN HFETs and MISFETs. GaN HFETs have demonstrated a transconductance of 45 mS/mm at 30 °C and 22 mS/mm at 350 °C. For a 1 μm gate length, an f_T of 8 GHz and f_{max} of 22 GHz were measured. Using a Si_3N_4 gate insulator, GaN MISFETs were fabricated and have exhibited a transconductance of 16 mS/mm at 30 °C and 11 mS/mm at 200 °C. For a 0.9 μm gate length, the MISFETs have an f_T and f_{max} of 5 and 9 GHz, respectively.

1. Introduction

Wide-bandgap semiconductors are of interest for high-temperature electronics and high-power microwave devices [1,2]. This interest stems primarily from the lower thermal generation rates and higher breakdown fields inherent in wide-bandgap materials. In addition, GaN-based devices can also be designed with heterojunctions for improved performance [3]. The dc and microwave characteristics of GaN HFETs have been recently reported [4,5]. In this paper, we present high transconductance HFETs, 350 °C HFET operation, and the demonstration of a GaN MISFET.

2. Material Growth and Device Fabrication

The GaN epitaxial layers used in this study were unintentionally doped and grown on basal-plane sapphire substrates by OMVPE [6]. A 40 nm AlN buffer layer was grown at a substrate temperature of 450°C, followed by a 3 μm undoped, highly-resistive GaN layer grown at 1050°C. A 6 nm AlN layer was then grown, followed by the growth of the 0.2 μm GaN channel. C-V measurements indicated that most of the carriers in the active channel are located near the GaN/AlN interface, as shown in Fig 1. The variable-temperature Hall-effect measurements shown in Fig. 2 indicated a constant mobility and sheet carrier concentration over the temperature range from 10 and 120 K. The sheet concentration obtained by integrating the concentration profile of Fig. 1 is in good agreement with the value obtained by the Hall measurements. These results are consistent with the existence of a 2DEG at the GaN/AlN interface.

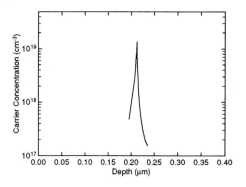

Fig. 1. Carrier concentration profile

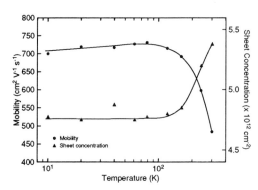

Fig. 2. Hall measurements on the FET wafer

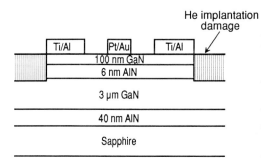

Fig. 3. HFET cross section

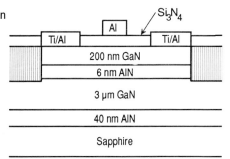

Fig. 4. MISFET cross section

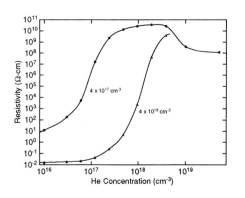

Fig. 5. Resistivity of GaN damaged by He implantation

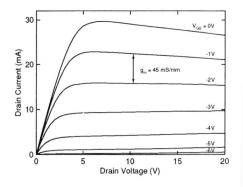

Fig. 6. HFET drain characteristics at 30 °C

The device cross sections of the HFET and the MISFET are shown in Figs. 3 and 4, respectively. Both devices were fabricated with a source-drain spacing of 5 μm and gate lengths of 0.7 to 2.0 μm. The total gate width is 150 μm. Alloyed Ti/Al was used to form the source and drain ohmic contacts [7]. For the HFETs, 1000 Å of GaN was removed by BCl_3 RIE before device processing. The gate metallization was Pt/Au (500/2000 Å). In the case of the MISFETs, an 800 Å Si_3N_4 gate insulator was deposited by plasma enhanced chemical vapor deposition and the gate metallization was 3000 Å of Al.

The devices were isolated using He-implantation-induced damage. Helium-implantation is capable of increasing the resistivity of n-type GaN material to 10^{10} Ω-cm. Shown in Fig. 5 is the resistivity as a function of He concentration for two GaN samples, one with a carrier concentration of 4×10^{17} and the other 4×10^{18} cm^{-3}. The higher carrier concentration sample requires a He concentration 10 times greater in order to produce a similar resistivity. For He concentrations exceeding 1×10^{19} cm^{-3}, the resistivity of the damaged GaN is stable to over 850 °C. The isolation properties of the He implanted GaN are superior to that of H implanted GaN, which has a maximum resistivity of 10^7 Ω-cm that anneals out at 300 °C.

3. Device Characteristics

The dc drain characteristics for HFETs with a gate length of 1 μm are shown in Fig 6. The HFETs have a pinch-off voltage of -6 V and a maximum g_m of 45 mS/mm. This g_m is significantly higher than that previously reported [4,5]. The Pt/Au Schottky barrier has an ideality factor of 1.3. As shown in Fig. 7, the HFETs were operational at a baseplate temperature of 350 °C with a g_m of 22 mS/mm. At this temperature, the gate leakage current begins to affect the ability to fully pinch off the device. The reduction in g_m at elevated temperatures is attributed to a reduction in electron mobility.

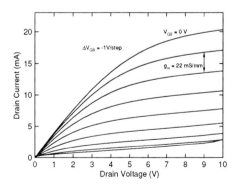

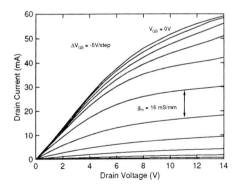

Fig. 7. HFET drain characteristics at 350 °C

Fig. 8. MISFET drain characteristics at 30 °C

The S-parameters of these devices were measured as a function of frequency and bias using on-wafer probing. A small-signal equivalent circuit was developed by matching the measured and calculated S-parameters. The circuit has the topology commonly used for FETs. Using the GaAs MESFET as a reference, the present GaN HFET has large drain and source resistances. These resistances are made up in roughly equal parts by the contact and access resistances. With modified channel doping and improved ohmic contact technology, both of these values can be lowered. For a 1.0 μm gate length device, the measured f_T and f_{max} values were 8 and 22 GHz, respectively. In terms of frequency performance, these initial results with GaN are competitive with those obtained with SiC MESFETs.

The drain characteristics of the GaN MISFETs are shown in Fig. 8. The MISFETs displayed a maximum g_m of 16 mS/mm and a pinch-off voltage of -50 V. The gate leakage current was less than 0.2 μA over the entire range of gate and drain biases. The relatively low g_m at low gate bias is attributed to interface traps and is consistent with a relatively large 6V hysteresis in the C-V characteristics of test MIS diodes. The measured f_T and f_{max} values were 5 and 9 GHz, respectively. The value of f_{max} is substantially better than that achieved with SiC MOSFETs. At 200 °C, the maximum g_m was 11 mS/mm. Device failure occurred while operating at a baseplate temperature of 300 °C.

4. Conclusions

GaN HFETs with a 1.0 μm gate length have a transconductance as high as 45 mS/mm and an f_T and f_{max} of 8 and 22 GHz, respectively. In addition, these HFETs are operational at a baseplate temperature of 350 °C. Finally, we have demonstrated a GaN-based MISFET with f_T and f_{max} values of 5 and 9 GHz, respectively. These results indicate that GaN devices show promise for high-temperature and high-frequency applications.

Acknowledgments

The authors thank J. B. Boos, P. E. Thompson, M. J. Yang, D. Park, L. Dobisz, R. Kaplan, and W. E. Carlos for valuable discussions. The n+ sample used in the He implantation study was provided by D. Wickenden of APL. One of us (LBR) acknowledges support from a NRC-NRL Cooperative Research Associateship. This work was supported by the Office of Naval Research.

References
[1] Driver M C, Hopkins R H, Brandt C D, Barrett D L, Burk A A, Clarke R C, Eldridge G W, Hobgood H M, McHugh J P, McMullin P G, Siergiej R R, and Sriram S 1993 *Proc. GaAs IC Symposium*, 19 - 22
[2] Palmour J W, Carter C H, Weitzel C E, and Nordquist K J 1994 *Diamond, SiC, and Nitride Wide-Bandgap Semiconductors, Materials Research Society Proceedings*, **339**
[3] Khan M A, Kuznia J N, Van Hove J M 1992 *Appl. Phys. Lett.*, **60** 3027-9
[4] Khan M A, Kuznia J N, Olson D T, Schaff W J, Burm J W, and Shur M S 1994 *Appl. Phys. Lett.*, **65** 1121-3
[5] Binari S C, Rowland L B, Kruppa W, Kelner G, Doverspike K, and Gaskill D K 1994 *Electronics Lett.* **30** 1248-9
[6] Rowland L B, Doverspike K, Gaskill D K, Freitas Jr J A 1994 *Diamond, SiC, and Nitride Wide-Bandgap Semiconductors Materials Research Society Proceedings* **339**
[7] Lin M E, Ma Z, Huang F Y, Fan Z F, Allen L H, and Morkoc H 1994 *Appl. Phys. Lett.* **64** 1003-5

Physics and Simulation of InGaAsP/InP Lasers

R. F. Kazarinov

AT&T Bell Laboratories, Murray Hill, NJ 07974

ABSTRACT

We review AT&T works on modeling and physics of communication lasers. They include a method for calculating the electronic states and optical properties of bulk and quantum well active region laser structures and comparison of modeling and experimental results. They also include two dimensional numerical simulation of carrier transport in laser structures, which allows calculation of the efficiency of injected carrier consumption by the active region and the dependence of the laser current on applied voltage. Calculations of quantum efficiency and threshold current for bulk InGaAsP lasers are supported by the experimental data.

1. INTRODUCTION

This paper is a review of some works done at Bell Laboratories on modeling of semiconductor lasers and related experimental studies aimed to verify model predictions. Naturally, we are interested in communication lasers. The list of these lasers includes analog lasers, uncooled lasers, pump and digital lasers. These devices have different designs, they face different technological challenges and different problems. The major requirement to analog lasers for CATV application is to provide a linear response of the output power to a modulation of an applied voltage. These are sophisticated multi-quantum-well (MQW), DFB devices. Uncooled lasers for Local Area Network applications should be low cost devices operating up to 85 C with a decent performance. Lasers for pumping of Erbium Doped Optical Fiber Amplifiers are high power devices (~50 mW). And digital lasers for long haul communication should have broad bandwidth, low chirp and low bit error rate. These are also DFB, MQW devices.

Driven by ever increasing demands to performance laser designs are getting more and more complex. With strain MQW structures replacing a simple bulk active layer in all types of lasers the number of laser structure parameters becomes too large to optimize device performance empirically. Therefore, Computer Aided Laser Design tools become increasingly important.

This paper is focused on experimental verification of physical assumptions the laser modeling is based on. We started this study with a simple Fabry-Perot, bulk active layer laser designed for uncooled application. The program we use for laser modeling consists of two parts. The first one, gain program, calculates electronic states and energy levels of a semiconductor material. It also calculates optical properties of laser structures as a function of injected carrier concentration. The second part describes carrier transport in lasers. It calculates distribution of carrier concentration and electrical potential in the device and the recombination current.

2. GAIN, MODELING AND VERIFICATION

The gain model we use has been developed by Gershoni, Henry and Baraff [1]. The electronic states and the energy levels of the semiconductor device are found, in principle, by solving the Schroedinger equation. This equation can be solved for the perfect spatially uniform semiconductor using a variety of well known techniques [2,3], but in a heterostructure or a quantum wire or dot, the crystal composition and/or strain varies from region to region and approximations are needed in order to solve it. Many such approximate methods are now well known and extensively used [4]. In the paper [1], the authors use Kane's method [2] and envelope functions [5]. The result of this approximation is that the Schroedinger equation is converted into a set of eight coupled differential equations for envelope functions. The equations are ordinary differential equations for quantum wells, and partial differential equations for quantum wires and quantum dots.

The gain model developed described above accounts for a realistic band structure. But, it regards electrons and holes as ideal Fermi gases. Some of carrier interaction effects, such as an enhancement of optical absorption above the energy gap caused by electron-hole attraction without carriers, can be put into the model. However, a reliable theory describing many body effects at actual levels of carrier concentration is not available. Therefore, it is important to verify experimentally whether a simplified gain model can give predictions with a reasonable accuracy.

Such verification has been carried out in reference [6] for a simple bulk, Fabry-Perot laser. Maximum modal gain Gmax and gain peak energy E(Gmax) are obtain in this work from measured subthreshold gain spectra as illustrated schematically in Fig. 1.

In [6] an accurate determination of of quasi-Fermi level separation $\mu = E_{Fn} - E_{Fp}$ is achieved by injection of a weak, single- frequency, ac probe signal into a laser cavity to determine conditions of material transparency [7] thereby allowing us to extract total modal gain G_{tot} at E(Gmax). Modal gain as a function of current is inferred by measurement of spectral fringe contrast [8], as is free carrier loss under transparency conditions.

In order to compare the experiment with results of modeling the energy gap of the active layer material was found by fitting E(Gmax) at threshold and the optical confinement factor was calculated to relate the material and modal gains.

Fig. 1 shows measured (solid curves) and calculated (dotted curves) modal gain G_{tot} as functions of electron-hole Fermi level separation μ at 25 and 85°C for a bulk, Fabry-Perot laser.

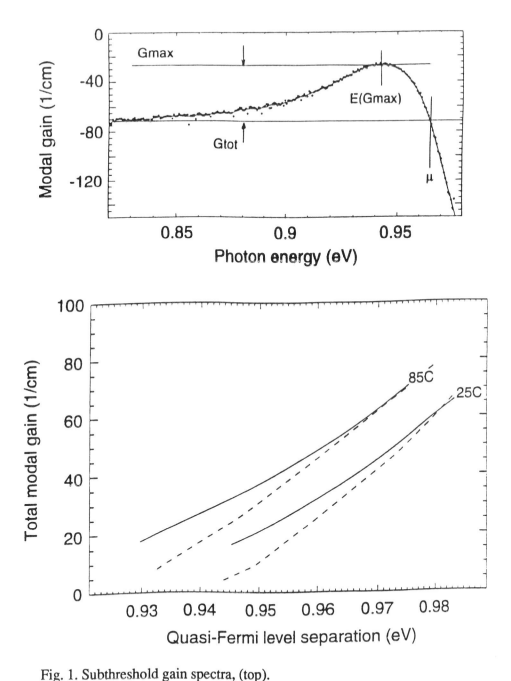

Fig. 1. Subthreshold gain spectra, (top).

Gain peak vs. quasi-Fermi level separation. Solid curves-experiment; dotted curves- theory

A significant difference between theory and experiment for μ below and at the energy gap is caused by spectral broadening due to disorder effects which are not included into the model. However, for values of μ approaching a threshold value calculated and measured G_{tot} are close. The discrepancy between the theoretical and the experimental μ at threshold corresponds to less than 5% of carrier concentration for both temperatures.

Discussed above relations between G_{tot}, E(Gmax) and μ are independent on the recombination and carrier loss processes. These processes determine the μ dependence on current. We address this issue in the next section.

3. CARRIER TRANSPORT AND RECOMBINATION

For numerical simulation of carrier transport and recombination processes we use the program PADRE which solves Poisson's and carrier transport equations for a semiconductor heterostructure in up to three spatial dimensions for an arbitrary doping distribution and arrangement of contacts. This program was modified to incorporate description of stimulated emission and supplemented with additional calculations of the optical properties of the laser. The program allows calculation of the efficiency of injected carrier consumption by the active region. It also calculates the current, carrier and potential distribution in a laser structure [9].

We applied this program to calculate the leakage current through the CMBH blocking structure [10] used in AT&T lasers for uncooled and analog applications. Our calculations have generally supported an observation that it can be made leak free. However, there is another mechanism of carrier loss which is related to thermionic emission of carriers out of the active layer. This effect depends exponentially on temperature and the magnitude of potential barriers confining carriers in the active region. Because electron and hole Fermi-levels are pinned above threshold, it may appear that these potential barriers are also fixed. In fact, these barriers depend on current and the physical reason for their reduction above threshold will be discussed below. The increase of carrier loss with increase in current manifests itself experimentally as a reduction of the laser slope efficiency η_{ext} caused by a reduction of the internal quantum efficiency η_i describing a fraction of carriers recombining in the active region.

Experimental verification of carrier loss due to thermionic emission has been done in ref.[10]. The devices studied in this paper are InGaAsP/InP Fabry-Perot capped- mesa-buried-heterostructure (CMBH) lasers whose basic design and gross electro-optic characteristics are described in reference [11].

Internal quantum efficiency, η_i, was determined though a measurement of the external quantum efficiency, η_{ext}, and the internal loss, α_i. Internal loss at threshold was estimated using the transparency method described in [7]. This method deduces the internal loss from a measurement of the spectral resonances at the transparency current. Transparency is determined at a specific wavelength with the aid of a probe laser.

In the absence of current shunt paths, carrier loss dominates the temperature dependence of η_i. The physical picture of carrier loss caused by the thermionic emission of electrons out of the active layer is described in [9]. It is related to the reduction of the

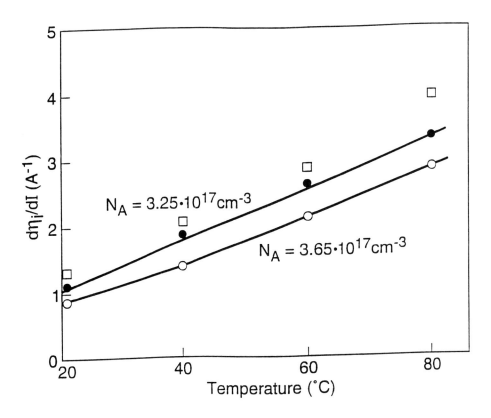

Fig. 2. Current derivatives of the internal quantum efficiency as a function of temperature. Open circles, squares and dots are the experimental values for lasers 1,2 and (3,4) respectively. Solid curves are the results of a theoretical fit with $N_A = 3.65*10^{-17} cm^{-3}$ and $N_A = 3.25*10^{-17} cm^{-3}$.

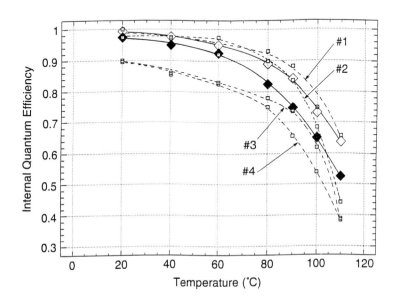

Fig. 3. Temperature dependence of internal quantum efficiency. Dashed curves represent experimental data for four lasers. Open and closed diamonds are the results of theoretical calculations for $N_A = 3.65*10^{17}$ cm^{-3} and $N_A = 3.25*10^{17}$ cm^{-3} respectively.

potential barrier confining electrons in the active layer due to the voltage drop across the heterojunction between the p- and active layers. The most important parameter determining this voltage drop (and therefore the carrier loss due to the thermionic emission) is the hole concentration in p-material adjacent to the active layer. This concentration is reduced compared to its nominal value due to Zn diffusion, and it is not well known. We find it by fitting the current derivative of the internal quantum efficiency $d\eta_i/dI$ obtained from the experiment to its calculated value at a fixed temperature. The results of fitting are shown in Fig. 2. We see that the Zn concentration $N_A = 3.25*10^{17} cm^{-3}$ is a good fit for lasers 3 and 4, and $N_A = 3.65*10^{17} cm^{-3}$ is a good fit for the laser 1. Fig. 2 also shows that the calculated dependence of $d\eta_i/dI$ on temperature is in good agreement with the experimental data.

These values of N_A were used to calculate the internal quantum efficiency η_i. Fig. 3 shows the results of this calculation (open and closed diamonds) and the experimental data. Both experimental data and calculation show a significant reduction of η_i above 80°C. Experimental curves for lasers 3 and 4 are shifted down compared to the theoretical one. We attribute this difference to leakage through the current blocking structure which has not been taken into account in calculations. We also note that the internal quantum efficiency of these lasers are significantly larger at high temperature than those studied in [12]. The calculated N_A for the lasers in [12] is about 30% smaller than obtained for devices studied in this paper. This is consistent with the theoretical conclusions of [9] indicating that increase in N_A should result in an improvement of η_i.

4. CONCLUSIONS

We review some AT&T works on physics of semiconductor lasers. They include a method for calculating the electronic states and optical properties of semiconductor quantum structures which is applicable to bulk, quantum wells, quantum wires and quantum dot lasers. They also include two dimensional numerical simulation of carrier transport in laser structures, which allows calculation of the efficiency of injected carrier consumption by the active region and the dependence of the laser current on applied voltage. Preliminary experimental works indicate that a semiconductor laser model neglecting most of many body effects can be reasonably accurate. Calculations of quantum efficiency and threshold current for bulk InGaAsP lasers are supported by the experimental data.

REFERENCES

[1] Gershoni, D., Henry, C. H. and Baraff, G. A. (1993) Calculation of the Optical Properties of Multidimensional Heterostructures: Application to the Modeling of Quaternary Quantum Well Lases, IEEE, J. Quant. Electron. vol. QE-29, pp. 2433-2450.

[2] Kane E. O. (1982) Energy Band Theory, in Handbook on Semiconductors, edited by W. Paul (North Holland, Amsterdam) Vol 1, 193-217.

[3] Callaway J (1964) Energy Band Theory, (Academic Press, New York).

[4] Bastard G and Blum J A (1986) Electronic States in Semiconductor Heterostructures, IEEE J. Quant. Electron., Vol QE-22, 1625-1644.

[5] Baraff G A and Gershoni D (1991) Eigenfunction-Expansion Method for Solving the Quantum Wire Problem: Formulation, Phys. Rev. B vol. 43, pp. 4011-4022.

[6] Ackerman D A, Morton P A, Kazarinov R F and Hybertsen M S, Comparison of direct gain measurements in semiconductor lasers to a Fermi gas model, to be published.

[7] Andrekson P A, Olsson N A, Tanbun-Ek T, Logan R A, Coblenz D, and Temkin H (1992) Electron. Lett., vol. 28, p. 171.

[8] Hakki B W and Paoli T I (1975) J. Appl. Phys., 46, 1299.

[9] Kazarinov R F and Pinto M R (1994) Carrier Transport in Laser Heterostructures, J. of Quant. Electron.

[10] Ketelsen L J P and Kazarinov R F, Carrier Loss in InGaAsP/InP Lases Grown by Hydride CVD, to be published.

[11] Zilko J L, Ketelsen L J P, Twu Y, Wilt D P, Napoltz S G, Blaha J P, Strege K E, Riggs V G, Van Haren D L, Leung S Y, Nitzshe P M, Long J A, Roxlo C B, Przyblek G, Lopata J P, Focht M W and Koszi L A (1989) Growth and characterization of high yield, reliable, high power high speed, InP/InGaAsP capped mesa buried heterostructure distributed feedback (CMBH-DFB) lasers, IEEE J. Quantum Electron. vol. 25, pp 2091-2095.

[12] Andrekson P A, Kazarinov R F, Olsson N A, Tanbun-Ek T, Logan R A, (1994) Effect of Thermionic Electron Emission from the Active Layer on the Internal Quantum Efficiency of InGaAsP Lasers Operating at 1.3 µm, IEEE J. Quantum Electron vol. 30, pp. 219-222.

Frequency Limitations of Quantum-Wire Lasers

Sandip Tiwari

IBM Research Division, T. J. Watson Research Center, P.O. Box 218, Yorktown Heights, NY 10598

Abstract. We show, based on experimental data and theoretical analysis, that quantum-wire lasers exhibit a substantial degradation in operable frequencies due to fundamental and practical reasons. Among the fundamental reasons is a substantial increase in apparent gain compression that arises from capture and emission processes and carrier heating. Among the practical reasons are transmission line effects in the presence of a large electric parasitic loading. Bandwidths in excess of 10 GHz have been achieved at sub-mA currents and bandwidths approaching 20 GHz are considered achievable at sub-mA currents.

1. Introduction

Recent experimental work on strained multi-well quantum-wire lasers[1][2] has demonstrated lower threshold currents (≤ 200 μA) and comparable K-factors (0.344) to those of simultaneously grown ridge lasers. However, the 3-dB response frequencies remain low (≈ 12 GHz at < 600 μA bias current) and significantly below the highest frequencies of 30 GHz[3] demonstrated for ridge lasers. The limit frequencies are also lower than those of ridge lasers fabricated and grown simultaneously with the quantum-wire lasers. This raises questions about limitations in the dynamic operation of the quantum-wire lasers. Here, we point out a number of fundamental and practical considerations that will limit the operational frequencies to below those postulated in earlier analysis of quantum-well and quantum-wire structures.[4][5] While we use the operation of quantum-wire lasers fabricated in a V-groove[2] for our discussion, several of the considerations apply to lasers with other designs.

2. Fundamental Limitations: Carrier Dynamics

Using conventional small-signal analysis[6] for ridge (4 × 1000 μm^2) and quantum-wire lasers (1000 μm)[1], key parameters have been obtained. The most striking changes occur in differential gain (g) and gain-compression coefficient (ϵ); the former increases from 2.1×10^{-15} cm^2 to 1.5×10^{-13} cm^2, while the latter increases from 2.54×10^{-17} cm^3

to 1.79×10^{-16} cm^3. The increase in the former, which is the basis of the high frequency predictions, is largely negated by the increase in the latter. To analyze the cause of this apparent increase in gain compression coefficient, we revisit the operation of quantum well lasers by considering the effect of capture[9] and emission processes.

Starting with the single-mode rate and band occupation equations,

$$\frac{dN_b}{dt} = \frac{I}{eVol} - \left(\frac{N_b}{\tau_{cap}} - \frac{N_w}{\tau_{em}}\right) - R(N_b), \tag{1}$$

$$\frac{dN_w}{dt} = \left(\frac{N_b}{\tau_{cap}} - \frac{N_w}{\tau_{em}}\right) - R(N_w) - v_{gr}\Gamma G(N,S)S, \tag{2}$$

$$\frac{dS}{dt} = -\frac{S}{\tau_{ph}} + \beta R(N_w) + v_{gr}\Gamma G(N,S)S, \tag{3}$$

$$\text{and } V(N_w) = \frac{E_g}{e} + V_T \ln\left\{\left[\exp\left(\frac{N_w}{N_c}\right) - 1\right]\left[\exp\left(\frac{N_w}{N_v}\right) - 1\right]\right\}, \tag{4}$$

where the subscripts b and w denotes barrier and well region, N, N_c and N_v are the carrier and band-edge conduction and valence band state densities, S is the photon density, R denotes spontaneous recombination, G is the gain, V_T is the thermal voltage, Γ is the confinement factor, and τ_{ph}, τ_{cap} and τ_{em} are the photon, carrier capture and carrier emission time constants, and using τ_{Nb} and τ_{Nw} for the differential lifetimes in the barrier and the well region, the small-signal solution for $\tilde{S}/\tilde{I}$, which is proportional to the modulation transfer function, is of the form

$$\frac{\tilde{S}}{\tilde{I}} = \frac{(1/eVolA)(\beta/\tau_{Nw} + v_{gr}\Gamma g\bar{S})}{D_1 + j\omega D_2 - \omega^2 - j\omega^3 D_4}, \tag{5}$$

where $A = 1 + \tau_{cap} \times (\tau_{Nb}^{-1} + \tau_{em}^{-1} + v_{gr}\Gamma\epsilon\bar{S})$ and the gain is linearized using $G(N,S) = \bar{G} + g\tilde{N} - \epsilon\tilde{S}$. We can write the constants and compare them to the conventional equations where capture and emission are ignored ($\tau_{cap} \to 0$ and $\tau_{em} \to \infty$). D_1, originally ω_r^2 - the square of relaxation oscillation frequency, is

$$D_1 = A^{-1}\left\{\frac{v_{gr}\Gamma\epsilon\bar{S}}{\tau_{Nw}} + \frac{\beta}{\tau_{Nw}\tau_{ph}} + \frac{v_{gr}\Gamma g\bar{S}}{\tau_{ph}} - \frac{v_{gr}\Gamma\epsilon\bar{S}\beta}{\tau_{Nw}} + \right.$$
$$\left. \frac{\tau_{cap}}{\tau_{Nb}}\left[v_{gr}\Gamma\epsilon\bar{S}\left(\frac{1}{\tau_{em}} - \frac{\beta}{\tau_{Nw}} - v_{gr}\Gamma g\bar{S}\right) + \frac{1}{\tau_{ph}}\left(\frac{\beta}{\tau_{Nw}} + v_{gr}\Gamma g\bar{S}\right)\right]\right\}. \tag{6}$$

D_2, originally γ - the damping factor, is

$$D_2 = A^{-1}\left\{v_{gr}\Gamma\epsilon\bar{S} + v_{gr}\Gamma g\bar{S} + \frac{1}{\tau_{Nw}} + \tau_{cap}\left[\frac{1}{\tau_{em}\tau_{Nb}} + \frac{1}{\tau_{ph}}\left(\frac{\beta}{\tau_{Nw}} + v_{gr}\Gamma g\bar{S}\right) + \right.\right.$$
$$\left.\left. v_{gr}\Gamma\epsilon\bar{S}\left(\frac{1}{\tau_{Nb}} + \frac{1}{\tau_{em}} - \frac{\beta}{\tau_{Nw}} - v_{gr}\Gamma g\bar{S}\right)\right]\right\}. \tag{7}$$

The coefficient for the third power in frequency, D_4 is τ_{cap}. These expressions show clearly the perturbation caused by finite capture times: (a) the response is directly attenuated by a fraction that measures the efficiency of the capture process vis-a-vis the barrier differential lifetime and the emission time-constant from the well and it accentuates directly the effect of gain compression non-linearities, (b) damping is accentuated and the relaxation oscillation frequency reduced by a factor that is again tied to the efficiency of the

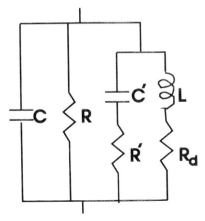

Figure 1. Electric equivalent circuit of the quantum-well laser in presence of non-zero capture and finite emission time-constants.

process that makes carriers available in the quantum well for direct recombination, and (c) an additional amplitude reduction effect tied directly to the capture process through the time constant $\omega\tau_{cap}$. As an estimate, the corner frequency for the 6-db/octave roll-off is 31 GHz for $\tau_{cap} = 5$ ps.

A useful way to look at this is through the electric equivalent circuit shown in Figure 1, which reproduces these equations. In this figure, the various elements are,

$$C = \frac{eVol\bar{N}}{mV_T}\left[1 + \tau_{cap}\left(\frac{1}{\tau_{Nb}} + \frac{1}{\tau_{Nw}} + \frac{1}{\tau_{em}} + v_{gr}\Gamma g\bar{S}\right)\right], \tag{8}$$

$$\frac{1}{R} = \frac{eVol\bar{N}}{mV_T}\left\{\frac{1}{\tau_{Nw}} + v_{gr}\Gamma g\bar{S} + \tau_{cap}\left[-\omega^2 + \frac{1}{\tau_{Nb}}\left(\frac{1}{\tau_{Nw}} + v_{gr}\Gamma g\bar{S}\right)\right]\right\}, \tag{9}$$

$$L = \frac{mV_T}{eVol\bar{N}}\frac{\tau_{ph}}{(\beta/\tau_{Nw} + v_{gr}\Gamma g\bar{S})(1+\tau_{cap}/\tau_{Nb})(1-v_{gr}\Gamma\epsilon\bar{S}\tau_{ph})}, \tag{10}$$

$$R_d = Lv_{gr}\Gamma\epsilon\bar{S}, \tag{11}$$

$$C' = \frac{eVol\bar{N}}{mV_T}\left(\frac{\beta}{\tau_{Nw}} + v_{gr}\Gamma g\bar{S}\right)(1 - v_{gr}\Gamma\epsilon\bar{S}\tau_{ph})\frac{\tau_{cap}}{v_{gr}\Gamma\epsilon\bar{S}\tau_{ph}}, \tag{12}$$

$$R' = \frac{1}{C'}\frac{1}{v_{gr}\Gamma\epsilon\bar{S}}. \tag{13}$$

Above and beyond the modifications introduced to the conventional equivalent circuit elements[7], the capture process also introduces additional lossiness through the series R-C network. Note that unlike the effect of this process, both parasitic capacitances such as that due to bonding and inter-level dielectrics[6] and transport[8] cause a dominant pole effect, and both allow laser design to circumvent them. Employing a gain-compression coefficient in analyzing experimental data based on unmodified equations leads to an anomalously large but physically meaningless parameter. The effect of capture/emission processes is folded into an apparent gain-compression coefficient which assumes a large value.

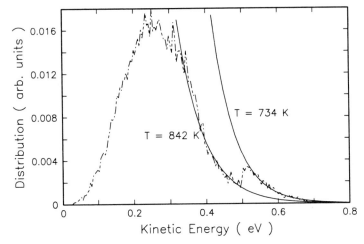

Figure 2. Carrier distribution in the first quantum-well of a triple quantum-well laser structure with a strong optical and electrical confinement.

An equally important effect of this capture and emission behavior is the large amount of carrier heating that occurs in a laser structure. Figure 2 shows the electron distribution using Monte Carlo calculations in a $Ga_{0.75}In_{0.25}As$ quantum-well with strong optical and electrical confinement that mimics the quantum-wire structure utilized in the experimental work. The carrier temperatures are very high and the consequence is a strong optical and electrical power dependent non-linearity. A time transient of this injection-capture-emission problem also shows clearly the large time constants involved in getting carriers into the well of a low-loss laser structure. Figure 3 shows the evolution of current in the triple quantum-well structure as a result of application of 0.05 V bias step. The time constant, in the presence of strong velocity overshoot that occurs in such structures at the high forward bias, can exceed 10 ps in poorly designed structures and causes an intrinsic corner frequency of 30 GHz.

3. Practical Limitations: Parasitics

The injection of the carriers into the quantum wire region of a structure grown in a V-groove occurs through carriers injected directly into the lasing region as well as laterally through the thinner quantum-well sloped wall regions. One significant advantage of this transport is that carriers to the reduced number of states in the quantum-wire regions are not only available through capture processes due to carriers being directly injected into the wire regions but also laterally from the confined states along the sloped wall where they can flow in over a significantly larger area. However, this has a significant parasitic effect. A large forward bias exists in the structure leading to a large capacitance along the sloped walls of the structure. The long cavity structure, inherently required due to the small gain volume, results in strong electrical transmission-line effect which is dominated by the inductive-capacitive loading. For a cavity of 200 μm long cavity, we estimate that the transmission-line effects will limit the 3-dB frequencies to below 20 GHz. An example

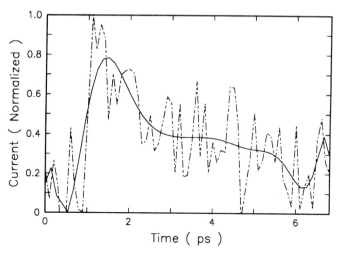

Figure 3. A time-transient due to a step voltage pulse of 0.05 V in the laser structure modeled in the previous figure.

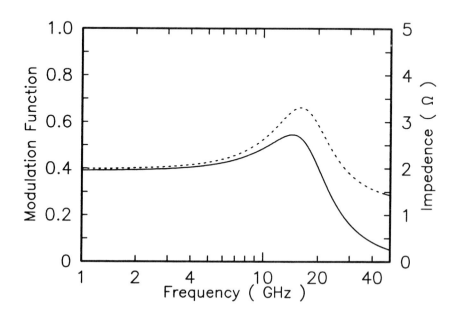

Figure 4. Modulation function assuming absence of any capture/emission effect in a 200 μm long laser cavity assuming the parasitics of experimental devices.

of the response function in a 200 μm long cavity is shown in Figure 4. The transmission-line effects limit the 3-dB frequency to approximately 22 GHz.

Conclusions

Fundamental and practical limitations constrain the usefulness of quantum-wire lasers to similar frequencies as their uni-directionally confined counterparts. Carrier dynamics in the confined regions cause reduction in the amplitude of the response and an amplified dispersive effect. The advantages of an increase in differential gain towards higher operating frequency are largely negated by increase in gain compression. The geometries that allow useful quantum-wires to be fabricated lead to a strongly loaded transmission line with additional frequency constraints.

Acknowledgments

Discussions and collaborations with Bob Davis, Massimo Fischetti, Dan Kuchta, Steve Laux, Keith Milkove, Dave Pettit, and Jerry Woodall have been crucial to the experimental and theoretical understanding summarized here.

References

[1] Tiwari S and Woodall J M 1994 *Appl. Phys. Lett.* **64** 2211–2213

[2] Tiwari S., Pettit G D, Milkove K R, Legoues F, Davis R J and Woodall J M, 1994 *Appl. Phys. Lett.* **64** 3536–3538

[3] Esquivias I, Weisser S, Schönfelder A, Ralston J D, Tasker P J, Larkins E C, Fleissner J, Benz W and Rosenzwig J 1993 *20th International Symposium on GaAs and Related Compounds*

[4] Arakawa Y and Yariv A 1986 *IEEE Trans. on Quantum Electron.* **22** 1887–1899

[5] Suemune I, Coldren L A, Yamanishi M and Kan Y 1988 *Appl. Phys. Lett.* **53** 1378–1380

[6] Olshansky R, Hill P, Lanzisera V and Powazinik W 1987 *IEEE Trans. on Quantum Electron.* **23** 1410–1418

[7] Katz J, Margalit S, Harder C, Wilt D and Yariv A 1981 *IEEE Trans. on Quantum Electron.* **17** 4–7

[8] Nagarajan R, Fukushima R, Ishikawa M, Bowers J E, Geels R S and Coldren L A 1992 *IEEE Photon. Techn. Lett.* **4** 121–123

[9] Kan S C, Vassilovski D, Wu Ta C, Kam Y L 1992 *Appl. Phys. Lett.* **61** 752–754

Fundamental Limits for Linearity of Cable TV Lasers

Vera B. Gorfinkel[a] and Serge Luryi[b, c]

[a]University of Kassel, Kassel, Germany
[b]AT&T Bell Laboratories, Murray Hill, NJ 07974
[c]*Present address:* Electrical Engineering Dept, SUNY at Stony Brook, NY 11794-2350

Abstract. We have considered nonlinear distortions in a subcarrier multiplexed optical transmission system arising from carrier heating effects in the source laser. We included two sources of carrier heating: free-carrier absorption of the coherent radiation in the cavity and the power flux into the active layer, associated with the input current. Simple analytic expressions have been derived for the intermodulation distortion in multi-channel single-mode lasers. For an 80-channel 60-540 MHz system with 3% optical modulation depth per channel, the composite second order distortion brought about by the hot-carrier effects is near the tolerance limit for CATV systems.

1. Introduction

Sub-carrier multiplexed (SCM) optical transmission systems, such as cable television (CATV) networks, require low-distortion distributed feedback (DFB) lasers. The amplitude modulation (AM) frequency division multiplexing (FDM) scheme, the preferred SCM technique in CATV networks, presents severe requirements to the linearity of the transmission. Typically, the composite second-order distortion must not exceed $-60\,\text{dBc}$. Channel capacity in current systems is about 40 channels, but future networks are expected to carry much larger number of channels [1]. Linearity specifications for an 80-100 channel system become extremely stringent.

The dominant intermodulation distortion (IMD) is commonly attributed to the nonlinearity of the light-current characteristics arising from a spatially inhomogeneous distribution of the optical field. Other sources of nonlinearity, such as the leakage current and the intrinsic (resonance) distortion are considered relatively small [2]. However, inhomogeneous field distribution is not fundamentally inherent to DFB lasers. Thus, lasers with a practically homogeneous field distribution have been obtained [3] by introducing two $\lambda/8$ phase shifts in the DFB grating. It is therefore reasonable to discuss *fundamental* limitations to the linearity of analog lasers without the spatial hole burning.

Two fundamental sources of nonlinearity have been discussed in the literature: the intrinsic (resonance) distortion [4] and the clipping of the multiplex signal at the laser threshold [5]. The clipping distortion has been extensively studied [6-9] and accurate formulae are available. As to the intrinsic distortion, it is sometimes claimed to be small in the CATV frequency range (≤ 600 MHz). However, since the intrinsic IMD decreases with the increasing resonance frequency f_r of the laser as $(f/f_r)^2$ [4], it becomes small only at sufficiently high bias currents. On the other hand, increasing the steady-state bias introduces additional nonlinearities that account for the observed minimum in the dependence of the CSO on the bias current [10]. Whether or not the intrinsic distortion is "small" depends on the position of the minimum, and therefore on the nature of the additional nonlinearity.

Presently, we consider another fundamental nonlinearity, arising from the *free-carrier absorption* of the coherent radiation in the laser cavity. Although by itself this effect is linear, it becomes a source of nonlinearity when one takes into account the *carrier heating* by the coherent radiation and the input current. Resulting from the very existence of a modulated optical signal, these effects cannot be avoided in any laser structure. Extending the standard model of laser dynamics to take into account the carrier heating effect, we derive simple analytic expressions both for the static nonlinearity of the light-current characteristics and for the CSO distortion in an arbitrary AM-FDM SCM system with any (not necessarily random) distribution of carrier phases. For an exemplary random-phase 80 channel 540 MHz system, based on a 1.5 μm InGaAsP DFB laser, we show that these fundamental nonlinearities alone, even without the inclusion of clipping, account for an IMD near the limit of acceptable.

2. Model

We describe the laser by the standard system of rate equations for carrier (n) and photon (S) densities in the active layer, combined with an energy balance equation for carriers:

$$\frac{dn}{dt} = J - gS - \frac{n}{\tau_{sp}} ; \tag{1a}$$

$$\frac{dS}{dt} = \Gamma S (g - \alpha n) - \frac{S}{\tau_{ph}} , \tag{1b}$$

$$\frac{3}{2} \frac{dT_e}{dt} = \frac{\alpha S \hbar \Omega}{2} + \frac{J \Delta}{2n} - \frac{3}{2} \frac{T_e - T}{\tau_\varepsilon} \tag{1c}$$

where $J \equiv I/eV_a$ is the electron flux per unit volume V_a of the active layer, I the pumping current, g the optical gain in the active layer, Ω the optical frequency, Γ the confinement factor for the radiation intensity, τ_{sp} the lifetime of carriers, τ_{ph} the photon lifetime in the cavity, T_e the carrier temperature in energy units, τ_ε the energy relaxation time, $\alpha > 0$ is the free carrier absorption rate and Δ is the kinetic energy per carrier injected into the active region.

We assume that the carrier ensembles are at all times described by a quasi-equilibrium distribution, parameterized by the effective temperature T_e, which we moreover assume to be the same for both electrons and holes. Validity of this approximation relies on the rapidity of the electron-electron and electron-hole scattering rates, compared to all other characteristic times in the problem, including energy relaxation. The gain function g is then of the form

$$g(T_e, n, \Omega) = g_{max} (f_e + f_h - 1) , \tag{2}$$

where f_e and f_h are the Fermi functions of electrons and holes, respectively, at energies selected by the incident photons $\hbar\Omega$.

We consider a multiple QW laser with the radiative transition between the heavy-hole and the lowest electron subbands near the fundamental absorption edge. In this case:

$$f_e(n_S, T_e) = 1 - e^{-\pi \hbar^2 n_S / m_e T_e} ; \tag{3a}$$

$$f_h(n_S, T_e) = 1 - e^{-\pi \hbar^2 n_S / m_h T_e} , \tag{3b}$$

where m_e and m_h are, respectively, the effective masses of electron and holes, and $n_S = n\, d_{QW}$ is the sheet carrier concentration in the QW. Assumed parameters of the laser at $T = 300\,K$ are listed in Table 1 below.

Table 1.   **Assumed Parameters (Multiple QW Laser)**

Symbol	Value	Comment
$\hbar\Omega$	0.8 eV	$\lambda = 1.55\,\mu m$
(m_e, m_h)	$(0.041, 0.5)\, m_0$	(electron, hole) eff. mass
g_{max}	$4.0 \cdot 10^3\,cm^{-1}$	Eq. (2)
d_{QW}	70 Å	7 wells
V_a	$12.25\,\mu m^3$	$250 \times 1 \times 0.05$
Γ	0.04	confinement factor
Δ	300 meV	$(\Delta E_G)_{eff}$
α	$0.19 \cdot 10^{-6}\,cm^3/s$	$\alpha_0 = 25\,cm^{-1}$ at $p = 10^{18}\,cm^{-3}$
τ_{ph}	3.34 ps	photon lifetime
τ_ε	2 ps	energy relaxation time
P	$\eta V_a\,\hbar\Omega S/\Gamma\tau_{ph}$	output optical power
η	0.5	$S \to P$ efficiency

Henry et al. [11] cite $\alpha_0 = 25\,cm^{-1}$ at $p = 10^{18}\,cm^{-3}$ and $\lambda = 1.6\,\mu m$. The energy Δ depends on the fraction of carriers that get into the active region by tunneling; in general, $\Delta \leq \Delta E_G$ between the cladding and the active layers. For InP cladding ($E_G = 1.35\,eV$) and 1.55 μm active layer, $\Delta E_G = 0.55\,eV$; we take a conservative estimate $\Delta = 0.3\,eV$. Experimental estimates for τ_ε vary widely; recently Hall et al. [13] found a time constant of ~ 1 ps in strained-layer quantum well 1.5 μm laser amplifiers. We take $\tau_\varepsilon = 2\,ps$, which is quite modest compared to calculations (see, e.g., [12]) that include the phonon bottleneck effect, resulting in a still longer τ_ε at high carrier concentrations. Our assumed values of Δ and τ_ε result in a steady-state overheating of $T_e - T \approx 4\,K$ at the laser threshold. The carrier lifetime $\tau_{sp}(n)$ will be assumed in the form [14]: $\tau_{sp}^{-1}(n) = A + Bn + Cn^2$, with the coefficients A, B, C chosen so as to produce $\tau_{sp} = 0.5\,ns$ at threshold.

3. Dynamic Characteristics

We shall denote the static values by the subscript 0, and time-dependent terms by an index $i = 1\ or\ 2$, with the increasing refinement of the approximation:

$$S = S_0 + S_1 + S_2\,; \qquad T_e = T_{e0} + T_{e1} + T_{e2}\,; \qquad n = n_0 + n_1 + n_2 \qquad (4)$$

The input current is of the form $J = J_0 + J_1$. The signal $S_1(t)$ is the first order modulated response of the laser. The pump current signal contains N channels at frequencies $\omega_{K_1} = \omega_0 + K_1 \Delta\omega$, $K_1 = 1, 2, \cdots, N$, viz.

$$J_1 = \mathcal{Re}\left[\sum_{K_1=1}^{N} a_{K_1} \exp(i\theta_{K_1}) \exp(i\omega_{K_1} t)\right] \equiv \sum_{K_1=1}^{N} \mathcal{Re}[J_{K_1} \exp(i\omega_{K_1} t)]\,, \qquad (5)$$

where $a_{K_1}(t)$ is the (real) modulated amplitude, $J_{K_1} \equiv a_{K_1} \exp(i\theta_{K_1})$, and θ_{K_1} is the phase of K_1-th channel signal. In a CSO testing setup the amplitudes are kept constant, $a_{K_1} \equiv a$, and the optical modulation depth is defined by $a \equiv \mathcal{OMD} \cdot (J_0 - J_{th})$.

The second order variations S_2, n_2 occur at frequencies $\omega_{K_2}^{(\pm)} = |\omega_{K_1} \pm \omega_{M_1}|$, where $K_1, M_1 = 1, 2, \cdots, N$. The set of channel numbers $\{K_1, M_1\}$ contributing a second order beat into a given channel K_2 will be denoted by $[K_2]$ (with the subsets $[K_2^{(+)}]$ and $[K_2^{(-)}]$ contributing the sum and the difference beats, respectively.)

Inasmuch as the energy relaxation is much faster (by two to three orders of magnitude) than the characteristic frequencies in the problem, $\tau_\varepsilon \ll \omega^{-1}$, we shall neglect the small frequency dependence of T_e and assume that T_e follows S and J instantaneously,

$$T_{ei} = \gamma_S S_i + \gamma_J J_i, \quad i = 1, 2, \tag{6}$$

where $\gamma_S \equiv \alpha \tau_\varepsilon \hbar\Omega/3$ and $\gamma_J \equiv \tau_\varepsilon \Delta/3\bar{n}$. Assuming that nonlinear effects are small, $n_2 \ll n_1$ and $S_2 \ll S_1$, we solve Eqs. (1) by successive approximations. A compact analytic expression obtains if we neglect compared to unity the following terms:

$$\frac{\omega}{\Gamma g_{th}} \approx \omega \tau_{ph} ; \tag{7a}$$

$$\frac{\omega^2 \tau_{ph}}{S_0 g'_n} \equiv \frac{\omega^2}{\omega_r^2} ; \quad \frac{\omega \tau_{sp}^{-1} \tau_{ph}}{S_0 g'_n} \equiv \frac{\omega \tau_{sp}^{-1}}{\omega_r^2} . \tag{7b}$$

These approximations are valid sufficiently far from the threshold, at frequencies much lower than the resonant frequency of the laser, defined by $2\pi f_r = \omega_r$, where

$$\omega_r^2 \equiv \frac{S_0 (g'_n - \alpha)}{\tau_{ph}} . \tag{8}$$

Under these conditions we find the following expression for the CSO distortion in the presence of carrier heating effects:

$$\frac{|S_{K_2}|}{S_{K_1}} = \frac{OMD}{4} \left[\left(\frac{\omega_{K_2}^2}{\omega_r^2} + \xi S_0 \right)^2 + \left(\frac{\omega_{K_2}}{\omega_r^2 \tau_{eff}} \right)^2 \right]^{1/2} |\Sigma[K_2]|, \tag{9}$$

where $\tau_{eff}^{-1} \equiv A + 2Bn_{th} + 3Cn_{th}^2$, and ξ is the static nonlinearity coefficient (zero-th order equations give $\xi \approx -g_{th}^2 \, d^2S_0/dJ_0^2$)

$$\xi \equiv 2\alpha(\tau_S + \tau_J) + 2g_{th}(B + 3Cn_{th})(\tau_S + \tau_J)^2, \tag{10}$$

where
$$\tau_S = \frac{\alpha n_{th}}{g_{th}} \frac{\tau_\varepsilon \cdot \hbar\Omega}{3T_{th}} ; \quad \tau_J = \frac{\tau_\varepsilon \cdot \Delta}{3T_{th}} . \tag{11}$$

The sums $\Sigma[K_2]$ are given by

$$\Sigma[K_2^{(+)}] = \sum_{K_1 > M_1}^{[K_2^{(+)}]} \exp[i(\theta_{K_1} + \theta_{M_1})] ; \tag{12a}$$

$$\Sigma[K_2^{(-)}] = \sum_{K_1 > M_1}^{[K_2^{(-)}]} \exp[i(\theta_{K_1} - \theta_{M_1})] . \tag{12b}$$

Note that all terms appearing in the sums (12) are independent of the first-order frequencies that combine in ω_{K_2} and depend only on the phases of the components. If the system were designed to control the subcarrier phases, it would be possible to reduce the CSO distortion by a suitable selection of the set $\{\theta_{K_1}\}$; thus, it is feasible to suppress the sum beats in high-frequency channels at the expense of an enhancement in difference beats which are dominant in low-frequency channels.

In a typical experimental arrangement, the CSO is measured with random phases θ_{K_1}, generated from a thermal source. Averaging $|\Sigma|$ over the random phases, we find

$$\overline{|\Sigma[K_2^{(+)}]|^2} = \sum_{K_1 > M_1}^{[K_2^{(+)}]} 1 \equiv \mathcal{N}[K_2^{(+)}] \; ; \qquad \overline{|\Sigma[K_2^{(-)}]|^2} = \sum_{K_1 > M_1}^{[K_2^{(-)}]} 1 \equiv \mathcal{N}[K_2^{(-)}] \; ;$$

The number of pairs $\mathcal{N}[K_2^{(-)}]$ contributing to difference beats decreases linearly from 70 at 60 MHz to 0 in channels with $\omega_{K_2}^{(-)} \geq 480$ MHz. For the sum beats, $\mathcal{N}[K_2^{(+)}]$ increases from 0 for $\omega_{K_2}^{(+)} \leq 120$ MHz to 35 at the high end.

Figure 1 shows the calculated CSO for an exemplary laser. Note that at high bias currents the low-frequency channel (difference beat) has a larger distortion than the high-frequency channel (sum beat), hence the curve crossing in Fig. 1. This phenomenon, often seen experimentally, occurs because the number of contributing channel pairs is twice larger for the low-frequency channel. Increasing current increases the resonant frequency and in the limit of $\omega_r \to \infty$ the only dependence on the channel frequency is contained in $\mathcal{N}[K_2]$. In this limit, the difference between the two curves tends to 3 dB.

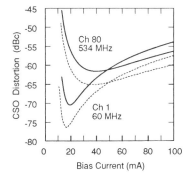

Fig. 1. Composite second order distortion in a 80-channel analog laser. Laser parameters as in Table 1, $OMD = 3\%$.
Carrier lifetime $\tau_{sp}(n_{th}) = 0.5$ ns results from $B \approx 3.8 \cdot 10^{-10}$ cm^3/s and $C \approx 1.5 \cdot 10^{-28}$ cm^6/s, assuming $A = 0$. For comparison, broken lines show the result at constant $\tau_{sp} = A^{-1} = 0.5$ ns. Note that τ_{eff} resulting from the variable $\tau_{sp}(n)$ is shorter than $\tau_{sp}(n_{th})$; in our example, $\tau_{eff} = 0.2$ ns.
The minimum in Ch.80 equals -61.6 dBc and occurs at $I \approx 40$ mA (when $f_r \approx 7$ GHz).

From Eq. (9) the minimum CSO distortion in a channel of frequency ω_{K_2} equals

$$\left[\frac{|S_{K_2}|}{S_{K_1}}\right]_{min} = \frac{OMD \cdot \omega_{K_2}}{2\sqrt{2}} \left[\frac{\tau_{ph}\,\xi}{g_n' - \alpha}\left(1 + \sqrt{1 + (\omega_{K_2}\tau_{eff})^{-2}}\right) \mathcal{N}[K_2]\right]^{1/2}.$$

We see that for a given number of channels, the hot-carrier distortion increases linearly with the OMD. This is in contrast to the clipping distortion, which is exponential [5-9] and hence must dominate at high enough OMD. However, for $OMD = 0.03$ which is the case we considered, the clipping distortion is immaterial (< 100 dBc for all channels).

Note that enhancement of g_n' by a factor of 2 decreases the minimum distortion by about 3 dB. Clearly, increasing g_n' is the key strategy for improving CATV lasers.

References

[1] M. Ishino, M. Kito, S. Yamane, K. Fujihara, N. Otsuka, H. Sato, and Y. Matsui, "1.3 µm strained-layer MQW-DFB laser with low noise and low distortion characteristics for 100-channel CATV transmission", IEDM'93 *Tech. Digest*, pp. 613-616 (1993).

[2] H. Yonetani, I. Ushijima, T. Takada, and K. Shima, "Transmission characteristics of DFB laser modules for analog applications", *IEEE J. Lightwave Technol.* **11**, pp. 147-153 (1993).

[3] U. Cebulla, J. Bouayad, H. Haisch, M. Klenk, G. Laube, H. P. Mayer, R. Weimann, and E. Zielinski, "1.55 µm strained layer multiple quantum well DFB-lasers with low chirp and low distortions for optical analog CATV distribution systems", CLEO'93 *Tech. Digest* **11**, pp. 226-227 (1993).

[4] T. E. Darcie, R. S. Tucker, and G. J. Sullivan, "Intermodulation and harmonic distortion in InGaAsP lasers", *Electron. Lett.* **21**, pp. 665-666 (1985).

[5] A. A. M. Saleh, "Fundamental limit on number of channels in subcarrier-multiplexed lightwave CATV system", *Electron. Lett.* **21**, pp. 776-77 (1989).

[6] T. E. Darcie, "Subcarrier multiplexing for lightwave neetworks and video distribution systems", *IEEE J. Select. Areas Commun.* **8**, pp. 1240-1248 (1990).

[7] K. Alameh and R. A. Minasian, "Ultimate limits of subcarrier multiplexed transmission", *Electron. Lett.* **27**, pp. 1260-1262 (1991).

[8] Q. Shi, R. S. Burroughs, and D. Lewis, "An alternative model for laser clipping-induced nonlinear distortion for analog lightwave CATV systems", *IEEE Photon. Technol. Lett.* **4**, pp. 784-787 (1992).

[9] N. J. Frigo, M. R. Phillips, and G. E. Bodeep, "Clipping distortion in lightwave CATV systems: models, simulations, and measurements", *IEEE J. Lightwave Technol.* **11**, pp. 138-146 (1993).

[10] H. Kawamura, K. Kamite, H. Yonetani, S. Ogita, H. Soda, and H. Ishikawa, "Effect of varying threshold gain on second-order intermodulation distortion in distributed feedback lasers", *Electron. Lett.* **26**, pp. 1720-1721 (1990).

[11] C. H. Henry, R. A. Logan, F. R. Merritt, and J. P. Luongo, "The effect of intervalence band absorption on the thermal behavior of InGaAsP lasers", *IEEE J. Quant. Electron.* **QE-19**, pp. 947-952 (1983).

[12] C. Y. Tsai, L. F. Eastman, anfd Y. H. Lo, "Hot carrier and hot phonon effects on high-speed quantum well lasers", *Appl. Phys. Lett.* **63**, pp. 3408-3410 (1993).

[13] K. L. Hall, G. Lenz, E. P. Ippen, U. Koren, and G. Raybon, "Carrier heating and spectral hole burning in strained-layer quantum-well laser amplifiers at 1.5 µm", *Appl. Phys. Lett.* **61**, pp. 2512-2514 (1992).

[14] G. P. Agrawal and N. K. Dutta, *Semiconductor Lasers*, 2nd edition, (Van Nostrand Reinhold, New York 1993) p. 124.

Mechanical stress in AlGaAs ridge lasers: its measurement and effect on the optical near field

P W Epperlein, G Hunziker, K Dätwyler, U Deutsch, H P Dietrich and D J Webb

IBM Research Division, Zurich Research Laboratory, CH-8803 Rüschlikon, Switzerland

Abstract. We report stress measurements on $Al_{0.04}Ga_{0.96}As/Al_{0.44}Ga_{0.56}As$ single quantum well laser diodes with 5-μm-wide ridge waveguides. Raman scattering and photoluminescence microprobe spectroscopy have been employed to measure the stress distribution in the ridge region on the cleaved, uncoated (110) mirror facets and to identify the stress sources. The camel hump-like stress profiles along the active layer show compressive stress with maximum amplitudes of 4 kbar near the ridge slopes and with reduced amplitudes close to the ridge center. Both the ridge-embedding Si_3N_4 layer and the top p-metallization introduce strain. Model experiments with externally applied stress of up to ≈ 1 kbar demonstrated the high sensitivity of the laser near field and threshold current to stress fields. They also allowed the measurement of low built-in stress amplitudes as small as 50 bar.

1. Introduction

The distribution of mechanical stress in semiconductor laser waveguide structures is a subject of great perennial interest, because stress may be of various practical importance for laser operation [1]. Thus, quantum well (QW) lasers with elastically built-in stress can have performance characteristics, including reliability, that are substantially superior to those of unstrained devices. On the other hand, stress may be the driving force for a variety of degradation phenomena [2] and may modify the laser waveguide modes. In this paper, we report stress measurements on $Al_{0.04}Ga_{0.96}As/Al_{0.44}Ga_{0.56}As$ single quantum well (SQW) graded-index separate-confinement heterostructure (GRINSCH) laser diodes with 5-μm-wide ridge waveguide structures. Raman scattering and photoluminescence (PL) spectroscopy have been employed to measure the stress distribution in the ridge region of cleaved (110) mirror facets with high spatial resolution (≈ 1 μm) at room temperature and to identify the stress sources. Results of model experiments with externally applied stress will also be discussed.

2. Experimental

$Al_{0.04}Ga_{0.96}As/Al_xGa_{1-x}As$ lasers were grown by molecular beam epitaxy on n^+-doped, (100) GaAs substrates with an 8-nm-thick $Al_{0.04}Ga_{0.96}As$ SQW sandwiched between 100-nm graded regions with linear Al concentration profiles ($x = 0.24 - 0.44$) and adjacent ≈ 1.8-μm-thick $Al_{0.44}Ga_{0.56}As$ cladding layers. $H_2SO_4 : H_2O_2 : H_2O$ etch solutions were used to form the ridge-waveguide stripes with typical widths of 5 μm along (110) directions on the (100) surface. The lateral slopes of the ridge are inclined and usually show a slight concave upward curvature. Si_3N_4 used for insulation was LF-plasma deposited with a thickness of $\approx$200 nm on the ridge region up to the upper edges, leaving a narrow stripe on top of the ridge free of the dielectric. A $\approx$400-nm-thick TiPtAu metallization overlaying the entire structure formed the top ohmic contact. The (110)-oriented facets of the 5-μm-wide and 750-μm-long lasers were formed by cleaving in air and were not coated. The laser devices were soldered junction-side-up on a heat sink. Details of laser fabrication have been described elsewhere [3].

Raman and photoluminescence (PL) microprobe spectroscopy have been used to characterize the stress on the ridge-waveguide facets. Raman scattering is known to have a detection limit of only $\approx$500 bar, and therefore is limited to areas with larger stress. The Raman microprobe spectra were measured on the (110) facets in air and at room temperature in the backscattering configuration by using the 457.9 nm line of an Ar^+ laser as excitation source with power densities of the order of $\leq$100 kW/cm^2 on the sample surface. This leads to temperature increases of only $\approx$10 K [4]. In this configuration the longitudinal-optical (LO) phonon mode is forbidden, but the transversal-optical (TO) mode is allowed, in accordance with the symmetry selection rules. This means that the spectra will show only the GaAs- and AlAs-like TO phonon modes when the probe laser spot hits the AlGaAs of the (110) mirrors [4]. We used the GaAs-like TO mode as the optical probe for sampling the stress distribution on the laser mirrors. The low self-heating of the probe laser has no detectable influence on the Raman phonon spectra and therefore will not impact the stress measurements. Spatial resolution was achieved by focusing the laser beam to a spot diameter of $\approx$1 μm. Positioning and aligning were realized by mounting the samples on a high-precision (0.1 μm) x, y, z-translator stage and observing the structure with a video camera and monitor under high magnification ($> 10^4 \times$). After appropriate dispersion in a 0.9-m double-grating spectrometer, the Raman-scattered light was detected by a cooled photomultiplier with GaAs cathode and single-photon counting electronics.

The micro-PL measurements on the laser mirrors were performed with the same spectroscopy setup. Here we used the $n = 1$ e-hh exciton transition in the active QW as the optical probe. This allowed a direct measurement of the relative lateral stress distribution in the active layer. Compared to Raman scattering, PL also has a higher sensitivity ($\approx 5\times$), which can be expressed from a practical point of view as the ratio between the pressure coefficient of the peak energy and the spectral line width of the transition involved.

3. Results and discussion

Raman spectra in Fig. 1 show the first-order Stokes GaAs-like TO-phonon mode detected in the AlGaAs several micrometers outside the ridge region at a peak position of

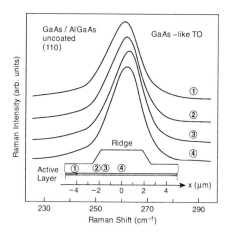

Figure 1. Raman spectra from a (110) AlGaAs ridge laser facet, (1) outside ridge region, (2) and (3) near ridge slopes and (4) at ridge center.

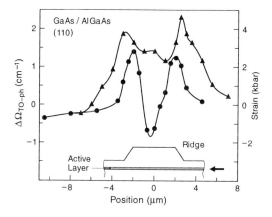

Figure 2. GaAs-like TO-phonon energy shift and corresponding strain profiles as measured along the active layer (see arrow in inset) of two different 5-μm-wide ridge lasers. Measurement points are connected by smoothed curves as a guide to the eye. Inset shows schematic diagram of ridge profile.

261 cm^{-1}. This point is far away from the strained region around the ridge and therefore represents the reference point. Figure 1 clearly exhibits a shift of this mode to higher energies when approaching the ridge along the active layer. Typical maximum shifts of 1.5 cm^{-1} can be found close to both ridge slopes. Inside the ridge the Raman mode shifts to lower energies with a minimum in the center.

Figure 2 depicts the mode shift profiles of two different laser devices. Both profiles are similar in overall shape, but are differently well-pronounced at the ridge slopes and ridge center. Using the known pressure dependence of the TO-mode energy shift [5], the experimental mode shift profiles were converted to stress profiles. The camel hump-like

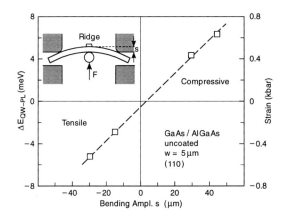

Figure 3. Dependence of the GRINSCH-QW $e-hh$ exciton energy shift ΔE and corresponding strain figures on the laser bar bending amplitude s. Size of squares indicates error bars for ΔE and s. Probe laser spot was positioned on (110) facet at center of ridge on active layer in the PL measurements. Inset shows schematic diagram of the three-point laser bar bending device.

stress profiles along the active layer show high compressive stress of up to 4 kbar near the ridge slopes. Usually, the compressive stress is reduced close to the ridge center, but sometimes low-tensile stress fields can be detected towards the center. The stress profiles strongly change in shape and strength from laser to laser (see Fig. 2). PL measurements led to the same camel hump-like stress profiles, but proved to be superior to Raman scattering because of their higher sensitivity (see Section 2).

Measurements of special samples with (i) just the ridge structure, (ii) the ridge structure with the overlaid insulating Si_3N_4 ridge-embedding layer, and (iii) the ridge structure with the top metallization layer revealed that both the nitride and metallization layer can introduce stress.

To investigate the effect of stress on laser operation, we performed model experiments in which stress can be applied externally by bending bars (a parallel array of laser chips) in a mechanical instrument (see inset of Fig. 3). The effective length of the bar in the bending experiment is 6 mm, the width is 750 μm, and thickness $\approx$150 μm. Figure 3 exhibits the achievable strain magnitudes. Here the relative QW-PL energy shift ΔE is plotted as a function of bending amplitude s with the probe laser spot positioned in the center of the ridge. The energy shift has been converted to stress amplitudes by using $\approx$10 meV/kbar for the pressure coefficient of the QW-PL energy [6]. Maximum bending amplitudes of $\approx$40 μm (fracture limit), corresponding to $\approx$0.6 kbar, have been achieved. As demonstrated in Fig. 3, the bending technique enables the measurement of built-in stress amplitudes (at $s = 0$) as small as 50 bar.

As shown in Fig. 4, the lateral near-field pattern is strongly affected by external stress. Here tensile stress was induced by the bending technique at the bar surface, where the lasers are located. As the two-dimensional transverse stress profile is unknown, the near field cannot be compared with theoretical results. Nevertheless, we were able to rule out some effects by modelling the laser structure [7]. Thus, the effect on the mode

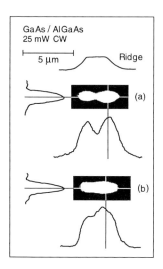

Figure 4. Near-field pattern of an AlGaAs ridge laser (a) without external strain and (b) with tensile strain (≈ -0.6 kbar) produced by the laser bar bending technique.

of the mismatch of the photoelastic coefficients between GaAs and Si_3N_4 could be ruled out. The refractive index step variation with ($\Delta p = -1$ kbar) and without external stress was estimated to be $\delta(\Delta n) \approx 0.014$ by using $\partial n/\partial p = -4 \times 10^{-6}$ bar^{-1} for GaAs [8] and $\partial n/\partial p = +1 \times 10^{-5}$ bar^{-1} for Si_3N_4 [9]. However, the intensity of the mode in these outer regions is so small that the near field remains unaffected. The presence

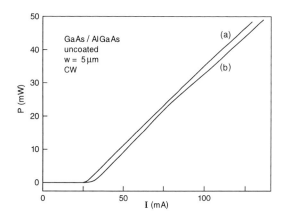

Figure 5. Light (P)-current (I) characteristic of the AlGaAs laser in Fig. 4 (a) without external stress and (b) with tensile stress (≈ -0.6 kbar) produced by the laser bar bending technique.

of a higher order lateral mode shown in the near-field pattern of Fig. 4a was confirmed by the modelling [7]. The modification of the near-field pattern due to the bending is accompanied by a modification of the differential efficiency around 25 mW (see also Fig. 5b). Our qualitative understanding of this is as follows: the lateral effective index profile is modified asymmetrically in the central regions of the ridge as a result of bending the bar. Therefore, the position and intensity distribution in the two lobes of the higher order mode are laterally shifted. This leads to an effective reduction of the width of the total intensity distribution (Fig. 4b), whereby the internal losses of the waveguide are modified and the differential quantum efficiency is changed (Fig. 5b).

A reversible increase of the threshold current of up to 15% was observed for tensile stress superposition in the bending experiment (Fig. 5b). As a consequence of the shift of the light-hole and the heavy-hole bands, this result can be expected for the transition from the usually compressive-strained laser to one that is more of tensile-strained nature.

These model experiments clearly demonstrated the high sensitivity of laser performance characteristics to mechanical stress fields.

4. Summary

Raman scattering and photoluminescence microprobe spectroscopy measurements on AlGaAs ridge lasers revealed strong compressive strain fields of up to 4 kbar near the ridge slopes. The center of the ridge is usually under reduced compressive or even tensile strain. The strain fields arise from the top p-metallization layer as well as from the ridge-embedding Si_3N_4 layer. Laser near-field pattern and light-current characteristics are highly sensitive to strain fields as demonstrated in model experiments by employing a three-point laser bar bending technique.

Acknowledgment

We would like to thank H.P. Meier for the MBE growth of the laser structures.

References

[1] For a review, see Thijs P J A 1992 *Proc. 13th IEEE Int. Semiconductor Laser Conf., Takamatsu, Japan* (Tokyo: Business Ctr. for Academic Societies Japan) p 2.
[2] For a review, see Ueda O 1988 *J. Electrochem. Soc.* **135** 11C.
[3] Harder C, Buchmann P and Meier H 1986 *Electron. Lett.* **22** 1081.
[4] Epperlein P W, Buchmann P and Jakubowicz A 1993 *Appl. Phys. Lett.* **62** 455.
[5] Trommer R, Mueller H, Cardona M and Vogl P 1980 *Phys. Rev. B* **21** 4869.
[6] Gell M A, Ninno D, Jaros M, Wolford D J, Kuech T F and Bradley J A 1987 *Phys. Rev. B* **35** 1196.
[7] Hunziker G and Harder C 1994 submitted to *Appl. Optics*.
[8] Samara G A 1983 *Phys. Rev. B* **27** 3494.
[9] Okada Y and Nakajima S 1991 *Appl. Phys. Lett.* **59** 1066.

Highly Efficient Light-Emitting Diodes with Microcavities

E. F. Schubert and N. E. J. Hunt

AT&T Bell Laboratories Murray Hill, NJ 07974

One-dimensional microcavities are optical resonators with coplanar reflectors separated by a distance on the order of the optical wavelength. Such structures quantize the energy of photons propagating along the optical axis of the cavity and thereby strongly modify the spontaneous emission properties of a photon-emitting medium inside of a microcavity. This report concerns semiconductor light-emitting diodes with the photon-emitting active region of the light-emitting diodes placed inside a microcavity. It is shown that these devices have strongly modified emission properties including experimental and theoretical emission efficiencies are more than a factor of 5 and 10 times higher than in conventional light-emitting diodes, respectively.

Fabry-Perot cavities are optical resonators defined by two coplanar reflectors separated by a macroscopic distance [1]. If this distance is small, namely one half or one single wavelength of operation, i.e., on the order of one μm for visible and near infrared light, the cavities are called optical *microcavities*. The density of optical modes within a microcavity is strongly modified as compared to the mode density in free space surrounding the cavity: The cavity allows for the propagation of optical waves only at the fundamental resonance mode and its harmonics along the optical axis of the Fabry-Perot microcavity. Modes other than the fundamental resonance and its harmonics cannot exist inside the cavity. The energy of photons confined to the microcavity is thus quantized, in near perfect analogy to the quantization of electron energy inside a quantum well potential.

Intriguing physics evolves, if an optically active medium, i.e. a photon-emitting medium, is placed between the two reflectors of a coplanar Fabry-Perot microcavity. This novel field of condensed matter research has, during the past few years, made possible a new generation of microresonator devices and the exploration of quantum electrodynamic phenomena [2,3]. Several microcavity geometries have been employed including planar (Fabry-Perot) microcavities, spherical resonators, disk-shaped resonators, and fully three-dimensional resonators [2-6]. The optically active materials inside the microcavity structures included optically active rare-earth atoms, semiconductors, and dyes [7-11].

It is the purpose of this publication to demonstrate that microcavity physics can be exploited in devices which have superior properties when compared with conventional devices. We show that a new generation of light-emitting diodes which employs microcavity effects has clearly improved characteristics as compared to conventional LEDs.

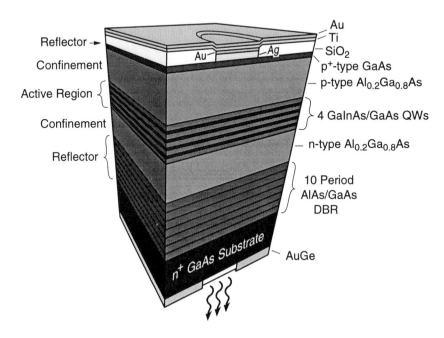

Fig. 1 Layer sequence and structure of the resonant-cavity light-emitting diode. The planar semiconductor microcavity is defined by a distributed Bragg reflector and a silver mirror. The circular silver reflector also serves as ohmic contact. Light emission occurs through the substrate coated with an anti-reflection film.

Of particular interest for the technological application of microcavity effects is the enhancement and suppression of radiation. This interest is based on the fact that optical emission rate in microcavities along the cavity axis can be higher under certain experimental conditions which will be defined below. To illustrate this effect we consider the semiconductor microcavity shown in Fig. 1, which consists of two reflectors namely a silver (Ag) mirror and a distributed Bragg reflector (DBR). The reflectivity of the semiconductor/silver interface is 96% for near-infrared light. The 90% reflective DBR consists of 10 pairs of an AlAs/GaAs multilayer structure with each layer being $\lambda/4$ thick, where λ is the resonance wavelength of the cavity. Since the DBR reflectivity is lower than the reflectivity of the silver mirror, photons will escape from the cavity mainly through the DBR. The active region of the cavity contains four $In_{0.15}Ga_{0.85}As$ quantum wells.

The enhancement of the emission intensity along the optical axis of the cavity can be calculated in terms of Fermi's Golden Rule [12]. At the resonance wavelength, the emission enhancement factor is given by [13]

$$G_e = \frac{\xi}{2} \frac{(1+\sqrt{R_2})^2 (1-R_1)}{(1-\sqrt{R_1 R_2})^2} \frac{\tau_{cav}}{\tau} \quad (1)$$

where R_1 and R_2 are the reflectivities of the mirrors and, since we assume $R_1 < R_2$, most light exits the cavity through the reflector with reflectivity R_1. The anti-node enhancement factor ξ has a value of 2, if the active region is located exactly at an antinode of the standing optical wave inside the cavity. ξ has a value of unity, if the active region is smeared out over many periods of the optical wave, and finally $\xi = 0$ if the active region is located exactly at an optical node. Calculations of the radiative lifetime τ_{cav} for the structure revealed that it is changed by less than 10% relative to the spontaneous lifetime τ, i.e. $\tau_{cav} \cong \tau$ [13,14]. Eq. (1) allows us to calculate the on-resonance enhancement and using $\xi = 1.7$ one obtains $G_e = 67$ for the structure shown in Fig. 1.

Rather than the enhancement at the resonance wavelength, the *total* enhancement integrated over all wavelength is relevant for practical devices. Exactly on resonance, the emission is enhanced along the axis of the cavity. However, off resonance, the emission is strongly suppressed. Considering that the natural emission spectrum of the active medium (without cavity) may be much broader than the cavity resonance, it is, *a priori*, not clear if the *integrated* emission is enhanced at all. To calculate the integrated enhancement, the spectral width of the cavity resonance and the spectral width of the natural emission spectrum must be determined. The former can be calculated from the uncertainty relation and the photon lifetime in the cavity, τ_{ph}, according to $\Delta E = \hbar/\tau_{ph}$ or

$$\frac{\Delta\lambda}{\lambda} = \frac{\lambda}{2nL_{cav}} \left[\frac{1-\sqrt{R_1 R_2}}{\pi(R_1 R_2)^{1/4}} \right] \quad (2)$$

where L_{cav} is the effective cavity length and n is the refractive index in the cavity. For typical reflectivities of $R_1 = 90\%$ and $R_2 = 96\%$ and an effective cavity length of $2.7(\lambda/n)$, the spectral width is $\Delta\lambda \cong 5$ nm at $\lambda = 900$ nm [13]. The theoretical width of the emission spectrum of bulk semiconductors is 1.8 kT [17], where k is the Boltzmann constant and T = 300K. The value of 1.8 kT corresponds to $\Delta\lambda_N = 31$ nm at an emission wavelength of 900 nm. Thus only a small part of the spectrum is strongly enhanced, whereas the rest of the spectrum is suppressed. The total enhancement/suppression can be calculated by assuming a gaussian natural emission spectrum [13]

$$G_{int} = G_e \sqrt{\pi \ln 2} \, (\Delta\lambda/\Delta\lambda_N) \quad (3)$$

For the above values for G_e, $\Delta\lambda$, and $\Delta\lambda_N$, we deduce an integrated enhancement of $G_{int} \cong 16$. The integrated enhancement can be quite different for different types of materials. Narrow atomic emission spectra can be enhanced by several orders of magnitude [7]. On the other hand, materials having broad emission spectra such as dyes or polymers [10] may not exhibit any integrated enhancement at all.

A significant difference exists between LEDs used for display purposes and for optical fiber communication. The figure of merit for communication LEDs is the *photon flux density* per unit solid angle. A high photon flux density allows one to couple much of the emitted light into an optical fiber. For display LEDs, the figure of merit is the *total power* emitted into the hemisphere of the observer, i.e., the total number of photons rather than the photon density is relevant. Several structures were proposed and realized to improve the efficiency of LEDs including spherical shapes [16], multiple surfaces [17], reflectors [18], and non-planar surfaces [19]. However, none of these concepts can fundamentally modify the photon flux density emanating from the active region, but are rather standard geometric optical constructions.

An LED structure which provides an enhanced photon flux density is the resonant-cavity light-emitting diode (RCLED) shown in Fig. 1. The RCLED is pn junction diode integrated with a microcavity. The fundamental mode of the cavity is in resonance with the light-emitting active region of the diode. The active region emits at 930 nm where all other materials used in the structure are transparent. Most light escapes from the cavity through the bottom reflector, since its reflectivity is lower than that of the top Ag reflector. The top silver reflector also serves as a non-alloyed ohmic contact to the heavily doped top p^+-type GaAs layer. The thicknesses of the two confinement layers are chosen in such a way to (i) have the fundamental mode of the cavity at 930 nm, and to (ii) place the active GaInAs region into an optical anti-node of the cavity. Light exits the structure through the substrate. The exit window is coated with an anti-reflection coating. AuGe surrounding the optical window provides an ohmic contact to the n^+-type GaAs substrate.

The figure of merit of LEDs used in optical fiber communication systems is the photon flux density emitted from the diode at a given current which, for a given wavelength, will be discussed in terms of µW/steradian. The optical power coupled into a fiber is directly proportional to the photon flux density.

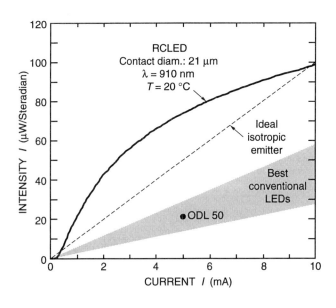

Fig. 2 Light intensity versus current for a planar RCLED emitting at 910 nm. Also shown is the theoretical curve of the ideal isotropic emitter which is an upper limit for all conventional LEDs. The shaded area indicates experimental intensities of the best conventional LEDs and of the ODL 50 which is a commercial product.

The intensity of the RCLED as a function of the injection current is shown in Fig. 2. For comparison, we also show the calculated intensity of an *ideal isotropic emitter* which is a hypothetical device. It is assumed to have a 100% internal quantum efficiency and to be clad by an a anti-reflection coating providing R = 0 for all wavelengths emitted from the active region. If the photon emission inside the semiconductor is isotropic, then the optical power per unit current per unit solid angle normal to the planar semiconductor surface is given by

$$\frac{P_{optical}}{\Omega} = \frac{1}{4\pi n^2} \frac{2\pi\hbar c}{e\lambda} \qquad (4)$$

where n is the refractive index of the semiconductor, c is the velocity of light, and λ is the emission wavelength in air. This equation is represented by the dashed line in Fig. 2. Note that neither the 100% internal quantum efficiency nor such a hypothetical anti-reflection coating can be reduced to practice for fundamental reasons. Therefore, the ideal isotropic emitter represents an upper limit for the intensity attainable with any conventional LED. In fact, the best conventional LEDs have intensities lower than the ideal isotropic emitter. Also included in Fig. 2 is the state-of-the art *ODL50* GaAs LED used for optical fiber communication [20]. All devices shown in Fig. 2 have *planar* light-emitting surfaces and no lensing is employed.

Figure 2 reveals that the RCLED provides unprecedented intensities in terms of both, absolute values as well as slope efficiencies. The slope efficiency is 7.3 times higher than the efficiency of the best conventional LEDs and 3.1 times higher than the calculated efficiency of the ideal isotropic emitter. At a current of 5 mA, the intensity of the RCLED is 3.3 times higher than that of the best conventional LEDs including the ODL50. The unprecedented efficiencies make the RCLED very promising for optical interconnect systems which will be discussed further in a subsequent section.

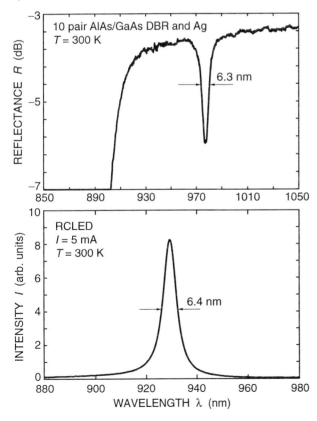

Fig. 3 Reflectivity spectrum (top) from a semiconductor microcavity and emission spectrum (bottom) from a microcavity structure. The similarity of the two spectra shows that optical emission is restricted to the fundamental cavity mode.

The RCLED presented here has been designed for an operating current of 5 mA and for usage with optical multimode fibers with a core diameter of 62.5 µm. At higher currents, the RCLED intensity starts to saturate as shown in Fig. 3. The saturation was shown to be due to band filling [21]. At operating voltages of 1.5V, the electrical power consumed by the devices is 7.5 mW. Less than 7.5 mW are converted to heat. Heating of linear arrays of devices separated by 2×62.5 µm = 125 µm is not a problem.

The reflection and emission properties of the RCLED are shown in Fig. 3. The reflection spectrum of the RCLED exhibits a highly reflective band for wavelengths >900 nm and a dip of the reflectivity at the cavity resonance. The spectral width of the cavity resonance is 6.3 nm. The emission spectrum of an electrically pumped device, shown in the lower part of Fig. 3, has nearly the same shape and width as the cavity resonance width.

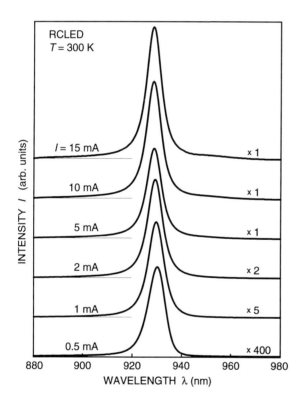

Fig. 4 Emission spectra of a resonant-cavity light-emitting diode at different excitation currents. The spectral shape is independent of the current indicating the absence of stimulated emission or superluminescence.

In conventional LEDs, the spectral characteristics of the devices reflect the thermal distribution of recombining electrons and holes in the conduction and valence band. The spectral characteristics of light emission from microcavities are as intriguing as they are complex [22]. However, restricting our considerations to the optical axis of the cavity, simplifies the cavity physics considerably. If we assume that the cavity resonance is much narrower than the natural emission spectrum of the semiconductor, then the on-resonance luminescence is enhanced whereas the off-resonance luminescence is suppressed. The on-axis emission spectrum should therefore reflect the enhancement, i.e. resonance spectrum of the cavity. The strong similarity of the emission and the reflectance spectrum shown in Fig. 3 confirm this conjecture.

Microcavities provide enhanced optical feedback through photons oscillating in the cavity. The enhanced photon density results in a higher probability of stimulated emission processes. To test the relevance of stimulated emission processes, we have measured the RCLED spectrum at different excitation intensities and these results are shown in Fig. 4. For injection currents ranging from 0.5 mA to 15 mA, the optical spectra are virtually the same indicating the insignificance of stimulated emission.

A design criterion for the RCLED is $(1 - R_1) \gg \alpha \ell_{active}$, where α and ℓ_{active} is the absorption coefficient and the thickness of the active region. This condition insures that the exit-mirror loss is much larger than self-absorption in the active region. If the above criterion were not fulfilled, the RCLED would not provide any intensity enhancement. On the other hand, laser oscillation requires that the round-trip gain exceeds the mirror loss. The maximum gain of a population-inverted structure cannot exceed the magnitude of the absorption in the structure, i.e. $|g| \leq |\alpha \ell_{active}|$ (23). Thus, the maximum achievable gain is *always* smaller than the mirror loss, irrespective of the pump level, which shows that laser oscillations cannot occur in RCLEDs.

Optical sources based on spontaneous and stimulated emission, i.e. LEDs and lasers, form the basis of all silica fiber communication. In practice, LEDs are preferred over lasers whenever possible. There are three major reasons for the preferability of LEDs. First, LEDs have much higher reliability which can exceed that of lasers by one order of magnitude [16,24]. Second, the temperature sensitivity, which is known to be governed by the well-known T_0-law in lasers, is smaller for LEDs [24]. Third, the simpler fabrication process of LEDs results in lower production cost. However, LEDs cannot be used in high bit rate (>1 Gbit) and long distance (>5 km) communication.

References

[1] C. Fabry and A. Perot, *Ann. Chim. Phys.* **16**, 115 (1899).

[2] Y. Yamamoto and R. E. Slusher *Physics Today*, 66 (June 1993).

[3] H. Yokoyama, *Science* **256**, 66 (1992).

[4] E. M. Purcell, *Phys. Rev.* **69**, 681 (1946).

[5] R. E. Slusher et al., *Appl. Phys. Lett.* **63**, 1310 (1993).

[6] S. John, *Physics Today*, **44**, No. 5, p. 32 (1991).

[7] E. F. Schubert et al., Appl. Phys. Lett. **61**, 1381 (1992).

[8] H. Yokoyama et al., *Appl. Phys. Lett.* **57**, 2814 (1990).

[9] E. F. Schubert, X. H. Wang, A. Y. Cho, L.-W. Tu, G. J. Zydzik, Appl. Phys. Lett. **60**, 921 (1992).

[10] F. De Martini, G. Innocenti, G. R. Jacobovitz, P. Mataloni, *Phys. Rev. Lett.* **59**, 2955 (1987).

[11] M. Suzuki, H. Yokoyama, S. D. Brorson, E. P. Ippen, *Appl. Phys. Lett.* **58**, 998 (1991).

[12] The Golden Rule is a fundamental result of time-dependent pertubation theory for harmonic pertubations. The rule allows one to calculate transition rates in quantum mechanical systems. See, for example: Merzbacher E. *Quantum Mechanics* (Wiley, New York, 2nd ed., 1970).

[13] N. E. J. Hunt *et al. Appl. Phys. Lett.* **63**, 2600 (1993).

[14] X. P. Feng, *Opt. Commun.* **83**, 162 (1991).

[15] See, for example, E. F. Schubert, *Doping in III-V Semiconductors* (Cambridge University Press, Cambridge, 1993).

[16] For review see R. H. Saul, T. P. Lee, C. A. Burrus, in *Semiconductors and Semimetals* Vol. 22, Part C p. 193 (Academic Press, Orlando, 1985).

[17] Hewlett-Packard, *Optoelectronics Components Catalogue*, 1994.

[18] T. Kato *et al., J. Cryst. Growth* **107**, 832 (1991).

[19] I. Schnitzer, E. Yablonovitch, C. Caneau, T. J. Gmitter, A. Scherer, *Appl. Phys. Lett.* **63**, 2174 (1993).

[20] Optical Data Link 50 or ODL50 is an AT&T product (AT&T part number 1261 AAC).

[21] N. E. J. Hunt, E. F. Schubert, D. L. Sivco, A. Y. Cho, G. J. Zydik, Electron. Lett. **28**, 2169 (1992).

[22] G. Björk, S. Machida, Y. Yamamoto, K. Igeta, *Phys. Rev.* A**44**, 669 (1991).

[23] Since the Einstein coefficients $B_{21} = B_{12}$, the gain in a population-inverted structure can not exceed the loss in a non-inverted structure.

[24] M. Fukuda *Reliability and Degradation of Semiconductor Lasers and LEDs* (Artech, Boston, 1991).

SiC-A³N alloys and wide band gap nitrides grown on SiC substrates

V A Dmitriev, K G Irvine, J A Edmond, G E Bulman, C H Carter, Jr.

Cree Research, Inc., Durham, USA

A S Zubrilov, I P Nikitina, V Nikolaev, A I Babanin, Yu V Melnik, E V Kalinina, V E Sizov

A F Ioffe Institute and Cree Research EED, St. Petersburg, Russia

Abstract. Single crystal GaN, GaN-AlN, GaN-InN, SiC-AlN and Si-C-Ga-Al-N layers were grown on SiC substrates. Crystal quality, film composition and optical properties were studied. The full width at half maximum (FWHM) of x-ray ω-scan rocking curves was 1.2 - 2 arcmin for GaN layers. The N_d-N_a concentration for undoped GaN layers was 10^{14} - 10^{16} cm^{-3}. The N_d-N_a concentration for n-doped GaN layers was controlled in the range 1×10^{17} - 4×10^{18} cm^{-3}, and N_a-N_d concentration for as-grown p-doped layers was in the range 1×10^{17} - 4×10^{18} cm^{-3}. GaN/SiC mesa-structures were fabricated using contact metallization, photolithography and reactive ion etching. The current-voltage (I-V) characteristics of n+GaN/n+SiC mesa-structures were linear with a resistivity of ~3 Ω (300 K). GaN pn junctions were also fabricated on SiC substrates.

1. Introduction

SiC is an attractive substrate for nitrides due to its well matched lattice parameters and high electrical conductivity [1-3]. We will present results of research on GaN, GaN-AlN, GaN-InN, SiC-AlN and Si-C-Ga-Al-N grown on SiC substrates. Layers were grown on the (0001)Si face of 6H-SiC manufactured at Cree Research. Crystal quality was investigated by x-ray diffraction, high energy electron diffraction (HEED) and transmission electron microscopy (TEM). Film composition was measured by Auger electron spectroscopy (AES) and secondary ion mass spectroscopy (SIMS). Optical properties were studied by photoluminescence (PL), cathodoluminescence (CL) and electroluminescence (EL).

2. GaN layers and pn structures

2.1. Surface and Crystal structure

The thickness of GaN layers grown on SiC ranged from 0.1 to 6 microns. Undoped and doped layers with a dopant concentration less than 10^{19} cm^{-3} had a smooth surface (Fig. 1). HEED measurements showed that the surface of the films was single crystal. The minimum FWHM of x-ray ω-scan rocking curves for the (0002) reflection was 1.2 arcmin. It is the most narrow ω scan FWHM for GaN layers reported to date. Results of TEM measurements showed that the GaN films consist of two layers, (1) a thin interface layer with a high dislocation density and (2) a top layer with a random dislocation distribution and lower dislocation density (Fig. 2).

Figure 1. Surface morphology of a GaN layer grown on 6H-SiC (thickness ~ 1.8 μm).

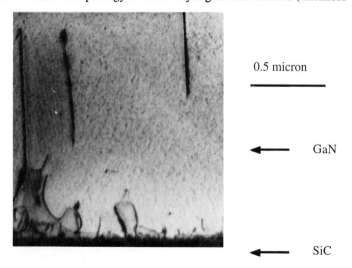

Figure 2. Cross-section TEM image of a GaN layer grown on 6H-SiC.

2.3. Optical properties and stimulated emission

Photoluminescence and CL of GaN layers grown on 6H-SiC substrates were investigated through a wide temperature range. The edge luminescence dominated the PL spectrum in the temperature range 77 - 900 K (Fig. 3,4). The temperature dependence of the GaN exitonic band gap was determined: $E(eV) = 3.56 - 5.6 \cdot 10^{-4} \cdot T$. Edge cavity stimulated emission from photopumped GaN layers was observed in the temperature range 77 - 450 K. It is the first observation of stimulated emission from any A^3N material at a temperature higher than 300 K. The wavelength and FWHM of the stimulated emission peak was 364 nm and 2 nm at 77 K; 374 nm and 3 nm at 300 K; and 386 nm and 7 - 9 nm at 450 K, respectively. Multipass stimulated emission with Fabry-Perot modes was observed from GaN for the first time. The FWHM of Fabry-Perot modes was ~0.2 nm (300 K).

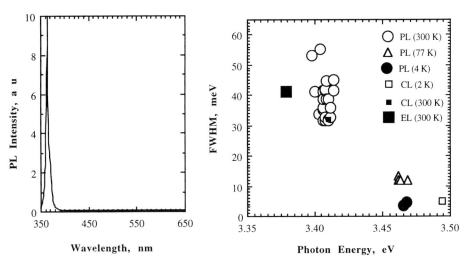

Figure 3. PL spectrum of a GaN layer at 300 K.

Figure 4. Wavelength and FWHM of PL, CL, and EL edge peak at various temperature.

2.4. Electrical properties and doping

The N_d-N_a concentration measured using a mercury probe was 3×10^{14} - 5×10^{16} cm^{-3} for undoped layers. The N_d-N_a concentration for n-doped layers was controlled in the range 1×10^{17} - 4×10^{18} cm^{-3}. The N_a-N_d concentration for p-doped as grown layers was in the range 1×10^{17} - 4×10^{18} cm^{-3}. n+GaN/n+SiC mesa-structures were fabricated using contact metallization, photolithography and reactive ion etching. The current-voltage (I-V) characteristics of the mesa-structures were linear with a resistivity of ~3 Ω (300 K).

2.5. pn structures and EL

GaN pn structures were grown and diodes with a 0.4x0.4 mm mesa structure were fabricated. Capacitance-voltage measurements showed a linear impurity distribution at the pn junction. The electroluminescence spectrum for this n^+GaN-p^+GaN structure is shown in Fig. 5.

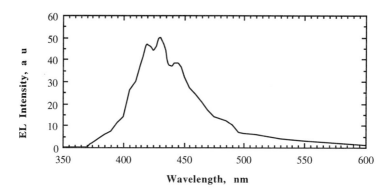

Figure 5. EL spectra of GaN n^+p^+ junction: I_f = 20 mA, 300 K.

3. GaAlN layers

GaN-AlN single crystal layers were grown with an AlN concentration from 2 to 25 mol. %. The layer thickness was 0.1 - 0.4 µm. The FWHM of the x-ray ω-scan (0002) rocking curve was 1.6 arcmin for $Ga_{0.95}Al_{0.05}N$, 4.7 arcmin for $Ga_{0.89}Al_{0.11}N$ and 6.5 arcmin for $Ga_{0.83}Al_{0.17}N$. These are the best x-ray rocking curves for GaAlN layers ever reported. Layers displayed sharp PL and CL edge peaks (Fig. 7). The FWHM of CL edge peaks were ~ 65 meV (300 K) for a $Ga_{0.95}Al_{0.05}N$ layer, and ~ 90 meV (80K) for a $Ga_{0.83}Al_{0.17}N$ layer.

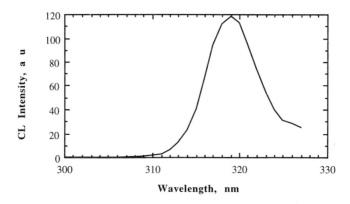

Figure 7. Edge CL peak of $Ga_{0.87}Al_{0.17}N$ layer (77 K).

4. GaInN layers and GaN/GaInN/GaN structures

Single crystal layers of GaN-InN alloys with an InN concentration up to 20 mol. % were grown (Fig. 8), as well as GaN-GaInN-GaN heterostructures (Fig. 9). The thickness of the layers ranged from 0.05 to 0.5 μm. Optical properties of the layers are under investigation.

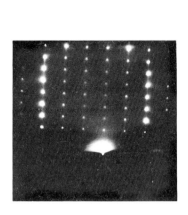
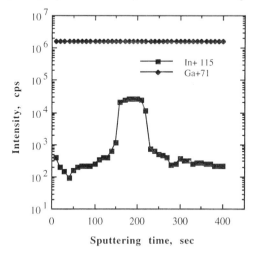

Figure 8. HEED pattern for InGaN layer.

Figure 9. SIMS depth profile of GaN/InGaN/GaN heterostructure.

5. SiC-based alloys

Single crystal SiC-AlN layers (AlN ~ 30 - 90 mol. %) were also grown (Fig. 10). For an AlN concentration of ~50 mol. % the minimum FWHM of the x-ray ω–2Θ rocking curve was 3.5 arcmin. Layers displayed CL in the UV and visible range. The minimum CL wavelength was ~ 230 nm (hν ~ 5.4 eV).

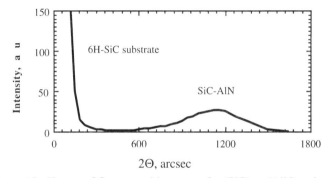

Figure 10. X-ray ω-2Θ scan rocking curve for $(SiC)_{0.2}$-$(AlN)_{0.8}$ layer.

In addition, monocrystalline Si-C-Al-Ga-N alloy layers were grown for the first time (Fig. 11). The single crystal structure of the alloy was verified by HEED. Cathodoluminescence showed several peaks in the UV and violet region with a minimum wavelength of ~ 300 nm. The photon energy of the UV peak decreased when the Si and C content in the film increased.

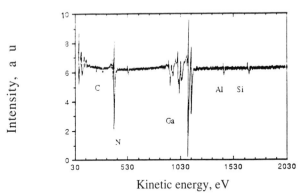

Figure 11. Auger electron spectra for a Si-C-Al-Ga-N layer

6. Summary

Single crystal layers of GaN, GaAlN, GaInN, SiC-AlN, and Si-C-Ga-Al-N alloys were grown on SiC substrates. The high quality of the grown layers was demonstrated. GaN layers had a FWHM of 1.2 - 2 arcmin as measured by double crystal x-ray ω-scan (0002) rocking curves. The N_d-N_a concentration was 10^{14} - 10^{16} cm^{-3} for undoped GaN layers and 1×10^{17}-4×10^{18} cm^{-3} for n-doped layers. The N_a-N_d concentration was 1×10^{17} - 4×10^{18} cm^{-3} for p doped as-grown layers. For the first time, stimulated emission was detected from photopumped GaN at a temperature higher than 300 K (up to 450 K).

Linear current-voltage characteristics with low resistance (R ~ 3 Ω) were observed in n$^+$GaN/n$^+$SiC mesa-structures proving that a vertical device geometry with a low resistance can be realized for GaN using SiC substrate. The first GaN pn junctions on SiC were made.

Acknowledgments. This work was supported in part by the Advanced Research Projects Agency and the Office of Naval Research.

Reference

[1] Strite S and Morkoc 1992 *J. Vac. Sci. Technol.* **B10** 1237-66

[2] Lin M E, Sverdlov B, Zhou G L, Morkoc H 1993 *Appl. Phys. Lett.* **62** 3479-81

[3] Weeks T W Jr., Kum D W, Carlson E, Perry W G, Ailey K S, and Davis R F 1994 *Transactions of the Second Int. High Temperature Electronics Conference* (Charlotte, North Carolina) **2** P-209 - 14

Effect of Atomic Hydrogen on 670nm AlGaInP Visible Laser

Won-Jin Choi, Ji-Ho Chang, Won-Taek Choi, Seung-Hee Kim, Jong-Seok Kim, Shi-Jong Leem, and Tae-Kyung Yoo

GoldStar Central Research Laboratory, 16 Woomyeon-Dong, Seocho-Gu, Seoul 137-140, Korea

Abstract : An origin of improvements in the characteristic of 670nm strained multi-quantum well AlGaInP visible lasers during accelerated aging test is investigated. During the test, the overall laser characteristics are improved considerably after a short-term continuous operation of 3mW at 50℃. From SIMS measurement, we find out that the atomic hydrogen in Zn doped AlGaInP cladding layer are dissociated from zinc-hydrogen complex and drifted into active region due to bias voltage and aging temperature. Because the dissociation of the atomic hydrogen results in the decrease of series resistance and the atomic hydrogen drifted into the active region passivating defects at AlGaInP/GaInP interfaces, it is believed that the atomic hydrogen plays a great role in the reliability of MOCVD-grown AlGaInP lasers.

1. Introduction

The AlGaInP visible semiconductor lasers have become one of the most important light sources for optical information systems. Hence, several investigations on reliability of AlGaInP lasers have been reported. In some of the reports, there were specific results showing that the characteristics of AlGaInP visible lasers were improved in the initial stage of accelerated aging test[1-2]. These phenomena, however, are not understood clearly yet. In this paper, we investigate the aging characteristics of AlGaInP lasers, especially for explaining improvements in the overall lasing characteristics during aging test.

2. Experiments

An index guided strained multi-quantum well (SMQW) separate confinement heterostructure AlGaInP laser was fabricated by three step low pressure metalorganic chemical vapor deposition (MOCVD). DEZn, and SiH_4 were used as dopant sources. Lattice mismatch in GaInP well was + 0.3%. The wavelength and threshold current of the laser with 250μm cavity length were 670nm and 35mA. respectively. An accelerated aging test was carried out for the laser in the automatic power control (APC) mode with 3mW output at 50℃. The end facets of the laser were not coated. For the purpose of investigating changes in each layer comprising laser structure before and after aging test, we fabricated a SMQW AlGaInP visible light emitting diode (LED) of the same double-hetero(DH) structure with the laser and operated it continuously in the same aging condition. The diode was later examined by secondary ion mass spectrometry (SIMS). In addition, we fabricated a specific laser, "the hydrogen-reduced laser", in which the concentration of hydrogen atoms is reduced, especially in p-type cladding layer, by the heat treatment in H_2 atmosphere.

3. Results and Discussion

Figure 1 shows the aging characteristics of the index guided SMQW AlGaInP visible laser during the initial stage of the accelerated aging test. The aging characteristic of AlGaAs laser, the 860nm liquid phase epitaxy (LPE)-grown index guided AlGaAs laser is included for comparison.

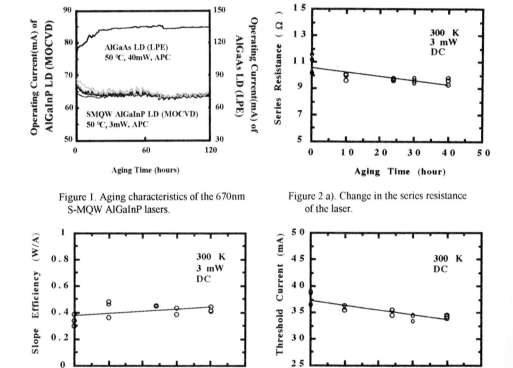

Figure 1. Aging characteristics of the 670nm S-MQW AlGaInP lasers.

Figure 2 a). Change in the series resistance of the laser.

Figure 2 b). Change in the slope efficiency of the laser.

Figure 2 c). Change in the threshold current of the laser.

Unlike LPE-grown index guided AlGaAs laser, the operating current of the AlGaInP visible laser decreased gradually for about 60 hours from the start of the aging test and then stabilized. This decrease in the operating current is explained from the results shown in Figure 2a), Figure 2b), and Figure 2c), in which changes in the series resistance, the slope efficiency, and the threshold current of the laser during the aging test are shown, respectively. The series resistance and threshold current of the laser are decreased and the slope efficiency is increased during the first 40 hours of the aging test. To compare with the initial values, the average series resistance and threshold current decrease about 8% and 8.5%, respectively, and the average slope efficiency increases about 24% after the first 40 hour aging. Putting all of these results together, it is apparent that the lasing characteristics of the AlGaInP laser are improved during the initial stage of the aging test.

Figure 3 shows atomic hydrogen depth profile of the LED by SIMS before and after the aging test. Before the aging test (thick solid line), the atomic hydrogen concentration in the p-AlGaInP layer is higher than those in both the active layer and the n-AlGaInP layer. The atomic hydrogen concentration of the active region is lowest among three layers. But, after the aging test (thin solid line) the atomic hydrogen concentration of p-AlGaInP layer reaches nearly the value of n-AlGaInP layer. This result clearly indicates that the atomic hydrogen in the Zn doped AlGaInP layer are dissociated from zinc-hydrogen complexes and drift into the active region by the bias voltage and the aging temperature[3-4].

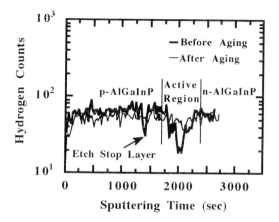

Figure 3. SIMS depth profiles of the atomic hydrogen in cladding and active layers of the LED before and after aging test.

From above results we suggest the following scenario. First, the free carrier concentration of each cladding layer, especially that of p-type cladding layer, which is more critical in determining the characteristics of the laser, increases as the concentration of the dopant-hydrogen complex decreases in each cladding layer. Thus, the series resistance of the laser decreases gradually and the electron overflow is reduced as the height of the hetero-barrier between p-cladding layer and the active region becomes larger[5]. Second, the atomic hydrogen drifts into the active region passivating both bulk and interfaces defects of the active region[6].

Figure 4 shows the aging characteristics of the hydrogen-reduced laser. The hydrogen-reduced laser was obtained by an annealing of the laser structure of the first step MOCVD growth in H_2 atmosphere[7]. The aging condition was the same as that for the results shown in Fig. 1. According to our suggestion, we expect that the degree of the improvement in the lasing characteristics of the hydrogen-reduced laser should be inferior to the laser made from an as-grown wafer. As expected, the operating current of the hydrogen-reduced laser increases gradually from the start of the aging test. This result strongly confirms our suggestions.

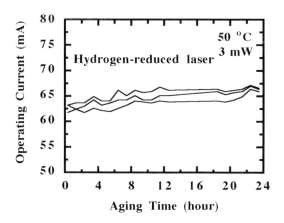

Figure 4. Aging characteristics of the annealed 670nm SMQW AlGaInP lasers.

4. Summary

An accelerated aging test has been carried out on the 670nm SMQW AlGaInP laser. The overall lasing characteristics of the laser have been shown to improve during the initial stage of the aging test. We suggest that this phenomenon is related to the behavior of the atomic hydrogen in the epitaxial layer of the laser on the basis of the results of SIMS measurement. Further confirmation of our suggestion is made by another aging test with the hydrogen-reduced lasers. In conclusion, we propose that the behavior of an atomic hydrogen should be considered seriously because of its importance on the reliability of the semiconductor laser.

Acknowledgment

We would like to thank Dr. Jongje Jung for his helpful discussion and Min Yang, Ju-Ok Seo, Min-Soo Noh, and Jae-Hak Lee for their experimental supports.

References

[1] K. Itaya, M. Ishikawa, H. Okuda, Y. Watanabe, K. Nitta, H. Shiozawa, and Y. Uematsu, Appl. Phys. Lett.**53**, 1363 (1988).
[2] M. Ishikawa, H. Okuda, K. Itaya, H. Shiozawa, and Y. Uematsu, Jpn. J. Appl. Phys. **28**, 1615 (1989).
[3] H. Y. Cho, S.-K. Min, K. J. Chang, and C. Lee, Phys. Rev. **B 44**, 13779 (1991).
[4] T. Zundel, and J. Weber, Phys. Rev. **B 39**, 13549 (1989).
[5] K. Itaya, M. Ishikawa, H. Shizawa, M. Suzuki, H. Sugawara, and G. Hatakoshi, Extended Abstract of the 22nd Int. Conf. on Solid State Devices and Materials, 1990, p. 565.
[6] S. M. Lord, G. Roos, and J. S. Harris, Jr., N. M. Johnson, J. Appl. Phys., **73**, 740(1993).
[7] H. Hamada, S. Honda, M. Shono, R. Hiroyama, K. Yodoshi, and T. Yamaguchi, Electron. Lett. **28**, 585 (1992).

Threshold Current Density Reduction by Annealing in 630-650nm GaInP Strained Single Quantum Well Lasers

Ichirou Nomura and Katsumi Kishino

Department of Electrical and Electronics Engineering, Sophia University
7-1 Kioi-cho, Chiyoda-ku, Tokyo 102, Japan

Abstract. Annealing effects on threshold current density (J_{th}) of GaInP strained single quantum well (SSQW) lasers were systematically investigated. After the annealing, remarkable reduction of the J_{th} of 630-650nm lasers was observed, accompanied by increase of p carrier density of the p-cladding layers. Especially, the J_{th} of 633-nm lasers was reduced from 850 A/cm^2 to 555 A/cm^2. The annealing effect on the J_{th} reduction can be theoretically explained by suppressing electron leakage currents, especially drift leakage current over p-side heterobarriers.

1. Introduction

AlGaInP red-light semiconductor lasers emitting in 600-nm wavelength range are crucial devices in realization of high-performance optical information processing systems, and many studies have been made on methods to improve the lasing properties.

For the AlGaInP lasers, especially with shorter-wavelengths (~633 nm), the electron leakage current over p-side heterobarriers is a dominant factor which deteriorates the lasing properties such as threshold current [1]. Thus, in order to improve the lasing properties, the leakage currents must be reduced.

The leakage current consists of two components, i.e., diffusion and drift leakage currents. These leakage currents can be reduced by pushing up the p-side heterobarriers. Especially, the drift leakage component, which is dominant in the leakage currents for the shorter-wavelength lasers [2], can be reduced by increasing electrical conductivity of the p-cladding layers, because of decrease in electric field induced in the cladding layer. Both pushing up the heterobarriers and increasing the electrical conductivity can be achieved by increasing p carrier density in p-cladding layers. Here, note annealing of the lasers, which is a very attractive method for increasing p carrier densities [3]. Furthermore, it was shown that the annealing is effective in reducing nonradiative recombination currents at the interfaces between active and cladding layers of GaInAs/GaAs strained quantum well lasers [4].

In this study, annealing effects on threshold current density (J_{th}) of GaInP strained single quantum well (SSQW) lasers [5,6] with Be-doped AlInP p-cladding layers grown by gas source molecular beam epitaxy (GSMBE) were systematically investigated. After annealing, remarkable reduction of the J_{th} of 630-650nm lasers was observed, accompanied by increase of p carrier density of the p-cladding layers. Especially, the J_{th}

of 633-nm lasers was reduced from 850 A/cm^2 to 555 A/cm^2. From these results and theoretical consideration of diffusion and drift leakage currents, it was shown that the annealing is very effective in suppressing the leakage currents, especially the drift leakage current, due to increased heterobarrier height and increased electrical conductivity of the p-cladding layers, both induced by the increased p carrier density.

2. Preparation of GaInP SSQW lasers

In this study, to fabricate GaInP strained single quantum well (SSQW) lasers, gas source molecular beam epitaxy (GSMBE) [7,8] was employed. In this system, group (Al,Ga,In) beams were supplied from solid sources, while the group V phosphorus and arsenic beams from cracking gas sources (i.e., using 100% PH$_3$ and AsH$_3$ gases, respectively) [8]. For n- and p-type dopants, Si and Be were used, respectively.

For the growth of GaInP and AlInP layers, the standard growth condition [7,8] were employed, that is, with the growth temperature of around 500 °C, growth rate of 0.86 μm/h, and PH$_3$ flow rate of 6.8 sccm (giving the corresponding V/III ratio of 6.6). Here, the typical substrate rotation speed during growth was 30 rpm. The lasers were grown on misoriented (toward [011] direction by 15°) n-GaAs substrates [1] mounted on molybdenum substrate blocks.

GaInP SSQW lasers consisted of GaInP strained quantum well active layers sandwiched between GaInP/AlInP short-period superlattice cladding (SLC) layers with 3-monolayer (ML) and 2-ML thickness combination. In the growth of the lasers, first n-GaAs (70 nm) and n-GaInP (10 nm) buffer layers were grown on GaAs substrates, followed by the sequential growth of SSQW laser layers in the following order: n-AlInP cladding layers (0.7 μm), undoped SLC layers (80 nm), undoped GaInP SSQW active layers (10 nm), undoped SLC layers (80 nm), p-AlInP cladding layers (0.7 μm), and p$^+$-GaInP cap layers (280 nm).

Here, the GaInP composition was switched abruptly at quantum active layer boundaries from lattice-matched to mismatched composition for introducing biaxial crystal strain. The composition change was successively performed without growth interruption by use of the novel shutter control method [5,6]. In this experiments, five types of the laser with different amounts of strain (i.e., -0.9, -0.5, 0, +0.5 and +1.1 %) were prepared.

After growth of the lasers, annealing was performed at 550 °C for 30 minutes in a N$_2$/H$_2$ (H$_2$ 5%) blended gas atmosphere. In order to investigate annealing effects, the lasing characteristics and photoluminescence (PL) properties of as-grown and annealed laser samples were evaluated.

3. Annealing effects

3.1. Threshold current density

For as-grown and annealed GaInP SSQW lasers, J_{th} values were evaluated as broad-contact-type lasers under the pulsed condition (repetition rate: 1 kHz, width: 100 ns).

Figure 1 shows threshold current density (J_{th}) at room temperature for annealed and as-grown lasers with cavity length of 1 mm as a function of lasing wavelength. The lasing wavelengths were 635, 642, 657, 691 and 708 nm for the as-grown lasers with corresponding strain amounts of -0.9, -0.5, 0, +0.5 and +1.1 %, respectively. In Fig.1, we notice that remarkable reduction in J_{th} value for shorter wavelength (630-650 nm) lasers was obtained by annealing, while for the longer wavelength (> 690 nm) lasers, no improvement in J_{th} was observed. Here, the lasing wavelength did not change by annealing for overall wavelength range (i.e., for any strain amounts). Especially, for 633-nm lasers with strain amount of -0.9 %, the J_{th} was reduced from 850 A/cm^2 to 555 A/cm^2.

3.2. Photoluminescence property

Photoluminescence (PL) spectra from GaInP SSQW active layers of the lasers were evaluated under 488-nm Ar$^+$ laser light irradiation with a power of 2 mW at room temperature after removing GaInP cap layers by chemical etching.

Figure 2 shows the PL peak intensity as a function of PL peak wavelength for as-grown and annealed GaInP SSQW lasers. The PL intensity for both the as-grown and annealed lasers increased with lengthening the lasing wavelength, because of enhanced carrier confinement into the active layer (i.e., of increased heterobarriers). The PL intensity for the annealed lasers was several times larger than one for the as-grown lasers. This probably shows that nonradiative recombination components at the interfaces between the active and the cladding layers were reduced by annealing.

We note here that the PL intensity for longer-wavelength (> 680 nm) lasers increased by the annealing as well as that for shorter-wavelength lasers. However, J_{th}

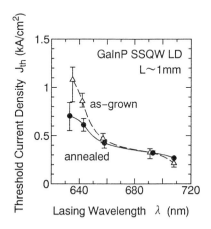
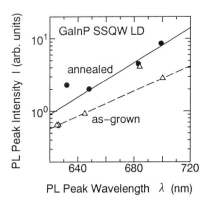

Fig.1 Threshold current density (J_{th}) of as-grown and annealed GaInP SSQW lasers as a function of lasing wavelength.

Fig.2 Photoluminescence (PL) peak intensity as a function of PL peak wavelength for as-grown and annealed GaInP SSQW lasers.

for the longer-wavelength lasers was not reduced by the annealing (see Fig.1). These results suggest that the increase of the PL intensity is not a main factor of the J_{th} reduction by annealing.

3.3. Temperature dependency of threshold current density

Dependencies of J_{th} on heatsink temperature for as-grown and annealed 633-nm GaInP SSQW lasers with strain amount of -0.9 % are shown in Fig.3, where the cavity length was 0.5 mm. The J_{th} value of annealed lasers was lower than that of as-grown lasers around room temperature, and the room-temperature characteristic temperature (T_0) was improved from 41 K to 52 K by annealing. On the other hand, in the low-temperature range (< 200 K), where the carrier leakage over heterobarriers can be negligible, the J_{th} values for both as-grown and annealed lasers were nearly the same.

Note that for short-wavelength (~ 633 nm) lasers with insufficient heterobarrier height, the carrier leakage over heterobarriers is a dominant factor of rapid increase in J_{th} at high temperature range (> 300 K) [1]. Thus, the reduced J_{th} in Fig.3 shows that the annealing is very effective in suppressing the carrier leakage currents in the J_{th}.

3.4. Diffusion and drift leakage currents

In this section, the annealing effect on leakage current reduction in GaInP SSQW lasers was theoretically investigated, taking diffusion and drift leakage currents into consideration.

The total electron leakage current density (J_L) with the diffusion and the drift leakage current components [2] is given as follows:

$$J_L = qD_n N_0 \left[\sqrt{\frac{1}{L_n^2} + \frac{1}{4z^2}} \operatorname{Coth} \sqrt{\frac{1}{L_n^2} + \frac{1}{4z^2}} x_p + \frac{1}{2z} \right]. \tag{1}$$

where, q, k and T are the electronic charge, Boltzmann's constant and the absolute temperature, respectively. x_p is the p-cladding layer thickness, and L_n is the minority electron diffusion length. D_n is the minority electron diffusion coefficient. N_0 is the density of minority electrons at the edge of the p-cladding layer. z means a length characteristic of drift leakage. These parameters (D_n, N_0, z) are described in ref.[2] in detail. Other parameters used in this investigation are summarized in Table 1.

p carrier density of Be-doped AlInP p-cladding layers was measured by C-V method, and it was increased from 1.7×10^{18} to 2.1×10^{18} cm^{-3} by the annealing. By increase of p carrier density, heterobarrier height on p-cladding layer side is enhanced, and electrical conductivity of the cladding layer increases. In this case, increase of the heterobarrier height of 6 meV and increase of the electrical conductivity from 0.82 to 1.0 $(\Omega \cdot cm)^{-1}$ were estimated from the increase of the p carrier density.

Taking the increased heterobarrier height and the increased electrical conductivity into account, the leakage current reduction amount by the annealing effect was calculated using Eq.(1). Here, the effective heterobarrier height (ΔE) before

Table 1. Parameters used in calculation of leakage currents.

Parameter		Value
Minority electron mobility in AlInP p-cladding layer	μ_n	100 cm^2/Vs
p (hole) carrier density in AlInP p-cladding layer	p	1.7×10^{18} cm^{-3} (as-grown)
		2.1×10^{18} cm^{-3} (annealed)
Majority hole mobility in AlInP p-cladding layer	μ_p	3 cm^2/Vs
Thickness of AlInP p-cladding layer	x_p	0.7 μm
Minority electron diffusion length in AlInP p-cladding layer	L_n	1 μm
Density of states X-electron effective mass in AlInP	m^*_n	0.6m_0

annealing were estimated for as-grown lasers by fitting leakage current components in the experimental J_{th} values to Eq.(1).

Figure 4 shows the reduction amounts of the leakage current density (J_L) calculated for two cases, i.e., considering both the diffusion and the drift leakage and only the diffusion leakage as a function of lasing wavelength, respectively. Reduction amounts of experimental J_{th} values by the annealing (see Fig.1) are also shown in Fig.4. The calculated reduction amount with both the diffusion and the drift leakage can explain the reduction amounts of J_{th}, while the calculation with only the diffusion leakage is not sufficient to explain the J_{th} reduction. From these results, we concluded that the annealing was very effective in suppressing the leakage current, especially the drift leakage current, due to increased heterobarrier height and increased electrical conductivity of the p-cladding layers, both induced by the increased p carrier density.

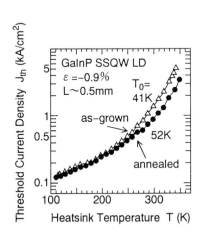
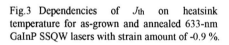

Fig.3 Dependencies of J_{th} on heatsink temperature for as-grown and annealed 633-nm GaInP SSQW lasers with strain amount of -0.9 %.

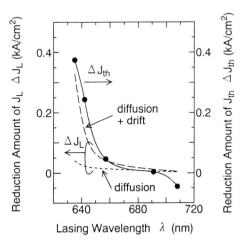

Fig.4 Calculated reduction amounts of leakage current density (J_L) and experimental reduction amounts of J_{th} by annealing for GaInP SSQW lasers as a function of lasing wavelength.

4. Conclusion

Annealing effects on J_{th} reduction of GaInP SSQW lasers with Be-doped AlInP p-cladding layers grown by GSMBE were systematically investigated. After annealing at 550 °C for 30 min, remarkable reduction of the J_{th} of 630-650nm lasers was observed, accompanied by increased p carrier density of p-cladding layers. The J_{th} of 633-nm lasers was reduced from 850 A/cm^2 to 555 A/cm^2. From these results and theoretical consideration including diffusion and drift leakage currents over p-side heterobarriers, it was concluded that the annealing was very effective in suppressing electron leakage currents, especially the drift leakage current, due to increased heterobarrier height and increased electrical conductivity of the p-cladding layers, both induced by the annealing.

References

[1] Kikuchi A., Kishino K. and Kaneko Y., 1991, *Jpn. J. Appl. Phys.* **30**, 3865-3872.
[2] Bour D. P., Treat D. W., Thornton R. L., Geels R. S. and Welch D. F., 1993, *IEEE J. Quantum Electron.* **QE-29**, 1337-1343.
[3] Hamada H., Honda S., Shono M., Hiroyama R., Yodoshi K. and Yamaguchi T., 1992, *Electron. Lett.* **28**, 585-587.
[4] Yamada N., Roos G. and Harris J. S., Jr., 1991, *Appl. Phys. Lett.* **59**, 1040-1042.
[5] Nomura I., Kishino K. and Kaneko Y., 1992, *Electron. Lett.* **28**, 851-853.
[6] Nomura I., Kishino K., Kikuchi A. and Kaneko Y., 1994, *Jpn. J. Appl. Phys.* **33**, 804-810.
[7] Kikuchi A., Kishino K. and Kaneko Y., 1989, *J. Appl. Phys.* **66**, 4557-4559.
[8] Kaneko Y., Nomura I., Kishino K. and Kikuchi A., 1993, *J. Appl. Phys.* **74**, 819-824.

Compositional Control of (Zn,Mg)(S,Se) Epilayers Grown By MBE for II/VI Blue Green Laser Diodes

M. Ringle, D. C. Grillo, J. Han, R.L. Gunshor, and G. C. Hua

School of Electrical Engineering, Purdue University
West Lafayette, Indiana 47907-1285 USA

A. V. Nurmikko

Division of Engineering and Department of Physics
Brown University, Providence, Rhode Island 02912 USA

Abstract: ZnMgSSe is a technologically important material for II-VI lasers; but there are unresolved difficulties in growing the material that our previous papers have described. Compositional drifting in ZnMgSSe and the resulting strain can have a detrimental effect on laser lifetime. By experimentally determining the temperature dependance of the sulfur incorporation, we suggest that the drifting is partially due to drift in the substrate temperature during growth. Improved temperature control using an infrared optical pyrometer corrected for thin film interference has resulted in better control of the composition of the ZnMgSSe films. Single peaked X-ray rocking curves of ZnMgSSe with FWHM of 24 arc seconds have been obtained.

1. Introduction

II-VI semiconductor lasers have received much attention in recent years. One of the more significant advances has been the introduction of the ZnMgSSe quaternary alloy [1]. This alloy has made it possible to design and build pseudomorphic lasers with separate optical and electrical confinement. It was this design, in conjunction with Zn(Se,Te) graded contacts to p-ZnSe, which resulted in room temperature, CW operation of II-VI lasers [2],[3]. However, these devices suffer from the problem of short lifetimes. III-V based devices demonstrated the need for the reduction of strain for successful operation [4]. Strain from lattice mismatch can help to form dark line defects which drastically shorten the lifetime of laser devices. Since it is likely that ZnMgSSe will be an integral part of II-VI laser design, it is essential to be able to accurately control the composition of ZnMgSSe to insure lattice matching to the GaAs substrate.

Much work has been done to improve the quality of ZnMgSSe. Electrically, the problem of nitrogen doping in ZnMgSSe has been recently addressed by Hall and photoionization cross-section measurements which suggest that p-type ZnMgSSe exhibits a behavior similar to that of the DX center in n-type AlGaAs [5]. However, less attention has been given to the issue of the epitaxial growth of the quaternary. Previously, we have reported the observation of multiple, narrow (10-30 arcseconds) peaks in x-ray rocking curves of ZnMgSSe, such as the one in Figure (1a), which can be explained by compositional drifting [6]. Figure (1b) is a theoretical x-ray rocking curve where the experimental x-ray data in Figure (1a) was iteratively fit using a computer simulation based on the Takagi-Taupin equations; the predicted drift in sulfur fraction is shown in Figure (3a). The problem of compositional drifting provided the motivation to understand what factors affect the incorporation of each element into the ZnMgSSe film, and detailed examination of the sulfur

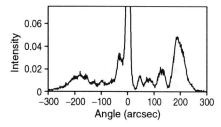
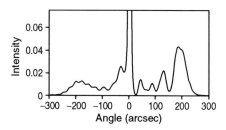

Figure 1(a). Experimentally measured x-ray rocking curve for a particular sample.

Figure 1(b). Simulated x-ray rocking curve assuming the sulfur concentration change verses thickness shown in figure 3a.

incorporation showed substrate temperature fluctuations to be a likely cause for much of the observed drift in composition. A procedure was developed to precisely monitor and control the substrate temperature to minimize the problem of compositional drifting in ZnMgSSe.

2. Experimental determination of sulfur incorporation

The films of this investigation were grown in a Perkin Elmer 430 MBE. The ZnMgSSe epilayers were grown on GaAs homoepilayers grown in a separate III-V chamber connected to the II-VI chamber by an ultra-high vacuum transfer tube. The ZnMgSSe was grown using Se(6N), Mg, and ZnS(6N). The Mg was commercial 4N purity material further refined by a two-step vacuum distillation process. Substrate temperatures varied from 230 to 310 °C, monitored by an infrared optical pyrometer. In order to measure the flux from each source, the chamber is equipped with a quartz crystal microbalance.

If the growth rate is known and we assume the epilayer is reasonably stoichiometric, then the rate of cation or anion incorporation (J_{inc}) can be calculated.

$$J_{inc} = 4R/(a^3)$$

R is the growth rate and a is the lattice constant. Growth rate was measured by TEM, SEM, selective etching, or by pyrometer oscillations [7],[8]. We define the fraction of the sulfur flux that is incorporated as K_s; the sticking coefficient.

$$K_s = J_{inc} x / J_{zns}$$

J_{zns} is the ZnS flux measured by the quartz crystal microbalance. The sulfur fraction is x and has been found by either x-ray diffraction (ZnSSe films only) or electron microprobe. Sulfur sticking coefficients were calculated for 23 ZnMgSSe and 14 ZnSSe epilayers grown under similar conditions (on the anion rich side of the transition between cation and anion stabilized growth as determined by RHEED). The sulfur sticking coefficient is plotted verses substrate temperature in Figure (2); the line is a guide for the eye. (We did not include those epilayers that were grown under highly Zn rich conditions in order to enhance the sulfur incorporation.) The scatter in the data is from three sources. One can expect normal errors in growth rate determination, etc. Also, in the past there were significant fluctuations in growth temperature before we improved our technique, so the temperature is only an approximate average for some growths. Finally, there is the fact that the quartz crystal microbalance is not exactly in the same location as the substrate growth position. Because the flux patterns from the ovens change with time as the ovens are used and recharged, the measured flux will not always be the same as the flux at the exact growth position. Even with the scatter, the data shows a clear trend. As the growth temperature drops from 300 to 260 °C, the sulfur sticking coefficient increases by a factor of 2 from 0.05 to 0.1. From 260 to 230 °C, the sticking coefficient increases by another factor of 2 from 0.1 to 0.2. From this data, a 5 °C change in substrate temperature will result in about a 10% change in both the sulfur sticking coefficient and the sulfur fraction because Zn, Mg, and Se do not have as large a variation in sticking coefficients for such small changes in temperature.

3. Temperature control

A standard thermocouple based temperature controller may not be sufficient to keep the substrate temperature constant. On the Perkin Elmer 430 system, the thermocouple is simply pressed against the back of the substrate holder. This does not make consistent contact, especially when the substrate is rotating to ensure spatially uniform quaternary film growth. Another serious problem comes from direct heating of the unshielded thermocouple by the heating element. As the heating element radiates power to the substrate holder, the thermocouple also is heated, resulting in false readings. Optical pyrometry can be used to avoid these problems. However, there is a complication to using an infrared optical pyrometer which comes from the interference between the light reflected from the substrate/epilayer interface and light reflected from epilayer/vacuum interface which causes an oscillation in pyrometer readings which have been observed in all films that we have grown: ZnSe,ZnSSe,ZnMgSSe,ZnCdSe, and ZnTe. Other groups have used this effect to measure growth rate once the index of refraction is known [7],[8]. We have had to quantitatively predict these oscillations in order to accurately control the substrate temperature because of their large amplitude (approximately 14 °C peak to peak at a growth temperature of 260 °C). The reason for this large oscillation is the large difference in refractive index between GaAs and ZnMgSSe which creates a large reflection at the epilayer/substrate interface. (The refractive index for GaAs is about 3.3 in the 2.0 to 2.6 µm wavelength range used by the pyrometer, where ZnSe and related materials have a refractive index of about 2.3.)

The problem of analysing the thin film interference from a single layer sandwiched between two layers of material of differing index of refraction can be solved using classical optics almost identical to that of the Fabry-Perot interferometer problem [9]. The optical power transmission (t) as a function of epilayer thickness (d) is expressed by the formula:

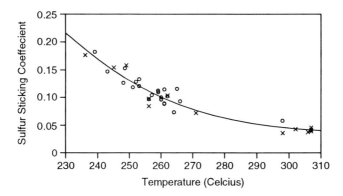

Figure 2. Sulfur sticking coefficient as a function of temperature.
(O) are ZnMgSSe growths, (X) are ZnSSe growths.

$$t(d)=(1-r_1^2)(1-r_2^2)/(1+r_1^2 r_2^2+2r_1 r_2 \cos(4\pi n_2 d/\lambda)) \quad (1)$$

where $r_1 = (n_1-n_2)/(n_1+n_2)$ and $r_2 = (n_2-1)/(n_2+1)$

The refractive index of the substrate is n_1 and n_2 is the refractive index of the epilayer. The wavelength is λ. In order to use this information to predict the oscillation, we must translate the optical intensity to a temperature reading. The pyrometer uses a relationship between the temperature (T) and the measured optical power (I) which can be expressed as:

$$I \propto \exp(-C/\lambda T) \quad (2)$$

where C is a constant equal to 14388 μm*K. By taking the ratio of optical power evaluated at two different thicknesses, where the real temperature is the same but the measured optical powers are different, the proportionality is removed from Eq.2 yielding in a exact equation.

$$I1/I2 = t1/t2 = \exp[-C/\lambda(1/T1-1/T2)] \quad (3)$$

The ratio of I1/I2 is the ratio of the optical transmissions expressed in Eq.1. If T1 is the pyrometer reading at the start of the film growth before thin film interference is a problem and T2 is the pyrometer reading at some point in the film growth, this formula can predict the pyrometer readings for a constant growth temperature by fixing T1 and solving for T2. The real growth temperature can be calculated by letting T2 be the pyrometer reading and solving for T1, the pyrometer reading without thin film interference.

The aforementioned model is not complete in that the experimental pyrometer readings show damped oscillations. The infrared radiation measured by the pyrometer has much less energy than the bandgap of the epilayer or the substrate, so the damping is not attributed to absorption by the semiconductor. The reason for the damped oscillations is that the pyrometer used in this experiment does not select a single wavelength, but passes a finite band extending from 2.0 μm to 2.6 μm. After integrating the optical transmission over the band of wavelengths and then calculating the temperature reading, the theoretical result does indeed exhibit damping due to the phase difference of each wavelength resulting in a cancellation of the oscillation as the film grows thicker. The actual rate of damping depends on the spectral response of the pyrometer. Experimental observations of the damping of pyrometer oscillations were used to estimate the spectral response.

Figure (3a) shows the sulfur fraction as a function of film thickness determined by the x-ray simulation for the ZnMgSSe film discussed in the introduction (Fig. 1a,b). For the same epilayer, the reconstructed growth temperature, calculated using the pyrometer oscillation model, is plotted in Figure (3b). The sulfur sticking coefficient using the x-ray simulation data

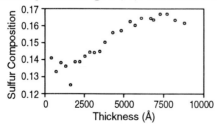

Figure 3(a). Sulfur concentration profile used to obtain figure 1(b).

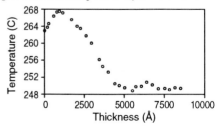

Figure 3(b). Growth temperature profile of the sample in figure 1.

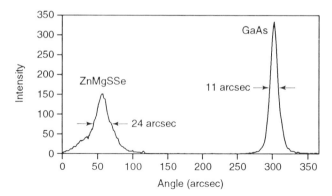

Figure 4. Experimentally measured x-ray rocking curve of a ZnMgSSe film growth with improved temperature control.

is .12 at 265 degrees and .145 at 250 degrees, which is in agreement with the data presented in Figure (2).

We have also applied the pyrometer oscillation model to a film growth to confirm that improved temperature control can result in constant epilayer composition. Figure (4) is a plot of the four crystal x-ray diffraction rocking curve of a growth with the improved temperature control. The FWHM is 24 arcseconds and there is only one peak, consistent with constant composition.

It is possible to anticipate the two main processes that change the amount of power needed to maintain constant temperature. The substrate temperature change due to changes in background temperature can be a major factor in growth using a compound ZnS source. For example, in the Perkin Elmer MBE, when the film is not being grown, the substrate holder is lowered out of the flux path of the ovens; the background temperature is the temperature of the liquid nitrogen cryoshrouds. When the substrate holder is raised into the flux paths, the approximately 900 °C ZnS source presents a much hotter background which can cause a 10 °C rise in substrate temperature over a 5 to 10 minute period. Therefore, we must initially drop the substrate heater power upon nucleation of the film. Also, deposition of the first micron of the film raises the emissivity of the substrate holder so that more power is needed to maintain a constant temperature. Because the emissivity change from a totally uncoated to a coated substrate holder is very significant, the GaAs homoepilayer is at least 1 μm thick to minimize the change in the emissivity of the substrate holder during the II-VI film growth.

Variation in substrate temperature is not the only cause of sulfur drifting. As mentioned earlier in reference to the flux measurements, the flux sources are not perfectly stable. There will be some drift in the flux from the ZnS oven with time that will affect the film's composition. However, this effect is less severe than the observed drifting due to substrate temperature fluctuations.

4. Conclusions

The technological importance of ZnMgSSe mandates that more effort be expended to optimize the quality of the material. We have described a means to experimentally determine the sticking coefficients of the elements as a function of growth parameters. These results suggested that proper substrate temperature control is crucial for accurate sulfur composition in ZnMgSSe. In order to improve temperature control, we have calculated the effect of thin film interference on optical pyrometer readings to allow the pyrometers use in accurate substrate temperature control. A useful side benefit is that the growth rate can be estimated from the pyrometer oscillation period. The pyrometer oscillation model was used to analyze a film to independently confirm the temperature dependance of the sulfur sticking coefficient. The improved temperature control was used to grow a 2 µm thick ZnMgSSe epilayer with constant composition indicated by x-ray diffraction rocking curves with FWHM of 24 arcseconds.

Acknowledgements

The authors would like to thank Professor Dave Dewitt for his useful comments on temperature control. This work was supported by ARPA/ONR URI grant 286-25043, NSF/MRG grant 9221390-DMR and AFOSR grant F49620-92-J-0440.

References

[1] H Okuyama, Y. Morinaga, T. Miyajima, F. Hiei, M. Ozawa and K. Akimoto, 1992 *Electron. Lett.* 28, 1798-9.
[2] N. Nakayama, S. Itoh, K. Okuyama, M. Ozawa, T. Ohata, K. Nakano, M. Ikeda, A. Ishibashi, and Y. Mori, 1993 *Electron. Lett.* 29, 2194-5.
[3] A. Salokatve, H. Jeon, J. Ding, M. Hovinen, A.V. Nurmikko, D.C. Grillo, L. He, J. Han, Y. Fan, M. Ringle, R.L. Gunshor, G.C. Hua, and N. Otsuka, 1993 *Electon. Lett.* 29, 2192-3.
[4] O. Ueda, 1988 *J. Electrochem. Soc.* 135, 11C-22C.
[5] J. Han, M. Ringle, Y. Fan, R.L. Gunshor, and A.V. Nurmikko, submitted to Appl. Phys. Lett. 1994.
[6] D.C. Grillo, Y. Fan, J. Han, L. He, R.L. Gunshor, A. Salokatve, M. Hagerott, H. Jeon, A.V. Nurmikko, G.C. Hua, and N. Otsuka, 1993 *Appl. Phys. Lett.* 63, 2723-5
[7] S. Wright, T. Jackson, and R. Marks, 1990 *J. Vac. Sci. Technol.* B 8, 288-92.
[8] A. SpringThorpe and A. Majeed, 1990 *J. Vac. Sci. Technol.* B 8, 266-70.
[9] E. Hecht, 1987, Optics (Addison-Wesley), 333-391.

Dark defects in II-VI blue-green laser diodes

G C Hua, D C Grillo, J Han, M D Ringle, Y Fan and R L Gunshor

School of Electrical Engineering, Purdue University, West Lafayette, IN 47907, USA

M Hovinen and A V Nurmikko

Division of Engineering and Department of Physics, Brown University, Providence, RI 02912, USA

N Otsuka

Department of Materials Science, Japan Advanced Institute of Science and Technology, Hokuriku, Tatsunokuchi-machi, Nomigun, Ishikawa 923-12, Japan

Abstract: The nature and origin of the nonluminescent dark defects responsible for the current rapid degradation of II-VI blue-green laser diodes based on the ZnCdSe/ZnSSe/ZnMgSSe pseudomorphic separate confinement heterostructure have been examined by transmission electron microscopy. The dark defects observed in the laser stripe region by electroluminescence microscopy have been identified to be dislocation networks developed at the quantum well region. The dislocation networks have been observed to be nucleated at dislocations originating from pairs of V-shaped stacking faults which are nucleated at or near the II-VI/GaAs interface.

1. Introduction

The development of blue-green lasers and light emitting diodes based on II-VI wide bandgap semiconductor quantum well structures received much attention due to their potential application in the areas including high density optical data storage and high brightness visual displays. Since the first blue-green lasers and light emitting diodes based on wide band gap II-VI semiconductor materials grown by molecular beam epitaxy (MBE) were realized in 1991 [1-3], progress has been made in obtaining low threshold current and voltage, high output power and long lifetime. Brief continuous-wave (CW) laser operation at room temperature has been reported for laser diodes based on the ZnCdSe/ZnSSe/ZnMgSSe separate confinement heterostructure (SCH) configuration and using the Zn (Se, Te) graded contact [4-6].

Understanding and subsequently eliminating the device degradation is an important task to further extend the device lifetime and make practical devices. Guha et al [7] were the first to report structural studies of degradation of blue-green light emitting devices based on the ZnCdSe/ZnSSe single confinement structure, although not operating as a laser. We previously reported our preliminary results of the microstructural study of a particular degraded gain-guided laser diode based on the ZnCdSe/ZnSSe/ZnMgSSe pseudomorphic SCH structure carried out by transmission electron microscopy (TEM) [8, 9]. In this paper, we describe our study on the degradation process, and the nature as well as the origin of the degradation dark defects in II-VI blue-green laser diodes based on the ZnCdSe/ZnSSe/ZnMgSSe pseudomorphic SCH structures. Nonluminescent dark defects observed in the laser stripe region have been identified to be patches of a dislocation network developed at the active quantum well (QW) region. By using TEM stereo microscopy, it has been found that the dislocation networks are nucleated at dislocations originating from pairs of V-shaped stacking faults, which are nucleated at or near the II-VI/GaAs interface.

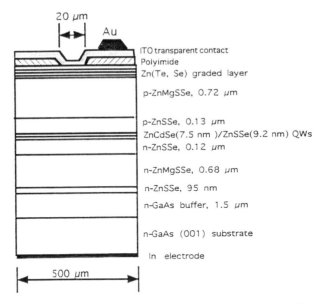

Fig. 1: Configuration of the gain-guided laser diodes mainly employed in this study.

2. Experimental procedures and results

Results described in this paper are mainly from gain-guided laser diodes which were fabricated from one particular three quantum well ZnCdSe/ZnSSe/ZnMgSSe pseudomorphic SCH structure, as shown in Fig.1. Transparent indium-tin-oxide (ITO) contacting material was deposited as the contact for the 20 μm wide and 1 mm long laser stripe of each laser diode. For degradation, the laser diodes were kept at a voltage of typically 10 - 15 V with a duty cycle of 4%. The voltages were chosen to be just above the threshold for lasing in the beginning of the tests. (The high threshold voltages were due to higher contact resistance of the ITO contacting material compared to the conventional metal contact.) The injected current was estimated to be 1.5 - 2 kA/cm^2. The devices lased typically for a few seconds at the beginning of the tests. Formation of triangular nonluminescent dark defects in the laser stripe region was observed by *in situ* EL microscopy performed through the transparent ITO contacting material. Figure 2 shows two typical triangular dark defects observed. As the triangular dark defects grew, they remained well bounded along two directions, forming two sharp edges as shown in Fig. 2.

Following degradation, the microstructure of one particular degraded diode was studied using plan-view TEM in a JEOL 2000 EX transmission electron microscope at 200 kV. The TEM sample was prepared by ion-milling the device from both the top and the bottom sides. Details about the sample preparation and TEM plan-view observation can be found in our previous paper [9]. Areas that are away from the laser stripe region, and therefore are non-degraded, were investigated first. Pairs of V-shaped stacking faults and associated threading dislocations were observed as the most commonly observed type of defects present in the structure. This type of complex of defects will be described in more detail later. In the degraded laser stripe region, patches of dislocation networks were observed.

Figure 3 is a representative TEM bright field image under normal two-beam condition showing a part of such a dislocation network. Weak-beam imaging indicated that the network consists mainly of dislocation dipoles elongated primarily along two directions [9]. It is quite evident that the directions along which the dipoles extend coincide with those of the fine dark features observed in Fig. 2. It is not clear at present whether the dipoles are of the interstitial

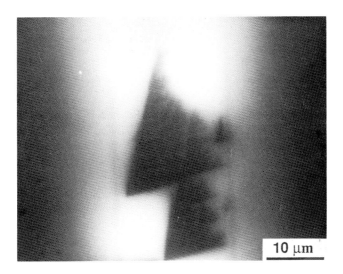

Fig. 2: Two triangular dark defects observed in the laser stripe region by electroluminescence microscopy after about two minutes of degradation.

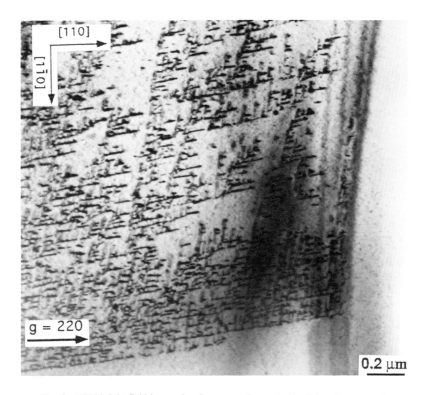

Fig. 3: TEM bright-field image showing a part of a patch of a dislocation network.

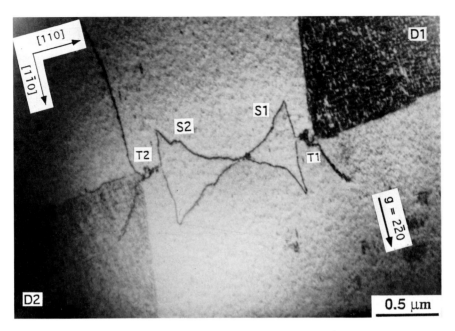

Fig. 4: Bright-field image showing two patches of dislocation networks (D1 and D2), which are nucleated at threading dislocations (T1 and T2) originating at a pair of V-shaped stacking faults (S1 and S2).

type or of the vacancy type. Three nearly vertical dark bands are seen in the right part of Fig.3. They have been identified to be the edges of the three QWs in the wedge shaped plan-view TEM sample. Only the area on the left of the dark bands includes the QWs. It is apparent that the dislocation network starts to show at the edges of the QWs, which indicates that the network is localized at the QW region.

Figures 4 is a bright field image showing the ends of two patches of dislocation networks (marked D1 and D2). D1 is the end of the same dislocation network as is shown in Fig. 3, while D2 is the end of another dislocation network. The two patches of dislocation networks are much denser at the end area, and they are well bounded along two directions, which are nearly the [340] and [3̄40] directions. Dislocation networks bounded along nearly the [430] and [43̄0] directions have also been observed. These directions are in agreement with the two initial sharp edges of the triangular dark defects observed by EL microscopy. (As can be seen in Fig. 2, as the dark defects grew, the angle between the two sharp edges bounding the triangular dark defects were slightly decreased.) We identify the dislocation networks with the triangular dark defects observed in our EL microscopy.

It is observed in Fig. 4 that each of the two dislocation networks originates from a dislocation (marked T1 or T2) which in turn starts from a pair of V-shaped stacking faults (S1 and S2). The contrast from the two stacking faults is absent due to the diffraction condition, while the partial dislocations bounding them are clearly observed [9]. As mentioned before, this type of defect complex, consisting of a pair of V-shaped stacking faults together with associated threading dislocations, is the most commonly observed defect present in the device structure (including the non-degraded areas that are as far away from the laser stripe as 0.25 mm). Figure 5 shows such a defect complex observed in the non-degraded area; the defect complexes are therefore believed to represent defects formed during the MBE growth. The density of these defects is estimated to be in the middle $10^5/cm^2$ range.

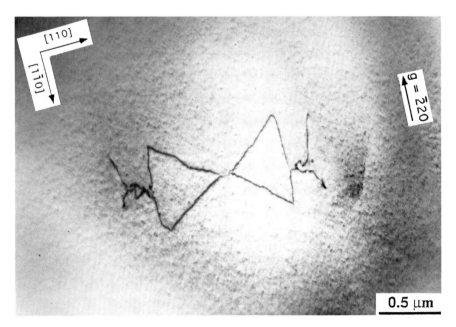

Fig. 5: Bright-field image showing a defect complex observed in the non-degraded area.

The three-dimensional arrangement of the defects shown in Fig. 4 has been investigated by using TEM stereo microscopy [9]. The result is schematically shown in Fig. 6. The two patches of dislocation networks reside at the QW region, as mentioned before. The two stacking faults appear to begin at or near the II-VI/GaAs interface and end at the n-ZnSSe/ n-ZnMgSSe interface. The two Shockley partial dislocations bounding each of the stacking faults unite at the n-ZnSSe / n-ZnMgSSe interface to form a perfect (or narrowly dissociated) dislocation having the Burgers vector b=1/2[110] and threading upwards. When the dislocation reaches the QW region, as can be seen in Fig. 4 (and also in Fig. 5), it appears that the dislocation starts to dissociate widely again, resulting in formation of a small stacking fault starting at that region. The defect finally becomes a dislocation half-loop at or near the p-ZnMgSSe/p-ZnSSe interface, and threads upwards. The detailed dislocation reaction near the p-ZnMgSSe/p-ZnSSe interface is not clear at present.

Our TEM microstructural study performed on other degraded laser diodes which were fabricated from another particular ZnCdSe/ZnSSe/ZnMgSSe SCH configuration also revealed that pairs of stacking faults nucleated at or near the II-VI/GaAs interface can be the nucleation source for dark defects. In that particular SCH structure, while the stacking faults ended at the n-ZnSSe/ n-ZnMgSSe interface, two separate dislocations have been observed to be generated from each stacking fault and to thread upwards, causing nucleation of dark defects at the quantum well region.

3. Conclusions

In summary, we have carried out degradation studies of blue-green laser diodes fabricated from the ZnCdSe/ZnSSe/ZnMgSSe SCH structures. Nonluminescent dark defects observed by EL microscopy in the laser stripe region have been identified to be patches of dislocation networks developed at the QW region. Complexes of defects starting from a pair of V-shaped stacking faults which are nucleated at or near the II-VI/GaAs interface have been observed as the most commonly observed defects present randomly throughout the SCH laser structures including non-degraded areas. These defect complexes are believed to represent defects

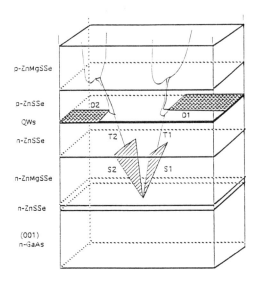

Fig.6: TEM stereo microscopy determined three-dimensional arrangement of the defects shown in Fig. 4.

formed during the MBE growth. The dislocation networks have been observed to be nucleated at such complexes of defects. The results suggest the importance of eliminating point defects as well as stacking faults and dislocations in the laser structures. Subsequent structures having modified growth of the quantum well region (primarily higher growth temperature to reduce point defects) exhibited a reduced tendency for the generation of dark defects, and resulted in the room temperature CW operation cited above [5, 6].

Acknowledgments:

This work has been supported by ARPA/ONR University Research Initiative grant number 28625043, AFOSR grant number F49620-92-J-0440, NSF/MRG grant number 9221390-DMR, and NSF grant number 9202957-DMR.

References:

[1] Haase M A , Qiu J, Depuydt J M and Cheng H 1991 *Appl. Phys. Lett.* **59** 1272-1274
[2] Jeon H, Ding J, Patterson W, Nurmikko A V, Xie W, Grillo D C, Kobayashi M and Gunshor R L 1991 *Appl. Phys. Lett.* **59** 3619-3621
[3] Xie W, Grillo D C, Gunshor R L, Kobayashi M, Jeon H, Ding J, Nurmikko A V, Hua G C and Otsuka N 1992 *Appl. Phys. Lett.* **60** 1999-2001
[4] Nakayama N, Itoh S, Ohata T, Nakano K, Okuyama H, Ozawa M, Ishibashi A, Ikeda M and Mori Y 1993 *Electron. Lett.* **29** 1488-1489
[5] Salokatve A, Jeon H, Ding J, Hovinen M, Nurmikko A V, Grillo D C, He L, Han J, Fan Y, Ringle M D, Gunshor R L, Hua G C and Otsuka N 1993 *Electron. Lett.* **29** 2192-2194
[6] Fan Y, Grillo D C, Ringle M D, Han J, He L, Gunshor R L, Salokatve A, Jeon H, Hovinen M, Nurmikko A V, Hua G C and Otsuka N 1994 *J. Vac. Sci. Technol.* **B12** 2480-2483
[7] Guha S, Depuydt J M, Haase M A, Qiu J and Cheng H 1993 *Appl. Phys. Lett.* **63** 3107-3109
[8] Hua G C, Otsuka N, Clark S M, Grillo D C, Fan Y, Han J, He L, Ringle M D, Gunshor R L, Hovinen M and Nurmikko A V 1994 *Electron Microscopy 1994* (Physique Les Ulis, France, ed. by Jouffrey B and Colliex C) Volume 2A 625-626
[9] Hua G C, Otsuka N, Grillo D C, Fan Y, Han J, Ringle M D, Gunshor R L, Hovinen M and Nurmikko A V 1994 *Appl. Phys. Lett.* **65** 1331-1333

Stimulated emission from GaN grown on SiC

A S Zubrilov, V I Nikolaev

A F Ioffe Institute and CREE Research EED, 26 Polytechnicheskaya Str., St. Petersburg, Russia

V A Dmitriev, K G Irvine, J A Edmond and C H Carter, Jr.

Cree Research, Inc., 2810 Meridian Parkway, Durham, NC 27713 USA

Abstract. Photoluminescence (PL) of GaN layers grown on 6H-SiC substrates was investigated in the temperature range 77 - 900 K. The edge luminescence was dominant in the PL spectrum throughout the entire temperature range. The temperature dependence of the GaN exitonic band gap was measured. The edge cavity stimulated emission from photopumped GaN layers was observed in the temperature range 77 - 450 K. It is the first observation of stimulated emission from any A^3N material above 300 K. The full width at half maximum (FWHM) of the stimulated emission peak was ~ 7 nm at 450 K. Multipass stimulated emission with Fabry-Perot modes was observed from GaN for the first time. The FWHM of Fabry-Perot modes was ~0.2 nm (300 K). We believe that microcracks created during the cleaving of the samples served as the Fabry-Perot cavity mirrors.

1. Introduction

The optical properties of GaN and its solid solutions with other nitrides is currently being investigated in many laboratories for application in electro-optical devices such as light emitting diodes and lasers. Stimulated emission from photopumped GaN and InGaN films (about 1 micron thick) grown on sapphire substrate has been reported for 77 K and 300 K [1-7]. SiC is a promising substrate for A^3N films because of its smaller lattice mismatch and high electrical conductivity [8-10]. In this paper we report the first observation of stimulated emission from a photopumped GaN layer grown on a SiC substrate.

2. Experiment

We investigated photoluminescence properties of GaN layers grown on 6H-SiC wafers. GaN layers about 1 micron thick were grown epitaxially on the (0001)Si face. The PL experiments were performed using a nitrogen laser (wavelength ~337.1 nm) with a peak power of 2 kW. The pulse duration was about 10 ns and repetition rate was up to 100 Hz. Samples were cleaved with a size of approximately 5 x 2 mm. The laser beam was focused on the sample surface yielding a pump density around 10 MW/cm^2 (laser beam diameter was about 100 microns). The angle of excitation on the sample face was ~90º. Emission spectra were detected in a direction perpendicular to the exciting beam (edge emission mode). The emission spectra was measured by a 0.6 m monochromator MDR-23 using a photo multiplier connected with dc digital voltmeter. The experiments were carried out in the temperature range from 77 K to 900 K.

3. Results

Generally PL spectra from GaN grown on SiC consisted of two main peaks: (1) an edge peak assigned to a bound exciton (wavelength of ~ 364 nm at 300 K), and (2) the so called "defect" peak at ~550 nm. In the entire temperature range 77 - 900 K the edge peak dominates the spectrum (Fig. 1). At 300 K, the FWHM of the edge peak was ~ 30 meV. From this data, the temperature dependence of the exciton band gap was determined: $E(eV) = 3.56 - 5.6 \cdot 10^{-4} \cdot T$ (Fig. 2). The GaN optical energy gap data measured by optical absorption by Leszczynski et al. [11] is also presented in Fig. 2.

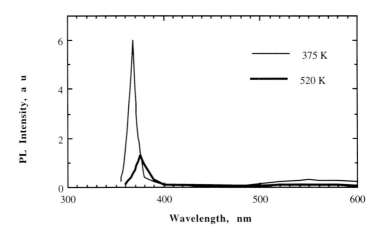

Figure 1. PL spectrum of GaN layers grown on SiC.

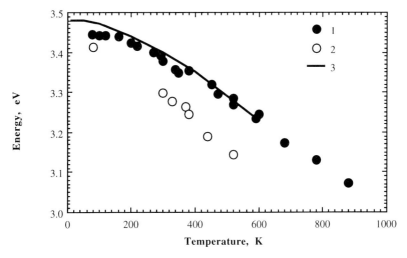

Figure 2. Temperature dependence of the excitonic band gap of GaN (1) and stimulated emission from GaN (2) grown on SiC, and optical energy gap (3) measured by optical absorption for GaN grown on sapphire [11].

Room temperature PL spectra of cleaved samples for various excitation levels are shown in Fig. 3. The PL output power at a peak wavelength of ~ 374 nm rose non-linearly with increasing laser excitation density (Fig. 4). Narrowing of the peak and its shift to lower photon energy were also observed. The minimum FWHM of the peak was ~ 3 nm. These features of the PL spectra are identical to features of stimulatied emission observed from photopumped GaN grown on sapphire [1].

Stimulated emission from cleaved GaN samples was observed in the temperature range from 77 to 450 K (Fig. 2). The stimulated emission peak at various temperatures is shown in Fig. 5. At 77 K, stimulated emission was observed at a wavelength of 364 nm. The intensity of the peak was about ten times higher than that at 300 K. At 450 K, the stimulated emission peak wavelength was ~ 386 nm. The FWHM of the peak was in the range of 7- 9 nm with an intensity approximately one order of magnitude lower than at room temperature. A very weak peak was detected at 520 K with a wavelength of ~ 395 nm and a FWHM equal to ~ 18 nm.

We also observed stimulated emission with Fabry-Perot cavity modes (Fig. 6). The FWHM of the Fabry-Perot modes were 0.2 - 0.5 nm and the distance between the fingers was ~ 0.7 nm (300 K). The cavity length was estimated from:

$$L = \frac{\lambda_o^2}{2n(\lambda_{m+1} - \lambda_m)},$$

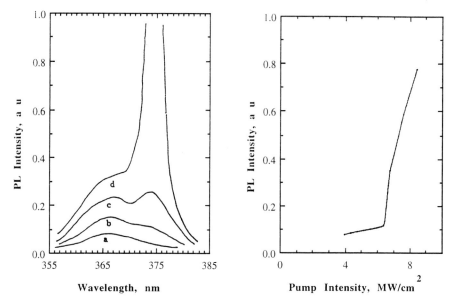

Figure 3. Room temperature PL spectra at various excitation intensities.

Figure 4. Dependence of stimulated emission signal on the pump power density (300 K)

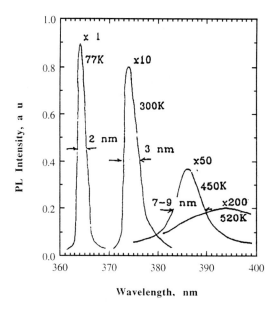

Figure 5. The stimulated emission peak at various temperature.

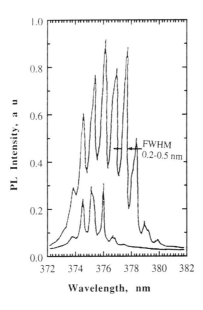

Figure 6. Multipass stimulated emission signal with Fabry-Perot cavity modes at two different pump densities (300 K).

where λ_m, λ_{m+1} are the wavelengths of neighbor Fabry-Perot fingers, n is the refractive index, and λ_o is the wavelength of the main emission peak. According to these calculations, L equals 0.03 - 0.04 mm. This value is much smaller than the sample size. Optical microscopy shows that the sample had micro cracks produced by cleaving. These cracks are parallel and the distance between cracks is 20 - 50 microns (Fig. 7). We believe that these cracks serve as cavity mirrors and generate the Fabry-Perot modes.

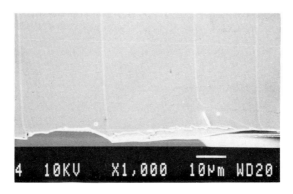

Figure 7. Photomicrograph of the cleaved GaN sample with cracks.

Estimation of the optical gain in the sample with a cavity was done in the same manner as that described by Hakki and Paoli [12]. The calculated net gain at threshold pumping is 100 cm^{-1}. Estimation of the cavity quality in this case yields a Q of 1800.

4. Summary

Photoluminescence of GaN grown on 6H-SiC substrates was investigated in the temperature range from 77 to 900 K. The temperature dependence of the excitonic band gap of GaN was measured for the entire temperature range. Stimulated emission from photopumped GaN layers was observed up to 450 K. It is the first observation of stimulated emission from GaN at a temperature higher than 300 K. The wavelength and FWHM of the stimulated emission peak were 364 nm and 2 nm at 77 K; 374 nm and 3 nm at 300 K; and 386 nm and 7 - 9 nm at 450 K, respectively.

Fabry-Perot cavity modes were observed in stimulated emission from GaN for the first time. The minimum estimated optical gain of the cavity was about 100 cm^{-1} with a Q of 1800. The FWHM of the Fabry-Perot modes was ~ 0.2 nm corresponding to a cavity length of ~ 0.03 - 0.04 mm. It is speculated that micro cracks produced during sample cleaving serve as cavity mirrors.

Acknowledgments. We appreciate the measurements performed by D.V. Tsvetkov. This work was supported in part by the Advanced Research Projects Agency and the Office of Naval Research.

References

[1] Dingler R, Shaklee K L, Leheny R F, 1971 *Appl. Phys. Lett.* **19** 5-7
[2] Catalano I M, Cingolani A, Ferrara M , Lugara M and Minafra A 1978 *Solid State Commun.* **25** 349-51
[3] Cingolam R, Ferrara M, and Lugara M 1986 *Solid State Commun.* **60** 705
[4] Amano H, Asahi T, Akasaki I, 1990 *Jap. J. Appl. Phys.* **29** L209-10
[5] Khan M A , Olson D T, Van Hove J M, and Kuznia J N 1991 *Appl. Phys. Lett.* **58** 1515-7
[6] Amano H, Tanaka T, Kunii Y, Kato K, Kim S T, and Akasaki I 1994 *Appl. Phys. Lett.* **64** 1377-9
[7] Yung K, Yee J, Koo J et al 1994 *Appl.. Phys. Lett.* **64** 1135-7
[8] Strite S and Morkoc 1992 *J. Vac. Sci. Technol.* **B10** 1237-66
[9] Lin M E, Sverdlov B, Zhou G L, and Morkoc 1993 *Appl. Phys. Lett.* **62** 3479-81
[10] Dmitriev V A, Irvine K, Edmond J A, and Carter C H 1994 *Technical Program with Abstracts, 1994 Electronic Materials Conference* A21
[11] Leszczynski M, Perlin P, Suski T, Teisseyre H, Grzegory I, Polowski S, Jun J 1994 *Transactions of the Second Int. High Temperature Electronics Conference* (Charlotte, North Carolina) **2** P-239 - 42
[12] Hakki B W and Paoli T L 1973 *J. Appl. Phys.* **44** 1299-306

High Speed Optoelectronics

Radhakrishnan Nagarajan, Kirk Giboney, *Rangchen Yu, Daniel Tauber, John Bowers and Mark Rodwell

Department of Electrical & Computer Engineering, University of California, Santa Barbara, CA 93106-9560
*Conductus Inc., 969 W. Maude Ave., Sunnyvale, CA 94086

Abstract. We review the state of the art in design and fabrication of high speed semiconductor lasers and photodetectors. For high speed operation, semiconductor lasers are designed for large differential gain, minimal carrier transport time, low device parasitics and optimum microwave signal transmission. We will present data on narrowband millimeter wave optical link at 35 GHz operating at 1.3 μm wavelength and a wideband cryogenic optical link with 30 GHz bandwidth operating at 1.55 μm. Conventional p-i-n photodetectors with larger than 100 GHz bandwidth are reviewed. An improvement over this is a traveling-wave geometry which is impedance-matched for increased bandwidth and bandwidth-efficiency product over a lumped-element design. Bandwidths higher than 170 GHz at an efficiency of 42% have been measured in these devices.

1. Introduction

High speed optoelectronic devices are essential components in the modern optical fiber communication systems, which are operating at increasing bit rates. Recently, an optical fiber transmission system using the NRZ (non-return to zero) format at 160 GBit/s (eight channel wavelength division multiplexed, with each channel at 20 GBit/s) has been demonstrated [1]. Another advancement in the transmission system technology is the use of optical solitons. Due to their unique properties, these soliton systems are capable of error-free transmission over *unlimited* distances [2]. Many of these systems are also being applied to solve the speed bottlenecks in short haul interconnections between high speed computing elements and control processes.

The integral part of all these high speed fiber optic links are the lasers, photodetectors, modulators, multiplexers, demultiplexers and switches. In this paper, we will review research in high speed lasers and photodetectors.

2. High Speed Lasers

2.1. High Speed Laser Design

To maximize the modulation bandwidth of a semiconductor laser one has to maximize the differential gain (quantum well active areas, which may also be strained or doped), minimize the carrier transport and capture time (especially in quantum well lasers) and minimize the device parasitics (optimized device structures). Close attention also has to be paid to the microwave design of the laser structure. Low threshold current and large output power are essential to obtain the large photon densities required without incurring thermal problems.

One of the effects of carrier transport is given in Fig. 1 which shows the experimental and theoretical variation in the 3 dB modulation bandwidth with separate confinement heterostructure (SCH) width at different power levels for a InGaAs/GaAs single quantum well laser. The optimum SCH width also corresponds roughly to the point at which the optical confinement factor is a maximum. For a narrow SCH, the bandwidth drops off due to decreasing optical confinement, resulting in a larger threshold gain and a lower differential gain. At larger SCH widths, the combination of a decreasing confinement factor and an increasing carrier transport time across the SCH limits the modulation bandwidth. At sufficiently high powers for wide SCH devices, a characteristic *drop* due to carrier transport appears in the modulation bandwidth curve [3].

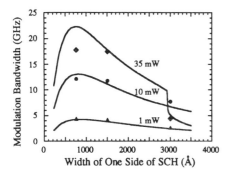
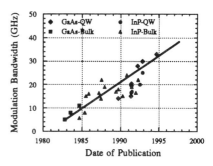

Fig. 1 Comparison of the calculated and measured variation in modulation bandwidth with SCH width.

Fig. 2 The trend in the CW 3 dB modulation bandwidth at room temperature reported for GaAs and InP high speed lasers.

As shown in Fig. 2, presently quantum well lasers have the largest small signal modulation bandwidths (around 30 GHz for InGaAs/GaAs and 25 GHz for InGaAsP/InP systems) [4,5]. The largest bandwidth lasers demonstrated to date also have p-doped active regions to further increase the differential gain and minimize the hole transport delay component, which is the major contributor to the total carrier transport and capture time.

2.2. Microwave Transmission Properties

Conventional high speed laser design assumes that the modulation current pumping the laser is uniform in amplitude and phase along the laser cavity. At very high frequencies this is invalid because the time delay from the feed point to the far end of the device can be a significant fraction of the modulation period. The distributed microwave nature of the laser under these conditions modifies the high frequency current injection and the modulation response [6].

Experimentally, the measured loss per unit length and the phase velocity as a function of frequency are shown as solid lines in Fig. 3. The striking features in the graphs are (1) the enormous loss, near 600 dB/cm at 40 GHz for the forward biased case and (2) the slow phase velocity, around one-tenth the speed of light in vacuum. The large loss has its origin in three sources. First, in a typical laser structure the thick n-doped cladding and substrate form a lossy ground plane for the transmission line. Because the skin depths in the semiconductor for the measured frequency range are significantly less than the 100 μm thickness of the n-cladding and n-substrate, almost all of the longitudinal current on the n-side is carried in the poor conductivity doped semiconductor, and not in the metal contact. In contrast, the 1 μm thick top p-cladding layer is thinner than the semiconductor skin depth and the longitudinal p-

side current will penetrate into the gold contact layer. The current should therefore be carried primarily in the high conductivity gold and this second source of loss should be low if the gold layer is thick enough. The forward biased junction is a third source of loss, because under forward bias a dissipative conduction path is opened that is not present at zero bias.

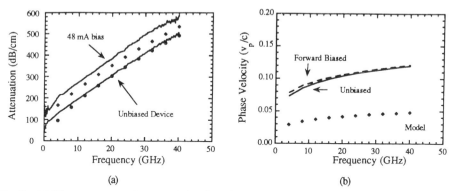

Fig. 3 (a) Microwave attenuation per unit length vs. frequency. (b) Phase velocity, normalized to the speed of light in vacuum, vs. frequency. The solid lines are measured values and the points are calculated.

The frequency dependent loss results in a frequency dependent current injection to the active region. For these set of devices, the 3 dB rolloff frequency occurs at approximately 25 GHz for a 300 µm long device. This is computed for a device that is fed at the center of the laser stripe. This is the optimal feeding condition since waves can propagate out in both directions from the center and maximally pump the device.

2.3. Room Temperature Narrowband Millimeter Wave Fiber Optic Links

The maximum direct modulation bandwidth in semiconductor lasers is presently limited to a little over 30 GHz. Millimeter wave modulation of semiconductor lasers at frequencies larger than this is of great interest for radar, satellite links, and electronic warfare systems. Due to their nature, millimeter wave systems will benefit from efficient narrow-band optical modulation techniques at frequencies above the intrinsic laser modulation bandwidth. One of the techniques that may used to implement a narrowband millimeter wave system is a resonantly enhanced external cavity or monolithically integrated multiple section laser [7].

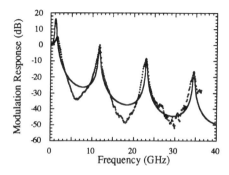
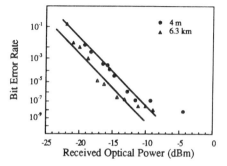

Fig. 4 Experimentally measured and theoretically modeled modulation response of an external cavity laser.

Fig. 5 Comparison of bit error rate as a function of received optical power for 40 MBit/s BPSK data.

Fig. 4 shows the experimental and modeled small signal response of the external cavity configuration operating at 1.3 μm wavelength. The optical feedback from the mirror or grating (for single optical mode operation) in the external cavity causes the modulation response of the system to be enhanced at frequencies corresponding to the multiples of the cavity round trip time. The resonant enhancement of the modulation response results in a series of frequency windows extending out into the millimeter wave range suitable for narrow band data transmission. The BER performance of the grating tuned external cavity laser transmission system for a 40 MBit/s BPSK digital data recovered from the 35 GHz subcarrier is shown in Fig. 5. The BER for the back to back case (4m) was always slightly worse than the transmission over 6.3 km. This shows that the system is not dispersion limited. There is a distinct error floor at around 10^{-9} BER that we believe is the result of the enhancement in the transmitter noise. The error floor can be lowered by increasing the signal power to the laser.

2.4. Cryogenic Millimeter Wave Fiber Optic Links

High speed cryogenic optical fiber links operating at and around 100 K are good candidates for the implementation of interconnects between high T_c superconducting integrated circuits and room temperature electronics. All the superior properties of optical fiber links like low loss, minimum crosstalk, and high bandwidth are available for use at these low temperatures.

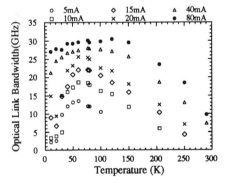
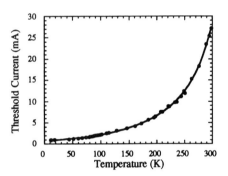

Fig. 6   Variation of fiber optic link bandwidth with temperature.

Fig. 7   Variation of laser threshold current with temperature.

Fig. 6 shows that the bandwidth of an optical link operating at 1.55 μm increases with temperature from about 10 GHz at room temperature to over 30 GHz at about 100 K. At the same time, as shown in Fig. 7, the threshold current is reduced from over 25 mA to about 1 mA. This reduction in threshold makes it possible for the low temperature electronics to directly drive the laser. The decrease in bandwidth below 100 K is due to carrier trapping effects leading to poor carrier transport in the device active region [8].

3. High Speed Photodetectors

3.1. Vertically Illuminated Photodiodes

The other vital component for the development of large-capacity optical fiber communications systems is the photodetector operating at 1.3 μm and 1.5 μm wavelengths. We have fabricated high-speed, high-efficiency GaInAs /InP p-i-n vertically-illuminated photodetectors

(VPDs) which are monolithically integrated with bias tees and matched resistors [9]. This is to enhance their frequency response because connections to bulk bias tees and various mismatched loads degrade the frequency response of high-speed photodetectors.

The integrated circuit design is shown together with external connections in Fig. 8. The bypass diode serves the function of a shunt capacitor. The matched source resistor greatly reduces reflections at the photodetector so frequency response variations are suppressed. This resistor also doubles the ideal RC bandwidth limitation since it cuts the net load presented to the photodiode in half. However, since the resistor forms a current divider with the load, the effective quantum efficiency is also cut in half. By implementing the bypass diode and matched resistor monolithically and close to the photodiode, these elements may be regarded as lumped over the design bandwidth.

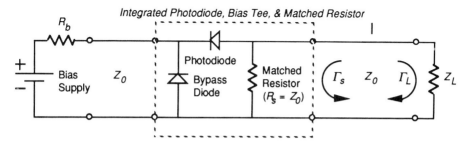

Fig. 8. Circuit diagram of integrated photodiode, bias tee, and matched resistor with external connections.

The photodiode design utilizes a GaInAs/InP double heterostructure to eliminate carrier diffusion, and graded bandgap layers at the hetero-interfaces to minimize carrier trapping. The time domain response of the photodiodes is measured by pump-probe electro-optic sampling, using the InP substrate as the electro-optic modulator. The measured pulse responses having full-width at half-maximum (FWHM) times as short as 3.0 ps for a discrete 2 x 2 μm^2 devices and 3.8 ps for 7 x 7 μm^2 photodiodes integrated with bias tees and matched resistors. The corresponding 3 dB electrical bandwidths from Fourier transforms are 110 and 108 GHz. Discrete devices were found to have 50% external quantum efficiency at 0.97 μm wavelength, and 32% at 1.3 μm. The bandwidth-efficiency product of these photodetectors is near the theoretical limit.

3.2. Traveling Wave Photodiodes

To further increase the bandwidth-efficiency product we developed the traveling-wave photodetector (TWPD). The TWPD is an in-plane illuminated device similar to a waveguide photodetector (WGPD). This type of device is not subject to the bandwidth-efficiency limit of VPDs because its absorption path is orthogonal to the carrier drift field. It is a strong candidate for high-speed communications systems because it is not inherently narrowband, it can operate over a wide temperature range, and its quantum efficiency can be near unity. Also, it is easily integrated with an optical preamplifier.

The trait that distinguishes a TWPD from a WGPD is that the photodetector structure is designed as a microwave transmission line with a characteristic impedance matched to the external circuit. This eliminates the RC time limitation dependence on device area without compromising the advantages of a WGPD. The TWPD is a fully distributed device, and the bandwidth limitation depends on the optical absorption coefficient and velocity mismatch [10].

For direct comparison, TWPDs, WGPDs, and VPDs are fabricated on the same wafer. Fabrication methods for the GaAs TWPDs are similar to those of the long-wavelength VPDs described above except that proton implants are used for isolation, rather than mesa etches. The GaAs/AlGaAs graded double heterostructure is grown by MBE with layer thicknesses such that a 1 µm wide TWPD has a calculated characteristic impedance of about 50 Ω. The ridge waveguide for the PD is defined by chlorine RIE, self aligned to the top ohmic contacts.

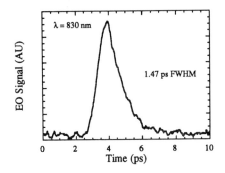

Fig. 9. (a) Measured pulse response and (b) derived electrical frequency response of a 1 µm x 7 µm TWPD.

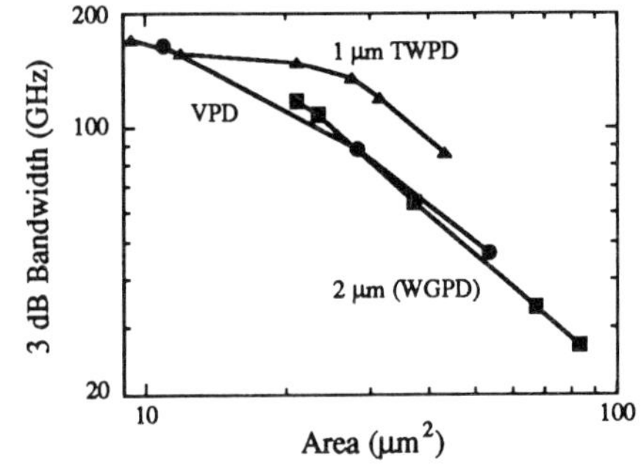

Fig. 10. 3-dB electrical bandwidths of various size in-plane and vertically-illuminated photo-detectors from the same wafer.

Fig. 9 shows the measured pulse response of a 1 x 7 µm^2 GaAs p-i-n TWPD having 1.47 ps FWHM [11]. The equivalent electrical frequency response gives a 3 dB bandwidth of 172 GHz, and significant conversion efficiency at much higher frequencies. The external quantum efficiency of 42% gives this device a 72 GHz bandwidth-efficiency product, the highest reported for any photodetector without gain. Evidence that the devices are indeed TWPDs is presented in Fig. 10, where the bandwidths of VPDs, 2 µm wide (WGPDs), and 1 µm wide TWPDs from the same wafer are plotted as functions of their area. The TWPDs have much higher bandwidths and efficiencies than the lumped devices of the same areas. The decrease in bandwidth for longer TWPDs is due to microwave loss rather than the lumped RC time limitation of the VPDs and WGPDs. TWPDs are best suited for communications systems requiring greater than 100 GHz bandwidth.

This work was funded by ARPA and Office of Naval Technology block program in Electro-Optics Technology.

References

[1] Chraplyvy, A.R., et al., 1994, *Conf. on Optical Fiber Comm. (OFC)*, paper PD19.
[2] Nakazawa, M., et al., 1993, *Electron. Lett.* **29** 729.
[3] Nagarajan, R., et al., 1992, *IEEE J. Quantum Electron.* **28** 1990.
[4] Weisser, S., et al., 1992, *Electron. Lett.* **28** 2141.
[5] Morton, P.A., et al., 1992, *Electron. Lett.* **28** 2156.
[6] Tauber, D.A., et al., 1994, *Appl. Phys. Lett.* **64** 1610.
[7] Nagarajan, R., et al., 1994, *IEEE J. Lightwave Technol.* **12** 127.
[8] Yu, R.C., et al., Aug. 1994, *Appl. Phys. Lett.* **65**.
[9] Wey, Y.G., et al., 1993, *IEEE Photon. Technol. Lett.* **5** 1310.
[10] Giboney, K.S., et al., 1992, *IEEE Photon. Technol. Lett.*, **4**, 1363.
[11] Giboney, K.S., et al., 1994, *52nd Annual Device Research Conf.*, paper VIA-9.

Silicon recrystallization effects in mirror coatings of high-power 980-nm InGaAs/AlGaAs lasers

P W Epperlein and M Gasser

IBM Research Division, Zurich Research Laboratory, CH-8803 Rüschlikon, Switzerland

Abstract. High-reflective PECVD Si/Si_3N_4 and IB Si/Al_2O_3 back facet coating stacks of MBE-grown, strained InGaAs/AlGaAs GRINSCH-SQW, 4-μm-wide ridge-waveguide lasers exhibit distinctive properties under laser output power. Raman microprobe spectra show that PECVD Si layers remain amorphous independent of exposure time and power applied, whereas IB Si layers rapidly recrystallize within hours and at lower power levels. Local mirror temperatures are $\approx 2\times$ higher in the IB stacks than in the PECVD stacks. The refractive index of IB Si changes from 3.85 to 3.45 at 980 nm according to a transition from amorphous to crystalline Si. This reduces the reflectivity of the stack from 95 to 91%, and increases the power transmitted by the back facet by a factor of 1.7. Model experiments employing a controlled local recrystallization by external Ar^+ laser irradiation showed that recrystallization has an impact on near-field patterns and on nonlinearities in the light-current characteristics.

1. Introduction

High-power, fundamental-mode, InGaAs/AlGaAs strained quantum well lasers play a dominant role in optical fiber communication systems as pump sources for Er^{3+}-doped fiber amplifiers [1]. This application requires devices with excellent electrical and optical properties and reliability. Front and back mirrors are known to be very sensitive elements in laser cavities. Using Raman microprobe spectroscopy, we have investigated the structural stability of Si layers in mirror coating stacks under optical power exposure. We have also characterized other distinctive features between Si layers deposited by plasma-enhanced chemical-vapor deposition (PECVD) and ion-beam (IB) sputtering, and in particular investigated the effect of Si recrystallization on the performance of the lasers. Additional results for model experiments employing a controlled local recrystallization by external Ar^+ laser irradiation will also be reported.

2. Experimental

Ridge waveguide, strained InGaAs/AlGaAs, single quantum well (SQW), graded-index separate-confinement heterostructure (GRINSCH) lasers grown by molecular beam epitaxy (MBE) have been used throughout these studies. The facets of the 750-μm-long and 4-μm-wide lasers are (110)-oriented and were formed by cleaving. After cleaving, the back facets were coated with high-reflective (95%) PECVD-Si/Si$_3$N$_4$ or IB-Si/Al$_2$O$_3$ 5-layer stacks. Single PECVD-Si$_3$N$_4$ or IB-Al$_2$O$_3$ layers, respectively, were deposited on the front facets to obtain a reflectivity of about 10%. The laser chips were soldered junction-side-up onto heat sinks for measurements.

The Raman microprobe spectra were measured on the (110) rear mirrors in air and at room temperature in the backscattering geometry by using the 457.9-nm line of an Ar$^+$ laser as excitation source. The probe beam was focused to a spot diameter of about 1 μm, and its low incident power of 2 mW heated the mirror surface by only about 20 K. The Raman signal was dispersed in a 0.9-m double-grating spectrometer and detected by a cooled photomultiplier with GaAs cathode and standard photon counting electronics. The spectra were recorded with a spectral resolution of 1.5 cm^{-1}.

Local temperatures on the back facets have been measured by Raman microprobe using the Stokes/anti-Stokes line intensity ratio of the Si phonon mode in the Si layers. Details of this evaluation method are given elsewhere [2].

3. Results and discussion

The Raman spectra of the as-deposited PECVD and IB Si layers show no sharp lines, but they do show a broad asymmetric peak near 480 cm^{-1} with a typical line width of 70 cm^{-1} (see Fig. 1a). The spectra strongly resemble those usually observed for amorphous silicon (a-Si) [3], and thus verify that the original Si layers were amorphous. The broad peak arises from the relaxation of the normal Raman selection rules for scattering from a crystal as a result of the loss of translational symmetry in the amorphous phase. However, under laser operation, the structure of PECVD- and IB-prepared a-Si layers was modified in quite a different manner. This is definitely shown in Raman spectra recorded from the optical near-field pattern of typically 3 μm$\times$1 μm. Thus PECVD a-Si remains amorphous, independent of exposure time and cw optical output power (measured up to 4 (230) mW emitted from the back (front) facet). This structural stability is known to be due to the large amount ($\approx$10 at.%) of atomic hydrogen [4] in the a-Si films prepared in a glow discharge decomposition of SiH$_4$. Whereas hydrogen is known to be essential for good electrical performance of a-Si devices by saturating randomly distributed dangling bonds, it protects the a-Si from recrystallization.

In contrast, hydrogen-free IB a-Si layers show an additional sharp line in the Raman spectra near 520 cm^{-1} after only brief laser operation at lower power levels. This line can be attributed to scattering from the threefold degenerate, $k = 0$, optical phonon of crystalline Si (c-Si) [3]. Figure 1 shows Stokes–Raman spectra obtained from the Si layers in IB-deposited back facet stacks at various laser operation times. It clearly exhibits an increase of the intensity of the c-Si phonon mode relative to that of the a-Si mode. After about 10 h at 200 mW, the c-Si peak intensity saturates in keeping with a nearly complete recrystallization of a-Si to c-Si. From the spectra in Fig. 1 we have deduced the volume

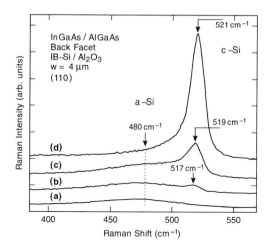

Figure 1. Raman scattering spectra for a-Si layers in IB-deposited Si/Al$_2$O$_3$ back facet coating stacks of InGaAs/AlGaAs lasers; (a) as deposited, (b) after 0.5 h, (c) 4 h and (d) 10 h operation at 4 (230) mW back (front) facet power emission.

fraction of crystallinity ρ in the two-phase system of Si microcrystallites embedded in the a-Si matrix. Thus $\rho \approx 5$, 30 and 100% after a laser operation of 0.5, 4 and 10 h, respectively, at 4 (230) mW back (front) facet power emission. The recrystallization strength is strongest within the optical near-field spot. A line scan perpendicular to the active layer towards the substrate showed that ρ decreases rapidly outside the near field, and goes to zero at about 5 µm away from the near field. Figure 1 also shows a slight shift of the c-Si phonon peak to higher energies with increasing ρ, which can be interpreted as a particle size effect [5]. According to [5] the crystallite size in the Si layers was estimated to be about 5 nm for a crystallized layer with $\rho \approx 5\%$ and > 10 nm for a layer with $\rho \approx 100\%$.

Different mirror heating is a further distinctive feature between the two types of high-reflective back facet coating stacks. Local mirror temperatures have been measured within ≈1 µm of the near-field spot from the Stokes/anti-Stokes intensity ratio of the a-Si Raman mode. Figure 2 exhibits actual temperature increases ΔT as a function of laser operating current I and optical power P plotted for IB-Si and PECVD-Si layers. The ΔT data are with respect to room temperature and allow for the low heating effect of the probe laser intensity. The power scale is valid for the IB-Si and PECVD-Si lasers, which have practically the same differential quantum efficiency. As is evident from Fig. 2, the temperatures are at least 2× higher in the IB-Si than in the PECVD-Si layer containing stacks. This is mainly due to the higher absorption coefficient α of IB a-Si, which is $\approx 8 \times 10^3$ cm^{-1} compared to ≈60 cm^{-1} for PECVD a-Si at 980 nm [6]. The highest temperatures measured in the IB-Si layers are significantly lower than those given in the literature for recrystallizing a-Si exclusively with thermal annealing treatments [7], and thus suggests a photothermal activation of the crystallization process in the IB a-Si layers.

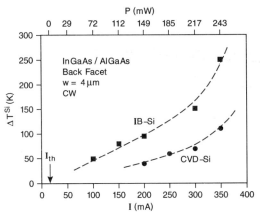

Figure 2. Measured temperature increases of IB- and CDV-deposited a-Si layers in back facet coating stacks of InGaAs/AlGaAs lasers as a function of current I and optical power P.

To quantify the effect of Si recrystallization on laser characteristics in a unique way, we produced the crystallization in well-controlled steps by irradiating the back facet coating in the near-field area with a focused Ar$^+$ laser beam. In this way a ρ-value of $\approx 100\%$ can be produced in IB a-Si in ≈ 2 h with 30 mW of the 457.9 nm line focused in a spot size of about 3 μm. A value of $\rho \approx 100\%$ increases the transmitted power at the back facet by a factor of 1.7 (see Fig. 3). This enhancement of power emission can be accounted for by a reflectivity decrease from 95% to 91% due to a transition of the

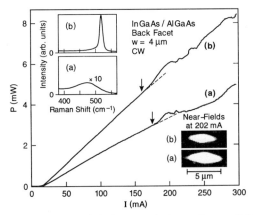

Figure 3. Light (P)-current (I) characteristics and near-field patterns taken from back facet; (a) back facet coating as deposited, (b) after controlled recrystallization of IB a-Si layers in back facet coating by Ar$^+$ irradiation. Inset shows corresponding Raman spectra (a) a-Si as deposited, (b) complete c-Si.

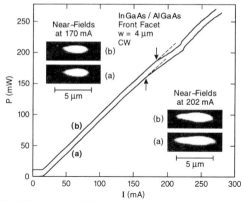

Figure 4. Light (*P*)-current (*I*) characteristics and near-field patterns taken from the non-manipulated front facet of same laser in Fig. 3 with IB a-Si layers in back facet coating (a) as deposited and (b) completely recrystallized.

refractive index from 3.85 (a-Si) to 3.45 (c-Si). An increase of the transmitted power caused by an absorption coefficient decreasing from 8×10^3 cm^{-1} (a-Si) to 5×10^2 cm^{-1} (c-Si) plays only a minor role (10%).

Figure 4 shows the corresponding *P–I* characteristics measured at the front facet. Of significance is the clear movement of a "kink", or nonlinearity, observed in the *P–I* curve upon Si recrystallization. Such a kink can be associated with a movement of the optical mode along the junction plane. This effect is known to cause, for example, a nonstable coupling of laser power into optical fibers. Finally, Figs. 3 and 4 also exhibit the effect of the recrystallization on the near fields, which remain unchanged in shape when recorded below the power onset value of the kink. In contrast, near-field patterns recorded above the kink shrink in lateral dimension.

4. Summary

We have demonstrated distinctive features between PECVD-Si and IB-Si layers in rear mirror coating stacks of 980-nm InGaAs/AlGaAs lasers under power exposure. PECVD-Si layers remain amorphous, whereas IB-Si layers rapidly recrystallize with time. Raman spectra proved to yield an unambiguous distinction between the amorphous and crystalline Si state. Local mirror temperatures are $\approx 2\times$ higher in the as-deposited IB than in the PECVD stacks due to a higher absorption coefficient of a-Si. The recrystallization of the a-Si layers in the back facet coating leads to a movement of the optical mode above a certain power level and shrinks the lateral near-field pattern.

Acknowledgment

We would like to acknowledge the support of the Laser Enterprise group. We are also grateful to C.S. Harder, G. Hunziker and P. Roentgen for useful discussions concerning mirror reflectivity and laser modes.

References

[1] Desurvire E 1990 *IEEE Photonic Tech. Lett.* **2** 208.
[2] Brugger H, Epperlein P W, Beeck S and Abstreiter G 1989 *Inst. Phys. Conf. Ser.* **106** 771.
[3] Tsang J C, Oehrlein G S, Haller I and Custer J S 1985 *Appl. Phys. Lett.* **46** 589.
[4] Mei P, Boyce J B, Hack M, Lujan R A, Johnson R I, Anderson G B, Fork D K and Ready S E 1994 *Appl. Phys. Lett.* **64** 1132.
[5] Tsu R 1981 *Defects in Semiconductors* (North-Holland Inc.) p 445.
[6] Webb D J *private communication.*
[7] Hartstein A, Tsang J C, DiMaria D J and Dong D W 1980 *Appl. Phys. Lett.* **36** 836.

Record Low Resistance Vertical Cavity Surface Emitting Lasers Grown by Molecular Beam Epitaxy Using Hybrid Mirror Approach

M. Hong, D. Vakhshoori, J. P. Mannaerts, and F. A. Thiel

AT&T Bell Laboratories, Murray Hill, NJ 07974

Abstract. Low resistance vertical-cavity surface emitting lasers employing a hybrid mirror as the top reflector have been grown by molecular beam epitaxy. A sinusoidal-like compositional grading between GaAs and $Al_{0.9}Ga_{0.1}As$ was used in both n- and p-type distributed Bragg reflector mirrors. Single crystal Au films which were *in-situ* deposited on top of the p-DBR to achieve low ohmic contact resistance of $10^{-7} \Omega\text{-cm}^2$ were part of the hybrid mirror. This novel combination of *in-situ* ohmic contact and sinusoidal grading in composition gave an extremely low total device resistance of $5 \times 10^{-5} \Omega\text{-cm}^2$. The threshold current density was 2 kA/cm^2.

1. Introduction

Vertical-cavity surface emitting lasers (SELs) are promising candidates for many optoelectronic applications such as optical communication and computer interconnections. One crucial factor of realizing SEL in a high-performance and reliable device is to decrease the high series resistance, which mainly comes from the p-type semiconductor distributed Bragg reflector (DBR) mirror, and may also be attributed to by a small contact area. The series resistance in the DBR mirror can be reduced when the barrier height at the heterojunctions between the two constituents of the mirror is lowered.

Experimentally during molecular beam epitaxial (MBE) growth, lowering the barrier height has been achieved by modifying the band discontinuity with inserting an AlGaAs region between GaAs and AlAs: This AlGaAs layer can be in different forms such as (1) one intermediate composition [1]; (2) two or more intermediate compositions [2]; (3) digital alloy linear grading [3], and (4) continuously compositional grading by variation of group III (Ga, Al) cell temperatures [4]. These approaches were used in the early stage of the research activity to lower the series resistance in SELs. More recently, much lower total resistances of $1.9 \times 10^{-4} \Omega\text{-cm}^2$ and $3.6 \times 10^{-4} \Omega\text{-cm}^2$ have been reported for $Al_{0.1}Ga_{0.9}As/Al_{0.9}Ga_{0.1}As$ mirrors [5,6] grown with a doping of $3 \times 10^{18} \text{cm}^{-3}$ and for $GaAs/Al_{0.67}Ga_{0.33}As$ mirrors [7] with a doping of $6 \times 10^{18} \text{cm}^{-3}$, respectively.

In this paper, we present MBE growth of an SEL device using a hybrid mirror approach combining semiconductor DBR and metal. The SEL with the hybrid p-mirrors has the advantage of requiring less epitaxial growth and reducing device resistance while maintaining high mirror reflectivity. The *in-situ* deposited Au was used to achieve both non-alloyed low ohmic contacts [8] and a smooth metal/GaAs interface [9].

2. Experimental

The growth was carried out in a multi-chamber molecular beam epitaxy (MBE) system which includes a solid source GaAs-based III-V chamber and an As-free metal chamber. The two deposition chambers are connected by transfer modules which are maintained with a 5×10^{-11} torr vacuum condition. Both n- and p-DBR mirrors were grown using the flux modulation of Al and Ga by simultaneously varying the cell temperatures of Al and Ga without any shutter operation. The compositional profile was determined by cell temperature controllers. Three $In_{0.2}Ga_{0.8}As$ quantum wells each 100 Å thick separated by 120 Å of GaAs barriers formed the active region. The gain peak was around 980 nm and the carrier confinement was provided by graded AlGaAs cladding.

The bottom mirror was a 26-pair n-type $\lambda/4$ distributed Bragg reflector (DBR) and the top one was a hybrid mirror combining a 16-pair p-type $\lambda/4$ DBR and a Au film. The first 3-pair p-DBRs close to the active region were doped with $3 \times 10^{18} cm^{-3}$ and the remaining DBRs were with $5 \times 10^{18} cm^{-3}$, except the top 500 Å which was doped heavily with Be of more than $10^{19} cm^{-3}$ for achieving low resistivity ohmic contacts. The SEL samples were then transferred to the metal chamber for *in-situ* deposition of Au, which was achieved using an effusion cell. The total thickness of the Au film is 300 Å.

After the MBE growth of SEL and *in-situ* Au, another Au film 2000 Å thick were *ex-situ* deposited on the samples in an e-beam evaporator. The devices were made by first ion-milling the metal to define the dot for ohmic contact. The SEL mesa was defined using wet chemical etching. The etching was stopped just before the active region.

3. Results and Discussion

An example is given here for the Ga cell temperature and flux for growth of one period of the DBR mirror: Figure 1 is a cell temperature as function of time produced by the temperature controller for one pair of DBR mirror. The input temperature profile was designed to give a sinusoidal composition-grading for the interface between the two constituents of the DBR mirror. In this work, these two components are chosen to be GaAs and $Al_{0.9}Ga_{0.1}As$. The design was based on the growth rates of GaAs and AlGaAs as

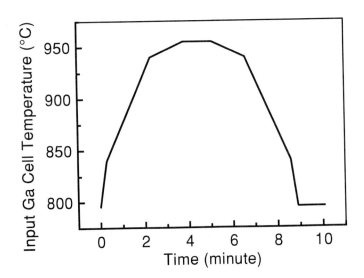

Fig. 1 Input Ga cell temperature as function of time for one pair of DBR. The period for this mirror is 10.1 minutes.

function of the cell temperatures as obtained using the reflection high energy electron diffraction (RHEED) oscillation. The actual Ga cell temperature shown in Fig. 2 follows the input temperature, but lags behind it by a few seconds. Moreover, during the cooling, actual temperature undershooting the input temperature by as much as 15°C was observed. Nevertheless, the undershooting repeats itself periodically. The Ga flux measured by a flux gauge near the substrate as shown in Fig. 3 indicates that the compositional profile is very much sinusoidal-like. A similar sinusoidal-like flux profile was also obtained for the Al cell.

The *in-situ* deposited Au is single crystal and epitaxially grown on GaAs with a substrate temperature of 100°C [9]. The Au/GaAs interface is very smooth with a roughness of 5.5 Å as determined from an X-ray reflectivity measurement [10]. The interfacial smoothness may also enhance the reflectivity of the hybrid mirror. Very low ohmic resistivity of 10^{-7} Ω-cm^2 was achieved for this type of non-alloyed contact [9]. Therefore, the ohmic contact resistance is negligible compared with the resistance from the p-DBR.

This novel combination of the *in-situ* non-alloyed ohmic contacts and a sinusoidal-like compositional grading profile between GaAs and $Al_{0.9}Ga_{0.1}As$ gave a total resistance of 5.0×10^{-5} Ω-cm^2. The total device resistance included the p- and n- ohmic metal contacts as well as the resistance from the p- and n- compositionally graded DBR mirrors. The SEL threshold current density was approximately 2 kA/cm^2 with a Fabry-Perot wavelength of 940 nm. By better alignment of the cavity Fabry-Perot wavelength with the gain peak, reduction in threshold current density is expected. We note that when the first 13-pair p-DBRs were doped with 3×10^{18} cm^{-3} and the remaining DBRs were doped with 5×10^{18} cm^{-3}, a total resistance was slightly increased to the value lower than 1.0×10^{-4} Ω-cm^2.

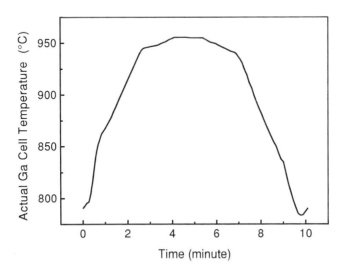

Fig 2. Actual Ga cell temperature as function of time for one pair of DBR. Note that the temperature undershooting near the 795°C.

4. Conclusion

The SEL structure using a hybrid mirror has the advantage of requiring less epitaxial growth and reducing device resistance without detrimental effect on the mirror reflectivity. The total resistivity of the SEL in the range of 10^{-5} $\Omega\text{-cm}^2$ has been demonstrated. We believed that further reduction in the SEL resistance is possible when the composition-grading and the doping profiles are optimized for the DBR mirrors. The continuous grading in composition for the growth of DBR mirrors first used in MBE [4] has now being successfully extended to metalorganic chemical vapor deposition (MOCVD) growth of SEL. Low threshold voltage SEL was achieved [11].

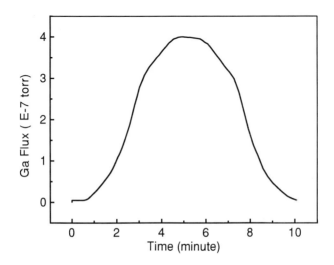

Fig 3. Ga flux as function of time for one pair of DBR. A sinusoidal-like curve is obtained.

References

[1] K. Tai, L. Yang, Y. H. Wang, J. D. Wynn, and A. Y. Cho, Appl. Phys. Lett. **56**, 2496 (1990).
[2] Y. H. Wang, R. J. Fischer, and A. Y. Cho, unpublished results.
[3] R. S. Geels, S. W. Corzine, and L. A. Coldren, IEEE J. Quantum Electron. **27**, 1359 (1991).
[4] M. Hong, J. P. Mannaerts, J. M. Hong, R. J. Fischer, K. Tai, J. Kwo, J. M. Vadenberg, Y. H. Wang, and J. K. Gamelin, J. Crystal Growth **111**, 1071 (1991).
[5] S. A. Chalmers, K. Lear, and K. P. Kileen, Appl. Phys. Lett. **62**, 1585 (1993).
[6] K. L. Lear, S. A. Chalmers, and K. P. Kileen, Electronics Letters **29**, 584 (1993).
[7] J. D. Walker, D. M. Kuchta, and J. S. Smith, Appl. Phys. Lett. **59**, 2079 (1991).
[8] M. Hong, D. Vakhshoori, J. P. Mannaerts, F. A. Thiel, and J. D. Wynn, J. Vac. Sci. Technol. **B12**, 1047 (1994).
[9] D. Y. Noh, Y. Hwu, H. K. Kim, and M. Hong, submitted to Phys. Rev. **B** rapid communications (1994).
[10] K. L. Lear, IEEE Photonics Technol. Lett. **6**, No. 9 (1994).

Effects of Impurity-Free and Impurity-Induced Disordering (IID) on the Optical Properties of GaAs/(Al,Ga)As Distributed Bragg Reflectors

P.D. Floyd and J.L. Merz

Department of Electrical and Computer Engineering, University of California, Santa Barbara
Santa Barbara, California 93106

Abstract. The variation of the reflectivity and stopband width of disordered GaAs/(Al,Ga)As distributed Bragg reflectors (DBRs) is studied by thermal annealing of semiconductor DBRs and modeled by simulating interdiffusion of Al and Ga at the interfaces within the DBR. Undoped mirrors show stability in their reflectivity and stopband width for annealing durations of up to 24 hours. Heavily Si doped mirrors show strong degradation for annealing durations as short as one hour. Using the model, excellent agreement with experimental results is found and interdiffusion coefficients can be estimated. The results show that important optical parameters are unchanged for short diffusion lengths, but drastically degrade for larger diffusion lengths. The results indicate the limits of thermal processing for device structures containing GaAs/(Al,Ga)As DBRs. The ability to accurately predict the effects of disordering on DBRs permits the proper design of vertical cavity structures containing DBRs, by specifically accounting for the effects of disordering.

1. Introduction

Vertical cavity optoelectronic devices such as lasers [1], [2] and modulators [3], [4] have been studied due to unique advantages of low threshold currents, low operating voltages, compactness and the ability to be fabricated into 2-D arrays. Semiconductor DBRs are used to form the cavities of such devices. High temperature processing, such as impurity diffusion for edge-emitting lasers fabricated by impurity-induced disordering (IID) [5], [6], bonding by atomic rearrangement, to join dissimilar semiconductors [7], Zn diffusion for low resistance mirrors and optical mode control [8] and regrowth over etched pillar vertical cavity lasers [9], can result in high performance optoelectronic devices. However, application of such high temperature techniques to vertical cavity devices requires understanding the effects caused by impurity-induced and impurity-free disordering on the optical properties of DBRs. Because DBRs are often doped with impurities known to enhance interdiffusion such as Si and Zn, their effect on the mirror during a high temperature process must be understood to insure proper optical cavity design.

In this work, we analyze the effects of impurity-induced and impurity-free disordering on important characteristics of DBRs such as reflectivity and stopband width. The variation of these parameters is calculated as a function of diffusion length and measured as a function of anneal duration.

2. Theory

To model the interdiffusion of Al and Ga in a DBR, we solve Fick's equation using a constant interdiffusion coefficient D_{Al-Ga} with the initial conditions of a square well Al profile and final conditions of a uniform alloy, along with periodic boundary conditions. [10] X_Z and X_B are the Al fraction in the quarter-wavelength layers of low Al content and the quarter-wavelength layers of high Al content, respectively. L_Z and L_B are the thicknesses of the layers and their sum the thickness of a single mirror period, T. The general solution for the Al profile in the mirror is given by

$$X_{Al}(z,t) = X_{Al}(z,\infty) + \frac{2}{\pi}(X_Z - X_B)\sum_1 \frac{1}{n}\cos(\pi n)\sin\left(\frac{\pi n L_Z}{T}\right)\cos\left(\frac{2\pi n z}{T}\right)\exp\left[-(\frac{2\pi}{T})^2 n L_D^2\right]$$

where L_D is the diffusion length given by $L_D^2 = D_{Al-Ga}t$. To calculate the reflectivity spectra, we use the transmission matrix method, including the wavelength dispersion of the (Al, Ga)As using the model of Afromowitz et. al. [11] and fixed losses for the layers, and break up the calculated Al profile into 1 nm slices. From the calculated reflectivity spectra, we extract the parameters of interest.

Figure 1(a) shows the calculated reflection spectra for a structure similar to the actual material used in experiments described later. The structure consists of an 18.5 period GaAs/AlAs DBR centered at 950 nm, capped with a one wavelength thick GaAs layer. This capping layer causes the structure to actually be a low finesse resonant cavity, with a Fabry-Perot mode at the center wavelength. The behavior of the average reflectivity (average of the reflectivity maxima within the stopband) and the stopband width of the spectra clearly depends on diffusion length. Figure 1(b) shows both parameters plotted as a function of L_D.

The sudden reduction of reflectivity and stopband width beyond a "threshold" diffusion length, as shown in the calculated curves, can be understood by considering extreme stages of interdiffusion. The early stages of interdiffusion will grade the interfaces in the DBR, but the maximum Al content in the high Al content layers and the maximum Ga content in the high Ga content layers does not change over most of the thickness of a quarter-wavelength layer. This implies that the index of refraction step between subsequent layers is little changed, therefore the reflectivity and other parameters vary little from those of a non-disordered DBR. As the interdiffusion proceeds and the diffusion length becomes a significant fraction of the quarter-wavelength thickness, the Al content in the high Al content layers and the Ga content in the high Ga content layers decreases over the entire length of a quarter-wavelength layer. This process reduces the index of refraction step between the layers, hence reducing the reflectivity and narrowing the stopband width of the DBR.

3. Experiment

Experiments were performed on Molecular Beam Epitaxy (MBE) grown, doped and undoped mirror samples. Undoped samples have 18.5 periods of quarter-wavelength GaAs/AlAs layers, and capped with a 1 wavelength GaAs cap layer. The DBR has a center wavelength of 950 nm. Because a one-wavelength layer was grown, rather than a 1.25 optical wavelength thick layer, the structures are actually low finesse resonant cavities with a Fabry-Perot mode at the center wavelength. Silicon doped ($\approx 10^{18}$ cm^{-3}), 20 period GaAs/AlAs DBRs with a center wavelength of 925 nm were used to study IID of the DBRs.

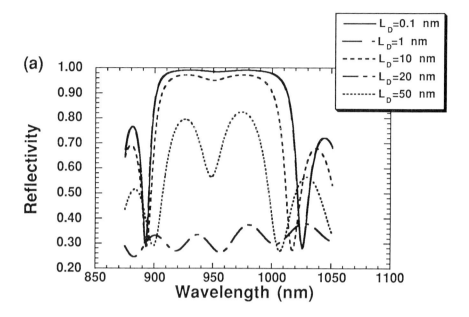

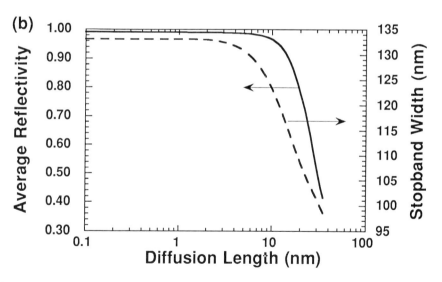

Figure 1. (a) Calculated reflectivity spectra for 18.5 period DBR with one optical wavelength GaAs capping layer centered at 950 nm. (b) Average reflectivity (average of the reflectivity maxima within the stopband) and stopband width as a function of diffusion length

SiO$_2$ was deposited on the samples by plasma-enhanced chemical vapor deposition (PECVD) at 270 °C. The samples were enclosed in a quartz ampoule along with elemental arsenic for overpressure. The thick GaAs cap layer is helpful in preventing impurity-induced disordering via Al-SiO$_2$ interaction. [12] Because we are interested in using DBRs in devices fabricated using IID by Si diffusion, we choose to anneal at a temperature, arsenic overpressure, and encapsulation condition similar to that used to fabricate diode lasers by Si-IID. [5,6] Therefore, the anneal is performed at 850°C with 0.5 Atm arsenic overpressure. After annealing, the dielectrics are stripped by CF$_4$ plasma and buffered hydrofluoric acid. Reflectivity spectra of samples with the best surface morphology were measured using a spectrophotometer, from which the mirror stopband width and reflectivity can be directly determined.

4. Experimental Results / Discussion

Figure 2(a) shows the reflection spectra of the undoped samples for various anneal durations. The behavior of the reflectivity and stopband width is similar to that of the calculated results of Figure 1. Figure 2(b) plots the stopband width and the average reflectivity (average of the reflectivity maxima within the stopband) as a function of anneal duration. Both parameters remain roughly constant for anneal durations < 10 hours and the reflectivity itself is stable for times ≤ 24 hours.

The doped DBRs exhibit rapid degradation for annealing durations of ≤ 1 hour. The reflectivity and stopband widths degrade in a manner similar to the undoped samples, but at a much accelerated rate, due to the silicon dopant which enhances the interdiffusion of Al and Ga.

Because the optical properties of the mirrors can be calculated as a function of L_D, an estimate of the interdiffusion coefficient, D_{Al-Ga}, for the doped and undoped structures is possible by using the stopband width as a fitting parameter and relating the annealing time to diffusion length using $L_D^2 = D_{Al-Ga} t$.

The undoped mirrors show $D_{Al-Ga} \approx 10^{-16} - 10^{-17}$ cm^2/s, which is in the range of values reported for undoped SiO$_2$ capped structures at identical annealing temperatures. [13], [14] Doped mirrors show much larger $D_{Al-Ga} \approx 4 \times 10^{-15}$ cm^2/s, similar to values measured by Mei et al. [15]

Comparing the results for undoped DBRs to results for undoped heterostructures investigated by other workers, the larger measured values of $D_{Al-Ga} \approx 1 \times 10^{-16}$ cm^2/s are from samples capped with electron-beam evaporated SiO$_2$ films. [13] The smaller measured values of $D_{Al-Ga} \approx 3 \times 10^{-18}$ cm^2/s come from undoped material capped with SiO$_2$ deposited by chemical vapor deposition (CVD) at 420 °C. [14]

The difference between the interdiffusion coefficient in the references and that found in this work may result from quality of the SiO$_2$ films, the least porous being the high temperature CVD film and the most porous, the electron-beam evaporated film. More porous films may allow easier outdiffusion of Ga from the GaAs DBR surface into the SiO$_2$ film, thereby enhancing the generation of column III vacancies necessary for interdiffusion. [13]

5. Conclusion

The experiments and calculations imply that moderate layer interdiffusion will not change the optical properties of a DBR, but for diffusion lengths or anneal durations greater than a threshold diffusion length or anneal duration, the optical properties degrade rapidly. These

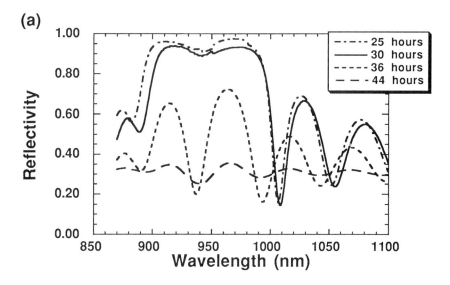

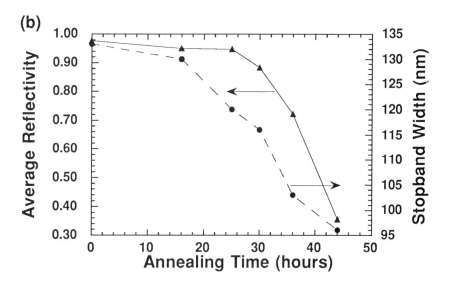

Figure 2. (a) Measured reflectivity spectra for undoped 18.5 GaAs/AlAs DBR with one optical wavelength thick GaAs capping layer sample, annealed at 850°C for different anneal durations. (b) Average reflectivity (average of the reflectivity maxima within the stopband) and stopband width as a function of annealing duration.

results indicate that to obtain a desired value of reflectivity after annealing, the number of mirror periods necessary will depend on the duration of the anneal and on the effective interdiffusion coefficient, which itself is directly dependent on doping of the DBR and the annealing temperature. The results imply that annealed, undoped DBRs should maintain their as-grown characteristics, due to the typically small D_{Al-Ga} of undoped structures. Si doped DBRs, widely used in vertical cavity optoelectronic devices, have much higher D_{Al-Ga}, and degrade even for short annealing durations. Structures must be designed with this in mind by either avoiding doping with interdiffusion enhancing impurities, reducing anneal times and temperatures, or adding periods to the DBR to achieve the desired reflectivity after annealing.

6. Acknowledgments

This work was supported by the ARPA Optoelectronics Technology Center (OTC), and by the NSF Center for Quantized Electronic Structures (QUEST), grant # DMR 91-20007. The authors would like to thank R.P. Mirin for MBE growth of material used in these experiments.

References

[1] Jack L. Jewell, J. P. Harbison, A. Scherer, Y.H. Lee, and L. T. Florez, *IEEE J. Quantum Electron.* **QE-27**, 1332 (1991).
[2] R.S. Geels and L. A. Coldren, *Appl. Phys. Lett.* **57**, 1605 (1990).
[3] K.-K. Law, R.H. Yan, J.L. Merz, L.A. Coldren, *Appl. Phys. Lett.* **56**, 1886 (1990).
[4] M. Whitehead, A. Rivers, G. Parry, and J.S. Roberts, *Electron. Lett.* **26**, 1588, (1990).
[5] W.X. Zou, T. Bowen, K-K Law, D.B. Young, and J.L. Merz, *IEEE Photon Technol. Lett.* **5**, 591 (1993).
[6] P.D. Floyd, C.P. Chao, K.-K. Law, and J.L. Merz, *IEEE Photon Technol. Lett.* **5**, 1261 (1993).
[7] J.J. Dudley, D.I. Babic, R. Mirin, L. Yang, B.I. Miller, R.J. Ram, T. Reynolds, E.L. Hu, and J.E. Bowers, *Appl. Phys. Lett.*, **64**, 1463 (1994).
[8] Y.J. Yang, T.G. Dziura, T. Bardin, S.C. Wang, and R. Fernandez, *Electron. Lett.* **28**, 274 (1992).
[9] C.J. Chang-Hasnain, Y.A. Wu, G.S. Li, G. Hasnain, K.D. Choquete, C. Caneau, and L.T. Florez, *Appl. Phys. Lett.* **63**, 1307 (1993).
[10] I. Szafranek, M. Szafranek, J.S. Major, Jr., B.T. Cunningham, L.J. Guido, N. Holonyak, Jr. and G.E. Stillman, *J. Electron. Mater.* **20**, 409 (1991).
[11] M.A. Afromowitz, *Solid State Commun.* **15**, 59 (1974).
[12] L.J. Guido, J.S. Major, Jr., J.E. Baker, W.E. Plano, N. Holonyak, Jr., K.C. Hsieh, and R.D. Burnham, *J. Appl. Phys.* **67**, 6813 (1990).
[13] J.D. Ralston, S. O'Brien, G.W. Wicks, and L.F. Eastman, *Appl. Phys. Lett.* **52**, 1511 (1988).
[14] L.J. Guido, N. Holonyak, Jr., K.C. Hsieh, R.W. Kaliski, W.E. Plano, R.D. Burnham, R.L. Thornton, J.E. Epler, and T.L. Paoli, *J. Appl. Phys.* **61**, 1372 (1987).
[15] P. Mei, H.W. Yoon, T. Venkatesan, S.A. Schwarz, and J.P. Harbison, *Appl. Phys. Lett.* **50**, 1823 (1987).

ID# Period Deviations in Distributed Bragg Reflectors: X-ray Diffraction and Optical Reflectivity Measurements

K. Matney,* M.S. Goorsky,* J. Tartaglia,[†] R. Rai,[†] and C. Parsons[†]

*University of California, Los Angeles, Department of Materials Science and Engineering, Los Angeles, CA 90024.

[†]Bandgap Technology Corporation, Broomfield, CO 80021.

Abstract In this study, high resolution double and triple axis x-ray diffractometry were employed to resolve superlattice deviations in an AlAs/GaAs distributed Bragg reflector. This work focused on determining whether superlattice imperfections were caused by interfacial roughness, compositionally graded interfaces, or changes in period thicknesses of a systematic or random format. The effects of such imperfections on the DBR optical properties are calculated and discussed. Double axis diffraction scans revealed the reflector to be of excellent crystalline quality, but degradation and/or splitting of the high order satellite peaks indicated a non-perfect periodicity. Comparison between experiment and dynamical simulations ruled out the possibility of compositionally graded interfaces and systematic period deviations. Triple axis reciprocal space mapping showed interference effects in the form of high order satellite peak splitting, and an absence of diffuse scatter from interfacial roughness, indicating that the perturbations were in the form of an abrupt but randomly deviated period defined by a standard deviation σ of about 1%-2%. The results were confirmed with transmission electron microscopy, however the standard deviation was lower (~0.5% -0.7%).

1. Introduction

Vertical cavity surface emitting laser (VCSEL) diodes show great promise for use in integrated optical and electronic devices. The VCSEL structure consists of an active p-i-n junction surrounded by distributed Bragg reflectors (DBRs). GaAs/AlAs superlattice (SL) mirrors represent a natural choice for DBRs due to superior material properties. Because the whole structure is epitaxially grown, control of crystal growth is of utmost importance. To provide enough gain for emission, the DBRs must be >99% reflective. Furthermore, the DBRs must be tuned to the emission wavelength of the active layer. In general, better than 1% accuracy must be obtained for both layer thickness and composition to realize a successful device.[1] For a two dimensional array of lasing diodes, this accuracy must extend laterally across the wafer as well. Such standards not only challenge existing growth methods, but materials characterization techniques as well.

High resolution x-ray diffraction (HRXRD) is a proven method for the study of strained layer superlattices.[2][3] The technique is a quick, accurate, non-destructive, and requires no sample preparation. The amount of information available with x-ray diffraction depends on the resolution and signal-to noise ratio (s-n) of the machine used; at least 4-5 orders of magnitude of diffracted intensity - to observe the typically weaker high order satellite peaks - are needed to resolve subtle perturbations in an actual SL.

Triple-axis diffraction (TAD) is becoming increasingly popular as a means to separate strain, tilt, and diffuse scatter (which may arise from rough interfaces).[4] In the presence of interface roughness, the superposition of x-rays scattered diffusely from each interface should be more pronounced in the high diffraction order satellite peaks. This effect can be analyzed in a TAD reciprocal space map (RSM).

The effects of several kinds of SL defects on diffraction measurements, and their distinctions, are considered in our study. These consist of compositionally graded interfaces, and period fluctuations of random and systematic formats. The effects of interface roughness are also discussed. Graded interfaces result from the diffusion of Al and Ga during growth, creating non-abrupt boundaries. Random deviation is described by individual layers (t_{GaAs} or t_{AlAs}) which fluctuate with a standard deviation (σ_{GaAs} or σ_{AlAs}), for this study. Systematic deviation, or drift, involves layer thicknesses becoming successively either smaller or larger in the growth direction, i.e. from t-σ to t+σ for increasingly larger layers.

All of these of the defects either lower the intensity of and/or broaden the higher order satellite peaks. A careful study, however, can differentiate between their respective effects. Determination of the type of period deviation may be crucial to device operation. For instance, the top and bottom mirrors may be completely detuned from one another as a result of systematic variation. The effect of layer thickness variation on the optical properties of VCSEL devices has been previously reported.[5] Through HRXRD, we obtained the SL defect structure (layer variation, etc.). We then can calculated, using transmission matrices, the optical properties of the device, such as the optical reflectivity and phase shift. We also compare the calculated results to measurements of optical reflectivity.

2. Experimental Procedure

In this study, we examined an MBE bottom DBR mirror structure, grown at ~580 °C. It consists of a 3000 Å buffer layer, followed by 15 periods of alternating AlAs and GaAs layers with nominal thicknesses of 828 Å and 693 Å, respectively. The substrate was rotated during growth to provide uniformity across the 2" (001) GaAs wafer. The HRXRD measurements were performed using a Bede D^3 diffractometer[6] (with an intensity range spanning ~5.5 orders of magnitude). More precise details of the experimental set-up may be found in an earlier paper.[7] (002), (004), and (006) scans were generated. X-ray diffraction curves were simulated with both commercial software[8] and an in-house computer program[9] utilizing dynamical x-ray diffraction theory. Triple axis diffraction measurements were performed with a silicon (111) 4-bounce channel-cut crystal between the sample and the detector. The diffracted intensity was mapped about the (004) peak. Conversion from diffractometer coordinates to reciprocal space coordinates followed the format of Iida and Kohra.[10]. Optical reflectivity and high resolution transmission electron microscopy (TEM) measurements utilized equipment described elsewhere.[11] Optical reflectivity and phase shift curves were simulated using the transmission matrix method, taking into account the affects of absorption and dispersion.[12]

3. Results and Discussion

Figure 1(a) shows the (004) diffraction peak. The presence of the high number of satellite peaks suggests a high crystalline quality of the DBR. An estimation of crystalline quality was attained from the FWHM of the substrate and zero order peaks which were found to be 15.4 sec and 16.8 sec, respectively for the (004) peak. These values, which are higher than those for a perfect sample, stem from strain in the DBR which introduces wafer curvature.[11] The radius of curvature in our sample was determined to be approximately 200 m.

While the low order satellite peaks ($m<5$, where m is the satellite order) provide SL properties such as average period thickness and composition, gaining information on the average individual thicknesses t_{GaAs} and t_{AlAs} requires the presence of higher order satellite peaks. The high order peaks provide an "envelope" function from which the individual layer thicknesses are obtained by matching the relative peak intensities.[13] Fig. 1 shows the (004) peak, with a corresponding simulation matching the envelope function. As seen in Fig. 1(a), the first ±5 satellite peaks are well matched, however the goodness of fit decreases with increasing peak order (satellites +14 - +22) and the experimental peaks are somewhat broader than the simulated ones. This is due to imperfections (possibly compositional transition layers, period deviations, etc.) in the SL itself. Since the high order satellite peaks represent the high order coefficients of the Fourier transform of the direct space SL structure, they are sensitive to small fluctuations in the layer thicknesses and interfacial grading.

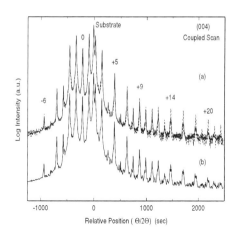

Fig. 1 - (a) Experimental HRXRD curve solid, and calculated curve with perfect SL structure (dashed). and (b) The Best-fit simulation, utilizing random period thickness deviations.

Comparison of the HRXRD data with dynamical simulation ruled out the possibility of interfacial grading greater than 20 Å. Adding graded interface layers to the calculated curves lower the satellite peak intensity, but do not cause broadening. Fig. 1(b) shows the best fit simulation for the (004) diffraction scan. This fit was obtained by introducing a randomly varying period, with the standard deviation of period thicknesses and layers given in Table I. The (006) and (002) scans (not shown) exhibited a similarly close fit between experimental and theoretical curves.

The effects of systematic vs. random deviation are highlighted in Fig. 3, for the (004) scan. The top curve shows the same layers used in the best fit simulation above, but with the layer order rearranged with the period becomes smaller with the growth direction. The effects of this arrangement on the diffraction scans, particularly the satellite peak shape, are clearly visible.

Table I. Structural parameters of the best fit simulations used in Figs. 1 and 2. The ± indicate the range of standard deviations needed for a noticeable degradation of fit.

Layer	Nominal	Experimental (avg.)	σ
Period	1521 Å	1575 Å	13.55 ± 2 Å
T_{AlAs}	828 Å	857 Å	13.05 ± 1 Å
T_{GaAs}	693 Å	718 Å	14.55 ± 1 Å

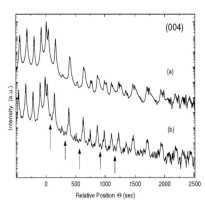

Fig. 2 - HRXRD simulations employing random(a), and systematic (b) period deviations (layers getting thinner with growth). The differences are mostly accounted for in the skewness of the SL peaks.

Another disparity between random and systematic period deviation is the existence of many fine "peaks" between satellite peaks in the former case (as shown by arrows in Fig. 3).[14] The appearance of these extra peaks is attributed to the overlapping of satellite peaks from periods of varying thicknesses. These peaks were also observed in the experimental scans, supporting the conclusion that a random variation is present. This variation is consistent with a study by Fernandez, et al [15], who demonstrated that random variations of layer growth rates were on the order of about 1%.

A triple axis RSM is shown in Fig. 3. The horizontal axis (q_z) of the RSM represents crystal strain and any associated interference effects. Crystal tilts produce intensities extending along the vertical axis (q_y). Diffuse scatter, arising from a decrease in the spatial coherence of the diffracted beam, would present itself as a broad central hill on the contour plot[16], centered about the reciprocal lattice point. Diffuse scatter may be caused by structural irregularities such as islanding between the AlAs/GaAs interfaces, or by misfit dislocations (which were below the detectable limit for this sample).[17] We

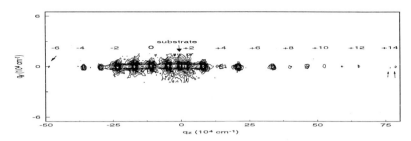

Fig. 3 - RSM about the (004) reciprocal lattice point.

measured a consistent FWHM (~11 sec) of the satellite peaks along the q_y axis, indicating a lack of notable interfacial roughness. Again, the FWHMs are slightly greater than predicted by dynamical theory due to wafer curvature. Also of note are the "extra" peaks between the SL peaks (due to period deviations) visible through the analyzer crystal (the +14 harmonic on the right edge of the picture is a doubly split peak, as is also observed in the double axis measurement.

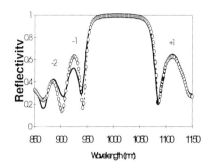

Fig. 4 - Measured (—) and calculated (ooo) optical reflectivity. The calculated value assumes the best-fit parameters from HRXRD.

Fig. 4 shows the optical reflectivity measurement, and a simulation with the best-fit HRXRD structure. The simulation fits nicely in the bandpass and low energy region, however the oscillatory fringes at shorter wavelengths are not matched. This may be due to the fact that a different area of the wafer was measured for each technique.

Figs. 5 and 6 show calculations of the optical reflectivity performance of the DBR with random (Fig. 5) and systematic (Fig. 6) deviations of the SL layer thicknesses. Both are compared to the calculated optical reflectivity for that of a perfect SL DBR. For the calculations, we used the same layer parameters as that in Fig. 2. Interestingly, random period deviations do not seem to greatly effect the DBR performance. Indeed, random deviations with layer thickness variations on the order of 30Å did not perceptibly change the reflectivity in the bandpass region. The calculated curve is much more effected, however, by a systematic period thickness deviation (Fig 6). Here, the critical bandpass region is shifted and lowered, which may be devastating to device performance. Thus, HRXRD measurements can distinguish the difference between systematic and random period deviations, which may then be applied to predict the DBR device performance.

High resolution TEM measurements were performed to measure the thicknesses of each layer. The average TEM thicknesses were 846 Å and 723 Å, for the AlAs and GaAs layers, respectively; these were found to have random variations; the standard deviation of layer thicknesses was found to be 5 Å and 4 Å, respectively; lower than the result from HRXRD. This discrepancy is partly due to the fact that different portions of the wafer were examined in each experiment. More importantly, HRXRD integrates a large physical area (compared to TEM) into the measurement. Thus, lateral variations in layer thicknesses may contribute to the "randomness" of the structure as seen in the x-ray diffraction scans.

4. Conclusion

In summary, high resolution x-ray techniques were utilized to obtain the true SL defect structure from proposed models. A systematic study showed that this particular SL did not exhibit interfacial grading or roughness, and superlattice period deviations were found to be randomly arranged with a standard deviation of about 1.5% to 2.0% for each quarter wave stack throughout the structure. The precision of this technique was exhibited by extremely close fits between experiment and dynamical x-ray diffraction theory. The results agreed with optical reflectivity measurements, and to some extent with TEM measurements, which confirmed that the layer thicknesses were randomly varying from layer to layer, although to a lesser extent than the HRXRD results.

Furthermore, the optical reflectivity was calculated for both randomly and systematically varying layer thicknesses; it was shown that systematic period deviation has a much greater effect on the optical properties. Additionally, the phase shift on reflection (not shown) is also much greater in the systematic arrangement. This phase shift can be calculated by directly applying the structural parameters derived by HRXRD. Thus it is shown that HRXRD can not only provide accurate structural information to improve the growth process, but can be used to predict DBR optical properties, such as reflectivity and the phase shift on reflection, as well. Future work will foacus on full VCSEL structures,

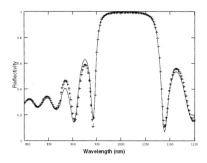
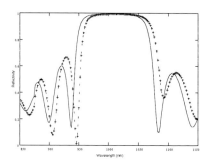

Fig. 5 - Calculated optical reflectivity curve for a perfect SL (+++) and a SL containing random period deviations.

Fig. 6 - Calculated DBR optical reflectivity curve for a perfect SL (+++) and a SL with a systematic period thickness deviation become smaller with growth.

References

1   J.L. Jewell, J.P. Harbison, A Scherer, Y.H. Lee and L.T. Florez, IEEE Jour. Quant. Elect. 27 (1991) 1332.
2   P.J. Wang, M.S. Goorsky, B.S. Meyerson, F.K. LeGoues, M.J. Tejwani, Appl. Phys. Letters 59 (1992) 814.
3   J.M. Vanndenberg, R.A. Hamm, M.B. Panish, H. Temkin, J. Appl. Phys. 62 (1987) 1278.
4   D.K. Bowen and M. Wormington, Advances in X-ray Analysis 35 (1993) 35.
5   J.P. Weber, K. Malloy, and S. Wang, IEEE Phot. Tech. Lett. 2 (1990) 162.
6   Bede Scientific Instruments Ltd., Lindsey Park, Bowburn, Durham DH6 5PF, UK.
7   K. Matney and M.S. Goorsky, submitted to Journal of Crystal Growth.
8   R.A.D.S. Program, Bede Scientific Instruments, Lindesy Park, Bowburn, Durham DH6 5PF, U.K.
9   K. Matney, unpublished.
10 A. Iida, K. Kohra, Phys. Stat. Sol. A 51 (1979) 533.
11 D.H. Christenson, J.G. Pellegrino, R.K. Hickernell, S.M. Crochiere, C.A. Parsons, and R.S. Rai, J. Appl. Phys., 72 (1992) 5982.
12 M. Born and E. Wolf, Principle of Optics (Pergamon, New York, 1970).
13 R. Fernandez, A. Harwit, and D. Kinell, J. Vac. Sci. Technol **B12** (1994) 1023.
14 M Quillec, L. Goldstein, G. Le Roux J. Burgeat, and J. Primot, J. Appl. Phys. 55 (1984) 2904.
15 P. Auvrey, M. Baudet, and A. Regreny, J. Appl. Phys., 62 (1987) 456.
16 B.K. Tanner, D.K. Bowen, J. Crystal Growth 126 (1992) 1.
17 M.S. Goorsky, M. Meshkinpour, D.C. Streit, and T. Block, to be pubished.

0.85-μm 8×8 Bottom-Surface-Emitting Laser Diode Arrays Grown on AlGaAs Substrates by MOCVD

Yoshitaka Kohama, Yoshitaka Ohiso, Chikara Amano, and Takashi Kurokawa

NTT Opto-electronics Laboratories, 3-1 Morinosato-Wakamiya, Atsugi-shi, Kanagawa 243-01, Japan

Abstract The first 0.85-μm bottom-emitting surface-emitting laser diodes (SELDs) were made by using novel 0.85-μm transparent substrates. Substrate cleaning and the introduction of a buffer layer were keys to acquiring excellent epilayers on AlGaAs. The laser characteristics of those SELDs are the same as those of SELDs grown on a GaAs substrate. We also fabricated 8×8 independently addressable SELD arrays that exhibited good lasing characteristics and uniform threshold current density.

1. Introduction

Surface-emitting laser diodes (SELDs) offer many advantages over conventional edge-emitting lasers. These advantages include a noncleaving fabrication process, a circular light output mode, high two-dimensional packing density for arrays, single longitudinal mode emission due to their short cavity length (about 300 nm), and wafer-scale testing [1]. 0.85-μm SELDs are more promising than 0.98-μm ones from the viewpoint of application to optical interconnection because sensitive Si or GaAs photodetectors and optical components such as lens, fibers, and PBS are readily available. Because GaAs substrates are opaque, all the SELDs reported in the 0.85-μm band were of the top-emitting-type, which emits light from the front surface. Bottom-emitting SELDs, however, like the conventional 0.98-μm ones that emit light from the bottom surface, have many advantages over top-emitting SELDs: thermal dissipation is better in a junction-down structure, the structure is optically accessible from both faces, and device fabrication is easier. Etching the substrate to emit lasing light, however, it is difficult and complicated [2]. This etching technique is not also suitable for an array structure.

In this paper, we report the first 0.85-μm bottom-emitting-type SELDs made by using metalorganic chemical vapor deposition (MOCVD) to grow SELDs on 0.85-μm transparent AlGaAs substrates. Most reported SELDs have been grown by molecular beam epitaxy (MBE); there have been only a few reports on SELDs grown by using MOCVD [3]. This is because MBE growth is thought to be better than MOCVD in terms of the layer thickness uniformity over the wafer, ease and accuracy of growth rate control by in-situ monitoring, and freedom from toxic gases like AsH_3 or SiH_4. The high productivity and throughput of MOCVD, however, make it desirable for growing SELDs.

2. Experimental

The SELDs were grown on n-type 35×35-mm AlGaAs substrates (Al content was about 0.1) by low-pressure MOCVD. Very thick AlGaAs liquid-phase-epitaxy-grown epilayers on GaAs substrates were used as the AlGaAs substrates [4]. Typical growth conditions were as follows: the pressure was 0.1 atmosphere, the growth temperature was 700-750 °C, and the growth rate was 3.2 µm/h except when growing certain parts of the active layer. This growth rate is three times that of conventional MBE, and the total growth time was greatly reduced: to about 4.5 hours. Figure 1 schematically shows the bottom-emitting SELD structure, which consists of a buffer layer and a p-i-n junction with bottom and top distributed Bragg reflectors (DBRs). A thin GaAs buffer layer was introduced on the AlGaAs substrate to improve the quality of epilayers. The Si-doped bottom DBR had 28.5 pairs of an $AlAs/Al_{0.15}Ga_{0.85}As$ optical-quarter-wave stack. The undoped spacer region consisted of six $Al_{0.2}Ga_{0.8}As/GaAs$ multiple-quantum-well layers as an active region sandwiched between two $Al_{0.3}Ga_{0.7}As$ layers. The C-doped top DBR consisted of 30 pairs of $AlAs/Al_{0.15}Ga_{0.85}As$ layers with a 10-nm-thick intermediate $Al_{0.5}Ga_{0.5}As$ layer. For array structures, the individual elements of a SELD array were isolated by using Cl_2 reactive ion etching and wet etching to etch just through the active layer. After the individual array elements were planarized using polyimide, a SiO_2 layer was selectively deposited to electrically isolate them. Gold was deposited to bond the individual elements and electrode pads. Finally, an anti-reflection coating layer, (SiO_2) was coated onto the back surface. The diameters of the individual elements were 16, 21, or 26 µm; the distance between the individual elements on a chip was 250 µm; and the unit array occupied on an area of 3×3 mm. The arrays were mounted on ceramic packages, and Au wires were bonded from the electrode pads to electrodes.

3. Results

AlGaAs substrates oxidize because the aluminum readily bonds with oxygen; and the epilayers we grew on AlGaAs by conventional MBE therefore had poor crystal quality. We found that to improve the surface morphology of epilayer on AlGaAs it was important to clean the substrate before growth and to introduce a GaAs buffer layer. To clean the substrate we etched it deeply with $H_2SO_4:H_2O_2:H_2O$ (4:1:1) and then loaded it into the MOCVD apparatus as soon as possible. As for the buffer layer, growing AlGaAs directly on an AlGaAs substrate resulted in morphology worse than that of AlGaAs grown on GaAs (Fig. 2). This is thought to be because the initial growth of AlGaAs readily becomes three-dimensional growth. Figure 3 shows how surface defect density and absorbance at 850 nm depend on the thickness of the GaAs buffer layer. The surface defect density of an epilayer grown in this study is one-tenth that of epilayers grown by conventional method. As the buffer layer gets thicker, the surface defect density drops dramatically but absorbance increases. A 100-nm-thick GaAs buffer layer was therefore introduced and its transmissivity at 0.85 µm was expected to be about 90%.

Figure 4 shows that the typical characteristics of light output and voltage as functions of injected current. The typical threshold current and threshold current density were 4.58 mA and 1.4 kA/cm^2 for elements 21 μm in diameter. The external quantum efficiency was 2.9 %. The optical output power was almost the same as that of the elements grown on GaAs substrates with the same total reflectivity, so the absorbency of the AlGaAs substrate has little effect on laser optical power. Figure 5 shows the dependence of the threshold current density and the threshold voltage on element diameter for elements on GaAs substrates (top-emitting type) and for elements on AlGaAs substrates (bottom-emitting type). By considering absorption in layers, the total reflectivity was adjusted to give the same value for both types of elements. When the element diameter is less than 20 μm, the threshold current density increases gradually because of surface recombination on the side wall of the element. In terms of the threshold current density, the lasing characteristics for AlGaAs elements were almost the same as those for GaAs elements. This indicates that the crystallinity of the epilayer on AlGaAs is good. The threshold voltage for AlGaAs was 3.1 V, which is 0.6 V higher than that for GaAs. This higher threshold is due to the larger number of pairs in the p-type DBR.

The variation of the threshold current across an 8×8 array is shown in Fig. 6. The diameter of the individual elements was 21 μm. They all performed as expected at room temperature under CW operation. The mean value ± the standard deviation of the threshold current density was 4.73 ± 0.25 mA, and the mean value ± the standard deviation of the lasing wavelength was 844.72 ± 1.70 nm. The deviation of the threshold current is almost as small as any ever reported [5]. Because the deviations in the threshold current come from process variations rather than fluctuations in the thicknesses of layers, we should get better uniformity when we improve process conditions.

4. Conclusion

We have demonstrated 0.85-μm SELDs grown on novel AlGaAs substrates. The keys to obtaining a high-quality epilayer are to clean the substrate and use a GaAs buffer layer. The characteristics of these SELDs were the same as those of SELDs made on a GaAs substrate. We also fabricated 8×8 independently addressable SELD arrays and obtained uniform laser array characteristics.

Acknowledgments

We thank Dr. Yoshihiro Imamura, Dr. Seiji Fukushima, and Dr. Yoshiaki Kadota for their support and encouragement throughout this work.

References

[1] Morgan R A, Chirovsky L M F, Focht M W, Guth G, Asom M T, Leibenguth R E, Robinson K C, Lee, Y H and Jewell J L 1991 *SPIE 1562: Devices for Optical Processing. SPIE Bellingham WA* 149-59.

[2] Koyama F, Kinoshita S and Iga K 1989 *Appl. Phys. Lett* **55** 221-2.

[3] Zhou P, Cheng J, Sun S Z, Zheng K, Armour E, Hains C, Hsin W, Myers D R and Vawter G A 1991 *IEEE Photon. Technol. Lett.* **3** 591-3.

[4] Kurata K, Ono Y, Morioka M, Ito K and Mori, M 1977 *IEEE J. Quantum Electron.* **QE-13**, 525-7.

[5] Vakhshoori D, Wynn J D, Zydzik G J and Leibenguth R E 1993 *Appl. Phys. Lett.* **62** 1718-20.

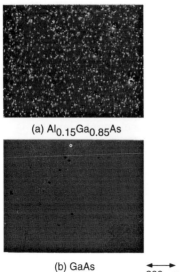

Fig. 2. Photographs of epilayer surfaces on two types of 100-nm-thick buffer layers.

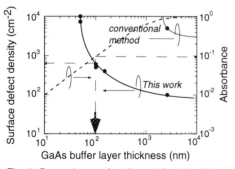

Fig. 3. Dependence of surface defect density and absorbance at 0.85 μm on GaAs buffer layer thickness.

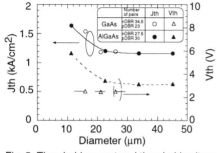

Fig. 5. Threshold current and threshold voltage as functions of mesa diameter.

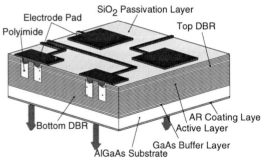

Fig. 1. Schematic diagram of the bottom-surface-emitting laser diode arrays grown on AlGaAs substrate.

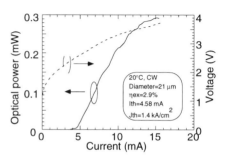

Fig. 4. Light-current characteristics for the bottom surface-emitting laser diode grown on AlGaAs substrate.

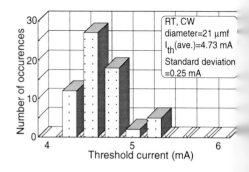

Fig. 6. Variations in the threshold current across an 8×8 array.

Vertical-cavity surface-emitting lasers with buried lateral current confinement

S. Rochus, T. Röhr, H. Kratzer, G. Böhm, W. Klein, G. Tränkle, and G. Weimann

Walter-Schottky-Institut, Technische Universität München, D-85748 Garching, Germany

ABSTRACT: Vertical-cavity surface-emitting lasers (VCSELs) with lateral current injection via the doped P-AlGaAs cladding layer have low series resistances around 100Ω and low threshold voltages around 2V. VCSEL diodes with three strained $In_{0.17}Ga_{0.83}As$ QWs in the cavity show cw threshold currents of only 1.2mA and optical output powers of 1mW. The threshold current and the laser efficiency are limited essentially by current spreading due to the lateral injection. Lateral current confinement is improved by introducing a 50nm thick n-type GaAs current blocking layer into the p-doped cladding. The laser diodes are structured by selective removal of this layer in the cavity and subsequent growth of the top Bragg mirror in a second MBE growth sequence. Threshold currents can be reduced by a factor between 2 and 5, depending on the aperture in the blocking layer.

1. Introduction

In the last years VCSELs found great interest for potential applications in optoelectronic systems. Whereas low cw threshold currents have been reported repeatedly, most of the devices have rather high series resistances and high operating voltages [e.g. 1,2]. Low series resistances and low lasing threshold voltages around 1.5V have been obtained by using doped, piecewise linearly graded distributed Bragg reflectors (DBRs) [3]. We, alternatively, achieved low series resistances of 50Ω and comparable threshold voltages by using lateral current injection via both, n and p-doped, cladding layers in the cavity [4,5].
Lateral contacting leads to current spreading outside the optical resonator. Furthermore it requires thick claddings and thus a long cavity, which reduces the longitudinal confinement factor. These two effects increase the threshold current and limit the laser output efficiency.
We improved the optical confinement by a reduction in cladding thickness and cavity length. Thin n-cladding layers and doped n-DBRs were used for the n-contact. Lateral injection for holes was retained, since the series resistance with vertical current injection is dominated by the multi-heterobarriers of p-type DBRs. Additionally, the current was confined in the optical resonator by inserting a n-GaAs blocking layer into the P-AlGaAs cladding in the contact regions [6,7,8].

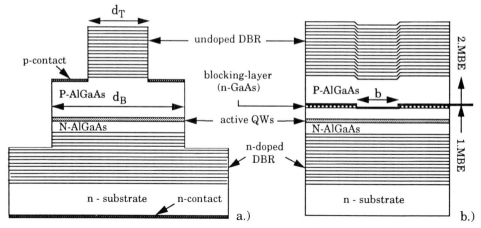

Fig.1: Vertical layer structure for resonators a.) without and b.) with current blocking layer. The solid horizontal line in b.) indicates the surface for the second MBE growth sequence.

2. VCSELs with lateral current injection

Fig.1a schematically shows VCSELs with lateral p-contacts, depicting epitaxial layer structure and lateral dimensions. Compared to our previous design [4,5], where both top and bottom cladding layers were contacted, these diodes have a lateral contact to the P-AlGaAs only, with electron injection from the substrate. This reduces the inner cavity length significantly from 12 to 7 half wavelengths, increasing the longitudinal confinement factor from 0.016 to 0.023. The active region, although asymmetrically placed, is still, however, at an antinode of the resonator.

The bottom n-doped DBR, with 26.5 pairs of λ/4-AlAs/GaAs stacks, is grown on a n-GaAs substrate with a nominal Si-concentration of $1 \cdot 10^{18} cm^{-3}$. The N-$Al_{0.35}Ga_{0.65}As$ cladding (Si: $1 \cdot 10^{18} cm^{-3}$) is 110nm thick, followed by the undoped active region with three strained $In_{0.17}Ga_{0.83}As$ QWs of 8nm width, separated by GaAs barriers and spacers of 10nm thickness. The top P-$Al_{0.35}Ga_{0.65}As$ cladding (Be: $1 \cdot 10^{18} cm^{-3}$) is 850nm thick to facilitate lateral contacting. The undoped DBR consists of 18 pairs of λ/4-AlAs/GaAs stacks (see Fig.1a).

Double mesa diodes, with varying diameters d_T and d_B, are structured by reactive ion etching (RIE) and selfadjusted ohmic contacting [4,5]. Whereas our previous diodes, with two lateral contacts, had differential series resistances below 50Ω, the injection over the n-DBR leads to increased resistances of about 100Ω for our present diodes (measured for $2 \cdot I_{th}$ with d_B and d_T 22 and 12μm).

Fig.2 shows cw light vs. current characteristics for different lateral dimensions. For top mesa diameters d_T smaller than 6μm single mode operation is observed. These diodes have threshold currents between 1.2 and 2mA. The highest external differential quantum efficiencies were measured to 9% for diodes with d_T between 8 and 12μm, with output powers saturating at approximately 1mW.

It is obvious, however, that in the design investigated here, the major part of the current flows outside the optical resonator. This unwanted current can be reduced by insertion of a blocking layer into the top cladding for lateral current confinement.

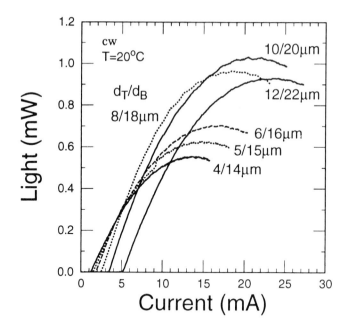

Fig.2: L-I-characteristics of one-sided laterally contacted VCSELs with different mesa diameters d_T/d_B in cw operation at 20°C.

3. VCSELs with buried blocking layers

The incorporation of an additional blocking layer into the resonator structure requires a two-step MBE growth [6,7,8]. In the first growth sequence the unchanged bottom DBR, the N-AlGaAs cladding, the active region, and a 100nm thick fraction of the P-AlGaAs cladding are grown, followed by a 50nm thick GaAs layer with $n=2 \cdot 10^{18} cm^{-3}$ forming a reversely biased blocking junction. Apertures with diameters b between 1 and 10 µm are opened in this layer by selective RIE with CCl_2F_2. In the second growth run the remaining 750nm of the P-AlGaAs cladding and the top DBR are grown. The optical resonator is thus essentially unchanged. The thickness of the GaAs blocking layer is limited to approximately 50nm to obtain planar growth without excessive facetting at the edges. Low growth temperatures of around 540°C further serve to reduce lateral dimensions of the higher indexed facets. MBE regrowth requires a GaAs growth surface for in-situ oxide desorption. Therefore GaAs is used instead of AlGaAs; the blocking layer actually consists of 50nm GaAs and two sacrificial few nm thick AlGaAs and GaAs layers, which make growth on a GaAs surface possible in the aperture, too [7,8]. For aperture diameters smaller than the top mesa (b < d_T, see Fig.1) current flow is thus effectively confined to the resonator. The processing of the diodes is identical with that for VCSELs without confinement layer [5].

Fig. 3 compares the output characteristics for VCSELs with and without blocking layer fabricated on the same wafer. VCSELs with lateral confinement show concentrated current flow, evident from the reduction in threshold current by factors between 2 and 5 depending on device dimensions. Moreover, the threshold current becomes independent on the mesa diameters and is now determined by the hole diameter in the blocking layer only. For hole diameters smaller than 6µm no laser emission is observed. Possible reasons are the non-

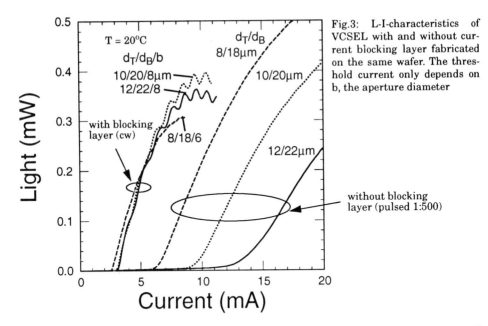

Fig.3: L-I-characteristics of VCSEL with and without current blocking layer fabricated on the same wafer. The threshold current only depends on b, the aperture diameter

radiative recombination at defects caused by the regrowth, dominating for small apertures, or an "unstable" resonator configuration due to the facetting. For the investigated wafer the emission wavelength for VCSELs without blocking layer is around 950nm, while the QW emission is at 970nm. These diodes, therefore, show only pulsed operation, and have threshold currents which are increased by a factor of 2.5 compared to the well matched resonators, described in the previous section. On the other hand, cw operation is observed from VCSELs with blocking layers at an emission wavelength of 942nm, in spite of an even larger mismatch in the emission wavelengths of the lasers and the QWs.

We gratefully acknowledge the financial support of the Bundesminister für Forschung und Technologie (BMFT) under contract BV 219/4.

References

[1] R.S. Geels, B.J. Thibeault, S.W. Corzine, J.W. Scott, and L.A. Coldren, 1993, IEEE J. Quant. Electron, vol. 29, 2977-2987
[2] T. Wipiejewski, K. Panzlaff, E. Zeeb, and K. J. Ebeling, 1992, Proc. 18th European Conf. Opt. Commun., Berlin, Germany, 903-906
[3] S. A. Chalmers, K. L. Lear, and K. P. Killeen, 1993, Appl. Phys. Lett., vol. 62, 1585-1587
[4] S. Rochus, C. Dill, G. Böhm, W.Klein, M. Walther, G. Tränkle, G. Weimann, Proc. LEOS'93, 6th Annual Meeting, San Jose, CA, 548-549
[5] S. Rochus, G. Böhm, W.Klein, G. Tränkle, G. Weimann, to be published
[6] T. J. Rogers, C. Lei, B. G. Streetman, and D. G. Deppe, 1993, J. Vac. Sci. Technol. B, vol. 11, 926 - 928
[7] T. Röhr, M. Walther, S. Rochus, G. Böhm, W. Klein, G. Tränkle, G.Weimann, 1993, Materials Science and Engineering, B21 , 153-156
[8] H. Kratzer, T. Röhr, S. Rochus, G. Böhm, A. Zrenner, G. Tränkle, and G. Weimann, to be published

MBE growth of low threshold laser structures employing short-period superlattices

H. Riechert, D. Bernklau, A. Milde, M. Schuster, H. Cerva, C. Hoyler and H.-D. Wolf

Siemens Corporate Research and Development, D-81730 München, Germany

Abstract. MBE growth of bulk-like layers of short-period $(AlAs)_n(GaAs)_m$ superlattices as replacement for alloy AlGaAs is reported. Structures with mean Al-contents up to 0.8 were characterised by high-resolution X-ray diffraction and by quantitatively evaluated high-resolution TEM. Sample periodicity, the profile of Al-content on an atomic scale and the influence of growth parameters are presented. Photoluminescence shows that optical quality comparable to that of the best alloy materials can be attained even at growth temperatures low enough to avoid Ga desorption (610°C). Employing superlattices grown in this way, high optical confinement InGaAs-SQW laser structures and vertical cavity lasers emitting at ~ 980 nm were realised with threshold current densities of 85 A/cm^2 and 400 A/cm^2, respectively.

1. Introduction

The MBE growth of AlGaAs with high optical quality - as it is essential for application in lasers - is usually performed at high growth temperatures. To our knowledge, in all reported (In)GaAs/AlGaAs lasers with threshold current densities below 100 A/cm^2, the AlGaAs layers were grown at ~ 700°C [e.g. 1]. Whereas the desorption of Ga occuring at these temperatures may aid in the desorption of unwanted oxygen [2], small variations in growth temperature will cause fluctuations in both composition and thickness of AlGaAs layers. Especially for vertical cavity surface emitting lasers (VCSELs) this is expected to limit the device yield substantially. Therefore the aim of this study was the MBE growth of low threshold lasers entirely in the regime where Ga-desorption is negligible (i.e. below 610°C).

One way to avoid elevated MBE growth temperatures for AlGaAs is the use of cracked, i.e. dimeric arsenic [3]. Alternatively, the replacement of alloy AlGaAs by short-period superlattices (SPS), i.e. by multilayers of the type $(AlAs)_n(GaAs)_m$ seemed very promising since quantum wells with thin SPS barriers are reported to show superior photoluminescence [4] and have been successfully employed in lasers [3].

Since very little is known about the growth and properties of bulk-like SPS material we have grown ~1 μm thick SPS layers and have analysed them by high-resolution X-ray diffraction (HRXRD), high-resolution transmission microscopy (HRTEM) and by photoluminescence (PL), if their bandgap is direct. Based on this, conventional lasers with high optical confinement and VCSELs were fabricated, replacing all bulk AlGaAs by SPS layers and employing low growth temperatures throughout.

2. Growth and characterisation of SPS layers

2.1. MBE growth

1.0 - 1.3 μm thick SPS layers were grown with a periodicity $n + m = 10$ monolayers. Thus 1 μm of material consists of ~ 360 periods of $(AlAs)_n(GaAs)_m$.

Growth was performed in a standard solid source MBE system with magnetically coupled, sinusoidally driven shutters [5]. The mean Al-contents $\bar{x}_{Al}$ were set around 0.3 and 0.8. GaAs and AlAs growth rates were 0.28 and 0.12 nm/sec and tetrameric As was always used. The ratio of beam equivalent pressures of As and Ga was kept at about 22.

Whereas previous experiments with the growth of $Al_xGa_{1-x}As$ with $x_{Al} \approx 0.8$ had shown that reasonable surface morphology could only be attained by growth above 720° C, we find that SPS layers with similar Al-content can be grown with excellent surface morphology over the whole range of growth temperatures investigated (580 - 680°C). Apparently, the intermittent smoothing of the growth front during the growth of the GaAs layers is sufficient to suppress any roughening. 680°C is the practical upper limit for controllable growth, since in the growth of SPS layers Ga desorption becomes noticeable (see below) at lower temperatures than in AlGaAs growth.

2.2. High-resolution X-ray diffraction and high-resolution TEM

HRXRD was performed and the measured satellite reflectivities were compared to those calculated with dynamical theory using angle dependent structure factors. This provides precise information about superlattice periodicity, mean lattice mismatch (and thereby about $\bar{x}_{Al}$) as well as the thicknesses of the GaAs and AlAs layers. The decay of the reflectivities of satellites of increasing order was described by a one-dimensional Debye-Waller factor, yielding the fluctuations in periodicity $<\Delta d^2>^{1/2}$ [6,7]. Results for five samples, grown at different temperatures in the range of low Al-contents, are given in table 1. All were grown with the same beam-flux settings and shutter timings, aimed at achieving a SPS with $n = 3$, $m = 7$. Superlattice peaks are clearly observed for all these samples, but we find that their

Table 1: Summary of results from X-ray diffraction and photoluminescence (2 K)

sample	grown at	X - ray diffraction				photoluminescence	
		d(AlAs)	d(GaAs)	$\bar{x}_{Al}$	$<\Delta d^2>^{1/2}$	$\bar{x}_{Al}$	FWHM
1	580 °C	0,892 nm	2.120 nm	0.296	0.29	0.264	7.3 meV
2	610° C	0.915 nm	2.122 nm	0.301	0.35	0.268	6.6 meV
3	640° C	0.924 nm	2.037 nm	0.312	0.61	0.283	8.0 meV
4	660° C	0.924 nm	1.987 nm	0.317	0.60	0.289	11.0 meV
5	680° C	0.939 nm	1.317 nm	0.416	0.66	0.392	22.0 meV

intensities vary systematically with growth temperature. Whereas the AlAs thickness is roughly the same, the GaAs thickness decreases significantly with growth temperature. Correspondingly, $\bar{x}_{Al}$ increases from 0.30 to 0.42. Similarly the thickness fluctuations increase by more than a factor of two with increasing growth temperature. Thus, the structurally most perfect SPS layers are obtained at low growth temperatures. For SPS layers with high Al-content this is even more essential, since such samples grown at or above 660°C do not show superlattice peaks.

HRTEM and quantitative evaluation of image contrast [8] were applied to find the profile of the Al-content along the growth direction. Unlike XRD which reveals fluctuations averaged over the whole sample thickness, this gives a local image of interface abruptness. The HRTEM image for a cleaved-edge specimen of sample 1 and the mean Al-contents (error ± 0.1) for two superlattice periods (averaged over a total of 14 cation rows at a thickness of 10 nm) are shown in fig. 1.

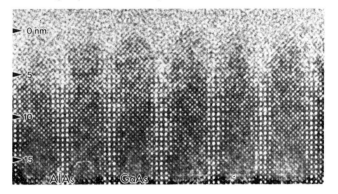

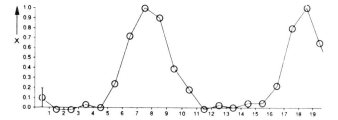

Fig. 1:

HRTEM image of a cleaved edge of sample 1. The local Al-contents shown in the lower part of the figure are the results of a quantitative evaluation of image contrast (see text). In this sample as well as in sample 5 a spreading of the interface over 2 - 3 cation layers is observed. Previously we have shown [8] that under similar growth conditions a spread over one intermediate layer can be achieved in double barrier quantum well structures. The larger spread in SPS layers is very likely due to a cumulation effect and to the fact that the GaAs layers (which smooth out roughness) are thinner in these samples.

2.3. Low temperature photoluminescence

Data from photoluminescence at 2 K are presented on the right hand side of table 1. The systematic variation in peak positions and widths with growth temperature corresponds well to our results from HRXRD (The systematic difference between PL and HRXRD for $\bar{x}_{Al}$ is due to the commonly used dependence between PL peak position and x_{Al}). HRXRD reveals that the peak shift between samples 1 and 2 is not due to Ga desorption but to a smaller AlAs layer thickness in sample 1. Thus it can be assumed that desorption of Ga becomes effective above 610° C, but we find that it is more pronounced than in the growth of AlGaAs. This is explained by the fact that the rate of Ga desorption from AlGaAs [9] is proportional to the surface coverage with Ga, which in a SPS is periodically very high.

With the exception of sample 5 which shows a strongly broadened line, all PL spectra consist of single, rather narrow lines and exhibit no second low energy peak. Such a rather wide second peak ("pair transition") is generally observed in all but the very best bulk AlGaAs samples and is associated with impurity-related transitions. Unless SPS samples behave differently in this respect, the absence of pair transitions can be taken as an indication of excellent optical quality. The linewidths we find (6.6 meV FWHM for sample 2) are larger than those for AlGaAs but still indicate a (mini-)band to (mini-)band transition. A more detailed study of the optical and electrical properties of some of our samples is given in [10].

3. Device results

3.1. Broad area SQW test lasers

The results from photoluminescence have prompted us to grow single quantum well (SQW) $In_{0.2}Ga_{0.8}As/Al_xGa_{1-x}As$ separate confinement laser structures employing SPSs for both the cladding layers (1μm thick, $\bar{x}_{Al} = 0.8$) and the waveguide layers (2x 115 nm thick, $\bar{x}_{Al} = 0.35$). Growth temperatures were 520° C for the active region and 610° C for the SPSs. The threshold current densities for broad area devices (100 x 800 μm^2) emitting at $\lambda \approx 980$ nm were as low as 85 A/cm^2, demonstrating that very low losses can be achieved with the growth scheme presented above.

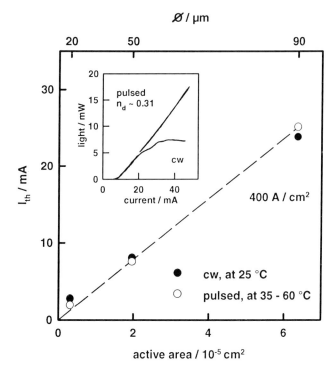

Fig. 2:

Threshold current I_{th} of VCSELs versus active area. Inset: Light vs. current characteristics (25° C) of a device with 50 μm diameter

3.2. Vertical cavity surface emitting lasers (VCSELs)

Low temperature growth and SPSs were also used for VCSEL structures. The DBRs (distributed bragg reflectors) consisted of 20 n-doped and 21 p-doped AlAs/GaAs pairs where the interfaces between AlAs und GaAs were pseudograded by 10 period superlattices of stepwise varying layer thicknesses. The doping concentrations were enhanced at the carrier injection sides.

Three InGaAs quantum wells were embedded in an SPS ($\bar{x}_{Al} = 0.35$). The emission wavelength of the VCSELs was about 975 nm.

Devices with cylinder-shaped mesas were RIE-etched from the p-side just through the active region. The threshold currents of VCSELs with different diameters are summarised in fig. 2. Threshold current densities of 400 A/cm² are evaluated for devices with diameters ≥ 50 μm. These results are among the best reported in the literature [11] and were achieved for VCSELs with comparatively few DBR mirror pairs. The low threshold values and high differential external quantum efficiencies ($\eta_d > 30$ %, uncorrected for absorption losses in the substrate, see e.g. inset of Fig. 4) indicate very low internal losses in our VCSEL structure.

An aging experiment at 40°C and 20 mA cw with the device corresponding to the inset of Fig. 2 did not show any significant degradation over a test duration exceeding 3000 h.

4. Conclusion

We have grown and analysed bulk-like short-period superlattices as a replacement of AlGaAs with high aluminum contents.

Smallest deviations from periodicity and excellent optical quality can be achieved at growth tempateratures low enough to avoid any desorption (~ 600° C). Employing this growth scheme, extremely low loss laser structures were obtained with maximum control over layer thickness and composition. This was demonstrated by SQW (InGaAs) lasers and 3 QW (InGaAs) VCSELs with threshold current densities of 85 and 400 A/cm^2, respectively.

This work was partly supported by the German Federal Ministry of Research and Technology. The authors alone are responsible for the contents.

5. References

[1] N. Chand, S.N.G. Chu, N.K. Dutta, J. Lopata, M. Geva, A.V. Syrbu, A.Z. Mereutza and V.P. Yakovlev 1994 IEEE J. Quantum Electronics **30** 424 - 440

[2] S.V. Iyer, H.P. Meier, S. Ovadia, C. Parks, D.J. Arent and W. Walter 1992 Appl. Phys. Lett. **60** 416 - 418

[3] C.T. Foxon, P. Blood, E.D. Fletcher, D. Hilton, P.J. Hulyer and M. Venig 1991 J. Crystal Growth **111** 1047 - 1051

[4] K. Fujiwara, A. Nakamura, Y. Tokuda, T. Makamaya and M. Hirai 1986 Appl. Phys. Lett. **49** 1193 - 1195

[5] Fisons Instruments (VG Semicon)

[6] M. Schuster and N. Herres 1993 Proceedings RöTo 1993 (Güntersberge) Arbeitskreis Röntgentopographie

[7] V. Holy, J. Kubena and K. Ploog 1990 phys. stat. sol. (b) **162** 347 - 361

[8] S. Thoma, H. Riechert, A. Mitwalsky and H. Oppolzer 1992 J. Crystal Growth **123** 287 - 305

[9] J.-P. Reithmaier, R.F. Broom and H.P. Meier 1992 Appl. Phys. Lett. **61** 1222 - 1224

[10] F.A.J.M. Driessen, H. Riechert, D. Bernklau, G.J. Bauhuis and L.J. Gilling 1994 22nd Int. Conf. on the Physics of Semiconductors (Vancouver)

[11] T.E. Sale, J. Woodhead, R. Grey and P.N. Robson 1992 IEEE Phot. Techn. Lett. **4** 1192 - 1194

Monolithically Integrated Optical Receivers and Transmitters

D. T. Nichols, W. S. Hobson, P. R. Berger, N. K. Dutta, P. R. Smith, J. Lopata, D. L. Sivco and A. Y. Cho

AT&T Bell Laboratories, 600 Mountain Avenue, Murray Hill, New Jersey 07974-0636 USA

Abstract. We present results on optoelectronic integrated circuits in both the GaAs and InP systems. The GaAs transmitters employed GaAs MESFETs integrated with an MQW-SCH ridge laser. These circuits operated with a bandwidth of 3.5 GHz. The GaAs receivers consisted of a p-i-n photodiode integrated with a GaAs MESFET in a simple single stage preamplifier configuration. The circuits exhibited gains of 17 dB and bandwidths of 4 GHz. The InP receivers employed an InGaAs/InAlAs photodiode integrated with a pseudomorphic MODFET. These circuits exhibited bandwidths as high as 10 GHz and sensitivities of -31.8 dBm, -26 dBm and -17 dBm at 1, 2 and 10 Gb/s, respectively.

1. Introduction

Substantial progress in the fabrication of optoelectronic integrated circuits (OEICs) has recently been made and there have been many reports in the literature which document greatly improved perfomance of photoreceivers [1], [2], integrated laser/modulators [3], [4], wavelength division multiplexers [5] and integrated laser/driver circuits [6]-[10]. Approaches to integration have included selective area growth and adaptation of the structure so that the laser/detector and transistor may utilize the same layer structure. For ease of fabrication and potentially greater reliability it is desirable to accomplish the integration with a single growth on a planar substrate.

The InP system is well-suited to long-haul communications applications because of the availability of low-loss fibers at long wavelengths and many of the OEICs which have been reported have concentrated in the 1.3 - 1.55 μm range. However, for short haul applications, such as LANs, it may also be desirable to employ GaAs-based circuits for improvement in high-temperature operation. The development of high-quality InGaP, along with the availability of highly selective wet etches for GaAs on InGaP and InGaP on GaAs make the GaAs/InGaP system particularly desirable for the fabrication of vertically integrated OEICs. In this paper, we present results on integrated photoreceivers and transmitters in the InP, GaAs/AlGaAs and GaAs/InGaP material systems.

2. Monolithically Integrated Photoreceiver Circuits

2.1. Gallium Arsenide-Based p-i-n/MESFET Photoreceivers

The GaAs p-i-n/MESFET structure was grown on semi-insulating GaAs and consisted of a 1000 Å layer of GaAs which was unintentionally doped, 1000 Å of $Al_{0.4}Ga_{0.6}As$, also unintentionally doped, an n-type channel layer consisting of 2300 Å of GaAs and a 3000 Å GaAs n-contact layer. This contact layer served as the ohmic contact layer for both the MESFET and the p-i-n diode. These layers were followed by the p-i-n structure which consisted of 2000 Å of n-type $Al_{0.4}Ga_{0.6}As$, a 6000 Å undoped GaAs absorption layer, a 2000 Å p-type $Al_{0.4}Ga_{0.6}As$ layer and a GaAs p-contact layer. Similar structures were also grown in which the $Al_{0.4}Ga_{0.6}As$ was replaced in favor of $Al_{0.6}Ga_{0.4}As$ and InGaP in order to facilitate selective wet chemical etching. Standard photo-lithographic techniques were then employed to selectively etch the p-i-n layers, define the FET mesas, define the ohmic contacts, gates, Ti resistors and interconnect metal. The MESFETs employed a π-gate configuration in order to reduce the gate resistance. The 1.0 μm gates were defined using optical lithography. The gate widths were varied (2x40, 2x75 and 2x100 μm wide.) A schematic of the receiver circuit is shown in the inset of Fig. 1(a).

The DC characteristics of isolated GaAs MESFETs were measured. The devices exhibited transconductances as high as 341 mS/mm with current densities (I_{ds}) of up to 537 mA/mm. Similarly, the DC characteristics of the isolated p-i-n photodiodes were also measured. The diodes had reverse breakdown voltages between 15 and 20 V and reverse leakage currents smaller than 10 nA when biased at -5 V. The RF characteristics of the MESFETs were measured between 100 MHz and 40 GHz. The FETs were biased for maximum transconductance and the devices were found to have cutoff frequencies as high as f_t = 11.3 GHz and f_{max} = 35.1 GHz.

High-speed characterization of the photodiodes and photoreceiver circuits was carried out between 130 MHz and 20 GHz. A vertical cavity surface emitting laser diode with a measured bandwidth of 6 GHz was modulated by the lightwave analyzer and its output was coupled to the photoreceivers via an optical fiber. The responses of the individual diodes and circuits were then measured. The diodes had diameters of 18 and 50 μm and were biased at -5 V. The smaller diodes had bandwidths which were limited by the response of the laser. The response of the photodiode of each receiver was used to calibrate the gain of the amplifier section. Typical characteristics of the receivers are shown in Fig. 1. The circuits exhibited 3-dB bandwidths as high as 4.5 GHz and flatband gains as high as 17 dB.

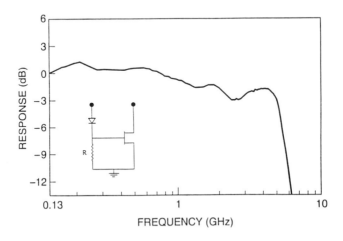

Fig. 1 RF response of the photoreceiver fabricated on GaAs (with a circuit schematic in the inset).

2.2. Indium Phosphide-Based p-i-n/MODFET Photoreceivers

The InP-based p-i-n/MODFET structure was processed in a similar manner, but with a few important differences. After, growing the p-i-n diode structure, selected areas of the wafer were etched, the remaining p-i-n structure was masked with oxide and the MODFET structure was regrown. The p-i-n photodiode was grown first and consisted of a superlattice buffer, a 0.4 μm n+ $In_{0.52}Al_{0.48}As$ ohmic contact layer, a 0.75μm i $In_{0.53}Ga_{0.47}As$ active region with 300 A i $In_{0.52}Al_{0.48}As$ setback layers on either side, a 0.4 μm p+ $In_{0.52}Al_{0.48}As$ p-contact layer and a 200 A p+ $In_{0.53}Ga_{0.47}As$ cap layer. The MODFET structure was comprised of a superlattice buffer, an i $In_{0.52}Al_{0.48}As$ buffer, a 400 A i $In_{0.53}Ga_{0.47}As$ layer, a 100 A strained i $In_{0.65}Ga_{0.35}As$ channel layer, a 50 A $In_{0.52}Al_{0.48}As$ spacer layer, a 150 A n+ $In_{0.52}Al_{0.48}As$ donor layer, a 400 A i $In_{0.52}Al_{0.48}As$ layer and a 200 A n+ $In_{0.53}Ga_{0.47}As$ cap layer. Subsequent processing steps were similar to those outlined for the GaAs devices.

DC characteristics of the regrown InP-based MODFETs were measured on the wafer and compared with as-grown MODFETs. The transconductances of the regrown MODFETs were as high as 495 mS/mm at current densities of up to 250 mA/mm. Compared to the as-grown MODFETs, the regrown devices suffered roughly a 10 % degradation in the DC performance due to regrowth. Likewise, the RF characteristics were evaluated for both the regrown and as-grown devices. The as-grown MODFETS had cutoff frequencies as high as f_t = 58 GHz and f_{max} = 67 GHz, while regrown MODFETs ranged up to f_t = 24 GHz and f_{max} = 51 GHz.

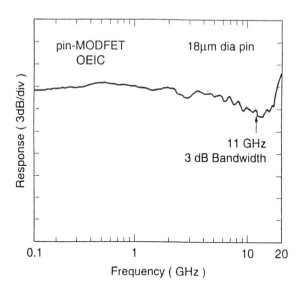

Fig. 2 RF response of the photoreceiver fabricated on InP.

The RF characteristics of the InP-based photoreceivers were then measured. The MODFETs were biased with Vds = 1.5 V and the photodiode reverse biased to -4.0 V. A frequency response characteristic of the p-i-n/MODFET is shown in Fig. 2. The measured 3-

dB bandwidth of the circuits ranged up to 10 GHz. The flatband gain for the circuits with 10 GHz bandwidth which have 2 X 40 μm gates was 17 dB. The maximum gain of 24.8 dB was observed in circuits with 2 X 100 μm gates. These circuits had bandwidths of 1.8 GHz. The receiver sensitivity was measured at 1 Gb/s using a 1.55 μm PRBS = 2^{15} - 1 NRZ signal and a plot the measured BER versus received optical power at a data of 2 Gb/s is shown in Figure 3. The sensitivities were found to be -31.8 dBm, -26 dBm and -17 dBm at 1, 2.5 and 10 Gb/s, respectively, at a BER of 10^{-9}.

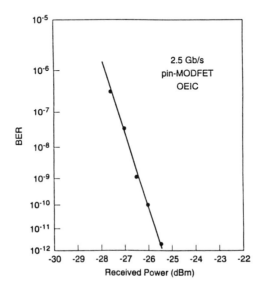

Fig. 3 Measured bit error rate as a function on received optical power at 2.5 Gb/s for the InP-based receiver circuits.

3. Integrated Optical Transmitter Circuits

The transmitter structure was grown via metalorganic chemical vapor deposition (MOCVD) and consisted of 1000 Å of unintentionally doped GaAs, 1000 Å of unintentionally doped $Al_{0.4}Ga_{0.6}As$, a 2300 Å n-type GaAs channel layer and a 600 Å GaAs n-contact layer. This contact layer served as the ohmic contact layer for both the laser and the MESFET. These layers were followed by the laser structure, which consisted of a 1.5 μm n-type $Al_{0.6}Ga_{0.4}As$ lower cladding layer, a 600 Å undoped $Al_{0.2}Ga_{0.8}As$ inner clad, an undoped multiquantum well active region consisting of 400 Å of GaAs, three 80 Å InGaAs quantum wells sandwiched between 200 Å GaAs barriers and 400 Å of GaAs, another 600 Å undoped $Al_{0.2}Ga_{0.8}As$ inner clad, a 1.5 μm p-type $Al_{0.6}Ga_{0.4}As$ upper cladding layer and a 1000 Å GaAs p+ ohmic contact layer. Similar structures were grown employing InGaP rather than AlGaAs. The laser was a 4 μm wide ridge waveguide structure. The n-contact layer for the laser and the MESFET was reached via wet chemical etching. After standard photolithographic processing, the devices were thinned down to a thickness of 150 μm and cleaved into 810 μm long bars for testing.

The isolated transistors exhibited transconductances as high as 138 mS/mm with current densities (I_{ds}) of up to 455 mA/mm. Similarly, the DC characteristics of isolated laser diodes were also measured. The lasers had threshold currents of 24 mA, which corresponds to a threshold current density of 700 A/cm^2, and were operated at powers of around 5 mW. The RF responses of the MESFETs and lasers were then measured. The FETs were found to have cutoff frequencies as high as f_t = 6.3 GHz and f_{max} = 8.5 GHz and the lasers exhibited cut-off frequencies of several gigahertz.

The high-frequency characteristics of the transmitter circuits were evaluated. A schematic of the circuit is shown in the inset of Fig. 4. The laser is turned on by applying a gate voltage sufficient to pinch off the MESFET in the branch of the circuit without the laser diode, typically about -5 V. The light output of the laser may be modulated by superimposing an RF signal on to this gate bias through a high-frequency bias network. A frequency response characteristic of the circuit is shown in Fig. 4. The circuits exhibited bandwidths as high as 3.5 GHz, believed to be limited by the frequency response of the lasers. This limitation is indicated by the presence in the response of the resonance peak of the laser, which would not be present if the response were limited by the MESFETs or circuit parasitics.

4. Summary

Monolithically integrated photoerceivers and transmitters have been fabricated on InP and GaAs. The GaAs-based receivers consisted of a p-i-n diode integrated with a single-stage MESFET amplifier and exhibited a bandwidth of 4.5 GHz. The InP-based receivers consisted of a p-i-n diode integrated with a single-stage MODFET amplifier and exhibited bandwidths as high as 10 GHz and sensitiviies as high as -31.8 dBm at 1 Gb/s. The transmitters consisted of a GaAs MQW ridge waveguide laser integrated with MESFETS in a differential pair configuration and gave modulation bandwidths as high as 3.5 GHz.

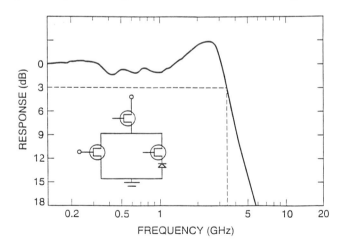

Fig. 4 RF response of the integrated transmitter circuit with a schematic in the inset.

References

[1] N. K. Dutta, J. Lopata, P. R. Berger, S. J. Wang, P. R. Smith D. L. Sivco and A. Y. Cho, Appl. Phys. Lett., **63**, 2115 (1993).
[2] J. S. Wang, C. G. Shih, W. H. Chang, J. R. Middleton, P. J. Apostolakis, and M. Feng, IEEE Photon. Technol. Lett., **5**, 316 (1993).
[3] I. Kotaka, K. Wakita, M. Okamoto, H. Asai and Y. Kondo, IEEE Photon. Technol. Lett., **5**, 61 (1993).
[4] N. K. Dutta, J. Lopata, R. Logan and T. Tanbun-Ek, Appl. Phys. Lett., **59**, 1676 (1991).
[5] J.-M. Verdeill, M. A. Newkirk, T. L. Koch, R. P. Gnall, U. Koren, B. I. Miller and B. Tell, IEEE Photon. Technol. Lett., **4**, 451 (1993).
[6] K.-Y. Liou, S. Chandrasekhar, A. G. Dentai, E. C. Burrows, G. J. Qua, C. H. Joyner and C. A. Burrus, IEEE Photon. Technol. Lett., **3**, 928 (1993).
[7] J. Shibata, I. Nakao, Y. Sasai, S. Kimura, N. Hase and H. Serizawa, Appl. Phys. Lett., **45**, 191 (1984).
[8] D. T. Nichols, J. Lopata, W. S. Hobson, N. K. Dutta, P. R. Berger, D. L. Sivco and A. Y. Cho, Electron. Lett., **30**, 490 (1994).
[9] E. Kuhn, C. Hache, D. Kaiser, G. Laube P. Speier, F. J. Tegude and K. Wunstel, in *Conf. Proc., Third Internat. Conf. Indium Phosphide and Related Materials,* (Cardiff, Wales, UK), pp. 419-422 (1991).
[10] Y. H. Lo, P. Grabbe, M. Z. Iqbal, R. Bhat, J. L. Gimlett, J. C. Young, P. S. D. Lin, A. S. Gozdz, M. A. Koza and T. P. Lee, IEEE Photon. Technol. Lett., **2**, 673 (1990).

Monolithically integrated microwave OEICs by selective OMVPE

H. Kanber, W. W. Lam, S. X. Bar, C. Beckham*, P. E. Norris**

Hughes Aircraft Company, GaAs Operations, Torrance, CA 90509
*EMCORE Corporation, Somerset NJ 08873
**NZ Applied Technologies, Cambridge, MA 02139

Abstract. We have combined selective fabrication technologies of selective organometallic vapor phase epitaxy and selective ion implantation to demonstrate monolithically integrated microwave optoelectronic circuits operating at 10 GHz. We report the microwave and optical performance of such circuits consisting of a microwave SAE grown oscillator, an X-band microwave switch, a transimpedance amplifier and a 0.83 μm optical detector, all monolithically fabricated on the same semi-insulating GaAs substrate. We also demonstrate selective area epitaxy of GaAs based HEMT structures for the first time.

1. Introduction

Selective fabrication technologies such as selective organometallic vapor phase epitaxy and selective ion implantation can be combined to demonstrate monolithically integrated microwave optoelectronic circuits operating at 10 GHz for advanced phased array radar systems applications such as smart skins. These combined technologies offer advantages over others such as selective etching. In this paper, we report microwave and optical performance of such circuits consisting of an X-band microwave selective area epitaxial (SAE) grown oscillator, an X-band microwave switch, a transimpedance amplifier and an optical detector at 0.83 μm all fabricated monolithically on the same semi-insulating GaAs substrate. Our results are the first demonstration of a planar SAE grown microwave circuit operating at 10 GHz and the first attempt at combined selective fabrication technologies to realize a multifunction complex RF-optical circuit at microwave frequencies.

Organometallic vapor phase epitaxy (OMVPE) has been the technique of choice for selective area epitaxy of GaAs based device structures [1-4]. SAE is a technique for controlling growth in the lateral dimensions as well as the vertical dimension. SAE can use a combination of photolithographically defined masks and the naturally occurring anisotropy in growth rates to achieve three dimensional patterned structures with a single epitaxial growth step. We have used this SAE feature to integrate SAE islands with planar ion implanted device structures side-by-side for a complex three dimensional circuit operating at 10 GHz.

Perfect selectivity is defined as no deposition on the masked region. We have shown [5,6] that perfect selectivity can be attained if the OMVPE growth conditions are selected such that differential desorption flux dominates the surface migration flux. AlGaAs can also be grown with 100% selectivity (up to x=0.35) by low pressure OMVPE using a Si_xN_y mask [7]. In the case of low pressure OMVPE of AlGaAs, we have found a range of growth rates and substrate temperatures which satisfy this condition. Many of the previous applications of SAE have been oriented toward optoelectronic devices; however, here we discuss the fabrication and characterization of microwave metal-semiconductor field-effect transistor (MESFET) devices, integrated with selective ion implanted devices.

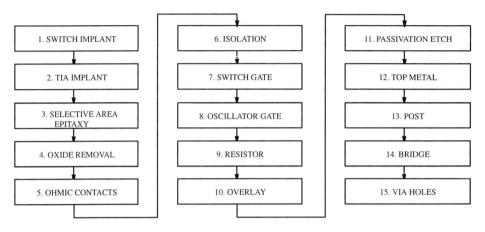

Figure 1.   Microwave OEIC fabrication process flowchart.

2. Experimental

The SAE growths by low pressure OMVPE were performed in an EMCORE GS3300 vertical, low-pressure rotating disk reactor on planar semi-insulating 76 mm GaAs substrates masked with SiO_2 or Si_xN_y masks. Growth conditions have been described previously [6]. The process sequence for fabrication of the monolithic microwave OEICs integrated 15 mask layers shown schematically in Figure 1. The active FET channel layers of the microwave switch and TIA circuits were first formed by selective ion implantation and capless annealing using standard FET processing technology. The preprocessed substrates were patterned for the third time for selective area epitaxy and the oscillator structures were epitaxially grown by OMVPE. After growth, the dielectric mask was partially removed and wafers continued processing through the subsequent process steps shown in Figure 1. The gate lengths used for the oscillator was 0.5 μm, 0.6 μm gates were used for the switch and 0.8 μm gates were used for the TIA, fabricated all by contact phtolithography. Both Pi-gate and T-gate geometries were investigated for optimum device performance.

3. Selective Area Epitaxy Results

3.1 X-Band Oscillator FET Structure

A low noise FET device structure was employed for the SAE grown X-Band oscillator. The structure consisted of a 0.2 μm thick undoped GaAs buffer and a 0.2 μm thick Si doped GaAs channel layer with an electron concentration of $3 \times 10^{17} cm^{-3}$. A 400 Å Si doped GaAs cap layer with $n=2 \times 10^{18} cm^{-3}$ was also added to improve contact resistance on the second MMIC lot. p-type buffer layers were obtained by optimizing the growth parameters and decreasing the V/III ratio to 15, leading to decreased leakage in the buffer layer and improved breakdown voltages up to −35 volts. Figure 2a shows a Pi 4 gate having dimensions of 0.5x300 μm fabricated on an SAE grown island for the oscillator, while Figure 2b shows a T-shaped GaAs FET island.

3.2 Selective Area Epitaxy of HEMT Structures

Low pressure OMVPE was employed to also selectively grow AlGaAs/GaAs HEMT structures using optimized growth conditions similar to the GaAs FET ones on Si_xN_y masked GaAs substrates. The traditional HEMT structure consisted of a 1 μm thick undoped GaAs buffer layer

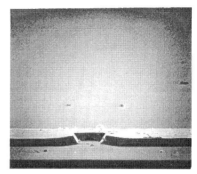

Pi 4 gate
0.5 x 300 µm

3 SAE 37 WAFER #65 (3")
T-shaped GaAs FET island

Figure 2. Morphology of oscillator FETs fabricated on selective area grown GaAs islands.

and an undoped 75Å thick AlGaAs spacer layer. The doped layers were made up of a 350 Å AlGaAs layer with a carrier concentration of $2.0 \times 10^{18} cm^{-3}$ and a 350 Å GaAs cap layer with a carrier concentration of $2.0 \times 10^{18} cm^{-3}$. Aluminum composition was 30%. The 77K sheet carrier concentration measured $1.2 \times 10^{12} cm^2$ and the mobility measured 45,000 cm^2/V-s on a blank GaAs substrate. The fast selective growth rate of 1600 Å/minute was also applied to demonstrate a 94 Å thin GaAs/AlGaAs quantum well measured by room temperature photoluminescense. Essentially 100% selectivity was obtained for the HEMT device structure, as seen by the smooth morphology in Figure 3 of this 1.0775 µm thick AlGaAs/GaAs structure.

Figure 3. SEM micrograph of selective area grown AℓGaAs/GaAs HEMT structure.

TABLE 1
FET DEVICE PERFORMANCE AFTER COMPLETION OF PROCESSING

Wafer	Idss (mA/mm)	Vpo (v)	G_m (mS/mm) (vgs = OV)	G/D Breakdown (v) (Ig = 100 µA)
2 x 125 µm Switch				
3D10	300-360	4.2-6	108-116	−13.1 to −14
3D12	340-400	4.2-6.3	116-121	−12.5 to −13.5
3D13	332-384	5-5.3	124-130	−11.4 to −13.5
4 x 75 µm Oscillator				
3D10	127-173	2.3-3	65-100	−35 to −39
3D12	147-207	2.1-2.6	75-95	−22 to −25
3D13	126-147	2.3-2.7	60-90	−33 to −37
120 µm TIA				
3D10	120-128	1.2-1.3	100-120	
3D13	180-200	2-2.3	80-90	12-14

4.0 Device Results

Three lots were fabricated using combined selective area and selective ion implant technologies. FET characteristics of the oscillator, switch and the TIA were optimized in Lot 3. Summaries of FET performance in terms of I_{dss}, pinch off voltage, dc transconductance G_m and gate-drain breakdown voltage for each type of FET are given in Table 1. FET characteristics of the SAE FETs were improved using p-type buffer layers and we obtained remarkably high breakdown voltages with the SAE FETs ranging from –26 to –39 volts. We measured good pinch-off behavior on the SAE FETs with the improved buffer layer.

5.0 RF and optical MMIC performance

5.1 X-Band Oscillator

The X-band oscillator was fabricated on the SAE grown FET islands using coplanar waveguide transmission line and positive feedback designs. Figure 4 shows a photograph of the X-band oscillator. Figure 5 shows the measured frequency spectrum of the oscillator. The broadband sweep indicated that the oscillator generates a single carrier frequency, while the narrow band sweep shows the noise performance of the oscillator near the carrier. The measured noise density is –100 dBc/Hz at 1-MHz offset from the carrier. The maximum output power is measured to be –2.8 dBm at 10.7 GHz. Similar results were measured on all three wafers from Lot 3.

5.2 X-Band Microwave Switch

The 10 GHz microwave switch was designed in coplanar waveguide transmission line and fabricated monolithically with the other MMICs using selective ion implantation technology. A single pole double throw design was utilized. The wafers were on-wafer probed and tested using the HP8510 network analyzer. The measured RF performance shows the switch worked well at X-band frequencies. The measured insertion loss is 1.5 dB at 10 GHz, while the measured return loss is better than 9 dB at 10 GHz as shown in Figure 6.

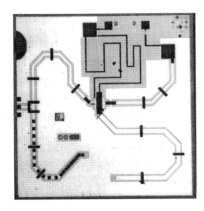

Figure 4. The X-band oscillator fabricated on SAE GaAs islands.

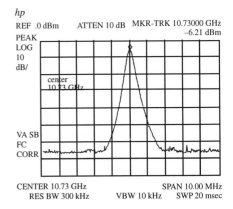

Figure 5. Measured X-band frequency spectrum of the SAE oscillator.

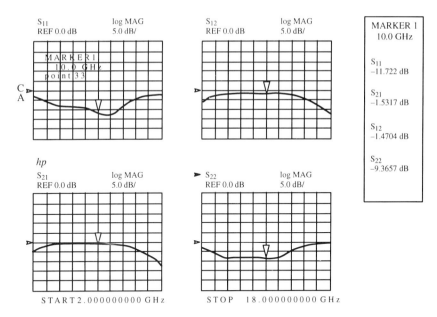

Figure 6. RF performance of the X-band microwave single pole double throw switch (from Port 1 to 2).

5.3 Transimpedance Amplifier

The transimpedance amplifier (TIA) basically contains two amplifiers in cascade and a level shifter. Both the open loop and feedback type of circuits were incorporated in the design. Figure 7 shows the input-to-output voltage transfer curve for the TIA. The off-state voltage is 0.2 V and the on-state voltage is 5.3 V. The transition from off-state to on-state is very sharp. The measured time response of the TIA was 20 ns, which was better than the desired response of less than 100 ns.

5.4 MSM Detector

The metal-semiconductor-metal (MSM) detector converts incident light into output current and is the optical component of the RF-optical complex multifunction circuit. We investigated four types of MSMs on our mask set. MSMs with anti-reflective coating MSMs without anti-reflective coating, MSMs with ion implantation and anti-reflective coating, and MSMs with I^2 and no anti-reflective coating. The dark current, responsivity and photocurrent from each type was measured at $\lambda=830$ nm. The highest photocurrent is measured on the MSM with ion implantation and anti-reflection coating, it also had dark current 2 orders lower than the other types. The measured responsivity of the detectors was 0.17 A/W as a function of incident optical power, as shown in Figure 8. Diode sizes were 40 by 60 microns and used small bias voltages between –2 to –3 V for operation.

6.0 Summary and conclusions

We have successfully demonstrated a monolithic complex RF-optical MMIC operating at 10 GHz for microwave applications. The X-band oscillator was fabricated by selective area epitaxy by low pressure OMVPE, the X-band switch the transimpedance amplifier and the optical MSM detector were fabricated by selective ion implantation technology. Excellent

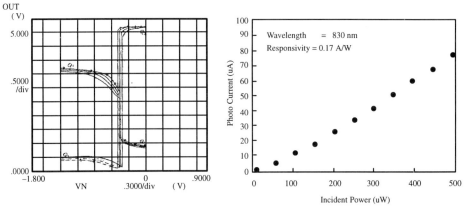

Figure 7. Measured performance and transfer curve for the transimpedance amplifier (TIA).

Figure 8. Optical responsivity of the MSM detector at 830 nm.

performance of each component was shown in this paper. Combined fabrication technologies and such component functionality has paved the path to the optimized design of a complex multifunction RF-optical OEIC chip which can operate at microwave frequencies.

7.0 Acknowledgements

The authors would like to thank various members of Hughes Gallium Arsenide Operations for their contributions. This work was supported by the U.S. Department of the Air Force, Electronic Technology Laboratory, Wright Patterson Air Force Base under Contract Number F33615-89-C-1147.

References

[1] Azoulay R, Bouadma N, Bouley J C and Dugrand L J 1981 Cryst Growth **66** 229

[2] Gale R, McClelland R, Fan J and Bozler C 1983 Inst Phys Conf Ser No. **65** 101

[3] Keuch T, Goorsky M, Tischler M, Palevski A, Solomon P, Potemski R, Tsai C, Lebens J and Vahala K 1991 J Cryst Growth **107** 116–128

[4] Azoulay R and Lugrand L 1991 Appl Phys Lett **58** 129

[5] Kanber H, Sokolich M, Bar S, Herman P, Beckham C and Walker D 1991 Inst Phys Conf Ser No. **120** 215

[6] Kanber H, Bar SX, Norris PE, Beckham C and Pacer M 1994 J Electronic Materials **23** 159–166

[7] Kamon K, Takagishi S and Mori H 1985 J Cryst Growth **73** 73

Optical refractometry with GaAs/AlGaAs interferometers

Hans P. Zappe

Paul Scherrer Institute, Badenerstrasse 569, 8048 Zurich, Switzerland

Abstract. Chemical, biological or environmental processes often give rise to a change of refractive index in a sensing film which is selective to certain species of interest; interferometry allows a precise measurement of these small index changes. The use of GaAs/AlGaAs-based interferometer structures for refractometry, the measuring of refractive index changes, is discussed. Device structures and performance of Fabry-Perot and Mach-Zehnder interferometers are studied for this application and initial experimental sensor results are presented.

1. Introduction

Optical sensors, particularly those employing fibers, have seen extensive development and have been applied to chemical, biological and environmental sensing [1]. Many of these sensors rely on the measurement of refractive index changes for their operation; either the presence of the medium to be sensed leads to a change in the refractive index near the sensor, or customized sensor membranes undergo an index shift selectively in response to a certain analyte.

Most of these optical sensing schemes rely on the coupling of an evanescent wave from a waveguide into the area to be sensed; the shift in real [2] or imaginary [3] refractive index, typically of magnitude between 10^{-6} to 10^{-4}, is then often determined by interferometric techniques. Waveguide integrated optic approaches to such sensors have employed glass [4], $LiNbO_3$ [5] or silica-based [6] substrates, to which a light source must be coupled by pig-tailing, butt-coupling or grating-coupling. III-V technology and devices can offer performance advantages for the fabrication of integrated optical sensors [7] since, through monolithic integration of laser light sources, completely integrated sensor chips may be fabricated. Monolithic integration should provide improvements in functionality and robustness with respect to alternative optical sensor configurations.

GaAs/AlGaAs-based interferometers are well established for use as optical switches [8]; we examine in this brief paper their use for measurement of refractive index changes in the off-chip environment. Fabry-Perot and Mach-Zehnder structures are studied in order to evaluate their performance in refractometry measurements. The use of standard material structures and conventional processing technology assures manufacturability and compatibility with other III-V applications.

2. Layer and device structures

To allow the monolithic integration of active optoelectronic components, such as lasers and electro-optic modulators, with a waveguide-based refractometer into a complete sensor system, a GaAs/AlGaAs layer structure compatible with these devices must be employed. The layer structure of Figure 1, a standard double heterostructure configuration, has proven to be suitable. The optical mode is guided in the waveguide core, with index n_g grown on a suitable cladding with index n_s. In the sensor recess, the upper cladding is selectively removed from the waveguide leaving an unprotected length of waveguide core. The recess brings the surface region, a sensing layer or a fluid to be sensed directly with the variable index n_c, close to the core so that changes in n_c directly affect the propagation of the waveguide mode. The purpose of refractometry in our case is then to determine the changes in the value of the refractive index of this surface layer.

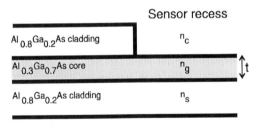

Figure 1 - GaAs/AlGaAs layer structure for refractive index sensing in the etched recess.

The interferometer structures of Figure 2 permit the translation of a surface index change, Δn_c, into an intensity change of an optical mode. Both the Fabry-Perot and Mach-Zehnder devices are waveguide-based, each with sensor recess of length L as shown in the figure. A change in the refractive index of the surface medium in this waveguide segment leads to a phase shift and therefore a modulation of the interferometer operating point; this modulation provides the measurable change in intensity.

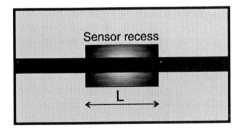

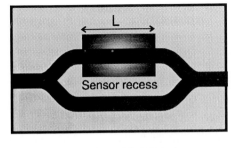

Figure 2 - Fabry-Perot and Mach-Zehnder interferometer structures showing the etched sensor recess.

3. Analysis

Examination of the behavior of a waveguide-based interferometer with respect to changes of the surface index will allow us to better understand the refractometric mechanisms and, more importantly, define the most sensitive structural parameters.

3.1. Effective index variations

The primary effect of a change in the index of the surface medium, Δn_c, is to promote a change in the effective index of the mode in the waveguide; determining the latter will allow us to calculate the interferometer behavior. Solving the asymmetric dielectric waveguide [9] in the recess of Figure 1, we obtain the variation of effective guide mode index, N_{eff}, as a function of surface index, n_c; the relationship, as shown in Figure 3, was calculated for the lowest order TE mode at a wavelength of 850 nm, with $n_g = 3.35$ and $n_s = 3.12$ and various core thicknesses, t. We see from the figure that low index values, $n_c < 2$, and large values of t lead to relatively small changes in N_{eff}.

This latter point may be seen more clearly in Figure 4 where we have plotted the sensitivity of effective index with respect to changes in the surface index, $\partial N_{eff} / \partial n_c$. We see clearly that, for index changes in the value range far from the waveguide index (3.35) and for core thicknesses far from cutoff, the sensitivity is low. Repeating the calculation for the lowest order TM mode shows that a marginally higher sensitivity can be achieved for this polarization. Most measurands of interest, primarily biological and chemical analytes, have index values in the range 1 to 1.5; aqueous solutions, for example, typically have values around 1.33. Sensitivity could be improved by replacing the III-V waveguide layers with materials of lower index [10], albeit at the expense of greater processing complexity.

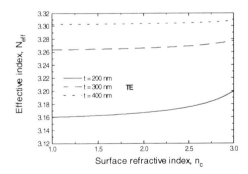

Figure 3 - Variation of effective refractive index with the index of a surface film in the sensor recess.

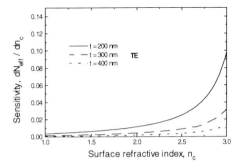

Figure 4 - Sensitivity of the change in refractive index with respect to changes in the index of the surface film.

3.2. Core thickness variations

From Figure 4, we saw that an increase in t resulted in a rapid decrease of sensitivity. By examining $\partial N_{eff} / \partial n_c$ as a function of t, as seen in Figure 5, we see that an optimum waveguide thickness can be determined. The characteristic was derived for small changes in n_c around a value of $n_c = 1.33$. We see that, due to the large index difference between the waveguide and the surface medium, the change in N_{eff} with respect to a change in n_c is limited to less than 0.01; the TM mode is again the more sensitive. It is apparent from the figure that an optimum waveguide thickness for TE polarized modes is 190 nm and for TM 240 nm. The peak sensitivities are 0.005 and 0.009 for TE and TM, respectively; we will see the implications for real sensor applications of these values in the next section.

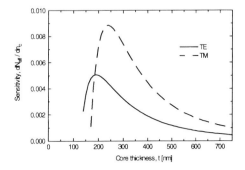

Figure 5 - Sensitivity to changes in a surface refractive index as a function of waveguide core thickness for TE and TM modes.

4. Interferometers

As we have just seen, changes in external refractive index give rise to small changes in the waveguide effective index for a III-V waveguide. The use of interferometer structures, however, allows us to translate these shifts into signals of easily measurable magnitude.

4.1. Fabry-Perot interferometer

Using the calculated change in effective index with change in index of the surface medium, we can predict the behavior of the Fabry-Perot waveguide interferometer of Figure 2. For an recess section of length L, the phase shift due to an index shift Δn_c at the surface of a waveguide with sensitivity S is given by

$$\delta = \frac{2\pi S \Delta n_c L}{\lambda}.$$

For a sensor recess length of $L = 1$ mm, we can use the above relation to determine the expected phase shift for the lowest order TE mode as n_c varies from 1 to 1.5. Plugging the value of δ directly into the expression for the transmission characteristic of a Fabry-Perot interferometer, ignoring optical absorption loss,

$$T_{FP} = \frac{(1-R)^2}{(1-R)^2 + 4R\sin^2\left(\frac{\delta}{2}\right)}$$

we obtain the behavior plotted in Figure 6. In the above expression, R is the facet reflectivity, which we take to be 31% for a typical GaAs-based structure.

The decrease in the period of the interference cycles with increasing index reflects the enhanced sensitivity of this system for cover media whose indices more closely match that of the waveguide. The relatively small change in N_{eff} (about 0.002) which results from the change in n_C of 0.5 does result in 3.5 interference fringes. A typical refractometry measurement, however, typically involves a much smaller change in index and is accomplished by monitoring small changes in transmitted intensity about a fixed bias point.

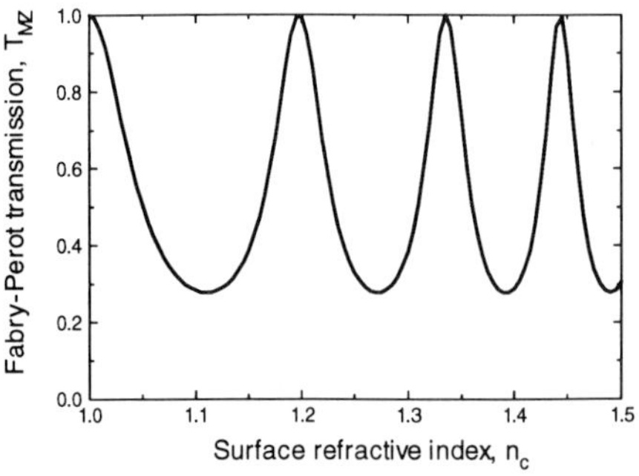

4.2. Mach-Zehnder interferometer

The same calculation can be carried out for the Mach-Zehnder interferometer, whose transmission characteristic, again ignoring waveguide loss, is given by

$$T_{MZ} = \frac{1}{2}\left[1+\cos\left(\frac{2\pi LS\Delta n_c}{\lambda}\right)\right].$$

The variation of T_{MZ} with the index of the surface medium is very similar to that for the Fabry-Perot resonator; in particular, the number of cycles

Figure 6 - Transmission characteristic of the Fabry-Perot interferometer as the surface refractive index varies from 1 to 1.5; TE lowest order mode.

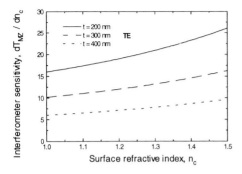

Figure 7 - Sensitivity of the Mach-Zehnder interferometer for small changes of surface refractive index around n_c.

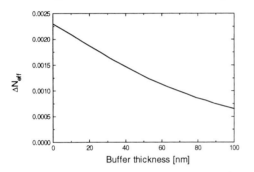

Figure 8 - Change in effective waveguide index between an air and an organic solvent-coated surface ($\Delta n_c = 0.5$) as a function of remaining cladding thickness after recess etch.

traversed as n_c varies from 1 to 1.5 is identical.

By examining the sensitivity of the Mach-Zehnder interferometer, namely the change in transmitted intensity with respect to a change in surface refractive index, $\partial T_{MZ}/\partial n_c$ for small Δn_c, we find the behavior of Figure 7. The sensitivity is calculated for the most sensitive region of the interference characteristic, namely halfway between maximum and minimum transmission ($\delta = m\pi/2$ for $m = 1,3,...$).

From this figure, we see that a sensitivity of about 22 is expected for $t = 200$ nm and $n_c = 1.33$, using TE polarization. If we define the minimum measurable output change of the interferometer to be 2 % (1/50 of the maximum modulation depth), this sensitivity implies a measurable index change of $\Delta n_c = 9 \times 10^{-4}$ per millimeter of sensor recess length. Since, for example, a change of glucose concentration in water from 0 to 1% gives rise to a refractive index shift of 1.4×10^{-3} [4], the device resolution we have determined would require a 3 mm long sensor recess to measure a 0.2 % concentration change of glucose, a sensitivity range with important medical relevance.

5. Fabrication issues

In order to demonstrate the compatibility of the above structures with conventional III-V-based optoelectronic processes, a recess etch which allows selective removal of the waveguide cladding in the sensor area has been developed. In this manner, the measurand of the surface medium, n_c, can be brought as close to the waveguide core as possible.

Using the layer structure of Figure 1, waveguides are defined by dry etch and completely covered with Si_3N_4. During the subsequent contact etch, nitride is removed over the sensor regions and thus acts as an etch mask for dry-etch removal of the cladding. The resulting structure allows easy definition of the sensor areas, the Si_3N_4-covered regions immune from any environmental variations at the surface. The recess is then formed by a further dry-etch step, resulting in the required uncovered waveguide region.

Experimental work has shown that complete removal of the upper cladding is essential for adequate performance; a remaining buffer layer between core and surface medium will rapidly reduce the sensitivity of the structure. This can be seen from Figure 8 which presents the results of a calculation of the change of the effective waveguide index as the surface index, n_c, changes from air ($n_c = 1.0$) to $n_c = 1.5$ (for example, an organic solvent) as a function of buffer thickness of this four-layer waveguide system. We see that as little as 60 nm of remaining cladding decreases ΔN_{eff}, and thus the sensitivity, by a factor of two. Thus adequate control of etch depth is essential for fabrication of the sensor recess.

The experimental response of a Mach-Zehnder interferometer with such a sensor recess is shown in Figure 9. The plot shows the transmitted interferometer intensity as a water film is slowly deposited on the surface; measurement is with a recess length of $L = 1$ mm and at a wavelength of 830 nm. Although, as mentioned above, a typical real-life sensor application involves much smaller index changes around a fixed bias point, this initial measurement demonstrates the response of the III-V-based interferometer sensor to changing surface conditions.

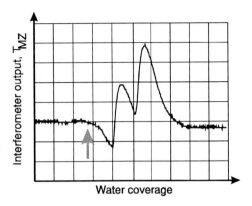

Figure 9 - Variation of the output intensity of a Mach-Zehnder interferometer as a water film is deposited on the sensor surface. At the arrow, a film of water begins to cover the sensor recess and completely covers the sensor when the signal stabilizes.

Conclusions

The use of purely III-V-based interferometers for measuring the change of environmental refractive indices has been discussed. The large index step between the GaAs/AlGaAs waveguides and the surface media, the latter typically with indices 1 to 2, limits the change in effective waveguide index achievable. The effective amplification of these small changes resulting from the use of interferometric devices, however, provides larger output signals and usable sensitivities with device sizes in the millimeter range. Further improvements in sensitivity are expected through the use of alternate waveguiding layers, such as dielectrics, in the device sensor area. Fabrication of the required structures is fully compatible with normal III-V processing techniques.

Acknowledgements

The author is grateful to Bernd Maisenhölder for sensor measurements and his colleagues Mike Gale, Bruno Graf, Daniel Hofstetter, Rino Kunz, and Jürgen Söchtig for technical assistance, useful discussions and generous support.

References

[1] Dakin J and Culshaw B 1988 *Optical Fiber Sensors* (Boston: Artech House)
[2] Fabricius N Gauglitz G and Ingenhoff J 1992 *Sensors and Actuators B*, 7 672-676
[3] Kang S-W Sasaki K and Minamitani 1993 *Applied Optics* 32 3544-3549
[4] Liu Y Hering P and Scully M O 1992 *Applied Physics B* 54 18-23
[5] Hinkov I and Hinkov V 1993 *IEEE Journal of Lightwave Technology* 11 554-559
[6] Heideman R G Kooyman R P H Greve J 1993 *Sensors and Actuators B* 10 209-217
[7] Zappe H P Arnot H E G and Kunz R E 1994 *Sensors and Actuators A*, 41 141-143
[8] Erman M Jarry P Gamonal R Autier P Chané J-P and Frijlink P 1988 *IEEE Journal of Lightwave Technology* 6 837-846
[9] Marcuse, D. 1974 *Theory of Dielectric Optical Waveguides* (Orlando: Academic)
[10] Zappe H P Arnot H E G and Kunz R E 1994 *Sensors and Materials* 6

Lateral npn-phototransistors with high gain and high spatial resolution fabricated by focused laser beam induced Zn doping of GaAs/AlGaAs quantum wells.

P. Baumgartner, C. Engel, G. Abstreiter, G. Böhm, G. Tränkle, and G. Weimann

Walter Schottky Institut, TU München, Am Coulombwall, D-85748 Garching, Germany

Abstract. We propose a novel kind of phototransistor with high gain and high spatial resolution. Lateral *npn*-transistors are fabricated by focused laser beam induced Zn doping of *n*-modulation doped GaAs/AlGaAs quantum well structures. This technique allows direct writing of lateral sub-μm sized *npn*-structures without damaging the sample. Spatially resolved photocurrent measurements show typical responsivities above 10^3 A/W and linewidths down to 605 nm. Direct measurement of the spatial derivative of a light distribution is possible simply by applying an AC bias voltage between emitter and collector.

1. Introduction

We have fabricated a new type of photodetector, a lateral *npn*-phototransistor, by focused laser beam-induced (FLB) thermal Zn doping of *n*-modulation doped GaAs/AlGaAs quantum well samples. The lateral device combines high gain with sub-μm spatial resolution. The recently introduced technique of local Zn-doping [1] preserves the high quality of the modulation doped samples and permit direct writing of arbitrary shaped *p*-doped lines. The peculiar interstitial-substitial diffusion mechanism of Zn in GaAs leads to an almost boxlike depth-profile of the Zn concentration. So it is possible to achieve a remote *p*-doping without introducing Zn-dopants into the quantum well layer. Due to model calculations, using a concentration and temperature dependent diffusion coefficient, nonlinear effects of Zn-diffusion restrict the width of the *p*-doped lines to less then 200 nm, which is well below the diffraction limited laser spot size of about $d_{FWHM} = 440$ nm [1-3]. We have already demonstrated the potential of this method by direct writing of in-plane-gate transistors, which make use of the good insulating properties of this lateral *npn*-structure [1].

In this paper we report on results obtained from single *p*-doped lines, which form the base of a lateral *npn*-transistor. Photogenerated holes supply the base current and control the electron current from the n-doped emitter region to the n-doped collector region, resulting in high current gain. The emitter- and collector-region of this device is formed by the two-dimensional electron gas in the GaAs quantum well. The light sensitive region is limited to the narrow p-type line, since only holes, created nearby the base, contribute to the base current. This leads to a phototransistor with high gain and high spatial resolution.

2. Fabrication method

The lateral npn-structures are fabricated from MBE-grown modulation doped GaAs/AlGaAs single quantum well structures. The two-dimensional electron gas, which forms 360 Å beneath the surface in the 80 Å thick GaAs quantum well, has an electron density of $n = 5.3 \cdot 10^{11}$ cm^{-2} and a mobility of $\mu_H = 6300$ cm^2/Vs at room temperature. The detailed layer sequence is described elsewhere [1].

A 10 μm wide mesa structure is prepared by standard photolithography, wet chemical etching and alloying of NiGeAu contact regions. The p-dopant is supplied by a 2000 Å thick SiO$_2$ film, containing about 14% Zn.

A microscope objective is used to focus the Ar$^+$ laser beam ($\lambda = 514.5$ nm) to a full width half maximum diameter of about $d_{FWHM} = 440$ nm [4]. Focused laser beam-induced doping is performed by local heating of the sample to temperatures of about T = 600°C. Nearly arbitrarily shaped lateral npn-structures can be written by moving the sample with a xyz-translation stage with an accuracy of 10 nm. Further important features of the setup are a sensitive autofocus system, ensuring a constant width of the laser spot, and the in situ measurement of the sample's reflectivity, to keep the absorbed laser power unchanged. We used absorbed laser intensities ranging from 4 to 10 mW and a writing speed of 7 nm/s for the fabrication of the phototransistors. The lateral npn-phototransistors were produced by writing a p-doped line across the mesa structures to insulate the n-contacts on both sides from each other.

3. Results

The barrier heights of the p-doped lines were estimated from temperature dependent current measurements. Figure 1 shows the current for an applied voltage of 100 mV as a function of temperature in an Arrhenius-plot. The measured straight lines confirm that the current is correlated to a thermionic emission across the potential barrier and follows an $I \sim e^{-E_B/kT}$ dependence. The potential barrier can be adjusted by the writing laser intensity and increases from $E_B \cong 200$ meV up to $E_B \cong 830$ meV when the intensity is increased from 4.1 mW to 10 mW. Figure 1 also reveals the good insulating properties of the lateral npn-structures, since the resistances of the highly p-doped samples reach the TΩ region at room temperature.

The schematic band diagram of a lateral npn-structure is depicted in the inset of figure 2. A small bias voltage U_{EC} is applied between the n-doped emitter and collector regions. Without illumination only a small thermionic current is able to flow over the p-doped barrier, as shown in the upper half of the inset. The thermionic dark current and the barrier height are determined by the p-doping concentration and consequently by the laser intensity used for local doping.

The lower half of the inset shows the corresponding band diagram under illumination. Photogenerated electron-hole pairs are separated by the high electric field in the space-charge region. The additional positive charge of holes accumulated in the p-doped base region reduces electrostatically the barrier height and leads to an increased thermionic current. Only electron-hole pairs, which are excited in the space charge region or within

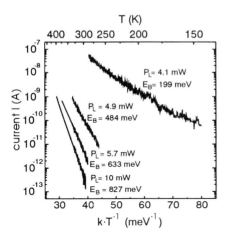

Figure 1: Temperature dependent dark current measurements of different lateral npn-structures. The barrier heights of the individual structures are estimated of the slope from the curve.

the holes diffusion length contribute to the base current. The spatial resolution of this detector is consequently dependent on width and shape of the doping profile, the corresponding depletion length and the diffusion length of the holes.

The expected good spatial resolution of the lateral npn-structures is confirmed in spatially resolved photocurrent measurements as shown in Figure 2. The linescans are recorded perpendicular to the p-doped line centered at the origin of the ordinate. A bias voltage of $U_{EC} = 50$ mV is applied to the sample with a barrier height of $E_B = 484$ meV. The narrower profile, measured at a wavelength of $\lambda = 514.5$ nm and a light intensity of $P_L = 220$ pW shows a width of only $d_{FWHM} = 605$ nm, including the width of the probing laser spot of about $d_{FWHM} = 440$ nm. The diffusion length for holes is apparently rather small. Strong radiative recombination reduces efficiently the lifetime of the minority charge carriers [5]. The spatially resolved experiments verify that one can achieve very narrow p-doped lines with laser-induced local Zn-doping.

The maximum photocurrent of about $I_{EC} = 210$ nA corresponds to a responsivity of roughly $R = 1000$ A/W and demonstrates the high internal gain of the lateral npn-transistor. For comparison the responsivity of an ideal photodiode is given by $R = e/\hbar\omega = 0.42$ A/W at the wavelength used.

The second wider photocurrent profile measured with an Argon ion laser pumped Ti:Sapphire laser tuned to a wavelength of $\lambda = 800$ nm has a width of $d_{FWHM} = 800$ nm. The corresponding photon energy of $E = 1.55$ eV is well below the AlGaAs absorption edge of the quantum well cladding layers. The increased width is understood by taking into account the enlarged diffraction limited laser spot diameter of about $d_{FWHM} = 680$ nm. The high sub-μm spatial resolution of this photodetector is therefore almost unaffected by the increased absorption length of the red light, since the photosensitive sheet is given by the quantum well layer.

The responsivity decreases compared to the excitation with green light, since the absorbing volume is restricted to the GaAs layers. To obtain the same photocurrent of $I_{EC} = 210$ nA a higher laser intensity of about $P_L = 1.3$ nW at $\lambda = 800$ nm is required.

The variation of the emitter-collector current I_{EC} as a function of light intensity is shown as a double logarithmic plot in figure 3 for lateral npn-structures with different barrier heights and an unstructured mesa as reference. The wavelength of the illuminating light was $\lambda = 514.5$ nm and the applied bias voltage $U_{EC} = 100$ mV. The different line

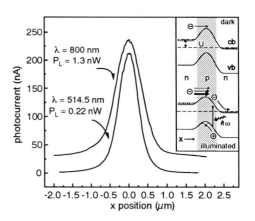

Figure 2: Spatially resolved photocurrent curve of a phototransistor structure at different exciting wavelengths. The upper curve is shifted by 25 nA for clarity. The inset shows schematically the principle of operation of the device.

segments correspond to measurements with different combinations of neutral density filters.

The current of the unstructured mesa is determined by the resistance of the two-dimensional electron gas which is about 25 kΩ for the given geometry. It is barely influenced by the illumination. Only at illumination levels above $P_L = 100\ \mu W$ the carrier concentration in the sample increases noticeably and the resistance decreases slightly.

The photocurrent curves of the npn-structures can be divided in three regions. At very weak illumination the current is dominated by the individual dark current of the structure and is insensitive to the illumination. At increased illumination levels the current is well controlled by the incident light intensity. Increasing the illumination further, the current saturates and reaches the reference values of the unstructured mesa, indicating a lateral flat band condition. Even the highly doped barriers can be completely eliminated by illumination. This proves once more, that laser-induced Zn doping does not degrade the sample's quality.

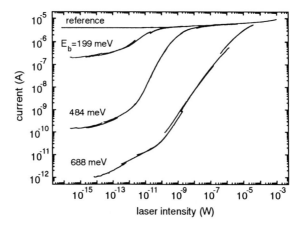

Figure 3: Influence of the incident light intensity on the photocurrent for structures with different potential barriers ($U_{EC} = 100\ mV$). The current in the reference mesa structure is also shown.

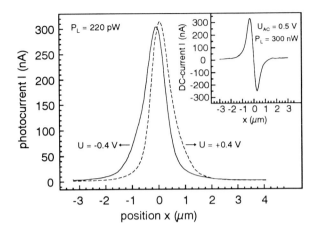

Figure 4: Two linescans measured with different applied bias voltages. The inset shows the spatially resolved DC photocurrent with an applied AC voltage.

The dark current, the responsivity and the sensitive region of a npn-structure is mainly determined by the Zn doping concentration and consequently by the barrier height. Besides the dark current also the responsivity drops down, e.g. from $R = 7 \cdot 10^5$ A/W down to $R = 0.7$ A/W when the potential barrier increases from $E_B = 200$ to $E_B = 633$ meV. The sensitive region is limited by the dark current on the one side and the saturation light intensity on the other side. The dynamic range of low doped phototransistors is indeed very small since these devices exhibit a large dark current and saturate at rather low light intensities. For example a 200 meV barrier is flattened out by a light intensity of only $P_L = 100$ pW. The samples with high potential barriers demonstrates in contrast phototransistor action over an intensity range of more than 5 decades at the cost of internal gain.

The dependence of the responsivities on the doping concentration can be compared with the current gain of conventional bipolar transistors. The emitter efficiency and accordingly the current gain is proportional to the ratio of the emitter to base doping concentration $h_{FE} \sim N_e/(N_b \cdot W)$ and increases with decreased base doping concentration N_b and with decreased base width W [6]. The responsivities of our lateral phototransistor structures increase strongly with decreased base doping, similar to conventional vertical transistors. The high obtainable responsivities confirms also the small lateral extension of the base W.

The current-voltage characteristics I_{EC} of the lateral phototransistors at various illumination levels resemble the corresponding current-voltage characteristics of a conventional bipolar transistor with different bias currents. The current saturates at voltages $U_{EC} \geq 1$ V and proofs the transistor action of this device.

Figure 4 shows two linescans across a phototransistor structure with higher applied voltages of $U_{EC} = +0.4$ V and $U_{EC} = -0.4$ V respectively. The lineshape becomes asymmetric with a tail located at the positive biased side of the structure. The maximum shifts in the same direction. If the applied voltage is inverted, the lineshape is basically reflected at the origin. The pn-diode from base to collector is reverse biased and the np-diode from emitter to base is forward biased. The main part of the applied voltage drops therefore at the reverse biased base to collector diode. The depletion length of this pn-junction increases consequently with the applied bias voltage whereas the depletion length of the forward biased emitter base np-junction decreases slightly. The changes in the observed photocurrent profiles are caused by the altered depletion length, since photogenerated electron-hole pairs are mainly separated in the space-charge regions. The collector and accordingly the longer tail of the lateral npn-structures are always located at the positive charged side of the structure.

The inset of figure 4 demonstrates a special application of this voltage dependent changes in the photocurrent lineshape. In this linescan we have recorded the DC current with an applied AC voltage of $U_{AC} = 0.5$ V. The AC components of the signal are suppressed by a RC lowpass filter. The measured lineshape is the spatial derivative of the photocurrent profile. We can understand this behaviour considering that the center of gravity of the photocurrent profile is always shifted towards the positive charged side of the device and that the direction of the photocurrent changes it's sign when the applied voltage is reversed. The conventional way to measure the spatial derivative of a light distribution is to modulate the position of the detector, and measure the AC component of the photocurrent with a lock-in-amplifier. With our proposed new measurement method the modulation of the position is obtained by applying an AC bias voltage U_{EC}. The lateral phototransistor structures change the sign of the photocurrent in phase with the applied voltage U_{EC} and contain therefore a "built in" lock-in amplifier.

4. Conclusion

In conclusion, we have fabricated a new kind of phototransistor with lateral npn-geometry by laser-induced local Zn doping of a GaAs/AlGaAs quantum well sample. The high gain of these transistors demonstrate the high quality of the fabrication method. The lateral sub-μm resolution and the ability to measure directly the spatial derivative of a light distribution are of substantial interest for detector systems with high spatial resolution.

Acknowledgment

We gratefully acknowledge the financial support by the Bundesministerium für Forschung und Technologie (Verbundprojekt DFE) and the Deutsche Forschungsgemeinschaft (SFB 348).

References

[1] P. Baumgartner, K. Brunner, G. Abstreiter, G. Böhm, G. Tränkle, and G. Weimann, Appl. Phys. Lett. **64**, 592 (1994).

[2] H. C. Casey, Jr. in *Atomic Diffusion in Semiconductors*, edited by D. Shaw (Plenum, New York, 1973), p. 351.

[3] M. Lax, Appl. Phys. Lett. **33**, 786 (1978).

[4] K. Brunner, U. Bockelmann, G. Abstreiter, M. Walther, G. Böhm, and G. Tränkle, Phys. Rev. Lett. **69**, 3216 (1992).

[5] R. K. Ahrenkiel, D. J. Dunlavy, H. C. Hamaker, R. T. Hayes, and H. Fardi, Appl. Phys. Lett. **49**, 725 (1986).

[6] S. M. Sze, *Physics of Semiconductor Devices*, 2nd Ed. John Wiley&Sons (New York 1981).

Ga_2O_3 films for insulator/III-V semiconductor interfaces

M. Passlack, M. Hong, E.F. Schubert, J.P. Mannaerts, W.S. Hobson, N. Moriya, J. Lopata, G.J. Zydzik

AT&T Bell Laboratories, 600 Mountain Avenue, Murray Hill, New Jersey 07974

Abstract. Film and interface properties of Ga_2O_3 films deposited by electron-beam evaporation from a high-purity single-crystal $Gd_3Ga_5O_{12}$ source are reported. The amorphous dielectric Ga_2O_3 films are stoichiometric, homogeneous and very uniform in thickness and refractive index over a 2 inch wafer. Optical properties including refractive index (n = 1.84 - 1.88 at 980 nm wavelength) and bandgap (4.4 eV) were found to be close or identical, respectively, to Ga_2O_3 bulk properties. Reflectivities as low as 10^{-5} for Ga_2O_3-GaAs structures and a small absorption coefficient (≈ 100 cm^{-1} at 980 nm) were measured. Dielectric properties include a static dielectric constant (9.9 - 10.2) identical to bulk Ga_2O_3 and dc breakdown fields up to 3.6 MV/cm. InGaAs/GaAs separate confinement heterostructure single quantum well (SCH-SQW) lasers emitting at 980 nm were fabricated with low reflectivity Ga_2O_3 front facet coatings. The peak power is limited by thermal rollover and catastrophic optical mirror damage (COMD) was not observed even at high optical power densities of 17 MW/cm^2 at the front facet. Finally, we show that ultra-high vacuum processing combining molecular beam epitaxy and Ga_2O_3 film deposition provides dielectric-GaAs structures with low interface recombination velocity/interface state density.

1. Introduction

Dielectric films find a wide range of applications from passivation of surface states in various types of conventional and low-dimensional devices to metal-insulator-semiconductor field-effect devices. The lack of stable dielectric films providing a low interface state density is still a serious drawback of III-V semiconductors [1, 2]. Different dielectric materials including Si_3N_4, SiO_x, Al_2O_3, and Ga_2O_x have been used in combination with dry, liquid, and photochemical semiconductor surface treatments [3 - 7]. Recently, Aydil et al. [8, 9] achieved passivation of surface states during a NH_3 or H_2 plasma treatment at room temperature by removal of excess As and As_2O_3 and subsequent formation of a Ga_2O_3 film (a few monolayers thick) on the GaAs surface. Furthermore, Ga_2O_3 films are chemically stable on GaAs as demonstrated by thermochemical phase diagrams [10]. This paper reports on film and interface properties of Ga_2O_3 films deposited by electron-beam evaporation from a high-purity single-crystal $Gd_3Ga_5O_{12}$ source. These Ga_2O_3 films have been employed for low reflectivity front facet coatings on InGaAs/GaAs SCH-SQW lasers emitting at 980 nm. Furthermore, we show that ultra-high vacuum processing combining molecular beam epitaxy and Ga_2O_3 film deposition provides dielectric-GaAs structures with low interface recombination velocity/interface state density.

2. Film fabrication

Electron-beam deposition imparts only a small amount of energy into the surface [11] and is compatible with standard semiconductor processing methods. Other Ga_2O_3 thin film fabrication techniques using oxidation of evaporated Ga in an O_2 plasma may cause severe surface damage prior to deposition and did not produce dielectric Ga_2O_3 films [6]. Splattering of Ga_2O_3, which occurs due to rapid sublimation during electron-beam heating, can be eliminated by using a single-crystal $Gd_3Ga_5O_{12}$ source for electron-beam evaporation [12]. Using this technique, the decrepitation of $Gd_3Ga_5O_{12}$ controls the evolution of Ga_2O_3 during electron-beam heating. The deposition rate is kept low at about 0.5 Å/s in order to maintain a high purity of the deposited Ga_2O_3 film.

3. Ga_2O_3 films deposited in a vacuum chamber

Stoichiometric, homogeneous, and uniform dielectric Ga_2O_3 amorphous films with thicknesses between 40 and 4000 Å on GaAs, Si and fused silica substrates have been fabricated in a vacuum chamber. The background pressure in the chamber was $1 - 2 \times 10^{-6}$ Torr. Both cases of no excess oxygen and introduction of additional oxygen into the evaporation chamber have been investigated. The maximum O_2 partial pressure was 2×10^{-4} Torr when additional O_2 was introduced.

3.1. Ga_2O_3 film properties

Thin film properties including stoichiometry, microstructure, phase, oxidation state, homogeneity, uniformity, and temperature integrity as well as optical and electrical characteristics have been investigated. An excellent uniformity and homogeneity of our Ga_2O_3 films over a 2 inch wafer was demonstrated by ellipsometry and Auger depth profiling. The Ga_2O_3 films are stoichiometric (Fig. 1) and amorphous as shown by Rutherford backscattering spectroscopy (RBS) and X-ray photoelectron spectroscopy (XPS) as well as transmission electron microscopy (TEM), respectively. The films maintain their integrity during annealing up to 800 °C on GaAs and up to 1200 °C on Si substrate. Optical properties including refractive

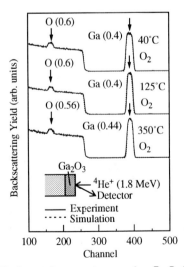

Fig. 1. Rutherford backscattering spectra measured on Ga_2O_3 films deposited with introduction of additional oxygen into the deposition chamber. An excellent agreement between simulated and experimental curves indicates homogeneous and stoichiometric films with 40 % Ga and 60 % O for moderate substrate temperatures.

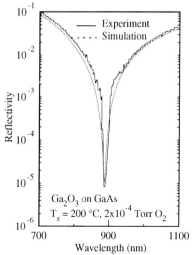

Fig. 2. Measured reflectivity vs wavelength of a Ga_2O_3-GaAs structure. The measured reflectivity minimum is $< 10^{-5}$ at 885 nm wavelength which is in excellent agreement with the simulation data. The film thickness is 1170 Å.

index (n = 1.841 - 1.885 at 980 nm wavelength) and bandgap (4.4 eV) are close or identical, respectively, to Ga_2O_3 bulk properties as shown by spectroscopic ellipsometry and optical transmission measurements. Reflectivities as low as 10^{-5} for Ga_2O_3-GaAs structures (Fig. 2) and a small absorption coefficient (≈ 100 cm^{-1} at 980 nm) insensitive to substrate deposition temperature were measured. The static dielectric constant of the films (9.9 - 10.2) is identical to bulk Ga_2O_3. The highest breakdown field measured was 3.6 MV/cm and it was found to degrade only by 6 % after exposure of the films to typical device processing temperatures up to 400 °C. A detailed discussion of thin film properties can be found in Ref. 13.

3.2. Ga_2O_3 optical coatings

Low reflectivity Ga_2O_3 front facet coatings on high power lasers are reported in Ref. 14. Ridge-waveguide InGaAs/GaAs SCH-SQW lasers emitting at 980 nm were fabricated. The laser structure was grown by organometallic vapor phase epitaxy (OMVPE) on n$^+$-GaAs substrate and consists of a 0.5 μm thick n$^+$-GaAs buffer layer, a 1 μm thick n-$In_{0.5}Ga_{0.5}P$ cladding layer, a 0.1 μm thick undoped GaAs optical confinement layer, a 7 nm undoped $In_{0.2}Ga_{0.8}As$ quantum well, a 0.1 μm thick undoped GaAs optical confinement layer, a 0.1 μm thick p-$In_{0.5}Ga_{0.5}P$ cladding layer, a 6 nm GaAs etch stop layer, a 1 μm thick p-$In_{0.5}Ga_{0.5}P$ cladding layer and a 0.2 μm p$^+$ GaAs cap layer. After processing the wafer using standard dielectric deposition and metallization techniques, a 5 μm thick heat spreading Au layer was deposited on the p-side of the 100 μm thick wafer by electrochemical plating. The wafer was subsequently cleaved into 750 μm long bars and the laser facets were coated. The low (1 %) and high (98 %) reflectivity facets were implemented by a Ga_2O_3 coating and by SiO_2/Si distributed Bragg reflectors, respectively. Finally, individual laser chips were bonded in a p-side down configuration on BeO submonts. As cleaved 6 μm x 750 μm SCH-SQW ridge-waveguide lasers have a threshold current of 14.8 mA and an external differential quantum efficiency of 27 % per facet. After coating, the external differential quantum efficiency at the front facet and the threshold current are 59 % and 20 mA, respectively. A high maximum

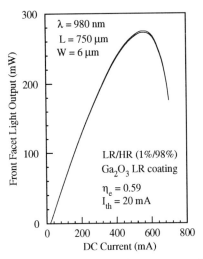

Fig. 3. Light output at the front facet under continuous wave conditions for a 6 μm x 750 μm SCH-SQW ridge waveguide laser at room temperature [14]. The optical power density at the front facet is 17 MW/cm² at the rollover point.

power density $P_D = P/(W\,d/\Gamma) = 17$ MW/cm² at the front facet at thermal rollover is obtained, where P, W, d, and Γ are the maximum output power at the front facet, the ridge width, the thickness of the active layer (quantum well), and the confinement factor (0.0245), respectively. More important, repeated cycling beyond thermal rollover produced no change in operating characteristics and the lasers did not exhibit catastrophic optical mirror damage. This is in contrast to results in the literature, which reported catastrophic optical mirror damage being the output power limiting factor at front facet power densities between 5 and 15 MW/cm² [15 - 17]. In pulsed mode operation, an optical power of 700 mW was measured at 1 A using a 1% duty cycle.

4 Integrated ultra-high vacuum processing

An integrated ultra-high vacuum processing system first proposed by Cho in 1984 [18] was used to combine molecular beam epitaxy and dielectric film deposition. Details of our in-situ processing system are described elsewhere [19]. Deposited dielectric-GaAs structures with low interface recombination velocity/interface state density are reported in Ref. 20 for the first time. A 1.5 μm thick GaAs n-type (2×10^{16} cm^{-3}) layer was grown by molecular beam epitaxy on a n$^+$ doped (100) GaAs substrate. Subsequently, the wafer was transferred into another chamber for deposition of the dielectric film. Finally, Ga_2O_3 films were deposited by electron-beam evaporation from a single-crystal $Gd_3Ga_5O_{12}$ source at substrate temperatures ranging from 0 °C to 500 °C. No optimization of the deposition process has been carried out. The Ga_2O_3-GaAs interface was investigated by photoluminescence measurements using an argon ion laser ($\lambda = 514.5$ nm) operated at an optical power density of 0.5 W/cm². Fig. 4 shows the photoluminescence data measured on an in-situ fabricated Ga_2O_3-GaAs sample. Compared to the photoluminescence signal of the bare surface, which is also shown in Fig. 4, the integrated photoluminescence intensity of the in-situ fabricated Ga_2O_3-GaAs sample is higher by a factor

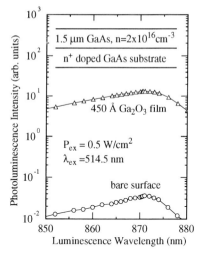

Fig. 4. Photoluminescence intensity vs wavelength measured for an in-situ fabricated Ga_2O_3-GaAs interface (triangles) and on a bare GaAs surface (circles) [20]. The inset shows the semiconductor layer structure before deposition of Ga_2O_3.

of 420. The integrated photoluminescence signal F, derived for a semiconductor absorption coefficient α which is much larger at the excitation wavelength than at the wavelength of the photoluminescence emission [21], is given for flatband condition by

$$F = \frac{(1-R)\,G_0 L^3}{4\pi\tau_r D}\left(\frac{\alpha L + SL/D}{\alpha L\,(\alpha L + 1)\,(1 + SL/D)}\right). \tag{1}$$

Here, L, G_0, τ_r, D, and S are the minority carrier diffusion length, the carrier generation rate at the surface, the bulk radiative lifetime, the minority carrier diffusion coefficient, the surface recombination velocity, and the sample reflectivity, respectively. According to Eq. (1), the surface recombination velocity at the Ga_2O_3-GaAs interface is less than 10^4 cm/s. The sample was also found to be stable during photoexcitation. Preliminary time resolved photoluminescence measurements indicated photoluminescence lifetimes in the ns-range. Capacitance-voltage measurements on metal-Ga_2O_3-GaAs structures demonstrated both accumulation and inversion.

5. Conclusions

Film and interface properties of Ga_2O_3 films deposited by electron-beam evaporation from a high-purity single-crystal $Gd_3Ga_5O_{12}$ source were reported. Thin film properties including stoichiometry, microstructure, phase, homogeneity, uniformity, and temperature integrity as well as optical and electrical characteristics make these films well suited for electronic and optoelectronic applications. Ultra-high vacuum processing combining molecular beam epitaxy and Ga_2O_3 film deposition provided deposited dielectric-GaAs structures with low interface recombination velocity/interface state density.

Acknowledgement

We would like to thank G.P. Schwartz, R.A. Gottscho, R.L. Opila, T.D. Harris, C.G. Bethea, L.C. Feldman, R. Becker, and R.J. Fischer for many useful discussions. Special thanks to S.N.G. Chu, K. Konstadinidis, U.K. Chakarabarti and R.L. Masaitis for TEM, XPS, spectroscopic ellipsometry and Auger spectroscopy measurements, respectively. M. Passlack gratefully acknowledges support by the Deutsche Forschungsgemeinschaft. We are also grateful to N.K. Dutta, A.Y. Cho and D.V. Lang for their support.

References

[1] C.W. Wilmsen, *Physics and Chemistry of III-V Compound Semiconductor Interfaces* (Plenum Press, New York, 1985)
[2] W.F. Croydon and E.H.C. Parker, *Dielectric Films on Gallium Arsenide* (Gordon and Breach, New York, 1981)
[3] F. Capasso and G.F. Williams, J.Electrochem. Soc. **129**, 821 (1982)
[4] S.D. Offsey, J.M. Woodall, A.C. Warren, P.D. Kirchner, T.I. Chapell, and G.D. Pettit, Appl. Phys. Lett. **48**, 477 (1986)
[5] C.J. Sandroff, R.N. Nottenburg, J.-C. Bischoff, and R. Bhat, Appl. Phys. Lett. **51**, 33 (1987)
[6] See, for example, A. Callegari, P.D. Hoh, D.A. Buchanan, and D. Lacey, Appl. Phys. Lett. **54**, 332 (1989)
[7] J.S. Herman, and F.L. Terry, Appl. Phys. Lett. **60**, 716 (1992)
[8] E.S. Aydil, K.P. Giapis, R.A. Gottscho, V.M. Donnelly, E. Yoon, J. Vac. Sci. Technol. **B11**, 195 (1993)
[9] E.S. Aydil, R.A. Gottscho, Materials Science Forum, **148 - 149**, 159, (1994)
[10] G.P. Schwartz,Thin Solid Films, **103**, 3, (1983)
[11] C.W. Wilmsen, *Physics and Chemistry of III-V Compound Semiconductor Interfaces* (Plenum Press, New York, 1985), p. 185
[12] S. Singh, L.G. van Uitert, and G.J. Zydzik, unpublished
[13] M. Passlack, E.F. Schubert,W.S. Hobson, M. Hong, N. Moriya, S.N.G. Chu, K. Konstadinidis, J.P. Mannaerts, M.L. Schnoes, G.J. Zydzik, unpublished
[14] M. Passlack, E.F. Schubert, C.G. Bethea, W.S. Hobson, J. Lopata, G.J. Zydzik, N.K. Dutta, unpublished
[15] S. Ishikawa, K.Fukagai, H. Chida, T. Miyazaki, H. Fujii, and K. Endo, IEEE J.Quantum Electron., **29**, 1936, (1993)
[16] J. S. Yoo, S.H. Lee, G.T. Park, Y.T. Ko, and T. Kim, J. Appl. Phys., **75**, 1840, (1994)
[17] Z.L. Liau, S.C. Palmateer, S.H. Groves, J.N. Walpole, and L.J. Missaggia, Appl. Phys. Lett., **60**, 6, 1992
[18] A.Y. Cho, Am. Chem. Soc. Symp. Series **290**, 118 (1985)
[19] M. Hong, K.D. Choquette, J.P. Mannaerts, L.H. Grober, R.S. Freund, D. Vakhshoori, S.N.G. Chu, H.S. Luftman, R.C. Wetzel, J. Electronics Materials, **23**, 625 (1994)
[20] M. Passlack, M. Hong, E.F. Schubert, J. Kwo, J.P. Mannaerts, S.N.G. Chu, N. Moriya, F.A. Thiel, unpublished
[21] H.B. Bebb and E.W. Williams, *Semiconductors and Semimetals*, vol. 8 (Academic, New York, 1972) p. 182

MOCVD GaInAsSb Heterostructure Materials for 2−4μm Photodetectors

Wei Guangyu, Peng Ruiwu and Wu Wei

Shanghai Institute of Metallurgy, Chinese Academy of Sciences, Shanghai 200050, P. R. China

The GaInAsSb epilayers were grown by metalorganic chemical vapor deposition (MOCVD). The n−type and p−type doping characteristics by using Te and Zn as dopants were investigated and, hence, various PN and PIN multilayer structural materials were prepared on GaSb substrates. GaInAsSb photodetectors were fabricated with room temperature detectivity of $D_\lambda^* = 5.04 \times 10^9 \mathrm{cmHz}^{1/2}/\mathrm{W}$.

1. Introduction

III−V antimony containing compound semiconductors are promising materials for long wavelength optoelectronics. The preparation of GaInAsSb materials and photodetectors are potential important in 2~4μm range. Several reports on the growth of these materials by LPE, MBE and MOCVD have been published[1~7]. Although MOCVD has many advantages compared with other growth techniques, a few work on n−type and p−type doping, growth of structural materials and fabrication of photodetectors by MOCVD has been reported. In this paper, the investigation on n−type and p−type doping characteristics for GaSb and GaInAsSb materials and the preparation of multilayer structures for 2−4μm photodetectors by MOCVD are presented.

2. Experimental

The epitaxial growth was carried out in a horizontal MOCVD system at atmospheric pressure using TMG, TMIn, TMSb and AsH_3 as the metalorganic sources and n−GaSb as the substrates. DETE and Zn as well as DEZn were used as n−type and p−type dopant respectively. The influence of growth procedure and parameters on layer properties were investigated in detail to obtain the following optimum conditions[6~8]: Tg=550~600℃, V/III=1,0~1.5, P_{III}=5~10×10^{15} atm, TMIn/III=0~0.3, TMSb/V=0.8~1.0, H_2=3.5 SLM.

The characterization of GaInAsSb multilayers was made by scanning electron microscope for surface and interface morphology, by electron probe microanalysis for solid compositions and by Van de Pauw technique for electrical properties. The PN GaInAsSb photodetectors were characterized by I−V curve and detectivity measurements.

3. Results and Discussions

In the previous work, p-GaInAsSb epilayers with background concentrations of about $10^{17} cm^{-3}$ were grown directly onto n-GaSb substrates to form PN junctions[2,9-11]. But it is well known that the lattice constant of P-GaInAsSb quaternary alloy is very difficult to be controlled precisely matched to n-GaSb substrate and high density of misfit dislocations will be created at the PN junction region, which are quite harmful to the preformance of PN photodiodes. Furthermore, the background concentrations of GaInAsSb epilayers are sensitively influenced by growth conditions and also difficult to be reproducibly controlled.

In this case, it is necessary to study the n-type doping processes in GaInAsSb epilayers for fabricating homo- and hetero-junction photodetectors. For n-type dopant, DETe was selected in the present work with the reduction of its strong memory effect in the reaction system. Fig. 1 shows the relationship between n-type carrier concentrations in Te-doped GaSb and GaInAsSb and DETe partial pressures in the vapor phase together with the results of Te-GaSb by other authors[12~14]. It can be seen that the electron concentrations of $n = 2 \times 10^{16} \sim 8 \times 10^{17} cm^{-3}$ and $n = 2 \times 10^{16} \sim 5 \times 10^{17} cm^{-3}$ have been obtained for GaSb and GaInAsSb alloys respectively and the incorporation chacteristics for Te in GaSb and GaInAsSb are almost the same, i.e, they all have a saturated concentration caused by native point defects in these epilayers during growth. The variation of Hall mobility at room temperature with electron concentration is shown in Fig. 2. The Hall mobilities of $\mu = 3,300 \sim 1,303 \ cm^2/V.s$ and $\mu = 880 \sim 200 \ cm^2/V.s$ were obtained for GaSb and GaInAsSb respctively, in

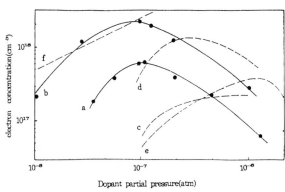

Fig 1 Variation of electron concentration at 300K with dopant partial pressure for GaSb and GaInAsSb epilayers grown on GaAs substrate. (a)DETe doped GaSb,present work; (b)DETe doped GaInAsSb,present work; (c)DMTe doped GaSb, F.Pascal ; (d)DETe doped GaSb, Nakamura ; (e)H₂Se doped GaSb,Nakamura ; (f)DETe doped GaAs, Collins .

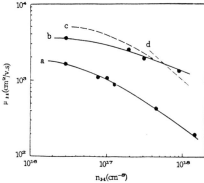

Fig 2 Variation of Hall mobility with carrier concentration for DETe doped GaSb and GaInAsSb epilayers grown on GaAs substrates. a)Te-GaSb,present work; b)Te-GaInAsSb, present work; c)Te-GaSb,F.Pascal ; d)Te-GaSb, Nakamura .

which the results of GaSb are comparable with that obtained in the literature by MBE and MOCVD[12~16], while that of GaInAsSb have not yet been reported.

3.2 P-type doping

In order to grow PN junction multistructures, p-type GaInAsSb over layers should be deposited onto n-GaSb layers. As mentioned above, DETe has strong memory effect and the background concentration of the alloy is difficult to control. Therefore, p-type doping investigations were studied in detail.

Since the p-type doping for Sb-compound is very scarce[17~18], we have employed several doping method, such as metallic Zinc, metalorganic DEZn and DEZn diluted in H_2[19]. Using metallic Zinc as a p-type dopant, the hole concentrations of $p = 3 \times 10^{16} \sim 1 \times 10^{18} cm^{-3}$ with Hall mobilities of $\mu = 880 \sim 200 cm^2 / V.s$ for GaSb and $p = 5 \times 10^{16} \sim 1 \times 10^{18} cm^{-3}$ with Hall mobilities of $\mu = 720 \sim 180 cm^2 / V.s$ for GaInAsSb have been obtained respectively, as shown in Fig. 3. In the case of using DEZn, GaSb and GAInAsSb epilayers with carrier concentrations as high as $p = 2.7 \times 10^{19} cm^{-3}$ and $p = 4 \times 10^{19} cm^{-3}$ were achieved. The later concentrations may be suitable not only for active layers, but also for top contact layers.

3.3 Multistructures and Photodetectors

Based on the above doping study, various types of PN and PIN multilayer GaInAsSb structural materials were grown on GaSb substrates. Each of these materials has a $0.5 \mu m$ p^+ or n^+ contact layer and a more than $2 \mu m$ GaSb buffer layer. The typical structural materials prepared are: p^+-GaInAsSb / p-GaInAsSb / n-GaSb(Sub), p-GaInAsSb / p^--GaInAsSb / n-GaSb / n-GaSb (Sub), p^+-GaInAsSb / p^--GaInAsSb / n^--GaInAsSb / n-GaSb / n-GaSb (Sub).

Using these structural materials, GaInAsSb photodetectors were fabricated[20]. The ohmic contacts were made by AuIn alloy on the p-type epilayers and AuGeNi alloy on

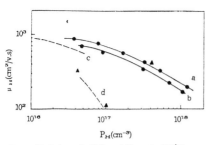

Fig 3 Variation of Hall mobility on carrier concentration for p-type doped and undoped GaSb and GaInAsSb. a)Zn-GaSb, present work, b)Zn-GaInAsSb,present work; c)undoped-GaSb, present work; d)undoped GaInAsSb,present work; ▲)Si-GaSb grown by MBE, T.M.Rossi.

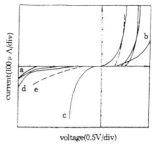

Fig 4 I -V characteristic curves for GaInAsSb photodiodes fabricated with different structural materials at room temperature.
a)p-GaInAsSb/n-GaInAsSb/n-GaSb,present work;
b)n-GaInAsSb/p-GaSb,present work;
c)p^+-GaInAsSb/p-GaInAsSb/n^--GaInAsSb/n^+-GaSb/ GaSb(PIN),present work;
d)p-GaInAsSb/n-GaSb,Xu chenmei (LPE);
e)p-GaInAsSb/n-GaSb,Bougnot (77k);

the n-type subsurates. Fig. 4 shows the typical I-V characteristic curve for these PN photodetectors at room temperature, compared with that obtained by other authors[2,21]. The reverse background voltage of −1.5V and dark current of 10nA were obtained. The detectivities at room temperature are almost in the range of $D_{bb}^* = 10^8 \sim 10^9 cmHz^{1/2}/W$ with a typical peak detectivity of $D_\lambda^* = 5.04 \times 10^9 cmHz^{1/2}/W$, which is comparable to that obtained in recent literatures[2,9~11]. The influence of substrate quality, solid composition in GaInAsSb, doping concentration and lattice imperfection at PN junction on detector performance has been investigated[22]. It is believed that the performance of GaInAsSb photodetectors made of doped epitaxial multistructures can be more effectively improved than that of undoped heterojunctions.

4. Conclusion

In this paper, the investigations on n-type and p-type doping characteristics for MOCVD GaSb and GaInAsSb epilayers have been reported and the multistructures with high electrical properties have been obtained. Using these PN and PIN GaInAsSb multistructural materials, 2~4μm photodetectors has been fabricated with the room temperature detectivity of $D_\lambda^* = 5.04 \times 10^9 cmHz^{1/2}/W$.

References

[1] Tournie E and Pitard F 1990 J. Appl. Phys. 68 5936.
[2] Xu C M, Peng R W and Wei G Y 1994 Rare Metals. 13 26.
[3] Chiu T H and Zyskind J L 1987 J. Electron. Mater 16 57.
[4] Cherng M J and Jen H R 1986 J. Cryst. Growth. 77 408.
[5] Giani A and Bougnot J 1991 Mat.Sci.Eng. B9 121.
[6] Wei G Y and Peng R W 1992 ICMOVPE-VI, USA.
[7] Wei G Y and Peng R W 1993 Rare Metals 12 104.
[8] Wei G Y and Peng R W 1994 J. Electron. Mater. 23 217.
[9] Srivastava A K and Dewinter J C 1986 Appl. Phys. Lett. 48 903.
[10] Bowers J E and Srivastava A K 1986 Electron. Lett. 22 137.
[11] Tournie E and Lazzari J L 1991 Electron. Lett. 27 1237.
[12] Pascal F and Delannoy F 1990 J. Electron. Mater. 19 187.
[13] Nakamura F and Taira K 1991 J. Cryst. Growth. 115 474.
[14] Collins D M and Miller J N 1982 J. Appl. Phys. 53 3010.
[15] Subbanna S and Tuttle G 1988 J. Electron, Mater. 17 297.
[16] Chen J F and Cho A Y 1991 J. Appl. Phys. 70 277.
[17] Rossi T M and Collins D A 1990 Appl. Phys. Lett. 57 2256.
[18] Su Y K and Kuan H 1993 J. Appl. Phys. 73 56.
[19] Wei G Y and Peng R W 1993 EWMOVPE-V, Sweden.
[20] Wei G Y 1994 Doctoral Disertation, Shanghai Institute of Metallurgy.
[21] Bougnot G and Delannoy F 1988 J. Electrochem. Soc. 135 1783.
[22] Peng R W and Wei G Y 1994 High. Tech. Lett. in Perss.

InGaAs/AlGaAs quantum well infrared photodetectors with 3-5 μm response

L.C. Lenchyshyn, H.C. Liu, M. Buchanan, and Z.R. Wasilewski

Institute for Microstructural Sciences, National Research Council
Ottawa, Ontario, Canada, K1A 0R6

Abstract: We demonstrate the detection of mid-wavelength infrared (~3-5 μm) radiation by intersubband transitions in multiple quantum well structures. These devices exploit the large conduction band offset between $In_xGa_{1-x}As$ and $Al_yGa_{1-y}As$ in order to obtain shorter wavelength detection than is available with the more conventional $Al_yGa_{1-y}As$/GaAs infrared photodetector technology. The peak detector response is shifted from 4.8 to 4.3 μm by increasing the indium fraction from x = 0.05 to x = 0.20, while simultaneously decreasing the well width to keep the first excited eigenstate near the top of the wells. These detectors can be integrated with $Al_yGa_{1-y}As$/GaAs long-wavelength (8-12 μm) infrared quantum well detectors for multi-band imaging applications.

1. Introduction

With the maturation of compound semiconductor growth and device fabrication, there is increasing interest in applications to infrared photodetector technology. The III-V material system in particular offers excellent uniformity and control, which is very attractive for use in detector arrays. The development so far (for a review see Levine [1]) has concentrated on $Al_yGa_{1-y}As$/GaAs quantum well infrared photodetectors (QWIPs), which are most suitable for the long-wavelength infrared atmospheric window (8-12 μm). However, for many applications the mid-wavelength infrared (3-5 μm) region offers better or complimentary information. In this paper we utilize III-V related compounds (InGaAs and AlGaAs) grown on GaAs substrates to achieve mid-wavelength infrared detection. These detectors can then be integrated with the more conventional long-wavelength infrared $Al_yGa_{1-y}As$/GaAs QWIPs for multi-band imaging applications.

2. Device Design

In a QWIP the infrared radiation is absorbed via intersubband transitons between the ground state (E_0) and the first excited state (E_1) of the quantum wells. In order to efficiently collect the photoexcited carriers, the excited state must be near the top of the barriers [2]. The detector wavelength response consists of a narrow symmetric peak centered at (E_1-E_0) in the case of a bound final state (bound-bound QWIP) or a broader response with a low energy threshold near (E_1- E_0) in the case of an extended final state (bound-continuum QWIP) [1]. The curve in Figure 1a shows the calculated energy level spacing, (E_1- E_0), for an $Al_yGa_{1-y}As$ /GaAs QWIP as a function of the Al alloy fraction (y), with the GaAs well width varied so as to keep the excited state in resonance with the top of the barrier. The shortest available wavelength is 5.3 μm, corresponding to y=0.45. Above 45% Al, the barrier band gap becomes indirect and the transport of the photoexcited carriers deteriorates due to Γ-X scattering. Such indirect band gap QWIPs have been found to exhibit unusually high dark currents [3]. As described by Schneider et al. [4], one means around this problem is the use of thin AlAs cladding layers which shift the QWIP response to shorter wavelength. However, questions remain on the understanding of such devices and the optimization of their design [5].

Because of the larger band offsets, shorter wavelengths can be obtained directly by switching to InGaAs wells and AlGaAs barriers. This is shown in Figure 1b, where the calculated energy level spacing (E_1-E_0) is plotted as a function of In fraction (x) for $In_xGa_{1-x}As$ wells and either GaAs (dashed curve) or $Al_{0.45}Ga_{0.55}As$ (solid curve) barriers. We approximated the $In_xGa_{1-x}As/Al_{0.45}Ga_{0.55}As$ conduction

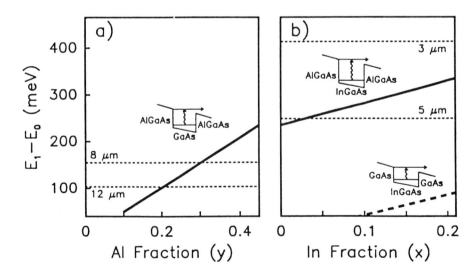

Figure 1: Dependence of calculated intersubband transition energies on a) Al fraction for $Al_yGa_{1-y}As$/GaAs and b) In fraction for $In_xGa_{1-x}As/Al_{0.45}Ga_{0.55}As$ (solid curve) and $In_xGa_{1-x}As$/GaAs (dashed curve).

band offset by adding the known band gap difference for GaAs/$Al_{0.45}Ga_{0.55}As$ to that for $In_xGa_{1-x}As$/GaAs. Again the calculations are shown for varying well width such that the excited state is at the top of the barriers. With GaAs barriers, the energy spacing is too small for mid-wavelength or even long-wavelength infrared detection and is used instead in the extended-wavelength region (11-18 μm) [1]. However, by combining $In_xGa_{1-x}As$ with $Al_{0.45}Ga_{0.55}As$ one can easily reach the mid-wavelength infrared.

Unfortunately, we are limited by the fact that the epitaxial InGaAs layers are strained and hence are subject to relaxation beyond an x-dependent critical thickness. The resulting dislocations may be expected to deteriorate the device performance. Very recently, Fiore et al. [6] applied the cladding approach to these materials, using very thin AlAs layers in an $In_{0.16}Ga_{0.84}As/Al_{0.37}Ga_{0.63}As$ QWIP to obtain 4.5 μm detection. Here we choose to study simply the InGaAs/AlGaAs devices first, without introducing cladding layers. However, we do decrease slightly the well thicknesses in the high x samples from the optimum values in order to avoid dislocations. This means that the excited state is relatively high in the continuum and the peak response will be shifted to somewhat higher wavelength from that shown in Figure 1b.

Our structures consist of 32 periods of $In_xGa_{1-x}As$ wells separated by 350 Å $Al_{0.45}Ga_{0.55}As$ barriers. We report on four samples, with indium fractions of x = 0.05, 0.10, 0.15, and 0.20 and corresponding well widths of 40, 37, 33, and 29 Å, respectively. (The calculated well thickness for resonance of the excited state with the top of the barrier is 40, 39, 38.5, 38 Å, for x = 0.05, 0.10, 0.15, and 0.20 respectively.) The wells are center delta-doped with Si to 5×10^{11} cm^{-2}. These multiple quantum well structures are sandwiched between 1.5×10^{18} cm^{-3} Si-doped contact layers (0.4 μm top contact thickness and 0.7 μm bottom contact thickness). The structures were grown by molecular beam epitaxy on semi-insulating GaAs substrates. Mesa structures of 120 x 120 μm were then fabricated by standard wet chemical etching and NiGeAu alloy contacts were made. Infrared light coupling to the multiple quantum well structures is provided by 45° edge facets. This is necessary because the optical electric field must have a component in the growth direction in order to induce an intersubband transition.

3. Results

The spectral response for the x = 0.05 and x = 0.20 QWIPs is shown in Figure 2 for a detector bias of -2 V applied to the top contact with respect to the bottom. These measurements were performed using a Fourier transform interferometer, with the detectors maintained at 82 K by mounting them in a liquid nitrogen cooled optical dewar. Similar spectra are observed for the x = 0.10 and x = 0.15 detectors, but shifted in wavelength so that they lie between the curves of Figure 2. The detector response peaks at wavelengths of 4.81, 4.60, 4.45, and 4.28 ± 0.02 μm for x = 0.05, 0.10, 0.15 and 0.20, respectively. The broad spectral response observed in Figure 2 ($\Delta\lambda/\lambda \sim 25\%$) is indicative of a bound-to-continuum

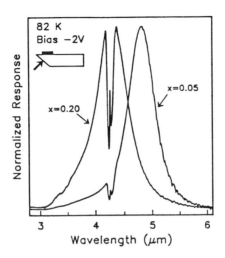

Figure 2: Spectral response at 82 K for $In_xGa_{1-x}As/Al_{0.45}Ga_{0.55}As$ quantum well infrared photodetectors with In fractions of x = 0.05 and x = 0.20, as indicated. The feature at 4.25 μm is due to CO_2 absorption in the infrared beam path.

QWIP detector [1]. The sharp dip in the spectra at 4.25 μm is caused by CO_2 absorption in the infrared beam path of the interferometer. Since this feature is of interest in some infrared imaging systems, these spectra clearly demonstrate the good response of our detectors for such applications.

Next we quantitatively investigate the magnitude of the mid-wavelength infrared response of the detectors. To perform this calibration we used a 1000 K black-body source, with the photocurrent collected by standard lock-in techniques. These responsivity measurements were taken at the peak detection wavelength for each sample, which was selected from the broad black-body spectrum by a narrow bandpass, tunable interference filter. The bias dependence of the peak responsivity is shown by the dashed curve in Figure 3 for the x = 0.05 detector for unpolarized light. Similar voltage dependences are observed for the other three QWIPs. The detectors are found to turn on at very low bias, again in agreement with the assignment as a bound-to-continuum QWIP. There is a small photovoltaic (zero bias) response, so that the photocurrent actually reaches zero at a positive bias of roughly 0.25 V. This indicates a very small built-in field in the devices [7]. Responsivities of the order of tens of mA/W are observed. This response is probably limited somewhat by our use of high Al fraction barriers (y ≈ 0.45) which, while ensuring a large conduction band offset, also have low mobilities for the transport of photoexcited electrons. Note however that our responsivities are comparable with other reports of InGaAs mid-wavelength infrared QWIP detectors [6,8].

A more illustrative figure of merit is the detectivity, particularily since these detectors have very low dark current (in comparison to long-wavelength QWIPs). The detectivity can be obtained from $D^* = R\sqrt{A}/\sqrt{4egI}$, where R is the measured

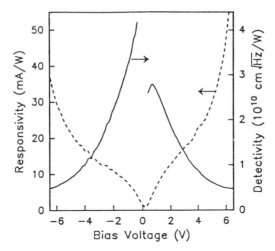

Figure 3: Peak (4.81 μm) responsivity (dashed curve) and detectivity (solid curve) versus detector bias for an $In_{0.05}Ga_{0.95}As/Al_{0.45}Ga_{0.55}As$ quantum well infrared photodetector at 82 K.

responsivity, A is the device area, g is the noise gain, and I is the measured dark current. The noise gain for a bound-to-continuum QWIP can be approximated by the photoconductive gain [9], which is then given by the expression $g = h\nu R/e\eta$ where η is the absorption efficiency and ν is the photon frequency. The absorption efficiency is found by converting room temperature Brewster angle absorption measurements to 82 K and 45° facet geometry. The resulting detectivity versus bias is shown in Figure 3 (solid curve) for the x = 0.05 detector. The detectivity is observed to be in excess of 10^{10} cm$\sqrt{Hz}$/W. The dark current increases more rapidly with bias than the responsivity, so that even though the responsivity is not saturated, the detectivity is found to drop to about 5×10^9 cm$\sqrt{Hz}$/W for |Vbias| > 5 V. Detectivities for the remaining three device structures are 0.7, 0.5, and 1.9 $\times 10^{10}$ cm$\sqrt{Hz}$/W for x=0.10, 0.15, and 0.20 respectively at the peak detection wavelengths and a bias voltage of -2 V. Therefore there appears to be very little degradation in the QWIP detectivities despite the possibility of relaxation in the high In fraction detectors. These detectivities are comparable with other detectors commonly in use in the mid-wavelength infrared.

In conclusion, InGaAs/AlGaAs QWIPs grown on GaAs have been shown to give good performance in the detection of mid-wavelength infrared radiation. At present, we are fabricating a second set of InGaAs/AlGaAs QWIPS with wider well widths to obtain better resonance of the excited state with the top of the barriers. In the event that the relaxation continues to pose no problem, this is expected to improve the peak detectivity and shift the response further into the mid-wavelength infrared. Further efforts will then be directed towards the integration of these detectors with long-wavelength QWIPs for multi-color applications.

ACKNOWLEDGEMENTS: We thank P. Chow-Chong, P. Marshall, J. Stapledon, and J. Thompson for device fabrication. This work was supported in part by DND DREV.

References

1. B.F. Levine, *J. Appl. Phys.* **74**, R1 (1993)
2. H.C. Liu, *J. Appl. Phys.* **73**, 3062 (1993).
3. B.F. Levine, S.D. Gunapala, and R.F. Kopf, *Appl. Phys. Lett.* **58**, 1551 (1991).
4. H. Schneider, F. Fuchs, B. Dischler, J.D. Ralston, and P. Koidl, *Appl. Phys. Lett.* **58**, 2234 (1991).
5. H.C. Liu, P.H. Wilson, M. Lamm, A.G. Steele, Z.R. Wasilewski, Jianmeng Li, M. Buchanan, and J.G. Simmons, *Appl. Phys. Lett.* **64**, 475 (1994).
6. A. Fiore, E. Rosencher, P. Bois, J. Nagle, and N. Laurent, *Appl. Phys. Lett.* **64**, 478 (1994).
7. H.C. Liu, Z.R. Wasilewski, M. Buchanan, and H. Chu, *Appl. Phys. Lett.* **63**, 761 (1993).
8. G. Hasnain, B.F. Levine, D.L Sivco, and A.Y. Cho, *Appl. Phys. Lett.* **56**, 770 (1990).
9. B. Xing, H.C. Liu, P.H. Wilson, M. Buchanan, Z.R. Wasilewski, and J.G. Simmons, *J.Appl. Phys.*, to be published in Aug. 1, 1994 issue.

Planar, low loss InGaAsP/InP Photoelastic waveguide devices

X. S. Jiang, Q. Z. Liu, L. S. Yu, B. Zhu, Z. F. Guan, S. A. Pappert, P. K. L. Yu, and S. S. Lau

Dept. of Electrical and Computer Engineering, University of California, San Diego
La Jolla, CA 92093-0407

Abstract. Two simple schemes are used to fabricate low loss planar photoelastic InP-based waveguides. The first scheme is to use metal semiconductor reactions between Ni and InP or InGaAs to form a stressor. The second scheme is to use rf sputtered refractory metal where W is sputtered onto a negative dc biased substrate. Stable, low loss planar photoelastic Ni/InP and W/InP-based waveguides have been obtained. An efficient planar photoelastic Franz-Keldysh electroabsorption waveguide modulator has been demonstrated for the first time with the WNi stressor. A little amount of Ni (5%) co-sputtered with W effectively enhances the stress in the WNi stressor and its thermal stability. WNi serves very well both as a stressor and as an electrode. Planar photoelastic quantum-confined-stark-effect (QCSE) waveguide modulator has been demonstrated. Optical beam splitters and couplers have also been achieved using this photoelastic effect.

1. Introduction

The study of photoelastic effect for planar waveguides in III-V compound semiconductors with a surface stressor dates back to mid 1970's[1, 2, 3]. This technique has not been explored extensively as compared to other waveguide fabrication techniques because of the lack of good control of the stress and inferior thermal stability in the stressor layers reported in earlier work. Recently we have developed two schemes to introduce controlled stable strains into GaAs/AlGaAs heterostructures[4]. In the first scheme which uses metal-semiconductor interfacial reaction, photoelastic waveguides are formed underneath a stripe window (a few μm wide) in an otherwise continuous Ni film in tension; in the second scheme which uses rf sputtered refractory metal film with dc bias at the substrate, photoelastic waveguides are formed underneath a stripe (a few μm wide) of W film under compression. Figs. 1 (a) and (b) schematically depicts these two schemes. In both schemes, a stress field exists in the semiconductor, which locally changes the refractive index and can provide desirable lateral optical confinement for waveguiding. For optical fiber communication application emphasizing the 1.3 and 1.5 μm wavelengths, waveguide devices based on InP/InGaAsP materials are of interest. In this study, InP-based planar photoelastic waveguides with both Ni and W stressors are fabricated. Thermal stability and optical losses of these waveguides have been studied. Planar photoelastic electroabsorption waveguide modulators based on either Franz-Keldysh effect or the quantum confined Stark effect (QCSE) have been fabricated. Photoelastic optical splitters and couplers have also been demonstrated.

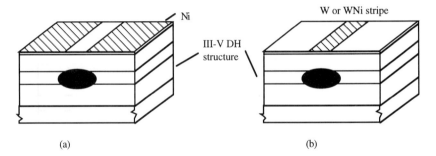

Figure 1. Schematics of the two schemes of photoelastic waveguides. Darkened elliptical spots represent the waveguide region under the stripe

2. Experimental

Material structures A and B for waveguides are schematically shown in Figures 2(a) and (b); while material structures for FKE modulators and QCSE modulators are shown in Figures 3(a) and (b) respectively. The samples are grown by low pressure OMVPE on n-InP substrate[5]. All epilayers of waveguide samples are undoped. After samples are cleaned chemically, they are etched in $HF:H_2O$ for 30 seconds to remove the native oxide before loading into an e-beam evaporator or a sputtering chamber.

As mentioned earlier, two schemes are used to fabricate the waveguides. In the first scheme, a photoresist lift-off technique is employed to form a 5 μm stripe window in an otherwise continuous Ni film. About 1000 Å Ni film is evaporated in a conventional e-beam system. In the second scheme, the photoresist lift-off technique is employed to form a 5 μm stripe of W film. About 950Å W film is rf sputtered onto the sample with a -220 V bias on the substrate. The background Ar pressure is 25 mTorr. After lift-off, the samples are lapped down and cleaved into bars for waveguide measurements. The cut-back method is used to measure the optical loss at 1.52 μm wavelength. Some samples are annealed in forming gas (85% N_2, 15% H_2) under various combinations of temperature and time to investigate the thermal stability of the stress induced in the structures.

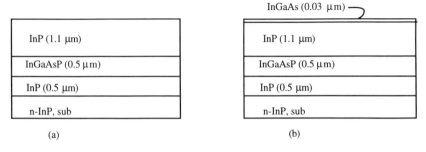

Figure 2. (a) Material structure A and (b) material structure B for waveguides

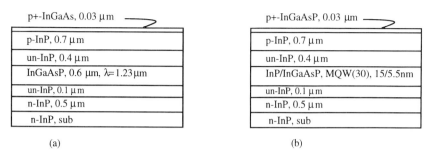

Figure 3. Schematic material structure of (a) FKE waveguide modaulator and (b) QCSE waveguide modulator

For FKE and QCSE waveguide modulators, the photoresist lift-off technique is employed to form a 5 μm stripe pattern with electrode pads. A little amount of Ni (~5%) is rf co-sputtered with W with -220V bias on the substrate at a background Ar pressure of 20 mTorr. This little amount of Ni has been shown to be effective in enhancing the stress in the WNi stressor and its thermal stability[4]. Sputtered WNi film is about ~1000Å. The samples are then loaded into an e-beam evaporator chamber. The ion implantation mask consists of about 58 Å Hf followed by 4520Å Au. Hf is to enhance the sticking of the Au layer to the WNi stressor. After lift-off, 200 kev O^+ implantation is performed at 510°C with a dose of 5×10^{13} cm^{-2} to provide electric isolation between devices. WNi stressor also becomes an ohmic contact to the p^+ layer of the samples during this high temperature implantation. Thereafter, the sample is lapped down and loaded into an e-beam evaporator. About 1000Å AuGe is evaporated onto the back side of the sample. After samples are annealed at 380°C for 3 min. using RTA to form the backside ohmic contact[6], they are cleaved into short bars for optical transmission versus bias voltage measurement.

3. Results and discussion

3.1. InP-based Planar low loss waveguide

Results of optical loss measurement and thermal stability study of photoelastic waveguides are summarized in Table 1. As shown in the table, Ni and W photoelastic planar waveguides have low propagation losses. Although they behave differently from GaAs based Ni and W photoelastic waveguides in term of thermal stability, they have good thermal stability except for the Ni/Structure B ones. We are especially interested in the success of the scheme of sputtered W stripe because the resulting waveguide structure can easily be extended for waveguide modulator with W stripe as an electrode.

3.2. Planar, photoelastic waveguide modulator

A typical optical transmission vs. bias voltage of a 830 μm long photoelastic FKE waveguide modulator is shown in Figure 4. The optical source is a 1.31 μm Nd:YAG laser. About 10 dB

Table 1

	Structure A		Structure B	
	Ni window	W stripe	Ni window	W stripe
As deposited	2.8 dB/cm (10 µm)	4.3 dB/cm (5.6 µm)	2.0 dB/cm (10 µm)	4.4 dB/cm (5.3 µm)
280°C, 30 min.	3.0 dB/cm (13.3 µm)	---	slab waveguide	---
300°C, 30 min.	---	3.5 dB/cm (7.5 µm)	---	(6 µm)
350°C, 30 min.	---	good lateral confinement	---	good lateral confinement
450°C, 30 min.	good lateral confinement	slab waveguide	---	---

Table 1. Summary of the result of photoelastic waveguides. The samples are annealed for 30 minutes at various temperatures. The values in dB/cm denote propagation loss, while the values in parenthesis denote lateral mode size, the dashed line means that the waveguides are not evaluated at that temperature.

extinction is achieved at a bias voltage of 3V. As far as we know, this is the first planar FKE photoelastic waveguide modulator with such a low driving voltage. The fiber-coupled insertion loss is -12dB for the TE mode and is -13 dB for the TM mode. It should be noted that during the implantation, the WNi stressor experiences a high temperature 510°C annealing for about 20 min.; it experiences further annealing at 380°C during the formation of the backside AuGe ohmic contact. The lateral light confinement of the modulator is still very good after all these high temperature processing steps. This indicates excellent thermal stability of the WNi stressor.

Optical transmission vs. bias voltage of QCSE photoelastic waveguide modulator is obtained using a 1.301 µm laser diode. For a 1100 µm long device, 10 dB extinction ratio is obtained at about 16 V. Further optimization is underway.

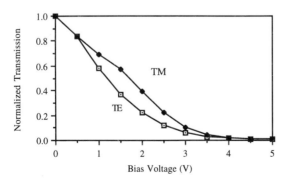

Figure 4. Optical transmission vs. Bias voltage of planar FKE waveguide modulator

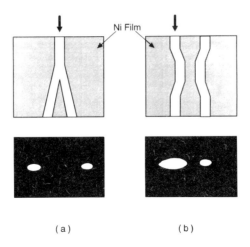

Figure 5. Schematic diagrams of (a) Y-junction splitter, and (b) directional coupler using Ni-window-stripe waveguides. The arrows point to the input facets. The displays beneath the diagrams are the respective optical output image.

3.3. Beam splitters and couplers

Preliminary planar photoelastic optical beam splitter and a coupler are fabricated with unannealed Ni-stripe InP/InGaAsP/InP waveguide. A typical near field of the optical output from the beam splitter as well as the coupler and the schematics of the metallization are shown in Figures 5(a) and (b) respectively. The width of the stripe opening is 2.5 μm. The angle of the Y junction splitter is 2°.

4. Summary

Planar, low loss InP-based photoelastic waveguides have been fabricated with two schemes developed for GaAs-based material systems. Although the post-annealing properties of the InP-based Ni photoelastic waveguides are different from those of GaAs-based Ni photoelastic waveguides due to different metal-semiconductor interfacial reactions, the Ni/InP/InGaAsP/InP waveguides have good thermal stability and low propagation losses. InP-based W waveguides have low losses and similar good thermal stability to that of GaAs-based W waveguides. An efficient (10 dB extinction at 3V for a 830 μm long modulator) planar FKE waveguide modulator at 1.3 μm has been demonstrated. Co-sputtered WNi film has been found to be ideal in serving both as an effective stressor with good thermal stability and as an electrode. A QCSE planar photoelastic waveguide modulator has been demonstrated for the first time. Planar photoelastic optical beam splitter and coupler have also been demonstrated in this material system.

Acknowledgment

The financial support from the National Science Foundation and from the Advanced Research Project Agency via the Office of Naval Research is appreciated.

Reference

[1] Yamamoto Y, Kamiya T and Yanai H 1975 *Applied Optics* **14** 322
[2] Campbell J C, Blum F A, Shaw D W and Lawley K L 1975 *Appl. Phys. Lett.* **27** 202
[3] Kirkby P A, Selway P R and Westbrook L D 1979 *J. Appl. Phys.* **50** 4567-79
[4] Liu Q Z, Deng F, Yu L S, Guan Z F, Pappert S A, Yu P K L, Lau S S, Redwing J, and Keuch T F to be submitted to *J. Appl. Phys.*
[5] Jiang X S, Clawson A R, Yu P K L 1992 *Journal of Crystal Growth* **124** 47-52
[6] Pappert S A, Xia W, Jiang X S, Guan Z F, Zhu B, Liu Q Z, Yu L S, Clawson A R, Yu P K L and Lau S S 1994 *J. Appl. Phys.* **75** 4532-61

A Planar Self-Aligned GaAlAs/GaAs HBT Technology Achieved by CBE Selective Base and Collector Contacts Regrowth

P.Launay, R.Driad, F.Alexandre, Ph.Legay, AM.Duchenois

FRANCE TELECOM, CNET-PAB, Laboratoire de Bagneux
196, Avenue Henri Ravera, BP 92225 BAGNEUX CEDEX, FRANCE

Abstract: A new planar and self-aligned GaAlAs/GaAs HBT technology based on a two selective Chemical Beam Epitaxy regrowth process is described. A C-doped ($> 10^{20}$ cm^{-3}) GaAs is selectively regrown on the extrinsic base layer providing both self-alignment and base resistance reduction. A Si-doped ($> 10^{19}$ cm^{-3}) GaAs/GaInAs bilayer is selectively regrown on the sub-collector layer. The final device is planar. Tungsten is used for both the base and collector ohmic contacts with a low resistivity (10^{-6} Ωcm^2). HBTs exhibit 40 GHz Ft and 30 GHz Fmax. The base regrowth process has been applied to the fabrication of multiplexers 2:1 and laser drivers, operating at 10 Gb/s.

1. Introduction

Self-alignment and device size reduction are the two main factors leading to very high values of Ft and Fmax. Much attention has been given to these points during the last decade leading to impressive performances [1]. In order to further increase HBTs high frequency performances, a selective base contact regrowth process has recently been proposed [2]. This process allows the fabrication of HBTs with thin intrinsic base layer for low base transit time and thick highly doped extrinsic base layer to reduce the base resistance.

Moreover, a planar HBT technology is also required to increase the interconnection reliability and integration scale, which can be obtained through selective collector contact deposition [3]. To obtained HBTs with a planar structure and high frequency performances, we have developed a new self-aligned HBT technology with two Selective Chemical Beam Epitaxy steps for the extrinsic base and collector contacts respectively. A highly C-doped GaAs regrown layer on the extrinsic base layer insures the emitter-base self-alignment, allows a reduction of the base access resistance and brings the opportunity to use tungsten as a low resistivity base ohmic contact. A Si-doped GaInAs/GaAs bilayer, selectively regrown on the sub-collector layer, also allows the use of tungsten as the collector ohmic contact. Emitter, base and collector ohmic contacts are on top of the device providing a full planarisation.

This paper describes this planar technology; DC and microwave performances are also discussed. Finally we report for the first time the demonstration of digital circuits (laser driver and multiplexer) achieved with the selective base regrowth process.

2. Planar HBT Technology

The HBT's basic epitaxial structure is grown by MOCVD and consists in Si-doped GaAs sub-collector ($2\ 10^{18}$ cm^{-3}, 500 nm) and collector ($2\ 10^{16}$ cm^{-3}, 450 nm) layers, a C-doped ($3\ 10^{19}$ cm^{-3}, 80 nm) base layer, a Si-doped ($5\ 10^{17}$ cm^{-3}, 250 nm) GaAlAs emitter layer, and finally a Si-doped GaAs ($2\ 10^{18}$ cm^{-3}, 150 nm) emitter contact layer.

The selective CBE extrinsic base regrowth proceeds as follows (Figure 1 A). First a thin Si_3N_4 film (100 nm) is deposited by PECVD on the wafer. Then the emitter mesa is patterned by contact printing followed by etching of the Si_3N_4 mask and etching of the emitter contact and emitter layers down to the extrinsic base layer. Finally Si_3N_4 spacers are fabricated on the mesa edges. So the emitter mesa is encapsulated by Si_3N_4 which insures an electrical isolation between the emitter layers and the base regrown layer (spacers) and acts as a mask for the selective growth process.

Prior to the selective regrowth process a specific surface preparation is made to insure the electrical conductivity between the base layer and the regrown layer. This preparation consists in a wet chemical etching of the base layer (10 nm) followed by a UV assisted ozone treatment which eliminates all surface contaminants and forms an oxide protecting the surface. This oxide is eliminated in the CBE system just before the regrowth process itself.Trimethylgallium and arsine are used to grow the heavily C-doped ($>10^{20}$ cm^{-3}) GaAs layer on the extrinsic base layer of the device. The growth selectivity is achieved for a growth temperature of 530 °C.

The selective CBE sub-collector regrowth is then achieved in a similar way (Figure 1 B). This regrown layer consists in a Si-doped ($5\ 10^{18}$ cm^{-3}) GaAs first layer with a Si-doped ($>10^{19}$ cm^{-3}) GaInAs thin layer on top.

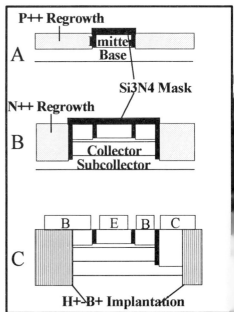

Figure 1: Planar HBT cross-section

Triethylgallium and arsine are used to grow the GaAs layer, trimethyl indium is additionally used to grow the GaInAs layer, solid Si is used as the dopant source. The growth selectivity is achieved for a growth temperature of 550 °C.

Deep boron and hydrogen implantation (2 µm) provides the device isolation. Tungsten is deposited for both the base and collector ohmic contacts, AuGeNi/Ag/Au is used for the emitter ohmic contact. When the process is completed the final device is planar, with all the ohmic contact on top of the device (Figure 1C).

3. Results and discussion

3.1. DC electrical characterisation

In this section a comparison will be made between four HBT processes with the same basic epitaxial structure: A standard double mesa one without epitaxial regrowth (Technology A), a second one with only the base contact regrowth (Technology B), a third one with only the collector contact regrowth (Technology C), and finally the one with both base and collector contacts regrowth (Technology D).

With the standard technology A, the specific contact resistances extracted from TLM measurements are about 10^{-6} Ωcm^{-2} for the three emitter, base (Mn/Au) and collector (AuGeNi/Ag/Au) contacts. With the N+ and P+ regrowth processes the base (W) and collector (W) specific contact resistances are improved to a value of 10^{-7} Ωcm^{-2}.

As we have mentioned elsewhere [4] the thermal treatment associated with the epitaxial regrowth induces a DC current gain degradation which is related to a decrease of the minority carrier lifetime in the base. We have plotted on figure 2 this DC current gain degradation (1-Hfe/Hfeo) as a function of the emitter-base junction area for the different technologies. Hfe stands for the DC current gain measured for a current density Jc of 10^3 Acm^{-2}, for the technologies with epitaxial regrowth (B, C, D), while Hfeo is the current gain without epitaxial regrowth (A).

As expected, the current gain degradation is more important when the regrowth temperature is increased. This degradation never exceeds 35 % for the P+ regrowth process which is achieved at 530 °C, and is lower than 50 % for the N+ regrowth process achieved at 550 °C.
The degradation is more severe, between 50 % and 80 %, when both base and collector regrowth processes are involved where two thermal treatments are cumulated.

This current gain degradation also increases when the emitter-base junction area is increased.

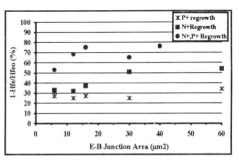

Figure 2: DC Current Gain Reduction

This can be particularly seen when the degradation ratio is high (Technology C, D), and can be correlated with the recombination current in the bulk of the base layer induced by recombination centers generated during the regrowth process [5]. Nevertheless DC current gain of 30 are obtained for the larger HBTs (2x20 µm^2 and 12x20 µm^2 for E-B and C-B junction area respectively) with the two regrowth processes. Further improvement to reduce the thermal budget, in particular the cleaning treatment (650 °C) used to eliminate the oxide prior to the regrowth process, is now under development. All the other characteristics are the same for all technologies. The emitter-base and collector-base reverse currents are around 10^{-8} Acm^{-2}, the emitter-base and collector-base ideality factors are 1.9 and 1.4 respectively.

On figure 3 we have reported the Gummel plot of an HBT with an emitter-base junction area of 40 µm^2 processed with both base and collector contacts regrowth. The maximum DC current gain is about 30. The corresponding Ic (Vce) curves are reported on figure 4.

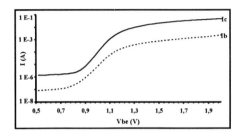

Figure 3: Gummel Plot (technology D)　　　**Figure 4**: Ic(Vce) curves (technology D)

3.2. Microwave characterisation

S-parameters were measured on wafer over the frequency range of 1 to 40 GHz using Cascade microwave probes and a Wiltron 360 network analyser. For each technology, we have reported on table 1 the microwave performances for the same HBT design which are still limited by a large collector-base junction area (12 x 20 µm^2). The cut-off frequencies are about the same value, except in the N+ regrowth process case where a high emitter resistance induces a decrease of Ft. As it can be noticed the base regrown process (technology B and D) induces a drastic base resistance decrease, and as a consequence the maximum oscillation frequency is increased. The base resistance decrease is obtained only when the base layer is partially etched off (10 nm) before the regrowth process. Otherwise no improvement in Rb and Fmax is observed. In this case, with the same Ft value (44 GHz), Rb is 49 Ω and Fmax 25 GHz which are about the same values than with a standard technology.

Technology	Re	Rb	Ft	Fmax
Standard (A)	9	55	42	21
P+ Regrowth (B)	11	11	38	29
N+ Regrowth (C)	24	64	30	15
N+ et P+ Regrowth (D)	10	18	44	31

Table 1: HBT microwave performances (E-B junction area: 60 μm^2, Vce=2V, Jc=50 000 Acm^{-2})

3.3. Digital circuits results

Digital circuits have been fabricated for the first time with the base regrown process. A laser driver, including an input buffer for impedance matching and a transimpedance wide band amplifier for input signal regeneration, operates at 10 Gb/s (see output eye diagram on figure 5), and multiplexers 2:1 which operate at 9 Gb/s. A high fabrication yield (90%) has been achieved for all these circuits, which demonstrates the potentiality of this new technology.

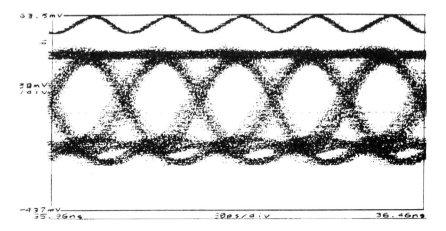

Figure 5: Laser Driver eye diagram (10 Gb/s)

4. Conclusion

In summary, we have reported the first successful demonstration of digital circuits, a laser driver and a multiplexer 2:1 which operate at 10 Gb/s, fabricated with a planar self-aligned GaAlAs/GaAs HBT technology based on selective CBE regrowth of the extrinsic base and sub-collector layers.

HBT's DC and microwave performances achieved with this new technology have been discussed. The results were compared to those obtained with a standard technology: an improvement in Fmax with the reduction of the base resistance obtained through the P+ regrowth process was shown. HBTs with an emitter-base junction area of 60 μm^2 exhibit Ft and Fmax of 44 GHz and 31 GHz respectively. These early results are very promising and demonstrate the potential of this planar technology. Improvement in the epitaxial process to reduce the current gain degradation and a new HBT design to reduce the base-collector junction area, are now underway to fabricate digital circuits and HBT with higher performances.

Acknowledgements

The authors would like to thank C.Besombes, F.Héliot, L. Bricard for wafer processing, JP. Medus and E.Wawrzynkowski for electrical measurements, and A.Scavennec for fruitful discussions.

References

[1] T.Sugiyama, Y.Kuriyama, M.Asaka, N.Iizuka, T.Kobayashi, M.Obara, 1994, *Jpn. J. Phys*, **33**, 786-789.
[2] H.Shimawaki, Y.Amamiya, N.Furuhata, K.Honjo, June 1993, *51st Annual Dev. Res. Conf.*, Santa-Barbara.
[3] K.Mitani, H.Masuda, K.Mochizuki, C.Kusano, 1992, *IEEE Elect. Dev. Lett.*, 13, N° 4, p 209-210.
[4] D.Zerguine, F.Alexandre, P.Launay, P.Legay, JL.Benchimol, R.Driad, 1993, *20th International Symposium on Gallium Arsenide and Related Compounds,* proceedings.
[5] R.Driad, F.Alexandre, JL.Benchimol, P.Launay, 1994, *this conference*.

Experimental Investigation of Low Frequency Noise Properties of AlGaAs/GaAs and GaInP/GaAs Heterojunction Bipolar Transistors

Jin-Ho Shin, Joon-Woo Lee, Young-Seok Seo, Young-Sik Kim and Bumman Kim

Department of Electronic and Electrical Engineering,
Pohang University of Science and Technology,
Hyojadong San-31, Pohang, Kyung-pook, 790-330, Korea

ABSTRACT : The intrinsic low frequency noise characteristics of AlGaAs/GaAs and GaInP/GaAs HBTs have been studied. HBTs with large emitter size of $120 \times 120 \mu m^2$ have been fabricated on MOCVD-grown abrupt junction emitter materials without undoped spacer layer. The leakage current of GaInP/GaAs HBTs is a little lower than that of AlGaAs/GaAs HBTs. However, AlGaAs/ GaAs HBTs have 10 ~ 25 dB lower noise level than GaInP/GaAs HBTs. For GaInP/GaAs HBTs, the base current noise power spectral densities are proportional to $\sim \exp(\frac{V_{BE}}{V_T})$ at a low current level(Ic $\leq$ 1mA) and saturate at a high current level. This is due to the gain-creeping effect of HBT originating from the persistent band discontinuity. Thus, the dominant noise generation process occurs at the base side of hetero-interface, which is very noisy. But AlGaAs/GaAs HBT noise source is the recombination at the base region.

1. Introduction

HBTs have excellent microwave performances such as high power, high gain, high cut-off frequency and high linearity. These features are very desirable for many applications at rf and microwave bands. The low frequency(LF) noise of HBT is also reported to be very low[1], but relatively little study has been made. Recent experimental works showed that the fluctuation of surface recombination velocity at the extrinsic base of HBTs is the dominant 1/f noise source of small size HBTs. However, there have been large deviations (10 ~ 20 dB) in 1/f noise data among the surface-passivated HBTs[1, 2]. More recently, W.J.Ho et.al reported that small feature size GaInP/GaAs HBTs have 20-30 dB lower LF noise than the comparable AlGaAs/GaAs HBTs[3]. However, the extrinsic base surface passivations of the HBTs may be different and this may be related to the improper conclusion. GaInP/GaAs HBTs are expected to have lower emitter base(E-B) space charge region(SCR) recombination 1/f noise or burst noise, because it does not have any DX centers. But the recombination noise is a minor one among the LF noise sources. Domain Costa and James S. Harris's experiments showed that AlGaAs/GaAs HBTs with reduced Al mole fraction have a lower burst noise by 4-5 dB but with a similar overall LF noise characteristics[2]. This data suggest that a detailed study is necessary to evaluate the intrinsic noise properties of AlGaAs/GaAs and GaInP/GaAs HBTs without having side effects such as surface recombination.

2. Device Structure

We have investigated the intrinsic LF noise properties of HBTs using large size emitter devices($120 \times 120 \mu m^2$). In order to reduce the LF noise, we chose a low leakage current structure, that is abrupt E-B junction without undoped spacer layer. This spacer layer enhances a recombination current[4], and the abrupt E-B heterojunction launcher injects electrons into the base with a high kinetic energy and the electrons transverse the interface region with high trap density very fast, resulting in reduced recombination. Both materials were grown by MOCVD. The emitter layer($Al_{0.3}Ga_{0.7}As$ or $Ga_{0.5}In_{0.5}P$) is doped to $2 \times 10^{17} cm^{-3}$. The GaAs base is 700 Å thick and is carbon doped to $3 \times 10^{19} cm^{-3}$. Detailed HBT layer structures are shown in Table 1. Using the materials, we have fabricated two identical MESA type HBTs and have characterized. Fig.1 (a) and (b) show the measured Gummel plots of AlGaAs/GaAs and GaInP/GaAs HBTs, respectively. The base current ideality factors indicate that AlGaAs/GaAs HBT has more SCR recombination component than GaInP/GaAs HBT. Current gain of AlGaAs/GaAs HBT is typically 60 and that of GaInP/GaAs HBT is 55. For both HBTs, emitter resistance (r_e) and base series resistance (r_b) are $3 \sim 5$ Ω and $100 \sim 200$ Ω, respectively. The current level at unity current gain of GaInP/GaAs HBTs is less than 100 pA, while that of AlGaAs/GaAs HBTs, less than 1 nA.

3. Low Frequency Noise

The LF base voltage noise (S_{Vb}) and total device voltage noise at collector(S_{Vc}) were measured from 1 Hz to 25 kHz using a low noise amplifier and a spectrum analyzer. The HBTs are in common emitter mode. The collector current noise spectra(S_{Vc}/R_c^2) and base current noise spectra(S_{Vb}/r_π^2) are deduced from the data at the collector current of 11 mA and are depicted in Fig.2. Here, Rc is a collector bias resistance and r_π is an effective base input resistance; i.e, $r_\pi = r_{\pi i} + \beta \times r_e$, β and $r_{\pi i}$ are differential current gain and base input resistance of intrinsic transistor, respectively. The spectra exhibit two distinctive regions, i.e, a white noise above 1 kHz (shot noise and thermal noise from the large base bias resistor) and a colored 1/f noise below 1 kHz. Over all the current range we measured, AlGaAs/GaAs HBTs have $10 \sim 25$ dB lower noise than GaInP/GaAs HBTs.

Table 1. HBT Layer Structures

Layer	AlGaAs/GaAs HBT			GaInP/GaAs HBT		
	Thickness (Å)	Doping (cm^{-3})	Composition	Thickness (Å)	Doping (cm^{-3})	Composition
Cap	400	10^{19}	$In_{0.5}Ga_{0.5}As$	400	10^{19}	$In_{0.5}Ga_{0.5}As$
Cap	400	10^{19}	$In_yGa_{1-y}As$ (y=0→0.5)	400	10^{19}	$In_yGa_{1-y}As$ (y=0→0.5)
Subemitter1	1300	4×10^{18}	GaAs	1500	5×10^{18}	GaAs
Subemitter2	500	5×10^{17}	$Al_xGa_{1-x}As$ (x=0.3→0)	300	10^{18}	$Ga_{0.5}In_{0.5}P$
Emitter	500	2×10^{17}	$Al_{0.3}Ga_{0.7}As$	700	2×10^{17}	$Ga_{0.5}In_{0.5}P$
Base	700	3×10^{19}	GaAs	700	3×10^{19}	GaAs
Collector	11000	2×10^{16}	GaAs	10000	2×10^{16}	GaAs
Subcollector	14000	4×10^{18}	GaAs	10000	5×10^{18}	GaAs

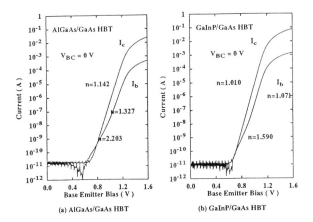

Fig. 1. Measured Gummel Plots

This feature can be qualitatively explained from the bias dependences of LF noises at f=10 Hz, shown in Fig.3. For AlGaAs/GaAs HBTs, $S_{Vc}/R_c^2 \sim I_C^{1.4} \sim \exp(\frac{V_{BE}}{V_T})^{1.2}$, and $S_{Vb}/r_\pi^2 \sim I_B^{1.6} \sim \exp(\frac{V_{BE}}{V_T})^{1.25}$, indicating that the dominant 1/f noise source is the recombination at the base and partly SCR recombination. For GaInP/GaAs HBTs, $S_{Vc}/R_c^2 \sim I_C^{1.2} \sim \exp(\frac{V_{BE}}{V_T})^{1.2}$ over all the measured current range but $S_{Vb}/r_\pi^2 \sim I_B \sim \exp(\frac{V_{BE}}{V_T})$ at a low bias current below 1 mA and S_{Vb}/r_π^2 saturates above 1mA. Since the recombination current in the band bending region of hetero-interface is proportional to $\exp(\frac{V_{BE}}{V_T})$ at a low current and saturates at a high current level, we postulate that the base noise in GaInP/GaAs HBT originates from the recombination at the potential well located at the neutral base side. The increase of collector current noise after saturation of base current noise is due to the emitter-to-collector current controlled by the collector potential. This anomalous noise phenomenon is related to the gain creeping effects of HBTs[5, 6].

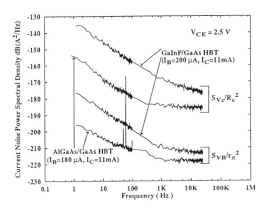

Fig. 2. Low Frequency Noise Spectra

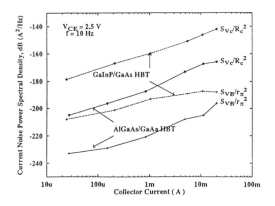

Fig. 3. Low Frequency Noise Spectra.vs. Collector Currents

4. Summary

In summary, our experimental results show that the intrinsic LF noise of AlGaAs/GaAs HBTs is lower by 10 ~ 25dB than that of GaInP/GaAs HBTs. The intrinsic noise of GaInP/GaAs HBT stems from the recombination at the potential well in the base side of hetero-interface and this recombination process is very noisy. On the contrary, AlGaAs/GaAs HBT noise source is the recombination at the neutral base.

References

[1] Hayama N., LeSage S.R., Madihian M. and Honjo K., 1988 IEEE MTT-S 679-682

[2] Costa D., and Harris J.S., 1992 IEEE Trans. on Electron Devices, 39 2383-2394

[3] Ho W.J., Chang M.F., Sailer A., Zampardi P., Deakin D., McDermott B. , Pierson R., Higgins J.A., and Waldrop J., 1993 IEEE Electron Device Lett ., 14 572-574

[4] Ito H., 1986 Japan. J. Appl. Phys., 25 1400-1404

[5] Day D.J., Jue S.C., Margittai A., and Houston P.A., 1989 IEEE Tans. Electron Devices, 36 1015-1019

[6] Jue S.C., Day D.J., Margittai A., and Svilans M., 1989 IEEE Trans . on Electron Devices, 36 1020-1025

Effect of carrier recombination at the emitter-base heterojunction on the performance of GaInP/GaAs and AlGaAs/GaAs heterojunction bipolar transistors

Z.H. Lu* and A. Majerfeld,
Department of Electrical and Computer Engineering, University of Colorado, Boulder,
CO 80309-0425. Tel/Fax (303 492-7164/2758)
* Presently at National Renewable Energy Laboratory, 1617 Cole Blvd., Golden, CO 80401

L.W. Yang, Martin Marietta Laboratories-Syracuse, Electronics Park, Syracuse, NY 13221

P.D. Wright, Presently at Motorola Inc., Phoenix Corporate Research Laboratories, Tempe, AZ 85284

ABSTRACT. In this work, carrier recombination at the emitter-base heterojunction is found to be the dominant mechanism to explain the distinctive DC characteristics of the GaInP/GaAs and AlGaAs/GaAs heterojunction bipolar transistors. An optical technique based on photoluminescence spectroscopy provides a non-destructive measure of the recombination strength and, thus, the carrier injection efficiency.

1. Introduction

It is well known [1] that the emitter-base heterojunction (E-B HJ) plays an essential role for heterojunction bipolar transistors (HBTs). However, the detailed mechanisms, through which the device performance is influenced by the E-B HJ, are not well understood. Effective characterization techniques that can be applied for wafer screening prior to processing are still lacking. In this work, the effect of interfacial recombination at the E-B HJ on the DC characteristics is studied both theoretically and experimentally. It is found that the DC current gain β is critically affected by the properties of the E-B HJ interface for compositionally graded AlGaAs/GaAs HBTs, whereas the dependence is less for abrupt GaInP/GaAs HBTs. A photovoltaic effect observed in photoluminescence (PL) spectroscopy was analyzed and applied to explain the device characteristics.

2. Experimental

The GaInP/GaAs and AlGaAs/GaAs HBT wafers analyzed in this work were grown by the MOVPE process with the C doped base of 10^{19}-10^{20} cm^{-3}. Microwave devices fabricated from the AlGaAs/GaAs wafers exhibit a cut off frequency, f_t =52 GHz, and a maximum frequency of oscillation, f_{max} = 86 GHz [2]. For the GaInP/GaAs wafers, emitter doping as low as 5x10^{16} cm^{-3} was used to reduce the junction capacitance while maintaining sufficient carrier injection because of the low conduction band offset. As a result, f_{max} values over 100 GHz were achieved [3]. Large area GaInP/GaAs HBTs with reduced emitter doping levels consistently show improved DC current gain, β, as compared to the best AlGaAs/GaAs structures. Optical measurements on these wafers were performed using a standard PL setup [4] in the temperature range of 10-300 K. The optical excitation power was varied from

0.01-5 W/cm². The n^+-GaAs contact layer was chemically removed to reveal optical emissions from the layers underneath it.

3. Results and Discussion

To understand the significant improvements in GaInP/GaAs structures, device characteristics were calculated using a modified SEDAN simulator [5] and known material parameters. The bandgap narrowing effect (BGNE) in the heavily doped GaAs base was included in the simulations using the experimental dependence [4]: $\Delta E_g = 2.4 \times 10^{-8} p^{1/3}$. The importance of the BGNE can be seen in the following results: for an AlGaAs/GaAs HBT with base doping of 10^{20} cm^{-3}, the experimental β is 20 while the simulated β with and without the BGNE is ~ 20 and 1.5, respectively, for a given collector current (800 A/cm²). The conduction band discontinuity for GaInP/GaAs was taken to be $\Delta E_c/\Delta E_g = 0.27$. A compositionally graded E-B HJ was assumed for the AlGaAs/GaAs structure while an abrupt HJ was assumed for the GaInP/GaAs structure, as shown in Fig. 1a. The choice of interfacial structure was mainly based on the difference in conduction band discontinuity in the two material systems and other experimental considerations. To understand the role of traps in the E-B HJ region, a thin layer with a thickness of t_r and short lifetimes contiguous to the base region was assumed. The lifetime in the Shockley-Reed-Hall (SRH) recombination process is related to the trap density N_t and trap cross-section σ by: $\tau = 1/N_t \sigma$. The calculated β as a function of t_r at several current density levels is shown in Fig. 1b. It is seen that for the AlGaAs structure β is reduced drastically as t_r increases up to about 50 Å but becomes near constant thereafter. The interface layer is likely related to the necessity of a growth interruption step between the heavily doped base and the AlGaAs emitter, during which oxygen and other impurity species are adsorbed on the surface and become efficient traps when Al is subsequently introduced. The sensitive dependence of β on the interface quality may explain the difficulty in achieving current gain reproducibility for high performance AlGaAs/GaAs HBTs.

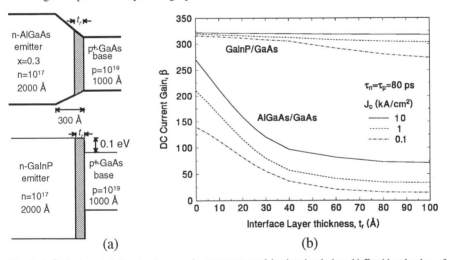

Fig. 1. a) Critical part of the structures and parameters used in the simulation. b) Rapid reduction of current gain β in AlGaAs/GaAs HBTs as the E-B interface layer thickness increases; a lifetime of 80 ps is assumed for both carriers within the interface layer to simulate the presence of a high density of traps. This effect is much reduced for abrupt GaInP/GaAs HBTs.

Our experimental data from GaInP/GaAs HBTs show high current gain even at very low current injection levels, which is consistent with most of the published results on similar structures. This behavior is generally attributed to the very low interface recombination velocity of this heterojunction. The calculation displayed in Fig. 1b shows that the current gain for GaInP/GaAs HBTs is much less sensitive to interfacial recombination than their AlGaAs/GaAs counterparts. In general, an abrupt HJ structure has intrinsically lower interfacial recombination than a graded one, which is explained as follows. The SRH recombination is proportional to the intrinsic carrier density n_i in the space charge region, which occurs mainly on the emitter side and is pushed towards the base under forward bias; compositional grading near this region leads to a larger n_i and, thus, more recombination. In an abrupt structure, on the contrary, n_i is small over the entire emitter region. In practice, an abrupt AlGaAs/GaAs junction is not favorable due to the significant reduction of the collector current associated with a large ΔE_c [6]. The small ΔE_c of the GaInP/GaAs heterojunction results in much of the advantages of this system for HBT applications.

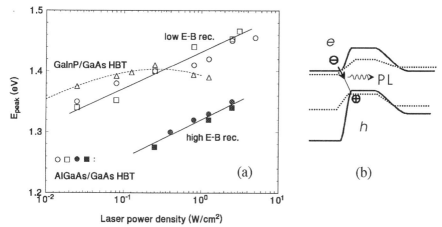

Fig. 2. a) Photovoltage as a function of laser power for four AlGaAs/GaAs and GaInP/GaAs HBTs; for clarity, results for only one GaInP HBT (Δ) are included. For the high quality AlGaAs HBTs and GaInP HBTs, η_{EB} = 1.07-1.1; for the two AlGaAs devices with high recombination currents, η_{EB} = 1.86-1.94. b) Optical transitions in the E-B HJ.

The sensitive dependence of current gain on the interface quality imposes a stringent requirement on a non-destructive technique to monitor the quality of the E-B HJ. Our theoretical analysis shows that under optical illumination, the recombination centers at the E-B junction also reduce the photovoltage across the E-B junction in a similar form to the observed reduction of carrier injection efficiency by a forward bias. The photovoltage is measured from the energy shift of a previously observed prominent radiative transition peak below the band edge of GaAs layers in PL spectroscopy [7]. In Fig. 2, the PL peak energy of this transition, which is essentially the photovoltage across the E-B junction, is plotted as a function of laser power. For AlGaAs/GaAs HBTs, the photovoltage-laser power dependence resembles the V-I curve of a diode. The quality of the E-B junction is assessed from the conventional I-V characteristics of the diode, where a near-unity ideality factor (η_{EB}) indicates low recombination and η_{EB} ~ 2 signals the presence of a large density of recombination centers. For the high quality AlGaAs/GaAs and GaInP/GaAs HBTs in the figure, η_{EB} =1.07-1.1; whereas for the two AlGaAs devices with high recombination currents, η_{EB} =1.86-1.94. It can be seen from the figure that for devices with a high density of recombination centers, the photovoltage is significantly reduced. This dependence can be used for a quantitative measure of the injection efficiency for the E-B junction. For GaInP/GaAs HBTs, all the devices measured in this work show large photovoltages, which is associated with low recombination currents in these devices. The photovoltage saturation of the GaInP/GaAs devices is in contrast with the variation observed for the AlGaAs/GaAs structure. The saturation and a slight reduction of the photovoltage result from the finite conduction band discontinuity, which in an abrupt HJ provides a potential barrier for electrons at large current densities.

4. Summary

Based on the experimental observation and numerical simulation of high performance GaInP/GaAs and AlGaAs/GaAs HBT structures, we conclude that the DC current gain of AlGaAs/GaAs HBTs is highly sensitive to the presence of traps at the graded E-B heterointerface; in contrast, the abrupt junction GaInP/GaAs HBTs are quite robust regarding interface quality if the lattice matching condition is carefully controlled. By comparing the analysis of the electrical characteristics of these devices and the PL spectroscopy study on the same wafers, we demonstrated that the photovoltage determined from a PL spectrum can be used as a sensitive indicator of the recombination current in the E-B junction. The difference between a graded and an abrupt HJ is clearly observed in the PL results.

5. Acknowledgments

We thank D.A. Suda, D.W. Winston and R. E. Hayes for making available to us the SEDAN simulator. We also wish to acknowledge the support of this work by the Colorado Advanced Technology Institute through the Colorado Advanced Materials Institute.

References

1. P.M. Asbeck, M-C. F. Chang, J.A. Higgins, N.H. Sheng, G.J. Sullivan, and K-C. Wang, *IEEE Trans. Electron Devices,* vol. ED-36, 2032 (1989).
2. L.W. Yang, P.D. Wright, P.R. Brusenback, S.K. Ko, A. Kaleta, Z.H. Lu, and A. Majerfeld. Electron. Lett. 27, 1145 (1991).
3. L.W. Yang, S.T. Fu, B.F. Clark, R.S. Brozovich, H.H. Lin, S.M. Liu, P.C. Chao, F. Ren, C.R. Abernathy, S.J. Pearton, J.R. Lothian, P.W. Wisk, and T.R. Fullowan, Proc. IEEE Cornell Conf. on Adv. Concepts in High Speed Semicon. Dev. & Circuits, p 43, 1993.
4. Z.H. Lu, M.C. Hanna and A. Majerfeld, *Appl. Phys. Lett.* vol. 64, 88, 1994.
5. D. Suda, Ph.D. Thesis, University of Colorado, Boulder, 1992.
6. H. Kroemer, *Proc. IEEE* 70, 13 (1982).
7. Z.H. Lu, M.C. Hanna, E.G. Oh, A. Majerfeld, P.D. Wright, and L.W. Yang, *Inst. Phys. Conf. Ser.* No. **120**, 335 (1991).

Energy Transport Modeling of HBTs Considering Composition-, Doping- and Energy-Dependence of Transport Parameters

A. Nakatani and K. Horio

Faculty of Systems Engineering, Shibaura Institute of Technology, 307 Fukasaku, Omiya 330, JAPAN

Abstract. A method of evaluating energy-transport parameters for devices with position-dependent band structure is proposed and successfully applied to the energy transport simulation of graded AlGaAs/GaAs HBTs. For several representative Al composition x and doping densities, parameters such as average electron mobility, energy relaxation time and upper valley fraction are evaluated (by a Monte Carlo method) as a function of electron energy. For the other x, these are determined by linear extraporation. The problem of this extraporation method is also addressed. Calculated cutoff frequency characteristics and electron velocity profiles are compared with those by using more simplified approaches, demonstrating the importance of using adequate parameters.

1. Introduction

AlGaAs/GaAs heterojunction bipolar transistors (HBTs) have received great interest for applications to high-speed and high-frequency devices. Since nonequilibrium carrier transport becomes important in the HBTs, carrier energy must be considered in the modeling of them. For this purpose, an energy transport model that uses hydrodynamic equations derived from the Boltzmann equation is attractive and has been applied to AlGaAs/GaAs HBTs (Horio and Yanai 1987, Horio et al. 1989, Azoff 1989, Tomizawa et al. 1990, Teeter et al. 1993) as well as GaAs MESFETs (Curtice and Yun 1981, Cook and Frey 1982 etc.) and Si MOSFETs (Fukuma and Uebbing 1984 etc.).

In the HBTs, material parameters should depend on material composition as well as carrier energy and doping densities. However, up to now, methods of giving parameters in the transport equations were too rude. In earlier works, a constant value of energy relaxation time was assumed regardless of electron energy. In another work (Azoff 1989), energy-dependence of transport parameters is included, but parameters for GaAs at a given doping density were also assumed for all x in $Al_xGa_{1-x}As$.

So, we have tried to give more reasonable transport parameters in the energy-transport simulation of AlGaAs/GaAs HBTs (Horio and Nakatani 1994). In this work, within an equivalent one-valley model where an upper and a lower valley are considered, we present a method of giving Al composition-, doping- and energy-dependent transport parameters. Some device characteristics simulated by using this method are also described briefly, and compared with those by the more simplified approaches.

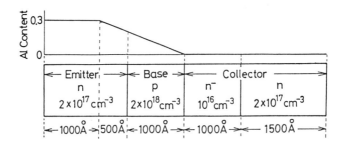

Figure 1: Test HBT structure analyzed in this study.

2. Device Structure and Basic Equations

We consider an npn^-n AlGaAs/GaAs HBT shown in Figure 1 where Al composition changes from 0.3 to 0 in the emitter and base regions. So, this is a graded band-gap base HBT. Of course, doping densities are different in respective regions. So, in general, transport parameters should be given as functions of Al composition, doping density and carrier energy.

Next we briefly describe the derivation of electron transport equations. Assuming paravolic energy bands, three conservation equations, that is, a particle conservation equation, a momentum conservation equation and an energy conservation equation are obtained by taking moments of the Boltzmann equation (Blotekjaer 1970). To make these equations tractable and applicable to the HBTs, the following assumptions are made (Horio et al. 1989). Firstly, we adopt a relaxation time approximation for collision terms. Secondly, we adopt an equivalent one-valley model. If we assume that electron drift energy is negligible small as compared to its thermal energy, average electron energy w_n can be written as

$$w_n = \frac{3}{2}kT_n + F_U \Delta_{LU} \qquad (1)$$

where T_n is electron temperature, F_U is upper valley fraction and Δ_{LU} is energy difference between upper and lower valleys. To treat HBTs, some factors are also considered. Carrier recombination is treated by SRH statistics, and an effective field acting on electrons which arises due to the position dependence of band structure is included. In the electron transport equations, parameters that should be given are electron mobility μ_n, energy relaxation time τ_w and upper valley fraction F_U.

For holes, we use drift-diffusion type equations. In addition to the transport equations, we include the Poisson's equation, which completes the basic equations for device simulation.

3. Transport Parameters

Here we describe methods of giving transport parameters such as electron mobility μ_n, energy relaxation time τ_w and upper valley fraction F_U. To do this, usually, a homogeneous bulk is assumed (Cook 1983). Then, the following two equations are obtained from the previous electron transport equations.

$$v_n = -\mu_n E \tag{2}$$

$$w_n = w_0 - q\tau_w v_n E \tag{3}$$

where E is electric field, v_n is average electron velocity and w_0 is an equilibrium value of w_n. Based on these two equations, one can obtain the values of μ_n and τ_w as describe below.

First, we describe conventional approaches to determine the transport parameters. After that, we present an approach made in this study. A simple way done first (Horio and Yanai 1987) is to assume that the energy relaxation time is independent of electron energy (and Al composition) and takes a constant value of 1 psec. Further, if an analytical expression of $\mu_n(E)$ is given, the relation between E and w_n can be obtained by combining Eqs.(2) and (3). Thus, μ_n as a function of w_n is obtained. As to upper valley fraction F_U, it is given analytically by assuming maxwellian distribution. This approach is simple, but may be a little too rude. Another approach is to use a Monte Carlo method for evaluating energy-dependence of transport parameters. By the Monte Carlo method, v_n, w_n and F_U are obtained as a function of electric field E. So, by using Eqs.(2) and (3), energy-dependent μ_n, τ_w and F_U are obtained. However, in a published work (Azoff 1989), data for GaAs at a given doping density are used for all x in $Al_xGa_{1-x}As$ and for all doping densities. Therefore, this approach also seems to be too rude.

So, in this work, we have tried to give transport parameters as functions of Al composition x, electron energy w_n and doping density N. To do this, we first evaluate μ_n, τ_w and F_U for $x = 0, 0.05, 0.1, 0.15, 0.2, 0.25$ and 0.3 by a Monte Carlo method. Doping densities are also varied. Once these parameters are available as fit curves or tables for a given doping density, parameters for any x between 0 and 0.3 can be obtained (as a function of w_n) by a linear extraporation method. That is, if x lies between x_1 and x_2, the parameter f is given by the following equation.

$$f(w_n) = f_1 + \frac{x - x_1}{x_2 - x_1}(f_2 - f_1) \tag{4}$$

where f_1 and f_2 are corresponding values of f for $x = x_1$ and x_2, respectively.

Next we show some examples of transport parameters estimated by a Monte Carlo method. Figure 2 shows electron mobility μ_n, energy relaxation time τ_w and upper valley fraction F_U as a parameter of Al composition x. The doping density is 2×10^{18} cm^{-3}. Here, we consider L-valley as upper valley and use L-valley's parameters for Δ_{LU} and upper-valley effective mass. We also treat a case somewhat considering X-valley's contribution artificially, but x and energy dependences of estimated parameters are essentially similar to those shown in Figure 2. As seen in Figure 2(b), τ_w depends on electron energy and decreases as x increases. So, to use a constant τ_w regardless of the values of x and electron energy seems to be an inadequate approach. As described before, once the figures like Figure 2 are obtained for a given doping density, the transport parameters for any x at any electron energy are obtained by the linear extraporation method. Of course, when the doping density becomes different, another figure is required. Furthermore, the parameters must be evaluated in every mesh point. However, this approach is simple in itself and can be easily applied to other devices.

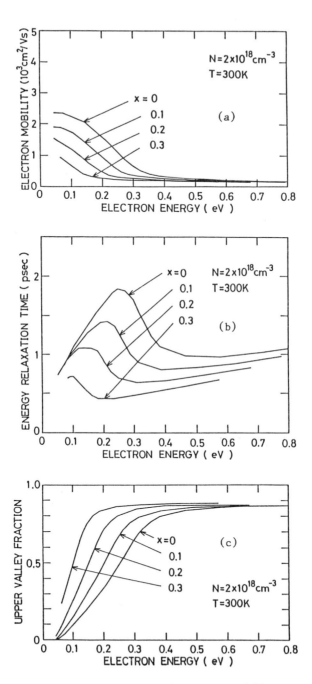

Figure 2: (a) electron mobility μ_n, (b) energy relaxation time τ_w, and (c) upper valley fraction F_U versus electron energy w_n curves as a parameter of x in $Al_xGa_{1-x}As$, calculated by a Monte Carlo method. The energy difference between upper and lower valleys Δ_{LU} and upper-valley effective mass m_U are set to $0.284 - 0.605x$ (eV) and $0.23m_0$, respectively.

It should be mentioned that there is one point to be taken care in the linear extraporation method when we determine electron energy at thermal equilibrium: w_{n0} for every x. Careless determination by $w_{n0} = w_{n1} + (x - x_1)(w_{n02} - w_{n01})/(x_2 - x_1)$ leads to a great fluctuation of electron temperature T_{n0} at zero bias. Instead, by combining an equation: $w_{n0} = (3/2)kT_{n0} + F_{U0}\Delta_{LU}$ (cf. Eq.(1)) and a relation between w_{n0} and F_{U0} (determined by linear extraporation: cf. Eq.(4)), we can obtain constant T_{n0} throughout the device.

4. Calculated Device Characteristics

Here we present some device characteristics calculated by the present model, and compare them with those obtained by other simplified approaches.

Figure 3 shows cutoff frequency f_T versus collector current density I_C curves of the HBT shown in Figure 1. The solid line is calculated by the present model. The dotted-and-dashed line is calculated by using GaAs parameters for all x in $Al_xGa_{1-x}As$, where the doping-density dependence of transport parameters is included. The dashed line is for constant τ_w method described before. For comparison, a case by drift-diffusion model is also shown by the dotted line, where the estimated f_T is much lower than those by considering energy transport effects. In the case using only GaAs parameters, f_T is estimated rather higher than that by the present model where Al-composition dependence of transport parameters is considered. On the other hand, in the case by constant τ_w method, f_T is estimated lower.

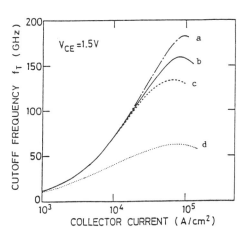
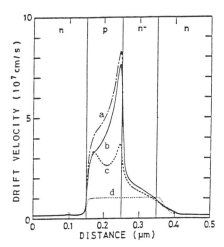

Figure 3: f_T versus I_C curves calculated by energy transport models (a, b, c) and by drift-diffusion model (d).
    a : Parameters for GaAs ($x=0$)
    b : x-dependent μ_n, τ_w and F_U
    c : Constant τ_w ($= 1$ psec)

Figure 4: Average electron velocity versus distance curves calculated by energy transport models (a, b, c) and by drift-diffusion model (d), corresponding to Figure 3. $I_C = 5 \times 10^4$ A/cm^2.

Figure 4 shows comparison of electron velocity profiles calculated by various methods. Collector current density is 5×10^4 A/cm^2. In cases considering energy transport effects, so-called velocity overshoot is observed in the base and n^--collector regions. It is clearly seen that in the case using GaAs parameters for the AlGaAs base, velocity overshoot in the base region is overestimated. Therefore, f_T is estimated rather higher as described before. On the other hand, in the case by constant τ_w method, the velocity overshoot is underestimated in the base region, resulting in the lower f_T. It should also be noted that in the case by constant τ_w method, the velocity profile deviates much from those obtained by using Monte Carlo parameters. So, this approach may be inappropriate for evaluating device performance.

From these we can say that to give adequate transport parameters is important for eatimating the device characteristics of AlGaAs/GaAs HBTs.

5. Conclusion

We have developed an energy-transport simulation method for AlGaAs/GaAs HBTs, which uses Al composition-, doping-, and energy-dependent transport parameters estimated by a Monte Carlo method. For representative Al composition x and doping densities, parameters such as average electron mobility, energy relaxation time and upper valley fraction are evaluated as a function of electron energy. For the other x, these are determined by linear extraporation. Calculated cutoff frequency characteristics and electron velocity profiles are compared with those by using more simplified approaches, demonstrating the importance of using adequate transport parameters. The presented approach can also be applied to other devices which have position-dependent band structures.

Acknowledgment

This work was supported by the Murata Science Foundation and by the Project Research Fund at the Shibaura Institute of Technology.

References

Azoff E M 1989 *IEEE Trans. Electron Devices* **36** 609.
Blotekjaer K 1970 *IEEE Trans. Electron Devices* **ED-17** 38.
Cook R K and Frey J 1982 *IEEE Trans. Electron Devices* **ED-29** 970.
Cook R K 1983 *IEEE Trans. Electron Devices* **ED-30** 1103.
Curtice W R and Yun Y H 1981 *IEEE Trans. Electron Devices* **ED-28** 954.
Fukuma M and Uebbing R 1984 *IEDM Tech. Dig.* p621.
Horio K and Yanai H 1987 *Proceedings of NASECODE V* p231.
Horio K, Iwatsu Y and Yanai H 1989 *IEEE Trans. Electron Devices* **36** 617.
Horio K and Nakatani A 1994 *Abstracts of NASECODE X* p3.
Teeter D A, East J R, Mains R K and Haddad G I 1993 *IEEE Trans. Electron Devices* **40** 837.
Tomizawa K and Pavlidis D 1990 *IEEE Trans. Electron Devices* **37** 519.

Inst. Phys. Conf. Ser. No 141: Chapter 6
Paper presented at Int. Symp. Compound Semicond., San Diego, 18–22 September 1994
© 1995 IOP Publishing Ltd

Measurement of the electron ionization coefficient at low electric fields in Heterojunction Bipolar Transistors

C. Canali
Facoltà di Ingegneria, Università di Modena, Via Campi 213, 41100 Modena, Italy

S. Chandrasekhar
AT&T Bell Laboratories, Crawford Hill Lab, Holmdel, NJ 07733, U. S. A.

F. Capasso, R. A. Hamm and R. Malik
AT&T Bell Laboratories, Murray Hill, NJ 07974, U.S.A.

C. Forzan, A. Neviani, L. Vendrame and E. Zanoni
Dipartimento di Elettronica e Informatica, Università di Padova, Via Gradenigo 6a, 35131 Padova, Italy

New data for the electron impact-ionization coefficient α_n in GaAs and $In_{0.53}Ga_{0.47}As$ are derived from measurements of the multiplication coefficient $M-1$ in Heterojunction Bipolar Transistors (HBT). We show that the intrinsic limiting factors of the $M-1$ measurement at low electric fields are the Early effect and the uncertainty in the electric field value in the collector depletion region. By adopting HBT's characterized by lightly doped collectors and heavily doped base regions we were able to extend previously available data on the impact-ionization coefficient of GaAs and $In_{0.53}Ga_{0.47}As$ by two orders of magnitude, down to $\alpha_n = 1\ cm^{-1}$. Our data on $In_{0.53}Ga_{0.47}As$ also confirm the theoretical results on the weak field dependence of α_n at low electric fields.

1. Introduction

Measurements of impact-ionization multiplication factor $M-1$, where M is the avalanche gain, at high reverse base-collector biases have proven to be a useful tool in the characterization of the breakdown behavior and current instabilities occurring at high electric fields in both homojunction and heterojunction bipolar transistors [1-6]. The extension to lower electric fields of the $M-1$ measurement may provide a deeper insight into the physics of the impact-ionization mechanism, such as information on the energy thresholds and changes in the dependence of the ionization coefficients on the electric field related to the density of states of the semiconductor [7,8]. The existing data of ionization coefficients in semiconductors are generally obtained from gain measurements in avalanche photodiodes [8,9], in which the primary photocurrent is generated by an optical beam incident on the device. The major disadvantage of this experimental configuration is the uncertainty in the determination of the primary photocurrent initiating impact ionization (base-line for the avalanche gain) [8], which becomes a critical source of error in low electric field conditions. In bipolar transistors instead, the avalanche generated current is superimposed on the base current, which is smaller than the injected current initiating multiplication by a factor equal to the current gain β of the transistor [10,11]. As a result, it can be measured more accurately than the collector current, permitting a high sensitivity measurement of the avalanche gain at low electric fields. Recently the electron ionization coefficient α_n of $In_{0.53}Ga_{0.47}As$ at low electric fields was measured using the avalanche induced variation of the collector current in InP/InGaAs HBT's [12]. Note however that the lowest reported values of α_n were $\simeq 10^2$ cm^{-1}.

In this work, we discuss the factors limiting the $M-1$ measurement at low electric fields in lightly-doped collector AlGaAs/GaAs and InP/InGaAs HBT's. The electron ionization coefficient α_n is extracted down to values as low as 1 cm^{-1} from the measured multiplication factor. Our data on α_n in $In_{0.53}Ga_{0.47}As$ also extend those in [12] by two orders of magnitude in the low field domain, demonstrating an abrupt reduction of the electric field dependence of α_n at field values below $2 \cdot 10^5$ Vcm^{-1}, in agreement with the theoretical predictions in [7].

2. Samples and experimental technique description

The n-p-n transistors adopted are: (a) AlGaAs/GaAs HBT's on GaAs, grown by Molecular Beam Epitaxy (MBE), with collector doping $N_C \simeq 3 \cdot 10^{15}$ cm^{-3} and width $W_C = 0.5$ μm, base doping $N_B \simeq 2 \cdot 10^{19}$ cm^{-3}, base thickness $W_B = 0.12$ μm and emitter area $A_E = 7.9 \cdot 10^{-5}$ cm^2 [6]; (b) InP/$In_{0.53}Ga_{0.47}As$ HBT's on InP, grown by Metal Organic Molecular Beam Epitaxy (MOMBE), characterized by an average collector doping $N_C \simeq 5 \cdot 10^{15}$ cm^{-3}, collector width $W_C = 0.85$ μm, base doping $N_B \simeq 4 \cdot 10^{19}$ cm^{-3}, base thickness $W_B = 0.05$ μm, and emitter area $A_E = 2.5 \cdot 10^{-5}$ cm^2. Multiplication factor measurements were performed with the device in common base configuration [1,6], using a HP 4145B semiconductor parameter analyzer.

When a bipolar transistor is biased in the active region at sufficiently high values of V_{CB}, electron-hole pairs are generated in the base-collector (BC) space-charge region due to impact ionization. Electrons injected into the collector (primary current I'_C) are accelerated by the reverse biased base-collector (BC) junction electric field, gaining sufficient energy to create electron-hole pairs by impact ionization. The electrons and holes created by this process are collected at the collector and base contacts respectively. Secondary electrons contribute to the total collector current by an amount $\Delta I_C = I_C - I'_C$, while the generated holes contribute a negative term $\Delta I_B = -\Delta I_C$ to the base current. The behavior of I_B as a function of V_{CB} in both types of devices is reported in Fig. 1. It is clearly visible the sign inversion of I_B due to the collection of impact-ionization generated holes at the base contact at $V_{CB} \simeq 8.5$ V in the AlGaAs/GaAs HBT's, and at $V_{CB} \simeq 4.5$ V in the InP/InGaAs ones. The I_B sign inversion occurs at the value of V_{CB} where $\beta \cdot (M-1) = 1$, which is almost constant in the range of I_E used in our measurements. ΔI_B can be experimentally obtained from:

$$\Delta I_B = I_B(V_{CB}) - I_{B0} \qquad (1)$$

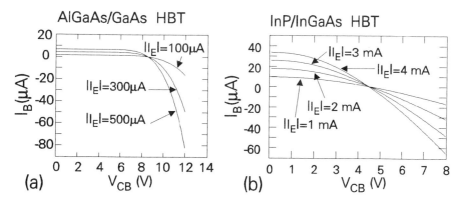

Figure 1: Base current I_B as a function of V_{CB} in an AlGaAs/GaAs HBT (a), and in an InP/InGaAs HBT (b). I_B changes its sign, due to the collection of impact-ionization generated holes at the base contact, at $V_{CB} \simeq 8.5\ V$ and $V_{CB} \simeq 4.5\ V$, respectively.

where I_{B0} is the base current without multiplication, assumed equal to I_B at $V_{CB} = 0\ V$. The multiplication factor M–1 can then be evaluated from the following equation [1,14]:

$$M - 1 = \frac{\Delta I_C}{I_C - \Delta I_C} = \frac{|\Delta I_B|}{I_C - |\Delta I_B|} \quad (2)$$

The ionization coefficient α_n can be obtained from the measured avalanche gain M through the relation [14]:

$$M = \frac{1}{1 - \int_0^{W_C} \alpha_n \exp\left[-\int_0^x (\alpha_n - \alpha_p)\, dx'\right] dx} \quad (3)$$

valid for electron-initiated multiplication, and derived under the assumption that α_n is a function of the local electric field. If the electric field is sufficiently low so that the M–1 factor is much smaller than unity ($10^{-5} \leq M - 1 \leq 10^{-2}$ in our measurement), and if α_n is approximately equal or greater than α_p, then the ionization events due to secondary carriers can be neglected [15], so that the term $\int_0^x (\alpha_n - \alpha_p)\, dx'$ is very close to zero. Furthermore, if the electric field is constant throughout the collector depletion region, eq. (3) reduces to:

$$M - 1 = \alpha_n(E) \cdot W_C \quad (4)$$

Although $\Delta I_C = |\Delta I_B|$, from an experimental point of view it is convenient to measure the avalanche current as a change in the base current, rather than a variation of the collector current, since the latter is superimposed on an injected carriers current I'_C, β times greater than I_{B0}. ΔI_B, as defined in eq. (1), actually yields the impact-ionization current only if the following conditions are verified: 1) the contribution to ΔI_B due to impact ionization is significantly larger than the contribution due to the Early effect; 2) the base-collector diode reverse current I_{CBO}, including the components due to surface generation-recombination and leakage, is negligible compared to ΔI_B. These assumptions are of major concern when eq. (1) is used in low field conditions.

To appreciate how parasitic contributions affect the base current at low electric fields, we report in Fig. 2 ΔI_B, as defined in eq. (1), measured at constant I_E and at constant V_{BE}; I_{CBO}, measured with the emitter floating, is also shown in Fig. 2. In both types of

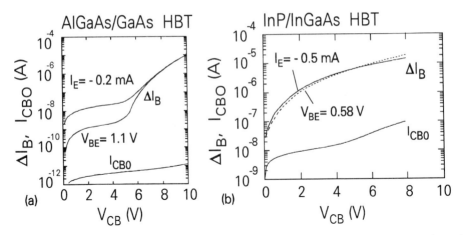

Figure 2: (a) $|\Delta I_B|$ versus V_{CB} measured at constant emitter current $I_E = -0.2\ mA$ and at constant emitter-base bias $V_{BE} = 1.1\ V$ in an AlGaAs/GaAs HBT, together with I_{CB0} measured with the emitter floating; (b) $|\Delta I_B|$ versus V_{CB} measured at constant emitter-base bias V_{BE} in an InP/InGaAs HBT, together with I_{CB0} measured with the emitter floating.

devices, ΔI_B is always much greater than I_{CB0}. As a consequence I_{CB0} does not affect impact-ionization measurements.

In the devices with GaAs collector, which is characterized by a lower ionization coefficient with respect to $In_{0.53}Ga_{0.47}As$, ΔI_B behaves differently at high and low V_{CB}. At high V_{CB} ($\geq 5.5\ V$), the increase of ΔI_B is dominated by impact-ionization-generated holes collected at the base contact [6]. At low V_{CB} ($\leq 4\ V$), the behavior of ΔI_B is dominated by the Early effect, which influences the base current through the modulation of the minority carrier profile in the base region. The base current is due to bulk recombination in the quasi-neutral base (I_{Br}), and to recombination in the base emitter space-charge region (I_{Bscr}) [16]. Free surface recombination is negligible in our samples due to the very low ratio of free surface to emitter contact area [6,17,18]. When the device is driven at constant V_{BE}, the concentration of minority carriers, $n_p(0)$, at the emitter edge of the neutral base region is kept constant, while the effective base width decreases as V_{CB} is increased. Since the total minority charge stored in the base is reduced, the base recombination current I_{Br} decreases, leading to a spurious contribution to $|\Delta I_B|$, Fig. 2(a). This contribution is even larger when the device is driven at constant I_E, since the injection level is externally fixed, so that the gradient of the electron density at the emitter edge of the neutral base is kept constant on increasing V_{CB}. Thus, the decrease in the effective base width must be accompanied by a reduction of the forward emitter-base bias, resulting in a decrease of the base-emitter space-charge recombination current I_{Bscr}. This, together with the reduction in the total minority charge stored in the base, leads to a contribution to $|\Delta I_B|$ which is greater than the one measured at constant V_{BE} by more than one order of magnitude, see Fig. 2(a). The influence of the Early effect on ΔI_B at low V_{CB} discussed above, sets the lower limit for the measurement of $M-1$ at low electric fields in the AlGaAs/GaAs HBT's. For this reason, in order to extract the ionization coefficient from $M-1$ data, it is convenient to drive the device at constant V_{BE}, rather than at constant I_E.

In the InP/InGaAs HBT's, due to the high value of α_n in $In_{0.53}Ga_{0.47}As$, ΔI_B is dominated by the collection of impact-ionization-generated holes even at low V_{CB} values. In fact, the curve of ΔI_B measured at constant V_{BE} and the one measured at constant I_E are very close, showing that the contribution of the Early effect is negligible.

3. Results and discussion

The values of α_n derived from the $M-1$ measured at constant V_{BE} using eq. (4) and eq. (2) are reported in Fig. 3(a) and 3(b) for GaAs and $In_{0.53}Ga_{0.47}As$, respectively. The α_n data have been attributed to the maximum value of the electric field E_{MAX} in

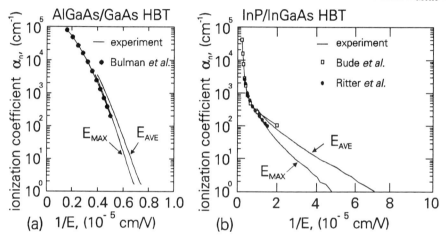

Figure 3: (a) Experimental data of the electron ionization coefficient α_n (dashed lines) in GaAs as extracted from the measured M-1 values. The experimental data for α_n in $\langle 100 \rangle$ GaAs obtained by Bulman et al. [9] are also reported (filled circles). (b) Experimental data of α_n (solid lines) in $In_{0.53}Ga_{0.47}As$ as extracted from the measured M-1 values. The experimental data for α_n obtained by Ritter et al. [12] (filled circles) and the theoretical calculation by Bude et al. [7] (open squares) are also reported.

the collector depletion region or to its average E_{AVE} to give an estimation of the error deriving from the slight variation of the electric field across the collector region. The previously available data for $\langle 100 \rangle$ GaAs [9] (filled circles) obtained on GaAs avalanche photodiodes are extended by almost two orders of magnitude in the low field domain. For $In_{0.53}Ga_{0.47}As$, a theoretical work by Bude et al. [7] predicted a weak electric field dependence of α_n below values of approximately $2 \cdot 10^5$ Vcm^{-1}, then observed by Ritter et al. [12] down to a minimum α_n value of 10^2 cm^{-1}. Despite the uncertainty due to the non-uniformity of the electric field, which sets the lower limit to the α_n values extracted, our measurement clearly demonstrates the reduction of the α_n field dependence down to a value of $\alpha_n = 1$ cm^{-1}. In the AlGaAs/GaAs devices, the lower limit of the α_n value extracted is set by the contribution of the Early effect to ΔI_B.

Since the distance over which non-equilibrium effects take place is short compared with the collector depletion width, the assumption of a local dependence of α_n on the electric field, on which eq. (3) is based, actually holds over most of the integration path. The same is not true for the highly-doped collectors, as discussed in ref. [6].

4. Conclusions

In conclusion, we have demonstrated that suitably designed HBT's and the measurement of $M-1$ in terms of ΔI_B rather than ΔI_C allow unprecedented sensitivity in measuring the impact-ionization coefficients at low electric fields. When the limit to the extension of the measurement to lower fields is set by the Early effect, as in our AlGaAs/GaAs HBT's, the limit can be reduced by driving the device at constant V_{BE} rather than at constant I_E. In particular, we were able to extend the measurement of α_n for $\langle 100 \rangle$ GaAs and $In_{0.53}Ga_{0.47}As$ down to 1 cm^{-1}. By increasing the width W_C of the collector depletion region, and by measuring devices with longer minority carrier lifetime in the base it should be possible to further increase the sensitivity of this technique. However, even if HBT's allow one to measure ionization coefficients with superior sensitivity at low electric field with respect to avalanche photodiodes, complementary n-p-n and p-n-p structures are required to measure α_n and α_p respectively.

References

[1] P.-F. Lu and T.-C. Chen, IEEE Trans. on Electron Devices, **36**, 1182, (1989).

[2] E. F. Crabbé, J. M. C. Stork, G. Baccarani, M. V. Fischetti and S. E. Laux, IEDM Tech. Dig., 463, (1990).

[3] E. Zanoni, E. F. Crabbé, J. M. C. Stork, P. Pavan, G. Verzellesi, L. Vendrame and C. Canali, IEEE Electron Device Letters, **14**, 69, (1993).

[4] K. Horio, Y. Iwatsu and H. Yanai, IEEE Trans. on Electron Devices, **36**, 617, (1989).

[5] J. J. Chen, G. B. Gao, J. I. Chyi and H. Morkoç, IEEE Trans. on Electron Devices, **36**, 2165, (1989).

[6] E. Zanoni, R. Malik, P. Pavan, J. Nagle, A. Paccagnella and C. Canali, IEEE Electron Device Letters, **13**, 253, (1992).

[7] J. Bude and K. Hess, J. Appl. Phys., **72**, 3554, (1992).

[8] F. Capasso, in *Semiconductors and semimetals*, edited by W. T. Tsang (Academic, Orlando, 1985), Vol. 22 D.

[9] G. E. Bulman, V. M. Robbins and G. E. Stillman, IEEE Trans. on Electron Devices, **32**, 2454, (1985).

[10] W. Maes, K. De Meyer and R. Van Overstraeten, Solid-State Electronics, **33**, 705, (1990).

[11] I. Takayanagi, K. Matsumoto and J. Nakamura, J. Appl. Phys., **72**, 1989, (1992).

[12] D. Ritter, R. A. Hamm, A. Feygenson and M. B. Panish, Appl. Phys. Lett. **60**, 3150, (1992).

[13] D. M. Kim, S. lee, M. I. Nathan, A. Gopinath, F. Williamson, K. Beyzavi, A. Ghiasi, Appl. Phys. Lett., **62**, 861, (1993).

[14] S. M. Sze, Physics of semiconductor devices, J. Wiley & sons, New York, U.S.A., (1981).

[15] This condition is satisfied in GaAs, where α_n/α_p is greater than unity, and increases with decreasing the electric field[9]. Note that, if α_n is instead sufficiently smaller than α_p, M–1 $\ll$ 1 does not imply that eq. (3) reduces to M$-1 = \int_0^{W_C} \alpha_n dx$.

[16] J. J. Liou, J. Appl. Phys., **69**, 3328, (1991).

[17] S. C. M. Ho, D. L. Pulfrey, IEEE Trans. on Electron Devices, **36**, 2173, (1989).

[18] W. Liu and J. S. Harris, IEEE Trans. on Electron Devices, **39**, 2726, (1992).

Electron Transport Mechanisms in Abrupt- and Graded-Base/Collector AlInAs/GaInAs/InP Double Heterostructure Bipolar Transistors

H. J. De Los Santos*, M. Hafizi, T. Liu, and D. B. Rensch

Hughes Research Laboratories, 3011 Malibu Canyon Road, Malibu, CA 90265
Hughes Space and Communications Company, El Segundo, CA, 90245*

Abstract. We present, for the first time, an analysis of the measured transport mechanisms operative in abrupt- and graded-base/collector AlInAs/GaInAs/InP double heterostructure bipolar transistors. We have found that the abrupt-BC structure gives rise to Schottky-Emission, Tunneling, Frenkel-Poole/Schottky, Thermionic, and Space-Charge-Limited Conduction current components, while only Schottky-Emission, Tunneling transport, and the usual high-current Ohmic behavior, are found in the graded-BC structure.

1. Introduction

The breakdown behavior of AlInAs/GaInAs heterostructure bipolar transistors can be significantly improved by replacing most of the collector with a wide gap material such as InP [1], [2]. Introducing the wide gap material, however, elicits *major* device design issues [3] in view of the impact the resulting base-collector (BC) conduction band step discontinuity has on important electrical parameters like, device current handling capability, saturation voltage, breakdown voltage, and microwave performance. Understanding of the transport mechanisms at work in AlInAs/GaInAs/InP double heterostructure bipolar transistors (DHBTs) is, therefore, crucial for elucidating design, optimization, and reliability issues pertaining the high-frequency/high-power operation of these devices and for the formulation of accurate physically-based analytical and numerical models. To this end, results on the transport analysis of fabricated abrupt- and graded-BC DHBT structures are presented.

2. Device Structures and Electrical Characteristics

The DHBT epitaxial structures were grown by our gas source MBE in a Varian GEN-II system. The layer design of the abrupt AlInAs/GaInAs/InP DHBT used for this study is shown in Figure 1. It consisted of a 0.7 mm GaInAs subcollector ($n^+=1 \times 10^{19}$ cm^{-3}) and a 0.75mm InP collector ($n=3 \times 10^{16}$ cm^{-3}) followed by a 50nm n-GaInAs spacer ($n=1 \times 10^{17}$ cm^{-3}) to allow the electrons to acquire sufficient energy to overcome the barrier. The base was 60nm thick and was heavily doped with Be ($p^+ = 6 \times 10^{19}$ cm^{-3}). The base-emitter

100 nm	GaInAs Contact	n = 1x10¹⁹ cm⁻³
70 nm	AlInAs Emitter Contact	n = 1x10¹⁹
120 nm	AlInAs Emitter	n = 8x10¹⁷
30 nm	Compositional Grade	n = 8x10¹⁷
10 nm	GaInAs Spacer	p = 2x10¹⁸
60 nm	GaInAs Base	p = 2x10¹⁹
20 nm	GaInAs Spacer	p = 5x10¹⁷
50 nm	GaInAs Spacer	n = 1x10¹⁷
750 nm	InP Collector	n = 3x10¹⁶
700 nm	GaInAs Subcollector	n = 1x10¹⁹
10 nm	GaInAs Buffer	Undoped
	InP Substrate	

Fig. 1 MBE-grown epitaxial layer design of abrupt base/collector AlIn/GaIn/InP DHBT

junction was compositionally graded over a distance of 30nm. A layer design similar to that of Figure 1, but with a 100nm compositionally graded superlattice inserted at the BC, describes the graded-BC DHBT structure [4]. Both abrupt- and graded-BC DHBT devices were fabricatedusing our triple mesa process [5].

3. Analysis of Transport Mechanisms

An assessment of transport mechanisms, in general, entails the interpretation of measured Current-Voltage and Current-Temperature characteristics in terms of the characteristics of fundamental transport processes, i.e., Schottky-Emission, Frenkel-Poole, Tunnel or Field Emission, Space-Charge-Limited Conduction, Ohmic Conduction, Ionic Conduction, Hopping Conduction, or Polaron Conduction [6], [7]. We have, therefore, analyzed the measured DC transport characteristics of abrupt- and graded-BC 6-finger power DHBTs
between 18°C and 150°C. From typical Gummel plots, shown in Figure 3 for a temperature of 50°C, we can immediately observe that a number of transport regimes are at work. Identification of these transport regimes has been carried out by expressing the collector current in terms of six major components [7, p. 496], namely,

$$I_C = I_{Schottky-emission} + I_{Frenkel-Poole} + I_{Tunneling} + I_{Thermionic} + I_{Space-charge} + I_{Ohmic} \quad (1)$$

$$I_{Schottky-emission} \propto T^2 \exp(+a\sqrt{V_{BE}}/T - q(\phi_B/kT)) \quad (2)$$

$$I_{Frenkel-Poole} \propto V_{BE} \exp(+2a\sqrt{V_{BE}}/T - q(\phi_B/kT)) \quad (3)$$

$$I_{Tunneling} \propto V_{BE}^2 \exp(-b/V_{BE}) \quad (4)$$

$$I_{Thermionic} \propto T^2 \exp(-q\phi_B/kT) \quad (5)$$

$$I_{Space-charge} \propto V_{BE}^2 \quad (6)$$

$$I_{Ohmic} \propto V_{BE} \exp(-c/T) \tag{7}$$

where, in the context of a base-collector junction, $a \equiv \sqrt{q/(4\pi\varepsilon_d W)}$, with ε_d being the depletion region permittivity, and W the width. ϕ_B represents the conduction band step discontinuity, and b and c are constants embodying the tunneling parameters, and the activation energy for ohmic conduction, respectively.

3.1 Abrupt-BC DHBT Transport Analysis

The changes in slope on the abrupt-BC DHBT's IC-VBE @ VCE=1V curve, Figure 2a, suggest the presence of at most six transport regimes [8]. Identification of the fundamental transport process operative in a given regime is done by re-plotting the data in question in such a way that a straight line is obtained [7]. Figures 3a-thru-3f identify the transport mechanisms operative in regions I-thru-VI of Figure 2a, respectively. They are found to be Schottky-Emission, Tunneling, Frenkel-Poole/Schottky, Thermionic, and Space-Charge-Limited Conduction. The extracted base/collector bandgap step discontinuity, ϕ_B, was approximately 0.26 eV.

3.2 Graded-BC DHBT Transport Analysis

Three transport regimes appear to be operative in the graded-BC DHBT, as can be deduced from Figure 2b. These have been found, Figure 4a-thru-4c, to be consistent with Schottky-Emission, Tunneling, and a rather weak Ohmic behavior with an activation energy of approximately 3.9 meV. The continuing presence of tunneling in the graded-BC structure, Figure 4b, dramatizes the partial success of the grading in eliminating the BC step discontinuity, and points to the need for further device optimization.

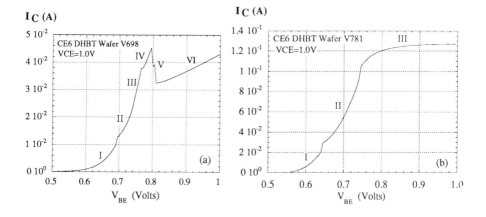

Fig. 2 Measured Gummel plots of 6-finger DHBTs: (a) abrupt base/collector, (b) graded base/collector

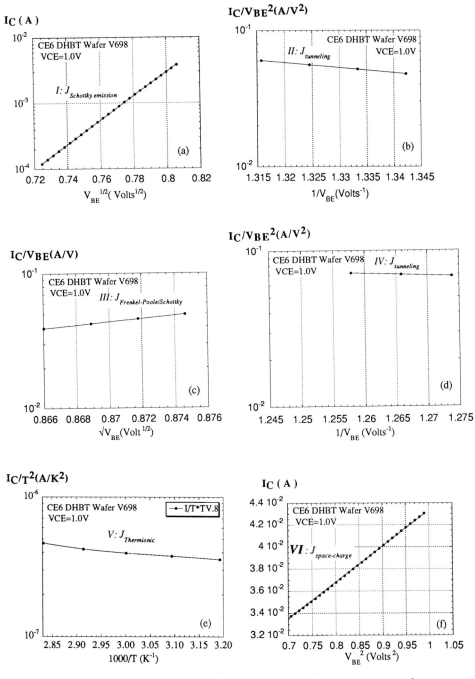

Fig. 3 Analysis of transport mechanisms in abrupt base/collector DHBT: (a) regime I, (b) regime II, (c) regime III, (d) regime IV, (e) regime V, (f) regime VI

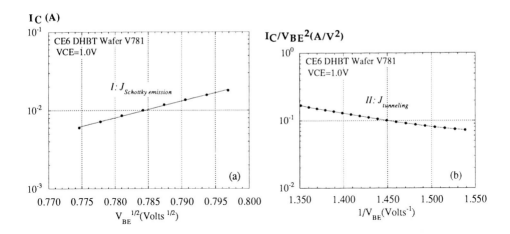
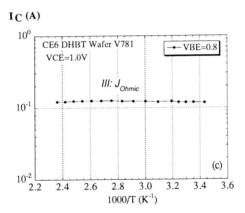

Fig. 4 Analysis of transport mechanisms in graded base/collector DHBT: (a) regime, (b) regime II, (c) regime III

4. Conclusions

We have analyzed the transport mechanisms operative in abrupt- and graded-BC AlInAs/GaInAs/InP DHBTs. Schottky-Emission, Tunneling, Frenkel-Poole/Schottky, Thermionic, and Space-Charge-Limited Conduction were found to be the mechanisms at work in the abrupt-BC DHBT, while only Schottky-Emission, Tunneling transport, and the usual high-current Ohmic behavior, were observed in the graded-BC DHBT. Despite the presence of tunneling in the graded-BC DHBT structure, which has revealed that the grading did not achieve a complete elimination of the step discontinuity, a stark contrast between the blocking effects of the abrupt- versus graded-BC DHBT structures under high-current conditions was observed. Our measured data displays, in a rather vivid fashion, the richness of physical phenomena underlying the operation of modern HBTs, and our analysis has clearly unveiled and demonstrated the important physics which an accurate formulation of analytical/numerical tools aimed at HBT device design must encompass [11]. These findings carry a high practical significance as they clearly show that the charge extraction mechanisms of the base/collector junction in DHBTs render the collector current a *non-exponential* function of VBE. This observation is highly relevant to power amplifier applications, as it implies a loss of efficiency, due to a reduced peak current capability [12]. In addition, the findings point to the detrimental effects of tunneling, $I_{Tunneling} \propto V_{BE}^2 \exp(-b/V_{BE})$, as a major component responsible for limiting the maximum attainable device efficiency. Finally, it is readily apparent that DHBTs exhibit multiple bias-dependent temperature behavior, which has serious implications for some applications, e.g. integrated circuits.

Acknowledgements

One of the authors (H.J.D.) would like to thank Dr. Paul Greiling for supporting this work at the Hughes Research Labs, and Dr. Mike Case, also of HRL, for making available his excellent computing resources. He also gratefully acknowledges the continuing support of Dr. Mark Shahriary and Mr. Robert Brunner, of Hughes Space and Communications Company.

References

[1] Feygenson A, Ritter D, Hamm R, Smith P R, Montgomery R, Yadvish R D, Temkin H, and Panish M B 1992 *Electronic Lett.* **28** 607-609
[2] Stanchina W E, Jensen J F, Metzger R A, Hafizi M E, and Rensch D B 1992 *Proc. 4th Intl Conf. on Indium Phosphide and Related Materials* (Newport, RI) 434-437
[3] Kurishima K, Nakajima H, Kobayashi T, Matsuoka Y, and Ishibashi T 1993 *Appl. Phys. Lett.* **62** 2372-2374
[4] Hafizi M, Liu T, Macdonald P A, Lui M, Chu P, Rensch D B, Stanchina W E, and Wu C S 1993 *IEEE IEDM Tech. Digest* 791-794
[5] Stanchina W E, Rensch D B, Jensen J F, Mishra U K, Kargodorian T V, Pierce M P, and Allen Y K 1990*Proc. State-of-the-Art Program Compound Semiconductors, Montreal,* Canada
[6] O'Dwyer J J 1973 *The Theory of Electrical Conduction and Breakdown in Solid Dielectrics* (Clarendon Press Oxford)
[7] Sze S M 1981 *Physics of Semiconductor Devices* 2nd Ed.(Wiley-Interscience)
[8] Mead C A 1962 *Phys. Rev.* **128** 2088-2093
[9] Chynoweth A G, Feldmann W L, and Logan R A 1961 *Phys. Rev.* **121** 684-694
[10] Forrest S R, Leheny R F, Nahory R E, and Pollack M A 1980 *Appl. Phys. Lett.* **37** 322-325
[11] Tanaka S and Lundstrom M S 1993 *IEEE IEDM Tech. Digest* 505-508
[12] Gao G, Morkoc H, and Chang M-C F 1992 *IEEE Trans. on Electron Dev.* **39** 1987-1997

Contactless HBT Equivalent Circuit Analysis Using the Modulation Frequency Dependence of Photoreflectance

Wojciech Krystek, H Qiang and Fred H Pollak

Physics Department and NY State Center for Advanced Technology in Ultrafast Photonic Materials and Applications, Brooklyn College of CUNY, Brooklyn, NY 11210 USA

and

Dwight C Streit and Michael Wojtowicz

TRW, Electronics and Technology Division, Redondo Beach, CA 90278 USA

Abstract. In this paper we report the contactless determination of the time constants of the equivalent circuits of both the GaAs collector and GaAlAs emitter portions of a GaAs/GaAlAs heterojunction bipolar transistor structure using the modulation frequency dependence (2 Hz $< f_m <$ 100 kHz) of the in-phase and quadrature photoreflectance signals from these spectrally separated regions. An analysis of the collector time constant reveals that the recombination mechanism in the collector-base region is dominated by the hole current and is due to midgap trap states. This experiment demonstrates a new area of application for this method for the characterization of these devices.

1. Introduction

The contactless electromodulation (EM) method of photoreflectance (PR) has proven to be a valuable experimental method to investigate the properties of semiconductors (bulk/thin film) [1,2], semiconductor microstructures [1,2] (quantum wells, quantum dots [3], heterojunctions), surfaces/interfaces [1,2] and actual device configurations (heterojunction bipolar transistors, pseudomorphic high electron mobility transistors, vertical cavity surface emitting lasers, etc.) [2,4]. Not only is PR contactless but it also requires no special mounting of the sample and hence can be performed nondestructively on wafer sized samples. Most of these investigations have used the sharp, derivative-like optical structures produced by this technique, including Franz-Keldysh oscillations (FKOs) [1,2]. However, since PR is an *ac* technique there is also important information in other modulation variables such as phase, modulation frequency (f_m), amplitude, etc. [2] in addition to the sharp, derivative-like spectral features. Photoreflectance is the optical response of the material to the modulating *ac* electric field and hence can be considered as a contactless optical impedance spectroscopy. However, very little work has been done in this area of modulation spectroscopy. The first report of EM as an optical impedance spectroscopy was by Silberstein et. al. [5], who demonstrated that electrolyte electroreflectance (EER) could be used to obtain the electrical impedance of a semiconductor-electrolyte interface. These authors measured the in-phase and quadrature EER signals as a function of f_m in the range 10 Hz $< f_m <$ 10 kHz. Subsequently a number of articles reported the evaluation of trap times by measuring the frequency response of PR signals from various semiconductors including temperature dependent measurements [2]. Also several investigators have made use of the phase of PR [2,6,7], including a recent study on interference effects from epitaxial layers [6]. Shen et. al. studied the time constants involved in PR from several surface/undoped/n^+-GaAs (001) structures by measuring the rise and fall times of the signals using a digital oscilloscope [8].

In this paper we report a room temperature PR study of the signals from the collector and emitter regions of a GaAs/GaAlAs heterojunction bipolar transistor (HBT) structure. The in-phase and quadrature signals from these spectrally isolated portions of the sample were evaluated in the frequency range 2 Hz $< f_m <$ 100 kHz. These measurements made it possible to determine the RC time constants, τ_1 and τ_2, of the equivalent circuits of the collector and emitter regions, respectively. An analysis of the former reveals that the recombination mechanism in the collector-base region is due to midgap trap states and is dominated by the hole current. This experiment demonstrates a new area of application for this method for the characterization of these devices.

2. Experimental Details

The GaAs/GaAlAs HBT sample used in this study was fabricated by MBE and had the following characteristics. The collector contact layer, 6000Å of n^+-GaAs ($\approx 5 \times 10^{18}$ cm^{-3}), was grown on an undoped liquid encapsulated Czochralski <001> semi-insulating substrate. This was followed by the collector layer of 7000Å of n^--GaAs ($\approx 7.5 \times 10^{15}$ cm^{-3}) and the 1400Å p^+-GaAs base ($\approx 1 \times 10^{19}$ cm^{-3}). The 1800Å n-GaAlAs ($\approx 5 \times 10^{17}$ cm^{-3}) emitter was graded in Al composition at the top and bottom over 300Å. The n^+-GaAs ($\approx 7 \times 10^{18}$ cm^{-3}) emitter contact layer was about 750Å thick. The dopants were Si and Be for the n- and p-type regions, respectively. The PR apparatus has been described in the literature [1,2]. The pump beam was the 670 nm line of a laser diode modulated by an acousto-optic modulator [9]. The intensities of the pump and probe beams were 600 μW/cm^2 and 100 μW/cm^2, respectively.

3. Experimental Results

Plotted in Fig. 1 is the PR spectrum at 300K from the collector and emitter regions of the sample at f_m = 500 Hz. The signals occur in the region of the direct band gaps of GaAs [E_0(GaAs)] and GaAlAs [E_0(GaAlAs)], respectively. From the position of the latter we can deduce an Al composition of 28%. Both spectra exhibit pronounced FKOs.

Displayed in Figs. 2 and 3 are the normalized in-phase (open circles) and quadrature (closed circles) amplitudes of the collector and emitter PR signals, respectively, as a function of modulating frequency in the range 2 Hz $< f_m <$ 100 kHz.

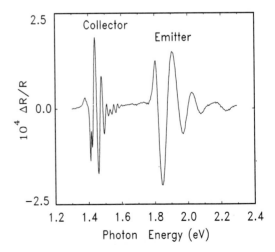

Fig. 1 Photoreflectance spectrum (in-phase component) in the region of E_0 of GaAs (collector) and GaAlAs (emitter) of a GaAs/GaAlAs HBT at 300K.

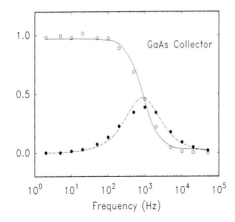 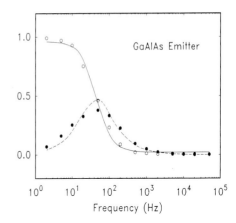

Fig. 2 In-phase (open circles)/quadrature (closed circles) signals vs f_m for the GaAs collector. The solid/dashed lines are fits to Eq. (2).

Fig. 3 In-phase (open circles)/quadrature (closed circles) PR signals vs f_m for the GaAlAs emitter. The solid/dashed lines are fits to Eq.(2).

4. Analysis

We have analyzed the frequency dependent data of Figs. 2 and 3 on the following basis. The photovoltage, $V_P(t)$, induced by the pump beam can be expressed as:

$$V_p(t) \propto \begin{cases} \dfrac{(e^x-1)}{2\sinh x} e^{-t/\tau} & -1/2f_m < t < 0 \\ 1 - \dfrac{(e^x-1)}{2\sinh x} e^{-t/\tau} & 0 < t < 1/2f_m \end{cases} \quad (1)$$

where $x = 1/2f_m\tau$ and τ is the characteristic time constant of the circuit. In Eq. (1) we have assumed equal rise and fall times for the circuit. By taking the Fourier transform of Eq.(1) it can be shown that the frequency dependence of the first harmonic in-phase and quadature components of the signal are given by:

$$\text{in-phase} = \frac{1}{[1+(2\pi f_m\tau)^2]}, \qquad \text{quadrature} = \frac{2\pi f_m\tau}{[1+(2\pi f_m\tau)^2]}. \quad (2)$$

Shown by the solid and dashed lines in Fig. 2 (GaAs collector) and Fig. 3 (GaAlAs emitter) are least-square fits of Eq.(2) to the frequency dependence of the in-phase and quadrature components of the PR signal, respectively. The obtained time constants are $\tau_1 = 1.8 \times 10^{-4}$ sec and $\tau_2 = 3.6 \times 10^{-3}$ sec, respectively. Note that the time constant of the emitter region is more than an order of magnitude greater than that of the collector portion.

In the collector-base *p-n* junction, assuming the saturation current is due to thermionic emission, the resistance times area (RA) can be written as [10]:

$$RA = (kT/q)(1/A^*T^2)\exp(E_b/kT), \qquad (3a)$$

where A^* is the Richardson constant and E_b is an effective barrier height for the saturation current. The capacitance per unit area (C/A) for an abrupt one-sided p-n junction is [10]:

$$C/A = (qn_c \kappa \epsilon_0 / 2V_{bi})^{1/2}, \qquad (3b)$$

with n_c being the n-type collector carrier concentration (7.5 x 10^{15} cm^{-3}), κ is the static dielectric constant (= 13 for GaAs), ϵ_0 is the permittivity of free space and V_{bi} is the built-in potential. For an ideal p-n junction V_{bi} can be expressed as [10]:

$$qV_{bi} = E_0 + kT \ln(n_c p_b / N_c P_v), \qquad (3c)$$

where p_b is the p-type carrier concentration in the base (≈ 1 x 10^{19} cm^{-3}) and N_c (4.7 x 10^{17} cm^{-3}) and P_v (7.0 x 10^{18} cm^{-3}) are the density-of-states of the conduction and valence bands, respectively. With $E_0 = 1.42$ eV we find $V_{bi} = 1.32$ V for the ideal case.

From Eqs. (3a) and (3b) the time constant, τ_1, is given by:

$$\tau_1 = RC = (kT/q)(1/A^*T^2)(qn_c \kappa \epsilon_0 / 2V_{bi})^{1/2} \exp(E_b/kT), \qquad (4a)$$

so that:

$$E_b = kT \ln[\tau_1 A^* T^2 / (kT/q)(qn_c \kappa \epsilon_0 / 2)^{1/2}] + (kT/2q) \ln V_{bi}. \qquad (4b)$$

The second term in Eq. (4b) is small compared to the first term.

With $\tau_1 = 1.8$ x 10^{-4} sec, $n_c = 7.5$ x 10^{15} cm^{-3} and assuming that the saturation current is due to the holes ($A^* = 74$ A/cm^{2}K^2 [10]) we find that $E_{bh} = 0.73$ eV for hole barrier height in the collector-base region.

Both the collector and emitter signals exhibit pronounced FKOs (see Fig. 1). The positions of the mth extrema in the FKOs are given by [1,2]:

$$m\pi = (4/3)[(2\mu_\parallel)^{1/2}(E_m - E_0)^{3/2}/q\hbar F] + \chi, \qquad (5)$$

where E_m is the photon energy of the mth extrema, F is the electric field, $\mu_\parallel$ is the reduced interband effective mass in the direction of $\vec{F}$ and χ is an arbitrary phase factor. Therefore, F can be obtained directly from the period of the FKOs if $\mu_\parallel$ is known. With $\mu_\parallel$ (in units of the free electron mass) = 0.055 (GaAs) and 0.073 (GaAlAs) we find F = 3.0 x 10^4 V/cm and 1.9 x 10^5 V/cm for the collector and emitter regions, respectively.

5. Discussion of Results

Because of the very short lifetimes the current flowing through the GaAs collector-base space charge region (SCR) under low bias is due primarily to Shockley-Read-Hall (SRH) recombination through deep traps in the SCR. The SHR recombination rate is given by [11]:

$$U_{SRH} = \frac{n'_c p'_c - n_i^2}{\tau_p\{n'_c + n_i\exp[(E_t - E_i)/kT]\} + \tau_n\{p'_c + n_i\exp[-(E_t - E_i)/kT]\}}, \quad (6a)$$

where $n'_c(p'_c)$ are the nonequilibrium electron (hole) concentration in the collector, n_i is the intrinsic carrier concentration, E_i is the Fermi level for intrinsic material, E_t is the trap level while τ_n and τ_p are the electron and hole lifetimes, respectively. For $\tau_n = \tau_p$, the recombination rate simplifies to:

$$U_{SRH} = \frac{n'_c p'_c - n_i^2}{\tau_n\{n'_c + p'_c + n_i\cosh[(E_t - E_i)/kT]\}}. \quad (6b)$$

This expression is maximized when $E_t = E_i$ and the Fermi energy level crosses the trap level, i.e., $E_f = E_t$. To reach the spatial location where $E_f = E_t$, the carriers must be thermionically emitted from the base and undepleted collector region. Since SRH recombination is a two carrier process, the carrier with the lowest emission current will provide the controlling current.

In the collector-base junction, the holes are emitted from the base while electrons emitted from the undepleted collector. Normally the collector region is non-degenerate so the potential barrier for electron emission into the deep trap is from the undepleted region conduction band to the conduction band energy where $E_f = E_t$. Depicted schematically in Fig. 4 is the HBT band diagram, in the region of the collector-base, showing the electron and hole barrier heights for SHR recombination in the collector. For mid-gap trapping levels, the majority of the recombination occurs where the Fermi level (E_f) crosses the intrinsic Fermi level (E_i).

Assuming that $E_t = E_i$, the electron barrier (E_{be}) in the collector is given by:

$$E_{be} = (E_c - E_i) - (E_c - E_f)|_{\text{undepleted collector}}, \quad (7a)$$

which can be rewritten as:

$$E_{be} = (E_0/2) - (kT/2)\ln(P_v/N_c) - kT\ln(N_c/n_c). \quad (7b)$$

Using the values of E_0, N_c, P_v and n_c listed above, we find that $E_{be} = 0.57$ eV.

For a non-degenerate base, the potential barrier for hole emission (E_{bh}) into the deep trap is from the base valence band energy to the valence band energy where $E_f = E_t$. This is given by:

$$E_{bh} = qV_{bi} - E_{be}, \quad (7c)$$

which yields $E_{bh} = 0.75$ eV. However, because of the heavy doping in the base the Fermi energy level is less than the valence band energy (see Fig. 4) and the holes are emitted from the Fermi level. In our HBT, this correction is 0.03 eV which yields $E_{bh} = 0.72$ eV. The hole barrier height is sufficiently larger than the electron barrier such that the hole current will limit the recombination process. Therefore, it is the hole barrier that is calculated from τ_1 and the thermionic emission current, i.e., 0.73 eV, which is in excellent agreement with the 0.72 eV value deduced above.

The FKOs from the collector region (see Fig. 1) also indicate a barrier height that is considerably less than qV_{bi} (1.32 eV) for the ideal case. From the relation between the field determined from the FKOs (3 x 10^4 V/cm) and the electron barrier height (E_{be}) [10]:

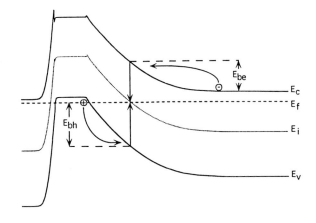

Fig. 4 HBT band diagram in the region of the collector-base showing the barrier heights for SHR recombination in the collector.

$$F^2 = (2n_c/\kappa\epsilon_0)(E_{be} - kT), \qquad (8)$$

we find that E_{be} = 0.46 eV (neglecting any photovoltage effects). This value is in reasonable agreement with that obtained from Eq. (7a).

6. Acknowledgements

The authors WK, HQ and FHP acknowledge the support of US Army Research Office contract #DAAL03-92-G-0189, NSF grant #DMR-9120363 and the New York State Science and Technology Foundation through its Centers for Advanced Technology program.

References

[1] See, for example, Glembocki O J and Shanabrook B V 1992 *Semiconductors and Semimetals*, **Vol. 36**, ed. Seiler D G and Littler C L (New York; Academic) p. 222-292 and references therein.
[2] See, for example, Pollak F H and Shen H 1993 *Materials Science and Engineering* **R10** 275-374 and references therein.
[3] Qiang H, Pollak F H, Tang Y-S, Wang P D, and Sotomayor Torres C M 1994 *Appl. Phys. Lett.* **64** 2830-2832.
[4] See, for example, Pollak F H, Qiang H, Yan D, Yin Y and Boccio V T 1993 *Photonics Spectra Magazine*, **Vol. 27**, Issue 8, August 1993, p. 78-84 and references therein; Qiang H, Yan D, Yin Y and Pollak F H 1993 *Asia-Pacific Engineering Journal, Part A: Electrical Engineering* **3** 167-198.
[5] Silberstein R P, Lyden J K, Tomkiewicz M and Pollak F H 1981 *J. Vac. Sci. Technol.* **19** 406-410.
[6] Lipsanen H K and Airaksinen V M 1993 *Appl. Phys. Lett.* **63** 2863-2865.
[7] Alperovich V L, Jaroshevich A S, Scheibler E H and Terehov A S 1994 *Solid State Electronics* **37** 657-660.
[8] Shen H, Dutta M, Lux R, Buchwald W, Fotiadis L and Sacks R N 1991 *Appl. Phys. Lett.* **59** 321-323.
[9] Shen H, Pollak F H, Woodall J M and Sacks R N 1990 *J. Electron. Materials* **19** 283-286.
[10] See, for example, Sze S M 1981 *Physics of Semiconductor Devices, 2nd edition* (New York; Wiley).
[11] See, for example, Tawari S 1992 *Compound Semiconductor Device Physics* (New York; Academic).

Threshold Tunable MESFET and Ultra-Fast Static RAM

P.J. Zampardi, S.M. Beccue, R.L. Pierson, W.J. Ho, J. Yu, M.F. Chang,
A. Sailor, and K.C. Wang

Rockwell Science Center
1049 Camino Dos Rios, Thousand Oaks CA 91360
(805) 373-4256 (TEL) (805) 373-4775 (FAX)

Abstract. This paper describes some of the unique features of a Metal Semiconductor Field Effect Transistor (MESFET) fabricated from the emitter epilayers of a heterojunction bipolar transistor (HBT). The addition of a backgate terminal allows such FETs have a tunable threshold voltage. The range of this tuning is large enough to allow a D-mode (normally on) device to be converted to an E-mode (normally off) device or to easily compensate for threshold voltage shifts with temperature. The effects of this backgate terminal on DC and RF performance of the MESFETs is reported. A 64-bit static random access memory with a read access time of 330-360 ps has been demonstrated in this Bipolar-FET (BiFET) technology and is also described.

1. Introduction

The integration of Metal-Semiconductor Field Effect Transistors (MESFETs) with Heterojunction Bipolar Transistors (HBTs) has significant impact on the design of high performance GaAs circuits and systems. The combination of the HBT's high switching speed, high drive capability, and low 1/f noise and the FET's low noise, high input impedance, high density and yield, offer designers a great deal of flexibility and new circuit opportunities [1,2]. In particular, the addition of MESFETs will provide high input impedance for sample-and-hold circuits, active loads (for low power applications), current sources and sinks, and low power static random access memories (SRAMs) with HBT driving circuitry and FET memory cells. The integrated FET characteristics are extremely important for many circuit applications: such as signal mixing and in threshold voltage matching for source coupled FET differential pairs. Moreover, a Direct Coupled Bipolar-FET (BiFET) Logic (DCBL), versus the traditional Direct Coupled FET Logic (DCFL), is now possible using D-FETs on GaAs for both pull-up and pull-down transistors, and electrically biasing the backgate of the pull-down device to transform its threshold voltage from depletion (negative V_t) into enhancement (positive V_t). The backgate can also serve a smart power shut-off gate for ultra-low power IC applications. This paper describes some of the unique properties of such FETs fabricated from the emitter layers of a heterojunction bipolar transistor (HBT).

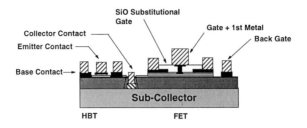

Figure 1. Schematic of HBT/FET Co-integration

2. BiFET FET Characteristics

The fabrication process by which these FETs were co-integrated with HBTs has been described previously [1,2]. A schematic diagram of the resulting HBT and FET is shown in Figure 1. As a result of this processing, there is direct access to the backgate of the FET via the base of the HBT layers. This is similar to the body-effect in CMOS, with the exception that each transistor can be individually controlled. This provides many interesting properties [3]. The conversion of a D-mode (normally on) FET to an E-mode (normally off) FET biasing the backgate potential is shown in Figure 2. This offers unprecedented fabrication tolerance to threshold voltage variations, as well as the ability to define both E and D mode devices with a single gate recess step. The required backgate voltage was found to be approximately 2.5 volts to obtain a threshold voltage of zero.

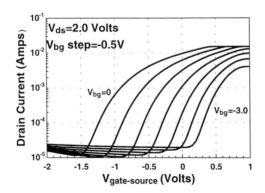

Figure 2. Demonstration of threshold control with backgate terminal

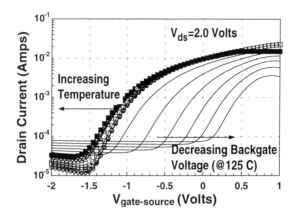

Figure 3. Temperature compensation with backgate terminal

The tuning range for the threshold voltage was found to be -1.2 to .6 volts for backgate voltages of 0 to -4.5 volts. Backgate control also allows the threshold voltage shifts due to temperature to be compensated for since increasing temperature shifts the threshold more negative and the backgate shifts it positive. This is demonstrated for FETs operating up to 125 °C in Figure 3. Various spacings for the source to backgate contact were investigated. No significant difference was noted in devices with spacings ranging from 2 to 8 microns. This allows the backgate contact to be removed from the source of the FET. This is critical in RAM applications were the pass elements are required to swap the source and drain for different modes of operation. We also note that this contact has little effect on the high frequency or noise properties of the FET. For sub-micron gate lengths defined using optical lithography and a substitutional gate process, F_ts of 39 GHz were observed. This is shown in Figure 4. The noise figure for these FETs was found to be 2 dB at 10 GHz with 8 dB associated gain.

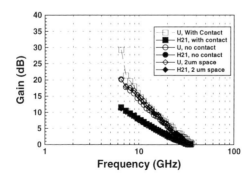

Figure 4. RF performance of FET with, without, and 2 μm removed p-contact

The RF and noise properties were independent of whether the backgate resides in the source of the FET or is removed by 2 μm.

3. BiFET RAM

As a demonstration of the utility of this technology, we have demonstrated a 64-bit SRAM with 330 ps best access time. This circuit takes maximum advantage of the high current drive capability of HBTs for the peripheral circuitry, and uses FETs and HBTs in the bit cells, to obtain lower power operation. A schematic for this SRAM is shown in Figure 5. The BiFET SRAM uses a conventional design in which six address lines, a WRITE/READ strobe, Din and Qout control all functions. The address lines are terminated with 50 Ohms to ground; however, ECL I/O levels may be used on all but the outputs. The outputs are complementary CML intended to be terminated with 50 Ohms to ground.

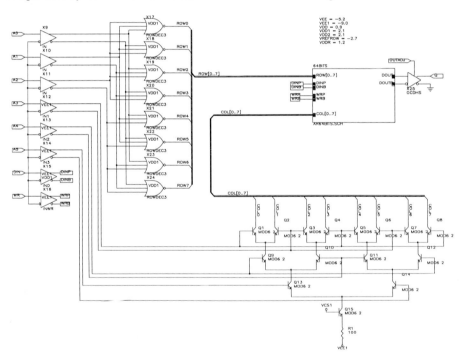

Figure 5. Schematic of HBT/FET SRAM

For testing purposes, all inputs were run at CML levels. The bit cell consists of 4 MESFETs and two HBTs. The two HBTs and two MESFETs form a cross-coupled latch,

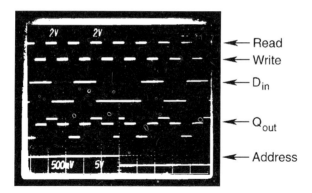

Figure 6. Operation of BiFET SRAM

while two MESFETs act as pass elements to complimentary bit lines. The pass elements have gates connected to word lines. All of the peripheral circuitry is implemented in HBT, with Schottky diodes used for row decoding. This design takes advantage of the high speed and excellent drive capability of the HBT for best access times. Operation of this SRAM is shown in Figure 6. Access time measurements were performed by selecting an arbitrary address line (A5) and toggling it at 1 GHz. This is a column select and ought to represent a worst case access time. The rows were selected to row 0, the furthest row from the sense amplifier. A 01 pattern was written (at DC by manually strobing the value); as the address toggles, a 01 pattern is observed on the outputs. The cable delays from the address line and Qout were matched. An HP54120 oscilloscope was used to capture the output and measure the delay, and is shown in Figure 7. Access times of 330-360 ps were observed. This is in good agreement with the simulated result of 336 ps based on SPICE simulation.

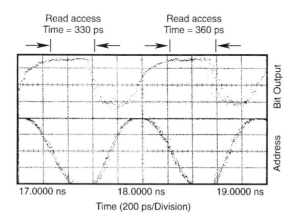

Figure 7. Read access time for SRAM

4. SUMMARY

We have demonstrated several interesting features that arise from fabricated FETs from the emitter layer of an HBT. The most important feature is the ability to tune the threshold voltage over a wide range allowing: E and D mode device to be fabricated from one gate recess, compensation for mismatch in V_t, and compensation for the shift of V_t with temperature. The addition of this extra terminal had very little effect on the RF performance of the FETs. In particular, the cut-off frequency and noise figure showed no noticeable degradation. To demonstrate the utility of this process, a 64-bit RAM was also demonstrated. Record access times of 330ps-360ps were achieved for this circuit.

5. ACKNOWLEDGMENTS

The authors acknowledge the support and encouragement of Dr. D.T. Cheung and Mr. Charles Young. Useful technical discussions with Dr. P.M. Asbeck are also gratefully acknowledged. We would also like to thank Randy Nubling for his careful reading of this paper. This work was partially supported by Air Force Wright Patterson Laboratory under contract #F33615-90-C-1505.

References

[1] Ho, Wu-Jing, M.F. Chang, S.M. Beccue, P.J. Zampardi, J. Yu, A. Sailer, R.L. Pierson, and K.C. Wang,"*A GaAs BiFET LSI Technology*", To be published in the Proceedings of the 1994 GaAs IC Symposium
[2] P.J. Zampardi, S.M. Beccue, K.D. Pedrotti, J. Yu, R.L. Pierson, W.J. Ho, M.F. Chang, and K.C. Wang, *"GaAs BiFET Process and Integrated Circuits"*, International Semiconductor Device Research Symposium, Virginia, Dec. 1-3, pp. 447-452, 1993.
[3] D. Cheskis, C. Chang, W. Ku, P. M. Asbeck, M.F. Chang, R. Pierson, A. Sailor, *"Co-integration of GaAlAs/GaAs HBTs and GaAs FETs with a Simple, Manufacturable Process"* IEDM Technical Digest, December 1992

Inst. Phys. Conf. Ser. No 141: Chapter 6
Paper presented at Int. Symp. Compound Semicond., San Diego, 18–22 September 1994
© 1995 IOP Publishing Ltd

An n-HJFET - pnp HBT process for complementary circuit applications

Primit Parikh, Kürşad Kızıloğlu, Mark Mondry, Prashant Chavarkar, Bernd Keller, Steven Denbaars and Umesh Mishra

Department of Electrical and Computer Engineering, University of California, Santa Barbara, CA 93106

ABSTRACT : We report a potential complementary circuit technology with pnp HBT's and n HJFET's. Both devices are fabricated from the same epitaxial layer structure, in the InAlAs-InGaAs on InP material system. This technology is centered around using the junction of the n-HJFET as the emitter for the pnp HBT. The HJFET has an I_{DSS} of 400 mA/mm with a peak g_m of 220 mS/mm. The maximum β of the HBT's is 200 at a collector current density of 16 kA/cm^2.

1. INTRODUCTION :

InP based heterojunction bipolar transistors (HBT's) and heterojunction field effect transistors (HJFET's) are finding extensive use in microwave as well as digital circuits due to their high electron mobility, peak velocity and conduction band discontinuity between $In_{0.52}Al_{0.48}As$ and $In_{0.53}Ga_{0.47}As$; all superior to the GaAs based system. The integration of FET's with HBT's is of great importance in the design of high performance integrated circuits due to the high current driving capability and the switching speed of the HBT coupled with the low noise, low power and high input impedance features of the FET . There are numerous applications for pnp devices in complementary npn -pnp circuits, such as the use of pnp devices as active loads, current mirrors and in complementary push-pull amplifiers with very low dc power dissipation and high linearity.
    Conventionally, high speed complementary technology has been in the form of npn-pnp HBT's or n and p channel HFET's [1,2]. A two dimensional electron gas base HBT has been reported [3]. We have fabricated n channel HJFET's and pnp HBT's from the same layer structure made in a single MBE growth. This technology is not limited by the relatively poor performance of p-channel FET's or regrowth of active regions. Our approach is based on using the p type junction gate of the n-HJFET as the emitter of the HBT. The n-channel is used as the base and the p buffer of the HJFET is used as the collector.

2. DEVICE STRUCTURE AND FABRICATION

The complete epitaxial layer structure common to both devices and the energy band diagram of the devices are shown in Fig. 1 and Fig. 2 respectively. A heavily doped p+ InGaAs cap is used for ohmic contact to the gate (emitter). The InGaAs spacer layer is 50 Å and the 500 Å channel consists of two delta dopings of 2.5 x 10^{12} cm^{-2} each. The substrate is p+ InP which is used to form backside contacts for the collector.

The HBT fabrication begins with PECVD deposition of SiO_2. The oxide is then patterned using the emitter mesa mask and etched using a CF_4 plasma. The base layer is exposed by a phosphoric acid solution. A second layer of oxide is then deposited and patterned to open windows for n+ InGaAs base contact regrowth by MOCVD (see Fig. 1). Simultaneously, source and drain layers are regrown for the FET.

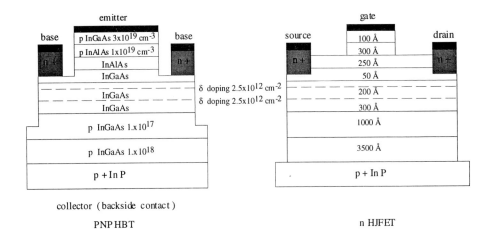

Fig 1. HJFET -HBT layer structure (not to scale)

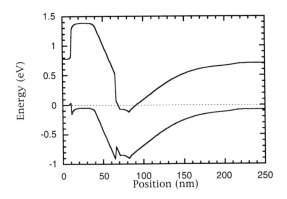

Fig 2. Energy band diagram for the HJFET-HBT structure at 0 V applied bias

Next, the base mesa is defined. An oxide layer is deposited in which windows are opened for base and emitter contacts. Final metallization of the base and emitter is completed with Ti/Pt/Au contact scheme. Finally, a backside contact is made for the collector. The FET fabrication has been described previously [4].

3. RESULTS AND DISCUSSIONS

The I-V characteristics of the HJFET are shown in Fig. 3a. The full channel current, I_{DSS},

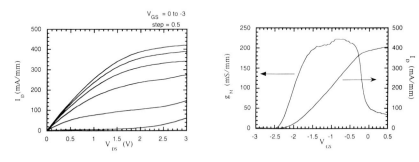

Fig 3a. HJFET I-V characteristics   Fig 3b. HJFET g_m and I_D vs V_{GS} (at V_{DS}=2.5V)

of the HJFET is 400 mA/mm. The device is fully pinched-off at a gate to source voltage of -3V. The peak g_m of the device is 220 mS/mm as seen in Fig. 3b. The two terminal diode breakdown voltage is 4 V and the turn on voltage is 0.8 V, both defined at 1 mA/mm of gate current. The source/drain regrown ohmic contact resistance is 0.45 Ω.mm.

For the HBT, the I-V characteristics for a device with emitter dimensions of 5 μm x 25 μm is shown in Fig 4.

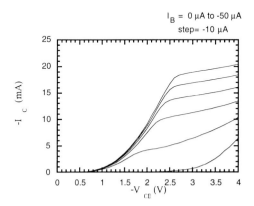

Fig 4. HBT common emitter I-V characteristics

The maximum current gain β of the HBT is 200 at an emitter current density of 16 kA/cm^2. The emitter base turn on voltage is around 1.5 V. This high value is attributed to a poor emitter contact (with improved epitaxial design for a non-alloyed p-contact, a specific contact resistivity of less than 1×10^{-6} $\Omega.cm^2$ is expected). The specific contact resistivity for the regrown base contact is 4.6×10^{-6} $\Omega.cm^2$. The early voltage of the devices is around 12 V which is typical of an InAlAs/InGaAs system with a narrow base [5]. The breakdown voltage for the base collector diode is around 3 V which explains the anomalous β values at lower base currents.

4. CONCLUSION

In summary, we have demonstrated components for a complementary technology featuring n-HJFET's and pnp-HBT's, fabricated from the same epilayer structure. All the processing steps of the two devices are compatible. HBT's with β values of 200 and HJFET's with g_m of 220 mS/mm have been obtained.

The primary advantage of this process is that it is easily integrable as we have both positive and negative transconductance devices from the same epitaxial layer . The gate and the emitter metals are at the same level which is ideal for circuit interconnects. The base emitter diode is same as the gate diode, facilitating matched voltages for linear operation. The InAlAs-InGaAs material system is attractive due to high electron mobility and velocity in InGaAs. With proper optimization of both the devices, this technology is well suited for the fabrication of high speed complementary integrated circuits.

REFERENCES

1. R.A. Kiehl et al., "High speed, low voltage, complementary FET circuit technology, Proc. of GaAs IC Symposium", p 101,1991.
2. D.B. Slater et al., "Monolithic integration of complementary HBT's by selective MOVPE", IEEE Elect. Dev.Lett., Vol 11, No.4, p 146, April 1990.
3. T. Usagawa et al., "A new two-dimensional electron gas base transistor (2DEG-HBT)", IEDM Tech. Digest, p 78, 1987.
4. J.B. Shealy et al., "High breakdown voltage InAlAs/GaInAs JHEMT's with regrown ohmic contacts", IEEE Elect.Dev.Lett.,Vol 14,No. 12, p 545, Dec 1993.
5. A. Miura et al., "InAlGaAs/InGaAs HBT", IEDM Tech. Digest, p 79, 1992.

Heterostructure Insulated-Gate-FET with Improved Gate Barrier Characteristics

R. Westphalen, K.-M. Lipka, J. Schneider* and E. Kohn

Department of Electron Devices and Circuits, *Department of Semiconductor Physics, University of Ulm, D-89069 Ulm, Germany

Abstract. A planar GaAs Insulated-Gate-HFET using an undoped MBE grown GaAs/AlAs insulating heterostructure layer between gate and channel has been investigated and compared with Schottky Gate FETs. The AlAs barrier layer in the IG-HFET represents an electron as well as a hole barrier between channel and gate, resulting in a reduced gate leakage. A reduced interface potential at the AlAs/channel interface of $\Phi_i=0.12eV$ is observed. This leads to a reduction of the space charge layer and an increase of the maximum open channel current as compared to the MESFET at identical $N_D \cdot$ t-product, whereas the breakdown voltage is $V_{DS,max}=36V$ (at $3 \times 10^{17} cm^{-3}$ channel doping) independent of the device configuration. 2W/mm RF power density has been calculated from DC-measurement.

1 Introduction

For GaAs based power FETs a high drain current has to be controlled by the gate source voltage in combination with a high drain source voltage. However, in GaAs MESFETs the maximum drain source voltage is inversely related to the maximum drain current due to the 2-dimensional avalanche breakdown of the gate drain space charge layer, resulting in a constant power density of approx. 1W/mm [1,2]. The forward bias is limited by the Schottky barrier height. The reverse breakdown is initiated by hole injection into the space charge region, subsequent electron hole pair generation, and eventually avalanche breakdown. This effect limits the gate drain breakdown and thus the power capability.

To overcome the power limitations, a number of attempts have been made, as in particular the Schottky drain [3], heterojunction gate barrier configurations [4,5] and MISFET structures [6,7]. Only the MISFET structure with a dielectric of high breakdown strength eliminates the GaAs related restrictions. However, GaAs based MISFETs suffer from a high density of insulated interface states and can only operate in the dynamic mode.

In this study an AlAs barrier was introduced to the SI-gate FET concept [8,9], forming an effective electron and hole barrier between the gate and the channel. The hole barrier in particular is extremely high with 1.8eV and hole injection to the gate should be totally suppressed. This will lead to reduced gate current and avalanche breakdown will be delayed. An improved gate drain breakdown characteristics results also in an improved drain source breakdown voltage. In combination with a low interface potential, supporting a high open

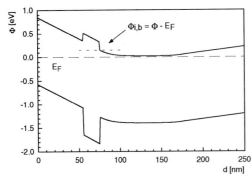

Figure 1: Calculated band structure of the GaAs/AlAs/GaAs:Si IG-HFET structure GaAs: 60nm, AlAs: 20nm, GaAs:Si 100nm, $n=3\times10^{17}cm^{-3}$, $\Phi_B=0.85eV$

channel current, high power densities can be obtained. Advantages of the AlAs barrier layer heterostructure configuration of the IG-HFET structure will be demonstrated by comparison with a conventional planar MESFET.

2 IG-HFET structure design

The IG-HFET contains an AlAs Barrier layer sandwiched in between the doped channel and the insulating GaAs gate layer. The top undoped GaAs layer serves as insulator in the SI-Gate-FET concept as well as surface passivation for the buried AlAs barrier (insert figure 2).

2.1 MIS-diode characteristics

Figure 1 shows the band diagram of an GaAs/AlAs/GaAs:Si heterostructure calculated by using CBAND [10,11]. The structure consist of a 60nm undoped GaAs layer, 20nm AlAs layer and 100nm channel with a Si doping of $3\times10^{17}cm^{-3}$. At the metal GaAs interface the conventional Schottky barrier potential of $\Phi_B=0.85eV$ was used.

At forward bias the AlAs barrier delay the electron injection into the Schottky contact increasing the maximum forward gate bias. At reverse gate bias the high barrier of the valance band of nearly 1.8eV suppresses hole injection into the gate contact.

The space charge layer of the IG-HFET underneath the gate is determined by the interface potential $\Phi_{i,b}$ of the AlAs/channel heterointerface (figure 1). It is drastically reduced compared to the Schottky barrier potential of $\Phi_B=0.85eV$ in a MESFET. This will allow to use nearly the entire $N_D \cdot t$-product of the channel for current transport. However, the transconductance is determined by the gate to channel separation.

2.2 Insulator/channel gate drain characteristics

In the planar structure the channel cross section between gate and drain is determined by the channel interface potential $\Phi_{i,s}$ related to the free surface potential. Calculating the band structure with the free surface potential of $\Phi_S=0.60eV$ at the SI-GaAs surface leads also to an essentially reduced depletion region. This also increases the usable

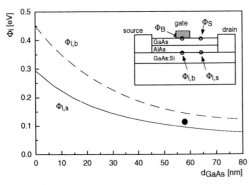

Figure 2: Calculated dependence of the thickness of the undoped GaAs layer on the channel interface potential Φ_i, structure see insert
$\Phi_{i,s}$ between gate and drain (solid line)
$\Phi_{i,b}$ below the Schottky Gate (dashed line)
point: $\Phi_{i,b}$ of the IG-HFET from CV-measurement

$N_D \cdot t$-product and thus the maximum open channel current, avoiding a parasitic current limiter as in the conventional MESFET. In the case of the MESFET, this can only be obtained by a recess structure, however, only a part of the channel $N_D \cdot t$-product can be used for current transport.

Figure 2 shows the dependence of the thickness of the undoped GaAs top layer d_{GaAs} on the channel interface potential $\Phi_{i,s}$ and $\Phi_{i,b}$ extracted from the calculation of the band structure. For the free AlAs surface, the GaAs potentials are used due to the lack of the exact value of the AlAs potential. The maximum reduction of the interface potential appears already for an insulating GaAs layers of 50nm thickness (figure 2). Therefore, in this study a 60nm layer were chosen. To take advantage of the reduced interface potentials using these heterostructure as a passivation and insulating layer, any free GaAs channel surface has to be avoided.

3 Device Structure and Fabrication

The MESFET and IG-HFET structures are grown in a Riber 32P MBE system on (100) SI-GaAs substrates on top of a 600nm thick GaAs Buffer layer under standard MBE conditions. Individual devices are insulated by wet chemical mesa etching.

In contrast to ohmic contact fabrication on most GaAs related devices it is not possible to alloy through the AlAs barrier layer. This would result in unacceptable high source and drain series resistances. To contact the channel directly it is necessary to open windows in the heterostructure by wet chemical selective etching. This will produce also lateral etching and result in unpassivated free surfaces adjacent to the contacts (figure 3a). To avoid this free channel surface, the contacts need to overlap onto the passivation layer. This is achieved by widening the resist structure with an oxygen plasma treatment before metal deposition (figure 3b). Ohmic contacts are made by evaporating and alloying Au/Ge/Ni and lift off (figure 4).

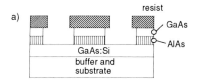

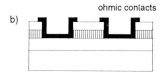

Figure 3: Sketch of overlapping ohmic contacts fabrication
a) after selective etching
b) after broadening the resist structure, metal deposition and lift off

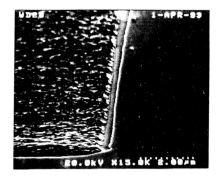

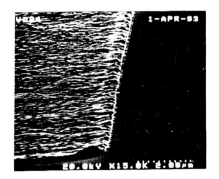

Figure 4: SEM picture of non overlapping (above) and overlapping ohmic contacts (below)

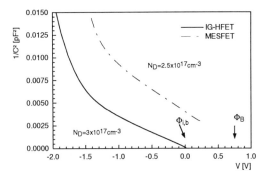

Figure 5: $1/C^2$ versus V for MESFET and IG-HFET (area of Schottky contact $100 \times 100 \mu m^2$)

Finally Ti/Pt/Au Schottky gate contacts are deposited by e-beam evaporation and lift-off without any gate recess. The gate length and width are $1\mu m$ and $50\mu m$. Drain source spacing is $4.5\mu m$.

4 Results

4.1 Gate barrier characteristic

As expected from the theoretical considerations the forward gate bias of a IG-HFET can be increased up to $V_{gs,max}=1.5V$, whereas the gate current density is still below $I_g=1mA/mm$. The improved MIS-diode characteristics is demonstrated by a reverse gate current below $I_g=5\mu A/mm$ before the drain source breakdown occurs.

The relevant barrier potential of the IG-HFET and the MESFET have been determined by CV-measurements. Form the $1/C^2$ versus voltage relationship an IG-HFET channel interface diffusion barrier potential of approx. 0.05V is extracted taking into account the bias independent insulator capacitance of the undoped GaAs and AlAs layers. This will result in a interfacial barrier height of $\Phi_{i,b}=0.12eV$ (point in figure 2). In Figure 5 the $1/C^2$ curve of the MESFET is also plotted.

4.2 AlAs/channel interface potential between gate and drain

To estimate the AlAs/channel interface potential in the region between gate and drain the maximum drain source current of the ungated IG-HFET structure is measured without and with the heterostructure layer. Removing the undoped GaAs layer reduces I_{DS} from 480mA/mm to 340mA/mm. The current differences allows to calculate an interface potential of $\Phi_{i,s}=0.22eV$ (assuming a saturated velocity of $1\times10^7 cm/s$). Etching off the AlAs layer has no further effect on the drain source current. This indicates a higher free surface potential of AlAs as compared to GaAs. However, the instability of AlAs exposed to air makes it difficult to determine the free surface potential exactly.

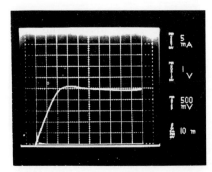

Figure 6: Two terminal saturation current with overlapping contacts

Supported by overlapping source and drain contacts a very uniform field distribution is obtained in the buried channel. This results in a constant saturation current for drain bias up to 9V (figure 6). In case of non overlapping contacts early breakdown and current instabilities are observed.

4.3 DC characteristics

The low potential of the insulator channel interface, is reflected in a clearly increased maximum open channel current of 450mA/mm (figure 7). In comparison, the MESFET delivered 240mA/mm. In order to exclude any difference of epitaxial growth between the channel of the MISFET and the channel of the MESFET, this results has been verified by selective removal of the insulating surface layer of the IG-HFET. Therefore, the increased current is due to an increased usable part of the $N_D \cdot t$-product of the channel, and although the current is increased similar drain source breakdown voltages are expected and indeed observed. For both devices drain source breakdown voltages of 36V are observed at pinch-off. However, in case of the MESFET a distinct leakage current is observed for drain source voltages larger than 28V. Although the threshold voltage V_{th} of the MESFET is 2V in comparison to $|V_{th}|=5.5V$ of the MISFET at high drain source voltages it has to be increased up to 4V to ensure low leakage current of the MESFET. The MISFET whereas shows a negligible leakage current up to 36V drain source voltage at constant gate bias of $|V_{th}|$. Therefore, in large signal applications lower non linear distortions are expected for the IG-HFET.

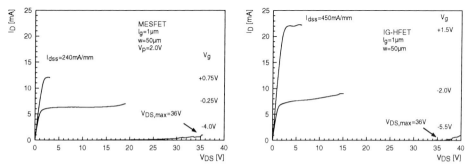

Figure 7: Output characteristics of MESFET and IG-HFET
($V_{p,MESFET}=2.0V$, to achieve $V_{DS,max}=36V$ in both devices Vg>4.0V)

In view of the power capability, the DC-output characteristics indicates a class A RF power density of 0.9W/mm for the MESFET. As expected, this value is close to the power limitation, predicted by the Schottky barrier depletion layer lateral spreading model of approx. 1W/mm. The Insulated-Gate-FET indicates that RF power density of 2W/mm is feasible, which would overcome the predicted power limit due to an improved current capability.

5 Summary

An AlAs barrier layer has been introduced in the well known SI-gate FET structure. This high barrier layer leads to a drastically improved Schottky barrier characteristic. In reverse bias gate current densities are below $I_g=5\mu A/mm$ before drain source breakdown occurs. For forward gate biases up to $V_g=+1.5V$, the gate leakage current is still in the limit of $I_g=1mA/mm$. The AlAs/channel interface potential is to be found approx. $\Phi_i=0.1eV-0.2eV$. Due to the reduction of the channel interface potential, in comparison to the MESFET, the space charge layer thickness is reduced and the maximum open channel current is increased. To exclude any parasitic current limiter any free surface of the channel has to be avoided by an overlapping drain source contact technology. The improved current capability results in an increased RF power density of 2W/mm, twice as high as the MESFET of identical drain source breakdown voltage of 36V. However the power density in the device structure is highest at medium current levels. Therefore the influence of any imperfections in the contact areas is most pronounced here. This technological detail is therefore most important to realise the full RF power potential.

Acknowledgement

Investigation supported by the ESPRIT project TAMPETS 6849 of the European Communities, which is in cooporation with the University of Lille, France and the University of Cardiff, UK.

References

[1] S.H. Wemple, W.C. Niehaus, H.M. Cox, J.V. Dilorenzo, W.O. Schlosser 1980 IEEE Trans. Electron Dev., **ED-27** 1013
[2] K. Hikosaka, Y. Hirachi, M. Abe 1986 IEEE Trans. Electron Dev. **ED-33** 583
[3] J.A. Calviello, P.R. Bie, R.J. Pomian and A. Capello 1985 IEEE Trans. Electron Dev. **ED-32** 2844
[4] W. Pletschen et al. 1993 Mat. Sci. Eng. **40** B21 304
[5] K.W. Eisenbeiser, J.R. East, J. Singh, W. Li and G.I. Haddad 1992 Electron Device Letter **EDL-13** 421
[6] J.M. Dortu, E. Kohn 1984 Inst. Phys. Conf. Series **74**. 563
[7] C.L. Chen et al. 1991 Electron Device Letter **EDL-12** 306
[8] H.M. Machksey, D.W. Shaw and W.R. Wissman 1976 Electronics Letters **EDL-12** 192
[9] N. Iwata, H, Mizutani, S. Ichikawa, A. Mochizuki and H. Hirayama 1991 Inst. Phys. Conf. Ser. **120** 119
[10] A.N. Khondker and A.F.M. Anwar 1987 Solid-State Electronics **30** 847
[11] M.C. Foisy 1990 "A Physical Model for the Bias Dependence of the Modulation-Doped Field-Effect Transistor's High-Frequency Performance" PH.D. Thesis Cornell University Ithaca NY

Strain-symmetrized $In_xGa_{1-x}As/In_yAl_{1-y}As$ HEMTs with extremely high 2DEG densities and mobilities

W. Klein, G. Böhm, H. Heiß, S. Kraus, D. Xu, R. Semerad,
G. Tränkle, and G. Weimann

Walter-Schottky-Institut, Technische Universität München, D-85748 Garching, Germany

ABSTRACT: Strain-symmetrized $In_xGa_{1-x}As/In_yAl_{1-y}As$ heterostructures with $0.53 \leq x \leq 0.74$ and $0.52 \geq y \geq 0.415$ and enhanced 2DEG densities and mobilities were grown by molecular beam epitaxy on InP substrates. The increased conduction band offset resulted in extremely high electron densities of 3.81×10^{12} cm^{-2}, with a 4.2 K mobility of 51100 cm^2/Vs for single-sided doping and highest ever densities of 6.70×10^{12} cm^{-2} for double-sided doping. The strain-symmetrization with stress compensating $In_yAl_{1-y}As$ barriers did not, however, allow an increase in the critical In content of the $In_xGa_{1-x}As$ QW.

1. Introduction

InGaAs/InAlAs high electron mobility transistors (HEMTs) on InP substrates are currently the most promising devices for high frequency amplification [1-5]. Device performance can be improved either by reducing the gate length [1,3] or by increasing the channel conductance [3,6-9]. The enhanced conduction band offset due to the increasing In content in the active channel leads to higher electron mobilities and saturation velocities. Numerous investigations have been made to maximize the In content in the channel [2,7,8], including stress compensated structures with opposite strain in quantum wells and barriers.

We investigated strain-symmetrized SQW structures with a homogeneously and compressively strained well and $In_yAl_{1-y}As$ under tensile strain for stress compensation. The essential advantages of these structures are a larger conduction band discontinuity due to the increased band gap in the $In_yAl_{1-y}As$ barrier, and a better pinch-off of the HEMTs due to improved electron confinement in the 12 nm wide QW. We demonstrate the optimization of 2DEG mobility and density in these structures with modified In contents in the barriers and the wells and, secondly, differing doping profiles. Optimized HEMT structures with single-sided and double-sided doping yielding 2DEG densities, measured at 4.2 K, as high as 3.81×10^{12} cm^{-2} and 6.70×10^{12} cm^{-2}, respectively.

2. MBE growth

All structures used here were grown by solid source MBE on (100) oriented InP:Fe substrates with growth rates around 1 μm/h, a substrate temperature of 530 °C and a V/III ratio (beam equivalent pressure) of 27 . The different compositions of the ternary layers were obtained by controlling the group III flux ratio.

Figure 1 schematically shows the epitaxial layer sequence of the investigated HEMT structure. The buffer consists of two 100 nm thick $In_{0.52}Al_{0.48}As$ layers and two $In_{0.53}Ga_{0.47}As/In_{0.52}Al_{0.48}As$ superlattices. The temperatures of the Ga effusion cell was decreased while growing the second $In_{0.52}Al_{0.48}As$ layer from the value necessary for lattice matched growth to the value appropriate for the strained QW. The In flux was kept constant, except for the growth of the lattice matched cap layer. As our MBE machine is equipped with two Al cells (and a single Ga and In cell each), we were able to grow the layer sequence shown in Fig. 1 without any growth interruption at the QW interfaces, having different Al fluxes at our disposal.

The investigated range in the In contents of the QWs and the barriers are given in Table I. The Si doping level in the $In_yAl_{1-y}As$ supply layer was 1.5×10^{19} cm^{-3} for sample M5409, M5424 and M5425 and 1.0×10^{19} cm^{-3} for all others. The measured transport properties reflect the influence of the parallel conduction in the cap layers: Samples M5406 - M5410 have doped caps of only 7 nm thickness, whereas the other samples have thicker caps (5 nm undoped InGaAs topped by 10 nm doped InGaAs). The doping in the cap layers is 3.0×10^{18} cm^{-3} in all samples.

The characterization of our HEMT structures was made by Hall measurements at 300, 77 and 4.2 K and Shubnikov-de Haas (SdH) measurements at 4.2 K to separate the 2DEG conductance from the parallel conductance in the cap layers. The thicknesses of the surface

InGaAs	$3 \cdot 10^{18}$ cm^{-3}	y=0.53	10 nm
InGaAs	-	y=0.53	5 nm
InAlAs	-	y=0.52	7 nm
InAlAs	-	y=0.52 - 0.415	3 nm
InAlAs	$1-1.5 \cdot 10^{19}$ cm^{-3}	y=0.52 - 0.415	12.5 nm
InAlAs	-	y=0.52 - 0.415	2 nm
InGaAs	-	y=0.53 - 0.74	12 nm
InAlAs	-	y=0.52 - 0.415	6 nm
InAlAs	$0-1.5 \cdot 10^{19}$ cm^{-3}	y=0.52 - 0.415	2 nm
InAlAs	-	y=0.52 - 0.415	16 nm
InAlAs	-	y=0.52	100 nm
SL			
InAlAs	-	y=0.52	100 nm
SL			
InP (:Fe) Substrat			

Figure 1: Schematic layer structure of a strain-symmetrized HEMT.

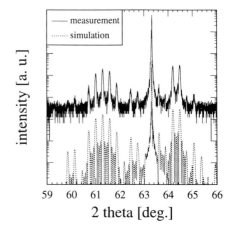

Figure 2: X-ray diffraction spectra (400) of a strain-symmetrized MQW structure (5 periods) with 1 μm $In_{0.52}Al_{0.48}As$ buffer layer on InP substrate (barrier: 23.8 nm $In_{0.53}Al_{0.47}As$, QW: 12.0 nm $In_{0.749}Ga_{0.251}As$; solid: measurement, dotted: dynamical scattering theory).

near layers were controlled by X-ray reflectivity measurements (XRR). Compositions of the strained layers were monitored by X-ray diffraction (XRD) on strain-symmetrized MQW structures, always grown in direct sequence with the HEMT structures.

3. Structural and transport properties

Figure 2 shows the XRD spectra of a strain-symmetrized $In_{0.74}Ga_{0.26}As/In_{0.415}Al_{0.585}As$ MQW structure. The nominal layer sequence is a 1 μm thick lattice matched $In_{0.52}Al_{0.48}As$ buffer layer, followed by 5 QWs, having 24 nm thick $In_{0.415}Al_{0.585}As$ barriers and 12 nm wide $In_{0.74}Ga_{0.26}As$ wells. The simulated spectrum using dynamical scattering theory fits the measured curve excellently, with only small deviations from the nominal values. The compositions and thicknesses from the simulation are: 23.8 nm $In_{0.419}Al_{0.581}As$, 12.0 nm $In_{0.749}Ga_{0.251}As$, 1 μm $In_{0.522}Al_{0.478}As$ buffer. Although we used high growth temperatures of 530 °C and a wide range of In contents from 0.415 to 0.74 in the QW and the barriers, the simulation indicates that there is no noticable In segregation at the interfaces, in contrast to observed segregation lengths of 3.5 nm at the interfaces of pseudomorphic $In_{0.74}Ga_{0.26}As/In_{0.52}Al_{0.48}As$ MQWs with lattice matched barriers, which were grown under identical growth conditions [13]. In segregation has also been reported for the growth of strained InGaAs layers on GaAs [10-12].

Table I: Structural parameters, mobilities and 2-DEG densities of strain-symmetrized $In_xGa_{1-x}As/In_yAl_{1-y}As$ HEMT structures with different In-content in the channel and the barriers (QW thickness: 12 nm)

Sample	LM	PM	PM	SS	SS	PM	SS	PM	SS
	ssd	ssd	ssd	ssd	ssd	ssd	ssd	dsd	dsd
	M5406	M5407	M5408	M5420	M5421	M5409	M5424	M5410	M5425
x ($In_xGa_{1-x}As$) QW	0.53	0.67	0.74	0.67	0.74	0.74	0.74	0.74	0.74
y ($In_yAl_{1-y}As$) Barrier	0.52	0.52	0.52	0.45	0.415	0.52	0.415	0.52	0.415
doping N_D [10^{19} cm^{-3}]	1.0	1.0	1.0	1.0	1.0	1.5	1.5	1.0	1.5
μ_{Hall}(300K) [cm^2/Vs]	9420	11390	12220	11040	12150	11090	11150	7450	7640
n_{Hall}(300K) [10^{12}cm^{-2}]	3.16	3.36	3.49	3.72	3.82	4.22	4.56	6.58	7.67
μ_{Hall}(77K) [cm^2/Vs]	30960	43000	50220	43150	50070	40700	46860	14160	16280
n_{Hall}(77K) [10^{12}cm^{-2}]	3.03	3.21	3.35	3.53	3.67	4.05	4.25	6.55	7.31
μ_{Hall}(4.2K) [cm^2/Vs]	33540	54940	62930	46820	-	54970	51080	16230	15060
n_{Hall}(4.2K) [10^{12}cm^{-2}]	2.79	2.97	3.18	3.79	-	3.64	4.30	6.07	6.92
$n_{1,SdH}$(4.2K) [10^{12}cm^{-2}]	2.55	2.68	2.74	2.81	-	3.05	3.12	3.84	4.15
$n_{2,SdH}$(4.2K) [10^{12}cm^{-2}]	0.26	0.38	0.45	0.46	-	0.61	0.69	2.20	2.55

LM=lattice matched QW and barriers; PM=pseudomorphic QW; SS=strain-symmetrized structure
ssd=single-sided doping in top barrier; dsd= double-sided doping in both barriers

The transport properties of all the HEMT structures investigated in the course of this work are given in Table I. Sample M5406 is a lattice matched reference sample with the QW width of 12 nm used in all our HEMT structures. The mobilities of this reference structure were measured to 9400 cm^2/Vs and 33500 cm^2/Vs at 300 and 4.2 K, respectively. These mobilities are clearly lower than those obtained on lattice matched structures with 32 nm wide QWs (10000 cm^2/Vs and 42000 cm^2/Vs) due to increased interface roughness scattering by the bottom barrier.

Using compressive strain only in the QWs, i.e with lattice matched barriers, increases the mobilities with rising In content from 11400 cm^2/Vs and 55000 cm^2/Vs for x = 0.67 (sample M5407) to 12200 cm^2/Vs and 63000 cm^2/Vs for x = 0.74 (sample M5408). Simultaneously there is an increase in the 4.2 K 2DEG densities from 2.81×10^{12} cm^{-2} to 3.06×10^{12} cm^{-2} and 3.19×10^{12} cm^{-2} (samples M5406, M5407 and M5408). SdH measurements clearly show two occupied subbands with densities n_1 and n_2. The intersubband separation in sample M5406 was found to be 108 meV by the peak separation in photoluminescence, agreeing well with the observed values of n_1 and n_2 using the calculated density of states.

On increasing the doping level to 1.5×10^{19} cm^{-3} in the supply layer, the total 2DEG density reaches 3.66×10^{12} cm^{-2}, while still maintaining a 4.2 K mobility of 55000 cm^2/Vs (sample M5409, lattice matched barriers, single-sided doping). Using In$_y$Al$_{1-y}$As barriers under tensile strain increases the carrier density even further. Sample M5424 is strain-symmetrized with x = 0.74 in the QW and y = 0.415 in the barriers, showing a total 2DEG density of 3.81×10^{12} cm^{-2} at 4.2 K. The mobilities are practically unchanged (see Table I).

A further increase of the In content in the QW to x = 0.78 with a corresponding decrease in the barrier to x = 0.34, however, resulted in a drastic decrease in mobility due to

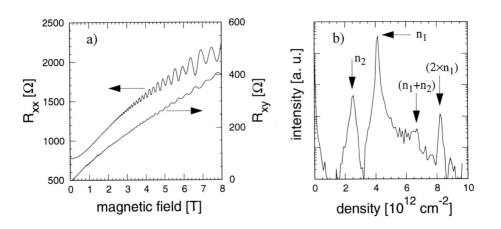

Figure 3: a) Shubnikov-de Haas measurement on a strain-symmetrized HEMT structure with double-sided doping (sample M5425); b) Fourier transformation of R$_{xx}$.

relaxation and the formation of misfit dislocations. Whereas it has been reported [8] that the critical In content of strained InGaAs QWs could be increased by the insertion of stress compensating layers, our results show that, under our growth conditions, the maximum In content is the same in pseudomorphic and strain-symmetrized structures.

Extremely high 2DEG densities were, however, obtained using double-sided doping from both - top and bottom - barriers. Total 2DEG densities as high as 6.04×10^{12} cm^{-2} and 6.70×10^{12} cm^{-2} were found by SdH measurements at 4.2 K in samples M5410 and M5425, respectively. The first sample is doped to 1.0×10^{19} cm^{-3} in the supply layer with a bottom spacer of 4 nm, the corresponding values for the latter are 1.5×10^{19} cm^{-3} and 6 nm. These 2DEG densities are, to our knowledge, the highest ever reported for InGaAs/InAlAs HEMTs. The marked reduction in mobility of double-sided doped structures in comparison to single-sided doping is due to Si segregation towards the channel [14]. So, for example in structure M5425 we found mobilities of 7600 cm^2/Vs and 15100 cm^2/Vs at 300 and 4.2 K, respectively. In spite of the low mobility and considerable parallel conduction in the cap layer we were able to determine clearly the separate densities of the two occupied subbands by SdH measurements and Fourier transformation, as shown in Fig. 3. This analysis yields densities of 4.15 and 2.55×10^{12} cm^{-2} for the first and the second subband in the QW.

4. High electron mobility transistors

HEMT devices with a gate length of 0.13 µm were fabricated from sample M5406 - M5408, using e-beam lithography and wet chemical selective gate recess with succinic acid. The increasing In content from x = 0.53 to 0.74 in the active QW (the barriers of these samples were lattice matched to the InP substrate) resulted in an increase in the extrinsic transconductance from 670 mS/mm to 1150 mS/mm, with a concurrent increase in the intrinsic values from 1000 mS/mm to 1600 mS/mm. The cut-off frequencies for the current gain were measured to 200 - 220 GHz for all devices.

5. Summary

Single-sided doped strain-symmetrized In$_{0.74}$Ga$_{0.26}$As/In$_{0.415}$Al$_{0.585}$As HEMT structures showed 2DEG densities as high as 3.81×10^{12} cm^{-2} and mobilities of 51100 cm^2/Vs at 4.2 K. Extremely high 2DEG densities of 6.70×10^{12} cm^{-2} were obtained using double-sided doping from both barriers. XRD of strain-symmetrized MQW structures showed that the segregation of In at the interfaces was significantly suppressed by strain-symmetrization.

6. Acknowledgement

We thank the German Federal Ministry of Research and Technology (BMFT) and Siemens AG for financial support under contract 01 BM 118/6.

References:

[1] Enoki T, Tomizawa M, Umeda Y and Ishii Y 1994 Jpn. J. Appl. Phys. 33 798-803
[2] Chough K B, Chang T Y, Feuer M D, Sauer N J and Lalevic B 1992 IEEE Electron Device Lett. 13 451-453
[3] Nguyen L D, Brown A S, Thompson M A and Jelloian L M 1992 IEEE Trans. Electron Devices 39 2007-2013
[4] Gueissaz F, Enoki T and Ishii Y 1993 Electron Lett. 29, 2222-2223
[5] Hwang K C, Ho P, Kao M Y, Fu S T, Liu J, Chao P C, Smith P M and Swanson A W 1994 Proc. of the 6th Intern. Conf. on Indium Phosphide and Rel. Mat. (Santa Barbara, California, USA, March 27-31, 1994) 18-20
[6] Tacano M, Sugiyama Y and Takeuchi Y 1991 Appl. Phys. Lett. 58 2420-2422
[7] Brown A S, Schmitz A E, Nguyen L D, Henige J A and Larson L E 1994 Proc. of the 6th Intern. Conf. on Indium Phosphide and Rel. Mat. (Santa Barbara, California, USA, March 27-31, 1994) 263-266
[8] Chin A, and Chang T Y 1990 J. Vac. Sci. Technol. B8 364-366
[9] Hong W-P, Ng G I, Bhattacharya P K, Pavlidis D and Willing S 1988 J. Appl. Phys. 64 1945-1949
[10] Toyoshima H, Niwa T, Yamazaki J and Okamoto A 1993 Appl. Phys. Lett. 63 821-823
[11] Kao Y C, Celii F G and Liu H Y 1993 J. Vac. Sci. Technol. B11 1023-1026
[12] Muraki K, Fukatsu S and Shiraki Y 1992 Appl. Phys. Lett. 61 557-559
[13] Klein W, Böhm G, Heiß H, Kraus S, Xu D, Semerad R, Tränkle G and Weimann G 1994 Proc. of the 8th Intern. Conf. on Molecular Beam Epitaxy (Osaka, Japan, Aug. 29 - Sep. 2, 1994), to be published in J. Crystal Growth
[14] Brown A S, Metzger R A, Henige J A, Nguyen L, Liu M and Wilson R G 1991 Appl. Phys. Lett. 59 3610-3612

POWER COMBINING OF DOUBLE BARRIER RESONANT TUNNELLING DIODES AT W-BAND.

R E Miles[†], D P Steenson[†], R D Pollard[†], J M Chamberlain[‡] and M Henini[‡]

[†] Department of Electronic and Electrical Engineering, University of Leeds, Leeds LS2 9JT, UK. [‡] Department of Physics, University of Nottingham, Nottingham NG7 2RD, UK.

Abstract. The double barrier resonant tunnelling diode has demonstrated the ability to produce power in the terahertz frequency region. This paper discusses device structures and circuit designs for combining a number of diodes in wave-guide so as to increase the power available at W-band (75 – 110 GHz). The devices are based on the GaAs/AlAs material system and have a planar configuration. Power is coupled into the wave-guide via an antenna structure connected to the devices. It is shown that owing to improved coupling the power available from two devices is greater than the sum of the individual powers.

1. Introduction

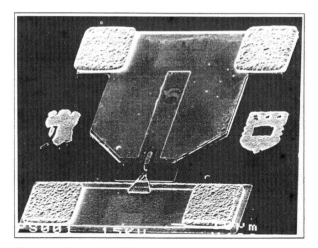

Figure 1 Planar DBRTD

Practical solid state sources developing power at frequencies much above 100 GHz at room temperature are currently non existent and this is one of the reasons why there is no large scale commercial exploitation of this part of the electromagnetic spectrum. The sources that do exist are either based on vacuum tube technology and are consequently bulky, fragile and require large and expensive power supplies or, if solid state, require cryogenic cooling [1]. The applications where such devices can be employed have therefore tended to be where cost is not a primary constraint, such as in astronomy. A solid state device capable of producing around 0.1 mW of power at terahertz frequencies would open up this part of the electromagnetic spectrum to other applications such as local area networks, high bandwidth communications, intelligent highways and aids to the disabled. An ideal solution would be an integrated technology [2] containing sources and other circuit components on a single chip.

While integrated circuits based on three-terminal devices are being developed to operate at frequencies up to 140 GHz and IMPATT diodes operating in harmonic mode at 200 GHz are available, the only other solid state source that will produce fundamental power at terahertz frequencies is the double barrier resonant tunnelling diode (DBRTD). This device, first suggested by Esaki in 1961 but realised in practice only in the last few years, has been shown to operate at over 700 GHz [3] and theoretical projections suggest it should be able to produce oscillations above 1 THz [4]. The device has thus lived up to expectations in terms of frequency of operation but only modest powers have been developed - typically falling from 20 mW at 2 GHz (using AlAs/InGaAs) [5] down to 0.3 µW at 712 GHz (using InAs/AlSb) [3]. An encouraging observation [3] is that a 50% DC to RF conversion gain has been observed at 2 GHz [5] - a figure that has not been approached in other two terminal devices such as IMPATTs and TEDs. Owing to technological difficulties, such as contacting the device and problems of device/circuit matching, it is believed that the powers so far obtained from DBRTD circuits are still well below what can be expected. The purpose of the work described here is therefore to develop the DBRTD structure to increase the power generation, to improve the device match to the circuit to optimise the power extracted and to investigate circuit techniques for combining the power from a number of devices to a useable level.

2. The Devices

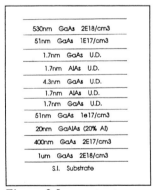

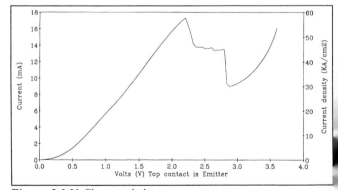

Figure 2 Layer structure.   **Figure 3** I-V Characteristic

The planar devices shown in figure 1 are fabricated from the GaAs/AlAs layer structure shown in figure 2. The layer structure has been designed both to maximise the peak-to-valley current ratio and to extend the negative differential resistance region over as large a voltage range as possible. This is achieved by inclusion of the 400 nm layer on the collector side of the double barrier. The DC current-voltage characteristics for the 30 µm^2 device of figure 1 are shown in figure 3 where the peak-to-valley ratio is seen to be 2:1 with a separation of 600 mV. From simple theory with optimum matching to the high frequency circuit this device should be capable of producing approximately 1 mW, providing the effects of the parasitics associated with the device are negligible at frequency of interest.

Figure 4 shows a typical planar device/antenna structure used to couple the power into the wave-guide. The antenna is deposited on a GaAs substrate with the DBRTD mounted adjacent to it and just below the W-band wave-guide slot. Details of a similar arrangement using whisker-contacted devices have been discussed elsewhere [6]; some results of this work are presented below for comparison purposes.

Figure 4 Device/antenna mounted in W-band wave-guide

3. Results

Figure 5a shows the spectrum of the signal generated by a single planar device mounted in the wave-guide. As can be seen the maximum power is 1.3 µW at 77 GHz. Figure 5b shows a comparable spectrum for two devices mounted in the waveguide where the power developed is 4 µW. The equivalent spectra for whiskered devices [6] are shown in figures 6a and 6b which indicate output powers of 2.6 µW and 12.6 µW respectively.

4. Discussion and Conclusion

Figure 5b shows that the power from two DBRTD planar devices operating at W-band can be combined to gain extra power. The combining action can be accomplished over a frequency range of at least 600 MHz by control of the individual bias voltages. The matching of the devices into the wave-guide by means of the antenna structure is crucial to obtaining power combining of these planar devices. In this work a number of different geometries have been investigated; the best observed output corresponds to less than 1% of what is theoretically possible so that much further optimisation is required.

The spectrum shown in figure 6b is similarly obtained when the DC bias to the two whiskered devices is individually adjusted to bring the signals together so that they injection-lock over a frequency range of ±300 MHz. It is again noted that the combined power from the two devices operating together is 12.6 µW i.e more than twice that produced by the devices operating individually. Although the whiskered device circuits in this case give rather better individual and power-combined performance, planar circuits discussed offer many advantages in that they can be scaled down for operation at higher frequencies. At such frequencies, because of the smaller dimensions, it is easier to fabricate devices and antennas in an integrated structure using a planar geometry.

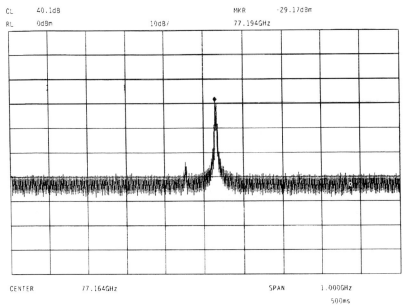

Figure 5a Output of a single planar device

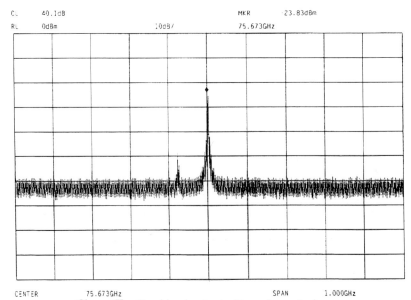

Figure 5b Combined output of two planar devices

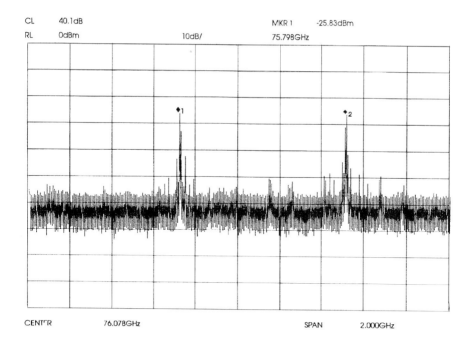

Figure 6a Output of two whiskered devices

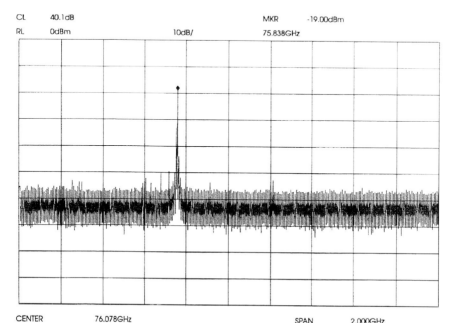

Figure 6b Combined output of two whiskered devices

5. Acknowledgements

The Engineering and Physical Science Research Council (formerly SERC) for financial support under grants GR/H-59497, GR/H-59503 and GR/H-96997. The EPSRC III-V Semiconductor processing facility at Sheffield University, especially Dr G Hill and Mr M Pate.

6. References

[1] Faist J, Capasso F, Sivco D L, Sirtori C, Hutchinson A L, Chu S N G and Cho A Y, 1994 *this conference*.

[2] Treen A J and Cronin N J 1993 *Proceedings of the 18th International Conference on Infrared and Millimeter Waves*, Proc. SPIE **2104** 470-471.

[3] Brown E R, Soderstrom J R, Parker C D, Mahoney L J, Molvar K M and McGill T C 1991 *Appl.Phys.Lett.* **58** 2291-2293

[4] Sollner T C L G, Brown E R, Goodhue W D and Le H Q 19876 *Appl. Phys. Lett.* **50** 322-324

[5] Javalagi S, Reddy V, Gullapalli, K and Neikirk D 1992 *Electronics Letters* **28** 1699-1701

[6] Steenson D P, Miles R E, Pollard R D, Chamberlain J M and Henini M 1994 *Proceedings of the Fifth International Symposium on Space Terahertz Technology*, Ann Arbor.

Selective molecular beam epitaxy for multifunction microwave integrated circuits

Dwight C. Streit, Donald K. Umemoto, Thomas R. Block, An-Chich Han, Michael Wojtowicz, Kevin Kobayashi, and Aaron K. Oki

TRW Electronic Systems and Technology Division
One Space Park, R6-2573, Redondo Beach, CA 90278 USA

Abstract. We report here the use of selective molecular beam epitaxy to produce microwave circuits that monolithically integrate multiple device technologies on the same chip. We have recently achieved several unique multifunction technology mixes using selective MBE. These include npn-pnp complementary HBT circuits such as push-pull amplifiers, and monolithic microwave HEMT-HBT integrated circuits such as 5-10 GHz HBT-regulated HEMT low-noise amplifiers.

1. Introduction

The integration of multiple device technologies on the same GaAs substrate is very attractive for multifunction microwave integrated circuits. The monolithic integration of npn and pnp heterojunction bipolar transistors (HBTs) allows complementary HBT circuits with improved efficiency and reduced power consumption. Active pnp loads can replace load resistors in HBT amplifier gain stages, reducing the required voltage supply. Push-pull amplifiers can be fabricated using complementary HBTs without the need for complex transformer circuits to implement 180° baluns. The monolithic integration of high electron mobility transistors (HEMTs) and HBTs allows high performance microwave circuits that combine the high frequency and low noise performance of HEMTs with the high linearity and low phase noise performance of HBTs.

Previous attempts at monolithic integration of HBTs with field-effect transistors have relied upon stacked epitaxial structures with the FET merged into the HBT collector [1] or emitter [2], or by using AlGaAs overgrowth combined with beryllium implantaion [3]. We have applied the selective molecular beam epitaxy technique to both npn-pnp HBT integration [4] and HEMT-HBT integration [5].

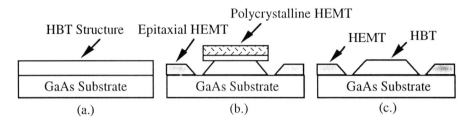

Figure 1. Selective MBE process flow schematic to achieve HEMT-HBT inetgration.

2. Fabrication

The selective MBE techniques described here are based on the conceptually simple use of silicon nitride definition and patterning of a previously grown epitaxial layer. The process used for HEMT-HBT material integration is shown in Figure 1. The npn GaAs-AlGaAs HBT structure is grown first. The wafer is removed from the MBE system and patterned with silicon nitride deposited by plasma-enhanced chemical vapor deposition. The silicon nitride and the HBT layer are etched to form HBT islands, and the next epitaxial layer is deposited. This layer can be a pnp HBT for complementary HBT integration, or HEMT for HEMT-HBT integration. Key to the success of this technique is the stability of the first layer when subjected to the thermal cycling associated with the regrowth process. We have found that HBTs optimized for high reliability [6] can survive the 600°C temperatures required for the regrowth process. However, pseudomorphic InGaAs HEMTs are subject to channel degradation when exposed to temperatures above 600°C [7], therefore the HBT layers are grown first.

Seperate merged processing flows were developed to allow the fabrication of the complementary HBTs or the HEMT-HBT integrated circuits without degrading the discrete device performance of either profile. In both the npn-pnp HBT and

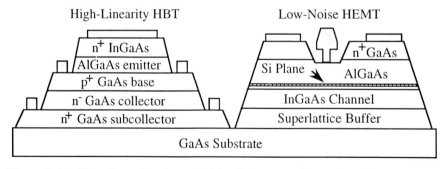

Figure 2. Profile schematic of monolithically integrated HEMT and HBT devices.

HEMT-HBT integration cases the additional process steps are minimized by using common ohmic metals and annealing steps wherever practical. In general, the additional complexity associated with the merged process flow is limited to less than three additonal masks levels, including the original silicon nitride mask step. The 0.2 μm HEMT gate is written by electron beam lithography, The HEMT surface is not affected by the selective MBE process, and we have found no degradation in the quality of the gate lithography due to the additional MBE growth steps. The devices are isolated within each island using oxygen ion implantation.

3. Device and circuit results.

In general the dc and microwave performance of discrete HBTs and HEMTs in both the complementary npn-pnp HBT and HEMT-HBT integration are equivalent to that achieved using our baseline single-technology process. The amount of base-dopant diffusion in the npn HBT during the HEMT regrowth is estimated to be about 25Å, in agreement with a 600°C beryllium diffusion coefficient of 2.6×10^{-17} cm^2/s.

The dc characteristics of devices reslting from the HEMT-HBT integration are shown in Figure 3. I-V curves for a 2x10 μm^2 HBT is shown in Figure 3(a). The device breakdown BV$_{ceo}$ is greter than 10V, Early voltage is greater than 500V, and ß>50 for Ic~4 mA. I-V curves for a 2-finger 80μm gate-width 0.2 μm gate-length HEMT device are shown in Figure 3(b). Transconductance is typical at 600 mS/mm and maximum drain current is ~600 mA/mm. Microwave performance is similiarly unaffected with fT = 22 and 70 GHz for the HBT and, HEMT devices, respectively.

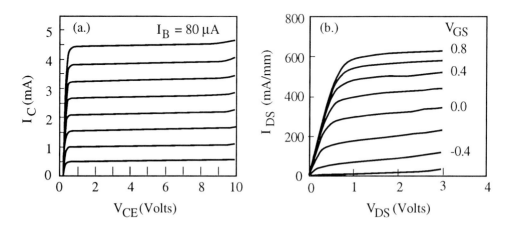

Figure 3. Current-voltage characteristics of monolithically integrated (a.) GaAs-AlGaAs HBT and (b.) InGaAs-GaAs pseudomorphic HEMT.

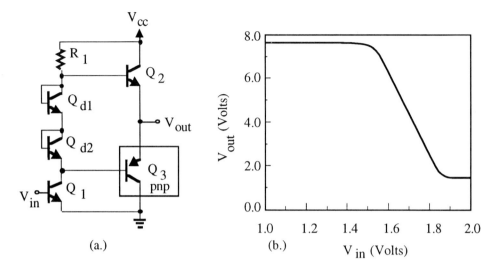

Figure 4. (a) Circuit diagram and (b) voltage transfer curve of basic complementary npn-pnp HBT push-pull amplifier.

The circuit and performance of a complementary HBT push-pull amplifier are shown in Figure 4. The saturated output power is 11 dBm at 2.5 GHz with a total power dissipation of 180 mW. This is the first demonstration of an MBE-grown monolithic push-pull complementary HBT amplifier. [8]

In conclusion, we have demonstrated the feasibility of using selective MBE to fabricate npn-pnp HBT and HEMT-HBT monolithic microwave integrated circuits.

References

[1] K. Itakura, *et al.*, IEDM Technical Digest, pp. 389-392, 1989.

[2] D. Cheskis, *et al.*, IEDM Technical Digest, pp. 91-94, 1992.

[3] J.Y. Yang, *et al.*, Proceedings IEEE GaAs IC Symposium, pp. 341-344, 1989.

[4] D.C. Streit, *et al.*, J. Vac. Sci. Tech. vol. B10, pp. 1020-1022, 1992.

[5] D.C. Streit, *et al.*, Proceedings IEEE GaAs IC Symposium, Philadelphia, 1994.

[6] D.C. Streit, *et al.*, J. Vac. Sci. Tech. vol. B10, pp. 853-855, 1992.

[7] D.C. Streit, *et al.*, Appl. Phys. Lett., vol. 58, pp. 2273-2275, 1991.

[8] K.W. Kobayashi, *et al.*, IEEE J. Solid-State Circuits, vol. 28, pp. 1011-1017, 1993.

High performance AlGaAs/InGaAs pseudomorphic HEMTs after Epitaxial Lift-Off

Y Baeyens† C Brys‡ J De Boeck, W De Raedt, B Nauwelaers†
G Borghs, P Demeester‡ and M Van Rossum

IMEC, Kapeldreef 75, B-3001 Leuven, Belgium

† KULeuven, ESAT-TELEMIC, Kardinaal Mercierlaan 94, B-3001 Leuven, Belgium

‡ RUG/INTEC-IMEC, St.-Pietersnieuwstraat 41, B-9000 Gent, Belgium

Abstract.
GaAs pseudomorphic HEMTs with a gatelength of 0.25 μm were lifted-off from the GaAs substrate and attached to a MgO-substrate. Only minor differences were observed in the DC and RF characteristics before and after the epitaxial lift-off (ELO). A high extrinsic transition frequency (95 GHz) and a slightly higher microwave gain (MAG=16dB @ 12GHz) were measured after transplantation. Also typical noise characteristics were measured (NF_{min} = 0.9dB @ 12 GHz), indicating that the high frequency noise was not affected by the ELO.

1. Introduction

Ever since the development of commercially viable products based on Silicon, III–V or any other materials technology, there has been a movement and a need to integrate the respective technologies to enhance and complement their relative strengths and weaknesses. Recently, the epitaxial lift-off (ELO) and transplantation of GaAs based FETs has gained a lot of attention as a cost-effective way to integrate in a planar process GaAs microwave MESFETs and HEMTs and other components based on different materials.

In the ELO-process, a sacrificial AlAs layer, buried between the GaAs substrate and the active device layer, is etched away using an extremely selective HF:DI (1:10) solution, leaving a very thin film (0.5 - 4 μm), which can then be transferred to a new host substrate, as was first demonstrated for GaAs solar cells in the late 70's by Konagai et al [1]. Since then, the ELO-step has been further optimised and demonstrated both for optoelectronic devices and for MESFETs and HEMTs [2, 3].

Recently, Young et al.[4] studied the effect of the ELO-step on the RF properties of conventional and pseudomorphic HEMTs and measured a consistent increase of 12 -

20% in the cut-off frequency f_T for lifted-off HEMTs with gatelengths varying between 0.8 and 1.2 µm. This change in f_T was attributed to an increase in the confined 2-DEG carrier concentration n_{so}, presumably caused by backgating effects [5].

2. Device fabrication

In our study, shorter devices with gatelengths between 0.25 and 1.0 µm were fabricated on a epitaxial layer consisting of a n^+ GaAs caplayer (40nm, $5.10^{18} cm^3$), a n^- $Al_{0.25}Ga_{0.75}As$ Schottky layer (30nm, $5.10^{17} cm^3$), a Si δ-doping of ($5.10^{12} cm^2$), a 5nm undoped $Al_{0.25}Ga_{0.75}As$ spacer layer and a 13 nm $In_{0.2}Ga_{0.8}As$ channel grown on top of an undoped GaAs buffer with a 50 nm AlAs release layer buried 2 µm below the heterojunction. Devices were fabricated using our standard PHEMT technology. [6]. Additionally, the HEMT devices were passivated using a PECVD deposited silicon nitride to protect the gate area. After passivation, the devices were fully DC and RF characterised.

Before performing ELO, the frontside of the wafer was coated with wax. After underetching the AlAs in HF:DI (1:5), the devices were transferred to the MgO-substrate. As an interlayer between the MgO and the GaAs HEMT, a 300nm polyimide layer was used to planarize the MgO substrate and to form a good interlayer for the thin film adhesion. This thickness is a good compromise between planarisation and heat sinking. To enhance process control and cleanliness, the ELO etching and subsequent transfer and deposition were performed "under water"[7]. Using an optical alignment procedure, a deposition accuracy of 5 µm can be achieved.

3. Experimental results and discussion

Different DC- and RF characteristics were measured before and after the ELO-step. As shown in fig. 1, the transplantation causes a small positive shift in the threshold

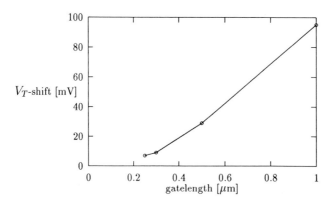

Figure 1. V_T-shift after ELO, as a function of the gatelength

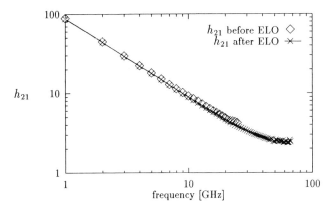

Figure 2. Measured h_{21}-characteristics of a 0.25μm PHEMT before ($\Diamond$ up to 25 GHz) and after (x up to 67 GHz) the ELO-step

voltage V_T. This shift is dependent on the gatelength of the device, the threshold voltage of short devices however is not affected by the transplantation.

The same applies to both the DC- and HF-transconductance g_m: for devices with a gatelength of 1 μm, a 10% increase of the transconductance is measured, a value comparable with the enhancement reported by Young [4]. For devices with shorter gates however, no real improvement is observed, which is an indication that the enhancement in g_m is indeed caused by a higher confined carrier concentration n_{so}, because this will affect especially the g_m of devices with longer gate lenghts, according to formula (1):

$$g_{max} = \frac{q\mu n_{s0}}{L\left(1 + (q\mu n_{s0} d_{eff}/\epsilon v_s L)^2\right)^{0.5}} \quad (1)$$

In this formula μ is the electron mobility, L the gate length, d_{eff} the effective 2-DEG to gate separation and v_s the carrier velocity. As can be seen from the formula, the g_m of shorter gate devices is less sensitive to a change in the carrier concentration.

The h_{21}-characteristics before and after ELO are compared in fig. 2. Again, no substantial degradation nor improvement can be noticed for the f_T-value of the 0.25μm PHEMTs after the transplantation. The measured extrinsic f_T, including all parasitics, is 95 GHz for a 150μm wide device. This is to our knowledge, the highest f_T-value reported for a FET after transplantation to another substrate.

When comparing the small-signal parameters, extracted before and after ELO, a small (5-10%) reduction of the output conductance g_{ds} is found for the 0.25μm PHEMTs, presumably due to a better carrier confinement. This reduction in g_{ds} causes a small increase of the microwave gain as plotted in fig. 3.

After ELO, the thermal noise of different devices was measured. Typical noise characteristics were measured: a minimal noise figure NF_{min} of 0.9dB at 12 GHz with an associated gain of 11dB, indicating that the high frequency noise was not affected by the ELO.

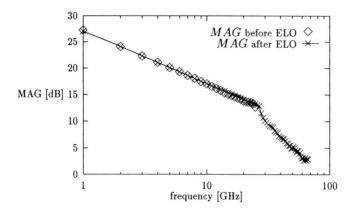

Figure 3. Measured Maximal Available Gain (MAG) of a 0.25μm PHEMT before (◇ up to 25 GHz) and after (x up to 67 GHz) the ELO-step

4. Conclusions

We demonstrated the succesfull lift-off of high-performant pseudomorphic HEMTs. No real degradation nor improvement of the HF- and noise characteristics was observed after ELO, indicating the potential of ELO-integration as an attractive alternative for bonding in millimetre wave systems.

Acknowledgments

This work was partly supported by the EC (BRA #6625 X-band SRO). Y. Baeyens and C. Brys thank the IWONL for financial support.

References

[1] Konagai M, Sugimoto M and Takaashi K 1978, *J. Crystal Growth* **45** 277–280

[2] Yablonovitch E, Hwang D, Gmitter T, Florez L and Harbison J 1990 *Appl. Phys. Lett.* **56** 2419–2421

[3] De Boeck J and Borghs G 1993, *J. Crystal Growth* **127** 85–92

[4] Young P G, Romanofsky R R, Alterovitz S A, Mena R A and Smith E D 1993 *IEEE Trans. on El. Dev.* **40** 3970–3976

[5] Mena R A, Schacham S E, Young P G, Haugland E J and Alterovitz S A 1993 *J. Appl. Phys.* **74** 1905–1909

[6] Zou G, De Raedt W, Van Hove M, Van Rossum M, Jin Y, Launois H 1992 *Microelectronic Engineering* **19** 321–324

[7] Demeester P, Pollentier I, De Dobbelaere P, Brys C and Van Daele P *Semicond. Sci. Technol.* **8** 1124–1135

Inst. Phys. Conf. Ser. No 141: Chapter 6
Paper presented at Int. Symp. Compound Semicond., San Diego, 18–22 September 1994
© 1995 IOP Publishing Ltd

Electron Transport in Doped InAs/InGaAs/AlInAs Quantum Wells and the Effect on FET Performance

John K. Zahurak[†‡], Agis A. Iliadis[‡], Stephen A. Rishton [†], and W. Ted Masselink[†*]

[†] IBM T. J. Watson Research Center, Yorktown Heights, N.Y. 10598
[‡] Dept. of Electrical Engineering, University of Maryland, College Park, Md. 20742
[*] Institute for Physics, Humboldt-University, Berlin

Abstract: We report the development of a novel, high-performance doped-channel heterojunction FET which incorporates a pseudomorphic, undoped InAs layer into the center of the uniformly-doped channel. Low-field mobility, peak electron velocity, and transistor performance have been studied as a function of InAs thickness. An optimal thickness of 30 Å significantly reduces ionized impurity scattering and enhances electron transport at all fields. Transistors based on the InAs quantum-well structures with 0.5μm gates yielded record (for doped-channel devices) room temperature extrinsic transconductances of 708 mS/mm; more than a 100% increase over the control samples with no InAs.

1. Introduction:

The InGaAs/AlInAs material system is well suited for high performance transistor fabrication due to its high electron mobilities, high peak electron velocity, large Γ-L valley separation, and large conduction band discontinuity at the heterojunction. Pseudomorphic InGaAs/AlInAs MODFETs grown on InP currently hold most FET performance records [1]. Modulation-doped structures are, however, limited in the number of carriers that can be transferred away from the donors, across the potential barrier, and into the channel. The gate voltage swing in MODFETs is also limited due to the formation of a parasitic MESFET in the AlInAs layer under positive bias conditions. Additionally, MODFETs suffer from trapping problems associated with doping the AlInAs layer [2,3]. These problems can be minimized by introducing the dopants directly into the channel. The drawback of doping the channel is a reduction in low-field electron mobilities. However, the importance of low-field mobility is expected to be minimized as device gate lengths shrink into the deep-sub micron regime and more of the channel is in the high-field transport regime [4]. At high fields, the effectiveness of ionized impurity scattering is reduced and phonon emission becomes the dominant scattering mechanism.

Although doped channel FETs have been fabricated with good results [5-8], a systematic study and optimization of design is needed for a better understanding of device operation. Our previous studies of DCHFETs grown lattice matched to InP showed that a 300 Å quantum well channel, either uniformly or center δ-doped, is the optimum size for both transport and transistor performance [9]. In the present work, a 300 Å uniformly-doped quantum well channel transistor is developed with a thin layer of undoped InAs inserted into the center of the channel. The operation of this device is examined, and the concept of quasi-modulation doping is developed.

Increasing the In mole fraction in the channel simultaneously decreases the electron effective mass while improving carrier confinement. Furthermore, by selectively introducing excess In into the channel, the wavefunction of the electrons can be shifted away from the position of the dopants. The reduced overlap between the electronic wavefunction and the parent donors causes a reduction in ionized impurity scattering, thereby enhancing transport and essentially creating a quasi-modulation doped structure.

2. Experimental Techniques:

All structures in this study were grown on InP:Fe substrates using gas-source MBE. Prior to growth, the native oxide was desorbed under a phosphorus flux to maintain surface stoichiometry. The composition of the layers was calibrated using x-ray analysis. Typical FWHM values are 20 arc seconds for the epitaxial layers and 13 arc seconds for the InP substrate. As shown in Figure 1, growth begins with a 400 Å AlInAs buffer layer. A thickness of 400 Å provides good carrier confinement while minimizing layer thickness. Next, a 300 Å, uniformly-doped (doping concentration of $8 \times 10^{17} cm^{-3}$) n-$In_{0.53}Ga_{0.47}As$ conducting channel was grown. The resulting sheet carrier concentration is $2.44 \times 10^{12} cm^{-2}$. Inserted into the center of the conducting channel is an undoped InAs layer with thickness ranging from 0 to 40 Å in 10 Å increments. Next, a 250 Å $Al_{0.49}In_{0.51}As$ gate layer was grown to facilitate Schottky barrier formation. The final layer is 50 Å of n+ InGaAs to facilitate ohmic contact formation.

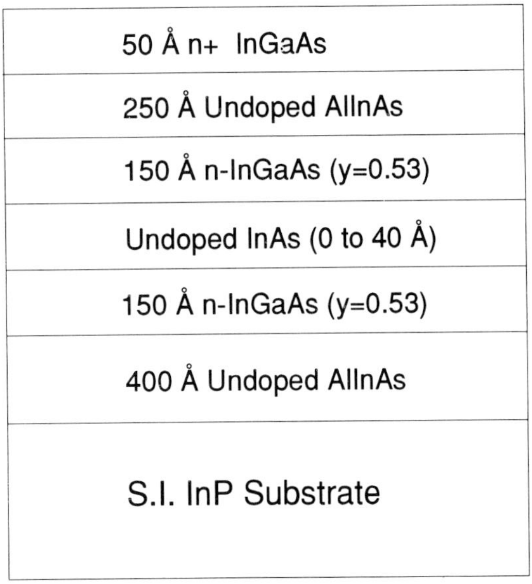

50 Å n+ InGaAs
250 Å Undoped AlInAs
150 Å n-InGaAs (y=0.53)
Undoped InAs (0 to 40 Å)
150 Å n-InGaAs (y=0.53)
400 Å Undoped AlInAs
S.I. InP Substrate

Figure 1. Cross section of the structures used to study electron transport and to fabricate both micron and sub-micron FETs.

Electron transport in the structures was characterized using low-field mobility measurements and microwave-heated carrier measurements. Low-field mobilities were measured as a function of temperature using the Hall effect measurement technique in the van der Pauw geometry. At higher electric fields, domain formation prevents accurate measurement of electron velocities using dc techniques. We therefore utilized a microwave-heated carrier technique that uses a large, 35GHz sinusoidal field superimposed on a comparatively small dc field to measure electron velocities at high fields. The ac-field serves only to heat the carriers while the dc-field drifts them. The measured dc conductivity as a function of applied ac field uniquely determines the electron velocity-field relationship [10].

Transistors with gate-lengths of 1.8 and 0.5μm were fabricated using optical and electron-beam lithography respectively. AuGe based ohmics were first deposited, lifted-off, and then annealed. To prevent lateral diffusion of the ohmic contacts in multi-layered structures, an optimized gradual, low-temperature alloying cycle was developed [11]. Typical contact resistances of $0.1\Omega-mm$ with no detectable lateral diffusion and excellent surface morphology were obtained. Next, mesa isolation was achieved using $H_3PO_4 : H_2O_2 : H_2O$ (1:1:38) wet-chemical etching. The gate fabrication process begins by removing the n+ InGaAs ohmic layer in the gate region using a selective citric acid etch [12]. Finally, Ti/Pt/Au gates were deposited and lifted-off.

3. Experimental Results:

Low-field mobilities were measured in the device structures at temperatures ranging from 12 to 300K. Figure 2 shows that the addition of a thin InAs layer into the channel causes an increase in mobility. Room temperature mobilities reach a peak value of 5329 cm^2/Vs for the 30 Å InAs sample: a 31% increase over the control sample (0 Å of InAs). The mobility of the sample with 40 Å of InAs decreased to 3453 cm^2/Vs indicating that the critical thickness for InAs in InGaAs is between 30 and 40 Å of InAs. Beyond this thickness, strain is accommodated by the formation of dislocations which impede electron transport. A critical thickness of 30 Å is in agreement with that measured by Eugster [13] and is 10 Å less than that reported by Akazaki [14] and Tournie [15]. As seen in

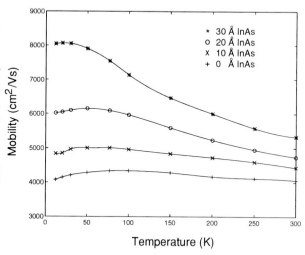

Figure 2. Electron mobility measured as a function of InAs thickness and temperature. The distinct increase in mobility with increasing InAs thickness and decreasing temperature indicates a substantial reduction in ionized impurity scattering in these structures.

Figure 2, the mobility of the 0-Å control sample has little dependence on temperature. Such a flat response of mobility with temperature is characteristic of heavily-doped channels: ionized impurity scattering dominates at low temperatures while phonon scattering dominates at higher temperatures. In contrast, we find that the mobility of the 30 Å InAs sample increases substantially with decreasing temperature indicating that ionized impurity scattering plays a less important role in this structure and therefore, that this structure is behaving in a quasi-modulation doped fashion. While a 30 Å well is too narrow to allow the electronic wavefunction to be bound in the well, the data demonstrates that the overlap of the electron wavefunction with the InAs is sufficient to avoid a significant portion of the ionized impurity scattering in the InGaAs layers.

Measurements at higher electric fields show a well defined peak velocity followed by a negative differential resistance region. Electron velocities at higher fields also increased with increasing InAs thickness. The peak velocity of the control sample is $1.4 \times 10^7 cm/s$ and increases to $1.63 \times 10^7 cm/s$ for the 30 Å sample. Again, the velocity of the 40 Å structure decreased to $1.35 \times 10^7 cm/s$. Correspondingly, the critical field (the electric field where the peak velocity occurs) decreased from 4.1 to 3.3 kV/cm respectively. These results are in

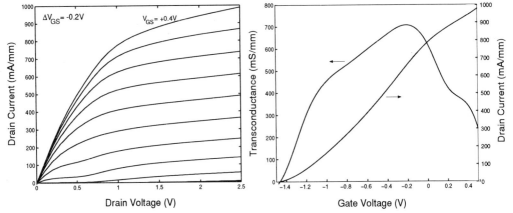

Figure 3. Typical output characteristics of a doped-channel HFET with 30Å of InAs inserted into the center of the uniformly-doped well.

Figure 4. Typical transconductance and drain current characteristics for a 0.5 μm Q-MODFET with 30 Å of InAs inserted into the center of the uniformly-doped well.

agreement with a model which predicts that the peak velocity is proportional to $\mu^{\frac{1}{2}}$ and the critical field is proportional to $\mu^{-\frac{1}{2}}$ [10].

With electron velocities at both low and high fields increasing with increasing InAs thickness (up to 30 Å), we observe a corresponding improvement in transistor performance. The 30 Å sample displayed the best results with an extrinsic transconductance of 708 mS/mm for the 0.5μm gate-length FETs and 420mS/mm for 1.8μm FETs. This extrinsic transconductance approaches the best results for any 0.5μm gate-length FET and it is the best result for any doped-channel FET. By comparison, these results are significantly better than those obtained from the control sample: transconductances of 315mS/mm (0.5μm) and 267mS/mm (1.8μm). Peak drain currents also increased substantially with increasing InAs concentration reaching 730mA/mm ($V_{GS} = 0V$, $V_{DS} = 2.5V$) for the 30 Å sample. At slightly positive gate voltages, drain currents of 990mA/mm are observed ($V_{GS} = +0.4V$ and $V_{DS} = 2.5V$). Figures 3 and 4 show the output and transfer characteristics of a typical transistor.

Output conductances in all of the InAs-containing FETs are higher than in the control FET increasing from 16 to 37mS/mm in the 0.5μm FETs. No dependence of output conductance on InAs thickness was observed in the InAs containing structures. Under zero gate bias conditions, the drain-source breakdown voltage decreases with increasing InAs thickness from 6.7 to 5.0 V for the control and the 30 Å InAs sample respectively. Drain-source breakdown voltages under pinch-off conditions are independent of InAs thickness and are in excess of 12V. The gate-drain breakdown voltage decreases with increasing InAs thickness from 8.0V to 6.5V for the control and the 30 Å sample respectively. Although it is difficult to make direct device comparisons with breakdown voltages reported elsewhere due to variations in device geometries, our breakdown voltages are much greater than those of InAs-Sb based FETs (approximately 1V for drain-source breakdown) [16] and are comparable to those of Bahl [17] (BV_{GS} of 9.6V for InGaAs/AlInAs HFETs with $2\mu m$ gate-lengths) and Hwang [18] (BV_{GD} of 10V on quarter-micron GaAs FETs). We therefore conclude that breakdown voltages are not a limitation in the use of our InAs based FETs.

Based on transconductance results and measured gate capacitances, we anticipate f_t of 50GHz. Furthermore, using various analytical FET models which approximate the velocity field curve using a two-piece model [19], we extract an average electron velocity in the channel of the $0.5\mu m$ FETs to be $1.6x10^7 cm/s$. This average velocity corresponds closely to the peak velocity we measure using the microwave-heated carrier technique indicating that already in FETs with $0.5\mu m$ gate-lengths, a significant fraction of the channel electrons travel with velocities in excess of the maximum steady-state velocity.

4. Conclusions:

We have demonstrated that the inclusion of a thin, undoped InAs layer into the center of a uniformly-doped channel reduces ionized impurity scattering and therefore improves electron transport at all electric fields. The optimal thickness of the InAs layer is 30 Å for all characteristics as shown in Figure 5. Beyond 30 Å, the critical thickness of the InAs/InGaAs system is exceeded and the strain is accommodated by the formation of dislocations which degrade performance. The 30 Å InAs structure set new performance records for doped-channel FETs. Electron velocities extracted from $0.5\mu m$ transistors indicate that the average electron velocity in the channel is the same as the steady-state peak electron velocity measured using the microwave-heated carrier technique. The correlation between

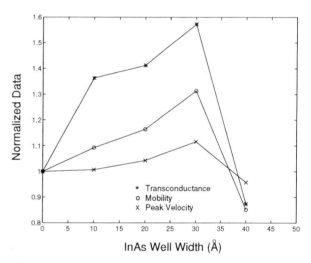

Figure 5. Transport and transistor performance as a function of InAs thickness. Clearly the optimal thickness of the InAs layer is close to 30 Å.

the average electron velocity and the peak velocity demonstrates that the most important velocity in sub-micron FETs is the materials steady-state peak velocity. Low-field mobilities are indirectly important through the square-root dependence of peak velocity on mobility. These results indicate that already in half-micron gate length FETs, a significant fraction of the channel electrons are traveling with velocities in excess of their steady-state peak velocities.

References

1. L.D. Nguyen, A.S. Brown, M.A. Thompson, and L.M. Jelloian, IEEE Transactions on Electron Devices, Vol. 39, pp. 2007-2014, 1992.

2. W-P. Hong, S. Dhar, P.K. Bhattacharaya, and A. Chin, Journal of Electronic Materials, Vol. 16, pp. 271-274, 1987.

3. K. Nakashima, S. Nojima, Y. Kawamura, and H. Asahi, Phys. Stat. Sol. (A), Vol. 103, pp. 511-516, 1987.

4. M.V. Fischetti and S.E. Laux, IEEE Trans. on Electron Devices, Vol. 38, No. 9, pp. 650-660, 1991.

5. H. Hida, Y. Tsukada, Y. Ogawa, H. Toyoshima, M. Fujii, K. Shibahara, M. Kohno, and T. Nozaki, IEEE Transactions on Electron Devices, Vol. 36, No. 10, pp. 2223-2229, 1989.

6. P.P. Ruden, M. Shur, A. Akinwande, J.C. Nohava, D.E. Grider, and J. Baek, IEEE Transactions on Electron Devices, Vol. 37, No. 10, pp. 2171-2175, 1990.

7. M. Matloubian, A.S. Brown, L.D. Nguyen, M. Melendes, L.E. Larson, M.J. Delaney, J.E. Pence, R.A. Rhodes, M.A. Thompson, and J.A. Henige, IEEE Electron Device Letters, Vol. 14, No. 4, pp. 188-189, 1993.

8. J. Dickmann, H. Daembkes, H. Nickel, W. Schlapp, and R. Losch, IEEE Electron Device Letter Vol. 12, No. 6, pp. 327-328, 1991.

9. J.K. Zahurak, A.A. Iliadis, S.A. Rishton, and W.T. Masselink, *Proceedings: IEEE/Cornell Conference on Advanced Concepts in High-Speed Semiconductor Devices and Circuits*, IEEE Publishing, New York, pp. 270-278, 1993.

10. W.T. Masselink, Semicond. Sci. Technol., Vol. 4, pp. 503-512, 1989.

11. J.K. Zahurak, A.A. Iliadis, S.A. Rishton, W.T. Masselink, and T. Neil, To be published.

12. M. Tong, K. Nummila, A. Ketterson, I. Adesida, C. Caneau, and R. Bhat, IEEE Electron Device Letters, Vol. 13, pp. 525-527, 1992.

13. C.C. Eugster, T.P. Broekaert, J.A. del Alamo, and C.G. Fonstad, IEEE Electron Device Letters, Vol. 12, pp. 707-709, 1991.

14. T. Akazaki, K. Arai, T. Enoki, and Y. Ishii, IEEE Electron Device Letters, Vol. 13, pp. 325-327, 1992.

15. E. Tournie, K.H. Ploog, and C. Alibert, Appl. Phys. Lett., Vol. 61, pp. 2808-2810, 1992.

16. K. Yoh, T. Moriuchi, and M. Inoue, IEEE Electron Device Letters, Vol. 11, pp. 526-528, 1990.

17. S.R. Bahl, B.R. Bennett, and J.A. del Alamo, IEEE Electron Device Letters, Vol. 14, pp. 22-24, 1992.

18. T. Hwang and M. Feng, IEEE Electron Device Letters, Vol. 13, pp. 445-447, 1992.

19. M.B. Das, W. Kopp, and H. Morkoc, IEEE Electron Device Letters, Vol. 5, pp. 446-449, 1984.

Very Low Contact Resistance to n+-InP Grown Using SiBr4

Michael T. Fresina, Steven L. Jackson and Gregory E. Stillman

Center for Compound Semiconductor Microelectronics and Department of Electrical and Computer Engineering, University of Illinois, Urbana, IL 61801

Abstract. In this paper we report on a study of metal contacts to heavily doped n-type InP layers. We have achieved very low specific contact resistances using both nonalloyed Ti/Pt/Au and alloyed AuGe/Ni/Au contacts to n+-InP. The InP layers were grown by GSMBE with Si doping levels of $N_D > 6 \times 10^{19}$ cm^{-3} using SiBr$_4$. The layers exhibited no surface morphology degradation. Average specific contact resistances of 6.09×10^{-8} and 5.42×10^{-8} Ω-cm^2 were measured for AuGe/Ni/Au and Ti/Pt/Au contacts, respectively.

1. Introduction

Ohmic contacts to n-type InP are very important for heterojunction bipolar transistors (HBTs) and optoelectronic devices based on this material. Considerable research has been done on alloyed AuGe/Ni/Au-based contacts to n-InP [1 – 6]. Reports of specific contact resistances, ρ_c, in the low 10^{-7} Ω-cm^2 range are common [1 – 5], and Fatemi and Weizer [6] reported a $\rho_c < 2 \times 10^{-8}$ Ω-cm^2 for a thin InP/Au/Ni contact. However, all of these reports are for contacts to moderately doped (10^{17} - 10^{18} cm^{-3}) n-InP.

There are fewer reports of ohmic, nonalloyed Ti/Pt/Au contacts to n-InP [7, 8]. Low resistance, non-alloyed contacts to n-InP are not as easily formed due to the absence of an alloyed dopant (i.e., Ge) and the inherent Schottky barrier between a Ti/Pt/Au contact and n-InP. Katz et al. [7] fabricated Ti/Pt contacts by rapid thermal annealing to InP doped n = 5×10^{18} cm^{-3} with $\rho_c = 8 \times 10^{-7}$ Ω-cm^2, which, to the best of our knowledge, is the lowest reported contact resistance for Ti/Pt/Au nonalloyed contacts on n-InP.

For InGaAs/InP HBTs, employing a very heavily doped n-InP emitter contacting layer would eliminate the need for the InGaAs layer that is generally required when using nonalloyed Ti/Pt/Au contacts, and would considerably simplify the growth and fabrication of these devices. We have recently reported on the heavy doping of InP ($N_D - N_A > 6 \times 10^{19}$ cm^{-3}) grown by GSMBE using SiBr$_4$ without surface morphology degradation [9, 10]. Nonalloyed Ti/Pt/Au and alloyed AuGe/Ni/Au contacts have been fabricated to this material. In this paper, we report on the optimization of the sintering/alloying conditions for these contacts and the variation of ρ_c with the Si doping level. Minimum contact resistances of 6.09×10^{-8} and 5.42×10^{-8} Ω-cm^2 have been measured for alloyed AuGe/Ni/Au and non-alloyed Ti/Pt/Au contacts on n+-InP.

2. Experimental Procedure

All of the samples used in this study were grown by gas source molecular beam epitaxy (GSMBE). The growth precursors were PH$_3$ and elemental In. The doping source was SiBr$_4$; the details of doping with this source have been discussed in references [9] and

[10]. It has been shown that SiBr$_4$ has an incorporation efficiency 10^4 times greater than Si$_2$H$_6$ and has 100% activation up to [Si] = 7×10^{19} cm^{-3} in InP with specular surface morphology. All samples used in this study were 0.25 or 0.50 μm thick epitaxial layers of n-type InP grown on a semi-insulating Fe-doped InP substrate.

Contact resistances were measured using the transmission line model (TLM) technique. Rectangular TLM patterns were defined on all samples using standard lift-off lithography. Samples were etched in 1 BOE : 1 H$_2$O to remove native oxides before being placed in the evaporator. Contacts composed of 150 Å Ti / 150 Å Pt / 1000 Å Au or 750 Å AuGe (eutectic, 12% Ge by weight) / 150 Å Ni / 1200 Å Au were deposited by electron-beam evaporation at a pressure of 8×10^{-7} Torr. To prevent current spreading during measurements, the TLM patterns were isolated by etching into the substrate using an HCl-based etch. Samples were then cleaved into multiple pieces for alloying/sintering on a hotplate in a forming gas atmosphere (~ 5 % H$_2$ in N$_2$). All of the TLM patterns used in this study had contacts 60 μm wide and 40 μm deep with approximate spacings of 2, 4, 9, 14, 19, and 24 μm. The actual spacings and mesa widths were measured on each sample with a scanning electron microscope.

The technique used to calculate the specific contact resistance from the TLM patterns has been developed by several authors and is described by Williams [11]. Measurements were made using a four-probe arrangement to eliminate probe and system resistances. The resistance was measured between each contact in a TLM pattern, and these values were plotted versus the contact spacing (see Figure 1). A linear fit to each TLM data set was obtained using the least-squares technique, and the fits all had a regression coefficient greater than 0.999. From the slope of this line, the sheet resistance, R_s, of the conductive layer was calculated from

$$R_s = slope \times w, \qquad (1)$$

where w is the width of the TLM pattern. The y-intercept of the linear fit is two times the contact resistance, R_c, and the specific contact resistance, ρ_c, is

$$\rho_c = \frac{R_c^2 \times w^2}{R_s}. \qquad (2)$$

3. Results and Analysis

The alloying time and temperature for the AuGe/Ni/Au contacts and the sintering temperature for the Ti/Pt/Au contacts on the most heavily doped InP (n = 6.0×10^{19} cm^{-3}) were first optimized. Initial investigation of the Ti/Pt/Au contact resistance versus sintering time showed that at any temperature, the minimum value was attained after a 1 min sinter. Specific contact resistance is plotted in Figure 2 for Ti/Pt/Au contacts sintered for 1 min at temperatures between 200 °C and 350 °C. Each data point is the average contact resistance measured from three sets of TLM patterns on that sample. The lowest average contact resistance, $\rho_c = 5.42 \times 10^{-8}$ Ω-cm^2, was measured on a sample which was sintered at 320 °C. To the best of our knowledge, this is the lowest specific contact resistance ever reported for Ti/Pt/Au contacts on n-type InP. Figure 1 shows the TLM data for the lowest single measurement from that sample. The contact resistance calculated from these data was 4.63×10^{-8} Ω-cm^2.

The dependence of the AuGe/Ni/Au contact resistance on alloying time and temperature is shown in Figure 3. The temperature was varied from 300 to 375 °C and the samples were alloyed between 1 and 10 min. Again, each data point is the average

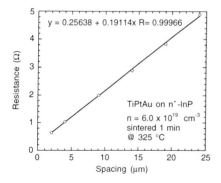

Figure 1. TLM data for lowest measured contact resistance of Ti/Pt/Au on InP.

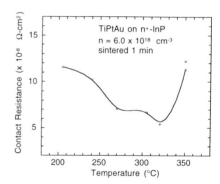

Figure 2. Sintering temperature optimization for Ti/Pt/Au contacts on n+-InP.

contact resistance measured from three sets of TLM patterns on that sample. From the data it appears that there may be a constant time/temperature product which results in the optimal alloying of the contacts over a significant temperature range. It does not appear that 305 °C is high enough to alloy the contact completely. An alloy time between 3 and 5 min at 320 °C provides a minimal contact resistance while longer times at that temperature appear to degrade the contact. At 350 and 375 °C, the minimum contact resistance is reached in less than a minute and then increases rapidly. A 3 min alloy at 320 °C was chosen as the optimal for this contacting system. The sample alloyed under these conditions had an average contact resistance of 6.09×10^{-8} Ω-cm^2. Only Fatemi and Weizer [6] have reported a lower AuGe/Ni-based contact resistance: $\rho_c < 2 \times 10^{-8}$ Ω-cm^2 for a thin Au/Ni contact.

The dependence of the contact resistance on the doping level of the InP was examined next. Figure 4 shows I-V curves for Ti/Pt/Au contacts on InP with various n-type doping levels. The data were taken by sweeping the voltage between two TLM contacts. The sample carrier concentrations for these traces are 7.5×10^{17}, 1.8×10^{18}, 5.8×10^{18}, 4.0×10^{19}, and 6.0×10^{19} cm^{-3}. From these data, it appears that only the contacts on the three most heavily doped samples are nonrectifying. These samples had average contact resistances of 5.01×10^{-5}, 2.90×10^{-7}, and 5.42×10^{-8} Ω-cm^2.

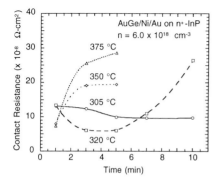

Figure 3. Alloying time and temperature optimization for AuGe/Ni/Au on n+-InP

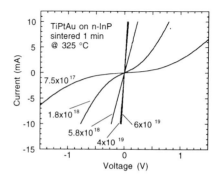

Figure 4. I-V characteristics of Ti/Pt/Au contacts on InP with varying doping levels.

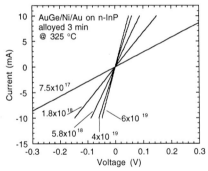

Figure 5. *I-V* characteristics of AuGe/Ni/Au contacts on n-InP with varying doping levels.

Figure 6. AuGe/Ni/Au specific contact resistance versus InP doping level.

The *I V* curves for AuGe/Ni/Au contacts on these same growth samples are shown in Figure 5. These contacts are nonrectifying for all doping levels because of the Ge doping of the InP during the alloying. The variation of the contact resistance with the InP doping level is plotted in Figure 6. As expected, ρ_c decreases as the doping level increases, although it appears that ρ_c may reach a minimum at very high doping levels.

4. Conclusion

Very low resistance alloyed and nonalloyed contacts to n-type InP have been fabricated using standard techniques. On InP with n = 6.0 × 10^{19} cm^{-3}, obtained using SiBr$_4$, minimum specific contact resistances of 6.09 × 10^{-8} Ω-cm^2 and 5.42 × 10^{-8} Ω-cm^2 were measured for alloyed AuGe/Ni/Au contacts and nonalloyed Ti/Pt/Au contacts, respectively. This is the lowest contact resistance reported for Ti/Pt/Au contacts on n-type InP. Low resistance, nonalloyed Ti/Pt/Au contacts to n-InP should make InP/InGaAs HBTs simpler to grow and fabricate thereby improving their viability.

Acknowledgments

The authors gratefully acknowledge helpful conversations with Dr. S.J. Prasad of Tektronix. This work was supported by the Joint Services Electronics Program (N00014-90-J-1270), SDIO/IST (DAAL 03-89-K-0080) administered by the Army Research Office, and by the National Science Foundation (NSF ECD 89-43166 and DMR 89-20538).

References

[1] D.G. Ivey, D. Wang, and D. Yang, *J. Electron. Mater.* **23**, 441 (1994).
[2] T. Clausen, O. Leistiko, I. Chorkendorff, and J. Larsen, *Thin Solid Films* **232**, 215 (1993).
[3] V.G. Weizer and N.S. Fatemi, *Appl. Phys. Lett.* **62**, 2731 (1993).
[4] J.A. Del Alamo and T. Mizutani, *Solid-St. Electron.* **31**, 1635 (1988).
[5] G. Bahir, J.L. Merz, J.R. Abelson, and T.W. Sigmon, *J. Electron. Mater.* **16**, 257 (1987).
[6] N.S. Fatemi and V.G. Weizer, *J. Appl. Phys.* **74**, 6740 (1993).
[7] A. Katz, B.E. Weir, S.N.G. Chu, P.M. Thomas, M. Soler, T. Boone, and W.C. Dautremont-Smith, *J. Appl. Phys.* **67**, 3872 (1990).
[8] A. Katz, B.E. Weir, and W.C. Dautremont-Smith, *J. Appl. Phys.* **68**, 1123 (1990).
[9] S.L. Jackson, S. Thomas, M.T. Fresina, D.A. Ahmari, J.E. Baker, and G.E. Stillman, *Proceedings of the 6th International Conference on InP and Related Materials*, Santa Barbara, CA (1994).
[10] S.L. Jackson, M.T. Fresina, J.E. Baker, and G.E. Stillman, *Appl. Phys. Lett.*, **64**, 2867 (1994).
[11] R.E. Williams, Gallium Arsenide Processing Techniques. (Artech House: Dedham, MA, 1984) p 241.

Ultra-high-speed HEMTs: progress towards the next generation

L D Nguyen

Hughes Research Laboratories, Malibu, CA 90265

Abstract. The push towards the next generation of ultra-high-speed High Electron Mobility Transistors (HEMTs) is once again in full swing. In just a few years, significant progress has been made, and it now appears that a 0.05-µm HEMT technology will soon become a reality. In this talk, I will review the status of this technology and discuss its potential impact on future systems.

1. Introduction

The successful development of the HEMT technology has given birth to a new class of ultra-high-speed transistors with cutoff frequencies (f_T, f_{max}) well over 100 GHz. Since the mid-1980s, HEMT technology has advanced at an incredible pace: a 4-fold increase in speed from 1984 to 1988 [1-4], and again a nearly 2-fold increase from 1988 to 1992 [5, 6]. Thanks to its higher speed, the HEMT has now replaced the GaAs FET for a wide range of microwave and millimeterwave low-noise applications. Today, there are tens of millions of HEMTs in operation around the world.

The need for lower noise figures of both the defense and commercial industries has brought the latest HEMT technologies to the market place. The 0.25-µm and sub-0.25-µm GaAs-based HEMTs offer the best combination of performance and cost and are now the work horse at X- and Ku-bands [7, 8]. At millimeterwave frequencies, a more advanced technology, the InP-based HEMT scaled to 0.10 µm, has emerged as the most attractive candidate for ultra-low-noise systems [9-11]. The InP HEMT offers significantly lower noise figures than the GaAs HEMT above 30 GHz, and is now the preferred technology for millimeterwave communications and radio astronomy.

Starting in 1989, the push for even higher performance HEMTs began in earnest. This time, the goal is to develop a 0.05-µm HEMT technology with significantly higher cutoff frequencies (>300 GHz) and lower noise figures. In this talk, I will review the status of this emerging technology and discuss its potential impact on future systems.

2. Scaling Rules

In the past decade, the III-V industry has successfully scaled the HEMT technology by a factor of 5, from 0.5 µm to 0.1 µm, by continually improving the device materials and lithography processes. Assuming that an aspect ratio of five (5) or greater can be maintained, it is straightforward to show that a properly-scaled HEMT must approximately obey a simple set of scaling rules, as listed in Table 1.

Table 1. Scaling rules for state-of-the-art HEMTs

Parameter	Scaling Rule	Uni
$C_{gs} + C_{gd}$	1.0	pF/mm
g_m	$100/L_g$	mS/mm
R_s	$< 2L_g$	$\Omega \cdot$mm
V_{ds}	$10L_g$	V
f_T	$20/L_g$	GHz
f_{max}	$2f_T$	GHz

where C_{gs}, L_g, R_s, and V_{ds} are the gate-to-source capacitance, gate length (in micrometers), source resistance, and drain-source operating voltage, respectively.

As L_g continues to shrink, however, it becomes increasingly more difficult to obey the scaling rules. First, it becomes very difficult to maintain an aspect ratio of five. At $L_g = 0.05$-μm, for example, the vertical dimension of a HEMT — the total thickness of the Schottky, spacer, and channel layers — must be reduced to approximately 100 Å! Second, for a given ambient temperature, there exists a minimum V_{ds} for the electrons to reach their maximum velocity. Empirically, this minimum V_{ds} appears to be a strong function of temperature, about 0.7 V at room temperature and 0.3 V at 18 K [11], but not of gate length. The 0.7-V minimum may also preclude the use of extremely narrow band gap materials, such as InAs or InSb, at room temperature!

3. Review of 0.05-μm HEMT technology

The real challenge in the development of a 0.05-μm HEMT technology is scaling its layer thicknesses. Despite the fact that modern MBE technology can produce materials with just a few monolayers, it remains a challenge to realize HEMTs with an active layer of only 100 Å. The main problems are thinning the Schottky and channel layers while maintaining low gate leakage current and achieving good ohmic contact. Table 2 summarizes these problems and offers a few potential solutions.

The implementation of the approaches in Table 2 has been a stepping stone towards the development of a 0.05-μm InP HEMT technology. In 1992, my colleagues and I reported a successful demonstration of both lattice-matched and pseudomorphic 0.05-μm InP HEMTs with f_T over 300 GHz [6]. We relied on a high-resolution 50-kV electron-beam lithography

Table 2. Scaling problems for 0.05-μm HEMT technology (see Refs. [6, 12])

Problem	Approach	Values Achieved
Schottky layer	Delta doping or low temperature growth	100 - 130 Å
Channel layer	Lattice-matched or pseudomorphic	150 - 200 Å
Reduce R_s & R_d	Self-aligned gate	0.1 $\Omega \cdot$mm
Reduce R_g	T-shaped gate	1000 Ω/mm
Ohmic contact	Non-alloyed ohmic contact	0.1 $\Omega \cdot$mm

system to define the 0.05-μm T-shaped gates, delta doping and low temperature growth to minimize the diffusion and surface segregation of the dopants, self-aligned gate technology to reduce R_s and R_d, and a low-temperature ohmic contact to achieve low resistance to the thin channel. In 1994, Enoki et al. reported a more refined 0.05-μm T-shaped gate process utilizing dielectric side wall and non-alloyed ohmic contact [12]. This approach is very promising and potentially could lead to a properly scaled 0.05-μm HEMT technology. Table 3 summarizes the results from both efforts and compares them with the scaled model.

A quick glance at Table 3 will reveal that these 0.05-μm HEMTs, despite their impressive g_m and f_T values, are not quite properly scaled. Their total gate capacitance, including a 0.1 to 0.2 pF/mm pad capacitance, is still too low. Their aspect ratio, as deduced from the capacitance value, is only about 3 (we want 5 or greater!). This low aspect ratio results in a high output conductance and consequently a low f_{max}. Future work must concentrate on the reduction of the Schottky layer (<80 Å), channel width (<75 Å), and output conductance (<60 mS/mm) to clearly demonstrate the potentials of a 0.05-μm HEMT technology.

4. Potential impact on future systems

The 0.05-μm HEMT technology is potentially an enabling technology for ultra-high-performance systems such as millimeterwave communications, optical communications with transmission speeds over 40 Gb/s, and ultra-low-noise cryogenic receivers. The key word here is *enabling*. There is little doubt that 0.05-μm HEMT technology will play a key role in these future systems, especially when the transmission speed approaches 100 Gb/s or when the operating frequency exceeds 100 GHz.

Table 3. Summary of Nguyen et al. and Enoki et al. 0.05-μm HEMTs

Parameter	Target	Nguyen et al. [6]	Enoki et al. [12]	Unit
$C_{gs} + C_{gd}$	1.0	0.80	0.7[*]	pF/mm
g_m	2000	1700	1280	mS/mm
R_s	0.1	0.1	0.1	Ω·mm
V_{ds}	0.5	0.8	0.8[**]	V
f_T	400 GHz	340	300	GHz
f_{max}	800 GHz	250	235	GHz

[*]Estimated from f_T & g_m values
[**]Estimated from I-V characteristics

References

[1] Camnitz et al. 1984 *IEDM Tech. Dig.*,360-363
[2] Lepore et al. 1988 *Electronic Lett.* **24** 364-366
[3] Moll et al. 1988 *IEEE Trans. Electron Devices* **35** 878-886
[4] Mishra et al. 1988 *IEDM Tech. Dig.* 180-183
[5] Mishra et al. 1989 *IEDM Tech. Dig.* 101-104
[6] Nguyen et al. 1992 *IEEE Trans. Electron Devices* **39** 2007-2014
[7] Henderson et al. 1986 *IEEE Electron Device Lett.* **7** 649-651
[8] Kawasaki et al. 1989 *IEEE MTT-S Sym. Dig.* 423
[9] Duh et al. 1990 *IEEE MTT-S Sym. Dig.* 595
[10] Chow et al. 1992 *IEEE MTT-S Sym. Dig.* 807
[11] Pospieszalski et al. 1994 *IEEE MTT-S Sym. Dig.* 1345
[12] Enoki et al. 1994 *Jpn. J. Appl. Phys.* **33** 798-803

External gate fringing capacitance model for optimizing a Y-shaped, recess-etched gate FET structure

M. Fukaishi, M. Tokushima, S. Wada, N. Matsuno and H. Hida

Microelectronics Research Laboratories, NEC Corporation
34 Miyukigaoka, Tsukuba, Ibaraki 305, Japan

Abstract. A simple yet accurate model is proposed for analyzing the gate fringing capacitance (C_f^{ext}) above the semiconductor in a Y-shaped, recess-etched gate FET. In this model, it is assumed that the capacitance value is determined by integrating the inverse of the minimum distance between the gate electrode and either the n^+-cap layer or the ohmic electrode with the smaller electrode area. Comparisons with two-dimensional simulations and experiments verify the high accuracy of the model. Also, the optimized gate structure is determined, in which C_f^{ext} is reduced to about half that of a conventional one, and its variation during the process is minimized.

1. Introduction

As gate lengths are reduced to sub-0.5 μm in the pursuit of higher-performance FETs, the gate fringing capacitance (C_f) contribution to the total gate capacitance increases. Since this C_f can ultimately limit the maximum obtainable FET performance, we must decrease it as much as possible. The C_f has two sources: yet, C_f in the semiconductor (C_f^{int}) and C_f above the semiconductor (C_f^{ext}). Many papers have reported simulating C_f^{int} dependence on gate-bias and on the recessed structure (Anholt 1991). C_f^{ext} hasn't been discussed quantitatively, while C_f^{ext} is significantly affected by gate structure parameters (Hida et al. 1993). In this paper, we focus on C_f^{ext}, propose an analytical model for determining gate structure parameters that dominate C_f^{ext}, and discuss C_f^{ext} dependence on these parameters to determine how we can reduce C_f^{ext}.

2. Analytical model

The capacitance between two electrodes is determined under the assumption that the electric field lines emanating from one electrode terminate at the nearest points on the other electrode. This method can be generalized as a way of deriving the capacitance value by integrating the inverse of the minimum distance between two electrodes with the area of the smaller electrode. The capacitance is given by

$$C = \varepsilon_{df}\varepsilon_0 \int\int \frac{1}{d(x,y)}dxdy,$$

where ε_{df} is the relative dielectric constant of the insulator between the electrodes, ε_0 is the dielectric constant of vacuum and $d(x,y)$ is the minimum distance between the two electrodes at the point (x,y) on the smaller electrode.

A cross-sectional view of the Y-shaped, recess-etched gate FET used in this C_f^{ext} analysis is shown in Fig. 1. C_f^{ext} is given by $C_f^{ext} = C_{f1} + C_{f21} + C_{f22} + C_{f3}$, where C_{f1}, C_{f21}, C_{f22}, and C_{f3} are the capacitances between the surface of the gate metal and either the n^+-cap layer or the ohmic electrode. We integrated the minimum distance with the surface of the n^+-cap layer or the ohmic electrode for C_{f1}, C_{f21}, and C_{f22}, and with the gate electrode surface for C_{f3}. The minimum distances used in these integrals are

$$d_1(x_1,\theta) = \begin{cases} L_r & (0 \leq x_1 \leq L_r(\frac{\sin\theta-1}{\cos\theta}) + H_g) \\ -x_1\cos\theta + L_r\sin\theta + H_g\cos\theta & (L_r(\frac{\sin\theta-1}{\cos\theta}) + H_g \leq x_1 \leq H_r) \end{cases}$$

$$d_{21}(x_{21},\theta) = \begin{cases} x_{21} + L_r & (0 \leq x_{21} \leq (H_g - H_r)\frac{\cos\theta}{1-\sin\theta}) \\ (x_{21}+L_r)\sin\theta + (H_g - H_r)\cos\theta & ((H_g - H_r)\frac{\cos\theta}{1-\sin\theta} \leq x_{21} \leq L_w - L_r) \end{cases}$$

$$d_{22}(x_{22},\theta) = \cos\theta(x_{22} + H_g - H_r + L_w\tan\theta - H_o) \quad (0 \leq x_{22} \leq H_o)$$

$$d_3(x_3,\theta) = x_3 + H_g - H_r + L_w\tan\theta - H_o \quad (0 \leq x_3 \leq H_{gw}),$$

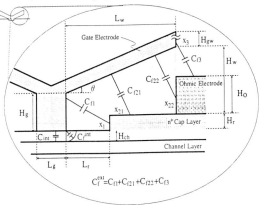

where x_1, x_{21}, x_{22}, and x_3 are the coordinates of the analysis points shown in Fig. 1. Here, θ is the angle between the Y-shaped gate wing and the substrate surface, L_r is the recess length, L_w is the Y-shaped gate wing length, H_g is the height of the "leg" part of the Y-shaped gate electrode, H_r is the recess height, H_o is the ohmic electrode height, and H_{gw} is the gate wing metal thickness.

Fig. 1. Cross-sectional view of the Y-shaped, recess-etched gate FET used in the external gate fringing capacitance analysis.

3. Results and discussions

To confirm the accuracy of the simplified C_f^{ext} model, comparisons with numerical simulations from an in-house two-dimensional electromagnetic field simulator and with experiments are made. Experimental results for the 0.25 μm Y-shaped, recessed gate n-$AlGaAs/InGaAs$ heterojunction FETs (Hida et al. 1993) were precisely extracted from measured S-parameter data using "TOUCHSTONE". The value of C_f^{ext} is estimated from the differences in the gate-to-source and drain capacitance values of the FETs before (C_{gSiO2}) and after (C_{gvac}) removing the SiO_2 film under the gate wings, while ensuring there were no changes in other equivalent circuit parameters.

Comparisons among analytical, numerical, and experimental results are shown in Fig. 2, which shows the total gate capacitance versus the angle θ and the height from the ohmic electrode to the gate wing. It is clear that good agreement with the numerical results is obtained for a wide range of θ.

To optimize the gate structure parameters, we analyzed C_f^{ext} dependence on these parameters in detail. As a result, we find that there are three dominant factors: the angle θ, the height H_g and the length L_r.

The C_f^{ext} dependence on θ is shown in Fig. 3, where H_g is used as a parameter. C_f^{ext} diverges at 205 nm or less in this analysis because the gate electrode touches the n^+-cap layer or the ohmic electrode in this gate structure. The C_f^{ext} dependence on θ is reduced by increasing H_g, and the minimum value of C_f^{ext} is independent of H_g. This reason is that the capacitance C_{f21}, which dominates θ dependence of C_f^{ext}, is constant because the electric field lines from the n^+-cap layer terminate on the "leg" part of the Y-shaped gate electrode as H_g increases.

It is assumed that the C_f^{ext} dependence on θ can be neglected, if C_{f21} variation is less than $0.01 fF/\mu m$ within 80 degrees. This assumption is represented by $\mid \partial C_f^{ext}/\partial \theta \mid \simeq \mid \partial C_{f21}/\partial \theta \mid \leq 0.0125 \times 10^{-15}$. Figure 4 shows $2\partial C_{f21}/\partial \theta$ versus H_g for different values of θ. The C_f^{ext} dependence on θ is ignored when $2\partial C_{f21}/\partial \theta$ is below the dashed line corresponding to the criterion mentioned above. As H_g increases, the minimum angle θ that allows us to ignore the dependence on θ decreases. If the height H_g is more than 250 nm, C_f^{ext} is independent of θ, where θ will be above about 20 degrees in our device process condition.

As L_r increases, C_f^{ext} dependence on θ doesn't vary, and the minimum value of C_f^{ext} is reduced, because C_{f1} and C_{f21} decrease in almost inverse proportion to L_r.

Based on these results, a Y-shaped gate structure with low C_f^{ext} and low C_f^{ext} dependence on the gate structure parameters is investigated and the results are shown in Fig. 5. L_r was expanded from 20 nm to 100 nm to reduce C_f^{ext} without large increase in the source and drain resistances and to obtain higher device performance. We also increased H_g from 65 nm to 250 nm to reduce C_f^{ext} dependence on the gate structure and

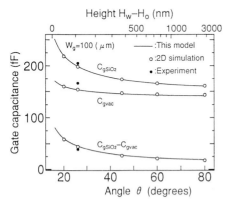
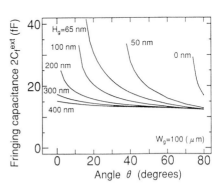

Fig. 2. Comparison of analytical (solid line), numerical (open circles) and experimental (closed circle) results for total gate capacitance as a function of the angle θ, and the height from the ohmic electrode to the gate wing.

Fig. 3. Analytical results of the total C_f^{ext} versus the angle θ with the height H_g.

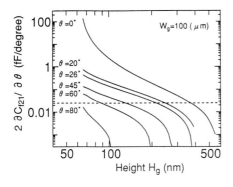 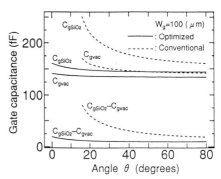

Fig. 4. Analytical results of the total partial differential of C_{f21} versus the height H_g for different θ.

Fig. 5. Comparison of analytical results of the total C_f^{ext} versus the angle θ between conventional (dashed lines) and optimized (solid lines) gate structure FETs with 0.25 μm gate length.

to obtain higher uniformity of device performance, where we consider the practical metal filling technology because it is difficult to fill in the gate opening (Wada et al. 1994) as H_g increases. The C_f^{ext} dependence on θ becomes negligible. The estimated C_f^{ext} value of this structure is about half that of a conventional structure.

Although the C_f^{ext} of the 0.25 μm FET has been discussed, this analytical model is also applicable to sub-0.25 μm gate FETs since C_f^{ext} is independent of gate length.

4. Conclusion

A simple yet accurate analytical model is proposed for analyzing the external gate fringing capacitance in Y-shaped, recessed gate FETs. By comparing the model with numerical simulations and experiments, its high accuracy is verified. Hence, this model will be a useful tool for optimizing gate structure parameters of fine gate-length FETs.

Acknowledgments

The authors would like to thank Mr. M. Ishikawa and Ms. T. Nakayama for their invaluable contributions to this device fabrication, Dr. N. Goto and Mr. H. Yano for their help with measurements and extraction of equivalent circuit parameters, and Messrs. T. Maeda, M. Fujii, Y. Suzuki, and K. Numata for mask design. They are also grateful to Drs. T. Nozaki, Y. Ohno, and K. Honjo for valuable advice, support and encouragement throughout the course of this work.

References

Anholt R *1991 Solid-State Electron.* **34** p 515

Hida H, Tokushima M, Maeda T, Ishikawa M, Fukaishi M, Numata K and Ohno Y *1993 GaAs IC Symp. Tech. Dig.* p 197

Wada S, Tokushima M, Fukaishi M, Matsuno N, Yano H and Hida H *1994 Device Res. Conf. Dig.* VIB-2

Electron Cyclotron Resonance Microwave Plasma Enhanced SiGe Oxidation and MOS Transistors

Pei-Wen Li, E. S. Yang, Y. F. Yang, and X. Li

Department of Electrical Engineering, Columbia University, New York, NY 10027

Abstract. A new low temperature process using electron cyclotron resonance (ECR) microwave plasma has been developed to grow (not deposit) a stoichiometric SiGe oxide directly on $Si_{1-x}Ge_x$ alloys. *In situ* x-ray photoelectron spectroscopy (XPS) measurements indicate that both Si and Ge are fully oxidized and Auger depth profiling reveals that no Ge segregation occurs at the oxide/SiGe interface or near the oxide surface. The ECR-grown SiGe oxides have fixed oxide charge density 1×10^{11} cm^{-2} and interface trap density 1×10^{11} cm$^{-2} \cdot$ eV^{-1}. The mean breakdown strength for the ECR oxide is 8-10 MV/cm. High quality 1 μm Al-gate $Si_{0.85}Ge_{0.15}$ pMOSFETs with an ECR gate oxide have also been fabricated. It is found that saturation transconductance increases from 48 mS/mm (300 K) to 60 mS/mm (77 K). Low field hole channel mobility is measured at 300 K and 77 K. The best experimental values are about a factor of two better than the corresponding silicon devices.

1. Introduction

SiGe based system has attracted a lot of attention because of its potential to make Si more versatile and functional in microelectronics and optoelectronics applications. To be compatible with the state-of-the art Si microelectronics fabrication technology, a device quality SiGe oxide is essential for surface passivation, interlayer isolation, and gate oxides for MOS devices. Besides, the SiGe-oxide system has very important applications in CMOS technology and could offer the possibility of direct improvements in packing density and speed for VLSI applications.

Due to the metastable nature of strained SiGe layers [1], the high temperature needed for conventional thermal oxidation results in the introduction of strain relief defects at the strain-layer/substrate interface and Ge segregation at the oxide/alloy interface [2]. The resulting Ge buildup at the oxide/SiGe interface and strain relaxation at the SiGe/Si interface render the heterostructure useless for electronic device applications. With the successful fabrication of SiGe heterojunction bipolar transistors, resonant tunneling diodes, and photodetectors comes the need for a low-temperature technique that produces high-quality SiGe oxide. ECR plasma oxidation is currently being investigated as an attractive alternative to replace thermal oxidation because it provides many advantages for low temperature oxidation.

We have used surface analysis tools such as *in situ* x-ray photoelectron spectroscopy (XPS) and *ex situ* Auger electron spectroscopy (AES) to investigate the chemical properties of the ECR-grown SiGe oxides and the interfacial chemistry between the ECR oxide and SiGe alloys. High frequency and quasi-static capacitance-voltage (C-V) measurements indicate that the interface states at the oxide/SiGe interface is quite low and the SiGe oxide exhibits a high breakdown field strength (10 MV/cm). The SiGe pMOSFETs exhibits high saturation transconductance and channel mobility both at 300 K and 77 K. This demonstrates the feasibility of making high speed CMOSFETs in SiGe/Si system for low power, high speed VLSI applications.

2. ECR plasma oxidation of SiGe

In ECR plasma processing, ion energy control is critical for controlling the properties, such as stress, index of refraction, and stoichiometry, of deposited thin film [3]. In this work, dc bias was used to examine the roles of electrons, ions, and neutrals.

2.1 Positive bias condition

The ECR-grown SiGe oxides with the best thickness uniformity and electrical characteristics were grown under positive bias conditions.

The electronic states of Si and Ge in ECR oxides can be determined by the binding energy shift with respect to the bulk Si and Ge signals. A systematic XPS study of the ECR oxidation of $Si_{0.5}Ge_{0.5}$ with a sample bias +22 V at room temperature was shown in Fig. 1. Spectra 1(a) shows a chemically cleaned $Si_{0.5}Ge_{0.5}$ surface with well defined Si2p and Ge3d substrate signals. After a 1 minute ECR oxygen plasma exposure, both Si and Ge are fully oxidized which are characterized by core-level chemical shifts of 3.8 eV for Si2p and 4 eV for Ge3d photoelectrons towards higher binding energies, respectively. As the exposure to the oxygen plasma increases, the Si2p signal positioned at 99.6 eV gradually decreases, while a signal at 103.4 eV corresponding to SiO_2 increases. The same observations were made with the Ge3d signals. The thermal stability of the SiGe-oxide was assessed by *in vacuo* (5×10^{-8} Torr) anneal at 450 °C for 1 hour or 600 °C for 30 minutes after ECR oxidation. No modification of the stoichiometry or chemical bonding of both Si and Ge was observed by XPS after the anneal, which indicates that the ECR-grown SiGe oxides are thermally stable enough to withstand further device processing. Different germanium contents of $Si_{1-x}Ge_x$ samples (x = 0.15, 0.2, 0.5) were studied. From XPS study it was found that both Si and Ge are simultaneously fully oxidized independent of the Ge content in $Si_{1-x}Ge_x$ alloys.

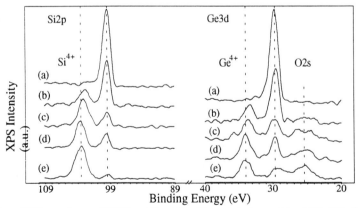

Figure 1   XPS spectrum of ECR oxidized $Si_{0.5}Ge_{0.5}$. (a) cleaned by 10% HF, (b) 1 min. (c) 1 hour, and (d) 3 hours ECR oxygen plasma exposure.

From the Auger depth profiling study, Ge was found to be uniformly distributed in the oxide and there is no Ge-rich layer formed between the oxide layer and the alloy, which is totally different from the results of thermal oxidation.

2.2 Floating and Negative bias conditions

Stoichiometric SiGe oxides could be grown by ECR oxygen plasma under floating or negatively biased conditions from the XPS and Auger depth profile studies. However, the quality of the oxides as determined by electrical measurements (C-V and I-V) are poor and not reproducible. The interface state density is larger than 10^{12} cm$^{-2} \cdot$ eV^{-1} and increases abruptly somewhere in the SiGe bandgap, which is reflected in pronounced shape distortion (flatten out) of the high-frequency C-V curve. The breakdown field of the ECR oxide is 3-7 MV/cm. This indicates that these bias regimes should not be used in a gate dielectric oxidation process.

3. SiGe MOS capacitors

To investigate the electrical properties of the ECR-grown SiGe oxides, Al-gate MOS capacitors were fabricated. High-frequency C-V and ramped bias I-V studies indicate that the electrical properties of the ECR oxides are sensitive to the growth temperature and substrate bias conditions during ECR process.

Figure 2 shows the high-frequency and quasi-static C-V curves of the ECR oxide grown at 15 °C with a substrate bias voltage of +4V. The fixed charge density Q_f for this film is -1.5×10^{11} cm^{-2} and the interface state density D_{it} is 1×10^{11} cm$^{-2} \cdot$ eV^{-1}. D_{it} decreases from 7×10^{11} cm$^{-2} \cdot$ eV^{-1} to 1×10^{11} cm$^{-2} \cdot$ eV^{-1} if the growth temperature decreases from 400 °C to 15 °C. This is in contrast with the prevailing view that a higher growth temperature is favored to produce a better oxide. This phenomenon is not fully understood yet, but it may be attributed to the reduced photon radiation damage at low temperature.

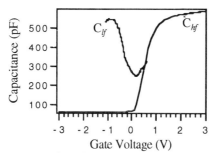

Figure 2 The high-frequency and quasi-static C-V curves of the ECR oxide grown at 15 °C with a substrate bias voltage of +4V.

The I-V characteristics of an 80 Å SiGe oxide is shown in Fig. 3. Breakdown field was defined as the field required to pass 1 µA through the defined gate. ECR oxides exhibit intrinsic breakdown behavior at approximately 8-10 MV/cm. Some of this spread in the mean breakdown field strength may be attributed to the slight film thickness nonuniformity and the lack of particle-free clean room conditions during wafer handling prior to and post-ECR processing. However, some of the spread in distribution is believed to be due to the residual radiation damage induced by the plasma.

4. SiGe pMOSFET

To prove that the ECR SiGe oxide is good enough as the gate dielectric in MOS devices, we have fabricated $Si_{0.85}Ge_{0.15}$ pMOSFETs. The starting material for the SiGe MOSFET consists of a 0.0038-0.008 Ω cm, n-type, (100) Si substrate on which a 1000 Å Si buffer layer and then a 1000 Å strained $Si_{0.85}Ge_{0.15}$ with doping concentration of 2×10^{17} cm^{-3} were grown by UHV/CVD at 500 °C. A 4000 Å field oxide (SiO_2) was deposited by PECVD and patterned to define the active area of the MOSFETs. The source and drain were formed by BF_2^+ implantation and activated by rapid thermal anneal. No additional implant was used for threshold-voltage adjustment. Subsequently, a 100 Å gate oxide was grown by ECR oxygen plasma at room temperature. Aluminum was used for drain, gate, and source contact, and a 15 minutes 400 °C forming gas anneal completed the fabrication.

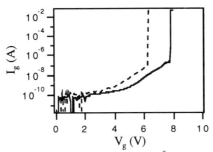

Figure 3 I-V characteristic of an MOS capacitor with 80 Å ECR-grown SiGe gate oxide.

The threshold voltage of the $Si_{0.85}Ge_{0.15}$ pMOSFET is -0.2 V at 300 K and -0.9 V at 77 K. Figure 4 shows the output characteristics of a 1 μm x 10 μm $Si_{0.85}Ge_{0.15}$ pMOSFET at 300 K and 77 K. The saturation transconductance increases from 48 mS/mm at 300 K to 60 mS/mm at 77 K. The external series resistance measured by Van der Paul technique is 2730 Ω-μm. The large series resistance is due to the large spacing (10 μm) between the source contact and the active gate. The intrinsic transconductance is estimated to be 55 mS/mm at 300 K and 72 mS/mm at 77 K. The replacement of the Al gate with a self-aligned polysilicon gate is expected to reduce the series resistance and thus to improve the device performance further.

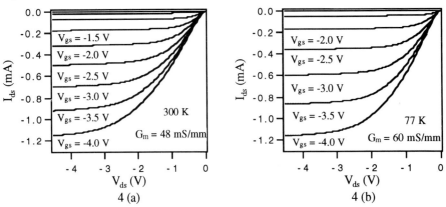

Figure 4 I-V characteristics of a 1 μm x 10 μm $Si_{0.85}Ge_{0.15}$ pMOSFET at (a) 300 K and (b) 77 K

At V_{ds} = -50 mV, the $Si_{0.85}Ge_{0.15}$ pMOSFETs have a subthreshold slope of 105 mV at 300 K and 40 mV/dec at 77 K. In comparison with Si devices, the subthreshold behavior is poor and the leakage current is high (~10^{-9} A at 300 K). Two mechanisms may contribute to the drain-source leakage current [4]: (1) reverse-bias drain junction leakage current of the drain junction, which could be caused by dislocations occurring during the implant anneal. (2) surface channel current that flows when the surface is weakly inverted. This might be contributed by the interface states ($D_{it} \sim 1 \times 10^{11}$ cm^{-2} · eV^{-1}) at the oxide/SiGe interface.

The low field hole mobilities $\mu_{fe} = \{(L/W)(g_m(V_{gs})/V_{ds}C_{ox}(V_{gs})\}$ at 300 K and 77 K for a 100 μm x 100 μm $Si_{0.85}Ge_{0.15}$ pMOSFET are shown in Fig. 5. The peak hole mobility is 167 cm^2/V-s at 300 K and increases to 530 cm^2/V-s at 77 K. For comparison, the Si pMOSFETs with channel doping concentration of 2.7×10^{17} cm^{-3} exhibit peak hole mobility of 100 cm^2/V-s at 300 K and 245 cm^2/V-s at 77 K [5]. The $Si_{0.85}Ge_{0.15}$ pMOSFETs exhibit as much as 67% at 300 K and 116% at 77 K larger peak hole mobility, compared to the Si pMOSFETs. The large increase in mobility at low temperature is not fully understood although a reduction of effective mass in the strained layer is a plausible explanation.

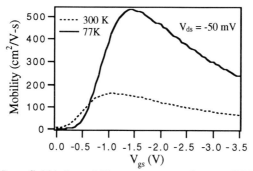

Figure 5 Low-field hole mobility versus gate voltage at 77 K and 300 K.

5. Summary

Stoichiometric SiGe oxide has been grown (not deposited) directly on SiGe alloys by low temperature ECR plasma oxidation. Both Si and Ge are fully oxidized and no Ge segregation occurs at the oxide/SiGe interface or near the oxide surface as observed from XPS and Auger depth profiling studies. C-V and I-V measurements indicate that a low growth temperature is favored producing a high quality SiGe oxide with a fixed charge density 1×10^{11} cm^{-2}, an interface state density 1×10^{11} cm^{-2} · eV^{-1}, and a breakdown field 8-10 MV cm^{-1}. 1 μm Al-gate $Si_{0.85}Ge_{0.15}$ pMOSFETs with the ECR-grown gate oxide have also been fabricated. The saturation transconductance increases from 48 mS/mm at 300 K to 60 mS/mm at 77 K with series resistance of 2730 Ω-m. The intrinsic transconductance is estimated to be 55 mS/mm at 300 K and 72 mS/mm at 77 K. The peak hole channel mobility is 167 cm^2/V-s at 300 K and increases to 530 cm^2/V-s at 77 K. These results show that the SiGe MOSFETs are potentially useful for applications in high speed electronics and optoelectronics.

Reference

[1] F. K. LeGoues, B. S. Meyerson, and J. Morar, *Phys. Rev. Lett.*, **22**, 2903 (1991)
[2] F. K. LeGoues, R. Rosenberg, T. Nguyen, F. Himpsel, and B. S. Meyerson, J. Appl. Phys. **65**, 1724 (1989)
[3] W. C. Dautremont-Smith, R. A. Gottscho, and R. J. Schutz, in *Semiconductor materials and Process Technology Handbook*, edited by G. E. McGuire (Noyes, NJ, 1988)

[4] W. M. Gosney, *IEEE Trans. Electron Devices*, **ED-19**, 213 (1972)
[5] S. Takagi, M. Iwase, and A. Toriumi, in *IEDM Tech. Dig.*, 398 (1988)

Non-Alloyed Contacts to GaAs Using Non-Stoichiometric GaAs

M.P. Patkar, J.M. Woodall, T.P. Chin, M.S. Lundstrom, and M.R. Melloch

School of Electrical Engineering, Purdue University, West Lafayette, IN 47907-1285

Abstract :
Non-alloyed ohmic contacts to p^{++} GaAs and n-GaAs were made by using low temperature molecular beam epitaxy of GaAs. Growth of p-type GaAs using a high concentration of Be followed by an anneal under As overpressure was used to achieve p^{++} GaAs with a low concentration of fast diffusing interstitial Be. Non-alloyed Ti/Au contacts to such p-type GaAs had specific contacts resistivities of about 10^{-7} Ω-cm^2. Such contacts should not have the stability problems that are usually caused by fast interstitial diffusion of Be. A low temperature grown layer of GaAs was also used to protect the high space charge density layer that exists at the surface of n-type GaAs heavily doped with Si. Ag and Ti/Au ohmic contacts to such structures showed specific contact resistivities of mid 10^{-7} Ω-cm^2.

1. Introduction

The dominant technology for forming ohmic contacts to n-GaAs consists of depositing a Au-Ge-Ni dopant/metal structure followed by an anneal (typically in the range of 400-500°C) to form an alloy. Ohmic contacts to p-GaAs are typically formed by depositing Au-Be or Au-Zn and alloying. Such alloyed contacts have resulted in specific contact resistivities on the order of 10^{-6} Ω-cm^2, but alloyed contacts are not good optical reflectors. Many devices (e.g. light emitting diodes, vertical surface emitting lasers, and solar cells) would benefit from an ohmic contact that would also be a good optical reflector. We have developed a technology for forming non-alloyed ohmic contacts to n- and p- GaAs.

2. Non-alloyed contacts to p^{++} GaAs

It is easy to make non-alloyed ohmic contacts to p^{++} GaAs, but for Be concentrations above $\approx 5 \times 10^{19}$ cm^{-3}, high concentrations of interstitial Be occur. Interstitial Be is a fast diffusing species that smears doping profiles. Also, it is a donor, which compensates the p-type doping. Low temperature growth has been used in the past to improve the Be doping efficiency in InGaAs devices [1]. In this work, we used low-temperature growth of GaAs with high Be concentrations followed by a subsequent anneal to drive interstitial Be onto acceptor sites. Our results demonstrate that this technique produces a higher p-type doping that lowers the specific contact resistivity and also improves the film's stability.

The ratio of Be atoms on acceptor sites to Be atoms on donor sites is given by [2]

$$\frac{N_A^-}{N_D^+} = K'(T) \times \frac{n_i^2}{p^2} \times \left(P_{As_2}\right)^{0.5} , \qquad (1)$$

where $K'(T)$ is a temperature-dependent constant, P_{As_2} is the arsenic dimer pressure, p is the hole concentration in GaAs, and n_i is the intrinsic carrier concentration in GaAs. Excess As is incorporated in the bulk of heavily doped p-GaAs grown at low temperature. According to Eq. (1), a subsequent anneal under As overpressure should result in interstitial Be moving to substitutional sites owing to the As overpressure caused by the excess As in the bulk. The MBE p-GaAs structures consisted of a GaAs film with a high Be concentration grown on a semi-insulating substrate. Three structures were grown; a p^{++}-GaAs layer grown at 600°C, a p^{++}-GaAs layer grown at 300°C, and a p^{++}-GaAs layer grown at 300°C that was subsequently annealed at 600°C under As overpressure. The results will be described in Sec. 5.

3. Non-alloyed contacts to n-GaAs

Achieving high n-type doping concentration in bulk GaAs is more difficult than achieving high p-type concentration. Silicon is a typical n-type dopant used in MBE of GaAs. The bulk doping density is limited to $N_D - N_A \approx 5 \times 10^{18}$ cm^{-3} in Si doped n-typed GaAs. The equilibrium ratio of donors to acceptors in n-GaAs with high Si concentration is given by [3],

$$\frac{N_D^+}{N_A^-} = K(T) \times \frac{n_i^2}{n^2} \times P_{As_2}, \qquad (2)$$

where $K(T)$ is a temperature-dependent constant incorporating the equilibrium arsenic pressure over GaAs, P_{As_2} is the arsenic dimer pressure, n is the concentration of electrons in GaAs, and n_i is the intrinsic carrier concentration in GaAs. During the MBE of n-GaAs, the electron concentration, n, at the surface is very near n_i because of Fermi level pinning. According to Eq. (2), this suggests that under a high As overpressure one should have a high concentration of Si atoms on donor (Ga) sites. The native oxide formation and subsequent native oxide etch that is done prior to metal evaporation, however, removes this high space charge density layer at the surface precluding the possibility of making ex-situ tunneling ohmic contacts. We choose to passivate the high space charge density layer at the surface with a layer of undoped, low temperature grown GaAs. Such a layer has a number of mid gap states owing to the presence of excess As in it. Ohmic contact should then be achieved due to the tunneling of carriers through the high space charge density region and due to defect assisted tunneling of carriers through the low temperature grown GaAs layer. (Metal contacts deposited on thick low temperature grown GaAs do exhibit ohmic behavior but show high specific contact resisitivities of about 10^{-3} Ω-cm^2 [4].)

For the n-GaAs structures, the heavily doped n-GaAs layer was grown at 600°C using As$_4$ instead of As$_2$. The V/III ratio was approximately equal to 2.0. A resistively heated Si filament was used to produce high fluxes of Si for the n-type doping of GaAs. The low temperature caps on the heavily doped n-GaAs layer were grown at 250°C. Films with low temperature caps with thicknesses ranging from 0 to 50 Å were grown. Figure 1 shows a schematic of the film structure.

4. Processing and Characterization:

Transmission line measurements were used to characterize the specific contact resistivities of the p-GaAs and the n-GaAs contact structures. The transmission lines were fabricated by using conventional photolithographic and wet etching techniques. Ag and Ti/Au were used as metals for the contacts to n-GaAs. Ti/Au contacts were used for making contacts to the p-GaAs contact structures.

Low temperature grown GaAs	2 nm
n-GaAs heavily doped with Si Si concentration of 1×10^{20} cm^{-3} $N_D - N_A = 5 \times 10^{18}$ cm^{-3}	100 nm
n-GaAs $N_D - N_A = 5 \times 10^{17}$ cm^{-3}	400 nm
Semi-insulating GaAs substrate	

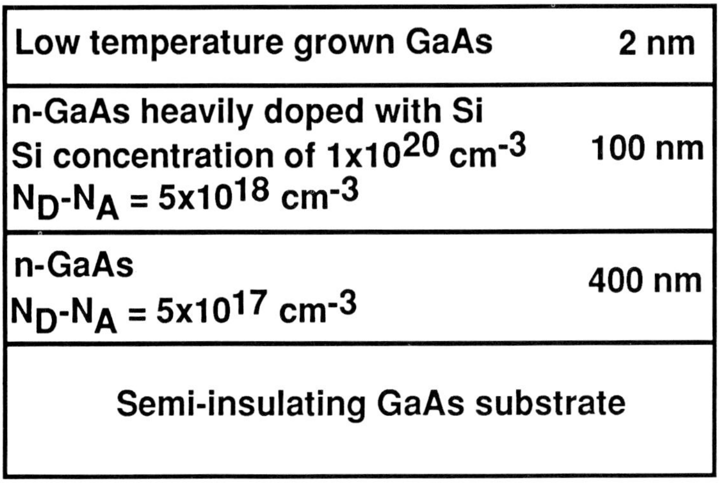

Fig. 1 Contact structure for n-GaAs contacts. The contact structure shows low temperature grown GaAs layer protecting the high space charge density layer.

5. Results

The specific contact resistivity for the contacts on p-type GaAs grown at 600°C and heavily doped with Be and the contacts on GaAs grown at 300°C were both about 10^{-8} Ω-cm^2. The p-GaAs sample grown at 300°C and annealed under As overpressure at 600°C exhibited a specific contact resistivity that was somewhat higher than the other two samples, however, the channel resistance for this sample was one-half of that of the other two samples. This suggests that the excess As in low temperature grown GaAs caused the interstitial Be to go onto acceptor sites due to the anneal under As overpressure thereby decreasing the channel sheet resistance. This results in high Be doping efficiency. Because it is on substitutional sites, the high concentration Be in such samples would not diffuse rapidly during subsequent high temperature processing steps. Out-diffusion of Be from a few monolayers of GaAs near the surface, during the 600°C anneal is a probable cause for the increased specific contact resistivity for the annealed sample.

A comparison of the measured I-V characteristics for Ag contacts to n-GaAs structures with and without the low temperature GaAs cap is shown in Fig. 2 [5]. As expected, the sample without a cap did not yield an ohmic contact. The high space charge density layer in the capped structures is protected by the cap layer thus allowing a tunneling ohmic contact to be formed. For the n-GaAs contact structures, specific contact resistivities of well below 10^{-6} Ω-cm^2 were obtained for the contact structures with cap thicknesses of 20, 30 and 50 Å. More details on the contact resistivities versus cap thickness will be reported elsewhere [5].

6. Conclusion

In conclusion, nonalloyed contacts to heavily Be-doped p-type GaAs gives specific contact resistivities of about 10^{-8} Ω-cm^2. The sheet resistance of low temperature grown, Be-doped ($\approx 10^{20}$ cm^{-3}) GaAs under As overpressure is lowered by annealing because excess As in the bulk causes interstitial Be to move onto acceptor sites. We have also developed a method

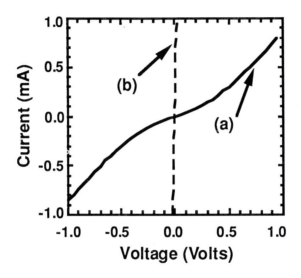

Fig. 2 I-V characteristics for Ag contacts. (a) without low temperature grown GaAs layer, (b) with low temperature grown GaAs layer [4].

for making non alloyed contacts to n-type GaAs by protecting the high space charge density layer, which naturally occurs on the growth surface, with a low temperature grown, undoped GaAs layer. Specific contact resistivities of well below 10^{-6} Ω-cm^2 were obtained.

References:

[1] R.A. Mertzger, M. Hafiji, W.E. Stanchina, T. Liu, R.G. Wilson, and L.G. McCray, *Appl. Phys. Lett.*, **63** (10), 1360 (1993).
[2] F.H. Pollack, P. Parayanthal, and J.M. Woodall, Solar Energy Research Inst., final report for subcontract ZL-4-03032-9, submitted August 1985.
[3] P.D. Kirchner, T.N. Jackson, G.D. Petitt, and J. M. Woodall, *Appl. Phys. Lett.*, **47** (1), 26 (1985).
[4] H. Yamamoto, Z-Q. Fang, and D.C. Look, *Appl. Phys. Lett.* **57** (15), 1537 (1990).
[5] M.P. Patkar, T.P. Chin, J.M. Woodall, M.S. Lundstrom, and M.R. Melloch, submitted to *Appl. Phys. Lett.*, July 1994.

Development of Refractory NiGe-based Ohmic Contacts to n-type GaAs

Takeo Oku, Hiroki Wakimoto, Masaki Furumai, Hidenori Ishikawa, Hirotaka R. Kawata, Akira Otsuki, and Masanori Murakami

Department of Materials Science and Engineering, Kyoto University, Sakyo-ku, Kyoto 606-01, Japan

Abstract Thermally stable, low resistance NiGe-based Ohmic contacts to n-type GaAs were developed by adding a small amount of a third element (Au, Ag, Al, Pd, or In) to the refractory NiGe compound contacts. These NiGe-based contacts had smooth surface and shallow diffusion depth, which are required for the future very large scale integration (VLSI) GaAs devices. The electrical properties were correlated with the microstructure by analyzing the interfacial structure using high-resolution electron microscopy, which enabled us to propose models for the current transport mechanism of the NiGe-based contacts. These models explained nicely the effect of the third elements on the electrical properties of the NiGe Ohmic contacts.

1. Introduction

The progress of the GaAs device technology is strongly dependent on the development of thermally stable, low resistance Ohmic contacts. Currently, AuGeNi Ohmic contacts have been extensively used in the manufacturing GaAs devices, because they provide low contact resistances (~0.1 Ωmm) and their reproducibility is excellent [1]. However, these contacts have poor thermal stability, rough surface and deep diffusion depth into the GaAs substrate, which prevent the AuGeNi contacts from using in the sub-micron devices. Previously, the cause of the thermal instability of the AuGeNi contacts was found to be due to formation of low melting point (370°C) β'-AuGa phases during contact formation [2], as schematically illustrated in Fig. 1(a). Based on this study, we believed that formation of high melting point (T_m) compounds would improve the thermal stability of the contacts, as illustrated in Fig. 1(b). In addition, formation of a heavy doping layer and/or an intermediate semiconducting layer which had low energy barrier height to the contact metals is essential to reduction of the contact resistances (R_c) [3]. The contacts which have interface structure of Fig. 1(b) is very attractive for the future submicron devices. In the present paper, new Ohmic contact metals which improve the poor contact properties of the alloyed Ohmic

2. Refractory NiGe Ohmic contacts

Recently NiGe Ohmic contacts which are compatible with the GaAs metal-semiconductor field-effect transistor (MESFET) fabrication process were prepared in our laboratory [4]. Samples with layered structure of GaAs/Ni/Ge were prepared by electron beam evaporation. (A slash "/" between two materials indicates deposition sequence). These samples were annealed in a conventional furnace or rapid thermal annealer in 5 % H_2/N_2 atmosphere. Ni(75nm)/Ge(90nm) contacts with 38 atomic % Ge prepared by annealing at 600℃ had smooth surface and yielded excellent thermal stability at 400℃ due to formation of NiGe compounds with high melting point of 850℃. This NiGe layer satisfies the condition of formation of high-T_m compound layer after annealing at elevated temperatures. However, the NiGe contacts yielded the contact resistance of 0.8 Ωmm, and reduction of the contact resistance by a factor of ~3 was mandatory to use these contacts in the sub-micron devices.

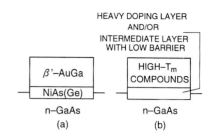

Fig. 1. Interface structures of (a) AuGeNi contacts and (b) contacts expected to have thermal stability.

3. Effects of third element addition to NiGe Ohmic contacts

3.1. Desirable third elements for NiGe contacts

Further studies to reduce the contact resistance without deteriorating the thermal stability of the NiGe contacts by adding a small amount of a third element have continued [5,6]. Such third elements should assist to increase the carrier concentration at the GaAs surface and/or form an intermediate semiconductor layer which has low energy barrier height to the contact metals.

To increase the carrier (donor) concentration in the GaAs surface, the third elements (M) must contribute to create Ga vacancies in the GaAs where the Ge atoms can diffuse in, providing donors to the GaAs. Thus, the binding energy of the element M with Ga should be stronger than that with As. These metals are Au, Pd and Pt [7]. The elements, which form solid solution with Ga at elevated temperatures would also create Ga vacancies in the GaAs substrate. Ag forms the Ag-Ga solid solution and an addition of a small amount of Ag to the NiGe is expected to reduce the R_c values of the NiGe contacts. To form an intermediate layer with low energy barrier, an addition of a small amount of In to the NiGe is desirable.

Indium is known to grow an epitaxial $In_xGa_{1-x}As$ layer on the GaAs substrate at temperatures above 300℃ [8]. In addition, an addition of Al to the NiGe is important to understand effect of barrier height on the R_c values. Al is expected to form an $Al_xGa_{1-x}As$ layer which has high energy barrier to contact metals [9]. This $Al_xGa_{1-x}As$ layer would increase the R_c values. The electrical properties and the microstructure of the NiGe contacts with a small amount of each third element are described.

3.2. NiGe contacts with third elements which increase donor concentrations

As an example of the NiGe Ohmic contacts with a small amount of third elements which increase the donor concentration at the GaAs surface, the results of the NiGe contacts added with a small amount of Au [denoted as NiGe(Au)] are described [6]. Figure 2 shows the dependence of R_c values on the Au layer thickness. The R_c value of the NiGe contact without Au was close to $\sim 2\,\Omega$mm after annealing at 500℃ for 5 sec. An addition of 3 nm thick Au to the NiGe contact reduced the R_c value down to $\sim 0.5\,\Omega$mm. The minimum R_c value of $\sim 0.2\,\Omega$mm was obtained by adding the Au layer with 5 nm thickness. R_c values were also reduced by addition of Ag or Pd, and the minimum R_c values of the NiGe contacts with Ag or Pd are shown in Fig. 2 [10,11]. The best R_c values were obtained by adding a 6 nm thick Ag or Pd layer to the NiGe.

However, NiGe(Al) contacts with layered structures of Ni/Al/Ge, Ni/Ge/Al, and Ge/Ni/Al (which were annealed at temperatures ranging from 400 to 700℃ for 5-240 sec) did not show Ohmic behavior [12].

Isothermal annealing at 400℃ after contact formation was performed for the NiGe(Au) contacts, which were prepared by annealing at 500℃ for 5 sec. The R_c values of the contact during isothermal annealing at 400℃ are almost constant for 3 h. The surface morphology of the NiGe contacts did not deteriorate by addition of small amounts of these elements. In order to investigate the correlation

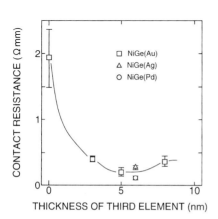

Fig. 2. The minimum contact resistances of NiGe contacts with third elements.

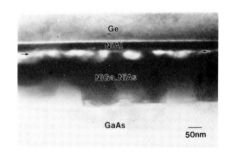

Fig. 3. Transmission electron micrograph of NiGe(Al) contact annealed at 700℃ for 5 sec.

Fig. 4. High-resolution image of NiGe(Au) contact annealed at 400℃.

between the electrical properties and microstructure, microstructural analysis by transmission electron microscopy was carried out. Figure 3 is a transmission electron micrograph of the NiGe(Al) contact annealed at 700℃ for 5 sec. No $Al_xGa_{1-x}As$ layer formed, and NiAl compounds formed by a reaction between Al and Ni. The NiAl layer which has a high melting point of 1640℃ prevents Ni from reacting with Ge. The NiAl layer also prevents Ge-doping in the GaAs substrate, which results that the NiGe(Al) contacts did not show Ohmic behavior. This result shows that a third element which prevents a reaction between Ni and Ge should not be selected.

Figure 4 is a high-resolution image of the interface of NiGe(Au) contacts prepared at 400℃ for 5 sec. The thickness of the regrown GaAs was measured to be about 20 nm, and the NiGe compound layer covers the regrown GaAs layer doped heavily with Ge. The carriers would transport through the n^+-GaAs/n^{++}-GaAs/metal interface by the tunneling mechanism. Another model to explain the low R_c values is that the R_c reduction is due to the reduction of the barrier height at the n^+-GaAs/metal interfaces by forming an epitaxial n^{++}-Ge(As) layer. (In Fig. 4, a contrast of the regrown GaAs layer near the NiGe (as indicated by white arrows) is different from that of the GaAs, which would be due to the existence of the epitaxially grown Ge layer.)

3.3. NiGe contacts with a third element which forms low barrier layer

The NiGe contacts with a small amount of In were prepared [5]. Dependence of the In layer thickness on the R_c values for Ni(60 nm)/In(x nm)/Ge(100 nm) contacts which were annealed at temperatures around 700℃ is shown in Fig. 5. The R_c value of the NiGe contact without In was 1.3 Ωmm. The R_c value decreased by a factor of about 4 by adding 3 nm thick In to the NiGe contact, and the lowest R_c value of 0.28 Ωmm was obtained.

A high-resolution image of the interface of the NiGe(In) contacts prepared at 700℃ is shown in Fig. 6. The GaAs surface is covered by the NiGe compounds. A "regrown" GaAs layer is formed

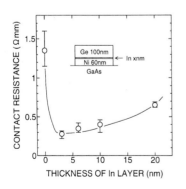

Fig. 5. Contact resistances of NiGe(In) contacts annealed at 600-700℃ for 5 s.

at the GaAs/NiGe interface. The regrown GaAs and $In_xGa_{1-x}As$ layers are measured to be about 50 nm in thickness, and the {111} twin-related structure is observed in these layers. The In composition (x) of the $In_xGa_{1-x}As$ is measured to be about 0.4 from the high-resolution images and the diffraction spots.

The present microstructural analysis indicates that the reduction of the R_c values by addition of a small amount of In to the NiGe is due to formation of $In_xGa_{1-x}As$ layers on the GaAs surface.

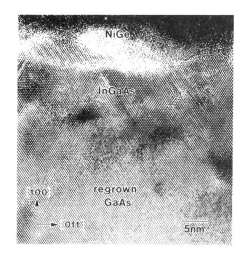

Fig. 6. High-resolution image of NiGe(In) contacts annealed at 700℃ for 5 sec.

4. Comparison among NiGe based Ohmic contacts

The properties of the NiGe contacts with or without third elements and the conventional AuGeNi contact are compared with an "ideal" contact by marking each property by a "+" or "-" sign in Table I where a "-" sign indicates the necessity of improvement of the property. (A "*" indicates that data are not available.)

Table I. The properties of ideal, AuGeNi, and NiGe-based contacts.

	Ideal contact	AuGeNi	NiGe	NiGe(Au)	NiGe(Ag)	NiGe(Al)	NiGe(Pd)	NiGe(In)
Contact resistance	Low	+	-	+	+	-	+	+
Sheet resistance	Low	+	+	+	+	+	+	+
Thermal stability	Excellent	-	+	+	-	*	-	+
Contact surface	Smooth	-	+	+	+	-	+	+
Diffusion depth	Shallow	-	+	+	+	-	+	+
Process window	Wide	+	-	-	+	-	-	+
Reproducibility	Good	+	+	-	+	-	+	+

Although the NiGe contacts had higher contact resistances than those of the AuGeNi contacts, the thermal stability at 400℃ and the surface morphology were significantly improved. The electrical properties were improved by adding a small amount (~5 nm) of third elements such as Au, Ag, Pd or In. However, deterioration of the thermal stability during annealing at 400℃ was observed when Ag or Pd was added to NiGe. A reason is believed to be due to formation of low melting PdGa compounds or Ag-Ga solid phase at the metal/GaAs interface. More detailed study is needed to correlate the microstructure and the thermal stability. Although the NiGe(In) contacts have all "+" in every category, our efforts will continue to search a third element which would further reduce the R_c values and improve the reproducibility of the contact fabrication.

5. Summary

In the present paper, recent progress of refractory NiGe-based Ohmic contacts to n-type GaAs was described. The future submicron GaAs devices require not only the good electrical properties but also the excellent thermal stability and smooth contact morphology. Such the contacts were prepared by depositing the metal layers which form refractory compounds and an intermediate layer with high carrier concentration and/or low energy barrier height to the metals. This contact is attractive for the future submicron GaAs MESFET devices.

References

1. Braslau N, Gunn J B and Staples J L, 1967 *Sol. Stat. Electron.* **10** 381-3.
2. Murakami M, Childs K D, Baker J M and Callegari A, 1986 *J. Vac. Sci. Technol.* **B4** 903-11.
3. Murakami M, 1990 *Mat. Sci. Rep.* **5** 273-317.
4. Tanahashi K, Takata H J, Otsuki A and Murakami M, 1992 *J. Appl. phys.* **72** 4183-90.
5. Oku T, Wakimoto H, Otsuki A and Murakami M, 1994 *J. Appl. Phys.* **75** 2522-9.
6. Kawata H R, Oku T, Otsuki A and Murakami M, 1994 *J. Appl. Phys.* **75** 2530-7.
7. Niessen A K, de Boer F R, Boom R, de Châtel P F, Mattens W C H and Miedema A R, 1983 *CALPHAD.* **7** 51-70.
8. Kim H J, Murakami M, Price W H, and Norcott M, 1990 *J. Appl. Phys.* **67** 4183-9.
9. Rideout V L, 1974 *Sol. Stat. Electron.* **17** 1107-8.
10. Wakimoto H, Oku T and Murakami M, to be submitted to *J. Appl. Phys.*
11. Oku T, Furumai M and Murakami M, to be submitted to *J. Appl. Phys.*
12. Wakimoto H, Ishikawa H, Oku T and Murakami M, unpublished.

Rapid thermal annealing effects on the structural integrity of InAlAs/GaAsSb HIGFET epilayers on InP

K.G. Merkel, C.L.A. Cerny[†], R.T. Lareau[‡], V.M. Bright, P.W. Yu[*], and F.L. Schuermeyer[†]

Air Force Institute of Technology, Department of Electrical and Computer Engineering
Wright-Patterson AFB (WPAFB), OH 45433-7765 USA

[†]Wright Laboratory, Solid State Electronics Directorate, WPAFB, OH 45433-7331 USA

[‡]U.S. Army Research Laboratory, Electronics and Power Sources Directorate
Fort Monmouth, NJ 07703-5601 USA

[*]Wright Laboratory, Materials Directorate (WL/MLPO), WPAFB, OH 45433-7707 USA

Abstract. The upper thermal limit for $In_{0.52}Al_{0.48}As/GaAs_{0.51}Sb_{0.49}$ heterojunction insulated-gate FET (HIGFET) epilayers is determined for both self-aligned gate (SAG) and recessed gate (RG) designs. Photoluminescence (PL), secondary ion mass spectroscopy (SIMS) and Auger electron spectroscopy (AES) are used to monitor degradation following rapid thermal annealing (RTA) at temperatures between 600 and 800 °C. Epilayer integrity is maintained in both the SAG and RG structures to 600 °C. Interfacial degradation following RTA at 700 °C is more severe for the SAG structure. After 800 °C RTA, the $GaAs_{0.51}Sb_{0.49}$ channel no longer exists in either sample due to Ga and Sb indiffusion, and In and Al outdiffusion to the sample surface. The SIMS and AES results correlate with the PL results and indicate rapid thermal processing temperature limits of 600 °C for the SAG HIGFET and 700 °C for the RG HIGFET.

1. Introduction

The $In_{0.52}Al_{0.48}As/GaAs_{0.51}Sb_{0.49}$ heterojunction, lattice matched to InP, is a novel III-V semiconductor system with superior hole confinement due to a high valence band offset [1,2]. This heterojunction has demonstrated initial promise in both the heterojunction bipolar transistor (HBT) [3,4] and heterostructure insulated-gate FET (HIGFET) [5]. Maturation of the InAlAs/GaAsSb HIGFET to complementary technologies requires progression to self-aligned gate (SAG) devices. The SAG process uses thermal cycling during a number of process steps including implanted dopant activation and ohmic contact alloying. Thermal cycling of the InAlAs/GaAsSb heterojunction in a HIGFET structure determines the allowable temperature limit for device fabrication. This report establishes the upper temperature limit for both SAG and recessed-gate (RG) epilayers and explains the sources of structural degradation.

2. Experiment

The two InAlAs/GaAsSb HIGFET structures used in this study were grown by molecular beam epitaxy on semi-insulating InP:Fe substrates using a Varian Gen-II system. These two structures are illustrated in Figs. 1(a) and (b). The thickness and doping concentration of the $GaAs_{0.51}Sb_{0.49}$ surface layer determine the HIGFET gate configuration. The SAG structure has a 50 Å thick, undoped, cap layer while the RG structure has a 500 Å thick, Be-doped, cap

(a)	
GaAsSb:i, 50 Å	
InAlAs:i, 300 Å	
GaAsSb:i, 200 Å	
InAlAs:i, 3000 Å	
InAlAs/InGaAs SL, 30/30 Å, 10 Per.	
Semi-Insulating InP Substrate	

(b)	
GaAsSb:Be, 1E19, 500 Å	
InAlAs:i, 300 Å	
GaAsSb:i, 200 Å	
InAlAs:i, 3000 Å	
InAlAs/InGaAs SL, 30/30 Å, 10 Per.	
Semi-Insulating InP Substrate	

Figure 1. Layer profiles of the two HIGFET structures. Surface layer doping and thickness determine HIGFET gate configuration. The $GaAs_{0.51}Sb_{0.49}$ surface layer thickness designates the sample name: (a) self-aligned gate (SAG, "50 Å" sample), and (b) recessed gate (RG, "500 Å" sample).

layer. The samples are referred to as the "50 Å" or "500 Å" sample based on the thickness of the $GaAs_{0.51}Sb_{0.49}$ surface layer. Both samples are otherwise the same with the following undoped epilayer thicknesses: 300 Å $In_{0.52}Al_{0.48}As$ gate, 200 Å $GaAs_{0.51}Sb_{0.49}$ channel, 3000 Å $In_{0.52}Al_{0.48}As$ buffer, and a 10 period $In_{0.52}Al_{0.48}As$(30 Å)/$In_{0.53}Ga_{0.47}As$(30 Å) superlattice buffer.

Following growth, the sample surfaces were covered with 1000 Å of Si_3N_4. The samples were subsequently subjected to rapid thermal annealing (RTA) in a Heatpulse 410 system in a forming gas (5% H_2, 95% Ar) ambient. The samples were placed inside a graphite ring, underneath a GaAs wafer during RTA in order to limit As sublimation from the $GaAs_{0.51}Sb_{0.49}$ surface layer. Separate wafer pieces were annealed at temperatures of 600, 700 and 800 °C for 10 seconds.

Following RTA, photoluminescence (PL) measurements were performed at a temperature of 4.2 K using an Ar laser source with 4888 Å and 5145 Å lines, and monitoring luminescence with a 1.2 m spectrometer. Secondary ion mass spectroscopy (SIMS) depth profiles were obtained using a Cameca IMS-3F with an O_2^+ primary ion beam, raster scanned over a 125 x 125 μm^2 area with an 8 keV impact energy. A Perkin-Elmer PHI 660 Scanning Auger Microanalyzer provided Auger electron spectroscopy (AES) depth profiles. Following removal of the Si_3N_4 cap, the samples were sputtered with a 20 nA, 4 keV Ar^+ primary ion beam. The ion beam was scanned over a 200 x 200 μm^2 analysis crater, and the AES data was collected from the crater center over an area approximately one-tenth of the total crater area.

3. Results and discussion

Photoluminescence measurements were used to determine the optical quality of the $GaAs_{0.51}Sb_{0.49}$ channel layer following the 10 second RTA cycles. Both the PL linewidth and peak energies were monitored with respect to RTA temperature. The PL results are shown in Figs. 2(a) and (b). Both sample types exhibited the same trends as the RTA temperature was increased to 700 °C - increasing linewidth and peak energy. Increasing linewidth indicates a microscopic loss of homogeneity in the channel width. Increase in the peak energy indicates a macroscopic narrowing of the channel width as the temperature increases. No luminescence was observed from the $GaAs_{0.51}Sb_{0.49}$ channel layer in either sample following 800 °C RTA, indicating catastrophic degradation of the gate-channel heterojunction for both samples.

The SIMS depth profiles are illustrated in Fig. 3(a) for the 50 Å sample. The $GaAs_{0.51}Sb_{0.49}$ and $In_{0.52}Al_{0.48}As$ layers are clearly delineated in the as-grown spectrum by following the Sb, In and Al element traces. The center of the $In_{0.52}Al_{0.48}As$ gate layer is loca-

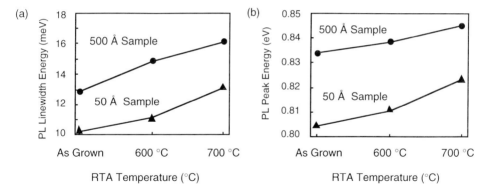

Figure 2. Photoluminescence energy versus RTA temperature: (a) linewidth energy and (b) peak energy. The 500 Å and 50 Å samples are represented by the circular and triangular data points respectively. Rapid thermal anneal time was 10 seconds at each temperature.

ted approximately 250 Å below the surface as evidenced by the first minimum in the Ga and Sb secondary ion yield and the peaks in the In and Al element profiles. Likewise, the center of the $GaAs_{0.51}Sb_{0.49}$ channel layer is positioned approximately 450 Å from the surface. Following annealing at 600 °C the profile resembled the as-grown case where the interfaces between the $GaAs_{0.51}Sb_{0.49}$ and $In_{0.52}Al_{0.48}As$ layers are well defined. At 700 °C considerable cross diffusion occurred as indicated by broadening of the As, Sb, In and Al element traces. Additionally, the Sb signal increased at depths below 600 Å, indicating diffusion into the $In_{0.52}Al_{0.48}As$ buffer layer. The $GaAs_{0.51}Sb_{0.49}$ channel layer was still detectable by slight dips in the In and Al signals at depths between 450 and 500 Å. The diffusion process continued at 800 °C as evidenced by a decrease in the slopes of the As, Al, In and Sb signals.

Figure 3(b) contains the SIMS spectra for the 500 Å sample. The thicker $GaAs_{0.51}Sb_{0.49}$ cap layer is observed by the flat minima in the In and Al signals near the surface in the as-grown profile. Little change in the elemental contours were observed following RTA at 600 and 700 °C RTA. Therefore, structural integrity of the InAlAs/GaAsSb heterojunction is maintained to 700 °C. After the 800 °C RTA the interfaces between the individual layers are no longer identifiable due to considerable cross diffusion of the elements. A comparison of the SIMS profiles in Figs. 3(a) and 3(b) indicates that the elemental diffusion is more severe at 700 °C for the 50 Å sample than the 500 Å sample. Also, both samples have no clearly definable InAlAs/GaAsSb heterojunction region following RTA at 800 °C.

Figure 4(a) illustrates the AES depth profiles for the 50 Å sample. Each spectrum demonstrates the depth profile for a single element. There are two spectra for the 700 °C case. One of the spectra was obtained in a specular, non-degraded surface region (open circle) while the other spectra was measured in a circular, degraded surface region approximately 100 µm in diameter (dark circle). The two different surface regions are henceforth referred to as the "field" and "circular" regions. The as-grown cap, gate and channel layers are easily delineated in each profile. Little diffusion was evidenced following 600 °C RTA. The AES profiles for the 700 °C case differ depending on whether AES occurred in the field or circular region. The AES profiles in the field region show little variation from the 600 °C case. However, in the

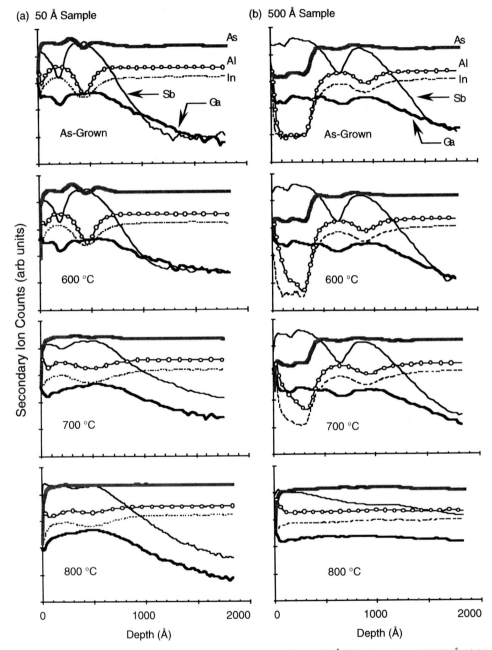

Figure 3. SIMS depth profiles for an InAlAs/GaAsSb HIGFET with a (a) 50 Å (left column) and (b) 500 Å (right column), GaAs$_{0.51}$Sb$_{0.49}$ cap layer. The SIMS spectra were measured as-deposited, and following rapid thermal annealing for 10 seconds at 600 °C, 700 °C and 800 °C.

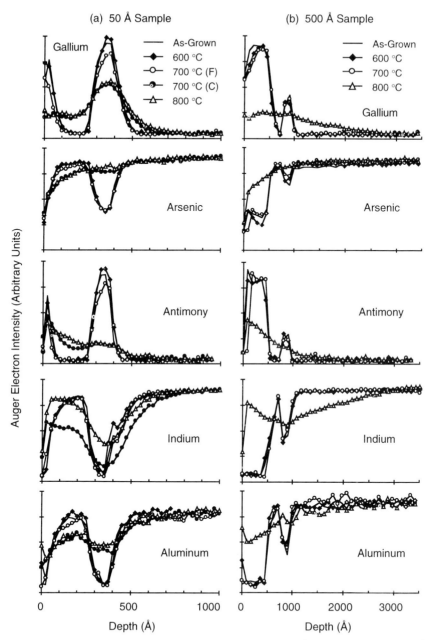

Figure 4. AES depth profiles for an InAlAs/GaAsSb HIGFET with a (a) 50 Å (left column) and (b) 500 Å (right column), GaAs$_{0.51}$Sb$_{0.49}$ cap layer. The AES profiles were measured as-grown, following RTA for 10 seconds at 600 °C, 700 °C (in both field and circular regions of the 50 Å sample) and 800 °C. The element profiles are Ga, As, Sb, In and Al in order from top to bottom.

circular region considerable diffusion occurred at 700 °C. AES profiles following 800 °C RTA are quite similar to those obtained in the circular region following 700 °C RTA.

Results of AES depth profiling on the 500 Å sample are shown in Fig. 4(b). These spectra indicated little diffusion at temperatures of 600 and 700 °C. These results are consistent with the observation of a specular surface on the 500 Å sample following RTA at 700 °C. Noticeable diffusion of elements was only detected following the 800 °C RTA.

A few insights are available through comparison of the PL, SIMS and AES results. First, RTA below 700 °C is dictated for the SAG HIGFET. Photoluminescence from the SAG HIGFET demonstrated a larger increase in linewidth and peak energy than the RG HIGFET as the RTA temperature was increased from 600 °C to 700 °C. Significant diffusion of elements from the $In_{0.52}Al_{0.48}As$ gate and buffer layers into the channel was subsequently detected by SIMS and AES. The AES depth profiles of the SAG HIGFET at 700 °C show these degradation regions are localized and were not detected by PL which used an excitation beam area much larger than the circular, degraded area. Second, the circular degradation regions in the SAG HIGFET structure appear to cause excesses of In, Al, and Sb at the surface, and a deficiency of Ga and As at the surface. Thus outdiffusion of Ga and As into the Si_3N_4 followed by Sb clustering appears to cause initial decomposition at the $GaAs_{0.51}Sb_{0.49}$ surface. Surface decomposition of the thin 50 Å cap layer versus the thicker 500 Å cap layer would create a greater disposition for outdiffusion from the $In_{0.52}Al_{0.48}As$ gate layer. Finally, while the thicker $GaAs_{0.51}Sb_{0.49}$ cap layer prevents epilayer degradation at 700 °C, it also serves as a source for greater indiffusion of cap layer elements at 800 °C. Both the SIMS and AES depth profiles showed deeper indiffusion of Ga from the 500 Å cap layer at 800 °C.

4. Conclusions

A combination of PL, SIMS and AES accurately monitor the structural integrity of InAlAs/GaAsSb HIGFET epilayers subjected to rapid thermal annealing. These results dictate maximum rapid thermal processing temperatures of 600 °C for self-aligned gate structures and 700 °C for recessed gate structures.

Acknowledgments

This effort was supported in part by a research grant from the Air Force Office of Scientific Research (AFOSR) monitored by Lt Col G.S. Pomrenke. Author P.W. Yu was supported under AFOSR contract No. F33615-86-C-1062 and is a National Research Council Associate. The authors thank C.E. Stutz and E.N. Taylor of Wright Laboratory, Solid State Electronics Directorate, and W. Rice of Wright State University for technical support.

References

[1] Martinez M.J., Scherer R.L., Schuermeyer F.L., Johnstone D.K., Stutz C.E. and Evans K.R., 1992 *Proc. 4th Intl. Conf. InP and Related Matls.*, 354-356.
[2] Stutz C.E., Evans K.R., Martinez M.J., Taylor E.N., Ehret J.E., Yu P.W. and Wie C.R., 1992 *J. Vac. Sci. Technol. B*, **10** 892-894.
[3] Sullivan G.J., Farley C.W., Ho W.J., Pierson R.L., Szwed M.K., Lind M.D. and Bernescut R.L., 1992 *J. Elect. Mat'ls*, **21** 1123-1125.
[4] Fathimulla A., Hier H. and Guitierrez D., 1994 *Tech. Dig. of the Dev. Res. Conf.*, IVB-6.
[5] Martinez M.J., Schuermeyer F.L. and Stutz C.E., 1993 *IEEE Elect. Dev. Lett.*, **14** 126-128.

A Novel Area-Variable Varactor Diode

Dong-Wook Kim, Jeong-Hwan Son, Song-Cheol Hong and Young-Se Kwon

Opto-Electronics Research Center
Department of Electrical Engineering, Korea Advanced Institute of Science and Technology 373-1 Kusong-dong, Yusong-gu, Taejon 305-701, Korea

Abstract : We propose a novel area-variable p-n varactor diode the depleted area of which changes sharply at a specific reverse bias voltage. The proposed diode can increase the capacitance ratio to the larger value than that of the other varactors and have various capacitance ratios on the same wafer. The diode can be fabricated by two ways. One is by etching away part of the top n layer, and the other is achieved by separating uncontacted n layer area from operating area by selective epitaxy. The fabrication process and quasistatic C-V characteristics of the diode are presented and the high frequency characteristics will be discussed.

1. Introduction

Varactor diodes with various capacitance-voltage characteristics have been extensively used in Monolithic Microwave Integrated Circuits (MMICs) as phase shift elements (Krafcsik et al. 1988), tunable bandpass filter elements(Chandler et al. 1993), voltage-controlled oscillators(Scott and Brehm 1982), and frequency multipliers(Hwu 1993). However, all the devices previously reported have the common limitations such that the devices fabricated on a wafer must have the same capacitance ratio (C_{max}/C_{min}) , which is determined by epitaxial structure, and are not easily integrated with other active devices. In this paper, we propose a novel area-variable p-n varactor diode the depleted area of which changes sharply at a specific reverse bias voltage. The proposed diode can increase the capacitance ratio to the larger value than that of the other varactors and have various capacitance ratios on the same wafer. The doping profile is not limited to a specified profile because its capacitance ratio is determined by top and bottom area ratio in the layout of the diode. Also highly nonlinear C-V characteristics can be obtained by the abrupt change of the depleted area. The diode can be fabricated by two ways. One is by etching away part of the top n layer of the diode and the other is achieved by separating uncontacted n layer area from operating area with triangular void formed by selective epitaxy.(Asai H. and Ando S. 1985, Kim C. T. et al. 1991)

2. The etched area-variable varactor diode

The concept of area-variable diode fabricated by etching away a part of the top n layer is shown in Fig. 1. The diode has the zero bias depletion depth, d_0, which does not cross the point, A, of etched area, and its capacitance value is proportional to the depletion width Wo. As the reverse bias voltage increases, the depletion depth crosses the point A and its effective depletion width is abruptly reduced to W1. Therefore we can have abrupt change of diode capacitance because the capacitance value of the diode is proportional to W1, which is much smaller than Wo.

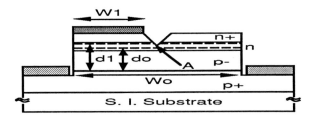

Figure 1. The cross sectional view of the etched area-variable varactor diode

    The etched area-variable varactor diode were fabricated using general MMIC process. The epilayer is arbitrarily chosen to be 500nm p^+-layer, 500nm p^--buffer layer, 200nm n-layer and 200nm n^+-layer grown by MBE. First, isolation is processed by chemical etching and SiO_2 sputtering. And then ohmic contact for cathode is formed by depositing AuGe/Ni/Au on the n^+-layer. And the etching to the middle of n-layer is performed to isolate the operating region with dummy region when the reverse bias voltage is fully applied. Next the etching to the p^+-layer and Au/AuZn/Au deposition for p-ohmic contact are performed. After ohmic metals are alloyed at 400 °C for 90 sec using rapid thermal annealing(RTA) system, Au is finally electroplated.

    Fig. 2 shows quasistatic C-V characteristics of large-area varactor diodes which have 9:1 and 16:1 area ratio respectively and are fabricated on the same wafer. The top n^+ contact area of both diodes are $1.75 \times 10^{-4} cm^2$ and the transition region of capacitance has about 2V span. The measured capacitance ratio is matched to the area ratio of top and bottom mask patterns. The quasistatic C-V characteristics of the small-area varactor diodes are also shown in Fig. 3. Fig.3 (a) shows the quasistatic C-V characteristics of area-variable varactor diode with 30:1 area ratio. The result shows that the capacitance ratio is not exactly matched to the area ratio. This is because the minimum capacitance value of the area-variable varactor diode is small and the parasitic capacitance is comparable to it. Fig. 3 (b) shows quasistatic C-V characteristics of area-variable varactor diode with 10:1 area ratio. The capacitance transition voltage is above 0V because n active layer is etched much more than other varactors mentioned above. So we can fix the capacitance transition voltage arbitrarily by adjusting the etching time of n active layer. The diode with floating top n^+ layer is expected to have the better high frequency characteristics. This is because the top n^+ layer decreases series resistance and results in high cutoff frequency. These structures are readily fabricated by utilizing anisotropic etching.

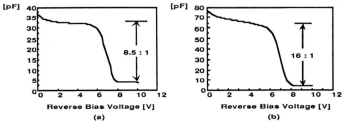

Figure 2. Large-area varactor quasistatic C-V characteristics (top contact area=$1.75 \times 10^{-4} cm^2$)
    (a) area-variable varactor diode with 9 : 1 area ratio
    (b) area-variable varactor diode with 16 : 1 area ratio

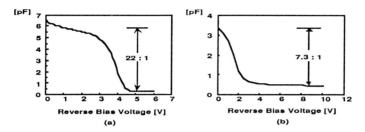

Figure 3. Small-area varactor quasistatic C-V characteristics
(a) area-variable varactor diode with 30:1 area ratio
(b) area-variable varactor diode with 10:1 area ratio

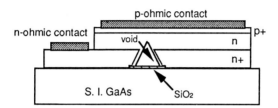

Figure 4. Schematic structure of the selectively grown area-variable varactor diode

3. The selectively grown area-variable varactor diode

A selectively-grown area-variable varactor diode is fabricated in a simple manner using selective metalorganic chemical vapor deposition(S-MOCVD) technique. It is experimentally confirmed that the geometrical shapes of the laterally grown facets during S-MOCVD growth depend on the growth condition, the crystal orientation of substrate, and the mask orientation. (Asai H. and Ando S. 1985, Kim C. T. et al. 1991) The proposed varactor diode, schematically illustrated in Fig. 4, features a trianguar void over SiO_2 stripe, $10° \sim 20°$ off from [T10] direction. The n^+-GaAs and n-GaAs layer right from void over SiO_2 stripe are isolated from the left n-contact metal when the distance between the top vertex of void and the p^+-GaAs layer is smaller than a depletion depth in n active layer. Thus, it is possible to make an abrupt change of the capacitance of the varactor diode. The device fabrication process is briefly described as follows.

After cleaning and etching an undoped semi-insulating GaAs (001) substrate with $H_2SO_4:H_2O_2:H_2O$ solution, the SiO_2 stripes, 20° off from [T10] direction, are formed on the substrate using e-beam evaporation and lift-off process. The width and thickness of SiO_2 stripes are 2.5um and 0.1um, respectively. The void height, H, can be experimentally expressed by

$$H = L / 2(\alpha \tan\theta + 0.71)$$

where L is the SiO_2 stripe width, α(1.25 for $\theta=20°$) is empirical constant, which is called the lateral growth parameter, and θ is the mask orientation angle. Expected void height from the above equation is 1um. 750nm, 5×10^{18} cm^{-3} doped n^+-GaAs layer, 550nm, 1×10^{17} cm^{-3} doped n-GaAs layer, and 100nm, 2×10^{19} cm^{-3} carbon-doped p^+-GaAs layer are grown at 650°C by atmospheric pressure-MOCVD system. Trimethygallium(TMG) and AsH_3 are used as source materials. N- and p-type dopant source materials are SiH_4 and CCl_4, respectively. During the growth, the void with height of 1um is formed over SiO_2 stripe. Using wet chemical etching with $H_2SO_4:H_2O_2:H_2O$ solution and lift-off process, the selectively grown area-variable varactor diode is fabricated as mesa-type. AuGe/Ni/Au and Au/AuZn/Au are used as n- and p-type ohmic metals, respectively, and alloyed at 430°C using RTA system.

The contrast of the quasistatic C-V characteristics of selectively grown area-variable p-n varactor diode with 2:1 area ratio and conventional p-n diode is shown in Fig. 5. The capacitance of the selectively grown p-n diode is reduced by the amount of capacitance isolated by void over SiO_2 stripe. Therefore we can make various C-V characteristics of varactor diode only with the height and position of void, i. e. the width and position of the SiO_2 stripe.

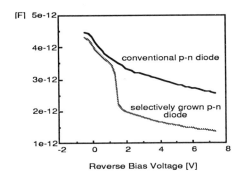

Figure 5. The contrast of the quasistatic C-V characteristics of the selectively grown area-variable p-n varactor diode with 2:1 area ratio and conventional p-n diode (top contact area=5×10^{-5} cm^2)

4. Conclusion

The novel area-variable p-n diodes the depleted area of which changes sharply at a specific reverse bias voltage have been presented. It is found that the proposed diode can increase the capacitance ratio to the larger value than other varactors and have various capacitance ratios on the same wafer. And the proposed diodes were fabricated by two ways of etching and selective MOCVD techniques and their capacitance ratios were well matched to the layout area ratio. It is believed that the further optimization of structure will give the better performance.

References

Krafcsik D. M., Imhoff S. A., Dawson D. E. and Conti A. L. 1988 IEEE MTT-36, 1938
Chandler S. R., Hunter I. C. and Gardiner J. G. 1993 IEEE Microwave and Guided Wave Lett. (3), 70
Scott B. N. and Brehm G. E. 1982 IEEE MTT-30, 2172
Hwu R. J. 1993 Microwave J. , 240
Asai H. and Ando S. 1985 J. Electrochem. Soc. (132), 2445
Kim C. T., Hong C. H. and Kwon Y. S. 1991 Jpn J. Appl. Phys. (30), 3828

Fabrication and Performance of In$_{0.53}$Ga$_{0.47}$As/AlAs Resonant Tunneling Diodes Overgrown on GaAs/AlGaAs Heterojunction Bipolar Transistors

K. B. Nichols, E. R. Brown, M. J. Manfra, and O. B. McMahon

Lincoln Laboratory, Massachusetts Institute of Technology
Lexington, Massachusetts 02173

P. J. Zampardi, R. L. Pierson, and K. C. Wang

Rockwell Science Center
Thousand Oaks, California 91360

Abstract. Resonant tunneling diodes (RTDs) have been fabricated in the pseudomorphic In$_{0.53}$Ga$_{0.47}$As/AlAs material system on GaAs substrates. The RTDs have peak-to-valley current ratios (PVCRs) of approximately 9 and peak voltages of approximately 1 V at room temperature. This PVCR is 50% greater than any reported RTD using the lattice-matched GaAs/AlGaAs or GaAs/AlAs material systems. When compared to an identical RTD structure grown lattice-matched on an InP substrate, these RTDs exhibit only a 20% reduction in PVCR with similar peak current and voltage. The RTD layers were overgrown on GaAs/AlGaAs heterojunction bipolar transistor (HBT) layers. The HBTs fabricated showed only a small degradation in their current-voltage characteristics

1. Introduction

Since the first observation of negative differential resistance in semiconductor resonant tunneling diodes (RTDs) [1], there has been great interest in using these devices in high-speed analog and digital circuits. Switches have been demonstrated with peak-to-valley rise times of 2 ps[2], and 6-10 ps[3]. The most promising digital applications would combine RTDs monolithically with conventional GaAs devices such as heterojunction bipolar transistors (HBTs), metal-semiconductor field effect transistors (MESFETs), and pseudomorphic high electron mobility transistors (pHEMTs) to create new switching devices, logic gates, and memory elements on GaAs substrates. A fundamental objective in the design of these devices is to maximize the peak-to-valley current ratio (PVCR) of the RTDs at the lowest possible peak voltage, while maintaining a high peak current and wide valley region. If the PVCR of the RTDs is high, these devices will have a very low static power dissipation. The major limitation to using conventional RTDs from the GaAs/AlAs or GaAs/AlGaAs material systems is their relatively low PVCRs. The best reported PVCR for these systems is 6 [4] while typically the PVCR is between 3 and 4 [5]. In order to circumvent this problem, we have looked at the pseudomorphic In$_{0.53}$Ga$_{0.47}$As/AlAs material system. When epitaxially deposited on latticed-matched InP substrates such that the AlAs barriers are pseudomorphic, In$_{0.53}$Ga$_{0.47}$As/AlAs RTDs have shown PVCRs as high as 28-30[6,7]. In this study we have fabricated In$_{0.53}$Ga$_{0.47}$As/AlAs RTDs on GaAs substrates and on GaAs HBT layers. GaAs substrates are preferable to InP substrates because of their lower cost, larger size, greater strength, and the maturity of HBT, MESFET, and pHEMT fabrication -- all important factors in integrated circuit fabrication.

2. RTD fabrication on GaAs substrates

Prior to studying RTDs grown on HBT wafers, we fabricated a control sample on an n-type InP substrate and a control sample on a semi-insulating GaAs substrate to determine the effects of the lattice mismatch on GaAs and to see if those effects could be mitigated. The RTD heterostructures were grown using a 3-in.-wafer compatible, solid-source, Intevac MOD GEN II, molecular beam epitaxy system. The InP substrates had a crystallographic orientation of (100), while the GaAs substrates had an orientation of either (100) or 2° off (100) toward (110). Reflection high-energy electron diffraction intensity oscillations were used to calibrate the growth rates. The $In_{0.53}Ga_{0.47}As$ growth rate was 1.08 μm/h and the AlAs growth rate was 0.31 μm/h with the substrate temperature at 480 °C. Growth interruptions of 25 s were used at the interfaces of the barriers and quantum well. The RTD structure consisted of a 55-Å-thick, undoped, $In_{0.53}Ga_{0.47}As$ quantum well with 18-Å-thick, undoped, pseudomorphic AlAs barriers, 400-Å-thick, lightly doped, $In_{0.53}Ga_{0.47}As$ cladding layers outside the barriers, and heavily doped $In_{0.53}Ga_{0.47}As$ ohmic contact layers outside the cladding layers. The overall RTD material profile is shown in Fig. 1.

The test samples consisted of the same $In_{0.53}Ga_{0.47}As$/AlAs RTD structure grown on GaAs substrates using various types of buffers between the substrate and the RTD layers. The buffer shown in Fig. 1 produced the best device results. It is a modified version of the buffer used for growing $In_xGa_{1-x}As/In_zAl_{1-z}As$ heterostructure field effect transistor material on GaAs [8]. The present buffer consists of a 1500-Å-thick, undoped, low-temperature-grown (LTG) $In_xGa_{1-x}As$ layer with x graded from 15% to 53%, and a 3000-Å-thick, undoped, LTG $In_{0.53}Ga_{0.47}As$ layer. The substrate temperature during the growth of these layers was 320 °C. A temperature ramp of 0.1 °C/s was used to continuously increase the temperature of the In cell from the initial value for the growth of $In_{0.15}Ga_{0.85}As$ to the final value for the growth of $In_{0.53}Ga_{0.47}As$. Thick $In_{0.53}Ga_{0.47}As$ layers, GaAs/$In_{0.53}Ga_{0.47}As$ superlattices with high and low arsenic fluxes, and LTG $In_{0.53}Ga_{0.47}As$ layers were also tried in this study but produced significantly inferior RTDs.

The processing of the wafers into diodes was completed by standard microfabrication steps, starting with the photolithography of square active areas. Evaporation and photoresist

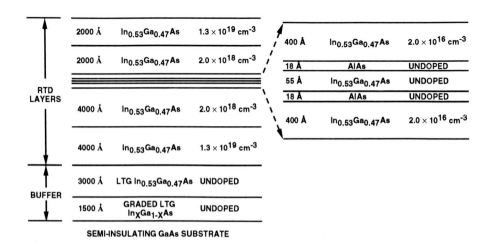

Figure 1. Schematic cross section of RTD and buffer layer structures.

lift-off were then used to define the ohmic contact metallization of Pd, Ge, and Au with thicknesses of 300, 400, and 3400 Å, respectively. The ohmic contacts, which are on $In_{0.53}Ga_{0.47}As$ having a carrier concentration of 1.3×10^{19} cm^{-2}, were not annealed or alloyed in any way. The diodes were isolated by wet chemical etching. Devices with active areas of approximately 49 and 9 μm² were produced on each wafer.

3. RTD performance

The current-voltage (I-V) characteristics of the control RTD on InP are shown in Fig. 2(a). This device had an active area of approximately 9 μm². When the top ohmic contact is biased negatively, the control RTD has a PVCR of 11.4, a peak voltage (V_P) of 0.95 V, and a peak current density (J_P) of 2.06×10^4 A/cm². This PVCR is substantially lower than the values of 28 to 30 seen by other workers[6,7]. However, the higher PVCRs were obtained at the expense of either significantly lower J_P or significantly higher V_P. The control RTD grown directly on GaAs has a PVCR of 4.0, a V_P of 1.05 V, and a J_P of 1.4×10^4 A/cm² using the same bias conditions. This significant degradation is attributed to the lattice mismatch with the GaAs substrate.

The I-V characteristics of a test RTD grown on a GaAs substrate using the best LTG buffer are shown in Fig. 2(b). This device had an active area of approximately 9 μm². When the top ohmic contact is biased negatively, the device has a PVCR of 9.3, a V_P of 1.00 V, and a J_P of 1.7×10^4 A/cm². The negative bias PVCR is 50% greater than any reported in RTDs made from the lattice-matched AlGaAs/GaAs or AlAs/GaAs material system[4]. When compared to the InP-substrate control RTD, these RTDs exhibit only a 20% reduction in PVCR, a 15% reduction in J_P, and a 5% increase in V_P.

A slightly modified RTD structure with the same buffer layer but thinner heavily doped $In_{0.53}Ga_{0.47}As$ ohmic contact layers above the tunneling layers was also fabricated. The thickness of the top two layers shown in Fig. 1 was 500 and 1000 Å, respectively, rather than 2000 Å each. The I-V characteristics of this RTD were nearly identical to the RTD shown in Fig. 2(b). When the top ohmic contact is biased negatively, the device has a PVCR of 9.3, a V_P of 1.00 V, and a J_P of 1.6×10^4 A/cm².

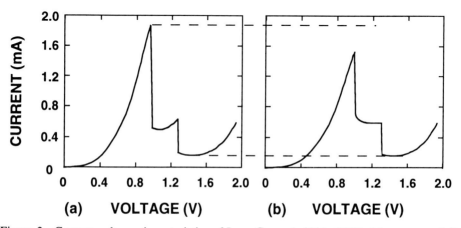

Figure 2. Current-voltage characteristics of $In_{0.53}Ga_{0.47}As$/AlAs RTDs (a) grown on InP and (b) grown on GaAs with an LTG buffer.

4. Transmission electron microscopy

To determine the effect of the LTG buffer, we used transmission electron microscopy (TEM) to observe the material quality in the RTD epitaxial layers with and without the LTG buffer. Fig. 3(a) shows a TEM cross section of the GaAs-substrate control RTD at 120,000 times magnification. A lower magnification view indicated that the ~4% lattice mismatch causes the formation of misfit dislocations at the substrate/epitaxial-layer interface that transform into threading dislocations and propagate into the RTD layers. From Fig. 3(a) we see that these dislocations penetrate into and through the $In_{0.53}Ga_{0.47}As/AlAs$ double-barrier structure. No dislocations were observed to propagate back into the GaAs substrate.

Fig. 3(b) shows a TEM cross section at the same magnification of the thinner test RTD described at the end of the previous section, which incorporated the LTG buffer between the RTD structure and the GaAs substrate. The density of dislocations in the $In_{0.53}Ga_{0.47}As/AlAs$ double-barrier structure in Fig. 3(b) is roughly an order of magnitude lower than in Fig. 3(a). The excellent RTD resulting from the use of the LTG buffer is undoubtedly a direct result of the lower dislocation density in the critical double-barrier region of the device. The LTG buffer appears to serve two functions. First, it accommodates the large lattice mismatch between the GaAs substrate and the $In_{0.53}Ga_{0.47}As$ RTD by the formation of misfit and threading dislocations. Second, it traps most of these dislocations in the LTG region and prevents their propagation into the RTD structure grown above it. Further TEM examination of the LTG buffer in this sample also supports these conclusions.

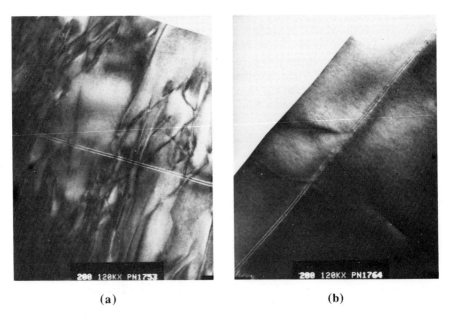

(a) (b)

Figure 3. Cross-sectional TEMs of $In_{0.53}Ga_{0.47}As/AlAs$ RTDs grown on GaAs substrates (a) without a buffer and (b) with an LTG buffer.

5. Integration of RTD and HBT

The primary goal of this study was to integrate high-quality RTDs made from the pseudomorphic $In_{0.53}Ga_{0.47}As/AlAs$ material system with commercially available HBT material grown on GaAs wafers. The HBTs were grown by organometallic chemical vapor deposition (OMCVD) on GaAs substrates oriented 2° off (100) toward (110). Therefore, we decided to first test the RTD structure on bare substrates having the same orientation as the HBT wafers. The RTD structure and LTG buffer layer were grown and fabricated simultaneously on sections of GaAs substrates with crystallographic orientations of both (100) and 2° off (100) toward (110). PVCRs of 9, V_P's of 1 V, and J_P's of ~2×10^4 A/cm^2 were measured for both orientations.

The same experiment was then performed with an on-axis GaAs substrate and an off-axis GaAs HBT wafer. For this experiment approximately 500 Å was removed from the top of the HBT structure by wet chemical etching before loading into the MBE system. This was required since the HBT wafers had not been kept continuously in an inert atmosphere after the OMCVD growth. Based on work done at Lincoln Laboratory, we believe that in the future HBT wafers kept in an inert atmosphere will not require this cleaning[9]. The measured PVCRs for both the GaAs substrate and the HBT wafer were 9 with V_P's of 1 V and J_P's of ~2×10^4 A/cm^2. From these results, we can conclude that the growth of a pseudomorphic $In_{0.53}Ga_{0.47}As/AlAs$ RTD is not affected by an off-axis orientation of 2° or by the epitaxial HBT layers on the substrate.

The second important issue was the effect of the RTD overgrowth on the HBT. To address this we cleaved two sections from the same HBT wafer. The first section went directly into fabrication and served as a control wafer. The second section had the RTD structure and LTG buffer grown on the HBT layers. The two sections were fabricated together such that the resulting HBTs had an emitter area of 72×72 μm. The Gummel plots of HBTs from the control wafer and the RTD/HBT wafer are shown in Fig. 4. The β was

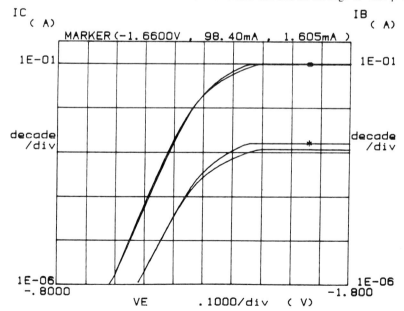

Figure 4. Gummel plots for a control HBT wafer and an RTD/HBT wafer (with marker).

88.5 for the control wafer and 61.5 for the RTD/HBT wafer. We believe that this difference in β is due to fabrication variations and does not indicate there was a degradation of the HBT in the RTD growth. Circuits combining these MBE-grown RTDs and OMCVD-grown HBTs are currently being fabricated since each building block appears to function properly.

6. Conclusion

In summary, RTDs with excellent characteristics have been fabricated in the pseudomorphic $In_{0.53}Ga_{0.47}As/AlAs$ material system on GaAs substrates. The RTDs have PVCRs of 9.3, peak voltages of 1 V, and a peak current density of 1.7×10^4 A/cm^2 at room temperature. This PVCR is 50% greater than any reported RTD using the lattice-matched AlGaAs/GaAs or AlAs/GaAs material systems. These RTDs exhibit only a 20% reduction in PVCR and a similar peak current and voltage compared to an identical RTD structure grown lattice-matched on an InP substrate.

The LTG $In_xGa_{1-x}As$ buffer layer was critical to producing these high-PVCR RTDs. Other buffers such as superlattices and thick InGaAs layers produced inferior RTDs. The LTG buffer appears to serve two functions. First, it accommodates the large lattice mismatch (~4%) between the GaAs substrate and the $In_{0.53}Ga_{0.47}As$ RTD by the formation of misfit and threading dislocations. Second, it traps these dislocations and prevents their propagation into the RTD structure grown above it.

These RTDs can be combined monolithically with conventional GaAs three-terminal devices such as HBTs, MESFETs, and pHEMTs to create high speed switching devices, logic gates, and memory elements on GaAs substrates. We have demonstrated the integration of these RTDs with HBT wafers grown by a commercial vendor using OMCVD. Because of the high PVCR of the RTD, digital circuits combining the RTD and HBT could have much lower static power dissipation than conventional HBT circuits.

Acknowledgments

This work was sponsored by the Advanced Research Projects Agency (ULTRA Program) and the Department of the Air Force. The authors acknowledge K. M. Molvar, L. J. Mahoney, and R. W. Murphy for their assistance in the fabrication of the RTDs, C. D. Parker and K. A. McIntosh for their assistance in the electrical characterization of the RTDs, and P. M. Nitishin for his assistance in transmission electron microscopy.

References

[1] L. L. Chang, L. Esaki, and R. Tsu, 1974 *Appl. Phys. Lett.* **24**, 593.
[2] J. F. Whitaker, G. A. Mourou, T. C. L. G. Sollner, and W. D. Goodhue, 1988 *Appl. Phys. Lett.* **53**, 385.
[3] S. K. Diamond, E. Ozbay, M. J. W. Rodwell, D. M. Bloom, Y. C. Pao, and J. S. Harris, Jr., 1989 *OSA Proceedings on Picosecond Electronics and Optoelectronics* (Washington D. C.: Optical Society of America), pp. 101-105.
[4] V. K. Reddy, A. J. Tsao, and D. P. Neikirk, 1990 *Electron. Lett.* **26**, 1742.
[5] E. R. Brown, 1992 *Hot Carriers in Semiconductor Nanostructures: Physics and Aplications*, 469-498.
[6] T. P. E. Broekaert, W. Lee, and C. G. Fonstad, 1988 *Appl. Phys. Lett.* **53**, 1545-1547.
[7] E. R. Brown, C. D. Parker, A. R. Calawa, M. J. Manfra, T. C. L. G. Sollner, C. L. Chen, S. W. Pang, and K. M. Molvar, 1990 *SPIE High-Speed Electronics and Device Scaling* **1288**, 122-135.
[8] D. E. Grider, S. E. Swirhun, D. H. Narum, A. I. Akinwande, T. E. Nohava, W. R. Stuart, P. Joslyn, and K. C. Hsieh, 1990 *J. Vac. Sci. Technol. B* **8**, 301-304.
[9] Private communication with G. W. Turner of MIT, Lincoln Laboratory.

Two-Dimensional Self-Consistent Modeling of InP/GaInAsP Lateral Current Injection Lasers

D.A. Suda[†], T. Makino[‡], and J.M. Xu[†]

[†]University of Toronto, Dept. of Electrical Engineering
Toronto, Ontario, M5S 1A4, Canada

[‡]Bell Northern Research
Ottawa, Ontario, Canada

Abstract – Lateral Current Injection (LCI) lasers have the potential to overcome many of the limitations inherent to conventional vertical injection structures, but until now there has been very little theoretical analysis and optimization of LCI laser designs. 2D self-consistent simulations of these devices have been carried out for the first time. Calculated characteristics are in good agreement with experimental results and offer explanations for the observed roll-off in differential quantum efficiency. In addition, we have explored other key design issues unique to LCI lasers such as the consequences of the non-uniform gain profile, ways to minimize this non-uniformity, and the effects limiting the maximum small-signal modulation frequency.

Lateral Current Injection (LCI) lasers have the potential to overcome many of the limitations inherent to conventional vertical injection lasers. In an LCI laser (Figure 1), current does not pass through the cladding layers or the substrate, allowing both these regions to be made of wide bandgap semi-insulating material for reduced parasitic loss and capacitance and for improved high-speed operation. In addition, the planar geometry of LCI lasers is well suited for OEIC applications.

While there has been a considerable amount of experimental effort devoted to these structures [1-6], there has until now been very little theoretical analysis of the device operation or design optimization. In this paper, we will present the first 2D analysis of some recently fabricated lasers. This will provide insight into the experimental results and the underlying physics and point the way to improved designs.

We first analyze a 1.55μm InP/GaInAsP LCI laser design which was recently reported in [6]. In this buried heterostructure, carriers are laterally injected from regrown n and p-doped InP regions into a 0.1μm thick GaInAsP active layer. We simulated this structure using our 2D self-consistent Finite Element Light Emitter Simulator (*FELES*) [7]. The predicted single mode operation, threshold current, and L-I characteristics were found to be in very good agreement with experiment, including the non-linearities at high injection (Figure 2).

A striking feature of both the measured and simulated results is a rapid decrease in differential quantum efficiency with increasing current. This leads to output power saturation at a relatively low level: only 10 mW at a current of 150mA. Speculations based on inspection have been made regarding the possible causes of this problem, including poor carrier confinement in the lateral direction, high non-radiative recombination at the regrowth interface, and the spatial mismatch between the gain peak and the peak of the optical mode. Contrary to these hypotheses, a series of simulations in which each of these possible detrimental effects was isolated and examined individually demonstrated that none of them could sufficiently account for the observed decrease in differential quantum efficiency.

Instead, the simulations reveal that this roll-off in quantum efficiency is mostly due to the activation of parallel conduction paths through the buffer and guiding layers (Figure 3). These layers form wide bandgap PIN diodes which are electrically in parallel with the active region. As the injected current increases, so does the bias across these parallel diodes, resulting in an exponential increase in leakage current and a corresponding decrease in quantum efficiency. With this leakage problem identified, we have been able to use *FELES* to develop designs which minimize this and other aforementioned problems while maintaining good optical and carrier confinement.

Another key design issue unique to LCI lasers is the non-uniform gain profile in the QWs. This problem, which can only be seen clearly through simulation, was found to originate from the disparity in mobilities between electrons and holes in the active region. The higher mobility electrons tend to "pile up" at the p-doped/active region interface, resulting in a highly asymmetric gain profile (Figure 4) which decreases the overlap of the optical mode and the gain peak and ultimately increases threshold and lowers efficiency.

A promising technique for minimizing this problem is to add doped layers above and below the undoped active region. Addition of these "current guiding" layers results in a design which is a lateral/vertical injection hybrid. Current is not only injected from the ends of the active region as in previous lateral injection designs but also along the length of the channel as is the case for vertical injection lasers. While the injection paths are now similar, the differences in fabrication and contacts remain. No current flows through the ridge or substrate so most of the benefits of lateral injection are retained. The use of these layers produces a more uniform gain profile (Figure 5). However, we also observed that they increase leakage. One can use a simulator like ours to study this trade-off and optimize the design.

Our analysis indicates that well designed LCI lasers should be comparable to vertical injection lasers in terms of threshold and quantum efficiency. Another concern which naturally arises is the dynamic performance. We find that this is an area where LCI lasers show the potential for large improvements over conventional designs. Since the substrate is semi-insulating, the effective area of the n and p contacts is small, and the separation between the contacts is relatively large, the capacitance of LCI lasers should be much lower than vertical injection designs. In confirmation of this, *FELES* simulations predict a zero bias capacitance of 0.8pF for the structure reported in [6], compared to the 0.5pF measured experimentally. While it is difficult to experimentally measure the capacitance under large forward bias, it was easily determined from *FELES* simulations and found to be approximately 15pF with the device biased at twice the threshold current. Combined with a low differential resistance (Figure 6), these numbers indicate that the modulation speed will be limited more by carrier transport times than R-C charging delays.

In summary, our analysis shows that LCI lasers have the potential to be low power, high speed components suitable for OEIC applications. While the performance of experimental devices produced until now have not yet been competitive with conventional vertical injection lasers, our simulations have led to an understanding of the problems encountered and indicate ways in which these problems can be overcome.

References
[1] A. Furuya, M. Makiuchi, O. Wada, and T. Fujii, *IEEE J. Quant. Elec.*, **24**, pp. 2448-2453, Dec. 1988.
[2] N. Yasuhira, I. Suemune, Y. Kan, and M. Yamanishi, *Appl. Phys. Lett.*, **56**, pp. 1391-1393, Apr. 1990.
[3] Y.Kawamura,A.Watatsuki,Y.Noguchi,and H.Iwamura,*IEEE Phot.Tech.Lett.*,**3**,pp.960-962, Nov. 1991.
[4] W. Zou, K. Law, J. Merz, R. Fu, and C. Hong, *Appl. Phys. Lett.*, **59**, pp. 3375-3377, 23 Dec. 1991.
[5] Y. Kawamura, Y. Noguchi, and H. Iwamura, *Electronics Lett.*, **29**, pp. 102-104, 7 Jan. 1993.
[6] K. Oe, Y. Noguchi, and C. Caneau, *IEEE Photonics Tech. Lett.*, **6**, pp. 479-481, Apr. 1994.
[7] G.L Tan, K. Lee, and J.M. Xu, *Jpn. J. Appl. Phys., part 1*, **32**, pp. 583-589, Jan. 1993.

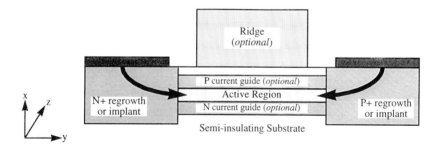

FIGURE 1. A typical LCI laser including optional ridge and current guiding layers. Arrows indicate direction of current injection from the contacts into the active region. The axes correspond to the directions used in the following figures.

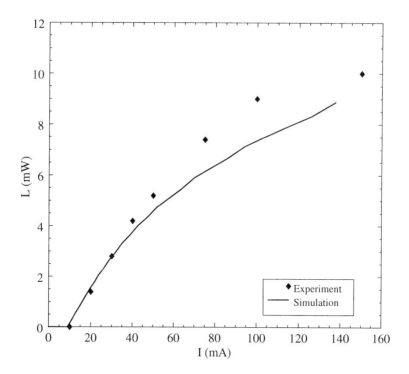

FIGURE 2. Experimental and simulated L-I curves for an LCI laser. The laser design and experimental data were taken from[6] and the simulation results were produced by *FELES* [7].

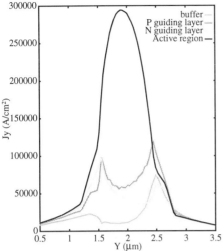

FIGURE 3. Simulated current densities in various cross-sections perpendicular to the growth direction of the LCI laser reported in [6]. N-type regrowth is present for Y<1.5μm and p-type for Y>2.5μm.

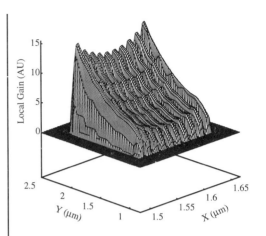

FIGURE 4. Simulated gain profile showing non-uniformity due to electron and hole mobility differences. In the plot, 10 QWs are contacted by an n-type regrowth at Y=1.0μm and a p-type regrowth at Y=2.2μm. The effects of spatial hole burning in X and Y are also visible.

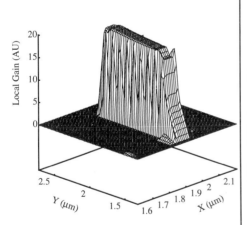

FIGURE 5. Simulated gain profile for an LCI laser with current guiding layers. A 0.1μm thick active region extends from the n-type regrowth at Y=1.5μm to the p-type regrowth at Y=2.5μm. The region above the active layer (X<1.8μm) is p-type and the region below (X>1.9μm) is n-type.

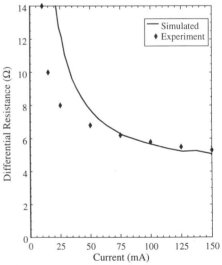

FIGURE 6. Simulated and experimental differential resistance for the LCI laser reported in [6]. The simulation assumes a 3.5Ω external contact resistance.

SimWindows: A New Simulator for Studying Quantum Well Optoelectronic Devices

D. W. Winston, R. E. Hayes

Department of Electrical and Computer Engineering and Optoelectronic Computing Systems Center, University of Colorado at Boulder, Campus Box 425, Boulder, CO 80309, USA

Abstract: A new one-dimensional device simulator, SimWindows, addresses the issues associated with complex optoelectronic devices. SimWindows solves the standard semiconductor equations, but includes thermionic emission current, quantum confinement, photon generation, and heat conduction. SimWindows is designed as a useful tool for analyzing optoelectronic devices. Results for a vertical cavity surface emitting laser are presented, and the effects of internal heating are discussed.

1. Introduction

Greater demands have been placed on the physical models incorporated into optoelectronic device simulators primarily because of two factors. These are the incorporation of complex structures such as distributed bragg reflectors (DBRs) and multiple quantum well (MQW) regions, and the need to understand the thermal issues associated with optoelectronic devices. The one-dimensional device simulator, SimWindows, addresses both of these issues while providing an easy interactive environment for evaluating optoelectronic device designs. SimWindows is being developed within the framework of a modified drift-diffusion model. Poisson's and the current continuity equations as well as a photon rate equation and a heat-flow equation are solved self-consistently. Currents result from the drift and diffusion of carriers as well as thermionic emission over heterojunction barriers. Quantum well electron and hole wave functions are coupled into these equations through the carrier concentrations and recombination rates. SimWindows is being applied to a variety of optoelectronic devices including vertical cavity surface emitting lasers (VCSELs) and QW solar cells. SimWindows runs on any IBM-compatible personal computer under the Microsoft Windows environment[1].

2. Models Description

One of the basic features of SimWindows is that it can easily apply different physics to different regions of the device. Regardless of the physics used, total charge, carrier currents, and total energy flow must be known within any region of the device. This is also true when applying different boundary conditions to the device. The particular boundary condition dictates the values of the charge, the currents, and the energy flow. Matching the conditions imposed by the various regions results in a self-consistent numerical solution.

An example of a specialized region is the quantum well where the physics associated with quantum confinement is applied independently from that of the bulk regions. At present,

---

[1]SimWindows is available via anonymous ftp from hopper.colorado.edu (128.138.248.150) in the pub/modeling/SimWindows directory.

the carrier concentrations are calculated using an approach similar to [1] where "bound" carriers are described by the step-like density of states, and "free" carriers are described by the 3D density of states. The thermionic emission approach of Grinberg [2] is used to model the carrier currents over the QW barriers. The energy flow resulting from the thermionic emission of carriers can also be determined using a similar approach. Electron and hole wave functions are incorporated into this model by considering the quantum well as a single node where the total number of carriers in the quantum well are calculated. The carriers are then distributed over a number of "subnodes" within the quantum well according to the appropriate wave function. This approach allows quantities to vary within the quantum well while only using a single node during the numerical iterations.

Thermal effects are also important in optoelectronic device simulation because of the heat dissipation problems associated with large arrays of optoelectronic devices. SimWindows addresses the thermal issues by solving the heat-flow equation, which yields the internal temperature profile of the device. The carriers and the lattice are assumed to be in local equilibrium and described by the same temperature. The heat-flow equation then balances the energy carried by the electrons and holes with the energy carried by the lattice. The heat flow between the device and the environment is controlled by setting the thermal conductance between the contacts and the environment to infinity (ideal heat sink), zero (ideal insulator), or an arbitrary value. Lateral heat flow is modeled by using a distributed thermal conductance along the length of the device.

3. Simulation Results

To demonstrate the functionality of SimWindows, results for a DBR VCSEL are presented. Figure 1 shows a band diagram for the VCSEL at threshold under isothermal conditions. The left DBR (top) consists of 18 periods of alternating $Al_{0.16}Ga_{0.84}As/AlAs$ layers p-typed doped between 2×10^{19} and 5×10^{18} cm^{-3}. The right DBR (bottom) consists of 29 periods of the same material n-type doped 3×10^{18} cm^{-3}. The active region consists of 3 80Å GaAs quantum wells with 80Å barriers [3].

One problem associated with VCSELs is the internal heating of the device caused by

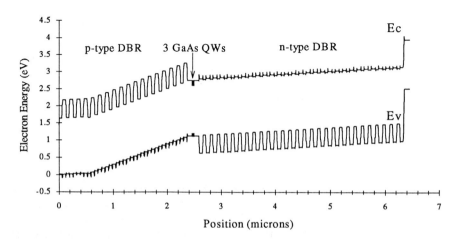

Figure 1. Band diagram of the VCSEL at threshold.

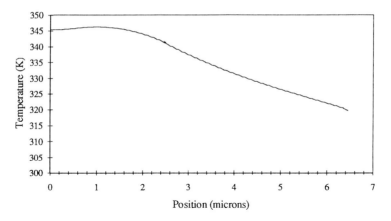

Figure 2. Temperature profile of the VCSEL in Figure 1 at four times the threshold current.

the resistive heating of the DBRs. The increase in temperature causes the resonant frequency of the cavity to shift away from the frequency of maximum gain, thus reducing the gain. It is desirable to design DBRs that reduce the resistance not only to reduce the threshold voltage, but also to reduce the internal heating. Figure 2 shows the temperature profile of the VCSEL at a current level approximately four times the threshold current. An insulating boundary was used at the left contact and an idealized thick substrate boundary was used for the right contact. The small ridges in the temperature are due to the varying thermal conductivity. Even at the modest light output of 0.5 mW and the given boundary conditions, the heating within the DBR is significant. The resistance of the DBR results from the thermionic emission barriers in the conduction band (for an n-type DBR) or the valence band (for a p-type DBR) which limit the carrier flow. Simulation results show that when currents are modeled as drift-diffusion currents only, then the DBR resistance is reduced by over an order of magnitude. To reduce the DBR resistance, it is necessary to either reduce the barrier height or assist tunneling through the barriers. Various schemes have been proposed to reduce the resistance of DBRs, and SimWindows provides the means for evaluating various DBR designs.

Figure 3 shows the light output power and the voltage across the VCSEL as a function of the drive current under both isothermal and non-isothermal conditions. From this figure the need for incorporating heat flow again becomes clear. Isothermal conditions can yield an overestimate of the optical output power. There are a number of physical effects which will be incorporated into future enhancements SimWindows. These include the band gap narrowing and the shift in the index of refraction with temperature. Including these effects will allow SimWindows to predict the further decrease in optical output power of a VCSEL at higher drive currents.

4. Summary

SimWindows is a new optoelectronic simulator which incorporates the features necessary for modeling a wide range of optoelectronic devices. In particular, SimWindows is being used to design and analyze VCSEL structures. Results for a DBR VCSEL were presented and the effect of resistive heating within the device structure was demonstrated. Including effects

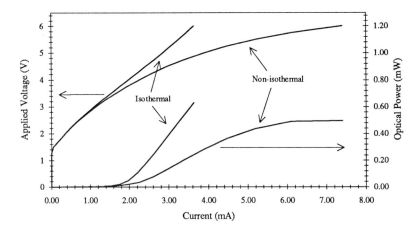

Figure 3. Light output and voltage vs. current under isothermal and non-isothermal conditions for the VCSEL in Figure 1.

such as quantum confinement and heat flow allows SimWindows to assist device designers in evaluating the performance of optoelectronic devices. SimWindows has been designed so that models can be improved and enhanced as necessary while also being an easy environment to use.

References

[1] J. Nelson, M. Paxman, K.W.J. Barnham, J.S. Roberts, and C. Button, *IEEE J. Quantum Electron.*, vol. 29, pp. 1460-8, 1993.
[2] A.A. Grinberg, *Physical Review B*, vol. 33, pp. 7256-8, 1986.
[3] H. Temkin, private communication.

Development of an Intelligent Heterostructure Material Parameter Database System

W. R. Frensley, R. C. Bowen, C. L. Fernando, and M. E. Mason

University of Texas at Dallas
P.O. Box 830688, EC-33
Richardson, Texas 75083-0688

ABSTRACT: Simulations of modern heterostructure devices are increasingly constrained by their need for a variety of material parameters. An intelligent database system has been developed to supply these parameters. The system stores expressions of arbitrary complexity, and organizes them in a flexible hierarchical fashion. Using empirical or theoretical rules, it is able to infer quantities that are not explicitly stored. This provides the flexibility required to deal with heterostructures containing semiconductor alloys.

1 Introduction

As heterostructure devices become more sophisticated, our ability to analyze, simulate, and design them is increasingly constrained by the volume of material parameter data required. Some of the devices of greatest current interest require the greatest diversity of material data. For example, to simulate the electronic and optical properties of a vertical-cavity laser with pseudomorphic layers requires elastic constants, energy-band data, deformation potentials, optical constants, and carrier transport parameters. Moreover, many of these numbers are not known by direct experimental measurements on the composition in question, but must be inferred from data on related materials. The data management problem in heterostructure technology is particularly compounded by the use of solid solutions whose compositions span a continuum. As a consequence, conventional database technologies [based upon the assumption that data are discrete entities (Parsaye et al., 1989)] are not particularly applicable to the problems posed by heterostructure technology.

In the course of developing interactive, general-purpose heterostructure and quantum device simulation programs (Frensley, 1991) we have squarely encountered the data management problem. The volume of data is proportional to an increasing number of materials of interest multiplied by an increasing number of physical effects to be modeled. As a consequence of interactions with the users of our simulation programs we have identified several features required of a heterostructure database system. First, no fixed data structures will be adequate in the long term. For example, the bandgap versus composition relations of most ternary alloys (in the compilation of Casey and Panish, 1978) are quadratic functions. The prominent exception is $Al_xGa_{1-x}As$, in which the direct bandgap is linear over part of the range and quadratic over the rest. The more recent results of Aspnes *et al.* (1986) find a cubic dependence over the entire composition range. In neither case would a simple array of three coefficients be adequate to define the bandgap of $Al_xGa_{1-x}As$, though such a data structure would work for other ternary alloys.

The second point is that the database must be modifiable by the users of any simulation programs which employ it. The reason for this requirement is not merely that experimental data continue to accumulate and must be added, but that the view of which data are "correct" can vary from laboratory to laboratory. For example, alloy compositions are controlled by epitaxial growth conditions. The epitaxy systems are calibrated by performing a series of growths at varying conditions, the bandgaps of the resulting samples are measured optically, and the compositions are then inferred from the "known" bandgap versus composition relation. If one uses a simulation program to optimize the design of a heterostructure device, it thus becomes much less important which of the published expressions is more accurate. What is important is to assure that the simulation program uses the same expression as that used to calibrate the epitaxial system on which the device will fabricated.

2 The Database Program

The design of the present system was determined largely by the what we view as the fundamental requirement of a heterostructure database: the need to store and evaluate algebraic expressions of arbitrary complexity. The software technology necessary to implement such a capability is well developed in the context of compiler design (Aho *et al.* 1986). Compilers necessarily construct dynamic data structures, and we found that this sort of software technology provided an elegant solution to the problem of constructing a heterostructure database system. Some aspects of the design were adapted from the program hoc described by Kernighan and Pike (1984).

The database management system, which we have called HeteroData, consists of a database in the form of one or more ASCII text files containing the information to be accessed, and a C code library which reads and interprets the text files, constructs a representation in memory, and provides functions for searching the database and evaluating the located entries. (This last step is necessarily a separate operation in the present system, because many entries depend quantitatively on such arguments as temperature and alloy mole fraction, which must first be specified.) The system is designed to be easily accessed by device simulation programs, but we have also implemented an interactive graphical interface for human users, illustrated in Fig. 1.

The use of text files for the nonvolatile representation of the data provides several advantages: such files may be readily moved between different types of computer systems by means such as electronic mail, and they may be constructed and modified using a simple text editor program. It also significantly simplifies the design of the system, as the code is only required to read and interpret the data files, not write or modify them. Modifications to the database are not expected to be frequent (as compared to searches),

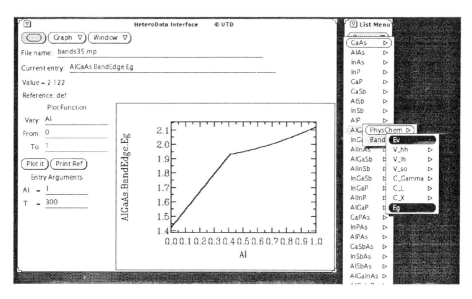

Figure 1: Screen image of the user interface to the heterostructure database system. The hierarchical organization is displayed using cascaded menus. The arguments of each located entry are displayed, and the functional dependencies may be automatically plotted.

so this approach is appropriate.

The data are arranged in a hierarchical organization, with groups of entries containing both individual data entries and other groups, as illustrated in Fig. 2. By convention, each material is represented by a group at the outermost level, which contains groups of conceptually related data such as physical and chemical constants, band-edge parameters, impurities, etc. These may be subdivided as logic dictates; for example, the band-edge groups contain groups representing individual bands, each of which includes energy, effective mass, and deformation potential entries. The structure of the hierarchy is determined by the use of nested delimiters within the text files. An entry is accessed by its "pathname" consisting of the sequence of group names and entry name separated by periods. Thus the static dielectric constant of GaAs is accessed by GaAs.PhysChem.epsilon_dc.

Each entry can consist of a constant or an algebraic expression of arbitrary complexity. An expression can include symbolic arguments whose value is set at the time the expression is evaluated. Arguments represent such quantities as temperature and mole fraction, and may be declared at any group level or for individual entries. The list of arguments defined for a particular entry can be obtained from that entry. By convention, alloys are identified by a string of all constituent elements and the mole fractions are denoted by the appropriate chemical symbol. Thus $Al_{0.3}Ga_{0.7}As$ would be specified by AlGaAs(Al=0.3, Ga=0.7). An alias mechanism permits one to define GaAlAs as another name for AlGaAs.

Expressions can also include previously defined entries. This permits the parameters of alloys to be expressed in terms of the parameters of the pure compounds, for example.

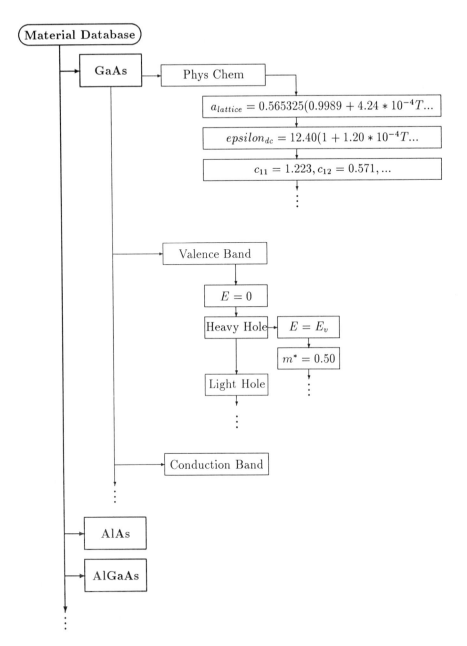

Figure 2: Schematic representation of the hierarchical organization of the database.

The requirement of previous definition places constraints on the order in which entries may appear, but it prevents the creation of circular definitions. Each entry definition is syntactically required to include reference string, which is intended to identify the source of the information. A function is provided to trace and display the entries (and their references) on which a given item depends, thus supplying the "family tree" of the information.

Special constructs are provided to permit a compact descriptions of common mathematical forms. These include polynomial and rational functions, specified by arrays of coefficients. A spline interpolation function is also included.

3   Data Organization and Rules for Inferring Unknown Parameters

It should be clear that the purpose of HeteroData is not merely to store numerical parameters, but to encode a potentially very complex network of interrelationships between parameters. There are two reasons for this rather ambitious approach. The first is to achieve what is known in database terminology as "data normalization" (Parsaye, et al., 1989), which in essence means that each independent piece of information should appear only once in the data files. This is addressed by such features as the use of previously defined entries in expressions, and the alias mechanism.

The second reason is the need to provide plausible estimates of parameters for which direct experimental measurements have not been performed, and which are thus not entered into the data files. This problem is addressed at two levels: a mechanism for specifying default expressions is designed into the software system, and the rules for inferring the value of each parameter must be determined from appropriate theoretical or empirical relationships. The HeteroData mechanism is the default expression, which can be associated with any group. An example is the default expression for AlGaAs, which describes linear interpolation on the composition:

```
? = Al * AlAs.? + (1.0 - Al) * GaAs.? ;
```

This expression is invoked in the data search procedure, if the requested item is not located directly. HeteroData then attempts to construct the entry by string substitution on the ? character, searching (in this case) for the corresponding AlAs and GaAs entries before reporting success or failure.

The task of determining the rules by which unknown quantities are to be inferred requires, in effect, a critical review of much of heterostructure physics. A central issue is the determination of heterojunction band alignments. Yu, McCaldin, and McGill (1992) have reviewed the status of experiment and theory for the band alignment problem, and the result is that there remain significant experimental uncertainties in the alignments of many of the heterojunctions of interest. In the context of such uncertainties one has a certain amount of freedom to impose simple mathematical or theoretical notions upon the data, but one should also provide for the possibility that a tightening of the experimental uncertainty will render such notions invalid.

Predicting unknown band alignments is quite simple if they can be obtained as energy differences on a universal, absolute energy scale. Most of the simpler theories of band alignment provide such a scale, but no case is the accuracy of the resulting predictions adequate for device simulations (Yu et al., 1992). One can, however, construct an empirical scale by the following procedure: Choose a reference material such as GaAs, and let its valence band edge be the reference energy. Use the experimental band alignments of heterojunctions with GaAs to define the absolute valence band energies of materials for which such heterojunction data are available, such as AlAs, GaP, and InAs. Other heterojunction data are then used to define the valence-band energies of GaSb and AlSb in terms of that of InAs, InP in terms of $In_{0.53}Ga_{0.47}As$, and so on, using the best known heterojunctions at each stage. This process is repeated until all materials of interest are included. This is the method presently implemented in HeteroData. The definitions are coded precisely as described, so that a revision to the GaAs/InAs heterojunction alignment would shift the absolute energies of InAs, GaSb and AlSb, but would not change the (relative) band alignment at the InAs/GaSb, InAs/AlSb or GaSb/AlSb heterojunctions.

One remaining issue is how the band offsets should be interpolated for heterojunctions involving alloys. The results of Batey and Wright (1985) demonstrate that in the $GaAs/Al_xGa_{1-x}As$ system, the valence bands vary linearly with composition. An extrapolation of these results produces a rule to the effect that the valence bands of alloys are always interpolated linearly, and any bowing is included in the absolute conduction-band energies.

The notion of a universal energy scale, to the extent that it is consistent with experimental data, is quite powerful in the present context. For example, the correlation between heterojunction band alignments and deep impurity levels observed by Langer and Heinrich (1985) provides a rule for inferring deep level energies in alloys and for impurity/semiconductor combinations that have not been experimentally reported. The rule is simply that the impurity-state energy is fixed in absolute energy.

4 Quaternary Alloys

Quaternary alloys such as $In_xGa_{1-x}As_yP_{1-y}$ and $Al_xGa_yIn_{1-x-y}As$ present a special problem for the HeteroData system. These materials are usually much less well-studied that the component ternary systems, so their properties must usually be inferred from the component ternaries. Note that there are two classes of quaternaries, of the form $A_xB_{1-x}C_yD_{1-y}$ (which we will refer to as the 2 × 2 case) and $A_xB_yC_{1-x-y}D$ (which we will call the 3 × 1 case). The interpolation formula for the 2 × 2 case is straightforward (Adachi, 1992):

$$P(A_xB_{1-x}C_yD_{1-y}) = \{x(1-x)[yP(A_xB_{1-x}C) + (1-y)P(A_xB_{1-x}D)] \\ + y(1-y)[xP(AC_yD_{1-y}) + (1-x)P(C_yD_{1-y})]\} \\ / [x(1-x) + y(1-y)], \quad (1)$$

where P represents some property of the semiconductor material.

For the 3 × 1 case we have derived an interpolation formula based upon a geometric model involving barycentric coordinates in an equilateral triangle (a well-known convention for plotting the properties of three-component mixtures). The resulting formula, presuming $x + y + z = 1$, is given by:

$$P(A_xB_yC_zD) = [xyP(A_{x+z/2}B_{y+z/2}D) + yzP(B_{y+x/2}C_{z+x/2}D) + P(C_{z+y/2}A_{x+y/2}D)] \\ / (xy + yz + zx). \quad (2)$$

Functions implementing these coefficients are built into the HeteroData system.

5 Needed Refinements

One significant question about the present organization of HeteroData concerns the validity of the universal energy scale assumption for heterostructures. This assumption is valid if the experimental band alignments can be shown to conform to the mathematical properties of an equivalence relation: symmetry, reflexivity, and transitivity. In the context of heterojunctions, symmetry means that there is no band discontinuity at the junction between a semiconductor and itself; since in this case there is no junction, this property is always satisfied. Reflexivity means that the same alignment is obtained if the ordering of the materials is reversed, and transitivity means that the band alignment at a junction AC is equal to the superposition of the alignments at junctions AB and BC (Frensley and Kroemer, 1977). It appears that transitivity is satisfied in those cases for which high-quality interfaces have been grown (Yu et al., 1992). However, there are experimental indications that reflexivity is violated to the level of a few hundredths of eV in some systems (Waldrop et al., 1992; Waterman et al., 1993). (The term "commutativity" has also been used in the literature to describe this test.) One should also note that the transitivity test is significantly complicated by strain effects at lattice-mismatched heterojunctions.

The implication of this for the HeteroData system is that it should be expanded to permit storage of information by interface, as well as by material. This is possible within the present software design, but it is not yet clear whether the added burden of searching the much larger space of interfaces would require a more specialized mechanism to acheive adequate performance.

6 Summary

The HeteroData system provides a framework within which the detailed properties of semiconductor heterostructures may be encoded and made available to both simulation programs and human users. The organization of any heterostructure database system is significantly complicated by the prevalence of semiconductor alloys whose composition spans a continuum. The mechanisms built into HeteroData provide the means to deal with this complication, and to estimate parameters for which experimental data are not available.

This work was supported in part by the Advanced Research Projects Agency, via subcontracts from Texas Instruments, monitored by the Air Force Wright Aeronautical Laboratory, and the University of California at Santa Barbara, monitored by the U. S. Army Research Office.

References

Adachi, S., 1992, *Physical Properties of III-V Semiconductor Compounds: InP, InAs, GaAs, GaP, InGaAs, and InGaAsP*. (Wiley, New York)

Aho, A. V., R. Sethi and J. D. Ullman, 1986, *Compilers: Principles, Techniques, and Tools*. (Addison-Wesley, Reading, Mass.)

Aspnes, D. E., S. M. Kelso, R. A. Logan and R. Bhat, 1985, *J. Appl. Phys.* **60**, pp. 754–67.

Batey, J. and S. L. Wright, 1985, *J. Appl. Phys.* **59**, 200.

Casey, H. C. Jr. and M. B. Panish, 1978, *Heterostructure Lasers. Part B: Materials and Operating Characteristics*. (Academic Press, Orlando) ch. 5.

Frensley, W. R., and H. Kroemer, 1977, *Phys. Rev. B* **16**, pp. 2642–52.

Frensley, W. R., 1991, in *Proc. 7th Internat. Conf. on Numerical Analysis of Semiconductor Devices and Integrated Circuits*, pp. 7–8.

Kernighan, B. W., and R. Pike, *The UNIX Programming Environment*. (Prentice-Hall, Englewood Cliffs, NJ) ch. 8.

Langer, J. M., and H. Heinrich, 1985, *Phys. Rev. Lett.* **55**, pp. 1414–7.

Parsaye, K., M. Chignell, S. Khoshafian and H. Wong, 1989. *Intelligent Databases*. (Wiley, New York)

Waldrop, J. R., G. J. Sullivan, R. W. Grant, E. A. Kraut, and W. A. Harrison, 1992, *J. Vac. Sci. Technol. B* **10**, pp. 1773–8.

Waterman, J. R., B. V. Shanabrook, R. J. Wagner, M. J. Yang, J. P. Davis and J. P. Omaggio, 1993, *Semicond. Sci. Technol.* **8**, pp. S106–111.

Yu, E. T., J. O. McCaldin and T. C. McGill, 1992, "Band Offsets in Semiconductor Heterostructures," in *Solid State Physics*, H. Ehrenreich and D. Turnbull, eds. (Academic Press, San Diego) vol. 46.

Steady-state, dynamic and spectral performance of microcavity surface-emitting strained quantum-well lasers

Igor Vurgaftman and Jasprit Singh

Solid State Electronics Laboratory, Department of Electrical Engineering and Computer Science, University of Michigan, Ann Arbor, MI 48109-2122, USA

Abstract. We examine the modifications in the spontaneous emission pattern of a microscopic cavity semiconductor laser in the surface-emitting geometry and their influence on the laser characteristics. We find that the threshold current density can be reduced to 60 A/cm^2 for a laser embedded in a cavity with a transverse dimension of 0.5 μm, while the -3dB modulation frequency exceeds 20 GHz for a moderate injected current density of 600 A/cm^2, allowing high-speed operation without laser heating that normally accompanies high current injections. Linewidths of several hundred MHz at moderate power levels are calculated for microcavity lasers resulting from a balance of the enhancement of the fraction of spontaneous emission coupling into the lasing mode and of the reduction of the linewidth enhancement factor in the strained active region.

1. Introduction

Spontaneous emission from excited states is an intrinsic quantum process that can be derived from thermodynamic considerations. It is crucial to laser operation in that spontaneously emitted photons fed back into the cavity from the mirrors provide a necessary source for initiating stimulated emission. Laser pioneers realizing the difficulty of constructing a cavity with all dimensions of the order of the optical wavelength proposed multimode cavities in which a large number of modes of the electromagnetic field may couple radiatively to the lasing transition in the active medium [1]. Mode selection in such a cavity is accomplished by the dispersion in the optical gain for photons traversing the cavity and/or in the cavity quality factor for the various modes. The disadvantage of multimode cavities is that intense pumping of the active medium is required to overcome the radiation loss into the modes whose gain and feedback parameters are suboptimal before a phase transition to an ordered state of stimulated emission into the lasing mode can take place.

Until very recently, all semiconductor lasers were based on the macroscopic cavities with the fraction of spontaneous emission into the lasing mode in the $10^{-4} - 10^{-6}$ range. Following the latest developments in the fabrication technology, a number of microcavity laser diodes have been proposed and fabricated. One of the most promising configurations is one based on the surface-emitting geometry, in which two high-reflectivity multilayer mirrors composed of quarter-wavelength stacks of materials with dissimilar refractive indices are constructed perpendicular to the substrate surface separated by an integral number of half-wavelengths. A standing-wave mode at the Bragg wavelength is formed in the one-dimensional electromagnetic bandgap with its maximum overlapping the active region placed midway between the mirrors [2] (see Fig. 1). If the transverse dimension of the cavity is brought down to the optical wavelength, the fraction of spontaneous emission into the lasing mode can be increased by several orders of magnitude. If the cavity can be made truly single-mode, then, disregarding cavity losses, *any* emitted photon induces subsequent stimulated emission and a transition to the ordered state. Therefore, the steady-state photon density in the cavity is linear in the pumping intensity, or injected current density.

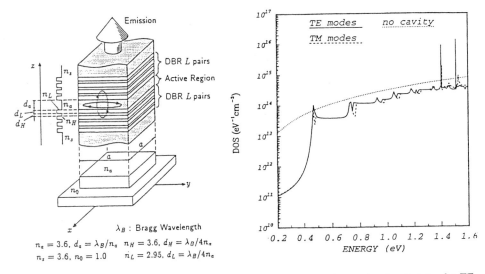

Fig. 1. The surface-emitting microscopic laser cavity. Adapted from Ref. 2.

Fig. 2. The photon density of states for TE and TM polarized modes in a 0.5 μm cavity and in absence of a cavity.

2. Results

2.1. Photon density of states

To quantify the qualitative discussion above, it is necessary to calculate the spontaneous emission rate for all the modes and use its value in the static and dynamic analysis of the laser operation. Insofar as spontaneous emission is a consequence of the coupling between the electromagnetic field constrained by the cavity boundary conditions and of the inverted transitions in the active medium, it requires a solution of wave equation in the laser cavity as well as of the bandstructure in the active layer. In the surface-emitting cavity with 15.5 pairs of AlAs/GaAs alternating stacks with a reflectivity of 0.991 at normal incidence, the former is most easily accomplished by separating the wave equation in the transverse and longitudinal coordinates. The longitudinal problem is complicated by the presence of a large number of layers and is treated by the transfer matrix technique [2]. The transverse solution is analytical for a square cross-section of the laser structure and given by linear combinations of standing waves. The final cavity modes are taken as crosspoints of the longitudinal and transverse modes. Three classes of modes are recovered: 1) resonant modes discussed above; 2) leaky modes beyond the 1D bandgap; 3) horizontally propagating modes. Of these, only the leaky modes form a continuum. Since the spontaneous emission rate is proportional to the photon density of states, the latter is found by assuming the Lorentzian profile for discrete modes and an appropriately normalized density-of-states function for continuous ones and integrating over all allowed cavity modes. The photon density of states in a 0.5 μm cavity is shown in Fig. 2. The interesting features are the formation of the staircase-like structure reminiscent of the electronic subbands in quantum wells, the emission resonances near the TE and TM modes, and the overall decrease in the emission rate due to insufficient filling of the photon wavevector space by discrete transverse modes.

2.2. Spontaneous emission rate

It has been shown that strained quantum-well lasers are capable of the best threshold-current and modulation performance of all currently practical semiconductor lasers. The valence-band electronic states of the quantum-well layer are calculated from a 4-band k·p method in conjunction with the deformation potential theory. The material gain spectrum is then found using the Fermi Golden Rule and assuming that the conduction-band effective mass is unaffected by the introduction of strain. Since light polarized perpendicular to the direction of epitaxial growth couples solely to electron-light hole transitions, the gain for this polarization is suppressed in the vicinity of the band edge. Combining the knowledge of the electronic states with the results for the photon density of states, the spontaneous emission rate due to electron-hole pair recombination is given by (in cgs units) [3]:

$$R_{sp} = \int d(\hbar\omega) \frac{4\pi^2 e^2 \hbar}{n^2 m_0^2 \hbar\omega} \frac{1}{2\pi^2} \sum_{n,m} \int dk \sum_{\sigma,\hat{\varepsilon}} |\hat{\varepsilon} \cdot \vec{P}_{nm}^{\sigma}(\mathbf{k})|^2$$
$$\times \rho_{\hat{\varepsilon}}(\hbar\omega) \delta(E_n^e(\mathbf{k}) - E_m^h(\mathbf{k}) - \hbar\omega)[f^e(E_n^e(\mathbf{k}))][1 - f^h(E_m^h(\mathbf{k}))], \qquad (1)$$

where $\rho_{\hat{\varepsilon}}$ is the (in general) polarization-dependent photon density of states, and $P_{nm}(\mathbf{k})$ is the momentum matrix element between the nth subband in the conduction band and the mth subband in the valence band for the in-plane wavevector $\mathbf{k}$. The resulting spontaneous emission rate is shown in Fig. 3 for TE and TM polarizations. While spontaneous emission is enhanced around the resonant peak, it is suppressed elsewhere. As a result, the radiative lifetime in the cavity increases.

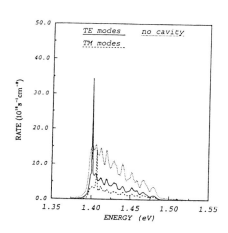

Fig. 3. The spontaneous emission rate for TE and TM polarizations in a 0.5 μm cavity and in absence of a cavity.

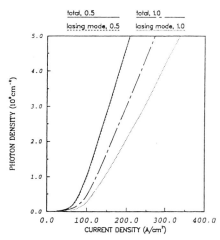

Fig. 4. The photon density vs. the current density for 0.5 and 1.0 μm cavities.

2.3. Steady-state performance

The steady-state solutions of the multimode rate equations are studied:

$$S_{\hat{\varepsilon},m} = \frac{R_{sp,\hat{\varepsilon},m}(n)}{(\frac{c}{n_g})[\alpha_c - \Gamma g(n, E_m)]}; \tag{2}$$

$$\frac{J}{e} = R_{sp}(n)[1 + \sum_{m,\hat{\varepsilon}} \frac{R_{sp,\hat{\varepsilon},m}(n)\Gamma g(n, E_m)}{R_{sp}(n)[\alpha_c - \Gamma g(n, E_m)]}]; \tag{3}$$

where $S_{\hat{\varepsilon},m}$ is the photon density per unit area in mode m of $\hat{\varepsilon}$ polarization, $R_{sp,\hat{\varepsilon},m}$ is the spontaneous emission rate into the mode, $R_{sp} = \sum_{\hat{\varepsilon},m} R_{sp,\hat{\varepsilon},m}$ is the total spontaneous emission rate, n is the carrier density, α_c is the total cavity loss, and Γ is the optical confinement factor assumed to be proportional to the quantum well width for narrow wells, and taken to be 1.25% in these calculations. The spontaneous emission coupling factor ($R_{sp,\hat{\varepsilon},m}/R_{sp}$) for the main mode is ≈ 0.1 for the 0.5 μm cavity decreasing to 0.05 for the 1.0 μm cavity. Only the resonant modes need be included in the calculation since any photons emitted into the leaky and propagating modes will immediately escape from the cavity.

The photon density as a function of the injected current density is shown in Fig. 4 for several transverse cavity widths. The curvature in the light-current characteristic persists even though the only excited resonant mode is the main mode owing to the photon losses in the leaky and propagating modes. The threshold current density defined as the extrapolated intersection of the light-current curve with zero photon density is found to be ≈ 60 A/cm^2 in the 0.5 μm cavity, 80 A/cm^2 in the 1.0 μm cavity, increasing to ≈ 200 A/cm^2 for the macroscopic cavity surface-emitting laser.

2.4. Large-signal modulation performance

The large-signal modulation performance of semiconductor lasers can be addressed by numerically integrating the rate equations with the current source turned on to a certain value at time 0. The multimode rate equations are as follows:

$$\frac{dS_{\hat{\varepsilon},m}}{dt} = [\Gamma g(n, E_m) - \alpha_c](\frac{c}{n_g})S_{\hat{\varepsilon},m} + R_{sp,\hat{\varepsilon},m}(n, E_m) \tag{4}$$

$$\frac{dn}{dt} = \frac{J}{e} - R_{sp}(n) - (\frac{c}{n_g})\sum_{\hat{\varepsilon},m} \Gamma g(n, E_m)S_{\hat{\varepsilon},m}, \tag{5}$$

where the symbols have been defined in the previous subsection. Following the switching event there is an interval (the turn-on time), in which the photon and electron densities undergo increase and fluctuations before settling to the steady-state values. The rate of increase of the electron density is limited by the injected current density, as can be seen from (5). Until the electron density reaches its transparency value, stimulated emission is nonexistent, and the photon density is constrained to remain close to zero. Therefore, the increase in the spontaneous emission coupling factor is not expected drastically to affect the turn-on delay time. However, in multimode lasers, mode competition following turn-on may increase the relaxation time by as much as an order of magnitude[4]. Since well-designed microcavity lasers are intrinsically single-mode, the relaxation time is given by the time in which the total photon density reaches its steady-state value.

Detailed numerical simulations show that the turn-on time for the 0.5 μm microcavity laser is close to 1 ns, while it may exceed 10 ns in 300 μm-long Fabry-Perot lasers, if the comparison is done in terms of the final output power. If the turn-on time is made in terms of the final current density value, the advantage of the microcavity laser is greater still owing to its small threshold current density.

2.5. Small-signal modulation performance

The small-signal frequency response of the 0.5µm-wide microcavity laser is shown in Fig. 5 as a function of the average injected current density and the total photon density. The microcavity laser is modulated by a sinusoidal waveform at current densities 2, 5 and 10 times that at threshold (120, 300 and 600 A/cm^2 respectively). As expected from the simple expressions for the relaxation oscillation frequency and the damping factor, both increase with a higher bias point. The resonance frequencies for the current densities 2, 5 and 10 times the threshold current density are 3.2, 7.5 and 14.0 GHz respectively. The corresponding damping factors are 5.7, 10.4 and 22.3 ns^{-1}. The -3dB modulation frequencies are 4.9, 11.6 and 21.5 GHz respectively. The resonance frequecies calculated for the microcavity lasers exceed those in 300 µm-long edge-emitting Fabry-Perot lasers with an identical active region [4]. The increase in the relaxation oscillation frequency may be attributed to a higher spontaneous emission coupling factor. It is particularly to be noted that a low threshold current density for the microcavity laser allows high modulation frequency results to be obtained at relatively low values of the injected current density. Therefore, the heating effects which set an upper limit to the current that can be injected into the lasing structure are expected to be much less in evidence in microcavity lasers. Note, however, that the present calculation ignores the effects of the particular design of the laser structure which may lead to additional heating, the effects of the device parasitics and the influence of gain compression at high photon densities which sets another upper limit on the maximum modulation frequency. Recent calculations of the former limit have shown, however, that it is much higher that the best modulation frequency results in semiconductor lasers [5].

2.6. Spectral linewidth

The modified Schawlow-Townes formula [1],[6] for the laser linewidth may be expressed in terms of the spontaneous emission rate into the lasing mode and the photon density in the cavity as follows:

$$\delta\omega_{ST} = \frac{\beta R_{sp}}{4\pi S}(1+\alpha^2), \qquad (6)$$

where α is the linewidth enhancement factor:

$$\alpha = \frac{d\chi_R(n)/dn}{d\chi_I(n)/dn}, \qquad (7)$$

where χ_R and χ_I are the real and imaginary parts of the susceptibility respectively.

Although the spontaneous emission rate into the lasing mode in microcavity surface-emitting lasers is obviously considerably increased, the linewidth enhancement factor is determined by the strained active layer and is found to have a low value of ≈ 1.25. Moreover, the *total* spontaneous emission rate is decreased, a phenomenon also reflected in the reduction of the threshold current density. Additionally, significantly lower current injections are needed to achieve comparable power outputs in microcavity lasers in contrast with conventional lasers. Detailed calculations show that for moderate injections the microcavity laser linewidth is of the order of several hundred MHz (see Fig. 6). This value shows that in spite of the appearance of proportionality between the spontaneous emission coupling factor and the laser linewidth in the Schawlow-Townes expression, tolerable linewidths may be obtained in microcavity lasers provided a strained quantum well is used as the active region.

3. Conclusions

In this paper, we have examined the steady-state, modulation and spectral performance characteristics of microcavity surface-emitting lasers with a strained quantum-well active region. The effect of the modified photon density of states on the light-current curve is to lower the threshold current density point. Although the latter can be decreased to 60 A/cm^2 in 0.5μm-wide structures, the features achievable with the current fabrication technology are insufficient to realize truly "thresholdless" lasers in the surface-emitting geometry. The large-signal modulation performance in microcavity lasers is also beneficially affected by the elimination of mode competition at the laser turn-on point. Small-signal modulation frequencies in excess of 20 GHz may be achieved at moderate current injections (600 A/cm^2). The spectral linewidth of microcavity lasers, although broadened as predicted by the Schawlow-Townes formula, can be kept to a moderate level of several hundred MHz by a proper design of the strained quantum-well active region.

Microcavity surface-emitting lasers represent an extremely attractive next step in the surface-emitting laser technology and the semiconductor laser technology in general by virtue of their superior cw and modulation performance. Experimental effort should be directed towards practical realizations of these lasers in high-density arrays with low power consumption.

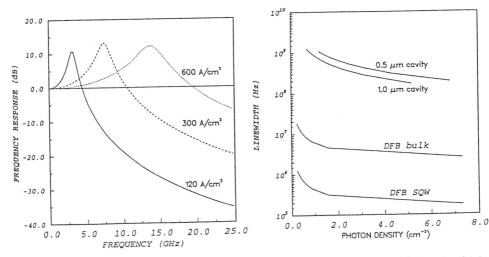

Fig. 5. The small-signal frequency response for current densities 2,5, and 10 times that at threshold (120, 300 and 600 A/cm^2 respectively).

Fig. 6. The spectral linewidth for 0.5 and 1.0 μm microcavities, bulk and strained-quantum-well layer DFB lasers.

References

[1] Schawlow A.L. and Townes C.H. 1958 *Phys. Rev.* **112** 1940-1949
[2] Baba T. *et al.* 1991 *IEEE J. Quant. Electron.* **QE-27** 1347-1358
[3] Vurgaftman I. and Singh J. 1994 *Appl. Phys. Lett.* **64** 1472-1474
[4] Lam Y., Loehr J.P. and Singh J. 1993 *IEEE J. Quant. Electron.* **QE-29** 42-50
[5] Lam Y. 1993 Ph.D. dissertation, Univ. of Michigan.
[6] Henry C.H. 1982 *IEEE J. Quant. Electron.* **QE-18** 259-264

Table-Based Monte Carlo Simulation of Electron, Phonon, and Photon Dynamics in Quantum Well Lasers

Muhammad A. Alam and Mark S. Lundstrom

1285 Electrical Engineering Building, Purdue University, West Lafayette, IN 47907

Abstract: A new, Monte Carlo based simulation technique is described and applications to quantum well lasers are presented. The technique treats the dynamics of 2D electrons as well as the phonons and photons to which they are coupled. It begins by pre-computing a transition matrix by Monte Carlo techniques. Initial conditions are then defined, and the system dynamics are simulated by simple matrix multiplication. The approach provides essentially the same physics as a Monte Carlo simulation, but treats a wider range of timescales and essentially eliminates statistical noise. Simulations which examine the influence of hot phonon effects on the performance of quantum well lasers are presented.

1. Introduction

Quantum well laser dynamics is currently a topic of considerable interest because the modulation bandwidth of quantum well lasers remains far below the initial predictions. Extensive experimental studies have probed various aspects of laser dynamics including carrier transport in the separate confinement region, carrier capture in the quantum wells, the relaxation dynamics of confined electrons, hot phonon effects, and photon propagation in the Fabry Perot cavity [1]. Phenomenological models with adjustable parameters are often used to interpret experimental results, but the parameters vary from model to model and a comprehensive and detailed understanding is still lacking. Monte Carlo studies have been reported [2], but the statistical noise and the difficulty of treating a wide range of time scales limits the number of problems that can be addressed. While the general features of the problem are now understood, a quantitative understanding, which is necessary for designing high-performance devices, is essential. In this talk, we describe a new, Monte Carlo based simulation technique and use it to examine the effects of hot phonons on quantum well laser dynamics.

2. Theory

To simulate the dynamics of electrons in quantum wells, we have developed a Transition Matrix Approach (TMA), which provides a semiclassical description of confined electrons [3]. The basic ideas are illustrated in Fig. 1. Momentum space for each subband is first discretized into bins in energy and angle as illustrated in Fig. 1. (Since the in-plane field is zero for this example, there is actually no need to resolve the angle in momentum space.) Consider a group of electrons located in bin j at time, t, as shown in the shaded box. At time $t + \Delta t$ these electrons will have made transitions to other bins in the same and other subbands. We can transform the Boltzmann equation into an equation relating the population of a bin in momentum space, $N_i(t + \Delta t)$, to the population of the bins at time, t, by forward differencing it in time and by integrating over the area of each bin to obtain

$$N_i(t + \Delta t) = \sum_j P_{ij} N_j(t), \quad (1)$$

where P_{ij} is the probability that an electron in bin j make a transition to bin i in a time Δt, and the sum over j runs over all the momentum space bins in all the subbands. It is readily shown that Eq. (1) applies only when $\Delta t \ll \tau$, where τ is the scattering rate. The transition matrix elements, P_{ij}, can be directly related to the scattering processes as described by $S(\mathbf{p}, \mathbf{p'})$, but we find it more convenient, however, to evaluate P_{ij} by direct Monte Carlo simulation. A large number of electrons is injected in bin j and tracked by 2D Monte Carlo simulation (e.g. [4]) for a time Δt. By noting the number of electrons in each bin at t = Δt, P_{ij} is evaluated by taking simple ratios. The transition matrix is pre-computed, stored, then used repeatedly for simulation. Initial conditions are defined, and electron dynamics are then simulated according to Eq. (1) by simple matrix multiplication. Because P_{ij} is evaluated once (or infrequently), the time-consuming Monte Carlo simulation is amortized over many simulations, and because a large number of electrons can be used to define P_{ij}, statistical noise is virtually eliminated.

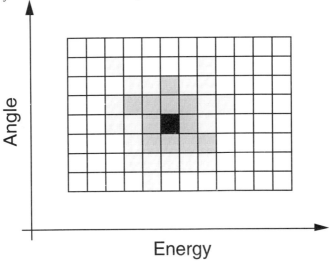

Fig. 1   Illustration of the transition matrix approach. Momentum space for each subband is divided into a finite number of bins. Assume that the electrons in bin i at time, t (the black bin above). At time t + Δt, electron scattering events will cause the distribution to spread. If N_i are injected in bin i at t = 0 and N_j appear in bin j at t = Δt, then the transition matrix element is P_{ji}. The transition matrix is precomputed by conventional Monte Carlo simulation then stored and used for subsequent simulations.

Nonlinear effects such as band filling, electron-electron scattering, and hot phonons cannot be treated with the simple technique as described above because these effects depend on knowing the evolving distribution carrier distribution. To illustrate how these effects are included, consider electron-electron scattering. The transition rate for electrons in state $\mathbf{p}$ scattering to $\mathbf{p'}$ from electrons initially at $\mathbf{p_0}$ which scatter to $\mathbf{p_0'}$ can be separated into two factors, one that is independent of the carrier distribution and one that accounts for the initial and final state occupancies and for screening and which depends on the carrier distribution. From the distribution-independent factor, we pre-compute by direct Monte Carlo simulation a transition matrix, P_{ijkl}, which give the probability that an electron in bin j make a transition to bin i by scattering from an electron in bin l which then makes a transition to bin k. While evaluating P_{ijkl}, we assume that the target electron is always present ($f_l = 1$) and that the final states are always empty ($f_i = f_k = 0$). Using the pre-computed electron-electron transition matrix, we then update the bin populations at $t = t + \Delta t$ according to

$$N_i(t+\Delta t) = P'_{ii}N_i(t) + \left(1 - f_i(t)\right) \sum_{j,k,l \neq i} P_{ijkl}\left[f_l(t)(1 - f_k(t))\right] N_j(t). \quad (2)$$

In Eq. (2), P'_{ii} is P_{ii} renormalized so that particles are conserved. Similar techniques are used to treat other nonlinear effects, such as bandfilling.

As carriers relax in energy, they emit optical phonons, and the resulting non-equilibrium phonon population has a strong influence on carrier relaxation [1,2]. The electron system is also coupled to the photon population in the laser cavity. Rate equations for both phonons and phonons can be discretized and solved by a similar transition matrix approach. Note that electron scattering by phonon absorption and emission produced recombination-generation terms for phonons, and optical absorption and emission produces similar recombination-generation terms for electrons.

Having outlined the basic approach for treat electron, phonon, and photon dynamics, we briefly describe the simulation process. We begin by discretizing momentum space for electrons and phonons and wavevector space for photons. Electron-phonon (P_{ij}), electron-electron (P_{ijkl}), and phonon generation-recombination transition matrices are precomputed by direct 2D Monte Carlo simulation. Initial conditions are defined, then the electron distribution is updated by splitting the effects of electron-phonon and electron-electron scattering. The electron distribution is first updated for phonon scattering using Eq. (1) adjusted for band filling, then the distribution is updated for electron-electron scattering by Eq. (2). This splitting of the scattering operator can be justified as long as the timestep Δt is much less than the average time between scattering events. After updating the electron population, the optical phonon and the photon populations are updated by solving similar procedures. The process continues at timesteps of Δt (typically 10-20 fs) until steady state is achieved.

3. Simulation Results

To illustrate the technique, we present dynamic phonon distributions in a GaAs quantum well laser. For this model simulation, the quantum well was 80 Å wide with a potential barrier of 0.600 eV. We treated electron-electron scattering with a non-equilibrium phonon population. We assumed a quasi-equilibrium distribution of holes and evaluated optical transitions using a 4x4 k · p valence bandstructure. Figure 2 displays the computed phonon distributions at three different times. In this particular simulation, the laser was biased in the amplifier mode with a carrier density of 3.21×10^{16} m^{-2}. When a step light pulse ($S = 8.5 \times 10^{15}$ m^{-2}) is introduced in the cavity, a spectral hole develops in the distribution function. To fill the spectral hole, electrons with large kinetic energy begin to relax from the top of the quantum well giving off small wavevector phonons. As the electrons continue to lose their energy, the phonons emitted begin to have larger wavevectors, so that the population at these wavevectors go up as can be seen in Fig. 2. Finally, a steady state distribution is reached when the rate of phonon loss and the rate of phonon generation balances each other. The most important point to note from this simulation is that the phonon distribution cannot be modeled with a single temperature which is often assumed in simple rate equation analysis [5].

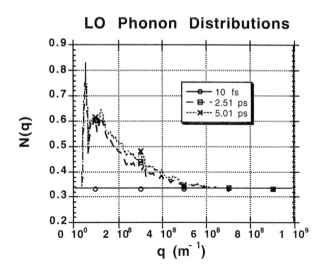

Fig. 2 Computed optical phonon distribution functions as a function of wavevector, q, at three different times after the introduction of an optical pulse.

4. Summary

In summary, we have described a new, Monte Carlo based simulation technique which is especially well-suited for quantum well lasers. By using pre-computed transition matrices, the time-consuming Monte Carlo simulations can be performed once, with high precision, then used repeatedly for simulations. The technique virtually eliminates statistical noise, treats both fast and slow phenomena, and provides an excellent computational efficiency. A key issue in the further development of the technique will be to encorporate an appropriate description of important many body effects [6]. State filling and broadening are already treated and a phenomenological bandgap renormalization can also be included, but the role of dynamic screening in the coupled phonon/electron system also needs to be addressed. With a careful consideration of such effects, the approach should provide a physically detailed treatment of carrier transport, capture, and relaxation in quantum well lasers.

References

[1] Shah J 1993 Editor., *Hot Carriers in Semiconductor Nanostructures: Physics and Applications*, (New York, Academic)

[2] Lam Y. and Singh J. 1993 *Appl. Phys. Lett.*, **63**, 1874

[3] Alam M A and Lundstrom M S 1994, Proc. of SPIE OE LASE 94, W.W. Chow and M. Osinski, Ed., 90

[4] Goodnick S M and Lugli P 1988 *Phys. Rev. B.*, **37**, 2578

[5] Tsai C Y, Eastman L F, and Lo Y H 1993 *Appl. Phys. Lett.* **63** 3408

[6] H. Huag and S.W. Koch, 1990 *Quantum Theory of the Optical and Electronics Properties of Semiconductors*, World Scientific, Singapore.

Tight-binding Calculation of Linear Optical Properties of $In_{0.53}Ga_{0.47}As$ Alloys and Heterostructures

V. Sankaran and K. W. Kim

Department of Electrical and Computer Engineering
North Carolina State University, Raleigh, NC 27695-7911.

G. J. Iafrate

U. S. Army Research Office, P.O.Box 12211
Research Triangle Park, NC 27709-2211.

Abstract

We present theoretical calculations of the dielectric function $\varepsilon(\omega)$ of $In_{0.53}Ga_{0.47}As/InP$ superlattices, using the empirical tight-binding (TB) description of electronic states. We demonstrate that the absorption spectra of the constituent bulk materials can be reproduced very well with an appropriate choice of values for optical matrix elements between (TB) basis functions. These parameters then yield the absorption spectra of superlattices. We have computed the imaginary part of the dielectric function, $\varepsilon_2(\omega)$, for ultrathin $(In_{0.53}Ga_{0.47}As)_n/(InP)_n$ superlattices, for $n = 1, 2$, and 3. It is found to be almost identical in all 3 cases. The dominant E_1 and E_2 peaks of $\varepsilon_2(\omega)$ of the superlattices are located in between the corresponding peaks of bulk $In_{0.53}Ga_{0.47}As$ and InP. $\varepsilon_2(\omega)$ is found to be identical for light polarized parallel or perpendicular to growth direction.

I Introduction

$In_xGa_{1-x}As$ alloys and their heterostructures with wider bandgap semiconductors such as InP have emerged as important materials for optoelectronic applications during the past two decades. Examples of such applications include photodetectors, LED's and double-heterostructure lasers for low-loss, low-dispersion optical fiber communication. Many of these applications require a knowledge of the dielectric function $\varepsilon(\omega)$ [1], which is a fundamental material parameter related to the energy band structure and optical matrix elements between valence and conduction band states. In bulk semiconductors $\varepsilon(\omega)$ is adequately modeled by analytic expressions derived using $\mathbf{k} \cdot \mathbf{p}$ theory. Here the dominant contribution to $\varepsilon(\omega)$ comes from interband critical points [2, 3]. The contribution from each critical point may be calculated using parabolic energy bands and a constant momentum matrix element.

The k · p theory, however, is not well suited to derive the dielectric function of superlattices for two reasons. Firstly, it describes the band structure of the host materials only around one symmetry point. Thus one cannot take into account, in a straightforward manner, important effects of band mixing in heterostructures. Secondly, using effective mass theory is unreliable for short-period superlattices.

Optical properties arise from interband transitions all over the Brillouin zone, and the ability to describe the band structure far from symmetry points is of great interest. Pseudopotential methods are reliable for such computations, but they require too much computation time. A viable alternative is the empirical tight binding (TB) method, which has been used extensively to calculate electronic properties of heterostructures. It has been demonstrated that the method yields accurate energy levels and wavefunctions of correct symmetry in bulk semiconductors as well as in superlattices (see, for example, refs. [4] and [5]). Recently, the TB description has been extended to calculate optical properties of of $GaAs/Al_xGa_{1-x}As$ heterostructures as well [6]. Just as the band structure is derived using a certain number of fitting parameters in the Hamiltonian, the optical properties can be derived with properly chosen values for matrix elements of the momentum operator between the atomic orbitals that make up the TB basis.

In this paper we report a study of the optical properties of $In_{0.53}Ga_{0.47}As/InP$ superlattices based on the TB method. We have fitted the optical matrix elements for GaAs, InAs and InP, and it will be shown that the resulting dielectric functions of these compounds as well as their alloys are in excellent agreement with the experimental spectra. The theory underlying our computation is discussed in the next section. The results of our computation for bulk materials as well as for short-period superlattices is presented in the following section.

II Theory

The imaginary part of the complex dielectric function $\varepsilon(\omega)$ characterizes optical absorption due to interband transitions, and is given by the following expression in the dipole approximation [7]:

$$\varepsilon_2(\omega) = \frac{4\pi^2 e^2}{m^2 \omega^2} \sum_{v,c} \int_{BZ} \frac{2\,d\mathbf{k}}{(2\pi)^3} |\beta \cdot \mathbf{P}_{cv}(\mathbf{k})|^2 \; \delta\left[E_c(\mathbf{k}) - E_v(\mathbf{k}) - \hbar\omega\right]. \quad (1)$$

The summation here is over filled valence bands and empty conduction bands indexed by v and c, respectively. e denotes the electronic charge, and m its mass. The incident radiation, of angular frequency ω, is polarized along the direction with unit vector β. The momentum matrix element appearing above is defined as:

$$\mathbf{P}_{cv}(\mathbf{k}) \equiv \int_{V_c} d\mathbf{r}\; \psi_c^*(\mathbf{k},\mathbf{r})\,(-i\hbar\nabla)\psi_v(\mathbf{k},\mathbf{r}). \quad (2)$$

The real part of $\varepsilon(\omega)$ may be calculated from the imaginary part through the Kramers-Kronig relation.

In order to calculate $\varepsilon_2(\omega)$, one requires a representation for the crystal electronic states. In the TB representation [8] the crystal eigenfunction $\psi_n(\mathbf{k},\mathbf{r})$ (with wavevector $\mathbf{k}$ and band index n) is built up from atomic-like states.

$$\psi_n(\mathbf{k},\mathbf{r}) = \sum_i \sum_\alpha c_{n;i,\alpha}(\mathbf{k}) \frac{1}{\sqrt{N}} \sum_a e^{i\mathbf{k}\cdot(\mathbf{R}_a+\mathbf{r}_i)} \xi_{i,\alpha}(\mathbf{r}-[\mathbf{R}_a+\mathbf{r}_i]) \qquad (3)$$

where a indexes Bravais lattice sites $\mathbf{R}_a$ of the crystal, i indexes ions within the primitive cell, at relative positions $\mathbf{r}_i$, $\xi_{i,\alpha}$ is an atomic orbital of the ith ion characterized by quantum numbers α (e.g. s, p_x, p_y, p_z), N is the number of primitive cells in the crystal, and $c_{n;i,\alpha}(\mathbf{k})$ are the expansion coefficients in this basis.

A basis of atomic orbitals consisting of s, p_x, p_y, p_z functions of both cations and anions is employed in our study. This choice is known to reproduce the band structure of most elemental and III-V compound semiconductors satisfactorily, provided one retains interactions up to second nearest neighbors. Including spin-orbit coupling, we work with a basis of 16 functions for the bulk materials. In the TB scheme the optical matrix elements between Bloch functions can be expressed in terms of a few parameters, which are the optical matrix elements between atomic orbitals [6]. Experimental data for $\varepsilon_2(\omega)$ in bulk materials may be used to determine optimized values for these parameters. We use the Gilat-Raubenheimer method [9] to perform the zone integration in eqn. 1, evaluating the optical matrix elements over a fine mesh within the Brillouin zone.

Equations 1- 3 apply to bulk materials as well as to infinite superlattices, as long as the corresponding primitive cells are chosen appropriately. The primitive cell of the superlattice should cover one period along the growth direction. It is essential to include interactions up to second nearest neighbors in order to derive superlattice eigenstates with correct symmetry [4, 5]. In deriving the electronic and optical properties of superlattices, we have assumed that the Hamiltonian and optical interaction parameters retain their corresponding bulk values in the interior of the regions comprising the superlattice. The parameters across an interface are taken to be the average of the corresponding values of the two materials.

III Results

We have fitted the parameters of the TB Hamiltonian to accurately reproduce the band structure, energy gaps and effective masses of various bulk semiconductors as determined by pseudopotential calculations [10] or experiments [11]. We have also optimized the optical parameters using experimental data for $\varepsilon_2(\omega)$ in bulk. The resulting dielectric functions of bulk InP and $In_{0.53}Ga_{0.47}As$ computed with our optimized parameters are compared with experimental data in figure 1. The data has been obtained from ref. [12] for InP, and from ref. [13] for the alloy. The overall agreement between experiment and theory is very good in both cases.

Figure 2a shows the computed spectra for short-period $(In_{0.53}Ga_{0.47}As)_n(InP)_n$ superlattices, for the cases $n = 1, 2$ and 3. The spectra are almost identical in all 3 cases. The dominant E_1 and E_2 peaks of $\varepsilon_2(\omega)$ of the superlattices are located in between the corresponding peaks of bulk $In_{0.53}Ga_{0.47}As$ and InP, as shown in Figure 2b. $\varepsilon_2(\omega)$ is also found to be identical for light polarized parallel or perpendicular to growth direction.

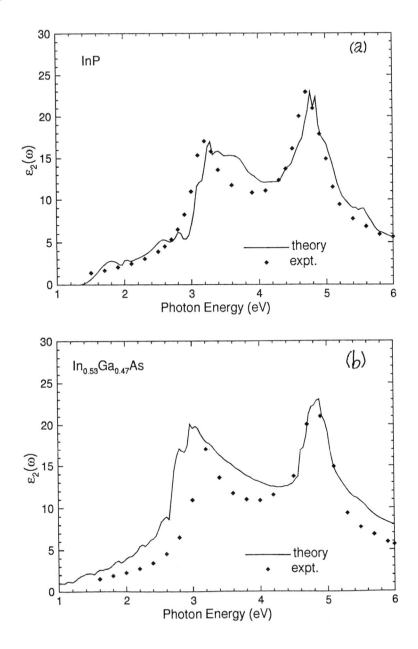

Figure 1: Dielectric functions of bulk semiconductors (a) InP and (b) $In_{0.53}Ga_{0.47}As$ compared with experiment.

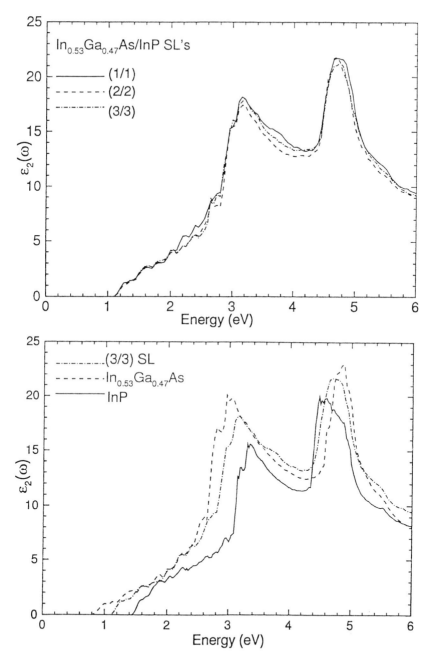

Figure 2: (a) Dielectric functions of ($In_{0.53}Ga_{0.47}As)_n(InP)_n$ superlattices for $n = 1, 2$, and 3; (b) Comparison of bulk and superlattice dielectric functions.

IV Conclusion

We have reported calculations of the dielectric function of $In_{0.53}Ga_{0.47}As/InP$ short-period superlattices using the TB method. The optical properties are derived using a few parameters, which determine the optical response of the constituent bulk semiconductors. We have optimized values of these parameters for several bulk materials with the resulting dielectric function in very good agreement with experiment. The computed $\varepsilon_2(\omega)$ of superlattices show features similar to those found in the spectra of the bulk constituents.

Our calculation is based on the full band structure, and not just the contribution from a few critical points. Such a detailed calculation will be useful in analyzing the contribution to $\varepsilon_2(\omega)$ from various regions of the superlattice Brillouin zone due to transitions between many zone-folded bands. The method itself is of interest for other material systems with components having indirect gaps, such as Si, Ge, AlSb, and GaP. It may be applied to quantum well structures also in a straightforward manner.

References

[1] H. C. Casey, Jr., and M. B. Panish. *Heterostructure Lasers.* Academic, New York, 1978.

[2] D. E. Aspnes. Modulation Spectroscopy/Electric Field Effects on the Dielectric Function of Semiconductors. In T. S. Moss, editor, *Handbook on Semiconductors*, chapter 4A, page 109. North-Holland Publishing Co., 1980. Vol.2, ed. M. Balkanski.

[3] S. Adachi. *Phys. Rev. B*, 39(17):12612, 1989.

[4] Y-T. Lu and L. J. Sham. *Phys. Rev. B*, 40(8):5567, 1989.

[5] J.-B. Xia and Y.-C. Chang. *Phys. Rev. B*, 42(3):1781, 1990.

[6] Y.-C. Chang and D. E. Aspnes. *Phys. Rev. B*, 41(17):12002, 1990.

[7] F. Bassani and G. P. Parravicini. *Electronic States and Optical Transitions in Solids.* Pergamon Press, 1975.

[8] J. C. Slater and G. F. Koster. *Phys. Rev.*, 94(6):1498, 1954.

[9] G. Gilat and L. J. Raubenheimer. *Phys. Rev.*, 144(2):390, 1966.

[10] J. R. Chelikowsky and M. L. Cohen. *Phys. Rev. B*, 14(2):556, 1976.

[11] O. Madelung, editor. *Landolt-Bornstein Numerical Data and Functional Relationships in Science and Technology*, volume 22a. Springer-Verlag, 1986.

[12] D. E. Aspnes and A. A. Studna. *Phys. Rev. B*, 27(2):985, 1983.

[13] S. M. Kelso, D. E. Aspnes, M. A. Pollack, and R. E. Nahory. *Phys. Rev. B*, 26(12):6669, 1982.

Resonant Tunneling Devices : Effect of Scattering

S.Datta, G.Klimeck*, R.K.Lake** and M.P.Anantram

School of Electrical Engineering, Purdue University, West Lafayette, IN 47907-1285.
Present Address : *University of Texas at Dallas, **Texas Instruments

Abstract : The non-equilibrium Green's function (NEGF) formalism provides a general framework for the simulation of quantum devices including electron-phonon and electron-electron interactions. We have used this approach to study the effect of phonon scattering on the dc and ac response of resonant tunneling diodes. In this talk we will discuss the basic results using a simple physical picture.

1. Introduction

Resonant tunneling diode is possibly the most well-known example of a 'quantum device' [1]. Such devices are commonly modeled under the assumption that electrons can traverse the active region coherently without any scattering. The current can then be computed from the expression

$$\hat{I} = \frac{2e}{h} \sum_n \int dE\, T_n(E) \left[f_1(E) - f_2(E) \right] \qquad (1)$$

where $f_1(E)$ and $f_2(E)$ are the Fermi functions for contacts '1' and '2' respectively. The index 'n' labels the different transverse modes and $T_n(E)$ represents the probability that an electron incident in mode 'n' with energy E transmits through the device from one contact to the other. The transmission probability, $T_n(E)$, is calculated by solving the Schrödinger equation for the appropriate device potential. This approach provides a fairly accurate description of the peak current, which is not affected significantly by scattering processes. The valley current, however, is strongly influenced by scattering processes and it is necessary to go beyond the coherent model in order to describe it accurately.

The non-equilibrium Green function (NEGF) formalism provides a general framework for the simulation of quantum devices, comparable to what the Boltzmann formalism provides for classical devices [2]. Recently we have applied this formalism to study the influence of scattering processes on resonant tunneling devices. The details have been described elsewhere [3-5] and will not be repeated here. Instead we will try to convey the main results in simple physical terms. To simplify our discussion we will not worry about different transverse modes and assume the structure to be one-dimensional. We will assume 'zero' temperature so that the Fermi functions appearing in eq.(1) can be replaced by step functions. Also we will assume that the bias is large enough that the injection of carriers from contact '2' can be neglected (see Figs.1 or 2). With these assumptions, eq.(1) reduces to

$$\hat{I} \approx \frac{2e}{h} \int_0^{\mu_1} dE\, T(E) \qquad (2)$$

2. Peak current

The current through a resonant level is maximum when it lies within the energy range of the incident electrons (Fig.1). Under these conditions, scattering processes have little effect on the current. Let us see why [6-9].
   If we neglect scattering then the current can be calculated from eq.(2). The transmission probability can be written approximately in the form of a Lorentzian function

$$T(E) \approx \frac{\Gamma_1 \Gamma_2}{(E - E_r)^2 + (\Gamma/2)^2} \qquad \text{where} \quad \Gamma = \Gamma_1 + \Gamma_2 \qquad (3)$$

The width of the Lorentzian (Γ) is determined by the sum of the escape rates (times $\hbar$) through barriers '1' and '2' respectively. These are obtained by multiplying the attempt frequency (times $\hbar$) by the transmission probabilities T_1 and T_2 through the two barriers : $\Gamma_1 = \hbar v\, T_1$ and $\Gamma_2 = \hbar v\, T_2$. Substituting eq.(3) into eq.(2) we obtain the peak current in the absence of scattering :

$$\hat{I} = \frac{2e}{\hbar} \frac{\Gamma_1 \Gamma_2}{\Gamma_1 + \Gamma_2} \qquad (4)$$

Scattering processes can be characterized by a parameter Γ_φ, which is related to the scattering time τ_φ by the relation $\Gamma_\varphi \equiv \hbar/\tau_\varphi$. If this quantity is much larger than the escape rates through the barriers (Γ_1 and Γ_2) then a significant fraction of the electrons will be scattered inside the well and get removed from the coherent stream. One way to model

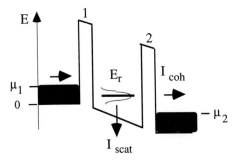

Fig.1. Condition for peak current : the resonant level E_r lies within the energy range of the incident electrons.

this loss of electrons is to introduce an imaginary component of the wavevector 'k' inside the well, causing the electronic wavefunction to decay. The transmission probability can still be written approximately in the form of a Lorentzian function, but the width is larger :

$$T(E) \approx \frac{\Gamma_1 \Gamma_2}{(E - E_r)^2 + (\Gamma/2)^2} \qquad \text{where} \quad \Gamma = \Gamma_1 + \Gamma_2 + \Gamma_\varphi \qquad (5)$$

Substituting into eq.(2) we obtain the coherent current :

$$I_{coh} = \frac{2e}{\hbar} \frac{\Gamma_1 \Gamma_2}{\Gamma_1 + \Gamma_2 + \Gamma_\varphi} \qquad (6)$$

But this is not the total current. The electrons that are scattered out of the coherent stream give rise to a scattered current which is given by an expression that looks just like that for the coherent current (see Eq.(6)) with Γ_2 replaced by Γ_φ :

$$I_{scat} = \frac{2e}{\hbar} \frac{\Gamma_1 \Gamma_\varphi}{\Gamma_1 + \Gamma_2 + \Gamma_\varphi} \qquad (7)$$

This result (known as the Breit-Wigner result) seems quite reasonable if we view the scattered current as a current that escapes into a fictitious third terminal having a coupling of Γ_φ [9]. The scattered current eventually exits from the device partially through barrier '1' and partially through barrier '2' in the ratio $\Gamma_1 : \Gamma_2$. The fraction coming out through '2' contributes to the net current and is referred to as the sequential current :

$$I_{seq} = \frac{\Gamma_2}{\Gamma_1 + \Gamma_2} I_{scat} = \frac{\Gamma_\varphi}{\Gamma_1 + \Gamma_2} I_{coh} \qquad (8)$$

The total current (coherent + sequential) is given by

$$I_{tot} = I_{coh} + I_{seq} = \frac{2e}{\hbar} \frac{\Gamma_1 \Gamma_2}{\Gamma_1 + \Gamma_2} \qquad (9)$$

This is *exactly the same* as the result obtained earlier (eq.(4)), neglecting scattering processes altogether ! Thus scattering has no effect on the peak current within this approximate model. More accurate calculations, however, do show small effects [10].

3. Valley current

The current through a resonant level is very small when it lies outside the energy range of the incident electrons (Fig.2). A number of authors have studied the influence of scattering processes on the valley current that flows under these conditions (see [11] and references therein]. In our simple model, we can calculate the coherent component simply by substituting eq.(5) into eq.(2), just as we did for the peak current (cf. eq.(6))

$$I_{coh} = \frac{2e}{\hbar} \Gamma_1 K \qquad \text{where} \quad K \approx \int_0^{\mu_1} \frac{\Gamma_2 \, dE}{(E - E_r)^2} \qquad (10)$$

assuming that the linewidth is much smaller than the energy difference, $E - E_r$, between the incident electrons and the resonant energy. Actually we should not be using the Lorentzian approximation far from resonance, but that only changes the precise value of K and does not affect the following arguments. The scattered current is obtained by replacing Γ_2 with Γ_φ, just as we did before :

$$I_{scat} = \frac{2e}{\hbar} \Gamma_1 g \quad \text{where} \quad g \approx \int_0^{\mu_1} \frac{\Gamma_\varphi \, dE}{(E - E_r)^2} \tag{11}$$

Earlier we had argued that the scattered current exits through the two barriers in the ratio $\Gamma_1 : \Gamma_2$ and the fraction coming out through '2' contributes to the sequential current. However, in the present case the scattered current must come entirely out of '2' since there are no states in '1' corresponding to E_r. We are assuming that the electrons scatter down in energy into the resonance by phonon emission and the temperature is low enough that they cannot climb out by absorbing phonons. Under these conditions the sequential current is simply equal to the scattered current so that the total valley current is given by

$$I_{tot} = I_{coh} + I_{scat} = \frac{2e}{\hbar} [K\Gamma_1 + g\Gamma_1] \tag{12}$$

But this cannot be right. Suppose we make barrier 2 so thick that Γ_2 (and hence 'K') is essentially zero. Eq.(12) still predicts a non-zero valley current equal to $(2e / \hbar) g\Gamma_1$! How can any current flow if electrons cannot transmit across barrier 2 ? What happens for very thick barriers is that the resonant level fills up, since it cannot empty anywhere. To include this effect we should modify Eq.(12) to write the valley current as

$$I_{tot} = \frac{2e}{\hbar} [K\Gamma_1 + g\Gamma_1 (1 - f_r)] \tag{13}$$

To complete the story we need to calculate the occupation factor f_r for the resonant level. This can be done quite simply using a sequential model as described below.

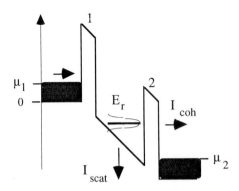

Fig.2. Condition for valley current : the resonant level E_r lies outside the energy range of the incident electrons.

4. Sequential model

In the sequential model the *peak current* is calculated by writing a rate equation for the currents I_1 and I_2 from the two terminals :

$$I_1 = \frac{2e}{\hbar} \Gamma_1 (f_1 - f_r) \quad \text{and} \quad I_2 = \frac{2e}{\hbar} \Gamma_2 (f_r - f_2) \tag{14}$$

Setting $f_1 = 1$, $f_2 = 0$ and $I_1 = I_2 \equiv I_P$, we obtain

$$f_r = \frac{\Gamma_1}{\Gamma_1 + \Gamma_2} \quad \text{and} \quad I_P = \frac{2e}{\hbar} \frac{\Gamma_1 \Gamma_2}{\Gamma_1 + \Gamma_2} \tag{15}$$

in agreement with our earlier results (see eqs.(4), (9)). We can use the same approach to describe the *valley current*, simply by modifying the rate equation as follows (cf. eq.(14)) :

$$I_1 = \frac{2e}{\hbar} g\Gamma_1 (f_1 - f_r) \quad \text{and} \quad I_2 = \frac{2e}{\hbar} \Gamma_2 (f_r - f_2) \tag{16}$$

All that we have done is to reduce the rate constant for barrier 1 from Γ_1 to $g\Gamma_1$, where the factor 'g' (<< 1) is determined by the electron-phonon coupling constant. This seems reasonable since an electron in order to get into the resonant level from terminal 1 must tunnel through barrier 1 *and* emit a phonon simultaneously. Setting $I_1 = I_2 \equiv I_V$ and $f_1 = 1$, $f_2 = 0$ we obtain (cf. eq.(15))

$$f_r = \frac{g\Gamma_1}{g\Gamma_1 + \Gamma_2} \quad \text{and} \quad I_V = \frac{2e}{\hbar} \frac{g\Gamma_1 \Gamma_2}{g\Gamma_1 + \Gamma_2} \tag{17}$$

Eq.(17) has also been derived from the NEGF formalism with suitable approximations and is found to describe the exact results obtained from a full numerical simulation fairly well [3].
Generally we assume that the valley current will be a small fraction of the peak current, independent of Γ_1 and Γ_2. But it is easy to see from Eqs.(15) and (17) that this is not true in general :

$$I_P \sim \frac{1}{\Gamma_1} + \frac{1}{\Gamma_2} \quad \text{and} \quad I_V \sim \frac{1}{g\Gamma_1} + \frac{1}{\Gamma_2}$$

For very asymmetric devices having $\Gamma_2 << g\Gamma_1$, the peak and valley currents can even become equal! This has been observed experimentally [12].

5. High frequency response

We have also used the NEGF formalism to study the ac response of resonant tunneling devices in the presence of inelastic scattering processes. The results can be understood fairly simply in terms of the sequential model. Under time-varying conditions we cannot assume that I_1 is equal to I_2. Instead, $I_1 - I_2 = 2e (df_r / dt)$, so that for the *peak* current we can use eq.(14) to write

$$\frac{df_r}{dt} + \frac{\Gamma_1+\Gamma_2}{\hbar} f_r = \frac{1}{\hbar}(\Gamma_1 f_1 + \Gamma_2 f_2) \qquad peak \qquad (19)$$

while for the *valley* current we can use eq.(16) to write

$$\frac{df_r}{dt} + \frac{g\Gamma_1+\Gamma_2}{\hbar} f_r = \frac{1}{\hbar}(\Gamma_1 f_1 + \Gamma_2 f_2) \qquad valley \qquad (20)$$

From eqs.(19) and (20) we would expect that if we modulate f_1 and/or f_2 through an ac signal applied across the contacts, the peak current should respond with a time constant of $(\Gamma_1 + \Gamma_2)$ while the valley current should respond with a time constant of $(g\Gamma_1 + \Gamma_2)$. This is exactly what we find from the NEGF formalism [5].

5. Concluding remarks

In this talk we have shown how the effect of scattering processes on the valley current can be understood in simple intuitive terms using rate equations based on a sequential model. In general for quantitative calculations one needs to go beyond simple models and include multiple levels, level broadening, non-zero temperature, transverse modes, details of scattering processes, band-structure etc. The NEGF formalism provides a rigorous approach suitable for this purpose (see [2,3] and references therein). Since it incorporates the physics of the interactions correctly from first principles, a device simulator based on this formalism should have significant predictive value. It should be mentioned, however, that in resonant tunneling devices with small cross-section, the electron-electron interactions can be very strong leading to Coulomb blockade and related effects [13]. The utility of the NEGF formalism in this transport regime is still unclear.

Acknowledgements

Parts of this article have been adapted from a forthcoming book by one of the authors [14]. This talk is based on work supported by the National Science Foundation (grant number ECS-9201446) and the Semiconductor Research Corporation (contract number 91-SJ-089).

References

[1] For a recent review see Liu H C and Sollner T C L G (1994) in *Semiconductors and semimetals* **41** eds. Kiehl R A and Sollner T C L G Academic Press.
[2] For a review of the work on uniform conductors see Mahan G D (1987) *Physics Reports* **145** 251.
[3] Lake R K, Klimeck G and Datta S (1993) *Phys.Rev. B* **47** 6427.
[4] Lake R K, Klimeck G, Anantram M P and Datta S (1993) *Phys.Rev. B* **48** 15132.
[5] Anantram M P and Datta S (unpublished).
[6] Stone A D and Lee P A (1985) *Phys. Rev. Lett.* **54** 1196.
[7] Jonson M and Grincwajg A (1987) *Appl. Phys. Lett.* **51** 1729.
[8] Weil T and Vinter B (1987) *Appl. Phys. Lett.* **50** 1281.
[9] Buttiker M (1988) *IBM J. Res. Develop.* **32** 63.
[10] Booker S M, Sheard F W and Toombs G A (1992) *Semicond. Sci. Technol.* **7** B439.
[11] Chevoir F and Vinter B (1993) *Phys.Rev.B* **47** 7260.
[12] Turley P J, Wallis C R, Teitsworth S W, Li W and Bhattacharya P K (1993) *Phys. Rev. B* **47** 12640.
[13] See Klimeck G, Lake R K, Datta S and Bryant G W *Phys. Rev. B* (to appear) and references therein.
[14] Datta S *Electronic Transport in Mesoscopic Systems, Chapter 6.* Cambridge University Press (to be published).

Quantum Cellular Automata: Computing with Quantum Dot Molecules

P D Tougaw and C S Lent

Department of Electrical Engineering, University of Notre Dame, Notre Dame, IN 46556 USA

Abstract. We present a new paradigm for computing—arrays of quantum dot molecules called quantum cellular automata (QCA). Such molecules are composed of coupled quantum dots sharing two electrons. Interaction between the electrons in a cell causes the charge density to be very highly polarized (aligned) along one of the two cell axes, and the polarization of one cell Coulombically induces a polarization in neighboring cells. Geometric arrangement of these cells allows the synthesis of wires, coplanar wire crossings, inverters, and a full range of logical devices including majority voting gates, programmable AND/OR gates, XOR gates, and full adders. Finally, we explore another realization of such cells, rings of small-capacitance metallic tunnel junctions, and show that they also have the potential to serve as the foundation of a QCA.

1. Introduction

While recent improvements in the size of microelectronic devices has led to improved speed and density of devices, the underlying paradigm of computation has remained largely unchanged. Cellular automata (CA) architectures have the potential to bring about revolutionary change in this basic computing paradigm. Such architectures also offer a natural match for the features of nanometer-scale quantum devices.

We propose a realization of such an architecture using two-electron quantum dot molecules.[1 - 3] A schematic of such a molecule, or cell, is shown in Fig. 1.

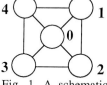

Fig. 1. A schematic of the five-site cell.

Fig. 2. The two bistable states of the QCA cell. Electron repulsion causes the cell to always polarize in one of these two states.

The Hamiltonian used to describe a cell includes on-site energies, tunneling between neighbors, on-site charging costs, and Coulombic interaction between each pair of sites:

$$H_0 = \sum_{i,\sigma} E_0 n_{i,\sigma} + \sum_{i>j,\sigma} t_{i,j}(a_{i,\sigma}^\dagger a_{j,\sigma} + a_{j,\sigma}^\dagger a_{i,\sigma}) + \sum_i E_Q n_{i,\uparrow} n_{i,\downarrow} + \sum_{i>j,\sigma,\sigma'} V_Q \frac{n_{i,\sigma} n_{j,\sigma'}}{|\vec{R}_i - \vec{R}_j|}$$

In the devices presented here, we use a "standard cell" for which the radius of the dots is 5 nm, the distance between neighboring dots is 20 nm, and the tunneling energy t between dots is 0.3 meV. Other parameters correspond to electrons in GaAs. Interaction between the electrons causes the cell to exhibit bistable behavior. The bistable states are shown in Fig. 2.

This bistable behavior makes it possible to encode binary information using the state of the cell. This state can be represented quantitatively by the cell polarization, which is a measure of which of the two bistable states the cell occupies. We define the polarization P as:

$$P \equiv \frac{(\rho_1 + \rho_3) - (\rho_2 + \rho_4)}{\rho_0 + \rho_1 + \rho_2 + \rho_3 + \rho_4}$$

2. Cell-cell response

In order for such a cell to be useful in a CA architecture, its state must interact strongly with the state of its neighbors. One measure of this is the cell-cell response function. To determine this response function, we consider the system of two cells shown in the insets to Fig. 3. We assume the polarization of cell 2 is fixed, and calculate the Coulombic energy at each site of cell 1 as a result of this fixed charge. The two-electron Schrödinger equation is then solved for cell 1. The resulting charge density and induced polarization of cell 1 can then be found for a number of values of P_2. The function $P_1(P_2)$ is called the cell-cell response and reveals how much influence one cell's state has over that of its neighbors.

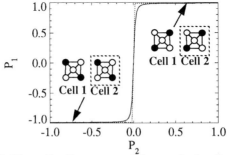

Fig. 3. The cell-cell response of the standard cell shown in figure 1. The bistable nature of the cell is apparent.

The response function shown in Fig. 3 is for a cell at zero temperature. The effect of nonzero temperature can be found directly by calculating the thermal expectation value for the charge densities on each site.[4] The results for the standard GaAs cell at several temperatures are shown in Fig. 4, while Fig. 5 shows the response of a cell with 2 nm near-neighbor spacing and the physical parameters corresponding to a vacuum. Such a molecular cell would work properly well above room temperature.

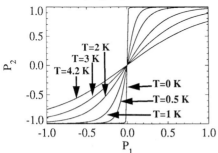 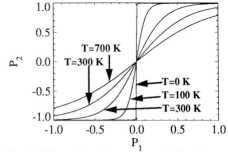

Fig. 4. The cell-cell response of the 20 nm GaAs cell at nonzero temperatures.

Fig. 5. The cell-cell response of the 2 nm vacuum cell at nonzero temperatures.

3. Lines of cells

The most fundamental QCA device is a line of cells, or binary wire. [5] Fig. 6 shows two such lines of standard cells, each being driven by a driver cell with fixed polarization at the left. The radius of each dot in the figure is proportional to the charge density on that site.

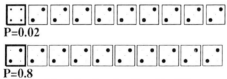

Fig. 6. QCA lines. Two lines driven by fixed cells with different polarizations.

Despite the fact that the driver cells are incompletely polarized, the polarization of each line rapidly saturates, propagating information about the driver polarization down the line. This saturation behavior will allow the line to recover from local decreases in polarization due to potential fluctuations in the fabrication parameters.

It should be noted that, unless stated otherwise, these figures are not simply schematic. The diameter of each of the dots is proportional to the charge density on the corresponding site, which is found by a self-consistent calculation of the ground state of the system shown.

4. Logic devices

While a binary wire is the most fundamental QCA structure, different geometric arrangements of cells can be used to perform other useful functions.[6] Fig. 7 shows how information in a binary wire propagates around a right angle corner, while Fig. 8 shows that the same signal can "fan out" to drive two similar wires.

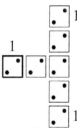

Fig. 7. Signal propagation around an abrupt corner.

Fig. 8. Fan-out of a binary wire to drive two similar wires.

In addition to propagation of QCA signals from one point to another, special arrangements of cells can be used to perform logic functions on the signal. Fig. 9 shows an arrangement of cells that can be used to invert the signal carried by a binary wire. The signal, coming from the left, is split into two parallel binary wires and then rejoined in a way that causes the cells to anti-align. Due to the geometrical symmetry of this arrangement, a 1 and a 0 will be inverted with the same reliability.

Fig. 10 shows a single central cell surrounded by four near neighbors. Three of the neighbors have fixed polarizations, while the fourth neighbor and the central cell are free to react to their charge densities. When the ground state of such a system is calculated, we find

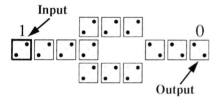

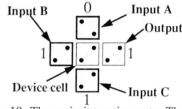

Fig. 9. The cellular arrangement used to invert signals.

Fig. 10. The majority voting gate. The central cell always matches a majority of the three fixed neighbors.

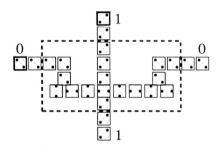
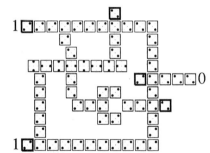

Fig. 11. The coplanar wire crossing.

Fig. 12. The XOR gate uses three reduced gates one inverter, and a coplanar wire crossing.

that the free cells always line up so as to match a majority of the driving neighbors. Therefore, we call such an arrangement of cells a majority voting gate.

While this novel function is quite useful by itself, its flexibility is seen when we single out one of the inputs and define it as the program line. If the program line is 1, then the result will be 1 unless both of the other two inputs are 0. Thus, the gate performs the OR function on the two non-program lines. Likewise, if the program line is 0, the gate performs the AND function on the other two lines. Such a gate is called a programmable AND/OR gate, since the program line determines the nature of the gate, and the other two lines are applied to the gate thus defined. Since the three inputs are indistinguishable, any of the three can serve as the program line.

Conventional electronic circuits cross wires by placing an insulator between them and physically passing one over the other in a second layer. Because of the cellular nature of the QCA architecture, two binary wires can be crossed in a plane with no interference. Such a coplanar wire crossing is shown in Fig. 11. By combining the reduced AND and OR gates, the coplanar wire crossing, and one inverter, we can make an XOR gate as shown in Fig. 12.

The one-bit full adder is the most complicated logical function synthesized thus far. By

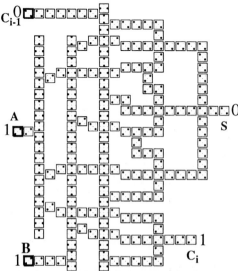

Fig. 13. The full adder implemented in the quantum cellular automata scheme.

using five majority voting gates in a novel way, we are able to design the entire device without using any reduced (simple AND or OR) gates. This greatly reduces the number of cells, and therefore the area needed to perform this function. Fig. 13 shows the result of a self-consistent simulation of the full adder design.

5. Metal electrode cells

The cells discussed thus far would typically be fabricated in semiconductor heterostructures using a depleted two-dimensional electron gas. An alternative realization can be seen in Fig. 14, which shows a ring of metallic tunnel junctions with very small capacitance.[7] The behavior of such a ring is controlled by Coulomb exclusion effects. The coupling between the metal "islands" and between cells is capacitive, not simply Coulombic. Such capacitive coupling allows better control than Coulombic interaction alone.

Each cell contains many conduction electrons, as well as two extra electrons provided by the ground substrate. The charge configuration which minimizes electrostatic energy is one in which electrons are on antipodal electrodes. This results in alignment of cell charge density in one of the two states shown in Fig. 15.

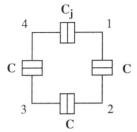

Fig. 14. A schematic diagram of the metal electrode cell.

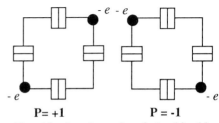

Fig. 15. A schematic of the bistable states of the cell caused by electron repulsion within the cell.

The Hamiltonian used to describe such a cell uses the capacitance matrix **C**, which describes the capacitance of the 4 electrodes as well as the nearest electrodes of each neighboring cell. The capacitance to ground of each electrode is also included:

$$H^k = \frac{1}{2} \sum_{\substack{i,j \in \\ \text{cell } k}} e^2 [\mathbf{C}^{-1}]_{i,j} \hat{n}_i \hat{n}_j + \sum_{\substack{i \in \text{cell } k \\ j \notin \text{cell } k}} e^2 [\mathbf{C}^{-1}]_{i,j} \hat{n}_i \hat{n}_j + \sum_{\substack{i,j \in \text{cell } k \\ i > j}} t_{i,j} (a_i^\dagger a_j + a_j^\dagger a_i)$$

Here the tunneling energy t, which is the same for all tunnel junctions, is expressed as a fraction of the charging energy of the junction.

$$t_{i,i+1} = t = \frac{\alpha e^2}{2C_j}$$

To find the cell-cell response function of these metal electrode cells, consider a single cell capacitively coupled to a neighboring cell as shown in Fig. 16. We fix the charge density in cell 2 and find the ground state of cell 1 as before. The charge densities on sites 5 and 6 couple capacitively to the charge in cell 1, inducing a polarization. When this is repeated for a number of values of P_2, the function $P_2(P_1)$ can be found. As shown in Fig. 17, the bistability apparent in such a plot indicates that this cell has the potential to serve as the basis of a QCA architecture.

Simulations of linear arrays of these metal cells shows behavior similar to that of the

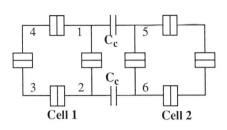
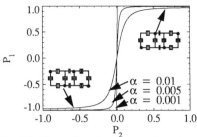

Fig. 16. Schematic diagram of two capacitively coupled cells. This arrangement is used to find the cell-cell response function.

Fig. 17. The cell-cell response function of the metal electrode cell for three values of tunneling coefficient.

quantum dot molecule cells. Fig. 18 shows two lines of metal cells being driven from the left by cells with different values of polarization. As before, the information in the driver cell is carried down the length of the binary wire, and a weakly polarized driver is still capable of driving the wire to full polarization within the span of a few cells.

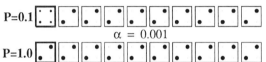

Fig. 18. Linear arrays of metal electrode cells transmit information in the same way as binary wires of quantum dot molecules.

5. Conclusions

We have demonstrated a new paradigm for computing with quantum dot molecules and discussed possible implementations for such an architecture. The complete set of logical functions can be constructed using this paradigm, and functions as complex as a one-bit full adder have been synthesized using these elements.

Acknowledgments

This work was supported in part by the Advanced Research Projects Agency, the Office of Naval Research and the Air Force Office of Scientific Research. This material is based in part upon work supported under a National Science Foundation Graduate Fellowship.

References

1. One-dimensional clocked cellular automata driven by quantum mechanical, but non-unitary, time evolution operators were first discussed by G. Grössing and A. Zeilinger in *Complex Syst.* **2**, 197 (1988) and G. Grössing and A. Zeilinger, *Physica D* **50**, 321 (1991). The system we examine here differs in several important respects but is motivated by similar goals. Our system is not clocked and undergoes unitary time evolution.
2. C. S. Lent, W. Porod, and P. D. Tougaw., *Appl. Phys. Lett.* **62**, 714 (1993).
3. C. S. Lent, P. D. Tougaw, W. Porod, and G. Bernstein, *Nanotechnology* **4**, 49 (1993).
4. P. D. Tougaw, C. S. Lent, and W. Porod, *J. Appl. Phys.* **74**, 3558 (1993).
5. C. S. Lent and P. D. Tougaw, *J. Appl. Phys.* **74**, 6227 (1993).
6. P. D. Tougaw and C. S. Lent, *J. Appl. Phys.* **75**, 1818 (1994).
7. C. S. Lent and P. D. Tougaw, *J. Appl. Phys.* **75**, 4077 (1994).

Quantum scattering states in open two-dimensional electronic systems and the local density of states *

Henry K. Harbury† and Wolfgang Porod

Dept. of Electrical Engineering, University of Notre Dame, Notre Dame, IN 46556

Abstract Recent experimental STM images [1-4] of the surfaces of Cu(111) and Au(111) have demonstrated the direct measurement of the local density of electronic states (LDOS) for the system of a two-dimensional electron gas (2DEG) in the vicinity of geometrically arranged scattering centers (Fe adatoms). We report on our numerical method and calculations of the LDOS for these "quantum corral" structures. The numerical technique implements non-reflecting boundary conditions on an artificial boundary enclosing the scattering domain and is compatible with standard finite element methods. Excellent agreement is achieved in a comparison of both the calculated energy spectra of the LDOS and the spatial variation of the LDOS with the recently reported experimental STM data [3,4]. Contrary to the results of the multiple scattering theory reported by Heller et al. [4], our results demonstrate that a completely coherent elastic scattering model in an open boundary system is sufficient to explain the reported experiments with comparable agreement.

1. Introduction

The detailed spatial variation of electronic scattering states in open and unconfined mesoscopic systems is of interest in both the asymptotic far-field regime and the near-field regime in the vicinity of the scattering potential. Scattering from defects or impurities in a two-dimensional electron gas (2DEG) is one such system of interest. The interaction of the electronic system with the scattering potential leads to spatial variations of both the local electrostatic field and the local density of states (LDOS). The effects of open boundary electronic scattering in a 2DEG have been demonstrated by recent scanning tunneling microscopy (STM) images of the spatial variation of the STM tip conductance due to electronic scattering of surface confined states from defects, impurities, and step edges on Cu(111) [1] at 4.2K and at room temperature on Au(111) [2]. These metallic surfaces are known to support a 2DEG through surface state confinement [5].
  Davis et al. [6], following the work of Tersoff and Hamman [7] and N.D.Lang [8], proposed a simple theory which relates the local density of surface electronic states (LDOS) to the differential conductance, dI/dV, of a scanning tunneling microscope (STM). The dI/dV of the junction at the energy $E = E_F + |e|V - E_0$, where E_F is the Fermi Energy and E_0 is the band edge, may be measured at a fixed bias potential, V, over the surface of the 2DEG system to obtain an image of the surface LDOS at energy E. The energy spectra of the differential conductance of the STM junction at a fixed sample point has also been reported [1-4]. Through atomic manipulation of surface adatoms of Fe, geometrically arranged scattering structures have been fabricated

[3-4]. The Fe adatoms are abrupt potential barriers to surface state electrons and act collectively as electronic "quantum corrals" in which the observed spatial variation of the LDOS is clearly demonstrated. The peaks in the measured energy spectra of the LDOS for circular shaped corrals fit remarkably well to the eigen-energies of a two-dimensional hard-wall quantum box [3]. This simple model, however, breaks down for the detailed spatial variation of the LDOS within the corral and completely neglects the LDOS outside.

Heller et al. [4] reported results for a multiple scattering theory approach to computing the two-dimensional LDOS in which the Fe adatoms act as purely s-wave type scatterers. For this model to achieve reasonable agreement with the experimental data, however, it was necessary to assert a significant inelastic absorption of the electronic flux incident on the scattering centers. The reported attenuation of this model was approximately 25% reflection, 25% transmission and 50% absorption. The authors speculate that the Fe adatoms act to couple the otherwise orthogonal surface electronic states to the bulk states of the metallic sample.

The method and results are presented for a purely elastic scattering calculation of the fully two-dimensional LDOS for several reported quantum corral structures. The LDOS results are compared to both the reported data and reported multiple scattering theory. These results demonstrate that an elastic scattering model is sufficient to achieve comparable agreement with both the reported energy spectra and the detailed spatial variation of the measured LDOS. The numerical technique implements non-reflecting boundary conditions on an artificial boundary enclosing the scattering domain and is compatible with standard finite element methods. The Hamiltonian formulated by this method is used to compute the LDOS in the scattering domain.

2. Solution Method

A method for implementing non-reflecting boundary conditions (NRBC) on an artificial boundary is presented. The authors have recently reported [9] an exact method which uses the known far-field solution for the asymptotic scattering state by matching the known partial-wave expansion to the near-field solution on the artificial domain boundary, and likewise matching the boundary normal derivatives. A more efficient but approximate NRBC method is presented herein. The approximate boundary condition is formulated from the application of operators which are, by construction, compatible with the outward Sommerfeld radiation condition in the far-field regime. This boundary condition is inserted into the surface integral term and results in only a local tridiagonally coupled boundary. A brief synopsis of the non-reflecting boundary condition will be presented and the reader is referred to the recent publication [9] by the authors for more details.

We initially formulate a method to solve the multidimensional, single-electron, effective-mass Schrödinger equation for the electronic scattering states, $\psi_E(\mathbf{r})$,

$$-\frac{\hbar^2}{2}\nabla \cdot \left[\frac{1}{m^*}\nabla \psi_E(\mathbf{r})\right] + V(\mathbf{r})\psi_E(\mathbf{r}) = E\psi_E(\mathbf{r}). \tag{1}$$

A two dimensional domain will be assumed for simplicity, although the method is easily generalized to three dimensional geometries. The scattering domain has an artificial boundary at a radius R_0 from the center of the scattering potential which encloses the solution domain, Ω_0. The potential is assumed constant outside the artificial domain boundary and is taken as the zero reference. An electronic flux incident upon the mesoscopic scattering region Ω_0, injected by some outside source, will be described by a plane wave of the form $\psi_{inc} = a\exp[i\mathbf{k}\cdot\mathbf{r}]$, where $k = |\mathbf{k}| = \sqrt{2m^*E/\hbar^2}$. In the subsequent application of the more general formulation to obtain the local density of states, only the part of the Hamiltonian which does not include incident wave terms will be needed.

The approximate non-reflecting boundary condition is based on the method of Bayliss and coworkers [10]. In solving Eq. 1 for the scattering states, a constraint on the artificial boundary is imposed which restricts the solutions to the class of wavefunctions which are compatible with the outward Sommerfeld radiation condition. The boundary constraint is exact in the limit of $R_0 \to \infty$ and has a radially decaying truncation error on the artificial boundary.

In the two-dimensional case, the scattered part of the wavefunction may be expanded in the form $\psi_{scatt} \propto e^{ikr}\sum_{j=0}^{\infty} f_j(\theta)/(kr)^{j+\frac{1}{2}}$. A set of linear operators, B_m, compatible with the outgoing Sommerfeld radiation condition are constructed such that on the artificial boundary each mth higher order term is eliminated when B_m operates on the far-field expansion, $B_m\psi_{scatt}|_\Gamma = \mathcal{O}\left(R_0^{-\frac{4m+1}{2}}\right)$. The set of operators are therefore defined as $B_0 = 1$, $B_m = (\partial_r - ik + \alpha_m/r)B_{m-1}$, where α_m is a constant chosen to eliminate the mth term of the expansion. In the two-dimensional case, the first two non-trivial operators are given by:

$$B_1 = (\partial_r - ik + \tfrac{1}{2r}) \quad ; \quad B_1\psi_{scatt}|_\Gamma = \mathcal{O}(R_0^{-5/2}) \tag{2.a}$$

$$B_2 = (\partial_r - ik + \tfrac{5}{2r})B_1 \quad ; \quad B_2\psi_{scatt}|_\Gamma = \mathcal{O}(R_0^{-9/2}). \tag{2.b}$$

Expanding $B_2\psi_{scatt}|_\Gamma$ and implicitly using the original Schrödinger equation, $\frac{\partial^2\psi}{\partial r^2} = -\frac{1}{r}\frac{\partial\psi}{\partial r} - \frac{1}{r^2}\frac{\partial^2\psi}{\partial\theta^2} - k^2\psi$, the result of operating with B_2 on ψ may be rearranged to obtain the boundary normal derivative of the wavefunction: $\frac{\partial\psi}{\partial r}\Big|_\Gamma = \{\frac{r}{2(1-ikr)}[B_2\psi_{inc} + \frac{1}{r^2}\frac{\partial^2\psi}{\partial\theta^2} - \frac{3\psi}{4r^2} + \frac{3ik}{r}\psi + 2k^2\psi]\}_\Gamma$. Similar to the derivation of the exact non-local NRBC [9], this boundary normal derivative may be used in the surface term resulting from the integration by parts of Eq. 1. The resulting formulation using the finite element approximation $\psi \approx \sum_i \psi_i u_i$, where u_i is the ith shape-function, recast into matrix notation has the form of the linear system: $[\mathbf{T} + \mathbf{V} + \mathbf{C}]\,\boldsymbol{\psi} = \mathbf{p}$, where $\mathbf{T}_{ij} = \frac{\hbar^2}{2}\int_{\Omega_0} \nabla u_i \cdot \frac{1}{m^*}\nabla u_j\, d\Omega$ and $\mathbf{V}_{ij} = \int_{\Omega_0} u_i[V-E]u_j\, d\Omega$ are the kinetic and potential energy matrix element, respectively, $\mathbf{C}_{ij} = \frac{\hbar^2}{2m^*}\frac{4}{3}\eta\int_0^{2\pi}\frac{\partial u_i}{\partial\theta}\Big|_\Gamma\frac{\partial u_j}{\partial\theta}\Big|_\Gamma d\theta + \frac{\hbar^2}{2m^*}(\eta - \xi - 2\gamma)\int_0^{2\pi} u_i\big|_\Gamma u_j\big|_\Gamma d\theta$, is the matrix element which results from the boundary integral, and the right-hand-side vector, $\mathbf{p}_i = \frac{\hbar^2}{2m^*}\{-\gamma\int_0^{2\pi} u_i(\cos\theta - 1)^2\psi_{inc}d\theta + \xi\int_0^{2\pi} u_i(\cos\theta - 1)\psi_{inc}d\theta + \eta\int_0^{2\pi} u_i\psi_{inc}d\theta\}$, contains the incident plane wave contribution where ψ_{inc} is the incident wave, and where $\gamma = \frac{(kR_0)^2}{2(1-ikR_0)}$, $\xi = \frac{3ikR_0}{2(1-ikR_0)}$, and $\eta = \frac{3}{4}\frac{1}{2(1-ikR_0)}$. Unlike the exact NRBC method, this formulation does not contain any embedded integral terms in $\mathbf{C}$ and results in a numerical formulation whose discretized boundary is only tridiagonally coupled. This local NRBC, however, has an approximation error due to the neglected terms of the far-field expansion. This error may be made less than the inherent discretization error in the domain by using a sufficiently large radius.

The local density of states (LDOS) is derived from the system Hamiltonian using the Green function formalism, $[E - H(\mathbf{r})]\,G(\mathbf{r},\mathbf{r}';E) = \delta(\mathbf{r} - \mathbf{r}')$, where G may be expressed as a sum over the discrete and an integral over the continuous spectrums, $G(\mathbf{r},\mathbf{r}';z) =$

$\sum_n \psi_n(\mathbf{r})\psi_n^*(\mathbf{r}')/(z-E_n) + \int dn \psi_n(\mathbf{r})\psi_n^*(\mathbf{r}')/(z-E_n) = 1/(z-H)$. The LDOS in the spectral representation may be written as

$$\begin{aligned}LDOS(\mathbf{r};E) &= \sum_n \delta(E-E_n)\psi_n(\mathbf{r})\psi_n^*(\mathbf{r}) + \int \delta(E-E_n)\psi_n(\mathbf{r})\psi_n^*(\mathbf{r})dn \\ &= \mp\frac{1}{\pi}\mathrm{Im}\left[G^\pm(\mathbf{r},\mathbf{r};E)\right],\end{aligned} \qquad (3)$$

where $G^\pm(\mathbf{r},\mathbf{r};E)$ indicates the analytic approach as $s \to 0$ of $z = E \pm is$. The numerical calculation of the Green function is performed using the matrices as defined in the previous section where $\mathbf{H} = \mathbf{T} + \mathbf{C} + \mathbf{V}$. The Green function is obtained through matrix inversion, $\mathbf{G} = [E\mathbf{M} - (\mathbf{T}+\mathbf{C}+\mathbf{V})]^{-1}$, where the matrix elements of $\mathbf{M}$ are given by $\mathbf{M}_{ij} = \int_{\Omega_0} u_i u_j d\Omega$. The LDOS is therefore obtained directly from the diagonal elements, $LDOS = -\frac{1}{\pi}\mathrm{Diag}\left(\mathrm{Im}\left[\mathbf{G}\right]\right)$.

3. Results

The numerical technique presented in Section 2 was used to compute the LDOS for several different scattering geometries. The modeled structures are based on the experimental corral geometries reported in Ref. [3-4], from which the experimental data on Cu(111) and the multiple scattering theory results shown in the following figures was adapted for comparison. Each numerical scattering domain was discretized by a non-uniform triangular mesh which extends beyond each Fe adatom corral. The Fe adatoms, discretized and modeled as hard-wall finite barriers of 1.52Å diameter, are centered about positions corresponding to sites equidistant from the 2.55Å spaced nearest-neighbor hexagonal Cu(111) sites. Careful placement of the adatom sites around the corral is necessary to obtain detailed agreement with the reported experimental spatial variation of the LDOS. The LDOS was computed for three different Cu(111) substrate domains: the 48 Fe atom 143Å diameter circular corral, the 60 Fe atom 177.4Å diameter circular corral, and the 76 Fe atom 141Å× 285Å oval "stadium" corral. The only two parameters which were varied were the electron effective-mass, m^*, and the barrier height, E_B, of the Fe atom hard-wall potential.

Plotted in Fig. 1 is the computed LDOS at the center of the 48 Fe atom circular corral. The reported STM differential conductance dI/dV, adapted from Ref [3], and the computed LDOS are plotted as a function of the STM bias potential. The shaded

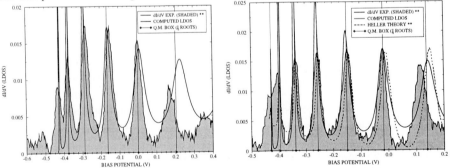

Figure 1. The LDOS and reported dI/dV measurements for a 48 Fe atom circular corral on Cu(111). ** Adapted from Ref. [3].

Figure 2. The LDOS and reported dI/dV measurements for a 60 Fe atom circular corral on Cu(111). ** Adapted from Ref. [4].

region identifies the reported experimental data, while the solid curve identifies the computed LDOS using the local NRBC method. The vertical lines, which are also marked on the axis, identify the zero angular momentum eigen-energies of the infinite hard-wall circular quantum box. The quantum box states were computed from the zeros of the Bessel function, J_0, using an effective mass $m^* = 0.38 \times m_0$. The elastic scattering calculation of the LDOS has excellent agreement with the experimental data. To achieve this agreement, an effective mass of $m^* = 0.345 \times m_0$ and an Fe barrier height of $E_B = 4.05$ eV was used. The lack of agreement of the high energy peak with the quantum box state has been attributed [3] to the motion of the Fe adatoms from their original positions, as well as possible finite barrier height effects of the Fe atoms. It is expected that the mislocation of the Fe atoms from their original corral sites would significantly change the interference behavior of the scattering states and thus change the energy spectra of the LDOS from that predicted by the NRBC calculation. The predicted low energy peaks are sharper than the experimental data, indicating the limit of the STM instrument broadening. The peak locations, however, show excellent agreement with the reported measurement.

Similarly plotted in Fig. 2 is the computed LDOS at the center of the 60 Fe atom circular corral. The reported STM differential conductance dI/dV, adapted from Ref. [4], and the computed LDOS are shown as a function of the STM bias potential. Also shown by the dashed curve is the reported multiple scattering theory result adapted from Ref. [4], while the vertical lines again represent the zero angular momentum eigen-energies of the hard-wall circular quantum box with $m^* = 0.38 \times m_0$. Again, the elastic scattering calculation of the LDOS has excellent agreement with both the peak widths and locations of the experimental data, and is comparable to the results of the alternative model of Heller *et al.* In this case an effective mass of $m^* = 0.3605 \times m_0$ and an Fe barrier height of $E_B = 3.871$ eV were used. The lowest peak is resolved as only a shoulder of the second peak in the reported experimental data. The downward shift of the high energy peak, presumably due to Fe atom displacement, is also seen.

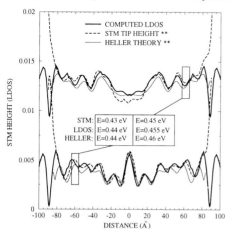

Figure 3. The spatial variation of the LDOS and reported dI/dV measurements across the diameter of the 60 Fe atom circular corral. ** Adapted from Ref. [4].

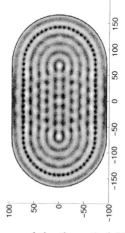

Figure 4. Image of the theoretical LDOS for the 76 Fe atom oval corral for $E = 0.45$ eV. Excellent agreement is achieved with the reported STM dI/dV image in Ref. [4].

To further demonstrate the detailed agreement of our model with the experimental data we compare the real space LDOS through the center of the 60 Fe atom circle. The LDOS at two energies (thick solid curve), offset for clarity, is plotted in Fig. 3 as a function of distance and is compared against both the experimental data (dashed line) and theoretical calculations (thin solid curve) adapted from Ref. [3]. The same effective mass and barrier height from Fig. 2 were used for the elastic scattering calculation of the LDOS. As was the case for the reported multiple scattering theory, a slightly higher energy was necessary to match the experimental data.

Finally, we demonstrate that our method reproduces the fully two dimensional image of the LDOS for the oval "stadium" shaped quantum corral reported by Heller et al. [4]. Shown in Fig. 4 is the contour shaded plot of the LDOS calculated using our elastic scattering model for an effective mass of $m^* = 0.321 \times m_0$ and an Fe barrier height of $E_B = 4.35$ eV. The calculation was performed for the same energy, $E = 0.45$ eV, as reported for the experimental measurement. The LDOS closely matches the reported image of the STM differential conductance and reproduces the fine details of the electron interference pattern. To achieve this level of agreement no energy shifting was necessary.

5. Conclusion

We have presented a method for solving the two-dimensional effective-mass Hamiltonian on open boundary domains for the local density of electronic state. The technique implements non-reflecting boundary conditions on an artificial boundary which are compatible with outward Sommerfeld radiation conditions. The LDOS was computed for the recently reported 48 Fe atom [3] and 60 Fe atom [4] circular quantum corrals and for the 76 Fe atom oval "stadium" corral [4]. Excellent agreement is achieved between our theoretical calculation and the experimental data. The results demonstrate that a purely elastic scattering model is sufficient to reproduce the details of the experimental STM data as opposed to the alternative multiple scattering theory of Heller et al.

References

[1] M. F. Crommie, C. P. Lutz, and D. M. Eigler, *Nature* **363**, 524 (1993).
[2] Y. Hasegawa and Ph. Avouris, *Phys. Rev. Lett.* **71**, 1071 (1993).
[3] M. F. Crommie, C. P. Lutz, and D. M. Eigler, *Science* **262**, 218 (1993). See also Graham P. Collins, *Physics Today* **46**, 17 (1993).
[4] E. J. Heller, M. F. Crommie, C. P. Lutz, and D. M. Eigler, *Nature* **369**, 464 (1994).
[5] S. D. Kevan and R. H. Gaylord, *Phys. Rev.* **B 36**, 5809 (1987).
[6] L. C. Davis, M. P. Everson, and R. C. Jaklevic, *Phys. Rev.* **B 43**, 3821 (1991).
[7] J. Tersoff and D. R. Hamman, *Phys. Rev.* **B 31**, 805 (1985).
[8] N. D. Lang, *Phys. Rev.* **B 34**, 5947 (1986).
[9] H. K. Harbury and W. Porod, *J. Appl. Phys.* **75**, 5142 (1994).
[10] A. Bayliss, M. Gunzburger, and E. Turkel, *SIAM J. Appl. Math.* **42**, 430 (1982).

* This work has been supported by the Office of Naval Research and the Air Force Office of Scientific Research.

† H.K.H. received support through a fellowship from the Center of Applied Mathematics of the University of Notre Dame.

Monte Carlo investigation of three-dimensional effects in sub-micron GaAs MESFETs

Shankar S. Pennathur and Stephen M. Goodnick

Department of Electrical and Computer Engineering
Center for Advanced Materials Research,
Oregon State University, Corvallis, OR 97331-3211

Abstract: We investigate three-dimensional effects in sub-micron GaAs MESFETs using PMC-3D, a parallel three-dimensional Monte Carlo device simulator. A realistic device structure is simulated to study the effects of scaling the length and width of the gate electrode of devices on the intrinsic transconductance and output impedance, computed using Monte Carlo transient statistics. Gate electrode overlap geometries and the resultant electric field variations for small gate widths could explain deviations from expected trends.

1. Introduction

Most device simulations still only consider two-dimensional device cross-sections, assuming a slow variation of electrical phenomena along the usually longer third (width) dimension [1, 2, 3]. The validity of such an assumption may not be universally true, especially given the increased integration densities of modern day commercial GaAs ICs. When widths of devices are scaled down progressively, realistic device simulations should consider not only the active device volume, but also the encompassing semiconductor geometries [4]. Furthermore, simple scaling rules for device parameters usually derived using elementary device simulation approaches such as the drift-diffusion technique start to lose validity [5]. Consequently, even the circuit simulation tools that often use such lumped parameter device models tend to be inaccurate when used to design large-scale integrated circuits. This work investigates the effects of the three-dimensional nature of small dimension devices using a sub-micron GaAs MESFET as an example. The effect of shrinking down the gate length and width on the frequency-dependent small-signal device parameters are computed using a three-dimensional Monte Carlo solver.

2. Monte Carlo device simulator

PMC-3D, a parallel Monte Carlo three dimensional device solver [6] is used to simulate MESFET geometries with different gate electrode dimensions. The Monte Carlo model for bulk GaAs includes all the pertinent scattering mechanisms and a three-valley conduction band. To compute the forces that accelerate the particles during statistically selected free-flights, Poisson's equation is solved periodically during the course of the simulation. The parallel implementation of the Poisson solver is based on an iterative method that uses an odd/even (or red/black) ordering of grid points with Chebyshev acceleration [7]. While Dirichlet boundary conditions are assumed for the three electrodes, Neumann boundary conditions are assumed elsewhere, except on the top surface wherein

Fermi-level pinning of the GaAs-air interface is assumed, resulting in a surface-potential of about 0.6 V [8]. The source and drain contacts are simulated by absorbing all the electrons that hit the electrodes and by injecting a sufficient number of electrons (according to a hemi-Maxwellian distribution) to maintain charge neutrality in the region adjacent to the electrodes. The Schottky gate (with a barrier height of 0.8 eV) is modeled by absorbing all the electrons that hit the gate. An electron that hits any other point on the boundary is reflected by reversing the component of velocity normal to the surface.

With a set of bias voltages applied to the electrodes, the current associated with an electrode is obtained using the time derivative of the equivalent integrated charge that is obtained by summing the displacement charge (due to temporal changes in the normal component of the electric field) at the electrode and the recorded net charge transfer (due to the flow of particles) across the electrode [8, 9].

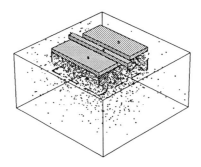

Figure 1: MESFET device structure used in this work.

The frequency-dependent small-signal parameters such as the transconductance g_m and the output impedance R_o are obtained simply by Fourier transforming the transient currents obtained at the electrodes by stepping the voltage of one electrode, while keeping the potential on the other electrodes constant [7, 8]. For example, by applying a step voltage to the *drain* electrode, the resultant drain current transient is recorded, and Fourier transformed to yield the output impedance R_o. Similarly, by stepping the *gate* voltage and tracking the drain current transient, the frequency-dependent device intrinsic transconductance g_m is obtained. While the minimum time step used in collecting the transient statistics determines the maximum frequency that can be obtained from the Fourier transform, the total time of the transient curve determines the frequency resolution attainable.

3. MESFET Device simulation

Owing to the 3-D nature of small-dimension devices, the incorporation of a realistic doping and geometry are necessary to simulate such devices. This implies that simulations need to consider not only the active device volume, but also the encompassing semiconductor geometries. In the structure simulated in this work, we consider a conventional MESFET geometry, that is surrounded by a semi-insulating substrate on all sides (refer to Fig. 1). The MESFET structure itself consists of gates 0.25-1.0 μm long, with drain-to-gate and source-to-gate distances of 0.25 μm each. The epi-layer is assumed to have a doping of 2×10^{17}/cm^{-3}, and the substrate a doping of 1×10^{15}/cm^{-3}. Gate widths between 10 μm and 40 μm are investigated.

3.1. Particle distribution

Figure 2 shows the particle distributions in steady-state for a 0.25 μm gate device, a gate bias of -0.2 V, and a drain bias of 2.0 V, for the planar structure used in this work. Also shown is the particle distribution of a recessed-gate structure, corresponding to a commercial MESFET device geometry ($L_g = 0.50$ μm, $V_d = 3.0$ V, $V_g = 0.0$ V). We see in both cases the formation of a depletion region under the gate, and the formation of a conducting channel. Velocity saturation is clearly responsible for saturation in device I-V characteristics, as opposed to channel pinch-off, for these short-channel devices. A considerable number of carriers stray into the lighter-doped substrate, thereby not participating in conducting current. While this is an effect that is not usually observable in device simulations of two-dimensional slices, it accounts for a slight reduction in the computed current in comparison with the values obtained using a more conventional structure.

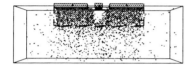

Figure 2: Particle distributions at steady-state for a planar device and a recessed-gate device.

3.2. Small-signal parameters

Figure 3a shows the variation of the DC transconductance with gate length for two different gate widths. While the transconductance scales with length in general, it saturates for very small gate lengths (below 0.25 μm). Furthermore, the transconductance does not scale linearly with width for a given gate length. This deviation from expected behavior is observed especially for smaller gate widths. For small width dimensions, due to the gate overlap region outside of the heavily doped epitaxial channel region, the electric fields are no longer strictly along the length of the channel. The resultant fringing fields could explain the deviation of of intrinsic transconductance from expected trends.

In Fig. 3b, the frequency-dependent voltage gain parameter associated with a MESFET device with a 10 μm wide gate is shown. The voltage gain is calculated as the product $g_m R_o$ of the frequency-dependent transconductance and the output impedance. A simulated unity voltage-gain frequency in excess of 100 GHz is obtained for both the 0.25 μm and 0.50 μm gate lengths.

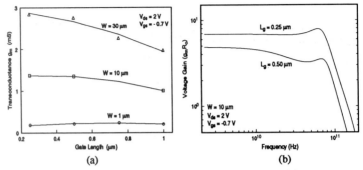

Figure 3: (a) DC transconductance variation with gate length. (b) Variation of voltage gain with frequency.

4. Conclusion

An investigation of the effects of scaling gate dimensions in a sub-micron MESFET reveals the increasing importance of three-dimensional effects for small-dimension devices. A rigorous study of such effects is crucial to the design of commercial GaAs integrated circuits.

References

[1] K. Yokoyama, M. Tomizawa, and A. Yoshii, *IEEE Electron Device Lett.*, Vol. EDL-6, **10**, 536, 1985.

[2] M. Tomizawa, K. Yokoyama, and A. Yoshii, *IEEE Trans. on CAD*, Vol. 7, **2**, 254, 1988.

[3] K. Throngnumchai, K. Asada, and T. Sugano, *IEEE Trans. on Electron Devices*, Vol. ED-33, **7**, 1005, 1986.

[4] P. Ciampolini, A. Pierantoni, and G. Baccarani, *IEEE Trans. on CAD*, **10**, 1141 (1991).

[5] K. Tomizawa, *Numerical simulation of submicron semiconductor devices*, Artech House, 1993.

[6] U. A. Ranawake, C. Huster, P. M. Lenders, and S. M. Goodnick, *IEEE Trans. on CAD*, Vol. 13, **6**, 712, 1994.

[7] R. W. Hockney and J. W. Eastwood, *Computer simulation using particles*, McGraw-Hill, 1981.

[8] C. Moglestue, *Monte Carlo simulation of semiconductor devices*, Chapman & Hall, 1993.

[9] C. Jacoboni and P. Lugli, *The Monte Carlo method for semiconductor device simulation*, Springer-Verlag, 1989.

Evaluation of electron ionization threshold energies for $Al_xGa_{1-x}As$ on a hypercube

V. Chandramouli and Christine M. Maziar

Microelectronics Research Center. BRC/MER 1.606E R9900, University of Texas at Austin, Austin, TX 78712.

Abstract. The impact ionization threshold energies for $Al_xGa_{1-x}As$ were evaluated over the entire first Brillouin zone using a 16 node Intel iPSC/860 hypercube. The bandstructures were evaluated using the empirical pseudopotential method and Anderson and Crowell's procedure was used to evaluate the ionization threshold energies. An efficient parallel sorting and searching procedure was used to obtain significant speedups in the calculation. Our results show that while the threshold energy for GaAs, $E_{th}(\mathbf{k})$, is reasonably isotropic over the Brillouin zone, it displays marked anisotropy for $Al_xGa_{1-x}As$ with aluminum mole fraction greater than 0.45 (corresponding to the transition from direct to indirect bandgap material). The results also show that while the use of a constant ionization threshold energy value for GaAs may be a good approximation in carrier transport calculations, it may be necessary to use wavevector dependent threshold energies for $Al_xGa_{1-x}As$. We believe that this is the first time that such a calculation has been performed over the entire first Brillouin zone in $Al_xGa_{1-x}As$.

1. Introduction

The importance of impact ionization in modern semiconductor devices has stimulated extensive research in recent years [1]. Many novel heterostructure devices containing AlGaAs layers have been proposed to increase the performance characteristics of devices like optical detectors, resonant tunneling diodes, electro-optic devices and bipolar transistors. Many of these devices employ thin layers, multi quantum wells (MQW) or superlattices [2]. Since high electric fields can exist in such thin layers, even for moderate applied voltages, it is important to study high energy processes like impact ionization in such material systems to optimize the device performance.

Impact ionization occurs when a carrier with sufficient energy excites an electron from the valence band to the conduction band. In the presence of a large electric field, the newly generated carriers may themselves acquire sufficient energy to initiate an impact ionization event. Early models for impact ionization were based on expressions that related a uniform electric field to the observed impact ionization rate of the material. Implicit in such models is the assumption that there exists a one-to-one correspondence between the local electric field and the carrier distribution function. This is clearly not the case for structures that exhibit electric fields varying rapidly in space. Because this process involves high energy electrons, accurate simulation of the phenomena requires detailed descriptions of the tail of the carrier distribution function and the energy band structure of the material. Microscopic models of the impact ionization phenomena are based on individual carrier energies. The minimum energy required by a carrier for impact ionization, consistent with energy and momentum conservation, is defined as the threshold energy. The threshold energy is a fundamental material parameter that appears directly in analytical expressions for the ionization rate. These expressions for ionization rates depend exponentially on the threshold energy. Hence small changes in the threshold energy can have a large effect on the calculated ionization rate. It is therefore important to have a good estimate of the threshold energy for any semiconductor material under consideration. The threshold energy is a difficult parameter to measure directly. In typical device simulation work, the threshold energy is used in conjunction with some model for the carrier distribution function (that itself is related to the electric fields found in the device structure). Because the distribution function is not well known, the treatment of the threshold energy as an adjustable fitting parameter has been deemed as somewhat justified [4]. Hence it is not surprising to find in the literature as many values for the threshold energy as there are analytical models for the ionization rates for a given semiconductor.

Since bandstructures of semiconductors take complicated forms at high energies, the calculation of threshold energies for ionization becomes a compute intensive problem. Parallel machines such as the Intel iPSC/860 hypercube are well suited to perform such computations due to the high degree of parallelism in this type of a problem. In this paper, we will describe our work in evaluating the threshold energies in $Al_xGa_{1-x}As$ using a 16 node Intel iPSC/860 hypercube.

2. Simulation Model

The threshold energy for impact ionization, E_{th} clearly depends on the bandstructure of the material. To evaluate $E_{th}(\mathbf{k})$ throughout the Brillouin zone, a realistic bandstructure must be used. Real bandstructures take complicated forms in the three dimensional $\mathbf{k}$ space. Hence determination of E_{th} from first principles is a difficult task and requires efficient algorithms. Anderson and Crowell [5] developed a method that simultaneously minimizes the energy of the initiating particle with respect to small changes to the wave vector of the final particles and conserves both the energy and momentum. Though this method can be restrictive under certain circumstances[6], it provides a straightforward approach to estimating the threshold energies throughout the Brillouin zone and produces good results for direct band gap materials like GaAs. Indirect bandgap materials, like AlAs, exhibit bands that are not as isotropic as those found in direct bandgap materials and hence it is very difficult to formulate a relationship between the directions of the final and initial states. Consider an initial electron '1' in the conduction band exciting an electron from the valence band as shown in Figure (1). The resulting electrons '2' and '3' reside in the conduction band. In this process, phonons may be either absorbed or emitted. For this process the total energy E and the total wave vector $\mathbf{k}$ of the resultant carriers can be expressed as

$$E = E_c(k_2) + E_c(k_3) - E_v(k_1) + \sum_{phonons} a_i \hbar \omega_\beta k_i \qquad (1)$$

and

$$\mathbf{k} = k_2 + k_3 - k_1 + \sum_{phonons} a_i k_i \qquad (2)$$

where E_c and E_v are the energies of the electron in the conduction and valence bands, respectively, with wave vector 'k', a_i is an integer representing the number of phonons participating in the event, $\omega_\beta k_i$ is the angular frequency for a β-phonon with wave vector k_i. For any given k, the minimization of the energy E requires phonon absorption processes only. This minimum energy for an electron (or hole) to create an electron-hole pair subject to the constraint that the energy and momentum is conserved can be evaluated using Lagrange's method of undetermined multipliers. Using this procedure, Anderson and Crowell showed that a necessary condition for an initiating particle to have a minimum energy consistent with pair

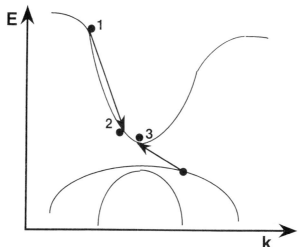

Figure 1: Schematic figure illustrating an impact ionization event. Here a high energy electron, '1', in the conduction band excites an electron from the valence band. The resulting electrons, '2' and '3' reside in the conduction band.

production is that all the resultant particles have the same group velocity. The group velocity $\vec{v}$, is given by:

$$\vec{v} = \frac{1}{\hbar} \nabla_k E(\mathbf{k}) \qquad (3)$$

where $E(\mathbf{k})$ is the dispersion relationship (bandstructure) of the semiconductor. When the dispersion relationship is known, graphical techniques can be used to determine the threshold energies. Due to the compute-intensive nature of this problem, the threshold energies have not been calculated in full three dimensions until recently. In this work, we generated the bandstructure for GaAs using the empirical pseudopotential method. The bandstructures for AlGaAs were generated using Vegard's law [3]. Since the experimental form factors needed for the pseudopotential calculations are not accurately known for $Al_xGa_{1-x}As$ for various mole fractions, we used linear interpolations to obtain the form factors. We then applied the graphical procedure as outlined in [5]. The initiating electron at wavevector $\mathbf{k}$ travels with along the direction $\mathbf{k}/|\mathbf{k}|$ [9]. Hence the threshold energy $E(k)$ is determined by the direction of the wavevector of the initiating electron. This procedure involves the evaluation of the slopes of the dispersion relationship at different wavevector positions. It also involves exhaustive searches within a given band for a given $dE(\mathbf{k})/d\mathbf{k}$. We found that by the selection and careful implementation of an efficient search technique, we can significantly speed up the computation of $E_{th}(\mathbf{k})$.

3. Parallel Implementation

Massively parallel processor (MPP) computing systems have very low 'cost per MFLOP' compared to conventional single-processor supercomputer systems. Combined with their ready scalability and ease of use, they are now an attractive choice for the implementation of parallel algorithms. We implemented a sorting/searching procedure on the Intel iPSC/860 hypercube. The sorting procedure used was a modified form of 'radix sort', which is commonly used in parallel applications, the search procedure was the standard 'binary-search' algorithm. Valiant [7] obtained a bound on the number of operations required to perform parallel searching. The complexity of this model is based on the number of computations involved and hence it provides a useful *lower bound* on the inherent computational complexity of the algorithm. Using this model it can be shown that the speedup achieved by the parallel searching algorithm 'S' is given by:

$$s = \frac{\lceil \log(n+1) \rceil}{\left\lceil \frac{\log(n+1)}{\log(P+1)} \right\rceil} \approx \log(P+1) \qquad (4)$$

where 'n' is the number of keys being searched and 'P' is the number of processors and $\lceil x \rceil$ is the 'ceil' function which returns the least integral value greater than or equal to 'x'. Hence the speedup achieved through parallelization is logarithmic only in the number of the processors used. Since this shared memory SIMD model can require as many as $\lceil \log(n+1)/\log(P+1) \rceil$ comparison steps, any model of parallel computation can require at least this many comparison steps in the worst case. From the above result, it is clear that it is not fruitful to speedup a single search on a parallel machine. Instead it is more interesting to consider speedup of several searches. In our model, we decomposed the problem so that searches on the bandstructure data are evenly divided among the various processors. We used the library function calls available on the iPSC/860 for communications between the processors and for sharing of the data. A significant speed up in the computation (five to six times) was observed on the parallel machine when compared to our simulations on an IBM RS6000/320.

4. Results and Conclusions

The results of our simulations are shown in Figures 2-5. Figure 2, in particular, provides a guide for relating the plots of threshold energy to positions in the Brillouin zone. The evolution of the threshold energy from an isotropic quantity in GaAs to one that is strongly anisotropic is readily observed. We can also see the gradual increase in anisotropy in the threshold energies in $Al_xGa_{1-x}As$ as the mole fraction gradually increases from 0.00 to 1.00. This can be explained by the fact that $Al_xGa_{1-x}As$ becomes an indirect gap material for x > 0.45. When the threshold energy is anisotropic over the Brillouin zone, the ionization cross section becomes a slowly increasing function of the energy above threshold. This is recognized as so-called 'soft' threshold [8]. From our results, it can be clearly seen that for $Al_xGa_{1-x}As$ (x > 0.45), the ionization threshold is 'soft' compared to GaAs. In summary, the impact ionization threshold energies were evaluated for $Al_xGa_{1-x}As$ on a 16 node, Intel iPSC/860 hypercube. Significant speedups were achieved in the calculation of this compute-intensive problem by using efficient parallel algorithms for searching. We believe that this is the first time that such a calculation has been performed over the entire first Brillouin zone in $Al_xGa_{1-x}As$.

Acknowledgments -We would like to thank Prof. Robert Van de Geijn in the Department of Computer Sciences, University of Texas at Austin for his suggestions for efficient parallel algorithms. This work was supported (in part) by the NSF Science and Technology Center Program, NSF grant CHE 8920120, the Texas Advanced Research Program (TARP), the Joint Services Electronics Program F49-620-92-C-0027 and through an Intel equipment grant.

References

[1] P J van Hall and E A E Zwaal 1993 *Superlattices and Microstructures* **13** 3 323
[2] S L Feng, J Krynicki, M Zazoui, J C Bourgoin, P Bois and E Rosencher 1993 *J. Appl. Phys.* **74** 1 341
[3] R E Nahory, M A Polack, W D Johnston and R L Barns 1978 *Appl. Phys. Lett.* **33** 7 659
[4] F Capasso, K Mohammed and A Y Cho 1985 *J. Vac. Sci. Tech.* B **3** 1245
[5] C L Anderson and C R Crowell 1972 *Phys. Rev.* B**5** 2267
[6] J Bude and K Hess 1992 *J. Appl. Phys.* **72** 8 3554
[7] L G Valiant 1975 *SIAM J. Comput.* **4** 348
[8] A R Beattie 1988 *Semicond. Sci. Tech.* **3** 48
[9] N Sano, T Aoki and A Yoshii 1989 *Appl. Phys. Lett.* **55** 14 1418

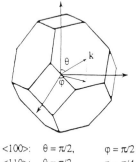

<100>:	θ = π/2,	φ = π/2
<110>:	θ = π/2,	φ = π/4
<111>:	θ = 54.74°,	φ = π/4

Figure 2: Schematic figure illustrating the definition of θ and φ inside the Brillouin zone.

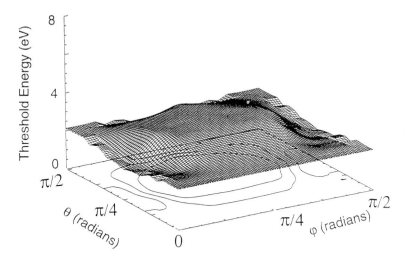

Figure 3: Electron ionization threshold energies for $Al_xGa_{1-x}As$ (x=0.0)

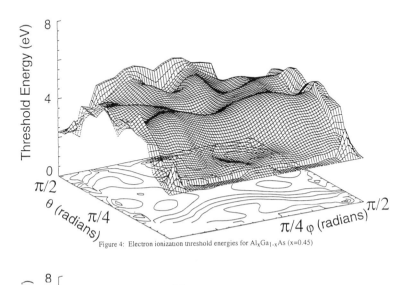
Figure 4: Electron ionization threshold energies for $Al_xGa_{1-x}As$ (x=0.45)

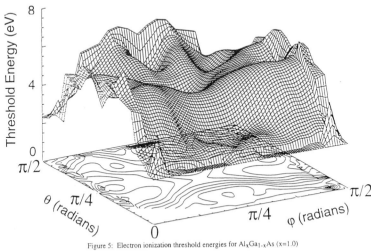
Figure 5: Electron ionization threshold energies for $Al_xGa_{1-x}As$ (x=1.0)

Ensemble Monte Carlo Calculation of the Hole Initiated Impact Ionization Rate in Bulk GaAs and Silicon Using a k- Dependent, Numerical Transition Rate Formulation

Ismail H. Oguzman, Yang Wang, Jan Kolnik, and Kevin F. Brennan

School of Electrical and Computer Engineering, Georgia Tech, Atlanta, GA USA 30332-0250

Abstract. The hole initiated impact ionization rate in bulk silicon and GaAs is calculated using a numerical formulation of the impact ionization transition rate incorporated into an ensemble Monte Carlo simulation. The transition rate is calculated from Fermi's golden rule using a two-body screened Coulomb interaction including a wavevector dependent dielectric function. It is found that the effective threshold for hole initiated ionization is relatively soft in both materials, that the split-off band dominates the ionization process in GaAs, and that no clear dominance by any one band is observed in silicon, though the rate out of the light hole band is greatest.

1. Introduction

The study of interband impact ionization is of great importance to understanding the operation of most semiconductor devices. In many structures, such as field effect transistors and most bipolar transistors, impact ionization can lead to deleterious effects greatly reducing the overall performance of these devices. In other structures, such as avalanche photodiodes, APDs, and IMPATT diodes the control of impact ionization is essential in producing an optimized device. Subsequently, the design and simulation of many important semiconductor structures requires an accurate description of impact ionization.

The early theories of impact ionization [1,2] provided a means of understanding the underlying physics of the process, but have limited usefulness for predicting the ionization rate in a device structure. Major strides in improving the theory of impact ionization were first taken by Shichijo and Hess [3] by incorporating the full details of the energy bandstructure into the calculation of the ionization rate. However, their method and those of others [4] used a simplified formula for the ionization transition rate, the Keldysh formula [5]. Though the Keldysh formula has been commonly used it has several major limitations. Among these are its failure to include any wavevector dependence of the ionization rate, its reliance on two fitting parameters, and its derivation based on parabolic energy bands.

Recently, there has been great progress in constructing an accurate theory of the impact ionization transition rate which overcomes the limitations of the Keldysh formula [6-16]. in most of these theories the transition rate is numerically evaluated enabling the inclusion of the full details of the bandstructure through a direct calculation of the transition rate without any parametrization. These theories have enabled a much greater understanding of the physics of impact ionization. To the authors' knowledge, the numerical theories have all concentrated on electron initiated ionization and no work has been done on hole initiated impact ionization. It is the purpose of this paper to present the first calculations of the hole initiated impact ionization rate in bulk silicon and GaAs using a numerical formulation of the transition rate.

2. Model description

The calculations are performed using an ensemble Monte Carlo simulator which includes the full details of the valence bands and the phonon scattering mechanisms. The model incorporates two advanced features beyond that typically used [3,4]. The first improvement is the inclusion of a completely numerical technique for calculating the phonon scattering rates, i.e., acoustic, nonpolar and polar optical, which accounts for the anisotropy and warping of the energy bands [17]. The second modification is the inclusion of a fully numerical treatment of the ionization transition rate including a wavevector dependent dielectric function. The dielectric function is determined using the model dielectric function of Levine and Louie [18]. The background carrier concentration is assumed to be 5.0×10^{16} cm^{-3}.

The transition rate is evaluated following the approach of Wang et al. [7,8] used previously for electron initiated ionization. The hole initiated ionization rate is determined for the heavy, light and split-off valence bands at each mesh point within a finely spaced, 1419 point, k-space grid spanning the reduced zone of the first Brillouin zone and is incorporated into the Monte Carlo simulator. By performing an additional averaging over energy, further information about the ionization transition rate can be gleaned. It is found that the energy dependent transition rates are comparable among all three valence bands at low energy in silicon. At higher energy, the transition rate within the split-off band is only slightly larger than the corresponding rate within the light hole band, but the rates within both of these bands are very much larger than that within the heavy hole band. In contrast, in GaAs, the energy dependent ionization rate within the split-off band is substantially larger than that in either the light or heavy hole bands. The rate within the heavy hole band in GaAs is particularly low.

3. Calculated results

The calculated hole initiated ionization rate as a function of inverse electric field along the (100) direction in bulk silicon is plotted in Figure 1. Experimental measurements [19-21] which comprise a representative set of data are plotted in Figure 1 as well. As can be seen from the figure, the calculations lie within the range of the experimental measurements and are in particularly good agreement with Overstraeten's results [19]. It is further found that most of the ionization events (between 60% and 50 % of the total) in silicon originate from within the light hole band at all of the applied fields considered. At higher electric field strengths, the contribution from the split-off band increases significantly, to ~35%, while that from the heavy hole band decreases , to ~10%.

The calculated hole impact ionization rate in bulk GaAs along the (100) direction as a function of inverse electric field is plotted in Figure 2 along with the experimental measurements of Bulman et al. [22]. Again fair agreement between the calculations and the experimental results are obtained. In GaAs, owing to the much higher transition rate within the split-off band, the vast majority of hole initiated events, from ~60% at 250 kV/cm to ~80% at 500 kV/cm, originate from within the split-off band. Less than 5% of the ionization events originate from within the heavy hole band at all electric fields in GaAs.

The nature of the effective threshold energy can be ascertained by calculating the quantum yield. The quantum yield is plotted as a function of hole injection energy in both GaAs and silicon in Figure 3. As can be seen from Figure 3, 100% hole ionization occurs at ~5 and ~4 eV in silicon and GaAs respectively. For both materials notice that the quantum yield is appreciably less than 1 at energies less than 5 and 4 eV respectively indicating that the

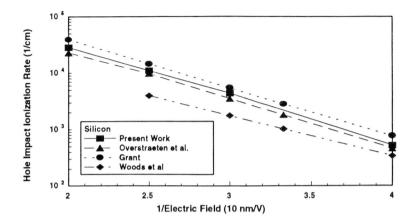

Figure 1: Hole impact ionization rate in silicon as a function of inverse electric field.

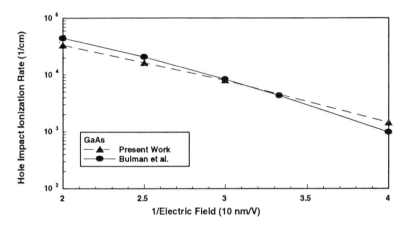

Figure 2: Hole impact ionization rate in GaAs as a function of inverse electric field.

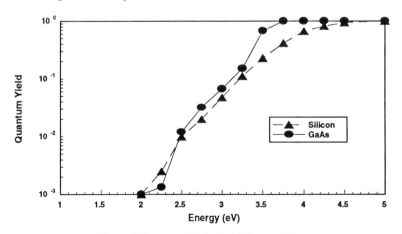

Figure 3: Quantum yield for both Silicon and GaAs.

holes do not necessarily ionize at these energies. This result implies that the effective hole ionization thresholds in both silicon and GaAs are relatively soft. Additionally, owing to the fact that the quantum yield saturates sooner in GaAs than in silicon, the threshold is apparently somewhat harder in GaAs than in silicon.

4. Conclusions

In this paper, the first calculations of hole initiated impact ionization in bulk GaAs and silicon using a complete numerical formulation of the transition rate have been presented. It is found that the effective hole ionization threshold energies in both materials are relatively soft. Comparison of the quantum yield data for holes to that for electron initiated ionization [23] shows that the threshold energy is relatively soft for both carrier species in these materials. Interestingly, the electron ionization threshold in GaAs is also somewhat harder than the electron ionization threshold in silicon as was determined in references 7 and 8. It is further found that the vast majority of ionization events originate from within the split-off valence band in GaAs while no band clearly dominates the ionization process in silicon.

Acknowledgements

This work was sponsored in part by the National Science Foundation through a PYI Award made to K. Brennan, through Bell Northern Research PYI matching grant I25A3J, by ARPA through contract to NASA, NAGW-2753 and by the National Science Foundation through grant E21-X63.

References

[1] Shockley W 1961 *Solid-State Electron.* **2** 35-67
[2] Wolff P A 1954 *Phys. Rev.* **95** 1415-1420
[3] Shichijo H and Hess K 1981 *Phys.Rev.B* **23** 4197-4207
[4] Fischetti M V and Laux S E 1988 *Phys. Rev. B* **38** 9721-9745
[5] Keldysh L V 1965 *Zh. Exp. Teor. Fiz.* **48** 1692-1707 (1965 *Sov. Phys. JETP* **21** 1135-1144)
[6] Bude J and Hess K 1992 *J. Appl. Phys.* **72** 3554-3561
[7] Wang Y and Brennan K F 1994 *J. Appl. Phys.* **75** 313-319
[8] Wang Y and Brennan K F 1994 *J. Appl. Phys.* **75** 974-981
[9] Quade W Rossi R and Jacoboni C 1992 Semicond. Sci. Technol. **7** B502-B505
[10] Yoder P D, Higman J, M Bude J and Hess K 1992 *Semicond. Sci. Technol.* **7** B357-B359
[11] Sano N and Yoshii A 1992 *Phys. Rev. B* **45** 4171-4180
[12] Kunikiyo T, Takenaka M, Kamakura Y, Yamaji M, Mizuno H, Morifuji M, Taniguchi K and Hamaguchi C 1994 *J. Appl. Phys.* **75**, 297-312
[13] Kamakura Y et al. 1994 *J. Appl. Phys.* **75**, 3500-3506
[14] Stobbe M, Konies A, Redmer R, Henk J and Schattke W 1991 *Phys. Rev. B* **44** 11105-11110
[15] Stobbe M, Redmer R and Schattke W 1994 *Phys. Rev. B* **49** 4494-4500
[16] Quade W, Schöll E and Rudan M 1993 *Solid-State Electron.* **36** 1493-1505
[17] Hinckley J M and Singh J 1990 *Phys. Rev. B* **41** 2912-2926
[18] Levine Z H and Louie S G 1982 *Phys. Rev. B* **25** 6310-6316
[19] Van Overstraeten R and De Man H 1970 *Solid-State Electron.* **13** 583-608
[20] Grant W N 1973 *Solid State Electron.* **16** 1189-1203
[21] Woods M H, Johnson W C and Lambert M A 1973 *Solid State Electron.* **16** 381-394
[22] Bulman G E, Robbins V M and Stillman G E 1985 *IEEE Trans. Electron Dev.* **ED-32** 2454-2466
[23] Kolník J, Wang Y, Oğuzman I H and Brennan K F 1994 *J. Appl. Phys.* **76** xxxx-xxxx

Analysis of a Heterojunction MOSFET Structure for Deep-Submicron Scaling

Scott A. Hareland, Al F. Tasch, and Christine M. Maziar

Microelectronics Research Center, PRC/MER 2.608B, MS R9950, University of Texas at Austin, Austin, TX 78712.

Abstract. As silicon-based MOSFETs used in CMOS ULSI circuits are scaled to physical gate lengths below 100nm, it becomes much more difficult to maintain the requirement for acceptable standby power because of the problem of increasing off-state leakage currents. In addition, conventional scaling techniques will not be able to offset device leakage currents indefinitely without seriously degrading other aspects of device performance. This paper focuses on a novel approach that makes use of a heterojunction barrier at the source/channel junction in order to control off-state leakage current and reduce the effects of drain-induced barrier lowering (DIBL). Results of two-dimensional numerical simulations of the new structure will be discussed and compared with conventional silicon MOSFET structures. The device will be illustrated with the Ge_xSi_{1-x}/Si heterojunction system, but additional material systems will be discussed.

1. Introduction

In deep submicron silicon MOSFETs for ULSI-generation circuits, it is becoming increasingly difficult to control the off-state leakage currents in the device without degrading other key device performance characteristics. This is one of the major obstacles to continued scaling of MOSFET devices [1]. With the demand for improved low power, portable electronics, the minimization of transistor off-state leakage is crucial for success. Physically, as devices are scaled into the deep submicron regime, the depletion regions formed at the source and drain sides of the device can overlap in the channel and allow the electric field from the drain to reach across the channel and lower the barrier to the source carriers, thereby causing an increase in the drain leakage current. This is a phenomena known as drain-induced barrier lowering (DIBL). Since the number of source carriers that are able to surmount the potential energy barrier is exponentially dependent on the barrier height, any amount of barrier lowering due to changes in the drain bias can greatly increase the total device leakage. In addition, because the drain is able to influence the off-state leakage current, the MOSFET is no longer exclusively a gate controlled transistor, and the electrical characteristics can deteriorate considerably, especially when channel lengths decrease to and below 100nm.

Conventionally, the control of the off-state and subthreshold characteristics of MOSFETs has been straightforward and has involved a combination of higher channel doping levels, thinner gate oxides, and shallower junction depths as the device dimensions were decreased. The primary purpose of these approaches was to maintain an adequate potential energy barrier between the source and the drain. These approaches were developed to minimize off-state leakage currents, while giving the gate, and not the drain, control of the channel potential. The higher channel doping concentration prevents the depletion regions from the source and the drain sides from overlapping in the channel and causing DIBL effects such as a lowering of the threshold voltage

(ΔV_T). However, with the reduction in total device area, these scaling methods are approaching limits, introducing additional physical effects that degrade other aspects of the device performance and pushing the edge of reliable manufacturing technology. Higher levels of channel doping degrade the inversion layer mobility and start to introduce quantum mechanical shifts that increase the threshold voltage [2]. In addition, it has recently been shown that significant statistical variations in the threshold voltage arise due to statistical variations in the number and location of the channel dopant atoms as the channel volume becomes very small. This can greatly complicate device, circuit, and system performance [3,4]. It is advantageous to maintain the channel doping as low as possible to avoid these effects. One example of a fundamental limit is the minimum gate oxide thickness that is widely accepted to be limited to about 3nm because of direct tunneling from the substrate to the gate.

In order to overcome these scaling limitations, a new structural concept has been introduced that makes use of a heterojunction to provide a potential energy barrier between the source and the drain that is not easily lowered by application of a drain bias [5]. Control of the off-state leakage of a device with the built-in band offset of a heterojunction provides an additional degree of freedom in designing deep submicron MOSFETs and helps avoid some of the problems with the traditional scaling approaches. Previous simulation results indicated that there was indeed a reduction in off-state leakage current when the heterojunction was used to enhance the total barrier height from the source to the drain. In this work, a discussion of heterojunction MOSFET scaling to ultra-small (<100nm) dimensions will be discussed and different structural options will be presented. Numerical simulations of these devices have been performed with ATLAS/BLAZE[6] from SILVACO using the heterojunction transport equations and a representative suite of MOSFET models, such as advanced inversion layer mobility and bandgap narrowing in the heavily doped source and drain regions.

2. Scaling Approaches

The basic failure of the homojunction MOSFET at extremely short gate lengths (50nm in these simulations) is readily seen in Figure 1. In this particular device (t_{ox}=4nm, x_j=20nm, N_{sub}=5x10^{16}cm^{-3} at the surface, N_{sub}=1x10^{18}cm^{-3} 10nm below the interface, and N_{sub}=10^{17}cm^{-3} in the bulk), the insufficient doping in the channel region is unable to prevent the lateral spread of the depletion regions from overlapping in the channel. This reduces the potential energy barrier between the source and drain and causes a large off-state leakage current to flow. As the drain bias is increased from -0.1 to -1.0V, the potential energy barrier is further lowered due to the drain electric field and the peak of the potential energy barrier is shifted towards the source. Since

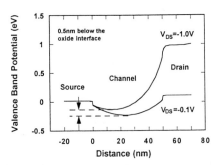 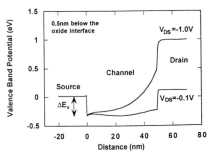

Fig. 1: Drain-Induced Barrier Lowering (DIBL) in a deep submicron (L_{eff}=50nm) MOSFET. The applied drain bias will increase the off-state leakage current and lower V_T.

Fig. 2: A heterojunction MOSFET similar in structure and doping to the conventional MOSFET but with $Ge_{0.4}Si_{0.6}$ source/drain regions. In this simulation, the total barrier height to holes near the surface is not influenced by the drain bias.

the amount of off-state leakage is exponentially dependent on the barrier height, this barrier lowering effect is undesirable in CMOS ULSI circuits. In a device such as this, unacceptably high values of off-state leakage would prevent this device from being used in low power applications and would make the threshold voltage and logic levels difficult to control.

In contrast to the conventional MOSFET, the heterojunction MOSFET (HJMOSFET) shown in Figure 2 has the same oxide thickness, junction depth, and doping parameters as the conventional MOSFET, but does not rely on the pn junction doping to provide the potential energy barrier that is necessary for acceptable subthreshold performance. In this case, the source/drain regions are $p^+Ge_{0.4}Si_{0.6}$ on an n-type Si substrate so the built-in band offset of the heterojunction provides a potential barrier of about 0.3eV and, most importantly, the barrier is not easily influenced by the bias applied to the drain side of the device. With such a barrier, the off-state leakage current is limited by thermionic emission over the barrier. Furthermore, as the bias is increased on the drain side, the total barrier height seen by the carriers in the source (holes in this case) is not significantly influenced by these electric fields. In this case, the off-state leakage current can be directly controlled by the choice of the heterojunction band offset, or a Schottky barrier workfunction difference on the semiconductor. This approach reduces the need for extremely high doping levels in the channel region that significantly compromise device performance at deep submicron dimensions. Figure 3 shows the off-state condition of two HJMOSFETs with different barrier heights. For the purposes of illustration in this figure, the built-in band offset has been raised by introducing a larger mole fraction of Ge into the source/drain material. In fabricated devices, when they become available, it will be necessary to observe constraints regarding the critical layer thickness of GeSi on Si due to the lattice mismatch of the two materials. It may be possible with sufficiently thin GeSi layers that mole fractions on the order of 0.4 to 0.5 Ge could potentially be achieved. If these issues are not observed, the benefits gained by the larger barrier height may be bypassed because of the large number of misfit dislocations that will increase the leakage current through interface defects. However, the role of defects and material properties in the transport of carriers across heterojunctions is not fully understood for GeSi/Si systems at this point and is an area of active research.

The use of strained Ge_xSi_{1-x} material in the source and drain regions on a Si substrate, limits hole injection and is only useful for p-channel MOSFETs. For n-channel devices, suitable material systems with band offsets in the conduction band are necessary. One potential candidate for NMOS devices is strained Si on relaxed GeSi since it exhibits a conduction band offset. An additional design constraint is the requirement that the band offset prevent the source carriers from entering the channel. One such potential system would be the use of Si source and drain regions

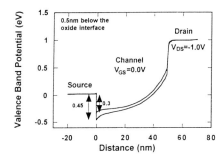

Fig. 3: Comparison of two heterojunction MOSFETs with different band offsets. Off-state leakage can be controlled by the size of the band offset, thereby bypassing the need for extremely high channel doping concentrations.

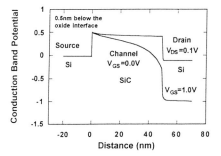

Fig. 4: Illustration of an n-channel heterojunction MOSFET using Si source/drains on a SiC epi-layer. The appropriate conduction band offset makes this one potential candidate for NMOS devices.

on a SiC epi-layer. Applying the Anderson electron affinity rule suggests that the conduction band offset (ΔE_c) is approximately 0.3-0.5eV. Because silicon has a larger electron affinity than SiC (approximately 4.1eV vs. 3.7eV for 6H-SiC), electrons in the source will be prevented from entering the SiC channel because of the conduction band offset. If SiC were used in the source/drain regions, the band offset would allow carriers to flow readily from the source into the channel. Figure 4 illustrates the barrier provided to electrons in this SiC/Si device.

The heterojunction MOSFETs illustrated above have a heterojunction at both the drain and the source ends of the device. For most practical purposes, the heterojunction really only needs to reside at the source end of the device because, with most device applications, the drain is usually biased slightly higher than the source, even when it is "off". For this reason, simulations were performed to examine the influence of the barrier to source carriers for both cases when the drain side of the device had a heterojunction barrier and the case where it was simply an extension of the material used in the channel with the appropriate doping. Figure 5 shows a double heterojunction MOSFET ($Ge_xSi_{1-x}/Si/Ge_xSi_{1-x}$) compared to a Ge_xSi_{1-x}/Si source HJMOSFET. From these simulations, it appears that the addition of the heterojunction at the drain does not play a significant role on the shape and height of the potential barrier at the source end of the device. In addition to the barrier at the source end, the device with a homojunction at the channel/drain interface may allow carriers to flow more easily in the on-state (i.e. when both the gate and the drain are biased appropriately). As will be shown, the presence of the heterojunction barrier at the drain end may reduce the drain current by hindering carriers from surmounting the small barrier near the drain due in part to the band bending near the drain heterojunction. For the purposes of reducing the overall device series resistance, it may be advantageous to eliminate the heterojunction at the drain, but this issue must be resolved with further simulations. In addition, the potential profile in the double heterojunction device is more localized near the drain. This may act as a "ballistic launching ramp" for carriers into the drain and away from the oxide. It should also be noted that by introducing an asymmetrical device, circuit design becomes more complicated. This fact may make the asymmetrical device undesirable for ULSI applications.

In some instances, the total barrier height to source carriers can be comprised of a barrier due to the heterojunction plus an additional barrier due to the doping profile between the source and the channel. Figure 6 shows a device with a larger channel doping level (uniform $10^{18} cm^{-3}$) to introduce an enhanced barrier height to source carriers near the interface. From an off-state

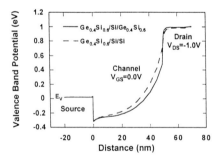

Fig. 5: Comparison of a double heterojunction (source and drain) MOSFET and a single heterojunction (source only) MOSFET. The total barrier height to source carriers is not influenced by the condition of the barrier at the drain side of the device. Trade-offs concerning series resistance, asymmetrical devices, and possible ballistic effects must be further investigated in this single heterojunction structure.

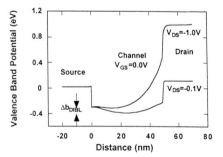

Fig. 6: A heterojunction MOSFET with a high channel doping concentration. The total barrier height to source carriers is composed of the heterojunction plus a barrier due to the doping. Even though the total barrier is higher, the barrier due to the doping is susceptible to drain-induced barrier lowering and high levels of channel doping are not desirable in deep submicron MOSFETs.

leakage point of view, this additional barrier further decreases the amount of off-state leakage current in these deep submicron MOSFETs. However, one of the advantages of the HJMOSFET illustrated in Figure 2 was the higher immunity to drain-induced barrier lowering because the barrier height was constant at the level of the heterojunction band offset, despite the applied drain bias. In this device, with higher N_{sub}, the barrier is susceptible to lowering by the drain electric field. Even though the total barrier height is larger and leakage currents should be lower, the ability of the drain to shift the threshold voltage and increase device leakage current must be considered for each particular device application. In addition, one reason for examining the HJMOSFET is its potential advantage in decreasing the total channel doping concentration that can lead to mobility reduction, increases in V_T due to quantum confinement effects, and statistical fluctuations due to the number and location of impurity atoms. It is advantageous to engineer the structure to provide the turn-off with the heterojunction barrier.

3. Drive Current Considerations

In addition to the need for low off-state leakage, high drive currents are necessary for improving circuit speed and fanout of devices. Conventional MOSFETs are based on the principle that a large gate bias will lower the potential energy barrier such that source carriers can be readily injected into the channel to form an inversion layer of minority carriers that are then swept to the drain by the lateral electric field. In the HJMOSFET, the existence of the heterojunction barrier is not expected to limit the drive current by thermionic emission because carriers are expected to be able to quantum mechanically tunnel through the thin barriers formed in the valence or conduction bands with the application of large gate biases.

Since the off-state leakage current is controlled by the height of the heterojunction band offset and the drive current is dependent on the barrier thickness, which is controlled by the strength of the electric field perpendicular to the interface, it is reasonable to conclude that the threshold voltage of the HJMOSFET can be controlled by choosing a gate with a suitable workfunction. In such a device, the gate electrode should be chosen to bend the valence or conduction bands sufficiently in order to permit significant tunneling at a specified gate voltage. Once carriers tunnel through the thin potential energy barrier, they will be swept to the drain by

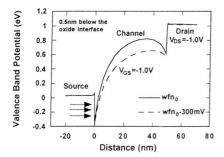
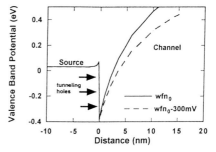

Fig. 7: Illustration of a heterojunction MOSFET under large gate and drain bias. The spike formed in the valence band is thin enough to allow quantum mechanical tunneling of carriers from the source to the channel. Threshold voltage of heterojunction MOSFETs should be controlled by using gate materials with appropriate workfunctions to shift the flatband voltage and allow the appropriate amount of tunneling to occur at a specified gate voltage. The heterojunction at the drain end may also alter the carrier transport.

Fig. 8: Closer view of the valence band spike near the source in a device with a large gate and drain bias. The higher workfunction (wfn_0) gate material makes the barrier thinner for a given gate voltage. The tunneling probability through such a barrier is exponentially dependent on the barrier thickness at a given kinetic energy (E). Therefore, the device with the larger workfunction would have a lower V_T.

the large electric fields. In this way, the reliance on the channel doping to control the threshold voltage is no longer necessary. Figure 7 shows the valence band potential when the drain and the gate are both biased at -1.0V in the GeSi/Si device and two different gate materials are used, one with a workfunction 300mV larger than the other. The band bending near the drain of the device indicates that, if the double heterojunction device is used, a larger drain series resistance may be encountered. More accurate simulations of carrier transport through this device are required for better understanding of this feature. A close-up view of the barrier (Figure 8) indicates that for this particular simulation, the valence band spike is only about 2-3nm thick at its widest point. Using the WKB approximation, a carrier in the source with kinetic energy E in the direction perpendicular to the barrier tunnels through a triangular potential with a probability of

$$P(E) = \exp\left(\frac{-4x_0}{3\hbar}\sqrt{2m^*(\Delta E_v - E)}\right) \quad (1)$$

where E is the kinetic energy, ΔE_v is the heterojunction barrier height, and x_0 is the effective barrier thickness for a carrier with energy E. Presently a tunneling current model is being implemented in a Beta version of ATLAS/BLAZE in order to self-consistently account for tunneling current through the use of physical models for carrier occupancy, density of states, tunneling probability, and carrier velocity. Once this model is fully implemented, more rigorous investigations of the on-state of the HJMOSFET can be investigated.

4. Conclusions

As MOSFET devices are scaled, it is important to consider alternative structures and scaling concepts as the conventional scaling limits for MOSFETs are approached. An overview of possible structural approaches for MOSFETs using heterojunctions has been introduced and scaling considerations have been discussed. Even though the primary illustration of this concept has been through the use of GeSi/Si heterojunction system, other material systems and Schottky barriers, such as WSi_2 or $TiSi_2$ on Si, have potential applications in this type of device structure. Schottky barrier devices may have the additional advantage of reducing the interconnect resistance by bonding directly to the source and drains of the MOSFET. Once more sophisticated models for the tunneling current and device transport at deep submicron dimensions become available, refinements to the proposed structures can be investigated.

Acknowledgments -This work was supported by the Semiconductor Research Corporation and the Texas Advanced Technology Program.

References

[1] J.M. Pimbley and J.D. Meindl 1989 *IEEE Trans. on Electron Devices* **36** 9 1711.
[2] M.J. vanDort et al. 1992 *IEEE Trans. on Elec Dev* **39** 4 932.
[3] Mizuno et al. 1993 *Symposium on VLSI Technology* 41.
[4] H.-S. Wong and Y. Taur 1993 *IEDM Technical Digest* **93** 705.
[5] S. A. Hareland et al. 1993 *Electronics Letters* **29** 1894
[6] SILVACO 1994 *ATLAS/BLAZE User's Manual Ver.2.0*

Full Scale Simulation of Quantum Devices with Interface Disorder and Randomness

J.P. Leburton, D. Jovanovic[1] and I. Adesida

Beckman Institute for Advanced Science and Technology, University of Illinois, Urbana, IL 61801

We present a three dimensional self-consistent numerical analysis of the transport properties in quantum dot nanostructures. It is shown that structural disorder in the form of interface and surface roughness gives rise to resonant features that strongly resemble the experimental data. We thereby demonstrate that disorder mechanisms in planar nanostructures have a profound influence on near-threshold carrier transport to the extent that they may dominate the effects of intentionally fabricated confinement features.

1. Introduction

Recent advances in semiconductor fabrication techniques have resulted in a large variety of new experimental devices exhibiting quantum effects due to high degrees of confinement. These devices have generated significant interest because they exhibit strong Coulomb blockade effect [2] and show promise for robust resonant tunneling applications. Although analytical [3,4] and numerical [5] models have successfully explained the conductance characteristics observed at cryogenic temperatures, several features of the experimental data have not been analyzed in detail and their explanation to date remains inconclusive [6,7,8]. Specifically, fluctuations in the conductance amplitudes with gate bias implies the presence of some type of disorder which influences the transport properties of the structure.

In order to provide a realistic model for the physical analysis and design of these structures, we have developed a comprehensive self-consistent simulation tool that merges the statistical and quantum mechanical aspects of the problem. The model takes into account the various regions of reduced dimensionality throughout a particular structure which can consist of 0D, 1D, 2D electron gases or bulk. High order effects such as exchange/correlation and interface disorder are also included in the model.

The purpose of the present paper is to analyze the transport characteristics of lateral quantum dot devices and assess the influence of structural disorder on their transport properties. Although, the theoretical investigation of ionized dopant randomness by Davies and Nixon [9] leaves no doubt of its importance on the electronic properties of quantum devices, we omit this disorder mechanism due to difficulties in capturing its essential features with a coarse computational mesh. Instead, we show that interface and surface roughness are sufficient to account for the particular resonant transport features observed in experimental devices. Our results can be generalized to other disorder mechanisms (i.e., ionized dopant randomization) which are not directly included in our model but which undoubtedly contribute to quantum transport in an analogous fashion.

2. Theoretical Background

We obtain the electronic properties of the system by solving Schrödinger equation with the iterative extraction orthogonalization method (IEOM) that propagates a set of eigenstates according to [5,10]

$$|\phi_n^i\rangle = e^{-\alpha \hat{H}}|\phi_n^{i-1}\rangle \quad , i^{th} \text{ iteration} \tag{1}$$

where $\hat{H}$ is the Hamiltonian and n indicates a particular eigenstate. The parameter α is selected so as to minimize the error in the Taylor expansion of the exponential operator $exp(-\alpha\hat{H})$. In general many iterations are required to maintain the accuracy of the Taylor expansion and eliminate all the error projections of the basis states $|m\rangle$ of $\hat{H}$ onto the initial guess state $|\phi_n^0\rangle$. Gram-Schmidt orthonormalization is performed over the entire set of states after each iteration to prevent any excited states from collapsing to the ground state. This procedure also eliminates all projections of states with $m < n$ such that after N_i iterations,

$$|\phi_n^{N_i}\rangle \simeq |n\rangle + \sum_{m=n+1} |m\rangle e^{-\alpha N_i (E_m - E_n)} \frac{\langle m|\phi_n^0\rangle}{\langle n|\phi_n^0\rangle}. \tag{2}$$

The repeated exponential scaling of the error projections and Gram-Schmidt orthonormalizations therefore eventually convert $|\phi_n^0\rangle$ into a pure basis state $|n\rangle$

In general, each application of the propagator scales with N_G and the Gram-Schmidt algorithm scales with $N_G N_E^2$ where N_E is the number of eigenenergies required in a particular simulation. If N_E is relatively small, as is often the case in nanostructures, $N_G N_E^2 \ll N_G^3$ and the IEOM shows a significant improvement over conventional eigenvalue solvers. The chief advantage offered by the rapid convergence of this method is the ability to accurately treat the dimensionality of each region in a quantum device out to an appropriate set of boundaries. By using a self-consistent scheme for the electrostatic Hartree potentials, we solve the 3D Schrödinger equation in the quantum dot (0D) regions, the 2D Schrödinger equation in the lead (1D) regions, and 1D solutions are obtained in the reservoir (2D) regions. Semiclassical solutions are used for carriers that do not exhibit confinement (holes) and regions that are not governed by quantum mechanics with the charge densities. In addition, exchange and correlation effects are self-consistently treated with the Kohn-Sham approach using Perdew-Zunger parameterization for the correlation potential [11]. Presently, the simulation is carried out in equilibrium although in principle, nonequilibrium solutions are possible for accurate evaluations of the quantum-mechanical current.

The computational mesh and boundary conditions are cast to accurately model the various material properties and overall 3D device geometry. A nonuniform mesh is used in the vertical direction (heterojunction) to allow refinement of the mesh width down to 5 Å to accurately model the disorder at the GaAs-$Al_{0.3}Ga_{0.7}$As interface. The disorder is created by allowing interpenetration of GaAs mesh "tiles" into the $Al_{0.3}Ga_{0.7}$As and vice versa. Mesh tiles of various lateral dimensions (200 Å× 200 Å, 800 Å× 800 Å, and 1600 Å× 1600 Å) are randomly assembled along the interface to create disordered interface clusters. Similarly, surface roughness is modeled by spatially varying the thickness of the surface boundary tiles over $\simeq 100 nm$. This approach results in a spatial variation of the Fermi level pinning boundary condition which subsequently generates long-range

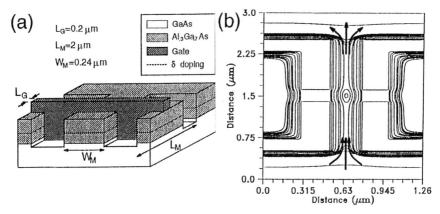

Figure 1: (a) Schematic configuration for the experimental planar thin-gated quantum dot device. (b) Potential contours calculated in a plane just below the $GaAs - Al_{0.3}Ga_{0.7}As$ interface. The arrows indicate the direction of current flow.

randomization of the confining potential. We note that both the disorder mechanisms in our model experience screening since they are treated self-consistently within our formalism.

A transfer matrix calculation is used to generate the transmission probability $|T(\epsilon)|^2$ over the relevant energy range. The finite temperature current, I, is finally obtained by inserting $|T(\epsilon)|^2$ into the standard tunneling current formula: [12]

$$I = \frac{2e^2}{h} \int d\epsilon \, |T(\epsilon)|^2 \, [f(\epsilon) - f(\epsilon + V_{DS})]. \qquad (3)$$

where e is the electron charge, h is Planck's constant, and $f(\epsilon)$ is the Fermi-Dirac distribution, and V_{DS} is the drain source voltage.

3. Experimental Device

The experimental device examined here is a thin-gated mesa-etched quantum-dot nanostructure. A cross-section of the device and a set of theoretical potential energy contours are shown in Fig. 1. Briefly, the heterostructure was grown by molecular beam epitaxy (MBE) on an undoped GaAs substrate. A 1 μm thick GaAs buffer layer was grown on the substrate followed by a 17.5 nm undoped $Al_{0.3}Ga_{0.7}As$ spacer, a Si δ-doping of 5×10^{12} cm^{-2}, a 20 nm layer of $Al_{0.3}Ga_{0.7}As$, and a 20 nm n$^+$-GaAs cap layer. Hall measurements on the sample etching indicated a sheet carrier density of $N_s = 4.7 \times 10^{11}$ cm^{-2} and a mobility of $\mu_s = 5.0 \times 10^5$ cm^2/Vs at T=4.2 K. A set of I-V characteristics of this device is shown in Fig. 2 for various drain-source biases, V_{DS}. The predominant features in the conductance are the independence of the threshold voltage from V_{DS} and the conductance peaks that persist beyond $V_{DS} = 1mV$. Both of these features point to resonant tunneling as the predominant mechanism for controlling the near turn-on I-V characteristics. However, the presence of resonant features is rare in the device population and, when present, exhibits a strong lack of uniformity in peak amplitude and periodicity from sample to sample.

4. Results and Discussion

Initially, we simulate the device in the absence of disorder to determine whether the perfect device potential alone is sufficient to account for the resonant tunneling features found in the experimental device characteristics. The adiabatic quasi-1D ground state eigenenergy and I-V characteristics of the perfect device are shown in Fig. 3. The adiabatic quasi-1D eigenenergy (Fig. 3(a)) clearly exhibits the presence of a quantum dot in the vicinity of the thin-gate and the formation of barriers in the quasi-1D channel which separate the quantum dot from the 2D contact regions. Examination of Fig. 3(a) reveals that the quasi-1D barriers are relatively flat and thick ($\simeq$500 nm). Although some degree of resonant tunneling should exist in the perfect device, the transport characteristics (Fig. 3(b)) show a complete lack of detectable structure in the device current. The explanation for the structure-free I-V characteristics therefore follows by comparing the relative magnitudes of the tunneling component of the current to the thermionic emission component. As V_G is increased, the long range influence of the gate lowers the quasi-1D barriers and therefore monotonically enhances the thermionic emission current at T=4.2 K. On the other hand, the tunneling current, which is superimposed on the thermionic emission current, has a magnitude proportional to the resonant linewidth Γ when $kT \gg \Gamma$. Since the broad barriers in the perfect device lead to a set of Γ which satisfy this condition, resonant tunneling contributes indistinguishably to the overall current and the transport characteristics in the VCA are completely dominated by thermionic emission. In the presence of disorder, weak localization in the barrier regions should significantly increase Γ for each resonance and thereby provide a discernible resonant tunneling contribution to the overall current. To quantitatively assess the influence of realistic interface and surface roughness, we have introduced these disorder mechanisms into the 3D device geometry as described in previous sections. Figure 4(a) shows the adiabatic eigenenergy corresponding to disordered structures. The disorder was produced by randomly assigning 800Å × 800Å × 5Å tiles at the heterointerface to either GaAs or $Al_{0.3}Ga_{0.7}As$ and randomly varying the position of the etched surface using 200Å × 200Å × 200Å boundary tiles. The influence of the disorder on the adiabatic eigenenergy is twofold. First, there is the vertical modulation of the potential which provides an adiabatic shift of the local eigenenergy. In addition, the transverse modulation, largely due to surface roughness, causes local fluctuations in the confinement strength which are manifest as peaks and valleys in the adiabatic quasi-1D eigenenergy. The impact of the disorder is immediately apparent on the theoretical I-V characteristic shown in Fig. 4(b) which

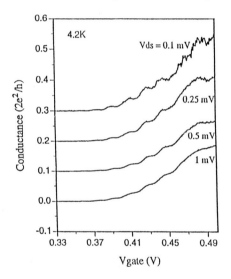

Figure 2: Experimental conductance characteristics at T = 4.2 K. The various curves are offset for clarity. (After Ref. 7)

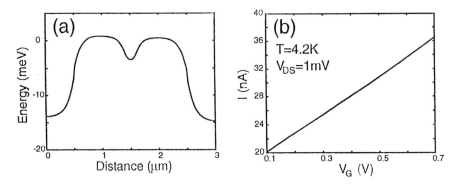

Figure 3: (a) Adiabatic quasi-1D eigenenergy calculated for the device geometry in the absence of disorder. (b) Transport characteristics at T = 4.2K.

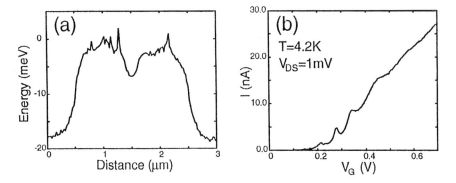

Figure 4: (a) Adiabatic quasi-1D eigenenergy arising from structural disorder randomly built from $800\text{Å} \times 800\text{Å} \times 5\text{Å}$ tiles at the $GaAs - Al_{0.3}Ga_{0.7}As$ interface. Surface disorder occurs over $200\text{Å} \times 200\text{Å} \times 200\text{Å}$ boundary tiles. (b) I-V characteristics.

exhibit strong resonant tunneling features near threshold. As V_G is swept to higher values and the quasi-1D barriers are lowered, the thermionic emission component of the current begins to dominate and the resonant features are gradually diminished. The effect of varying the size of the interface clusters is shown in Fig. 5. We have also simulated coarser interface disorder which showed the least structure in the adiabatic quasi-1D eigenenergy. The primary disorder mechanism for this case was surface roughness which, owing to long-range screening, results in a more gradual variation of the adiabatic eigenenergy, and was apparent in very subtle resonant features in the conductance characteristic. These data suggest that interface irregularities on the order of $500\text{Å} - 1500\text{Å}$ are the most likely explanation for the resonant tunneling features found in the experimental device. Fig. 5 shows the influence of temperature on the conductance The curve at T=0.5 K shows the strongest structure in the conductance and most closely resembles the bare transmission coefficient through the device. As the temperature is increased, the thermal broadening softens the resonant behavior and modifies the threshold voltage due to the onset of thermionic emission. Some resonant features remain up to T=10 K indicating the relative robustness of the disorder-induced resonant tunneling. We should emphasize the disorder-induced near-threshold resonant behavior we have investigated has been repeatedly observed in planar modulation doped nanostructures.

5. Conclusions

In this investigation, we have demonstrated that disorder mechanisms play a prominent role in determining near-threshold conductance characteristics in quantum dot nanostructures for $T \leq 10K$. While our simulations indicate that interface and surface roughness alone can explain anomalous resonant tunneling features, other disorder mechanisms (i.e., disorder in the impurity layer) are undoubtedly present in fabricated devices and contribute to the general trends we have observed. As fabrication techniques improve and lead to higher temperature operation, disorder-induced potential fluctuations will become insignificant thereby enabling a practical resonant tunneling device technology.

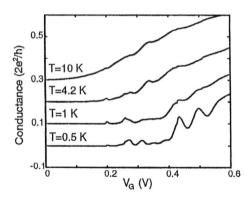

Figure 5: Theoretical conductance characteristics for the same configuration as in Fig. 4 with temperature as a parameter for $V_{DS} = 0.1mV$.

Acknowledgements

This work was supported by NSF Grant # ECS 91-20641 and the Joint Services Electronics Program. We would like to thank Ms. C. Willms and F. Register for clerical assistance.

References

[1] Present address: Central Research Laboratory, Texas Instruments Inc., P.O. Box 655936, MS134, Dallas, TX 75265.
[2] U. Meirav, M. A. Kastner, and S. J. Wind, Phys. Rev. Lett. **65**, 771 (1990). U. Meirav, P.L. McEuen, M.A. Kastner, E.B. Foxman, A. Kumar, and S.J. Wind, Z. Phys. B **85**, 357 (1991).
[3] C. W. J. Beenakker, Phys. Rev. B **44**, 1646 (1991).
[4] Y. Meir, N. S. Wingreen, and P. A. Lee, Phys. Rev. Lett. **66**, 3048 (1991).
[5] D. Jovanovic and J. P. Leburton, Phys. Rev. B **49**, 7474 (1994).
[6] Y. Wang and S. Y. Chou, Appl. Phys. Lett. **63**, 2257 (1993).
[7] H. Chang, R. Grunbacher, D. Jovanovic, J. P. Leburton and I Adesida, J. Appl. Phys. (In Press).
[8] S. Washburn, A. B. Fowler, H. Schmid, and D. Kern, Phys. Rev. B **38**, 1554 (1988).
[9] J. H. Davies and J. A. Nixon, Phys. Rev. B **39**, 3423 (1989). M. J. Laughton, J. R. Barker, J. A. Nixon and J. H. Davies, Phys. Rev. B **44**, 1150 (1991).
[10] R. Kosloff and H. Tal-Ezer, Chem. Phys. Lett. **127**, 223 (1986).
[11] J. P. Perdew and A. Zunger, Phys. Rev. B **23**, p. 5048, (1981).
[12] R. Landauer, Philos. Mag. **21**, 863 (1970).

Progress in self-assembled quantum dots of In$_x$Ga$_{1-x}$As on GaAs

D. Leonard, G. Medeiros-Ribeiro, H. Drexler[†], K. Pond,
W. Hansen[†], J. P. Kotthaus[†], and P. M. Petroff

Department of Materials and the center for Quantized Electronic Structures (QUEST)
University of California, Santa Barbara, CA 93106

[†] at the Ludwig-Maximilians-Universitat in Munich, Germany

The growth of self-assembling InAs quantum dots, as formed during deposition on (100)GaAs until the onset of the Stranski-Krastanow growth mode transition, is examined in detail. Atomic force microscopy has revealed a tendency toward uniformity only at the initial stages of the nucleation of the InAs islands. Coulomb energies are observed to be smaller than quantum confinement energies, allowing density of states spectroscopy of distinct s-like and p-like energy levels. Two fold orbital degeneracy is lifted in p-like levels in the presence of a large perpendicular magnetic field. Structural, optical and electronic measurements corroborated atomic-like behavior of these quantum dots. Although fundamental studies of quantum dots formed with this method continue, we discuss the ultimate uniformity with respect to the requirements for common applications.

1. INTRODUCTION

Epitaxial islands which form as a result of lattice mismatch strain have recently obtained the focus of many of those interested in fabricating quantum structures[1]. Without any pre-growth or post-growth processing steps, quantum dots with high optical[2] and electrical[3] activity have been created by embedding these islands in a suitable cladding material. Molecular beam epitaxy[1] and metal organic vapor phase epitaxy[4] have been employed in this fashion to produce arrays of strained quantum dots. The idea of using three dimensional growth to fabricate quantum structures had already received some attention[5]. Upon the discovery of very uniformly sized quantum dots[6] with high photoluminescence efficiency[7], current scientific interest has increased.

Mainly two experimental methods of creating confinement in the self-assembled dots (SAD) have emerged. Notzel *et al* used a remarkable post deposition annealing to produce surprisingly regular features of 70 to 210 nm.[8] We previously used direct cladding of naturally formed strain-coherent islands to create random arrays of uniform 20 nm InAs quantum dots. Although laterally disorganized at present, it is unclear for which direct applications of quantum dots this arrangement is a requirement. Using the latter technique, the quantum dots which are formed fulfill the size requirement for observation of quantum confinement effects for both electrons and holes.

Before the advent of this approach, quantum dots were by and large created by electrostatic confinement in GaAs using a suitably patterned surface topology and gate electrodes.[9] In these devices, typical lateral confinement lengths are 100 nm and larger and thus Coulomb charging energies often dominate quantization energies when individual electrons are added to the structure.[10] Other schemes which attempt to confine electrons and holes in all three spatial dimensions by suitable growth and pattering techniques have been able to produce complex and intriguing luminescence spectra [11] but so far have not yet resulted in objects which allow controlled occupation of atomic-like levels.

Using layers of SAD in an appropriately designed GaAs/AlAs metal-insulator-semiconductor-field-effect-transistor (MISFET), such quantum dots have been realized[12]. Evidence of discrete density-of-states (DOS) of the quantum levels was observed in capacitance-voltage traces, and quantum confinement energies larger than the expected Coulomb charging energies were extracted. Elsewhere photoluminescence excitation experiments further suggested a peaked DOS in these quantum dots[13] and similar energy level spacings of the SAD were measured. In both techniques very large numbers of dots were probed ($>10^6$). Energy level positions of SAD are expected to vary greatly with respect to changes in size, composition or strain of SAD. Consequently broadened spectral features have been obtained with both of these techniques, thought to arise mostly from SAD height variations since this is the direction of strongest confinement. Recently however photoluminescence from well defined numbers of quantum dots demonstrated that the observed broad spectra are in fact made up of individual peaks with discrete and very narrow line widths.[14] Higher spectral resolution revealed photoluminescence from a single SAD.[15]

In these proceedings we review the current state of understanding of the structural properties of SAD, and demonstrate further DOS spectroscopy in InAs SAD embedded in GaAs. Atom-like behavior of these high quality arrays of quantum dots is demonstrated. From recent research in this area arise fundamental questions regarding the implementation of quantum dots into useful devices.

2. EXPERIMENTAL DETAILS

SAD layers studied here were grown by molecular beam epitaxy (MBE) with the deposition of InAs directly on a 0.5 μm (100) GaAs buffer layer including a GaAs/AlAs (2nm / 2nm) short period superlattice. The substrate pyrometer reading during InAs deposition was 530 °C, the temperature at which the c(4x4) GaAs reconstruction begins to give way to the (2x4) reconstruction. The As rich reconstruction was verified by reflection high energy electron diffraction (RHEED). A variation in the deposited In coverage was implemented in the sample A by continuously sweeping a Ta shadow mask across the wafer during InAs deposition. Variation in InAs coverage of sample B were obtained from nonuniformities of the In flux due to the source geometry in the Varian Gen II MBE machine used. Substrate rotation was not used only during InAs deposition. Under a constant As4 beam equivalent pressure of 7×10^{-6} Torr, cycles of In were deposited corresponding to an average growth rate of 0.01 ML per second. In this way we emulate equilibrium surface conditions as closely as possible.

The In flux was calibrated with RHEED oscillations at 530 °C by subtracting the Ga flux from the In + Ga flux. After formation of SAD, samples were rapidly cooled to 350° C and removed from the growth chamber for AFM measurement. Samples were stored in a vacuum desiccator less than one day before AFM measurements were completed in ambient conditions. This produces a thin oxide layer which does not significantly effect AFM measurement in the range of interest. A SiN tip from a Nanoscope III AFM with a 200 μm cantilever was kept in direct contact with the sample surface for measurement. AFM images shown here were flattened and planefit with the Nanoscope III software.

For capacitance measurements, the structure included a highly doped back contact layer 30 nm below the SAD layer. The SAD layers for these measurements were immediately covered with a thin GaAs spacer layer, a GaAs/AlAs (2nm / 2nm x 7 periods) short period superlattice (SPS), and a GaAs cap layer. The devices are completed by alloying In contacts to the back contact layer and evaporating a Ni/Cr gate electrode.

3. STRUCTURAL PROPERTIES

Underlying an understanding of the potential confinement in a SAD is a better understanding of the geometry of the InAs, as formed at the Stranski-Krastanow growth mode transition. The shape of a SAD effects not only the geometry of the potential confinement, but of the strain in and around the SAD. This has the effect of raising or lowering the bandgap respectively. To this end several techniques have been utilized including AFM[16] and TEM in cross-section and plan view[17]. TEM showed a radial symmetric strain field which extends 30 to 50 nm into the surrounding lattice. Radial changes in the strain are still not well understood. This strain was also observed in cross-sectional TEM micrographs, preventing the acquisition of accurate chemical composition and size information.

The dependence of InAs SAD density and diameter on growth parameters such as substrate temperature and As flux have been documented.[1] However, the most significant changes in SAD are associated with the total InAs coverage. In particular for sample A shown in figure 1, density of the SAD abruptly increases from zero to a value of 1×10^{10} cm^{-2} between regions separated by only 1 mm on the

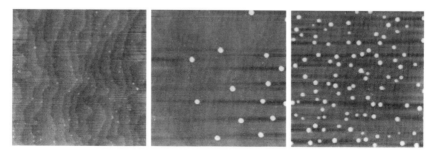

Figure 1: 1μm x 1μm atomic force microscopy images of various coverage of InAs on GaAs. Self-assembled quantum dots in (b) are more uniform than with higher InAs coverage in (c).

wafer. We therefore deduce that the strain, induced by increasing the total amount of deposited InAs, is a more critical growth parameter for the tuning of SAD size and density. To ascertain the exact coverage dependence of SAD density, as well as other structural variations, an additional type of sample was grown. On sample B changes in SAD were produced across a two inch wafer by the InAs flux variations from the In effusion cell. 5 μm^2 measured areas were spaced by 0.7 mm on the two inch wafer, with samples of higher density coming from areas in closest proximity to the In effusion cell. Using well known predictions of the flux variations[2] for our source-sample geometry we estimate that 0.7 mm represents only a 0.01 ML change in the amount of deposited InAs. SAD density increases monotonically from 8×10^6 cm^{-2} to 2.5×10^8 cm^{-2} with increased coverage of only 0.021 ML. The increasing numbers of SAD allow greater relief of the strain which increases with further InAs coverage. In figure 2 we plot the SAD density versus the estimated total InAs coverage on sample B.

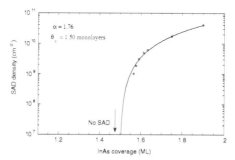

Figure 2: Self-assembled dot density versus estimated InAs coverage.

The SAD density is essentially zero until a certain critical coverage at which the value increases sharply. The data are fit with a function of the form $\rho_{SAD} = \rho_0 (\Theta - \Theta_c)^\alpha$. In this case ρ_{SAD} is the SAD density and Θ is the estimated InAs coverage. This functional behavior is that of a phase transition, with SAD density representing an order parameter. The critical coverage Θ_c, the exponent α, and the normalization density ρ_0 were obtained from a least squares fit shown by the solid line. From the best fit to the data, we extract a ρ_0 of 2×10^{11} cm^{-2}, an exponent of $\alpha = 1.76$, and a critical InAs coverage of 1.50 ML. With the AFM, we are able to observe the initial response of the surface to increasing mismatch strain in the InAs epitaxial layer. Consequently, we report a value of the critical thickness lower than expected.

A statistical analysis of many such images has been performed with the use of a computer program which records diameter and height of SAD from AFM images. Histograms of the SAD height and diameter are created from several nearby 1 µm x 1 µm AFM images corresponding to the estimated coverage of 1.6, 1.65, 1.75 and 1.9 ML. SAD size uniformity, reported as standard deviation in AFM data, of ±10% in height and ±7% in diameter was observed at the initial stages of formation (figure 1b), after which the uniformity degraded. The mean diameter decreased from nearly 30 nm to below 20 nm at higher InAs coverage. The additional strain induced by further coverage with InAs evidently causes no further increase in the diameter or height of SAD. Instead, the SAD form at a certain size (~ 30 nm diameter) after which only an increase in the nucleated density of SAD takes place. It is likely that the greater amount of nucleation leads to the SAD size reduction through In surface diffusion away from already formed SAD. At a density of approximately 4×10^{10} cm^{-2}, large relaxed InAs islands form, and the SAD density does not increase further.

It is rather unexpected that after the initial nucleation, further InAs coverage does not produce larger SAD, but leads to increasing numbers of SAD. This suggests an energy barrier to SAD growth. This energy barrier to continued SAD growth may simply be the energy barrier to the formation of a misfit dislocation. This could explain the tendency toward size uniformization of the SAD. Additionally a strong correlation must exist between SAD formation and the nature of nucleation on the starting surface. The selection of nucleation sites by SAD can be observed directly by AFM. In figure 1 (a)-(c), monolayer steps were observed that indicated preferential SAD nucleation at step edges. It is unknown if the initial nucleation (figure 1 (b)) occurs at kinks into the step edge or if the initial nucleation process causes depletion of the nearby terrace. In any case these images indicate that surface steps play a large role in determining SAD nucleation, and therefore a large role in determining SAD structural properties. A more detailed study of the growth mechanisms of InAs SAD can be found in reference[20].

4. CAPACITANCE SPECTROSCOPY

Observations of SAD with quantum-sizes and good size uniformity must be accompanied with measurement of their electronic and optical properties for this is the true test of quantum dot like behavior. Complex and interesting photoluminescence and cathodoluminescence spectra were presented elsewhere[21]. Here we use a MISFET structure described earlier to directly sweep the Fermi level across the DOS in SAD. At certain forward gate biases electrons can be injected into the quantum dots by tunneling through the 30nm of undoped GaAs between the back contact layer and the SAD layer. Typical capacitance traces measured at 4.2 K are shown in figure 3, illustrating the charging characteristics of SAD in a surface normal magnetic field B.

At small forward gate bias we measure the capacitance between the back contact layer and the gate. At much larger bias (>0.6 V), electrons are transferred directly into a 2 dimensional electron gas formed near the GaAs interface with the GaAs/AlAs SPS. The two maxima in the magnetocapacitance trace at intermediate gate voltages are due to charging the lowest quantum states of the SAD with electrons. The gate voltage can be converted into an energy scale by a simple relation $V_g - V_{tz} = (N-1/2)e/C + 2E_n/e$. V_{tz} denotes the voltage at which the Fermi level in the InGaAs dots is equal to the quantization energy in the growth direction E_z. The first term on the r.h.s. of the equation is the Coulomb charging energy required to charge N electrons into the dot capacitance C and the second term reflects the lateral quantization energy En in the quantum dot with the factor of 2 reflecting that the gate potential difference is reduced by about a factor of 2 at the SAD plane. Calculations of the charging energy $e^2/2C$ using the self capacitance $C = 4\varepsilon\varepsilon_0 d$ of a 20nm disk in GaAs yield a value of about 17.5 meV. This Coulomb energy is difficult to observe in the figure because of broadening of the overall measurement due to variations in the ground state energies of different dots, estimated to be 25 mV. For other MISFET geometries, a clear Coulomb splitting of the capacitance of the ground state peak is observed.[22]

Assuming spin degeneracy, $2E_0/e + 3e/2C$ is thus needed to fully charge the ground state of the SAD. Interpreting the next maximum of the capacitance as reflecting the charging of the four-fold degenerate p-like E_1 state of the SAD we expect a gate voltage difference of $\Delta V_g = 3\cdot 17.5$ mV + $2(E_1 - E_0)/e$ between the two peaks at B = 0. The value $\Delta V_g = 145$ mV measured from capacitance gives $E_1 - E_0 = 46$ meV, a value consistent with far infrared absorption data reported elsewhere[23] from which is extracted $E_1 - E_0 = 41$ meV.

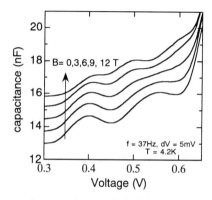

Figure 3: Capacitance-voltage traces in perpendicular 0,3,6,9, and 12 T magnetic fields. This density of states spectroscopy shows charging in the quantum states of the self-assembled quantum dots.

The magnetic field dependence reflects the atom-like behavior of SAD. The lowest capacitance maximum shifts only slightly for B fields used. However the upper maximum splits into two distinct peaks one decreasing and one increasing in gate voltage with increasing B field. We model the lowest quantum states to arise from lateral quantization only, since the confinement is much stronger in the growth direction. We assume a parabolic lateral confinement potential which yields an energy level structure in which the s-like E_0 level has two fold spin degeneracy and in which the p-like E_1 level has a two fold orbital and a two fold spin degeneracy[24]. With the magnitudes of the B field used spin splitting is not observed, whereas the orbital degeneracy will be lifted and the expected energy difference is $(E_1 - E_0)/\hbar = \omega_\pm = \sqrt{\hbar\omega_0^2 + (\omega_c/2)^2} \pm \omega_c/2$ where $\omega_c = eB/m^*$ is the cyclotron frequency and $\hbar\omega_0 = E_1 - E_0$ at B = 0 T. This behavior is observed in the experiment. The magnetic field dependence of the measured splitting is well described by m* = 0.07m0. This is close to the conduction electron mass of GaAs, possibly due to the extention of the SAD electronic wave function into the surrounding GaAs. Nevertheless, this value correlates well with m* measured with far infrared absorption.

5. DISCUSSION AND CONCLUSIONS

These simple experiments with SAD reveal atom-like behavior, and provide a new method of exploring a vast body of quantum dot theory. Unlike previous attempts to create artificial atoms[ashoori], Coulomb blockade energies here are smaller than quantum confinement energies, presenting SAD as a unique tool for understanding the behavior of *true* quantum dots. We have integrated layers of SAD into common device structures, showing them to be ready for practical implementation. For practical applications, however, this technological capability now requires attention to a fundamental problem in quantum dot fabrication, that of creating quantum dots with extreme uniformity. Energy level positions in quantum dots varies quadratically with its dimensions. Therefore stochastic variations of just a few atoms could significantly broaden spectral features of an ensemble of various SAD. Although uniform beyond previous expectations, clearly the SAD size variations observed here exceed this necessary uniformity. Current work is being directed at devices which supercede this uniformity restriction. We emphasize that although SAD dimensions are such that size fluctuations blur the effects of quantum confinement, the achieved uniformity remains sufficient to observe and differentiate quantum and Coulombic effects.

ACKNOWLEDGEMENTS

This research was supported by the National Science Foundation Science and Technology Center for Research in Quantum Structures (QUEST, DMR #91-20007) and by the Air Force Office of Scientific Research (AFOSR, #F49620-92-J-0124). We also acknowledge the MRL for use of its AFM facility. G. Medeiros-Ribeiro would like to acknowledge financial support from the Brazilian agency CNPq.

REFERENCES

1. D. Leonard, *et al*, Appl. Phys. Lett. **63**, 3203 (1993), or J. M. Moisson, *et al*, Appl. Phys. Lett. **64**, 196 (1994).
2. G. Wang, *et al*, Appl. Phys. Lett. **64**, 2815 (1994), or R. Nötzel, *et al*, Jpn. J. Appl. Phys. **33**, L275 (1994).
3. H. Drexler, *et al*, submitted to Phys. Rev. Lett.; G. Medeiros-Ribeiro, *et al*, submitted to Appl. Phys. Lett.
4. J. Ahopelto, *et al*, Jpn. J. Appl. Phys. **32**, L32 (1993); R. Nötzel, *et al*, Jpn. J. Appl. Phys. **33**, L275 (1994);
5. L. Goldstein, *et al*, Appl. Phys. Lett. **47**, 1099 (1985); J. Ahopelto, *et al*, Jpn. J. Appl. Phys. **32**, L32 (1993); M. Tabuchi, *et al*, *Science and Technology of Mesoscopic Sctructures*, edited by S. Namba, *et al* (Springer, Tokyo, 1992), P. 328.
6. D. Leonard, *et al*, Appl. Phys. Lett. **63**, 3203 (1993); J. M. Moisson, *et al*, Appl. Phys. Lett. **64**, 196 (1994); D. Leonard, *et al*, Phys. Rev. B, to be published Oct. 15, (1994).
7. D. Leonard, *et al*, Appl. Phys. Lett. **63**, 3203 (1993).
8. R. Nötzel, *et al*, Jpn. J. Appl. Phys. **33**, L275 (1994).
9. D. Heitmann and J. P. Kotthaus, Physics Today **46**, 56 (1993); M. A. Kastner, Physics Today **46**, 24 (1993).
10. R. C. Ashoori, *et al*, Phys. Rev. Lett. **71**, 613 (1993).
11. Y. Arakawa, Solid State Electron. (in press); K. Brunner, *et al*, Phys. Rev. Lett. **69**, 3216 (1992); L. Katsikas, *et al*, Chem. Phys. Lett. **172**, 201 (1990).
12. H. Drexler, *et al*, submitted to Phys. Rev. Lett.
13. D. Leonard, *et al*, J. Vac. Sci. Technol. B, in press.
14. S. Fafard, *et al*, Phys Rev. B Rap. Comm., in press.
15. J. Y. Marzin, *et al*, Phys. Rev. Lett. **73**, 716 (1994).
16. D. Leonard, *et al*, Phys. Rev. B, scheduled to be published Oct. 15, (1994); J. M. Moisson, *et al*, Appl. Phys. Lett. **64**, 196 (1994).
17. S. Guha, *et al*, Appl. Phys. Lett. **57**, 2110 (1990); D. Leonard, *et al*, Appl. Phys. Lett. **63**, 3203 (1993)
18. D. Leonard, *et al*, Journal of Vacuum Science and Technology B **12**, 1063 (1994).
19. D. A. Herman and H. Sitter, *Molecular Beam Epitaxy* (Springer-Verlag, Berlin, 1989).
20. D. Leonard, *et al*, Phys. Rev. B, scheduled to be published Oct. 15, (1994).
21. P. M. Petroff, *et al*, submitted 7th Int. Conf. Superlatt., Microstruct. and Microdev., Banff, Canada 1994.
22. G. Medeiros-Ribeiro, et al, submitted to Appl. Phys. Lett.
23. H. Drexler, *et al*, submitted to Phys. Rev. Lett.
24. V. Fock, Z. Phys. **47**, 446 (1928).

Controlled Nucleation of InAs Clusters on (100) GaAs Substrates by Electron Beam Lithography

J. W. Sleight, R. E. Welser, L. J. Guido, M. Amman, and M. A. Reed

Center for Microelectronic Materials and Structures
Yale University, Box 208284, New Haven, CT 06520, USA

Abstract. We have developed a combined electron beam lithography / epitaxial growth procedure for controlling the nucleation of InAs clusters on (100) GaAs. A sacrificial AlGaAs layer was employed to avoid interactions due to surface contamination effects. Atomic force microscopy studies of the surface morphology after electron beam lithography and after InAs growth, as well as scanning electron micrographs and Auger analysis of the surface will be presented.

1. Introduction

The fabrication of low dimensional semiconductor structures usually involves some type of lithographic pattern transfer strategy, such as dry-etching of topographic features or selective deposition of material through a patterned mask. Almost always, these post-growth fabrication techniques damage the structure by creating electrical and/or optical defects. [1,2] We have developed a combined electron beam lithography / epitaxial growth procedure for controlling the nucleation of InAs clusters on (100) GaAs substrates. This technique provides sub-micron lateral resolution and does not require pattern transfer.

2. Procedure

The starting material, except where noted, is (100) silicon-doped GaAs which has in some cases been pre-coated with either a 60 Å or 1000 Å layer of $Al_{0.6}Ga_{0.4}As$ via low-pressure metalorganic chemical vapor deposition. The purpose of the $Al_{0.6}Ga_{0.4}As$ layer is to eliminate hydrocarbon contamination of the GaAs surface during the subsequent electron beam (e-beam) exposure.

The e-beam exposures are performed at an accelerating potential of 40keV with a beam current of 100nA and an exposure dose that varies between 10 and 100 C/cm^2. Several different types of patterns are written using the e-beam, including square lattices of 0.5 μm pixels (variable pixel spacing), lines (variable width and separation), and more complex combinations of lines. The base pressure of the scanning electron microscope during e-beam writing is 2×10^{-7} Torr, and a liquid nitrogen cold trap is used to reduce any residual hydrocarbon contamination from the diffusion pump.

After e-beam exposure, the sacrificial $Al_{0.6}Ga_{0.4}As$ layer is etched away in a 1:1 mixture of hot (~75°C) $HCl:H_2O$ in preparation for growth. The InAs clusters are grown using a low-pressure horizontal geometry metalorganic chemical vapor deposition (MOCVD) reactor with an H_2 carrier gas flow of 12 SLM and a base pressure of 100 Torr. The indium and arsenic are supplied by TMIn (trimethylindium) and AsH_3. After the hot HCl etch, samples are rinsed in de-ionized water and then loaded into the MOCVD reactor. The samples are heated by RF induction to 700°C in an AsH_3 overpressure for 6 minutes, which is sufficient to establish a surface ready for epitaxial growth, i.e., the native oxide has been removed and the rough GaAs surface has been converted into a series of atomic terraces and steps, verified by atomic force microscopy.

By varying the flow rates for AsH_3 and TMIn and the substrate temperature, the InAs crystallite density can be controllably varied from ~10^5 cm^{-2} (far above the substrate etch pit density of 250 cm^{-2}) to ~10^9 cm^{-2}, high enough for quasi-two-dimensional InAs growth. Previous work [3] has examined this process in detail. Under typical growth conditions used in this study, the growth rate is approximately 1.5Å/s at a substrate temperature of 650°C. The island density is approximately constant for growth times in the range of 5 to 60 seconds (i.e., a saturation value is quickly reached), and decreases after that due to coalescence of the islands. The surface morphology has been studied by many techniques, including Normarski optical microscopy, scanning electron microscopy (SEM), electron microprobe analysis (EMA), Auger analysis, and atomic force microscopy (AFM).

3. Results

3.1 Analysis of Surface Morphology After E-beam Exposure

AFM studies were performed to see how the e-beam exposure modified the sample surface morphology. Figure 1 shows an AFM image of a sample with a 1000Å $Al_{0.6}Ga_{0.4}As$ cap layer. There are three important regions: (1) the dark background region which was not exposed to the e-beam, (2) the lighter region near the line structure, and (3) the bright line structure. The surface of region 1 is not changed during the process as expected, and high resolution AFM shows atomic terracing. Region 2 is approximately 15Å higher than the background and the line in region 3 is 30Å higher than the background. The magnitude of this deformation is e-beam dosage dependent. High resolution AFM reveals atomic terracing is still present in these regions, as can be seen in Figure 2. Region 3 corresponds to where the e-beam dosage was actually swept, and the width (~1μm) corresponds to the lithographic width of the feature. The width of region 2 is approximately 10μm, consistent with the length scale for energy loss of 40 keV electrons, suggesting that this feature is caused by the electrons damaging the lattice beneath the surface.

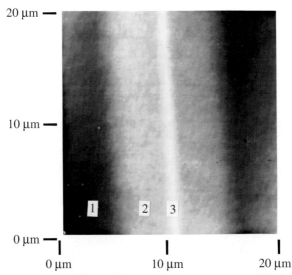

Figure 1. AFM image of an e-beam written line on (100) GaAs. Total height variation is 82Å.

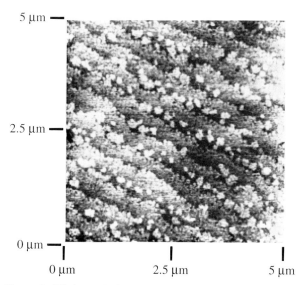

Figure 2. High resolution AFM image showing atomic terracing. The total height variation is 11Å.

3.2 InAs Surface

Scanning electron and atomic force microscopies were used to study the topology of the InAs clusters. Figures 3 and 4 are high magnification AFM and low magnification SEM images, respectively. Note the alignment of the clusters to the GaAs substrate orientation, and the uniformity in size (±25%).

Figure 3. AFM image of InAs clusters on (100) GaAs. The height of these clusters is ~4000Å and the width of the base is ~5000Å.

Figures 5-7 show SEM images of the InAs growth on substrates which received e-beam patterning. In all cases, a 60Å thick sacrificial $Al_{0.6}Ga_{0.4}As$ layer was used. Figure 5 shows an InAs grid fabricated with a 2 μm center to center spacing (InAs clusters are white and the GaAs substrate is the black background). Figure 6 shows a more complex grid of InAs clusters, demonstrating anti-alignment of the clusters to e-beam exposure. The pattern consists of 0.5μm pixels, with variable pixel to pixel spacings. This figure shows the effect that varying the pixel spacing has upon the cluster growth. Figure 7 shows lines of clusters that were confined to grow between written e-beam lines.

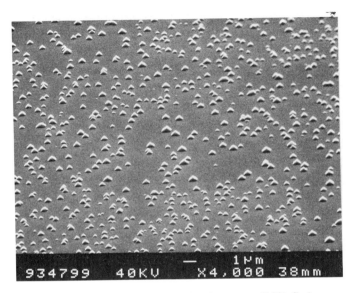

Figure 4. An SEM image of InAs clusters on (100) GaAs.

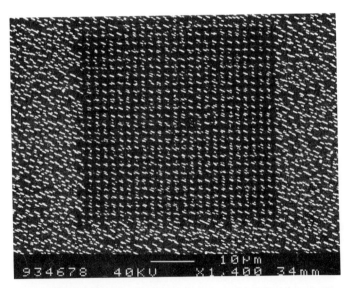

Figure 5. An SEM image showing InAs clusters (white) forming an array due to the electron beam patterning.

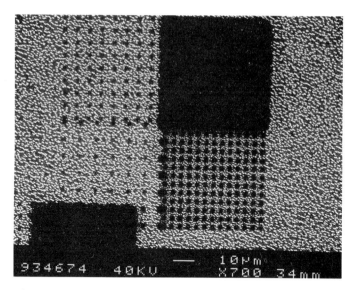

Figure 6. An SEM image showing that the InAs clusters do not nucleate in the e-beam exposed regions. Pixel spacings are 2 μm (upper right), 4 μm (lower right), 6 μm (upper left), and 8 μm (lower left); a second 2 μm spaced lattice is partially seen in the far lower left, overlapping the 8 μm lattice.

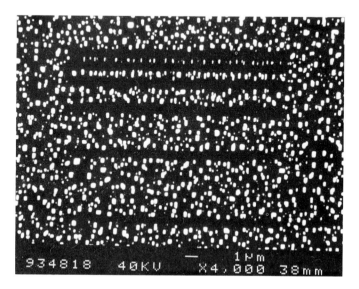

Figure 7. A series of lines written using the e-beam. Note the thin ~0.5 μm line of InAs clusters

3.3 Auger and Electron Microprobe Results

In order to better understand the nature of this process, we investigated these structures with Auger and electron microprobe measurements. These results are on samples that had a sacrificial $Al_{0.6}Ga_{0.4}As$ layer. The following was determined:

(1) In the areas between InAs clusters, far away from e-beam exposure an In signal is observed, implying an InAs wetting layer of ~2 monolayers forms. No trace of aluminum is found in this region, and as this region is sputtered, the indium signal decreases.

(2) In regions with moderate electron beam dosages (<10 C/cm^2), no aluminum or indium signals are observed for the samples with a 60 Å $Al_{0.6}Ga_{0.4}As$ layer.

(3) On other samples that had very high (~100 C/cm^2) electron beam dosages, traces of aluminum, oxygen, and indium are seen for samples with the 60 Å $Al_{0.6}Ga_{0.4}As$ layer. With sputtering, the indium signal disappears first, then the aluminum and oxygen signals disappear together. For samples with a 1000Å $Al_{0.6}Ga_{0.4}As$ layer, an indium signal corresponding to a wetting layer is seen, though no nucleation occurs. No aluminum signal is seen in this case.

(4) In all cases, a similar initial carbon signal is present, but it is the first to disappear with sputtering. This is most likely due to organics absorbed onto the sample when exposed to ambient conditions.

From this we conclude that for the 60 Å $Al_{0.6}Ga_{0.4}As$ cap layer samples, moderate e-beam dosages inhibit the InAs cluster growth. At high e-beam doses, it appears that an oxidized $Al_{0.6}Ga_{0.4}As$ layer becomes resistant to the etch, and InAs can grow on top of the oxidized $Al_{0.6}Ga_{0.4}As$ layer. It is still unclear why this layer becomes resistant to etching. Samples with the much thicker 1000Å $Al_{0.6}Ga_{0.4}As$ top layer appear to fully etch, even at high dosages. InAs cluster growth is still inhibited in the e-beam written area, though a wetting layer of InAs is forming.

4. Conclusions

We have demonstrated the use of electron beam lithography to control the nucleation of InAs clusters on (100) GaAs surfaces. A variety of patterns have been achieved using this method. The surface morphology of the substrates after the e-beam lithography, but prior to growth, is also examined and shows a distinct relationship to both the electron beam size and the length scale for energy loss of the 40 keV electrons in GaAs. Auger analysis confirms that InAs growth is inhibited in regions that receive approximately 10 C/cm^2 for the 60 Å $Al_{0.6}Ga_{0.4}As$ case, though this becomes more complicated for higher doses and thicker $Al_{0.6}Ga_{0.4}As$ layers.

Acknowledgments

The authors gratefully acknowledge the financial support of the National Science Foundation via the Presidential Fellowship Program (Grant # ECS-9253760) and DMR-9112497, the NASA Graduate Student Researchers Program (Grant # NGT-50832), and the Office of Naval Research N00014-91-J-1561.

References

[1] H. Qiang, F.H. Pollack, Y.-S. Tang, P.D. Wang, and C.M. Sotomayor Torres, *Appl. Phys. Lett.*, **64** (21), 2830 (1994).

[2] R. Cheung, Y.H. Lee, C.M. Knoedler, K.Y. Lee, T.P. Smith, III, and D.P. Kern, *Appl. Phys. Lett.*, **54** (21), 2130 (1989).

[3] *Unpublished Data*, R.E. Welser, L.J. Guido, J.W. Sleight, and M. Amman.

# Quantum Interference Effects in Finite Antidot Lattices

R. Schuster *, K. Ensslin *, D. Wharam *, V. Dolgopolov *, J. P. Kotthaus *
G. Böhm+, W. Klein+, G. Tränkle+, and G. Weimann+

* Sektion Physik, LMU München, Geschw. Scholl Platz 1, D-80539 München, Germany
+Walter Schottky Institut, TU München, D-85748 Garching, Germany

Antidot superlattices are fabricated on AlGaAs/GaAs heterostructures by electron beam lithography followed by a carefully tuned wet etching step. We study electron transport in an antidot lattice with a finite (9x9) number of periods. At low temperatures, T=30mK, the phase coherence length of the electrons exceeds the size of the total system (2.4 µm). At low-carrier densities and high magnetic fields the electrons travel phase coherently over the entire system in magnetic edge channels and interfere with each other under the influence of the geometry of the antidot potential landscape. The resistance shows pronounced Aharonov-Bohm-type oscillations with a characteristic area defined by the circumference of the antidots.

1 Introduction

Antidot superlattices represent a model system to study electronic transport through a periodic system [1-4]. Starting from a high mobility two-dimensional electron gas a periodic array of potential pillars can be fabricated by various technological means. So far most experiments have been performed on systems whose extent is much larger than the electron mean free path l_e as well as the phase coherence length L_ϕ. The experimental observations are successfully described by classical dynamics which neglects the phase of the electrons [5]. If the magnetic field is such that the classical cyclotron diameter fits around a single antidot or a group of antidots a maximum in the magnetoresistance arises.

In order to investigate phase coherence effects in antidot systems we fabricate an array of 9x9 antidots surrounded by a square geometry (see inset of Fig. 1). For very low temperatures T<100mK where electron-electron scattering is significantly reduced both L_ϕ and l_e may exceed the size of the system [6]. The electrons now carry a phase and an amplitude and therefore can interfere with each other.

2 Fabrication process

The fabrication process starts from a $GaAs / Al_xGa_{1-x}As$ heterostructure which contains a two-dimensional electrons gas 65nm below the surface. Its electron density is $n_s = 3 \cdot 10^{15} m^{-2}$ and the elastic mean free path is $l_e = 8\ \mu m$. A Hall bar is defined by wet etching and provided with Ohmic contacts (AuGe/Ni). The pattern is produced by electron beam lithography and transferred onto the sample by a carefully tuned wet eching step. The square geometry around the finite antidot lattice has point contact-like openings at its corners as contacts to the system. This geometry is written in the same step with the electron beam as

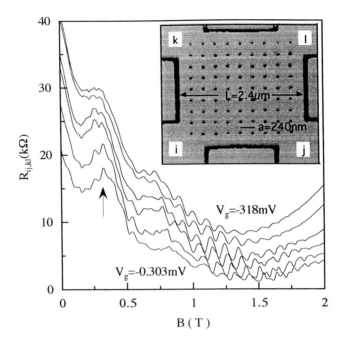

Figure 1: Magnetoresistance traces of a finite antidot lattice at T= 30 mK at very low carrier densities. The arrow indicates the classical commensurability maximum. The inset shows an image taken with an atomic force microscope of a wet etched surface of a GaAs/AlGaAs- heterostructure. Ohmic contacts are made to the corners of the square indicated by i,j,k and l.

the antidot lattice. That procedure provides an inherently good alignment of the two structures. For small structures the etch rate depends sensitively on the size of the features. In making the width of the bars that define the square confining geometry larger than the diameter of the antidots it is guaranteed that the finite lattice is decoupled from the outside 2DEG before the antidot potential is actually formed in the 2DEG. The inset of Fig. 1 shows an atomic force microscope image of a finite antidot lattice with a system size $L = 2.4 \mu m$ and a period $a = 240 nm$. Each antidot is well developed and the variation in size is remarkably small. The whole structure is covered by a gate metal which allows to change the Fermi energy in the system. The sample is cooled in a dilution refrigerator down to temperatures of 30 mK. Typical four-terminal-measurements of the resistance $R_{ij,kl}(B) = (U_k - U_l) / I$ are made by passing a current I through the contact i and j and measuring the voltage drop across the other two contacts (k and l).

3 Experimental results

In a recent publication we investigated the regime of high carrier densities [7]. In this case the classical commensurability oscillations are superimposed by strong reproducible fluctuations in the magnetoresistance that die out at larger fields where the cyclotron

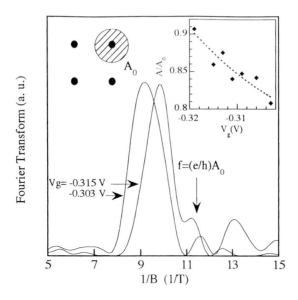

Figure 2: Fourier transform calculated on the experimentally obtained magnetoresistance traces (Fig. 1). The right inset shows the dependence of the AB-frequency on the applied gate voltage. The frequencies approach a value which is given by the circular area as indicated in the left inset.

diameter becomes smaller than the lattice period. The Fourier analysis reveals an Aharonov-Bohm effect related to the area enclosed by the classical cyclotron orbit around groups of antidots. Here we focus on the regime of low-carrier densities and magnetic fields around filling factor ν=2. The system now resembles a shallow lake (Fermi sea) between the antidot pillars.

In the quantum Hall regime the electrons travel along the boundary of the confining geometry in magnetic edge channels and therefore do not feel the presence of the antidot lattice. If the Fermi Energy E_F is lowered by a very negative gate voltage the constrictions between the wall and the antidots and between two antidots become very narrow. This leads to a coupling of the edge channels along the boundaries and the edge channels which travel along the circumference of the antidots. Figure 1 presents the magnetoresistance $R_{ij,kl}(B)$ for different negative gate voltages. At low magnetic fields $B = 0.3T$ the resistance peak corresponding to a classical orbit around a single antidot can be seen. In the minimum of the magnetoresistance related to filling factor ν=2 (0.8T ≤ B ≤ 2.3T) highly periodic Aharonov-Bohm (AB)-oscillations with a period $\Delta B = 100mT$ are observed. The characteristic area in this case is defined by the circumference of the antidots. The role of the classically pinned trajectories is now taken over by quantum mechanical edge states. In single quantum dots such AB-oscillations have been observed where the electrons are confined to edge states close to the perimeter of the dot [8].

At high magnetic fields the location of the edge channels is determined by their guiding center energy $E_G = E_F - \hbar\omega_c(n + 1/2)$ where ω_c is the cyclotron frequency and n is the

Landau level index. The electrons follow the equipotential lines $V(x,y) = E_G$ around the antidots. During one revolution they accumulate a phase $\Delta\varphi = 2\pi\phi / \phi_0$ where $\phi_0 = h/e$ is the elementary flux quantum and $\phi = A(E_G)B$ is the flux through the area enclosed by the edge channels around the antidots. The oscillations are periodic in B corresponding to the addition of one flux quantum through the enclosed area $A(E_G)$.

In order to get a quantitative understanding of the oscillations we Fourier transform the magnetoresistance as displayed in Fig. 2. If the window is chosen around the minimum corresponding to ν=2 pronounced maxima in the Fourier transforms are found. The maxima shift to higher frequencies $f = (e/h)A(E_G)$ for more negative gate voltages. This is in agreement with the simple picture that with decreasing carrier density the antidots size increases. Consequently, the area given by the circumference of the antidots and the corresponding frequency of the AB-oscillations increases. The inset in Fig. 2 shows the frequency as a function of the applied gate voltage. The frequencies approach a value $f = (e/h)A_0 \approx 11.5T^{-1}$ which is determined by a circular area $A_0 = \pi(a/2)^2$ with the diameter equal to the lattice period a. If the gate voltage is lowered further the edge channels are reflected at the barriers between the antidots. The oscillations disappear if the temperature is raised above T= 900mK.

In summary, we have reported on a cross-over from classical trajectories to quantum mechanical edge states in a finite antidot lattice. The characteristic area for the AB-oscillations is given by the circumference of the antidots.

Acknowledgements

It is a pleasure to thank M. Wendel for help with the atomic force microscope. We acknowledge financial support by Deutsche Forschungsgemeinschaft and the Esprit Basic Research Action.

References:

[1] H. Fang, R. Zeller, and P. J. Stiles, Appl. Phys. Lett. **55**, 1433 (1989).
[2] K. Ensslin and P. M. Petroff, Phys. Rev. **B41**, 12307 (1990).
[3] A. Lorke, J. P. Kotthaus, and K. Ploog, Superlattices and Microstructures **9**, 103 (1991).
[4] D. Weiss, M. L. Roukes, A. Menschig, P. Grambow, K. v. Klitzing, and G. Weimann, Phys. Rev. Lett. **66**, 2790 (1991).
[5] R. Fleischmann, T. Geisel, R. Ketzmerick, Phys. Rev. Lett. **68**, 1367 (1992).
[6] A. Jacoby, U. Sivan, C.P. Umbach, and J.M. Jong, Phys. Rev. Lett. **66**, 1938 (1991).
[7] R. Schuster, K. Ensslin, D. Wharam, S. Kühn, J. P. Kotthaus, G. Böhm, W. Klein, G. Tränkle, and G. Weimann, Phys. Rev. **B49**, 8510 (1994).
[8] B. J. van Wees, L. P. Kouwenhoven, C. J. P. M. Harmans, J. G. Williamson, C. E. Timmering, M. E. I. Broekhaart, C. T. Foxon, and J. J. Harris, Phys. Rev. Lett. **62**, 2523 (1989).

Interference of ballistic quantum electron waves in electronic stub tuners

P. Debray, J. Blanchet, R. Akis*, P. Vasilopoulos*, and J. Nagle+

Service de Physique de l'Etat Condensé and Centre National de la Recherche Scientifique, Centre d'Etudes de Saclay, 91191 Gif-sur-Yvette, France
*Department of Physics, Concordia University, Montreal, Quebec, Canada, H3G1M8
+Laboratoire Central de Recherche, Thomson-CSF, Route Départementale 128, 91401 Orsay, France

Abstract. We have measured the conductance G of electronic stub tuners (ESTs) and quantum wires (QWs) in the ballistic regime as a function of an applied gate voltage V_g, which changes the stub length and wire width. For QWs we observe quantized conductance plateaus. The transmission pattern of ESTs as a function of V_g is oscillatory in nature. We attribute this to interference effect between electron waves reflected from the stub and those in the main wire. Confirmation of this interpretation is obtained from measurements of G in an applied magnetic field B. This work presents unequivocal experimental evidence for interference of ballistic quantum electron waves.

1. Introduction

In quantum transport, electrons are described by their quantum mechanical wave functions. When transport is ballistic, the electron wave retains its phase coherence and is subject to interference effects common to all wave phenomena. Though various concepts [1-2] have been put forward for experimental observation of this effect, as of now, very few attempts [3-4] have been made to experimentally observe quantum interference in the ballistic regime. Moreover, the reults of these attempts are inconclusive. *In this work, we report unambiguous experimental evidence for interference of ballistic quantum electron waves.* The experimental device we have chosen is the electronic stub tuner (EST)

2. Quantum electron transport in electronic stub tuner (EST)

An EST can be viewed as the mesoscopic analog of an electromagnetic stub tuner that can be tuned on a femtosecond scale. A simple EST is a quantum wire which has the form of an inverted T and thus consists of a quantum wire to which is attached a lateral branch or stub, the length of which can be controlled by the voltage V_g of a Schottky gate (Fig. 1). *When the total length of the EST is smaller than all the characteristic transport scattering lengths, and its width is comparable to the Fermi wavelength, electron transport through it is ballistic.* One can also consider an EST of a special shape in the form of a cross or double stub connected to quantum wires on each side (Fig. 1, inset). To distinguish the two different types of ESTs, we designate the single-stub EST as SEST and the double-stub EST as DEST.

The conductance G in the fundamental mode of a SEST at zero temperature has been theoretically shown [5] to be a periodic function of the stub length. Here we consider the case when all the dimensions W, H, and W' (Fig. 1) of a SEST or DEST change simultaneously and in the same linear manner with V_g. One can then calculate the conductance as a function of one parameter W, remembering that H and W' change simultaneously as W changes. We have calculated the conductance of SESTs and DESTs, for the case W = W', using the transfer matrix technique [5] and assuming *square-well confinement*. The elements of the transfer matrix are obtained by mode-matching. The conductance G is evaluated from the Landauer formula and at zero temperature is given by:

$$G(E) = (2e^2/h) \sum_{mn} |c_n^{right}|^2 (\alpha_n / \alpha_m).$$

The summation is over all propagating channels. C_n^{right} is the outgoing amplitude, α_n is the wave number for the nth channel, and E the Fermi energy. Fig. 2 shows the calculated conductance of a DEST at zero temperature as a function of W for the dimensions (see below) of our experimental sample. The influence of finite temperature is to make the minima shallower, by lifting them to higher G values as W increases. Inspection of Fig. 2 shows that the expected plateaus of a QW are split into a series of transmission minima. Similar results have been obtained for SESTs. *The physical picture of the origin of these minima is the quantum interference between electron waves reflected from the stub and those in the main wire.*

3. Results and discussion

The QWs, SESTs, and DESTs used in this study were fabricated from high-mobility ($\mu \cong 110$ m^2/Vs at 4.2 K) AlGaAs/GaAs modulation-doped (Si) heterostructures grown by MBE with low electron concentration (n = 3.1 × 10^{15} m^{-2}). The electron mean free path is 10 µm. The lithographic geometric dimensions of the devices were : L = 550, H = 300, and W = W' = 250 nm (Fig. 1). With the device gates grounded, measurements of Shubnikov-de Haas oscillations at 70 mK gave a Fermi energy E$_F$ of 8.9 meV. The conductance G of the devices was measured as a function of the split-gate voltage V_g at 70 mK in the absence and presence of a magnetic field B applied perpendicular to the device plane. To measure G, an ac voltage of 64 µV at 7Hz was applied onto the drain contact of the device and the current flowing through the device and out the source contact (a virtual ground) was monitored by a current-to-voltage preamplifier, the output of which was fed into a lockin amplifier, tuned to the frequency of the drain-source voltage.

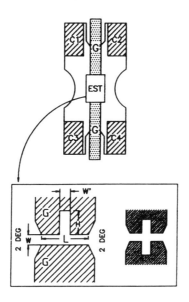

Fig. 1. Schematic diagram of a sample chip with EST devices. C's represent ohmic contacts and G's are Schottky gates.

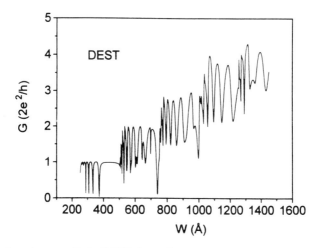

Fig. 2. Calculated conductance G of a DEST of experimental sample dimensions at zero temperature as a function of wire width W.

Figure 3 shows the conductance G of a QW and a SEST as a function V_g in zero magnetic field. The QW shows conductance steps with plateaus at close to, but lower than integral values of ($2e^2/h$). The oscillatory nature of the conductance between the plateaus is reproducible and can be attributed to scattering from disorder; specifically to backscattering and resonant tunneling through single impurities in the Q1D channel. The observed recovery of exact quantization and the improvement in the sharpness of the steps in a magnetic field of 1 T (Fig. 3, inset) are consistent with this interpretation. *The conductance of the SEST, contrary to the QW, shows no plateaus or quantization and is oscillatory in V_g.* The oscillations are reproducible. In Fig. 4, we show the conductance G of a DEST in zero and applied magnetic field. As for the SEST, G of the DEST shows oscillations and does not exhibit quantized conductance steps. The oscillatory pattern of the conductance is found to evolve with applied magnetic field until for B = 1T the oscillations disappear and well-defined, quantized conductance steps appear. We have obtained similar results for the SEST as well. Measurements on other SESTs and DESTs confirm this general pattern, except that the pinch-off voltage is found to depend on the specific sample.

For the QW, we find the midpoints of the plateaus to be equispaced in V_g (Fig. 2 does not show all the observed plateaus) with a spacing ΔV_P = 29 mV, which is the plateau width. From this we construct the conductance steps for the ideal QW (Fig. 2, dotted). The nonzero value of G below the first step in the experimental curve is due to smearing of the subband threshold caused by quantum tunneling. This results in a shift ΔV_S = 12 mV in the effective pinch-off voltage. Assuming that ΔV_S remains the same for all the devices, we can superimpose the ideal QW quantized steps (dotted in Fig. 3) on the experimental G curves of SEST and DEST. Since the midpoint of a plateau corresponds to the situation when the Fermi level is in the middle between two subbands, from a knowledge of E_F and assuming square-well confinement one can compute the wire widths for values of V_g corresponding to the midpoints of the plateaus. We find a linear relationship between V_g and W with dW/dV_g = 0.84 (nm/mV). One can then convert the V_g scales of Figs. 3 and 4 into W scale and compare the experimental results with theoretical predictions.

We attribute the observed conductance oscillations for SEST and DEST to quantum interference between electron waves. However, we note that the observed oscillatoty pattern of the conductance looks more like an envelope of the calculated ones. A smearing of the zero-temperature theoretical oscillations may result due to finite temperature effect. However, at 70 mK this effect should be negligible. It should be noted that the theoretical conductance pattern of Fig. 2 was computed on the basis of rectangular stubs with sharp wall boundaries which is unrealistic for electrostatic confinement. In fact, the real shape of the stubs can deviate substantially from the ideal case. Recent theoretical calculations [6] on the shape dependence of G of SESTs show that the oscillation period increases as the stub shape changes from rectangular to Lorentzian to right-triangle, and can be of the order of 100

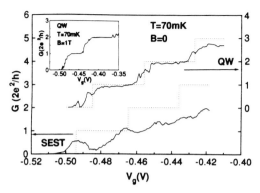

Fig. 3. Measured conductance G of a quantum wire (QW) and a SEST at 70 mK as a function of gate voltage V_g. The dotted lines show ideal QW steps.

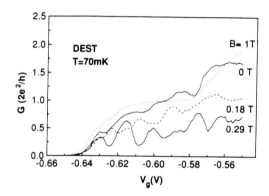

Fig. 4. Measured conductance G of a DEST as a function of gate voltage V_g at 70 mK for different values of applied magnetic field.

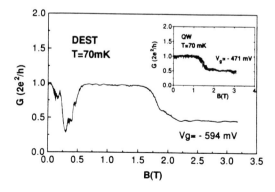

Fig. 5. Measured conductance G of a DEST in the fundamental mode at 70 mK as a function of magnetic field. The inset shows data for a quantum wire (QW).

Å. A one-to-one correspondence between theoretical predictions and experimental results can not, therefore, be expected. Our interpretation of the experimental results in terms of quantum interference of electron waves, as outlined above, is substantiated by the results of measurements of G in magnetic field B (Fig. 4). For low values, the field introduces an additional local phase factor. The oscillation pattern $G(V_g)$ of ESTs is thus expected to be different for different values of B, in agreement with observation. This evolution has not been observed for QWs. When B is strong enough, the current is carried by the edge states and interference between electron waves can not take place. In the absence of interference, one gets back the quantized conductance steps of QWs (Fig. 4).

A more convincing evidence for quantum interference between elctron waves in SESTs and DESTs is obtained from measurements of G as a function of B for fixed V_g (Fig. 5). To do this, the split-gate voltage was so chosen that transport was in the fundamental mode. The G of the DEST shows a pronounced structured dip in the range where B is not strong enough to produce edge states. This happens because the field modifies the interference between electron waves. Hence as B changes one can expect a structure in the conductance. The structured dip is not observed for the QW (Fig. 5, inset) in which there is no interference effect. Indeed recent calculation of G [7] for ESTs in the fundamental mode as a function of magnetic field shows a conductance dip similar to the one observed experimentally.

4. Acknowledgments

The authors wish to thank A. Kumar, M. Rooks, and D. Carter for sample preparation. The work of R A and P V was supported by NSERC grant No. OGPIN028.

5. References

[1] Capasso F and Datta S 1990 Phys. Today **43** 74-82
[2] Joe Y S and Ulloa S E 1993 Phys. Rev. B **47** 9948-51[1]
[3] Wu J C and Wybourne M N 1991 Appl. Phys. Lett. **59** 102-4
[4] Aihara K, Yamamoto M and Mizutani T 1992 Jpn. J. Appl. Phys. **31** L916-9
[5] Wu H, Sprung D W L, Martorell J and Klarsfeld S 1991 Phys. Rev. B **44** 6351-60 a nd references therein
[6] Akis R, Vasilopoulos P, Blanchet J and Debray P 1994 *22nd Intl. Conf. on the Physics of Semiconductors*, Aug.15-19, Vancouver, Canada
[7] Akis R unpublished

Efficient modulation of carrier density in novel double quantum well structure by low optical pump power

M. Rüfenacht, H. Akiyama, S. Tsujino, Y. Kadoya, and H. Sakaki
Research Center for Advanced Science and Technology (RCAST), University of Tokyo, 4-6-1, Komaba, Meguro-ku, Tokyo, 153 Japan

Abstract. We demonstrate that the density of carriers can be efficiently modulated by low cw pump power I_{EXC} when a novel GaAs/GaAlAs double quantum well structure is used to separate photoexcited electron-hole pairs into different quantum wells. A separated carrier density $N_{SC}=1.4*10^{11}$ carriers cm^{-2} was found at a pump power $I_{EXC}=250$mWcm^{-2}. This was studied via the photoluminescence peak position which is very sensitive to the carrier density. Detailed measurements revealed the efficiency of carrier separation $\varepsilon_{CS} \approx 20\%$ and the separated carrier lifetime $\tau_{SC}=2.3\mu s$.

1. Introduction

Efficient spatial separation of carriers in double quantum well (DQW) structures is expected to open doors to various devices. This phenomenon can be used to modulate efficiently the carrier concentration by optical excitation or to control the intensity and wavelength of photoluminescence from biased DQWs. Numerous attempts have been made to control this process by using such structures as nipi-structure [1], type-II GaAs/AlAs quantum wells [2] or coupled DQW [3]. We investigate here the carrier separation process in a novel double quantum well structure to show that high efficiency of carrier separation and long lifetime of separated carriers are simultaneously achieved.

This novel DQW structure consists of two GaAs wells surrounded by AlAs external barrier (EB) and separated by a low GaAlAs middle barrier (MB) (inset Fig. 2). The middle barrier has been chosen wide enough to avoid tunneling between the two wells and acts also as an absorbing layer. Electron-hole pairs optically generated inside MB can be separated by a perpendicular electric field E_z and relax into a respective well as carriers are to be stopped by the external barrier.

The concentration of separated carriers (SC) for a fixed pump power depends on two factors, namely (1) the efficiency of carrier separation ε_{CS} which is defined as the number of spatially separated electron-hole pairs divided by the total number of electron-hole pairs; (2) the lifetime τ_{SC} of SC. Mechanisms of carrier separation in this structure are revealed here by a series of photoluminescence (PL) measurements. From the dependence of integrated PL intensity on applied bias, ε_{CS} is estimated. We then point out that the peak position of PL depends on the concentration N_{SC} of SC as the field-induced Stark effect is reduced by the screening of the field by spatially separated electron-hole pairs. To analyze the data, we calculate by a self-consistent Hartree approximation the shift of the PL peak as a function of carrier density and bias and compare the result with experiments to determine N_{SC} as a function of pump power. Finally, the time resolved PL study is done to determine the peak position as a function of time, from which the lifetime of SC is estimated.

2. Sample structure and experimental set-up

The sample used in these measurements consisted of fifteen periods of DQW grown by molecular beam epitaxy on a semi-insulating GaAs substrate. The DQW consists of two GaAs quantum wells QW-A and QW-B (70Å and 50Å, resp.) separated by a 200Å $Ga_{0.66}Al_{0.34}As$ middle barrier (MB). These wells are surrounded by a 100Å-thick AlAs external barrier (EB) (cf. inset Fig. 2). The well region was clad between an n^+-doped layer and a 100Å GaAs top-layer. Ohmic contact to the doped layer was taken by annealing InSn contact after an etching of 4000Å on part of the sample. An electric field perpendicular to the layers was applied by using a Schottky contact, formed by evaporating a 100Å-thick semi-transparent gold film on the non-etched part of the sample. PL measurements were carried out at 77K with Tungsten lamp excitation for low power measurement and with Ar^+ laser excitation for higher pump power. Titanium-Sapphire laser was used as a probe in the time resolved measurements. The Ohmic contacts were opaquely covered during the measurements.

3. Experimental results

In the following, we discuss the photoluminescence of the wide well (QW-A). Figure 1 shows the integrated PL intensity of QW-A as a function of bias. Monochromated light from Tungsten lamp is used for excitation with the wavelengths λ_{EXC}=680nm (1.82eV) and 550nm (2.25eV). In case I (λ_{EXC}=680nm) where the photon energy is lower than MB gap, the carriers are directly excited inside the wells and no carrier separation is expected. In case II (λ_{EXC}=550nm) where the carriers are excited above MB, the direct separation of carrier can occur. In a reverse bias condition, holes will relax to QW-A, and electrons to QW-B. Hence, QW-A has a majority of holes and its PL intensity will be governed by its number of electron. Note that the PL intensity (point A in Fig. 1) exhibits a drastic quenching when an electric field is applied (point B and C) which is not seen in case I. This quenching reflects the sudden onset of the carrier separation which decreases the number of electron relaxing in QW-A. As PL intensity at B results from the recombination of non-separated EH pairs, the ratio $(I_A-I_B)/I_A$ gives us a measure ε_{CS} of the efficiency of carrier separation, where I_A and I_B are the PL intensity at point A and B. In Fig. 1, we have $\varepsilon_{CS} \approx 57\%$. The increase of PL intensity at larger reverse bias (point D, Fig. 1) seen in case I and II is not related with carrier separation. This phenomenon will be discussed elsewhere.

A series of PL study was done next by using an Ar^+ laser excitation with pump powers ranging from 1.9mWcm^{-2} to 5Wcm^{-2}. First, ε_{CS} was studied as a function of pump power

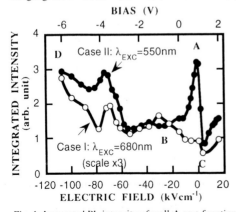

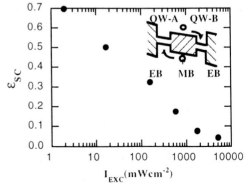

Fig. 1: Integrated PL intensity of well A as a function of electric field at two excitation wavelengths.

Fig. 2: ε_{SC} as a function of I_{EXC}. The inset shows the structure under a reverse bias condition.

I_{EXC}. As shown in Fig. 2, ε_{CS} is found to decrease rapidly when I_{EXC} is increased. This suggests that photogenerated carriers in MB are separated less efficiently due to the weakening of the field in MB. This weakening results from the opposite electric field created by the separated carriers population.

The peak position of the PL spectra of QW-A was studied and is plotted in Fig. 3 as a function of corrected bias voltage at different pump power. The bias was estimated by considering the potential drop due to the current through a contact resistance of $1k\Omega$. Note that the field-induced Stark shift decreases at high pump power. This decrease comes from the screening of the external field by separated carriers (SC) and reflects an increase in the concentration N_{SC} [5]. To interpret this data, we have calculated self-consistently the PL peak shift as a function of bias and carrier concentration. Results are shown in Fig. 4 for several electric fields. By comparing Fig. 4 with the data of Fig. 3, the concentration N_{SC} of SC was determined as a function of pump power I_{EXC}. As shown in Fig. 5, N_{SC} is not proportional to the pump power but increases slower. It is clear that N_{SC} is given by

$$N_{SC} = C \, \varepsilon_{SC} \, \tau_{SC} \, I_{EXC} \qquad (1)$$

where C is a proportionality constant that can be approximated as

$$C \approx T \exp(-\alpha \, d_{well}) (h\nu)^{-1} \approx 6.5 \ast 10^{14} mJ^{-1} \qquad (2)$$

with T the transmission of the gold layer, α the absorption coefficient of GaAs, d_{well} the total width of the wells and $h\nu$ the excitation photon energy. If we substitute into (1) the carrier concentration $N_{SC} = 1.4 \ast 10^{11}$ cm^{-2} at pump power $I_{EXC} = 250$mWcm^{-2}, $\varepsilon_{SC} = 20\%$ and $C = 6.5 \ast 10^{14}$mJ^{-1}, we find that $\tau_{SC} = 4.3\mu s$. If the lifetime τ_{SC} is constant, N_{SC} must be proportional to $\varepsilon_{SC} \ast I_{EXC}$. In Fig. 5, the dashed curve shows $\varepsilon_{SC} \ast I_{EXC}$ estimated from the values of Fig. 2. The good agreement with the curve for N_{SC} confirms that the slow increase of N_{SC} can be ascribed to the decrease of ε_{SC}. The highest carrier concentration found at -4V is $N_{SC} = 2.2 \ast 10^{11}$ carriers cm^{-2} for a pump power $I_{EXC} = 5$Wcm^{-2}.

In our calculation of the Stark shift, many-body effect was not considered. As the quenching of exciton compensates approximately the band-gap renormalization for carrier concentration lower than $3 \ast 10^{11}$ carriers cm^{-2} [6,7], Hartree approximation should be a good approximation in our case. Many-body effect gives an additional shift to the low energy side when carrier density increases. Hence the carrier concentration is underestimated.

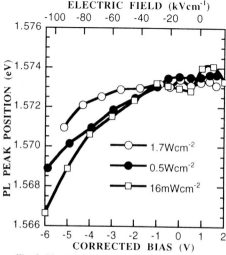

Fig. 3: PL peak position as a function of corrected bias for three excitation intensities.

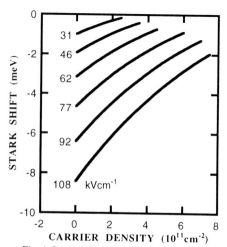

Fig. 4: Stark shift under different electric fields calculated as a function of carrier concentration.

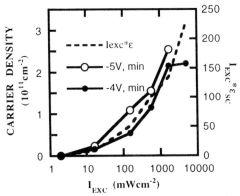
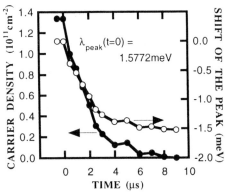

Fig. 5: $\varepsilon_{SC} \cdot I_{EXC}$ (dashed line) and carrier concentration measured at -4V and -5V (solid lines) as a function of excitation intensity

Fig. 6: Shift of the peak and carrier concentration at -6V as a function of time after the pump is turned off.

As the peak position changes sensitively with N_{SC}, we can estimate the lifetime τ_{SC} of the SC from the time variation of the peak position. For this experiment, carriers are generated above the MB by 30μs-long Ar+ laser pulse while the PL was measured using a weak Ti-Sapphire laser excitation at 724nm (1.71eV). In Fig. 6 the peak position of the photoluminescence spectra is plotted as a function of time after the Ar+ laser is turned off at pump power I_{EXC}=250mWcm^{-2} and bias U=-6V. Note that the Stark shift recovers, from which we find a lifetime τ_{SC}=2.3μs. This is consistent with the above estimation of τ_{SC}.

4. Conclusion

We have demonstrated that the spatial separation of photogenerated carriers in a novel quantum well structure results in an efficient modulation of carrier density N_{SC} with N_{SC} being $1.1 \ast 10^{11}$ cm^{-2} at a low pump power I_{EXC}=155mWcm^{-2}. When a high carrier density N_{SC}=2.5*10^{11} carriers cm^{-2} is reached, however, the efficiency ε_{SC} of spatial separation is reduced. A long lifetime of separated carriers τ_{SC}=2.3μs was found. An improvement of the sample quality and structure, such as the use of a higher and thicker external barrier, will further enhance τ_{SC} and ε_{SC} and will allow even more efficient modulation of the separated carrier density N_{SC}.

Helpful discussion with Professor G. Fasol and Professor T. Matsusue are gratefully acknowledged. One of the authors (M.R.) acknowledges partial financial support from the Japan Society for the Promotion of Science.

References

[1] G. Hasnain, G.H. Döhler, J.R. Whinnery, J.N. Miller, and A. Dienes, Appl. Phys. Lett. **49**(20), 1357-1359 (1986).
[2] P. Dawson, I. Galbraith, A.I. Kucharska, and C.T. Foxon, Appl. Phys. Lett. **58**(25), 2889-2891 (1991).
[3] A. Alexandrou, J.A. Kash, E.E. Mendez, M. Zachau, J.M. Hong, T. Fukuzawa and Y. Hase, Phys. Rev. B **42**(14), 9225-9228 (1990).
[4] A.R. Bonnefoi, D.H. Chow, T.C. McGill, R.D. Burnham, and F.A. Ponce, J. Vac. Sci. Tech. B **4**(4), 988-995 (1986).
[5] I. Galbraith, P. Dawson, and C.T. Foxon, Phys. Rev. B **45**(23), 13499-13508 (1991).
[6] G.E.W. Bauer and T. Ando, J. Phys. C **19**, 1537-1550 (1986).
[7] H. Yoshimura, G. E. W. Bauer, and H. Sakaki, Phys. Rev. B **38**(15), 10791-10797 (1988).

Resonant Tunneling in Semi-metal/Semiconductor, ErAs/(Al,Ga)As, Heterostructures

Kai Zhang, D.E. Brehmer, S.J. Allen, Jr. and C.J. Palmstrøm*

Quantum Institute, University of California, Santa Barbara, CA 93106
*Bellcore, Redbank, NJ 07701

Abstract. Resonant tunneling appears in GaAs/AlAs/ErAs/AlAs/GaAs structures fabricated by MBE on (211)B, (311)B, (511)B and (100) semi-insulating substrates and ErAs thicknesses of 5, 10 and 15 monolayers. The resonant channel is essentially temperature independent, it dominates the background current and exhibits differential negative resistance only below 150 K. The exact nature of the resonant tunneling channel is made clear by magneto-tunneling experiments. Magnetic fields below 8 Tesla are sufficient to produce an anomalously large splitting of the resonant channel. This result is consistent with the known, large, Er 4f exchange enhanced, splitting of the conduction band in ErAs and implies that the resonant channel is a quantized electron state.

I. Introduction

Hybrid metal or semimetal/semiconductor heterostructures have promising applications for three terminal quantum devices and other novel metal/semiconductor devices, such as metal base transistors and random distributed Schottky diode arrays buried in a semiconductor [1]. The key concerns for synthesis of such hybrid structures have been the difference in crystal structures, thermal stability and growth compatibility [2]. Rare-earth monoarsenides (REAs), such as ErAs, have been shown to be excellent choices as thermodynamically stable compounds for use in epitaxial semi-metal/semiconductor heterostructures with III-V semiconductors [3]. Previous work [3] demonstrated that pinhole-free ErAs thin films down to 3 ml can be epitaxially grown on GaAs. Transport measurements [4] of relatively thick, (200Å), ErAs films buried in GaAs confirmed semimetallic conduction in this material as predicted earlier [5]. Unlike other semimetals, ErAs shows no semimetal-semiconductor transition as the thickness is reduced. Also, the differences in material properties of ErAs and GaAs allow reliable and reproducible selective contact to the thinnest of ErAs layers [1]. This makes ErAs quite appealing for active metal film applications. This hybrid structure is essentially a building block for metal/semiconductor electronic devices. With the exception of some early, tentative results on NiAl[6], little has been known about vertical transport in such a structure.

In this paper, we present our experimental work on the vertical electrical transport through ErAs/III-V semiconductor heterostructures which exhibit clear and reproducible resonant tunneling through a semi-metallic quantum well. Most striking is the anomalously large magnetic field dependence which is caused by the exchange interaction between the quantum states and the localized moments on the Er ions.

II. Experimental

The test devices used in this experiment have a double barrier resonant tunneling diode (RTD) structure. It consists of a thin ErAs well (5~15 ml) sandwiched between 20Å AlAs double barriers, with a 100Å GaAs spacer layer and n+GaAs electrode on each side, as schematically shown in figure 1. The devices were grown on GaAs semi-insulating substrates in a Varian Gen II solid source molecular beam epitaxy system. Previous work [7] showed that the electrical properties of ErAs remain unchanged as the growth temperature varies from 420°C to 580°C, while RHEED oscillation was observed only at the growth temperatures ≤ 450°C [8]. The lowest temperature for which RHEED oscillation are observed has not been determined. However, the overgrowth of GaAs (or AlAs) proceeds in a 3D island growth mode due to the poor wettability of GaAs on ErAs. Based on these earlier results and our recent growth study, the ErAs well and the overgrown layers were grown at 450°C and 550°C respectively in this experiment. Substrates with different surface orientation, (100), (511)B, (311)B and (211)B, were used to examine the effects of orientation on the overgrowth and transport properties.

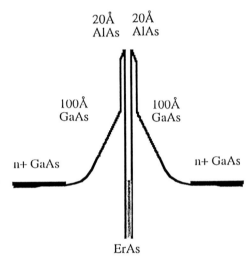

Figure 1. Schematic conduction band diagram of an ErAs RTD. The hatched region denotes well filling.

The device was formed by mesa etching. The bottom and top ohmic contacts were provided by alloyed Ni/AuGe/Ni/Au and non-alloyed Ti/Au, respectively. The devices were characterized by dc I~V measurement at different temperatures. Magnetotunneling measurement at low temperatures (>4.5K) was carried out to study the nature of the observed resonant tunneling.

III. Results and Discussion

The dc electrical measurement of devices indicated negative differential resistivity (NDR) at temperatures below 150K. The I~V curve measured at 12K for a 10ml RTD grown on (311)B substrate is shown in figure 2. Typical of all of the samples, the I~V curve is not symmetric. The NDR appeared only at the forward bias (bottom contact negative), even though the structure was symmetric. This is believed to be due to the difference in crystalline quality

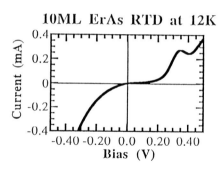

Figure 2. This typical I~V curve for an ErAs RTD shows NDR when the top contact is positive.

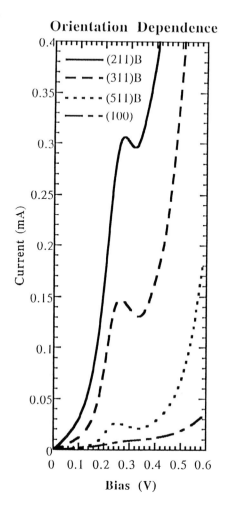

Figure 3. The strong orientation dependence of the background tunneling current correlates well with variation of the schottky barrier height

between upper and lower ErAs/AlAs interfaces. Earlier work [1] pointed out that the overgrowth of GaAs (or AlAs) on ErAs is the key issue to synthesis of this hybrid structure. The difficulty with the overgrowth arises mainly from the poor wettability of GaAs on ErAs and the difference in symmetry of crystalline structures between ErAs (NaCl structure) and GaAs (zincblende). It is the poor wettability that results in the 3D island growth of GaAs(AlAs) on ErAs, as strongly suggested by previous cross-section TEM studies of the ErAs/GaAs interfaces [9]. A spotty RHEED pattern usually appears immediately after the overgrowth initiates.

The wetting of GaAs (or AlAs) on ErAs may be improved either using a surfactant or growing on the substrate with a low energy surface. In this experiment, differently oriented GaAs substrates, (100), (511)B, (311)B and (211)B, were used. The preliminary results indicated that the overgrowth is improved by growing on (311)B, as indicated by a streaky RHEED pattern during the overgrowth and smooth surface morphology. However, no attempt has been made in this paper to correlate the quality of overgrown layers to the measured transport properties because of the complicated nature of the quantum mechanical transport in an ErAs/(Al, Ga)As heterostructure.

Figure 3 shows the I~V curves measured at 4.5K for devices grown on substrates with different surface orientation. The tunneling current exhibited a strong orientation dependence. This is consistent with previous results [10]. That is, the Schottky barrier height of ErAs on GaAs varies with the GaAs surface orientation, decreasing almost linearly from (100) to (111)B oriented surfaces.

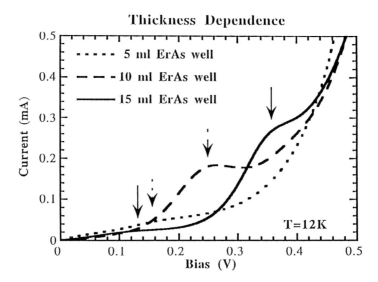

Figure 4. Anomalous behavior of the resonant channel seems to indicate that the tunnel current is through confined hole states in the well.

Since ErAs is a semimetal, either quantized electron or hole states in the well could serve as a resonant channel. On the other hand, localized defect states at the ErAs/AlAs interface, particularly at the overgrown interface, could also provide a channel for resonant tunneling. To clarify whether the resonant channel is a quantized state in the ErAs well or a localized defect state at the interface, RTDs with ErAs thicknesses of 5, 10, and 15 ml were fabricated on (311)B GaAs substrates. The I~V curves measured at 12K are shown in figure 4. They show what seems to be an anomalous dependence on the ErAs well thickness. This significant thickness dependence implies that the resonant channel is a quantized state in the ErAs well, although this data is not sufficient to clarify whether it is an electron or hole state. Since very different behavior is expected for electron and hole states in a magnetic field [11], magneto-tunneling measurements were made. As described below, these results make it clear that the resonant channel is a quantized electron state.

The magneto-tunneling measurement was carried out at a fixed temperature of 4.5K while varying the field and at a fixed field of 8T while varying the temperature. The external magnetic field (B) was parallel to the

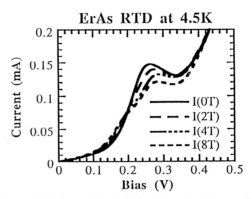

Figure 5. Variation of the I~V characteristic with magnetic fields from 0 to 8 Tesla.

electron tunneling direction (i.e., growth direction). The measured I~V characteristics at different B fields up to 8 Tesla are shown in figure 5. Plotting dI/dV against V reveals that the resonant channel starts splitting at a field of ~1.5T, and saturates above 4T, as shown in figure 6. This observed splitting is consistent with the known, large, Er 4f exchange enhanced, splitting of the conduction band in ErAs. Splittings taken from this data and data measured as a function of temperature at B=8T are plotted versus B/T in figure 7. Except for an offset near the origin, this data behaves qualitatively like the Brillouin function expected

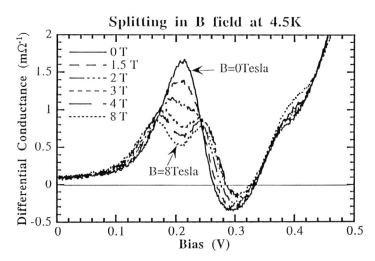

Figure 6. The deduced dI/dV~V curve clearly shows a huge splitting in the resonant level. Opening up abruptly between 1.5 and 2.5 Tesla. The splitting is nearly saturated above 4T.

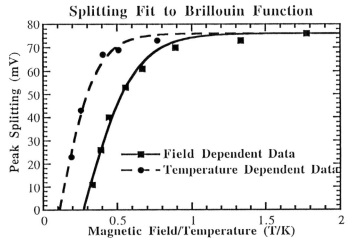

Figure 7. Fitting the splitting to the shape of the Brillouin function gives g=9.1 and 12.5 for the field dependent data and the temperature dependent data respectively.

for alignment and saturation of localized spins in a magnetic field. For the temperature dependent data this offset may be due to demagnetization effects. The field dependent data has an additional offset that has been determined to arise from the necessity to overcome the intrinsic antiferromagnetic interaction between Er spins.

Fitting the shape of this data to the Brillouin function, we obtain g=9.1 and 12.5 from the field data and the temperature data respectively. This compares well with that derived from magnetoresistance measurements of g=10.6 [12] and from neutron scattering in ErP (g=11.2) and ErSb (g=14.0) [13]. From the magnitude of the saturated level splitting, ~40meV, we estimate the exchange interaction between the 4f electrons and the conduction electrons to be Γ=3.4eV•Å^3 which agrees well with that known for Erbium metal, Γ=3.3eV•Å^3.

IV. Summary

We have found resonant tunneling through confined states in thin ErAs semimetal quantum wells to be robust, with NDR persisting to 150K. A strong dependence of the background current on orientation is accounted for by variation of the Schottky barrier height. The strength of the current through the resonant channel also has a strong orientation dependence but no attempt has yet been made to account for it due to the complex nature of the wavefunction matching at the ErAs/GaAs interface which must occur for the transmitted electrons. An anomolously large splitting in a magnetic field is exhibited by the resonant channel which has been determined to be through electron subbands in the ErAs quantum well.

This work is supported by the ONR, the AFOSR and the NFS. The authors greatly appreciate Art Gossard, Herb Kroemer and John English for their help which has made it possible to grow ErAs at UCSB.

References

[1] S.J. Allen, D.E. Brehmer and C.J. Palmstrøm, "Rare Earth Doped Semiconductors", edited by G.S. Pomrenke, P.B. Klein and D.W. Langer, MRS Sym. Proc. 301, p.307 (1993).
[2] T. Sands, C.J. Palmstrøm, J.P Harbison, V.G. Keramidas, N. Tabatabaie, T.L. Cheeks, R. Ramesh and Y. Silberberg, Mat. Sci. Rep. 5, 99 (1990).
[3] C. J. Palmstrøm, K.C. Garrison, S. Mounier, T. Sands, C.L Schwartz, N. Tabatabaie, S.J. Allen, Jr., H.L. Gilchrist, and P. F. Miceli, J. Vac. Sci. Technol. B7, 747 (1989).
[4] S.J. Allen, Jr., F. DeRosa, C.J. Palmstrøm, and A. Zrenner, Phys. Rev. B43, 9599 (1991).
[5] A. Hasegawa and A. Yanase, J. Phys. Soc. Jpn. 42, 492 (1977).
[6] N. Tabatabaie, T. Sands, J.P. Harbison. H.L. Gilchrist and V.G. Keramidas, Appl. Phys. Lett. 53, 2528 (1988).
[7] J.D. Ralston, H. Ennen, P. Wennekers, P. Hiesinger, N. Herres, J. Schneider, H.D. Muller, W. Rothemund, F. Fuchs, J. Schmalzlin and K. Thonke, J. Electronic Mater. 19, 555 (1990).
[8] C.J. Palmstrøm (unpublished)
[9] J.G. Zhu, C.J. Palmstrøm and C.B. Carter, Mat. Sci.Proc. V198, 177 (1990).
[10] C.J. Palmstrøm, T.L. Cheeks, H.L. Gilchrist, J.G Zhu, C.B. Carter, B.J. Wilkens and R. Martin, J. Vac. Sci. Technol. A10, 1946 (1992).
[11] A.G. Petukhov, W.R.L. Lambrecht and B. Segall, Phys. Rev. Lett. (in press).
[12] S.J. Allen, Jr., N.Tabatabaie, C.J. Palmstrøm, G.W. Hull, T. Sands, F. DeRosa, H.L. Gilchrist and K.C. Garrison, Phys. Rev. Lett. 62, 2309 (1989).
[13] H.R. Child, M.K. Wilkinson, J.W. Cable, W.C. Koehler and E.O. Wollan, Phys. Rev. 131, 922 (1963).

Hole charging effects in Sb-based resonant tunneling structures

J.N. Schulman and D.H. Chow

Hughes Research Laboratories, Malibu, CA 90265

Abstract. The InAs/AlGaSb double barrier structure exhibits normal conduction band resonant tunneling for high aluminum concentrations. However, the presence of quantum wells for holes in the AlGaSb barrier layers for 40% to 60% aluminum concentrations allows the collection of charge which has a strong influence on the potential profile, and thus the current-voltage characteristic. Here we show results of calculations that demonstrate a strong voltage compression effect in which the expected current-voltage curve, including the tunneling resonance, is shifted to much lower voltages. We also discuss possible causes of the current controlled S-shaped negative resistance observed experimentally.

1. Introduction

The type-II band alignment found in Sb-based tunnel structures is well known to produce unique tunneling behavior due to the proximity in energy of the valence band maximum of one component with the conduction band minimum of another. This same band proximity can lead to situations in which a significant amount of charge can transfer among the layers of the structure. By contrast, the InAs/AlGaSb double barrier structure has a band diagram in which only normal resonant tunneling might be expected. Although it is a Type-II band alignment, it has staggered offsets in which the conduction band of any given component is above the valence bands of the others (Figure 1). Interband tunneling is therefore not expected, and it is also not obvious whether charge transfer would take place. However, experiments have shown unusual current-voltage characteristics which may indeed be demonstrating charging effects [1]. It has been conjectured that the charging that occurs is mainly due to holes occupying the hole quantum well formed by the AlGaSb valence band offset relative to InAs. Although normally unoccupied, a moderate applied bias pulls down the collector InAs conduction band far enough to expose the well and allow electrons to tunnel out (holes tunneling in). The dependence of this effect on the Ga content (it goes away for both alloy endpoints) is strong evidence for this. Although there is as yet no explanation of the anomalous characteristics observed, our calculations duplicate some aspects of experiment, especially a strong compression along the voltage axis.

2. Calculation

The calculation was based on a two-band tight-binding model for the tunneling transmission coefficient [2]. The emitter and collector band bending was included by solving Poisson's equation. The calculation is done as a function of the internal voltage drop across the well and barriers only, with the band bending added in separately. The two familiar effects of the

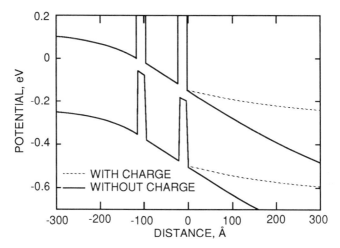

Figure 1. Band profile of InAs/AlGaSb double barrier structure with and without inclusion of hole charge in AlGaSb collector barrier.

band bending are that the emitter side accumulation region increases the charge density on the incoming side, and that the collector side depletion region increases the voltage drop there. This second aspect is the one strongly influenced by the accumulation of holes in the collector side AlGaSb barrier.

The charge due to the resonant tunneling electrons was calculated, but its electrostatic feedback on the potential profile was not included. Instead, focus was on the charge in the AlGaSb hole well, which was included self-consistently. The main effect is to supply positive charge that reduces the electric field that would otherwise decay gradually into the collector cladding region with a length scale dependent on the collector doping. In general, the amount of charge in the hole well increases as the Fermi level in the collector is lowered relative to the quantum confined hole level in the barrier. This allows more and more of the exposed electrons to tunnel out, but the exact processes are unknown and may involve a trade-off with other charge flowing in. For example, for larger applied fields the triangular barrier formed next to the AlGaSb layers by the InAs well valence band becomes quite steep. This allows the holes to tunnel out and thus limits their accumulation. For purposes of estimation we ignored this extra current and simply assumed that the Fermi level in the well equaled that in the collector. This is appropriate for smaller biases, but breaks down when the fields become large.

3. Results

Figure 1 demonstrates the effect of the hole well charge by plotting the calculated potential profile of the double barriers with and without the charge included. The plot is for a 25 monolayer (ML) InAs well and 6 ML $Al_{0.5}Ga_{0.5}Sb$ barriers. The temperature is 300 K and the carrier concentration is 10^{17} cm^{-3} on both sides. The heavy hole confinement energy is estimated to be 0.112 eV relative to the AlGaSb valence band edge. The internal voltage drop (across the double barriers only) is 0.15 V in both cases. This compares to the internal voltage at resonance which is calculated to be 0.22 V for this structure. The total voltage drop (Fermi level to Fermi level) is 0.39 V with the hole charge included, and 0.90 V without the charge. It can be seen that the collector depletion layer voltage drop is greatly reduced with the hole charge included.

The current-voltage curve is correspondingly compressed to smaller voltages. Figure 2 shows the room temperature current versus total voltage for the same two cases. The peak voltage is near 0.5 V, when the charge is included. There is one experimental comparison that is available. Measurements on a standard InAs/AlSb double barrier diode with layer thicknesses of 6/25/6 ML, 50 Å 2×10^{16} cm^2 spacers and 250 Å 10^{17} cm^2 cladding layers before highly doped contact layers show a voltage at peak current close to 1.0 volts. This is to be compared with an 8/25/8 ML InAs/Al$_{0.5}$Ga$_{0.5}$Sb device with 100 Å 2×10^{16} cm^2 spacers and 500 Å 10^{17} cm^2 cladding layers which exhibits normal resonant tunneling. Its peak voltage was measured to be just 0.48 volts, despite the larger spacer layers.

Figure 3 shows how the band bending outside the barrier on the collector side varies as a function of the internal voltage due to the influence of the confined hole charge. The magnitudes when the charge is included are much smaller, tens of millivolts instead of the tenths of volts which is the case without the charge. For small applied fields the hole charge is not yet built up and the band bending increases linearly. As the field increases, the holes screen the collector field more and more effectively. As mentioned above, this would be limited at large fields as the holes tunnel out through the through the triangular InAs well potential.

4. Conclusion

Hole trapping in the AlGaSb barriers has a strong effect on the current-voltage characteristics. This is already apparent in the voltage compression of the resonant tunneling peak. Additional ingredients are needed to produce the S-shaped curves previously reported [1].

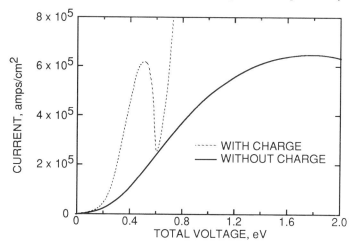

Figure 2. Current vs. voltage characteristics of InAs/AlGaSb double barrier structure with and without inclusion of hole charge in collector AlGaSb barrier.

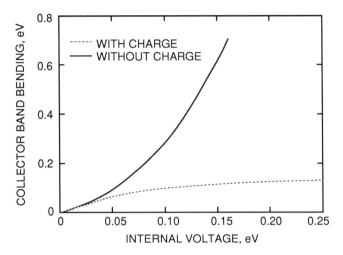

Figure 3. Collector side band bending vs. voltage drop across well and double only, for InAs/AlGaSb double structure including hole charge in AlGaSb collector barrier.

There are two other sources of charge in this structure: the hole charge in the emitter AlGaSb barrier, and the charge due to the electrons tunneling through the resonance. The magnitudes of these charges are all voltage dependent which in turn influences the potential profile. It would not be surprising if the interplay between these charges produces novel multiple solutions for the current-voltage characteristics.

References

[1]   Chow D H and Schulman J N 1994 *Appl. Phys. Lett.* **64** 76-8
[2]   Schulman J N and Waldner M 1988 *J. Appl. Phys.* **63** 2859-61

Resonant interband tunneling quantum functional device

S. Tehrani, J. Shen, H. Goronkin, G. Kramer, R. Tsui, and T. X. Zhu

Motorola Inc., Phoenix Corporate Research Laboratories
2100 E. Elliot Rd., Tempe, AZ 85284 USA

Abstract. Resonant interband tunneling based on InAs/AlSb/GaSb is used to demonstrate a SRAM unit cell and a three terminal device. The bistability and the switching principles for SRAM unit cell are shown and several key issues involving the applications of this device are also discussed. A planar three-terminal Resonant Interband Tunneling Field Effect Transistor (RITFET) has been fabricated. It is a resonant interband tunnel diode (RITD) combined with a pseudomorphic HFET in which the current is controlled by a Schottky gate. The three terminal device has current-voltage characteristics that exhibit pronounced negative differential resistance and a hard pinch-off at room temperature, thus making it a suitable candidate for the utilization of quantum functional devices in low-power applications.

1. Introduction

In CMOS technology, leakage current tunneling through the gate oxide and drain induced barrier lowering will eventually create obstacles to further scaling. An alternate technology that can make use of quantum effects to produce multifunctionality will become necessary in the near future. We are utilizing interband resonant tunneling in antimonide-based devices to develop devices having inceased functionality. In this way, one device can replace many conventional devices with the result so that circuit functionality can be substantially increased without additional interconnect metalization or semiconductor real estate.

The need for high-speed, high-density static RAM's (SRAM's) for enhancing the system performance of computers, microcomputers, workstations, terminals, etc. is rapidly increasing. While significant improvements in SRAM's can be achieved by scaling device dimensions and by cell layout innovations [1], a fast, compact, scalable single device SRAM is an attractive possibility. Resonant tunneling diode (RTD) based SRAM's [2] are potential candidates for such future generation memory devices when conventional devices reach the scaling limit in the sub $0.1\mu m$ regime.

Three-terminal quantum devices have been studied by a number of research groups. Some examples are the Resonant Hot Electron Transistor [3], the Resonant Tunneling Bipolar Transistor [4,5], a three-terminal device using a double barrier resonant tunnel diode in which the quantum well is modulated by a p^+ region from the sides [6], a resonant interband tunneling transistor in which AlGaAs/GaAs layers are grown on the etched sidewall of a GaAs p^+-i-n^+ stack [7] in which the gate potential controls a surface channel through which interband tunneling occurs, and a resonant tunneling transistor which combines a resonant tunneling diode and a MESFET [8].

We have taken advantage of the superior peak-to-valley ratios (P/V) in the type-II InAs/AlSb/GaSb material system [9] to fabricate Resonant Interband Tunneling Diode (RITD)-based devices. In this paper, we report on a single device SRAM and a planar three-terminal Resonant Interband Tunneling Field Effect Transistor (RITFET).

2. Static Read Access Memory (SRAM)

When two RTD's are connected in series and biased appropriately, the middle node is bistable which can be utilized for SRAM applications. Although it is straight forward to make a single memory cell with two RTD's, the addressing scheme must be capable of utilizing the speed and compactness of the RTD's. Mori et al. [2] have proposed and demonstrated a scheme that uses a tunnel diode (TD) connected to the middle node (Fig. 1). The tunnel diode current is used to read and write to the memory cell. A large peak-to-valley (P/V) current ratio is crucial for high density memory applications because the standby power consumption is proportional to the valley current and the speed is proportional to the peak current. Thus, we utilized high P/V ratios in the Resonant Interband Tunneling Diodes (RITD's) to demonstrate the SRAM.

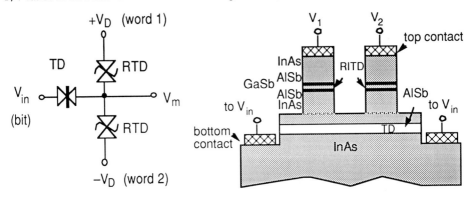

Fig. 1. Equivalent circuits of RTD SRAM's. The cell is selected when a differential voltage between V_D and V_{in} are applied.

Fig. 2. The layer structure of an SRAM cell based on RITDs in the InAs/AlSb/GaSb material system.

the cross-section of an SRAM cell is shown in Fig. 2. The epitaxial layers are grown on a GaAs substrate with a buffer layer containing a 10-period GaSb/AlSb superlattice for smoothing. The subsequent active layers are: n+ (Si, 2×10^{18} cm^{-3}) InAs (~5000Å), AlSb (1000Å), n+ InAs (~3000Å), n-(Si, 2×10^{16} cm^{-3}) InAs (~500Å), undoped InAs (100Å), AlSb (25Å), GaSb (65Å), AlSb (25Å), undoped InAs (100Å), n- InAs (500Å), and n+ InAs (2500Å).

To fabricate the SRAM's, the following procedures are used: The top contacts are defined and Ni/Ge/Au ohmic contacts are evaporated and lifted off. The RITD mesa is selectively etched into the bottom RITD InAs layer. The AlSb tunneling diode layer and part of the InAs layer is etched to define and evaporate the bottom contct layer, mesas are formed and the structure is passivated using Si_3Ni_4. Contact via holes are opened and interconnect metal is evaporated.

Electron transport through the resonant tunneling diodes in our SRAM cell is by interband tunneling. At resonance, electrons in the InAs conduction band tunnel through valence-band states in the GaSb. When off resonance, the InAs electrons are blocked by the bandgap of GaSb. Because of this bandgap blocking, the P/V ratio is expected to be much larger than in type-I GaAs/GaAlAs or InGaAs/InAlAs material systems. We have achieved room temperature P/V ratios as high as 30 [9]. Typical values range from 15-25. Current through the tunneling diode is used to switch the bistable point in the memory cell. To select an individual cell, a differential voltage is applied to the bit line and the word lines of the cell.

The current-voltage characteristics of two RITD's in series are plotted in Fig. 3. The P/V ratio in these wafers is around 18 at room temperature. The memory characteristic of an

SRAM cell is shown in Fig. 4. The high and low states are represented by two voltages (+0.14 and –0.14V) at the middle node. Switching from one state to the other occurs when current through the TD exceeds the peak current of the RITD. The switching voltages depend on the threshold voltage of the TD. A significant difference between the SRAM based on the InAs/AlSb/GaSb RITD's and the one based on the InGaAs/InAlAs RITD's [2] is the output voltage in the SRAM in the middle node. The RITD-based SRAM uses lower DC bias because of the smaller valley voltage (about half that of an RTD).

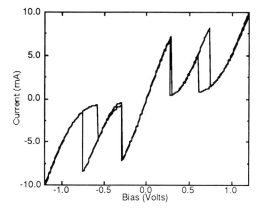
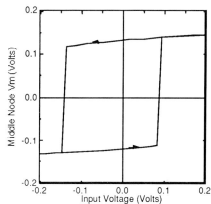

Fig. 3. I-V characteristics of two RITD's in series with the P/V ratio of 18 at 300K. The valley voltage of 0.3 Volts determines the operating point of the SRAM cell. Hysteresis appeared between the forward and reverse voltage scans.

Fig. 4. Demonstrated input-output characteristics of an SRAM cell. In this demonstration, the two diodes are biased at ±0.14V. The vertical axis is middle node voltage Vm.

We have estimated the optimal density of the RITD-based SRAM's and compared it to the currently most compact CMOS designs [1]. The RITD-based SRAM can save a factor of two in cell area and approximately a factor of four when the diodes are stacked on top of each other. The key parameter that needs to be improved is the P/V current ratio. The stand-by power consumption is limited by the valley current. The valley current density in the present devices is in the order of 10 A/cm^2 for a 1μm^2 diode, a 64 Mbit SRAM will consume approximately 6.4 W at 1 Volt. To retain or improve the access speed of the SRAM, a large drive current is necessary. The drive current is limited by the peak current of the RITDs. Again, for a diode size of 1μm^2 and a drive current of 1mA, the peak current density is 10^5 A/cm^2. Thus, depending on the applications, the P/V current ratio of the RITD needs to be improved by several orders of magnitude before it can be used in ULSI memories.

3. *Resonant Interband Tunneling FET (RITFET)*

The RITFET consists of a resonant interband tunneling diode (RITD) integrated with an AlGaAs/InGaAs pseudomorphic HFET in which the current is controlled by a Schottky gate. This structure takes advantage of the high peak-to-valley (P/V) current ratios that can be achieved in an antimonide-based RITD [9] and the excellent gate characteristics and superior modulation control of the InGaAs channel HFET. The current-voltage characteristics have pronounced negative differential resistance (NDR) and a hard pinch-off.

Fig. 5 shows the schematic cross-section of the device structure. The epitaxial layers are grown by MBE on a semi-insulating GaAs substrate. First, a buffer region with a GaAs/AlAs superlattice is grown. The HFET portion of the structure, with a pseudomorphic $In_{0.15}Ga_{0.85}As$ channel and a n^+ –GaAs cap, is then grown. Next, a step-graded InGaAs region is grown to accomodate the lattice mismatch between the n^+ –GaAs cap and the InAs bottom contact layer of the RITD. The layer sequence of the RITD has been reported previously [9], with the exception that the double barrier region is symmetric with a 6.5 nm GaSb quantum well and two AlSb barriers each 2.5 nm thick.

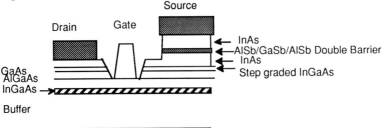

Fig. 5. Schematic cross section of the Resonant Interband Tunneling FET.

The device fabrication process begins with the evaporation of source contact metal which is used as a mask to etch into the InAs bottom contact layer of the RITD. The mesa isolation pattern is then defined and the structure is etched down to the GaAs in the buffer region. The drain contact is defined on the bottom InAs layer and evaporated. The gate is patterned, and the InAs bottom contact layer of the RITD, the InGaAs step-graded region and the GaAs cap layer are removed selectively down to the AlGaAs layer. Finally, Ti/Al gate metal layers are evaporated and lifted.

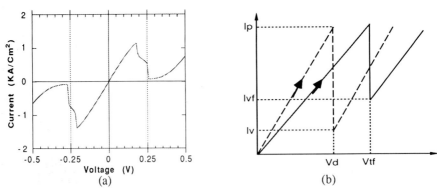

Fig. 6. (a) Experimental I–V characteristics of a RITD at room temperature. (b) Illustration of the operation principle of a RITFET (in forward direction mode) as a combination of a RITD and a FET channel.

The current-voltage characteristics of a RITD with a diameter of 44 μm are shown in Fig.6a. At room temperature, a peak current density, I_p, of 1.4 KA/cm^2 and a P/V current ratio, I_p/I_v, of 18 were achieved. No hysterisis was observed in these devices.

The I_{ds}–V_{ds} characteristics of the RITFET are shown in Fig. 7. NDR is clearly exhibited as well as solid pinch-off and low output conductance. In RITFETs the Schottky gate modulates the potential on the channel side of the RITD. The combination of this potential and the drain voltage will determine the potential across the RITD. The P/V current ratios observed in the RITFET are much lower than those observed in the corresponding RITDs. This is because the RITD is in series with the InGaAs channel of finite resistance, R_{ch}. Depending on whether the applied bias on the RITFET, V_{ds}, is swept from zero (forward direction, as in Fig. 7a) or to zero (backward direction, as in Fig. 7b), the observed NDR region will occur at a higher voltage, V_{tf}, or a lower voltage, V_{tb}.

Let us first examine the forward direction case, as illustrated in Fig. 6b. The dashed curve is the I–V characteristic of a RITD. The solid curve is the I–V characteristic of a RITD with an additional extrinsic series resistance analagous to the channel resistance of a RITFET. The peak current of the RITFET is the same as that of the RITD, I_p, while the observed valley current of the RITFET, I_{vf}, is much higher than that of the RITD, I_v. The onset voltage (for NDR) of the RITFET is given by

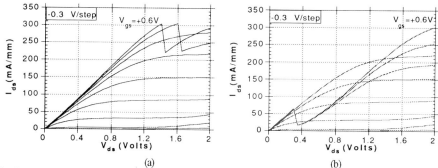

Fig. 7. I_{ds}–V_{ds} characteristics of the RITFET at room temperature. The gate length is 1.2 µm, the device width 100 µm, and the first sweep corresponds to a gate voltage of +0.6 V. The change in gate voltage was -0.3 V/step. V_{ds} was swept (a) from 0 V to 2.0 V, and (b) from 2.0 V to 0 V.

$$V_{tf} = I_p \times R_{di} + I_p \times R_{ch} = V_d + I_p \times R_{ch},$$

where V_d corresponds to the internal onset voltage of the RITD. Similarly, assuming the RITD has the same differential resistance, R_{di}, on both sides of the NDR region, the observed valley current of the RITFET, I_{vf}, can be derived as

$$I_{vf} = (I_v \times R_{di} + I_p \times R_{ch})/(R_{di} + R_{ch}).$$

In other words, when the value of R_{ch} is very high, the observed P/V ratio is low, since $I_{vf} \approx I_p$. On the other hand, when $R_{ch} \ll R_{di}$, the high intrinsic P/V current ratio of the RITD will appear in the RITFET.

Fig. 7b shows the I_{ds}–V_{ds} characteristics for a RITFET biased from 2.0 V to 0 V. In this case, the observed onset voltage V_{tb} is closer to the internal valley voltage V_d of the RITD,

$$V_{tb} = V_d + I_v \times R_{ch}.$$

Although the valley current of the RITFET is the same as that of the RITD, I_v, the observed peak current of the RITFET, I_{pb}, is decresed from I_p as given by

$$I_{pb} = (I_p \times R_{di} + I_v \times R_{ch})/(R_{di} + R_{ch}).$$

The drain current as a function of the gate voltage is shown in Fig. 8. The drain was biased from 2.0 V to 1.2 V in 0.2 V steps. The first four sweeps (2.0 V to 1.4 V) showed clear

NDR while none was observed for drain voltages below 1.4 V. The gate voltage was swept from -2.0 V to 0.8 V in Fig. 8a, and from 0.8 V to -2.0 V in Fig. 8b.

It can be seen that in order to take full advantage of the high P/V current ratios of the RITD in the RITFET structures, the following criteria have to be met. 1) The saturation current in the channel has to be larger than or equal to the peak current of the RITD, that is, $I_{dsat} \geq I_p$, 2) The internal resistance of the RITD has to be larger than the channel resistance, that is, $R_{di} > R_{ch}$.

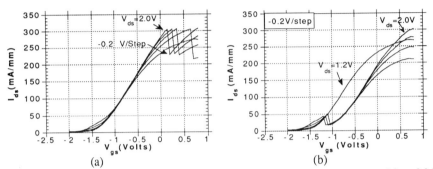

Fig. 8. I_{ds}–V_{gs} characteristics of the RITFET at room temperature. The drain voltage was steped from 2.0 V to 1.2 V in 0.2 V steps. (a) V_{gs} was swept from -2.0 V to 0.8 V, and (b) from 0.8 V to -2.0 V.

Summary

We have fabricated RITD-based SRAM's in the InAs/AlSb/GaSb material system. The bistability and switching principles are demonstrated. A planar three-terminal Resonant Interband Tunneling Field Effect Transistor (RITFET) has been fabricated. The device has current-voltage characteristics that exhibit pronounced negative differential resistance and a hard pinch-off.

Acknowledgment-This work was partially performed under the management of FED (the R&D Association for Future Electron Devices) as a part of the R&D of Basic Technology for Future Industries supported by NEDO (New Energy and Industrial Technology Development Organization, Japan).

References

[1] K. Sasaki, K. Ueda, K. Takasugi, H. Toyoshima, K. Ishibashi, T. Yamanaka, N. Hashimoto, and N. Ohki, IEEE J. Solid-State Circuits, 28, 1125, (1993).
[2] T. Mori, S. Muto, H. Tamura, and N. Yokoyama, Extended Abstracts of the 1993 International Conference on Solid State Devices and Materials, Makuhari, 1074 (1993).
[3] N. Yokoyama, K. Imamura, S. Muto, S. Hiyamizu, H. Nishi, Jpn. J. Appl. Phys. 24, L853 (1985).
[4] F. Capasso, S. Sen, A. C. Gossard, A. L. Hutchinson, J. H. English, IEEE Electron. Dev. Lett. EDL-7, 573 (1986).
[5] A.C. Seabaugh, A.H. Taddiken, E.A. Beam III, J.N. Randall, Y.-C. Kao, B. Newell, IEDM 93-149 (1993).
[6] K. Maezawa, T. Mizutani, Jpn. J. Appl. Phys. 32, 42 (1993).
[7] T. Uemura and T. Baba, Jpn. J. Appl. Phys. 33, L207 (1994).
[8] A. R. Bonnefoi, T. C. McGill, and R. D. Burnham, IEEE Electron Device Lett., EDL-6, 636 (1985).
[9] S. Tehrani, J. Shen, H. Goronkin, G. Kramer, M. Hoogstra, and T. X. Zhu, Int. Symp. on GaAs and Related Compounds, Freiburg, Inst. Phys. Conf. Ser. No. 136, 209, (1994).

Investigation of magneto-tunneling processes in InAs/AlSb/GaSb based resonant interband tunneling structures

Harald Obloh, Johannes Schmitz, John D. Ralston

Fraunhofer Institut für Angewandte Festkörperphysik,
Tullastraße 72, 79108 Freiburg, Germany

Abstract: AlAs monolayers grown adjacent to the AlSb barriers in InAs/AlSb/GaSb/AlSb/InAs resonant interband tunneling diodes have recently been proposed in order to improve the peak-to-valley ratio of these devices. We report the effects of high magnetic fields applied parallel and perpendicular to the tunnel current. Clearly-resolved oscillatory behaviour in the I(V) characteristics with the magnetic field parallel to the tunneling current can be attributed to resonant tunneling via Landau level states of the hole states in the GaSb well. The magnetotunneling data reveal much better confinement of the hole states in the GaSb well in tunneling structures with AlAs-hole-barriers as compared to devices with no hole barriers.

1. Introduction

Quantum mechanical tunneling in semiconductors and negative differential resistance (NDR) were first demonstrated in forward biased Esaki diodes [1]. Later, due to tremendous improvements in epitaxial sample growth, resonant tunneling diodes (RTD) based on a variety of III-V semiconductores have been grown and extensively analysed. During the past few years, the nearly lattice matched InAs/GaSb/AlSb material system has attracted much attention because it offers a large degree of flexibility for growing tunneling devices that can provide both high speed and low excess current operation. In particular, the type II character of the band offset (broken gap structure) and effective band-gap blocking allows a new class of tunnel structures known as resonant interband tunneling diode (RITD), which can be considered as a combination of a resonant tunneling diode and an Esaki diode.

The InAs/AlSb/GaSb/AlSb/InAs RITD is a promising candidate of this new class of tunnel structures. In these structures, resonant tunneling occurs when the occupied electron states in the InAs align with the unoccupied hole states in the GaSb quantum well. The tunneling is blocked when the electron states align with the bandgap of GaSb, resulting in the valley-current. In order to reduce the valley-current and to improve the peak-to-valley current ratio, hole barriers have been inserted by replacing a portion of the AlSb barrier with two monolayers of AlAs [2]. Negative differential resistance with peak-to-valley current ratios up to 30:1 and peak current densities in the order of kA/cm^2 have been achieved at room temperature. Recently, resonant Raman scattering experiments [3] performed on InAs/AlSb quantum wells with monolayers of AlAs deposited at the interfaces have shown that instead of binary AlAs interfaces and AlSb barriers, a strong exchange among the As and Sb atoms occurs resulting in pseudo-ternary $AlSb_{1-y}As_x$ barriers.

In this paper we report on resonant magneto-tunneling experiments performed on two series of InAs/AlSb/GaSb/AlSb/InAs RITDs. One series did not contain hole barriers whereas the other one was grown with AlAs-monolayers at the heterointerfaces.

2. Samples

All samples were grown on undoped GaAs substrates in a Riber 2300 MBE system. Growth temperatures ranged from 460 to 580°C, as measured by a thermocouple calibrated to the 640°C GaAs congruent sublimation temparature. Molecular beams of Sb_2 and As_2 were used, the latter being supplied by a valved cracker effusion cell in order to suppress the unintentional incorporation of As from the As background into the antimonide layers [4,5]. Reflection high-energy electron diffraction (RHEED) was used both for the calibration of growth rates and for in-situ surface monitoring during growth. To accomodate the misfit between the GaAs substrate and the GaSb, InAs and AlSb layers, a thick buffer layer was grown, consisting of a 100 nm thick AlSb nucleation layer, a 1 µm thick AlSb layer and a 10-period smoothing superlattice (2.5 nm GaSb/ 2.5 nm AlSb) [4].

The RITD structure consists of 1 µm of heavily Si-doped InAs (n = 10^{18} cm^{-3}), 50 nm of lightly Si-doped InAs (n = 10^{16} cm^{-3}), 10 nm InAs, the double barrier layer sequence (5ML AlSb, 6.5 nm GaSb, 8ML AlSb), 10 nm InAs, 50 nm of lightly Si-doped InAs (n = 10^{16} cm^{-3}) and 250 nm of heavily Si-doped InAs (n = 10^{18} cm^{-3}). The growth parameters were similar to those reported in [2]. In addition, a series of RITDs containing AlAs monolayers adjacent to the AlSb barriers was grown. The structure of these samples was the same as above except that monolayers of AlAs were grown before and after the GaSb layer and before the upper 10 nm undoped InAs layer. The AlSb layer thicknesses were reduced appropriately in order to maintain the original number of barrier monolayers as in the RITD structures without AlAs monolayers.

Mesa diodes were fabricated using contact photolithography and a standard H_3PO_4:H_2O_2:H_2O wet etch, with the mesa dimensions ranging from 30x30 µm² up to 160x160 µm². For ohmic contacts we used nonalloyed Au/Ge/Ni layers.

3. Results and discussion

At room temperature the smallest diodes (30x30 µm²) showed the same overall curve shape as reported before [2]. The best peak-to-valley current ratio of about 20:1 at 300 K was achieved on a device containing nominally 'AlAs-hole barriers', whereas samples without 'AlAs-hole barriers' showed peak-to-valley current ratios of less than 11:1 at room temperature.

At 4.2 K peak-to-valley current ratios of 65:1 on samples with hole barriers have been found compared to peak-to-valley current ratios of about 30:1 for the best sample without AlAs monolayers. Fig. 1a shows the I(V)-characteristic of a 30x30 µm² sample with hole barriers. It is clearly seen that the I(V)-characteristic is asymmetric. When the device is negatively biased, electrons tunnel into the quantum well through the thinner barrier and

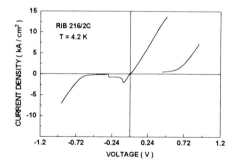
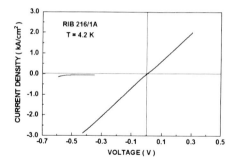

Fig. 1a: I(V)-characteristic of a 30x30 μm² RITD grown with AlAs-monolayers.

Fig. 1b: I(V)-characteristic of a 60x60 μm² RITD grown with AlAs-monolayers

tunnel out through the thicker barrier, which has a much lower transmission coefficient. This behaviour gives rise to an enhanced electronic space charge build-up and an intrinsic bistability in the I(V) characteristics [6]. Under positive bias, on the other hand, the charge build-up is relatively small, leading to the observed asymmetry between the forward or backward bias directions. The device had a peak-to-valley current ratio of about 60:1 for negative bias-voltage and of 25:1 for positive bias-voltage. In general, the valley current for positive bias-voltages was much higher than that for negative bias-voltages for all samples.

Because of the large current densities found in the RITD structures, ohmic heating due to contact and parasitic series resistances completely alters the I(V)-characteristic on larger mesas. Fig. 1b shows the I(V)-characteristic for a diode having the same structure as in Fig. 1a but with a mesa of 60x60 μm². In this case a negative differential resistance was found only for negative bias voltages, and showed no bistability. Small hysteresis effects between upwards and downwards-tracing curves have been found to be due to parasitic series resistances. The peak-to-valley current ratio of the device of Fig. 1b was 65:1 at 4.2 K. For small bias-voltages ($|V_b| < 100$ mV) the magneto-transport data did not show any dependency on the mesa area. Therefore, magneto-tunneling experiments have been peformed on samples with a 60x60 μm² mesa.

The I(V) characteristics are significantly modified in the transverse field B⊥J geometry. In devices with narrow quantum wells the peak current in the I(V)-characteristic is shifted to higher voltages whereas the valley-current remains mostly unaffected (Fig. 2). The increase of the peak voltage position can be understood qualitatively in terms of an increase in the effective height of the potential barrier due to a diamagnetic term caused by the external magnetic field.

Semiclassically, in a magnetic field applied parallel to the x-axis an electron tunneling in the z direction will have an additional in-plane wavevector component k_y'. For a system with approximately isotropic and parabolic band structure it has been shown that due to the k_1-conservation the additional voltage $V_B = -c\Delta E$ that must be applied to the barrier in order to offset the effect of B and maintain a constant tunneling current is given by $V_B = \beta(m^*) B^2$ where β depends on the effective mass m^* [7,8]. In the present system the electrons are tunneling into a complex hole band structure in the GaSb quantum well. By plotting the voltage positions V_{peak} of the peaks in the I(V) characteristics as a function of the in-plane

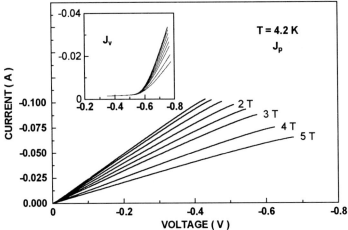

Fig. 2: 'Peak' (J_p) and 'valley'-current (J_v) of a RITD with hole barriers for different magnetic fields in the B⊥J configuration.

magnetic field B, the form of the in-plane energy dispersion curves $\varepsilon(k_\parallel)$ can be mapped out [9]. Up to 5 Tesla V_B was found to depend quadratically on the external magnetic field B. This is quite reasonable because for the devices investigated here a magnetic field of 5 Tesla means that only states very close to the Γ-point are probed.

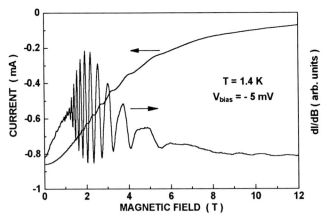

Fig. 3: I(B) and dI/dB-curve of a RITD in the B∥J configuration biased at V_{bias} = -5mV

When a quantising magnetic field is applied parallel to the tunnel current (B∥J) the electron states in the emitter and the hole states in the quantum well become quantized into Landau levels. Resonant tunneling already takes place at arbitrarily small applied voltages; but due to energy and momentum conservation rules the current decreases with increasing magnetic field superimposed by an oscillating structure, periodic in 1/B [10]. The general behaviour of the magnetotunneling current and the magnetotunneling oscillations (dI/dB) in the B ∥ J configuration are shown in Fig. 3. The bias-voltage was -5 mV. When monitoring the current in the perpendicular configuration (B⊥J), all oscillations are found to disappear, which indicates that they are 2D in character. Moreover, tilted field experiments show a characteristic frequency shift, given by $B_F(\alpha) = B_F / \cos(\alpha)$, where B_F is the so called fundamental field or inverse period $B_F = [\Delta(1/B)]^{-1}$ and α denotes the angle between the external field and the direction of the tunnel current. Due to the heavy doping in the contact layers, no two-dimensional electron gas is formed in the emitter accumulation layer, which means that the 2D behaviour reflects the 2D hole gas in the quantum well.

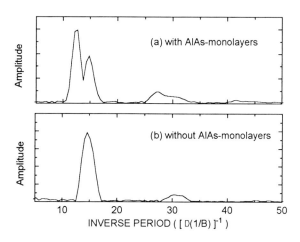

Fig. 4: FFT-spectrum of the dI/dB(1/B)-curves of RITDs with (a) and without (b) hole barriers.

A fast Fourier transform analysis of the dI/dB-data plotted against the inverse magnetic field 1/B revealed further information. As shown in Fig. 4b the Fourier spectrum of the dI/dB(1/B)-curve measured on a RITD without AlAs-monolayers consists of a single peak corresponding to the inverse period or fundamental field B_F. The first harmonic appears as a small hump. This spectrum has been found on all InAs/AlSb/GaSb/AlSb/InAs tunnel structures having no AlAs-monolayers incorporated. In sharp contrast, the Fourier spectra of the tunnel diodes grown with AlAs-monolayers and exhibiting large peak-to-valley current ratios showed a double peak structure (Fig. 4a). The experimental data indicate that the double peak structure is strongly correlated to the confinement of the hole states in the GaSb quantum well, although a complete explanation for this behaviour is still lacking. Strain induced field effects may play an important role as well as mixing between heavy and light hole states [11]. Further experiments and theoretical calculations are necessary in order to fully understand the above behaviour.

4. Acknowledgment

The authors would like to thank H. Goronkin for helpful discussions and assistance in preparing part of the diodes. We would also like to thank G. Bihlmann, J. Fleißner and K. Sambeth for sample preparation and technical support. This work was supported by the Bundesminister für Forschung und Technologie.

5. References

[1] Esaki L 1958 Phys. Rev. **109** 603
[2] Tehrani S, Shen J, Goronkin H, Kramer G, Hoogstra M and Zhu T X 1993 Int. Phys. Conf. Ser. **136**: Int. Symp. on GaAs and Related Compounds Freiburg 1993 pp. 209-214
[3] Wagner J, Schmitz J, Behr D, Ralston J D and Koidl P 1994 Appl. Phys. Lett. to appear
[4] Schmitz J, Wagner J, Maier M, Obloh H, Hiesinger P, Koidl P and Ralston J D 1993 Int. Phys. Conf. Ser. **136**: Int. Symp. on GaAs and Related Compounds Freiburg 1993 pp. 367-372
[5] Wagner J, Schmitz J, Ralston J D and Koidl P 1994 Appl. Phys. Lett. **64**, 82
[6] Leadbeater M L, Sheard F W, Eaves L 1991 Semicond. Sci. Technol. **6** 1021-1024
[7] Ben Amor S, Martin K P, Rascol J J L, Higgins R J, Torabi A, Harris H M and Summers C J (1988) Appl. Phys. Lett. **53** 2540
[8] Eaves L, Stevens K W H and Sheard F W 1986 Springer Proceedings in Physics 13: The Physics and Fabrication of Microstructures and Microdevices, ed. Kelly M J and Weisbuch C 343
[9] Hayden R K, Eaves L, Henini M, Valadares E C, Kühn O, Maude D K, Portal J C, Takamasu T, Miura N and Ekenberg U 1994 Semicond. Sci.Technol. **9** 298
[10] Leadbeater M L, Eaves L, Simmonds P E, Toombs G A, Sheard F W, Claxton P A, Hill G and Pate M A 1988 Solid-State Electronics **31** 707-710
[11] Kiledjian M S, Schulman J N and Wang K L 1992 Surface Science **267** 405

Inst. Phys. Conf. Ser. No 141: Chapter 8
Paper presented at Int. Symp. Compound Semicond., San Diego, 18–22 September 1994

DC and transient simulation of resonant tunneling devices in *NDR-SPICE*

P. Mazumder, J. P. Sun, S. Mohan and G. I. Haddad

Center for High Frequency Microelectronics, Department of Electrical Engineering and Computer Science, The University of Michigan, Ann Arbor, MI 48109, Phone: (313) 763 2107

Abstract. Efficient circuit simulation models for several different resonant tunneling devices have been developed and incorporated into the $NDR - SPICE$ circuit simulator. The models are table-based, in order to simplify the extraction of parameters from device measurements or from quantum device simulation. In addition, new convergence algorithms have been developed to overcome the numerical problems caused by table-based models and the negative differential resistance characteristics of the devices. DC and transient models for RTTs and RTDs have been implemented on the basis of device simulation and measurements. Innovative circuits have been designed using these models and RTD/HBT circuits have been built to validate simulation results.

1. Introduction

New circuit simulation models for Resonant Tunneling Diodes (RTDs), Multiple-peak Resonant Tunneling Diodes (MRTDs), Resonant Tunneling Transistors (RTTs) and Resonant tunneling Hot Electron Transistors (RHETs) have been incorporated into the $NDR - SPICE$ circuit simulator as internal models. While previous modeling efforts have been based on the macro-modeling capabilities of SPICE, these models are built-in, making the simulation more efficient and, more importantly, allowing the incorporation of special convergence routines to overcome the numerical problems caused by piece-wise linear and negative differential resistance (NDR) characteristics.

The new RTD model has better convergence properties and typical simulations with the new RTD model are up to five times faster than simulations using previously published external RTD models. It is a piecewise-linear model and the extraction of model parameters from measured data is very simple. The model has been validated by measurements on fabricated devices and by the verification of circuit operation against circuit simulation.

A new state-based simulation model for MRTDs has been developed and implemented. This model accounts for the presence of multiple stable states for given ranges of the applied voltage and produces accurate results for circuits where a macro-model

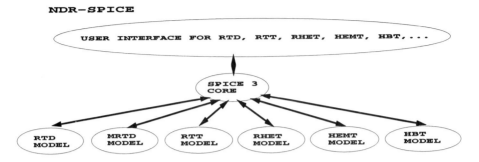

Figure 1. NDR-SPICE enhancements to the basic SPICE software

consisting of series connected RTDs produces results that are not in accordance with observed results.

An efficient lookup-table model with special convergence routines has been developed for RTTs. The tables are computed off-line from quantum device simulations, where the Schrödinger and the Poisson equations are solved self-consistently for the device, to provide the most accurate values while avoiding excessive run-times for circuit simulation. The table-driven model is again supported by convergence routines that force the Newton-Raphson iterations of SPICE to converge exactly when the actual circuit would converge to a stable solution. An empirical model for RHETs has also been incorporated into NDR-SPICE. This model is based on observed RHET characteristics. Fig. 1 shows how the basic SPICE software has been enhanced in NDR-SPICE.

The $NDR-SPICE$ circuit simulator has been used successfully to design several circuits containing RTDs, RTTs, RHETs, HBTs, HEMTs, and other conventional devices. These include conventional binary circuits, analog-digital converters, nanopipelined logic [1] and multivalued logic/memory circuits [2] and simulation results have been validated by measurements.

2. RTD Modeling

Accurate and computationally efficient models of RTDs have been developed and incorporated into NDR-SPICE. In the past, detailed models of RTDs, involving the solution of the Schrödinger equation, have been developed [3]. The main drawback of these models is the excessive computational time required. When a circuit containing several tens of RTDs has to be simulated, these detailed models are not suitable. Simple equivalent circuit models for RTDs have been developed by various authors [4]. These models take the form of SPICE subcircuit or function elements. A SPICE subcircuit for a single RTD consists of about 10 resistors, switches and voltage/current sources to produce a piecewise-linear approximation of the RTD characteristic. Discontinuous derivatives in the piecewise-linear approximation lead to convergence problems in the simulation algorithm. The new RTD model presented here is a SPICE internal model replacing the subcircuit of 10 elements by a single nonlinear element. While a piecewise-continuous approximation of the RTD characteristic is still used, additional algorithms have been added to detect convergence problems in the simulation algorithm and to force convergence in cases where the actual circuit would converge.

2.1. The Internal RTD Model

Linearized constitutive equations for the RTD have been incorporated into the $NDR-SPICE$ code so that an RTD may, for example, be specified as:

```
MODEL RTD NDR TYPE=2 IPEAK=35mA VPEAK=0.7V IVALLEY=1mA VVALLEY=0.75V
```

Using a Taylor series expansion, the RTD may be replaced at each Newton-Raphson iteration [5] by a circuit consisting of a current source in parallel with a conductance (see Fig. 2a). The values of the current source and conductance are as shown below:

$$I_0 = \begin{cases} 0, & 0 < V \leq V_{PEAK} \\ I_{PEAK} + b \cdot V_{PEAK}, & V_{PEAK} < V \leq V_{VALLEY} \\ I_{VALLEY} - c \cdot V_{VALLEY}, & V > V_{VALLEY} \end{cases} \quad (1)$$

$$g = \begin{cases} a, & 0 < V \leq V_{PEAK} \\ -b, & V_{PEAK} < V \leq V_{VALLEY} \\ c, & V > V_{VALLEY}. \end{cases} \quad (2)$$

This internal model results in one call to an RTD sub-routine in each iteration in the simulation loop [5], while the external model results in calls to current-source, resistance and switch sub-routines, each of which is of the same complexity as the RTD sub-routine. Hence the internal model is more efficient. The internal model has two other important features: 1) it is general in the sense that each segment in the piecewise characteristic may be replaced by a quadratic or some other differentiable function if required, as in the case of the 4-peak RTD model described below, and 2) it can be used to overcome the inherent limitations of certain $SPICE$ algorithms to make the simulation converge.

This RTD model has been implemented in $NDR-SPICE$ and the piecewise-linear characteristics are seen to be in good agreement with measured characteristics (see Fig. 2b). The complete internal RTD model includes a voltage-dependent capacitance in parallel with the current source.

A multiple-peak RTD is formed by connecting several single-peak RTDs in series. Elementary analysis of the resulting circuit shows that it has multiple stable states and it acts as a multi-state memory element. A new MRTD model that takes into account this state information, has been incorporated into NDR-SPICE.

3. RTT Modeling

The Bound-State Resonant Tunneling Transistor [6] is a three-terminal device whose NDR characteristics can be controlled by the voltage applied to the base terminal. The

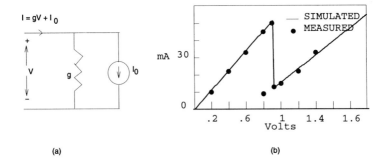

Figure 2. a) RTD equivalent circuit model b) Simulated characteristics of an RTD along with measured values

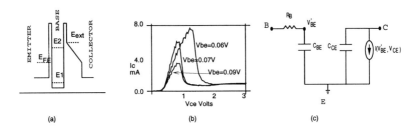

Figure 3. a) Band diagram of BSRTT; b) Collector current I_C as a function of V_{CE} for the BSRTT calculated with self-consistency c) circuit simulation model

I-V characteristics of this device were obtained by the self-consistent solution of the Schrödinger and the Poisson equations for various bias points. The Poisson equation relates the electrostatic potential to the charge distribution, in which fixed charge distribution represents ionized impurities (doping profile of the device) and mobile charge distribution represents electron transfer (space charge effects). Since the carrier density responds to the same electrostatic potential it generates, we have a self-consistent problem which entails simultaneous solution of the Schrödinger equation and the Poisson equation. Additional treatment must also be considered for charge in the quantum well to account for the quantized effects. The self-consistent calculations for the device models can be divided into three parts and are referred to as the Thomas-Fermi calculation for the self-consistent potential, the quantum-well calculation for space charge in the quantum well(s), and the traveling wave calculation for current density. Readers are referred to [7] for more details. The calculated I-V characteristics for a BSRTT are shown in Fig. 3, along with the circuit simulation model. The current source and capacitance values are obtained from quantum simulation, and the entire model is implemented as a SPICE internal model.

As in the case of the RTD model, the piecewise-linear table-driven RTT model has convergence problems due to the discontinuous derivatives and the NDR characteristic itself. The new convergence algorithm keeps track of the state of the RTT in all previous iterations during transient or DC analysis to determine when the simulation is oscillating, when the simulation should be forced to converge, and where the new operating point

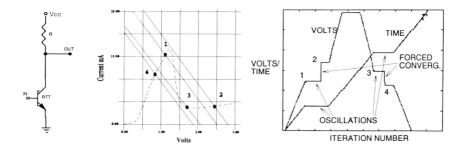

Figure 4. RTT circuit, load line, convergence problem and solution

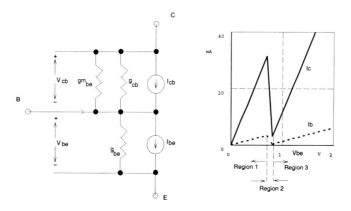

Figure 5. RHET equivalent circuit and simulated characteristics

should be. Fig. 4 shows a simple RTT circuit, its load line, and the simulation results when the supply voltage is first ramped up and then ramped down. When the operating point jumps from (1) to (2) on the loadline diagram, the simulation begins to oscillate around point (1) until the convergence algorithm forces the operating point to (2).

4. RHET Modeling

An empirical model for RHETs has been incorporated into NDR-SPICE. The RHET equivalent circuit shown in Fig. 5 contains a nonlinear current source element to model the NDR characteristic in the base-emitter region, and controlled current source and transconductance elements in the collector-base region to simulate the RHET characteristics.

5. Applications

Several different circuit applications have been developed with the help of the above simulation models. Ultrafast, compact, self-latching circuits using RTTs [8] and compact multiple-valued logic circuits using a combination of RHETs, RTDs and HBTs [2] have been designed and simulated in NDR-SPICE. The RTD+HBT circuits of [2] and self-latching, pseudo–bistable [8] circuits using RTDs and HBTs have been built to verify the correctness of the simulation.

Acknowledgements

This work was supported by the U.S. Army Research Office under the URI program contract DAAL 03-92-G-0109 and by ARPA under contract DAAH04-93-G-0242

References

[1] Mohan S, Mazumder P and Haddad G I 1994 Ultra-fast pipelined arithmetic using quantum electronic devices, *IEE Proc. E* **141(2)**, 104–110.

[2] Chan H L, Mohan S, Chen W L, Mazumder P and Haddad G I 1994 Ultrafast, compact multiple-valued multiplexers using quantum electronic devices, *Proceedings of the Government Microcircuit Applications Conference, San Diego*.

[3] Mains R K and Haddad G I 1988 Time-dependent modeling of resonant tunneling diodes from direct solution of the Schrödinger equation, *J. Appl. Phys.* **64(7)**, 3564–3569

[4] Kuo T H, Lin H C, Anandakrishnan U, Potter R C and Shupe D 1989 Large-signal resonant tunneling diode model for spice3 simulation, *International Electron Devices Meeting Technical Digest* pp. 567–570

[5] McCalla W J 1988 *Fundamentals of Computer-Aided Circuit Simulation*, (Norwell:Kluwer Academic)

[6] Haddad G I, Mains R K, Reddy U K, and East J R 1989 A proposed narrow band-gap base transistor structure, *Superlattices and Microstructures* **5(3)**, 437–441.

[7] Sun J P 1993 Modeling of semiconductor quantum devices and its applications *Ph.D. thesis* University of Michigan, Ann Arbor

[8] Mohan S, Mazumder P, Mains R K, Sun J P and Haddad G I 1993 Logic design based on negative differential resistance characteristics of quantum electronic devices, *IEE Proc. G* **140(6)**, 383–391.

Resonant tunneling through zero-dimensional Si-impurity states in GaAs/AlAs single-barrier structures

Hiroyuki Fukuyama and Takao Waho

NTT LSI Laboratories,
3-1, Morinosato Wakamiya, Atsugi, Kanagawa 243-01, Japan

Abstract. Tunneling current through a Si-planar-doped AlAs barrier is studied, showing excess current through this single barrier, which is not observed when the barrier is undoped. The dI/dV–V characteristics indicate that the excess current is due to resonant tunneling through the zero-dimensional Si-impurity states formed in the AlAs barrier. The impurity states are probably related to the X valley in the AlAs barrier.

1. Introduction

Compound semiconductor heterostructures have provided various quantum structures. One of the most promising is the double-barrier structure in which current flows through the resonant state confined vertically. This phenomenon has shown its high potential for realizing ultra-fast [1] and new-function [2, 3] devices. Recently, resonant tunneling through the states, confined laterally as well as vertically, is attracting much interest because lateral confinement is expected to exhibit new tunneling features. In several experimental studies, electron-beam lithography or focused-ion-beam techniques have been used to achieve lateral confinement [4, 5, 6]. In these works, current oscillations corresponding to energy levels of the confined states have been observed in the I–V characteristics. These oscillations were, however, rather small since lateral confinement is not as strong as vertical confinement. This is due to limitations in fabricating lateral structures; the lateral size which we can control is usually one order of magnitude larger than the controllable vertical size.

An alternative approach to achieving lateral confinement is to use the spherical Coulomb potential related to an ionized impurity. The possibility of resonant-tunneling current through the impurity states was first suggested for a gated double-barrier resonant tunneling diode with an undoped well layer [7]. This type of resonant-tunneling current was actually observed around the onset of the tunneling current in a double-barrier structure with intentionally doped Si impurities in its well [8]. Because impurity states are determined by material properties and free from lithography-induced fluctuations, tunnel characteristics reproducibility is expected. In addition, we can expect to achieve high resonant-tunneling current density by increasing the impurity density, which is important for application.

In this study, we employed GaAs/AlAs *single*-barrier structures in which Si impurities were introduced into the AlAs barrier layer. The I–V and dI/dV–V characteristics were observed at 4.2 K, showing excess current due to tunneling through zero-dimensional Si-impurity states in the AlAs barrier. This is the first time this has been observed to the best of our knowledge. These states are probably related to the X valley in the AlAs barrier.

2. Experimental

Si-planar-doped GaAs/AlAs single-barrier structures were grown on an n$^+$-GaAs substrate by molecular beam epitaxy (MBE); the details are shown in Tables 1 and 2. First, we grew a 300-nm n-GaAs buffer layer. This layer was doped with 5×10^{17}-cm^{-3} Si. We then grew a 5-nm undoped GaAs spacer layer and the AlAs single-barrier layer, either 3-, 4- or 5-nm thick. The barrier was Si-planar-doped to 5×10^{10} cm^{-2} at its center. Next, we grew a 1.4-nm undoped GaAs spacer layer and a 270-nm n-GaAs layer with 5×10^{17}-cm^{-3} Si. Finally, a 50-nm n$^+$-GaAs layer with 1×10^{19}-cm^{-3} Si was grown as a cap layer. We also grew single-barrier structures with an undoped barrier as reference samples. The barrier-layer thickness in the reference samples was 4 and 5 nm.

The single-barrier diodes were fabricated by conventional photolithography, wet etching, and metallization. Mesa structures were formed by wet etching to make 20 × 20-

Table 1. Epitaxial layer structure for the Si-planar-doped single-barrier diodes.

Material	Si concentration (cm^{-3})	Thickness (nm)
GaAs	1×10^{19}	50
GaAs	5×10^{17}	270
GaAs	undoped	1.4
AlAs	N_S†	L_b†
GaAs	undoped	5.0
GaAs	5×10^{17}	300
n$^+$-GaAs substrate		

† N_S represents the planar-doped Si concentration in the barrier layer and L_b is the barrier thickness.

Table 2. Planar-doped Si concentration in the barrier layer and thickness of the barrier layer of the single-barrier diodes.

Si concentration (N_S) (cm^{-2})	Thickness (L_b) (nm)
5×10^{10}	3, 4, 5
undoped	4, 5

μm² diodes. AuGe/Ni/Ti/Pt/Au was deposited both on the top of the mesa structures and on the back of the wafer, and then annealed to form ohmic contacts.

We measured the I–V and dI/dV–V characteristics of the diodes at 4.2 K. The dI/dV–V characteristics were obtained by applying low-frequency (ω) modulation to the dc voltage to measure the ω component of the current density with lock-in amplifiers.

3. Results and Discussion

The I–V characteristics of the single-barrier diodes with a Si-planar-doped barrier and the reference diodes with an undoped barrier are shown in Figure 1 by solid and broken lines, respectively. For the undoped barrier samples, the current density abruptly increased above 230 mV for the 4-nm diode and above 180 mV for the 5-nm diode. This is because a considerable number of electrons in the emitter transfer to the X valley in the AlAs barrier above these voltages and tunnel through the X valley [9]. For the Si-planar-doped barrier samples, excess current was clearly observed between 100 mV and 300 mV for the 4-nm diode, and between 50 mV and 250 mV for the 5-nm diode. This shows that electrons in the emitter tunnel through additional paths in these bias-voltage regions. Impurity states formed by planar-doped Si in the AlAs barrier must be responsible for the excess current because this was *not* observed in the undoped reference samples.

In the samples we employed, the average distance between impurities is estimated to be 45 nm, much larger than the size of the impurity state which is roughly estimated at 6 nm from the Bohr radius ($4\pi\varepsilon\hbar^2/m^*e^2$) in AlAs. In other words, the Si concentration is too low to form a two-dimensional state, which is expected when the donor impurity is highly planar-doped [10]. Therefore, the observed excess current possibly

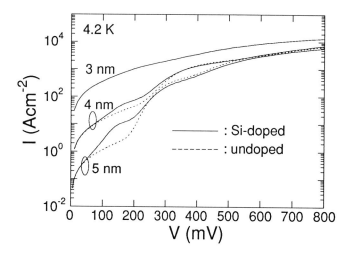

Figure 1. I–V characteristics of single-barrier diodes. Solid lines are for the Si-doped samples and broken lines are for the reference samples. Excess current in the I–V characteristics of the Si-doped samples is evident around 200 mV.

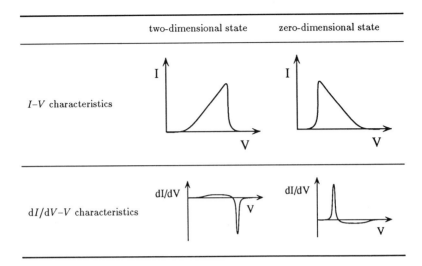

Figure 2. Schematic view of the I–V and dI/dV–V characteristics of the tunneling current via a two-dimensional or zero-dimensional state. A peak in the dI/dV–V characteristics should be observed for a zero-dimensional state, whereas a notch would be observed for a two-dimensional state.

flows via zero-dimensional states formed by ionized impurities. According to a theoretical calculation [11], the resonant-tunneling current through a zero-dimensional state increases abruptly when its level coincides with the emitter Fermi level and gradually decreases until the level coincides with the emitter band edge. This results in the peak in the dI/dV–V characteristics as seen in Figure 2. In contrast, the current through a two-dimensional state increases gradually after its level coincides with the emitter Fermi level and decreases abruptly when the impurity level coincides with the emitter conduction band edge. This results in a notch in the dI/dV–V characteristics. As shown in Figure 3, what we observed in the dI/dV–V characteristics was a peak instead of a notch, which corresponds to excess current. This leads to the conclusion that the excess current observed in the Si-doped sample is resonant-tunneling current flowing through zero-dimensional impurity states.

As shown in Figure 1, the excess current was found just below the bias voltage, above which the current via the X valley of the AlAs barrier was observed in the I–V characteristics of the undoped reference samples. This suggests that the impurity states are related to the X valley. In GaAs/AlAs/GaAs structures, the X-valley minima formed single quantum wells in contrast to the Γ-valley minima. If the impurity states are formed in this AlAs X-valley well, their energy level rises with decreasing AlAs X-valley well width (L_b) due to stronger confinement in the GaAs X-valley walls. Therefore, we can determine whether or not the impurity states are related to the X valley by observing the AlAs layer-thickness dependence of the peak voltage in the dI/dV–V characteristics. This dependence is illustrated in Figure 3. The peak voltage increases as L_b decreases, indicating that the energy level of the impurity states rises with decreasing barrier thickness L_b. Therefore, the impurity states are probably related to the X valley, as shown in Figure 4. According to the hydrogen-atom model, the binding energy of

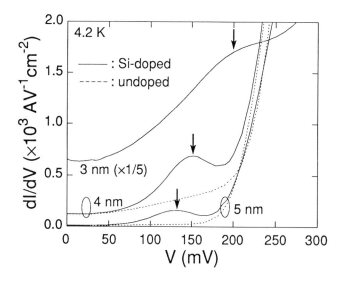

Figure 3. dI/dV–V characteristics of single-barrier diodes. Solid lines are for the Si-doped samples and broken lines are for the reference samples. We found a peak instead of a notch corresponding to the observed excess current. It was also observed that the peak shifted to a higher voltage as the barrier became thicker.

the impurity state $(m^* e^4/32\pi^2\varepsilon^2\hbar^2)$ related to the bulk-AlAs X valley is 35 meV. This indicates that the peak voltage in the dI/dV–V characteristics is about 70 mV lower than the voltage at which the current density abruptly increases due to electrons flowing

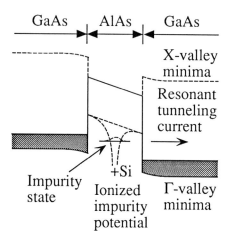

Figure 4. Schematic view of the band structure of the Si-doped GaAs/AlAs single-barrier diode and the proposed impurity state formed in the X-valley well.

via the X valley in the AlAs barrier. This agrees well with our experimental results.

4. Conclusion

We studied the I–V characteristics of Si-planar-doped GaAs/AlAs single-barrier structures, and found excess current which was not observed in undoped single-barrier structures. This excess current was determined to be due to resonant tunneling through zero-dimensional impurity states based on a peak observed in the dI/dV–V characteristics. The dependence of the peak voltage in the dI/dV–V characteristics on the barrier thickness indicates that the impurity states are related to the X valley in the AlAs barrier.

Acknowledgments

The authors would like to thank T. Ishikawa for MBE growth. They are also grateful to T. Mizutani for his continuous encouragement.

References

[1] Brown E R, Söderström J R, Parker C D, Mahoney L J, Molvar K M and McGill T C 1991 *Appl. Phys. Lett.* **58** 2291–2293
[2] Maezawa K and Mizutani T 1993 *Japan. J. Appl. Phys.* **32** L42–L44
[3] Akeyoshi T, Maezawa K and Mizutani T 1994 *Japan. J. Appl. Phys.* **33** 794–797
[4] Reed M A, Randall J N, Aggarwal R J, Matyi R J, Moore T M and Wetsel A E 1988 *Phys. Rev. Lett.* **60** 535–537
[5] Tarucha S, Hirayama Y, Saku T and Kimura T 1990 *Phys. Rev.* B **41** 5459–5462
[6] Tarucha S, Tokura Y and Hirayama Y 1991 *Phys. Rev.* B **44** 13815–13818
[7] Dellow M W, Beton P H, Langerak C J G M, Foster T J, Main P C, Eaves L, Henini M, Beaumont S P and Wilkinson C D W 1992 *Phys. Rev. Lett.* **68** 1754–1757
[8] Fukuyama H, Waho T and Mizutani T 1993 *Japan. J. Appl. Phys.* **32** 570–573
[9] Landheer D, Liu H C, Buchanan M and Stoner R 1989 *Appl. Phys. Lett.* **54** 1784–1786
[10] Schubert E F and Ploog K 1985 *Japan. J. Appl. Phys.* **24** L608–L610
[11] Liu H C and Aers G C 1988 *Solid State Commun.* **67** 1131–1133; 1989 *J. Appl. Phys.* **65** 4908–4914

Quantum dot single-photon and electron-hole turnstile devices

A. Imamoḡlu (*) and Y. Yamamoto (**)

*Department of Electrical and Computer Engineering, University of California, Santa Barbara, CA 93106
**E. L. Ginzton Laboratory, Stanford University, Stanford, CA 94305

Abstract. We describe a quantum dot turnstile device that is based on a double-barrier mesoscopic $p-i-n$ heterojunction driven by an alternating voltage source. The device suppresses the quantum fluctuations usually associated with electron and hole injection processes using Coulomb blockade and quantum confinement effects. Provided that the radiative recombination takes place in a short time-scale, it is possible to generate heralded single-photon states without the need for a high-impedance current source. Since the frequency of the alternating voltage source determines the repetition rate of the single-photon states and the magnitude of the junction current, the present scheme promises high precision photon-flux and current standards.

1 Introduction

There is an increasing interest in semiconductor light emitting diodes (LED) and lasers due to their applications in advanced optical communication systems and laser spectroscopy. The fundamental noise arising from the quantum nature of the generated optical fields is fast becoming a limiting factor in many of these applications. The quantum noise suppression in light generation is therefore interesting not only from a conceptual but also a practical point of view. One of the distinguishing factors of semiconductor lasers is that they are pumped by injection currents, which are not shot-noise limited as a result of the Coulomb interaction and Fermi exclusion principle. In contrast, optical pumping used in other practically significant laser sources, is a Poisson point process, and the generated laser light is in a (shot-noise limited) coherent state. The realization and experimental demonstration [1] that a constant-current pumped semiconductor laser generates a number-phase squeezed state with reduced intensity noise has had a great impact, partly because of the rather simple and fundamental nature of the noise-suppression mechanism.

Here, we propose a turnstile device that generates heralded single-photon states, which represent the ultimate limit in quantum control and intensity noise suppression in the light generation process in semiconductors. In a heralded single-photon stream, *one and only one* photon is generated (detected) at arbitrarily short time-intervals, and the successive photon generation (detection) events are separated by a *deterministic dwell-time*. The transition from the macroscopic squeezing regime mentioned previously to the strict regulation of the photon emission events in heralded single-photon states is achieved by utilizing the Coulomb blockade induced correlations in mesoscopic $P-I-i-I-N$ heterojunctions.

In the Coulomb blockade regime defined by $e^2/C \gg kT$, where C and T are the junction capacitance and temperature, respectively, the correlations betweeen the successive electrons (holes) are strong enough to suppress the randomness introduced by thermal fluctuations. There has been considerable interest in single electron charging effects in ultrasmall metal-insulator-metal tunnel junctions [2]. Mesoscopic semiconductor quantum well and dot structures offer a wealth of new phenomena, as both Coulomb blockade and quantum confinement effects are simultaneously important [3]. Until recently, both theoretical and experimental efforts have been focused on the modification of the current-voltage characteristics in constant-voltage driven double-barrier $n-I-i-I-n$ resonant tunneling structures.

We have previously analyzed the properties of light generated by mesoscopic $P-I-i-I-N$ AlGaAs-GaAs heterojunctions. We initially considered a constant-current driven mesoscopic ($e^2/2C \gg kT$) junction and simulated the dynamics using a semiclassical Monte Carlo method, where electrons and holes are treated as classical particles subjected to quantum jump processes such as resonant tunneling and radiative recombination. The simulation results showed that the electron injection and the single-photon generation times can be regulated with an accuracy limited only by the reciprocal peak resonant tunneling and radiative recombination rates [4]. The requirement of high-impedance constant-current operation however, is a major limitation in the mesoscopic regime due to the effects of the electromagnetic environment, in which the junction is embedded.

The ac-voltage driven single-photon turnstile device that we describe here removes the high-impedance requirement by utilizing Coulomb blockade and quantum confinement effects simultaneously, for both electrons and holes [5]. This turnstile device generates heralded single-photon states, where the deterministic dwell-time between successive single-photon emission events is given by the period of the ac-voltage source, and the uncertainty in the photon emission times is given (ultimately) by the (much shorter) radiative recombination time. Similar turnstile devices for single electrons have already been demonstrated [6, 7]: More specifically, it has been shown experimentally that the generated current is to a high degree determined by the frequency f of the applied source, through the relation $I = ef$, as each ac-voltage cycle contributes to one and only one electron transfer. These single electron turnstile devices as well as ours, can therefore be used as a frequency-based current standard.

2 Turnstile device

Figure 1 shows the energy band diagram of the turnstile device that we propose: If the junction voltage $V_j(t)$ is well below the built-in potential V_{bi} ($V_{bi} - V_j(t) \gg kT$), the carrier transport in such a structure takes place by resonant tunneling of electrons and holes through the undoped I_n and $I_p - AlGaAs$ *barrier* layers, respectively. The injected electron-hole pairs then recombine radiatively in the $i-GaAs$ layer. We assume that the width of the $i-GaAs$ quantum well is small enough that the energy separation of the quantized subbands well exceed the single electron (hole) charging energy of the *GaAs Coulomb island* and that resonant tunneling into a single conduction (valence) subband need to be considered. The resonant tunneling of an electron or a hole is allowed only when the junction voltage is such that

$$E_{fn} - e^2/2C_{ni} \geq E_{res,e} \geq E_{nc} - e^2/2C_{ni} \quad (electrons) \;, \tag{1}$$

and

$$E_{fp} + e^2/2C_{pi} \leq E_{res,h} \leq E_{pv} + e^2/2C_{pi} \quad (holes) \;. \tag{2}$$

Here, $E_{res,e}$ ($E_{res,h}$) is the energy of the electron (hole) resonant subband of the $i-GaAs$ quantum well (or dot); E_{nc} and E_{pv} are the energies of the conduction and valence bands in the $n-$ and $p-$ type layers, respectively; and, E_{fn} and E_{fp} are the Fermi energies in the corresponding layers. C_{ni} and C_{pi} are the capacitances of the $N - I_n - i$ and $P - I_p - i$ regions, respectively. The energies in Eq. (1-b) are determined by the applied junction voltage $V_j(t) = V_o + v(t)$, where $v(t) = 0$ ($0 \leq t < T_{ac}/2$); and $v(t) = \Delta V$ ($T_{ac}/2 \leq t < T_{ac} = f_{ac}^{-1}$). The impurity concentrations on both N and P sides should be small enough that $E_{fn} - E_{nc} \simeq e^2/C_{ni}$ and $E_{pv} - E_{fp} \simeq e^2/C_{pi}$, since we want to be able to turn the tunneling of a particular carrier on and off by applying a voltage pulse whose magnitude is on the order of (but larger than) the single-charge charging energy.

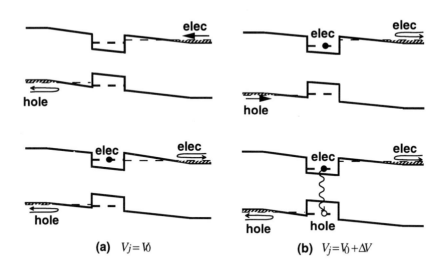

Figure 1: The energy-band diagram of the $p-i_p-i-i_n-n$ $AlGaAs-GaAs$ heterojunction, with (a) and without (b) the applied voltage pulse. For $V_j = V_o$ ($V_j = V_o + \Delta V$), the Fermi-energy of the N (P) type $AlGaAs$ layer is at least $e^2/2C_{ni}$ ($e^2/2C_{pi}$) higher (lower) than the energy of the quantum-well electron (hole) subband.

The dc-bias voltage (V_o) is chosen so that the electron tunneling is resonantly enhanced when $v(t) = 0$. The applied square voltage pulses ($v(t) = \Delta V$) enable the resonant hole tunneling, while blocking electron tunneling due to the second inequality in Eq. (1), i.e. by quantum confinement. A second tunneling event of the same carrier during the time interval where the junction voltage remains unchanged is blocked by Coulomb blockade. Therefore, only one electron and one hole can tunnel into the $i-GaAs$ layer within a single cycle of the applied ac-voltage. Assuming that the radiative recombination occurs in a time-scale short compared to the period of the ac-voltage, a single photon is generated in each cycle with a probability approaching unity. The emission of heralded single-photons into a single direction requires in addition, that the heterostructure be embedded in a micropost-cavity or photonic band-gap structure.

Figure 2 shows a semiclassical Monte Carlo simulation of the turnstile device operation. The period of the ac-voltage is chosen such that both the electron and hole tunneling occurs with very high probability during the time-intervals in which they are allowed. The photon emission events follow the hole tunneling in a very short time interval.

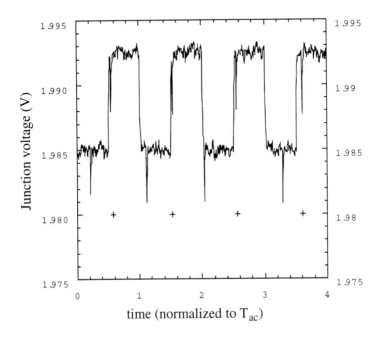

Figure 2: The junction voltage as a function of time shown together with the accompanying photon emission events (denoted by '+'). The *voltage spikes* correspond to electron or hole tunneling events.

The spectrum of the generated photon stream exhibits a squeezed background noise extending to $min[\Gamma_{tunn,h}, \Gamma_{rad}]/2\pi$, together with nonstochastic peaks at integer multiples of f_{ac}, indicating regulation with the *clock* of the ac-voltage source. Prior to this work, the squeezed background noise was predicted and demonstrated only for $p-n$ junctions driven by a high-impedance constant-current source [4]. The second-order correlation function of the photon stream obtained by inverse Fourier transforming the spectrum, exhibits peaks at integer multiples of T_{ac} with negligible amplitude in between the peaks [4]. Physically, this indicates that following a photon emission event at $\tau = 0$, the probability of emitting others at the integer multiples of T_{ac} is enhanced, whereas in between, it is strongly suppressed. The generated light field has accurate information on the photon emission times: In contrast, each photo-emission time is almost completely random for a photon-number or number-phase squeezed state [1].

3 Conclusion

If realized, the work proposed here could start a new subfield of quantum optics and mesoscopic physics, namely *single photonics*. The control of single-electron to single-photon (and vice-versa) conversion processes could have important applications in optoelectronics and optical communication systems.

From a practical perspective, reliable information encoding at the single-photon level is very attractive since it represents one of the fundamental limits in information transfer. In an ideal single-photon turnstile device, every full ac-voltage cycle generates one and only one photon and the successive cycles are completely independent. It is therefore possible to modulate the applied voltage pulses such that every bit of electrical information can be encoded: Presence or absence of a photon in a given time-window corresponds to the information in the voltage pulse.

References

[1] W. H. Richardson, S. Machida, and Y. Yamamoto, Phys. Rev. Lett. **66**, 2867-2870 (1991).

[2] D. V. Averin and K. K. Likharev, J. Low Temp. Phys. **62**, 345-373 (1986).

[3] A. Groshev, Phys. Rev. B **42**, 5895-5898 (1990); C. W. J. Beenakker, Phys. Rev. B **44**, 1646-1656 (1991); P. Gueret, N. Blanc, R. Germann, and H. Rothuizen, Phys. Rev. Lett. **68**, 1896-1899 (1992).

[4] A. Imamoglu, Y. Yamamoto, and P. Solomon, Phys. Rev. B **46**, 9555-9563 (1992); A. Imamoglu and Y. Yamamoto, Phys. Rev. B **46**, 15982-15991 (1992); A. Imamoglu and Y. Yamamoto, Phys. Rev. Lett. **70**, 3327-3330 (1993).

[5] A. Imamoglu and Y. Yamamoto, Phys. Rev. Lett. **72**, 210-213 (1994).

Overview and Status of the Quantum Functional Devices Project

Shigeo Okayama, Ichiro Ishida, Nobuo Aoi, and Shuku Maeda

Research & Development Association for Future Electron Devices
Fukide Building No.2, 4-1-21 Toranomon, Minato-ku, Tokyo 105, Japan

Abstract. The Quantum Functional Devices (QFD) Project is being carried out under the Industrial Science and Technology Frontier (ISTF) Program of the Ministry of International Trade and Industry (MITI) in Japan. The QFD project started in fiscal year 1991 as a ten-year project to establish basic technologies for microelectronics devices with low power consumption, high speed operation, and multifunctions by controlling quantum effects. Six companies including an American company are joining this project through the Research & Development Association for Future Electron Devices (FED).

1. Introduction

The rapid progress of semiconductor integrated circuit (IC) technology has brought great success to the field of microelectronics which is playing an important role as a first priority of infrastructure systems.

 The strong demand for smaller and faster information processing systems with lower power consumption has been boundless and has driven the research and development of new devices and their concepts down to a mesoscopic realm, and will reach a single electron level ultimately. The presence or absence of individual electrons, as well as their behavior, becomes significant for devices spanning only tens or hundreds of nanometers comparable to several kinds of characteristic lengths (wavelength, coherent length, mean free path, etc.) of electrons with low dimensional structures. At this point, new operation principles based on quantum effects are urgently required for advanced devices.

 The QFD project aims at establishment of basic technologies for developing such innovative microelectronic devices which operate by controlling quantum phenomena at practical temperatures and can be integrated with low power consumption, high speed operation, and multifunctions.

This project started in fiscal year 1991 after the completion of two earlier projects "Superlattice Devices" and "Three Dimensional Integrated Circuits".

This paper describes organization, schedule, and outline of the QFD project and proposed QFDs in this project.

2. Organization

The Agency of Industrial Science and Technology (AIST) of MITI assigned the Electrotechnical Laboratory (ETL; national laboratory) and FED as the research and development (R&D) organizations for the QFD project. FED is under contract with the New Energy and Industrial Technology Development Organization (NEDO) for promotion of this project. The QFD project budget is about seven hundred million yen for fiscal 1994.

FED subcontracted with six companies (NEC Corp., Motorola Inc., Matsushita Electric Industrial Co. Ltd., Fujitsu Ltd., Sony Corp., and Hitachi, Ltd.) for six subthemes and three universities (The University of Tokyo, Osaka University, and Kyushu University). The subthemes are introduced respectively in section 4.

FED has organized two technical committees, "Investigation and Survey Committee" and "Technology Prediction Committee," to promote this project aggressively. The former committee handles the latest R&D results and strategies of the QFD project with researchers of the subcontracted companies, ETL, and the three universities. The trend of quantum electronics and related fields are surveyed in the latter committee organized by researchers of fourteen FED member companies, ETL, and five universities. Furthermore, MITI has organized "Promotion Committee" and "Evaluation Committee" to set R&D targets and evaluate results of the QFD project.

FED has surveyed worldwide research activities on quantum function devices with the help of distinguished foreign researchers. They attend the "Investigation and Survey Committee" meeting once a year as advisors.

3. Schedule

The QFD project is divided into three phases, i.e., Phase 1 (1991–1994), Phase 2 (1995–1997), and Phase 3 (1998–2000). During Phase 1, basic structures of the devices proposed in the subthemes are being defined and basic technologies for design, fabrication, and characterization of device elements are being developed. In Phase 2, prototype devices will be fabricated and their performances evaluated. Technological means for integration of the device elements will also be investigated. During Phase 3, device performance such as operation speed, dissipation power, operation margin, and operating temperature will be advanced and integration technologies developed.

Progress of the R&D on each subtheme has to be evaluated according to flexible long-term targets even if some research results do not fit short-term targets. Any kind of research related to quantum effects including some possible device applications can be studied in the Phase 1 of this project. In the Phase 2, however, there are any possibilities that the current subthemes are selected and/or new subthemes are adopted from the view point of device feasibility.

4. QFDs of the six subthemes

4.1. Tunneling-control functional device (NEC Corp.)

A three-terminal tunnel device, named surface tunnel transistor (STT), has been studied by NEC Corp. Figure 1 shows a device image of the STT. The STT has an n+/i/p+ diode structure including highly degenerate regions with an insulated gate over the i-region. The interband tunnel current between the source and drain is controlled by a gate voltage. The gate voltage induces a two-dimensional electron gas (2DEG) at the surface of the i-region. The width of the tunnel barrier depends on the gate voltage. Therefore, the gate voltage modulates the tunnel current between the 2DEG and the p+-region. Their two-dimensional device-analysis shows that shrinkage of the gate length even to a few-ten nanometers does not affect the shape of the tunneling barrier in the vicinity of a drain region. Negative differential resistance characteristics of the STT were controlled at room temperature. The current density and peak-to-valley (P/V) current ratio were enlarged by an n+ channel inserted layer between a gate insulation layer and the i-region with an advanced interface-cleaning technique in MBE regrowth process.

4.2. Quantized-band-coupling multi-functional device (Motorola Inc.)

Motorola Inc. is studying a quantum multi-function transistor (QMT) as one candidate of Quantized-band-coupling multi-function device. The device image of the QMT is shown in Fig.2. Two channels coupled by resonant tunneling effects occurring between their quantum wells are stacked with multilayered structures. An InAs/GaSb/AlSb material system is chosen to obtain high P/V current ratios. The upper channel acts as a gate for the lower channel. They established a baseline epitaxial growth capability for antimonides and demonstrated the state of the art of a double barrier type II resonant interband tunneling diode (RITD) using MBE system. The P/V current ratio was improved from 15 to 30 (at room temperature) by inserting two monolayers of AlAs between the GaSb and InAs layers as hole barriers.

4.3. Resonant electron–transfer device (Matsushita Electric Industrial Co. Ltd.)

Figure 3 shows one unit structure of the Resonant electron–transfer device which consists of silicon quantum wires and silicon quantum dots. They are coupled with each other through a tunneling barrier of a silicon oxide insertion layer. High–speed electrons are transferred in the quantum wires and dots. Matsushita Electric fabricated the silicon quantum wires with a minimum width of 10 nm with a surface smoothness of about 3 nm using an anisotropic etching technique. Current oscillation in the wires due to one–dimensional structures was observed. Quantum dots with a size of 40 nm and silicon oxide double–barrier structures were also fabricated.

4.4. Quantized energy level memory device (Fujitsu Ltd.)

Ultrafine structure memory cells have been studied by Fujitsu. A static random access memory (SRAM) cell using a double–emitter resonant– tunneling hot electron transistor (RHET) was fabricated experimentally by conventional technology and performance of the cell operation in a 2 X 2 cell array was confirmed. A cross sectional structure of a one–bit cell is shown in Fig.4. The cell size can be shrunk to $0.2\ \mu\mathrm{m} \times 0.3\ \mu\mathrm{m}$ finally by developing an epitaxial growth technique for semiconductors on metals. A schematic image of a 4 X 2 cell array is shown in Fig.5. Fujitsu clarified that a multi–emitter RHET (MERHET) has a logic function itself. Using MERHETs for peripheral circuits, Gigabit–plus memory systems can be realized.

4.5. Coupled quantum dots device (Sony Corp.)

Sony has investigated the coupled quantum dots system (CQDS) consisting of quantum dots distributed in an array on a substrate as shown in Fig.6. Because the strength of quantum wave coupling between dots depends on their space, it is expected that the dynamics of the electrons in the CQDS will exhibit features corresponding to the configurations of the dots. A single electron tunneling between two dots controls the states of two dots. Patterns of electron distribution in the CQDS can find possible application in information processing. Statistical properties of the CQDS were calculated theoretically. An interface technology and an architecture suitable for the CQDS are major issues to be considered.

4.6. Quantum-wave structure functional device (Hitachi, Ltd.)

A Quantum-wave structure functional device has been studied by Hitachi. Phase modulation of a quantum wave such as polaritons (coexisting wave of electrons, holes, and photons) controlled by an electric field is utilized in this device. Multi-quantum wire waveguides with control gates are formed on a substrate as shown in Fig.7. This device processes a large number of information channels simultaneously in an ultra high-speed such as a massive parallel processing device. GaAs and InAs crystal whiskers were studied first for quantum wires as shown in Fig.8. The size of the crystal whisker is controlled by of crystal growth conditions. Crystals grow from Au particles seeded on a substrate. The direction of whisker growth depends on the crystal direction on the substrate's surface. Hitachi has also studied quantum wires growing laterally because a lateral structure of quantum wires is preferable to a perpendicular structure for the device application.

5. Summary

The significance of the QFD project has become more obvious recently due to the recession and also to the trend of recognizing the importance of fundamental research. Activities of this challenging project cover far-reaching fields of microelectronics based on quantum effects, from theory to experiment, from design to fabrication, and from materials to systems. The QFD project mission is not only theoretical and experimental research but also application research for devices using quantum effects. At this point, intense exchange of information among different disciplines is needed to identify practical uses for quantum effects.

This project has been promoted in cooperation with many foreign researchers and an American company as well as Japanese researchers. We aim to make every effort to pursue the internationalization of the QFD project systems.

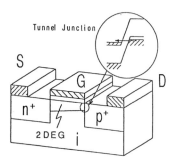

Fig.1. Schematic view of the STT with band diagram of the PN junction between 2DEG and the drain.

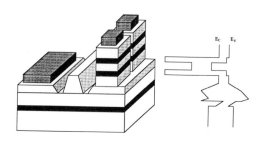

Fig.2. The Resonant Interband Tunnel Multi-Emitter HeterostructureTransistor with AlSb barriers, GaSb quantum well and InGaAs channel.

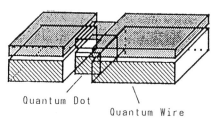

Fig.3. Unit structure of the Resonant Electron-Transfer Device.

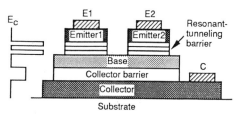

Fig.4. The Double-emitter RHET with onductionband diagram for one bit memory cell.

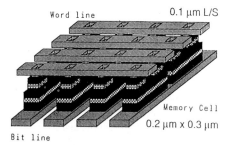

Fig.5. 4X2 memory cell array for the Gigabit-plus SRAM.

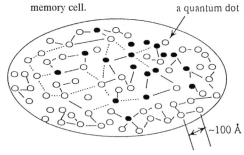

Fig.6. Schematic view of the CQDS.

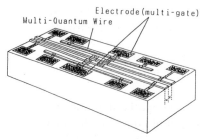

Fig.7. Device image of the Quantum Wave Structure Functional Device.

Fig. 8. SEM image of the crystal whiskers.

/ *Paper presented at Int. Symp. Compound Semicond., San Diego, 18–22 September 1994*

Overview and Status of the European Framework/ESPRIT programmes

Steven P. Beaumont

Nanoelectronics Research Centre, Department of Electronics and Electrical Engineering, University of Glasgow, Glasgow G12 8QQ, Scotland, UK

Abstract. The funding of nanostructure research by the European Union is discussed from the point of view of present arrangements and future plans.

1. Introduction

This contribution is written from the perspective of a participant in European Union (EU) nanostructure research and development activities and is based on my own knowledge and experience: it is not an official pronouncement from Brussels. With this proviso I plan to describe the present state of EU funding for nanostructure research and indicate what might happen in future, trying to explain how and why the European situation differs from Japan and the USA.

The EU (hitherto the European Community or EC) has no counterpart to the focused programmes concerned with nanoelectronics and nanophotonics described in the companion papers in this session: by which I mean there is no 'budget' for nanostructure research as such. That is not to say, however, that such programmes do not exist within the nations making up the EU: in the UK, for example, the now-completed Low Dimensional Structures and Devices Initiative of the Science and Engineering Research Council made a considerable investment in this area. Indeed, the objectives and funding rules for EU projects are predicated on nationally-funded research. EU funding is intended to support research which *requires* a trans-national approach, perhaps because facilities and know-how are unavailable in a single country. This concept of *subsidiarity*, now enshrined in the Maastricht treaty, is likely to be applied even more stringently in future. Therefore, the EU provides largely marginal funding to support labour, consumables and travel between members of consortia to make use of resources and expertise at different sites: the capital equipment needed for most nanostructure research, whether for growth, fabrication or characterisation, is in almost all cases assumed to exist, having been funded at the national level.

Most of the current EU support for nanostructure research has come from the programme known as ESPRIT. Its objective is to enhance the EU's competitiveness in Information Technology. Most of the research is 'pre-competitive' and is carried out by consortia comprised largely of industrial partners. Recognising the need to support somewhat longer-term research, but nevertheless with IT relevance as a selection criterion, ESPRIT Basic Research Actions were established. It is from the Basic Research programme that most nanostructure projects have been supported.

Support at a lower level has also come from the EU Human Capital and Mobility programme (HCM). This is intended to promote training, access to specialist resources and the dispersal of skills in short supply around the countries of the Union. Research training is at

Project	Status*	Topics
MONOFAST	C	Use of nanotechnology in yield-predictable, right-first-time MMIC fabrication
NANSDEV, LATMIC	C	Quantum transport, nanostructure spectroscopy and nanofabrication
NANOFET	C	Ultimate miniaturisation of HEMTs
LATMIC II	P	Quantum transport and single electronics in semiconductors
NANOPT	P	Fabrication and characterisation of quantum wires and dots for optical device applications
SETRON	P	Single electron systems in metals and semiconductors
SOLDES	P	Self organised growth of quantum dots using zeolites and precipitation techniques
PARTNER	P	Resonant tunneling devices for optical applications
PRONANO	P	Ultimate nanofabrication by probe techniques, including STM

Table 1: Past and present projects in Nanoelectronics and Nano-optics
(* C = completed P = in progress)

postdoctoral level and the project cannot be undertaken in the student's home country. Some projects involve research in nanostructures, but proposals are submitted on an ad-hoc basis without co-ordination.

2. Support for nanostructure research under ESPRIT

ESPRIT Basic Research provides three types of support, all of which are represented within the nanostructures discipline. They are:

Projects or Actions. Consortia (generally with four or more partners) are contracted to carry out a specified research programme. Funds can be used to support labour, consumables, equipment, travel and overheads. Within Basic Research, projects are reviewed annually and contracts normally last from two to three years. Calls for new proposals are made from time to time accompanied by a workplace which sets out broadly the areas which Basic Research is prepared to support. 'Ultimate Miniaturisation' typifies one of the themes which embraces nanostructure research. Projects are selected for their relevance to the workplan, but the main criteria are technical and scientific excellence, the calibre of the consortium and the feasibility of the management plans. No formal budget is set for any one theme, although notional targets may be in the minds of the Brussels project officers. As a consequence of these rather loose arrangements, the outcome of a Call is unpredictable.

Table 1 lists all the current projects involving nanostructures and semiconductor nanotechnology and indicates the theme of each. Further details can be obtained from the official EU Synopses. But it is clear from this list that with few exceptions the projects are science-based and likely to be exploited only in the very long term. The contrast with the managed programmes described in the companion papers could not be greater.

Working groups. Consortia are funded at a relatively low level to pursue their mutual interests by holding regular workshops and brief laboratory visits. Working groups can be large: QUANTECS (Quantised Electronics) has 10 partners, including two from EFTA (European Free Trade Association) countries. EFTA partners are allowed to participate in ESPRIT, but national governments have to meet the costs of involvement. QUANTECS embraces work on quantum transport, spectroscopy of nansotructures, single electronics, microcavity devices and fabrication technology. Working groups often carry out a basic level of work to demonstrate the feasibility of promising lines of research preparatory to the submission of a full project proposal. Two new proposals (SETRON and SMILES) were awarded to members of the working group in the last ESPRIT round.

Networks of Excellence. The objective of a network of excellence is to formalise links between centres ('nodes') of expertise throughout the EU with a view to co-ordinating and promoting a broad area of research, and providing a point of contact for the outside world, including the European Commission itself. One of the networks is concerned with nanostructures. PHANTOMS (Physics and Technology of Mesoscopic Systems) now has 62 members and associate members, of which 8 are industrial partners and 3 are in Eastern European countries. Full members only are allowed to receive funding from the network. These funds allow members to send representatives to network meetings, and make short visits to other network members to carry out experiments, or discuss results. Funding provides for travel and subsistence only, not for the other costs of research. PHANTOMS maintains a database of publications and information on its partners, so that anyone searching for a particular expertise or capability can readily establish its availability within the network. A newsletter is sent regularly to worldwide contacts. Each year, PHANTOMS organises a workshop for its members and some invited speakers to discuss recent results.

ESPRIT funding is also available to support international collaboration by projects, working groups and networks. These schemes are negotiated on a country-by-country basis. At present agreements exist with the NSF (USA) and the NRC (Canada). Under the NSF scheme the QUANTECS working group has a funded partnership with QUEST at Santa Barbara which allows for the exchange of personnel and the organisation of a workshop. PHANTOMS and a consortium of Canadian institutes, lead by the NRL in Ottawa, have established the ECAMI project. This has a formal workplan from which topics for research are selected on an annual basis.

3. Framework IV

The arrangements described above applied during the third round of ESPRIT, which has now terminated. From early 1995, EU research will be covered by Framework IV, a Framework being an umbrella scheme which defines the budget, strategy and principal themes for research and associated activities.

Framework IV is intended to ring some changes in the objectives and mechanisms for R&D funding. Most important is to attempt to improve the rate of exploitation of EU funded research. Although research will remain pre-competitive, consortia will be required to define and commit to an exploitation strategy for their projects, clearly implying that the nature of the research will be much more closely associated with commercial realities. Another manifestation of change of emphasis is the plan to establish a number of 'Focused Clusters'.

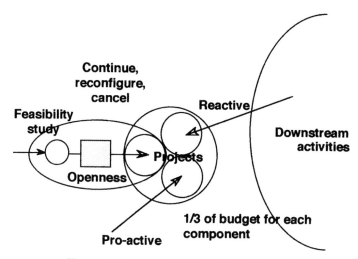

Figure 1: Structure of IT Long Term Research

These will involve suppliers and users of R&D and may cover a number of technologies as required by their goals. A Focused Cluster will be responsible for progressing its activities through the R&D phase into exploitation, by managing its resources according to the state of progress.

These changes in mainstream research projects have inevitable consequences for Basic Research. If Basic Research remains a novel science programme, judged by the publications it creates rather than the likelihood that the work will be exploited, there will be few opportunities to follow progress through to real applications. On the other hand, if Basic Research becomes too closely associated with mainstream projects, support for really long term research will fade.

The Commissions proposals for structuring Basic Research, renamed Long Term Research (LTR), try to balance these conflicts. Under the proposals, LTR will be divided into three modes:

(1) Responsive mode. These projects will carry out research closely connected with the perceived longer term requirements of mainstream (downstream) research. The Call for Proposals for this mode will parallel a main Call.

(2) Pro-Active mode. Two generic, long term themes will be defined and proposals solicited. Currently, the themes suggested are : handling the electronic information explosion: and miniaturisation beyond $0.1\mu m$ CMOS. The second topic evidently covers much of the present nanoelectronics programme.

(3) Open mode. This mode will allow any IT-relevant proposal to be considered. Acceptable projects will go through a 6 month feasibility study at which point they may be allowed to continue, be restructured or closed.

This structure is represented diagrammatically in Figure 1.

Clearly these arrangements address some of the concerns about the application of long-term research funds. Responsive mode funding allows downstream projects to be supported with upstream research. The pro-active mode encourages a longer term vision and also provides some focus, which was lacking in earlier rounds of Basic Research. And the open mode allows for novelty.

Each mode is to be allocated one third of the total available budget. So to judge the impact of the new structure on nanoelectronics research, it is necessary to examine the available funds and the specified topics for downstream research.

Objective
Small size, low cost, high functionality
High speed, low power mm- and μ-wave
Integration technologies, including optical & electrical interconnects and packaging
Design methodologies and tools
Devices & integration technologies for advanced peripherals and storage systems, communications, optical computers and microsystems
Effective manufacturability for small & large volume production
Fast turnround on ASICS

Table 2: Objectives of Silicon & Compound Semiconductor Research Activities

Tables 2 and 3 show the principle themes for mainstream semiconductor research activities, and the budget for the IT sector, respectively. Upstream work on high-speed electronics, high functionality, manufacturability and optical interconnections provide coverage for some of the topics I have classified under nanoelectronics: for example, ultrashort gate FETs, resonant tunneling devices and microcavity optical emitters could all fit these objectives. Significant support for nanoelectronics might also come from the 'Beyond $0.1\mu m$ CMOS' pro-active theme: it would not be unreasonable for one third of the budget for this theme to be devoted to nanoelectronics, giving around $13M over the lifetime of the Framework. Coupled with support from the responsive mode, and novel ideas in nanoelectronics contributing to the open theme, I believe that the funding level established in Framework III is likely to be

Area	%	Budget (M$US)
Software	22%	523
Multimedia technologies		
Technologies for components and subsystems	23%	544
Technologies for business processes	20%	471
Integration in manufacturing		
Open microprocessor systems	25%	595
High performance computing		
Long Term Research	10%	237

Table 3: Breakdown of effort in the Information Technology theme

sustainable in Framework IV. The sums are, however, extremely small in comparison with the Japanese Quantum Function Devices and the US Ultra programmes. To make best use of this funding requires more direction to be given to the nanoelectronics programme and a mission-oriented approach. Identification of priorities based on a thorough evaluation of options will be critical. The PHANTOMS network should be in a good position to provide this co-ordination and identification of the most promising lines of research, though it will have to take a stronger leadership role than hitherto whilst retaining the support of its predominantly science-based membership. If LTR carries out its recommendation that selection and monitoring of projects be carried out by Networks, then PHANTOMS' authority to construct and manage a programme will be greatly strengthened.

4. Conclusions

ESPRIT Basic Research has provided a modest level of funding over the past six or so years to establish a trans-national EU effort in nanoelectronics. The work is largely science-based, aimed at understanding physical phenomena upon which might form the basis for new types of device. In contrast to the Ultra and Quantum Functional Device programmes, EU Nanoelectronics has developed from the bottom up, rather than from the top, down. The programme could be threatened by policies to be adopted in Framework IV, but I believe that good co-ordination and promotion, together with a modest change of emphasis towards shorter-term objectives should see a sustained level of funding for nanoelectronics in the course of the next four years.

5. Acknowledgements

I am very grateful to the following for information and advice: Dr David Worsnip, EPSRC UK: Dr Ian Eddison, DTI UK: Professor Malcolm McLennan, University of Glasgow EU Office: Professor Peter van Daele, University of Gent. The views expressed in this paper are my own.

Subject Index

2DEG mobility   323
2DEG mobility   167
(311)A GaAs   149
6H SiC   415

AC response   775
Abrupt, graded   645
Accelerated aging   503
Admittance spectroscopy   411
AlN buffer layer   101
AlAs   35, 667
AlAs/GaAs   29
AlAs/GaAs heterointerface   345
AlAs/InGaAs/InAs   45
AlGaAs   57, 633, 797
AlGaAs substrates   559
AlGaAs/GaAs   81
AlGaAs/GaAs HEMT   579
AlGaInP   503
AlGaN (aluminium gallium nitride)   419, 497
AlSb   335
Ambipolar diffusivity   369
Amphoteric   339
Andreev reflections   7
Annealing effects   265
Annealing   275
Antimonide   17
Antimony   851
A-plane sapphire   101
Artificial control   149
As-termination of Si   93
Atom exchange   345
Atomic force microscopy   359
Atomic hydrogen   503
Auger recombination   243

Ballistic   835
Band alignment   751
Band bending   143
Band discontinuity   149
Bandgap narrowing   253
Base doping   253
BiFET   657
Blocking layer   563
Breakdown   301
Buffer thickness   101

Capture time-constant   471
Carbon doping   265, 269
Carrier heating   471
Carrier recombination   629
Carrier separation   841
Carrier transport   463
CATV (cable TV)   477
Characterisation   23, 567
Charge build-up   227
Chemical beam epitaxy   63, 69, 265
Chlorine-based plasma   329
Circuit simulation   867
Collector-base   651
Communication lasers   477
Compensation   269, 425
Complementary   663
Complementary HBT   685
Compositional drifting   513
Contact resistance   699
Corner frequencies   625
Coulomb blockade   781, 879
Critical layer thickness   819
Cross-hatching   87
Cryogenic optical links   531
Crystal growth   373

Current blocking layers   189
Cutoff frequency   633

δ-doping   149
Database system   751
DBR (Distributed Bragg reflector)   553
Deep levels   411
Defects   351, 405
Degradation of carrier density   323
Density of states   253
Device degradation   519
Device simulation   747, 793
Dielectric function   769
Diffusion   259, 339
Diode   851
Dislocations   519
Disorder   813
Disordering   547
Distributed Bragg reflector (DBR)   543, 489, 547
Doped-channel   693
Doping   399
Double heterojunction   645
Double quantum wells   841
Double   851
Driving force   161
Dry etching   237
Dual gate   183

ECR   711
ECR etching   329
ECR plasma   131
Effective mass   233, 415
Electrical and structure properties   295
Electrical characterizations   301
Electrical/optical properties   307
Electron beam lithography   237
Electron transport   645
Emission   525
Energy transport model   633
Epitaxial growth   339
Epitaxial lift-off   689
Epitaxial overgrowth   737
Epitaxy   81, 373, 619
Equivalent circuit analysis   651
ESPRIT   891

Etching   291
European funding   891
Exceed critical thickness   167
Excitons   233, 351

Facetting properties   177
FET   183, 667, 693, 707
Field effect transistors (MESFET)   373, 459
Field screening   227
Finite element modelling   177
Flow modulation epitaxy   155
Focused ion beams   359
Framework   891
Future electron devices (FED)   885

GaAlAs/GaAs HBT   625
GaAs (gallium arsenide)   183, 265, 269, 291, 295, 383, 619, 639, 667, 707
GaAs (III) A   339
GaAs/AlGaAs   227
GaAsSb   727
GaInAsP epitaxy   173
GaInP (gallium-indium phosphide)   63, 207, 195
GaN (gallium nitride)   101, 107, 119, 131, 137, 219, 425, 459, 497, 525
GaP   63, 287
$GaP_{1-x}N_x$ 97
GaSb   335
Gain-compression   471
Gallium (indium arsenide)   155
Gallium oxide   597
Gas source molecular beam epitaxy   319
Gate capacitance   707
Gate fringing capacitance   707
Graded buffer layers   87
Green's function   775
Growth   23
Growth interruption   97
Growth kinetics   97
Growth mechanics   137
Growth modeling   137
GSMBE   161

Harmonic balance   419
Heat flow   747
HEMT   673, 685, 703
Heterointerface   351, 355
Heterojunction   75, 807
Heterojunction bipolar transistors (HBT)   629, 633, 619, 639, 651, 663, 685, 737
Heterojunction field effect transistors (HJFETs)   419, 663
Heterostructure transistors   689
Heterostructures   149, 195, 307, 673, 751
HIGFET   727
High field transport   803
High gain   591
High power   1
High spatial resolution   591
High speed   703
High speed lasers   531
High speed photodetectors   531
High temperature   1, 383, 459
Highly lattice-mismatched heteroepitaxy   93
Hole transport   803
Homoepitaxy SiC   443
Hopping conduction   301
HRTEM (picture of GaN)   125
Hydrogen   287
Hydrogen azide   131
Hydrogen-reduced   503

ICs   455
IG-HFET   667
Impact ionization   639, 797, 803
Impurities   259
Impurity state   873
Impurity-free disordering   547
Impurity-induced disordering   547
In situ control   35
In situ deposition   543
In situ ellipsometry   41
In situ fabrication   597
In situ monitoring   35
In situ observation   345
In situ process monitoring   29

InAlAs   727
InAlAs-InGaAs   673
InAs   23, 35, 693, 825
InAs/Alsb quantum wells   7
$In_xGa_{1-x}As/InP$ 769
InAs/GaAs SL   167
InAs/GaSb/AlSb   855, 861
InAsSb   13, 243
Indium gallium nitride   497
Infrared   607
Infrared emitters   13
InGaAs   13, 35, 639, 693, 737
InGaAs on GaAs   87
InGaAs/AlGaAs   607
InGaAs/GaAs   69
InGaAs/InAlAs   313
InGaAsP layered structures   369
InGaAsP/InP compound semiconductor   613
InGaP/GaAs HBT   625
InGaP/InGaAs quantum wells   161
InGeAs/AlGaAs lasers   537
InP   35, 63, 155, 329, 663, 699, 703
InP/InGaAs   75
InSb on Si   93
Insulated gate barrier   667
Insulator-semiconductor interface   597
Integrated circuits   679
Integrated optics   585
Interband tunneling   855
Interdiffusion   547
Interface   161, 291
Interface perfection   51
Interface states   281
Interface traps   629
Interference   835
Interferometers   585
Intermodulation distortion   477
Intrinsic 1/f noise   625
Inverted interface   57

JFET   389

Large area   119
Laser   51, 399, 471, 483, 567, 573, 743, 765

Laser light scattering 45
Laser linewidth 759
Laser near-field 483
Lateral bipolar transistor 591
Lateral confinement 563
Lateral current injection (LCI) 743
Lateral injection 563
Lattice mismatch 143
Lattice tilt 87
Lifetime 405
Light-emitting diodes (LED) 489
Linearly-graded buffer 313
Local doping 591
Local-field effects 787
Local vibrational mode 287
Logic 455
Low energy ion scattering 345
Low frequency noise 625
Low ion energy 329
Low leakage 383
Low temperature GaAs 301
Low temperature grown (LTG) 737
LP-MOVPE 189
LP-MOVPE 23
LT-GaAs 717

Magneto-photoluminescence 243
Magnetotransport 319
MBE 81, 131, 339
Mechanical stress 483
Mechanisms 645
Mercury cadmium telluride 41
Merged technology 689
MESFET 383, 389, 657
Mesoscopic systems 879
Metal semiconductor reaction 613
Microcavity 489
Microcavity lasers 759
Micropipes 377
Microroughness 355
Microstructure 207
Microwave 459
Microwave devices 579
Microwave technology 689
Millimeter wave 703
Millimeter wave systems 531

Minority carrier mobility 253
Mirror coatings 537
Mirror temperatures 537
MITI's project 885
Mobility 319, 415, 425
MOCVD (metal-organic chemical vapor deposition) 13, 41, 107, 119, 177, 213, 559
Modeling 743
Modulation doped structure 319
Modulation frequency dependence 651
Molecular beam epitaxial growth (MBE) 29, 45, 51, 173, 237, 295, 355, 534, 567, 819
MOMBE 17
Monitoring and control 41
Monolithic microwave optoelectronic integrated circuits 579
Monte Carlo 137, 765
Monte Carlo simulation 793
MOS 449
MOSFET 377, 449, 711, 807
MOVPE 81, 97, 217, 629
Multiple quantum well 69
Multiplication factor 639
Multi-quantum well 607
Multi-wafer 119
Multiwavelength ellipsometry 41

Nanocrystal 143, 825
Nanoelectronic architectures 781
Nanoelectronic devices 781
Nanoelectronics 787, 891
Nanostructure 813
Native defects 259, 269, 275
Natural superlattice 213
Negative differential resistance 247
NiGe 721
Niobium contacts 7
Nitride 131, 497
Nitrogen desorption 97
NMOS 455
Non-alloyed ohmic contacts 717
Non-planar epitaxy 177
Non-planar selective epitaxy 189
Non-volatile memory 395

Novel group V precursors   63
N-type GaAs   721
Nucleation   825

ODCR   415
ODMR   405
OEICs   573
Ohmic contact   721
OMVPE   75
OMVPE growth   195
Optical absorption   769
Optical constants   35, 363
Optical properties   173
Optical pyrometer   513
Optical reflectivity   553
Ordering   195, 207
Oxidation   449

Parallel computer applications   793
Passivation   287
Phase separation   213
Photocurrent   233
Photodetector   591, 607
Photoelastic effects   613
Photoluminescence   107, 195, 207, 221, 237, 359, 629, 841
Photoluminescence spectroscopy   483
Photopumped   525
Photoreceivers   573
Photoreflectance   651
Planar device structure   613
Planar devices   679
Plasma-assisted epitaxy   93
Plasma-CVD   323
Plasma pretreatments   281
PL/PLZ   355
Polarization dependence   237
Post growth heat treatment   125
Power combining   679
Power device   1, 437
Precipitation   269
Pressure   275
Process control   29
Proximity effect   7

Quantum corral   787

Quantum device   855
Quantum dots   781, 813, 819, 879
Quantum functional devices   885
Quantum interference   831
Quantum optics   879
Quantum simulation   787
Quantum structures   831
Quantum well   29, 221, 227, 233, 355, 747, 765
Quantum wire   471
Quantum wire and dot   885

RAM   395, 657
Raman spectroscopy   483, 537
Rapid thermal annealing   727
Reactive ion etching   307
Recombination   369, 405
Refractometry   585
Refractory contact   721
Regrowth   291
Resonant cavity   489
Resonant tunneling   45, 775, 855, 861, 873
Resonant tunneling devices   867
Resonant tunneling diodes (RTDs)   17, 35, 679, 737
Restoration of carrier density   323
RHEED   167
Rotating disk   119

Scaling   807
Scattering   775
Schottky rectifier   437
Screw dislocation   335
Selective combined technologies of OMVPE and ion implantation   579
Selective epitaxy   183
Selective MBE   685
Selective organometallic vapor phase epitaxy   579
Selenium treatment   143
Self-assembly   819
Self-consistent modeling   813
Semiconductors   573
Semiconductor devices   619
Semiconductor laser   463, 597, 519

Semiconductors, II-VI   259
Semiconductors, III-V   259, 597
Semi-insulating   411
Sensors   585
Sequential resonant tunneling   247
Shadow masking   177
Shubnikov de Haas   313
SiBr$_4$ 699
SiC/6H-SiC system   443
Si-doping   217
SiGe-HBT   253
Silicon carbide (SiC)   1, 373, 377, 389, 395, 399, 405, 411, 437, 449, 445, 497
Silicon germanium (SiGe)   319, 711
Si-InP   189
SiN film   323
Si-PBH LD   189
Silicon recrystallization   537
Simulation   765, 807, 813
Single electronics   781
Single-barrier structure   873
Solid phosphorus   173
Spectroscopic ellipsometry   35, 363
Spectroscopy   233, 287
SPICE   867
Spiral   335
Spontaneous emission   489
Stacking faults   519
Stark shift   841
Static induction transistors (SIT)   373
Step-controlled epitaxy   437
Sticking coefficient   513
Stimulated   525
Strain   75
Strain compensation   161
Strain, electronic effects   243
Strain relaxation   819
Strained layer epitaxy   45
Strained layer superlattice (SLS)   13
Strained layers   51, 673
Strained quantum well lasers   759
Structural study of SiC   443
Structure   313
Stub   835
Submicron MESFETs   793

Substrates   377
Superconductivity   7
Superlattice   75, 155, 553, 567, 769, 831
Surface   227, 335
Surface segregation   57
Surface state density   449
Surface states   221
Surface-emitting laser   563, 759
Surfactant   51, 57
Synchrotron radiation photoelectron spectroscopy   143

TEM   101, 167
TEM characterization   443
TEM study SiC/6H-SiC   443
Terahertz   679
Thermal properties of 1-S interface   281
Thin layers   363
Three dimensional effects   793
Threshold energies   797
Thyristor   377
Tight-binding   769
TMAA   17
Transient gratings   369
Transistor   645
Transmission electron microscopy (TEM)   345, 519
Transmitters   573
Transport   351, 425, 693
Transport in multiquantum wells   247
Trap levels characterisation   281
Travelling wave photodiodes   531
Tris-dimethylaminoarsenic   69
Tunneling   851
Tunneling effect   885
Two-dimensional electron gas   313
Two-dimensional electronic system   787
Two-step growth   93
Type-II structures   351

Unified charge control model   419

Valley current   775

Valved cracking cells   173
Varactor diode   217
VCSEL (vertical cavity surface emitting laser)   553, 559, 775
Velocity overshoot   633
Vertical   183
Vertical-cavity laser   563
Vertical-cavity surface emitting laser (SEL)   543
Visible laser diode   503

Waveguide modulator   613
Wide band gap   107, 137
Wurtzite GaN   125, 425

X valley   873
X-ray   107
X-ray diffraction   553

Y/T-shaped gate   707

ZnSe   275

Index

Abstreiter G 591
Adesida I 813
Agarwal A K 373
Agawa K 149
Ager III J W 269
Ahmed S 399
Akis R 835
Akita H 437
Akiyama H 841
Alam Md A 765
Alexandre F 265, 619
Allen Jr S J 845
Allen S 389
Amano C 559
Amman M 825
Anand S 29
Anantram M P 775
Antonelli A 351
Aoi N 885
Armour E A 177
Astratov V N 227
Augustine G 373

Babanin A I 497
Bacalzo F T 351
Baeyens Y 689
Baillargeon J N 173
Bajaj K K 351
Ballal A K 443
Bar S X 579
Barbero C J 399
Baucom K C 13
Baumgartner P 591
Beaumont S P 891
Beccue S M 657
Beckham C 579

Behet M 23
Benchimol J L 265
Berger P R 573
Bernklau D 567
Bertness K A 207
Besabathina D P 443
Betko J 295
Bhat I B 41
Bhattacharya P 237
Biefeld R M 13, 243
Binari S C 101, 459
Blanchet J 835
Block T R 685
Boeck J De 689
Böhm G 563, 591, 673, 831
Bone D J 359
Borghs G 689
Bowen R C 751
Bowers J 531
Bowser M I 281
Brandt C D 373
Brar B 335
Brehmer D 845
Brennan K F 803
Bright V M 727
Brock T 237
Brown E R 737
Brys C 689
Buchanan M 607
Bugge F 217
Bulman G E 497
Burk A A 373

Canali C 639
Capasso F 639
Carius R 81

Carter Jr C H   377, 389, 395, 497, 525
Celii F G   35, 363
Cerny C L A   727
Cerva H   567
Cha S S   213
Chamberlain J M   679
Chandramouli V   797
Chandrasekhar S   639
Chang J-H   503
Chang M F   657
Chavarkar P   663
Chen A L   275
Chen J   313
Chen R   351
Chen W M   405, 415
Cheng K Y   173
Chern C S   119
Chin T P   717
Chindalore G L   449
Cho A Y   173, 573
Choi W-J   503
Choi W-T   503
Choo H R   189
Chow D H   851
Clarke R C   373
Clawson A R   75
Coldren L A   291
Cooper Jr J A   395, 449, 455
Crofts R J   383
Cronk D   17

Daiminger F   221
Damian J   119
Datta S   775
Dätwyler K   483
Dawson L R   243
De Los Santos H J   645
DeLong M C   207
Debray P   835
Demeester P   689
Denbaars S   663
Deutsch U   483
Dieker Ch   295
Dietrich H B   459
Dietrich H P   483
Dmitriev V A   431, 497, 525

Dolgopolov V   831
Dong H K   69
Doverspike K   101, 425, 459
Drexler H   819
Dreybrodt J   221
Driad R   265
Driad R   619
Driver M C   373
Droopad R   29
Duchenois A M   619
Duncan W M   35
Dutta N K   573

Eberl K   87
Edmond J A   497, 525
Eijkemans T J   307
Emerson D T   155
Engel C   591
English J H   335
Ensslin K   831
Epperlein P W   483, 537
Evans K R   57
Evwaraye A O   411

Fan Y   519
Fekade K   443
Feng Y H   383
Fernando C L   751
Fernández J M   319
Floyd P D   547
Follstaedt D M   13
Forchel A   221
Förster A   81, 295
Forzan C   639
Freitas Jr J A   101
Frensley W R   751
Fresina M T   699
Fu L P   351
Fujita K   339
Fukaishi M   707
Fukuyama H   873
Furdyna J   275
Furumai M   721

Gaskill D K   101, 425, 459
Gasser M   537

Gedridge Jr R W   63
Gelmont B   419
George T   167
Giboney K   531
Gilliland G D   351
Glass R C   373
Goldman R S   313
Goodnick S M   793
Goorsky M S   553
Gorfinkel V B   477
Goronkin H   855
Götz W K   125, 131
Gravé I   247
Griffin J A   167
Grillo D C   513, 519
Gruhle A   253
Grunthaner F   167
Guan Z F   613
Guido L J   825
Gunshor R L   513, 519

Haddad G I   867
Hafizi M   645
Haller E E   269, 275, 287
Hamadeh H   23
Hamm R A   639
Han A-C   685
Han J   513, 519
Hansen W   819
Hanson C M   75
Harbury H K   787
Hardtdegen H   81, 183
Hareland S A   807
Hariu T   93
Harris Jr J S   125, 131, 359
Hashimoto Y   149
Hattori R   329
Häusler K   87
Hayes R E   747
He P   29, 41
Heiß H   673
Heime K   23
Henini M   679
Hersee S D   177
Herzinger C M   363
Heymann P   217

Hida H   707
Hirai M   339
Ho I H   195, 207
Ho W J   657
Hobgood H M   373
Hobson W S   573, 597
Hollfelder M   81
Hong C-H   107
Hong M   543, 597
Hong S-C   733
Hopkins R H   373
Horio K   633
Hovinen M   519
Hoyler C   567
Hsu T C   207
Hu E L   7, 291
Hu J   657
Hua G C   513, 519
Hunt N E J   489
Hunziker G   483
Hwang C-Y   119
Hwu R J   383

Iafrate G J   769
Ikoma T   149
Iliadis A A   693
Imamoğlu A   879
Ingefield C E   207
Irvine K G   497, 525
Irvine K   443
Ishida I   885
Ishihara O   329
Ishikawa H   721
Ito R   97
Itoh A   437

Jackson S L   699
Jang S   437
Janzén E   405, 415
Jeon H I   213
Jiang X S   613
Johnson G M   395
Johnson N M   287
Johs B   29, 35, 41
Jones A C   63
Jovanovic D   813

Joyce B A 319
Juhel M 265
Juodkazis S 369
Jusserand B 265

Kızıloğlu K 663
Kadoya Y 841
Kalinina E V 497
Kanber H 579
Kao Y-C 35, 363
Karczewski G 275
Karimov O Z 227
Kaspi R 57
Kavanagh K L 313
Kawata H R 721
Kazarinov R F 463
Keller B 663
Kelner G 459
Khan A 419
Kibbel H 253
Kim B 625
Kim C W 63
Kim D K 213
Kim D W 355
Kim D-W 733
Kim H M 189
Kim J S 189
Kim J-S 503
Kim K W 769
Kim S-H 503
Kim W S 355
Kim Y M 355
Kim Y-S 625
Kimoto T 437
Kishino K 507
Klein W 563, 673, 831
Klem J 351
Klimeck G 775
Ko H S 355
Ko K 237
Kobayashi K 685
Kohama Y 559
Kohn E 667
Kolnik J 803
König U 253
Konstantinov A O 415

Kordina O 405, 415
Kordoš P 183, 295
Kotera N 233
Kotthaus J P 819, 831
Kramer G 17, 855
Kratzer H 563
Kraus S 673
Kroemer H 7
Krupper W 459
Krystek W 651
Kuklovský S 295
Kuo C-H 29
Kurokawa T 559
Kurtz S R 13, 243
Kwon Y-S 733

Lake R K 775
Lam W W 579
Lange J 81
Langen W 183
Lareau R T 727
Lau S S 613
Launay P 265, 619
LeRoux G 265
Leburton J P 813
Lee B T 213
Lee H J 213
Lee H 125, 131, 359
Lee J-W 625
Lee S W 189
Lee Y H 213
Leem S-J 503
Legay Ph 619
Lenchyshyn L C 607
Lent C S 781
Leonard D 335, 819
Levola T 29
Lewis J H 167
Li N Y 69
Li P-W 711
Li X 711
Liang S 119
Lim K Y 213
Lipka K-M 667
Lipkin L A 377, 395, 449
Liu H C 607

Liu H   119
Liu Q Z   613
Liu T   645
Lopata J   573, 597
Lu Z H   629
Lundstrom M S   717, 765
Luo H   275
Luo J K   301
Luo J S   201
Luryi S   477
Lüth H   81, 183

Maeda F   143
Maeda S   885
Majerfeld A   629
Makino T   743
Malik R   639
Manfra M J   737
Mannaerts J P   543, 597
Maracas G N   29
Mason M E   751
Masselink W T   693
Mathine D L   29
Matney K   553
Matsumura A   319
Matsunami H   437
Matsuno N   707
Mayo W E   119
Maziar C M   797, 807
Mazumder P   867
McCluskey M D   287
McHugh J P   373
McMahon O B   737
McMullin P G   373
Meideros-Ribeiro G   819
Melloch M R   395, 449, 455, 717
Melnik Yu V   497
Merkel K G   727
Merz J L   547
Milde A   567
Miles R E   679
Miller J N   313
Mishra U   663
Mitchel W C   411
Miyakuni S   329
Miyoshi S   97

Mizutani T   323
Mohan S   867
Moll A J   269
Molnar R   113
Mondry M   663
Monemar B   405, 415
Moore K   389
Morgan D V   301
Moriya N   597
Moulin D   23
Moustakas T   113
Mui D S L   291
Murakami M   721
Murthy S D   41

Nagarajan R   531
Nagle J   835
Nakata S   323
Nakatani A   633
Nauwelaers B   689
Neudeck P G   1
Neviani A   639
Nguyen C   7
Nguyen L D   703
Nichols D T   573
Nichols K B   737
Nikitina I P   431, 497
Nikolaev V I   525
Nikolaev V   497
Nomura I   507
Nordquist K J   389
Norris P E   113, 597
Nurmikko A V   513, 519

Oberman D B   125, 131
Obloh H   861
Oguzman I H   803
Oh D K   189
Ohba H   93
Ohiso Y   559
Ohnishi H   339
Okayama S   885
Oki A K   685
Oku T   721
Olson J M   201, 207
Onabe K   97

Oshima M 143
Otsuka N 519
Otsuki A 721

Pajot B 265
Palmer J E 345
Palmour J W 377, 389, 395, 449
Palmstrom C J 845
Pan J 449, 455
Pang S W 237
Pappert S A 613
Parikh P 663
Park C 189
Park H M 189
Park S-C 189
Parsons C 553
Passlack M 597
Patkar M P 717
Pavlidis D 107, 137
Pearah P J 173
Pease R F W 359
Peng R 603
Pennathur S S 793
Pereira R 307
Petrauskas M 369
Petroff P M 291, 819
Piao Z S 213
Pierson R L 657, 737
Pittal S 41
Ploog K H 51
Pollak F H 651
Pollard R D 679
Pond K 819
Porod W 787
Prasad R S 319

Qian W 101
Qiang H 651

Raafat T 81
Raedt W De 689
Rahbi R 265
Rai R 553
Ralson J D 861
Razumovsky O 113
Reed M A 825

Reithmaier J P 221
Rensch D B 645
Reynolds D C 57
Rhee S J 355
Richter E 217
Riechert H 567
Ringle M D 513, 519
Rishton S A 693
Robbins V M 313
Rochus S 563
Rode D L 425
Rodwell M 531
Röhr T 563
Roth M 411
Rowland L B 101, 425, 459
Rüders F 295
Rüfenacht M 841
Ryu H H 63

Sadwick L P 63, 383
Sailor A 657
Saito T 149
Saitoh T 345
Sakaki S H 841
Sakamoto N 149
Salagaj T 119
Salamanca-Riba L G 443
Sankaran V 769
Sato K 329
Schmitz J 861
Schneider J 667
Schneider P 23
Schubert E F 489, 597
Schuermeyer F L 727
Schulman J N 851
Schüppen A 253
Schuster M 567
Schuster R 831
Semerad R 673
Seo Y-S 625
Sermage B 265
Shakouri A 247
Shealy J R 155
Shen J 855
Shenoy J N 449
Shin J-H 625

Shin M W   419
Shin Y K   213
Shiraki Y   97
Shiralagi K   17
Shur M S   419
Shuttlewood J   237
Siergiej R R   373
Sigmon T W   399
Singh J   137, 237, 759
Singh R   113
Sivco D L   573
Sizov V E   497
Skowronski M   101
Sleight J W   825
Smith P R   573
Smith S R   411
Smith T J   373
Snyder P G   363
Sokolich M   383
Solomon J   411
Son J-H   733
Son N T   405, 415
Sörman E   405, 415
Spencer M G   167, 443
Sriram S   373
Stall R A   119
Steenson D P   679
Stillman G E   699
Strand T A   291
Streit D C   651, 685
Stringfellow G B   63, 195, 207
Su L C   195, 207
Suda D A   743
Suh E-K   213
Sun J P   867
Sun S Z   177
Swanson J G   281

Tamura M   345
Tanaka K   233
Tartaglia J   553
Tasch A F   807
Tauber D   531
Taylor E N   57
Taylor P C   207
Tehrani S   855

Theodore N D   17
Theron D   301
Thibeault B J   291
Thiel F A   173, 543
Thomas H   301
Thornton T J   319
Tiwari S   471
Tokushima M   707
Tougaw P D   781
Tournié E   51
Tränkle G   563, 591, 673, 831
Trew R J   419
Tsui R   17, 855
Tsujino S   841
Tsvetkov V F   377
Tu C W   69, 161

Umemoto D K   685
Ungermanns Chr   81
Urushidani T   437

Vaitkus J   369
Vakhshoori D   543
Van Es C M   307
Van Hove M   307
Van Rossum M   307, 689
Vasilopoulos P   835
Vendrame L   639
Vlasov Yu A   227
Vurgaftman I   237, 759

Wada S   707
Wagner J   265
Waho T   873
Wakimoto H   721
Walker J   287
Walukiewicz W   259, 269, 275
Wang K C   657, 737
Wang K   107, 137
Wang Q H   281
Wang Y   395, 803
Wasilewski Z R   607
Watanabe T   339
Watanabe Y   143
Webb D J   483
Wei G   603

Weimann G 563, 591, 673, 831
Weitzel C E 389
Welser R E 825
Westphalen R 667
Westwood D 301
Weyers M 217
Wharam D 831
Whittingham L 155
Wieder H H 313
Willander M 369
Williams K 359
Williams R H 301
Winston D W 747
Woitok J 23
Wojtowicz M 651, 685
Wolf H-D 567
Wolford D J 351
Wolter J H 307
Woo J C 355
Woodall J M 717
Woollam J A 29, 41
Wright P D 629
Wu W 603

Xie M H 319
Xie W 395, 455
Xu D 673
Xu J M 743
Xu Y 247

Yaguchi H 97
Yamamoto M 323
Yamamoto Y 879
Yan C H 161
Yang E S 711
Yang L W 629
Yang Y F 711
Yariv A 247
Yoo T-K 503
Yu K M 269
Yu L S 613
Yu P K L 613
Yu P W 727
Yu R 531

Zahurak J K 693
Zampardi P J 657, 737
Zanoni E 639
Zappe H P 585
Zhang D P 167
Zhang J 319
Zhang K 845
Zhang X 319
Zhao J 113
Zhu B 613
Zhu L D 113
Zhu T X 855
Zubrilov A S 497, 525
Zydzik G J 597